# CLASSZONE.COM

Looking for ways to integrate the Web into your curriculum?

ClassZone, McDougal Littell's textbook-companion Web site, is the solution! Online teaching support for you and engaging, interactive content for your students!

ClassZone is your online guide to *Earth Science*. You'll discover:
- Online resources that correlate to text chapters
- Inquiry-based explorations of key concepts to engage students
- Visuals and animations that support student learning
- Real world connections to science through links to local data and issues
- Online Lesson Planner to plan either a specific lesson or the whole year
- Laboratory support to ensure successful classroom experiences

Log on to ClassZone at
www.classzone.com

With the purchase of *Earth Science* you have immediate access to ClassZone.

Teacher Access Code

MCDQ8RPUH5XH3

Use this code to create your own user name and password. Then access both teacher and student resources.

Teacher's Edition

EARTH SCIENCE

Nancy E. Spaulding, Samuel N. Namowitz

McDougal Littell
A HOUGHTON MIFFLIN COMPANY
Evanston, Illinois • Boston • Dallas

Copyright ©2003 by McDougal Littell Inc. All rights reserved.

No part of this work may be reproduced or transmitted in any form or by any means, electronic or mechanical, including photocopy and recording, or by any information storage or retrieval system without the prior written permission of McDougal Littell Inc. unless such copying is expressly permitted by federal copyright law. Address inquiries to Manager, Rights and Permissions, McDougal Littell Inc., P.O. Box 1667, Evanston, IL 60204.

ISBN: 0-618-18739-1   1 2 3 4 5 6 7 8 9 DWO 05 04 03 02

Internet Web Site: http://www.mcdougallittell.com

McDougal Littell

EARTH SCIENCE
Teacher's Edition

TABLE OF CONTENTS

Program Overview	T4
Accessible Content	T6
Real-World Connections	T7
Integrated Technology	T8
Program Components	
Print Resources	T10
Technology Resources	T12
Teacher's Edition Features	
Chapter Planning Guide	T14
Point-of-Use Support	T16
Reading and Learning Science from Textbooks	T18
Pacing Guide	T20
Master Materials List	T24
Safety During Activities	T26
ClassZone	T28
Annotated *Pupil's Edition*	i

©2003 by McDougal Littell Inc. All rights reserved.

Choose the Right Environment for Your Classroom

Earth Science provides a wide range of teaching options and the flexibility to choose the resources that meet the needs of your classroom. Reliable content, opportunities for inquiry-based learning, and online resources help students understand the concepts of earth science.

Pupil's Edition

Accessible content and visuals, hands-on activities, and active investigations engage students in the exploration of earth science. Point-of-use references guide students to online resources that extend the content of the text.

Teacher's Edition

The new wrap-around *Teacher's Edition* provides a wealth of material to help you reach a broad range of students. Point-of-use features include strategies for teaching reading, differentiating instruction, and incorporating national standards.

Print Resources

Extensive print resources provide a variety of teaching and learning support, including teaching manuals for online and in-text labs and investigations, assessment options, lesson plans, and more.

Technology Resources

Fully integrated Internet resources and corresponding CD-ROMs include projects funded by NSF and provide solid, easy-to-use activities and teaching materials that support the content of the text.

Accessible content
STUDENT-FRIENDLY PRESENTATION

Reading strategies, paired diagrams, and visualizations complement the strong visual appeal and the dynamic, interactive content, making the program accessible to a wide variety of students.

KEY IDEA
This feature introduces each section of the text and helps students focus on key concepts.

Magma and Erupted Materials

Kilauea, a volcano on the island of Hawaii, has spewed molten rock for decades. In contrast, Mount St. Helens exploded violently in 1980 after more than a century of being a quiet, snow-capped mountain. Differences in the volcanic activity at these two sites result partly from differences in the magmas that rise to the surface there.

Types of Magma

Silica, a principal ingredient in all magmas, determines a magma's **viscosity**, or resistance to flow. Magmas high in silica resist flow, whereas magmas with lower silica content flow more easily. Basaltic magmas contain the least silica; as a result, they flow most easily. Andesitic and rhyolitic magmas contain more silica and are more resistant to flow.

Magmas also contain gases—mainly water vapor and carbon dioxide. The gases, dissolved in the magmas at the depths where they form, may bubble out of solution as the magmas rise. Most gases escape easily from basaltic magmas. When these magmas reach the surface, any remaining gases usually produce relatively harmless fountains and floods. Gases in more viscous andesitic and rhyolitic magmas, however, cannot escape as easily. As such magmas rise, the gases expand and propel the magma rapidly to the surface. The result can be an explosive eruption of gas and debris, such as the one at Mount St. Helens.

Each type of magma tends to form at specific locations. Basaltic magmas form at rifts and at oceanic hot spots. Andesitic magmas tend to form at subduction boundaries. Rhyolitic magmas generally form where hot spots underlie continental plates.

9.2
KEY IDEA
The composition of magma largely determines how explosive a volcanic eruption will be.

KEY VOCABULARY
- viscosity
- lava
- pahoehoe
- aa
- pillow lava
- pyroclastic material
- pyroclastic flow

KEY VOCABULARY
Vocabulary terms at the beginning of each section help students prepare to read.

SUMMARY Characteristics of Magma

	Basaltic Magma	Andesitic Magma	Rhyolitic Magma
Silica content	Least (about 50%)	Intermediate (about 60%)	Most (about 70%)
Gas content	Least	Intermediate	Most
Viscosity	Least viscous	Intermediate	Most viscous
Type of eruption	Rarely explosive	Sometimes explosive	Usually explosive
Melting temperature	Highest	Intermediate	Lowest
Location	Rifts, oceanic hot spots	Subduction boundaries	Continental hot spots

Mount St. Helens   Yellowstone caldera

Chapter 9 Volcanoes 199

GRAPHIC ORGANIZERS
Graphs and charts throughout the text provide engaging visuals that help students organize and understand chapter content.

Real-world connections
RELEVANT CONCEPTS

A variety of hands-on labs and investigations, as well as features in the text, help students experience earth science as vital, engaging, and relevant.

SCIENCE & SOCIETY
Real-world features raise students' awareness of environmental concerns and solutions.

SCIENCE & Society

Predicting Eruptions, Saving Lives

Consider two of the deadliest eruptions of the past century:

- **1902** An eruption of Mount Pelée on Martinique kills nearly 30,000 people.
- **1985** Lahars triggered by an eruption of Colombia's Nevado del Ruiz volcano sweep through the town of Armero, killing over 20,000 people.

Can such tragedies be prevented? Is it possible to predict a volcanic eruption in time to save lives?

DEVASTATION The Nevado del Ruiz, in extensive damage a

When Mount Pinatubo, a volcano in the Philippines, rumbled to life in April 1991, scientists quickly set up a station to monitor its activity. They also collected samples of the gas escaping from the summit of the mountain. Toward the end of May, they detected an increase in sulfur dioxide emissions, which led them to conclude that magma was rising. In early June, they observed that the side of the volcano was swelling. Meanwhile, seismic activity was increasing. All evidence suggested that a large eruption was about to occur.

From June 9 through June 15, Mount Pinatubo erupted in a series of increasingly violent explosions. Fortunately, scientists and public officials had worked together to set up a plan for educating and alerting the population. Some 58,000 people had been evacuated from the areas predicted to be in the greatest danger. Eventually, as many as 200,000 people were forced to evacuate their homes. The evacuees were spared from the dangerous lahars the eruption produced. Thousands of lives had been saved

from wh
most de
20th cer

Exte

SCIENCE NOTEBOOK
Consider ways in which predictions made by scientists promote safety. Explain how scientific predictions might affect your life.

INVESTIGATIONS
classzone.com
How Fast Do Gases from Volcanic Eruptions Travel?
Use satellite images of Mount Pinatubo to calculate the speed of volcanic ash and gases moving through the atmosphere.
Keycode: ES0906

SCIENCE NOTEBOOK
Writing and research assignments help students apply earth science concepts to their lives.

INVESTIGATIONS
Internet icons appear throughout the text and direct teachers and students to *ClassZone*'s relevant online resources. Investigations, visualizations, and other activities extend and enhance the content of the text.

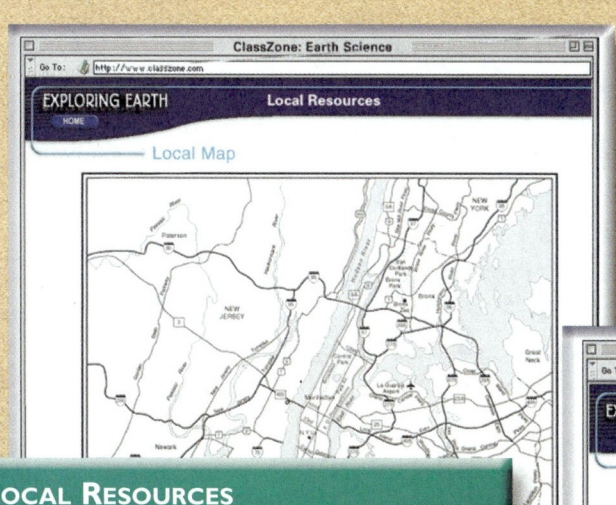

LOCAL RESOURCES
Create real-world connections through links to local data and issues.

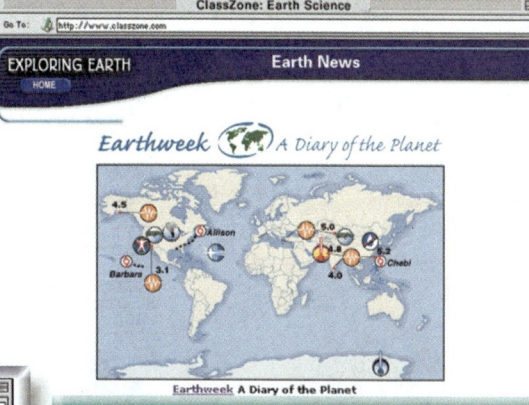

EARTH NEWS
Discover current events that illustrate earth science phenomena.

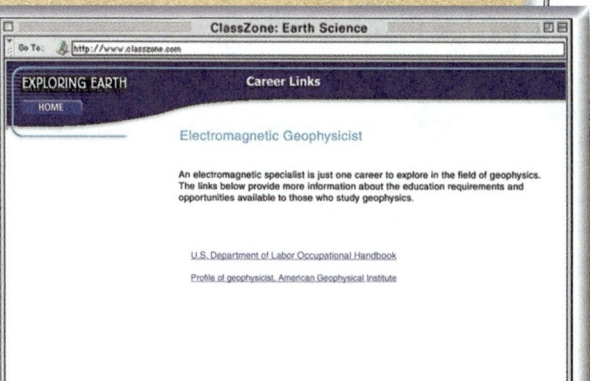

CAREER LINKS
Introduce your students to career opportunities in earth science.

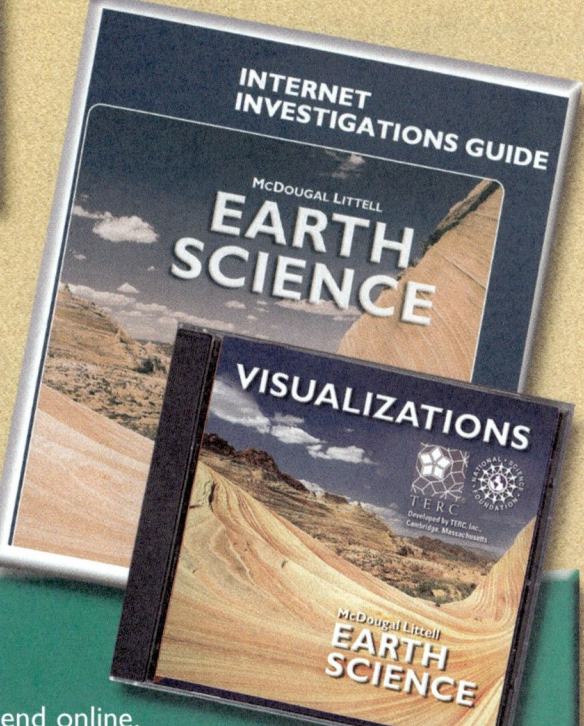

INTERNET INVESTIGATIONS GUIDE
Provide guidance to students as they learn to use Internet resources. These lesson plans help you prepare labs and evaluate the time students spend online.

VISUALIZATIONS CD-ROM
Extend and enhance instruction with a wealth of engaging visualizations. Created for classrooms without Internet access, this CD-ROM contains the animations, video clips, and images that are available online.

Program components
PRINT ANCILLARIES

From lab manuals to assessment booklets and transparencies, *Earth Science* provides a wide range of flexible teaching options to help you create an accessible and engaging classroom environment.

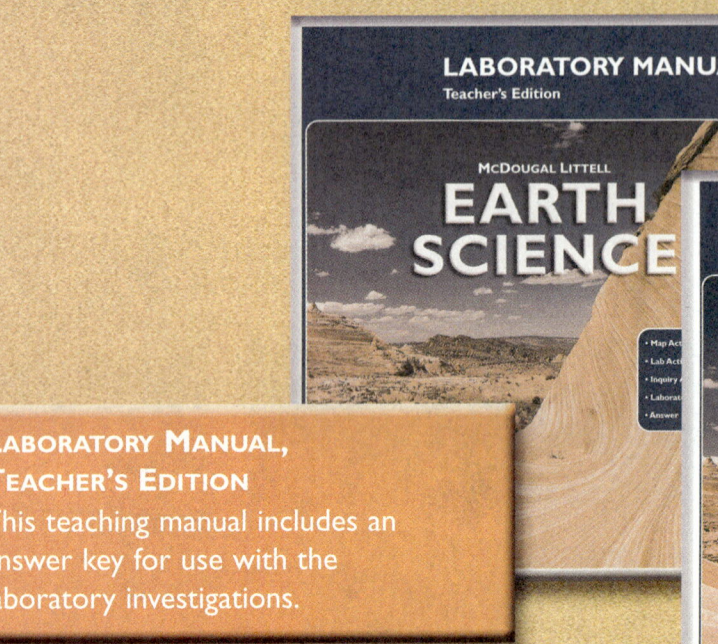

LABORATORY MANUAL, TEACHER'S EDITION
This teaching manual includes an answer key for use with the laboratory investigations.

GUIDE TO EARTH SCIENCE IN URBAN ENVIRONMENTS
This guide provides strategies and activities that help connect the concepts of earth science to the lives of urban students. TERC developed this guide with funding from NSF.

INTERNET INVESTIGATIONS GUIDE
This NSF-funded companion to the online Investigations guides students' use of Web resources and helps you evaluate the time students spend online.

READING STUDY GUIDE
Review questions, visuals, and easy-to-read chapter summaries help students prepare for tests and check comprehension.

Formal Assessment
This booklet contains tests for each unit and chapter, along with an answer key.

Spanish Vocabulary and Summaries
These copymasters provide Spanish translations of chapter summaries and key vocabulary terms.

Alternative Assessment
This booklet provides performance assessment tasks and rubrics for each chapter.

Lesson Plans
Lesson plans for each section of the text are customized for state-level standards.

Laboratory Manual, Pupil's Edition
Inquiry-based activities engage students in forming and testing hypotheses, designing experiments, and gathering and interpreting data.

Teaching Transparencies
These diagrams, infographics, and maps visually support the content of the text and include images from NASA and other government agencies.

Program components
INTEGRATED TECHNOLOGY RESOURCES

State-of-the-art technology resources reinforce and extend the content of the text and allow students to interact with the concepts of earth science. These resources save you valuable preparation time and provide teaching support to help you customize lessons.

CLASSZONE.COM
This online guide to *Earth Science* provides a variety of resources, including the Online Lesson Planner, chapter overviews, a student research tutorial, and access to Exploring Earth.

EXPLORING EARTH
Up-to-date, relevant resources connect the content of the textbook with real-world science. Engaging features include Investigations, Visualizations, Local Resources, Data Center, Earth News, and Career Links. The NSF funded TERC's development of this innovative Web site.

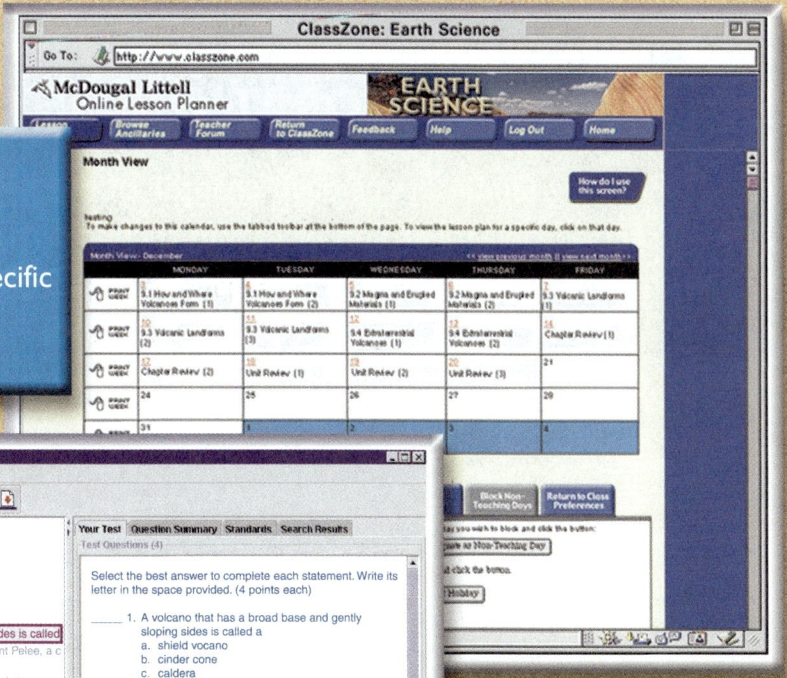

ONLINE LESSON PLANNER
This Internet planning tool provides all of the teacher's resources in one convenient, online format. Plan a specific lesson, or plan the whole year using preset or customized plans.

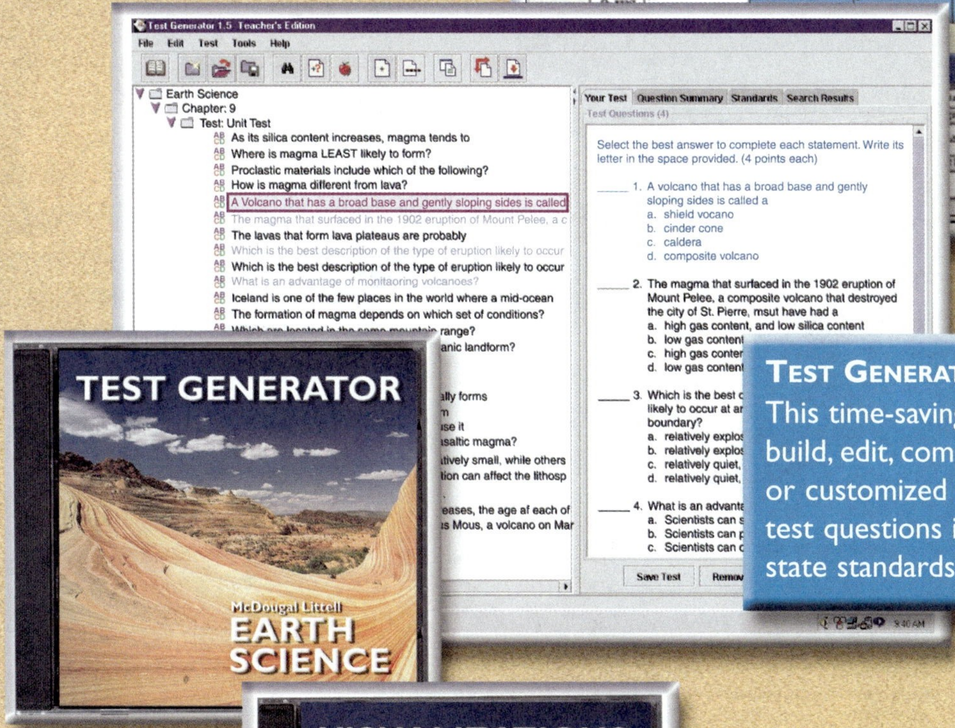

TEST GENERATOR CD-ROM
This time-saving tool helps you easily build, edit, compose, and print pre-made or customized tests. The database of test questions is correlated to multiple state standards.

VISUALIZATIONS CD-ROM
This CD-ROM contains over 90 of the animations, video clips, and images that are available through Exploring Earth. The NSF funded TERC's development of this CD-ROM.

ELECTRONIC TEACHER TOOLS CD-ROM
This CD-ROM includes all of the *Earth Science* teaching resources, organized by chapter in a convenient PDF format.

The new wrap-around Teacher's Edition
YOUR GUIDE TO THE EXPLORATION OF EARTH SCIENCE

A convenient, two-page Planning Guide opens each chapter of the *Earth Science Teacher's Edition*. This overview of section objectives and print and technology resources makes it easy for you to choose the materials that meet the diverse needs of your students.

PRINT RESOURCES
This organized view of chapter resources helps you choose from the available teaching options and take full advantage of the flexibility the program offers.

SECTION OBJECTIVES
This centralized outline of objectives helps you plan lessons and organize teaching materials ahead of time.

ASSESSMENT
Alternative and formal assessment options throughout the program support a variety of testing requirements.

CHAPTER 9 PLANNING GUIDE

Section Objectives		Print Resources
		Laboratory Manual, pp. 53–54
		Internet Investigations Guide, pp. 35–37
		Guide to Earth Science in Urban Environments, pp. 50–52
		Spanish Vocabulary and Summaries, pp. 21–22
		Formal Assessment pp. 25–27
		Alternative Assessment pp. 17–18
Section 9.1 How and Where Volcanoes Occur, pp. 194–198	1. A... re... in... 2. E... th...	
Section 9.2 Magma and Erupted Materials, pp. 199–201	1. I... and locate where each one forms. 2. Compare characteristics of magma and lava. 3. Describe pyroclastic materials.	Reading Study Guide, p. 29
Section 9.3 Volcanic Landforms, pp. 202–204	1. Compare and contrast landforms that result from volcanic activity. 2. Hypothesize how predicting eruptions and emergency preparedness can reduce loss of life and loss of property.	Lesson Plans, p. 38 Reading Study Guide, p. 30
Section 9.4 Extraterrestrial Volcanism, pp. 206–207	1. Classify volcanic activity on other planets and moons. 2. Analyze the tectonics behind extraterrestrial volcanic activity.	Lesson Plans, p. 39 Reading Study Guide, p. 31

INVESTIGATIONS
CLASSZONE.COM

Section 9.1 INVESTIGATION
How Are Volcanoes Related to Plate Tectonics?
Computer time: 90 minutes / Additional time: 90 minutes
Students use 3-D maps to plot areas of volcanic activity.

Section 9.3 INVESTIGATION
How Fast Do Gases from Volcanic Eruptions Travel?
Computer time: 30 minutes / Additional time: 30 minutes
Students use satellite images of Mount Pinatubo to calculate speed of ash and gases moving through atmosphere.

T14

TECHNOLOGY/ONLINE RESOURCES
These references allow you to preview integrated technology and online resources.

Classroom Management

Gather these materials for minilabs and laboratory activities.

Minilab, p. 200
   3 disposable cups
   water
   vegetable oil
   stop watch
   drinking straws (clear plastic)
   corn syrup or maple syrup
   eyedropper

Map Activity, pp.
   Map of plate b
   Colored pencil

CLASSROOM MANAGEMENT
An overview of lab materials helps you prepare for the mini-labs and laboratory activities featured in the chapter.

Online Resources

Electronic Teacher Tools
Visualizations CD-ROM
Classzone.com
Online Lesson Planner
Test Generator

Visualizations CD-ROM
Classzone.com
   ES0901 Investigation, ES0902 Visualization
   ES0903 Visualization, ES0904 Visualization

Visualizations CD-ROM
Classzone.com
   ES0905 Visualization

Classzone.com
   ES0906 Investigation

Chapter Review **INVESTIGATION**
Is It Safe to Live Near Volcanoes?
Computer time: 90 minutes / Additional time: 90 minutes
Students analyze data from Mount St. Helens and develop an evacuation plan for the area around Mt. Rainier.

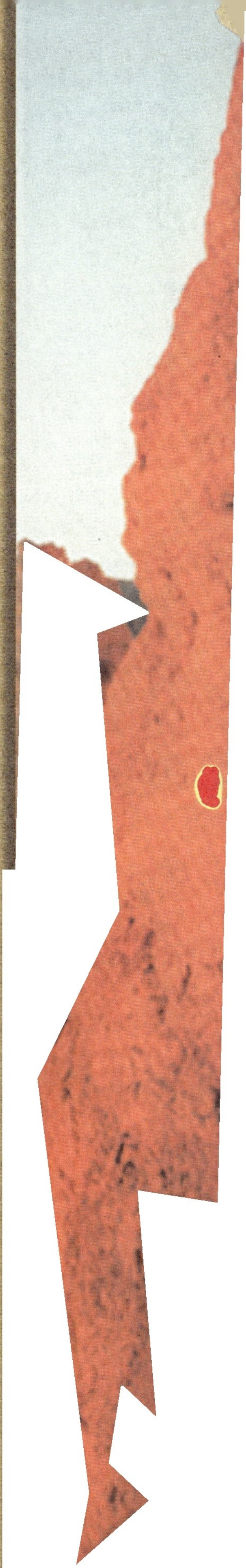

Chapter 9 Volcanoes   **192B**

The new wrap-around Teacher's Edition
TEACHING SUPPORT AT THE POINT OF USE

The *Earth Science Teacher's Edition* places a wealth of information and practical teaching suggestions at your fingertips. The wrap-around edition features strategies for teaching reading, differentiating instruction, and addressing national and state standards.

FOCUS
Clearly spelled-out objectives that incorporate the concepts and skills identified in the National Science Standards help focus student learning and define the purpose of the lesson.

INSTRUCT
Additional support allows you to customize lesson plans according to students' needs.

CHAPTER 9 SECTION 1

FOCUS

Objectives
1. Analyze how magma forms as a result of plate motion and interaction.
2. Explain why plate boundaries are the sites of most volcanic activity.

Set a Purpose
Have students read this section to find the answer to the question "How and where do volcanoes form?"

INSTRUCT

More about...

Magma Formation
As temperatures within the upper mantle increase, the rock begins to melt, forming small, isolated pockets of magma. These pockets rise through fractures or natural conduits. The magma may then collect in weakened patches of the crust, called magma chambers. These act as holding tanks and may be just a few miles beneath Earth's surface. With pressure rising in the magma chamber, magma may be forced up through weak spots in the lithosphere to the surface and a volcano forms.

9.1
KEY IDEA
Volcanoes form where magma reaches Earth's surface.

KEY VOCABULARY
- volcano
- hot spot

INVESTIGATIONS CLASSZONE.COM
How Are Volcanoes Related to Plate Tectonics? Plot volcano locations. Examine maps and 3D diagrams to explore volcanism at plate boundaries.
Keycode: ES0901

VOLCANO The diagram (inset) shows magma rising inside a volcano. Rising magma can lead to an explosive eruption like this one at Alaska's Mt. Augustine in 1986.

How and Where Volcanoes Form
One of the most dramatic activities associated with plate tectonics is the eruption of a volcano. The term **volcano** refers both to the opening in Earth's crust through which molten rock, gases, and ash erupt and to the landform that develops around this opening.

Magma Formation
A volcanic eruption occurs when magma—molten rock that has formed deep within Earth—rises to the surface. Most of the asthenosphere is solid because of the pressure exerted by the lithosphere above it. This pressure raises the melting temperatures of materials in the asthenosphere. Yet for magma to form, some of these materials must melt. The following three conditions allow magma to form:

- A decrease in pressure can lower the melting temperatures of materials in the asthenosphere. Such a decrease takes place along the rift valley at a mid-ocean ridge, where the lithosphere is thinner and exerts less pressure.
- An increase in temperature can cause materials in the asthenosphere to melt. Such an increase occurs at a hot spot.
- An increase in the amount of water in the asthenosphere can lower the melting temperatures of materials there. Such an increase occurs at subduction boundaries.

Conditions at both divergent and subduction boundaries are ideal for magma formation. Most volcanoes are found along mid-ocean ridges, where plates are moving apart, and at subduction boundaries, where plates are being forced under other plates.

Once magma forms, it tends to rise to the surface because its density is lower than that of the solid materials surrounding it. The characteristics of a magma and the rates at which it rises depend upon the amount of silica it contains.

DIFFERENTIATING INSTRUCTION

Hands-On Demonstration Why does magma rise? A substance that is less dense than the material around it rises to the top. Show students three materials: a cork, a glass marble, and vegetable oil. Have students predict what will happen when all three materials are placed in a container. Place the cork and the marble in the bottom of a clear plastic container. Fill the container with oil. As the cork rises, ask students which material is less dense. **the cork** Explain that magma is less dense, thus it rises to the surface.

T16

At Subduction Boundaries

As you may recall from Chapter 8, subduction boundaries are places where lithospheric plates collide. The diagrams below show the two processes by which volcanoes form: first, when an oceanic plate is forced beneath a continental plate, and second, when one oceanic plate is forced beneath another oceanic plate.

At a subduction boundary, volcanoes always form on the overriding plate. Where an oceanic plate collides with a continental plate, the volcanoes form on the overriding continental plate. The Cascade Range, which extends along the Pacific coast from northern California to British Columbia, Canada, is made up of volcanoes that have formed at an oceanic-continental subduction boundary.

Where two oceanic plates collide, the volcanoes create a chain of volcanic islands, called a volcanic island arc, on the overriding plate. The Mariana Islands in the Pacific Ocean are an example of a volcanic island arc formed at such a subduction boundary.

VISUALIZATIONS
CLASSZONE.COM
Observe an animation of volcanism at a subduction zone.
Keycode: ES0902

Volcanic Activity at a Subduction Boundary
BETWEEN AN OCEANIC PLATE AND A CONTINENTAL PLATE

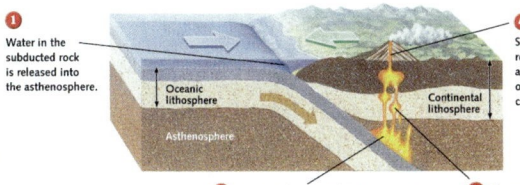

1. Water in the subducted rock is released into the asthenosphere.
2. The water lowers melting temperatures of materials in the asthenosphere, leading to the formation of magma.
3. The magma is less dense than its surroundings, so it rises.
4. Some of the magma reaches Earth's surface, and volcanoes form on the overriding continental plate.

BETWEEN OCEANIC PLATES
The process by which magma forms at an oceanic-oceanic subduction boundary is similar to the process at an oceanic-continental boundary. Notice that the difference between the two processes occurs at step 4.

4. Magma that reaches Earth's surface is underwater. Thus, an arc of volcanic islands forms underwater on the overriding oceanic plate.

READING AND LEARNING SCIENCE FROM TEXTBOOKS

Active, engaged readers make the best learners. Researchers have found that successful readers connect text information with what they already know. These readers build associations among ideas, create visual images of what they are reading, and continually refine their interpretations as they gather more information.

Encouraging active, engaged reading is difficult for teachers when they find that many students have learned to ignore their textbooks. This is often due to texts that are too difficult conceptually and are too dense in information. It is possible, however, for teachers and students to find texts that meet their needs.

Supporting Readers

Often we see students who don't have a foundation of information and experience on which to build. If there are not strong connections, experiences, or foundations, then the amount of support and elaboration provided by the text and teachers is critical. That support and elaboration can be provided in a variety of ways.

Personal Connections

For some students, personal stories and real-life connections can bring science to life. Taking a personal experience and relating it to a scientific concept can deepen students' learning. Studies of these techniques have generally shown significant increases in learning. You'll find a variety of hands-on labs and investigations in **Earth Science,** as well as *Science & Society* features to help make personal connections.

Learning Styles

We know that students have different styles of constructing meaning—some rely heavily on verbal input and discussion, some need visual support, and some need to make notes and drawings or create graphic organizers as they learn. For others, creating opportunities to challenge ideas and make comparisons helps them learn. Posing problems, comparing and contrasting ideas, and presenting different points of view on issues stimulates deeper engagement and leads to more learning than does a simple exposition of factual information.

Visual Information

Visual information is a resource that many readers rely on heavily when reading unfamiliar material. When text content is unfamiliar, the visual materials accompanying the text can provide strong support for the creation of meaning, especially in expository material. Pictures and illustrations create a visual context and help students more clearly understand scientific concepts. **Earth Science** has a wealth of visuals and graphic organizers throughout the text to help students visualize and organize information.

Strategies for Reading Science

There are specific strategies that you can encourage your students to use to learn science information and make sense of important scientific concepts. Some strategies are most effective when used before reading, while others are most effective during or after reading. At any stage of reading, having your students use a **Science Notebook** is valuable.

What is a Science Notebook?

In this textbook, you will find many references to the Science Notebook—consider it a tool for learning. You can instruct your students to turn any spiral notebook or binder with loose-leaf pages into their own Science Notebook.

- A place where they can record thoughts and observations
- A space for them to reflect what they are learning as they read

Sometimes you may want to assign activities for your students to keep in their notebook. Also, the text guides students to record items in their notebooks.

Your students can use their Science Notebook

- for keeping notes
- for organizing information
- for writing assignments
- for recording answers to section and chapter reviews in one place to help them prepare for tests

More importantly, the practice of keeping a Science Notebook enables your students to deepen their understanding of important scientific concepts. Thinking and writing about ideas helps your students understand what they are reading.

Before Your Students Read

Preview. Have your students skim through the section they are about to read. Have them read the headings to help them identify the main ideas. Have them look closely at diagrams, illustrations and photographs.

After you have your students preview the section, suggest that they do one or more of the following activities in their Science Notebook:

- List what they already know about the topic presented in the section.
- List any predictions or expectations they may have about what they might learn in the section.
- List what they want to find out about the topic.
- Formulate questions about the contents coming up in the topic.

Set a purpose. Each section begins with a statement that presents the Key Idea. Suggest to your students that they use the Key Idea to set a purpose for reading. For example, the Key Idea of a section might be "The composition of magma affects how explosive a volcano eruption will be." Already, from reading this statement, your students can figure out two things: (1) there must be a variety of magmas and (2) some volcanic eruptions are more explosive than others. Your students can then turn that Key Idea into a question that they can think about as they are reading the section.

While Your Students Read

Look for design features. Have your students pay attention to any text that is set into a box or treated differently.

Use text organizers. Chapters and section titles, headings and subheadings all provide a framework to help your students identify and organize main ideas.

Develop vocabulary. Have your students look up words that they do not know. Also, have them keep vocabulary words in their Science Notebooks.

Connect. As your students read a section, have them make connections between what they are reading and what happens in their day-to-day lives.

Record thoughts. Have your students use their Science Notebooks to record thoughts and track information they find interesting or important

After Your Students Have Read

Review. Have your students look over the section they've read.

Summarize. Using their Science Notebook, ask your students to answer the question they posed in *Setting a Purpose*.

Reflect. Ask your students to reflect on what they learned in the section they read.

Clarify. Ask your students to clarify new information by sharing their new knowledge with someone else or in a small group setting.

Pacing Guide

The following pacing guide shows how the chapters in *Earth Science* can be adapted to fit your specific course needs. The guide is based on 170 days of instruction for a traditional schedule or 85 days of instruction for a block schedule.

Unit and Chapter	Traditional Schedule (Days)	Block Schedule (Days)
Unit 1 Investigating Earth		
Chapter 1 Earth as a System		
1.1 A New View of Earth	1	.5
1.2 The Earth's System's Four Spheres	1	.5
1.3 Cycles Within the System	2	1
Laboratory Activity	1	.5
Chapter 2 The Nature of Science		
2.1 The Scientist's Mind	1	.5
2.2 Scientific Methods of Inquiry	1	.5
2.3 Scientist's Tools	1	.5
Laboratory Activity	1	.5
Chapter 3 Models of Earth		
3.1 Modeling Planet Earth	1	.5
3.2 Mapmaking and Technology	1	.5
3.3 Topographic Maps	2	1
Map Activity	1	.5
Total Days for Unit	**14**	**7**
Unit 2 Earth's Matter		
Chapter 4 Earth's Structure and Motion		
4.1 Earth's Formation	2	1
4.2 Earth's Rotation	2	1
4.3 Earth's Revolution	1	.5
Laboratory Activity	1	.5
Chapter 5 Atoms to Minerals		
5.1 Matter and Atoms	1	.5
5.2 Composition and Structure of Minerals	2	1
5.3 Identifying Minerals	1	.5
5.4 Mineral Groups	1	.5
Laboratory Activity	1	.5
Chapter 6 Rocks		
6.1 How Rocks Form	1	.5
6.2 Igneous Rocks	2	1
6.3 Sedimentary Rocks	2	1
6.4 Metamorphic Rocks	1	.5
Laboratory Activity	1	.5
Chapter 7 Resources and the Environment		
7.1 Mineral Resources	1	.5
7.2 Energy Resources	2	1
7.3 Environmental Issues	1	.5
Laboratory Activity	1	.5
Total Days for Unit	**24**	**12**

Unit and Chapter	Traditional Schedule (Days)	Block Schedule (Days)
Unit 3 Dynamic Earth		
Chapter 8 Plate Tectonics		
8.1 What Is Plate Tectonics?	1	.5
8.2 Types of Plate Boundaries	2	1
8.3 Causes of Plate Movements	1	.5
8.4 Plate Movements and Continental Growth	1	.5
Map Activity	1	.5
Chapter 9 Volcanoes		
9.1 How and Where Volcanoes Occur	2	1
9.2 Magma and Erupted Materials	1	.5
9.3 Volcanic Landforms	1	.5
9.4 Extraterrestrial Volcanoes	1	.5
Map Activity	1	.5
Chapter 10 Earthquakes		
10.1 How and Where Earthquakes Occur	1	.5
10.2 Locating and Measuring Earthquakes	2	1
10.3 Earthquake Hazards	1	.5
10.4 Studying Earth's Interior	1	.5
Laboratory Activity	1	.5
Chapter 11 Mountain Building		
11.1 Where Mountains Form	1	.5
11.2 How Mountains Form	2	1
11.3 Types of Mountains	1	.5
Map Activity	1	.5
Total Days for Unit	**23**	**11.5**
Unit 4 Earth's Changing Surface		
Chapter 12 Weathering, Soil, and Erosion		
12.1 Weathering	2	1
12.2 Soil	1	.5
12.3 Mass Movements and Erosion	1	.5
12.4 Soil as a Resource	1	.5
Laboratory	1	.5
Chapter 13 Surface Water		
13.1 Streams and Rivers	1	.5
13.2 Stream Erosion and Deposition	2	1
13.3 River Valleys	1	.5
13.4 Flood Plains and Floods	1	.5
Map Activity	1	.5
Chapter 14 Groundwater		
14.1 Water in the Ground	2	1
14.2 Conserving Groundwater	1	.5
14.3 Groundwater and Geology	1	.5
Laboratory Activity	1	.5
Chapter 15 Glaciers		
15.1 What is a Glacier?	1	.5
15.2 Glacial Movement and Erosion	2	1
15.3 Glacial Deposits	2	1
15.4 Ice Ages	1	.5
Laboratory Activity	1	.5

Unit and Chapter	Traditional Schedule (Days)	Block Schedule (Days)
Chapter 16 Wind, Waves, and Currents		
16.1 Wind as an Agent of Change	1	.5
16.2 Waves in the Sea	2	1
16.3 Shoreline Features	1	.5
Laboratory Activity	1	1
Total Days for Unit	**29**	**15**
Unit 5 Weather and Atmosphere		
Chapter 17 Atmosphere		
17.1 The Atmosphere in Balance	1	.5
17.2 Heat and the Atmosphere	2	1
17.3 Local Temperature Variations	1	.5
17.4. Human Impact on the Atmosphere	1	.5
Laboratory Activity	1	.5
Chapter 18 Water in the Atmosphere		
18.1 Humidity and Condensation	2	1
18.2 Clouds	2	1
18.3 Precipitation	1	.5
Laboratory Activity	1	.5
Chapter 19 The Atmosphere in Motion		
19.1 Air Pressure and Wind	2	1
19.2 Factors Affecting Wind	1	.5
19.3 Global Wind Patterns	1	.5
19.4 Continental and Local Winds	1	.5
Laboratory Activity	1	.5
Chapter 20 Weather		
20.1 Air Masses and Weather	1	.5
20.2 Fronts and Lows	2	1
20.3 Thunderstorms and Tornadoes	1	.5
20.4 Hurricanes and Winter Storms	1	.5
20.5 Forecasting Weather	1	.5
Map Activity	1	.5
Chapter 21 Climate and Climate Change		
21.1 What is Climate?	1	.5
21.2 Climate Zones	2	1
21.3 Climate Change	1	.5
Laboratory Activity	1	.5
Total Days for Unit	**30**	**15**
Unit 6 Earth's Oceans		
Chapter 22 The Water Planet		
22.1 Oceanography	1	.5
22.2 Properties of Water	1	.5
22.3 Properties of Ocean Water	1	.5
22.4 Ocean Life	2	1
Laboratory	1	.5

Unit and Chapter	Traditional Schedule (Days)	Block Schedule (Days)
Chapter 23 The Ocean Floor		
23.1 Studying the Ocean Floor	1	.5
23.2 The Continental Margin	1	.5
23.3 The Ocean Basin	2	1
23.4 Ocean Floor Sediments	1	.5
Laboratory	1	.5
Chapter 24 The Moving Ocean		
24.1 Surface Currents	1	.5
24.2 Currents Under the Surface	2	.5
24.3 Tides	1	.5
Laboratory	1	.5
Total Days for Unit	**17**	**8**
Unit 7 Space		
Chapter 25 Earth's Moon		
25.1 Origin and Properties of the Moon	1	.5
25.2 The Moon's Motions and Phases	2	1
Laboratory Activity	1	.5
Chapter 26 The Sun and the Solar System		
26.1 The Sun's Size, Heat, and Structure	2	1
26.2 Observing the Solar System: A History	1	.5
Laboratory Activity	1	.5
Chapter 27 The Planets and the Solar System		
27.1 The Inner Planets	2	1
27.2 The Outer Planets	2	1
27.3 Planetary Satellites	1	.5
27.4 Solar-System Debris	1	.5
Laboratory Activity	1	.5
Chapter 28 Stars and Galaxies		
28.1 A Closer Look at Light	2	1
28.2 Stars and Their Characteristics	2	1
28.3 Life Cycle of Stars	1	.5
28.4 Galaxies and the Universe	1	1
Laboratory Activity	1	.5
Total Days for Unit	**22**	**11.5**
Unit 8 Earth's History		
Chapter 29 Studying the Past		
29.1 Fossils	1	.5
29.2 Relative Time	2	1
29.3 Absolute Time	1	.5
Laboratory Activity	1	.5
Chapter 30 View's of Earth's Past		
30.1 The Geological Time Scale	2	.5
30.2 The Precambrian	1	.5
30.3 The Mesozoic	1	.5
30.4 Earth's Recent History	1	.5
Laboratory Activity	1	.5
Total Days for Unit	**11**	**5.0**

Master Materials List

Quantities are based on a class size of 24 students and the following assumptions:
- All mini-labs are done individually.
- The activities for Chapters 3, 4, 6, 9, 11, 13, 14, 20, 23, 27, 28, and 29 are done individually.
- The activities for Chapters 1, 2, 5, 7, 10, 12, 15, 16, and 25 are done in groups of 2.
- The activities for Chapters 17, 18, 21, 22, 24, 26, and 30 are done in groups of 4.

Equipment: materials ordered through an equipment catalog and that do not need to be replaced each year.

Supplies: materials that are available locally or in the school, and/or materials that need to be replaced each year.

Samples: materials unique to earth science such as rock and mineral specimens, fossils, and sediments.

Equipment	Lab	Quantity
anemometer	19	6
aquarium	24	6
balance, triple-beam	1, 2, 5, 18, 25	12
bar magnet	mini-4	24
barometer	19	6
beaker		
50-mL	1	12
100-mL	16, 24	24
250-mL	5, 12, 17, 18, mini-14	30
500-mL	22, mini-5	24
large, clear	mini-7	24
block		
cubic	1	12
cylindrical	1	12
rectangular	1	12
wooden, 5 × 10 cm	16	12
bowl, 2-qt. mixing	15	12
buzzer	mini-28	8
C-clamp	5	12
calculator	2, 5, 14, 26, mini-25	24
compass, safety	4	24
container		
clear	mini-12	48
clear, rectangular	16	12
clear, tall	mini-13	24
film (plastic)	2	156
dropper, medicine	18, mini-9	24
funnel	mini-7	24
graduated cylinder		
50-mL	18	6
100-mL	1, mini-14	24
hole punch	7	12
jar, 1-qt. glass	21	18
lab apron	12, 18, 21	24
lamp, desk or clip-on	17	6
large pebble	15	72

Equipment	Lab	Quantity
meter stick	25, 26	12
oven mitts	21	6 pairs
pan		
clear rectangular	mini-16	24
not glass, at least 25 × 30 × 7.5 cm	25	12
shallow	mini-17	24
probe		
pH	18	6
temperature	24	6
rain gauge	19	6
ring clamp	17	18
ring stand	17	6
rinse bottle	22	6
round dish	mini-24	24
safety goggles	12, 18, 25	24 pairs
salinometer	22	6
scrubbing brush, small	18	6
shovel, small	30	6
sieve or sifter	25	12
silk cloth or fake fur	mini-22	24 pieces
spring scale	mini-5	24
stopwatch	12, 16, 17, mini-9	24
thermometer		
Celsius	7, 12, 17, 19, 21, mini-18, mini-21	30
long	24	30
short	mini-17	24
wet-bulb	19	6
tongs	18, 21	6 pairs
trowel, small hand	30	6
tubing, plastic	mini-7	24 lengths
wind vane or flag	19	6

Supplies	Lab	Quantity
alfalfa sprouts	18	6 plugs
aluminum foil	7, 23	2 rolls
bag, clear plastic	mini-1	24
bag of colored beads	mini-30	24
balloon	28, mini-22	24
box		
cardboard	10	24
cardboard, at least 60 × 60 × 30 cm	30	6
plastic	15	12
brick	7	12
card stock	mini-4	24
cardboard	24	6 pieces
colored pencils	2, 3, 7, 9, 13, 20, 21, 23, 24, 25, 27, 29, mini-24, 27	12 boxes
concrete		
piece	18	24
coarsely ground	17	750 mL
cornstarch	15	12 boxes
corn syrup	mini-9	3 bottles
cotton gauze	mini-18	24 wads
cup		
paper	7, mini-9, mini-19	144
Styrofoam, medium	24	24
Styrofoam, small	mini-6, mini-17	24
dry ice	21	6 pieces
dye, 4 colors	24	6 each
effervescent antacid tablet	12	60
flour	25	12 bags
ice cube	15	12
index card, 5 × 7 in.	15	24
iron filings	mini-4	6 oz.
latex gloves	18	24 pairs
marbles	10	240
marker		
fine-tip	16, 21, 23, 24, 26	24
thick, 2 colors	mini-8	48
metal washer	mini-2	24
metric ruler, flexible	1, 2, 3, 4, 6, 10, 18, 22, 23, 24, 25, 26, 28, 29, mini-20	24
mini-marshmallows	10	480
modeling clay	22, 23, mini-6, mini-7, mini-11	12 bricks
paint		
dry tempera	25	1 jar
dull black	7, 17	1 can
dull white	17	1 can
shiny silver	17	1 can
paintbrush	30	6
paper, tracing	11, 13, 20, 29	120
paper circle, small	mini-24	400
paper clip	5, 13, 20, 28	36
paper fan	mini-18	24
pH indicator strips	18	48
plaster of Paris	mini-6	10 lbs.

Supplies	Lab	Quantity
plastic wrap	7	2 rolls
potted plant	mini-1	24
protractor	4, mini-2	24
push pin	mini-19	24
puzzle pieces from 2 or more sets	30	72
rubber band	5, 10, mini-18	48
rubber cement	7	12 jars
salt	22, 24	6 boxes
scissors	7, 10, 22, 24, mini-8, mini-12	24 pairs
shoebox	7, 23	24
shoebox lid	mini-8	12
soup can	17	18
spherical objects of varying size and weight	25, 26	60
spoon	15, mini-13	24
stapler	10, mini-19	6
steel nail	18	48
steel wool pad	18, mini-12	48
stiff cardboard strip	mini-19	48
stirrer, coffee	10, 23, mini-6	480
straw, drinking	22, mini-2, mini-7, mini-9, mini-24	48
string	3, 5, 28, 30, mini-2, mini-5, mini-26	6 balls
sulfuric acid, 0.1 M	18	1350 mL
tape		
clear	24, mini-2, mini-20	2 rolls
masking	7, 16, 18, 21, 23, 24, 30, mini-1, mini-4, mini-28	6 rolls
tennis ball	mini-26, mini-28	24
thumbtack	30	40
toothpick, wooden	15	60
towel	15	12
vegetable oil	mini-9	3 bottles
vegetable oil spray	22	6 cans
wire	mini-11	1 roll

Samples	Lab	Quantity
calcite	5	6
chalcopyrite	5	6
chips, wet limestone	1	240
clay, dry sediment	mini-13	1 pail
galena	5	6
gravel	16, mini-7, mini-13, mini-14	1 pail
gypsum	5	6
magnetite	5	6
mineral sample	mini-5	24
pyrite	5	6
sand	16, 30, mini-7, mini-13	6 pails
soil	17, 18, 30	6 pails
sulfur	5	6

SAFETY DURING ACTIVITIES

The activities in McDougal Littell Earth Science are designed to actively involve students in the processes of science. Each activity has been developed with safety as a major concern. The activities use readily accessible materials and can be easily performed by high school earth science students.

Laboratory safety can only be achieved by constant vigilance on the part of both teachers and students. Be alert to potential danger. Some activities use chemicals, sharp instruments, glassware, electric circuits, and fire as a source of heat. By your continuous insistence on the use of precautions, you can prevent mishaps.

General Safety Procedures

Safety Rules General safety rules for you and your students are summarized on pages xxii–xxiii. The same safety information can be found in the *Laboratory Manual* for McDougal Littell *Earth Science*.

Safety symbols are used to flag specific caution statements within an activity. Each statement provides a brief explanation of the potential hazard. All safety symbols are listed on page xxiii. Also, be sure to read the additional safety information provided in the margins of the *Teacher's Edition*.

Facilities It is your responsibility to provide a safe environment for laboratory work. The laboratory should be equipped with proper ventilation. Check your work area on a regular basis. Equipment and materials should be properly stored. Hazardous materials should never be left exposed. Never keep food or drink in the classroom or laboratory, or in a refrigerator that is used for chemicals or biological samples. Your laboratory, classroom, and storage facilities should be locked at all times when not under your direct supervision.

Safety Equipment Check smoke detectors, safety showers, and eyewash facilities before each activity. Know the location and operation of all master shut-offs for any utilities used.

Activity Preparation Before handling equipment and materials, thoroughly familiarize yourself with the necessary precautions. Read and follow the instructions in equipment manuals and material safety data sheets, and on labels for chemicals. You should always perform each activity before assigning it to your students. Always wear required safety equipment when performing demonstrations for your students and when showing them the proper use of equipment and materials.

Time Allotment Limit the size of the group working on an activity to a number that you can safely supervise without causing confusion and accidents. Be sure to allow sufficient time for students to perform each activity and to clean up afterward. Students hurrying to finish their work often become careless and inattentive to safety procedures.

Waste Disposal Dispose of dangerous waste chemicals and materials as prescribed by appropriate standards and the laws of your community. Provide separate labeled waste receptacles for certain chemicals as called out in the margins of this *Teacher's Edition;* also provide labeled receptacles for broken glass, waste paper, and used matches.

Accident Policies In case of accidents, know your school's policy and procedures. Make accident reports promptly, accurately, and completely.

Using Chemicals Safely

Review the safety rules listed on pages xxii–xxiii regarding chemical use. Both you and your students should wear plastic gloves when working with potentially dangerous chemicals. Instruct students never to smell a chemical directly. Instead tell students to waft the vapor toward the nose by fanning with their hands. When diluting acid with water, always add acid to water. Once a chemical has been removed from its stock bottle, discard the chemical properly. Do not return any unused chemical to its stock bottle for future use. Equip chemical storage shelves with ledges to prevent the slipping or sliding of stock bottles.

Using Streak Plates Safely

Examine streak plates for chips or cracks before dispensing to your students. Caution students to always place streak plates on the table during use, because a streak plate can break in the hand and cause serious cuts.

Using Glassware Safely

Broken Glass Examine all glassware for chips or cracks before dispensing to your students. Avoid glass breakage by directing your students to keep flasks, bottles, beakers, and other equipment away from the edges of tables or lab benches. If glass breakage does occur, use a dustpan and brush, not your fingers, to pick up broken glass.

Storage Make sure all glassware is clean before returning it to its storage location. Shelves in your classroom used for glassware storage should also be equipped with ledges.

Using Heat and Fire Safely

Be especially cautious when dealing with fire hazards. Review the cautions listed on pages xxii–xxiii regarding heat and fire safety. Familiarize students with the school's fire regulations and evacuation procedures, as well as with the location and use of first-aid supplies and equipment such as fire extinguishers and fire blankets.

Demonstrate the correct procedure for lighting a Bunsen burner. When lighting a Bunsen burner, students should wear safety goggles and a lab apron; long hair should be confined; and long, floppy sleeves should be rolled up. Instruct students to immediately turn off the gas if the burner does not light or if the flame goes out. Instruct your students never to reach over an open flame or other heat source. Remind students to keep books and papers as far away from a flame as possible.

Give students only Pyrex glassware to use when heating substances. When students are heating test tubes, always supervise closely to ensure that students point the test tube opening away from themselves and their classmates. Before using any flammable chemical, be sure there are no ignition sources in the room.

Using Electrical Equipment Safely

All devices using 110–115 volt AC power should be equipped with three-wire cords and three-prong plugs. Each socket should be three-holed, be protected with a ground fault interrupter (GFI), and be checked for correct polarity (polarity testing devices display a coded sequence of red and green lights). Instruct students to switch off equipment before plugging it into the socket and to turn equipment off and unplug it after an activity is completed. Use 1.5–volt or 9–volt "dry-cell" batteries as direct current sources; do not use 6– or 12–volt automobile storage batteries. Wiring hookups should not be made, or altered, except when both the power switch is off and the apparatus has been disconnected from its AC or DC power supply. Cover battery terminals with insulating tape before storing the batteries.

Student Safety Contract

The following student safety contract was adapted from one developed by the National Science Teachers Association. Before being allowed to work on activities, students should sign the safety contract. However, students must be thoroughly familiar with the rules for proper laboratory behavior and safety precautions before signing.

I will
- follow all instructions given by the teacher
- protect eyes, face, hands, and body while conducting class and laboratory activities
- carry out good housekeeping practices
- know the location and proper use of first-aid and fire-fighting equipment
- conduct myself in a responsible manner at all times.

I,_____ have read and agree to abide by the safety regulations set forth above and also any additional printed instructions provided by the teacher and/or the district. I further agree to follow all other written and oral instructions given to me by the teacher in charge.

SIGNATURE:
DATE:

Integrate the Web into your curriculum

ClassZone

Visit **classzone.com**, McDougal Littell's textbook-companion Web site, and experience state-of-the-art teaching support and engaging, interactive content!

ClassZone is your online guide to McDougal Littell's *Earth Science*.

- **Exploring Earth** provides access to a variety of resources, correlated to the chapters of the textbook. The NSF funded TERC's creation of these materials.
  - ◆ Investigations
  - ◆ Visualizations
  - ◆ Local Resources
  - ◆ Data Center
  - ◆ Earth News
  - ◆ Career Links
- **Online Lesson Planner** offers lesson planning support and teacher's resource materials in one convenient location.
- **Research Zone** includes a tutorial that helps students learn to conduct online research.

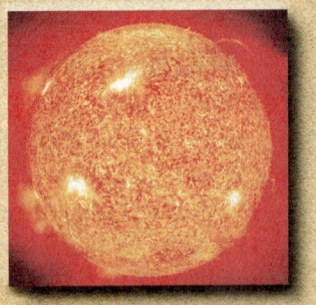

Teacher Access Code
MCDQ8RPUH5XH3

Use this code to create your own user name and password. You will be able to access all student and teacher-only resources.

EARTH SCIENCE

Nancy E. Spaulding · Samuel N. Namowitz

McDougal Littell
A HOUGHTON MIFFLIN COMPANY
Evanston, Illinois · Boston · Dallas

Copyright ©2003 by McDougal Littell Inc. All rights reserved.

No part of this work may be reproduced or transmitted in any form or by any means, electronic or mechanical, including photocopy and recording, or by any information storage or retrieval system without the prior written permission of McDougal Littell Inc. unless such copying is expressly permitted by federal copyright law. Address inquiries to Manager, Rights and Permissions, McDougal Littell Inc., P.O. Box 1667, Evanston, IL 60204.

ISBN: 0-618-11550-1   1 2 3 4 5 6 7 8 9 DWO 05 04 03 02 01

Internet Web Site: http://www.mcdougallittell.com

AUTHORS and PROGRAM CONSULTANTS

Authors

Nancy E. Spaulding
Retired Earth Science Teacher
Elmira Free Academy
Elmira, NY

Samuel N. Namowitz
Former Principal and Earth Science Teacher
Charles Evans Hughes High School
New York, NY

Content Consultants

Eric D. Carlson, Ph.D.
Astronomer Emeritus
Adler Planetarium and Astronomical Museum
Chicago, IL

Duncan M. FitzGerald, Ph.D.
Sedimentologist
Department of Earth Sciences
Boston University
Boston, MA

Nancy W. Hinman, Ph.D.
Associate Professor
Department of Geology
The University of Montana
Missoula, MT

Richard D. Norris, Ph.D.
Paleobiologist
Woods Hole Oceanographic Institution
Woods Hole, MA

Patricia Pauley, Ph.D.
Meteorologist
Naval Research Laboratory
Monterey, CA

Isabelle Sacramento Grilo
Lecturer
Department of Geological Sciences
San Diego State University
San Diego, CA

James Schmitt
Associate Professor of Geology
Department of Earth Sciences
Montana State University
Bozeman, MT

Brian J. Skinner, Ph.D.
Eugene Higgins Professor of Geology and Geophysics
Yale University
New Haven, CT

Teacher Reviewers

Timothy Goldpenny, M.S. Ed.
Earth Science Teacher
Barker Central School
Barker, NY

Margaret Anne Holzer, M.A.T.
Earth Science Teacher
Chatham High School
Chatham, NJ

Kenneth F. Nossavage, M.S. Ed.
Earth Science Teacher
Niagara Falls High School
Niagara Falls, NY

Bernadette Tomaselli, M.S.
Earth Science Teacher
Lancaster High School
Lancaster, NY

John Wichowsky
Earth Science Teacher
Newfane High School
Newfane, NY

CONTENTS OVERVIEW

UNIT 1 Investigating Earth — 1
- CHAPTER 1 Earth as a System — 2
- CHAPTER 2 The Nature of Science — 24
- CHAPTER 3 Models of Earth — 42

UNIT 2 Earth's Matter — 66
- CHAPTER 4 Earth's Structure and Motion — 68
- CHAPTER 5 Atoms to Minerals — 88
- CHAPTER 6 Rocks — 116
- CHAPTER 7 Resources and the Environment — 142

UNIT 3 Dynamic Earth — 168
- CHAPTER 8 Plate Tectonics — 170
- CHAPTER 9 Volcanoes — 192
- CHAPTER 10 Earthquakes — 212
- CHAPTER 11 Mountain Building — 234

UNIT 4 Earth's Changing Surface — 254
- CHAPTER 12 Weathering, Soil, and Erosion — 256
- CHAPTER 13 Surface Water — 278
- CHAPTER 14 Groundwater — 298
- CHAPTER 15 Glaciers — 316
- CHAPTER 16 Wind, Waves, and Currents — 338

UNIT 5 Atmosphere and Weather — 362

CHAPTER 17	ATMOSPHERE	364
CHAPTER 18	WATER IN THE ATMOSPHERE	388
CHAPTER 19	THE ATMOSPHERE IN MOTION	412
CHAPTER 20	WEATHER	434
CHAPTER 21	CLIMATE AND CLIMATE CHANGE	464

UNIT 6 Earth's Oceans — 486

CHAPTER 22	THE WATER PLANET	488
CHAPTER 23	THE OCEAN FLOOR	508
CHAPTER 24	THE MOVING OCEAN	530

UNIT 7 Space — 552

CHAPTER 25	EARTH'S MOON	554
CHAPTER 26	THE SUN AND THE SOLAR SYSTEM	570
CHAPTER 27	THE PLANETS AND THE SOLAR SYSTEM	586
CHAPTER 28	STARS AND GALAXIES	610

UNIT 8 Earth's History — 644

| CHAPTER 29 | STUDYING THE PAST | 646 |
| CHAPTER 30 | VIEWS OF EARTH'S PAST | 664 |

TABLE OF CONTENTS

UNIT 1 Investigating Earth — 1

CHAPTER 1 EARTH AS A SYSTEM — 2
- 1.1 A New View of Earth — 4
- 1.2 The Earth System's Four Spheres — 8
- 1.3 Cycles and the Earth — 13
- Laboratory Activity — 20
- Chapter Review — 22

Rocks in geosphere p. 9

CHAPTER 2 THE NATURE OF SCIENCE — 24
- 2.1 The Scientist's Mind — 26
- 2.2 Scientific Methods of Inquiry — 29
- 2.3 Scientists' Tools — 35
- Laboratory Activity — 38
- Chapter Review — 40

Studying rainforests p. 24

CHAPTER 3 MODELS OF EARTH — 42
- 3.1 Modeling the Planet — 44
- 3.2 Mapmaking and Technology — 48
- 3.3 Topographic Maps — 53
- Map Activity — 58
- Chapter Review — 60

Satellite mapping p. 49

Unit Feature The Earth System and the Environment: An Old Idea Becomes New Again — 62

Unit Investigation

Standardized Test Practice — 64

UNIT 2 Earth's Matter — 66

CHAPTER 4 EARTH'S STRUCTURE AND MOTION — 68
- 4.1 Earth's Formation — 70
- 4.2 Earth's Rotation — 75
- 4.3 Earth's Revolution — 80
- *Laboratory Activity* — 84
- *Chapter Review* — 86

Earth seen by satellite p. 68

CHAPTER 5 ATOMS TO MINERALS — 88
- 5.1 Matter and Atoms — 90
- 5.2 Composition and Structure of Minerals — 96
- 5.3 Identifying Minerals — 104
- 5.4 Mineral Groups — 108
- *Laboratory Activity* — 112
- *Chapter Review* — 114

Microscopic rock structure p. 88

CHAPTER 6 ROCKS — 116
- 6.1 How Rocks Form — 118
- 6.2 Igneous Rocks — 121
- 6.3 Sedimentary Rocks — 127
- 6.4 Metamorphic Rocks — 133
- *Laboratory Activity* — 138
- *Chapter Review* — 140

Limestone cliffs of Dover p. 129

CHAPTER 7 RESOURCES AND THE ENVIRONMENT — 142
- 7.1 Mineral Resources — 144
- 7.2 Energy Resources — 148
- 7.3 Environmental Issues — 154
- *Laboratory Activity* — 160
- *Chapter Review* — 162

Unit Feature The Earth System and the Environment: Recycling Cents and Sensibility — 164

Unit Investigation

Standardized Test Practice — 166

Harnessing solar energy p. 157

UNIT 3 — Dynamic Earth — 168

CHAPTER 8 PLATE TECTONICS — 170
- 8.1 What Is Plate Tectonics? — 172
- 8.2 Types of Plate Boundaries — 176
- 8.3 Causes of Plate Movement — 180
- 8.4 Plate Movements and Continental Growth — 182
- Map Activity — 188
- Chapter Review — 190

Mid-Atlantic Ridge p. 175

CHAPTER 9 VOLCANOES — 192
- 9.1 How and Where Volcanoes Occur — 194
- 9.2 Magma and Erupted Materials — 199
- 9.3 Volcanic Landforms — 202
- 9.4 Extraterrestrial Volcanoes — 206
- Map Activity — 208
- Chapter Review — 210

Volcano exploration p. 192

CHAPTER 10 EARTHQUAKES — 212
- 10.1 How and Where Earthquakes Occur — 214
- 10.2 Locating and Measuring Earthquakes — 217
- 10.3 Earthquake Hazards — 222
- 10.4 Studying Earth's Interior — 228
- Laboratory Activity — 230
- Chapter Review — 232

Earthquake aftermath p. 212

CHAPTER 11 MOUNTAIN BUILDING — 234
- 11.1 Where Mountains Form — 236
- 11.2 How Mountains Form — 238
- 11.3 Types of Mountains — 243
- Map Activity — 246
- Chapter Review — 248

Measuring Mount Everest p. 242

Unit Feature The Earth System and the Environment: When Volcanoes Breathe Life — 250

Unit Investigation

Standardized Test Practice — 252

UNIT 4 Earth's Changing Surface 254

CHAPTER 12 WEATHERING, SOIL, AND EROSION 256
    12.1 Weathering 258
    12.2 Soil 264
    12.3 Mass Movements and Erosion 268
    12.4 Soil as a Resource 271
    Laboratory Activity 274
    Chapter Review 276

Weathered rock p. 256

CHAPTER 13 SURFACE WATER 278
    13.1 Streams and Rivers 280
    13.2 Stream Erosion and Deposition 283
    13.3 River Valleys 287
    13.4 Flood Plains and Floods 290
    Map Activity 294
    Chapter Review 296

Studying marsh water p. 282

CHAPTER 14 GROUNDWATER 298
    14.1 Water in the Ground 300
    14.2 Conserving Groundwater 306
    14.3 Groundwater and Geology 309
    Laboratory Activity 312
    Chapter Review 314

Karst topography p. 298

CHAPTER 15 GLACIERS 316
    15.1 What Is a Glacier? 318
    15.2 Glacial Movement and Erosion 321
    15.3 Glacial Deposits 326
    15.4 Ice Ages 330
    Laboratory Activity 334
    Chapter Review 336

Icebergs off Antarctica p. 316

CHAPTER 16 WIND, WAVES, AND CURRENTS 338
    16.1 Wind as an Agent of Change 340
    16.2 Waves in the Sea 344
    16.3 Shoreline Features 349
    Laboratory Activity 354
    Chapter Review 356

Unit Feature The Earth System and the Environment: Harnessing Earth's Water 358
Unit Investigation
Standardized Test Practice 360

Sea arch p. 338

UNIT 5 Atmosphere and Weather 362

CHAPTER 17 ATMOSPHERE 364
   17.1 The Atmosphere in Balance 366
   17.2 Heat and the Atmosphere 369
   17.3 Local Temperature Variations 374
   17.4 Human Impact on the Atmosphere 378
   Laboratory Activity 384
   Chapter Review 386

Photosynthesis releases oxygen p. 366

CHAPTER 18 WATER IN THE ATMOSPHERE 388
   18.1 Humidity and Condensation 390
   18.2 Clouds 396
   18.3 Precipitation 402
   Laboratory Activity 408
   Chapter Review 410

Seeding clouds p. 406

CHAPTER 19 THE ATMOSPHERE IN MOTION 412
   19.1 Air Pressure and Wind 414
   19.2 Factors Affecting Winds 419
   19.3 Global Wind Patterns 422
   19.4 Continental and Local Winds 427
   Laboratory Activity 430
   Chapter Review 432

Summer monsoon rains p. 428

CHAPTER 20 WEATHER 434
   20.1 Air Massses and Weather 436
   20.2 Fronts and Lows 439
   20.3 Thunderstorms and Tornadoes 445
   20.4 Hurricanes and Winter Storms 450
   20.5 Forecasting Weather 455
   Map Activity 460
   Chapter Review 462

Lightning p. 446

CHAPTER 21 CLIMATE AND CLIMATE CHANGE 464
   21.1 What Is Climate? 466
   21.2 Climate Zones 469
   21.3 Climate Change 474
   Laboratory Activity 478
   Chapter Review 480

Unit Feature The Earth System and the Environment:
Climate and Civilization 482

Unit Investigation

Standardized Test Practice 484

Highland climate p. 473

UNIT 6 Earth's Oceans — 486

CHAPTER 22 THE WATER PLANET — 488
- 22.1 Oceanography — 490
- 22.2 Properties of Water — 492
- 22.3 Properties of Ocean Water — 495
- 22.4 Ocean Life — 499
- Laboratory Activity — 504
- Chapter Review — 506

Diatoms p. 499

CHAPTER 23 THE OCEAN FLOOR — 508
- 23.1 Studying the Ocean Floor — 510
- 23.2 The Continental Margin — 514
- 23.3 The Ocean Basin — 517
- 23.4 Ocean Floor Sediments — 523
- Laboratory Activity — 526
- Chapter Review — 528

Submarine canyon p. 508

CHAPTER 24 THE MOVING OCEAN — 530
- 24.1 Surface Currents — 532
- 24.2 Currents Under the Surface — 536
- 24.3 Tides — 541
- Laboratory Activity — 544
- Chapter Review — 546

Bay of Fundy low tide p. 543

Unit Feature The Earth System and the Environment: Ocean Effects on Weather and Climate — 548

Unit Investigation

Standardized Test Practice — 550

UNIT 7 Space — 552

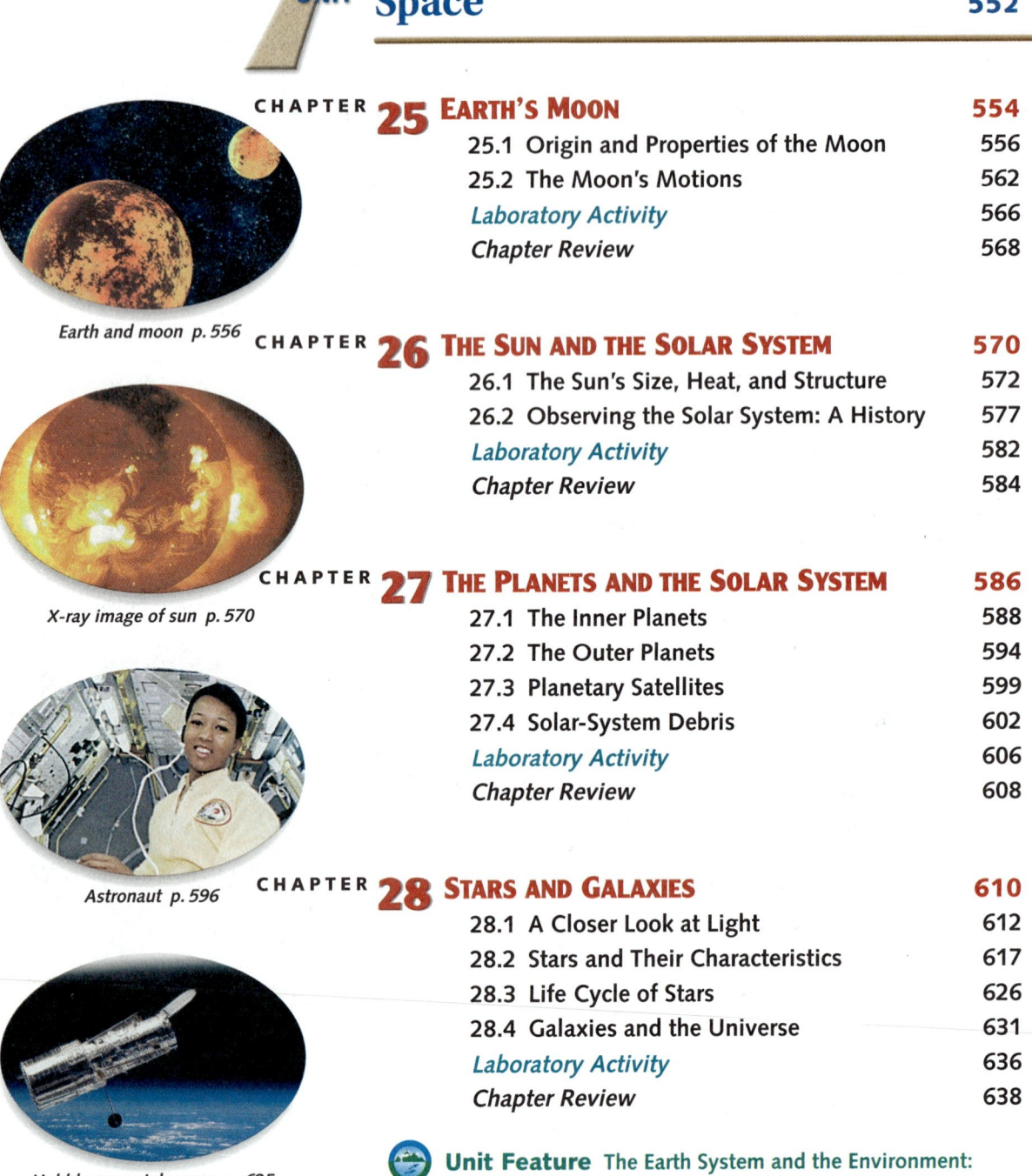

Earth and moon p. 556

X-ray image of sun p. 570

Astronaut p. 596

Hubble space telescope p. 625

CHAPTER 25 EARTH'S MOON — 554
- 25.1 Origin and Properties of the Moon — 556
- 25.2 The Moon's Motions — 562
- Laboratory Activity — 566
- Chapter Review — 568

CHAPTER 26 THE SUN AND THE SOLAR SYSTEM — 570
- 26.1 The Sun's Size, Heat, and Structure — 572
- 26.2 Observing the Solar System: A History — 577
- Laboratory Activity — 582
- Chapter Review — 584

CHAPTER 27 THE PLANETS AND THE SOLAR SYSTEM — 586
- 27.1 The Inner Planets — 588
- 27.2 The Outer Planets — 594
- 27.3 Planetary Satellites — 599
- 27.4 Solar-System Debris — 602
- Laboratory Activity — 606
- Chapter Review — 608

CHAPTER 28 STARS AND GALAXIES — 610
- 28.1 A Closer Look at Light — 612
- 28.2 Stars and Their Characteristics — 617
- 28.3 Life Cycle of Stars — 626
- 28.4 Galaxies and the Universe — 631
- Laboratory Activity — 636
- Chapter Review — 638

Unit Feature The Earth System and the Environment: Life in Extreme — 640

Unit Investigation

Standardized Test Practice — 642

UNIT 8 Earth's History — 644

CHAPTER 29 STUDYING THE PAST — 646
- 29.1 Fossils — 648
- 29.2 Relative Time — 650
- 29.3 Absolute Time — 656
- Laboratory Activity — 660
- Chapter Review — 662

Paleontologists excavate rhinoceros p. 646

CHAPTER 30 VIEWS OF EARTH'S PAST — 664
- 30.1 The Geological Time Scale — 666
- 30.2 The Precambrian and Paleozoic — 673
- 30.3 The Mesozoic — 678
- 30.4 Earth's Recent History — 681
- Laboratory Activity — 686
- Chapter Review — 688

Unit Feature The Earth System and the Environment: Mass Extinction — 690

Unit Investigation

Standardized Test Practice — 692

Appendices — 694
  Appendix A: Reference Tables — 696
  Appendix B: Atlas — 708
  Appendix C: Skills Handbook — 719

Glossary — 737
Index — 756
Acknowledgments — 768

Galapagos tortoise p. 671

xiii

FEATURES

The Earth System and the Environment

An Old Idea Becomes New Again	62
Recycling Cents and Sensibility	164
When Volcanoes Breathe Life	250
Harnessing Earth's Water	358
Climate and Civilization	482
Ocean Effects on Weather and Climate	548
Life in Extreme	640
Mass Extinction	690

Tropical growth in England p. 548

"Junk" art p. 165

Science and Technology

Linking Information and Location	52
Measuring Time	79
A New Form of Carbon Creates a New Ball Game ...	103
What Good Is a Rock?	132
Harnessing the Power of the Sun	157
Shifting Ground: Tracking Plate Movements	184
Earthquakes: Technology to the Rescue	227
Measuring Mount Everest	242
Rivers of Ice	329
New Ideas Help Fight Air Pollution	383
Extreme Science: Flying into the Storm	454
The NEPTUNE Project	522
One Small Step into the Cosmos	561
Seeking the Red Planet's Secrets	593
Telescopes: Windows to the Universe	625
Ancient Rocks Provide Clues	655

Astronomical and atomic clocks p. 79

Science and Society

Using Landscaping to Save Water	19
Scientific Literacy: What's the Big Deal?	34
Predicting Eruptions, Saving Lives	205
Preserving Rain Forest Topsoil: The Environment	267
Removing the Edwards Dam	293
The Shrinking of the High Plains Aquifer	308
Disappearing Beaches	348
Who Will Stop the (Acid) Rain?	407
Trade Winds	426
El Niño	468
Troubled Waters: Coral Reefs	503
Upwelling and Anchovies off Peru	540
Changing the Face of Science	581
Biodiversity	672

Rescuing sea turtles p. 672

Rainforest ecologists p. 12

Careers

Tropical Rainforest Ecologist	12
Land Surveyor	51
Marble Sculptor	136
Recycling Technician	158
Electromagnetic Geophysicist	187
Earthquake Engineer	221
Hydrologist	282
Wastewater Treatment Plant Operator	301
Environmental Consultant	368
Air Force Meteorologist	400
Weather Observer	456
Marine Biologist	500
Research Vessel Officer	524
Solar Physicist	576
Mission Specialist	596
Science Writer	622
Natural History Museum Curator	670

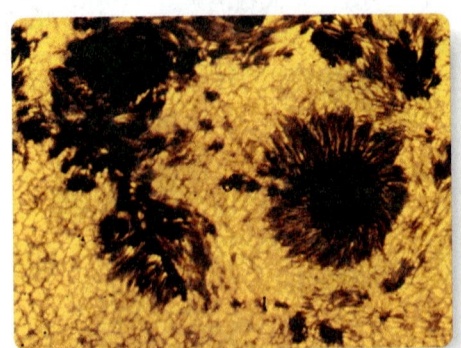
Sunspots p. 574

Scientific Thinking

Analyze pp. 196, 292, 382
Apply pp. 340, 459
Communicate pp. 196, 229, 332, 525
Compare/Contrast pp. 153, 417
Design an Experiment p. 73
Develop Models p. 577
Explain p. 534
Hypothesize pp. 124, 473, 674, 683
Infer/Deduce pp. 6, 50, 394, 598, 627, 671
Interpret pp. 239, 428, 653
Investigate p. 261
Observe p. 17
Predict pp. 32, 492, 498
Using Mathematics p. 619

Lab Activities

The Density of Earth Materials	20
Making Inferences From Observations	38
Eratosthenes and Earth's Circumference	84
Specific Gravity and Mineral Identification	112
Studying Rocks in Thin Section	138
Passive and Active Solar Heating	160
"Earthquake Engineering"	230
Chemical Weathering and Temperature	274
Water Budgets	312
Modeling Glacial Movement	334
Beach Erosion and Deposition	354
Absorption and Radiation of Heat	384
Effects of Acid Rain	408
Correlating Weather Variables	430
Observing Greenhouse Gases	478
Ocean Water and Fresh Water	504
Mapping an Unknown Surface	526
Oceanic Water Masses	544
Making Impact Craters	566
Scale Model of the Solar System	582
Galilean Moons of Jupiter	606
Expansion of the Universe	636
Deciphering Tree Rings	660
Fossil Excavation	686

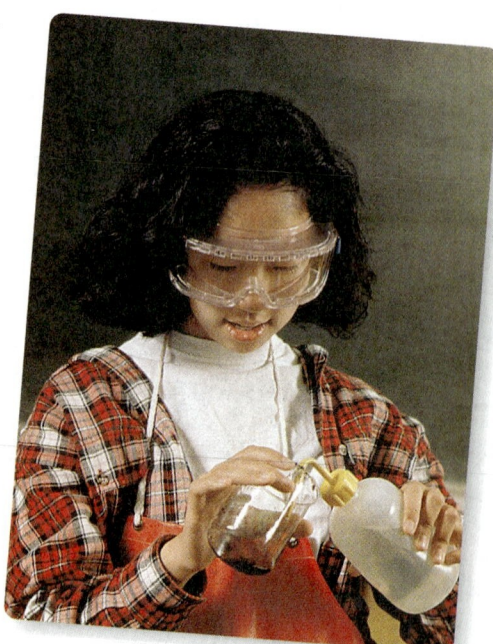
Effects of acid rain p. 408

Map Activities

The Roads to Harris Hill	58
Nuclear Sites and Plate Boundaries	188
Where Are Active Volcanoes?	208
Folded Mountain Range	246
Stream Divides and River Systems	294
Severe Storms	460

Mini Labs

How the Biosphere Affects the Hydrosphere	11
Making a Sextant	36
Drawing a Profile	56
Modeling Earth's Magnetic Field	74
Measuring Specific Gravity	107
Metamorphic Molds	130
Oil Well	150
Ocean Floor Magnetism	175
Modeling Viscosity	200
Interpreting a Travel-Time Graph	219
Modeling a Fault	240
Surface Area and Chemical Weathering	263
Modeling River Sediments	285
Measuring Porosity	302
Modeling Glacier Formation	320
Making Waves	345
Changes of State	370
Measuring Humidity	393
Observing Air in Motion	423
Graphing a Front	441
Classroom Microclimates	472
Observing Water's Polarity In Action	493
Continental Margin	516
Simulating Ocean Currents	537
Weights on the Moon and Earth	557
Orbital Forces	578
Design a Martian Calendar	592
Simulate the Doppler Effect	616
A Decay Path	657
Continental Isolation's Effect on Species	680

Dew on spider web p. 393

TECHNOLOGY

Explore Earth Science *on the* Internet

Extend your learning with a wealth of Internet resources. The Exploring Earth Web site offers links to scientific data, interactive visualizations, Earth science news, and information on Earth science in your region. The Internet resources have been selected to correspond with topics in the book.

DATA CENTERS   Links to scientific information

Data centers help you explore Earth science concepts with links to current and archived information. You can find out about recent volcanic eruptions, use an online mineral guide, or explore the latest mission to Mars.

VISUALIZATIONS   New ways to see Earth

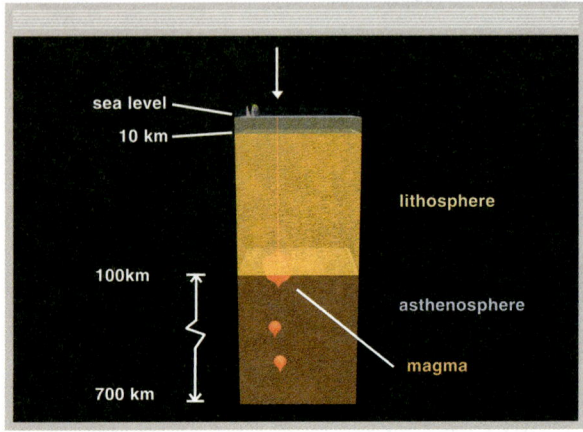

Interactive visualizations can help you learn core concepts of Earth science. You can explore Earth science as a dynamic process through animations, satellite images, video, and photographs.

EARTH NEWS   Current events

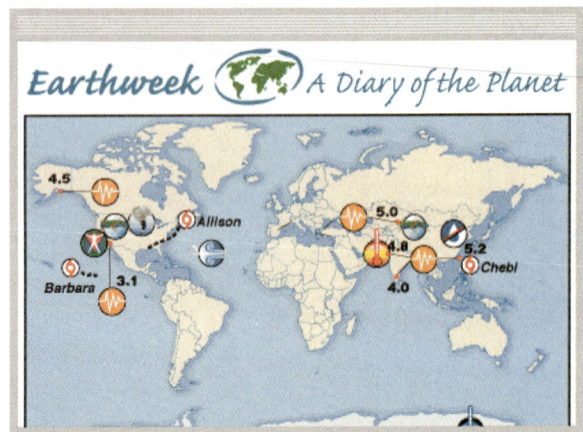

Earth news can help you stay current on events related to Earth science. You can click on an interactive map of Earth to get updates on important regional events.

LOCAL RESOURCES   Earth science in your region

Local resources help you learn about Earth science in your area. You can discover more about where you live through satellite images, geologic maps, water data, and weather information.

Direct Links from Your Textbook

Internet resources referenced in the textbook correspond to concepts as they are introduced. The Exploring Earth Web site has a carefully selected set of resources, and custom-built visualizations and investigations. There is no need for wide-range searches to find the information.

GETTING TO THE WEB SITE

Using a browser, enter "ClassZone.com" and select *Earth Science*. Once on the Earth Science page, click on *Exploring Earth* to access all of the online resources.

TWO EASY WAYS TO FIND WHAT YOU NEED:

1. **Use the keycode**

   - In your textbook, you will find a keycode for each direct reference to the Internet.
   - On the Exploring Earth Web site, enter the keycode in the search box. You will jump directly to the referenced material.

2. **Navigate using buttons and menus**

   - Every chapter begins with an overview of the chapter's Web resources, such as Data Centers, Earth News, and Local Resources.
   - Go to the Exploring Earth Web site and use the buttons at the top of the screen to select a resource. You will see a list of chapters for which material is available. You can then browse through the resources and select what you need.

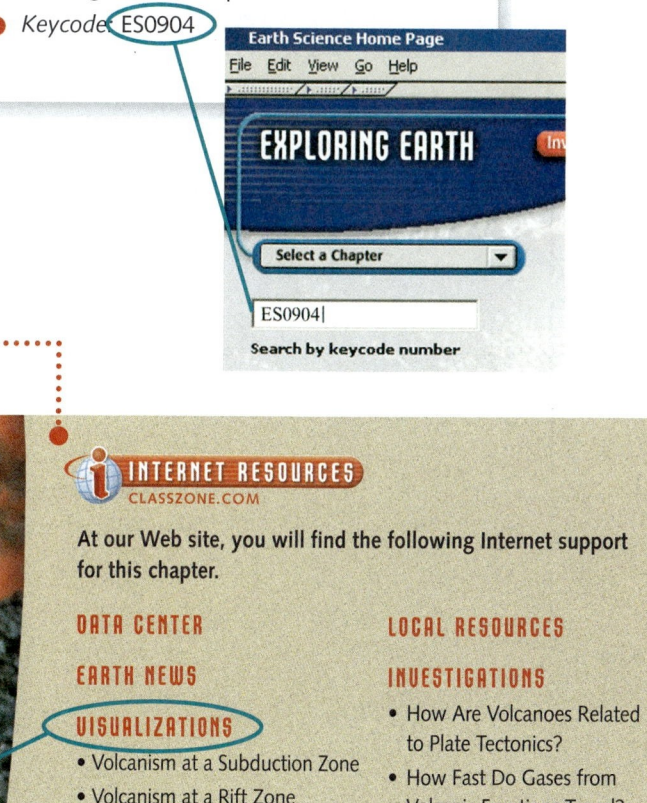

VISUALIZATIONS
CLASSZONE.COM
Observe an animation of the volcanic islands forming over a hot spot.
Keycode: ES0904

INTERNET RESOURCES
CLASSZONE.COM
At our Web site, you will find the following Internet support for this chapter.

DATA CENTER

EARTH NEWS

VISUALIZATIONS
- Volcanism at a Subduction Zone
- Volcanism at a Rift Zone

LOCAL RESOURCES

INVESTIGATIONS
- How Are Volcanoes Related to Plate Tectonics?
- How Fast Do Gases from Volcanic Eruptions Travel?
- Is It Safe To Live Near a Volcano?

Internet Investigations

Investigations are interactive, structured activities on the Web that help you learn more about Earth science concepts. Each investigation has instructions right on the Web page to walk you through the site. Using images and information, these instructions can help you answer the investigation question.

There are Investigation references and questions in the textbook that correspond to important topics and concepts. You will navigate to a specific reference on the Web site (see previous page), and answer the question by working through the activity. Each investigation has a worksheet for you to fill out as you work through the question.

HERE ARE THREE SAMPLES OF INTERNET INVESTIGATIONS

INVESTIGATIONS

How Are Volcanoes Related to Plate Tectonics?

In this investigation, you will explore the relationship between volcanism and two different types of plate boundaries.

- Plot volcano locations on an interactive map of the Pacific Northwest to see the patterns they form.

- Predict the locations of plate boundaries in the region and draw them on the map, like the example at the bottom right.

- View 3-D diagrams that help you see how volcanoes are related to plate tectonics.

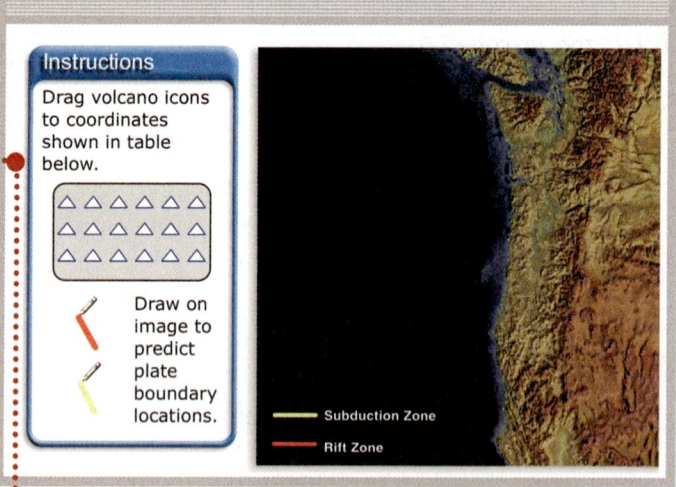

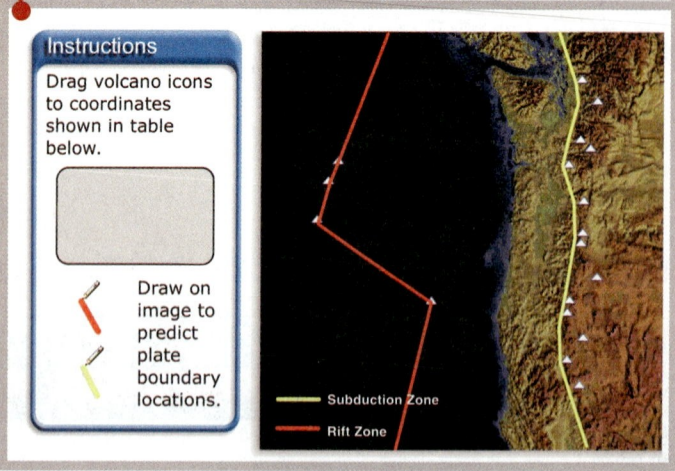

INVESTIGATIONS

What Time Is It?

For this investigation, you will learn how the apparent motion of the sun can be used to track time.

- View a rotating globe, like the one at the left, and flat maps that show where it is light and dark on Earth. Use natural time indicators like sunrise, sunset, and local noon to understand global time.

- Overlay time zones to tell what time it is in different parts of the world.

INVESTIGATIONS

How Does a Mid-latitude Low Develop into a Storm System?

In this investigation, you will examine the most common type of weather system in the United States, a low-pressure system associated with the boundary between a warm air mass and a cold air mass.

- Track a dramatic winter storm from its early stages through its climax. The images at the right show the storm first over the southeastern United States and then farther north later the same day.

- Analyze weather data and satellite images of this low-pressure system that dumped over two feet of snow along much of the eastern seaboard of the United States.

SAFETY in the EARTH SCIENCE LABORATORY

The Earth science laboratory is a place where discovery leads to knowledge and understanding. It is also a place where caution is essential for your safety and the safety of others. You can avoid accidents by knowing and practicing safe laboratory techniques. The following guidelines present some basic rules for safe laboratory work. Your teacher will provide additional rules specific to your laboratory setting. Read the rules thoroughly and always apply them in the laboratory.

1. Do not touch materials or equipment until instructed to do so.
2. Locate and learn to use all laboratory safety equipment.
3. Never eat, drink, chew gum, or apply cosmetics in the laboratory. Do not store food or beverages in the lab area.
4. Never smell a chemical directly; instead, fan the vapor toward your nose with your hand. Never taste a chemical. Never taste a mineral sample unless instructed to do so by your teacher.
5. Never force or twist glass tubing into a stopper. Protect both hands with cloth pads and use an appropriate lubricant when inserting or removing glass tubing. Use a streak plate or glass plate on a flat surface, not in your hand.
6. Never reach over a flame or a heat source. Keep hair and clothing away from flames.
7. Never do an investigation without your teacher's supervision. Never perform your own experiment without your teacher's permission.
8. Read all parts of an investigation before you begin work. Follow your teacher's directions completely. Ask for help if you do not understand.
9. Never run, push, play, or fool around in the laboratory.
10. Keep your work area clean and uncluttered. Store books, backpacks, purses, and other items in designated areas. Keep glassware and containers of chemicals away from the edge of the lab bench.
11. Report all accidents to your teacher immediately. Do not touch broken, cracked, or chipped glassware.
12. Turn off electric equipment, water, and gas when not in use.
13. Pay attention to safety CAUTIONS. Wear safety goggles and a lab apron whenever you use heat, chemicals, solutions, glassware, or other dangerous materials.
14. Dress properly for the laboratory. Do not wear loose-fitting sleeves, bulky outerwear, dangling jewelry, or open-toed shoes. Tie back long hair and tuck in ties and scarves when you use heat or chemicals.

15. Never touch a hot object with your bare hands. Use a clamp, tongs, or heat-resistant gloves when handling hot objects.

16. When heating a test tube, point it away from yourself and others. Use sturdy tongs or test-tube holders. Do not reach over a flame. Do not heat a sealed container.

17. Use care in handling electrical equipment. Do not touch electrical equipment with wet hands or use it near water. Check for frayed cords, broken wires, and loose connections. Use only equipment with three-prong plugs. Make sure cords do not dangle from worktables. Disconnect an electrical appliance from an outlet by pulling the plug, not the cord.

18. Use flammable chemicals only after ensuring that there are no flames anywhere in the laboratory.

19. Use care when working with chemicals. Keep all chemicals away from your face and off your skin. Keep your hands away from your face while working with chemicals. If a chemical gets in your eyes or on your skin or clothing, wash off the chemical immediately with plenty of water. Tell your teacher what happened.

20. Do not return any unused chemical to its bottle. Follow your teacher's directions for the disposal of chemicals, used matches, filter papers, and other materials. Do not discard solid chemicals, chemical solutions, matches, papers, or any such substances in the sink.

21. Always clean your lab equipment and workspace after you finish an investigation. Always wash your hands with soap and water before leaving the laboratory.

Safety in the Laboratory

Throughout the investigations in this program, you will see symbols relating to safe laboratory procedures. These symbols and their meanings are shown below. Study the information and become familiar with all CAUTIONS in an investigation before you begin your work.

Wear safety goggles. Investigation involves chemicals, hot materials, lab burners, or the risk of broken glass.

Wear a lab apron. Investigation involves liquids, chemicals, hot materials, open flame, or the risk of broken glass.

Danger of cuts exists. Investigation involves scissors, wire cutters, pins, or other sharp instruments. Handle with care.

Danger of fire or explosion. Investigation involves hot plates, lab burners, lighted matches, or flammable liquids with explosive vapors.

Danger of burns, frostbite, or skin irritation; use protective gloves. Investigation involves hot materials, dry ice, or corrosive or irritating chemicals. In case of contact, notify your teacher immediately.

Danger of electrical shock. Investigation involves use of electrical equipment such as electric lamps or hot plates.

The triangle alerts you to additional specific safety procedures in an investigation. Always discuss safety CAUTIONS with your teacher before you begin work.

UNIT 1

UNIT OVERVIEW

Hold up the book and ask students what they see as the central image of the two opening pages. Most will probably say Earth; some may say home. The entire planet is our home. As large and complex as Earth is, its many different parts interact. Point out the clouds that stretch from the Pacific Ocean across North America to the Atlantic. This unit introduces students to Earth as a system of four interrelated spheres: the atmosphere, the geosphere, the hydrosphere, and the biosphere. As part of the biosphere, each student can affect the rest of Earth by a lifetime of choices.

VISUAL TEACHING

Chapter 1 Earth as a System

The graphic illustrates the water cycle, one example of how Earth's spheres interact. Water—in solid, gas, and liquid form—makes up the hydrosphere. It exists as water vapor within the atmosphere and is cycled back to Earth as precipitation. In the geosphere, water can take the form of snow, surface water (oceans, streams, lakes), or groundwater. Water is also a critical part of the biosphere, sustaining all life and supporting ecosystems. Students will learn about cycles that are part of the Earth system.

Answer: Both matter and energy are exchanged as various parts of Earth interact. This happens with water in the water cycle. It also happens with living organisms, as they take in nutrients from Earth and process them.

UNIT 1 Investigating Earth

The Earth System Viewed from space, the beauty and interdependent nature of the Earth's system is clear. You may not realize it, but you are an integral part of the Earth system.

How do the choices you make affect the rest of Earth?

Earth as a System CHAPTER 1

How do the many parts of Earth interact and affect one another?

CONNECTIONS TO . . .

Prior Knowledge Students have been observing Earth and its phenomena for many years. Unit 1 introduces them to the concept of viewing Earth as a system. It explains how Earth's different spheres—the atmosphere, hydrosphere, geosphere, and biosphere—interact. This unit reviews some of the methods and tools of scientific inquiry, and describes how they are used in Earth system science.

Unifying Themes Earth is a *system* of interacting spheres that follow basic principles of *order*. As students study the chapters in Unit 1, point out the interactions of Earth's spheres and the processes scientists use, such as making models, to help them understand Earth.

UNIT 1

CHAPTER 2
The Nature of Science

Why do scientists pursue knowledge?

CHAPTER 3
Models of Earth

What can maps and satellite images, such as the one of Death Valley above, tell us about Earth?

Chapter 2 The Nature of Science

Evan Forde, an American scientist with the National Oceanic and Atmospheric Administration (NOAA), uses instruments to make quantitative observations. The data he collects is used to understand Earth's past, assess present conditions, and consider what might happen in the future. Students will learn of the methods scientists use to answer questions about Earth.

Answer: Scientists seek to understand the workings of the physical world. They may focus on events such as earthquakes, study changes in climate, or assess Earth resources.

Chapter 3 Models of Earth

The Landsat satellite recorded this image of Death Valley, a desert in California. The bright red streak down the middle of the valley indicates extensive salt deposits, suggesting that this region once supported a large salty river that has since evaporated. Death Valley is the hottest and driest spot in the United States. At one time, it held the record for being the hottest place on Earth, with a record-setting temperature of 134°F (56°C) recorded on July 10, 1913. Students will learn how scientists use satellites to map and study Earth.

Answer: Maps and satellite images allow scientists to study large-scale phenomena, such as patterns of vegetation or variations in topography.

CONNECTIONS TO . . .

National Science Education Standards
- Students will come to understand that matter moves through the Earth system by way of geochemical cycles.
- Students will understand that the Earth system has internal and external sources of energy. Movement of matter within the Earth system is driven by these energy sources.
- Students will understand that interactions among Earth's spheres cause the Earth system to evolve continuously.

Unit Features

The Earth System and the Environment
An Old Idea Becomes New Again, pp. 62–63

How Can Getting Farther Away from Earth Help Us See It More Clearly? Explore how satellites gather information about Earth.
Keycode: ESU101

Standardized Test Practice, pp. 64–65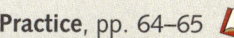

Chapter-Level Resources			Laboratory Manual, pp. 1–2 Internet Investigations Guide, pp. T7–T8, 7–8 Guide to Earth Science in Urban Environments, pp. 5–12, 13–20 Spanish Vocabulary and Summaries, pp. 1–2 Formal Assessment, pp. 1–3 Alternative Assessment, pp. 1–2
Section 1.1 A New View of Earth, pp. 4–7		1. Describe how scientists view Earth today. 2. Compare and contrast open and closed systems. 3. Explain the significance of Earth as a closed system.	Lesson Plans, p. 1 Reading Study Guide, p. 1
Section 1.2 The Earth System's Four Spheres, pp. 8–12		1. Describe the four spheres of the Earth system. 2. Explain how the spheres interact and change.	Lesson Plans, p. 2 Reading Study Guide, p. 2
Section 1.3 Cycles and the Earth, pp.13–19		1. Describe characteristics of the water, carbon, and energy cycles. 2. Analyze how humans interact with the water, carbon, and energy cycles.	Lesson Plans, p. 3 Reading Study Guide, p. 3 Teaching Transparency 1

INVESTIGATIONS
CLASSZONE.COM

Section 1.2 INVESTIGATION
How Are Earth's Spheres Interacting?
Computer time: 45 minutes / Additional time: 45 minutes
Students identify evidence of energy and materials being transferred between spheres.

Chapter Review INVESTIGATION
How Do Interactions between Earth's Spheres Vary Regionally?
Computer time: 45 minutes / Additional time: 45 minutes
Students compare and contrast interactions between spheres in the local region with other regions.

Electronic Teacher Tools
Visualizations CD-ROM
Classzone.com
Online Lesson Planner
Test Generator

Visualizations CD-ROM
Classzone.com
ES0101 Visualization

Visualizations CD-ROM
Classzone.com
ES0102 Visualization, ES0103 Investigation, ES0104 Career

Visualizations CD-ROM
Classzone.com
ES0105 Visualization, ES0106 Visualization, ES0107 Data Center

Classroom Management

Gather these materials for minilab and laboratory activities.

Minilab, p. 11
clear plastic bag
potted plant
tape

Lab Activity, pp. 20–21
balance
cubic block
metric ruler
Appendix C Area and Volume, page 732
rectangular block
cylindrical block
100-mL graduated cylinder
50-mL beaker
wet limestone chips
paper towels
Laboratory Manual, Teacher's Edition
  Lab Sheet 1, p. 137

CHAPTER 1

Earth as a System

This photograph displays all four spheres of the Earth system.

What is the Earth system, and where do we humans fit in?

INTRODUCE
Build Interest
Before sharing background information about the photograph with students, allow time for their questions and comments.

Ask questions like the following:
- What living things do you see? **trees, flowers, other plants**
- Where can you see water in the photograph? **water in the lake and snow on the mountain**
- Where is there water that you cannot see? **water vapor in the air**
- Where do you see rocks? **along the edge of the lake, the mountain**

Respond
- What is the Earth system, and where do we humans fit in? **The Earth system encompasses the interaction of all matter and energy within the geosphere, atmosphere, hydrosphere, and biosphere. As humans, we belong to the biosphere, but our activities affect all the spheres of the Earth system.**

About the Photograph
Earth's four spheres are visible in the photograph. The trees, shrubs, and flowers are part of the biosphere. Earth's sky is part of the atmosphere. The hydrosphere includes the water in the lake and the snow and ice on the mountaintop. The mountain itself and the soil from which the plants grow are part of the geosphere.

BEYOND THE TEXTBOOK

Print Resources
- Reading Study Guide pp. 1–3
- Transparency 1
- Formal Assessment, pp. 1–3
- Laboratory Manual, pp. 1–2
- Alternative Assessment, pp. 1–2
- Spanish Vocabulary and Summaries, pp. 1–2
- Internet Investigations Guide, pp. T7–T8, 7–8
- Guide to Earth Science in Urban Environments, pp. 5–12, 13–20

Technology Resources
- Visualizations CD-ROM
- Classzone.com
- Online Lesson Planner
- Electronic Teacher Tools
- Test Generator

2 Unit 1 Investigating Earth

CHAPTER 1

PREVIEW

▶ **FOCUS QUESTIONS** In this chapter you will study the Earth system and learn about the key questions below.

Section 1 What is Earth system science?

Section 2 What are the Earth system's four spheres, and how do they affect one another?

Section 3 What are cycles and how do they work?

▶ **READING STRATEGY**

PREVIEW

Before you begin to read, examine the headings, images, and captions in this chapter. Write your predictions about the chapter's contents in your science notebook.

INTERNET RESOURCES
CLASSZONE.COM

At our Web site, you will find the following Internet support for this chapter.

DATA CENTER
EARTH NEWS
VISUALIZATIONS
- New Perspective of Earth
- Earth's Spheres
- Water Cycle
- Carbon Cycle

LOCAL RESOURCES
CAREERS
INVESTIGATIONS
- How are Earth's Spheres Interacting?
- How Do Interactions between Earth's Spheres Vary Regionally?

CHAPTER 1

PREPARE

Focus Questions

Find out what knowledge students have about Earth as a system. For each focus question, have students consider the section title and then develop two or three additional questions for each section. For example, Section 1: How do scientists study Earth? Why is it called a system?

Review Topics

Students may need to review these concepts:

Life Science
- Water, heat, and light are essential for the maintenance of life.
- Ecology is the study of how organisms behave and interact with their environments.

Reading Strategy

Model the Strategy: To help students preview each section, use a version of the following: "I see the title of the first section is 'A New View of Earth.' I wonder what a 'new view' could mean? Below there is a photograph of Earth from the moon. That's a different perspective. It shows Earth as a whole. Also, many of the red headings are about Earth as a system. I think the 'new view' means viewing Earth as a system—as a whole rather than as unconnected parts." Have students continue to use this strategy for the remainder of the chapter.

BEYOND THE TEXTBOOK

INTERNET RESOURCES
CLASSZONE.COM

INVESTIGATIONS
- How Are Earth's Spheres Interacting?
- How Do Interactions between Earth's Spheres Vary Regionally?

VISUALIZATIONS
- Animation zooms in on Earth from beyond the moon
- Models of Earth's spheres
- Images and animations of a raindrop traveling through the water cycle
- Animation showing evidence of the carbon cycle

DATA CENTER
EARTH NEWS
LOCAL RESOURCES
CAREERS

Online Lesson Planner

CHAPTER 1 SECTION 1

FOCUS

Objectives
1. Describe how scientists view Earth today.
2. Compare and contrast open and closed systems.
3. Explain the significance of Earth as an essentially closed system.

Set a Purpose
Read aloud the key idea. Then have students read the section to answer the question "How do scientists describe Earth?"

INSTRUCT

More about...

The First View of Earth
On August 23, 1966, *Lunar Orbiter I* was on its 16th orbit around the moon when it took a photograph of the moon's surface. In the photograph's background was a crescent view of Earth. The view excited scientists because until then no one knew what Earth looked like as a planet. The photograph showed that Earth is a small planet consisting of blue waters, brown and green landmasses, and swirling white clouds.

1.1

KEY IDEA
Scientists and others are beginning to view Earth as a system of interconnected spheres, not a collection of unrelated parts.

KEY VOCABULARY
- Earth system science
- model
- system
- closed system
- open system

VISUALIZATIONS
CLASSZONE.COM
Examine Earth from a new perspective.
Keycode: ES0101

A New View of Earth

Earth scientists study the solid earth, the atmosphere, the oceans, and the stars and planets. Recently scientists have begun to study Earth as a whole. The change in view point is due to a new way of looking at Earth and to a better understanding of how solids, liquids, and gases interact.

Past Perceptions Meet New Issues

For decades, scientists studied Earth as a collection of parts. Some scientists devoted themselves to studying how mountains form, while others concentrated on the life forms found on those mountains, and still others examined the icy peaks. This approach led to many important discoveries about the planet. During the late 20th century, however, scientists and other concerned individuals came up against the limits that arise when regarding Earth science as a collection of specialties, with each specialty independent of the other. Environmental incidents at the time demonstrated that the various parts of Earth were connected and that they interacted; thus, these parts could not be understood fully in isolation.

During the 1970s, scientists realized that chlorofluorocarbons (KLAWR-oh-FLUR-oh-KAHR-buhnz)— or CFCs, chemicals used in everyday items like hair spray and air conditioners—were floating up into the atmosphere where they reacted with the atmospheric ozone. Ozone is a form of oxygen consisting of three atoms; the oxygen we breathe consists of two atoms. Ozone forms a layer around Earth that absorbs ultraviolet radiation from the sun. So many CFCs had been released that they were changing the three-atom ozone to two-atom oxygen, thus damaging the ozone layer and increasing the risk of exposure to harmful solar radiation. A high degree of exposure to the ultraviolet portion of solar radiation can cause skin cancer.

EARTH'S BEAUTY and fragility are especially striking in this view of our planet from the moon.

DIFFERENTIATING INSTRUCTION

Reading Support Tell students that as they read the text, they can use the headings as clues to finding the main ideas. Model how to use the heading "Past Perceptions Meet New Issues" to determine the main idea. Tell students that the information under the heading must relate to it, so the main idea is probably about the past and the present. Then have students read the text under the heading, and ask them to identify the main idea. In the past, scientists tended to study Earth in parts, but now many scientists recognize the importance of studying Earth as a whole. Explain to students that finding the main idea under each heading can help them identify the important points of a section.
Use Reading Study Guide, p. 1.

4 Unit 1 Investigating Earth

Nations around the world cooperated to ban the manufacture and use of CFCs, but the ozone layer will take years to recover. Human beings began to realize the huge impact they could have on the world around them and to recognize that human activities are one of the most powerful forces of change on Earth. Scientists had the unprecedented chance to see Earth from space, through satellite imaging and space exploration, and also from within the deep sea. These views led human beings—including earth scientists—to realize that Earth is more than a collection of parts. This new view of the earth is called **Earth system science.**

The Rise of Earth System Science

Earth system science has come about, in part, because advances in technology allow scientists to see Earth in new ways. Satellites let them see Earth from space. Deep-diving submersibles allow them to explore the ocean deeps. Powerful computers and computer programs called *Geographic Information Systems* (GIS) help scientists to interpret great amounts of data from various sources to see how Earth changes over time.

With the information from these new tools, scientists can create complex pictures of how processes work and from these pictures formulate models that can explain unseen details. A **model** is a representation of an object, a process, or a phenomenon. Before a house is built, architects create a model, a small cardboard version that shows where the windows will go, how many rooms there are, and so on. The model isn't exactly like the real house (there is no running water or heat, for example), but it provides vital information nonetheless. The models that scientists build may show physical features, such as globes or maps; explain how an element moves through the environment; or show how something works, especially when something is too large or too small to observe directly.

What Is a System?

A **system** is a kind of model. A system can be defined as a part of the universe that can be studied separately. In addition to this quality of being able to be studied separately, system models also include time as a variable; that is, we now measure things in real time and make models to describe what happened in the past and what might happen in the future. Scientists often describe the natural world by using models such as open and closed systems.

In a **closed system,** energy, such as sunlight, may enter, but matter doesn't enter or leave. An example of a closed system would be a sealed glass jar of tea. Energy, in the form of light, can enter. Energy can leave, too, as heat passes out through the glass walls of the jar. But if the jar is tipped over, the tea stays in. As a closed system, the jar of tea doesn't share matter with its surroundings.

An example of a closed system is a deep-sea submersible that carries human researchers to great depths. The submersible has been designed and constructed in a way that minimizes the loss of energy and matter to the surrounding ocean. Air remains inside the craft, and water is kept out, protecting the researchers inside.

SATELLITES such as Landsat, shown here in a laboratory before its launch, have given scientists a new perspective of Earth.

A CLOSED SYSTEM A closed jar of tea exchanges only energy with its surroundings.

Examine Earth from a new perspective.
Keycode: ES0101

Have students speculate how the area in which they live might look from space.

Visualizations CD-ROM

More about...

Models
Scientists develop models from information that is available at a particular time, but information can change as scientists learn more about the world around them. When this happens, scientists may have to modify a model or replace it with a new one. The changing nature of models is at the core of the nature of science—an ever-changing body of knowledge that is refined and augmented to over time.

VISUAL TEACHING
Discussion

Read aloud the caption to the image at the bottom of the page and then ask: What part of the illustration represents a system? the jar of tea What makes the jar of tea a system? It can be studied separately from the rest of the surroundings. Why is the jar of tea considered to be a closed system? It exchanges energy with its surroundings but not matter.

Developing English Proficiency As students read the text, have them make a list of unfamiliar words in their science journal or notebook. Encourage them to try to figure out the meanings of the words from context. After reading, have students find the definitions of the words in the dictionary and add them to their list of words. Then ask students to reread the sentences with the unfamiliar words and restate what they read in their own words.
Use Spanish Vocabulary and Summaries, pp. 1–2.

In an **open system,** the system and its surroundings freely exchange both energy and matter. For example, the jar of iced tea would become an open system if the lid of the jar were to be removed, allowing some of the tea to evaporate. If a lemon slice were added, new matter would enter the system.

An island is a natural example of an open system. Sunlight (energy) strikes the island, warming it. The island returns some of this energy to the atmosphere in the form of heat. Rain (matter) falls on the island. The rain runs off, possibly carrying sediment (another type of matter) into the waters around it. The island exchanges matter and energy with its surroundings.

AN OPEN SYSTEM Open, a jar of tea exchanges both energy and matter with its surroundings.

Earth system science is a means by which all the flows of energy and matter in and out of an open system are measured and related to all the other open systems.

A Planetary System

Scientists consider Earth to be an essentially closed system. Earth receives radiant energy from the Sun and reflects much of it back into space (an exchange of energy). Earth is not an entirely closed system, however. The atmosphere loses hydrogen atoms to space and meteorites enter the atmosphere from space. Overall, however, Earth's mass remains constant.

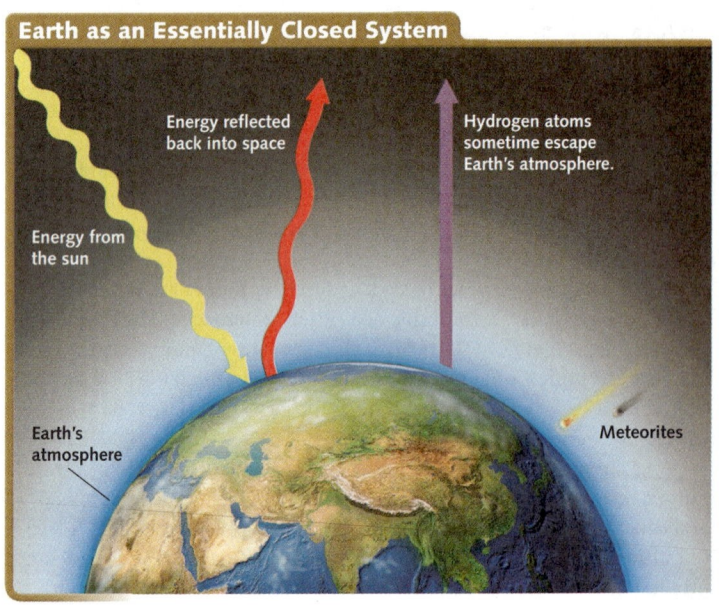

Earth as an Essentially Closed System

One of the consequences of Earth being an essentially closed system is that the planet's resources are finite: new matter is not going to form, and whatever matter exists on Earth is not going to go away.

Nature, Science, and Human Policy

Humankind's growing realization of its impact on the world has led to some important changes in the way human beings live. You may be aware of some of these changes in your daily life; for example, your school may recycle paper, plastics, and other materials.

Recycling is just one way of reducing humankind's impact on Earth. Scientists are seeking new and safer ways of providing the products people want and need. You will read about many of these projects throughout this textbook in the "Science and Society" and "Science and Technology" features.

It is not always easy, however, to balance human needs and wants with long-term concerns about other parts of the environment. Sometimes it is hard to visualize the impact people have on the planet and its resources because Earth seems so large and its resources so vast. Such appearances are deceiving. For example, as cities in the southwestern United States grow, there is an increase in the demand for essential needs—water, food, shelter, and power. How can enough water be found to accommodate the demands of a growing population in one area without taking water away from people in other areas or from the wilderness? What steps can be taken to lessen demand? How can water supplies be safeguarded?

Because of the integrated approach of Earth system science, issues such as water supply can be studied in depth from various points of view. Other concerns we face as 21st-century citizens include global warming, water and air pollution, fresh water conservation, declining variety among animal and plant species, and the already-mentioned ozone layer. Understanding these issues and making decisions about them requires an understanding of Earth, of its processes, and of science itself.

HUMAN ACTIVITY such as this development in Riverside County, California, impacts the Earth system.

1.1 Section Review

1. What technological advances led to the rise of Earth system science?
2. What is a system?
3. Compare and contrast an open system and a closed system. Use a Venn diagram.
4. Why do scientists consider Earth an essentially closed system?
5. **CRITICAL THINKING** As the human population expands, what might be some of the important issues policy makers face?
6. **PAIRED ACTIVITY** Build a model of a system and identify the features that make it open or closed.

CHAPTER 1 SECTION 1

ASSESS

1. satellites, deep-diving submersibles, computers, computer programs
2. A system is a part of the universe that can be studied separately.
3. The Venn diagram should show that energy is exchanged in both an open system and a closed system. Matter is also exchanged in an open system. No matter is exchanged in a closed system.
4. Very little matter is exchanged.
5. Answers may include overcrowding, insufficient resources, disease, and pollution.
6. Models will vary. Features of an open system should include an exchange of both energy and matter. Features of a closed system should include an exchange of energy only.

If students miss . . .

Question 1 Have students reread the first paragraph under "The Rise of Earth System Science." (p. 5)

Question 2 Reteach "What Is a System?" (pp. 5–6) Ask: In what way is the jar of tea in the illustration a system? **It can be studied separately from its surroundings.**

Question 3 Have students compare and contrast the illustrations of the open and closed systems on pages 5 and 6.

Question 4 Use the diagram at the bottom of page 6 to explain why Earth is an essentially closed system.

Question 5 Reteach "Nature, Science, and Human Policy." (p. 7) Ask students to consider what the basic needs of humans are.

Question 6 Review open and closed systems. (pp. 5–6) Focus on the exchange of matter and energy.

FOCUS

Objectives
1. Describe the four spheres of the Earth system.
2. Explain how the spheres interact and change.

Set a Purpose
Have students read this section to answer the question "What are Earth's four spheres, and how do they interact with one another?"

INSTRUCT

VISUAL TEACHING
Discussion
Have students use the illustration of the four spheres to predict what each sphere consists of. Then have them read about the four spheres to find out if their predictions were correct. Ask: What sphere or spheres is water vapor a part of? **atmosphere and hydrosphere** Is all of the geosphere solid? Explain. **No; melted rock, such as lava erupted from volcanoes, is part of the geosphere.** Which sphere is the largest? **the geosphere**

1.2

KEY IDEA
The Earth system consists of four spheres that all affect one another: the atmosphere, the geosphere, the hydrosphere, and the biosphere.

KEY VOCABULARY
- atmosphere
- geosphere
- hydrosphere
- biosphere

VISUALIZATIONS
CLASSZONE.COM
Observe a visual model of Earth's spheres.
Keycode: ES0102

The Earth System's Four Spheres

Earth is unique in the solar system: It is the only planet currently known to support life. This capacity is due to the interactions among its four spheres: the atmosphere, the geosphere, the hydrosphere, and the biosphere. Energy moves back and forth among the spheres within the Earth system.

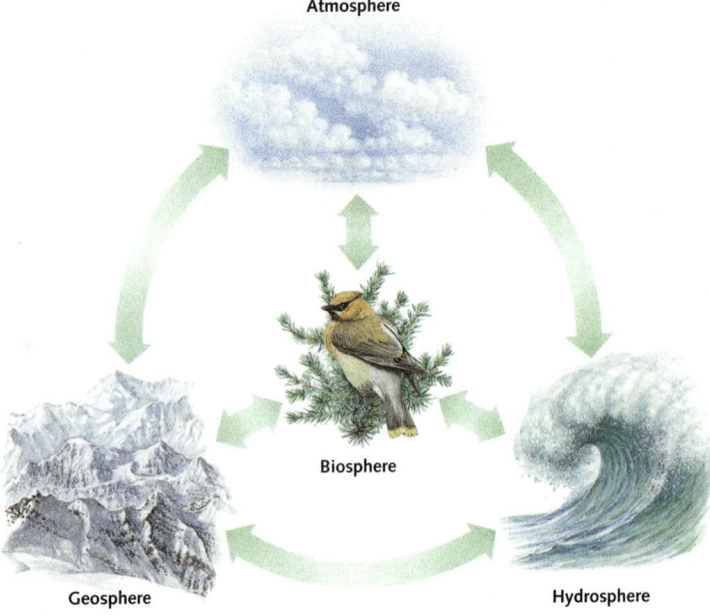

THE FOUR SPHERES of the Earth system interact continually, each affecting the others.

The Atmosphere

The gaseous envelope surrounding Earth is the **atmosphere.** Composed of a mixture of gases, the atmosphere provides living things with oxygen and carbon dioxide. Other gases in the atmosphere absorb and alter rays from the sun, blocking potentially harmful radiation. Changing amounts of water vapor in the atmosphere result in areas of high or low humidity and influence the formation of clouds. Varying areas of atmospheric pressure cause winds. Storms occur when water vapor, air pressure, and energy interact.

Earth's atmosphere is unique in the solar system; it is the only one that contains oxygen. The Moon has no atmosphere. The atmospheres of planets like Venus and Jupiter are thick and filled with gases that would be poisonous to most life forms as we know them. The mix of gases in Earth's atmosphere supports the life forms that thrive on Earth's surface.

THE ATMOSPHERE is a gaseous envelope surrounding Earth.

8 Unit 1 Investigating Earth

Reading Support Help students identify cause-and-effect relationships. Have them make a two-column chart, labeling the columns *Cause* and *Effect,* respectively. Tell students to record in the chart any cause-and-effect relationships that they come across as they read. As an example, point out the phrase "is due to" from the first paragraph. Explain that the phrase signals a cause-and-effect relationship. Then ask: What cause does this phrase refer to? **interactions among Earth's four spheres** What is the effect? **Earth supports life.** After students complete the section, have them share the cause-and-effect relationships they identified.
Use Reading Study Guide, p. 2.

8 Unit 1 Investigating Earth

The Geosphere

The earth itself—the rocks, the mountains, the beaches, and all the other physical features of the planet, except water—makes up the **geosphere.** The geosphere includes the ocean basins and the rock layers beneath your feet, including those that you cannot see. These layers include the mantle that is the source of the lava that pours from volcanoes, and the core that forms the planet's center and generates Earth's magnetic field.

Mineral resources, such as iron and copper, are mined from the geosphere. The stone and concrete used in building materials come from the geosphere. Cities and homes are perched on the geosphere—in some places, more precariously than in others.

Although it happens slowly, the geosphere is ever changing. Volcanic eruptions form new land. Mountains are uplifted and eroded. And Earth's continents are in slow but constant motion.

VOCABULARY STRATEGY

The names of the four spheres include prefixes with Greek roots. *Atmo-* means vapor; *geo-*, earth; *hydro-*, water; and *bio-*, life. By analyzing prefixes, you can often figure out the meaning of unfamiliar terms.

VISUALIZATIONS
CLASSZONE.COM

Observe a visual model of Earth's spheres.
Keycode: ES0102

Visualizations CD-ROM

More about...

The Hydrosphere
The atmosphere above the United States holds an average of 151,000 billion liters of water at any given time. This water is in the form of water vapor, liquid water, and ice. Water is the only substance on Earth found naturally in all three states.

THE GEOSPHERE includes rocks, such as these in Utah, and other physical features of the planet, except water.

The Hydrosphere

The **hydrosphere** contains all the water in the Earth system, including the water in the oceans, lakes, and rivers, and groundwater. The hydrosphere also includes the water locked up in ice and snow at the poles and in high mountains.

Most of Earth's water is salty. Only about 3 percent of the hydrosphere consists of fresh water, and most of that water (70 percent) is frozen, in the form of glacial ice. The remaining fresh water is found in groundwater, in lakes, as soil moisture, as water vapor, and as river water. Of all the water in the hydrosphere, only about one half of one percent is useable fresh water.

All the water on Earth is continually recycled. The water you drank this morning may have irrigated a field last year, flowed through aqueducts during the time of the Roman Empire, or washed around the feet of a dinosaur standing at a river's edge millions of years ago.

THE HYDROSPHERE includes all of Earth's water.

SECTION 2

VISUAL TEACHING
Discussion
Use the photograph on page 11 as a basis of discussion of interactions among the spheres. Ask: Which spheres can you see in the photograph? **all of them** How did each sphere help produce the rock formations? **atmosphere—wind eroded land; hydrosphere—rain and flowing water eroded land; biosphere—plant roots broke up rock; geosphere—earthquake or volcanic activity shaped land**

More about...

The Biosphere
The area in which living things can be found on Earth extends from a few kilometers into Earth's atmosphere to the bottom of the oceans. Although this may seem like a very large area, in relation to the entire Earth system, it occupies only a small space. In fact, if you reduced Earth to the size of an apple, the biosphere would occupy only an area as large as the apple's peel.

INVESTIGATIONS CLASSZONE.COM

How are Earth's Spheres Interacting? Examine images of Earth from the ground and from space. Identify evidence of energy and materials being transferred between spheres.
Keycode: ES0103

The Biosphere
The interactions of the geosphere, atmosphere, and hydrosphere gave rise to the conditions that support life. The living things on the planet compose the **biosphere**.

The biosphere includes all living things, from single-celled protozoans to redwood trees to people. Human activities have changed the face of the geosphere, had an impact on the oceans, and altered the atmosphere. We are living proof that one sphere can affect another.

THE BIOSPHERE encompasses all living things, including these bison crossing the Yellowstone River in Yellowstone National Park, Wyoming.

Interactions Among the Spheres
The four spheres of the Earth system are constantly moving, changing, and interacting. The wind blows. Rocks form, break down, and form again. Animals and plants live, grow, die, and decompose. Water moves from the rivers to the sea.

These actions do not occur in isolation. The four spheres are interacting every day, all around you:

- Volcanoes (geosphere) erupt, sending ash and gases into the air (atmosphere) and sending lava and ash down onto surrounding forests (biosphere) and human habitations (biosphere).
- Plants (biosphere) draw carbon dioxide from the air (atmosphere) and water from the hydrosphere; they release oxygen and water vapor, and then may be eaten by animals (another part of the biosphere).
- Hurricanes (atmosphere) sweep across the ocean (hydrosphere) and onto the land (geosphere), damaging the dwellings of people (biosphere) who live along the coast.
- Human beings (biosphere) drill wells into Earth's crust (geosphere) to draw out groundwater (hydrosphere) for drinking and irrigation of crops (biosphere).

Support for Visually Impaired Students Viewing the references to Visualizations and Investigations throughout this book may be difficult for some students. To assist them, produce printed copies of key images from the activities and enlarge them, either by photocopying or by using an overhead projector.

How Interactions Change the Spheres

Throughout history, the interactions among the spheres have changed the Earth system and the individual spheres. An interaction can take one of many forms. It can be a single event, such as the eruption of a volcano or a flood. It can be a temporary change, such as a cold snap that kills a fruit crop. Or it can be an ongoing, steady process, such as erosion.

Some changes can take thousands or millions of years. Erosion, which you will read more about in Unit 4, illustrates the effect of the hydrosphere and the atmosphere on the geosphere. Over time, water breaks down rock, and running water and ice carry it away. Wind can transport fine grains of rock great distances. These changes are all part of the rock cycle, which you will read about in Chapter 6.

Early in Earth's history, the atmosphere contained a high percentage of carbon dioxide and relatively little free oxygen. When tiny organisms and, later, plants evolved, they began to photosynthesize carbon dioxide into oxygen. Over time, photosynthesis resulted in an increase in the amount of oxygen available in the atmosphere; and when land animals evolved, they breathed this oxygen. Similarly, the cooling of the atmosphere that took place during the great Ice Age led to the formation of glacial ice from the liquid water on the planet; some plants, animals, and insects died, while others adapted to the change and survived.

Interactions also cause changes on a more personal scale, especially when they involve the biosphere. Extended periods of stormy or cold weather can disrupt bird nesting seasons, cutting them short or eliminating them completely. For example, the population of lesser snow geese fell from 400,000 to less than 50,000 during a 10-year period when storms kept them from nesting on Wrangel Island in the Arctic Ocean.

Human beings, too, are affected by these interactions. Storms can damage homes and crops. Periods of high rainfall may lead to bountiful harvests, while droughts may cause crop failures. Human beings also alter the other spheres. Since the industrial revolution—beginning in the late 18th century—the amount of carbon dioxide in the atmosphere has increased, which has implications not only for human beings, but for all life on Earth.

INTERACTIONS among the spheres have resulted in these remarkable rock formations in Arizona.

25-Minute Mini LAB

How the Biosphere Affects the Hydrosphere

Materials
- clear plastic bag
- potted plant
- tape

Procedure
1. Place the clear plastic bag over the plant.
2. Wrap the tape around the open end of the bag and the plant's stem to secure the bag.

Tape

3. Two days later, observe the contents of the bag.

Analysis
Describe what you see in the bag. What do you think is happening in the bag? How does this connect the hydrosphere and the biosphere?

INVESTIGATIONS
CLASSZONE.COM

How are Earth's Spheres Interacting?
Keycode: ES0103

Provide students with copies of the illustration of the four spheres on page 8. Students can write their evidence for interaction among the spheres on the appropriate part of the illustration.

Use Internet Investigations Guide, pp. T7, 7.

MINILAB

How the Biosphere Affects the Hydrosphere

Materials
- Use a leafy green plant, ideally one that needs frequent watering. The plant should have only green leaves on its stem.
- The soil should be moist.
- Instead of potted plants, the class could use a shrub or tree. Each student would place a bag around a different group of leaves at the end of a stem.

Analysis Answers
- Drops of water should appear on the inner surface of the bag.
- The plant takes in water from the soil and releases water vapor from its leaves.
- Water is part of the hydrosphere, and the plant is part of the biosphere.

For proper behavior in a laboratory setting, refer students to Safety in the Earth Science Laboratory on pages xxii–xxiii.

Developing English Proficiency List the Minilab materials on the board. Identify the materials by displaying each and pointing to the appropriate word on the board as you say it aloud. Then read aloud each step in the Procedure and demonstrate the action. To illustrate the two days in step 3, point to the two days on a calendar.

Find out more about careers in ecology.
Keycode: ES0104

ASSESS

① The atmosphere is a gaseous envelope surrounding Earth.

② No; the geosphere changes due to volcanic eruptions, uplifting, erosion, and movement of continents.

③ oceans, lakes, rivers, groundwater, ice, snow, glaciers, and water vapor

④ **Sample Response:** drink water; pollute rivers and lakes

⑤ **Sample Response:** An increase in rainfall might affect plant growth (biosphere) and cause flooding that erodes the land (geosphere).

⑥ **Sample Response:** Rain (hydrosphere) would fall from the sky (atmosphere). Strong winds (atmosphere) might knock down tree branches (biosphere). Plants (biosphere) would take in some rain through their roots. The soil (geosphere) would absorb some rain.

CAREER

Tropical Rainforest Ecologist

Imagine that your laboratory was located 50 meters above ground, surrounded by the canopy of a tropical rainforest. You might hang suspended from a crane that reaches from the ground far below into the treetops. Your research subjects would be exotic plants and animals few people have seen and even fewer have studied. For some rainforest ecologists, such working conditions are routine. These scientists study rainforest plant and animal species by observing them in their native habitats. A critical part of their job is to assess the impact of development on rainforest plants and wildlife. Rainforest ecologists also work with governments and local people to protect rainforest ecosystems by establishing wildlife preserves. For some of these scientists, teaching the public about the importance of conservation and the preservation of native wildlife is especially satisfying.

A rainforest ecologist usually holds at least a bachelor's degree in ecology, biology, or another related science. A master's or a doctoral degree may be required for ecologists who work for some research or academic institutions. An interest in wildlife conservation and an enjoyment of working outdoors also come in handy. Organizational and research skills, the ability to use computer databases, and readiness to work as part of a team are also essential. ■

ECOLOGISTS working in the rainforests of Sri Lanka and in other rainforests around the world are continually discovering new species of life.

Find out more about careers in ecology.
Keycode: ES0104

While the Earth's four spheres can be and have been studied separately, it is important to keep in mind how they interact. Changes to one sphere have profound effects on one or more of the others. The interconnected nature of the four spheres is well illustrated by the cycles that occur among them, as shown in Section 1.3.

1.2 Section Review

① What is the atmosphere?

② Is the geosphere static and unchanging? Explain your answer.

③ Which features on Earth make up the hydrosphere?

④ Name two ways in which human beings affect the atmosphere, geosphere, or hydrosphere.

⑤ **CRITICAL THINKING** How might an increase in rainfall in an area affect the geosphere and the biosphere?

⑥ **WRITING** Describe the interactions among the spheres that would occur during a spring rainstorm in your area.

If students miss . . .

Question 1 Have students reread the information about the atmosphere. (p. 8) Ask: What is the atmosphere composed of? Where is it located?

Question 2 Reteach "The Geosphere." (p. 9) Have students find examples in the text of how the geosphere changes.

Question 3 Work with students to brainstorm a list of features on Earth that contain water.

Question 4 Reteach "Interactions Among the Spheres." (p. 10) Ask students to speculate how humans might affect each sphere.

Question 5 Review the parts of the geosphere and the biosphere. (pp. 9–10)

Question 6 Have students describe what happens in their area during a rainstorm. Ask them to identify the interactions of the spheres in each event they describe.

Cycles and the Earth

A **cycle** is a sequence of events that repeats. Some cycles repeat over relatively short periods. Other cycles may occur over millions of years. On Earth, the water cycle, carbon cycle, and energy cycle all work together to maintain a dynamic planet.

The Water Cycle

Of the many cycles that happen on Earth, the **water cycle** is the one you probably notice most, especially when it rains or snows.

Most of Earth's water is in liquid form in the oceans. When the sun shines on the ocean, or on any body of water, it causes some of the water to evaporate, thus becoming water vapor, the gaseous phase of water. As water vapor rises into the atmosphere, it cools; and when water vapor cools, it condenses to form clouds. As the clouds cool further, rain begins to fall, or, depending on the temperature, it snows.

The rain falls to Earth. If rain flows over the ground and into a body of water, it is called *runoff*. Runoff flows into streams and rivers and eventually back to the sea. Water can also soak into the ground where it might be stored as groundwater in the small spaces between particles of soil, sand, and rock. It moves slowly, but eventually flows back to the ocean.

Some of the water evaporates again quickly or is breathed out by the leaves of plants in a process called **transpiration.** Scientists who study the hydrosphere call these processes **evapotranspiration.**

When water returns to the oceans, one turn of the water cycle is complete. The energy of the Sun drives the water cycle. All Earth's water, whether fresh or salt, has moved through this water cycle millions of times, over millions of years. Water is never created or destroyed—only changed.

1.3

KEY IDEA
The water cycle, the carbon cycle, and the energy cycle all involve interactions among the four spheres of Earth.

KEY VOCABULARY
- cycle
- water cycle
- evapotranspiration
- carbon cycle
- energy cycle
- solar energy
- geothermal energy
- tidal energy

VISUALIZATIONS
CLASSZONE.COM

Observe a raindrop traveling through various paths of the water cycle.
Keycode: ES0105

WATER cycles through the Earth system in solid, liquid, and gas forms, but the total amount of water remains relatively constant.

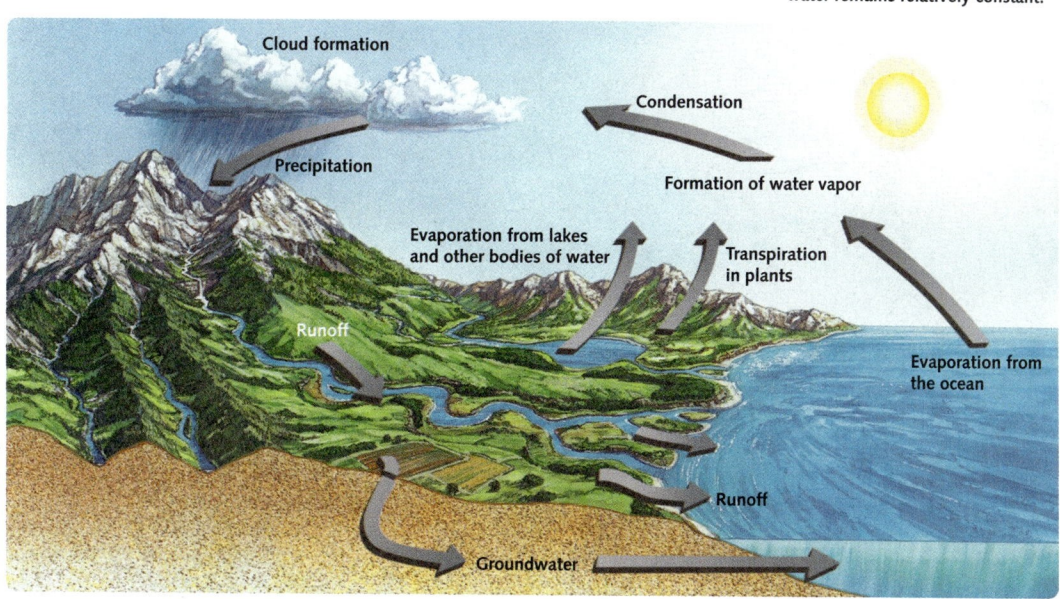

Chapter 1 Earth as a System 13

FOCUS

Objectives
1. Describe characteristics of the water, carbon, and energy cycles.
2. Analyze how humans interact with the water, carbon, and energy cycles.

Set a Purpose
Have a volunteer read aloud the headings in this section. Then have students form a question to set a purpose for reading. For example, "What are the water cycle, the carbon cycle, and the energy cycle?"

INSTRUCT

More about...

The Water Cycle
The atmosphere contains about 0.001 percent of the total volume of water on Earth. About 1.5×10^{18} gallons or 2.3×10^{15} liters of water evaporate from the land into the air each day.
Use Transparency 1.

VISUALIZATIONS
CLASSZONE.COM

Observe a raindrop traveling through various paths of the water cycle.
Keycode: ES0105

 Visualizations CD-ROM

Hands-On Demonstration Model the water cycle. Place a small, clear cup in the center of a large, clear bowl. Pour a small amount of water into the bowl, being careful not to get any in the cup. Cover the bowl with plastic wrap and seal it with a rubber band or string. Place a small weight in the center of the plastic wrap over the cup. Place the setup in the sun. After a few hours, students should observe water inside the cup. Ask them to explain what happened. The heat of the sun evaporated the water in the bowl. The water vapor condensed on the cool plastic and formed droplets. The droplets fell into the cup.

Chapter 1 Earth as a System 13

VISUAL TEACHING

Discussion

Have students read the captions in the illustration. Then ask: Why is the movement of carbon on Earth described as a cycle? **Carbon continuously moves through the Earth system as it is changed from one form to another.** Where is carbon found? **methane, carbon dioxide, seashells, plant and animal tissue, fossil fuels** How is carbon dioxide removed from the atmosphere? **Plants take in carbon dioxide during photosynthesis; carbon dioxide dissolves in ocean waters.**

VOCABULARY STRATEGY

When you encounter a long term such as *biogeochemical,* look for the smaller words or word parts within it. In biogeochemical you will find *bio-* (life), *geo-* (earth) and *chemical.* A biogeochemical cycle is one that involves Earth, chemistry, and life forms.

The Carbon Cycle

In a biogeochemical cycle, a chemical element or compound is changed as it moves through the Earth system. The **carbon cycle** is a biogeochemical cycle involving the element carbon (C).

Carbon has been called the building block of life. It is present in all organic material and in materials, such as coal and oil, that are derived from once-living things. You may think of carbon as primarily a solid, but it can also form gases, such as methane (CH_4) and carbon dioxide (CO_2).

Carbon enters the atmosphere in several ways as carbon dioxide. Living things, such as animals, breathe it out. Organisms that break down decaying organic matter give off carbon. When carbon-based materials burn, such as when trees are consumed in a forest fire, releases carbon dioxide into the atmosphere. Volcanic eruptions release carbon dioxide from inside Earth. Carbon dioxide also diffuses out of the ocean waters in which it is dissolved.

Carbon is also removed from the atmosphere. Plants remove carbon dioxide from the atmosphere during photosynthesis, and then release oxygen. The carbon is stored in their tissues as carbohydrates and passed on to animals that eat the plants.

Phytoplankton also play an important part in the carbon cycle. Like land plants, these tiny oceanic plants undergo photosynthesis. During photosynthesis, they take in carbon, in the form of carbon dioxide, and then release oxygen, which diffuses out of the water.

A FOREST FIRE, such as this one in Yellowstone National Park, returns stored carbon to the atmosphere.

Volcanoes add CO_2 to the atmosphere.

Forest fires add CO_2 to the atmosphere.

Carbon enters the ocean as CO_2 which is used by phytoplankton in photosynthesis.

Carbon dioxide dissolves in the ocean and the carbon is converted into carbon compounds.

When carbon compounds break down, CO_2 escapes into the atmosphere.

Humans and other animals exhale CO_2.

Reading Support Model for students how to develop a learning log that details their questions and ideas as they read. For example, pause after reading "The Carbon Cycle" and ask "How many plants would it take to form one piece of coal?" Tell students that when they finish reading, they should go back over their questions to see if they can add answers or additional information to their logs.
Use Reading Study Guide, p. 3.

When the phytoplankton die, they sink to the bottom of the ocean, taking with them the carbon stored in their cells. Sediment covers the dead phytoplankton and the carbon within them enters storage. A place that stores carbon, such as the ocean, is called a *carbon sink*.

The action of the ocean waves dissolves carbon dioxide in seawater; there, it is converted into bicarbonate and carbonate compounds, such as calcium carbonate (or lime), the major component of seashells. Through wave action and through photosynthesis by phytoplankton, the ocean removes about 40 percent of the carbon that is released into the atmosphere by the burning of fossil fuels.

Once removed from the atmosphere, carbon can be stored for long periods. In some cases, it is stored only for the life span of the plant or animal whose tissues contain it. When this plant or animal dies, its tissues are decomposed by bacteria. The decomposition process changes the carbon compounds in the tissues into carbon dioxide and methane.

Some parts of the carbon cycle happen quickly, such as when organisms live, die, and decay in a short time. Other parts of the cycle take longer, such as the formation of fossil fuels. If a plant or animal dies and seeps into a low-oxygen environment, such as marsh land, the carbon in its tissues—given enough time—can change into fossil fuels such as coal and oil. When burned, these fossil fuels release the stored carbon back into the atmosphere. Carbon, like water, is never destroyed; it is only changed from one form to another.

Observe an animation showing evidence of the carbon cycle.
Keycode: ES0106

PLANTS, such as these in the temperate rainforest on Washington's Olympic Peninsula, store absorbed carbon in their tissue.

CHAPTER 1 SECTION 3

Observe an animation showing evidence of the carbon cycle.
Keycode: ES0106
Visualizations CD-ROM

VISUAL TEACHING
Discussion
Discuss the forms that carbon assumes throughout the carbon cycle. Point to a particular object in the illustration and ask: In what form is carbon found here? **Possible examples include the fossil fuel in the truck and car; the carbohydrates in the plants and animals; the carbon dioxide in the atmosphere and dissolved in the ocean.**

Burning fossil fuels releases carbon back into the atmosphere.

Plants remove CO_2 from the atmosphere.

Carbon from tissue of decomposing plants and animals is converted to CO_2 and methane. Fossil fuels come from the carbon compounds of decomposed tissue.

DIFFERENTIATING INSTRUCTION

Hands-On Demonstration Demonstrate how carbon is stored in some ocean animals. You will need natural chalk (calcium carbonate), not dustless chalk. Explain that long ago, some ocean animals used the carbon dioxide dissolved in the water to make their shells. When the animals died, their shells sank to the ocean bottom and formed chalk. Crush a piece of chalk and put it in a glass beaker. Add vinegar and have students observe the gas that forms. Explain that the gas is carbon dioxide, which contains the same carbon that was stored in the chalk from ancient times.

⚠ *Exercise caution when working with glassware. Remind students that safety is everyone's responsibility. See Safety in the Earth Science Laboratory on pages xxii–xxiii.*

CHAPTER 1 SECTION 3

VISUAL TEACHING
Discussion
Review the summary chart with students. Ask: Which is the greatest source of energy? **the sun** Which type of energy is produced by the moon? **tidal energy** Which type of energy originates within Earth? **geothermal energy**

More about...

Geothermal Energy
Near Los Alamos, New Mexico, scientists have drilled parallel wells 3500 meters into the ground to heat water. At that depth, the temperature of the rock is about 230°C. Scientists use extremely high pressure to pump water down one of the wells. This causes the rock to crack, forming fissures that connect the two wells. The water that is forced down the well is heated, turns to steam, travels through the fissures to the second well, and is pumped back to the surface. Researchers are trying to perfect this technology so that the steam can be used to generate electricity without polluting the environment.

VOCABULARY STRATEGY
Applying what you know about the prefix *geo-*, what is the meaning of *geothermal*?

The Energy Cycle

The water and carbon cycles are like wheels, with water and carbon continually moving back and forth between the four great spheres. The **energy cycle** is different. It is more like a scale you would use in the laboratory. When you measure materials on the scale, you're usually looking for the two sides of the scale to balance; what is on one side should equal what is on the other. The Earth system works in a similar way in regard to energy.

The amount of energy that enters the system should equal the energy that is removed. If the planet were to take in more energy than it releases, the climate would get warmer. If it released more energy than it gained, the climate would turn cooler. Because of the balance-scale nature of Earth's energy cycle, scientists also call it *Earth's energy budget*. There are three main sources of energy for the Earth's energy budget: solar energy, geothermal energy, and tidal energy.

Most of the energy that enters the Earth system (99.895 percent) comes from the sun. **Solar energy** drives the winds, the ocean currents, and waves. It is also the source of the energy that causes rocks to weather, forming soil. You will read more about this in Chapter 17.

A much smaller part of the energy budget (0.013 percent) originates from within Earth, as **geothermal energy.** Geothermal energy drives the movement of Earth's oceans and continents; powers volcanoes, geysers, and earthquakes; and plays an important part in the rock cycle.

The third and smallest part of the energy budget (0.002 percent) is **tidal energy.** Tidal energy results from the Moon's pull on Earth's oceans. Although small when compared to solar energy, tidal energy is powerful enough to slow down Earth's rotation, acting as a brake as Earth moves through the tidal "bulge."

SUMMARY Sources of Earth's Energy

	Sun	Earth	Tide
Type of energy	Solar	Geothermal	Tidal
Percentage of total energy	99.895	.013	.002
Where does it come from?	Nuclear fusion reactions in the Sun	Decay of radioactive materials	Pull of the Moon on Earth's oceans

---

DIFFERENTIATING INSTRUCTION

Developing English Proficiency Explain that *geo* means "earth" and *thermal* means "heat." *Geothermal* means "heat from the earth." Refer students to the Word Roots and Prefixes section of the Glossary. (p. 737) Read the entry for *Geo-*. Have students look up the definitions of the sample words—*geology* and *geocentric*—in a dictionary. Point out the references to earth in the definitions. Then have students look up other words that begin with *geo-* and relate them to earth.

To maintain the balance of the energy budget, this incoming energy must go somewhere. About 40 percent of it is reflected back into space without being changed. Different areas of Earth's surface and different types of clouds reflect varying amounts of energy.

The remaining solar energy, along with the tidal energy and the geothermal energy, is used within the Earth system, where the energy may be used to bring about evaporation and precipitation in the water cycle. It may be changed to wave and wind energy or be converted to heat energy and radiated back into space. Some of the energy is stored in water and ice, in plants, and even in sedimentary rocks. The great reservoirs of fossil fuels, coal, gas and oil, were formed from dead plants—and so fossil fuel energy is in reality old solar energy.

As the energy moves through the Earth system, it is changed. With every change, a little bit of energy is converted to heat and is lost to the cycle. According to a basic principle of physics, energy can never be completely recycled. This degradation of energy is an important difference between the energy budget and the carbon and water cycles.

The Laws of Thermodynamics

Although it does not behave like carbon and water, which are both forms of matter, energy follows certain predictable rules that explain what it will do. These rules are called the laws of thermodynamics. Thermodynamics is a branch of physics that deals with how heat energy is converted into other forms of energy. The laws of thermodynamics deal with how energy flows.

The first law of thermodynamics states that energy can never be created or destroyed, only changed from one form to another. This can happen many times. Solar energy can be stored in plants, which die and eventually become fossil fuels. Fossil fuels can be burned at an electric power plant to generate electricity, which then powers a light bulb.

The second law of thermodynamics states that when energy changes, it is converted from a more useful, more concentrated form to a less useful, less concentrated form. This means that, unlike water, which can turn from ice to water to vapor and back without harm, energy can never be recycled completely. Some energy will always be lost, usually as heat. As the solar energy moves through the changes described above, it is gradually degraded into less useful forms.

The Effects of Earth's Surface

Earth's surface is not uniform. It is covered by oceans, deserts, grasslands, forests, cities, and glaciers. These different parts of the earth reflect solar energy at various rates. The percentage of energy that is reflected without being changed is called the *albedo*.

A forest has a low albedo, reflecting between 5 and 10 percent of the energy that reaches it. A field of freshly fallen snow has a considerably higher albedo than a forest, reflecting 80 to 90 percent of the energy that reaches it from the sun. Desert areas fall between the two, reflecting about half of the energy back into space.

Scientific Thinking

OBSERVE
What human-influenced developments in the atmosphere are changing the balance of Earth's energy budget? What is being done to offset those effects?

Scientific Thinking

OBSERVE
Students might consider the effects of ozone depletion, burning fossil fuels, deforestation, and development.

More about...

Energy Loss
As energy moves through a food chain, the total energy transfer from one level to the next is only about 10 percent. The other 90 percent of energy is lost because an organism doesn't eat all its prey, and not all the food that is eaten is digested. Also, each organism uses energy for its own life functions, so that energy is not available to the organisms above it in the food chain.

Challenge Activity Have small groups of students research the use of solar energy to generate electricity and heat homes. Have them investigate methods for generating and storing solar energy. Ask groups to present recommendations on the usefulness of solar energy.

CRITICAL THINKING

Have students analyze the albedo in their area. Ask: What was the area like before people settled? Did the people change the area's albedo? If so, how?

ASSESS

1. Water evaporates from the ocean to form water vapor, which rises and cools. As it cools, it condenses and forms clouds. As the clouds cool, precipitation falls to Earth's surface, where it runs off or seeps into the ground and eventually makes its way back to the ocean.

2. Plants remove carbon dioxide from the atmosphere during photosynthesis. They store the carbon in their tissues as carbohydrates. When animals eat plants, the carbon in the plant tissues passes to the animals. When plants and animals die, bacteria decompose them, producing carbon dioxide in the process.

3. As water and carbon move through the Earth system and change form, their total amounts do not change.

4. Models should be labeled with explanations.

5. With fewer trees, more energy will be reflected to the atmosphere, so the albedo will increase. More carbon dioxide will enter the atmosphere due to the increased use of fossil fuels in the city and the decrease in plant growth.

Although albedo varies across the Earth's surface, the planet's overall albedo is about 30 percent. In contrast, the Moon's albedo is about 11 percent; the albedo of cloud-covered Venus is about 75 percent.

An area's albedo can change. Thus, the amount of energy that is absorbed or reflected changes as well, and that can alter the energy budget of Earth. Some areas may have a thick snow or ice cover during the winter that melts in the spring. During the winter, this area would reflect more of the sun's energy, and thus have a higher albedo, than it would during the rest of the year.

If a green meadow is plowed for crop planting, its albedo will change. If a forest is cut down and replaced with houses, that area will reflect more energy back into the atmosphere. If former farmland is allowed to go fallow and reforests, its albedo changes as well. The alterations in the energy cycle brought about by transformations to the landscape are further examples of how each part of the Earth system affects other parts.

Human Activity and the Cycles

Like the spheres of the Earth system, the water, carbon, and energy cycles do not exist in isolation. The models presented in this section become more complicated when you consider the interactions among them. The water, carbon, and energy cycles are also affected by human activity.

By changing the albedo of an area, human beings can alter the amount of solar energy it absorbs or reflects, which in turn can affect the energy budget. By burning more fossil fuel, humans can put more carbon dioxide into the air—and by preserving and planting tracts of vegetation, they can remove carbon dioxide. By damming rivers and withdrawing groundwater, humans can lengthen the time it takes a water molecule to move through the water cycle. Every action has an effect, like the ripples that spread out on a pond's surface after you throw in a pebble. Earth system science is working to find out just how far those ripples spread.

REFORESTATION These jack pines growing in a clearcut area in Huron National Forest in Michigan are part of a reforestation project, a human activity that positively affects the carbon cycle.

1.3 Section Review

1. Describe the water cycle.
2. Summarize the carbon cycle, starting and ending with carbon dioxide in the atmosphere.
3. Why is the energy budget more appropriately described as a balance while the water and carbon cycles are more accurately described as circles?
4. **BIOLOGY** Draw a model showing how one of the cycles interacts with living things. Show relationships not mentioned in the text.
5. **CRITICAL THINKING** Humans cut down a large tract of forest, burn the cut trees, and turn the land into a city. Predict what changes are going to result, based on what you've learned about the carbon cycle, energy budget, and albedo.

If students miss . . .

Question 1 Have students trace the movement of water in the illustration on page 13.
Question 2 Refer students to the illustration of the carbon cycle on pages 14 and 15 and read the captions aloud.
Question 3 Reteach "The Energy Cycle." (pp. 16–17) Ask: How does energy enter and leave the Earth system?
Question 4 Review with students that during photosynthesis, plants use the sun's energy, carbon dioxide, and water to make carbon compounds.
Question 5 Reteach "The Effects of Earth's Surface" (pp. 17–18) and have students reread the captions for the illustration of the carbon cycle. (pp. 14–15)

SCIENCE & Society

Using Landscaping to Save Water

In the United States, the average household uses about half of its water on plants that enhance the landscaping around the house.

Can a new landscaping style save time while saving water?

Do you spend time on the weekend helping with your home's landscaping? As you water, weed, and mow, do you think, "There's got to be a better way"?

On average, 40 to 50 percent of the water consumption of a typical American household goes to water its outdoor plants and lawn. In areas such as Albuquerque, New Mexico, and Phoenix, Arizona, that need to conserve water, local officials and conservation groups are teaching residents the benefits of low-water landscaping.

Low-water landscaping makes use of native plants, which are acclimated to the rainfall and temperature conditions of the area in which they live. As a result, they don't need to

NATIVE CACTI are included as part of the landscaping on this Sedona, Arizona, property.

be watered as much as nonnative species. Some of these plants, such as cacti, are especially well adapted to arid climates because they do not transpire as much water as other plants.

Around the country, horticulturists (specialists who work with plants) are trying to find, develop, and make available plant species that are attractive and yet require little water.

Low-water landscaping does not require homeowners to replace all their plants or to give up their lawns entirely. It saves water (and time) because the plants are carefully chosen and need less water than grass does. It also involves improving the soil so it can absorb and retain more moisture, making smaller plots of thirsty lawn, and making efficient use of irrigation.

Extension

SCIENCE NOTEBOOK

Consider how the water cycle in an arid environment might differ from one in a wetter environment. What are some characteristics of desert plants that enable them to survive for longer periods without water? Write a description in your science notebook.

DATA CENTER CLASSZONE.COM

Learn more about the plants and techniques that the Xeriscape method uses.
Keycode: ES0107

LANDSCAPING around this house in Miami, Florida, includes native plants.

PURPOSE

To gain experience with lab equipment and procedures, including making measurements, computing, and graphing

MATERIALS

Provide students with a copy of Lab Sheet 1. A copymaster is provided on page 137 of the teacher's laboratory manual. The cube, cylinder, and block should be of the same material. Limestone chips are sometimes called marble chips. Soak them before use.

PROCEDURE

❶, ❻ You might want to copy the Density Charts for students in advance.

❷, ❸, ❹ You may want to write the following equations on the board:

$$\text{density} = \frac{\text{mass}}{\text{volume}}$$

volume of cube or rectangle:
volume = length × width × height

volume of cylinder with radius r:
volume = π × radius² × height

❺ Accepted densities for objects can be given, measured by you, or determined by class consensus.

❽ Soaking the chips in advance prevents distortion of the measurements by water absorption. Splashing water out of the cylinder may change the measurements.

SKILLS AND OBJECTIVES

- **Measure** the mass and volume of several objects.
- **Compute** and **compare** the density of these objects.
- **Graph** and **interpret** data.

MATERIALS

- balance
- cubic block
- metric ruler
- Appendix C Area and Volume, page 728
- rectangular block
- cylindrical block
- 100-mL graduated cylinder
- 50-mL beaker
- wet limestone chips
- paper towels

The Density of Earth Materials

Density, a measure of the amount of material (mass) in a given space (volume), is expressed as the ratio $D = m/V$. Differences in the densities of Earth's matter are the basis of many common processes. Wind, ocean currents, and plate tectonics are all driven by differences in density.

Even careful measurements of density contain some error. The **percent error** is the amount by which any measurement differs from the accepted value. Percent error is determined by using the following formula:

$$\text{Percent error} = \frac{\text{Calculated measurement} - \text{accepted measurement}}{\text{accepted measurement}} \times 100$$

In this lab, you will measure both mass and volume for regularly and irregularly shaped objects, and use these data to calculate densities.

Procedure

Part A: Finding the Density of Regular Solids

❶ Use the balance to measure the mass of the cube to the nearest tenth of a gram. Copy Table 1 and record the mass.

Table 1: Density Chart for Regular Solids							
Object	Mass (g)	Length (cm)	Width (cm)	Height (cm)	Volume (cm)	Density (g/cm³)	Percent Error
cube			—	—			
block							
cylinder			—				

**Round off all values to the nearest tenth of the given unit.

❷ Measure and record the dimensions of the cube. Using the formula on page 728, compute and record the cube's volume.

❸ Calculate the density of the cube using its mass and volume.

❹ Repeat Steps 1–3 using the rectangular block and the cylinder.

❺ Obtain the accepted value for the density of each object from your teacher. Use this value to calculate the percent error for each object.

Part B: Finding the Density of Irregular Solids

❻ Pour about 50 mL water into a 100-mL graduated cylinder. Copy Table 2 and use it to record the measurements. Record the water's volume to the nearest 0.1 mL. Leave the water in the cylinder.

❼ Find the mass of the cylinder + water. Record the mass.

Table 2: Density Chart for Irregular Solids

Solid	Mass Cylinder + Water (g)	Mass Cylinder + Water + Chips (cm)	Mass Chips (g)	Volume Water (mL)	Volume Water + Chips (cm)	Volume Chips (mL)	Density Chips (mL)
5 chips							
10 chips							

**Round off all values to the nearest tenth of the given unit.

8 Use the 50-mL beaker to obtain a sample of wet limestone chips. Select 5 chips and blot them with a paper towel. Slide the chips into the cylinder, without splashing any water. Record the volume of the water plus the chips. Leave the chips in the cylinder.

9 Subtract the volume you calculated in Step 6 from the volume in Step 8. Record this value as the volume of the 5 chips.

10 Find the mass of the cylinder containing the water and the 5 chips.

11 Subtract the mass of the cylinder and water from the mass of the cylinder, water, and 5 chips. Record your result as the mass of the 5 chips.

12 Use the formula $D = m/V$ to determine the density of the 5 limestone chips. Record the density in the table.

13 Repeat Steps 8–12 three times using a total of 10, 15, and 20 chips.

14 Copy the graph to the right and plot the data for the mass and volume of the chips. Draw a best-fit line for your data.

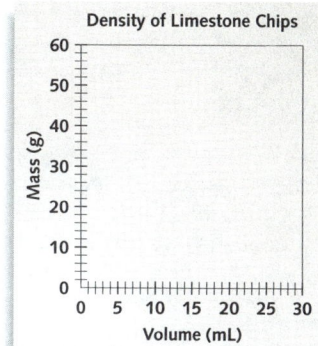

Analysis and Conclusions

1. Why are the densities of the regular solids about the same?

2. If one block were cut in two pieces, what would be the density of each piece? Why?

3. Why do the points on the graph nearly form a straight line? What does that say about the ratio of mass to volume? (Hint: Note that the slope of your line equals the density.)

4. On the graph, draw and label a line for a material with a density of 1.0 g/mL (slope equals m/V or 1.0 g/mL). Where is this line relative to the line for the chips? Explain.

5. On the graph, draw and label a line for a material with a density of 4.0 g/mL. Where is this line relative to the line for the chips? Explain.

6. What are some potential sources of error in this investigation?

ANALYSIS AND CONCLUSIONS

1. The densities are the same because all three objects are made of the same material.

2. The density of each piece would be the same as the density of the original piece because the material is still the same.

3. All the limestone chips are made of the same material with the same density, so they have the same ratio of mass to volume. Therefore, the graph is a straight line.

4. The line representing a material with a density of 1.0 g/mL is below (or to the right of) the limestone line because the material is less dense than the limestone chips.

5. The line representing a material with a density of 4.0 g/mL is above (or to the left of) the limestone line because the material is more dense than the limestone chips.

6. measuring the mass inaccurately, rounding off numbers, excess water remaining on the chips, splashing water out of the cylinder

Refer students to Safety in the Earth Science Laboratory on pages xxii–xxiii.

Chapter 1 Review

Summary of Key Ideas

1.1 Technological advances and environmental incidents have encouraged scientists to look at the earth as a system. A system is a kind of a model that allows scientists to study a process or phenomenon with time as a variable. Because it exchanges little matter with its surroundings, Earth is an essentially closed system.

1.2 The Earth system includes four spheres: the hydrosphere, atmosphere, geosphere, and biosphere. The atmosphere is the gaseous envelope surrounding Earth. The geosphere consists of all the physical features on Earth except water. The earth's water makes up the hydrosphere. Living things, including plants, animals and people, make up the biosphere. The spheres interact and change constantly.

1.3 All water on Earth is continually moving through the water cycle, which includes evaporation, transpiration (evapotranspiration), and precipitation. A biogeochemical cycle involves the movement of an element, such as carbon, through the four spheres of the Earth system. The Earth system has three energy sources: solar energy, geothermal energy, and tidal energy.

Key Vocabulary

atmosphere (p. 8)
biosphere (p. 10)
carbon cycle (p. 14)
closed system (p. 5)
cycle (p. 13)
Earth system science (p. 5)
energy cycle (p. 16)
evapotranspiration (p. 13)
geosphere (p. 8)
geothermal energy (p. 16)
hydrosphere (p. 8)
model (p. 5)
open system (p. 6)
solar energy (p. 16)
system (p. 5)
tidal energy (p. 16)
water cycle (p. 13)

Vocabulary Review

Answers (left column)
1. carbon cycle
2. biosphere
3. solar energy
4. model
5. Earth system science
6. cycle
7. hydrosphere

Concept Review

8. Water in the atmosphere (water vapor) condenses and comes to Earth's surface in the form of precipitation (snow, rain). Liquid water may collect underground as groundwater or runoff aboveground, into rivers and lakes, eventually moving toward the oceans. Some surface water evaporates, forming water vapor that can condense into clouds to become precipitation.

9. Plants take in carbon dioxide from the atmosphere during the process of photosynthesis, making carbohydrates and releasing oxygen. The carbon-based carbohydrates serve as a food source for animals.

10. Advances in technology allow scientists to study many aspects of Earth, integrating information about the atmosphere, biosphere, geosphere, and hydrosphere into a systems model.

11. The geosphere encompasses all the physical features of Earth and the rock materials they are made of. It includes external features, such as mountains and beaches, as well as internal features of Earth's layers.

12. Water vapor comes from evaporation of bodies of water, thus, it is part of the hydrosphere. However, it is located in the atmosphere, thus, it can be seen as part of the atmosphere.

13. Water Cycle: evapotranspiration, condensation, runoff;
    Carbon Cycle: biogeochemical, carbon sinks, photosynthesis;
    both: continuous.

Vocabulary Review

Write the term from the key vocabulary list that best completes the sentence.

1. An example of a biogeochemical cycle is the ___?___.
2. Human beings are part of the ___?___, one of the four spheres of Earth.
3. The largest source of energy for the Earth system is ___?___.
4. To understand how something works, scientists may create a(n) ___?___ of it.
5. The name given to Earth science that looks at the planet as a whole is called ___?___.
6. A(n) ___?___ is a sequence of events that repeats.
7. A river and a pond are both part of the ___?___.

Concept Review

8. Describe how water moves through the water cycle.
9. What role do plants play in the carbon cycle?
10. Describe the factors that have led to the rise of Earth system science.
11. What is included in the geosphere?
12. Explain how water vapor could be considered part of the atmosphere as well as part of the hydrosphere.
13. **Graphic Organizer** Copy and complete the Venn diagram below by adding the following terms in the appropriate places: biogeochemical; evapotranspiration; carbon sinks; condensation photosynthesis; runoff; continuous.

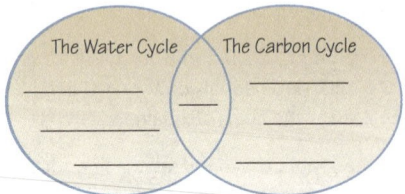

Critical Thinking

14. In a closed system, energy can enter but matter does not enter or leave. In an open system, the system and its surroundings exchange both energy and matter. Earth is considered a closed system because energy enters from the sun but very little matter either enters or leaves Earth.

15. Since energy entering the system of the planet would be heat energy, and there is less coming in than going out, the planet would probably be cooling down.

16. Many Earth processes take place over thousands and even millions of years. This needs to be considered when studying causes, effects, and interactions of spheres within the Earth system.

Critical Thinking

14. **Compare** How do an open system and a closed system differ? Why is Earth essentially a closed system?
15. **Predict** What would happen to a planet if the amount of energy coming into its system was less than the energy that was leaving its system?
16. **Analyze** A system can be defined as a model that includes time as a variable. Explain the effect of time on a system model.
17. **Infer** Is it possible for a planet to be a completely closed system? Why or why not?
18. **Communicate** Sometimes, human beings are considered the fifth sphere of Earth, the "androsphere." Do you agree with this separate classification? Why or why not?

Internet Extension

How Do Interactions among Earth's Spheres Vary Regionally? Compare and contrast interactions between Earth's spheres in your region with other regions of the world.
Keycode: ES0108

Writing About the Earth System

 SCIENCE NOTEBOOK Describe a time when you have experienced interactions among the hydrosphere, geosphere, atmosphere, and biosphere. The experience can be a common, everyday occurrence.

Interpreting Graphs

Scientists use diagrams such as this one to compare various aspects of the water cycle. This graph depicts three variables: evaporation, precipitation, and the amount of water stored as groundwater, shown as the level of the groundwater table.

19. What is the average precipitation for the year?
20. When does the groundwater table begin to rise? Why do you think this occurs?
21. What relationship exists between precipitation and the groundwater table?
22. Why does the groundwater table begin to decline in May?
23. Why do you think the groundwater table does not rise quickly in February, although there is a large increase in the amount of precipitation?

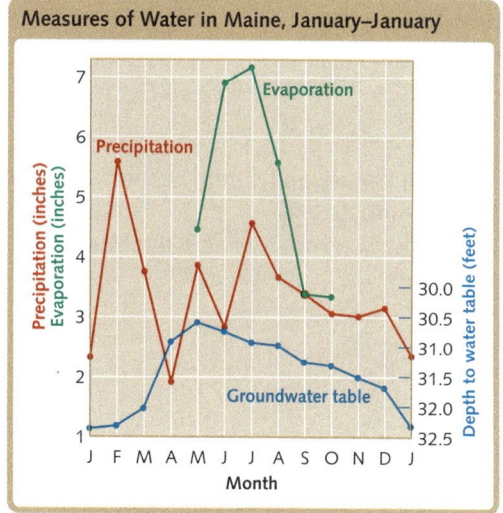

Measures of Water in Maine, January–January

Chapter-Level Resources		Laboratory Manual, pp. 3–6 Internet Investigations Guide, pp. T9–T10, 9–10 Spanish Vocabulary and Summaries, pp. 3–4 Formal Assessment, pp. 4–6 Alternative Assessment, pp. 3–4
Section 2.1 The Scientist's Mind, pp. 26–28	1. Identify possible similarities and differences among scientists who study Earth. 2. Describe the qualities of scientific thinking.	Lesson Plans, p. 4 Reading Study Guide, p. 4
Section 2.2 Scientific Methods of Inquiry, pp. 29–34	1. Explain the importance of scientific inquiry and peer review. 2. Explain the differences among a hypothesis, a theory, and a law.	Lesson Plans, p. 5 Reading Study Guide, p. 5
Section 2.3 Scientists' Tools, pp. 35–37	1. Describe simple and complex tools that Earth scientists use. 2. Explain how computers and satellites have advanced Earth science.	Lesson Plans, p. 6 Reading Study Guide, p. 6

CLASSZONE.COM

Section 2.2 INVESTIGATION
How Might a Scientist Investigate Annual Patterns of Fires?
Computer time: 45 minutes / Additional time: 45 minutes
Students examine images to look for conditions that influence fires.

Chapter Review INVESTIGATION
How Might You Investigate Scientific Phenomena?
Computer time: 45 minutes / Additional time: 45 minutes
Students examine image sets and develop a plan for conducting an investigation.

Electronic Teacher Tools
Visualizations CD-ROM
Classzone.com
Online Lesson Planner
Test Generator

Classzone.com
ES0201 Investigation, ES0202 Data Center

Visualizations CD-ROM
Classzone.com
ES0203 Visualization, ES0204 Visualization

Classroom Management

Gather these materials for minilab and laboratory activities.

Minilab, p. 36
- string
- protractor
- straw
- metal washer
- tape

Lab Activity, pp. 38–39
- balance
- 13 plastic film canisters lettered A–M
- graph paper
- ruler
- calculator (optional)
- colored pencil (optional)
- Laboratory Manual, Teacher's Edition Lab Sheet 2, p. 138

CHAPTER 2

The Nature of Science

This researcher in the rainforest of Borneo is studying the food supply of orangutans.

How is she like other scientists?

INTRODUCE

Build Interest

Before sharing background information about the photograph with students, allow time for their questions and comments.

Ask questions like the following:

- Why is this person hanging from a tree? *She is studying orangutans, which live in trees.*
- What do you think she will do with the yellow bag? *perhaps collect samples of food left by the orangutans*
- Where is Borneo? *It is a large island in the East Indies, near the Philippines.*

Respond

- How is she like other scientists? *She is making observations and gathering evidence to find answers to questions.*

About the Photograph

Researchers must climb trees to study orangutans because that is where these endangered animals live—in the treetops of Borneo's and Sumatra's rainforests. Orangutans rarely come down from the trees because almost all the food they eat, including fruits, leaves, and insects, can be found in the treetops. Researchers have built platforms in trees where they can observe the orangutans' feeding habits.

BEYOND THE TEXTBOOK

Print Resources
- Reading Study Guide, pp. 4–6
- Formal Assessment, pp. 4–6
- Laboratory Manual, pp. 3–6
- Alternative Assessment, pp. 3–4
- Spanish Vocabulary and Summaries, pp. 3–4
- Internet Investigations Guide, pp. T9–T10, 9–10

Technology Resources
- Visualizations CD-ROM
- Classzone.com
- Online Lesson Planner
- Electronic Teacher Tools
- Test Generator

24 Unit 1 Investigating Earth

CHAPTER 2

PREVIEW

▶ **FOCUS QUESTIONS** In this chapter you will study science and scientists and learn more about the key questions below.

Section 1 Where do scientists come from, and how do they think?
Section 2 How do scientists approach problems?
Section 3 What tools do earth scientists use?

▶ **READING STRATEGY**

SET A PURPOSE

Before you read, write each focus question at the top of a separate page in your science notebook. As you read, write information about each focus question on the appropriate page of your notebook.

INTERNET RESOURCES
CLASSZONE.COM

At our Web site, you will find the following Internet support:

DATA CENTER
EARTH NEWS
VISUALIZATIONS
- Technology Facilitating Discovery
- Geographic Information System (GIS)

LOCAL RESOURCES
INVESTIGATIONS
- How Might a Scientist Investigate Annual Patterns of Fires?
- How Might You Investigate Scientific Phenomena?

CHAPTER 2

PREPARE

Focus Questions
Find out what knowledge students have about the nature of science. For each focus question, have students consider the section title and then develop two or three additional questions for each section. For example, Section 1: What kind of person becomes a scientist? Do scientists think differently than other people?

Review Topics
In addition to the earth science topics listed, students may need to review this concept:

Mathematics
- Angles and distance can be used to determine height.

Reading Strategy
Model the Strategy: Tell students they will be using the focus questions to set a purpose for reading each section. Model how they would read to answer the question. For example, "For Section 1, I'll write that the three scientists came from different places and had different kinds of lives. So that must mean that scientists come from varied backgrounds. But they must think in similar ways. I will write down the things that the scientists have in common and try to explain what scientific thinking is." Have students continue to use this strategy for the rest of the chapter.

BEYOND THE TEXTBOOK

INTERNET RESOURCES
CLASSZONE.COM

INVESTIGATIONS
- How Might a Scientist Investigate Annual Patterns of Fires?
- How Might You Investigate Scientific Phenomena?

VISUALIZATIONS
- Images showing an example of how technology facilitated discovery
- Maps that demonstrate the usefulness of a Geographic Information System (GIS)

DATA CENTER
EARTH NEWS
LOCAL RESOURCES
CAREERS

Online Lesson Planner

CHAPTER 2 SECTION 1

▶ FOCUS

Objectives
1. Identify possible similarities and differences among scientists who study Earth.
2. Describe the qualities of scientific thinking.

Set a Purpose
Have students read the section to answer the question "How do scientists observe the world and solve problems?"

▶ INSTRUCT

DISCUSSION
Ask students what they know about scientists. What qualities do they think a scientist should have? Then have students compare and contrast the lives of the three scientists. Ask: When did the three scientists first become interested in science? **childhood** What quality do you think they share? **Sample Response:** **curiosity** How do the three scientists differ? **They are interested in different areas of Earth science; they have different jobs; they came from different backgrounds.**

2.1

KEY IDEA
Although they come from varied backgrounds, scientists share a way of observing the world and solving problems.

KEY VOCABULARY
- evidence
- hypothesis
- technology

The Scientist's Mind

Every technological device humans use and every scientific discovery from which they benefit begins in the mind of a person: a scientist. But where do these scientists come from? How do they think? How do they work?

Three Scientists, Three Individuals

A boy from Brooklyn, New York, named Carl Sagan began to wonder what the stars were. He asked other people, but he didn't get an answer that satisfied him. So he went to the library and asked for a book about stars. The librarian gave him one—but it was about movie stars. Sagan was embarrassed that she had misunderstood, but he was not too embarrassed to ask again. That time, she gave him a book about the right kind of stars. It answered some of his questions but started him thinking about many more. Carl Sagan became a noted astronomer, Pulitzer Prize-winning author, and popularizer of science through books and television shows.

CARL SAGAN did much to dispel the notion that science was for only a select few.

While growing up in a small mountain town outside Caracas, the capital of Venezuela, Enriqueta Barrera used to explore the tropical forest and a small river near her home. On her walks, she collected specimens of plants, rocks, and small animals that she studied at home, first with a magnifying glass and later with a microscope her father bought her when she was seven years old.

A high school course sparked Barrera's interest in geology, and she received her doctoral degree from Case Western Reserve University in Cleveland, Ohio. Currently, she is the director of the Geology and Paleontology Program at the National Science Foundation, which funds study of the history of life and the evolution of Earth's environments. Her recent work investigates what conditions were like just prior to the extinction of the dinosaurs 65 million years ago.

In seventh grade, Evan B. Forde's science teacher set up a candle in the front of the classroom. As their midterm exam, he asked his students to write as many observations about the burning candle as they could. Forde finished the exam with more than 100 observations on his paper.

DIFFERENTIATING INSTRUCTION

Reading Support Have students look for connections and similarities between the scientists' lives and their own lives. For example, ask if students would be able to make as many observations about a burning candle as Forde did. As they read, have them think about how they are similar to each scientist.
Use Reading Study Guide, p. 4.

Developing English Proficiency Often the names given to the tools of science are based in the language of ancient Greece. *Scope* (from the Greek *skopein*, to examine) is an instrument for viewing: microscope, telescope. *Meter* (from the Greek *metron*, a measure) is an instrument for measuring: magnetometer, gravimeter. A *bathyscaph*, which is a free-diving, deep-sea vessel, derives from the Greek *bathus* (deep) and *skephos* (boat).
Use Spanish Vocabularies and Summaries, pp. 3–4.

ENRIQUETA BARRERA became interested in geology during high school.

While Evan was growing up in Miami, his parents, both teachers, encouraged him to ask questions and to find the answers himself. He received his graduate degree from Columbia University after working for the National Oceanic and Atmospheric Administration (NOAA) as an oceanographer. He became the first African American to conduct research dives in the submersibles *Nekton Gamma, Alvin,* and *Johnson Sea Link*.

Today, Forde is working with NOAA to make the data from a satellite remote-sensing system more useful to meteorologists who forecast tropical storms. He feels observation is one of his great strengths as a scientist. "If you have enough detail," he says, "sometimes you can make correlations that might otherwise be missed." And he wonders if his powers of observation are a cause or an effect: "Am I a good scientist because I'm a keen observer? Or is it the other way around? Maybe I was born to be a scientist."

Three different people, three different stories, and yet all became earth scientists. Although they came from varying backgrounds, Carl Sagan, Enriqueta Barrera, and Evan B. Forde were each drawn to science when they were young. Although each faced obstacles, they all became respected scientists.

Different Lives, Common Goals

Some scientists wear lab coats, but many wear blue jeans. Some work in laboratories, but many (especially in the earth sciences) work in the field, diving deep into the ocean in bathyscaphs and searching for rocks in the desert or mountains.

Some scientists teach, while others focus on research. They work for universities and private corporations. Some study the universe; some study objects that can be seen only with a microscope—or, in the case of particle physicists, that really can't be seen at all.

Given all these differences, do scientists have anything in common? Yes—just like Sagan, Barrera, and Forde, they share what might best be called the "scientist's mind." It's a way of looking at the world with both logic and a sense of wonder.

EVAN B. FORDE is the first African-American to conduct research dives in the *Nekton Gamma, Johnson Sea Link,* and *Alvin,* shown here.

CHAPTER 2 SECTION 1

More about...

Stephen Hawking
Stephen Hawking is a famous scientist who developed new ideas about the universe. His work in theoretical physics did much to popularize the big bang theory and brought forth new ideas about black holes. He wrote a best-seller called *A Brief History of Time,* which helped explain the basic laws that govern the universe and detailed the birth of Earth and the solar system. Hawking did most of this work in a wheelchair. He has ALS (amyotrophic lateral sclerosis), or Lou Gehrig's disease, a progressive neurological disease.

DIFFERENTIATING INSTRUCTION

Hands-On Demonstration Challenge students to take Evan B. Forde's midterm exam. Place a lighted candle where students can see it. Have them list as many observations about the burning candle as they can. After students have had a chance to compare their observations, ask them to describe the experience. Was it difficult to do? What did they notice most? Did they have to think creatively to come up with observations?

⚠ *Exercise caution when handling the candle. Remind students that safety is everyone's responsibility.*

CHAPTER 2 SECTION 1

CRITICAL THINKING

Evan B. Forde thinks that being able to make keen observations is one of his great strengths as a scientist. Ask: Why is making observations important to the study of science? **Sample Response:** Observations tell what is happening, can provide answers to questions, and give details needed to make connections between ideas.

ASSESS

1. **Sample Response:** Carl Sagan was curious about the stars. Enriqueta Barrera was interested in nature. Evan B. Forde was a keen observer.

2. Students might list any two of the following characteristics: inquisitive, observant, creative, skeptical, cooperative.

3. Scientists need to question ideas and prove or disprove them in order to understand the natural world better.

4. Students should choose technology that uses scientific knowledge to help meet human needs. For example, television technology uses knowledge of electromagnetic waves. Space technology uses knowledge of forces.

VOCABULARY STRATEGY

The word *skeptical* has its roots in the Greek word *skeptesthai,* which means "to examine." In ancient Greece, the Skeptics were a group of philosophers who believed nothing can be known for certain.

Qualities of Scientific Thinking

Scientists always seem to be asking questions. "Could there be life on other planets?" "What strange creatures live in the deep sea?" "How does human activity affect the atmosphere?" These and thousands more questions are being researched right now by scientists all over the world.

In the course of their careers, some scientists investigate many questions. Some seek the answer to only one. However many questions they ask, though, scientists are methodical in the way they go about finding the answers. Sometimes scientists must overcome great obstacles in their search for answers. A scientist who wants to study distant galaxies or Earth's core cannot physically go there. Scientists must be inventive as they figure out ways to get the information they need.

Scientists are observant. They ask questions about what they observe. They predict what might cause the phenomena they observe by using **evidence** that can be measured and tested to verify their prediction.

Scientists are creative. Drawing upon what they know, they form a **hypothesis.** A hypothesis is a tentative explanation for an observation or phenomenon. Then they think of ways to test their explanations.

Scientists are skeptical. They question long-held assumptions and try to prove or disprove ideas. As they do, they form hypotheses, analyze what is known, experiment to test their hypotheses, and interpret the results using their knowledge, mathematics, and **technology.** Where science's goal is to understand the natural world, technology's goal is to modify the world to meet human needs. For example, science may look for the answer to the question "How do plants grow?" Technology might use the answer to that question to figure out how to grow more plants per acre of land and thus provide more food for people.

Because no one can know everything, no matter how long he or she has studied or how smart he or she is, scientists must work together. Through cooperation, scientists share their knowledge by publishing scientific papers, speaking at conferences, and joining professional societies with other scientists.

Scientists approach questions in a methodical way, which you will read more about in the next section. What they learn adds to what we know and can also be built on by other scientists in the future.

2.1 Section Review

1. Describe the unique characteristics of Carl Sagan, Enriqueta Barrera, and Evan Forde that assisted them in their scientific careers.
2. Name two characteristics of scientific thinking that all scientists share.
3. Explain why it is important for scientists to be skeptical.
4. **COOPERATIVE LEARNING** Develop a presentation on two ways technology has built on scientific knowledge.

MONITOR AND RETEACH

If students miss . . .

Question 1 Have students reread "Three Scientists, Three Individuals." (pp. 26–27) Ask: What characteristics do each of these individuals have that make them good scientists?

Question 2 Reteach "Qualities of Scientific Thinking." (p. 28) Have students list adjectives that are used to describe scientists.

Question 3 Have students read the vocabulary strategy on page 28 and use *skeptical* in a sentence of their own.

Question 4 Review the example of technology described on page 28. Ask: "What technology was created to help grow more plants?" **irrigation systems, grow lights, fertilizers, pesticides**

Scientific Methods of Inquiry

Creativity plays a large role in science. So does logic. While it's important to see things in a new way, it's equally important to test those new ideas rigorously. Testing ideas with experiments is the key to much of science.

How Scientists Approach Questions

What is **scientific inquiry**? Although it is often presented as a universal series of fixed steps, that doesn't capture how complex scientific study can be. Not every investigation leads to new knowledge. Not every question can best be answered by doing an experiment; in some cases, observation is the best way to gather data. Sometimes the evidence that is uncovered leads the scientist to explore entirely different issues. However, scientific inquiry generally involves observing, asking questions, forming a hypothesis, gathering data, testing the hypothesis, and sharing what has been learned.

Applying this Approach

A look at the approach taken by one scientist can illustrate how scientists approach questions.

Geologist Stephan Custer of Montana State University studies the hydrosphere. One day local residents came to him with a problem. Their wells were running dry, and they thought a new housing subdivision that had been built nearby was responsible.

Dr. Custer decided to design an experiment to check the water table, the level below which the ground is saturated with water. You will read more about water tables and groundwater in Chapter 14.

He decided the best way to figure out what was going on was to collect data about the water table's level over time. So he devised an experiment that involved measuring the water table at various wells at different times of the year. Data from one well appears in the graph below.

2.2

KEY IDEA

Scientists approach problems methodically, and their ideas are tested by themselves and by their peers.

KEY VOCABULARY

- scientific inquiry
- peer review
- theory
- law

How Might a Scientist Investigate Annual Patterns of Fires?
Examine satellite images to look for conditions that influence fires.
Keycode: ES0201

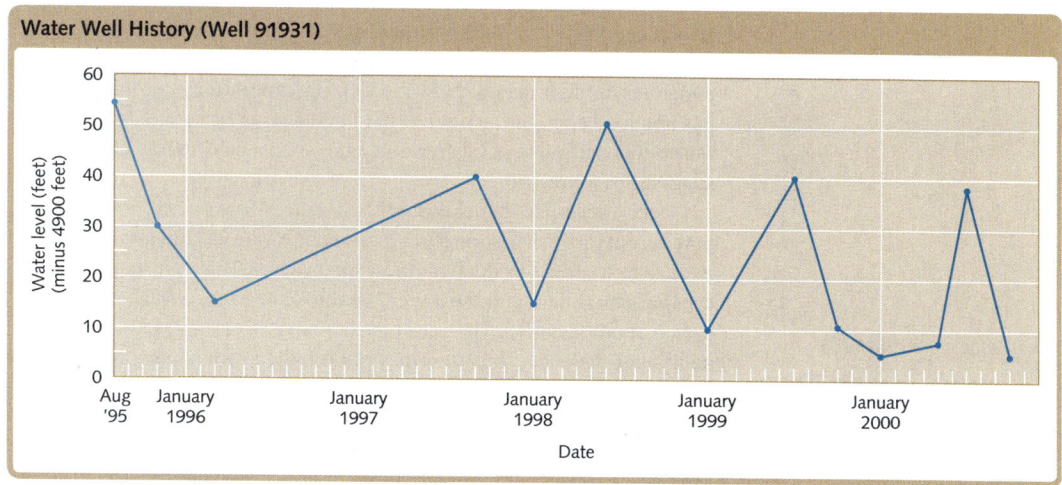

CHAPTER 2 SECTION 2

More about...

A "Modern" Scientific Approach

The Greek philosopher Aristotle (384–322 B.C.) was one of the first to look to natural (not supernatural) causes to explain earthly phenomena, such as earthquakes. Sometimes described as a natural philosopher, Aristotle based his ideas upon logic and observations. When he could not directly observe an object of study, such as in astronomy, he relied on basic principles, taken mostly from physics. The scholars of the Roman Empire used Aristotle's work as their chief guide to nature. Although many of Aristotle's explanations were eventually proved wrong, his idea of a science based on observation still prevails today.

IN THE FIELD Stephan Custer prepares to measure a well's water level.

VOCABULARY STRATEGY

Your *peers* are your equals. To be reviewed by your peers means that your equals are looking over your work. A scientist's peers would be other scientists.

When Dr. Custer analyzed the data he and his graduate students had gathered, he noticed something startling: the level of the water table varied as much as 12 meters over the course of the year. Dr. Custer knew that too little rain fell in the area to account for such a large variation. So he formed a hypothesis to account for what he had seen. His hypothesis was that the water table was recharged (filled up) by spring runoff from melting snow in the nearby mountains.

To see if his hypothesis was correct, Dr. Custer investigated how much water from one mountain stream actually made it to the local river. Measurements showed that except for one day of the year, none of the stream's water made it to the river. Instead, it all soaked into the ground and raised the water table's level.

By repeating these measurements and collecting additional data, Dr. Custer has been able to observe the changes in the water table over a period of four years. He has found that the level, on average, has declined. That leads him to the next question: what is the cause of the decline?

Because Dr. Custer approached the question about the water table logically and methodically, he was able to gather evidence about changes in the water table before conflicts arose between area residents.

Peer Review and Scientific Journals

The end of an experiment is not the end of the scientific process. The scientist's new knowledge needs to be shared with and tested by other scientists. After completing experiments, scientists write scientific papers about their findings. A scientific paper explains what the scientist observed; how he or she gathered, analyzed and interpreted data; and what he or she thinks the results mean.

To share their knowledge, scientists publish their papers in a scientific journal. Worldwide, thousands of scientific journals are published every year. Some are widely read, such as the journals *Science* and *Nature*. Others are meant for smaller, more specialized audiences.

Scientific journals differ from magazines because of **peer review.** As part of peer review, the editors at a scientific journal ask experts in the field to read the papers scientists have submitted. If the experts agree that a paper has merit, it may be published. Or the experts may feel the scientist has not done enough work or has not proved his or her point. In that case, the editors decline to publish the paper or send it back to the scientist with suggestions for revision.

Once a scientist's paper is published, other scientists will read it. Some may accept the explanation. Others may not. Some will be interested enough to want to test the hypothesis themselves. They may repeat the original experiment and see if they can duplicate its results. They may design another experiment to test the hypothesis, or they may search for additional evidence. Then they may publish their own papers, supporting or disproving the original hypothesis.

This gathering of new evidence and suggesting of new hypotheses constantly changes science. As hypotheses are disproved, others are proposed to take their place. Scientists test the new hypotheses in turn.

DIFFERENTIATING INSTRUCTION

Developing English Proficiency Refer students to the vocabulary strategy on this page, then help clarify the meanings of *peer*. Point out that as a noun, the word *peer* means "an equal." In the term *peer review*, the word *peer* is used as an adjective that describes the word *review*. Tell students that *peer* can also be a verb. As a verb, *peer* means "to look at something carefully" or "to come partially into view." For example, "The girl peered at the drawing in the book" and "The sun peered through the clouds." Finally, tell students that the word *pier*, a homophone, has the same pronunciation as *peer* but a different spelling and meaning. A pier is a platform that reaches out from the land into water.

A good example of how science changes is scientists' hypotheses about how the moon formed. Several ideas have been advanced and disproved; you can read about them on pages 32 and 33.

The Importance of Testing Ideas

Scientists need to be willing to ask questions and experiment. They also need to be willing to change their minds about their hypothesis if the evidence they gather does not support it.

Do you think you could change your mind like that? It takes a lot of courage to let go of an idea, especially if you have worked hard to find evidence that supports it. But new evidence is being found all the time, and scientists need to be flexible. Scientific ideas are tested and retested every day. Some advances that look especially promising fall short when tested. In 1989, a team of researchers from the University of Utah announced that they had succeeded in producing atomic fusion at room temperature (a phenomenon called "cold fusion").

Fusion is the process in which atomic nuclei combine to create new elements. For example, the nuclei of hydrogen atoms fuse to become helium. Along with producing a new element, fusion releases tremendous amounts of energy; it is fusion that fuels the stars. And unlike fission (which breaks up atoms to release energy), fusion leaves much less radioactive residue.

Why isn't fusion in common use? The problem is that it happens only at very high temperatures and pressures. Hydrogen bombs use fusion, but the resulting release of energy is uncontrolled. Many scientists have dreamed of creating controlled fusion at lower temperatures as a clean source of immense amounts of energy.

The Utah researchers were excited when they thought they had successfully produced a fusion reaction at room temperature. However, rather than waiting for their results to be published in a journal and tested by their peers, they announced their results to the media. The media was also excited by the possibility of cold fusion, which could solve humanity's energy problems.

A FAILED ATTEMPT Researchers around the world tried to re-create the University of Utah team's cold fusion experiment with devices like this one. None of the attempts succeeded.

CHAPTER 2 SECTION 2

More about...

Scientific Testing
The natural philosophers of ancient Greece thought that all matter was made up of a combination of four basic elements: earth, air, fire, and water. For example, Aristotle explained the cause of earthquakes in terms of air trying to escape from Earth. These ideas were based on logic and observation, two important components of scientific inquiry. Scientific testing, which would eventually prove the ancient Greeks wrong about earthquakes, was not to become part of scientific inquiry until the 17th century. It was then that a new breed of natural philosopher, led by Francis Bacon, decided that observation and logic were not enough. These scientists insisted that explanations for natural phenomena had to be tested in order to prove their validity.

DIFFERENTIATING INSTRUCTION

Challenge Activity Have students use scientific inquiry to solve an everyday problem, such as how to keep beverage cans or bottles chilled in summer without using a refrigerator. They should develop a hypothesis and a plan for performing an experiment, do the experiment, and provide an explanation for the results of the experiment. Students might present the steps they followed and their findings in the form of a scientific paper and then exchange their papers with classmates for peer review.

CHAPTER 2 SECTION 2

VISUAL THINKING
Discussion
Have students read aloud each caption. Be sure they understand how each illustration relates to the description in the caption. Review how each of the first three hypotheses were disproved, making sure students understand how the information in the caption disproves the hypothesis. For example, a lack of iron on the moon disproves the fission hypothesis because if the moon had spun off from Earth, you would expect it to have a similar composition as Earth.

Extend
Ask: What type of information might disprove the currently accepted impact hypothesis?
Sample Response: New data might show that Earth could not have remained intact after such a large collision.

Scientific Thinking

PREDICT
Students might mention that valuable resources, like people and money, would be taken away from more viable research.

DEVELOPMENT OF A THEORY

FISSION This hypothesis proposed that the moon had once been part of Earth. Early in its formation, Earth was spinning so fast that a large piece of it tore away and flew off into space.

FISSION WAS DISPROVED because it could not explain why the moon lacks iron, a common element on Earth.

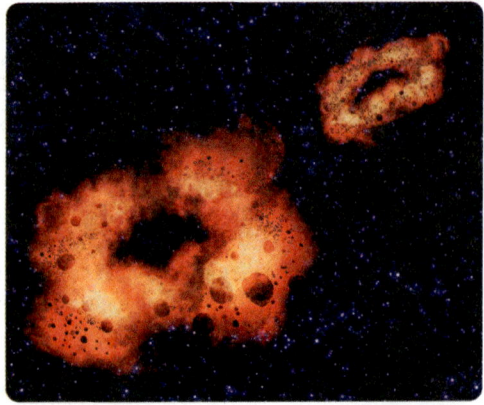

CO-FORMATION This hypothesis proposed that the moon and Earth formed at about the same time out of the rocky debris ring that circled the sun.

CO-FORMATION WAS DISPROVED when analysis of the energy and momentum required showed that they could not have formed this way.

The story appeared on TV screens and in newspapers all over the world.

But then the news came from their peers. Other scientists had tried to duplicate the experiment—and failed. The failure of cold fusion was a disappointment. If it had worked, it would have been a tremendous benefit to humankind. But fortunately, other scientists tested the idea before hopes got too high. The story of cold fusion shows the value of testing and retesting experimental results.

Science depends on inquiry—asking questions—and experimentation—testing to find evidence. Testability separates scientific knowledge from nonscientific knowledge. Many beliefs people hold cannot be tested. Scientific explanations must be supportable by evidence and be able to be tested in the natural world.

Scientific Theories and Laws

When a hypothesis has been thoroughly tested and retested successfully, it may be considered a **theory**. A theory is an explanation for observable events or facts for which no exception has been found. A theory has passed the scrutiny of many people and seems to be the best explanation for an observed phenomenon, based on the available information.

One of earth science's best-known theories is the big bang theory. This theory states that the entire universe was once packed into a very hot, very dense sphere. The sphere exploded, forming a huge, expanding cloud. As parts of the cloud cooled, they condensed into galaxies. Right now, this theory is the best scientific explanation we have for the origins of the

Scientific Thinking

PREDICT
What do you think might have happened if people had taken the researchers' word that cold fusion worked?

32 Unit 1 Investigating Earth

DIFFERENTIATING INSTRUCTION

Support for Visually Impaired Students Have students with good eyesight describe the details in the illustrations to students who are visually impaired. They should include a description of the compositions of the two bodies, the shapes of the forms, and the implied actions. Also have students read the captions aloud.

CAPTURE This hypothesis proposed that the moon was a celestial body on a path that took it near Earth, and Earth's gravitation captured the moon, drawing it into orbit.

CAPTURE WAS DISPROVED when lunar rocks were found to have the same isotope composition as Earth rocks do.

IMPACT This hypothesis proposes that during Earth's formation, a planet-sized object struck it. The impact destroyed the other planet and released debris, which eventually coalesced into the moon we know today. This theory is currently accepted among scientists.

universe. Scientists have found evidence to support the theory, including cosmic background microwave radiation that is an echo of the big bang. However, it is still possible that someone will find evidence that disproves the big bang theory. Theories rise and fall from favor as new information comes to light.

Scientific **laws** are different from theories. They are generalizations about how the natural world behaves under certain conditions. No exceptions to these generalizations have ever been found; for example, no object that has been dropped has ever fallen up. This upholds the law of gravity. Humans don't create the laws; they discover them. Theories grow from laws, but they do not become laws after further testing.

2.2 Section Review

1. What is a hypothesis? What steps are necessary before a hypothesis can be formulated?
2. What is peer review? What is its role in scientific inquiry?
3. Explain the value of testing and retesting ideas scientifically.
4. **CRITICAL THINKING** How have you applied the method of scientific inquiry—formulating a question, gathering data, and forming a hypothesis—in your own life?
5. **PAIRED ACTIVITY** Prepare a presentation to explain the difference between a theory and a law.

CHAPTER 2 SECTION 2

Learn more about how science affects your everyday life.
Keycode: ES0202

Have students describe ways in which they use science every day.

CRITICAL THINKING

Over the years, computers have become smaller, faster, and more powerful. Ask students to predict future features and uses of computers. **Computers might be voice-activated, thus no longer requiring keyboards and mouse pads. All books, magazines, and newspapers might be accessed via the Internet only. Tiny computers might be implanted in humans to monitor their vital signs, such as heart rate and body temperature. Computers might "drive" cars and other vehicles.**

Science Notebook
Students might list the Internet, cell phones, DVDs, portable CD players, MP3 players, digital cameras, animal cloning, the Human Genome Project, and the International Space Station.

SCIENCE & Society

Scientific Literacy: What's the Big Deal?

You may think that science doesn't have much impact on your daily life, but every time you turn on a computer or listen to a CD, you are enjoying the fruits of scientific research, the advancement of technology.

What changes lie ahead? How can you make sure you aren't left behind?

PERSONAL COMPUTERS today are less expensive, faster, easier to use and are becoming more and more a part of our everyday lives.

Do the adults in your life ask you for help when they are trying to set the VCR to record or when the computer won't work? Maybe you wonder why adults seem so hopeless when it comes to technology.

Although many adults are perfectly capable of debugging their computers and recording a TV show successfully, other adults aren't so lucky. Technology has changed so fast that they just can't keep up.

To put the changes in perspective, remember that Steven Jobs introduced the Apple II, the first personal computer marketed for home use, in 1977. Since then, computers have become part of many aspects of life, and the Internet has

THE ENIAC was the first general purpose computer. First activated in 1946, it weighed 30 tons and contained 18,000 vacuum tubes.

experienced explosive growth.

Now consider when your parents were born. How old were they when the Apple II was introduced?

Learning can be easier when you're younger, and that's part of the reason your parents turn to you for help with the technological devices around your house. For you, they're second nature because you've grown up with them, but for many adults, they're still new and a little strange.

The rate of technological development isn't likely to slow down. How are you going to keep up with each technological breakthrough reported in the media or hyped by the marketplace? How will you know

THE LISA was a desktop computer introduced by Apple in 1983 for home or office use. Its high price ($10,000) prevented it from becoming a market success.

whether you should be concerned about advances in biotechnology? How will you be able to evaluate national issues such as financing a new space mission?

The best way is to make sure you understand the science behind these technological advances. You don't need to get a Ph.D. You should, however, learn as much about science and technology as you can. Even after you graduate, read. Keep learning.

After all, when you're an adult, you don't want your kids to roll their eyes at you, do you? ■

Extension

SCIENCE NOTEBOOK
What scientific and technological advances have occurred since you were born? Make a list of ways things have changed during your lifetime.

Learn more about how science affects your everyday life.
Keycode: ES0202

Scientists' Tools

Scientists use a variety of tools in their inquiries. Some, like computers and satellites, are complex and may cost billions of dollars. Others are simple and may cost just a few pennies each, such as bags to hold specimens.

Tools to Study the Earth and Ocean

One of the geologist's most basic and simplest tools is a rock hammer. It is used (sometimes with a rock chisel) to split rocks or to crack specimens, or samples, loose from rock faces. Geologists use soil augers and earth drills to sample material beneath the ground's surface. The geologists collect their specimens in cloth or plastic sample bags and take them to the laboratory for further study.

In the laboratory, geologists may use a crusher to pulverize a specimen for chemical analysis. They may use a hand lens to examine it closely or a spectroscope to analyze its composition. They may measure it with a caliper, an L-shaped instrument that measures irregular objects by determining the distance between its two short arms. Using sieves and a sieve shaker, a geologist can quickly sort a bag of mixed sediments by size.

Geologists also use laser range finders for measuring distance, clinometers for measuring the slope of inclines, magnetometers for measuring Earth's magnetic field, and gravimeters to measure Earth's gravity field. Using a seismograph, geologists can determine the composition of the ground beneath their feet by measuring how fast vibrations travel through it from a source such as a dynamite blast, a sledgehammer blow, or an earthquake. Marine geologists and oceanographers employ tools, such as dredges and gravity corers, specially suited for ocean research. You will read more about these tools in Chapters 22 and 23.

RESEARCH VESSELS, such as the *Edwin Link* pictured here, serve as ocean-going laboratories for marine scientists.

2.3

KEY IDEA
Scientists use both simple and complex tools in their inquiries.

Examine a case of how technology facilitated discovery.
Keycode: ES0203

CHAPTER 2 SECTION 3

FOCUS

Objectives
1. Describe simple and complex tools that Earth scientists use.
2. Explain how computers and satellites have advanced Earth science.

Set a Purpose
Have students read the key idea. Then have them read the section to answer the question "What simple and complex tools do Earth scientists use?"

INSTRUCT

More about...

Research Tools
Before 1950, so little was known about the ocean that most of the information was contained in one book, *The Oceans,* by H. Sverdrup, M. Johnson, and R. Fleming. Since 1950, more than 1000 research ships have been built. Probably the greatest impact on ocean study has been the development of shipboard computers.

Examine a case of how technology facilitated discovery.
Keycode: ES0203

 Visualizations CD-ROM

DIFFERENTIATING INSTRUCTION

Reading Support Model for students how to preview the red headings as a way of organizing information as they read. For example, you might say, "All the headings are about tools. Some tools are for studying Earth and the ocean, some are for studying the sky and stars, and some have many uses. As I read, I'll find out which tools are used to study each." Suggest that students write the headings on a sheet of paper and then list the different tools that are discussed under each heading.
Use Reading Study Guide, p. 6.

CHAPTER 2 SECTION 3

MINILAB
Making a Sextant

Materials
- The string should be at least 50 cm long.
- Any small, heavy object can be used in place of the metal washer.

Management
- Students should hold the the long end of the straw.

Analysis Answers
- Yes, using trigonometry: (angle) = height of object from observer's eye ÷ horizontal distance to object. Add the height of the observer's eye to determine the height of the object from the ground.
- Early seafarers could use a sextant to record the position of stars relative to the horizon. This information was used to determine latitude and longitude. For example, the angle of the North Star measured from the horizon gives a ship's present latitude north of the equator.

More about...

Hubble Space Telescope
The Hubble Space Telescope was deployed in 1990. Scientists soon realized that it was not operating correctly; the images transmitted were extremely blurred. The cause was a minute flaw in the telescope's main mirror. Adjustments were made via computer, but the problem was not corrected until 1993, when astronauts were able to fix it manually.

GEOLOGY A creepmeter monitors creep, the downslope movement of soil.

OCEANOGRAPHY Oceanographers need their equipment to meet the special demands of exploring ocean environments.

25-Minute Mini LAB

Making a Sextant

Materials
- string
- protractor
- straw
- metal washer
- tape

Procedure

1. Tie the washer to one end of the string. Fasten the other end to the protractor's center.
2. Tape the straw at the center mark so that one end hangs a little beyond the 90° mark.
3. Look through the straw at a tree. Let the string hang freely. When the string stops swinging, press the string in place and record the angle.

Analysis
Could you determine the height of a tree using the sextant? Why would a sextant have been important for early seafarers?

Tools to Study the Sky and Stars

Meteorologists use a variety of instruments to study weather and climate. These include the thermometer, which measures temperature; the aerovane, which measures wind speed and direction; and the barometer, which measures air pressure. Weather radar and satellite pictures give meteorologists a view of Earth's weather patterns that was unavailable to them just a few decades ago.

Astronomers use telescopes to study the stars and planets. Reflecting telescopes use mirrors to focus light waves into images, while refracting telescopes use lenses. Radio telescopes pick up radio waves rather than light waves. The Hubble Space Telescope, an orbiting reflecting telescope, is providing astronomers with images of objects never before seen. Some telescopes are portable and affordable enough for amateur stargazers. Others, such as the Arecibo radio telescope in Puerto Rico, take up about three acres of land.

Tools with Many Uses

Not all of the Earth scientist's tools developed for purely scientific reasons. The telescope, for example, was originally designed for seafarers. Also originally used by seafarers was the sextant. The sextant is an instrument that measures the angle between two points, such as the horizon and a star. By determining the altitude of a star viewed through a sextant and comparing it with altitudes the star was known to have at various latitudes, a ship's navigator could find the latitude of the ship.

Two tools in particular have revolutionized Earth science: the computer and the satellite. With computers, earth scientists can now analyze larger amounts of data than ever before, in a fraction of the time it took in the past. Computers can generate graphs, charts, and images that make data easier to interpret and understand. Among the most exciting uses of computer

---

DIFFERENTIATING INSTRUCTION

Developing English Proficiency Demonstrate making the sextant rather than having students rely on the written directions. List the materials on the board. Point to each word as you refer to it in your directions. As you complete each step, have students describe the process in their own words.

METEOROLOGY Meteorologists use both computers and pencil and paper to do their work.

ASTRONOMY Telescopes, such as the Anglo-Australian Telescope in Sliding Spring, Australia (shown here) are among the largest tools of Earth science.

technology are geographic information systems, which you will read more about in the next chapter. During the last few decades, computer technology has advanced rapidly, and the computer power that once filled a room now fits in a pocket. More portable computers can go with scientists into the field, instead of staying behind in the laboratory. In the field, computers allow scientists to analyze data and to connect to other computers.

Like computers, satellites are useful to earth scientists in all fields. Oceanographers can use satellite data to map the sea floor. Meteorologists use satellite images to study weather patterns. False-color satellite images can tell geologists where to find different rock types. They have dramatically changed mapmaking, as you will learn in the next chapter.

From rock hammers to satellites, earth scientists have an amazing array of tools to choose from as they do their research. The most important tool, however, is the mind—the ability to wonder why, to ask questions and seek answers. Such inquiry has led us to invent and construct the most advanced tools to tell us more about our planet Earth and the worlds and stars that surround us.

VISUALIZATIONS
CLASSZONE.COM
Observe some products of a Geographic Information System (GIS).
Keycode: ES0204

2.3 Section Review

1. Describe how tools help scientists in their work.
2. **WRITING** Earth scientists must take notes in the field. Practice your note-taking skills by observing an animal for five minutes and noting its activities, or by writing a description of a rock sample.
3. **CRITICAL THINKING** How are computers and satellites changing the nature of science?
4. **PAIRED ACTIVITY** Make a poster about one of the scientific tools in your science laboratory, describing its appearance and function.

Chapter 2 The Nature of Science 37

CHAPTER 2 SECTION 3

VISUALIZATIONS
CLASSZONE.COM

Observe some products of a Geographic Information System (GIS).
Keycode: ES0204

Have students describe how GIS works and explain why it is a helpful tool.

Visualizations CD-ROM

ASSESS

1. **Sample Response:** Tools allow scientists access to more information.
2. Students should record what they see, hear, smell, and feel.
3. Computers allow scientists to analyze larger amounts of data in a fraction of the time it took in the past. Satellites provide new views of Earth—both its surface and the ocean floor.
4. Posters should clearly explain what the tool looks like and how the tool is used in science.

MONITOR AND RETEACH

If students miss . . .

Question 1 Have students make a list of the tools discussed in this section and describe their functions.

Question 2 Discuss the kinds of things students might look for as they observe the animal or rock sample. Stress the importance of writing detailed notes when recording observations.

Question 3 Reteach "Tools with Many Uses." (pp. 36–37) Have students give examples of how computers and satellites have helped scientists in each of the following branches of Earth science: geology, oceanography, meteorology, and astronomy.

Question 4 Display tools from the science laboratory. If available, provide instruction manuals or science catalogs for students to read about the functions of the tools.

CHAPTER 2 ACTIVITY

PURPOSE
To measure the masses of several objects and use the measurements to make inferences about those objects

MATERIALS
Provide students with a copy of Lab Sheet 2. A copymaster is provided on page 138 of the teacher's laboratory manual.

Opaque-plastic film canisters are ideal to use for this lab. Before class, fill 12 canisters with 1 to 12 objects each. Leave one canister empty. Objects could be pennies, identical buttons, or identical beads. You may want to seal the canisters with tape. Label the canisters randomly A to M.

PROCEDURE

1. Emphasize that students should not open the canisters. You may wish to have students order the canisters by their approximate weight before using the balance.
2. Students can easily find the empty canister by shaking the containers.
4. A more complex variation of this lab would be to fill the 12 canisters with 1 to 25 objects each.

CHAPTER 2 LAB Activity

Making Inferences from Observations

SKILLS AND OBJECTIVES
- **Measure** and **record** the masses of several small canisters.
- **Interpret** and **graph** the data resulting from those mass measurements.

MATERIALS
- balance
- 13 plastic film canisters lettered A–M
- graph paper
- ruler
- calculator (optional)
- colored pencil (optional)

How do scientists study structures they cannot see? For example, people cannot directly observe Earth's core and mantle. Likewise, they cannot directly see objects in deep space. Instead, scientists use indirect observations to make inferences about such structures.

In this activity, you will use a balance to collect data from which you will make inferences about the contents of several canisters. The canisters all hold identical objects, but in varying quantity.

Procedure

DO NOT OPEN THE CANISTERS.

1. Determine the masses of each of the 13 plastic film canisters to the nearest tenth of a gram. Copy Table 1 to record your values under "Mass of canister." Be careful to record the mass under the correct letter.

Table 1:	Canisters Arranged by Letter												
	A	B	C	D	E	F	G	H	I	J	K	L	M
Mass of canister (g)													
Mass of empty canister (g)													
Mass of canister contents (g)													
Number of objects in canister													

2. Find the canister that contains no objects and designate it as the mass of an empty canister. Write this value in each column on the line labeled "Mass of empty canister."

3. Subtract the mass of the empty canister from each of the canister masses to determine the total mass of the objects within each canister. Record the values under "Mass of canister contents."

4. If the canister with the smallest mass is empty, then the canister with the next smallest mass must contain only one object, the next smallest two objects, and so forth. Using this reasoning, infer the number of objects in each canister. Write your prediction for the number of objects that each canister contains in Table 1 under "Number of objects in canister."

5. Copy Table 2 to arrange your data by mass.

Table 2:	Canisters Arranged by Content												
	0	1	2	3	4	5	6	7	8	9	10	11	12
Number of objects in canister													
Mass of canister contents (g)													
Canister letter													

6. Plot the data from Table 2 on the graph paper, using the diagram on the right as a guide. The *x*-axis will be "Number of objects in canister" and the *y*-axis will be "Mass of canister contents." You may use a colored pencil so that the plotted points show clearly on the graph. Label each point with the letter of the canister. Connect your plotted points with a straight best-fit line. Note: If the points do not fall in a relatively straight line, check the masses and calculations for the canisters that seem incorrect.

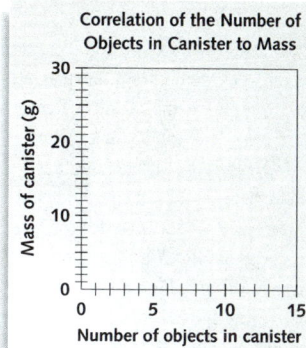

Analysis and Conclusions

1. Identify one property that is the same for all of the canisters. Also identify one property that is different for all of the canisters. Note: Color, shape, volume, and mass are examples of properties.

2. For the objects inside the canisters, identify one property that can be inferred about each of them. Also identify two properties that cannot be known about any of them without opening the canisters.

3. For each of the following, determine if the graph for the modified activity would be a straight line. Explain your answers.

   a. The number of objects in each canister continues to vary, but the type of object used is different.

   b. The number of objects in each canister continues to vary, but no two canisters contains the same kind of object.

   c. Each canister holds the same number and type of object.

   d. Each canister holds the same number of objects, but no two canisters hold the same type of object.

4. What are some other methods you could use to infer more properties of the objects in the canisters? Hint: Remember how you found the empty canister.

CHAPTER 2 ACTIVITY

ANALYSIS AND CONCLUSIONS

1. same properties for all canisters: shape, color, size, volume; different properties for all canisters: mass, density, weight

2. properties that can be inferred: qualitative shape (round, flat), qualitative number (none, one, many), total mass of objects; properties that cannot be inferred: color, individual volume

3. a. Yes; the line would still be linear.
   b. No; there would be no correlation between the number of objects in the canister and the mass of the canister.
   c. No; the graph would be a single point.
   d. Yes; the graph would be a vertical line.

4. **Sample Response:** The sound the objects make when the canister is rattled could give more information on the shape of the objects (round or with edges) or the kind of material the objects are made of (soft or hard).

Refer students to Safety in the Earth Science Laboratory on pages xxii–xxiii.

CHAPTER 2 REVIEW

Summary of Key Ideas

2.1 Scientists come from many different backgrounds, but in their work they use many common tools, techniques, and habits of mind. The qualities of scientific thinking include asking questions, seeking evidence, forming hypotheses, testing hypotheses, being skeptical, and working cooperatively. Technology plays an important role in applying scientific discoveries to everyday life.

2.2 Doing science is a complex process that does not proceed neatly from one revelation to the next. When scientists investigate questions, they state the question, gather evidence, form a hypothesis, and test the hypothesis. Scientists publish their results in scientific journals, which provides the opportunity for their peers to review their work. Testing ideas is vital to science. When a hypothesis has been thoroughly tested, it may be considered a theory, an explanation for observable events for which no exceptions have been noted. Scientific laws are generalizations about the natural world and how it behaves.

2.3 Scientists use a variety of tools, both simple and complex, as they study the earth, ocean, sky, and stars.

Key Vocabulary

evidence (p. 28)
hypothesis (p. 28)
law (p. 33)
peer review (p. 30)
technology (p. 28)
theory (p. 32)

Vocabulary Review

Explain the difference between the terms in each pair.

1. theory, law
2. hypothesis, theory
3. science, technology
4. cooperation, peer review
5. evidence, inquiry
6. hypothesis, question

Concept Review

7. What are some qualities that scientists share?
8. How do scientists support their hypotheses?
9. Describe the way in which a scientist might approach a question.
10. Why is it important for scientists to be skeptical?
11. Describe how peer review works.
12. Explain why it is important to test scientific ideas.
13. Can a scientific theory become a scientific law? Why or why not?
14. Describe some of the tools Earth scientists use as they work.
15. **Graphic Organizer** Copy and complete the concept map below by adding characteristics that are important to looking at the world scientifically.

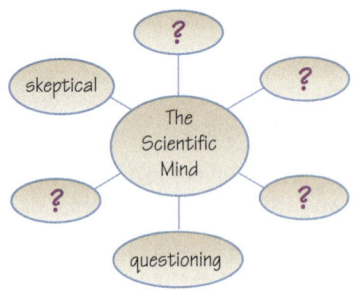

Vocabulary Review (Answers)

1. A theory is an explanation based on available information; a theory may change as new information is found. A law is a generalization to which no exception has been found; a law does not change.
2. A hypothesis is a temporary explanation for an observation or a phenomenon. A theory is a carefully tested hypothesis that seems to be the best explanation for the phenomenon.
3. The goal of science is to understand the natural world. Technology's goal is to modify the world to meet human needs.
4. Cooperation is the sharing of information by scientists. A peer review is the review of a scientific paper by experts to determine whether the paper has merit.
5. Inquiry is the process of asking questions. Evidence is information that can be measured and tested to verify a prediction.
6. A question seeks an answer. A hypothesis is a temporary answer to a question.

Concept Review (Answers)

7. Scientists have strong powers of observation and use logic. Most scientists have a sense of wonder, are creative, and are skeptical.
8. Scientists test their hypotheses by gathering, analyzing, and interpreting data.
9. Scientists are methodical in their approach and inventive in figuring out ways to get information.
10. Scientists need to question assumptions and try to prove or disprove ideas.
11. When a scientist submits a scientific paper to a journal, the editor gives the paper to experts in the field to determine whether the paper has merit or needs more work.
12. Testing scientific ideas leads to evidence that proves or disproves the ideas.
13. Theories grow from scientific laws, but they do not become laws. Theories may change as more information is found, but laws have no exceptions and do not change.
14. magnifying glass, microscope, telescope, submersibles, satellites
15. creative, inventive, methodical, observant

Critical Thinking

16. **Observe** Think about Evan Forde's seventh-grade experience, when his science teacher asked him to write down observations about a candle. Take a common object and challenge yourself: How many observations about the object can you make? Remember that details are the key.

17. **Communicate** Without naming the object you observed in item 16 above, write a description based on your observations. Include enough detail so someone can guess what it is. As Evan Forde says, "Write your description as if an alien from another planet were going to read it."

18. **Predict** What might happen if scientific discoveries were not thoroughly tested through peer review? Explain your thinking.

Internet Extension

How Might You Investigate Scientific Phenomena? Examine sets of satellite images. Develop a plan for conducting an investigation.

Keycode: ES0205

Writing About the Earth System

SCIENCE NOTEBOOK If you could meet any of the scientists described in this chapter, or another scientist (past or present) you have heard of, what would you like to ask him or her? Make a list of questions you would ask your chosen scientist.

Interpreting Graphs

As you saw in Section 2.2, it is vital for scientists to be able to interpret graphs and draw useful information from them. The graph shown is similar to the one Dr. Stephan Custer used in his study of well water levels. Each bullet on the graph represents the result of a water level test.

19. How many times was the water level tested in Study 1? in Study 2?
20. During Study 1, by how much did the water level decline between the first measurement and the final measurement?
21. During Study 2, by how much did the water level decline between the first measurement and the final measurement?
22. Given both studies, for how many weeks were water levels monitored?
23. What must have taken place between the measurement taken near March 1, 1995, and the one taken September 1, 1995?
24. If a third study were made, what do you think a graph made from its data would look like? Explain your thinking.

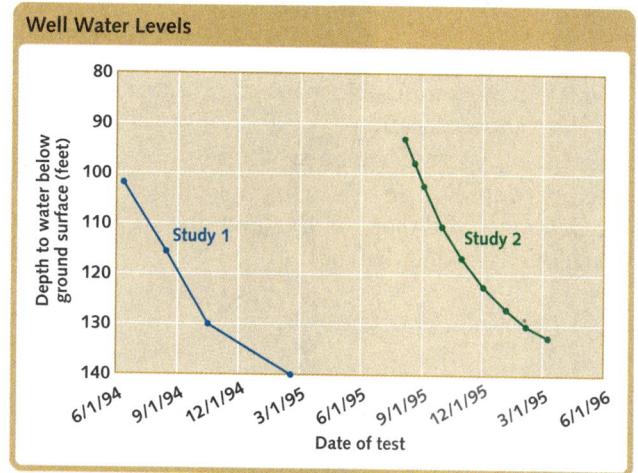

Chapter 2 Earth as a System 41

CHAPTER 2 REVIEW

How Might You Investigate Scientific Phenomena?
Keycode: ES0205

Use Internet Investigations Guide, pp. T10, 10.

CRITICAL THINKING

16. **Sample Response:** Observations of a pencil: long and narrow, wooden, yellow paint, sharp point, carbon tip, rippled metal band, eraser tip, lettering, number 2

17. **Sample Response:** This object is long, narrow, and thin. It is made of wood that is painted yellow. It has an inner core of carbon. The wood and carbon are sharpened to a point. It has a number 2 near the end, next to a rippled band of metal. There is a pink eraser on the tip.

18. The scientific discoveries might include errors. These would lead to even more errors as people used the erroneous information.

INTERPRETING GRAPHS

19. Study 1: 4 times; Study 2: 9 times
20. about 40 feet
21. about 40 feet
22. about 60 weeks
23. There must have been a lot of rain.
24. If it covered the same months, it would probably look the same as the other two studies. There are probably periods of rainy months and dry months in each season of the year. These would be roughly the same from year to year.

WRITING ABOUT SCIENCE

Writing About the Earth System

Sample Responses: Students might ask Carl Sagan if he would like to visit another planet. They might ask Enriqueta Barrera how changing conditions killed off the dinosaurs. They might ask Evan B. Ford what phenomena he observed during his dives.

CHAPTER 3 PLANNING GUIDE: MODELS OF EARTH

	Section Objectives	Print Resources
Chapter-Level Resources		Laboratory Manual, pp. 7–12 Internet Investigations Guide, pp. T11–T13, 11–13 Guide to Earth Science in Urban Environments, pp. 13–20, 21–28 Spanish Vocabulary and Summaries, pp. 5–6 Formal Assessment, pp. 7–9 Alternative Assessment, pp. 5–6
Section 3.1 Modeling the Planet, pp. 44–47	1. Explain how maps are models of Earth. 2. Compare and contrast three types of map projections. 3. Use latitude and longitude to describe a location on Earth.	Lesson Plans, p. 7 Reading Study Guide, p. 7 Teaching Transparency 2
Section 3.2 Mapmaking and Technology, pp. 48–52	1. Relate the history of mapmaking. 2. Describe the roles of satellites and computers in making and using maps.	Lesson Plans, p. 8 Reading Study Guide, p. 8
Section 3.3 Topographic Maps, pp. 53–57	1. Explain how topographic maps use contour lines to show elevation. 2. Describe topographic map symbols. 3. Demonstrate how to use topographic maps to determine the shape of the land, the flow of rivers, and distance.	Lesson Plans, p. 9 Reading Study Guide, p. 9 Teaching Transparency 3

 INVESTIGATIONS
CLASSZONE.COM

Section 3.1 INVESTIGATION
How Do Map Projections Distort Earth's Surface?
Computer time: 45 minutes / Additional time: 45 minutes Students compare and contrast area, distance, and shape across several map projections.

Section 3.1 INVESTIGATION
How Do Latitude and Longitude Coordinates Help Us See Patterns on Earth? Computer time: 45 minutes / Additional time: 45 minutes. Students plot latitude and longitude data to interpret geographic patterns.

Technology/Online Resources

Electronic Teacher Tools
Visualizations CD-ROM
Classzone.com
Online Lesson Planner
Test Generator

Visualizations CD-ROM
Classzone.com
  ES0301 Investigation, ES0302 Visualization, ES0303 Investigation

Visualizations CD-ROM
Classzone.com
  ES0304 Visualization, ES0305 Career, ES0306 Data Center

Classroom Management

Gather these materials for minilab and map activities.

Minilab, p. 56
  graph paper

Map Activity, pp. 58–59
  7.5′ quadrangle topographic maps for Big Flats, Elmira, Horseheads, and Seeley Creek, NY
  colored pencils
  string
  ruler

Chapter Review INVESTIGATION

How Are Landforms Represented on Flat Maps?
Computer time: 45 minutes / Additional time: 45 minutes
Students examine topographic maps and 3-D models to learn how contour lines represent Earth's surface.

CHAPTER 3

Earth's Models

This false-color image of Death Valley provides information that cannot be seen in a photograph. Technological advances such as satellite imagery give us a new view of Earth.

Why do we make maps, and how do they help us?

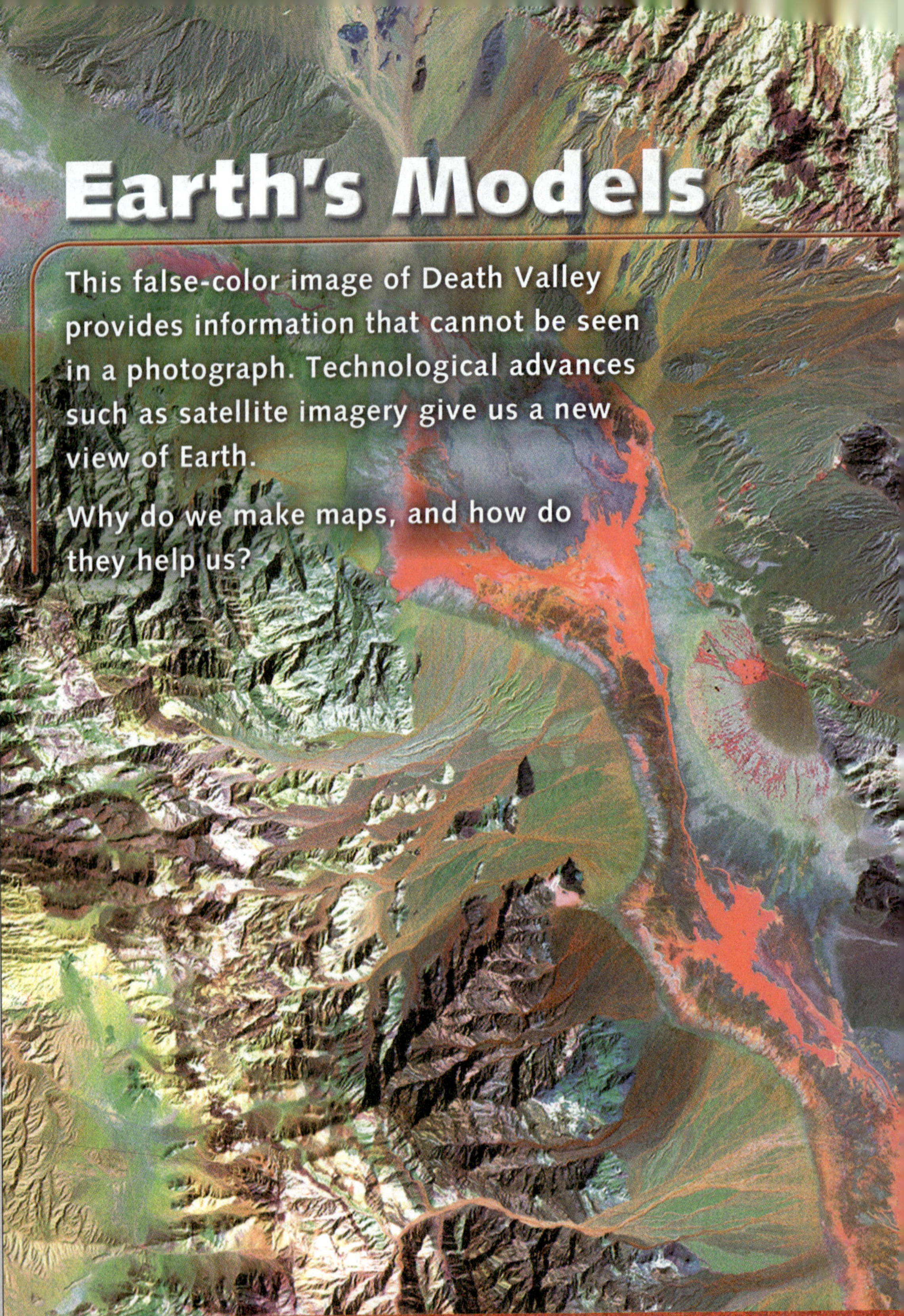

INTRODUCE

Build Interest

Before sharing background information about the satellite image with students, allow time for their questions and comments. Ask questions like the following:

- Can you identify any landforms on this false-color image? **Students might identify areas that look like streambeds, ridges, and valleys.**
- What do you think the colors on the map represent? **Answers might include types of vegetation, types of soil, or types of surface sediments.**
- Who might find false-color maps useful? **geologists, miners, ranchers, farmers, land-use planners, water scientists**

Respond

- Why do we make maps, and how do they help us? **They help us organize information on features of Earth's surface, both natural and human-made. They help us locate places and resources. They enable us to see patterns relating natural phenomena to physical features.**

About the Photograph

This false-color image identifies surface deposits and vegetation. For example, the bright red streak down the middle of the valley indicates extensive salt deposits. The magenta at higher elevations denotes substantial plant life.

BEYOND THE TEXTBOOK

Print Resources
- Reading Study Guide, pp. 7–9
- Transparencies 2, 3
- Formal Assessment, pp. 7–9
- Laboratory Manual, pp. 7–12
- Alternative Assessment, pp. 5–6
- Spanish Vocabulary and Summaries, pp. 5–6
- Internet Investigations Guide, pp. T11–T13, 11–13
- Guide to Earth Science in Urban Environments, pp. 13–20, 21–28

Technology Resources
- Visualizations CD-ROM
- Classzone.com
- Online Lesson Planner
- Electronic Teacher Tools
- Test Generator

42   Unit 1   Investigating Earth

CHAPTER 3

PREVIEW

▶ **FOCUS QUESTIONS** In this chapter you will study maps and mapmaking and learn more about the key questions below.

Section 1 How do maps model Earth?
Section 2 How has technology changed mapmaking?
Section 3 What are topographic maps and what information do they contain?

▶ **REVIEW TOPICS** As you investigate maps and mapmaking, you will need to use information from earlier chapters.

• satellites as tools of Earth scientists (p. 36)

▶ **READING STRATEGY**

USE TEXT AIDS

Terms that you will use to talk about models of Earth are in boldface type. Scientific terms are also in the glossary beginning on page 737. Be sure you understand the meaning of these terms.

INTERNET RESOURCES
CLASSZONE.COM

At our Web site, you will find the following Internet support for this chapter.

DATA CENTER
EARTH NEWS
VISUALIZATIONS
• One Place at Many Scales
• Many Representations of One Place

LOCAL RESOURCES
CAREERS
INVESTIGATIONS
• How Do Map Projections Distort Earth's Surface?
• How Do Latitude and Longitude Coordinates Help Us See Patterns on Earth?
• How Are Landforms Represented on Flat Maps?

Chapter 3 Models of Earth 43

CHAPTER 3

PREPARE

Focus Questions
Find out what knowledge students have about maps. For each focus question, have students consider the section title and then develop two or three additional questions for each section. For example, Section 1: What are some different types of maps? Section 2: How do people make maps?

Review Topics
In addition to the earth science topics listed, students may need to review this concept:

Physical Science
• Electromagnetic energy travels in waves of different wavelengths. The full range of these waves is the electromagnetic spectrum. The spectrum ranges from gamma rays and X-rays, which have some of the shortest wavelengths, to radio waves and microwaves, with some of the longest.

Reading Strategy

Model the Strategy: Point out the list of Key Vocabulary terms on page 44. Have a student read the first word, review its definition and use in the text, and say an original sentence using the word. Write the sentence on the board, and ask other students in the class to evaluate the sentence. Was the term used correctly? If not, how would they change the sentence? Have students continue this strategy for every vocabulary term before starting each section of the chapter.

BEYOND THE TEXTBOOK

INTERNET RESOURCES
CLASSZONE.COM

INVESTIGATIONS
• How Do Map Projections Distort Earth's Surface?
• How Do Latitude and Longitude Coordinates Help Us See Patterns on Earth?
• How Are Landforms Represented on Flat Maps?

VISUALIZATIONS
• Maps of one place at many scales
• Maps, aerial photographs, and other representations of a single place

DATA CENTER
EARTH NEWS
LOCAL RESOURCES
CAREERS

Online Lesson Planner

Chapter 3 Earth's Models 43

CHAPTER 3 SECTION 1

FOCUS

Objectives
1. Explain how maps are models of Earth.
2. Compare and contrast three types of map projections.
3. Use latitude and longitude to describe a location on Earth.

Set a Purpose
Read the title of the section and discuss with students their understanding of what a model is. Ask them to describe any models of Earth they have seen or used. Then have students read to answer the question "How are models of Earth made and used?"

INSTRUCT

More about...

Map Accuracy
Some investigators believe that a mapping error led to the disappearance of aviator Amelia Earhart in 1937, during her attempt to fly around the world. The map she carried placed Howland Island, in the South Pacific, 7 miles northwest of its true location. Armed with faulty coordinates, Earhart and her navigator vanished, never reaching the island.

3.1

KEY IDEA
Maps are models of Earth that share special features to make them as accurate and useful as possible.

KEY VOCABULARY
- map
- cartographer
- projection
- hemisphere
- equator
- prime meridian
- latitude
- longitude
- map scale

Modeling the Planet

Just as some tools are better suited to particular tasks, some models of Earth are better suited to certain uses than are others. Globes are three-dimensional representations of Earth, which makes them accurate models of our three-dimensional planet.

However, globes are not convenient. You cannot fold up a globe and put it in your pocket. And because the globe is a model of the whole Earth, it is not very handy for finding your way across town.

Maps

Sometimes, a map is a better tool. A **map** is a flat, two-dimensional representation of Earth's surface. But because it is two dimensional, a map cannot depict the three-dimensional Earth as accurately as a globe can. All maps distort Earth's features to a degree. However, **cartographers** (mapmakers) have devised a number of tools that help a map user relate the two-dimensional image on the map to the three-dimensional world.

Map Projections

To reduce distortion, cartographers draw world maps in different ways called **projections**. Three common projections are described below.

Mercator Projection

The Mercator projection depicts Earth as if a large cylinder of paper had been wrapped around the planet, and the outlines of the continents had been projected onto the paper. This projection shows the whole world, except the extreme polar regions, on one continuous map.

However, a Mercator projection distorts the areas near the poles. On a Mercator map, Greenland looks like it is almost twice as big as the continental United States, when in reality it is just under one-quarter the size. Nevertheless, Mercator projections are useful because they show true directions as straight lines, and thus are a great boon to navigators at sea.

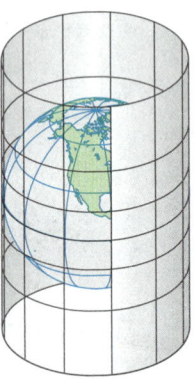 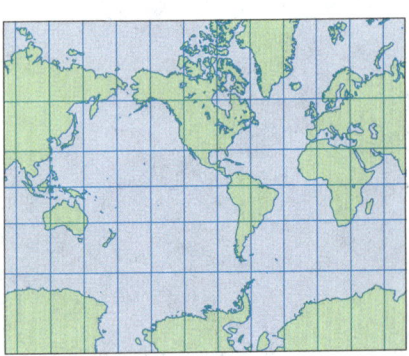

A MERCATOR PROJECTION shows true direction in straight lines but distorts distance near the poles.

44 Unit 1 Investigating Earth

DIFFERENTIATING INSTRUCTION

Reading Support Ask students what they do when they read a word they are not familiar with. As a class, brainstorm a list of strategies for dealing with this challenge. Strategies might include looking for clues to the meaning of the word in the sentence, using the dictionary to look up the word, or using illustrations to help infer the meaning. As an example, demonstrate strategies students can use to find the meaning of *cylinder* on page 44. Point out clues to the word's meaning found in the description "cylinder of paper had been wrapped around the planet" and in the diagram.
Use Reading Study Guide, p. 7.

Gnomonic Projection

Another projection is called the gnomonic, or planar, projection. This kind of projection is made as if a sheet of paper had been laid on a point on Earth's surface, often one of the poles. Although it accurately shows the shortest distance between two points, the gnomonic projection distorts land mass shapes away from the center point. The gnomonic projection is useful in planning ocean or air voyages and for depicting the polar areas.

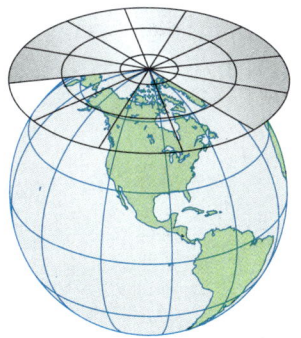 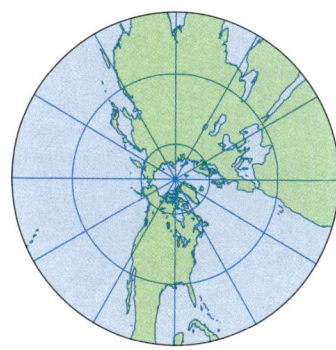

A GNOMONIC PROJECTION can be used to plot the shortest distance between two points, but land masses are distorted away from the center point.

Polyconic Projection

The polyconic, or conic, projection is made as if a cone of paper had been wrapped around Earth. The lines of latitude and longitude are curved slightly, and landforms have their true shape and size in relation to each other. The polyconic projection is especially useful for mapping large areas of land that fall in the middle latitudes, such as the United States.

The shortcoming of all three of these projections is that they most accurately depict the land at the point where the "invisible sheet of paper" touches the sphere; the more distant a landform is from this point, the more distorted it is likely to be.

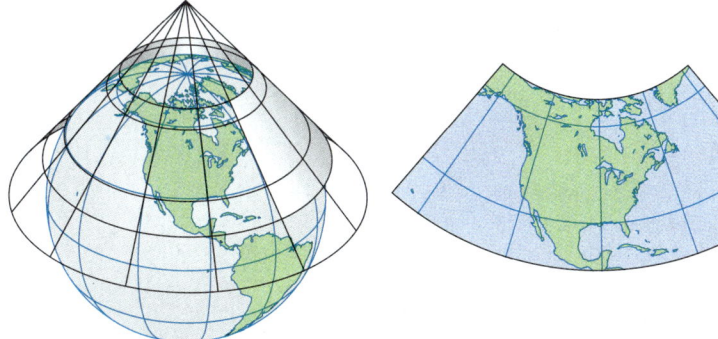

A POLYCONIC PROJECTION is especially useful for mapping large areas of land that fall in the middle latitudes, such as the continental United States.

INVESTIGATIONS
CLASSZONE.COM

How Do Map Projections Distort Earth's Surface? Compare and contrast area, distance, and shape across several map projections.
Keycode: ES0301

CHAPTER 3 SECTION 1

INVESTIGATIONS
CLASSZONE.COM

How Do Map Projections Distort Earth's Surface?
Keycode: ES0301

As a class, discuss the investigation and have students make predictions about how scale influences map projections.

📖 *Use Internet Investigations Guide, pp. T11, 11.*

VISUAL TEACHING
Discussion

Have students contrast the three map projection diagrams. In each diagram, have them identify where the invisible sheet of paper touches the sphere. Make certain they understand that these areas represent where land is most accurately depicted on the resulting map. Point out the grid system on the globes and use its varying appearance on the three maps as an indicator of distortion.

Use Transparency 2.

DIFFERENTIATING INSTRUCTION

Hands-On Demonstration With black marker, draw a grid and simple continent outline on a transparency. Shine a flashlight through the transparency so that it projects the grid on a sheet of paper. Relate this projection to a map projection. Then, using a globe and a large sheet of paper, model the placement of the invisible sheet of paper relative to the globe in the diagrams on pages 44 and 45. Tell students to imagine that the globe is transparent like the transparency and a light is shining at the center of the globe so that an image of a grid and continents would be projected on the paper.

CHAPTER 3 SECTION 1

DISCUSSION

To help students remember the further divisions of degrees of latitude and longitude, relate them to the further divisions of time in hours:

Latitude and Longitude
1 degree = 60 minutes
1 minute = 60 seconds

Time
1 hour = 60 minutes
1 minute = 60 seconds

More about...

Latitude and Longitude
The distance covered by one degree of latitude equals 1/360 of the circumference of Earth (about 40,000 km), or approximately 111 km. One minute of latitude equals approximately 1.85 km, and one second equals about 0.031 km. The distance covered by one degree of longitude varies with the position on Earth. At the equator, one degree equals about 111 km. As you move from the equator toward the poles where the meridians meet, this distance decreases.

VOCABULARY STRATEGY
The prefix *hemi* comes from Greek and means "half." A hemisphere is "half a sphere."

VISUALIZATIONS
CLASSZONE.COM
Observe one place at many scales.
Keycode: ES0302

Latitude and Longitude

The locations of places on Earth are pinpointed through a grid of imaginary lines that cartographers have placed over the planet. This grid, called the graticule, divides up the planet and gives each point a unique and unchangeable "address."

First, Earth is divided into halves, or **hemispheres.** Because Earth is spherical, the hemispheres can be north and south or east and west. The **equator** is the imaginary line dividing the Northern and Southern Hemispheres. The Eastern and Western hemispheres are divided by the **prime meridian** and, opposite it on the globe, the 180° meridian.

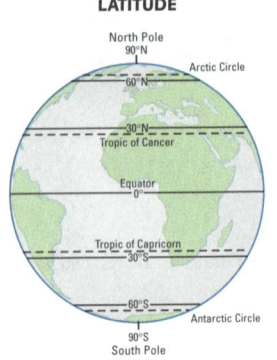
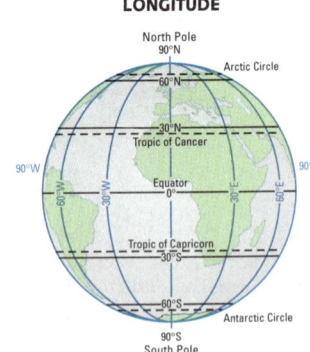

The equator is a **latitude** line. Latitude lines, or parallels, circle the world from east to west parallel to the equator. Latitude is measured in degrees north or south of the equator. The equator has a latitude of 0°. The North Pole's latitude is 90° north; the South Pole's, 90° south.

The prime meridian is a **longitude** line. Longitude lines, or meridians, are half-circles that run from north to south between the poles. Longitude is measured in degrees east or west of the prime meridian, which passes through Greenwich, England, and has the value of 0° longitude. Longitude lines can be numbered from 0° to 180° east or west. The United States lies between 65° and 125° west longitude. India lies on the 80° east longitude line.

Each degree of latitude or longitude is further divided into 60 minutes (written ′) and each minute into 60 seconds (written ″). This system allows you to pinpoint a location on Earth precisely.

Map Scales

How do you know the size of the area you're looking at on a map? Is the body of water in the middle of the map a pond, a lake, or a sea? Without some idea of the size of the landforms a map is depicting, the map is useless; there is no way to judge size or distance.

That is why **map scales** are vital. The scale of a map tells you how the map's features compare in size with Earth's surface. This comparison is usually shown as a ratio, such as one inch on the map equals one mile on the earth. A map can be a large-scale map, which shows a small area of the earth's surface, or a small-scale map, which shows a large area.

DIFFERENTIATING INSTRUCTION

Challenge Activity Have students express the following map scales in words and figures: 1:1,000 and 1:125,000. **1 unit of measure on the map equals 1,000 or 125,000 of the same (or other designated) unit on Earth.** Have them assume a scale based on centimeters, then ask: On a map of each scale, how many kilometers on Earth do 2 centimeters on the map represent? **2 cm at 1:1,000 equals 2000 cm or 0.02 km; 2 cm at 1:125,000 equals 250,000 cm or 2.5 km** Which map scale shows more of Earth's surface per centimeter? **1:125,000** On a map of which scale would your town appear larger? **1:1000**

A map scale can be expressed in several ways. It may be shown in words and figures, such as "1 inch to 50 statute miles." That means that one inch on the map equals 50 miles on earth. It may be shown graphically as a line divided into segments of equal length and then marked in units of measurement (feet, miles, or kilometers). It may be shown as a fraction, such as 1:100,000; that means one unit on the map equals one hundred thousand units on Earth. If the unit is a centimeter, then one centimeter on the map is equal to 100,000 centimeters—or one kilometer—on Earth.

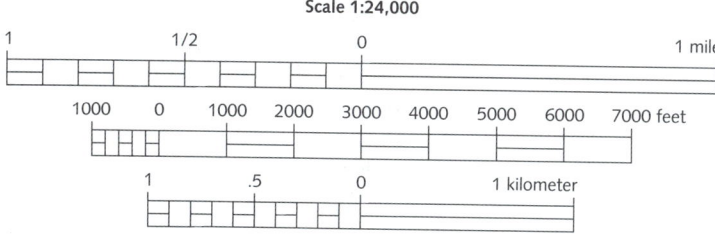

MAP SCALES can be depicted as words, figures, or graphics.

Map Orientation

When you're reading a map, you have to orient yourself to which directions the top, bottom, left, and right represent. In the past, mapmakers did not follow any particular rules when they drew their maps. On some, east is at the top of the page; on others, south is. These variations are confusing today because most modern maps follow the convention of having north at the top, south at the bottom, east on the right, and west on the left. In polar view maps, the pole is in the center. If there is any doubt about a map's orientation, look for the compass rose, an arrow indicating direction, somewhere on the map.

 INVESTIGATIONS
CLASSZONE.COM

How Do Latitude and Longitude Coordinates Help Us See Patterns on Earth? Plot latitude and longitude data on a global map to interpret geographic patterns.
Keycode: ES0303

3.1 Section Review

1. Discuss the advantages and disadvantages of the Mercator, gnomonic, and polyconic projections.
2. Explain the function of latitude and longitude.
3. **PAIRED ACTIVITY** Working with a partner, compare the landforms depicted on a globe and on a map. Which landforms seem most distorted by the map? Explain why they are.
4. **CRITICAL THINKING** What methods do you use to navigate the area in which you live? What advantages do your methods have? Would navigating by latitude and longitude work as well? Why or why not?
5. **GEOGRAPHY** What is the latitude and longitude of your area?
6. **MATHEMATICS** A map scale is essentially a ratio of a unit of distance on the map to a unit of distance on Earth. Calculate the scale used to make a classroom globe.

CHAPTER 3 SECTION 1

Observe one place at many scales.
Keycode: ES0302

Visualizations CD-ROM

How Do Latitude and Longitude Coordinates Help Us See Patterns on Earth?
Keycode: ES0303

Have students work in pairs.

Use Internet Investigations Guide, pp. T12, 12.

ASSESS

1. Mercator—advantages: shows the whole world on one continuous map, shows true directions as straight lines; disadvantage: distorts the areas near poles; Gnomonic—advantage: shows the shortest distance between two points; disadvantage: distorts landmass shapes away from center point; Polyconic—advantage: shows landforms with their true shape and size in relation to each other; Disadvantage of all projections—all distort landforms to some degree.

2. Latitude and longitude form a grid that gives each place on Earth a unique and unchangeable "address."

3. Landforms near the map edges are most distorted. On map projections, the more distant a landform is from the point where the invisible sheet of paper touches the sphere, the more distorted it is likely to be.

4. Answers will vary. Latitude and longitude would not likely work well because over small distances, you would need to refer to degrees, minutes, and seconds, which most maps are not detailed enough to show.

5. Responses will vary with location. Look for detailed coordinates of degrees, minutes, and seconds.

6. Students' scales can be expressed in words and figures or as a fraction.

MONITOR AND RETEACH

If students miss . . .

Question 1 Reteach "Map Projections." (pp. 44–45)

Question 2 Reteach "Latitude and Longitude." (p. 46) Ask: What do latitude and longitude help you do?

Question 3 Review the disadvantages of the three types of map projections shown on pages 44 and 45.

Question 4 Reteach "Latitude and Longitude."

Question 5 Refer students to the United States map on pages 708 and 709.

Question 6 Have students determine the actual distance between two locations and then use a sheet of paper and ruler to measure the distance between the two on the globe.

CHAPTER 3 SECTION 2

▶ FOCUS

Objectives
1. Relate the history of mapmaking.
2. Describe the roles of satellites and computers in making and using maps.

Set a Purpose
Read aloud the key idea. Then have students read this section to answer the question "How has technology influenced the making and use of maps?"

▶ INSTRUCT

> **More about...**
>
> **Mapmaking**
> The science of mapmaking—cartography—dates back to when nomadic humans became settled farmers and the ownership of land became significant. The oldest known map consists of a small clay tablet that shows the boundaries of an estate in Mesopotamia. It dates to about 2500 B.C.

3.2

KEY IDEA
Technological advances have profoundly changed the way maps are made and used.

KEY VOCABULARY
- radar
- false-color image

Mapmaking and Technology

Mapmaking is an ancient and important art. One of the first known maps was made on a clay tablet about 2300 B.C. in ancient Sumeria. Peoples around the world have long recognized how vital maps are. Today, technology is changing traditional methods of making maps.

How Cartographers Traditionally Worked

The United States Geological Survey (USGS) has mapped the entire United States. The USGS created its first map in 1879, the same year it was founded by Congress for the purpose of the "classification of the public lands." Most of the USGS's early work involved conducting ground surveys of the West. The 19th-century cartographers used a classic mapping technique called plane-table surveying.

The plane table was a drawing board with a sighting device mounted on a tripod. Carrying the plane table, the cartographer would climb to the area's highest vantage point and draw a map of whatever features he could see. The process was time-consuming and required great skill and, in many situations, great courage, as the cartographers braved difficult conditions, including bone-chilling weather, bandits, and horse thieves. Despite the drawbacks, plane-table surveying remained the main method of mapping until the 1940s, when a new technology arrived.

How Technology Has Changed Mapmaking

The long-standing method of making maps changed tremendously when it became possible to take photographs from airplanes. A photograph can show a cartographer exactly what a land area looks like. By combining the information from photographs and from surveys, cartographers made more accurate maps than they could make with only survey data. Cartographers could accurately copy the contours of the land from photographs onto paper.

The invention of **radar** gave cartographers another tool. If the day was cloudy, aerial photography was useless because the surface was hidden from view. Radar, however, could "see" right through the clouds.

A radar device sends out a radio signal, which consists of electromagnetic waves, a pattern of magnetic and electric energy. The radio signal bounces off any object it hits, such as a mountain peak, and is reflected back to the radar device that transmitted it. Because the radar used in mapping is directed sideways from the aircraft, not straight down, several signals come back instead of only one. Computers process the radar data and turn them into images of Earth.

AERIAL PHOTOGRAPHY In this 1950s photograph, a Coast Guardsman prepares to take pictures during a flight.

48 Unit 1 Investigating Earth

DIFFERENTIATING INSTRUCTION

Developing English Proficiency Students may become bogged down by unfamiliar technical terms used in this section. Encourage them to raise their hands when you use unfamiliar terms or to point out such terms in their textbooks. If possible, show students photographs of plane-table surveying and radar equipment and examples of false-color and high-resolution images so they can make visual associations to technical terminology.
Use Spanish Vocabulary and Summaries, pp. 5–6.

Satellite Technology

In the late 20th century, mapmaking technology changed again, as satellites trained their electronic eyes on Earth. In 1972, the National Aeronautics and Space Administration (NASA) launched the first Landsat satellite. The photographs produced from data gathered by Landsat 1 and its successors showed Earth in a way it had never been seen before.

Satellites use sensors to detect changes in the wavelength of light that is reflected from Earth's surface. They send this information to computers on Earth, which process it into images.

The images are not necessarily true-to-life color pictures because natural colors make perceiving the different wavelengths difficult. A **false-color image,** made up of colors such as reds, yellows, and purples, can be more useful. In a false-color image of Niagara Falls, the calm water upstream might be black while the waterfall's turbulent water is bright blue.

In addition to their value in mapmaking, these satellite images give other earth scientists a new window on the world. In 1999, a team of scientists from the American Museum of Natural History used Landsat images to find a new site of dinosaur and early mammal fossils in the Gobi Desert. The images helped them pinpoint areas of sedimentary rocks and low vegetation—both signs of a site that might hold fossils.

In 1999, NASA launched another satellite, the Earth Observing System (EOS). Unlike Landsat, EOS does not provide high-resolution images of Earth. Instead, it provides precise measurements of Earth's processes, such as air–land and air–sea exchanges of energy, carbon, and water.

Observe many representations of a single place.
Keycode: ES0304

Observe many representations of a single place.
Keycode: ES0304

Have students identify which representations they would find most useful in hiking around and navigating the roads around the Grand Canyon.

Visualizations CD-ROM

Visual Teaching

Interpret Diagram
The diagram shows how satellites and computers work together to produce images of Earth's surface. Ask: What device detects and records data from Earth's surface? **a scanner on a satellite** In what form does the data travel? **as electromagnetic waves** What happens to the data once it is recorded? **Computers convert the data to code, then to pixels, and finally to usable images.**

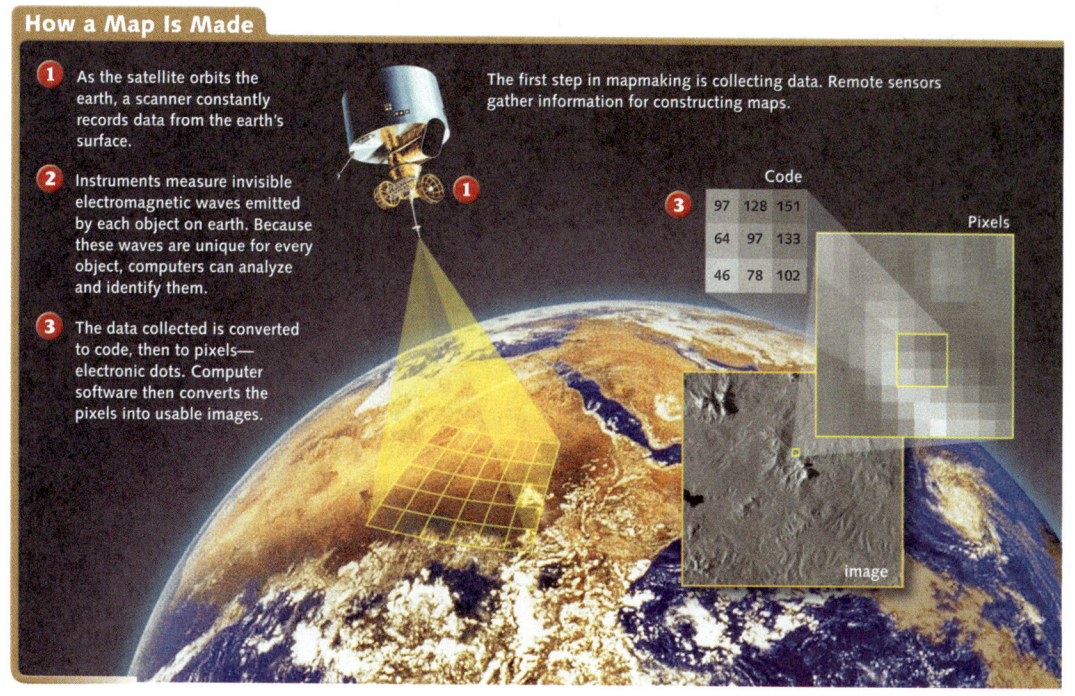

How a Map Is Made

1. As the satellite orbits the earth, a scanner constantly records data from the earth's surface.
2. Instruments measure invisible electromagnetic waves emitted by each object on earth. Because these waves are unique for every object, computers can analyze and identify them.
3. The data collected is converted to code, then to pixels—electronic dots. Computer software then converts the pixels into usable images.

The first step in mapmaking is collecting data. Remote sensors gather information for constructing maps.

DIFFERENTIATING INSTRUCTION

Reading Support Have students make a chart to organize the information in this section. In the first column, they should label the rows *Type of technology, What it is used for, How it works,* and *Other information.* Then in the second column, they should record information about each technology discussed in the section.
Use Reading Study Guide, p. 8.

Challenge Activity Have students use the Internet to find and print out two false-color images of Earth's surface. They should research what the various colors represent and present their images and findings in a poster. The poster should include a guide or key to interpreting the false colors and labels that explain parts of the images. Remind students to include the locations of the images.

CHAPTER 3 SECTION 2

More about...

The Global Positioning System
GPS receivers, once the tools only of government agencies and scientific researchers, can now be purchased by anyone. These are compact devices that can be handheld or installed in automobiles. GPS receivers can be used as interactive maps that graphically point to the user's location relative to roads, bodies of water, and other landmarks. GPS receivers can also act as odometers, tracking how far an individual has traveled, or speedometers, determining how fast an individual has moved. Since they require unobstructed communication with satellites, individual GPS receivers generally operate only when used outdoors.

Scientific Thinking

INFER
Students may suggest that by considering the source locations of different kinds of data, researchers will be able to discern information not available when looking at just the data alone.

Computer Technology

Satellites would not be much use without something to decode and make images out of the data they collect. Computers that are smaller, more portable, and more powerful than their predecessors have changed both how maps are made and used. Maps are not just on paper anymore.

The Global Positioning System

The power of computers is combined with the power of satellites in the Global Positioning System (GPS). Started as a tool for the military, GPS allows people to determine their position, speed, and time anywhere in the world, at any time of day, in any weather.

GPS consists of three segments: space, control, and user. The space segment of GPS is made up of 24 satellites that follow circular orbits above the earth's surface. At any given time, 5 to 8 satellites are in view from any part of the world. The control segment of GPS is a master control station near Colorado Springs, Colorado, along with three ground antennas and five monitoring stations located around the world. The control segment tracks the satellites, making sure they are in their proper orbits and sending them orbit-correction instructions as needed. The user segment of GPS consists of the receivers that allow people to use the system.

Each of the 24 GPS satellites continuously broadcasts its location and the exact time. The broadcasts act as reference points. The user's receiver picks up broadcasts from at least four satellites and uses the information to calculate the user's location, speed, and time. Because it uses satellite signals, GPS is not affected by bad weather. It can be used on land, at sea, and in the air.

GPS is used for many purposes besides finding out where a user is. At the University of West Florida, researchers use GPS information to study coastline erosion caused by hurricanes. An international bank uses GPS information to verify the time of business trades that take place between London and New York. NASA, with the help of local high school earth science students, has been using GPS to make precise measurements to assess the earthquake hazards in Alaska. Some people are using GPS as the key to a game called geocaching, in which one group hides a cache (a collection of objects, such as a log book and a disposable camera), and other groups try to find it by using its GPS coordinates.

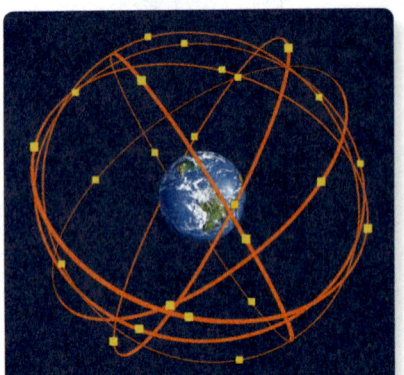

GPS SATELLITES A web of satellites that serve as beacons transmit signals that are unaffected by bad weather.

Scientific Thinking

INFER
Why would having data linked to their geographic origin be particularly valuable?

Geographic Information Systems

By combining data from satellite images, statistical surveys, and traditional land surveys, geographic information systems (GIS) let cartographers make maps unparalleled in their usefulness. A GIS is a computer system that can assemble, store, manipulate, and display data identified by location. To learn more about GIS, turn to the Science and Technology feature on page 52.

DIFFERENTIATING INSTRUCTION

Hands-On Demonstration If possible, obtain an individual GPS receiver and demonstrate its use. Show students how to use the device to find their current position relative to local landmarks. Let them examine any instructional materials packaged with the receiver to discover additional uses for it. Review with students the role that satellites play in the functioning of the device. Then brainstorm a class list of ways people might use individual GPS receivers.

CAREER

Land Surveyor

In many business transactions that involve land, the land surveyor's role is essential. Land surveyors collect information about a plot of land, including its boundaries and the artificial and natural structures on it. They then use maps and charts to communicate the findings to clients.

Like cartographers, surveyors use innovative technology, such as satellite positioning systems, laser measuring devices, and computer models. Their strong analytical skills are very useful for interpreting figures and data. Since surveyors must visit sites to make their assessments, they have frequent opportunities to travel. Because of the nature of their work, many land surveyors work well independently and are willing to brave all weather conditions.

A bachelor's degree in a field such as computer science, physics, geology, or engineering is helpful, as is computer proficiency. Besides having an analytical mind, many surveyors find the ability to communicate well with others an asset in dealing with clients.

Most surveying firms provide their employees with additional training and allow them to use expensive equipment. Employment opportunities can be found in private surveying companies, local governments, or civil-engineering firms. ∎

LAND SURVEYORS must use a variety of equipment while on the job to obtain the necessary information about a plot of land.

Learn more about a career as a land surveyor.
Keycode: ES0305

New Ways to Print and Produce Maps

Not only have computers changed the way maps are used, they have also changed the way maps are physically created. In the past, all the lines of each map had to be scribed by hand onto a special plastic sheet covered with a soft coating called scribecoat. A separate sheet was used for each color. The sheets were used to create printing plates that were put on the printing presses to print the maps. The process was time-consuming, as well as expensive.

The USGS creates new maps on computer. Now, updating the maps will be considerably easier than it was in the past. Printing plates will be made from the digital information, a faster and less-costly process than has been possible before.

3.2 Section Review

1. How did aerial photography and radar change mapmaking?
2. How do satellites and computers work together to produce images of Earth?
3. **CRITICAL THINKING** What other uses can you imagine for GPS?
4. **PAIRED ACTIVITY** With a partner, try drawing a map the way a plane-table surveyor did. What is easy about making a map this way? What is difficult?

CHAPTER 3 SECTION 2

Learn more about a career as a land surveyor.
Keycode: ES0305

ASSESS

1. Aerial photography allowed cartographers to study photographs of land areas and determine the contours of the land from the photographs, which made mapmaking more accurate. Radar allowed cartographers to obtain images of Earth in any weather.

2. Satellites collect and record data from Earth's surface. Computers then convert the data to a code that is used to produce images.

3. **Sample Response:** locate people and vehicles when they are lost, help someone find a route to a destination, and investigate scientific phenomena such as ocean currents

4. **Sample Response:** It is easy to identify and draw landmarks seen in person, but it is difficult to accurately map distances and reach rugged locations.

MONITOR AND RETEACH

If students miss . . .

Question 1 Reteach "How Technology Has Changed Mapmaking." (p. 48) Have students contrast mapmaking before and after the development of aerial photography.

Question 2 Use the diagram on page 49 to discuss how satellites and computers work together to produce images of Earth.

Question 3 Review with students how GPS works, as described on page 50. Have the class brainstorm possible uses of GPS.

Question 4 Have students reread "How Cartographers Traditionally Worked." (p. 48) If students will be mapping outdoors, preapprove their proposed mapping locations to minimize safety issues.

CHAPTER 3 SECTION 2

Make your own map with an online GIS.
Keycode: ES0306

Science and Technology

Using GIS for Preservation
Scientists use GIS to supply important environmental data to decision makers. For example, government scientists used GIS to help officials predict how much damage a mine would do to a forest. They combined data on the amount and location of the proposed mining with a map showing landforms. GIS created a three-dimensional view of the area, showing how the mine would affect the landforms and forest.

Science Notebook
Students might suggest linking a street map to data on crime occurrence to determine where to increase police protection, linking the incidence of cancer to the locations of industrial facilities to see if there is a connection between the occurrence of the disease and the facilities, and linking a street map to census data showing the addresses of seniors to determine where a new bus service should be routed.

SCIENCE & Technology

Linking Information and Location

Geographic Information Systems let scientists and others look at geographic data in a new, more useful way than other methods.

As we link geography and knowledge more closely, what more can we learn about our world?

Although sometimes called computer mapping, Geographic Information Systems (GIS) involve more than getting driving directions off the Internet. GIS maps can illustrate the effects of herbicides in Vietnamese wetlands, track excess nitrogen in the Mississippi River basin, and pinpoint cancer incidences on Cape Cod, Massachusetts. Scientists, public-health officials, city planners, natural-resources managers, and businesses can use GIS to find patterns and relationships that would go unnoticed if they were not linked to specific places.

By combining layers of data from various sources, GIS mapmakers can create custom maps. While paper maps show all the features in an area, such as highways, railroads, and bodies of water, GIS technology allows users to select the information that will be shown on a map according to the users' specific needs.

For example, a GIS map can show the population density of an area, which might help planners determine where a new hospital needs to be built. A biologist might use GIS to determine where the highest concentrations of certain waterfowl species make their nests. If they relate this information to wetland preservation, they can evaluate the success of wetland restoration efforts.

By combining the power of computer data with the usefulness of maps, GIS technology is helping people in a variety of professions to look at the world in a different way and to understand it better than ever before. ■

Extension

Can you think of other ways of linking data to specific geographic locations that might prove useful?

Make your own map with an on-line GIS.
Keycode: ES0306

BY USING GIS technology to combine a terrain map, a land use map, and a base map, developers were able to determine the best sites for a new airport.

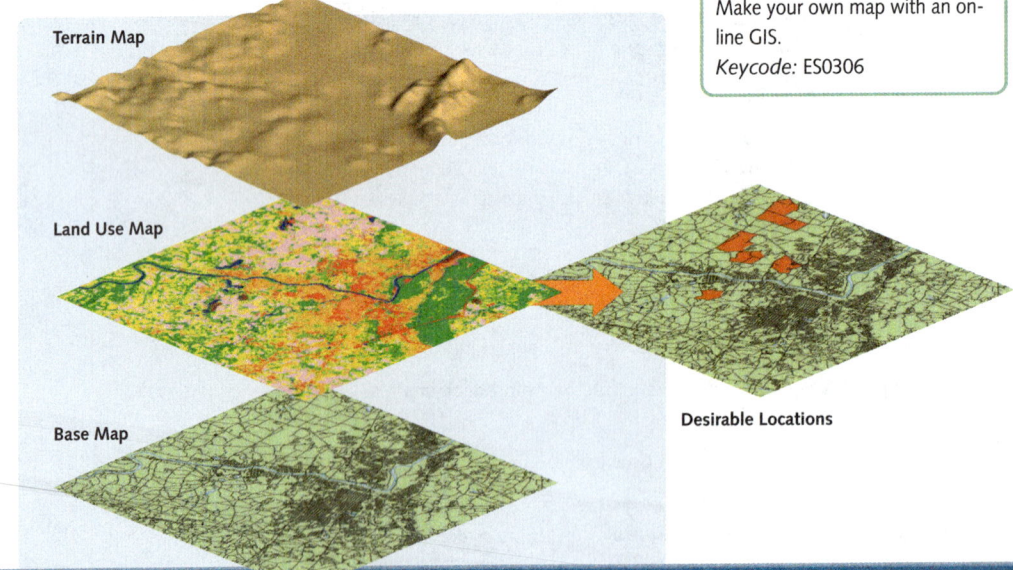

Topographic Maps

One of the most frequently used types of maps in earth science is the topographic map, which depicts the features of an area through the use of contour lines.

Qualities of Topographic Maps

A **topographic map** shows the **topography** (tuh-PAHG-ruh-fee) of an area, that is, its natural and human-made surface features. It indicates landforms such as cliffs, mountains, and beaches; bodies of water; and streets and buildings. To show landforms, maps must show the relief (the highs and lows) of Earth's surface. Relief can be shown in many ways, such as by shading or coloring. Topographic maps are unique and especially useful because they use contour lines to show the relief.

The United States Geological Survey (USGS) has mapped the entire United States in more than 53,000 maps. The USGS updates map information regularly, using information from aerial photography and field checks by volunteer members of the Earth Science Corps. Because making topographic maps is so expensive, most revisions do not involve radical changes. Instead, they involve updating boundary changes or adding features (such as new areas of city growth) that have appeared since the last revision.

A common USGS topographic map is the 7.5-minute map. It shows an area called a quadrangle that spans 7.5 minutes of latitude and 7.5 minutes of longitude. It has a scale of 1:24,000. Because one inch represents 2000 feet, these 7.5-minute maps give detailed information about the areas they depict. Some maps even indicate individual houses.

Topographic maps of Alaska are different from those for the continental United States. Called 15-minute maps, they cover areas of 15 minutes of latitude and 20 to 36 minutes of longitude. USGS 15-minute maps have a scale of 1:63,360 (one inch on this map equals one mile).

HIKERS, such as this woman exploring Mount Baxter in New York's Adirondack Mountains, are among the people who find topographic maps invaluable.

3.3

KEY IDEA
Topographic maps use contour lines to show the natural features of an area and the features created by people.

KEY VOCABULARY
- topographic map
- topography
- contour lines
- contour interval
- slope
- magnetic declination

CHAPTER 3 SECTION 3

FOCUS

Objectives
1. Explain how topographic maps use contour lines to show elevation.
2. Describe topographic map symbols.
3. Demonstrate how to use topographic maps to determine the shape of the land, the flow of rivers, and distance.

Set a Purpose
Have students read the section to answer the question "How do topographic maps use contour lines to show the features of an area?"

INSTRUCT

CRITICAL THINKING
Ask students how hikers might find topographic maps useful. Hikers can use topographic maps to find their way relative to land features, avoid bodies of water and too-steep trails, and estimate how far they have to travel.

DIFFERENTIATING INSTRUCTION

Reading Support Explain that making outlines can help students organize and summarize in a clear and logical manner what they have read. Then demonstrate the outlining process using the blue and red headings, key idea, and key vocabulary as guides. On the board, begin an outline of the section, such as the following:

I. Topographic Maps
   A. Qualities of Topographic Maps
      1. show natural and human-made features
      2. use contour lines to show slope and elevation

Have students suggest what should come next in the outline. **Possible Response:** B. Topographic Map Symbols
Use Reading Study Guide, p. 9.

CHAPTER 3 SECTION 3

VISUAL TEACHING

Discussion

Have students find contour lines, index contours, and contour intervals on the topographic map on pages 54 and 55. Ask: What is the elevation of the index contour on Long Island? **30 feet** Which is higher in elevation—Falmouth Foreside or West Gorham? **West Gorham**

Extend

Ask students to relate any experiences they have had using topographic maps. Discuss how they depended on the maps and how topographic maps were more useful than other kinds of maps in those circumstances.

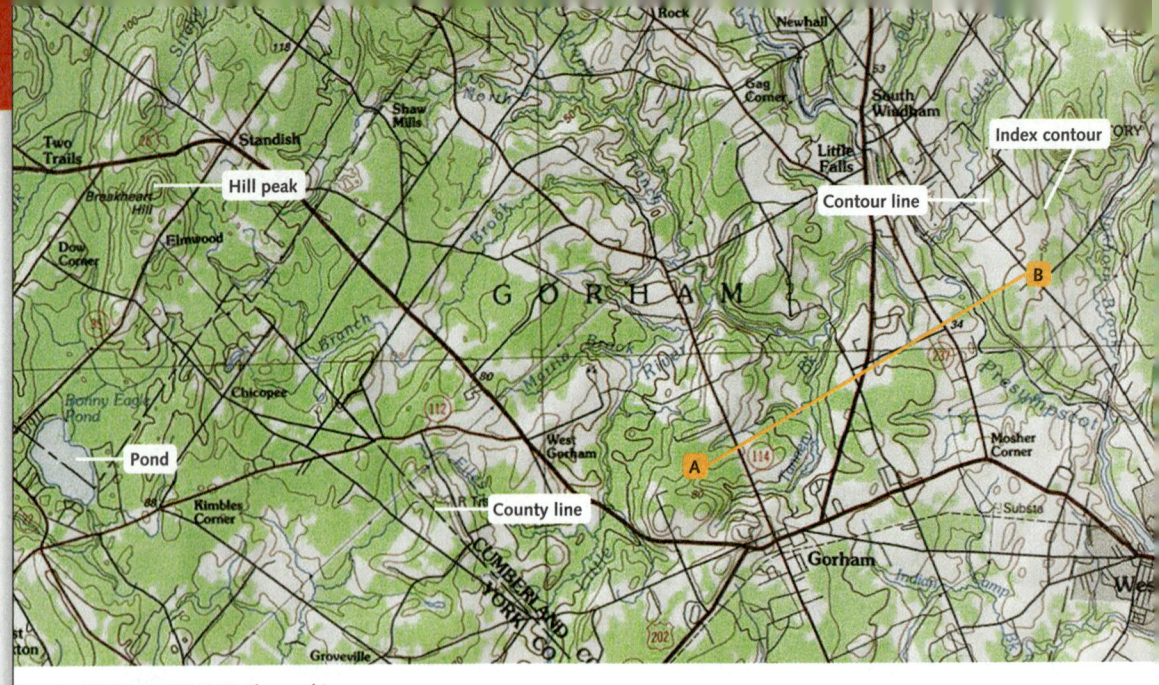

A TOPOGRAPHIC MAP The graphic above is a section of the topographic map of the Portland, Maine, area. The city is shown in pink. You will use the line marked "A–B" in the minilab.

Contour Lines

Unlike road maps, topographic maps show the shape of the earth's surface. By reading a topographic map, you can get a mental picture of the shape of the area it depicts if you know how to interpret the contour lines.

Contour lines are the narrow brown lines on topographic maps that indicate the landscape's elevation, or height above sea level. Contour lines are drawn to connect points of the same elevation in an area. They are the key to understanding what you are looking at on a topographic map.

On a contour map of the Maine seacoast, the first contour line is found at zero feet above sea level (at sea level). The next contour line appears 10 feet above the first one. That means that all the features that this line intersects are 10 feet above sea level. Anything below the line is between zero and 10 feet above sea level, and anything above it is more than 10 feet above sea level. The difference in elevation between two consecutive contour lines is called the **contour interval.**

The line above the 10-foot line indicates features at 20 feet; the next line above that indicates 30 feet, and so on, in 10-foot increments. Most of the lines on a topographic map are not marked, but every fourth or fifth interval will be a heavy line, similar to the one you see at sea level, with a number on it (100, 200, etc.), that serves as a reference point (these are called index contours). Not all maps begin at sea level, of course, and some have intervals greater than 20 feet. However, by looking for prominent index contours, you will be able to judge the elevation of the area the map depicts.

Some maps include information on the depths of the water they show; they are called topographic-bathymetric maps. Like the contour lines that show elevations on the land, blue lines called isobaths show the form of the land beneath the water's surface.

54 Unit 1 Investigating Earth

DIFFERENTIATING INSTRUCTION

Hands-On Demonstration Using a photocopier, enlarge a section of a topographic map that includes contour lines, particularly index contours. If possible, make the copy into a transparency and use it to demonstrate how to use contour lines to find elevation and the contour interval. Demonstrate how to determine the elevation of a point on an unmarked contour line and how to estimate the elevation of a point between two contour lines. For example, a point halfway between the 40-foot and 60-foot contours would have an estimated elevation of 50 feet.

54 Unit 1 Investigating Earth

Slope and Elevation

The distance between one contour line and the next shows you the **slope,** or steepness, of the landscape. If the contour lines are very far apart, the landscape is relatively flat. If they are close together, it is steep. To determine the slope of an area, divide the change in elevation by the distance covered. The figure that results will be the average slope.

For example, say you want to find the slope of a mountainside. The point you start measuring is at 1400 feet; the summit is at 3165 feet. The distance you're measuring is two and a half miles.

Subtract 1400 from 3165 to get 1765 feet (the change in elevation). Divide 1765 feet by 2.5 miles, and you get an average slope of 706 feet per mile. The actual slope may be different if some parts of the mountainside are steeper than others.

In mapmaking, surveyors find the exact elevation of many points in the map area. These may be shown on the map in a number of ways. A bench mark is a location whose exact elevation is known. At the survey location, it is noted on a brass or aluminum plate, which is permanently set into the ground. On a map, bench marks are shown by the abbreviation "BM" that appear next to a cross, an X, or a triangle. Numbers next to the abbreviation and symbol give the elevation to the nearest foot.

Spot elevations are the elevations of road forks, hilltops, lake surfaces, and other points of special interest. These points are usually shown on the map by small crosses. Numbers giving elevations checked by surveyors are printed in black. Unchecked elevations are printed in brown. Water depths are shown in blue, with blue lines called isobaths indicating the shape of the land below the water.

COAST OF MAINE The topography of Maine can be complex, as this photograph taken near Portland shows.

CHAPTER 3 SECTION 3

DISCUSSION
Refer students to the key to topographic map symbols on page 697. Ask them what a bench mark is and have them describe the bench marks listed on the key. A bench mark is a location whose exact elevation is known. On the key, bench marks are the abbreviation BM and/or an X followed by numbers that represent elevation. Then refer students to the crossed symbol north of Falmouth on the map on pages 54 and 55. Have them use the key to determine what the symbol represents. a quarry or gravel pit

---

DIFFERENTIATING INSTRUCTION

Developing English Proficiency Write the words *bathymetric* and *isobath* on the board. Circle the common root *bath* in each word. Explain that these words are derived from the Greek word *bathys*, meaning "deep" or *bathos*, meaning "depth." Point out the prefix *iso-*, which comes from the Greek *isos*, meaning "equal." Thus, an isobath is a line of *equal depth* in a body of water, and bathymetric maps show *depths* of water. As a memory tool, students might associate *bathymetric* and *isobath* with water in a bathtub.

CHAPTER 3 SECTION 3

MINILAB
Drawing a Profile
Management
- Point out the location of line AB on the topographic map on pages 54 and 55.
- Remind students to locate index contour lines to help determine the height of each contour that crosses the line.

Analysis Answers
- Some students might use straight lines because the heights between the points are unknown. Others might use curved lines because such lines show more realistic changes in height.
- You cannot determine with certainty the elevation of a point between contour lines except to say that it is between the elevations of the two adjacent contour lines.

VISUAL TEACHING
Discussion
As an example of a depression contour, point out the depression isobath in the lower right corner of the map on pages 54 and 55. For help in identifying natural and human-made features on the map, refer students to the key to topographic map symbols on page 697. Then have students locate Mill Brook on the map. Ask them to determine the direction in which the brook flows and explain their answer. Since the contour lines crossing the brook point toward the north, or top of the map, to the brook's source, Mill Brook must flow south, or toward the bottom of the map.

25-Minute Mini LAB
Drawing a Profile
Materials
- graph paper

Procedure
1. The map on pages 54–55 has a line from A to B marked on it. Place the bottom edge of the graph paper on that line.
2. At each point where the line crosses a contour, make a mark on the edge of the paper. Record the height of each contour next to its mark.
3. Raise the point at the highest contour close to the top of the page. Raise the other points to their proper places in relation to the highest point.
4. Connect the points to complete your profile.

Analysis
Did you connect the points with straight lines or curved lines? Why? Is there any way to tell what the elevation is between contour lines? Explain your answer.

Understanding Topographic Map Symbols
By understanding the unique symbols topographic maps use, you can tell human-made features, such as buildings, from natural ones. You can tell water from land, and you can visualize landforms.

Contour lines may form circles or ovals. These indicate hills, ridges, or mountains. The peak of a hill or mountain may also include a number, which indicates its elevation. The distance between the contour lines will give you an understanding of the slope of the land.

You may find a circle or oval with small lines jutting inward from the main contour line. This symbol indicates a depression in the landscape, and it is called a depression contour. The crater of a dead volcano would be shown by a depression contour.

Bodies of water are indicated in blue on topographic maps. If you know how to read the contour lines, you can figure out which way rivers flow by using a topographic map. When a contour line crosses a river, it forms a point back toward the river's source. On the map on pages 54–55, you can see how the rivers run down toward the sea.

Human-made features, as well as natural landforms, are depicted in black on topographic maps. Roadways, railway lines, bridges, gas and water tanks, dams, and individual buildings are often indicated. Thickly settled regions are depicted in pink.

Maps on which areas have been revised indicate these updated areas in purple. Refer to pages 712–713 for more information about symbols.

Topographic maps have a special kind of compass rose that shows the magnetic declination of the area pictured. Compasses will point to magnetic north, not to true north. The **magnetic declination** is the angle by which a compass needle will vary from true north. Each topographic map gives the angle of magnetic declination for the area it represents.

Approximate Mean Declination, 2001

MAGNETIC DECLINATION This symbol shows the angle by which a compass point varies from true north.

Using Topographic Maps
Topographic maps can be used to determine the shape of the land, the direction of the flow of rivers (as mentioned above), and distance.

It is easy to determine the shape of the landforms depicted on a topographic map. To do this, you make a profile of the land. Wherever the line meets a contour, the exact height above sea level is known. Plotting these points on a vertical scale results in a profile.

A profile is most easily made by placing the bottom edge of a sheet of paper on top of the line to be followed. At each point where the line crosses a contour, make a mark on the edge of the paper. Record the height of the contour next to its mark on the paper. When all points are marked, use the vertical scale to raise each point to its proper height. (Plotting is easier when graph paper is used.) A vertical scale is usually stretched out

---

DIFFERENTIATING INSTRUCTION

Support for Visually Impaired Students Team visually impaired students with partners with good eyesight to do the Minilab. Partners can read the maps and create the profiles while visually impaired students review the procedures for making profiles and answer the analysis questions. Some students may benefit from having a magnifying glass and task lighting available when reading the map on pages 54 and 55. If possible, provide these items as well as an enlarged photocopy of the map for their use.

compared to the horizontal scale. This is to make the differences in elevation more visible. An example of a vertical scale is $\frac{1}{8}$ inch = 20 feet. Of course, it is important to keep the points the same horizontal distance apart as they were on the map. Once the elevated points have been plotted, they are joined to make the profile.

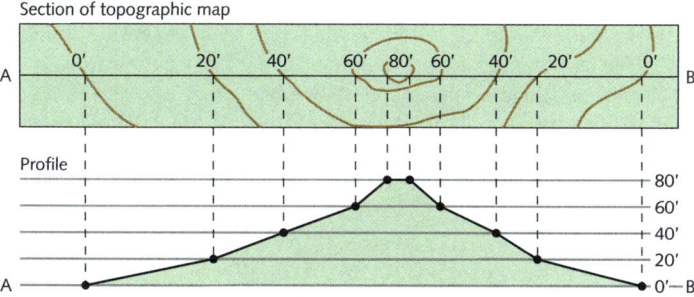

PROFILE DIAGRAM By transferring information from a topographic map to another sheet of paper, it is possible to the draw a landform's profile, or shape.

To measure distance on a topographic map, you will use the map scale. If the scale of a map is given in words, distances on the map may be measured with a ruler. When a graphic scale is printed on the map, the distance between two points can be marked off with a straightedge, such as the edge of a sheet of paper. A piece of string may also be used. The marked straightedge is held against the scale for reading. Zigzag distances along roads or rivers may be marked off one after the other on the edge of a sheet of paper before measuring against the graphic scale.

Topographic maps are useful in a variety of ways. City planners use them to plan public works projects. Engineers use them when they plan civil engineering projects, such as roads. Researchers use them as they search for deposits of coal, oil, and natural gas. And everyday people use them too, as they hike, camp, orienteer, and otherwise enjoy the outdoors.

3.3 Section Review

1. What do contour lines indicate on a topographic map?
2. What would the topography of an area be like if the contour lines were close together? If they were far apart?
3. Describe how to measure distance using a topographic map.
4. What human-made features might appear on a topographic map?
5. **CRITICAL THINKING** Why do you think USGS cartographers made 15-minute maps for Alaska, instead of 7.5-minute maps?
6. **MATHEMATICS** Determine the slope of an area, given the following: Start measuring at 700 feet. Measure a distance of four miles, and end at an elevation of 2432 feet.
7. **PAIRED ACTIVITY** Working with a partner, select part of the topographic map of the Portland, Maine, area, and create a profile.

CHAPTER 3 SECTION 3

VISUAL TEACHING
Discussion
Suggest that students think of each point on the profile as a coordinate plotted on *x*- and *y*-axes. The *x*-axis (horizontal) represents the position on the map. The *y*-axis (vertical) represents elevation.

Use Transparency 3.

ASSESS

1. elevation, or height above sea level
2. If the contour lines were close together, the topography would be steep. If the lines were far apart, the topography would be flat.
3. Use the map scale. If the scale is given in words, you can measure distance with a ruler. If there is a graphic scale, you can mark off the distance with a straightedge and hold it to the scale for reading.
4. Human-made features that appear on a topographic map include buildings, roadways, bridges, railway lines, gas and water tanks, and dams.
5. Because Alaska is so large, 15-minute maps, which cover more area than 7.5-minute maps, are needed to effectively cover it.
6. $(2432 - 700) \div 4 = 433$ feet per mile
7. Students' profiles will vary but should show the shape of the land and the elevations at various points along the land.

MONITOR AND RETEACH

If students miss . . .
Question 1 Reteach "Contour Lines." (p. 54) Have students identify contour lines on the map on pages 54 and 55.
Question 2 Point out on the map on pages 54 and 55 where the topography is steep and where it is flat.
Question 3 Have students use the map scale on page 55 to estimate the shortest distance between Cow Island and Little Chebeague Island. **about 1.8 miles, or 2.9 kilometers**
Question 4 Have students review the map symbols on page 697.
Question 5 Reteach "Qualities of Topographic Maps." (p. 53) Have students use a U.S. map to contrast the size of Alaska with that of other states.
Question 6 Review the equation slope = change in elevation ÷ distance.
Question 7 Discuss how the profile diagram on page 57 was generated.

CHAPTER 3 ACTIVITY

PURPOSE
To use a topographic map to determine characteristics of an area's landforms and roads

MATERIALS
If the required topographic maps are not available at a local library, they can be purchased or ordered at a map store or directly from the U.S. Geological Survey.

PROCEDURE

❶ Make sure students mark their strings with pencil, not pen, in order to avoid damaging the maps.

❹ West Hill Road extends from the Elmira map onto the Seeley Creek map.

Tell students to note that two roads seem to be labeled West Hill Road. The path up Harris Hill is the one that goes northwest out of Elmira and follows the left fork after passing the Elmira Reservoir.

Students should also note that the 1200-foot contour crosses West Hill Road three times. Students may or may not realize that this is not an important factor in their average slope calculations, but they will have to make a decision about how to mark their strings.

Answers for Slope Table

Harris Hill Road South: 1.5 mi; 400 ft/mi
Harris Hill Road North: 0.8 mi; 750 ft/mi
West Hill Road: 3.0 mi; 200 ft/mi

ANALYSIS AND CONCLUSIONS

1. 42° 7′ 30″ N latitude; 76° 52′ 30″ W longitude

2. about 5.8 miles × 8.7 miles

3. greatest: Harris Hill Road-North; smallest: West Hill Road

---

CHAPTER 3 MAP Activity

The Roads to Harris Hill

Harris Hill is known throughout the world for its sailplane activity. Winds that sweep up the steep north-facing slope of the hill make it an ideal location for launching sailplanes and soaring to higher elevations.

In this activity, we will consider three of the routes that may be used to climb Harris Hill. They are West Hill Road from Elmira, NY; Harris Hill Road from Route 352; and Harris Hill Road from County Route 64. You will investigate the elevation changes that occur along these three roads as well as some other elevation aspects of the area.

One way to study elevation changes is to determine the average slope, or gradient, of the land surface. To calculate the average slope, divide the change in elevation by the distance between the two locations:

$$\text{average slope} = \frac{\Delta \text{ elevation}}{\Delta \text{ distance}} \quad (\Delta = \text{"change"})$$

SKILLS AND OBJECTIVES
- *Interpret* a topographic map.
- *Determine* distance using a map's scale.
- *Calculate* the average slope for several distances.

MATERIALS
- 7.5′ quadrangle topographic maps for Big Flats, Elmira, Horseheads, and Seeley Creek, NY
- colored pencils
- string
- ruler

Arrange maps as shown.

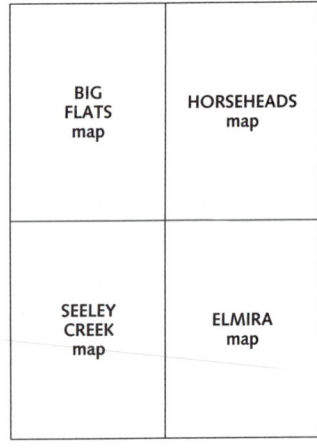

Procedure

❶ Identify where Harris Hill Road crosses the 1000-foot contour line near Harris Hill Manor on the Seeley Creek topographic map. Follow the road north until it crosses the 1600-foot contour. Lay a piece of string along this route. Mark the 1000-foot contour on the string in pencil. Then mark the string every time the road rises 100 feet in elevation until it reaches the 1600-foot contour.

❷ Use the string and the scale of the topographic map to determine the horizontal distance between the 1000-foot and the 1600-foot contours to the nearest tenth of a mile. Record your answer on a copy of the Slope Table.

❸ Identify where the northern end of Harris Hill Road crosses the 1000-foot contour on the Big Flats map. Trace its path southward until it crosses the 1600-foot contour. Lay a second piece of string along this route, and mark each 100-foot change in elevation, as you did in Step 1. Determine the horizontal distance between the 1000-foot and 1600-foot marks. Record this distance on the Slope Table.

❹ Identify where West Hill Road crosses the 1000-foot elevation mark near Elmira. Follow this route westward, taking the southern branch where the road forks, toward the top of Harris Hill. Use string to mark 100-foot changes in elevation and find the distance from the 1000-foot to the 1600-foot elevation. Record the distance on the Slope Table.

❺ Calculate the average slope of each road between the 1000-foot and 1600-foot elevations by using the formula at the top of the page. Record your answer on the Slope Table in feet per mile.

58 Unit 1 Investigating Earth

Analysis and Conclusions

1. What is the latitude and longitude of the point where the four topographic maps meet? Give your answers to the nearest second of latitude and longitude.

2. Each topographic map covers an area 7.5 minutes longitude by 7.5 minutes latitude. Choose one of the maps, and use the scale to determine the dimensions of the map to the nearest tenth of a mile.

3. Which of the three roads has the greatest average slope between the 1000-foot and 1600-foot elevations? Which road has the smallest average slope?

4. Use the maps or the string to examine the spacing of the contour lines on each of the three roads. For which road are the contour lines the greatest distance apart? For which road are the contour lines closest together?

5. What is the relationship between the spacing on the contour lines and the average slope?

6. On the Seeley Creek map, find the steep slope to the east of Sing Sing Creek and west of the hill with a height of 1648 feet. What is the average slope here in feet per mile? Show your work.

7. If you wanted to ride a bicycle up to the Harris Hill Picnic Area, where would be the best place to start: Harris Hill Manor, Lowes Pond, or Woodlawn Cemetery? Explain your reasoning.

8. Determine, to the best of your knowledge, the highest elevations on these features: Harris Hill, Hawes Hill, Northcrest Road, the Elmira Reservoir, Vanderhoff Road, and the WSYE Radio Tower.

9. Why is Hawley Hill a good location for a radio tower?

10. Why is the Harris Hill Gliderport well-situated for flying sailplanes?

11. In which general direction does the water in the creek flow, and how can you tell?

12. If you were to build a house within 5 miles of the Harris Hill Gliderport, where would you want to build? Explain your decision using information from the topographic maps, including locations of important structures, roads, and scenery.

Slope Table		
Road	Distance (mi)	Slope (ft/mi)
Harris Hill Road South		
Harris Hill Road North		
West Hill Road		

Find a topographic map of your local area.
Keycode: ES0308

CHAPTER 3 ACTIVITY

Find a topographic map of your local area.
Keycode: ES0308

(continued)

4. greatest distance: West Hill Road; closest together: Harris Hill Road-North

5. The closer together the contour lines are, the greater the average slope. The farther apart the contour lines are, the smaller the average slope.

6. Slope is about 720 feet in height for each distance along the ground of about 0.2 mile, or about 3600 ft/mi.

7. Woodlawn Cemetery, because the slope of the road from there is shallower than the slope of the road from the other two locations.

8. Harris Hill: 1752 feet; Hawes Hill, 1610–1620 feet; Northcrest Road: about 1640 feet; Elmira Reservoir: 1065 feet; Vanderhoff Road: 1540–1550 feet; WSYE Radio Tower: 1709 feet

9. It is the highest hill in the area and is easily accessible.

10. The Harris Hill Gliderport is on one of the highest hills in the area. The hill has a relatively flat and broad top for takeoff and landing areas. The top of the hill is also located very close to a steep slope that provides updrafts for sailplanes.

11. When contour lines cross a creek, they form a V directed upstream. This stream flows in a southerly direction, because the Vs point north and the land elevation around the creek decreases to the south.

12. Answers should indicate students' preferences and needs concerning available space for a structure; average slope of land on which the house is to be built; access to roads; closeness to desirable natural features such as creeks or forests; distance from undesirable features such as the gliderport, power lines, or major highways.

CHAPTER 3 REVIEW

Vocabulary Review

1. topographic map
2. false-color image
3. equator
4. map scale
5. latitude

Concept Review

6. A map, which is two-dimensional, cannot accurately depict Earth's three-dimensional features. Thus, all maps distort Earth's features to some degree.

7. Latitude describes location in degrees north or south of the equator. Longitude describes location in degrees east or west of the prime meridian. Each place on Earth can be described by the intersection of a line of latitude and a line of longitude.

8. A Mercator (cylindrical) projection depicts Earth as if a large cylinder were wrapped around the planet and the outlines of the continents were projected onto the paper. It shows true direction in straight lines but distorts polar areas. A gnomonic (planar) projection depicts Earth as if a larger plane were placed at a point on Earth's surface and the outlines of the continents were projected onto the plane. It can be used to plot the shortest distance between two points, but landmasses are distorted away from the center point.

9. Map scales tell you the sizes of the areas depicted by maps.

10. The Global Positioning System consists of 24 satellites in various orbits around Earth. The satellites continuously broadcast their locations and the exact time. Users on Earth pick up the signals of at least four of the satellites with a receiver. The receiver uses the signals to determine the user's location, speed (if the user is moving), and the time.

11. You can get a mental picture of the landscape the map depicts by using the contour lines to determine elevation as well as the shape and slope of landforms.

12. In the late 19th century, cartographers surveyed the ground and drew maps with pen and paper. The ability to take photographs from airplanes made it possible to use aerial photographs of the ground instead. Radar allowed aerial photographs to be taken even in cloudy weather. Instead of making maps largely by hand, cartographers now use computers and satellite technology to make maps largely through mechanical means. Scanners carried by satellites gather data from Earth's surface, which computers interpret to create maps.

CHAPTER 3 REVIEW

Summary of Key Ideas

3.1 A flat map of a curved surface is distorted. Different map projections are used to minimize distortion of shape, distance, or direction. On a map horizontal lines show latitude, positions north and south of the equator. Vertical lines show longitude, positions east and west of the prime meridian. Map scales compare the size of the map with Earth's surface.

3.2 The use of remote-sensing methods allows mapmakers to produce accurate maps of many places on Earth. With the help of computers cartographers use remote-sensing data to make detailed maps. Images produced using data from satellites are used in many areas of science and research.

3.3 Topographic maps indicate elevation and average slope by using contour lines. Colors are used to indicate various features on topographic maps, including water and human-made structures.

Key Vocabulary

cartographer (p. 44)
contour interval (p. 54)
contour lines (p. 53)
equator (p. 46)
false-color image (p. 49)
hemisphere (p. 46)
latitude (p. 46)
longitude (p. 46)
magnetic declination (p. 56)
map (p. 44)
map scale (p. 46)
prime meridian (p. 46)
projection (p. 44)
radar (p. 48)
slope (p. 55)
topographic map (p. 53)
topography (p. 53)

Vocabulary Review

Write the term from the key vocabulary list that best completes the sentence.

1. An area's natural and human-made surface features are shown on a(n) ___?___, which uses contour lines.

2. To perceive the various wavelengths of light used to create an image, scientists may use a(n) ___?___.

3. The Northern and Southern Hemispheres are divided by an imaginary line called the ___?___.

4. To determine how the map's features compare in size with Earth's surface, look at the ___?___.

5. Circling the world from east to west and running parallel to the equator are ___?___ lines.

Concept Review

6. What are the disadvantages of using a map as a model of Earth?

7. How do latitude and longitude describe location?

8. Describe the differences between a Mercator projection and a gnomonic projection.

9. How do map scales help you use maps?

10. What is the Global Positioning System?

11. What information can you get from the contour lines on a topographic map?

12. What are some ways technology has changed the way that maps are created and updated?

13. **Graphic Organizer** Copy and complete the concept map below.

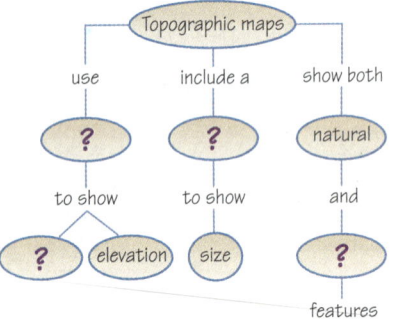

13. Topographic maps use *contour lines* to show *slope* and elevation. Topographic maps include a *scale* to show size. Topographic maps show both natural and *human-made* features.

Critical Thinking

14. Infer Unlike photography, radar is not affected by cloud cover. When would it be helpful to use radar for mapping?

15. Apply When would it be useful to find the slope of an area?

16. Compare Describe a situation in which a globe would be the better model of Earth. Describe another situation in which a map would be the better model.

17. Infer Small-scale maps show a larger land area than large-scale maps do. Which would be more useful in locating a hiking trail? Explain your reasoning.

Internet Extension

How Are Landforms Represented on Flat Maps?
Examine topographic maps and interact with 3-D models to understand how contour lines represent Earth's surface.
Keycode: ES0307

Writing About the Earth System

SCIENCE NOTEBOOK Mapmaking changed rapidly during the second half of the last century. Describe two ways in which new mapping technology has affected our understanding of the hydrosphere and the atmosphere.

Interpreting Graphs

The profile below was drawn from a topographic map. Use the profile to answer the following questions.

18. What is the difference in elevation between point B and point A?

19. What is the map distance between point B and point A?

20. Determine the average slope in feet per mile between point A and point B.

21. Determine the average slope in feet per mile between point C and point D.

22. The relief of an area is the difference between the highest and lowest point. What is the total relief of the profile?

23. The vertical exaggeration of a profile is the amount by which the vertical scale is expanded compared to the true scale. The horizontal scale in feet is the true scale. Compare the vertical scale with the true scale. What is the vertical exaggeration of this profile?

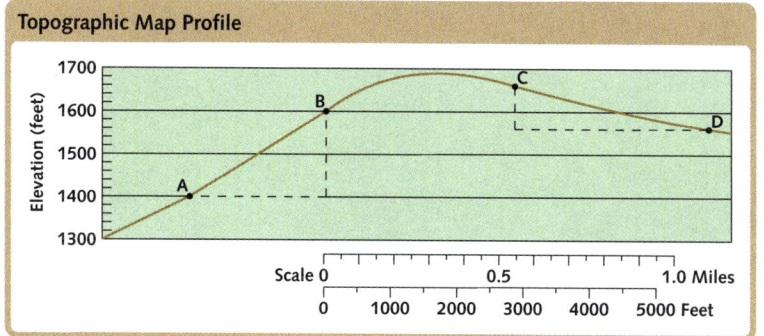

CHAPTER 3 REVIEW

How Are Landforms Represented on Flat Maps?
Keycode: ES0307

Use Internet Investigations Guide, pp. T13, 13.

CRITICAL THINKING

14. Radar is useful when it is cloudy, stormy, or dark.

15. Sample Response: when building roads or other structures on a site; when trying to determine potential for runoff and erosion in an area

16. Sample Response: A globe would be a better model to use when you want to see how continents are located in relation to one another. A map is a better model to use when you want to find your way within a small area, such as a city.

17. A large-scale map would be better because it would show a smaller area, such as an area containing a hiking trail, in greater detail.

INTERPRETING GRAPHS

18. 200 feet
19. about 2000 feet, or about 0.38 miles
20. about 526 feet per mile
21. about 182 feet per mile
22. about 400 feet
23. The vertical scale is exaggerated between 6 and 7 times, as compared to the horizontal scale.

WRITING ABOUT SCIENCE

Writing About the Earth System

Sample Response: determining pollution hotspots in the atmosphere and in bodies of water; tracking exchanges of energy, carbon, or water between the air, land, and bodies of water; mapping the ocean floor

UNIT 1 FEATURE

Focus

Objective
Explain how the environment will benefit if people consider Earth as a system.

Set a Purpose
Have students read to find answers to the question "What are some ways people have affected the Earth system?" Humans can have a negative effect on the environment. For example, chemical pesticides developed to control crop-damaging pests have also harmed birds and fish. Dams and building developments disturb wildlife habitats, and the burning of fossil fuels has increased air pollution. Humans can also work to correct these problems through their study of Earth systems, allowing people to take advantage of Earth resources in a responsible way.

UNIT 1 The EARTH SYSTEM

An Old Idea Becomes New Again

Although it came to the forefront during the last decade of the 20th century, Earth system science has roots that stretch back many decades.

Earth system science is not a new idea. The idea that Earth is a whole can be found in many cultures. In Western civilization, the Earth system has been acknowledged just recently—even though its scientific roots can be traced back to the 18th century.

At that time, James Hutton (1726–1797) realized that Earth was like a body consisting of many interconnected systems working together. Known as the founder of modern geology, Hutton was originally trained as a medical doctor. His medical training led him to see parallels between the human body and Earth. He thought a better name for geology would be *geophysiology*—the study of Earth's body. Despite Hutton's stature in the scientific community, his suggestion to consider Earth as a whole went largely ignored. Scientists continued to work within their disciplines, each one isolated from the other.

In the United States, during the mid-20th century, technology boomed as scientific discoveries made during World War II were adapted for civilian and industrial uses. For example, wartime chemicals found new life as pesticides, and portable radios were developed from military transistor radios. Houses, roads, dams, and other development projects sprung up around the country. Few people gave much thought to what effect such changes were having on the environment.

That changed in 1962, with the publication of a book called *Silent Spring*. Its author, a biologist named Rachel Carson (1907–1964), had earlier

published several informative and engaging books introducing readers to the wonders of nature—*The Edge of the Sea* and *Under the Sea-Wind*.

In *Silent Spring,* Carson presented evidence that the heavy use of pesticides, including the widely-used and highly toxic chemical DDT, had unforeseen

62 Unit 1 Investigating Earth

DIFFERENTIATING INSTRUCTION

Reading Support Tell students that identifying key ideas is critical to learning, and writing headings is one way to identify key ideas. Ask students to write headings for "An Old Idea Becomes New Again" and to indicate where to place them. As an example, tell students that you might write the heading "Parallels: Earth and The Human Body" between the first and second paragraphs. Give students time to write headings for the entire article. Then ask them to share the headings they wrote with the class. Write the headings on the board and discuss which ones best identify the article's key ideas. Point out that the headings can serve to organize an outline, which students can complete by adding supporting details.

and the ENVIRONMENT

effects on other life forms. DDT killed not only the crop-destroying insects, but birds and fish as well. Once it entered the environment, DDT spread its toxins throughout the food chain.

With her book's publication, Carson found herself at the center in a storm of controversy. The pesticide-industry dismissed her scientific evidence. Despite the opposition, *Silent Spring* became an influential book, leading to the ban on DDT in the United States. Carson is often called the founder of today's ecology movement. Rachel Carson's life might best be summed up in a single sentence: One person can make a difference. What can you do in your life to make a difference for yourself, for others, for the environment?

Human beings have seen the planet from space only for a few decades. Images and studies have driven home the realization that Earth, although it may seem vast and rich with resources, is in reality a small and fragile planet.

The biosphere, the realm of living things, is a just thin layer on the surface of Earth and of particular interest to Earth system scientists. Among the new ways of looking at Earth is the Gaia Hypothesis, proposed by Lynn Margulis and James E. Lovelock in the 1970s. It suggests that living things not only have a greater impact on the evolution of earth than previously thought, but also that living things cooperate to maintain the Earth system. Like *Silent Spring,* the Gaia Hypothesis has met with controversy. Changes in science do not come easily, and full acceptance of Earth system science may be years away, even though, around the world, organizations such as NASA have embraced this new way of looking at Earth.

At times it is difficult for people to see beyond their immediate needs and consider the future. You may have done this yourself; it's a very human thing to do. Growing environmental crises, however, make it plain that we can no longer live this way. Native American philosophy urges that for every decision made or action taken, we consider its impact on the lives of those seven generations from now. Would you say that decision-makers today follow that philosophy?

DDT has been linked to thinning eggshells and population decline in some bird species, such as the peregrine falcon shown at left.

UNIT INVESTIGATION
CLASSZONE.COM

How Can Getting Farther Away from Earth Help Us See It More Clearly? Instruments on orbiting satellites collect an incredible array of information about our planet every day. Explore how sensors gather information such as the amount of water vapor in the air or the shape of the sea floor. Plan to view Earth from a specific satellite. Make a prediction about what you might see.
Keycode: ESU101

UNIT 1 FEATURE

INVESTIGATIONS
CLASSZONE.COM

How Can Getting Farther Away from Earth Help Us See It More Clearly?
Keycode: ESU101

Use Internet Investigations Guide, pp. T14–T15, 14–15.

Investigation Photograph: The computer-generated image shows a portion of the Mid-Atlantic Ridge.

UNIT 1 ASSESSMENT

UNIT 1 STANDARDIZED TEST Practice

ANSWERS

1. d, The sun provides 99.895 percent of the energy in the Earth system.
2. b, The other terms are tools for mapmaking.
3. c, A skeptical approach ensures that scientific ideas get a thorough review.
4. d, A hypothesis is only a tentative explanation for observations or phenomenon.
5. a, A map scale relates distance on a map to distance on Earth.
6. d, A theory has passed scientific scrutiny and seems to be the best explanation based on available information.
7. b, These spheres are air, living things, water, and Earth itself, respectively.
8. b, Editors at scientific journals ask experts in the field to read the papers to judge the merit of the work.
9. b, Slope, or steepness, is the change in elevation with distance.
10. a, Most topographic maps of the United States show an area that spans 7.5 minutes of latitude and 7.5 minutes of longitude.
11. Core-drilling allows for analysis of a long swath of underground materials, giving information about the composition, layering, and depth of geologic features.
12. If water resource managers know the depth to which a water-saturated sandstone layer extends, for example, they can better calculate the amount of groundwater the aquifer contains. This gives them a better idea as to the rate at which water can be removed.

Directions (1–10): For *each* statement or question, select the word or expression that, of those given, best completes the statement or answers the question. Record your answer on a separate answer sheet.

1. The largest source of energy for the Earth system is
   (a) bioelectric energy
   (c) tidal energy
   (b) geothermal energy
   (d) solar energy

2. Cartographers make
   (a) satellites
   (c) surveys
   (b) maps
   (d) false-color images

3. It is important for a scientist to be skeptical and question the work of other scientists because
   (a) scientists need to share their knowledge
   (b) scientists need to develop new technology
   (c) scientists must be able to identify faulty reasoning
   (d) scientists want to publicize their findings quickly

4. A hypothesis has been tested and withstood the test. Which of the following is true?
   (a) It is sure to fail the next test.
   (b) It will never be disproved.
   (c) It becomes a scientific law.
   (d) It may someday be disproved.

5. What does a map scale tell you?
   (a) how the map's features compare in size with Earth's surface
   (b) what sort of projection the cartographer used
   (c) how far apart the map's contour lines are
   (d) how to find magnetic north

6. An explanation for observable phenomena for which no exception has been noted is called
   (a) a hypothesis
   (c) an experiment
   (b) a scientific law
   (d) a theory

7. The Earth system's four spheres are the
   (a) biosphere, hydrosphere, androsphere, and geosphere
   (b) atmosphere, biosphere, hydrosphere, and geosphere
   (c) mesosphere, photosphere, hydrosphere, and biosphere
   (d) lithosphere, asthenosphere, chromosphere, and geosphere

8. The process by which a scientist's findings are shared with and examined by other scientists is called
   (a) experimental theory
   (b) peer review
   (c) skeptical inquiry
   (d) hypothesis formation

9. For a landscape, the slope of an area on a topographic map is the
   (a) elevation
   (b) steepness
   (c) aridness
   (d) altitude

10. Most topographic maps of the United States are
    (a) 7.5-minute maps
    (b) 15-minute maps
    (c) 24-second maps
    (d) 20-minute maps

UNIT 1 ASSESSMENT

Directions (11–21): Record your answers on a separate answer sheet. Some questions may require the use of material in the Appendix.
Base your answers to questions 11 through 13 on the newspaper article below.

Science on the Beach in North Carolina

Scientists working with the the U.S. Geological Survey (USGS) are drilling holes at Kure Beach in North Carolina this summer to find out just where the water is—and how much of it is underground. It is the first step in a statewide program to describe the subsurface geology of the area's coastal plain. The goal of the drilling project is to study the relationship between groundwater reserves and geological features.

The research involves drilling a 1,500-foot-deep hole in the earth, bringing up an intact core of underground sediment and rocks for analysis, and installing a probe in bedrock to monitor seismic activity. Water resource managers will use the results from this study to make better decisions about the use of groundwater.

"Most geologists used to assume that the geology of the Coastal Plain was quite simple, like a stack of blankets on a bed," says USGS scientist Robert E. Weems. Similar studies in South Carolina and Virginia, however, revealed a pattern of earth layers "more like a patchwork quilt than a stack of blankets" that resulted in part from sea level changes. Weems explains, "This pattern makes understanding aquifers much more challenging than was previously thought. There is every reason to believe that we will find the same kinds of patterns in North Carolina."

11. Why would the scientists chose to use core-drilling as their tool in this study? Explain.

12. How does knowing an area's geology help water resource managers make better decisions about the use of groundwater? Give an example.

13. Describe the evidence that the scientist has to support the belief that the earth layers in the area will be like "a patchwork quilt."

Base your answers to questions 14 through 16 on the topographic map on pages 54–55 in your book.

14. Briefly describe the topography of the land depicted on this map.

15. Name two natural features depicted on the map, and two artificial features (created by people). Explain how you used the map symbols to identify them.

16. Explain what is unusual about a special kind of contour that appears just west of Chapboard Island, and estimate its depth.

Base your answers to questions 17 through 21 on the topographic map of the island below.

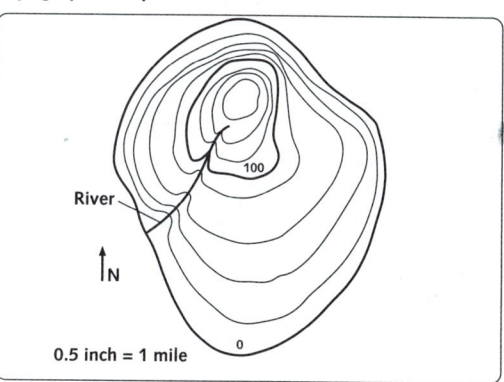

17. What is the contour interval of this map?

18. In which direction does the river flow?

19. Which side of the island would be most challenging to a hiker? Explain.

20. Approximate the elevation of this island's highest point.

21. Approximate the width and length of the island. Measure east to west and north to south.

Standardized Test Practice 65

(continued)

13. Studies in bordering coastal states, Virginia to the north and South Carolina to the south, revealed the "patchwork" pattern. It is reasonable to expect that North Carolina, situated between the two, will exhibit a similar subsurface pattern.

14. gently rolling topography, for the most part, with several river gorges

15. Answers will vary. The map shows many hills, a pond, several rivers, islands, and river gorges for natural features, which contour lines help to identify. Among the artificial features are highways, roads, and a gravel pit, all of which are identified by symbols found in the Appendix.

16. The contour lines used to mark the topography of the sea bottom are isobaths. They measure depth below sea level. To the west of Chapboard Island there is a depression contour that goes to a depth of 12 feet. Short hatchmarks come from the surrounding isobath, pointing toward the center of the depression.

17. 20 feet

18. southwest

19. Northern side; it is the steepest side, as shown by the closely spaced contour lines.

20. approximately 170 feet high

21. width (E-W): approximately $3\frac{1}{4}$ miles; length (N-S): approximately 4 miles

UNIT 2

Unit Overview

Earth can be studied as a system in which the matter and energy of one sphere interact with the energy and matter of the others. This can be done at a large scale, as suggested by the titles of Chapters 4 and 7, or at a small scale, as suggested by the titles of Chapters 5 and 6. Understanding Earth at both levels allows scientists to relate the individual properties of Earth materials to global phenomena such as climate variations or the location and accessibility of Earth resources.

Visual Teaching

Chapter 4 Earth's Structure and Motion

The time-lapse photograph of Alaska's midnight sun shows that the sun does not always set below the horizon. Students will learn how Earth's movement and tilt create seasonal effects.

Answer: During the summer, Earth's $23\frac{1}{2}°$ tilt prevents the sun from setting below the horizon in areas above the Arctic Circle ($66\frac{1}{2}°$ N).

Chapter 5 Atoms to Minerals

Naturally occurring diamonds formed as carbon atoms were exposed to great pressure, deep within Earth. These gemstones were forced to the surface millions of years ago as molten rock was propelled through Earth's crust by superheated gases. Students will learn how the arrangements of atoms in rock-forming minerals determine their physical properties.

UNIT 2 Earth's Matter

CHAPTER 4
Earth's Structure and Motion

Elapsed-time photography captures the midnight sun over a site in Alaska. What causes the midnight sun?

The Earth System Earth's matter includes a soaring rock cliff and a small but eye-catching gem. Other parts of Earth, such as a tiny atom or Earth's interior, may be hidden or cannot be seen with the unaided eye.

What can understanding Earth at all levels, large and small, tell us about how its parts relate?

CONNECTIONS TO . . .

Prior Knowledge In Unit 1, students were introduced to the concept of Earth as a system of interacting spheres. In this unit, students will begin to see how the interactions of these spheres create the rocks and minerals they encounter daily and how human activities can affect the availability of these and other resources. Units 3 and 4 further explain the processes responsible for the formation and transformation of rocks and minerals.

Unifying Themes Rocks and minerals can be *organized* into groups based on their characteristics and how they formed. As students study the chapters in Unit 2, point out how scientists have *organized* rocks and minerals and how this *organization* makes Earth's matter easier to study and understand.

UNIT 2

CHAPTER 5
Atoms to Minerals

Why is a diamond so different—and more valuable—than other minerals?

CHAPTER 6
Rocks

How could tiny sea creatures form the huge rock cliffs of Dover, England?

CHAPTER 7
Resources and the Environment

This dam in Turkey uses water to generate electricity. How does our use of such resources affect the environment?

(continued)

Answer: Diamonds are highly valued as a rare gem with brilliant luster and perfect cleavage. Diamond is also a very hard crystal, making it valuable in industrial uses.

Chapter 6 Rocks
The White Cliffs of Dover on the southeast coast of England are made of chalk. The chalk formed from ocean sediment laid down millions of years ago. Students will learn how sedimentary rock, and other types of rock, form.

Answer: As tiny sea creatures called plankton died, their shells sank to the ocean floor, forming layers of sediment. Cemented together, these layers eventually formed thick layers of chalk that have since been raised above sea level.

Chapter 7 Resources and the Environment
The Ataturk Dam in Turkey is one of the largest hydroelectric dams in the world. It uses running water from the Euphrates River to produce electricity for millions of people. Students will learn about how humans make use of Earth resources and how that affects the environment.

Answer: Using water to produce electricity conserves natural resources. However, hydroelectric dams cause great changes in the environment, such as flooding dry areas and limiting the flow of water to areas downstream.

CONNECTIONS TO . . .

National Science Education Standards
- Students will come to understand how Earth and the other planets in the solar system formed.
- Students will understand that the primary sources of Earth's internal heat are gravitational energy from Earth's formation and the decay of radioactive isotopes.
- Students will develop an understanding of the movement of matter in the Earth system. Specifically, they will learn how matter changes form as it moves through the rock cycle.

Unit Features
The Earth System and the Environment
Recycling Cents and Sensibility, pp. 164–165

Paper or Plastic—Which Type of Bag Is Best for the Environment? Evaluate the environmental effects of producing and disposing of paper and plastic bags.
Keycode: ESU201

Standardized Test Practice, pp. 166–167

CHAPTER 4 PLANNING GUIDE: EARTH'S STRUCTURE AND MOTION

	Section Objectives	Print Resources
Chapter-Level Resources		Laboratory Manual, pp. 13–16 Internet Investigations Guide, pp. T18–T19, 18–19 Guide to Earth Science in Urban Environments, pp. 45–52 Spanish Vocabulary and Summaries, pp. 7–8 Formal Assessment, pp. 10–12 Alternative Assessment, pp. 7–8
Section 4.1 Earth's Formation, pp. 70–74	1. Explain how most scientists explain the formation of our solar system. 2. Describe Earth's size and shape and the arrangement of its layers. 3. List three sources of Earth's internal heat. 4. Describe Earth's magnetic field.	Lesson Plans, p. 10 Reading Study Guide, p. 10 Teaching Transparency 4
Section 4.2 Earth's Rotation, pp. 75–79	1. Give evidence of Earth's rotation. 2. Relate Earth's rotation to the day-night cycle and the time zones.	Lesson Plans, p. 11 Reading Study Guide, p. 11
Section 4.3 Earth's Revolution, pp. 80–83	1. Give evidence for Earth's revolution around the sun. 2. Describe Earth's path and rate of revolution. 3. Explain why seasons occur.	Lesson Plans, p. 12 Reading Study Guide, p. 12 Teaching Transparency 5

INVESTIGATIONS
CLASSZONE.COM

Section 4.1 INVESTIGATION
How Do We Know about Layers Deep within Earth?
Computer time: 45 minutes / Additional time: 45 minutes
Students use animations of earthquake waves moving through model planets to predict Earth's interior.

Section 4.2 INVESTIGATION
What Time Is It?
Computer time: 45 minutes / Additional time: 45 minutes
Students view maps of Earth showing where it is light and dark and interpret them to tell time.

Technology/Online Resources

Electronic Teacher Tools
Visualizations CD-ROM
Classzone.com
Online Lesson Planner
Test Generator

Visualizations CD-ROM
Classzone.com
  ES0401 Visualization, ES0402 Investigation

Visualizations CD-ROM
Classzone.com
  ES0403 Visualization, ES0404 Visualization,
  ES0405 Investigation, ES0406 Data Center

Visualizations CD-ROM
Classzone.com
  ES0407 Visualization, ES0408 Visualization

Classroom Management

Gather these materials for minilab and laboratory activities.

Minilab, p. 74
  bar magnet
  iron filings
  card stock
  tape

Lab Activity, pp. 84–85
  blank sheet of paper
  safety compass
  protractor
  flexible metric ruler

CHAPTER 4

Earth's Structure and Motion

Each day, Earth's structure and its motions have profound effects on us.

What materials make up our home planet?

INTRODUCE
Build Interest
Before sharing background information about the photograph with students, allow time for their questions and comments.

Ask questions like the following:
- What is shown in the photograph? **Earth and its moon**
- What can you tell about Earth based on this photograph and your own knowledge? **Earth is a solid sphere with water on its surface and has significant cloud cover.**
- Where is the sun in relation to Earth and the moon in the photograph? **on the right side of Earth and the moon**

Respond
- What materials make up our home planet? **Answers may include land, water, and air in addition to various specific elements and compounds.**

About the Photograph

This enhanced image was produced from images of Earth and the moon taken by the Near Earth Asteroid Rendezvous (NEAR) spacecraft on January 23, 1998, as the spacecraft traveled to asteroid 433 Eros. Both Earth and the moon are viewed from above their south poles. The large white area in Earth's center is Antarctica.

BEYOND THE TEXTBOOK

Print Resources
- Reading Study Guide, pp. 10–12
- Transparencies 4, 5
- Formal Assessment, pp. 10–12
- Laboratory Manual, pp. 13–16
- Alternative Assessment, pp. 7–8
- Spanish Vocabulary and Summaries pp. 7–8
- Internet Investigations Guide, pp. T18–T19, 18–19
- Guide to Earth Science in Urban Environments, pp. 45–52

Technology Resources
- Visualizations CD-ROM
- Classzone.com
- Online Lesson Planner
- Electronic Teacher Tools
- Test Generator

68 Unit 2 Earth's Matter

CHAPTER 4

PREVIEW

▶ **FOCUS QUESTIONS** In this chapter you will study Earth's structure and motion and learn more about the questions below.

Section 1 How was Earth formed, and what are some characteristics of its structure?

Section 2 What is rotation and what are its effects?

Section 3 What is revolution and what are its effects?

▶ **REVIEW TOPICS** As you investigate Earth's structure and motion, you will need to use information from earlier chapters.

- characteristics of the geosphere (p. 9)
- models of Earth (p. 44)
- longitude and latitude (p. 46)

▶ **READING STRATEGY**

SET A PURPOSE

Read the key questions listed above. Before you begin reading Chapter 4, write a sentence or two in your science notebook that identifies a purpose for reading each section.

INTERNET RESOURCES
CLASSZONE.COM

At our Web site, you will find the following Internet support for this chapter.

DATA CENTER
EARTH NEWS
VISUALIZATIONS
- Origin of the Solar System
- Evidence of Earth Rotating about an Axis
- Earth's Daily Rotation
- Night Sky over a Year
- Earth's Revolution around the Sun

LOCAL RESOURCES
INVESTIGATIONS
- How Do We Know about Layers Deep within Earth?
- What Time Is It?

Chapter 4 Earth's Structure and Motion 69

CHAPTER 4

PREPARE

Focus Questions
Find out what knowledge students have about Earth's structure and its motion. For each focus question, have students consider the section title and then develop two or three additional questions for each section. For example, Section 2: How do we know Earth rotates? How does Earth's rotation affect our lives?

Review Topics
In addition to the earth science topics listed, students may need to review these concepts:

Physical Science
- Kelvin is the SI unit of temperature; zero K is called *absolute zero*, the coldest possible temperature.
- Almost all matter expands as it gets hotter and contracts when it cools.

Reading Strategy
Model the Strategy: To help students set a purpose for each section, use a version of the following: "Let's read the focus questions aloud and then write a sentence or two that identifies a purpose for reading each section. For example, possible purposes for studying section 1 could include, 'Learning about how Earth was formed helps explain the present physical structure of the planet that I live on. Knowing about Earth's structure, in turn, helps me to understand events such as earthquakes and volcanic eruptions.'" Have students continue to use this strategy for the remainder of the chapter.

BEYOND THE TEXTBOOK

INTERNET RESOURCES
CLASSZONE.COM

INVESTIGATIONS
- How Do We Know about Layers Deep within Earth?
- What Time Is It?

VISUALIZATIONS
- Animation of the origin of the solar system
- Animations of evidence that Earth turns about an axis
- Animation of Earth's daily rotation
- Images of the night sky from the same location over a year
- Animation of Earth's yearly revolution around the sun

DATA CENTER
EARTH NEWS
LOCAL RESOURCES
CAREERS

Online Lesson Planner

Chapter 4 Earth's Structure and Motion 69

CHAPTER 4 SECTION 1

Focus

Objectives
1. Explain how most scientists explain the formation of our solar system.
2. Describe Earth's size and shape and the arrangement of its layers.
3. List three sources of Earth's internal heat.
4. Describe Earth's magnetic field.

Set a Purpose
Have students read this section to find the answer to the question "How did Earth develop into the planet we know today?"

Instruct

More about...

The Shape of Earth
The reason Earth has a spherical shape can be understood in the context of Newton's law of gravitation. All the particles of Earth's matter are pulled toward the center of gravity. The natural response to the maximum possible concentration of particles around a center is a sphere. Every part of the sphere's surface is of equal distance from the center, resulting in a state of equilibrium.

4.1

Key Idea
Earth formed from a whirling cloud of gas and debris into a multilayered sphere, which has since been losing heat.

Key Vocabulary
- inner core
- outer core
- mantle
- crust
- lithosphere
- asthenosphere
- magnetic field

VISUALIZATIONS
CLASSZONE.COM
Observe an animation showing the origin of the solar system.
Keycode: ES0401

Earth's Formation

Physical and chemical processes change our planet every day. Earth as you know it today is a result of events that have transpired over billions of years.

Origin of the Solar System

The most widely accepted model of the formation of our solar system is called the nebular hypothesis. It suggests that about 4.6 billion years ago a great cloud of gas and dust was rotating slowly in space. The cloud was at least 10 billion kilometers in diameter. As time passed, the cloud shrank under the pull of its own gravity. As it shrank, its rate of rotation increased. Most of the material in the rotating cloud gathered around its center. The compression of this material made its interior so hot that a powerful reaction called *hydrogen fusion* occurred. At this time, the star we now know as our sun was born.

About 10 percent of the material in the cloud formed a great platelike disk surrounding the sun and extending far into space. Frictional, electromagnetic, and gravitational forces within the disk caused most of its mass to condense, forming solid particles of ice and rock. The condensed particles in the spinning cloud eventually combined into larger bodies called *planetesimals*, shown third from the left in the illustration below. Keep in mind that the illustration of this model of the solar system formation is not drawn to scale.

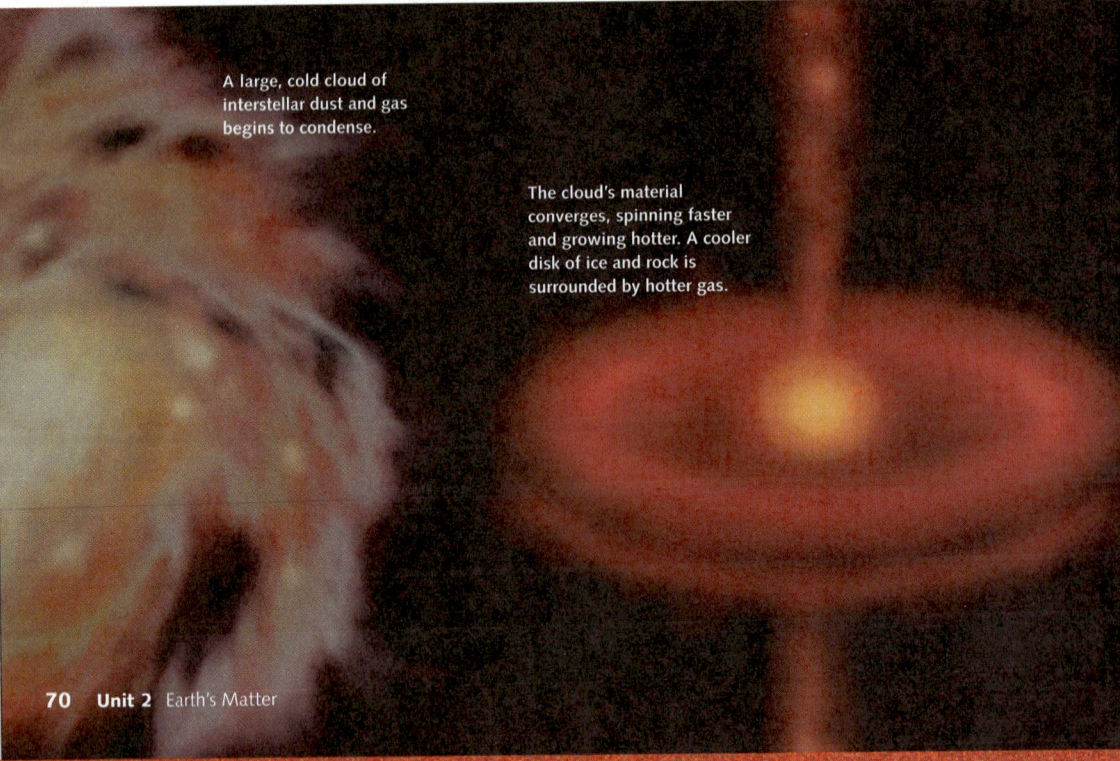

A large, cold cloud of interstellar dust and gas begins to condense.

The cloud's material converges, spinning faster and growing hotter. A cooler disk of ice and rock is surrounded by hotter gas.

70 Unit 2 Earth's Matter

DIFFERENTIATING INSTRUCTION

Reading Support Have students organize the information in this section in the form of a weblike concept map. Get students started by writing *Earth* on the board as the central theme. Extend five arms from this theme and ask students to look at the red heading to determine how to label the ends of the arms (*Origin, Size and Shape, Interior, Heat,* and *Magnetic Field*). Have students copy this partial concept map and complete it by listing the important facts related to each heading on lines emanating from the heading.
Use Reading Study Guide, p. 10.

Earth's Size and Shape

The planetesimals continued to compress and spin, sometimes colliding with each other and other objects in space. Eventually these planetesimals developed into planets and moons. Of these new objects, the third closest to the sun became Earth. The spinning motion of the young Earth caused it to form into a sphere that bulges in the center. Such a shape is called an *oblate spheroid* (AHB-layt SFEER-oyd).

While scientists cannot directly observe the events that led to Earth's formation, they can directly observe Earth's shape. Many photographs of Earth have been taken from space. The photographs show that Earth is spherical. It is not, however, a perfect shape.

Scientists discovered that Earth is not a perfect sphere by measuring the weight of an object at several places on Earth's surface. The weight of an object, in newtons, is the force with which gravity pulls the object toward Earth's center. The farther away an object is from Earth's center, the lighter it is. Conversely, the closer an object is to Earth's center, the heavier it is. Careful measurements show that an object that weighs 195 newtons at the North Pole or the South Pole weighs 194 newtons at the equator. This difference in weight is evidence that the object is nearer to Earth's center at the poles than at the equator. If Earth were a perfect sphere, the distance to its center would be the same at the poles as it is at the equator. In addition, an object's weight in newtons would be the same at any given point on the planet's surface. Of course, you would have to account for elevation, because an object at sea level is heavier than it is at the top of Mount Everest.

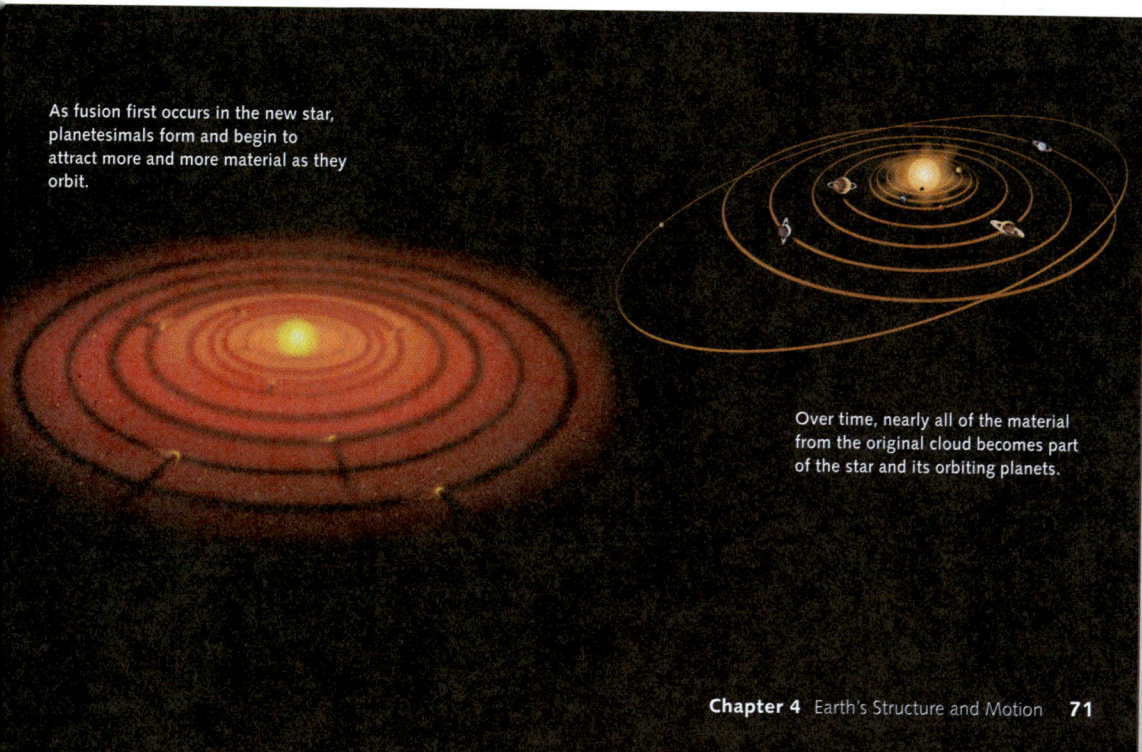

As fusion first occurs in the new star, planetesimals form and begin to attract more and more material as they orbit.

Over time, nearly all of the material from the original cloud becomes part of the star and its orbiting planets.

CHAPTER 4 SECTION 1

VISUALIZATIONS
CLASSZONE.COM

Observe an animation showing the origin of the solar system.
Keycode: ES0401

Have students make a flowchart showing the sequence of steps involved in the formation of the solar system.

Visualizations CD-ROM

VISUAL TEACHING
Discussion

Have students read the captions. Ask: What causes the cloud of interstellar dust and gas to condense? **forces of gravity, friction, and electromagnetism** In addition to forming a star and its orbiting planets, what other objects were formed from the material of the original cloud? **Answers may include moons, asteroids, and comets.**

DIFFERENTIATING INSTRUCTION

Developing English Proficiency
Have students study the illustration showing the stages in the formation of our solar system. Ask them to describe each stage in their own words. Then read aloud the captions that describe the stages. Discuss with students the meanings of the more difficult words, such as *interstellar*, *condense*, *converges*, *fusion*, *planetesimal*, and *orbit*. Have students write their own captions for the illustration.
Use Spanish Vocabulary and Summaries, pp. 7–8.

CHAPTER 4 SECTION 1

Visual Teaching

Interpret Diagram

Ask: Which of Earth's layers make up the lithosphere? **the crust and mantle** The asthenosphere? **the mantle** How do the lithosphere and the asthenosphere differ? **The lithosphere is a rigid solid, whereas the asthenosphere is a solid with liquid properties.**

Use Transparency 4.

CRITICAL THINKING

Have students brainstorm why 35 percent of Earth is made up of the element iron by weight, whereas Earth's crust is comprised of about 5 percent iron. **Iron, being a dense metal, sank to form most of the inner core during the early melting stage of Earth's history, while the lighter elements floated toward the surface forming Earth's crust.**

More about...

Earth's Core
Based on seismic data from earthquakes, scientists have inferred that the inner core is a 2400-kilometer-wide sphere consisting of solid iron and nickel that rotates in an outer core of molten iron and nickel. Together, the inner and outer core make up a sphere about the size of Mars.

VOCABULARY STRATEGY

The words *lithosphere* and *asthenosphere* both contain the root *sphere*, referring to the round shape of Earth. In Greek, *lithos* means "stone," while *asthenes* means "weak." So the lithosphere and the asthenosphere are respectively the "stony-sphere" and the "weak-sphere."

The total surface area of Earth is about 510 million square kilometers, equivalent to about 55 continental United States of Americas. Of this area, about 149 million square kilometers lie above sea level as continents and islands. Oceans cover the remaining 361 million square kilometers. In other words, about 29 percent of Earth's surface is dry land, while about 71 percent is covered by water.

Earth's Interior

According to the nebular hypothesis, the original surface of Earth looked much as the moon's surface does today. Earth was probably composed of the same type of material from its surface all the way to its center.

When objects collide, energy from the collision is converted into heat. In its early history, Earth frequently collided with material left over from the formation of the solar system. These impacts helped cause Earth to grow hot enough that heavy elements such as iron and nickel melted.

Recall that the density of a substance is the amount of mass of that substance that occupies a particular volume of that substance. If two liquids of different densities are mixed, over time the liquids will separate, with the denser liquid settling on the bottom. In the same way, the material composing Earth gradually separated into several layers, with denser material located toward the center.

At Earth's center is an **inner core** composed of solid iron and nickel. Surrounding the inner core is an **outer core** composed of iron and nickel in a liquid state. Around the core is the thickest of Earth's layers, called the **mantle**. The mantle is composed mostly of compounds rich in iron,

Earth's Interior

Material in Earth's interior is structured in zones of density. The most dense materials are at the inner core; the least dense materials, at the crust. Earth's mantle is the middle zone—what remains after heavier materials sank, and lighter materials rose to the surface.

DIFFERENTIATING INSTRUCTION

Support for Visually Impaired Students While teaching about Earth's interior, have a textured model available. Using a large polystyrene ball to represent Earth, cut out a large wedge to reveal Earth's interior. Represent Earth's layers by gluing onto the interior small objects of different textures, such as different types of pasta. Indicate the increasing density of the inner layers by gluing the pieces of pasta closer together.

Hands-On Demonstration Half fill a jar with water and salad oil. Cover the jar with a lid and shake the jar. Students should observe that the water sinks to the bottom of the jar, and the oil forms an upper layer. Ask: Which liquid is denser? **water** Relate this separation to the way material composing Earth separated into layers.

⚠ *Exercise caution when handling glassware. Remind students that safety is everyone's responsibility.*

SUMMARY: Characteristics of Earth's Layers

	State	Depth from surface (km)	Temperature (K)
Inner core	solid	6371	approximately 6000
Outer core	liquid	5150	3700–5500 (increases with depth)
Mantle	solid with liquid properties	2890	1500–3200 (increases with depth)
Crust	solid	0–65	<1000 (increases 10–30K/km of depth)

silicon, and magnesium. Although the mantle is solid, high pressures and temperatures cause it to behave as a liquid in some ways. Surrounding the mantle is the **crust,** a thin, rigid layer of lighter rocks that includes Earth's surface.

Earth's near-surface layers are further classified by their material properties. The crust and the uppermost portion of the mantle together make up the **lithosphere.** The more rigid material of the lithosphere floats upon a thin, slushlike layer of the mantle called the **asthenosphere.**

Compared to Earth's major layers, the crust has the smallest mass and volume. However, the crust is that part of the geosphere with which humans have direct contact, and it is the only place where life has been found. As you know from Chapter 1, we rely on the geosphere to provide the materials we need to build cities and grow crops. Although we do not have direct contact with the asthenosphere, it also affects our environment. As you will learn in Chapter 8, this part of the mantle is thought to be responsible for the movements of Earth's crust known as plate tectonics.

Earth's Heat

Events that gave rise to the formation of Earth generated heat. Some of the heat that caused Earth's layers to form came from meteorite impacts, and some arose as the weight of overlying materials caused compression in Earth's interior. Heat was also generated by the decay of radioactive isotopes, elements that release heat as they disintegrate into more stable forms.

Since its original heating, Earth has been slowly losing heat. The amount of heat loss varies from place to place, for the following reasons:

1. Some rocks lose heat more quickly than others.
2. The thickness of the crustal rock varies from place to place.
3. The percentage of radioactive materials in rocks varies.

You may have noticed that on a warm summer day, an underground cave remains cool. Deep caves stay about the same temperature all year because neither the sun's warmth nor the winter's cold can penetrate there.

INVESTIGATIONS
CLASSZONE.COM

How Do We Know about Layers Deep Within Earth? View animations of earthquake waves moving through model planets. Use the information to predict what Earth's interior is like.
Keycode: ES0402

Scientific Thinking

DESIGN AN EXPERIMENT

It is difficult and expensive to gather information about Earth's crust, let alone its inner layers. For example, the deepest underground gold mines reach depths of only about 4 kilometers, whereas Earth's crust is, on average, about 65 kilometers deep. In such situations, scientists often design experiments to help them test their hypotheses.

How would you go about building a working model of Earth's interior? What equipment and materials would you need? Think about ways a computer might help you test an hypothesis of Earth's structure.

CHAPTER 4 SECTION 1

INVESTIGATIONS
CLASSZONE.COM

How Do We Know about Layers Deep within Earth?
Keycode: ES0402

Use a computer and projector to view in class the animations of earthquake waves. Then assign as homework the data analysis and prediction portions of the investigation.

 Use Internet Investigations Guide, pp. T18, 18.

Scientific Thinking

DESIGN AN EXPERIMENT
You might have students complete the investigation above before they begin working on their models.

More about...

Investigating Earth's Interior
Scientists have been able to drill only about 2 km into Earth's crust. But Japan is building a ship that scientists hope will be able to drill all the way through 6 km of thin ocean crust, providing the first undisturbed samples of mantle material. Equipment required for such deep drilling may not be available until 2012. Drilling itself would take two years.

DIFFERENTIATING INSTRUCTION

Challenge Activity Have students research the relationship between Earth's internal heat due to disintegrating radioactive isotopes and the methods by which Earth cools itself. Students could identify the major radioactive elements involved (uranium, thorium, and potassium) and discuss the two methods of cooling (conduction and convection).

CHAPTER 4 SECTION 1

MINILAB
Modeling Earth's Magnetic Field

Management
Have a container available in which students can put the filings when finished.

⚠ *Students should take care not to inhale or ingest the filings or get them in their eyes.*

As an alternative, you can do the Minilab as a demonstration, using an overhead projector.

Analysis Answers
- The iron filings align in curved lines surrounding the bar magnet, from pole to pole.
- Sketches should show the poles of the bar magnet tilted about 11° away from Earth's geographic poles.

For proper behavior in a laboratory setting, refer students to Safety in the Earth Science Laboratory on pages xxii–xxiii.

ASSESS

1. In addition to photographs of Earth taken from space, evidence of Earth's oblate spheroidal shape comes from measuring the weight of an object at several places on Earth's surface. The farther away an object is from Earth's center, the lighter it is, and vice versa.

2. crust, mantle, outer core, inner core

3. Three sources: meteorite impacts, the weight of overlying materials causing compression in Earth's interior, and the decay of radioactive isotopes (which still produces heat).

4. Earth's interior layers would be at much lower temperatures. This would result in more solid, dense interior layers.

5. From Earth's crust to the center of the inner core is (6371 km − 65 km) 6306 km. Assuming 30K increase every km, then 30K × 6306 km = 189,180K at the center, compared to the actual inner-core temperature of 6000K.

10-Minute Mini LAB

Modeling Earth's Magnetic Field

Materials
- bar magnet
- iron filings
- card stock
- tape

Procedure
1. Tape the magnet beneath the center of a piece of white card stock.
2. Sprinkle the iron filings onto the card stock above the magnet.
3. Tap the card stock to allow the filings to align.
4. Draw what you see.

Analysis
Describe the alignment of the iron filings. Earth's magnetic field is similar to that of a bar magnet. Sketch Earth's outline over your drawing of the filing pattern, and mark the location of the poles.

Below a depth of 70 meters, however, ground temperatures begin to increase. Underground temperatures have been measured at tunnels, mines, oil wells, and water wells. While the rate of temperature increase varies from one location to another, the average rate of increase in the outer crust is about 1°C for every 40 meters of depth. Evidence suggests that the temperature increase becomes more gradual beneath the first 1000 meters of Earth's crust.

Earth's Magnetic Field

You may have noticed that a compass needle always points north. In fact, the compass needle aligns itself along the lines of force that make up Earth's **magnetic field.** The magnetic north pole is the equivalent of the attracting, or positive, end of a bar magnet, so it attracts your compass needle. On the other hand, the magnetic south pole is like the negative end of a bar magnet, so it repels the compass needle.

To visualize Earth's magnetic field, imagine a bar magnet lying inside Earth with each end pointing toward one of Earth's poles. Now imagine that the ends of the magnet are tilted about 11° away from the poles. Earth's magnetic field is the resulting lines of force that loop from one end of the bar magnet to the other. The 11° tilt explains why the magnetic north pole and the geographic north pole are not in exactly the same place.

Although scientists do not fully understand the origin of Earth's magnetic field, many support a hypothesis first developed in the 1900s. The hypothesis credits Earth's magnetic field to the movement of fluid in the outer core. An electric current is generated when liquid iron moves across an already existing, but weak, magnetic field. The electric current produces a magnetic field that, with the fluid motion, produces yet another magnetic field. Together, these fields create Earth's strong magnetic field.

4.1 Section Review

1. What evidence is there for Earth's spheroidal shape?
2. Describe the arrangement of Earth's layers, starting with the crust and moving inward.
3. Describe three sources of Earth's internal heat. Which internal process is still producing heat?
4. **CRITICAL THINKING** Describe how Earth would be different today if it contained no radioactive material. What would the consequences be for the Earth's interior layers?
5. **MATHEMATICS** Assume that the crust temperature increases by 30K for each kilometer below Earth's surface, where the temperature is 300K. Calculate what the temperature would be at Earth's center if the rate of increase were the same below the crust as in it. Compare your result with the actual inner-core temperature.

MONITOR AND RETEACH

If students miss . . .

Question 1 Reteach "Earth's Size and Shape." (p. 71) Have students explain the significance of weight in determining Earth's shape.

Question 2 Use the diagram of Earth's interior on page 72 to review the layers of Earth.

Question 3 Reteach "Earth's Heat." (pp. 73–74) Students can make a concept map that summarizes heat loss and generation on Earth, and temperatures in different parts of Earth.

Question 4 Reteach "Earth's Heat." (pp. 73–74) Ask: What do radioactive isotopes release as they disintegrate? heat

Question 5 Refer students to the chart on page 73 to help them answer the question.

Earth's Rotation

The spinning motion that helped form primitive Earth still influences our planet and our lives today. Earth completes one whole turn around its axis about every 24 hours. This spinning of Earth around its axis is called **rotation**.

Evidence for Rotation

One remarkable piece of evidence for Earth's rotation was built by physicist Jean Foucault in 1851. By attaching an iron sphere to a very long wire, Foucault constructed a pendulum that was 20 stories high. Physicists of the time knew that once a pendulum is set in motion, its direction of swing would not change. Foucault, however, observed that the direction of swing of his pendulum seemed to change. Each hour it shifted about 11° in a clockwise direction. After eight hours the pendulum was swinging at a right angle to its starting direction. Because the pendulum itself could not have changed its direction of swing, Foucault concluded that the shift he saw was caused by Earth's turning beneath his pendulum. The Foucault pendulum is now a famous demonstration of Earth's rotation.

More evidence of Earth's rotation can be seen by observing wind. If Earth did not rotate, winds would blow along straight paths from areas of high pressure to areas of low pressure. Because of Earth's rotation, winds appear to be turned, or deflected. In the Northern Hemisphere, winds are deflected to their right relative to Earth's surface. In the Southern Hemisphere, winds are deflected to their left. This apparent deflection is called the Coriolis effect. Any substance or object moving freely above or within Earth's surface is subject to the Coriolis effect. You will study the Coriolis effect in Chapter 19.

4.2

KEY IDEA
Earth rotates on its axis once approximately every 24 hours, resulting in day and night and providing the basis for time zones.

KEY VOCABULARY
- rotation
- standard time zones
- time meridian
- prime meridian
- International Date Line

VISUALIZATIONS
CLASSZONE.COM
Observe evidence of Earth turning about an axis.
Keycode: ES0403

This pendulum in the Cumberland Museum, Nashville, Tennessee, is suspended from a wire far above the heads of the observers.

CHAPTER 4 SECTION 2

FOCUS

Objectives
1. Give evidence for Earth's rotation.
2. Relate Earth's rotation to the day-night cycle and the time zones.

Set a Purpose
Read aloud the key idea. Then have students read this section to find the answer to "How does Earth's rotation result in day and night and affect the time zones?"

INSTRUCT

More about...

Earth's Rotation
Since Earth formed, its rate of rotation has gradually decreased. Earth is losing kinetic energy due to friction, caused in part by tides and galactic space dust. Scientists estimate that Earth's rotation is slowing 2.2 seconds every 100,000 years. The slowing of Earth's rotation allows gravity to pull the planet into an increasingly more-perfect sphere.

VISUALIZATIONS
CLASSZONE.COM
Observe evidence of Earth turning about an axis.
Keycode: ES0403
Visualizations CD-ROM

DIFFERENTIATING INSTRUCTION

Reading Support Have students begin to outline this section by first carefully reading the entire section. Model the process for students by writing on the board the red headings from the section. Label each heading with a separate Roman numeral. Then have students pick out key words or phrases that represent the main ideas under each heading and write these words or phrases separately under the numbered headings, using capital letters before each. Students should add details under these words or phrases, writing a number before each detail. Tell them to check the outline by reading the section again. They can make changes as needed.
Use Reading Study Guide, p. 11.

CHAPTER 4 SECTION 2

Visual Teaching

Interpret Diagrams
Refer students to the diagram at the far right, which shows Earth's speed of rotation at various locations. Point out that Miami, Florida, is located between the latitudes of Boston, Massachusetts, and the equator. Ask: What can you infer about Earth's speed of rotation in Miami? **It is greater than 1200 kilometers per hour but less than 1670 kilometers per hour.** Explain why Earth's speed of rotation changes with latitude. **Earth's circumference changes with latitude, but its rate of rotation is constant. Thus, the distance traveled per unit time changes with latitude.**

CRITICAL THINKING

Discuss the possible effects of Earth rotating at a rate much slower than once every 24 hours. Have students consider changes in the living and nonliving aspects of the environment, including people's lifestyles. **Longer days and nights would likely result in more extreme temperatures, resulting in hotter days and colder nights. Native vegetation might change. Basic animal behaviors, such as wake-sleep patterns, that are triggered by the day-night cycle would probably change. Animals exhibiting cyclic behavior—such as migration, estivation, and hibernation—would be affected. Human activities, including work and school schedules, might change.**

Axis and Rate of Rotation

Like the other planets in our solar system, Earth rotates as it travels around the sun. Recall that Earth's axis of rotation is an imaginary straight line through Earth between the North Pole and the South Pole. When Earth rotates, it turns around this axis. Earth's orbit, or path around the sun, lies within an imaginary flat surface called an orbital plane. As shown below, the axis of rotation is not perpendicular to Earth's orbital plane; that is, the two do not make a right angle. The axis is slightly tilted, making an angle of 23.5° with the perpendicular.

At present, Earth's axis points toward Polaris, the North Star. The tilt of Earth's axis stays the same throughout the year. This consistency in Earth's tilt is called *parallelism*. Because of parallelism, the North Star always appears at the same angle above the horizon in the Northern Hemisphere.

Earth's Axis

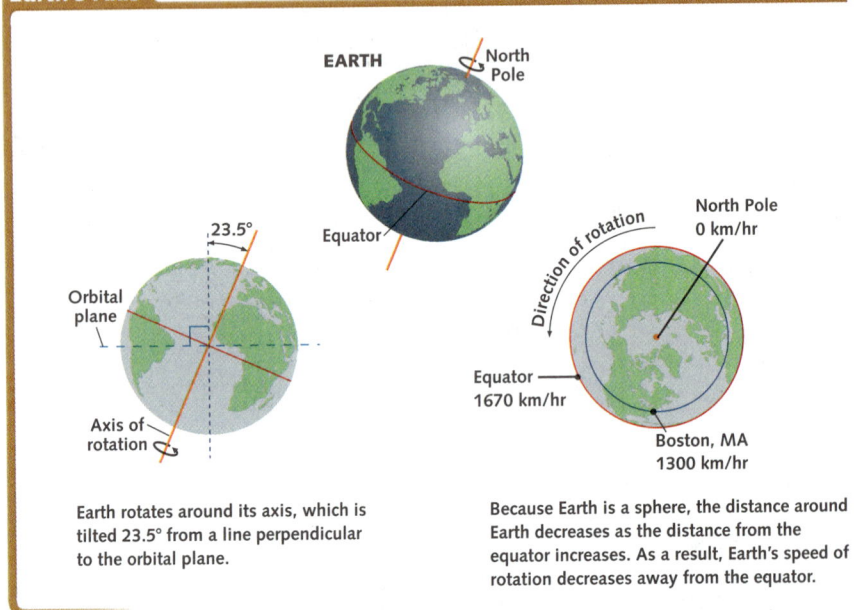

Earth rotates around its axis, which is tilted 23.5° from a line perpendicular to the orbital plane.

Because Earth is a sphere, the distance around Earth decreases as the distance from the equator increases. As a result, Earth's speed of rotation decreases away from the equator.

Explore a model of Earth's daily rotation.
Keycode: ES0404

Earth makes one complete turn equal to a rotation of 360° approximately every 24 hours. That means that every location on Earth's surface rotates at a rate of 15° per hour. However, because of Earth's spheroidal shape, the speed of rotation varies from point to point. While rate of rotation is measured in degrees per unit of time, speed of rotation is measured in distance per unit of time. The distance traveled in one rotation varies by latitude. At the equator, one rotation equals the Earth's circumference, or 40,074 kilometers. Therefore, points on the equator rotate at a speed of about 1670 kilometers per hour. At the latitude of Boston, Massachusetts, the speed of rotation is only about 1300 kilometers per hour. Near the poles, the speed of rotation is almost 0 kilometers per hour, because the poles are on the axis of rotation.

76 Unit 2 Earth's Matter

DIFFERENTIATING INSTRUCTION

Hands-On Demonstration Show how Earth's speed of rotation increases toward the equator. Rotate a globe a quarter turn using two lines of longitude as beginning and ending reference points. Show students that the distance between those lines of longitude at the equator is greater than at the Arctic Circle, yet the globe moved a quarter turn in the same amount of time at all places. Therefore, the globe had to move faster at the equator to cover its distance than it had to move at the Arctic Circle to cover *its* distance.

76 Unit 2 Earth's Matter

Effects of Rotation

The behavior of a Foucault pendulum and the Coriolis effect are both a result of Earth's rotation. Another effect of Earth's rotation is the daily change from day to night. From the standpoint of the North Pole, Earth rotates counterclockwise. Thus, the sun appears to rise in the east and set in the west. Only half of Earth receives sunlight at any given time. If Earth did not rotate, the half facing the sun would have constant light, while the other half would have perpetual dark.

Measuring Time

One day, 24 hours, is the approximate time it takes Earth to rotate once on its axis. For centuries, people figured the time of day by the sun's position in the sky. Each day, the sun rises on the eastern horizon, seems to move in an arc across the sky, and sets below the western horizon. Solar noon occurs when the sun is at the highest position on this arc.

Because of Earth's rotation, of course, solar noon does not occur at the same time everywhere. Instead, it moves westward at a rate of about 15° each hour, or 1° every four minutes. Consider New York City, located at longitude 74° W, and Philadelphia, near longitude 75° W. Because of the 1° difference in longitude, solar noon occurs in New York City about four minutes before it occurs in Philadelphia.

Standard Time Zones

The problem of having different solar times in nearby communities was solved through the development of time zones. As shown, 24 worldwide **standard time zones** were developed, each 15° of longitude wide.

The basis for time zones is the rate at which the sun appears to move across the sky. Each standard time zone is roughly centered on a line of longitude exactly divisible by 15°, called a **time meridian.** All areas within a time zone keep the same clock time. Clock time is the average solar time at that zone's time meridian.

INVESTIGATIONS
CLASSZONE.COM
What Time Is It? View maps of Earth showing where it is light and dark. Interpret them to tell what time it is in different parts of the world.
Keycode: ES0405

International Time

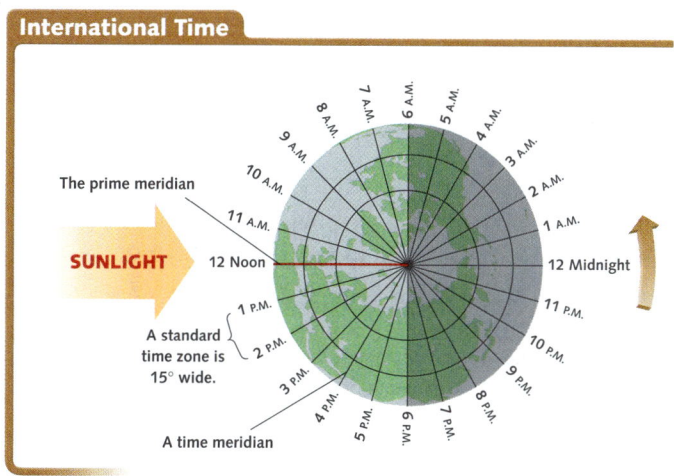

CHAPTER 4 SECTION 2

VISUALIZATIONS
CLASSZONE.COM

Explore a model of Earth's daily rotation.
Keycode: ES0404

Visualizations CD-ROM

INVESTIGATIONS
CLASSZONE.COM

What Time Is It?
Keycode: ES0405
Have students work in pairs.

Use Internet Investigations Guide, pp. T19, 19.

Visual Teaching

Interpret Diagram
Have students turn their book so that the prime meridian in the diagram is oriented vertically. Ask: Which direction relative to the prime meridian is east—to the right or the left? **right** Then, using the diagram as a guide, have students give examples of cities that experience solar noon before and after their area does.

DIFFERENTIATING INSTRUCTION

Hands-On Demonstration Use a globe and a flashlight to demonstrate the effects of Earth's rotation. Have students discover why the sun appears to rise in the east and set in the west by turning the globe counterclockwise relative to the North Pole while shining a light on the globe. Then ask a volunteer to locate New York City and Philadelphia on the globe and demonstrate why solar noon occurs in New York City before it occurs in Philadelphia.

CHAPTER 4 SECTION 2

ASSESS

1. One piece of evidence for Earth's rotation is Foucault's pendulum, which was observed to shift about 11° every hour clockwise, resulting in the conclusion that the shift was caused by Earth's turning beneath the pendulum. A second piece of evidence is found in observing winds. If Earth did not rotate, winds would blow along straight paths from areas of high pressure to areas of low pressure. However, winds appear to be deflected to the right in the Northern Hemisphere and to the left in the Southern Hemisphere.

2. Earth's circumference is greater at the equator than at the poles.

3. The basis for time zones is the rate at which the sun appears to move across the sky. Each standard time zone is roughly centered on a line of longitude exactly divisible by 15°, called a time meridian. All areas within a time zone keep the same clock time.

4. Earth rotates counterclockwise from the reference point of the North Pole. Thus, the east side of a time zone turns away from the sun before the west side does.

5. You would arrive in Seattle at 1:00 P.M., December 14, in Tokyo at 4:00 P.M. on Monday, December 15, in Baghdad at 8:00 P.M. on Monday, December 15, and finally in London at 3:00 A.M. on Tuesday, December 16.

The starting point for the standard time zones is an arbitrary longitude line called the **prime meridian**, which passes through Greenwich, England. Travelers moving westward from Greenwich move their clocks back to earlier times, while those moving eastward change to later times. When it is 10 A.M. in Greenwich (longitude 0°), it is 11 A.M. in Rome (longitude 15° E), 5 A.M. in Philadelphia (longitude 75° W), and 3 A.M. in Denver (longitude 105° W). People working for international businesses regularly place telephone calls across several time zones. They must keep these time differences in mind when scheduling their calls.

In theory, each standard time zone should be exactly 15° wide. On land, however, such exactness is not always convenient. For example, having a time-zone boundary cut right through a city could be confusing. Because of this, time-zone boundaries on land are seldom straight lines. Instead, they shift east or west to meet the needs of the people living in the area.

The International Date Line

Travelers going completely around the world gain or lose time at each time zone until they have gained or lost an entire day. How can travelers know where to change from one date to another? An imaginary line called the **International Date Line** represents the longitude at which the date changes. Upon crossing the date line, which goes through the Pacific Ocean, travelers change not their watches but their calendars. For travelers moving westward, the date is one day later; for eastward travelers, it is one day earlier. When travel agents don't keep these changes in mind, a traveler may miss a connecting flight or lose a hotel reservation.

The International Date Line lies within a time zone. Locations on either side of the date line within the same time zone keep the same time, but the western half is one day ahead of the eastern half. For much of each day, the continental United States is one day behind eastern Asia.

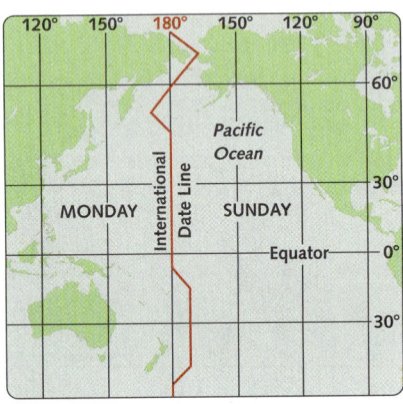

THE INTERNATIONAL DATE LINE follows the 180th meridian, varying from it only where necessary to avoid land.

4.2 Section Review

1. Describe two pieces of evidence for Earth's rotation.
2. Why is the speed of Earth's rotation different at the equator than it is at the poles?
3. Explain how time zones are determined.
4. **CRITICAL THINKING** In any given time zone, it gets dark a little earlier on the eastern side than on the western side. Why?
5. **WRITING** Imagine that you are on a flight leaving New York City at 6 A.M. on Sunday, December 14, and passing through Seattle, Tokyo, Baghdad, and London. Supposing it takes 10 hours to fly between cities, write a travel diary that includes the time and date you reach each city.

MONITOR AND RETEACH

If students miss . . .

Question 1 Use a pendulum to help explain Foucault's demonstration of Earth's rotation described on page 75.

Question 2 Use a globe and tape measure to compare the circumference of Earth at various latitudes.

Question 3 Point to the circle in the diagram on page 77. Remind students that there are 360° in a circle. Have them divide 360° by 15° per hour (rate at which Earth's surface rotates) to understand why there are 24 worldwide time zones.

Question 4 Use a globe to show why the eastern side of a time zone gets dark earlier than the western side.

Question 5 On an overhead projector, display a drawing of a clock and a world time-zone map that includes the International Date Line. Help students use both graphics to answer the question.

SCIENCE & Technology

Measuring Time

The way we tell time has changed significantly over the years. As technology improves, measuring time becomes increasingly precise.

How do you know time is passing? Is there more than one way to measure time?

ON A SUNDIAL, as the sun moves across the sky, the shadow cast by the center of the sundial moves along the perimeter scale, indicating the time of day.

In ancient times, when most activities were limited to daylight, people used shadow-casting instruments—such as gnomons, obelisks, and sundials—or water clocks to measure the passage of time. For over 5000 years there wasn't much change in how people marked time. It wasn't until improvements were made in machinery and gear making that telling time changed significantly.

By the mid-1300s, Europeans were building large mechanical, weight-driven clocks housed in towers for public reference. In 1656, Dutch astronomer Christiaan Huygens invented the pendulum clock, the first to accurately count seconds. Although driven by weights, the clock was regulated by the naturally periodic swinging motion of a pendulum. Before this invention, even the most efficient clocks could be off by 20 minutes per day!

The quartz clock, invented in 1927, was so precise that it helped scientists find irregularities in the rate of Earth's rotation. When placed in alternating electric fields, quartz crystals vibrate at a very regular frequency. These vibrations are used to regulate the clock precisely. Most wristwatches and clocks today are quartz clocks.

In nature, an atom of any chemical element emits electromagnetic radiation at a unique frequency. In 1957, the first atomic clock was made. Scientists based the design of the clock on the exposure of the element cesium to radiation. Radiation causes the cesium atom to release energy at a constant frequency, thus creating the basis for increments of time. By 1967, cesium's frequency became the new international unit of time.

One second is now defined as 9,192,631,770 cycles of the cesium atom's frequency, equal to an average second of Earth's rotation time. ■

Extension

SCIENCE NOTEBOOK

In addition to relying on clocks, people through the ages have used calendars to mark the passage of time. Research and report on the types of yearly calendars in use today or on a calendar used by an ancient culture.

Learn more about how time is measured.
Keycode: ES0406

ASTRONOMICAL CLOCKS like this one in Prague, the Czech Republic, are beautiful but are less efficient timekeepers than atomic clocks.

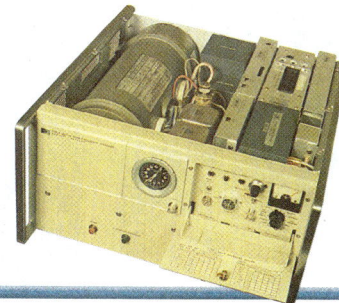

ATOMIC CLOCKS like this one use the vibrations of cesium atoms to mark time.

CHAPTER 4 SECTION 2

Learn more about how time is measured.
Keycode: ES0406

Science and Technology

A New Standard of Time?
Scientists at the National Institute of Standards and Technology in Boulder, Colorado, have built a mercury-based atomic clock that skips only one second every 30 billion years. By comparison, a cesium-based clock skips one second every 30 million years. The mercury atomic clock is kept in pace by a single mercury ion that vibrates at a constant cadence. The clock counts time by the femtosecond (a millionth-billionth of a second). Vibrations of the mercury ion are used to "tune" the light from a laser oscillating a million billion times a second. The mercury atomic clock, however, still needs refining if it is to replace the cesium-based clock as the international standard of time.

Science Notebook
Calendars used today include the Gregorian, Islamic, Hebrew, and Chinese calendars. Calendars used by ancient cultures include the ancient Egyptian, Julian (Roman), and Mayan calendars.

CHAPTER 4 SECTION 3

FOCUS

Objectives
1. Give evidence for Earth's revolution around the sun.
2. Describe Earth's path and rate of revolution.
3. Explain why seasons occur.

Set a Purpose
Have students read the section to find the answer to "How does Earth's revolution cause the seasons?"

INSTRUCT

More about...

Earth's Revolution
Earth revolves around the sun as a result of the combined effects of the centrifugal force generated by Earth's inertia as it orbits the sun and the inward pull of the sun's gravitational force.

4.3

KEY IDEA
Earth revolves around the sun in an elliptical orbit, causing seasonal variations.

KEY VOCABULARY
- revolution
- parallax
- summer solstice
- winter solstice
- vernal equinox
- autumnal equinox

VISUALIZATIONS
CLASSZONE.COM
Observe the view of the night sky from the same location over a year.
Keycode: ES0407

Viewed from New York City after 11 P.M. in January, the Big Dipper appears on the opposite side of the sky than when viewed at the same time of night six months later.

Earth's Revolution

Rotation is one type of motion characteristic of Earth. Another motion is **revolution**, the movement of Earth in its orbit around the sun.

Evidence for Revolution

For centuries, people gazing at the stars have observed evidence of Earth's revolution. Although groups of stars called *constellations* are visible every clear night, their positions in the sky appear to change as Earth rotates and revolves. Constellations which are visible on a winter evening are either in a different place in the summer night sky, or they are not visible at all. The shifting position of Earth in its orbit around the sun causes such changes in our view of the constellations.

As Earth moves in its orbit, nearby stars appear to shift position when compared to distant stars. This apparent shift in position is called **parallax.** Among stars, parallax cannot be detected by eye, but it can be measured with precise instruments. You can see the effect of parallax for yourself. Hold a pencil upright at arm's length and notice what happens when you look at the pencil with first one eye alone and then the other eye alone. Viewing the pencil from the two different positions produces the same apparent shift as seen in nearby stars when Earth is in two different positions. If Earth did not orbit the sun, no shift would occur. Therefore, the parallax of nearby stars is evidence of Earth's revolution.

Path and Rate of Revolution

The direction of Earth's revolution is the same as its direction of rotation, that is, counterclockwise when viewed from above the North Pole. Like the orbits of the other planets, Earth's orbit is an ellipse with the sun located at one focus. Because the orbit is elliptical, the distance between Earth and the sun changes throughout the year. The average distance is about

January

July

80 Unit 2 Earth's Matter

DIFFERENTIATING INSTRUCTION

Reading Support Have students preview the diagram at the bottom of page 81 before reading the text under the heading "Path and Rate of Revolution." Then as students read the text, have them stop after each main point and refer to the diagram for clarification or reinforcement. For example, after reading the first sentence, students can follow the direction of the arrows in the diagram to confirm that Earth revolves around the sun in a counterclockwise direction as viewed from the North Pole. After reading the second sentence, they can view the shape of Earth's orbit in the diagram. Have students continue using the visual aid in this manner as they read.

Use Reading Study Guide, p. 12.
Use Transparency 5.

80 Unit 2 Earth's Matter

Parallax

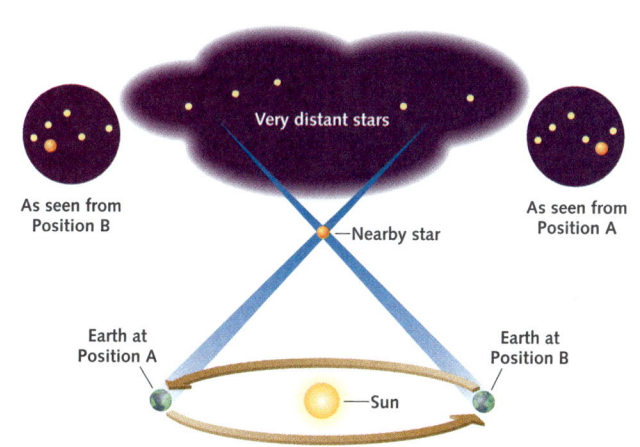

As Earth revolves around the sun, nearby stars appear to shift position. This shift, called parallax, is evidence of Earth's revolution. The diagram above shows two views of stars seen from Earth six months apart. Notice that the nearby star in the center seems to shift position relative to the very distant stars beyond it.

150 million kilometers. The sun is about 2.4 million kilometers from the center of the orbit. At perihelion, its point nearest the sun, Earth is about 147.6 million kilometers from the sun. Perihelion occurs on or about January 2. At aphelion, its point farthest from the sun, Earth is about 152.4 million kilometers from the sun. Aphelion occurs on or about July 4.

Earth makes one revolution around the sun every 365.24 days, defining the length of one year. Because one orbit represents a journey of 360°, Earth's rate of revolution around the sun is very close to 1° per day.

While Earth's rotation makes the sun appear to move across the sky once every day, Earth's revolution around the sun causes the sun's apparent path across the sky to change throughout the year.

VISUALIZATIONS CLASSZONE.COM

Explore a model of Earth's yearly revolution around the sun.
Keycode: ES0408

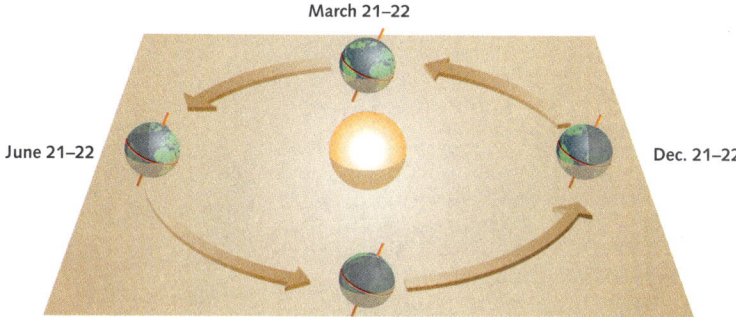

ORBITAL PLANE Earth's orbit around the sun defines the orbital plane. Earth remains tilted 23.5° from the vertical at all times during its orbit.

CHAPTER 4 SECTION 3

VISUALIZATIONS CLASSZONE.COM

Observe the view of the night sky from the same location over a year.
Keycode: ES0407

- Visualizations CD-ROM

VISUALIZATIONS CLASSZONE.COM

Explore a model of Earth's yearly revolution around the sun.
Keycode: ES0408

- Visualization CD-ROM

Visual Teaching

Interpret Diagram

As students study the diagram of parallax, point out that as Earth moves from position A to position B, the nearby (orange) star in the circle farther from the right will appear to shift to the left until it reaches the position shown in the circle to the left. Ask: How will the nearby star appear to shift as Earth moves from position B to position A? **The star will appear to shift to the right.** If Earth were always located at position A, how would the position of the star change? **The star would not appear to shift.**

Use Transparency 5.

DIFFERENTIATING INSTRUCTION

Hands-On Demonstration Have students speculate about how the distance of a star from Earth affects its parallax. **The closer a star is to Earth, the greater its parallax.** To illustrate this principle, have students repeat the pencil demonstration described on page 80, holding the pencil at a close distance and then farther away. **The pencil will appear to shift more when students hold it closer.**

CHAPTER 4 SECTION 3

Visual Teaching

Interpret Diagrams

Have students compare the tilts of the Northern Hemisphere and the Southern Hemisphere in the diagrams on this page. Ask: **When it is summer in the Northern Hemisphere, how is it tilted relative to the sun?** *toward the sun* **How is the Northern Hemisphere tilted in the winter?** *away from the sun* Lead students to recognize that because of the tilt of Earth's axis, it is summer in the Northern Hemisphere when it is winter in the Southern Hemisphere, and vice versa. Ask: **What is the first day of winter in the Southern Hemisphere?** *on or about June 21*

Use Transparency 5.

More about...

Solstices and Equinoxes

In different years, the exact times and dates of the solstices and equinoxes are not the same. This is due to leap years as well as small variations in Earth's orbit from year to year.

In describing the sun's position in the sky, astronomers refer to the point directly above the observer as the *zenith*. The angular distance between the horizon and the sun's position at any given time is called its *altitude*. When the sun is at the zenith, its altitude is 90°. When it is on the horizon, its altitude is 0°. For locations in the United States (except Hawaii), the sun is always below the zenith.

Effects of Revolution and Tilt

Effects of Earth's revolution include the seasons and variation in the length of days and nights. In addition to revolution, the tilt of Earth's axis relative to its plane of orbit has a profound effect on Earth. At almost any given time, one hemisphere is tilted toward the sun, as the other is tilted away. The hemisphere tilted toward the sun receives more direct sunlight and thus has warmer temperatures and longer days. The hemisphere tilted away from the sun receives indirect sunlight. That hemisphere has cooler temperatures and shorter days.

The changes in hours of daylight and in temperature caused by revolution and tilt lead to the yearly change of seasons at middle latitudes. If Earth's axis were perpendicular to its plane of orbit, seasons would not occur. In addition, every place on Earth's surface would experience 12 hours of daylight and 12 hours of darkness every day. On the other hand, if Earth's axis were tilted more than 23.5°, each hemisphere would experience hotter summers and colder winters.

The first day of summer in the Northern Hemisphere occurs on or about June 21 each year. This day has the longest daylight period, because the sun's path in the sky is longer and higher than at any other time of the year. The point at which this daily increase stops is the **summer solstice** (SOHL-stihs). At the summer solstice the Northern Hemisphere is at its maximum tilt toward the sun. Because this tilt is equal to 23.5°, the sun is straight overhead at locations along the latitude line of 23.5°N. This latitude line is called the Tropic of Cancer.

On the first day of summer, every point on Earth within 23.5° of the North Pole experiences 24 hours of daylight. The boundary of this region, at latitude 66.5° N, is the Arctic Circle. On June 21 in the Southern Hemisphere, every point south of the Antarctic Circle (latitude 66.5° S) experiences 24 hours of darkness.

EFFECT OF TILT Earth's Northern Hemisphere receives the most sunlight at the summer solstice (left) and the least sunlight at the winter solstice (right). Direct sunlight falls on the Tropic of Cancer at the summer solstice and on the Tropic of Capricorn at the winter solstice.

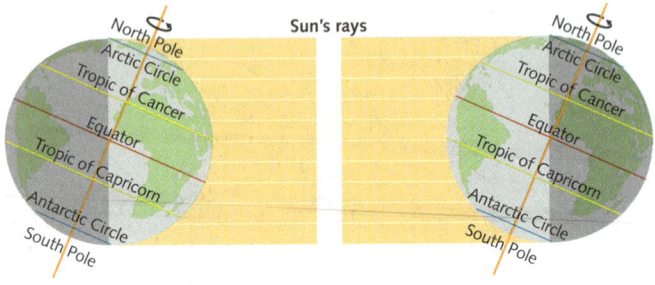

82 Unit 2 Earth's Matter

DIFFERENTIATING INSTRUCTION

Hands-On Demonstration Ask two students to model Earth's revolution around the sun, using the diagram on page 81 as a guide. The student representing Earth can hold a globe as he or she orbits the sun in an ellipse, while the student representing the sun continually shines a flashlight onto the globe. As Earth is revolving, have students point out its position at the summer solstice, winter solstice, vernal equinox, and autumnal equinox. Explain that while the flashlight shines in one direction, the sun shines in all directions.

Winter begins in the Northern Hemisphere on or about December 21. This is the **winter solstice,** the shortest day of the year, when the sun follows its lowest and shortest path across the sky. On this day, the Northern Hemisphere is at its maximum tilt away from the sun, while the Southern Hemisphere is at its maximum tilt toward the sun. The sun is straight overhead at the Tropic of Capricorn, which is at latitude 23.5° S. Daytime and nighttime conditions on December 21 are the opposite of those on June 21. On December 21, every point north of the Arctic Circle experiences 24 hours of darkness while every point south of the Antarctic Circle has 24 hours of daylight.

There are two days each year, midway between the solstices, when neither hemisphere tilts toward the sun. On these days, daytime and nighttime are equal in length all over the world. Each of these days, therefore, is known as an equinox (EE-kwuh-NAHKS). The **vernal equinox** occurs on or around March 21. The **autumnal equinox** is on or near September 22. The sun is overhead at the equator at noon on these dates.

VOCABULARY STRATEGY

Solstice comes from the Latin words *sol* (meaning "sun") and *stitium* (meaning "stoppage"). *Equinox* comes from the Latin words *aequalis* (meaning "equal") and *nox* (meaning "night"). So a solstice is a "sun stop," while an equinox is a time when night and day are of equal length.

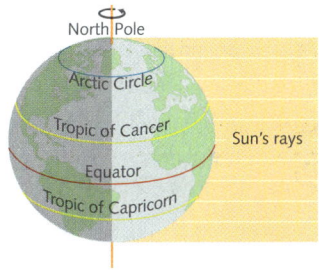

EQUINOX At an equinox, each of Earth's hemispheres receives equal amounts of sunlight. Direct sunlight falls on the equator.

The equinoxes also mark periods of long twilight at the poles. On March 21, the sun rises above the horizon at the North Pole for the first time in six months. The sun then remains visible at the North Pole for the next six months, while at the South Pole there are six months of darkness. On September 22, a six-month period of darkness begins at the North Pole, while that at the South Pole ends.

4.3 Section Review

1. What evidence is there for Earth's revolution around the sun?
2. Describe the shape of Earth's orbit and explain how Earth's position relative to the sun changes as Earth revolves.
3. Name the solstices and equinoxes and the dates on which they occur.
4. **CRITICAL THINKING** How would the solstices and equinoxes change if the Earth's orbit were circular instead of elliptical? Explain how a circular orbit would affect seasonal changes.

Chapter 4 Earth's Structure and Motion

CHAPTER 4 SECTION 3

ASSESS

1. Evidence of Earth's revolution around the sun includes the changing positions of constellations in the night sky in winter and summer and the apparent shift in position over time of nearby stars when compared to the positions of distant stars.
2. Earth's orbit is a counterclockwise ellipse (when viewed from the North Pole) with the sun located at one focus. Because the orbit is elliptical, the distance between Earth and the sun changes throughout the year, ranging from 147.6 million km around January 2 to 152.4 million km around July 4.
3. In the Northern Hemisphere, the summer solstice occurs on or about June 21, the winter solstice occurs on or about December 21, the vernal equinox occurs on or about March 21, and the autumnal equinox occurs on or about September 22.
4. Remind students that, although the path of Earth's revolution is elliptical, it is very nearly circular. The solstices and equinoxes largely result from the tilt of Earth's axis, not from the path of its revolution. Thus, a circular orbit, instead of an elliptical one, would not change the solstices and equinoxes. In addition, a circular orbit would not affect seasonal changes because the distance between Earth and the sun does not affect seasonal temperatures. This is evidenced by the fact that when the Northern Hemisphere is experiencing winter, Earth is nearest the sun, and when the Northern Hemisphere is experiencing summer, Earth is farthest from the sun.

MONITOR AND RETEACH

If students miss . . .

Question 1 Reteach "Evidence for Revolution." (p. 80)
Question 2 Have students use the diagram on page 81 to write three statements that describe Earth's orbit around the sun.
Question 3 Reteach "Effects of Revolution and Tilt." (pp. 82–83) Discuss seasonal changes that occur at the middle latitudes and the dates that mark the first day of summer, autumn, winter, and spring in the Northern Hemisphere.

Question 4 Have students draw a diagram similar to the diagram on page 81 that shows Earth moving around the sun in a circular path instead of an elliptical one. Students should compare Earth's positions at the solstices and equinoxes in their diagram with its positions in the diagram on page 81.

CHAPTER 4 ACTIVITY

PURPOSE
To explore the use of mathematics in earth science, using Eratosthenes' calculation of Earth's circumference as an example

MATERIALS
If a flexible metric ruler is not available, students can use string to measure the lengths of the arcs and then use a nonflexible metric ruler to measure the string lengths.

PROCEDURE
❶ This lab assumes that Earth is a sphere.

❺ Review metric units. For example, 1 cm is equal to 10 mm.

❼ Algebraic manipulations of the equations may need to be reviewed.

Students can calculate the percentage difference between their answers to questions 1 and 2. Assuming that the answer to question 2 is more accurate based on the circumference formula, the percent difference can be calculated by using:

$|C$ from #1 $- C$ from #2$| =$ Difference, where C is the circumference and $|x|$ is the absolute value of x

$$\frac{\text{Difference}}{C \text{ from } \#2} \times 100 = \%$$

❽ angle GFH = 60°

❾ Review the geometric concept of alternate interior angles of parallel lines.
angle IFH = 30°

CHAPTER 4 LAB Activity

Eratosthenes and Earth's Circumference

SKILLS AND OBJECTIVES
- **Model** Eratosthenes' Earth.
- **Measure** and **record** data from the model.
- **Compute** and **interpret** the data.

MATERIALS
- blank sheet of paper
- safety compass
- protractor
- flexible metric ruler

More than 2,000 years ago, a man named Eratosthenes made a surprisingly accurate estimation of Earth's circumference. He did this by using some simple geometric relationships. In this activity, you will use Eratosthenes' method to determine the circumference of a circle.

Procedure

Part A: Finding a Circle's Circumference

❶ Visually locate a point as close to the center of a blank sheet of paper as possible. Mark the point and label it point C. With a safety compass, draw a large circle around point C.

❷ Use the ruler to draw a straight line in any direction from point C to the edge of the paper. Label the point where the straight line intersects the circle point A. The line connecting points A and C is called line AC.

❸ Place a protractor along line AC so that its center is on point C. Mark off an angle between 15° and 50°. Make a copy of Table 1 and record the angle.

❹ Complete the angle by drawing a straight line from point C through the point you have marked off with the protractor to the edge of the paper. Label the point where this straight line intersects the circle point B. The line connecting points B and C is called line BC.

❺ Set the flexible metric ruler on edge and bend it to follow the circumference of the circle. Use the curved ruler to measure, to the nearest tenth of a centimeter, the length of the circle from point A to point B (an arc). Record the value in your copy of the Table.

❻ Measure the length from point C to point A, to the nearest tenth of a centimeter. Record this value in Table 1. Line AC is the radius of the circle.

❼ Answer Questions 1 and 2 under Analysis and Conclusions.

Part B: Finding Earth's Circumference

❽ Eratosthenes used careful observations of the sun's rays to find Earth's circumference. The figure on the next page shows the sun's rays striking two locations at Earth's surface, E and F. At point E, the sun is straight overhead and strikes the surface at a 90° angle. At point F, the sun's rays strike the surface at angle GFH. Measure angle GFH and record in a copy of Table 2 on the next page.

Table 1
Angle used (°) = _____
Length of arc AB (cm) = _____ (distance along circle)
Length of line AC (cm) = _____ (radius of circle)

84 Unit 2 Earth's Matter

⑨ Because the sun is far away, Eratosthenes considered its rays to be parallel. Based on geometric principles, this parallelism means that the angle IFH equals the angle between E and F at Earth's center. To find IFH, subtract GFH from IFG. Note that angle IFG is a right angle (90°).

⑩ Using the curved ruler technique from Part A, measure the distance from point E to F along the segment of Earth's surface shown, to the nearest tenth of a centimeter. Record the arc length in your copy of Table 2.

⑪ Use the scale 1 centimeter = 1,800 kilometers to convert the length of arc EF to kilometers. Record this value in Table 2.

⑫ Answer Questions 3 and 4 under Analysis and Conclusions.

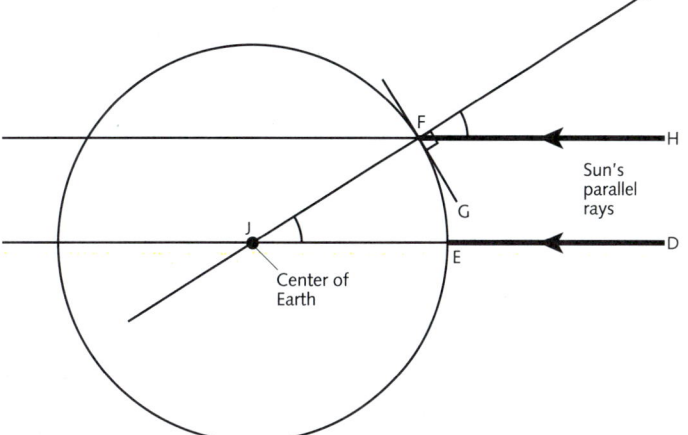

Table 2
Angle GFH (°) =	
Angle IFH (°) =	
Measured length, arc EF =	
(distance along circle in cm)	
Distance, E to F (km) =	

Analysis and Conclusions

Part A: Finding a Circle's Circumference

1. The length of arc AB has the same relationship to the circumference of the circle as the angle used has to the whole circle (360°). Use the equation to the right to find the circumference of the circle.

2. Using the standard formula $C = 2\pi r$ (C = circumference, r = radius), determine the circumference of your circle again. (Use $\pi = 3.14$.) Compare with your answer to Question 1.

$$\frac{\text{arc AB}}{\text{circumference}} = \frac{\text{angle used}}{360°}$$

Part B: Finding Earth's Circumference

3. Calculate Earth's circumference using the equation to the right.

4. Earth's actual circumference is approximately 40,000 km. Determine the percentage by which your answer differs from Earth's actual circumference using the equation to the right.

$$\frac{\text{distance EF}}{\text{circumference}} = \frac{\text{angle IFH}}{360°}$$

$$\% = \frac{\text{actual circ.} - \text{ans.}}{\text{actual circ.}} \times 100$$

Chapter 4 Earth's Structure and Motion 85

CHAPTER 4 ACTIVITY

(continued)

⑩ arc EF = 1.8 cm

⑪ distance EF = 3060 km

ANALYSIS AND CONCLUSIONS

1. Answers will vary, depending on the sizes of students' drawings. Students should perform a simple substitution of their values for *arc AB* and *angle used* and then solve for *circumference*.

2. Using the line segment AC as the radius, circumference (*C*) = 2 × 3.14 × length of segment AC. *C* is in centimeters. Answers to questions 1 and 2 should be very similar, within 1 to 2 percent.

3. angle IFH = 30°;
   distance EF = 3240 km;
   circumference = 38,880 km

4. Students should substitute their answer from question 3 into the equation, and use 40,000 km for Earth's actual circumference, giving a result of about 2.8% error. Discuss the uses and methods of error analysis.

Refer students to Safety in the Earth Science Laboratory on pages xxii–xxiii.

CHAPTER 4 REVIEW

Summary of Key Ideas

4.1 Earth formed about 4.6 billion years ago from a whirling cloud of dust and gas. It developed layers as it cooled and dense material sank to its center. Meteorite impacts, the weight of overlying material, and the decay of radioactive isotopes caused Earth to heat up soon after its formation. Since then, Earth has been losing heat. Earth has a magnetic field.

4.2 Earth makes one complete turn on its axis about every 24 hours. Its axis of rotation is tilted with respect to Earth's orbital plane. Effects of this rotation include the Coriolis Effect, Foucault pendulum behavior, day and night, and sunrise and sunset. Earth is divided into 24 worldwide standard time zones that begin at the prime meridian.

4.3 Earth revolves around the sun in an elliptical orbit with the sun as one focus. Combined with Earth's tilt, revolution causes seasonal changes. The summer and winter solstices are the longest and shortest days of the year, respectively. On the vernal and autumnal equinoxes, day and night are of equal lengths.

Key Vocabulary

asthenosphere (p. 73)
autumnal equinox (p. 83)
crust (p. 73)
inner core (p. 72)
International Date Line (p. 78)
lithosphere (p. 83)
magnetic field (p. 74)
mantle (p. 72)
outer core (p. 72)
parallax (p. 80)
prime meridian (p. 78)
revolution (p. 80)
rotation (p. 75)
standard time zones (p. 77)
summer solstice (p. 82)
time meridian (p. 77)
vernal equinox (p. 83)
winter solstice (p. 83)

Vocabulary Review

Explain the difference between the terms in each pair.

1. inner core, outer core
2. mantle, crust
3. time meridian, prime meridian
4. rotation, revolution
5. summer solstice, winter solstice
6. vernal equinox, autumnal equinox

Write the term from the key vocabulary list that best completes the sentence.

7. A compass needle points northward due to Earth's ____?____.
8. Earth is divided into 24 worldwide ____?____, each 15° wide.
9. A traveler crossing the ____?____ will either gain or lose a day.
10. The apparent shift in the position of nearby stars when compared to distant stars is ____?____.

Concept Review

11. In five or more steps, summarize the major events described by the nebular hypothesis.
12. Describe Earth's shape and surface area. Include any available numerical data.
13. List evidence for the rotation of Earth. What is one effect of rotation?
14. **Graphic Organizer** Copy and complete the concept map below.

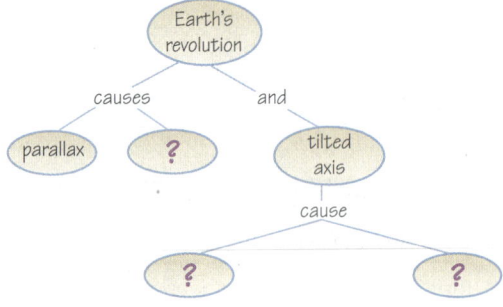

Vocabulary Review

1. inner core: located at Earth's center, composed of solid iron and nickel; outer core: surrounds inner core, composed of iron and nickel in a liquid state
2. mantle: surrounds Earth's core, thickest of Earth's layers, composed mostly of compounds rich in iron, silicon, and magnesium, behaves as a liquid in some ways; crust: a thin, rigid layer of lighter rocks covering the mantle that includes Earth's surface
3. time meridian: a line of longitude exactly divisible by 15°, the approximate center of a standard time zone; prime meridian: an arbitrary longitude line that passes through Greenwich, England and serves as the starting point for the standard time zones
4. rotation: the spinning movement of Earth around its axis; revolution: the movement of Earth in its orbit around the sun
5. summer solstice: the point in Earth's orbit when the sun's path in the sky is longest and highest, resulting in the longest day of the year (first day of summer); winter solstice: the point in Earth's orbit when the sun's path is at its shortest and lowest, resulting in the shortest day of the year (first day of winter)
6. Both occur midway between the solstices, when neither hemisphere tilts toward the sun. The vernal equinox occurs on or around March 21 and marks the beginning of six months of daylight at the North Pole; the autumnal equinox is on or near September 22 and marks the start of a six-month period of darkness at the North Pole.
7. magnetic field
8. standard time zones
9. International Date Line
10. parallax

Concept Review

11. (1) A great cloud of gas and dust rotated slowly in space. (2) The cloud shrank under the pull of its own gravity. (3) As the cloud shrank, its rate of rotation increased and material gathered around its center. (4) The compression of this material heated its interior, causing hydrogen fusion and the sun's formation. (5) A small percent of the material formed a platelike disk around the sun that eventually formed into planets and moons.
12. Earth is an imperfect sphere. It has a total surface area of about 510 million km^2. Of this area, about 149 million km^2 (about 29 percent) lie above sea level as continents and islands. Oceans cover the remaining 361 million km^2 (about 71 percent).
13. Evidence of rotation: a Foucault pendulum shifts clockwise about 11° every hour, indicating that Earth turns beneath the pendulum; the Coriolis effect, the apparent deflection of winds that would, on a stationary planet, blow along straight paths. Effect of rotation: the behavior of a Foucault pendulum, the Coriolis effect, or the daily change from day to night.

Critical Thinking

15. Infer What has happened to Earth's temperature over time? If this trend continues, what might eventually happen to the temperatures of Earth's inner and outer cores, its mantle, and its crust?

16. Draw Conclusions During a total eclipse of the sun, the moon's shadow falls on Earth as a dark circle. What does the shape of the shadow suggest about the shape of the moon?

17. Predict Earth experiences an apparent force away from its center of rotation. This force is greatest at the equator. Jupiter rotates more rapidly than Earth. Predict the shape of Jupiter. Explain your reasoning.

18. Infer How would increasing the tilt of Earth's axis affect the amount of daylight throughout the year? If Earth's axis were tilted at 33.5° instead of 23.5°, where would the Tropics of Cancer and Capricorn and the Arctic and Antarctic circles be located?

Interpreting Graphs

The apparent size or diameter of an object in the sky depends on your distance from the object. Apparent diameters are measured in degrees. The width of your fist viewed at arm's length has an apparent diameter of about 10°. The apparent diameter of the sun, a much larger object, is less than 1°. The graph records the apparent diameter of the sun when viewed from Earth over the course of a year.

19. In which season is the sun's apparent diameter largest? Is Earth closer to or farther from the sun at that time?

20. In which season is the sun's apparent diameter smallest? Is Earth closer to or farther from the sun at that time?

21. Based on your previous answers, explain how you know that distance from the sun is not a cause of seasons.

22. Use the graph to determine the months during which Earth is at aphelion and at perihelion.

Internet Extension

Learn more about methods of keeping time.
Keycode: ES0409

Writing About the Earth System

SCIENCE NOTEBOOK Over the course of the school year, record observations in your notebook about local seasonal changes, including observations about the changing amount of daylight. Using information about your local weather, describe the impact of seasons on the atmosphere and the biosphere.

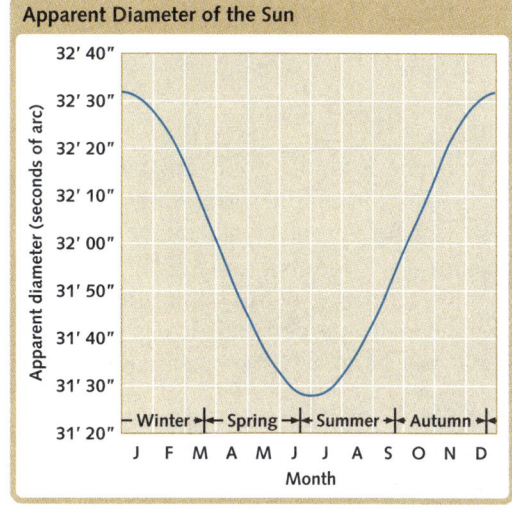

CHAPTER 4 REVIEW

Learn more about methods of keeping time.
Keycode: ES0409

(continued)

14. Earth revolution causes parallax and *changing positions of constellations*. Earth's revolution and tilted axis cause seasons and variations in length of days and nights.

Critical Thinking

15. Earth's temperature has been decreasing over time. If this trend continues, Earth's inner and outer cores, mantle, and crust might eventually cool to the same temperature.

16. The moon is spherical.

17. Jupiter is spherical and probably more oblate than Earth because it spins more rapidly.

18. An increased tilt would result in longer periods of daylight during the summer and longer nights during the winter. The Tropic of Cancer would be at 33.5° N, the Tropic of Capricorn at 33.5° S. The Arctic Circle would be 56.5° N (33.5° from the North Pole), and the Antarctic Circle at 56.5° S.

Interpreting Graphs

19. winter; closer to

20. summer; farther from

21. Winter occurs in the Northern Hemisphere when Earth is closest to the sun, and summer occurs when Earth is farthest from the hot sun.

22. The graph shows aphelion at the end of June or the beginning of July; perihelion at the end of December or the beginning of January.

WRITING ABOUT SCIENCE

Writing About the Earth System

Responses will vary, depending on where students live. The impact of the seasons includes increasingly shorter days and colder weather as summer turns into winter and increasingly hotter weather and longer days as winter changes into summer. Summer precipitation may be evidenced as high humidity and rain, whereas winter precipitation may be exhibited as snow, hail, or sleet. Animals, including birds and large mammals, might migrate to warmer weather during winter months, while other animals, including bears, may hibernate. Over hot summers, animals might estivate, or be active during the night hours when the temperature drops.

CHAPTER

PLANNING GUIDE ATOMS TO MINERALS

	Section Objectives	Print Resources
Chapter-Level Resources		Laboratory Manual, pp. 17–26 Internet Investigations Guide, pp. T20–T21, 20–21 Spanish Vocabulary and Summaries, pp. 9–10 Formal Assessment, pp. 13–15 Alternative Assessment, pp. 9–10
Section 5.1 Matter and Atoms, pp. 90–95	1. Identify the characteristics of matter. 2. Compare the particles that make up atoms of elements. 3. Describe the three types of chemical bonds.	Lesson Plans, p. 13 Reading Study Guide, p. 13
Section 5.2 Composition and Structure of Minerals, pp. 96–103	1. Identify the characteristics of minerals. 2. Explain how minerals form. 3. List the physical characteristics of minerals that are influenced by their crystalline structure.	Lesson Plans, p. 14 Reading Study Guide, p. 14 Teaching Transparency 6
Section 5.3 Identifying Minerals, pp. 104–107	1. Identify rock-forming minerals by inspection, using physical properties such as color, luster, and crystal shape. 2. Identify rock-forming minerals using simple tests that identify both physical and chemical properties, such as streak, cleavage, hardness, and specific gravity.	Lesson Plans, p. 15 Reading Study Guide, p. 15
Section 5.4 Mineral Groups, pp. 108–111	1. Describe the properties of the most common minerals, silicates and carbonates. 2. Describe tests used to identify mineral groups.	Lesson Plans, p. 16 Reading Study Guide, p. 16

 INVESTIGATIONS
CLASSZONE.COM

Section 5.1 INVESTIGATION
How Many Protons, Neutrons, and Electrons Are in Common Elements? Computer time: 45 minutes / Additional time: 45 Students build model atoms, ions, and isotopes from subatomic particles.

Chapter Review INVESTIGATION
How Do Crystals Grow?
Computer time: 45 minutes / Additional time: 45 minutes
Students select variable conditions and see their effects on crystal growth.

Technology/Online Resources

Electronic Teacher Tools
Visualizations CD-ROM
Classzone.com
Online Lesson Planner
Test Generator

Visualizations CD-ROM
Classzone.com
  ES0501 Investigation, ES0502 Visualization

Visualizations CD-ROM
Classzone.com
  ES0503 Visualization, ES0504 Visualization

Visualizations CD-ROM

Visualizations CD-ROM
Classzone.com
  ES0505 Visualization

Classroom Management

Gather these materials for minilab and laboratory activities.

Minilab, p. 107
  beaker
  water
  string
  mineral sample
  spring scale

Lab Activity, pp. 112–113
  mineral kit containing 3 unnamed minerals
  triple-beam balance
  C-clamp
  250-mL beaker
  water
  string
  paper clip
  rubber band
  calculator

CHAPTER 5

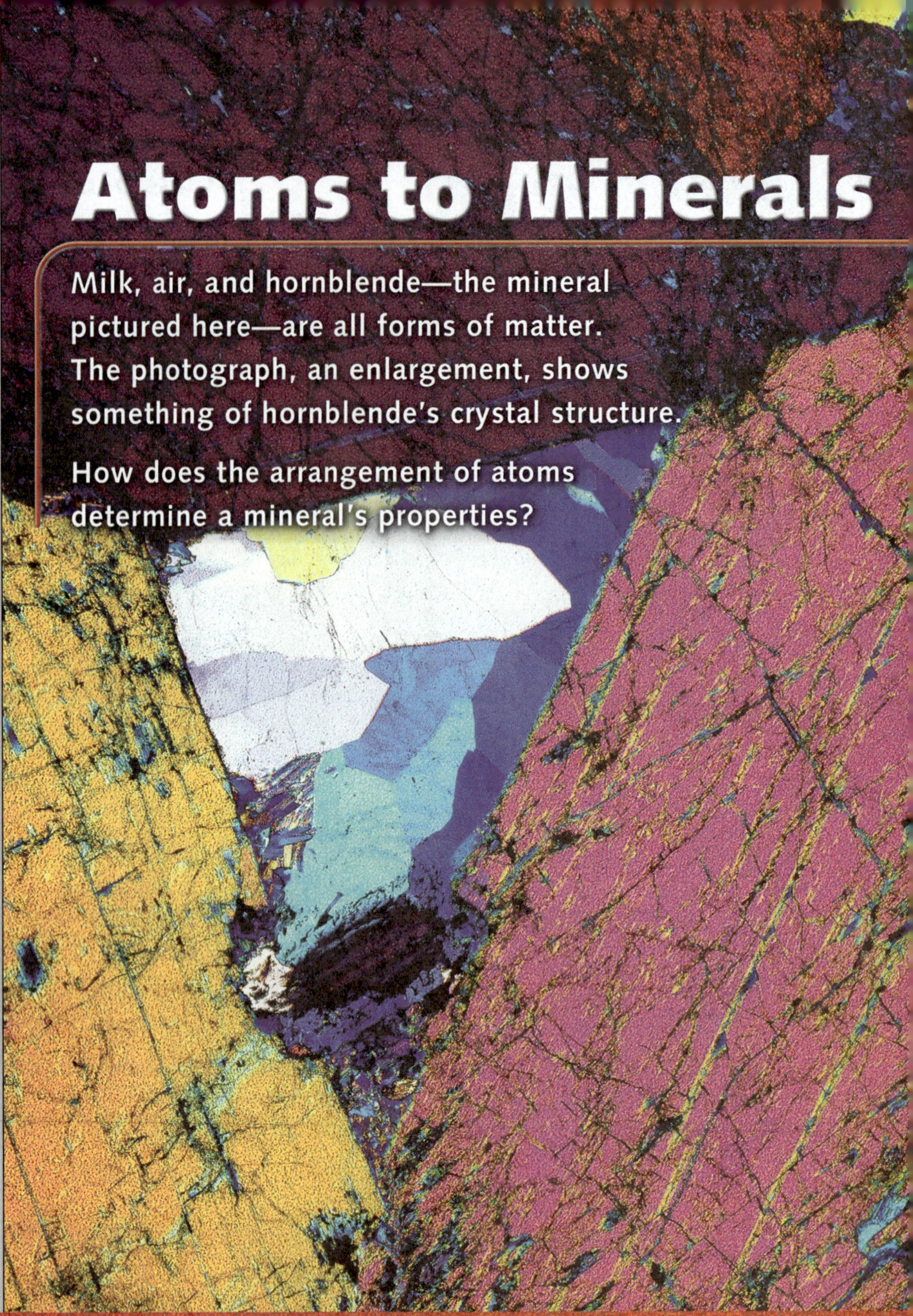

INTRODUCE
Build Interest
Before sharing background information about the photograph with students, allow time for their questions and comments.

Ask questions such as the following:

- What is matter? *anything that has mass and volume*
- What is a crystal? *a regular geometric solid with smooth surfaces*
- What property of hornblende does this photograph show? *has a tendency to split*

Respond
- How does the arrangement of atoms determine a mineral's properties? *An orderly arrangement of atoms results in the crystalline shape of a mineral. Such a structure determines whether the mineral will split in particular directions or how dense it is.*

About the Photograph
This is a micrograph of a thin section of hornblende, a silicate mineral. Hornblende and other silicates make up 95 percent of Earth's crust. The basic building block of most minerals, including hornblende, is a pyramid-shaped unit consisting of a silicon atom surrounded by four oxygen atoms. Minerals differ because of the other types of atoms they contain and the way in which the silica units are joined.

Atoms to Minerals

Milk, air, and hornblende—the mineral pictured here—are all forms of matter. The photograph, an enlargement, shows something of hornblende's crystal structure.

How does the arrangement of atoms determine a mineral's properties?

BEYOND THE TEXTBOOK

Print Resources
- Reading Study Guide, pp. 13–16
- Transparency 6
- Formal Assessment, pp. 13–15
- Laboratory Manual, pp. 17–26
- Alternative Assessment, pp. 9–10
- Spanish Vocabulary and Summaries pp. 9–10
- Internet Investigations Guide, pp. T20–21, 20–21

Technology Resources
- Visualizations CD-ROM
- Classzone.com
- Online Lesson Planner
- Electronic Teacher Tools
- Test Generator

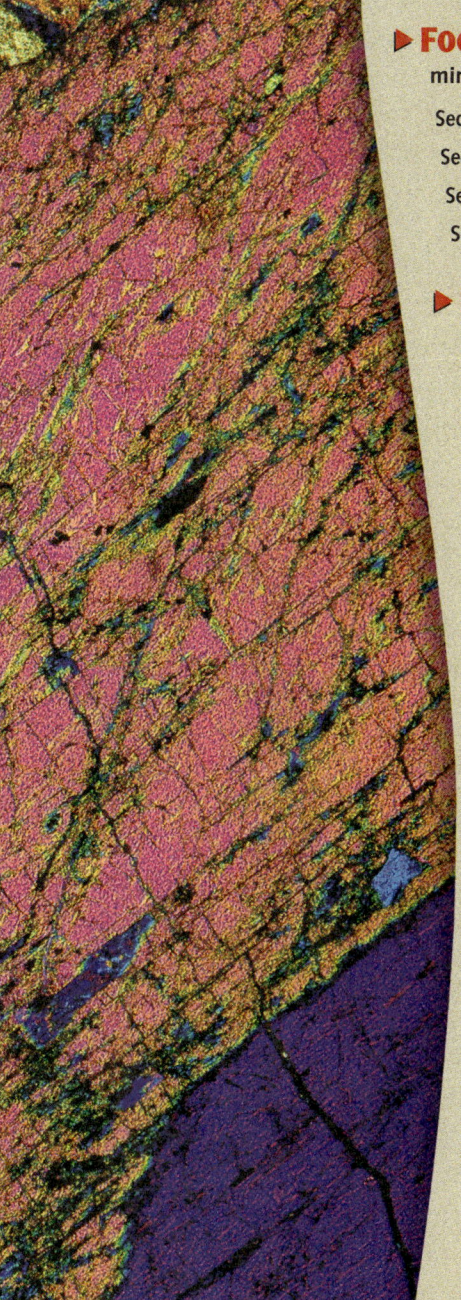

CHAPTER 5

PREVIEW

▶ **FOCUS QUESTIONS** In this chapter you will study atoms and minerals and learn more about the key questions below.

Section 1 What is the relationship of atoms to matter?
Section 2 What are minerals and how do they form?
Section 3 How can minerals be identified?
Section 4 What are common minerals and how are they classified?

▶ **REVIEW TOPICS** As you investigate atoms and minerals, you will need to use information from earlier chapters.

- geosphere (p. 9)
- Earth's structure (p. 72)
- crust (p. 73)
- lithosphere (p. 73)
- Earth's heat (p. 73)

▶ **READING STRATEGY**

QUESTION

Before you begin reading, scan the chapter, noting the photographs, illustrations, and headings. Develop a list of three or four questions that you would like to answer through your reading.

INTERNET RESOURCES
CLASSZONE.COM

At our Web site, you will find the following Internet support for this chapter.

DATA CENTER
EARTH NEWS
VISUALIZATIONS
- Molecules in Minerals
- Common Molecules
- Buckminsterfullerene
- Common Objects Made of Minerals

LOCAL RESOURCES
INVESTIGATIONS
- How Many Protons, Neutrons, and Electrons Are in Common Elements?
- How Do Crystals Grow?

Chapter 5 Atoms to Minerals **89**

CHAPTER 5

PREPARE

Focus Questions
Find out what knowledge students have about minerals. For each focus question, have students consider the section title and then develop two or three additional questions for each section. For example, Section 3: How are minerals unique? What properties are used to identify minerals?

Review Topics
In addition to the earth science topics listed, students may need to review these concepts:

Physical Science
- The arrangement and types of atoms give matter its properties.
- Most atoms tend to be stable when they have eight electrons in their outer shell.
- Radioactivity is the emission of high-energy radiation or particles from the nucleus of a radioactive atom.

Reading Strategy
Model the Strategy: To help students develop questions for each section, use a version of the following: "For the first section, I see that the title is 'Matter and Atoms.' Looking at the illustration on page 90, I can think of a couple questions that the section might answer: 'What is matter? How does an atom relate to matter?' Other questions might actually be found in the headings, such as 'What is a mineral?' on page 96 of the second section." Have students continue to use this strategy for the remainder of the chapter.

BEYOND THE TEXTBOOK

INTERNET RESOURCES
CLASSZONE.COM

INVESTIGATIONS
- How Many Protons, Neutrons, and Electrons Are in Common Elements?
- How Do Crystals Grow?

VISUALIZATIONS
- Images showing the arrangement of molecules in minerals
- Animation of 3-D models of common molecules

- A model of buckminsterfullerene
- Maps of mineral distribution and images of common objects made of those minerals

DATA CENTER
EARTH NEWS
LOCAL RESOURCES
CAREERS

Online Lesson Planner

CHAPTER 5 SECTION 1

▶ FOCUS

Objectives
1. Identify the characteristics of matter.
2. Compare the particles that make up atoms of elements.
3. Describe the three types of chemical bonds.

Set a Purpose
Have students read the key idea and discuss their understanding of what matter and atoms are. Tell students to read this section to find the answer to "What is the relationship between matter, elements, and atoms?"

▶ INSTRUCT

VISUAL TEACHING
Discussion
The arrangement of atoms making up a substance is important in determining the substance's properties. Have students note in the middle diagram the orderly arrangement of the carbon atoms making up the diamond. Tell students that these atoms are held together by strong bonds, which hold the atoms in a three-dimensional network. This structure makes diamond the hardest substance known. Graphite also is made of carbon atoms, but the atoms, which are arranged in sheets, are held together by weak bonds. The result is that graphite is soft and slippery.

5.1

KEY IDEA
Ordinary matter is composed of elements that can be broken down into particles called atoms.

KEY VOCABULARY
- element
- atomic number
- isotope
- mass number
- compound
- molecule
- ion
- metal
- nonmetal

Matter and Atoms

Telling the difference between sand, water, and diamonds is fairly simple. Each of these substances, like the thousands of other materials that occur naturally on Earth, has distinctive characteristics. Yet as different as they may seem, these materials have much in common. Each is a kind of matter.

Matter

What is matter? Matter is anything that has mass and volume. Mass is a measure of the amount of material in an object or a substance. Mass is often discussed in terms of weight, but mass and weight are not the same. Weight is a measure of the force of gravity acting on an object or substance. Where the force of gravity is very weak, as in outer space, an object may be virtually weightless, but its mass remains the same. Volume is the amount of space taken up by an object or a substance. All the materials found on Earth—diamonds, for example—have mass and volume and therefore are matter.

Ordinary matter is composed of elements. An **element** is a substance that cannot be broken into simpler substances by ordinary chemical means. You may already be familiar with the names of many elements. Oxygen and nitrogen are elements in the atmosphere; gold, silver, and iron are familiar metallic elements. Each element has a symbol as well as a name. In some cases, the symbol is the first letter or two of the element's name. For example, the symbol for hydrogen is H, and the symbol for helium is He. A few elements' symbols are based on their Latin names. For example, the symbol for gold is Au, from the Latin word for gold, *aurum*.

Of what do elements consist? About 200 years ago, the English chemist John Dalton formulated the modern particle model—the concept that each element is made up of tiny particles, all alike, called *atoms*. Dalton described an atom as the smallest part of an element that has all the element's properties. Today, scientists know that atoms are very complex and consist of still smaller particles.

This sample shows a "rough" or uncut diamond in its rock matrix.

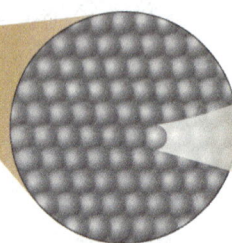

A diamond is made up of uniformly arranged carbon atoms.

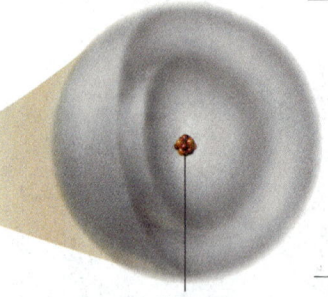

A carbon atom is made up of a nucleus surrounded by an electron cloud. The nucleus, in which most of the atom's mass is concentrated, consists of protons and neutrons.

90 Unit 2 Earth's Matter

DIFFERENTIATING INSTRUCTION

Reading Support Key vocabulary words represent many of the important concepts in this chapter. Have students write the vocabulary words for each section in their science notebooks. As they find the words in the text, have them write definitions in their own words. After reading the section, they can adjust the definitions as necessary.
Use Reading Study Guide, p. 13.

Developing English Proficiency Help students understand the relationship between an element's name and its symbol. Write the name of at least 20 elements on the board, including those mentioned in the text. Then list the symbol next to each name and underline the letter or letters of that symbol in the name. For symbols derived from the element's Latin name, have students recognize that the symbol does not make sense with the English name. Then write the Latin name and underline the appropriate letter(s).
Use Spanish Vocabulary and Summaries, pp. 9–10.

90 Unit 2 Earth's Matter

Structure of an Atom

An atom is essentially composed of negatively charged particles moving at high speed about a central nucleus (plural *nuclei*). Each negatively charged particle is called an *electron*. The moving electrons create a "cloud" of negative charge around the nucleus. To make atomic structure easier to understand, however, diagrams of atoms often show electrons orbiting nuclei like planets orbiting the sun.

The Nucleus

The nucleus of an atom contains protons and neutrons. Each proton carries a positive charge that is exactly equal to an electron's negative charge. A neutron (NOO-TRAHN) carries no charge. In its normal state, a neutral atom has an equal number of electrons and protons. The number of protons in the nucleus is the atom's **atomic number** and is also equal to the number of electrons in the atom's electron cloud. The number of protons and electrons in an atom determines the atom's properties.

Atoms are tiny and consist mainly of empty space. An atom of iron, for example, is about 25 ten-millionths of a meter in diameter. Most of an iron atom, however, is empty space between the nucleus and its surrounding electrons.

Although the nucleus is small, it contains the more massive atomic particles. A proton is 1836 times more massive than an electron, and a neutron is slightly more massive than a proton. More than 99.9 percent of the mass of an atom is in its nucleus.

The simplest and lightest of all atoms is that of hydrogen (H). The nucleus of the most common form of hydrogen has one proton and no neutrons. One electron circles the nucleus.

The next lightest element is helium (He). The nucleus of a typical helium atom contains two protons and two neutrons. The two electrons in its electron cloud balance the two positively charged protons in the nucleus.

One of the most important elements found on Earth is carbon (C). The most common form of the carbon atom has six protons and six neutrons in its nucleus. The electron cloud contains six electrons. In atoms that have more than two electrons, the electron cloud is divided into energy levels, as described below.

Electrons and Energy Levels

As the number of electrons in atoms increases, more energy levels are needed to hold them. The greatest number of energy levels in any atom is seven. Each level can hold only a specific number of electrons. For example, the innermost level never holds more than 2 electrons. Other levels hold as many as 32 electrons. Atoms of uranium (U), the heaviest of the natural elements, have 92 electrons distributed in seven energy levels containing 2, 8, 18, 32, 21, 9, and 2 electrons, counting innermost to outermost.

INVESTIGATIONS CLASSZONE.COM

How Many Protons, Neutrons, and Electrons Are in Common Elements? Build model atoms, ions, and isotopes from subatomic particles.
Keycode: ES0501

VOCABULARY STRATEGY

The word *nucleus* is from a Latin word, meaning "kernel."

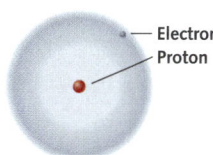

HYDROGEN (H) — Electron, Proton

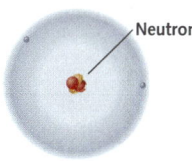

HELIUM (He) — Neutron

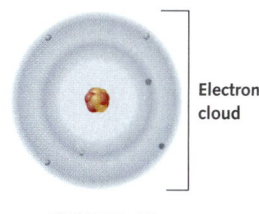

CARBON (C) — Electron cloud

CHAPTER 5 SECTION 1

INVESTIGATIONS CLASSZONE.COM

How Many Protons, Neutrons, and Electrons Are in Common Elements?
Keycode: ES0501

📖 Use Internet Investigations Guide, pp. T20, 20.

VISUAL TEACHING
Discussion

Have students analyze the three diagrams representing hydrogen, helium, and carbon. Ask: What are the electrical charges of the three particles that make up these atoms? **protons—positive, electrons—negative, neutrons—no charge.** What is the overall charge of the nucleus of each atom and of the atom as a whole? **Hydrogen nucleus is +1, helium nucleus is +2, and carbon nucleus is +6. All three atoms have no overall charge.**

More about...

Electrons and Energy Levels
The row number of an element in the periodic table (pp. 698–699) indicates how many energy levels its atoms have. For example, potassium (K), is in row 4 and has four energy levels. As the number of energy levels increases, the radius of an atom also increases. For example, the bromine atom (Br) is much larger than the chlorine atom (Cl).

DIFFERENTIATING INSTRUCTION

Hands-On Demonstration Demonstrate to students that matter, while it must have mass and volume, need not be visible. Use a balance to find and record the mass of a deflated balloon. Inflate the balloon, then find its mass again. Ask students to account for the increased mass of the inflated balloon. Guide students in recognizing that the air in the balloon, although invisible, has mass and takes up space and, therefore, is matter.

Support for Visually Impaired Students Enlarge the illustrations of the hydrogen, helium, and carbon atoms to help students with limited vision. You also might need to trace over parts of the illustrations to make them darker. The electrons may be particularly difficult to distinguish.

CHAPTER 5 SECTION 1

Visual Teaching

Interpret Diagrams

Mention to students that the box containing information on carbon represents one "cell" of the periodic table. Point out the atomic number (6) and have students compare it to the number of protons in the nucleus and electrons in the electron cloud of the carbon atom. Then ask: What important characteristic of a carbon atom results from this equal number of protons and electrons? **The atom is neutral.**

Extend

Have students locate carbon in the periodic table on pages 698–699. Ask: In which row and group (column) of the periodic table does it appear? **row 2, group 14 or 4A** Then have them use the information in the periodic table as a reference to find the symbol, atomic number, atomic mass, and the number of protons, electrons, and neutrons of potassium and fluorine. **Potassium—symbol K, atomic number 19, atomic mass 39.098, 19 protons, 19 electrons, and 20 neutrons. Fluorine—symbol F, atomic number 9, atomic mass 18.998, 9 protons, 9 electrons, and 10 neutrons.**

Classifying Atoms

The periodic table, a tool used to organize information about the elements, appears on pages 698–699 of the Appendix. Of the more than 100 known elements listed there, 92 occur naturally on Earth in significant amounts. The rest are synthetic elements produced by scientists. In each row of the periodic table, elements are listed from left to right in order of increasing numbers of protons. The vertical columns, called *groups,* comprise elements that have similar chemical properties. The location of an element in the table allows you to predict how it will react with other elements.

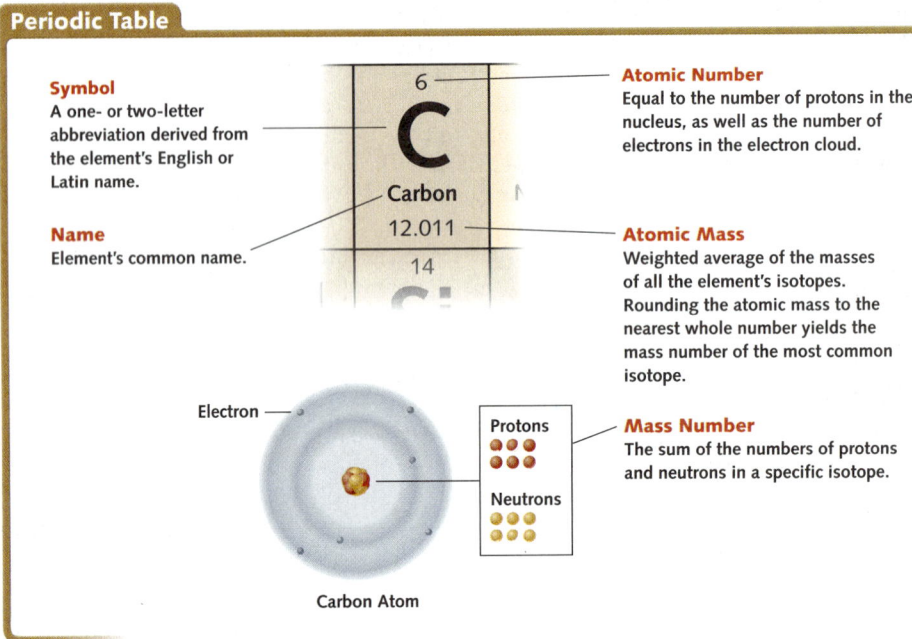

Periodic Table

Symbol A one- or two-letter abbreviation derived from the element's English or Latin name.

Name Element's common name.

Atomic Number Equal to the number of protons in the nucleus, as well as the number of electrons in the electron cloud.

Atomic Mass Weighted average of the masses of all the element's isotopes. Rounding the atomic mass to the nearest whole number yields the mass number of the most common isotope.

Mass Number The sum of the numbers of protons and neutrons in a specific isotope.

Carbon Atom

VOCABULARY STRATEGY

The word *isotope* is of Greek origin. *Iso-* comes from the word *īsos,* which means "equal," and *tope* comes from *topos,* which means "place."

The identity of an atom depends only on the number of protons, not on the number of neutrons. Many elements have atoms with the same number of protons but different numbers of neutrons. Atoms of an element that have different masses are **isotopes** (EYE-suh-TOHPS). The **mass number** of any given isotope is the sum of the numbers of protons and neutrons in that isotope. An atomic mass is a proportional average of the naturally occurring masses of an element's isotopes. This is why an atomic mass often includes a fractional part. For example, the average mass of carbon atoms is 12.011 atomic mass units (amu).

By learning how to interpret data in the periodic table, you can determine the number of particles in any element's atoms. For example, potassium's atomic number is 19, and the mass number of its most common isotope is 39. The potassium nucleus, therefore, contains 19 protons. By subtracting 19 from 39, you can find the number of neutrons; potassium has 20. How many

---

DIFFERENTIATING INSTRUCTION

Developing English Proficiency As students read "Classifying Atoms," they might have difficulty understanding these phrases: *weighted average, proportional average,* and *fractional part.* The phrases *weighted average* and *proportional average* mean the same thing. The average value used for atomic mass in the periodic table takes into account that there is a far greater number of carbon-12 atoms in existence (98.9%) than there are carbon-13 or carbon-14 atoms. The *fractional part* referred to in text is the decimal used to express this weighted average, the atomic mass shown in the illustration Periodic Table.

electrons surround the nucleus? Because the number of electrons in a neutral atom equals the number of protons, the answer is 19 electrons.

Hydrogen has three isotopes. The most common has one proton and no neutrons in its nucleus. Its atomic number is 1 and its mass number is also 1. However, hydrogen has a second isotope with one proton and one neutron in its nucleus. Its atomic number is 1, but its mass number is 2. This isotope is known as heavy hydrogen or deuterium (doo-TEER-ee-uhm). It is much less common than ordinary hydrogen. The third isotope of hydrogen is even rarer. Known as tritium (TRIHT-ee-uhm), it has a mass number of 3. Its nucleus contains one proton and two neutrons.

You may have heard of carbon–14, a "heavy" isotope of carbon. Each ordinary atom of carbon–12 has a nucleus consisting of six protons and six neutrons. A carbon–14 atom, with a mass of 14, has a nucleus with six protons and eight neutrons.

Uranium has several isotopes, all of which have 92 protons in the nucleus. The most commonly occurring isotope, Uranium–238, has 146 neutrons in its nucleus; Uranium–235, however, has only 143 neutrons.

Bonding of Atoms

Most substances on Earth are not pure elements but rather compounds. A **compound** is a substance that contains atoms of two or more elements that are chemically combined.

Understanding more about the energy levels in atoms can help you see why elements combine to form compounds. Atoms tend to be most stable when their outermost energy level is filled with electrons. For example, atoms of helium, neon, and the other elements in the right-most column of the periodic table naturally have their outermost energy level filled. For this reason, these elements are very stable and do not readily combine with other elements to form compounds. In contrast, the atoms of most other elements tend to fill their outermost energy level by gaining, losing, or sharing electrons with other atoms. The gain, loss, or sharing of electrons forms a chemical bond that holds atoms together. There are three main types of bonds: (1) covalent bonds, (2) ionic bonds, and (3) metallic bonds.

COVALENT BONDS

Some compounds form when atoms share electrons. The type of bond formed by sharing electrons is called a *covalent bond*. Two or more atoms held together by covalent bonds form a **molecule.** For example, in a molecule of water, two atoms of hydrogen and one atom of oxygen share electrons. In a molecule of carbon dioxide, one carbon atom and two oxygen atoms share electrons.

In some cases, atoms of the same element also form covalent bonds with each other to form molecules. For example, the two most abundant gases in Earth's atmosphere, nitrogen and oxygen, both form such molecules. Nitrogen naturally occurs as molecules of two atoms each (N_2), and oxygen also most commonly exists as molecules of two atoms (O_2).

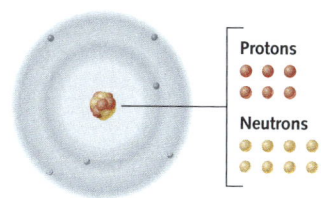

CARBON 14 is an isotope of carbon, with two more neutrons than ordinary carbon.

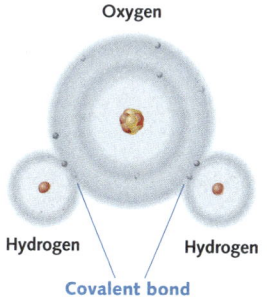

COVALENT BOND The two hydrogen atoms and one oxygen atom of a water molecule are held together by covalent bonds.

CHAPTER 5 SECTION 1

VISUAL TEACHING
Discussion

Have students compare the number of neutrons shown in the carbon 14 atom on page 93 to the number of neutrons in the carbon atom on page 92. Ask what effect this increase in the number of neutrons has on the atomic number of carbon and why. none; The atomic number is equal to the number of protons, which doesn't change. Ask what the number of neutrons affects. atomic mass Ask which form of carbon is most prevalent. carbon-12; Note that the atomic mass is very close to the whole number 12; 98.9% of all carbon is carbon-12.

DISCUSSION

Students may think of a chemical bond as a mechanical link between atoms. The model kits used in classrooms may reinforce this misconception because they typically use pieces of molded plastic that lock into place. However, the attachment in a chemical bond is very different. Emphasize that a chemical bond is produced by the attractive force that exists between the protons of one atom and the electrons of another.

DIFFERENTIATING INSTRUCTION

Support for Visually Impaired Students Have pairs of students use common materials, such as gumdrops and toothpicks, to construct atomic models of different isotopes of hydrogen, including hydrogen-1, deuterium, and tritium. Have students label the protons, electrons, and neutrons in each isotope. These models supplement the text description and will be helpful for review.

CHAPTER 5 SECTION 1

Visual Teaching

Interpret Diagram
Tell students that both sodium and chlorine in their elemental forms are poisonous. Have students look at the pictures of sodium and chlorine and note other characteristics of each. **sodium—soft, silvery white solid; chlorine—greenish yellow gas** Then have students note the characteristics of the sodium chloride. **white, hard crystal** Remind students that sodium chloride is a substance they probably eat every day. Emphasize that when a new substance forms during a chemical reaction, it has its own set of properties that are different from those of the reactants.

VISUALIZATIONS
CLASSZONE.COM
Observe the arrangement of molecules in minerals.
Keycode: ES0502

IONIC BONDS

Other compounds are held together by the force of electrical attraction between atoms that have lost or gained electrons. In the neutral state, atoms have equal numbers of protons and electrons. If an atom gains one or more electrons, it becomes negatively charged. But if an atom loses one or more electrons, it becomes positively charged. Such a charged atom is called an **ion** (EYE-uhn). Groups of atoms may also form ions.

Because opposite charges attract each other, ions of opposite charges bond to form compounds. The force of attraction, or ionic bond, between the oppositely charged ions holds them together. Ionic bonds are common in many minerals. For example, positively charged sodium ions are bonded with negatively charged chlorine ions in the compound sodium chloride, or table salt. Sodium chloride is found in nature as the mineral halite.

How do atoms lose or gain electrons to form ions? Consider the element sodium. You can see from the diagram below that the sodium atom's outer energy level contains only one electron. A stable, or nonreactive, element would have its outer energy level completely filled with electrons. Sodium reacts with many elements that cause it to lose this outer electron. Chlorine is reactive for the opposite reason. Its atom's outer energy level is short one electron, so it reacts with other elements to gain an electron. When sodium and chlorine react, sodium atoms lose their outer electrons to chlorine atoms' outer energy level.

A **metal** is an element that loses electrons easily to form positive ions. Among the metallic elements are aluminum, sodium, potassium, calcium, zinc, gold, silver, iron, copper, and lead. Metals dominate the periodic table. Ionic bonds do not form between metals.

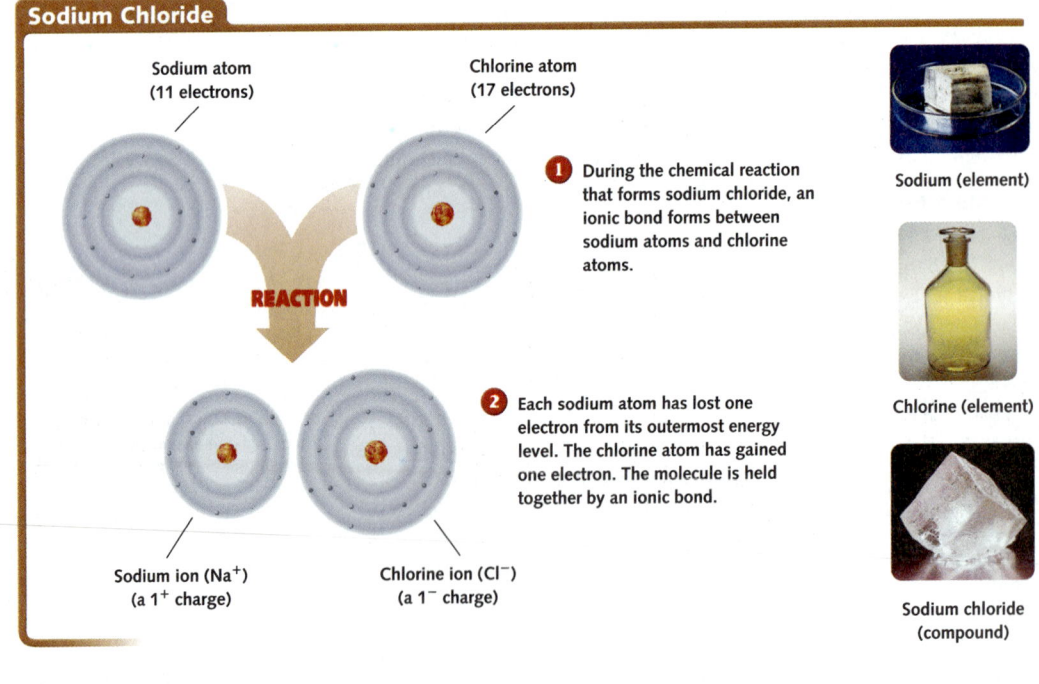

Sodium Chloride

Sodium atom (11 electrons)
Chlorine atom (17 electrons)

REACTION

1. During the chemical reaction that forms sodium chloride, an ionic bond forms between sodium atoms and chlorine atoms.

2. Each sodium atom has lost one electron from its outermost energy level. The chlorine atom has gained one electron. The molecule is held together by an ionic bond.

Sodium ion (Na^+) (a 1^+ charge)
Chlorine ion (Cl^-) (a 1^- charge)

Sodium (element)
Chlorine (element)
Sodium chloride (compound)

DIFFERENTIATING INSTRUCTION

Hands-On Demonstration Place two small magnets on an overhead projector. Tell students that the negatively charged side of one magnet is facing the positively charged face of the other. Ask: What will happen as the two magnets come close to each other? Test their predictions by moving the magnets toward each other. Use the reactions of opposite charges as an analogy for the attractive force between ions that have opposite electrical charges. Further this discussion by facing the ends of the magnets with the same charge toward each other. Have students predict what will happen. Ask: What types of elements—metals or nonmetals—are represented by the positively and negatively charged ends of the magnets? **The positively charged end is a metal; the negatively charged end is a nonmetal.**

An element that gains electrons easily to form a negative ion is classified as a **nonmetal.** Although ionic bonds form most readily between metals and nonmetals, compounds do form from ions of different nonmetals. Nonmetals include nitrogen, oxygen, fluorine, chlorine, phosphorus, and sulfur. Nonmetals predominate the right side of the periodic table.

METALLIC BONDS

The bonds that form between metal atoms have different characteristics from bonds that form between metal and nonmetal atoms. In pure metallic minerals such as iron, the iron atoms exist as positively charged ions. Rather than moving distinctly around one nucleus, the outer electrons move freely around all the positive ions. This arrangement of electrons is sometimes described as a *sea of electrons*. The free movement of electrons around metal ions accounts for some typical properties of pure metals, such as high conductivity.

Compounds and Mixtures

A compound can have properties entirely unlike those of the elements of which it is made. For example, water is certainly different from either hydrogen or oxygen. At room temperature, hydrogen and oxygen are gases, while water is a liquid. Another example is table salt, which is a compound of the elements sodium and chlorine. Both sodium and chlorine are poisonous. Yet when these two elements react chemically, they form table salt, a compound most people can eat safely with their food.

Compounds should not be confused with mixtures. In a mixture, the individual elements or compounds keep their own properties and can be present in any proportion. Most mixtures can be separated by physical means. Salt water is an example of a mixture. It can be separated by evaporating the water. The elements in a compound, however, can be separated only by chemical means. For example, water can be decomposed into hydrogen and oxygen by breaking the chemical bonds with a strong electric current.

5.1 Section Review

1. What are the characteristics of matter?
2. Compare and contrast a proton and a neutron. How are they alike? How are they different?
3. Name and describe the three models of chemical bonds.
4. **APPLICATION** Suppose you are given two liquids and told that one is a compound and the other is a mixture. How might you determine which was which? Describe one method you could use.
5. **MATHEMATICS** Use the periodic table on pages 698–699 to determine the numbers of protons, neutrons, and electrons in an atom of silver (Ag).

CHAPTER 5 SECTION 1

Observe the arrangement of molecules in minerals.
Keycode: ES0502

 Visualizations CD-ROM

ASSESS

1. Matter has mass and volume.
2. Differences—Proton is positively charged; neutron has no charge. Neutron has more mass than proton. Number of protons defines the atom; number of neutrons usually does not affect the atom's identity or reactivity. Similarities—Both are subatomic particles found in the nucleus.
3. Ionic bond—Force of attraction between oppositely charged ions holds them together; electron pair is dominantly possessed by one atom over the other; often occurs between a positive metal ion and a negative nonmetal ion. Metallic bond—forms between metal ions; outer electrons move freely around all the positive ions. Covalent bond—forms when two elements share electrons.
4. (a) Check for cloudiness; if it is cloudy, the substance is a mixture. (b) If material settles out of the liquid, it is a mixture. (c) Filter it through filter paper; compounds will not leave sediment behind.
5. The atomic number for silver is 47, so silver has 47 protons and 47 electrons. Rounding off the atomic mass (107.868) gives the mass number of its most common isotope: 108. The number of neutrons in this isotope is equal to 108 minus 47: 61 neutrons.

MONITOR AND RETEACH

If students miss . . .

Question 1 Reteach "Matter." (p. 90) Ask: Air takes up space and has mass, but you can't see it. Is air matter?

Question 2 Reteach atomic structure using the illustrations of hydrogen, helium, and carbon atoms. (p. 91) Have students make a Venn diagram to compare and contrast protons and neutrons.

Question 3 Reteach "Bonding of Atoms." (pp. 93–95) Have students compare what is happening in the illustrations that show the formation of sodium chloride (p. 94) and of water (p. 93).

Question 4 Have students define *mixture* and *compound*. (p. 95) Then have them list ways to identify each.

Question 5 Reteach "Classifying Atoms." (pp. 92–93). Have students use the periodic table to determine the number of protons, neutrons, and electrons in the atoms of other elements.

CHAPTER 5 SECTION 2

FOCUS

Objectives
1. Identify the characteristics of minerals.
2. Explain how minerals form.
3. List the physical characteristics of minerals that are influenced by their crystalline structure.

Set a Purpose
Have students read this section to find the answer to "What is a mineral, and how do minerals form?"

INSTRUCT

More about...

Mineral Composition
The chemical composition of a specific mineral is not always identical in different samples. In some minerals, one element can substitute for another having the same charge and being of similar size. For example, in olivine (Fe_2SiO_4), iron can be replaced by magnesium without changing the mineral itself, resulting in Mg_2SiO_4. In this case, both iron and magnesium have a charge of 2+. Some forms of olivine may contain both iron and magnesium, while other forms contain trace amounts of copper, nickel, cobalt, manganese, and other elements as ionic substitutes for iron and magnesium.

5.2 Composition and Structure of Minerals

KEY IDEA
A mineral is a naturally occurring element or compound that is inorganic and crystalline in structure.

KEY VOCABULARY
- mineral
- crystal
- silicate
- silica tetrahedron
- cleavage

What Is a Mineral?

All the rocks and minerals of Earth's crust consist of elements. How are minerals different from other forms of matter? A **mineral** has the following characteristics:

1. It occurs naturally.
2. It is solid.
3. It has a definite chemical composition.
4. Its atoms are arranged in an orderly pattern.
5. It is inorganic (was never alive).

There are about 4000 known minerals, and each satisfies the conditions listed above. You may already be familiar with the minerals quartz, halite (rock salt), gold, and diamond. Many materials found on Earth are not minerals, however. Water at room temperature is not a mineral because it is not a solid. Coal is not a mineral because it is made from plant remains, it lacks a definite composition, and its atoms are not arranged in an orderly way. (Although they are produced by living things, the shells of such marine animals as clams are composed of minerals.)

Of all the elements in Earth's crust, a mere eight make up 98.5 percent of the crust's total mass. These elements, which are the ones most common in minerals, are listed in the table below. More than 90 percent of the minerals in Earth's crust are compounds containing oxygen and silicon, the two most abundant elements.

Most minerals are compounds. Quartz is a compound of silicon and oxygen. The mineral galena is a compound of lead and sulfur. A few minerals, however, consist of single elements and are called native elements. Examples are silver (Ag), copper (Cu), sulfur (S), and diamond (C). Often different types of minerals (compounds and native elements) are found mixed together. Such mixtures of minerals are called rocks.

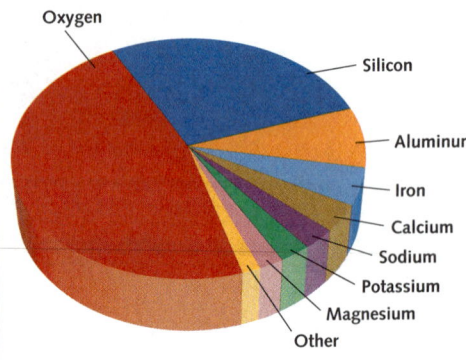

COMMON ELEMENTS The circle graph (below) and the chart (right) show the eight most common elements in Earth's crust by mass. Compare the amount of oxygen and silicon to the other elements.

Common Elements of Earth's Crust

Name	Element Symbol	Percent by Mass
Oxygen	O	46.6
Silicon	Si	27.7
Aluminum	Al	8.1
Iron	Fe	5.0
Calcium	Ca	3.6
Sodium	Na	2.8
Potassium	K	2.6
Magnesium	Mg	2.1
Other	–	1.5

DIFFERENTIATING INSTRUCTION

Reading Support Direct students to the numbered list of five characteristics of minerals on page 96. Model how the information in this list tells not only what a mineral is but what it isn't. For example, ask: Is diamond a mineral? Is synthetic diamond a mineral? A naturally occurring diamond is a mineral (a native element), a synthetic diamond is not. Scientists consider minerals to be the building blocks of rock, and in that context a synthetic mineral makes no sense. Is ice a mineral? is water? In a technical sense, ice meets all the criteria of a mineral, while water does not. Ask: Is ice likely to be found in a rock? no, because of the temperatures under which rocks form Ask students about the minerals listed in dietary supplements. No; these single elements (calcium, sodium, iron, for example) may be part of a mineral's composition, copper even occurs in Earth's crust as a native element, however these minerals are not rock-forming minerals.

Use Reading Study Guide, p. 14.

How Minerals Form

Minerals can form in several ways. Many minerals form out of molten rock, or magma, in which atoms, molecules, and ions can move freely. As magma cools, the atoms, molecules, and ions move closer together and form chemical bonds that create compounds.

Many kinds of minerals can form out of a single magma mass. The types of minerals that form depend in part on the types and amounts of elements that are present in the magma. The rate at which the magma cools determines the size of the mineral grains that form.

Mineral-forming processes can be gradual and lengthy. Some types of minerals form as water containing dissolved ions slowly evaporates. For example, halite forms when water evaporates from a solution of salt and water. As the water molecules evaporate, sodium and chlorine ions bond to form the mineral halite. New minerals also form when existing minerals are transformed by heat, pressure, or chemical action.

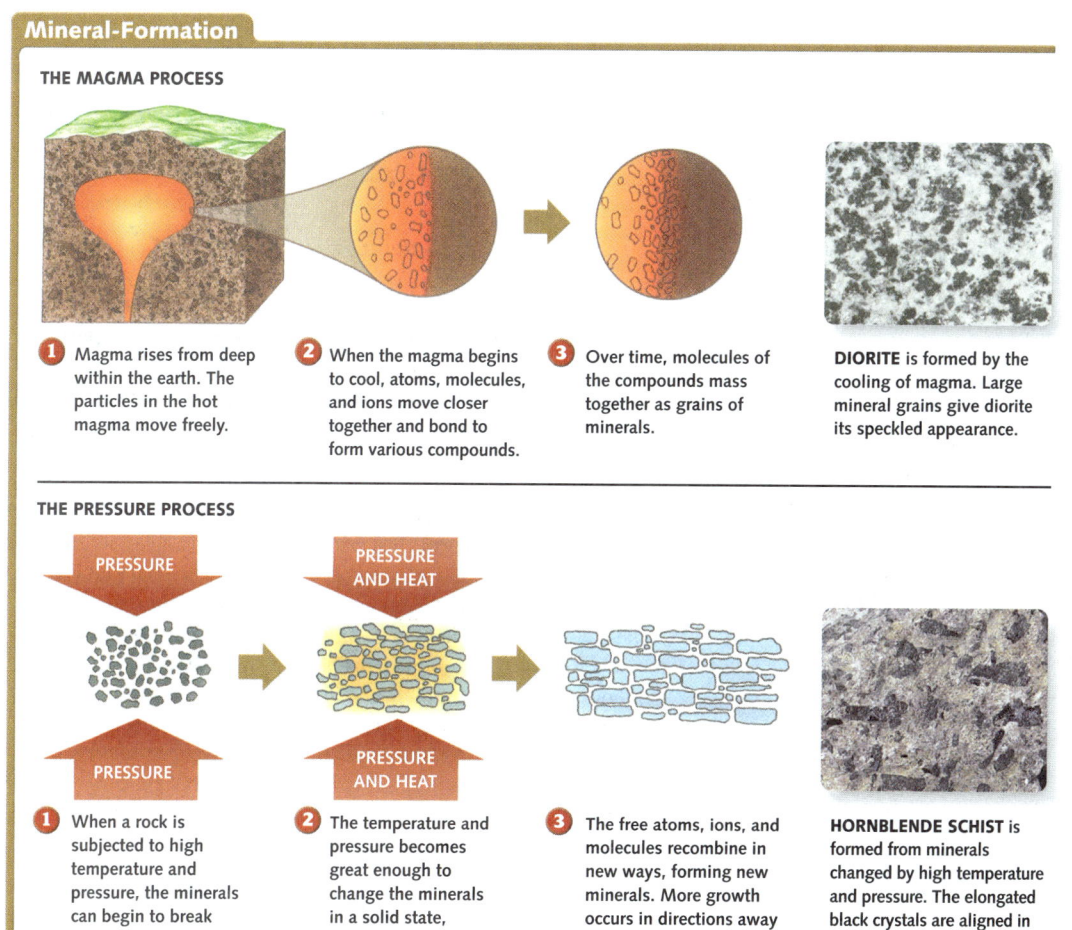

Mineral-Formation

THE MAGMA PROCESS

1. Magma rises from deep within the earth. The particles in the hot magma move freely.
2. When the magma begins to cool, atoms, molecules, and ions move closer together and bond to form various compounds.
3. Over time, molecules of the compounds mass together as grains of minerals.

DIORITE is formed by the cooling of magma. Large mineral grains give diorite its speckled appearance.

THE PRESSURE PROCESS

1. When a rock is subjected to high temperature and pressure, the minerals can begin to break down chemically.
2. The temperature and pressure becomes great enough to change the minerals in a solid state, without melting them.
3. The free atoms, ions, and molecules recombine in new ways, forming new minerals. More growth occurs in directions away from the pressure.

HORNBLENDE SCHIST is formed from minerals changed by high temperature and pressure. The elongated black crystals are aligned in one direction.

Chapter 5 Atoms to Minerals   97

CHAPTER 5 SECTION 2

Visual Teaching

Interpret Diagrams

It is important that students understand that the two mineral-forming processes diagrammed can occur during the same event, depending on how high temperatures go. Once melting occurs, the pressure process gives way to the magma process. Have students compare the two processes, using the illustration. The magma process involves molten rock that forms minerals as magma rising toward Earth's surface begins to cool. In the pressure process, minerals form in a solid state, as high pressures and temperatures cause rock material to break down and reform. Direct students' attention to part 3 of the magma process and ask: Why do the compounds mass together as the magma cools? The temperatures at which minerals crystallize vary, causing different minerals to separate into grains at different times.

Extend

Ask: How does the rate at which magma cools affect the size of the grains? If magma cools rapidly, the minerals do not have time to form large grains, and the minerals solidify into smaller grains.

DIFFERENTIATING INSTRUCTION

Developing English Proficiency Point out that some words can take on very distinct and restrictive meanings when used by a scientist. The word *mineral* is a good example. To a nutritionist, a mineral is an inorganic substance (calcium, sodium, zinc) mostly found in plant or animal food sources. A miner may think of coal or uranium as a mineral. None of these materials would meet the earth scientist's definition of *mineral*. Because dictionary definitions are often generalized, a scientific dictionary may be needed to obtain a scientifically accurate definition.

CHAPTER 5 SECTION 2

CRITICAL THINKING

After students have had a chance to review the text on page 98 and the chart on page 99, ask: What do the axes shown in the diagrams of crystal systems represent? **The minerals "grow" along these axes as they form. These axes, then, help to define the mineral's overall shape, assuming they have the space to grow freely.** Point out the photograph of quartz on page 98. Ask: How does the shape of the quartz crystals conform to the hexagonal crystal system shown on page 99? **The hexagonal shape is visible, especially along the sides of the more fully formed, upright crystals.** Ask: In what way do these crystals not conform to the shape of the crystal system shown in the chart? **The ends are pointed, not squared off as in the diagram.** Ask: Why is that? **The shape of the crystal system represents the parameters within which the crystal will form, but since minerals grow over time and often within confined spaces, they do not necessarily "fill out" to that "complete" form.**

Structure of Minerals

The orderly arrangement of atoms in a mineral is often apparent in the mineral's shape. You may be most familiar with minerals in the form of beautiful crystals. A **crystal** is a regular geometric solid with smooth surfaces, called crystal faces. By definition, all minerals have crystalline structures—that is, regular, orderly arrangements of atoms.

Crystal Structure

What is the crystalline structure of halite, commonly known as rock salt? Recall that halite is composed of positively charged sodium ions bonded ionically to negatively charged chlorine ions. In the resulting pattern, illustrated below, each sodium ion is surrounded by six chlorine ions, and each chlorine ion is surrounded by six sodium ions. This pattern is repeated throughout the mineral. As a result of this arrangement of ions, halite typically occurs in crystals that have a cubic shape.

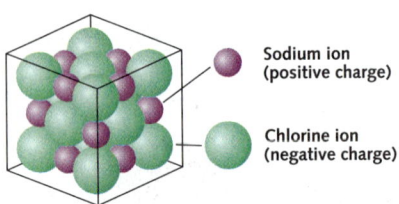

CRYSTAL STRUCTURE OF SALT

As in halite, it is the orderly arrangement of ions, molecules, or atoms in every mineral that determines the shape of its crystals. Each type of mineral has its own crystal form. For example, quartz, consisting of silicon and oxygen atoms bonded together, may form long six-sided crystals. All quartz crystals, from microscopic to hand-size or larger, have the same six-sided shape.

The angle at which crystal faces meet is characteristic for each type of mineral and can be used to help identify the mineral. Halite's crystal faces meet at right angles. In quartz, the crystal faces meet at angles of 120°. Why don't all crystalline substances have crystal faces? If space is limited when a mineral is forming, there may not be enough room for crystals with well-developed faces to grow. The mineral simply fills the available space. Such a mineral is still crystalline, but crystal faces are not visible.

Although there are thousands of different types of minerals, their crystals have only six basic types of shapes. Imaginary lines called crystallographic axes are used to distinguish the six systems of crystal shapes. Each axis connects the centers of a pair of opposite faces, passing through the center of the crystal. In crystals of the cubic system, including halite crystals, three axes equal in length intersect at right angles.

HEXAGONAL CRYSTALS
Under the right conditions, quartz crystals will exhibit their characteristic hexagonal shape.

DIFFERENTIATING INSTRUCTION

Developing English Proficiency The word *oblique* is used in the illustration of crystal systems to describe a type of angle: any angle under 180° that is not a 90° angle. Oblique angles can be either acute (less than 90°) or obtuse (over 90° but under 180°). All three of these words also are used in English as adjectives that do relate, in a general way, to their mathematical meanings. *Oblique* describes something that is indirect or not straightforward, for example an oblique remark. *Acute* is used to describe something that is sharp or severe, for example an acute pain or shortage. *Obtuse* is used to describe something or someone lacking sharpness, and so can mean insensitive or stupid.

Six Crystal Systems

Cubic System

Three axes of equal length intersect at 90° angles.
Examples: sylvite, halite

Sylvite

Orthorhombic System

Three axes, each of a different length, intersect at 90° angles.
Examples: topaz, sulfur

Topaz

Tetragonal System

Three axes intersect at 90° angles. Two of the axes are equal in length.
Examples: wulfenite, zircon

Wulfenite

Triclinic System
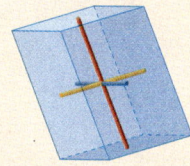
Three axes, each of a different length, intersect at oblique angles.
Examples: kyanite, turquoise

Kyanite

Hexagonal System
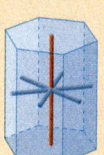
Three axes of equal length intersect at 60° angles. The fourth, vertical axis can be longer or shorter than the other axes.
Examples: emerald, graphite

Emerald (a form of beryl)

Monoclinic System
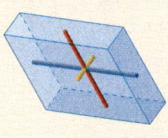
Three axes, each of a different length, intersect, with two of the intersections at 90° angles and the third at an oblique angle.
Examples: gypsum, mica

Gypsum

Axes of equal length are shown in the same color.

VOCABULARY STRATEGY
Prefixes in words that describe geometric shapes can often give you clues to the words' meanings. For example, *mono-* means "one" and *clin* means "incline." Therefore, *monoclinic* means "having one axis at an inclined (oblique) angle." In *tetragonal*, *tetra-* means "four," and *gonal* means "angle." Each face of a tetragonal shape has four angles.

CHAPTER 5 SECTION 2

VISUAL TEACHING
Discussion
Review with students the terms *90° angle* and *oblique angle*. **Any two axes that intersect so as to create equal angles, create angles of 90° and are said to be perpendicular; any angle less than 180° that is not perpendicular is an oblique angle.** Have students cover the last two columns of the chart and identify where axes in each system meet at 90° angles and at oblique angles. **90° angles are shown in yellow.** Ask: How do the angles affect the overall crystal-system shape? **Perpendicular angles alone create cubes or rectangles. When oblique angles are part of the system (triclinic, hexagonal, monoclinic), they create parallel sides at a slant.**

DIFFERENTIATING INSTRUCTION

Support for Visually Impaired Students Pair visually impaired students with sighted students, and have them use the sticks from organic chemistry modeling kits to model the axes of different crystal systems. Make sure students understand that the end of each stick represents the center of the plane that forms one of the faces of the crystal system.

Challenge Activity Provide pairs of students with an unknown mineral. Tell them that they have until the end of the chapter to identify the mineral, using what they learn in the chapter to do so. Provide reference materials to help with this challenge. Then have students display their mineral along with a card identifying it and giving information about its uses and properties.

CHAPTER 5 SECTION 2

More about...

Tetrahedrons
In 1947, R. Buckminster Fuller (1895–1983) patented an architectural structure called the geodesic dome, which is based on the tetrahedron. The dome is constructed of a network of interconnected tetrahedrons made of lightweight material. This design evenly distributes stress across the entire structure. The geodesic dome, free of interior structural supports, exhibits a high strength-to-weight ratio. Buildings that employ this architecture include the Climatron in St. Louis, Missouri, the Astrodome in Houston, Texas, and the United States Pavilion built for Expo 67 in Montreal, Canada.

VISUAL TEACHING
Discussion
Have students compare the ball-and-stick model of the silica tetrahedron to the space-filled model. Ask how the models differ. *The ball-and-stick model shows the bonds and bond angles, as viewed from the side. The space-filled model shows how much of the space within the compound is filled by each atom; it is viewed from below.*

Use Transparency 6.

VISUALIZATIONS CLASSZONE.COM
Examine 3-D models of common molecules.
Keycode: ES0503

SILICA TETRAHEDRON
The arrangement of silica tetrahedra determines many properties of silicate minerals, including cleavage. Several arrangements are shown in the table on page 101. For all but the first arrangement, oxygen atoms are shared by adjacent tetrahedra.

Silicates

Silicon and oxygen are the two most abundant elements in Earth's crust, so it is not surprising that most minerals contain these elements. Minerals that are compounds including silicon and oxygen are called **silicates.** A silicate may also contain one or more metallic elements, such as aluminum or iron. A few silicates do not contain metal. For example, quartz is composed only of oxygen and silicon. More than 90 percent of the minerals in Earth's crust are silicates.

The basic building block of a silicate is the **silica tetrahedron,** consisting of four oxygen atoms packed closely around a silicon atom. This unit is named for its shape. As shown in the model below, imaginary lines connecting the four oxygen atoms form a geometric figure called a tetrahedron. A silica tetrahedron is held together by covalent bonds between the silicon atom and the oxygen atoms. All silicates are composed of these tetrahedra, although the tetrahedra may be arranged in various ways. The table on the next page shows how silicates are classified according to the different arrangements of tetrahedral units. Note that the metals contained in silicate compounds are not considered in the classification.

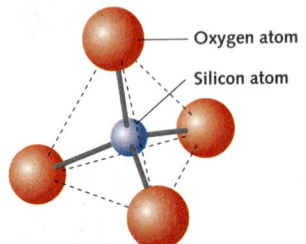

BALL-AND-STICK MODEL **SPACE-FILLED MODEL**

Crystal Structure and Physical Properties

As a result of their crystalline structures, minerals are solid. The atoms, ions, and molecules in minerals are closely packed, bound by strong chemical bonds. An increase in temperature, however, may weaken the bonds between particles. At high temperatures, minerals melt, becoming liquids made up of loose groups of particles. At even higher temperatures, minerals vaporize, becoming gases in which individual particles are far apart. The temperatures at which a mineral melts and vaporizes are characteristic of the mineral and can sometimes be used to differentiate two different minerals of similar appearance.

Crystal structure also determines a mineral's **cleavage,** or tendency to split in particular directions. The planes along which the mineral splits correspond to planes of weak bonds between the atoms, ions, or molecules of the mineral. Halite splits into cubes between layers of ions. Quartz, with its strong network of atoms, does not split along any plane.

DIFFERENTIATING INSTRUCTION

Hands-On Demonstration Have students use a microscope or magnifying lens to observe grains of table salt. Ask: What properties can you see with and without magnification?

⚠ *Exercise caution when working with microscope slides. Remind students that safety is everyone's responsibility.*

Support for Physically Impaired Students If students have difficulty with manual dexterity, mount magnifying lenses on ring stands so that students can look at salt and other crystals. Place the samples in large, flat boxes so that students can more easily move the boxes.

Molecular Structures of Some Common Silicate Minerals

	Structure	Cleavage	Mineral
Olivine Group	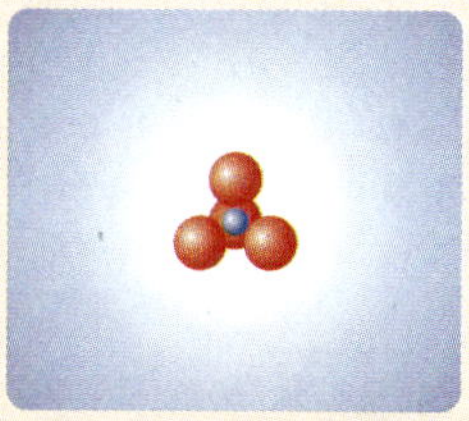	Olivine has no cleavage.	 Olivine
Beryl	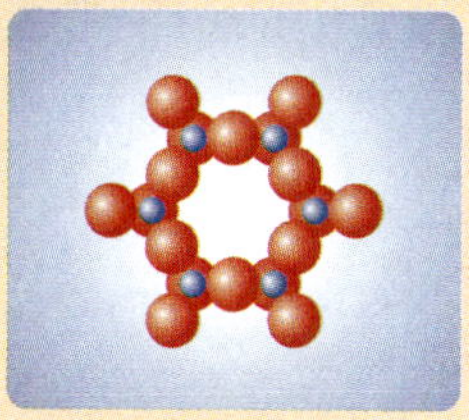	Beryl has one imperfect cleavage.	 Beryl
Pyroxene Group	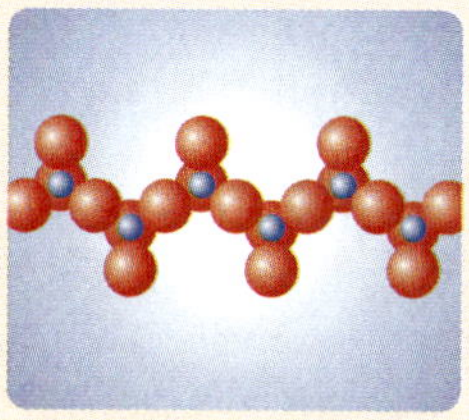	Diopside has two perfect cleavages, at close to 90° angles.	 Diopside
Amphibole Group	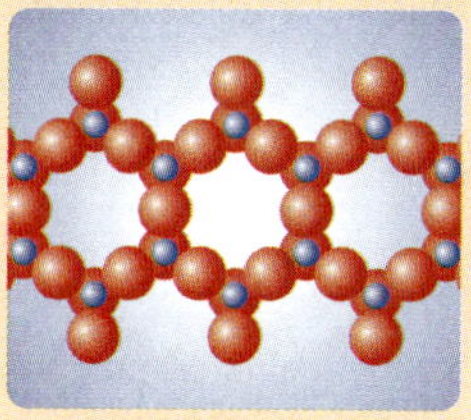	Tremolite has one perfect and one imperfect cleavage at close to 60° and 120° angles.	 Tremolite
Mica Group	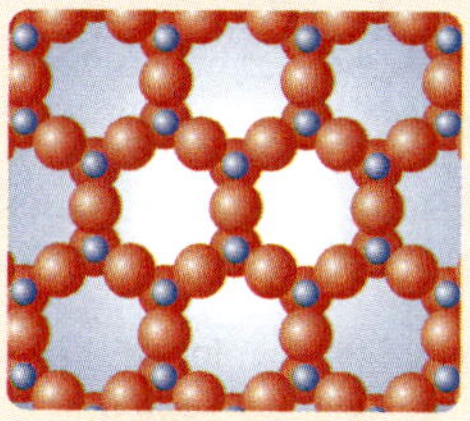	Micas exhibit perfect cleavage in one direction.	 Muscovite Mica
Feldspar Group	Too complex to draw	Microcline feldspar has two good cleavages, at 90° angles.	 Microcline Feldspar

CHAPTER 5 SECTION 2

Examine 3-D models of common molecules.
Keycode: ES0503

 Visualizations CD-ROM

VISUAL TEACHING

Discussion

Make sure students understand that the silica tetrahedron (shown in this illustration as a space-filled model) is the basic building block of minerals. Point out how in the more complex structures shown, the tetrahedra are bound together by a shared oxygen atom (red). As a mineral forms, more silica tetrahedra are layered on. The weaker the bonds between the layers of the tetrahedra, the more likely the mineral is to cleave along those layers or planes. A mineral's cleavage is classified based on the quality of surfaces produced and the ease of cleaving. *Perfect* cleavage results in extremely high quality, smooth surfaces. *Fair* to *poor* cleavages are irregular and result in the mineral easily breaking along directions other than the cleavage planes.

Extend

Make sure students understand that the crystal faces that result from mineral growth are distinct from planes of cleavage. However, since cleavage does follow the symmetry of a mineral, planes of cleavage are often parallel to crystal faces.

Use Transparency 6.

DIFFERENTIATING INSTRUCTION

Reading Support Make sure students understand the meaning of the words *imperfect* and *perfect* as they relate to cleavage. Then have them prepare molecular structure cards for the structures shown. On one side of an index card, they should list the group name; on the other side they can draw the structure and give information about cleavage.

CHAPTER 5 SECTION 2

ASSESS

1. naturally occurring; solid at room temperature; have a definite chemical composition; atoms of molecules are arranged in orderly patterns; have inorganic origins

2. Magma process—Magma rises and begins to cool; atoms, molecules, and ions move closer together and bond to form compounds, which mass together over time; example is diorite. Evaporation—As water evaporates, ions crystallize together; example is halite. Pressure process—High temperatures and pressure break down minerals in rock and they recrystallize in solid state, rearranging free atoms, ions, and molecules into new minerals; example is hornblende schist.

3. Answers should include three of the following: mineral's solid form, cleavage, hardness, and density.

4. The hardness of a mineral depends on the arrangement of its atoms. When carbon atoms are arranged in an interlocking network of silica tetrahedra, the result is a very hard, very strong mineral: diamond. When they are arranged in sheets of hexagonal networks, the result is a very soft, slippery mineral: graphite.

5. metals to which they bond; temperatures at which they form; pressures at which they form; speed at which the formation takes place

6. Answers will vary based on the region. Remind students that minerals in their region may have formed far away and been transported by water or glaciers.

The hardness of a mineral also depends on the arrangement of its ions, atoms, or molecules and on the strength of the chemical bonds between them. A good example of the relationship between hardness and crystalline structure is the two minerals that consist of the element carbon. When carbon atoms are arranged in a tetrahedral network, the result is diamond, the hardest natural mineral. Yet when carbon atoms are arranged in sheets of hexagonal networks, the result is graphite, a very soft mineral that flakes easily.

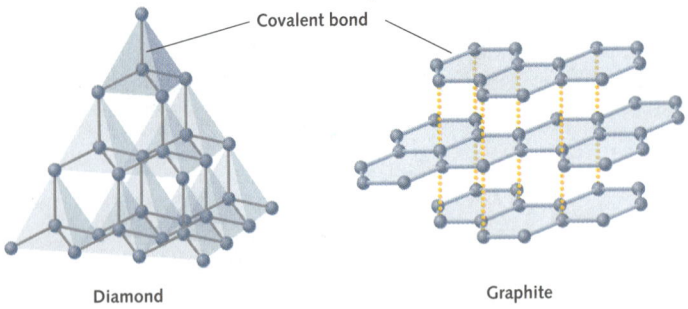

CARBON STRUCTURES

Recall that the density of a material is the ratio of its mass to its volume. Density depends not only on the masses of the atoms in the mineral but also on how they are arranged. For example, although both graphite and diamond are made of carbon atoms, the density of graphite is about 2.2 g/cm^3, whereas that of diamond is 3.5 g/cm^3.

5.2 Section Review

1. List the five characteristics of a mineral.
2. Describe two ways in which minerals are formed. Include an example of each process.
3. List three physical characteristics of a mineral that are influenced by its crystalline structure.
4. Both diamond and graphite are composed purely of carbon atoms. Explain why diamond is so much harder than graphite.
5. **CRITICAL THINKING** Silicate minerals have a variety of crystalline structures, even though they are made from the same building blocks—silica tetrahedra. What factors do you think affect the structures of silicate minerals as they are forming?
6. **GEOGRAPHY** What types of minerals are common in the region where you live? Speculate on the processes by which they might have formed.

MONITOR AND RETEACH

If students miss . . .

Question 1 Review the characteristics of minerals. (p. 96) Ask students to explain each characteristic.

Question 2 Have students review mineral formation including the diagrams. (p. 97) Have pairs of students retell each process in their own words.

Question 3 Have students reread "Crystal Structure and Physical Properties." (pp. 100–102) Have them outline what they read.

Question 4 Refer students to the Carbon Structures illustration (p. 102) and have them compare the crystalline structure of diamond and graphite.

Question 5 Have students reread "How Minerals Form" (p. 97) and "Silicates." (p. 100) Tell students to note any information describing factors that affect the crystalline structure of minerals.

Question 6 Have students research information about regional minerals at the library or on the Internet.

SCIENCE & Technology

A New Form of Carbon Creates a New Ball Game for Scientists

Until recently, scientists knew of only two forms of pure carbon: graphite and diamond. In 1985, however, a form of carbon with molecules resembling soccer balls bounced onto the scene. This new form of carbon didn't look like or behave like graphite or diamond.

Why do such discoveries excite scientists? In what ways can research on the unseen—atoms and molecules, for example—affect our daily lives?

Sometimes scientists make discoveries that surprise even themselves. Researchers trying to reproduce some unusual deep-space carbon chains placed graphite in a helium-filled chamber and then vaporized it with a laser. The result was a collection of sphere-shaped molecules of pure carbon. Most of these molecules were made up of 60 atoms of carbon. Each resembled a soccer ball, having 20 hexagonal sides and 12 pentagonal sides. Scientists named the substance buckminsterfullerene—and called its molecules buckyballs, in honor of R. Buckminster Fuller. A buckyball resembles one of the geodesic domes designed by this American architect.

The unusual electrical, structural, and chemical properties of buckminsterfullerene excited scientists. In the first years after its discovery, scientists proposed using it as an ingredient in substances ranging from photocopier toner to rocket fuel. But in tests it didn't always produce the desired results. Furthermore, buckminsterfullerene isn't cheap to produce. In October 2000 it cost more than $200 per ounce.

A more important breakthrough may have come in 1991, when a scientist manufacturing buckyballs instead produced "buckytubes," or nanotubes. Measuring one fifty-thousandth the width of a human hair, a nanotube consists of a layer of rolled-up graphite, usually sealed at each end with half a buckyball.

Nanotubes have remarkable electrical properties and are 100 times stronger than steel. These attributes, combined with their microscopic size, make nanotubes ideal for new technology. They can be used as the muscles of tiny robotic machines, and they can store energy generated by ocean waves or fuel cells. Nanotubes

GEODESIC DOMES like this one, which buckyballs resemble, were designed by architect R. Buckminster Fuller.

can transmit the photons for high-resolution, ultra-thin display screens and lead to the production of smaller and more powerful transistors for future computers. ■

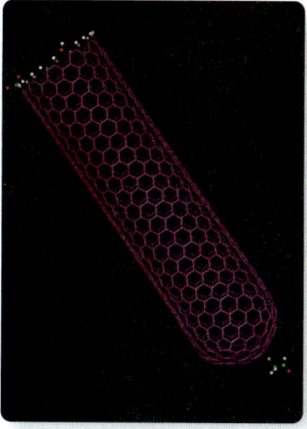

NANOTUBES, another form of carbon, are 100 times stronger than steel.

Extension

SCIENCE NOTEBOOK

Do some additional research on the current studies and uses of buckyballs and nanotubes. Since their discovery, what are some ways scientists have used them? Try to think of some practical applications on your own and write them in your science notebook.

Examine a model of a buckminsterfullerene.
Keycode: ES0504

CHAPTER 5 SECTION 2

Examine a model of a buckminsterfullerene.
Keycode: ES0504

Have students locate the hexagonal and pentagonal sides of the model molecule.

 Visualizations CD-ROM

Science and Technology

Buckyballs and Medicine
A unique feature of a buckyball is its hollow ball-like structure. Some scientists hypothesize that the buckyball might be used to deliver drugs at a molecular level. Such a process, for example, might be used in the treatment of cancer. Individual radioactive molecules could potentially be encased within and delivered via buckyballs to the exact location of the cancer within the body. This might eliminate the need for radiation therapy, in which patients are bombarded with relatively large quantities of low-level radiation.

Science Notebook
Scientists have proposed using buckyballs in lubricants and as catalysts, superconductors, medicine-delivery devices, and components of membranes. Nanotubes might be used in electronics, machinery, and epoxy composite materials.

CHAPTER 5 SECTION 3

▶ FOCUS

Objectives
1. Identify rock-forming minerals by inspection, using physical properties such as color, luster, and crystal shape.
2. Identify rock-forming minerals using simple tests that identify both physical and chemical properties, such as streak, cleavage, hardness, and specific gravity.

Set a Purpose
Read the key idea aloud. Have students read this section to answer the question "How can you use physical and chemical properties to identify minerals?"

▶ INSTRUCT

More about...

Rock-Forming Minerals
Geologists often examine rocks and minerals to deduce the processes and events that occurred in the past. For example, the kinds of minerals found in some volcanic rocks might give evidence of the eruptions that brought molten rock, at temperatures perhaps as high as 1000°C, to the surface of Earth. Also, minerals found in granite might give evidence that it was crystallized deep in Earth's crust under conditions that formed mountains such as the Himalayas.

5.3

KEY IDEA
Minerals can be identified by physical and chemical properties that include color, luster, crystal shape, streak, cleavage, fracture, hardness, specific gravity, and reaction to an acid.

KEY VOCABULARY
- mineralogy
- rock-forming mineral
- luster
- streak
- fracture
- specific gravity

Identifying Minerals

Mineralogy is the study of minerals and their properties. Many minerals can be identified and classified by inspecting them visually and performing simple tests to determine their properties.

Rock-Forming Minerals

Of the nearly 4000 known minerals, only about 30 are at all common. Most minerals, including gold and diamond, are rare. Among the most commonly found minerals are quartz, feldspar, mica, and calcite. These and the other minerals that make up most of the rocks in Earth's crust are called **rock-forming minerals.** Most rock-forming minerals are silicates.

Identifying rock-forming minerals can be difficult. A rock often consists of several minerals, some of which may be present in small amounts or well-mixed with other minerals. Sometimes a rock must be broken up in order to retrieve a mineral sample for identification. Although even very small mineral grains can usually be identified, identification is easier with a hand specimen, a sample that is large enough to be readily observed and manipulated. The method commonly used to identify a hand specimen of a mineral is the examination of the mineral's physical properties. Tests may also be performed to identify chemical properties of the mineral. However, the simplest way to identify a mineral is by means of physical tests and inspection.

Identifying Minerals by Inspection

A field guide to minerals is a useful tool, because it lists properties such as color, luster, and crystal shape, as shown below. Visual inspection of a mineral may reveal such properties, but they must be considered together. Rarely is a mineral identified by a single property.

Characteristics of Some Common Minerals

Mineral	Galena	Sulfur	Hornblende	Zircon
Crystal shape				
Color	lead or silver-gray; may have bluish tint	bright yellow as crystals; pale yellow as powder	black to dark green	brown, red, yellow, green, blue, black, or colorless
Luster	metallic to dull	glassy to earthy	glassy to dull	adamantine

104 Unit 2 Earth's Matter

DIFFERENTIATING INSTRUCTION

Reading Support Help students understand and remember what they read by finding the main idea of a paragraph. Use the second paragraph of "Rock-Forming Minerals." Ask: What is the main thing the writer is trying to tell about rock-forming minerals in this paragraph? *that they can be difficult to identify; The rest of this section will describe different techniques a scientist can use to identify minerals.* Tell students to use a similar strategy when reading, particularly when they have difficulty understanding.
Use Reading Study Guide, p. 15.

Color is the most easily observed mineral property. Some minerals have distinctive colors that help identify them. For example, cinnabar, an ore of mercury, is red. Malachite (MAL-uh-KYT), an ore of copper, is green. Color, however, is the least useful property for mineral identification. One reason is that many minerals have similar colors. Also, impurities can turn colorless minerals into colored minerals. Although pure quartz is colorless or white, a small amount of iron gives quartz a purple color, and trace amounts of titanium produce pink quartz. Another reason not to rely on color alone is that some minerals can change color in various circumstances. For example, brown-bronze bornite, a copper ore, turns purple when exposed to air.

The **luster** of a mineral is the way the mineral shines in light. Luster can be described as either metallic or nonmetallic. Minerals with metallic luster, such as galena and pyrite, shine like polished metal. A mineral that does not shine like a metal has a nonmetallic luster. There are several categories of nonmetallic lusters. For example, quartz has a vitreous (VIHT-ree-uhs) luster, like shiny glass. Mica has a pearly luster. The hard, brilliant luster of diamond is called adamantine (AD-uh-MAN-TEEN). Other terms that are used to describe luster are *greasy, oily, dull,* and *earthy.*

Although a mineral's crystal shape can help one identify it, recall from Section 5.2 that crystal faces do not form if space is limited during mineral formation. The mineral grains in most rocks are so small and imperfect that flat-faced, regularly shaped crystals are difficult to find.

Testing Mineral Specimens

Some mineral properties are not revealed by inspection and must be determined by simple physical tests. The streak, cleavage, hardness, and specific gravity of a mineral can be tested with a hand specimen.

The **streak** of a mineral is the color of its powder. The streak is obtained by rubbing the mineral on an unglazed white tile, called a streak plate. In many cases, the streak will not be the same color as the mineral. For example, iron-containing pyrite is brass yellow, yet its streak is always greenish black. Hematite, another iron-containing mineral, can be brown, red, or silver. Its streak, however, is always reddish brown. Although the color of a mineral may vary, the color of its streak rarely does. As a rule, the streak of a metallic mineral is at least as dark as the specimen. The streaks of nonmetallic minerals are usually colorless or white.

You may recall from Section 5.2 that the cleavage of a mineral is its tendency to split easily along flat surfaces. Mica splits very easily in one direction, into thin sheets. Mica is said to have one perfect cleavage. Feldspar splits readily in two directions, always at or near right angles. Feldspar is therefore said to have two good cleavages. Cleavage is useful for identifying minerals because cleavage surfaces can be observed even on tiny mineral grains.

Not all minerals exhibit cleavage. When minerals break in directions other than along cleavage surfaces, they are said to exhibit **fracture.** The mineral quartz and the rock obsidian show conchoidal (kahng-KOYD-uhl), or shell-like, fracture. The fracture surface is smooth and curved like the inside of a clam shell. Splintery fracture leaves a jagged surface with sharp

> **More about...**
>
> **Cleavage**
> Minerals characterized by strong covalent bonding throughout their structures will show little or no cleavage. Minerals that have within their structures planes of weaker ionic bonds will tend to break along those planes when struck, creating well-defined cleavage. Ask students to identify a mineral with strong bonding: quartz, see photograph on page 98 with weak bonds: mica, photograph shown below

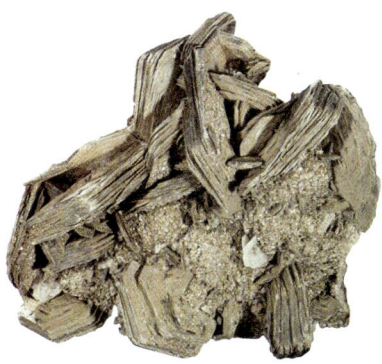

CLEAVAGE TEST Mica splits well in only one direction.

DIFFERENTIATING INSTRUCTION

Reading Support Tell students that when they come to an unfamiliar word in their reading, they can use context clues to help them figure out the meaning. Model the strategy, using this sentence from the first paragraph on page 105: *Also, impurities can turn colorless minerals into colored minerals.* Say: "As I read the sentence, I'm not sure what *impurities* means. Maybe I can get clues to its meaning if I read on. I see in the next sentence that pure quartz is colorless or white, but iron can make the mineral purple. I think an impurity must be something that changes a mineral's appearance but it's still the same mineral. That makes sense in the sentence." Have students make a list of unfamiliar words as they read and use context clues to figure out their meanings whenever possible.

CHAPTER 5 SECTION 3

VISUAL TEACHING
Discussion
Have students use the Mohs Scale of Hardness to answer the following questions: Which mineral would be harder—one with a hardness of 3.4 or one with a hardness of 8.6? **mineral with a hardness of 8.6** If you wanted to engrave a quartz crystal, would you use topaz or potassium feldspar? **topaz** If a mineral can scratch quartz but not diamond, in what range must its hardness rating fall? **between 8 and 9**

CRITICAL THINKING
Ask: Why is graphite added to motor oil in automobiles? **Because it is a soft mineral, graphite is used to lubricate the moving parts of a motor. It can also be used to lubricate other moving parts, such as in doors and locks.**

---

edges, as in native copper. Uneven or irregular fracture leaves a generally rough surface. The garnets show this type of fracture.

The hardness of a mineral is its resistance to being scratched. Diamond is the hardest of all minerals. It will scratch any other mineral against which it is rubbed. Talc is the softest of all minerals. All other minerals scratch talc.

The mineralogist Friedrich Mohs devised a numeric scale that is used to express the hardnesses of minerals. In this scale, ten well-known minerals are assigned numbers from 1 to 10. They range from the softest mineral, talc, to the hardest mineral, diamond. One limitation of Mohs' scale is that the increase in hardness at each successive step is not uniform. For example, Diamond, number 10, is several times harder than corundum, number 9.

You can determine the approximate hardness of any common mineral by using your fingernail, a copper penny, a small glass plate, and a steel file. Just see whether the mineral scratches or is scratched by each item. If, for example, the mineral scratches the glass plate but is scratched by the steel file, its hardness is between 5.5 and 6.5.

Mohs Scale of Hardness

Rating	Reference Mineral	Reference Tool (approximate value)
1	talc	
2	gypsum	fingernail (2.5)
3	calcite	copper penny (3.5)
4	fluorite	
5	apatite	glass plate (5.5)
6	potassium feldspar	steel file (6.5)
7	quartz	
8	topaz	
9	corundum	
10	diamond	

Remember that hardness and brittleness are different properties. Glass is a brittle substance that breaks easily when dropped. Glass, however, is harder (more resistant to scratching) than copper and many other metals.

Specific gravity is another property that is helpful in identifying a mineral. A mineral's **specific gravity** is the ratio of its weight to the weight of an equal volume of water. In other words, the specific gravity of a mineral tells you how many times denser the mineral is than water. You may recall that density is the ratio between a substance's mass and its volume.

Because minerals are denser than water, their specific gravities are greater than 1. Nonmetallic minerals, such as quartz, have specific gravities less than those of ore minerals and native metals. Some metallic minerals, such as hematite and magnetite, have specific gravities of about 5. Other metallic minerals are much denser. For example, gold has a specific gravity of about 19.3.

DIFFERENTIATING INSTRUCTION

Reading Support To help students remember the Mohs Scale of Hardness, have them create a mnemonic device, using the first letter of each reference mineral. Have volunteers share their mnemonics with the class.

To understand how specific gravity is determined, it is important to be familiar with the concept of buoyancy. Buoyancy is the tendency of an object to float in water, due to the difference in densities between the water and the object. Archimedes' Principle states that an object will weigh less when it is in water than when it is in air and that this difference is equal to the weight of the displaced water. Because a mineral is solid, a mineral sample will displace an amount of water equal to its own volume. By weighing the sample in air and again underwater, you can determine the weight of a volume of water that is equal to the volume of the mineral sample.

In addition to physical tests, some minerals can be identified by means of chemical tests. Calcite, the principal mineral in limestone and marble, is calcium carbonate, $CaCO_3$. When a drop of hydrochloric acid is placed on calcite, the drop of acid fizzes. The released bubbles are carbon dioxide gas. Other minerals also react with acid but not as readily.

Special Properties of Minerals

Some minerals have unusual characteristics that can be confirmed by inspection. For example, a variety of the mineral calcite, Iceland spar, splits light rays that pass through it. This property, called double refraction, causes a single object to appear as two when it is viewed through a transparent specimen of Iceland spar. Other minerals, such as fluorite and calcite, are fluorescent, appearing to glow when viewed under ultraviolet light. Some samples of the minerals willemite and sphalerite continue to glow after the ultraviolet light is turned off. They are phosphorescent.

Minerals that have unique properties are often easy to identify. For example, halite (rock salt) can be identified by its salty taste. Magnetite, an iron ore, is attracted by a magnet.

Some rare minerals, such as the uranium minerals carnotite and uraninite (yu-RAY-nuh-NYT), are radioactive. They give off subatomic particles that can be detected by a Geiger (GY-guhr) counter. Exposure to radioactive minerals can be dangerous to living organisms, including human beings.

5.3 Section Review

1. What type of compounds are most rock-forming minerals?
2. Explain why streak is a useful property for identifying minerals.
3. The hardness of a mineral is found to be between 9 and 10 on the Mohs scale. Can you accurately state that the mineral's hardness is 9.5? Why or why not?
4. **CRITICAL THINKING** You are given a sample of an unknown mineral. You determine that the mineral's streak is white and its specific gravity is 2.8. Is the mineral most likely to be a metallic mineral or a nonmetallic mineral? Explain your answer.
5. **PHYSICS** What is the difference between a mineral's density and its specific gravity?

25-Minute Mini LAB

Measuring Specific Gravity

Materials
- beaker
- water
- string
- mineral sample
- spring scale

Procedure

1. Fill the beaker $\frac{3}{4}$ full of water. Tie one end of the string around the mineral. Tie the other end to the scale's hook.

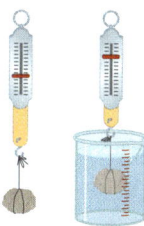

2. Hold the scale so that the sample hangs freely. Measure and record the mass in grams (M_1).
3. Lower the mineral into the beaker so that it is completely covered by water. Do not let the sample touch the bottom or sides of the beaker. Record the mass (M_2).
4. ($M_1 - M_2$) is the mass of the water displaced by the mineral. Calculate the specific gravity using the equation $\frac{(M_1)}{(M_1 - M_2)}$.

Analysis

How might a larger sample change your results? The specific gravity of water is 1. Pure gold has a specific gravity of about 19. Higher numbers indicate higher densities. Compare the density of your sample with those of water and gold.

Chapter 5 Atoms to Minerals **107**

CHAPTER 5 SECTION 4

▶ FOCUS

Objectives
1. Describe the properties of the most common minerals, silicates and carbonates.
2. Describe tests used to identify mineral groups.

Set a Purpose
Read aloud the key idea. Have students read this section to answer the question "How do the major mineral groups differ?"

▶ INSTRUCT

> **More about...**
>
> **Common Minerals**
> Scientists have identified approximately 4,000 minerals. Although most are found in Earth's crust, a few have been identified in meteorites and rocks brought back from the moon. Only 12 elements appear in Earth's crust in amounts greater than 0.1 percent. Rather than forming unique minerals, the remaining scarce elements tend to be found in ionic substitution. Two elements, oxygen and silicon, make up more than 70 percent of Earth's crust.

5.4

KEY IDEA
The most common minerals in Earth's crust are silicates and carbonates.

KEY VOCABULARY
- carbonate
- oxide
- sulfide

Mineral Groups

You can find minerals virtually everywhere. In fact, it is possible that you walk on or pass by samples of common minerals every day. Some minerals are abundant, and many also have important uses.

Major Silicates

More than 90 percent of the minerals in Earth's crust are members of the silicate family. A silicate is a compound of silicon, oxygen, and usually one or more metallic elements, such as aluminum or iron. In all silicates, the basic building block is the silica tetrahedron, consisting of four oxygen atoms bonded to a central silicon atom. Silicates are classified by the ways the tetrahedra are linked together—in chains, for example, or in single sheets. A common example of a mineral containing these silica tetrahedra is quartz.

Quartz

Quartz is made entirely of tightly bound silica tetrahedra. Quartz has the chemical formula SiO_2. Its chemical name, silicon dioxide, indicates that in quartz there are two oxygen atoms for each silicon atom.

Inspection of a quartz sample reveals a glassy or greasy luster. Pure quartz is colorless or white; colored varieties include rose quartz, amethyst, and smoky quartz. All varieties of quartz exhibit either conchoidal or irregular fracture. Quartz's hardness of 7 on the Mohs scale is a common property used for identifying it, since it is the hardest of the common minerals.

The hardness and optical properties of quartz make it ideal for certain industrial uses. It is commonly used in watch movements, prisms, heat lamps, lenses, glass, and paints. The crystal quartz is a semiprecious gemstone often worn as jewelry.

Quartz is found in granite and is a significant component of many other types of rocks. For example, quartzite consists almost entirely of quartz. Most sands consist mainly of grains of quartz.

Although the silica tetrahedra that make up quartz consist of Earth's two most common elements—oxygen and silicon—quartz is only the second most abundant mineral in Earth's crust. The most abundant family of minerals is the feldspars.

QUARTZ, such as the amethyst shown here, is often used in jewelry.

SMOKY QUARTZ

CITRINE

108 Unit 2 Earth's Matter

DIFFERENTIATING INSTRUCTION

Reading Support Have students begin a chart with these headings: *Mineral Group*, *Characteristics*, and *Importance*. As they read, have them fill in the chart with information from the text. *Use Reading Study Guide, p. 16.*

Feldspars

The feldspar family of minerals makes up about 60 percent of Earth's crust. Despite notable differences, all feldspars share three features: two directions of cleavage, a hardness of 6, and a pearly luster.

How does the molecular structure of feldspar differ from that of quartz? In feldspar, aluminum atoms replace some of the silicon atoms in the silica tetrahedra. This replacement creates a net electrical charge in these tetrahedra, which is balanced by the addition of other metals, such as potassium, sodium, or calcium. On the basis of the types of additional metal elements, feldspars are classified into two major groups—the potassium feldspars and the sodium-calcium feldspars.

The most common potassium feldspar is orthoclase (AWR-thuh-KLAYS). Typically, orthoclase is light-colored—pink or salmon. The two cleavage surfaces meet at right angles. Like quartz, orthoclase is most commonly found in granite.

The sodium-calcium feldspars are called plagioclase (PLAY-jee-uh-KLAYS) feldspars. Two examples are albite and labradorite. The plagioclase feldspars range in color from white to gray. The two cleavage surfaces meet at slightly less than a right angle. One cleavage surface is often marked by fine parallel lines, called striations (stry-AY-shuhnz).

Feldspars are important rock-forming minerals. Feldspars are also important economically. Minerals in this family are used in the manufacture of such everyday products as glass and ceramics.

ORTHOCLASE FELDSPAR

PLAGIOCLASE FELDSPAR

Other Silicates

Minerals in the pyroxene (py-RAHK-SEEN) family occur widely. Pyroxenes have cleavage surfaces that meet nearly at right angles. Augite (AW-JYT) is the most common member of the pyroxene family. Augite is also an example of a ferromagnesian silicate. These silicates can belong to almost any of the silicate families, but they all contain iron and magnesium. Ferromagnesian silicates are always dark in color. Augite has two good cleavages and a hardness between 5 and 6.

Minerals in the mica (MY-kuh) family are soft silicates. With a hardness of 2.5 and perfect cleavage, they can be easily picked out of rocks such as granite and gneiss. All micas form flat crystals that cleave in only one direction to form thin sheets or flakes. Muscovite, also known as white mica, is silvery white. Biotite is a dark brown or black mica.

VISUALIZATIONS CLASSZONE.COM

Observe common objects made of minerals.
Keycode: ES0505

VOCABULARY STRATEGY

The prefix *ferro-* in the word *ferromagnesian* comes from the Latin word for iron, *ferrum.*

CHAPTER 5 SECTION 4

VISUALIZATIONS CLASSZONE.COM

Observe common objects made of minerals.
Keycode: ES0505

Visualizations CD-ROM

More about...

Silicate Minerals
Silicate minerals include asbestos, although asbestos is not a specific mineral. Instead, asbestos is the name for a group of naturally occurring silicate minerals that can be separated into fibers, including several minerals and hydrous silicates of magnesium. Emeralds also are silicates, a variety of the mineral beryl. Beryl is a colorless, transparent silicate. Small amounts of chromium or vanadium provide the deep green color of emerald. Other gem varieties of beryl include the greenish blue aquamarine and the pink morganite.

DIFFERENTIATING INSTRUCTION

Support for Visually Impaired Students Provide students with samples of minerals from the different groups. Choose at least some samples in which students can observe differences using the sense of touch. For examples, quartz would be very hard and have a relatively smooth surface. Feldspar would be less smooth and feel "pearly." Mica would also be a good choice because of its distinctive feel.

More about...

Calcium Carbonate
Calcium carbonate, which is found in limestone, can play an important although incidental role in helping scientists in their studies of human ancestry. Scientists rely on DNA preserved in ancient human remains to help establish that ancestry. Mummies offer one rich source of ancient human DNA. Another source, oddly enough, are bodies that for one reason or another had the grave misfortune long ago to land in a peat bog. The conditions of the peat bog are ideal for preserving human remains except that the tannic acid commonly found there can destroy the DNA. However, if limestone is nearby, calcium carbonate from the rock will leach into the bog and act to neutralize the tannic acid. The DNA will be preserved.

Micas are useful in both sheet and powdered forms. Among their many applications, they are used in electronic insulators, paints, plastics, rubber, and roofing.

The amphibole (AM-fuh-BOHL) minerals are complex silicates that form long, needlelike crystals. The most common amphibole is hornblende. A ferromagnesian silicate, hornblende can be shiny dark green, brown, or black. Its hardness ranges from 5 to 6. Hornblende has two good cleavages that meet at oblique angles. It is found in igneous and metamorphic rocks.

Although the minerals mentioned above are the most common silicates, two others—kaolinite and the olivine group—have important practical uses. Named for its olive-green color, olivine is a ferromagnesian silicate. It has a hardness of 6.5 and a glassy, shell-like fracture. Gem-quality olivine used to make jewelry is called peridot (PEHR-ih-DAHT). Some meteorites contain olivine.

Kaolinite is an aluminum silicate formed by the weathering of feldspars and other silicates. Pure kaolinite is white and has a hardness of about 2 and perfect cleavage in one direction. Kaolin, a clay composed primarily of kaolinite, is sometimes called china clay and is commonly used in ceramics, paints, and fiberglass products.

Carbonates

Although most widely found minerals are silicates, several nonsilicate minerals are also common. One nonsilicate group is the carbonates. A **carbonate** mineral is made of negatively charged carbonate ions bonded to positive metal ions. Each carbonate ion is made up of one carbon atom covalently bonded to three oxygen atoms.

The rocks limestone and marble, often used in construction, consist almost entirely of carbonate minerals. Limestone is also used in the manufacture of everyday products ranging from paper to medicines.

CALCITE Calcium carbonate, or calcite (KAL-SYT), is the most common carbonate mineral. Usually colorless or white, calcite has a hardness of 3 and can be scratched with a knife. Calcite has three perfect cleavages that meet at oblique angles. Its cleavages give it a very strong tendency to break into flat-sided rhombohedra when dropped or struck. You can use an acid test to identify calcite easily. The chemical formula of calcite is $CaCO_3$.

CALCITE

DOLOMITE Calcium magnesium carbonate, or dolomite (DOH-luh-MYT), has a hardness of 3.5 to 4. Its chemical formula is $CaMg(CO_3)_2$. Like calcite, it cleaves into rhombohedra. Dolomite does not readily bubble when hydrochloric acid is dropped on it. Dolomite occurs as coarse or fine grains in dolomitic limestones.

DOLOMITE

Oxides and Sulfides

Some minerals contain significant amounts of the element iron. Although these minerals are not as common as the silicates or carbonates, they have economic importance. Iron-containing minerals are used in industry to make steel, magnets, and car parts. They are also used in such consumer goods as medicines, cosmetics, plastics, and paints.

DIFFERENTIATING INSTRUCTION

Challenge Activity Have students bring their own mineral samples to class. Working in groups, they can develop keys and identify groups of unknown mineral samples. Each group determines the minerals' properties by using their senses, available tools, and simple tests. After identifying the minerals, have students use reference books and/or the Internet to find the chemical formulas of the minerals.

In these iron minerals, the iron is usually combined with either oxygen or sulfur. An **oxide** (AHK-SYD) is a mineral consisting of a metal element combined with oxygen. A **sulfide** (SUHL-FYD) is a mineral consisting of a metal element combined with sulfur.

HEMATITE Hematite, the most common iron oxide, is usually red. Its properties include an earthy luster and uneven fracture. Some hematite samples have a silvery metallic luster that makes them attractive as gemstones. However, all hematite samples, regardless of their apparent color, leave a red-brown streak on a streak plate. Hematite has a hardness of 5 to 6 on the Mohs scale.

MAGNETITE Magnetite is a black iron oxide. Its name refers to the fact that it is attracted to a magnet. Lodestone is a variety of magnetite that is itself a natural magnet. Because of this unique property, the first compass needles were made from lodestone. Magnetite has a hardness of 5.5 to 6.5.

PYRITE Pyrite, an iron sulfide, is the most common sulfide mineral. Its color ranges from pale brass to golden yellow. Its hardness is about 6. Pyrite often occurs in 6- or 12-sided crystals. Because it is sometimes mistaken for gold, pyrite is commonly referred to as fool's gold.

HEMATITE

PYRITE

MAGNETITE

5.4 Section Review

1. Name one example of a silicate mineral and one example of an oxide mineral. Describe the distinguishing characteristics of each mineral.

2. How can you use cleavage to distinguish between orthoclase and plagioclase feldspars?

3. **CRITICAL THINKING** Suppose that you are given hand specimens of the carbonates calcite and dolomite. Describe two tests that would allow you to identify the samples.

4. **ECONOMICS** According to the principle of supply and demand, the price you pay for a product is determined both by its availability and by consumer demand for the product. On the basis of this principle, which minerals do you think have a higher value, iron-containing minerals or feldspars? Explain your reasoning.

CHAPTER 5 SECTION 4

More about...

Lodestone
Lodestone is relatively rare and only found in certain areas of the world, including Siberia, South Africa, and the island of Elba in the Mediterranean Sea. A strong magnetic field is required to turn magnetite into lodestone, such as occurs when magnetite is struck by lightning. Magnetite in Earth's crust can also become lodestone as hot magma containing magnetite cools in the presence of Earth's magnetic field.

ASSESS

1. Silicate minerals include quartz, orthoclase feldspar, plagioclase feldspars (such as albite and labradorite), pyroxenes (such as augite), micas, amphiboles (such as hornblende), olivine, and kaolinate. Oxides include hematite and magnetite. Descriptions of each should include color, hardness, fracture, cleavage, luster, and uses.

2. The two cleavages of orthoclase feldspars meet at a right angles; the two cleavages of plagioclase feldspars meet at slightly less than a right angle. Plagioclase feldspars have striations running along the cleavage surface; orthoclase feldspars do not.

3. Hardness test—Dolomite has a higher ranking than calcite. Acid test—Calcite will bubble when acid is applied, but dolomite will not.

4. Iron minerals would tend to have a higher value. Feldspars are much more widely found than iron-containing minerals. Iron oxides and iron sulfide are used much more in heavy industry.

MONITOR AND RETEACH

If students miss . . .

Question 1 Reteach silicates and oxides. (pp. 108–111) Have students review the pictures of different minerals and identify characteristics that can be used to distinguish each.

Question 2 Have students reread the information about orthoclase and plagioclase feldspars. (p. 109) Ask: How do the cleavage surfaces of the two differ?

Question 3 Have students skim the text and list characteristics of each mineral. (p. 110) Ask: Which tests would you use to determine each characteristic?

Question 4 Reteach iron-containing minerals and feldspars. (pp. 109–110) Ask: What is meant by the law of supply and demand?

CHAPTER 5 ACTIVITY

PURPOSE
To examine the usefulness of specific gravity measurements in identifying unknown minerals

MATERIALS
The balance should be calibrated with the paper clip and rubber band hanging from the string.

PROCEDURE

❶ The Minilab on page 107 also deals with specific gravity. If students did the Minilab, review the procedure and students' results.

❶ Discuss error analysis (pp. 724–725). Review the procedure, step by step, and discuss where errors might occur. Then discuss how each error might influence the measurements and results. Errors can be systematic (having to do with limitations of instruments) or procedural (errors by the experimenter).

❼ Students may be systematically low on measurements. This may be due to impurities in the mineral samples and the submerging of the rubber band and paper clip.

❽ Review Archimedes' Principle, and discuss the difference between density and specific gravity.

ANALYSIS AND CONCLUSIONS

1. Tests may include recognizing color, luster, or crystal shape.

CHAPTER 5 LAB Activity

Specific Gravity and Mineral Identification

SKILLS AND OBJECTIVES
- **Measure** data and **calculate** the specific gravity of several minerals.
- **Identify** minerals using their specific gravities.

MATERIALS
- mineral kit containing 3 unnamed minerals
- triple-beam balance
- C-clamp
- 250-mL beaker
- water
- string
- paper clip
- rubber band
- calculator

It is often quite difficult to distinguish one mineral from another. For example, albite and oligoclase, both plagioclase feldspars, are similar in appearance and in most other properties. One property that does slightly differ between these two minerals is specific gravity: 2.62 for albite and 2.65 for oligoclase. Careful measurements of the specific gravity of each mineral make it possible to tell the two apart. In this activity, you will determine the specific gravities of three unnamed minerals and then use these values to identify each mineral.

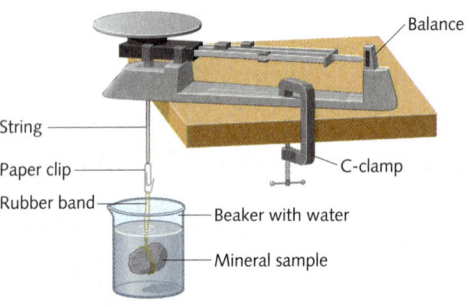

Procedure

❶ Construct a data table for recording the data based on Table 1. Add two more columns for Minerals 2 and 3.

❷ Set the balance on the lab table so that the pan hangs over the edge of the table. Attach the balance to the tabletop with a C-clamp.

❸ Look under the balance pan and locate the metal bar that hangs down from the pan. Attach a loop of string to this metal bar. Tie the string securely so that the loop of string hangs below the balance.

❹ Securely wrap a rubber band around the first mineral sample so that you can suspend the sample by holding onto one of the rubber band's loops without the sample falling out.

❺ Make a hook out of a paper clip by unbending it as shown in the diagram above. Attach one end to the free loop of the rubber band. Hook the other end to the loop of string suspended from the balance pan. Your mineral sample should now hang freely from the bottom of the pan.

❻ Find the mass of the sample by adjusting the sliding masses on the balance. Record the result in the row "Mass in air."

Table 1

Property	Mineral 1
Mass in air (g)	
Mass in water (g)	
Loss of mass in water (g)	
Specific gravity (calculated)	
Name of mineral	
Specific gravity (accepted value)	

112 Unit 2 Earth's Matter

7. Fill a 250-mL beaker with about 150 mL of water. With the mineral attached to the balance, raise the beaker of water under the mineral until the mineral is totally submerged in the water. Do not allow the mineral to rest on the bottom of the beaker. Measure the mass and record the measurement under "Mass in water."

8. Find the difference between the mass of the mineral in water and the mass in air. Record your result as "Loss of mass in water." Use the formula on the right to calculate the specific gravity of your sample.

9. Repeat Steps 3–7 for the other two minerals in your kit.

10. Use the accepted values for specific gravity in Table 2 to identify each of your mineral samples. Record the mineral names and actual values for specific gravity in your table.

Specific Gravity (SpG)

$$(SpG) = \frac{\text{Mass in air}}{\text{Mass in air} - \text{Mass in water}}$$

Table 2
Specific Gravities of Selected Minerals

Mineral	Specific Gravity
Sulfur	2.1
Gypsum	2.3
Calcite	2.7
Chalcopyrite	4.2
Pyrite	5.0
Magnetite	5.2
Galena	7.5

Analysis and Conclusions

1. Were you able to identify any of the minerals in your mineral kit by inspection prior to determining their specific gravities? If so, describe which tests you used.

2. Were you able to use specific gravity alone in identifying the minerals in your kit?

3. What sources of error in the experiment might account for any differences from the accepted value of specific gravity for each mineral?

4. If you held a sample of sulfur in one hand and an equal-sized sample of galena in the other hand, which would feel heavier? Be sure to use the words *specific gravity* in your response.

5. Why does the mass determined by the scale change when you suspend the mineral in water?

6. When identifying a valuable gemstone, why is specific gravity more likely to be used than mineral tests such as streak or hardness?

7. Why would the procedure for this investigation not be useful in trying to determine the specific gravity of halite?

8. Explain how a liquid with a specific gravity of 5.1 can be used to distinguish pyrite from magnetite.

Learn more about minerals and their properties.
Keycode: ES0507

CHAPTER 5 ACTIVITY

Learn more about minerals and their properties.
Keycode: ES0507

(continued)

2. Depending on the accuracy of the results and the purity of the samples, it may be possible to identify the minerals based on specific gravity alone.

3. Sources of error may include measurement error, the use of an impure material, having the mineral rest on the bottom of the beaker, having the rubber band and part of the paper clip submerged with the mineral sample, and not having the balance calibrated.

4. The galena would feel heavier because it has a higher specific gravity.

5. The mineral weighs less when submerged in water because of the buoyant effect of water. Archimedes' Principle states that the apparent loss in weight of the mineral when weighed in the water compared to air is equal to the weight of the water displaced by the mineral.

6. Testing for streak or hardness often involves damaging the mineral, which is undesirable for a gemstone. When performed accurately, specific gravity can also be a precise indicator.

7. Halite, or rock salt, dissolves in water.

8. Pyrite, with a specific gravity of 5.0, will float in the liquid, while magnetite, with a specific gravity of 5.2, will sink.

Refer students to Safety in the Earth Science Laboratory on pages xxii–xxiii.

CHAPTER 5 REVIEW

VOCABULARY REVIEW

1. Atomic number is the number of protons in the nucleus of an atom of a given element; mass number is the sum of the number of protons and neutrons in an isotope of that element.
2. A mineral is a naturally occurring inorganic solid with a definite chemical composition of one or more elements in an orderly arrangement of atoms; an element is the basic material from which all matter is formed, a substance that cannot be broken into simpler substances by ordinary chemical means.
3. Cleavage is a mineral's tendency to split along distinct planes where bonds are weak; fracture is a pattern of breakage not defined by cleavage planes.
4. A compound is a substance that contains atoms of two or more elements that are chemically combined; a molecule is the smallest unit of a compound.
5. A metal is an element that loses electrons easily to form positive ions; a nonmetal is an element that gains electrons easily to form negative ions.
6. ions
7. isotopes
8. crystal
9. silicates
10. oxides, sulfides

CONCEPT REVIEW

11. Knowing that an element loses electrons easily and forms positive ions helps identify it as a metal. Similarly, elements that easily gain electrons and form negative ions are classified as nonmetals.
12. Minerals may form through the cooling of magma, the evaporation of water, and the action of heat, pressure, or chemical action.
13. Minerals are solid because of their crystal structure. Crystalline structure can determine the temperature at which a mineral melts or vaporizes. It can also determine cleavage and hardness.

14. A mineral can be identified by inspection (color, luster, crystal shape) or through tests (streak, fracture, hardness, specific gravity) or by identification of special properties. The simplest method is inspection. Because color and crystal shape can be so variable, and not all minerals have unique properties, students may suggest the use of the simple tests first.

Summary of Key Ideas

5.1 Matter is made up of atoms. An atom is the smallest part of an element that has all the element's properties. Two or more chemically bound elements may form a compound. A metal is an element that easily loses electrons to form positive ions. A nonmetal easily gains electrons, forming negative ions.

5.2 A mineral is a naturally occurring, inorganic solid with a definite chemical composition and orderly atomic arrangement. Minerals may be either elements or compounds. Atomic structure determines a mineral's properties. Most of Earth's crust consists of silicate minerals.

5.3 A mineral is identified by its properties. Simple inspection reveals a mineral's color, luster, and crystal shape. Simple tests reveal a mineral's streak, cleavage, fracture, and hardness.

5.4 Silicates and carbonates are the most common minerals in Earth's crust. Less common but economically important minerals include iron-rich oxides and sulfides.

KEY VOCABULARY

atomic number (p. 91)
carbonate (p. 110)
cleavage (p. 100)
compound (p. 93)
crystal (p. 98)
element (p. 90)
fracture (p. 105)
ion (p. 94)
isotope (p. 92)
luster (p. 105)
mass number (p. 92)
metal (p. 94)
mineral (p. 96)
mineralogy (p. 104)
molecule (p. 93)
nonmetal (p. 95)
oxide (p. 111)
rock-forming mineral (p. 104)
silica tetrahedron (p. 100)
silicate (p. 100)
specific gravity (p. 106)
streak (p. 105)
sulfide (p. 111)

Vocabulary Review

Explain the difference between the terms in each pair.

1. atomic number, mass number
2. mineral, element
3. cleavage, fracture
4. compound, molecule
5. metal, nonmetal

Write the term from the key vocabulary list that best completes the sentence.

6. Atoms that have lost or gained electrons and are in a charged condition are ___?___.
7. Atoms of the same element having different sums of protons and neutrons are called ___?___.
8. A regular, geometric solid with smooth surfaces is called a ___?___.
9. Compound minerals made up of oxygen and silicon are called ___?___.
10. Types of minerals that contain large amounts of the element iron are ___?___ and ___?___.

Concept Review

11. How does knowing about ions help with the classification of elements?
12. List three ways minerals can form.
13. What physical properties are determined by a mineral's crystal structure?
14. List the methods you would use to identify a mineral. Describe the method that you would use first.
15. **Graphic Organizer** Copy the Venn diagram below. Organize the following items in the diagram: atomic particle; located in nucleus; has no charge; has positive charge; determines atom's atomic number; most massive atomic particle.

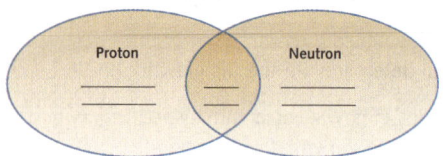

15. Proton: has positive charge; determines atom's atomic number
Proton and Neutron: atomic particle; located in nucleus
Neutron: has no charge; most massive atomic particle

Critical Thinking

16. **Draw Conclusions** The element mercury (Hg) is a metal. It is solid at temperatures of about −39° C (−38° F), has a definite composition, is inorganic, and lacks a crystalline structure. Ice (H_2O), on the other hand, is often considered a mineral. Based on the information given, do you think mercury is a mineral? Why or why not?

17. **Hypothesize** Perfect, naturally formed crystals are relatively hard to find. If you were to manufacture perfectly-shaped synthetic crystals, in what type of environment would you expect to produce them?

18. **Compare** Graphite and diamond are both made of the element carbon. Compare the conditions needed to form graphite with the conditions needed to form diamond. Why do you think graphite is more common?

Internet Extension

How Do Crystals Grow? Select conditions such as temperature and pressure and see their effects on crystal growth.
Keycode: ES0506

Writing About the Earth System

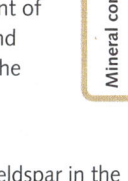 **SCIENCE NOTEBOOK** While minerals are commonly associated with Earth's crust, minerals are an essential part of hydrological and biological cycles. For example, dissolved minerals are used by plants and animals to carry out life processes. Investigate two minerals that are used by plants or animals. Determine how plants or animals obtain the minerals. Describe the roles of the minerals in biological functions.

Interpreting Charts

Seven rock-forming minerals make up the four rocks listed across the top of the chart shown. Copy the 0–100% scale on the left of the chart onto the edge of a piece of paper. Use your scale to read the percent of a mineral in a rock. For example, to find the percent of quartz in tonalite, slide your scale to the vertical line marked *tonalite*. Place the zero of your scale at the line between quartz and plagioclase feldspar. Read the percent of quartz at the line between quartz and potassium feldspar. About 35% of the tonalite's volume is quartz.

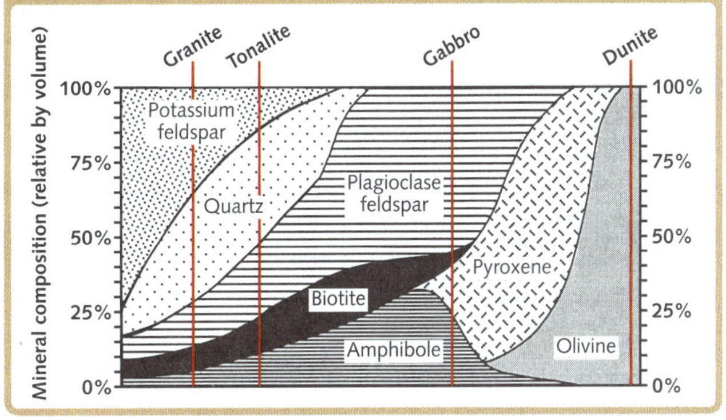

19. Which rock contains pyroxene?
20. Find the percent of potassium feldspar in the tonalite.
21. Identify the rock containing 40% potassium feldspar, 35% quartz, 12% plagioclase feldspar, 8% biotite, and 5% amphibole.
22. Describe the composition of the gabbro.

CHAPTER 5 REVIEW

How Do Crystals Grow?
Keycode: ES0506

 Use Internet Investigations Guide, pp. T21, 21.

CRITICAL THINKING

16. A mineral is a naturally occurring inorganic solid with a definite chemical composition and a crystalline structure (ordered atomic pattern). Although a temperature of −39°C is relatively rare, it does occur naturally. However, mercury's lack of a crystalline structure usually precludes it from being classified as a mineral.

17. Answers will vary, but students may recognize the importance of having the proper environment for crystal formation, including the proper amounts of heat and pressure, as well as the presence of the appropriate elements.

18. Diamond is a network of strongly bonded carbon atoms compared to graphite's sheets of weakly bonded carbon atoms. More energy would be required to produce the structure of diamond, which suggests that it forms under conditions of extreme heat and pressure found deep within Earth's crust. The fact that graphite is more commonly found suggests it forms under less extreme conditions.

INTERPRETING CHARTS

19. gabbro
20. approximately 15%
21. granite
22. 55% plagioclase feldspar, 25% amphibole, 5% biotite, 5% pyroxine

---

WRITING ABOUT SCIENCE

Writing About the Earth System

Minerals mentioned might include calcium, phosphorus, potassium, sodium, and iron. Animals ingest compounds of calcium, found in milk, eggs, and green leafy vegetables. Calcium is used to build strong bones and teeth, aids in blood clotting, and is involved in muscle and nerve activity. Potassium, found in bananas, potatoes, nuts, and meat, aids in the balance of water in cells, as well as nerve-impulse conduction in animals. Plants take up potassium compounds to maintain proper water content of stems and leaves. Iron compounds, also found in soil, are taken up by plants and are needed for healthy growth. Iron compounds carry oxygen in hemoglobin in red blood cells. A lack of iron may result in anemia.

CHAPTER 6 PLANNING GUIDE ROCKS

	Section Objectives	Print Resources
Chapter-Level Resources		Laboratory Manual, pp. 27–32 Internet Investigations Guide, pp. T22–T24, 22–24 Guide to Earth Science in Urban Environments, pp. 5–12, 45–52 Spanish Vocabulary and Summaries, pp. 11–12 Formal Assessment, pp. 16–18 Alternative Assessment, pp. 11–12
Section 6.1 How Rocks Form, pp. 118–120	1. Differentiate among the three major types of rocks. 2. Compare and contrast the processes in the rock cycle.	Lesson Plans, p. 17 Reading Study Guide, p. 17 Teaching Transparency 7
Section 6.2 Igneous Rocks, pp. 121–126	1. Distinguish between intrusive and extrusive igneous rocks and how they form. 2. Contrast the types of plutons that form as the result of intrusive igneous activity.	Lesson Plans, p. 18 Reading Study Guide, p. 18
Section 6.3 Sedimentary Rocks, pp. 127–132	1. Distinguish among the three types of sedimentary rocks and how they form. 2. Discuss different features typical of sedimentary rocks.	Lesson Plans, p. 19 Reading Study Guide, p. 19
Section 6.4 Metamorphic Rocks, pp. 133–137	1. Explain the processes involved in the formation of metamorphic rocks. 2. Differentiate among different kinds of metamorphic rocks.	Lesson Plans, p. 20 Reading Study Guide, p. 20

INVESTIGATIONS

CLASSZONE.COM

Section 6.1 INVESTIGATION
How Do Rocks Undergo Change?
Computer time: 45 minutes / Additional time: 45 minutes
Students follow a rock through various paths of the rock cycle.

Section 6.2 INVESTIGATION
How Do Igneous Rocks Form?
Computer time: 45 minutes / Additional time: 45 minutes
Students watch igneous minerals crystallize, examine rocks, and classify how the rocks formed.

Technology/Online Resources

Electronic Teacher Tools
Visualizations CD-ROM
Classzone.com
Online Lesson Planner
Test Generator

Visualizations CD-ROM
Classzone.com
  ES0601 Visualization, ES0602 Investigation

Classzone.com
  ES0603 Investigation

Visualizations CD-ROM
Classzone.com
  ES0604 Visualization, ES0605 Visualization, ES0606 Data Center

Visualizations CD-ROM
Classzone.com
  ES0607 Visualization, ES0608 Career

Classroom Management

Gather these materials for minilab and laboratory activities.

Minilab, p. 130
  object to make mold
  modeling clay
  plaster of Paris
  water
  small cup
  stirring straw

Lab Activity, pp. 138–139
  metric ruler

Chapter Review INVESTIGATION
What Kind of Rock Is This?
Computer time: 45 minutes / Additional time: 45 minutes
Students use a key to help determine a type of rock.

CHAPTER 6

Rocks

The rock in this outcrop is quartzite, a rock that is often used for building because it is very durable.

How did the processes that formed the quartzite contribute to its durability?

INTRODUCE
Build Interest

Before sharing background information about the photograph with students, allow time for their questions and comments.

Ask questions like the following:

- What is distinctive about the rock in this picture? **brightly colored, has distinct layers, appears to have been cut away**
- What does the shape of this outcropping of rock suggest to you? **the rock was quarried, probably long ago**

Respond

- How did the processes that formed the quartzite contribute to its durability? **Sand laid down in an ancient sea was, over time, transformed into sandstone, and then quartzite. Sand is made up of grains of quartz, a very hard mineral. In quartzite, these grains become densely packed, resulting in a very durable rock.**

About the Photograph

The Sioux quartzite, shown here in an old quarry at Blue Mounds State Park in Minnesota, was formed over $1\frac{1}{2}$ billion years ago. Its color ranges from pink to purple, due to the presence of iron oxide. Against a setting sun, the rock appeared blue to settlers moving west, and so was called "The Blue Mound."

BEYOND THE TEXTBOOK

Print Resources
- Reading Study Guide, pp. 17–20
- Transparency 7
- Formal Assessment, pp. 16–18
- Laboratory Manual, pp. 27–32
- Alternative Assessment, pp. 11–12
- Spanish Vocabulary and Summaries, pp. 11–12
- Internet Investigations Guide, pp. T22–T24, 22–24
- Guide to Earth Science in Urban Environments, pp. 5–12, 45–52

Technology Resources
- Visualizations CD-ROM
- Classzone.com
- Online Lesson Planner
- Electronic Teacher Tools
- Test Generator

116 Unit 2 Earth's Matter

CHAPTER 6

PREVIEW

▶ **FOCUS QUESTIONS** In this chapter you will study rocks and learn more about the key questions listed below.

Section 1 What is a rock, and what is the rock cycle?
Section 2 How do igneous rocks form and how are they classified?
Section 3 How do the different types of sedimentary rocks form?
Section 4 How do metamorphic rocks form and what are their characteristics?

▶ **REVIEW TOPICS** As you investigate rocks, you will need to use information from earlier chapters.

- mineral (p. 90)
- crystal (p. 92)
- the relationship among element, mineral, and rock (p. 90)
- silicates (p. 93)
- rock-forming minerals (p. 98)

▶ **READING STRATEGY**

CONNECT

Note the rocks you see and use every day. Make connections between your experience and what you are reading. Especially note information in the text that seems to contradict your experience.

INTERNET RESOURCES
CLASSZONE.COM

At our Web site, you will find the following Internet support:

DATA CENTER
EARTH NEWS
VISUALIZATIONS
- Many Views of Rocks
- Sediment Deposition
- Clastic Sedimentary Rock Formation
- Metamorphic Rock Formation

LOCAL RESOURCES
CAREERS
INVESTIGATIONS
- How Do Rocks Undergo Change?
- How Do Igneous Rocks Form?
- What Kind of Rock Is This?

Chapter 6 Rocks 117

CHAPTER 6

PREPARE

Focus Questions
Find out what knowledge students have about rocks. For each focus question, have students consider the section title and then develop two or three additional questions for each section. For example, Section 4: How are sedimentary rocks different from metamorphic rocks?

Review Topics
In addition to the earth science topics listed, students may need to review these concepts:

Physical Science
- In a solution, the substance that is dissolved is the solute, and the substance that does the dissolving is the solvent.
- A saturated solution contains all the solute it can hold at a certain temperature.

Reading Strategy
Model the Strategy: Have students work in groups to create a list of everyday uses of rocks. Encourage them to include objects that do not necessarily look like rocks, such as paving materials and jewelry. Remind students that rocks are used as writing tools (chalk), scouring powder (pumice), building material (marble), and fuels (coal). Share their lists in class. Students might connect their reading to their experience by examining building materials or roadside stones for grain size. They might not be able to connect the rock features they read about with the rock they see around them. For example, sedimentary rock may not show stratification if the sample is small or the outcrop belongs to a single bed.

BEYOND THE TEXTBOOK

INVESTIGATIONS
- How Do Rocks Undergo Change?
- How Do Igneous Rocks Form?
- What Kind of Rock Is This?

VISUALIZATIONS
- Images of rocks from satellite views to microscopic views
- Animation of how sediments are deposited
- Animation of clastic sedimentary rocks forming
- An animation of metamorphic rocks forming

DATA CENTER
EARTH NEWS
LOCAL RESOURCES
CAREERS

Online Lesson Planner

Chapter 6 Rocks 117

CHAPTER 6 SECTION 1

FOCUS

Objectives
1. Differentiate among the three major types of rocks.
2. Compare and contrast the processes in the rock cycle.

Set a Purpose
Have students read this section to find the answer to "What processes result in the formation of rocks?"

INSTRUCT

More about...

Rock Classification
One can use standard criteria to subdivide each group of rocks—igneous, sedimentary, and metamorphic. Igneous rocks are classified according to composition and texture. Sedimentary rocks can be classified based on particle size and the mode of formation. Metamorphic rocks are classified according to texture.

VISUAL TEACHING

Discussion
People often use the terms *rock* and *mineral* synonymously. As students examine the photographs, stress that rocks are made up of one or more minerals. Review from Chapter 5 that a mineral is a naturally occurring, inorganic solid with a specific chemical composition and a definite internal structure.

6.1

KEY IDEA
Three major types of rocks are formed, broken down, and reformed in a recurring process called the rock cycle.

KEY VOCABULARY
- rock
- igneous
- magma
- sedimentary
- sediment
- metamorphic
- rock cycle

VOCABULARY STRATEGY
The word "igneous" comes from *ignis*, the Latin word for fire. The Latin word for "the act of settling," *sedimentum*, is the origin of the word "sedimentary." "Metamorphic" is based on the Greek word *metamorphoun*, which means "to transform."

VISUALIZATIONS CLASSZONE.COM
Examine rocks from a satellite view and zoom in to a microscopic view. *Keycode:* ES0601

GRANITES vary in composition, but the minerals shown here are often present. Due to differences in their formation, a hand sample of a mineral will look different from a crystal of the same mineral that is part of a rock.

118 Unit 2 Earth's Matter

How Rocks Form

If you could dig a hole straight down through Earth's crust, what would you see? At first, you might find layers of soft dirt, sand, or clay. Eventually you would find the sturdy base on which we live, the solid material called rock.

What Is a Rock?

An understanding of Earth's processes requires knowledge about rocks and how they form. In general, a **rock** is a group of minerals bound together. Rocks can consist largely of one mineral or of several different minerals in varying quantities. In the granite rock shown below, four minerals are readily visible and identifiable; other minerals may be present in much smaller amounts. Other types of granites may be composed only of feldspar and quartz.

Rocks are found in Earth's crust and mantle. The rocks of the mantle are seldom seen at the surface and are largely similar. The crust, however, contains many different types of rocks. These rocks can be classified according to the processes by which they are formed:

- **Igneous** (IHG-nee-uhs) rocks are formed by the cooling and hardening of hot, molten rock, or **magma,** from inside Earth.
- **Sedimentary** (SEHD-uh-MEHN-tuh-ree) rocks are formed by the compaction and cementing of layers of sediments. **Sediments** are materials such as rock fragments, plant and animal remains, or minerals that settle out of solution onto lake and ocean bottoms.
- **Metamorphic** (MEHT-uh-MOR-fihk) rocks are formed by the effect of heat and pressure on other rocks.

Hornblende, Feldspar, Mica, Quartz

DIFFERENTIATING INSTRUCTION

Challenge Activity Provide students with hand lenses and hand specimens of granite like the one shown in the photograph on this page, a gneiss that contains hornblende, feldspar, mica, and quartz, and a sandstone composed of these same minerals. Have students contrast the textures of the rocks and attempt to classify each. Granite is an igneous rock whose minerals lend it a coarse-grained texture. Gneiss is a metamorphic rock with a foliated (banded) texture. Sandstone is a sedimentary rock whose texture can be described as grainy.

118 Unit 2 Earth's Matter

The Rock Cycle

Rocks form from other rocks. Classifying the rocks of the crust according to their origins shows how closely related they are. The **rock cycle** is the repeated series of events by which rock gradually and continually changes from one type to another. The diagram of the rock cycle below illustrates a simplified model of this continuous process of rock formation and change.

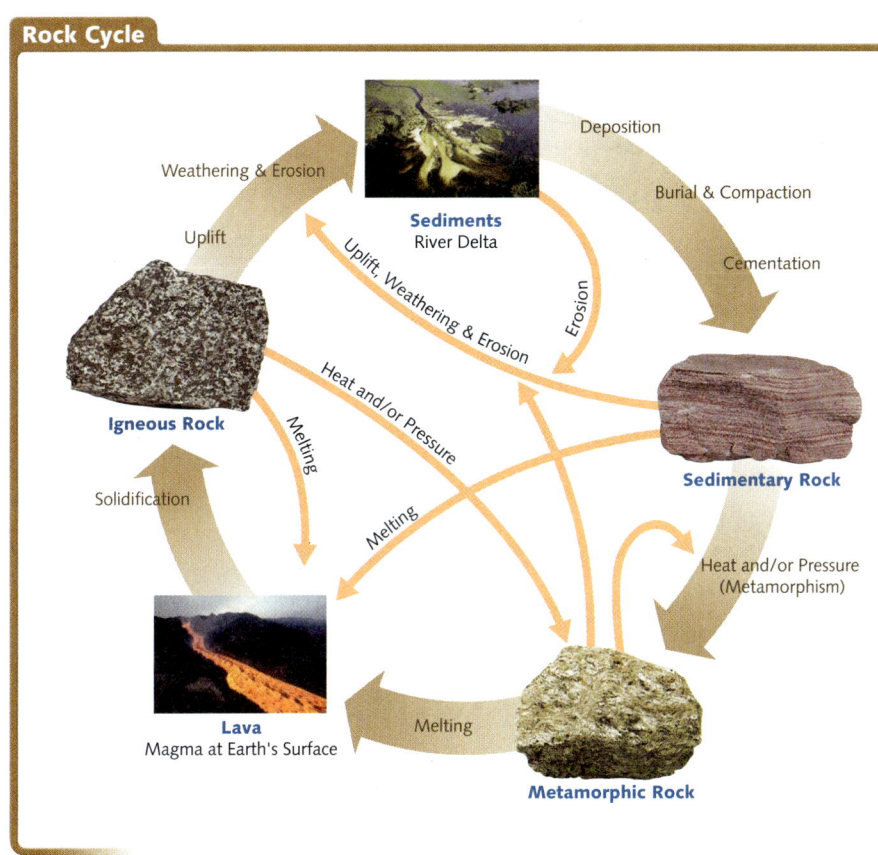

Rock Cycle

Magma from deep below Earth's surface is the source of all Earth's rocks. When this molten rock approaches or reaches the surface, it cools and solidifies to become igneous rock.

Once at the surface, igneous rocks are slowly broken down by weathering and erosion, forming sediments. As later sediments accumulate on top of earlier layers, the buried sediments begin to compact and cement together. In this way, the sediments become sedimentary rocks.

Over time, these sedimentary rocks are buried beneath other sediments or are caught in movements of Earth's crust that expose them to high

---

CHAPTER 6 SECTION 1

INVESTIGATIONS
CLASSZONE.COM

How Do Rocks Undergo Change?
Follow a rock through various paths of the rock cycle.
Keycode: ES0602

VISUALIZATIONS
CLASSZONE.COM

Examine rocks from a satellite view and zoom in to a microscopic view.
Keycode: ES0601

- Visualizations CD-ROM

INVESTIGATIONS
CLASSZONE.COM

How Do Rocks Undergo Change?
Keycode: ES0602

Have students work in pairs to do this investigation.

- Use Internet Investigations Guide, pp. T22, 22.

Visual Teaching

Interpret Diagram
As students refer to the diagram, ask: Can an igneous rock become another igneous rock? Explain. **Yes; if igneous rock is subjected to enough heat and pressure, it will melt to form magma. When this molten material solidifies, igneous rock forms.**

Compare and contrast the processes involved in the formation of metamorphic and igneous rocks. **Both involve subjecting rocks to increased temperature and pressure. Unlike igneous activity, no melting occurs during metamorphism.** Is there a beginning or an end to this cycle? **This cycle is a series of events that repeat many times, and as such, has no true beginning or end.**

Use Transparency 7.

Chapter 6 Rocks 119

DIFFERENTIATING INSTRUCTION

Reading Support Point out that students can use diagrams to summarize a large amount of text. Have students carefully study the diagram of the rock cycle and the accompanying text. Ask them to identify the main processes in this geologic cycle. **The main processes are overprinted on the wide arrows.**
Use Reading Study Guide, p. 17.

Support for Visually Impaired Students Pair visually impaired students with sighted students. Have pairs make a tactile rock cycle. Have them use actual rock samples, yarn for the arrows, and labels with enlarged type to recreate the rock cycle. If a Braille label maker is available, students can use it to make the diagram labels.

CHAPTER 6 SECTION 1

ASSESS

1. Igneous rocks form from magma or lava. Sedimentary rocks form from sediments. Metamorphic rocks form from other rocks—sedimentary, igneous, or metamorphic.

2. Both result from Earth processes, rocks are made up of minerals.

3. Magma cools to form igneous rock. Igneous rock is uplifted, weathered and eroded. Resulting sediments are buried, cemented, and compacted to form sedimentary rock. Sedimentary rock exposed to heat and pressure becomes metamorphic rock. These rocks melt when forced deep beneath the surface, becoming magma. Variations may include: igneous rock exposed to heat and pressure forms metamorphic rock; metamorphic rock or sedimentary rock that is uplifted, weathered, and eroded may be compacted and cemented into sedimentary rock.

4. Both require high temperatures. Igneous rock forms from molten rock (magma). Metamorphic rock forms under extreme heat and/or pressure, but does not melt.

5. Sedimentary rock exposed to heat and pressure forms metamorphic rock. Metamorphic rock exposed to heat and pressure forms other metamorphic rock. Metamorphic rock melts to form magma, which hardens to form igneous rock. Igneous rock is changed by heat and pressure into metamorphic rock. Through uplift, weathering, and erosion, metamorphic rock may become sediments, which may form sedimentary rock.

---

temperatures and pressures. In either case, heat, pressure, or both cause sedimentary rocks to acquire new characteristics, thus becoming metamorphic rocks.

Sometimes crustal movements force rocks deep into Earth. The rocks may become so hot that they melt into magma. The magma may then harden into igneous rocks to complete the rock cycle.

This description of the rock cycle is a simplified model of rock formation and change. This model is useful because it presents a clear "map" of the basic processes, the transformation of rock from one form to another. A complete model of the rock cycle would be very complex and include many variations on and departures from the basic cycle the model illustrates. That is, the transformation of rock does not always move from igneous to sedimentary to metamorphic rock. For example, an igneous rock may become a metamorphic rock without first becoming sediments and then sedimentary rock.

The rock cycle illustrated on the previous page shows several possible variations from the model. For example, igneous rock may melt back into magma before it can become any other type of rock. A sedimentary rock may weather back into sediment without first becoming a metamorphic rock. Some metamorphic rocks change into different metamorphic rocks instead of melting into magma.

Some sediments are created completely outside the rock cycle by means other than weathering. For example, sometimes sediments drop out of, or precipitate from, chemical solutions. Also, the bodies of some organisms have hard, mineral-rich shells or similar structures. When these organisms die, the structures can break down to form sediments.

6.1 Section Review

1. Using the information in this section copy and complete the concept map at right.

2. Distinguish between a rock and a mineral. How are they similar? How are they different?

3. Describe the most direct cycle of relationships between the three major rock types. Then create a chart of the other ways that rocks change. First list the rock, then the process it undergoes, and finally the result.

4. **CRITICAL THINKING** According to the information in this section, what are the similarities and differences between igneous rocks and metamorphic rocks?

5. **PAIRED ACTIVITY** Create a concept map that shows at least three different ways in which metamorphic rock can be formed and changed.

MONITOR AND RETEACH

If students miss . . .

Question 1 Reteach "What Is a Rock?" (p. 118) Have students summarize in their own words the bulleted information about different types of rocks.

Question 2 Review the photographs on page 118. Have students note that the minerals shown are found in the rock. Lead students to conclude that minerals are the building blocks of rocks.

Question 3 Use Transparency 6 to go through the rock cycle diagram. (p. 119) Explain the interrelationships represented by the cycle arrows.

Question 4 Reteach "The Rock Cycle." (pp. 119–120) Stress that melting is only involved in the formation of igneous rocks, not metamorphic rocks.

Question 5 Reteach the processes involved in the rock cycle. (pp. 119–120) Have students explain each process in their own words.

Igneous Rocks

Although all igneous rocks begin as magma, they have a variety of compositions and textures. Their differences are the result of variations in the composition of the magma and in the formation process.

Igneous Rock Formation

As with other rocks, igneous rocks are classified by their mineral composition and texture. Some igneous rocks form from volcanic ash. Most form directly from magma. The location of the magma determines the rate at which it cools, which determines the texture of the resulting rocks. Igneous rocks formed from underground magma are called intrusive igneous rocks. Those formed at Earth's surface are called extrusive igneous rocks.

The Starting Material

Magma may be classified as felsic, mafic, or an intermediate form. **Felsic** (FEHL-sihk) magma is thick and slow-moving. It contains large amounts of silica (SiO_2) and smaller amounts of the elements calcium, iron, and magnesium. This magma typically hardens into rocks of light-colored silicate minerals such as quartz and orthoclase feldspar.

Compared with felsic magma, **mafic** (MAF-ihk) magma is hotter, thinner, and more fluid, containing large amounts of iron and magnesium and much lower amounts of silica. Rocks formed from mafic magma usually contain large amounts of dark silicate minerals such as hornblende, augite, and biotite.

Underground Magma

Magma may harden slowly or quickly. The rate of cooling and the texture of the rock that forms depend on where the cooling occurs. Magma trapped deep in Earth's crust hardens very slowly to form intrusive igneous rocks. Evidence suggests that massive bodies of intrusive rock may take thousands of years to cool underground. Intrusive rocks appear at Earth's surface when they are uplifted and the overlying rock is worn away.

LAVA This lava is magma that poured onto Earth's surface during a volcanic eruption. It will harden to form extrusive igneous rock. Magma that hardens below the surface forms intrusive igneous rocks.

6.2

KEY IDEA

Igneous rocks form from magma in distinct ways and can be classified based on mineral composition and texture.

KEY VOCABULARY
- felsic
- mafic
- pluton
- batholith

CHAPTER 6 SECTION 2

FOCUS

Objectives
1. Distinguish between intrusive and extrusive igneous rocks and how they form.
2. Contrast the types of plutons that form as the result of intrusive igneous activity.

Set a Purpose
Have students read this section to find the answer to "How do igneous rocks form and how do they differ in composition?"

INSTRUCT

> **More about...**
>
> **Mineral Crystallization**
> Minerals crystallize from cooling magma in a specific order known as Bowen's reaction series. Those with the highest melting temperatures crystallize first. The series has two parts: the discontinuous series and the continuous series. Mafic minerals crystallize in the discontinuous series: olivine, pyroxene, amphibole, biotite, potassium feldspar, muscovite, and quartz. In the continuous series, calcium-rich feldspars crystallize first, followed by both calcium-rich and sodium-rich feldspars, then finally sodium-rich feldspars.

Chapter 6 Rocks 121

DIFFERENTIATING INSTRUCTION

Reading Support Have students identify the various adjectives used on this page to describe felsic and mafic magmas. *felsic magma—thick, slow-moving, silica-rich, light-colored; mafic magma—hot, thin, fluid, silica-poor, dark-colored* Have students make tables that use the adjectives to summarize magma properties. As you teach this section, have students add pertinent information to their tables.
Use Reading Study Guide, p. 18.

CHAPTER 6 SECTION 2

VISUAL TEACHING
Discussion
Refer students to the diagram on this page. Ask: Which of the rocks shown is an intrusive igneous rock? Support your choice. **Gabbro is an intrusive igneous rock as shown by its coarse-grained texture.** How do such rocks form? **Gabbros form when magma slowly cools deep beneath Earth's surface.** Explain how the extrusive rock shown forms. **Basalt forms along mid-ocean ridges (right side of diagram) as magma flows onto Earth's surface and quickly cools. Basalt also forms when lava erupts from volcanic mountains (left side of diagram).** From which type of magma do basalts and gabbros form? **Rocks in the gabbro family are dark in color and thus form from mafic magmas.**

> **VOCABULARY STRATEGY**
> The words *intrusive* and *extrusive* share a word root that means "push" or "thrust." The prefix *ex-* changes the meaning to "pushed out." The prefix *in-* alters the meaning to "pushed between or within."

Intrusive rocks have a coarse texture because the magma from which they formed cooled very slowly. Slowly cooling magma remains liquid for a long time, and its atoms move about quickly and freely. As you may recall from Section 5.2, large, well-defined crystals form only when space is available for crystal growth. The longer the magma stays liquid, the longer the atoms are free to move and the larger the crystals become. Such intrusive rocks have a granular, or coarse-grained, texture.

At the Surface
When magma pours onto Earth's surface during a volcanic eruption, it is called lava, and that cooled, hardened lava is known as volcanic rock or extrusive igneous rock. Extrusive rock hardens rapidly, sometimes within a few hours or days. However, it can take years for large lava flows to cool and harden completely. You will learn more about lava in chapter 9.

The particles within rapidly cooling magma have little time to move around. Crystals have such a short time to form that extrusive rocks may have only microscopic crystals or no crystals at all. Extrusive rocks with tiny crystals have a fine-grained texture. Those without crystals are smooth as glass and so are said to have a glassy texture.

Sometimes underground magma begins cooling slowly, as it would during the formation of intrusive rocks, but then is suddenly forced to the surface. During the period of slow cooling, some large crystals form. However, once the partially crystallized magma arrives at the surface, the remaining liquid cools quickly. Microscopic crystals form around the larger, original crystals. The result is porphyry (POR-fuh-ree), an igneous rock in which large crystals are surrounded by a fine-grained mass of rock.

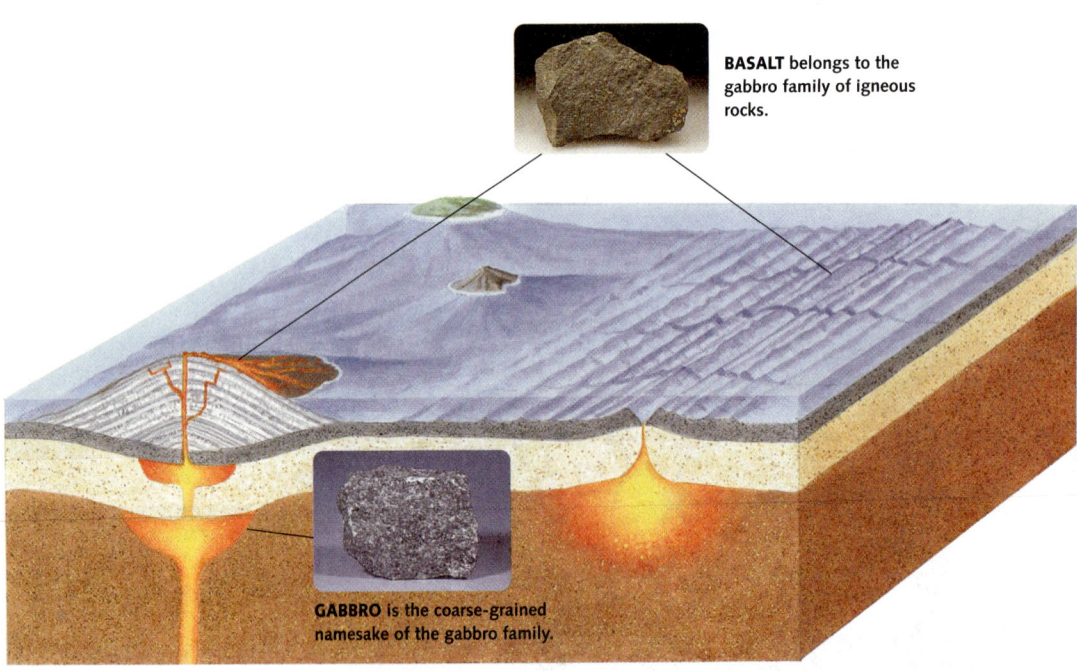

BASALT belongs to the gabbro family of igneous rocks.

GABBRO is the coarse-grained namesake of the gabbro family.

122 Unit 2 Earth's Matter

DIFFERENTIATING INSTRUCTION

Support for Hearing-Impaired Students Write on index cards in large letters the names of the various rocks referred to in this section. As you refer to each type of rock, point to the card.

Developing English Proficiency Help students differentiate between extrusive and intrusive igneous rocks. Write *Exit* on the board and ask if anyone has seen this word above a door. Lead students to realize that an exit is a passage to the outside. Relate this meaning to the meaning of the word *extrusive* as it refers to rocks. Then ask in which direction people go if they go *in* the doorway of a building. **toward the inside** Relate this meaning to the term *intrusive*. Use *Spanish Vocabulary and Summaries*, pp. 11–12.

Magma also reaches Earth's surface when expelled forcefully into the air as ash. The ash settles onto Earth's surface and, if present in large enough amounts, may eventually be buried and compressed into a rock called tuff. Although tuff originates from volcanic particles, some scientists consider it a transitional rock, sharing features of both igneous and sedimentary rocks.

Igneous Rock Descriptions

Igneous rocks are grouped into families according to mineral composition. A family may include intrusive and extrusive rocks. Where an igneous rock forms determines its texture, so each family may have coarse-grained, fine-grained, and glassy members. Specific igneous rocks can be recognized by their mineral composition and texture.

Granite Family

Rocks in the granite family form from felsic magmas. These rocks are usually coarse-grained because their slow-rising, "sticky" parent magmas tend to cool slowly underground. Members of this family typically contain quartz, felspar (orthoclase, plagioclase, or both), mica, and hornblende.

Granite, for which this family is named, is one of the coarsest-grained rocks in the family. Because granites often contain large amounts of light-colored feldspar, the color of this mineral usually determines the color of the granite. Granite usually ranges from white or gray to pink.

Granite is an intrusive igneous rock. It appears at the surface only after uplifting, or after erosion has removed thousands of meters of overlying rocks. Granite is a very common continental igneous rock found in many mountainous areas across the country.

How Do Igneous Rocks Form? Watch animations of igneous minerals crystallizing. Examine images of real rocks and classify how each one formed.
Keycode: ES0603

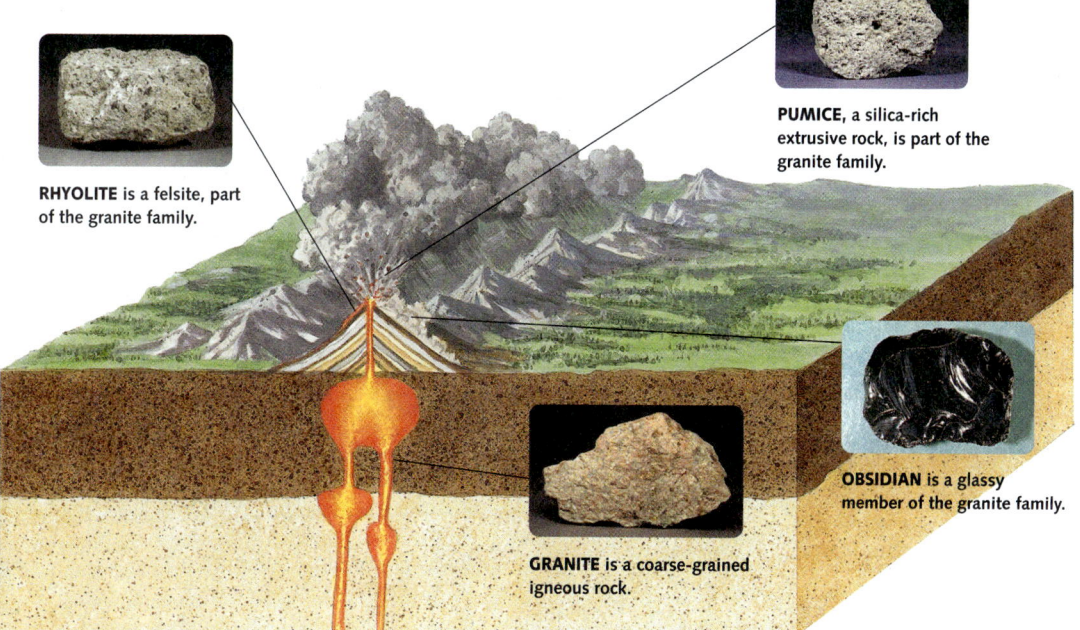

RHYOLITE is a felsite, part of the granite family.

PUMICE, a silica-rich extrusive rock, is part of the granite family.

OBSIDIAN is a glassy member of the granite family.

GRANITE is a coarse-grained igneous rock.

CHAPTER 6 SECTION 2

INVESTIGATIONS
CLASSZONE.COM

How Do Igneous Rocks Form?
Keycode: ES0603

- As students view the animations, have them note with short numbered steps the processes involved in the formation of each type of rock.
- If actual samples of the rocks shown in the photographs are available, have them displayed for students to see.

📖 *Use Internet Investigations Guide, pp. T23, 23.*

VISUAL TEACHING
Discussion

Have students examine the images on pages 122 and 123. Ask them to compare and contrast rocks of the granite family. **all except obsidian are light-colored; pumice, obsidian, rhyolite are fine-grained, extrusive; granite is coarse-grained, intrusive** Have students identify where rhyolite forms. **at the surface, at a volcano** Ask how this location could influence the rock's texture. **rapidly cooling magma such as this tends to form microscopic crystals or no crystals, resulting in fine-grained rock**

CRITICAL THINKING

Tell students that most obsidian has the same composition as rhyolite. Ask: Why then are the two rocks such different colors? **Tiny amounts of iron oxides throughout obsidian give the rock its dark color.** Why do you think obsidian appears opaque? **As a result of its fast rate of cooling, the grains that make up obsidian are so abundant that the rock appears opaque.**

DIFFERENTIATING INSTRUCTION

Reading Support Tell students that identifying cause-and-effect relationships can help them remember information throughout the chapter. Tell them that a cause is anything that makes something happen. The effect is the thing that happens. Model identifying cause and effect, using this sentence from the first paragraph of "Granite Family": *These rocks are usually coarse-grained because their slow-rising, "sticky" parent magmas tend to cool slowly underground.* Point out that the word *because* gives a clue of cause and effect. Ask: In this sentence, what causes something to happen? **slow-rising, sticky parent magmas tend to cool slowly underground** What happens because of this event? **rocks are usually coarse-grained** Tell students to look for other cause-and-effect relationships as they read.

CHAPTER 6 SECTION 2

Scientific Thinking

HYPOTHESIZE

Sample response: Basalt is an igneous rock, which forms at high temperatures. The Mesopotamians possibly collected the silt, packed it into blocklike forms, applied pressure to remove much of the water, and allowed the sediment to dry in the hot sun.

Scientific Thinking

HYPOTHESIZE

In an ancient Mesopotamian site in modern-day Iraq, archaeologists discovered rectangular slabs of rock that looked very much like basalt. However, the scientists realized that the ancient Mesopotamians had made the rock out of river silt, possibly because they lacked stone for building or grinding grain.

Based on what you know about basalt formation, propose a process that the Mesopotamians might have used to change river silt into flat, rectangular, basalt-like slabs.

VISUAL TEACHING

Discussion

Refer students to the summary table. Ask: **What is the intrusive equivalent of andesite?** diorite **Which three igneous rocks are composed mostly of olivine and pyroxene?** peridotite, dunite, pyroxenite **Which rock is most like diabase in texture and composition?** gabbro **Which rocks have glassy textures?** obsidian and basalt glass **What is the extrusive equivalent of granite?** rhyolite

CRITICAL THINKING

Inform students that viscosity is a measure of a material's resistance to flow. An increase in silica increases the viscosity of magma. Ask: **Which type of magma—felsic, intermediate, mafic, or ultramafic—has the highest viscosity?** felsic magma because of its high silica content

---

Obsidian, a volcanic rock with a glassy texture, is moderately hard (about 5 on the Mohs scale) and brittle with conchoidal fracture. Obsidian's chemical composition resembles that of granite and other light-colored rocks, so it is considered a member of the granite family. However, obsidian is usually dark brown or black due to tiny amounts of dark-colored iron oxides scattered throughout the rock.

Pumice is formed from silica-rich lava that hardened as steam and other gases bubbled out of it. It resembles a sponge because of its many holes and air pockets. It is often light enough to float on water.

The granite family also includes felsite, the general name for any light-colored, fine-grained rock. Rhyolite, a fine-grained rock that ranges from light gray to pink, is a common example.

Gabbro Family

The gabbro family consists of mafic rocks. They are dark in color and denser than rocks in the granite family. The dark minerals pyroxene and olivine, as well as plagioclase feldspar, are the most plentiful minerals in a gabbro rock. Other minerals often found in gabbros are amphibole and biotite. Gabbro, this family's namesake, is a coarse-grained rock, very dark in color.

The most common rock in the gabbro family is basalt. It has a composition similar to that of gabbro, but it is fine-grained. Basalt is typically dark gray or black and is the igneous rock that makes up the ocean floor. On land, basalt is the most common rock formed from lava flows.

Other members of the gabbro family include diabase, basalt glass, and scoria. Diabase has a composition identical to gabbro. Its texture is finer than gabbro but coarser than basalt. Basalt glass resembles obsidian but is mafic. Scoria, like pumice, is full of holes. However, scoria is made of denser minerals, and because its holes are commonly larger, it is unlikely to float.

Diorite Family

Members of the diorite family have an intermediate composition that is neither felsic nor mafic but has characteristics of both. Their colors tend to be medium grays and greens—darker than the granites and lighter than the gabbros. Diorite, a coarse-grained member of the family, has less quartz than granite and less plagioclase feldspar than gabbro. Andesite is a fine-grained member of the diorite family.

SUMMARY	Igneous Rocks				
Texture	**Chemical Composition**				
	felsic	felsic-intermediate	intermediate	mafic	ultramafic
coarse-grained	granite	granodiorite	diorite	gabbro	peridotite, dunite, pyroxenite
				diabase	
fine-grained	rhyolite		andesite	basalt	
glassy	obsidian			basalt glass	
porous	most pumice			scoria	

DIFFERENTIATING INSTRUCTION

Developing English Proficiency Pair students learning English with proficient English speakers. Provide each pair with hand lenses and labeled hand specimens of each of the rocks in the table. Have students sketch each specimen and list its characteristics under the drawing. Characteristics should include color, texture, mineral composition, and type of rock (intrusive or extrusive). Have volunteers share their drawings and characteristics.

Other Igneous Rocks

Igneous rock compositions do not occur as distinct categories but as a range, loosely separated by gradual boundaries. For example, granodiorite is a coarse-grained rock with a transitional composition between that of granite and diorite. Some rocks, including coarse-grained, dark, and dense pyroxenite, dunite, and peridotite, have a chemical composition described as ultramafic. Ultramafic rocks consist chiefly of mafic minerals, the ferromagnesian silicates olivine and pyroxene. Scientists hypothesize that these rocks are similar to those of Earth's mantle.

Igneous Intrusions

Volcanoes give only a hint of the amount and the activity of the magma that exists below Earth's surface. Forces deep within Earth may push magma into fractures in the bedrock. Sometimes magma is squeezed between rock layers, forcing the overlying rocks upward to form domes. Great masses of magma may solidify far below the surface to form the cores of mountains.

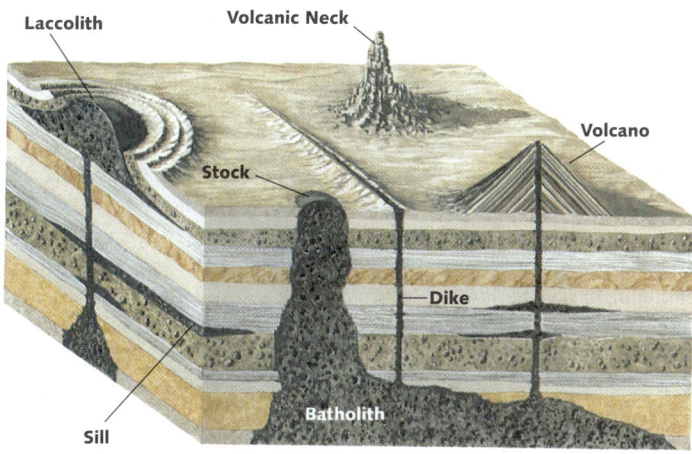

TYPES OF IGNEOUS FORMATIONS

In general, any igneous intrusion—a rock mass that forms when magma cools inside Earth's interior—may be called a **pluton.** Dikes, sills, laccoliths, and volcanic necks are sometimes called plutons. However, some scientists identify only the largest, thickest intrusions as plutons. A pluton reaches Earth's surface only after uplift, weathering, or both take place.

As the diagram above indicates, dikes and sills are sheets of magma intruded into previously formed rock. A dike is a sheet of igneous rock that cuts across rock layers vertically or at a steep angle. Dikes form when magma intrudes into angled cracks. Dikes may be hundreds of kilometers long and can range from about a centimeter to many meters thick. They are common in regions of volcanic activity.

As you can see above, a sill is a sheet of igneous rock that lies parallel to the layers it intrudes. A sill forms when magma is forced between, not across, rock layers. Sills can be hundreds of meters thick and many

SILL The dark rock layer above is a sill, a type of igneous intrusion, located in Big Bend National Park, Texas.

CHAPTER 6 SECTION 2

VISUAL TEACHING
Discussion

Ask these questions about the block diagram: What are the largest plutons called? **batholiths** How does a dike differ from a sill? **A dike cuts across rock layers, while a sill intrudes parallel to existing rocks.** How are dikes and sills similar? **Both are relatively small, sheetlike plutons.** How do you think a volcanic neck forms? **as the result of erosion**

CRITICAL THINKING

Have students use their knowledge of igneous rocks and the block diagram to answer the following questions: What type of texture do the rocks that make up plutons have? Explain your choice. **Plutons are coarse-grained intrusive rocks because they form as magma cools slowly beneath the surface.** Why do some scientists classify laccoliths as sills? **Laccoliths are parallel to the rocks into which they intrude and thus can be classified as sills.** How do plutons change the surrounding country rock? **Igneous intrusions often cause metamorphism of surrounding country rock.**

DIFFERENTIATING INSTRUCTION

Reading Support Tell students that identifying the main idea of a paragraph can help them understand and remember what they read. Have them read the first paragraph under "Igneous Intrusions." Then ask: What is the main idea of the paragraph? **Volcanoes represent only part of the amount of and activity of magma underground.** Then have students identify details from the paragraph that support the main idea. **Forces within Earth may push magma into fractures of the bedrock; magma may be squeezed between rock layers; magma may solidify far below the surface.**

CHAPTER 6 SECTION 2

More about...

Batholiths
The Sierra Nevada batholith formed tens of millions of years ago when two tectonic plates converged. Partial melting of the subducted plate generated magma, some of which lithified to form the batholith. Uplift and erosion have exposed the pluton at Earth's surface.

ASSESS

1. Both—types of molten rock, contain silica, solidify to form igneous rocks; felsic—thick, slow-moving, more silica, less calcium, iron, and magnesium, forms light-colored rocks; mafic—hotter, thinner, more fluid, less silica, more iron and magnesium, forms dark-colored rocks.

2. Magma's cooling rate affects rock texture. Slow cooling results in larger mineral grains. Rapid cooling results in fine-grained rock.

3. granodiorite, peridotite, dunite, and pyroxenite; their composition

4. A thicker felsic magma might not flow between rock layers as easily as a thinner, more fluid mafic magma, and might tend to push upward to form a dome instead of spreading out to form a sill.

5. obsidian; because of its brittleness, texture, and conchoidal fracture

6. low amounts; thin, fluid mafic magma is characterized by low amounts of silica

BATHOLITH The Sierra Nevada in California is part of a batholith, the largest of the igneous intrusions.

kilometers long. The exposed sill that makes up the Palisades along the Hudson River in New York is about 80 kilometers long.

Some magmas are stiff and do not flow easily. Instead of spreading into sheets, these magmas can bulge upward to form domed masses called laccoliths. The rock layers above a laccolith are also pushed upward to form a dome. (See photograph on page 125.) Laccoliths exist in the Henry Mountains in Utah.

As an inactive volcano erodes, a volcanic neck may be exposed. A volcanic neck is the central plug of hardened magma left after the volcanic material around it has worn away.

Batholiths, the largest of all plutons, form the cores of many of Earth's mountain ranges. They are usually made of either granite or granodiorite and can span tens of thousands of square kilometers. A batholith is exposed through uplift and erosion of the overlying rock layers. A small batholith, in which less than 100 square kilometers is exposed at the surface, is called a stock. The largest batholith in North America forms the core of the Coast Range of southern Alaska and British Columbia, Canada.

6.2 Section Review

1. Create a Venn diagram that illustrates the differences and similarities of felsic and mafic magmas.

2. Why do igneous rocks have different textures?

3. What types of igneous rocks do not belong to any of the three major igneous rock families? What traits set them apart?

4. Describe the factors that might cause magma to form a laccolith instead of a sill.

5. **CRITICAL THINKING** Stone arrowheads are made by chipping smooth, curved pieces from a rock to form a sharply pointed triangle with razor-sharp edges. Which rock, obsidian or gabbro, would make a better arrowhead? Why?

6. **CHEMISTRY** A sample of magma flows very quickly. Would you expect it to contain high or low amounts of silica? Why?

126 Unit 2 Earth's Matter

MONITOR AND RETEACH

If students miss...

Question 1 Reteach "The Starting Material." (p. 121) Have students list characteristics of felsic and mafic magmas.
Question 2 Have students review the photographs on pages 122 and 123. Point out that rocks that form at the surface are fine-grained. Rocks that form at depth are coarse-grained.
Question 3 Use the table on page 124 to differentiate among the three major igneous rock families and rocks that do not fit a distinct family.
Question 4 Reteach igneous intrusions. (pp. 125–126) Demonstrate how the viscosity of a substance such as honey, which can be used to model magma, changes with temperature.
Question 5 Review the photographs of gabbro (p. 122) and obsidian. (p.123) If possible, have students observe hand specimens of each.
Question 6 Reteach "The Starting Material." (p. 121) Have students compare and contrast the two types of magma.

Sedimentary Rocks

Although Earth's crust consists primarily of igneous rock, most of the crust's surface is covered by sedimentary rock.

Formation of Sedimentary Rocks

In simple terms, sedimentary rock forms through the compacting and cementing of layers of sediment. However, sedimentary rocks are classified by three basic formation processes.

Clastic Rocks

Clastic sedimentary rocks are formed from fragments of other rock. The fragments come from the weathering of igneous, metamorphic, and sedimentary rocks. The fragments may be the size of pebbles, gravel, grains of sand, tiny particles of silt, or microscopic flakes of clay.

The formation of clastic rocks begins with the movement and relocation of the fragments. The majority of these rock fragments are collected and moved by running water. When carried by a stream or river, the fragments become smooth and rounded from rubbing against one another and the stream bed. The farther the particles are carried, the more rounded they become. When a stream flows into a body of water, such as a lake or an ocean, it typically slows down and drops all but the smallest particles, as illustrated below. Waves and currents may then redistribute the sediment over great distances.

Larger pebbles and gravels are often the first to drop out and settle in the shallow water near shore. Next to settle are the smaller sands and finally, in calm water, the silts and clays. The sorting process is not always

6.3

KEY IDEA
The three major types of sedimentary rocks are categorized by the way they form and by their distinctive features.

KEY VOCABULARY
- cementation
- stratification
- fossil

 VISUALIZATIONS
CLASSZONE.COM
Observe how sediments are deposited.
Keycode: ES0604

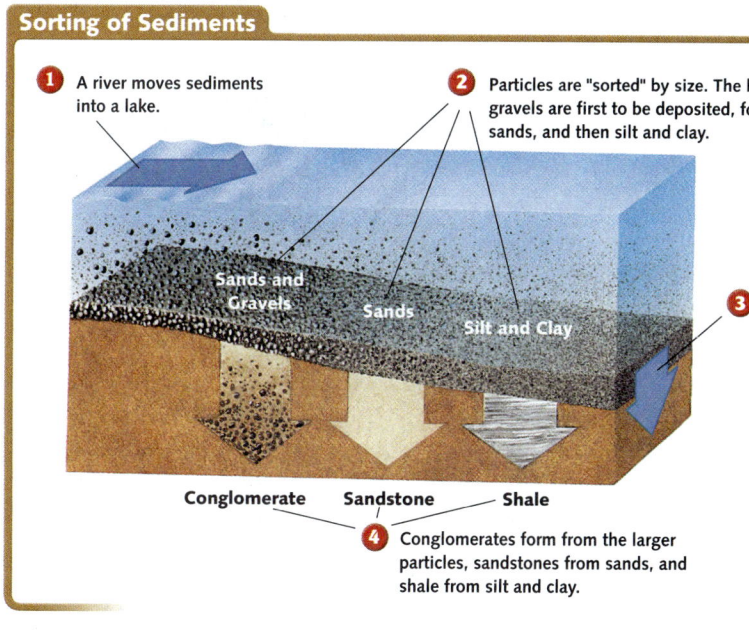

Sorting of Sediments

1. A river moves sediments into a lake.
2. Particles are "sorted" by size. The largest gravels are first to be deposited, followed by sands, and then silt and clay.
3. Over time, the sediments are buried and compacted. The particles may become cemented together.
4. Conglomerates form from the larger particles, sandstones from sands, and shale from silt and clay.

Conglomerate — Sandstone — Shale

Chapter 6 Rocks **127**

CHAPTER 6 SECTION 3

FOCUS

Objectives
1. Distinguish among the three types of sedimentary rocks and how they form.
2. Discuss different features typical of sedimentary rocks.

Set a Purpose
Have students read to find the answer to "What are sedimentary rocks and how do the different kinds differ from one another?"

INSTRUCT

Visual Teaching

Interpret Diagram

Ask: According to the diagram, what types of sediments make up conglomerates? **sands and gravels** Which sediments consist of the smallest particles? **silt and clay**

Extend

Contrast the textures of conglomerates and shales. **Conglomerates are coarse-grained; shales are fine-grained.**

 VISUALIZATIONS
CLASSZONE.COM

Observe how sediments are deposited.
Keycode: ES0604

🔘 Visualizations CD-ROM

DIFFERENTIATING INSTRUCTION

Hands-On Demonstration Half fill a large jar with water. Have students observe as you slowly add two tablespoons each of brightly colored aquarium gravel, aquarium sand (which is a different color from the gravel) and baking powder. Gently turn the jar over several times and have students observe the settling rates. The gravel is the first to reach the bottom of the jar, followed by the sand, and finally the baking powder.

⚠ *Exercise caution when handling the glass jar. Remind students that safety is everyone's responsibility.*

Reading Support Previewing the heads in the section helps prepare students for reading by organizing their thoughts and questions. Have students skim the blue heads to find the three types of sedimentary rocks. Then ask: What can you find out by reading the black subheads on pages 130 and 131? **the features of sedimentary rocks**

Use Reading Study Guide p. 19.

Chapter 6 Rocks **127**

CHAPTER 6 SECTION 3

VISUAL TEACHING
Discussion
Reinforce understanding of the formation of clastic rocks. Ask: What happens to sediments carried by a stream when the stream loses energy? **They are deposited.** How do sediments deposited near the mouth of the stream differ from those carried out into the standing body of water? **largest sediments deposited near mouth of stream; smaller sediments carried farther into body of water** What occurs during compaction? **Sediments are pressed together due to the overlying sediments.** What is cementation? **process during which loose sediments become bound together by natural cements**

More about...
Porosity and Permeability
Porosity is the percentage of the total volume of rock or sediment that consists of pore spaces. Permeability is the property of a material that allows fluids or gases to pass through it. Many sandstones are both porous and permeable, making them excellent petroleum and natural gas reservoirs. Shales, as a result of their low porosities and permeabilities, are often the cap rocks in such reservoirs that prevent hydrocarbons from migrating.

Observe an animation of clastic sedimentary rocks forming.
Keycode: ES0605

perfect. Sand is sometimes found mixed with pebbles and gravels in shallow water or with silts and clays in deeper water.

Loose sediments become solid clastic rock when particles of sediments become cemented. Ocean water, lake water, and groundwater all contain natural cements in the form of dissolved minerals. These natural cements include silica (SiO_2), calcite ($CaCO_3$), iron oxide (such as hematite, Fe_2O_3), and clay minerals. When minerals fill the spaces between sand grains, pebbles, or other rock particles, they bind the fragments together in a process called **cementation.** Over time, cement transforms loose sediments into bound sedimentary rock. The type of cement influences a rock's color. Rocks with silica and calcite tend to be gray or white, while iron-based cements help form red, brown, or rust-colored rocks. In some sedimentary rocks, though, the pressure of overlying sediments enables fine sediments to stick together without cement.

Conglomerates are the coarsest clastic rocks. A typical conglomerate is a cemented mix of rounded fragments (typically pebbles and sand grains) that were deposited in rough water. Quartz is a common ingredient because it is so durable. Breccia (BREHCH-ee-uh) is a conglomerate made of angular fragments surrounded by finer grains.

Sandstones made of quartz sand grains are rough, gritty, and durable, if they are well cemented. Although cement and fine particles tend to fill the space between such fragments, up to 30 percent of a sandstone may be unfilled space. As a result, sandstones are typically both porous (filled with small holes) and permeable (capable of passing water through it).

Silts and clays (usually tiny flakes of clay minerals) form shale. The spaces between the clay particles in shale are so small that it is virtually impermeable. Shales are smooth, soft, and easily broken.

SUMMARY — Clastic Sedimentary Rocks

Conglomerate	Sandstone	Shale
large fragments ranging from gravel to boulders	particles up to about the size of the head of a pin	sediments about the size of flour particles
any mineral composition	commonly quartz grains	clay minerals such as kaolinite
porosity and permeability depend on the degree of cementation	often highly porous, often highly permeable	not very porous, nearly impermeable

---

DIFFERENTIATING INSTRUCTION

Challenge Activity Provide students with hand lenses and specimens of conglomerate, sandstone, and shale. Have them identify the sediments in each, as well as the cement in the conglomerate and the sandstone. Challenge students to list the steps involved in the formation of each sample. **Answers should be specific to each of the rocks. Sample Response:** Clastic sedimentary rocks form when sediments are deposited. Compaction occurs over time as more sediments are deposited. Cements fill in the spaces among sediments, binding them to form rocks.

Chemical Rock

Sea, lake, swamp, and underground waters often contain dissolved minerals. Chemical sediments form when these minerals precipitate, or fall out of solution. Precipitation can occur through evaporation or through chemical action—for example, dissolved ions combining to form new minerals. Common chemical sedimentary rocks are rock salt, rock gypsum, and some limestones.

Rock salt, or halite, occurs in thick layers in many parts of the world. Rock gypsum occurs in layers or as nearly pure veins of the mineral gypsum. Although they are uncommon, limestones of chemical origin may form when tiny grains of calcite are deposited on the bottoms of seas or lakes. These limestones are often gray or tan, compact and dense, and smooth to the touch.

ROCK SALT FLAT at Bonneville, Utah

ORGANIC LIMESTONE CLIFFS in Dover, England

Organic Rock

An organic sedimentary rock forms from sediments consisting of the remains of plants and animals. Common organic sedimentary rocks are limestone and coal. Although coal consists primarily of carbon, it is formed from the fossilized remains of plants. The formation of coal is discussed in more detail in Chapter 7.

Organically formed limestones contain the mineral calcite. Limestone formation begins when water dissolves calcite out of rocks on land and carries it in the form of calcium ions to an ocean or lake. There, certain organisms use the ions to produce calcium carbonate shells or other support structures. Clams, corals, and some algae are just a few of these organisms, many of which live in shallow water near ocean shores. When the organisms die, their calcium-rich remains pile up on the ocean floor. The type of sedimentary rock formed from these sediments depends on the type of structures and the environment in which it forms. For example, the rock coquina develops when masses of whole or mostly whole shells are cemented together by minerals in ocean water. More often, waves break the shells into fragments. In time these fragments may become cemented into limestones. Limestones that form near shore may contain large amounts of clay. Those that form farther from shore may be almost pure calcite.

CHAPTER 6 SECTION 3

VISUALIZATIONS
CLASSZONE.COM

Observe an animation of clastic sedimentary rock forming.
Keycode: ES0605

Visualizations CD-ROM

DISCUSSION

Help students distinguish among the processes that result in the formation of chemical and organic sedimentary rocks. Ask: How do chemical sediments form during precipitation? **Minerals drop out of a solution.** What happens during evaporation? **A liquid changes into a gas.** How do minerals form during this process? **If the liquid contains dissolved solids, minerals will form as the liquid evaporates.** How do organic limestones form? **Certain organisms use calcium to produce calcium carbonate. When the organisms die, structures and shells of calcium carbonate accumulate and eventually form limestone.** Which process formed the rock salt flat in the photograph? **precipitation**

DIFFERENTIATING INSTRUCTION

Hands-On Demonstration Students may have difficulty understanding that water contains dissolved minerals. Make "hard water" by dissolving a little sodium bicarbonate (baking soda) and calcium chloride in water. Then freeze the solution into ice cubes. Place the ice cubes in hot water. As they melt, a layer of fluffy calcium carbonate will form at the bottom of the glass. Lead students to see that the "minerals" settled out of the hard water as its temperature increased.
⚠ *Exercise caution in handling calcium chloride.*

Reading Support Have students reread pages 127–129 and make a list of processes involved in the formation of sedimentary rocks. Then have them use context clues and the Glossary to write a brief definition next to each term. Challenge students to use each term correctly in a sentence.

CHAPTER 6 SECTION 3

MINILAB
Metamorphic Molds

Management
Choose an object with surface features or an identifiable shape.

Demonstrate two of three deformations (stretching, squeezing, and shearing). The original form should remain detectable.

Analysis Answers
Ideally features of the original form will be identifiable, though elongated, compressed, or twisted. The order of deformation may be identified if only two were made. This exercise models deformation of a fossil. The actual process requires intense pressure and stress over millions of years.

For proper behavior in a laboratory setting, refer students to Safety in the Earth Science Laboratory on pages xxii–xxiii.

More about...

Stratification
Rapid deposition can result in stratification called graded bedding. The sediments in such a bed are graded in size, coarse at the bottom, fine at the top.

25-Minute Mini LAB

Metamorphic Molds

Materials
- object to make mold
- modeling clay
- plaster of Paris
- water
- small cup
- stirring straw

Procedure
1. Press the object into a rectangular piece of clay so that plaster can be poured in the depression.
2. Deform the mold, noting the steps you take and the order in which they were done.
3. Mix plaster and water together until a runny mixture is obtained.
4. Pour plaster into the mold and wait 5–10 minutes for it to set.

Analysis
Can you recognize the orginal object from the resulting cast and detect the steps of the deformation? Trade casts with another person. Can you recognize this object? Can you determine individual steps of deformation? Describe how this exercise is similar to and different from metamorphic rock formation.

Features of Sedimentary Rocks

The single most characteristic feature of sedimentary rocks is **stratification**, the arrangement of visible layers.

STRATIFICATION A change in the type of sediment being laid down in one place results in the formation of a new rock layer. For example, when sand is deposited on top of clay, a layer of sandstone may form on top of a layer of shale. In this way, sedimentary rocks become stratified.

A bedding plane, the line between layers, separates each rock layer, or bed. Bedding planes are usually horizontal, but cross-bedding can occur when a river deposits sediments at an angle on sandbars, or when wind deposits beds at an angle on dunes.

Stratification occurs for a number of reasons. The river that brings sediment to an ocean or lake may break off and pick up new types of rock. A river in flood may carry larger amounts and different types of sediments. The river may carry the sediments farther out to sea than it normally does. Or it may drop its sediments closer to shore.

STRATIFICATION

FOSSIL

FOSSILS Sedimentary rock often contains fossils. A **fossil** is the remains, impression, or any other evidence of a plant or animal preserved in rock. Fossilization occurs when a dead plant or animal is buried by sediments that gradually turn to rock. The soft parts of the animals and plants usually decay, but the hard parts may become fossils. You will learn more about fossils in Chapter 29.

Splitting rock layers may reveal impressions in the rock, left when a shell or a skeleton was pressed into soft sediments. Plant impressions occur less frequently, but can be found in rocks formed from swamp sediments.

RIPPLE MARKS AND MUD CRACKS Two relatively common features of sedimentary rock are ripple marks and mud cracks. Ripple marks are sand patterns formed by the action of winds, streams, waves, or currents. You may have seen fresh ripple marks at a sandy beach or stream bed. Such ripple marks can be preserved when the sand becomes sandstone. Many sandstones show ripple marks on the surface of a bedding plane.

You may have seen fresh mud cracks where a muddy road or puddle of water has dried out after a rain. Mud cracks in sedimentary rock develop

---

MONITOR AND RETEACH

If students miss . . .

Question 1 Have students study the Sorting of Sediments illustration. (p.127) Have them explain in their own words the processes shown.
Question 2 Reteach the formation of sedimentary rocks. (pp. 127–129) Assist students with making a table similar to the one at the bottom of page 128 that includes chemical and organic rocks and their properties.

Question 3 Reteach the features of sedimentary rocks, using the photographs on pages 130 and 131. If possible, provide students with actual examples of each of the features presented.
Question 4 Reteach "Nodules, Concretions, and Geodes." (p. 131) Provide students with several different types of geodes that have been cut and polished. Have them note the ample space where crystals grow.

when deposits of wet clay dry and contract. The cracks are filled with different sediments and fossilize when the clay becomes shale rock.

RIPPLE MARKS

GEODE

NODULES, CONCRETIONS, AND GEODES Limestones and chalk often contain hard lumps of fine-grained silica called nodules. Whitish, brown, or gray nodules are called chert; darker varieties are sometimes called flint. Stone Age humans made many tools and weapons of chert and flint. The uniform fine grain and conchoidal fracture of these rocks made them easy to shape and to form sharp edges.

Round, solid masses of calcium carbonate called concretions often occur in layers of shale. Nodules and concretions probably form when minerals in a solution precipitate around a shell fragment or other impurity in the clay sediments.

Limestones sometimes contain spheres of silica rock called geodes (JEE-ohds). The interiors may be lined or completely filled with crystals—often quartz or calcite. Some geologists hypothesize that geodes form when groundwater creates cavities in limestone. Minerals in the groundwater then concentrate in the cavities, where they grow as crystals. Others propose that geodes form where older concretions have dissolved.

6.3 Section Review

1. From where a stream runs into a lake and out to the middle of that lake, in what order would you most likely find the three basic types of clastic sedimentary rock?
2. Create a chart that shows the differences and similarities between sedimentary rocks of chemical origin and those of organic origin.
3. List and describe three features of sedimentary rocks. Which feature is common to nearly every type of sedimentary rock?
4. **CRITICAL THINKING** Why is it possible to find large, well-formed crystals inside geodes?
5. **BIOLOGY** What type of environmental change might stop the development of an organic sedimentary rock? Explain.

CHAPTER 6 SECTION 3

ASSESS

1. conglomerate, sandstone, shale
2. Both are nonclastic rocks that form from sediments. Chemical rocks form when minerals precipitate from solution; organic rocks form from the carbon- or calcium-rich remains of plants or animals that settle in a freshwater or saltwater environment.
3. Stratification, the layering of sedimentary rock, is found in nearly all sedimentary rocks. Ripple marks are patterns formed by water or wind. Mud cracks form when deposits of wet clay dry and contract, fill with sediments, and then fossilize. Students may also list and describe fossils, nodules, concretions, or geodes.
4. The geode forms from cavities in limestone, thus there is room for crystal growth.
5. Any change that severely limits the number of organisms—for example, a drop or rise in ocean level, temperature, or salinity—could slow or stop the formation of organic rock.

MONITOR AND RETEACH

Question 5 Have students read "Organic Rock." (p. 129) Then discuss an example of how organic rock forms, focusing on the source and environment of the organic materials.

CHAPTER 6 SECTION 3

Learn more about how people use rocks.
Keycode: ES0606

Science and Technology

Obsidian in Aztec Culture

The Aztecs used obsidian probably more than any other culture. With no extensive knowledge of tool-making metallurgy, the Aztecs used obsidian for a variety of hunting tools such as arrowheads and spearheads, for cutting tools such as knives, and for weapons such as swords. Obsidian sword blades were extremely sharp weapons, but because the rock is brittle and cannot be reshaped, sword blades had to be replaced often. Some Aztecs wore highly polished obsidian as jewelry or used it for mirrors. Obsidian played a role in Aztec rituals and spiritual beliefs as well. After their conquest of the Aztec empire, the Spanish adopted obsidian for many uses, including blades for shaving. Obsidian remained important in Mexican technology until the late 1700s.

Science Notebook
Iron, steel, and diamond are among the materials that have replaced obsidian for cutting, splitting, and shredding.

SCIENCE & Technology

What Good Is a Rock?

Rocks exist in an astonishing variety. Their origins, the way in which they form, their chemical composition—these and other factors lend rocks a wide range of physical properties.

Which physical properties of rocks have humans used to their benefit?

Imagine a surgeon telling you she will use stone-age technology to operate on your eye. You might be tempted to run until you learn that she is referring to obsidian, the sharpest material available for scalpels—nearly 500 times sharper than surgical steel. Obsidian, an igneous rock, has been valued for its sharp edges since prehistoric times.

Obsidian, flint, chert, and chalcedony—all quartz-based igneous rocks with extremely fine crystals—have been prized for thousands of years by flint knappers. Flint knapping is the process of creating chipped stone tools such as knife blades, scrapers, arrowheads, and spear points. A small group of flint knappers carries on the tradition today by striking these rocks with harder rocks to create precise shapes and edges.

Because it is so sharp, obsidian was a valuable commodity and was widely traded during the Stone Age, at least as far back as 10,000 years ago. Ancient obsidian tools have been discovered at distantly scattered sites, even though there are actually very few sources of obsidian in the world. Scientists have learned to analyze trace elements in obsidian artifacts to match them to source areas. They have found that obsidian artifacts in Israel originated in Greece and other sites thousands of kilometers away. Ancient obsidian tools found in Alberta, Canada, have been traced to sources in Wyoming.

OBSIDIAN, an igneous rock that has sharp edges, is shown here in its natural form.

The study of obsidian artifacts allows archaeologists to gain a real sense of sophisticated interactions between distant communities of prehistoric people. ■

Extension

SCIENCE NOTEBOOK

Consider the ways in which sharp-edged stone tools might have been used in prehistoric times. Which materials seem to have replaced stone for these uses today? Write your answers in your science notebook.

Learn more about how people use rocks.
Keycode: ES0606

TOOLS AND WEAPONS, such as these knives and projectile-point artifacts, have been crafted from obsidian since prehistoric times.

Metamorphic Rocks

Metamorphic rocks are formed from preexisting rocks called **parent rocks.** As a result, a metamorphic rock often resembles its parent rock. Any differences between the two are a result of the metamorphic process the parent rock undergoes.

Metamorphic Processes

The process by which a rock's structure is changed by pressure, heat, and moisture is **metamorphism.** The pressure and heat can originate from Earth's internal heat, the weight of overlying rock, and the deformation of rock as mountains build. A metamorphosed, or changed, rock may have a chemical composition, texture, or internal structure that differs from that of the parent rock. Minerals may be enlarged or re-formed, or new minerals may appear. As shown below, pressure may force grains closer together, making the rock more dense and less porous.

6.4

KEY IDEA
Metamorphic rocks form when natural forces, such as heat and pressure, alter existing rocks.

KEY VOCABULARY
- parent rock
- metamorphism
- deform

Metamorphic Conglomerate

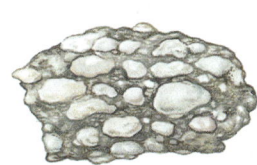
1 A conglomerate is a sedimentary rock consisting of pebbles and rounded rock fragments cemented together.

PRESSURE

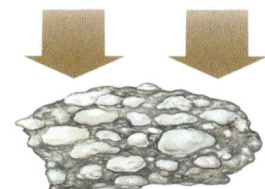
2 As the conglomerate undergoes metamorphism under high pressure, it flattens, stretches, and becomes more compact.

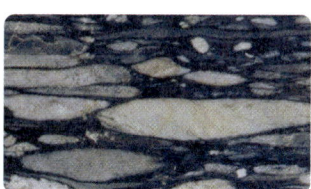
3 This is a photograph of a conglomerate that has undergone metamorphism. Note its compressed rock fragments and pebbles.

There are two basic types of metamorphism: regional and local. Regional metamorphism forms most of the metamorphic rock of Earth's crust; it occurs over very large areas. Local metamorphism occurs in much smaller, more distinct areas.

Regional Metamorphism

Regional metamorphism can occur during mountain-building movements of the crust, when large areas of rock change form after exposure to intense heat and pressure. As mountains form, deeply buried rocks are subjected to high heat and pressure. Due to Earth's internal heat, temperature increases with depth. The high pressure comes from the great weight of overlying rock. When the pressure is greater in one direction, minerals in the rock tend to align in layers. Hot liquids and gases in the deep rocks can help speed up the process.

CHAPTER 6 SECTION 4

FOCUS

Objectives
1. Explain the processes involved in the formation of metamorphic rocks.
2. Differentiate among different types of metamorphic rocks.

Set a Purpose
Have students read this section to find the answer to "What types of processes form metamorphic rocks, and how do these rocks differ from one another?"

INSTRUCT

Visual Teaching

Interpret Diagrams
Ask: How do the pebbles in the conglomerate differ from those in the metamorphosed rock? The pebbles in the conglomerate are rounded; in metamorphosed rock, elongated. The pebbles are closer together in metamorphic rock. What do you think happens to the matrix of a rock that is exposed to high pressure? The matrix becomes more dense.

Extend
Challenge students to draw "before"/"after" pictures of a coarse-grained granite that has been metamorphosed as the result of intense pressure. before—see photograph, p. 118; after—see photographs, p. 137

DIFFERENTIATING INSTRUCTION

Hands-On Demonstration Have volunteers help you embed rice grains into brightly colored modeling clay. Then cut the model rock in half. Have students describe the orientation of the long axes of the rice grains, which represent minerals. randomly distributed Apply moderate pressure to each half of the model rock by pressing down on the clay. Cut the model rocks in half again and have students describe the general orientation of the grains. perpendicular to the direction of force (pressure)

CHAPTER 6 SECTION 4

Visual Teaching

Interpret Diagram
Refer students to the diagram. Ask: What is responsible for the increase in temperature during contact metamorphism? **a magma chamber** What is the dark gray area in the figure? **the metamorphosed rock (hornfels)** How do temperatures and pressures change with distance from the magma body? **They decrease.**

Extend
Ask: What kinds of changes can occur during contact metamorphism? Explain. **changes in texture resulting from the pressure exerted on the parent rock by the magma chamber; changes in mineral composition caused by hot liquids and gases from the magma entering the intruded rock**

DISCUSSION

Help students differentiate among the three types of metamorphism. Ask: What are the three types of metamorphism? **regional, contact, and deformational** What temperature and pressure conditions are associated with regional metamorphism? **high temperature, high pressure** With contact metamorphism? **high temperature, low pressure** With deformational metamorphism? **low temperature, high pressure**

Observe an animation of metamorphic rocks forming.
Keycode: ES0607

The degree of regional metamorphism is influenced by the amount of heat, pressure, and fluids or gases to which the rock is exposed. Extremely high temperature and pressure can produce extreme metamorphism. Little metamorphism occurs when temperature and pressure are very low. Metamorphism occurs slowly when the parent rock is dry, and as a result fewer changes occur. These factors may occur in any combination. For example, a mass of rock may first experience moderate temperatures and high pressure and later be exposed to high temperature and low pressure.

Local Metamorphism

The two types of local metamorphism are called contact and deformational. Contact metamorphism occurs when hot magma moves into rock, heating and changing it. Hot liquids and gases from the magma may also enter the intruded rock and react with its minerals.

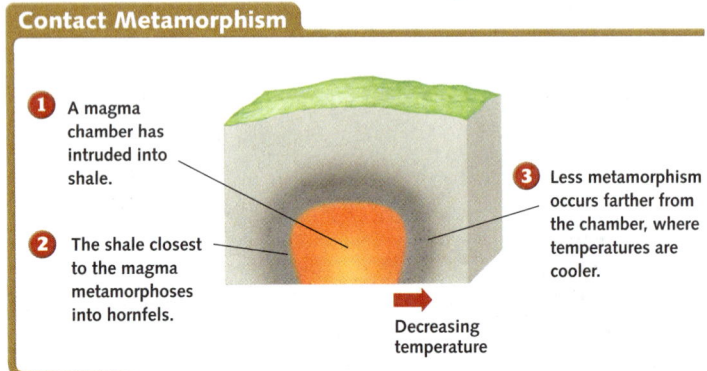

Contact Metamorphism
1. A magma chamber has intruded into shale.
2. The shale closest to the magma metamorphoses into hornfels.
3. Less metamorphism occurs farther from the chamber, where temperatures are cooler.

Decreasing temperature

Compared to regional metamorphism, contact metamorphism causes fewer changes in the rock and affects much less rock. The size of the affected area depends on the temperature of the magma and whether gases and fluids are present, but the area is rarely wider than one hundred meters. Shale that undergoes contact metamorphism may become hornfels, a dense, hard, and fine-grained rock.

Deformational metamorphism occurs at relatively low temperatures and high pressure caused by stress and friction, most often at faults where rock masses pass each other. As the masses move, heat from the friction, stress, and pressure cause the rock to **deform,** or change shape. The altered rocks usually have the same mineral composition of the rock around them but show changes in structure and texture.

Metamorphic Rock Descriptions

The descriptions and identifications of metamorphic rocks are often based on the parent rock, mineral content, and texture. Foliation, the tendency of a rock to form bands of minerals or split along parallel layers, can also help identify and classify metamorphic rocks.

DIFFERENTIATING INSTRUCTION

Reading Support Reiterate that metamorphism involves changes in rock texture, composition, or both as the result of changes in temperature and/or pressure and sometimes as the result of the presence of fluids. Help students make a table to summarize the parameters associated with regional, contact, and deformational metamorphism. **Information should be summarized as follows: regional metamorphism—high pressure, high temperature/changes in texture and composition; contact metamorphism—low pressure, high temperature/changes in composition; deformational metamorphism—high pressure, low temperature/changes in texture.**

Use Reading Study Guide, p. 20.

The properties of the metamorphic rock quartzite depend on the properties of its parent rock, sandstone. Some sandstones are made almost entirely of quartz sand grains. Quartzite consists largely of quartz crystal. Under heat and pressure, the spaces between the sandstone's grains are filled, and the rock recrystallizes into quartzite. Quartzite is a dense, nonfoliated, uniformly crystalline rock that is harder to break apart than the parent sandstone. An interesting type of quartzite called quartzite conglomerate is a metamorphosed conglomerate. Its parent rock was most likely a conglomerate composed largely of quartz fragments. Metamorphism changed the conglomerate into a "metaconglomerate."

Sandstone

Quartzite

QUARTZITE Sandstone (left) is the parent rock of quartzite (right). The sandstone consists of cemented quartz particles, such as the sand shown to the right of the sandstone.

Marble is a metamorphosed limestone with a relatively simple mineral content. When parent limestones consist almost entirely of the mineral calcite, the resulting marbles are also almost entirely calcite. Impurities in marble tend to appear as wavy streaks and color changes. These impurities are often introduced by fluids and gases flowing through the metamorphosing parent rock. Marbles are usually even-grained with a sugarlike texture.

Shells

Limestone

Marble

SEASHELLS (left), formed by marine animals out of carbon and calcium ions in seawater, form the sedimentary rock limestone (center). When it undergoes metamorphosis, limestone becomes marble (right).

When shale undergoes regional metamorphism, many changes in physical properties can occur. The rock becomes denser, and its grain size may increase. The elements recombine to form new minerals not found in shale, such as mica; this can cause foliation. Foliation occurs when pressure squeezes the flakes of mica into parallel layers, sometimes causing the new rock to split easily along the layers into thin, leaflike flakes.

The lowest degree of regionally metamorphosed shale is slate. In slate, the foliation layers are thinly spaced. The rock is usually very fine-grained and composed mainly of clay and quartz with microscopic flakes of

---

CHAPTER 6 SECTION 4

Observe an animation of metamorphic rocks forming.
Keycode: ES0607
Visualizations CD-ROM

VISUAL TEACHING
Discussion
Use the photographs on this page to discuss the following questions: What is the parent rock of quartzite? **sandstone** Does the metamorphism of sandstone to quartzite involve a compositional change, a textural change, or both? **textural** What is the parent rock of marble? **limestone** Does the change from limestone to marble involve changes in composition, texture, or both? **texture** Describe the texture of the metamorphic rocks shown on this page. **Quartzite is nonfoliated and uniformly crystalline. Marble is even-grained with a sugarlike texture.**

DIFFERENTIATING INSTRUCTION

Reading Support Have students reread page 135 to find the following related words: *metamorphic, metamorphosed, metamorphism, metamorphosing,* and *metamorphosis*. Have students use a dictionary to find the meaning of the Greek words *meta* and *morphe*, from which the words are formed. **meta**—"after"; **morphe**—"form" Relate these meanings to the meanings of the geologic terms. Have students define each term and use it correctly in a sentence.

CHAPTER 6 SECTION 4

VISUAL TEACHING
Discussion

Use the photographs on this page as the basis for the following questions: Which of the rocks shown in the photographs has undergone the highest degree of metamorphism? **schist** Which rocks most closely resemble each other? Why? **shale and slate because of the degree of metamorphism undergone by the shale** How does the slate differ from the phyllite? **The phyllite has pronounced foliation and is shinier than the slate.** What mineral gives slate and phyllite their somewhat shiny appearances? **mica** Other than changes in luster and color, what is the most obvious change that has occurred during the metamorphosis of a shale to a schist? **a change in texture, that is, an increase in grain size, or a change in mineral composition**

More about...

Metamorphosed Shales
Index minerals distinguish the degree of metamorphism undergone by a rock. In shale, low-grade minerals include chlorite and muscovite (white mica). Intermediate-grade metamorphic minerals include muscovite, biotite (black mica), garnet, staurolite, and kyanite. High-grade metamorphic minerals include biotite, garnet, and sillimanite.

Shale | Slate | Phyllite | Schist

SHALE (left) is sedimentary rock. Through the process of metamorphism, it becomes slate (center left), phyllite (center right), and schist (right).

muscovite or chlorite. These micas form in layers that produce slate's characteristic cleavage. When compared with its parent rock, slate is slightly shinier and has a flatter, less irregular surface.

With more intense metamorphism, the rock phyllite forms. Phyllite has pronounced foliation and a shiny surface. This shine is due in part to the increase in size of the mica grains. The tiny grains of mica are now large enough to be detected with the unaided eye.

Further metamorphism, usually the high temperatures and pressures of regional metamorphism, produces schist. The parent rocks include shales and basalts, so there are several varieties of schist. Schists are usually named for their most conspicuous mineral. For example, mica schist, talc schist, and hornblende schist are named for their most

CAREER

Marble Sculptor

Some of history's most celebrated artists, such as Michelangelo, were marble sculptors. Today, marble sculptors carry on this respected art form just using power hand tools, slabs of marble, and their artistic creativity. Depending on their style, some sculptors focus on classic subjects, such as human figures or religious themes, while others create modern, abstract works. Once completed, the sculptures are then exhibited and sold in art galleries. Sculptures are sometimes commissioned by patrons such as the government, schools, private foundations, churches, or individuals.

Being a successful sculptor requires imagination, fine motor skills, a ready supply of marble, and above all, patience. Because self-expression is so important to a marble sculptor, many artists are largely self-taught.

While a degree is not necessary, some find that a degree in fine arts is helpful in acquiring a sense of art history. Some sculptors also occasionally attend workshops, sculpting classes, and conferences to hone their skills and to share their work with colleagues. Part of a sculptor's training can come in the form of travel. As they visit cultural centers such as Carrara, Italy, which is famous for its sculpting tradition, sculptors gain a sense of what makes a true masterpiece. ∎

THIS MARBLE SCULPTOR in Tuscany, Italy, is creating a likeness of the human form.

Learn more about a career in sculpting.
Keycode: ES0608

DIFFERENTIATING INSTRUCTION

Challenge Activity Provide students with hand lenses and unlabeled hand samples of shale, slate, phyllite, and schist. Challenge students to identify the parent rock and then to arrange the metamorphic rocks from the lowest grade to the highest grade. Also have students identify as many minerals as they can in each specimen.

observable minerals. Schists are flaky rocks in which the foliation layers are easily seen. These rocks tend to be very large-grained and often contain large crystals of minerals not found in the parent rock.

Gneiss is one of the most highly metamorphosed of the metamorphic rocks. The various forms of gneiss metamorphose from a number of parent rocks, including shale, granite, and conglomerate. It is often difficult to determine the parent rock of a gneiss, so its name usually reflects its chemical composition. For example, the gneiss shown below is called a granite gneiss. It contains many of the same minerals found in granite.

Granite

Gneiss

GRANITE (left) is one of the parent rocks that can become gneiss (right).

Gneiss has the coarsest foliation of all the metamorphic rocks. Its minerals are arranged in wavy parallel bands, usually no more than a few centimeters thick. Bands of light-colored minerals such as quartz and feldspar alternate with dark minerals such as hornblende or biotite.

6.4 Section Review

1. What factors cause metamorphism? Which of those factors is most important for each type of metamorphism (regional, contact, and deformational)?
2. Name two examples of nonfoliated metamorphic rocks. Explain why they do not exhibit foliation.
3. Describe the metamorphism of shale and the rock sequence involved in that metamorphism.
4. **CRITICAL THINKING** Compare the formation processes of igneous, sedimentary, and metamorphic rocks. In general, which process can be the most direct, requiring the fewest number of steps? Which can be the most complex? Explain.
5. **VISUAL ARTS** Marble is a very popular stone for artists to use when carving sculptures. What properties of marble might make it good for sculpting?

CHAPTER 6 SECTION 4

Learn more about a career in sculpting.
Keycode: ES0608

ASSESS

1. Exposure to heat, pressure, or both; moisture (hot liquid or gas) can affect the process. Regional metamorphism: intense pressure and heat; contact metamorphism: heat; deformational metamorphism: high pressure, lower temperatures

2. quartzite and marble; their parent rocks, quartz sandstone and calcite limestone, respectively, generally consist of one, or mostly one, mineral that results in these uniform, even-grained rocks

3. Through regional metamorphism, shale is exposed to intense pressure and heat. From lowest to highest degree of metamorphism, shale becomes slate, phyllite, then schist. Through contact metamorphism, heat changes shale to hornfels.

4. Igneous rock formation is simple: the solidification of magma. Sedimentary rock formation is complex: uplift, weathering, erosion, deposition, burial, compaction, and cementation. Metamorphism can be complex—a parent rock undergoes many changes in chemical composition, texture, or internal structure.

5. even-grained, fine texture, few impurities

MONITOR AND RETEACH

If students miss . . .

Question 1 Reteach metamorphic processes. (pp. 133–134) Have students summarize the information in a table.
Question 2 Reteach "Metamorphic Rock Descriptions." (pp. 134–135) If possible, have students examine samples of these rocks.
Question 3 Reteach "Metamorphic Rock Descriptions." (pp. 135–136) Use the photographs at the top of page 136 to compare and contrast the changes that take place in the metamorphosis of a shale.
Question 4 Review the rock cycle. (p. 119) Redefine, if necessary, the major processes that are overprinted on the wide arrows.
Question 5 Reteach "Metamorphic Rock Descriptions." (p. 135) Have samples of different colored marbles available for students to examine with hand lenses. Remind students that calcite has a hardness of about 3 on the Mohs scale of hardness.

CHAPTER 6 ACTIVITY

PURPOSE
To identify the minerals contained in model rock thin sections and to classify the rocks based on their composition

PROCEDURE
1, 2. Rock A contains approximately 40% orthoclase feldspar, 30% quartz, and 10% each of plagioclase feldspar, biotite mica, and amphibole.

3. Rock B contains approximately 40% plagioclase feldspar, 30% amphibole, and 10% each of biotite mica, pyroxene, and quartz.

4. 4 cm

5. approximately 0.5 cm; Accept any reasonable value as long as the average is calculated correctly.

ANALYSIS AND CONCLUSIONS

1. Rock C is sedimentary, as shown by its rounded quartz grains that are cemented together. Rocks A and B are igneous, as indicated by their grains' interlocking pattern. Rocks D and E are metamorphic, as shown by their grains' linear pattern.

2. The actual size of an average grain is calculated to be approximately 0.06 cm, using the equation included on the next page. Accept any reasonable value as long as students have correctly performed the calculation.

CHAPTER 6 LAB Activity

Studying Rocks in Thin Section

SKILLS AND OBJECTIVES
- **Observe** and **interpret** several diagrams of rock thin sections.
- **Classify** and **identify** which minerals and rocks the diagrams represent.

MATERIALS
- metric ruler

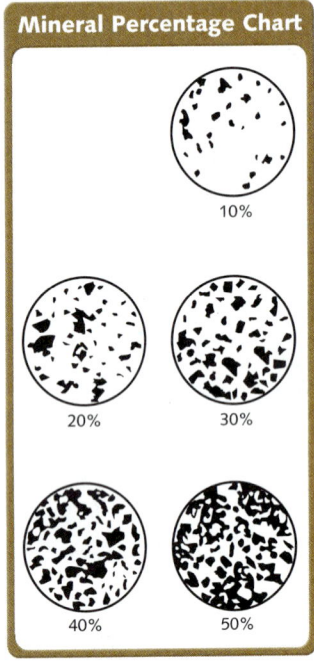

Mineral Percentage Chart (10%, 20%, 30%, 40%, 50%)

Have you ever tried to look through a rock? In addition to looking at hand-held rock samples, sometimes geologists need to see through the rock in order to study it. To do this, they make thin sections—slices of rock so thin that light actually passes through them! Geologists then use microscopes to analyze these sections. The magnified view allows scientists to see minerals that are not large enough to view in the handheld sample and to identify more easily the larger minerals that can be seen.

In this activity, you will look at some diagrams of thin sections. You will study minerals found in rocks and identify rocks containing such minerals.

Procedure

1. Look at the diagram of Rock A on the next page. Use the key to determine and list the name of each mineral found in Rock A.

2. Use the chart at left to estimate the percent of one mineral present in Rock A. Record the data on a separate sheet. Repeat for each of the minerals in Rock A. Your values should total 100%.

3. Repeat Steps 1 and 2 for Rock B.

4. Using the metric ruler, measure the diameter of the circular diagram for Rock C. Record your measurement.

5. Look at the mineral grains in Rock C. Measure the widths in any direction across five different mineral grains. Record your data. Calculate an average width for the grains.

Analysis and Conclusions

1. Grain or crystal size can provide clues to rock types. Typically, sedimentary grains are rounded and cemented together. Igneous grains fit together in a jigsaw-puzzle fashion. Metamorphic grains exhibit linear patterns and foliation. Which of Rocks A to E is sedimentary? Igneous? Metamorphic? Explain your answers.

2. Determine the actual size of an average grain of Rock C. The actual diameter of the rock sample shown in each diagram is 0.5 centimeters. Use your average grain diameter data and the magnified diameter you recorded. Show your work.

3. Using your answer to Question 2, determine the name of the average grain size in Rock C. The diameter of clay-size grains is less than 0.0004 cm, silt 0.0004 to 0.006 cm, sand 0.006 to 0.2 cm, and pebbles greater than 6.4 cm. Which kind of sedimentary rock is Rock C? Explain your answer.

4. In Rock A, which two minerals did you estimate make up over 50% of the rock? Which minerals make up the remainder of the rock?

5. In Rock B, which two minerals did you estimate make up over 50% of the rock? Which minerals make up the remainder of the rock?

6. Igneous rocks are commonly grouped into mafic rocks and felsic rocks, based on their chemical composition. Mafic rocks are dark in color because they contain many dark minerals such as amphibole, pyroxene, olivine, and biotite. Felsic rocks are light in color and tend to contain minerals such as quartz and orthoclase feldspar. Based on these definitions, is Rock A mafic or felsic? Rock B? Explain your answers.

7. Compare the diagrams for Rock C and Rock D. Look at each rock's texture in its crystal size, shape, orientation, and contact points with other crystals. How do the textures differ?

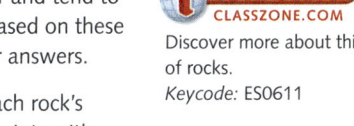

Discover more about thin sections of rocks.
Keycode: ES0611

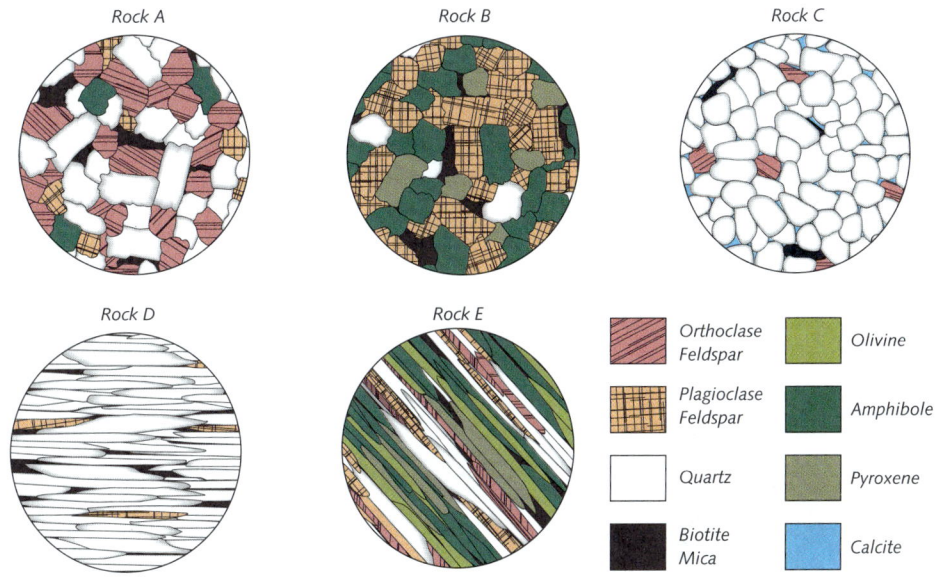

CHAPTER 6 ACTIVITY

Discover more about thin sections of rocks.
Keycode: ES0611

(continued)

0.5 cm (actual diameter of rock sample)/4.0 cm (measured diameter of rock sample from Step 4) = X cm (actual average diameter of grain)/0.5 cm (calculated average diameter of grain from Step 5)
$X = 0.06$ cm

3. The average grain size is 0.06 cm, which falls within the range of grain sizes classified as sand. Therefore, Rock C is a sandstone.

4. Orthoclase feldspar and quartz make up over 50% of Rock A. Plagioclase feldspar, biotite mica, and amphibole make up the other 50% of the rock.

5. Plagioclase feldspar and amphibole make up over 50% of Rock B. Pyroxene, biotite mica, and quartz make up the rest of the rock.

6. Rock A is felsic, because it contains mainly light-colored minerals such as quartz and orthoclase. Rock B is mafic as shown by the presence of mainly dark-colored minerals such as amphibole, pyroxene, and biotite.

7. The grains that make up Rock C are rounded and cemented together randomly. The grains that make up Rock D are elongated and fit together in a linear pattern.

Refer students to Safety in the Earth Science Laboratory on pages xxii–xxiii.

CHAPTER 6 REVIEW

Summary of Key Ideas

6.1 Igneous, sedimentary, and metamorphic rocks are formed, broken down, and reformed in a recurring process called the rock cycle.

6.2 Igneous rocks form from magma deep in the crust or from lava at Earth's surface. Igneous rocks are grouped into families by mineral composition and texture. Igneous rock texture depends mainly on the rate at which magma cools. Felsic magmas form light-colored, silica-rich rocks. Mafic magmas form dark-colored, rocks rich in iron and magnesiums.

6.3 Sedimentary rocks often occur in layers formed when different sediments are deposited on top of each other. Sedimentary rocks are grouped by the type of sediment from which they form: clastic, chemical, or organic. Clastic sediments are often sorted by water action before pressure and mineral cements turn them into rock. Fossils, ripple marks, mud cracks, nodules, concretions, and geodes are sedimentary rock features.

6.4 Metamorphic rocks form when heat or pressure or both alter parent rocks. A metamorphic rock may be described and identified according to its parent rock, mineral composition, and texture.

Key Vocabulary

batholith (p. 126)
cementation (p. 128)
deform (p. 134)
felsic (p. 121)
fossil (p. 130)
igneous (p. 118)
mafic (p. 121)
magma (p. 118)
metamorphic (p. 118)
metamorphism (p. 133)
parent rock (p. 133)
pluton (p. 125)
rock (p. 118)
rock cycle (p. 119)
sediment (p. 118)
sedimentary (p. 118)
stratification (p. 130)

Vocabulary Review

Explain the difference between the terms in each pair.

1. pluton, batholith
2. felsic magma, mafic magma
3. cementation, stratification

Write the term from the key vocabulary list that best completes the sentence.

4. Rocks classified as clastic, chemical, or organic are ___?___ rocks.
5. Evidence of an organism preserved in rock is called a ___?___.
6. If sedimentary or igneous rocks are subjected to heat and pressure, they become ___?___ rocks.
7. In the rock cycle, igneous rocks weather to form ___?___, which becomes sedimentary rocks.

Concept Review

8. Is it possible for rocks in the rock cycle to "skip" becoming sedimentary rock? Explain.
9. How does the cooling rate of magma affect the formation of igneous rock?
10. Why are fossils more likely to be formed in shale and sandstone than in conglomerate?
11. What characteristic might help you distinguish between a sandstone and a tuff?
12. **Graphic Organizer** Copy and complete the concept map below.

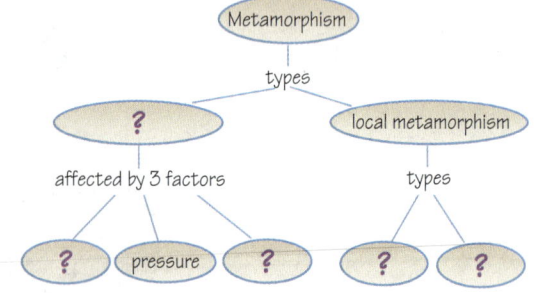

Critical Thinking

13. **Compare** Use the terms *felsic, mafic, plutonic,* and *volcanic* to compare each pair of igneous rocks: granite and gabbro, granite and rhyolite, gabbro and basalt, rhyolite and basalt.

14. **Apply** A newspaper reports that the fossil of an ancient plant has been found in a layer of gneiss. What can you conclude about the accuracy of the report, and why?

15. **Hypothesize** How might the locations of different types of sedimentary rocks help you map the boundary of an ancient inland sea?

16. **Infer** Prehistoric tools such as scrapers and knives are often made of chert and flint. What other type of rock has similar characteristics and might make similar tools?

Interpreting Graphs

The graph shows a classification system for sandstones made of varying amounts of three minerals: kaolin, feldspar, and quartz. Each corner shows a rock made up of 100% of the named mineral. The side opposite the same corner shows rocks with 0% of that mineral but 100% of the other two minerals. Determine the composition of a sandstone by finding the amount of each mineral in that sandstone. For example, to find the amount of feldspar in the sandstone associated with point X, start at the "0% Feldspar" side. Count the red lines from this side to point X. There are five lines. Each line is 10%; point X is on the 50% feldspar line. Likewise, point X is on the 30% kaolin line, and the 20% quartz line. It also falls within the graywacke area. Thus, point X shows a graywacke with 50% feldspar, 30% kaolin, and 20% quartz.

17. What is the composition and name of the sandstone at point Y?
18. What is the composition and name of the sandstone at point Z?
19. A sandstone contains 40% quartz, 30% feldspar, and 30% kaolin. What type of sandstone is it?

Internet Extension

What Kind of Rock Is This? Use a key to help you determine the type of rock you have found.
Keycode: ES0610

Writing About the Earth System

SCIENCE NOTEBOOK Research to find information about rocks in your region. Using the information, identify the types of rocks near you and explain how parts of the Earth system interacted to form the rocks. Describe what changes have helped shape the geologic features you see today.

20. Suppose a particular sandstone contains 95% quartz and 5% other minerals. What type of sandstone is it?

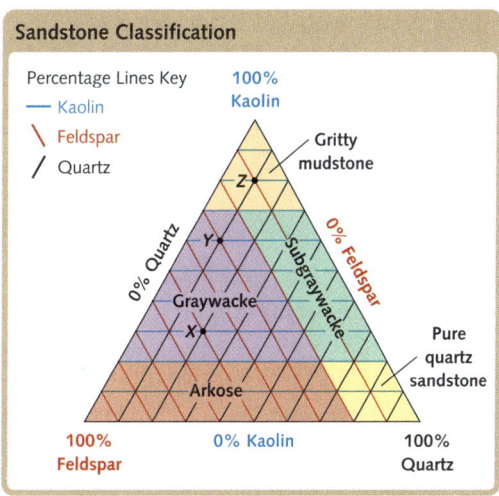

CHAPTER 6 REVIEW

What Kind of Rock Is This?
Keycode: ES0610
Review the rock key in class.
Use Internet Investigations Guide, pp. T24, 24.

CRITICAL THINKING

13. Granite is a plutonic felsic rock; gabbro is a plutonic mafic rock. Granite is a plutonic felsic rock; rhyolite is a volcanic felsic rock. Gabbro is a plutonic mafic rock; basalt is a volcanic mafic rock. Rhyolite is a felsic volcanic rock; basalt is a volcanic mafic rock.

14. The report is inaccurate. The heat and pressure involved in the formation of gneiss, a metamorphic rock, would destroy any organic material in the parent rock.

15. Coarse-grained rocks form in relatively shallow waters. Sands are deposited seaward of coarser deposits but landward of finer sediments. Because of their size, silts and clays can be carried out into deep waters.

16. The igneous rock obsidian, because of its hardness, conchoidal fracture, and fine grains, might make similar tools.

INTERPRETING GRAPHS

17. The sandstone at point Y is a graywacke made of 10% quartz, 60% kaolin, and 30% feldspar.
18. The sandstone at point Z is a gritty mudstone made of 80% kaolin, 10% quartz, and 10% feldspar.
19. graywacke
20. pure quartz sandstone

WRITING ABOUT SCIENCE

Writing About the Earth System

Your state geological survey and the United States Geological Survey are good sources of geologic maps, which show the ages and types of rocks at the surface. Students should also use field guides to identify the rocks and minerals in your area. Changes should include many of the processes involved in the rock cycle and, if applicable, glaciation and tectonism.

CHAPTER 7

PLANNING GUIDE: RESOURCES AND THE ENVIRONMENT

	Section Objectives	Print Resources
Chapter-Level Resources		Laboratory Manual, pp. 33–34 Internet Investigations Guide, pp. T25–T27, 25–27 Spanish Vocabulary and Summaries, pp. 13–14 Formal Assessment, pp. 19–21 Alternative Assessment, pp. 13–14
Section 7.1 Mineral Resources, pp. 144–147	1. Distinguish between renewable and nonrenewable resources. 2. Explain how the availability and use of minerals determine how long mineral reserves will last.	Lesson Plans, p. 21 Reading Study Guide, p. 21
Section 7.2 Energy Resources, pp. 148–153	1. Identify renewable and nonrenewable energy resources. 2. Explain how fossil fuels form. 3. Describe how humans use renewable and nonrenewable energy resources to meet their energy needs.	Lesson Plans, p. 22 Reading Study Guide, p. 22
Section 7.3 Environmental Issues, pp. 154–159	1. Describe how the use of nonrenewable and renewable resources affect the environment. 2. Explain how humans can slow the depletion of resources.	Lesson Plans, p. 23 Reading Study Guide, p. 23

INVESTIGATIONS
CLASSZONE.COM

Section 7.3 INVESTIGATION
What Happens When an Oil Spill Occurs?
Computer time: 45 minutes / Additional time: 45 minutes
Students examine the affects of an oil spill on the surrounding environment.

Section 7.3 INVESTIGATION
Why Is This Place Protected?
Computer time: 45 minutes / Additional time: 45 minutes
Students explore parks and monuments then decide if another place is worth protecting.

Technology/Online Resources

Electronic Teacher Tools
Visualizations CD-ROM
Classzone.com
Online Lesson Planner
Test Generator

Visualizations CD-ROM
Classzone.com
ES0701 Visualization, ES0702 Visualization

Classzone.com
ES0703 Investigation, ES0704 Data Center, ES0705 Investigation, ES0706 Career

Classroom Management

Gather these materials for minilab and laboratory activities.

Minilab, p. 150
- gravel
- clear beaker
- 2 straws
- modeling clay
- sand
- plastic tubing
- funnel
- water

Lab Activity, pp. 160–161
- 2 shoeboxes of the same size
- 1 brick
- 2 thermometers
- 1 sheet of aluminum foil
- 2 paper cups
- 2 pieces of plastic wrap
- masking tape
- black paint
- rubber cement
- 1 sheet of graph paper
- scissors
- hole punch
- colored pencils
- Laboratory Manual, Teacher's Edition
  Lab Sheet 7, p. 139

Chapter Review INVESTIGATION
What Environmental Changes Can We See with Satellites?
Computer time: 45 minutes / Additional time: 45 minutes
Students use images to investigate environmental change through time.

Chapter 7 Resources and the Environment

CHAPTER 7

Resources and the Environment

The air that moves the blades of these windmills near Palm Springs, California, is one of Earth's many resources.

What are Earth's resources, and how does the way we use them affect Earth's environment?

INTRODUCE
Build Interest
Before sharing background information about the photograph with students, allow time for their questions and comments.

Ask questions like the following:
- Describe the landscape you see. **large, open area**
- Why do you think windmills would be located in a place like this? **The open landscape allows wind to blow through and turn the windmill blades.**
- What do you think these windmills do? **produce electric power**

Respond
- What are Earth's resources, and how does the way we use them affect Earth's environment? **The resources include mineral resources, energy resources, living resources, air, water, sunlight, and soil. The extraction and use of some resources can lead to pollution and other damage to the environment.**

About the Photograph
The wind farm near Palm Springs, California, is located on the San Gorgonio Mountain Pass in the San Bernardino Mountains and consists of more than 4000 windmills. They require an average wind speed of 21 kilometers per hour and produce enough electric power for Palm Springs and the entire Coachella Valley.

BEYOND THE TEXTBOOK

Print Resources
- Reading Study Guide, pp. 21–23
- Formal Assessment, pp. 19–21
- Laboratory Manual, pp. 33–34
- Alternative Assessment, pp. 13–14
- Spanish Vocabulary and Summaries, pp. 13–14
- Internet Investigations Guide, pp. T25–T27, 25–27

Technology Resources
- Visualizations CD-ROM
- Classzone.com
- Online Lesson Planner
- Electronic Teacher Tools
- Test Generator

CHAPTER 7

PREVIEW

▶ **FOCUS QUESTIONS** In this chapter you will study Earth's resources and environment and learn more about the key questions below.

Section 1 What types of resources are part of Earth's environment, and how are they important to humans?

Section 2 What are nonrenewable and renewable energy resources?

Section 3 How does the use of Earth's resources affect Earth's environment?

▶ **REVIEW TOPICS** As you investigate Earth's resources and environment, you will need to use information from earlier chapters.

- interactions and the four spheres (p. 11)
- uranium and isotopes (p. 92)
- mineral (p. 96)
- rock-forming minerals (p. 104)
- mineral groups (pp. 108–111)

▶ **READING STRATEGY**

SET A PURPOSE

Read the key questions listed above. Before you begin reading Chapter 7, write a sentence or two in your science notebook that identifies a purpose for reading each section. Consider what questions about Earth's resources and environment you would like answered by your reading.

INTERNET RESOURCES
CLASSZONE.COM

At our Web site, you will find the following Internet support:

DATA CENTER
EARTH NEWS
VISUALIZATIONS
- Coal Formation
- Nuclear Fission

LOCAL RESOURCES
CAREERS
INVESTIGATIONS
- What Happens When an Oil Spill Occurs?
- Why Is this Place Protected?
- What Environmental Changes Can We See with Satellites?

Chapter 7 Resources and the Environment 143

CHAPTER 7

PREPARE

Focus Questions
Find out what knowledge students have about Earth's resources and the environment. For each focus question, have students consider the section title and then develop two or three additional questions for each section. For example, Section 2: Why are some resources nonrenewable? How can we make them last longer?

Review Topics
In addition to the earth science topics listed, students may need to review this concept:

Physical Science
- Nuclear energy can be produced by nuclear fission (the splitting of atoms) or nuclear fusion (the joining, or fusing, of atoms). Nuclear fission, used in the production of electric power, is the only type of nuclear power that we have learned to control and use.

Reading Strategy
Model the Strategy: Have students work in pairs to brainstorm additional questions that come to mind as they read the key questions. Model how certain words or ideas in a key question might spark a related question. For example, the second half of the question for Section 1 might prompt the question "Which resources do I use each day?" Have students write the answers to the questions in their notebooks when they find them in the text. If the text does not answer a particular question, encourage students to do additional research.

BEYOND THE TEXTBOOK

INTERNET RESOURCES
CLASSZONE.COM

INVESTIGATIONS
- What Happens When an Oil Spill Occurs?
- Why Is This Place Protected?
- What Environmental Changes Can We See with Satellites?

VISUALIZATIONS
- Animation of how coal forms
- Animation of nuclear fission

DATA CENTER
EARTH NEWS
LOCAL RESOURCES
CAREERS

Online Lesson Planner

Chapter 7 Resources and the Environment 143

7.1 Mineral Resources

You've probably heard that it's important to "protect Earth's environment," but what does that phrase mean? Earth's **environment** includes all of the resources, influences, and conditions near Earth's surface. We can make responsible decisions about the use of Earth's resources if we first know the types of resources available and how quickly they are used and renewed.

Renewable or Nonrenewable?

Some of Earth's most important resources, including air, water, land, and sunlight, are basic to life. Other resources have become critical to the world economy only since the 19th century. These include energy resources such as coal and oil and raw materials such as minerals and metal ores.

Earth's resources are classified as renewable and nonrenewable. A **renewable resource** is one that can be replaced in nature at a rate close to its rate of use. Examples are oxygen in the air, trees in a forest, food grown in the soil, and energy from the sun. See the table below.

A **nonrenewable resource** exists in a fixed amount or is used up faster than it can be replaced in nature. Nonrenewable resources include geological resources such as oil, and the metals, nonmetals, and other energy sources listed in the table on the following page. Some geological resources can be reused. Most, especially those used as energy resources, are destroyed once they are used.

Included among nonrenewable resources are the mineral resources mined from Earth's surface. Each United States citizen consumes, on average, about 40,000 pounds of new minerals a year. Minerals provide, for example, stone and cement for building, silicon for fiber optics and computer parts, fertilizers for farming, and aluminum for cars and trucks.

RENEWABLE RESOURCES Sunlight and trees are renewable resources.

SUMMARY: Some Common U.S. Renewable Resources

Resource	Type	Some Common Uses
air	nonmineral	respiration in organisms, generation of electricity
water	nonmineral	drinking, growth of crops, generation of electricity, transport
crops	nonmineral	food for humans and livestock, production of fabrics
forests	nonmineral	production of paper, building materials, medicines
sunlight	nonmineral	growth of crops and other plants, home heating, generation of electricity
soil	mixture of minerals and organic matter, water, air, and live organisms	growth of crops, foundation for buildings
ice	mineral based	food storage and preparation, recreation, medical and industrial uses

144 Unit 2 Earth's Matter

Earth's Minerals

In Chapter 5 you learned that chemical elements can be divided into two general categories, the metals and the nonmetals. Of the two, the metallic elements are of the greater economic importance.

All economically important metallic elements are obtained from minerals. Some of these elements, such as gold and silver, commonly occur as native metals—uncompounded with other elements—and can easily be separated from the rock surrounding them. Often, however, metals are found chemically combined with other elements. A metal must be chemically separated from other elements before it can be used. In either case, the element is often only a small part of the rock in which it occurs.

Rock that contains enough of a metallic element to make separation profitable is called an ore. Iron ore and copper ore are examples of rocks from which metallic elements (iron and copper) can be removed. The valuable mineral, the metallic element, is called an **ore mineral.** The rest of the rock is known as gangue (gang). Quartz, feldspar, and calcite are common gangue minerals.

There are important mineral resources other than those from which metals are extracted. Unlike ores, many of these resources are used in the forms in which they come out of the ground. Others are separated from surrounding materials by means of simple physical processes. Such resources include sand, gravel, building stone, rock salt, talc, and graphite.

SUMMARY: Some Common U.S. Nonrenewable Resources

Resource	Type	Some Common Uses
coal	nonmineral	generation of electricity
petroleum	nonmineral	production of gasoline, kerosene, fuel oil, lubricants, plastics, fertilizers, dyes, and medicines
natural gas	nonmineral	fuel, production of plastics, fertilizers, dyes, and medicines
sand, gravel, building stone, crushed stone	mineral based	construction of roads and buildings
salt	mineral based	production of chemicals, clearing road ice, food preservation
talc	mineral based	talcum powder, filler in paints and rubber
graphite	mineral based	dry cells, lubricants
sulfur	mineral based	soil conditioner, fungicides, production of sulfuric acid
gypsum	mineral based	plasterboard, plaster products
uranium	mineral based	generation of electricity, medical and industrial uses
phosphate rock, potash, nitrates	mineral based	fertilizers

NONRENEWABLE RESOURCES
Coal, shown being mined at a Pennsylvania strip mine, is used faster than it can be replaced in nature.

CHAPTER 7 SECTION 1

VISUAL TEACHING

Discussion
Have students use the table on page 144 to list the renewable resources used to generate electricity. **air, water, and sunlight** Ask: How can air be used to generate electricity? **The air, in the form of wind, turns windmills, which results in the generation of electricity.** Have students review the table on this page. Ask how the exhaustion of petroleum and natural-gas resources would affect the production of plastics. **It would require finding a substitute for plastics or a different way to produce plastics.**

Extend
Have students list the nonrenewable resources in the table that they have used in the past 24 hours. Use the fact that students have probably used all or most of the resources as a basis for discussion of the importance of the resources.

DIFFERENTIATING INSTRUCTION

Reading Support After students read pages 144 and 145, have them clarify their understanding by making a concept map that compares renewable and nonrenewable resources. The concept map should include definitions of the terms *renewable resource* and *nonrenewable resource* and give several examples of each type. Students' concept maps might classify the examples as nonmineral or mineral based.
Use Reading Study Guide, p. 21.

CHAPTER 7 SECTION 1

Visual Teaching

Interpret Diagram

Have students use the map to list the major mineral deposits in the United States. copper, lead, iron, phosphate, sulfur, and potash Ask: Which minerals are concentrated on the Gulf Coast of the United States? phosphate, sulfur, and potash

Extend

Ask: On which continent does the map show no major mineral deposits? Antarctica Can you conclude from this that Antarctica has no valuable minerals? no Explain your answer. Existing mineral deposits in Antarctica may be unknown because Antarctica has not been explored as extensively as the other continents.

CRITICAL THINKING

Discuss how the distribution of mineral resources on Earth might affect the economies of different nations. Nations with abundant mineral resources have the potential to gain wealth by exporting resources to other nations. Nations with few resources have to spend money to import what they need. Ask students if the lack of mineral reserves is necessarily an indicator of a poor economy. Ask if they can find a nation that lacks mineral reserves but still supports a global economy. Japan

Supply and Demand

Stores operate on the principle of supply and demand. That is, they try to provide supplies of products to meet the needs (demand) of the shoppers who want to buy the products. Similarly, the use of mineral resources is a matter of supply and demand. Unlike manufactured goods, however, minerals are not replaced as fast as or faster than they are used.

Availability of Minerals

Estimating the world's supply of a particular resource and the amount of it that should be used is a complex and controversial process. Reliable estimates of the total available amounts of minerals are hard to make because the entire world has not been explored for each resource. Most estimates of minerals' available supplies refer to reserves. A mineral **reserve** is the known deposits of a mineral in ores that are worth mining. The cost of mining and processing a mineral must also be considered. A high demand often means that an expensive mining operation can still be profitable.

Minerals are nonrenewable. Knowing the size of the reserve of a particular mineral and the rate at which it is being used makes it possible to estimate how long the reserve will last. The more of a mineral we demand, the faster it will be used up.

The map below shows the locations of some major mineral deposits. As you know from earlier chapters in this unit, Earth's mantle and crust are composed mostly of lighter minerals. However, metals exist in the crust in

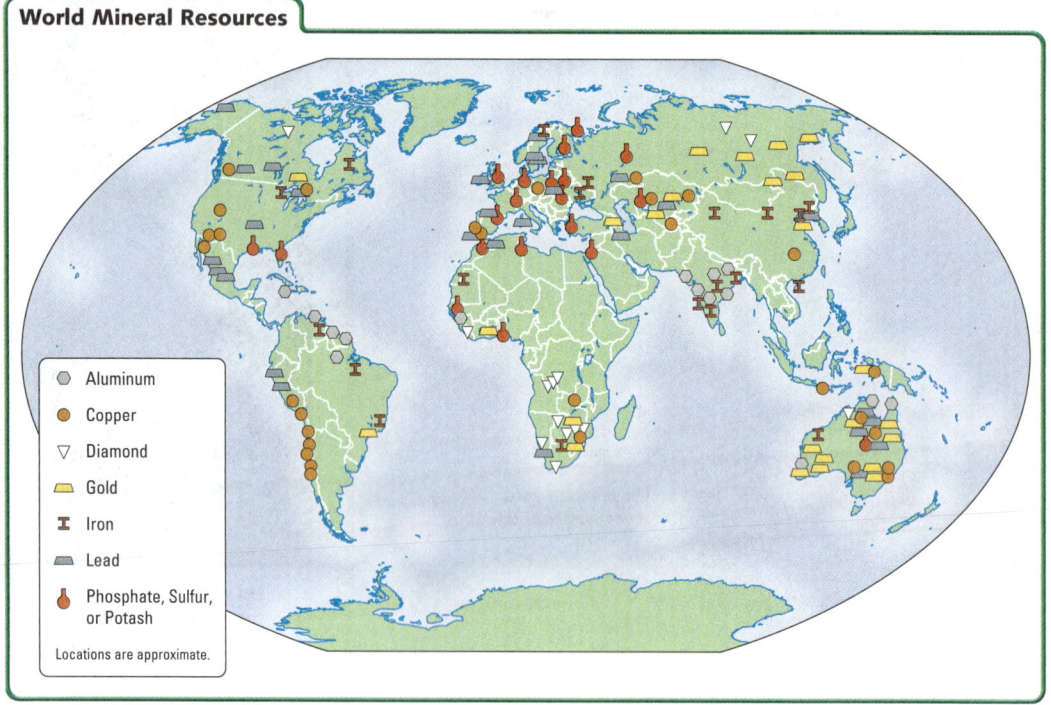

World Mineral Resources

- Aluminum
- Copper
- Diamond
- Gold
- Iron
- Lead
- Phosphate, Sulfur, or Potash

Locations are approximate.

146   Unit 2   Earth's Matter

DIFFERENTIATING INSTRUCTION

Support for Visually Impaired Students Some students may have difficulty seeing the small symbols on the map on this page. If possible, make an enlarged color photocopy of the map. Review the map to make sure students understand the symbols in the key. To practice using the map, ask students to point to the areas where there are large concentrations of copper. Then ask them to locate large diamond and gold deposits.

localized concentrations. It is not always easy to find and remove these metals. Ores deep in the ground are usually removed in underground mines reached by tunnels. Ores close to the surface are removed by digging great holes, called open-pit mines. Both types of mining can be costly, requiring the use of expensive machines and technologies as well as the labor and knowledge of many people.

Use of Minerals

The United States has some of the world's largest deposits of mineral resources. However, it is also one of the greatest consumers of these resources. If present rates of use continue, the world reserves of these and other elements could be used up within the next 60 years. Some metals, such as platinum, magnesium, cobalt, chromium, tin, and nickel, are scarce in the United States, and must be entirely imported.

Why are metals so desirable? Societies around the world depend on metals in a variety of ways. The metal iron, alloyed with elements such as carbon, nickel, chromium, or tungsten, is essential to steel production. Steel is used to make skyscrapers, bridges, tunnels, ships, planes, and trains. Steel also is used to make objects as small as pins, as well as utensils and tools. Another metal, copper, is used in electrical wiring and is combined with zinc to make brass. Zinc is also used as a protective coating to prevent rust. Cans, cookware, and bicycle frames are made with aluminum. Car batteries and the protective shielding around radioactive materials are made with lead.

Nonmetals such as sand, gravel, and crushed stone are taken from quarries (small open-pit mines). The construction industry has many uses for these resources, including the making of concrete and the construction of asphalt-gravel roofs. Phosphate rock, potash, and nitrates are used in fertilizers. All are mined or produced in the United States.

Because mineral resources are nonrenewable, all consumers must plan for the day when these resources disappear. If we can reduce the demand for them, we will increase the length of time that they will be available. A reduction in demand will also help reduce the creation of hazardous wastes, thus making Earth's environment safer for living organisms, including humans.

7.1 Section Review

1. Distinguish between a renewable and a nonrenewable resource.
2. What characteristics make a mineral an ore mineral?
3. In what ways is a mineral reserve different from a mineral resource?
4. **CRITICAL THINKING** Review the uses of mineral resources. Identify three such resources that you encounter daily and describe how life in your community might change if those mineral reserves were exhausted.

Chapter 7 Resources and the Environment 147

CHAPTER 7 SECTION 1

More about...

Mineral Uses
Diamond and gold are minerals usually associated with jewelry, but they have other uses as well. Diamond is the hardest natural substance known. Some industrial diamonds are made into cutting tools to shape and cut metals used to assemble airplanes, cars, and engines. Gold, by contrast, is a very soft metal, but it conducts electricity well, resists corrosion, and can easily be shaped into long, thin wires without breaking (ductile). For these reasons, gold is used in electrical equipment such as computers, radios, and televisions.

ASSESS

1. A renewable resource can be replaced in nature at a rate close to its rate of use. A nonrenewable resource exists in a fixed amount or is used up faster than it can be replaced in nature.

2. A mineral is an ore mineral if it contains a valuable metallic element that can profitably be removed from it.

3. A mineral reserve is the known deposits of a mineral resource in ores that can be mined profitably.

4. **Sample Response:** Gravel is used to make concrete. If gravel were not available, sidewalks, stairs, and roads would have to be made of materials other than concrete. Phosphate that is used to make fertilizers helps food crops grow. Without phosphate, food might not be as plentiful and might increase in price. Zinc is used as a protective coating to prevent rust. Without zinc, structures that contain iron would rust and corrode more easily.

MONITOR AND RETEACH

If students miss . . .

Question 1 Review the definitions of *renewable resource* and *nonrenewable resource.* (p. 144)

Question 2 Reteach "Earth's Minerals." (p. 145) Ask students which is more likely to be an ore: a 100-kg rock with 10 percent gold embedded in it or 10 percent iron. **10 percent gold, because gold is less abundant, so more highly valued**

Question 3 Refer students to the map on page 146. Lead them to understand that the map does not show all the places where minerals exist on Earth. Rather, it shows the locations of major known mineral deposits.

Question 4 Reteach "Use of Minerals." (p. 147) Have students list objects that contain certain mineral resources.

CHAPTER 7 SECTION 2

▶ Focus

Objectives
1. Identify renewable and nonrenewable energy resources.
2. Explain how fossil fuels form.
3. Describe how humans use renewable and nonrenewable energy resources to meet their energy needs.

Set a Purpose
Read aloud the key idea. Then have students read the section to learn the answer to the question "How do humans rely on renewable and nonrenewable energy sources to meet their energy needs?"

▶ Instruct

Visual Teaching

Interpret Diagram
Have students study the map on this page. Ask: Are non-renewable energy resources evenly distributed across the United States? *no* Have students compare this map with the one on pages 708 and 709. Ask: Where are most coal deposits located? *in the Appalachian and Rocky Mountain areas and in parts of the Midwest and Great Plains* What other nonrenewable energy resources are usually located near coal deposits? *gas and oil*

7.2

KEY IDEA
Humans depend on a variety of energy resources, both renewable and nonrenewable, to meet their energy needs.

KEY VOCABULARY
- fossil fuel

Energy Resources

Water, wind, and even humans can supply energy for work. Fuels also provide energy. In the past, the major source of energy was wood, a fuel burned to provide heat and light. However, the sources of energy and the demand for it have changed dramatically in the past 150 years.

Nonrenewable Energy Resources

Today the world's use of energy is greater than ever. About 7 percent of that energy comes from renewable sources. The rest comes from nonrenewable sources of energy, such as coal, petroleum, natural gas, and nuclear fission.

Fossil Fuels

In Chapter 6 you learned that a fossil is evidence of life preserved in rock. Coal, petroleum, and natural gas are called **fossil fuels** because they formed from the remains of organisms that lived millions of years ago. The burning of coal, petroleum, and natural gas, like that of other fuels, releases the energy stored in them. Fossil fuels are nonrenewable because they are being used up millions of times faster than they are forming. In the United States today, coal is used primarily in power plants to generate electricity, but it is also important in the manufacture of steel and as a raw material in chemical processes. Deep coal deposits are worked in underground mines. Coal in shallow deposits is dug up in surface mines called strip mines.

Most coal is considered an organic sedimentary rock formed from materials such as ferns, mosses and parts of trees. These and all other organic materials includes the elements carbon, hydrogen, and oxygen.

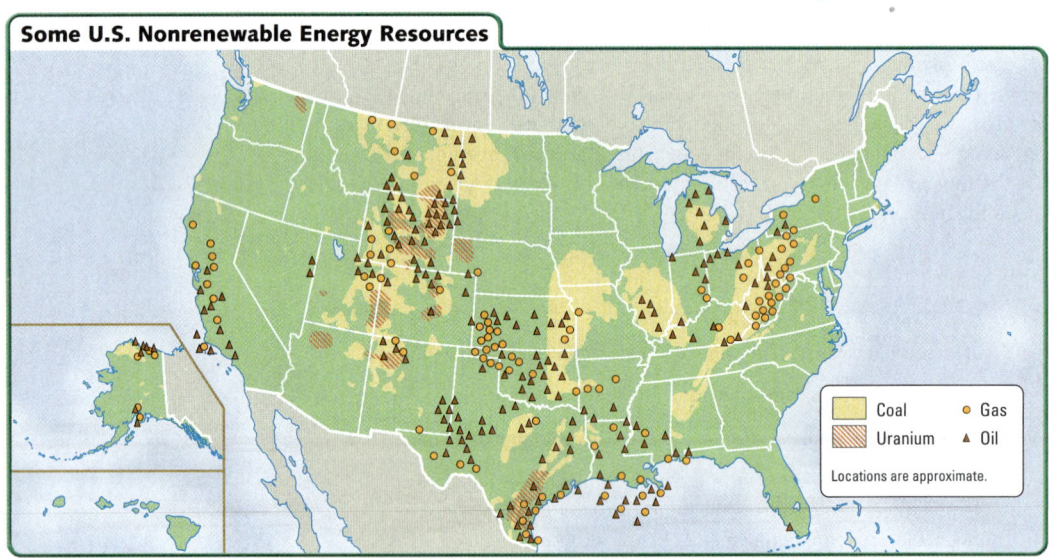

Some U.S. Nonrenewable Energy Resources

148 Unit 2 Earth's Matter

DIFFERENTIATING INSTRUCTION

Reading Support Have students preview this section by looking at the heads, subheads, photographs, and illustrations. Have students write three to five questions they would like answered based on what they previewed. Then tell them to look for and write the answers to their questions as they read the section. Use their answered questions as the basis for review. Use their unanswered questions as the basis for discussion and further investigation.
Use Reading Study Guide, p. 22.

148 Unit 2 Earth's Matter

When organic remains are buried in swamp waters—often under sand or clay—they decay slowly, gradually losing hydrogen and oxygen. The physical properties of the sediment change as it ages and is compacted.

The lightly compressed mass of plant remains is called peat. Over time, peat is compressed and more hydrogen and oxygen are lost. Eventually, a soft brown coal called *lignite* forms. Some types of lignite contain about 40 percent carbon. After millions of years of compression, lignite may become bituminous, or soft, coal. Bituminous coal is up to 85 percent carbon. Through regional metamorphism, bituminous coal may become anthracite, or hard coal. Anthracite is about 90 to 95 percent carbon. The higher the percentage of carbon, the greater the amount of energy released when the coal is burned.

Formation of Coal

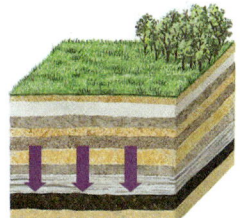
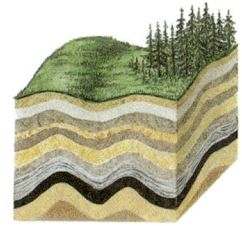

1. Swamp matter decays and is compacted over time, becoming peat. Close examination of peat can reveal leaves, sticks, and twigs.

2. Sedimentation and compaction of the peat force out water and gases. The resulting lignite has a higher carbon content than peat.

3. Over millions of years, more sediments cover the lignite. It becomes harder, sedimentary bituminous coal.

4. Anthracite, a metamorphic coal, takes the longest to form and has been exposed to the most heat and pressure.

PEAT

LIGNITE

BITUMINOUS COAL

ANTHRACITE

Unlike coal, petroleum is a liquid. However, both are sedimentary materials of organic origin. Petroleum, or oil, is mainly a mixture of liquid hydrocarbons (compounds of hydrogen and carbon). Petroleum is recovered by drilling wells into oil-bearing rock. Natural gas, a mixture of methane and other hydrocarbon gases, often occurs with petroleum—although it may exist in great deposits by itself. The pressure of natural gas overlying oil deposits helps bring the oil to the surface. The drilling must be carefully controlled, or the high pressure causes wasteful gushers. Even with modern technology, only about 60 percent of the oil in a given well can be pumped out of it.

VOCABULARY STRATEGY
The word *petroleum* (puh-TROH-lee-uhm) comes from the Latin words *pctra*, meaning "rock," and *oleum*, meaning "oil."

CHAPTER 7 SECTION 2

MINILAB
Oil Well

Management
Have paper towels on hand for cleaning up water spills. This Minilab can also be used with Chapters 29 and 30.

Materials
Any clear container with sides taller than 17 cm and a diameter of about 15 cm (such as a 2-L plastic bottle with its top cut off) will work. Use one-quarter-inch bore tubing (plastic or rubber). Be sure the funnels and tubing are sterile.

Procedure
An alternative to students using their mouths in step 3 is hooking up a hose to a pressurized air faucet or a balloon pump.

Analysis Answers
- When the funnel is held below the gravel, the water forms a reservoir in the funnel. When the funnel is raised, the water flows down into the tubing, straw, and gravel. As air and water are added, the pressure forces water to come out of the other straw.
- Traditional methods of drilling for oil (such as cable-tool drilling or rotary drilling) rely on hydrostatic pressure. Once a path is drilled to the oil rock, pressure exerted on the oil from overlying rock layers forces the oil up to the surface. However, some modern experimental methods rely on a system similar to that modeled in this Minilab. One method is to inject carbon dioxide into the ground to force out remaining oil from a long-established oil field.

For proper behavior in a laboratory setting, refer students to the section on Safety in the Earth Science Laboratory, pages xxii–xxiii.

25-Minute Mini LAB
Oil Well

Materials
- gravel
- clear beaker
- 2 straws
- modeling clay
- sand
- plastic tubing
- funnel
- water

Procedure

1. Place 10 cm of gravel in the bottom of the beaker. Stand the straws inside the beaker, on opposite sides. Pack 1 cm of modeling-clay to seal around each straw and the beaker's sides. Add 6 cm of sand on top of the clay. Use more clay to seal the tubing and funnel to one of the straws.

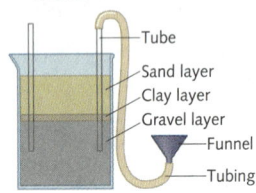

2. Fill the funnel with water while holding it below the gravel's level. Observe the flow of water as you slowly raise the funnel above the sand, adding more water as needed.

3. Blow directly into the funnel.

Analysis
Explain your observations. On the basis of your results, describe methods that can be used to pump oil from below ground.

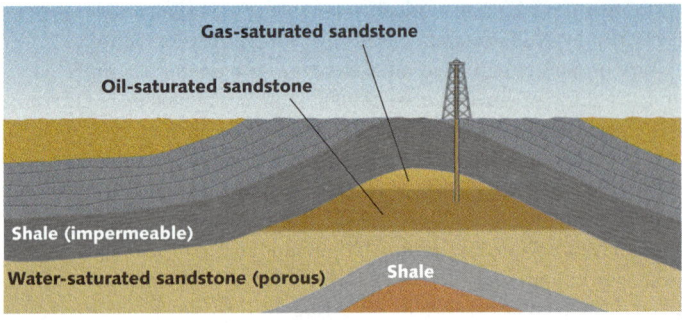

OIL AND GAS WELL

Petroleum is thought to have been formed by slow chemical changes in organic materials buried under sand and clay in shallow coastal waters. As the sediments were compacted, liquid and gaseous hydrocarbons were forced into pores and cracks of nearby sandstones or limestones. The pores and other spaces may have been filled with seawater; but the lighter petroleum rose above the water, and the natural gas collected above the petroleum.

Why haven't the petroleum and gas escaped from the rock during the millions of years that followed their formation? Probably large amounts have. The deposits found today are ones that were sealed underground by layers of virtually impermeable rock, such as shale, in structures called oil traps. The illustration above shows the most common type of trap—an anticline, or upfold.

Other sources of fossil fuels include oil shales and tar sands. When heated, oil shales release a petroleum vapor that can be condensed into liquid oil. The spaces between the grains of tar sands are filled with what may be the dried residue of petroleum. Oil can be removed from these sands. The amount of oil in Earth's oil shales and tar sands is estimated to be 50 percent greater than Earth's remaining oil reserves. However, the recovery processes for both are too expensive at present for either to compete with oil from wells. Coal is currently the least expensive fossil fuel.

Uranium

The metallic element most commonly used as a fuel is uranium. When atoms of a certain isotope of uranium are hit with neutrons, a reaction called atomic fission occurs, releasing much energy. The newly released neutrons cause the fission of other nuclei. This process, modeled on page 151, takes place in the nuclear reactor of a nuclear power plant. A coolant pumped through the reactor is heated by the energy released by fission. The hot coolant is used to convert water to steam. The steam moves turbines, which generate electricity.

Uranium is recovered primarily from the black mineral uraninite and the yellow mineral carnotite. The fission of one gram of uranium releases as much energy as the burning of nearly 3 tons of coal or 14 barrels of oil. Today, uranium is the fifth most important source of energy in the world, behind oil, natural gas, coal, and water power.

DIFFERENTIATING INSTRUCTION

Support for Physically Impaired Students Have students do the Minilab with a partner. Some students will be able to help set up and perform the activity. Others might be better able to participate by making observations or taking notes or recording observations on audiotape. Encourage partners to discuss their results and work together to present their observations.

Nuclear Power

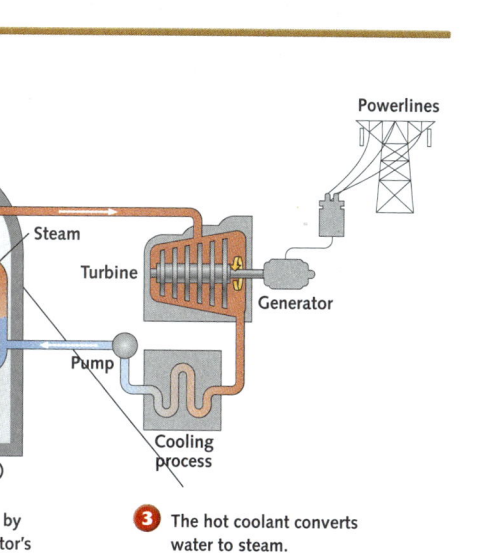

1. Fission occurs when the nucleus of a uranium isotope absorbs a neutron. Lighter elements form and much energy is released.
2. The energy produced by fission heats the reactor's coolant.
3. The hot coolant converts water to steam. The steam is used to generate electricity.

Renewable Energy Resources

Four of the most widely used sources of renewable energy are water, wind, the sun, and geothermal energy.

Water

The major use of water power today is to produce electricity. Electricity so produced is called hydroelectric power. Water power is the most efficient way of producing electricity because the turbines that power the electrical generators are turned directly by moving water. When coal or uranium is used, the energy they produce must first be used to convert water to steam. It is the steam that turns the turbines.

Tides, the rise and fall of Earth's oceans, can also be used to generate electricity. Water is trapped behind a barrier at high tide and then slowly released at low tide. As the water drops to the lower level, it spins a turbine. Currently, several large-scale tidal power plants are in operation. One has been operating in France since 1966.

HYDROELECTRIC POWER The Dallas Dam in Oregon generates electricity by using the force of the Columbia River's moving water to power turbines.

Observe an animation of nuclear fission.
Keycode: ES0702

Chapter 7 Resources and the Environment 151

CHAPTER 7 SECTION 2

Visual Teaching
Discussion
Tell students that the advantages of solar power include the generation of energy without causing pollution and the use of an energy resource that is renewable. But there are disadvantages as well. Ask students if they can see one disadvantage in the photograph of the Natatorium. **The thermal and photovoltaic panels take up a lot of space.** Where might this be a particular problem? **in cities** How might this problem be solved? **The panels could be set up outside the cities, using power lines to transmit the electricity produced to the cities.**

More about...

Geothermal Energy
The United States is one of the world's major geothermal energy users, with California in the forefront of development of this energy resource. California's geothermal plants produce 40 percent of the world's geothermally generated electricity. The Geysers, north of San Francisco, is one of only two places in the world where hot, dry steam is used directly to spin the blades of an electricity-generating turbine. The Geysers plant produces more than enough electricity to supply the entire city of San Francisco.

WIND POWER This windmill farm in Palm Springs, California, generates electricity by capturing the power of the wind with hundreds of windmills.

SOLAR POWER The Natatorium has harnessed solar power since it was built as the swimming facility for the 1996 Olympics in Atlanta. Its solar energy system uses thermal panels (black) to absorb the sun's warmth, and photovoltaic panels (gray) to convert sunlight into electricity.

Wind
Wind power—the force of moving air—can be captured by a windmill, which can be used to generate electricity. The amount of power produced can vary greatly, depending on the speed of the wind, the length of the windmill's blades, and the efficiency of the windmill. In order for wind power to provide electricity to large numbers of homes, windmill farms must be built. A windmill farm is a vast array of up to several hundred windmills. Wind farms in the United States are located in Hawaii, California, and New Hampshire. Occasionally, homeowners in remote areas use individual windmills to power small electrical generators.

Wind power is not the most efficient way of producing electricity, but better windmills are being developed. Since 1991, the number of windmills in California has dropped, yet the amount of electricity generated has remained about the same. In 1999, California windmills produced over 1650 megawatts of electricity, about the same amount produced by a medium-sized nuclear power plant.

The Sun
Solar power—energy from the sun—can be used to provide heat and electricity. "Passive" or "active" systems may be used. A building with a passive solar heating system is designed to collect and store solar energy. For example, a specialized window might let light into a house but not let heat escape. An outside wall might be made of a material that heats easily in sunshine and then transfers its heat to the inside of the house.

An active solar heating system has three parts. A solar collector facing the sun absorbs heat, which is transferred to a storage area. There the heat energy is stored until it is needed. Finally, a system distributes the heat throughout the building. The same system can also be used to heat water.

Most active solar heating systems are privately owned. Engineers, however, have developed solar power plants that can deliver electricity to thousands of homes. Throughout the day, mirrors focus sunlight on a large collector, heating molten salt to over 500°C. The salt stores heat so well it can make steam to turn turbines even at night or during cloudy days.

Solar cells, also known as photovoltaic cells, have been used to generate electricity in spacecraft since the start of the space age. These cells convert light into electricity. Advances in the design of solar cells may eventually lead to power plants that can produce millions of watts of electrical power.

Geothermal Energy
Heat from Earth's interior is called geothermal energy. In some areas, highly pressurized steam rises naturally out of deep hot rock. Most geothermal sources, however, are drilled and controlled like oil wells. The geothermal energy is converted to electrical energy when steam or hot water from below Earth's surface is piped into a power plant to run a generator. Hot water from geothermal areas can also be piped into homes for heating and cooking.

DIFFERENTIATING INSTRUCTION

Hands-On Demonstration Compare how well different materials can store heat from the sun. Place equal masses of water, vegetable oil, sand, and soil in four separate metal cans lined with black construction paper. Leave the cans in direct sunlight for at least an hour. Then record the temperatures of the materials. Remove the cans from the sunlight, and record the temperatures of the materials every 3 minutes for 15 minutes. Have students study the temperature data. Ask: Which material stores heat from solar energy the best? How do you know? **The material that shows the slowest decrease in temperature over time is the material that stores heat the best. Typically water stores heat the best.**

GEOTHERMAL POWER PLANT
Geothermal resources provide direct heating and generate electricity. This power plant in Wairakei, New Zealand, has operated since the 1950's.

Areas of volcanic activity have the greatest potential as geothermal energy sources. In the United States, geothermal sources generate enough energy to provide electricity for nearly 3.5 million homes. The Geysers, California, is the largest source of geothermal power in the world. As of 1999, geothermal power plants in 22 countries around the world provided electricity to over 60 million people.

Because hot rock can be found in all parts of Earth's crust at various depths, scientists have experimented with ways to use geothermal energy at any location. The United States, France, Australia, and Japan have experimented with "hot-dry-rock" projects. In such a project, cold water is pumped into an underground reservoir in hot rock. The water warms underground, then is pumped back to the surface to generate electricity and heat homes.

Ground-source heat pumps are another direct use of geothermal energy. Regardless of climate, Earth's ground temperature remains constant about 150 centimeters below the surface. As a result, the ground is warmer than the air in the winter and cooler in the summer. In winter, a heat pump moves fluid through pipes in the ground, transferring heat from the Earth to the fluid. The warm fluid is then circulated through a house, warming it. Similarly, the ground can cool the house in the summer. Ground pumps will work in nearly all parts of the United States.

7.2 Section Review

1. List four nonrenewable energy resources used today. Which one is not a fossil fuel?
2. What is a renewable energy resource? Name two and tell why they are considered renewable.
3. **CRITICAL THINKING** Compare peat and the three types of coal. Which might be preferred as a fuel for power plants? Why?
4. **GEOGRAPHY** Locate Iceland on a world map and on the plate tectonic map in the Appendix. Most of that country's energy comes from geothermal sources. What other countries might make similar use of geothermal energy? What geological characteristics or features must be present in these countries?

Scientific Thinking

COMPARE AND CONTRAST

As some power plants use coal or uranium to generate electricity, a waste-to-energy plant burns solid wastes. Although pollution from the resulting ash is a concern, this method not only provides energy but gets rid of mountains of solid wastes that would otherwise be buried in landfills. Landfills can also cause pollution and take up valuable space.

What risks and benefits might a small, densely populated country consider when choosing among energy sources; traditional fuels, such as coal, oil, and uranium; and renewable resources such as solid wastes as means of producing energy?

CHAPTER 7 SECTION 2

Scientific Thinking

COMPARE AND CONTRAST
Considerations might include which energy resources are available in the country; how much space is available (windmill farms, for example, take up a lot of space); and the amount of pollution or waste that would be produced.

ASSESS

1. coal, petroleum, natural gas, and uranium; uranium

2. a source of energy replaced in nature at a rate close to its rate of use; renewable energy resources include: water—constantly recycled between Earth and the atmosphere; wind—available as long as air moves; sunlight—available as long as the sun shines; geothermal energy—continual heat from Earth's interior

3. Peat is the softest and least compacted. Lignite, bituminous coal, and anthracite are progressively more compact, harder, and contain more carbon. Anthracite would be preferred because it releases the most energy when burned.

4. **Sample Response:** the United States, Colombia, Mexico, and Japan; geological features associated with hot rock underground, such as volcanoes and geysers

---

MONITOR AND RETEACH

If students miss . . .

Question 1 Review the definition of *fossil fuel*. (p. 148)

Question 2 Review "Renewable Energy Resources." (pp. 151–153) Ask students: Why aren't fossil fuels renewable resources? **They cannot be replaced in nature as fast as they are used.**

Question 3 Review the formation of coal. (p. 149) Have students compare the percentages of carbon in peat, lignite, bituminous coal, and anthracite.

Question 4 Show students the map of earthquake and volcanic activity on Transparency 8. Explain that in these areas, hot rock is more likely to be found near Earth's surface.

CHAPTER 7 SECTION 3

FOCUS

Objectives
1. Describe how the use of nonrenewable and renewable resources affects the environment.
2. Explain how humans can slow the depletion of resources.

Set a Purpose
Have students read the section to find the answer to the question "What environmental risks are associated with different energy resources?"

INSTRUCT

VISUAL TEACHING
Discussion
Tell students to use the information in the diagram on the bottom of pages 154 and 155 to create a table that lists one advantage and one disadvantage of using each type of energy resource shown. Have students use their tables to decide which energy resources they think are best to use where they live and to give reasons for their choices. Discuss students' ideas. They may conclude that no single source will meet their area's energy needs and that a mix of different sources in particular percentages may be best.

7.3

KEY IDEA
Human use of Earth's resources affects the living and nonliving parts of the environment.

KEY VOCABULARY
- conservation
- recycle

Environmental Issues

Human beings' use of Earth's resources often damages the environment that enables us to live. If Earth's environment is severely damaged, life as we know it will be changed.

Risks and Disadvantages

Our current use of nonrenewable resources can pollute land, water, and air and contribute to global warming. While renewable resources cause less pollution, they also have disadvantages.

Mining for Minerals

Surface mining can remove tons of soil, ore, or rock, often creating a rocky waste that supports little life. Although a renewable resource, topsoil is renewed naturally at a rate of only a few inches every thousand years. Strip-mined land can take decades to recover.

A landscape can be destroyed when mining leaves hills barren, levels mountains, or forms enormous craters. Mining produces huge piles of waste rock. Water that collects in open pits or runs off from piles of waste rock can be dangerous. Surface compounds in the waste can react with the water to form sulfuric acid. Ore processing can contaminate the waste chemically. Heavy metals dangerous to living things can be weathered out of waste rock. All these substances can pollute water, damaging or killing life in streams and lakes.

Nonrenewable Energy

One problem with the use of nuclear reactors is that fission produces dangerously radioactive by-products. Radioactive waste must be stored away from living things for thousands of years. No satisfactory way of safely

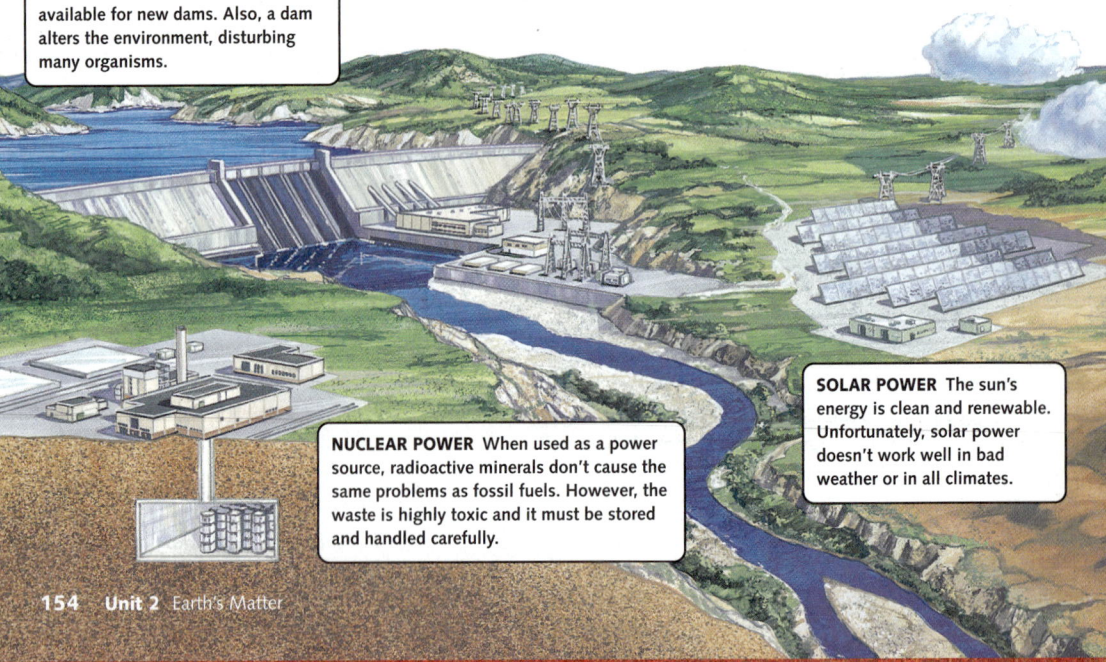

HYDROELECTRIC DAM Water is a renewable and nonpolluting energy resource. However, few sites are available for new dams. Also, a dam alters the environment, disturbing many organisms.

NUCLEAR POWER When used as a power source, radioactive minerals don't cause the same problems as fossil fuels. However, the waste is highly toxic and it must be stored and handled carefully.

SOLAR POWER The sun's energy is clean and renewable. Unfortunately, solar power doesn't work well in bad weather or in all climates.

154 Unit 2 Earth's Matter

DIFFERENTIATING INSTRUCTION

Reading Support Tell students that as they read, certain words and phrases are clues to the kind of information coming up. For example, point out that each caption on pages 154 and 155 gives at least one advantage and one disadvantage of using the energy resource described. Certain words clue the reader when a disadvantage is being described. Model this situation by reading the caption on page 154 for the hydroelectric dam. Ask: What are the advantages of using hydroelectric dams? Water is renewable and nonpolluting. What are the disadvantages? few new sites are available; changes the environment and disturbs organisms What word tells you that a disadvantage is coming up? however Continue with the other captions. Students should find the transition words *however*, *unfortunately*, *although*, and *yet*.
Use Reading Study Guide, p. 23.

storing or disposing of nuclear waste has been found. Also, accidents can happen. In 1986, workers caused a deadly accident at the Chernobyl nuclear power plant in the Soviet Union. Thirty-two people were killed immediately, thousands became ill, and millions of acres were contaminated with radiation. Fortunately, mechanical safeguards and extensive training of nuclear technicians make such accidents rare.

The burning of fossil fuels, especially coal and oil, releases pollutants into the air. Some of these pollutants can irritate the nose, throat, and lungs. Others contribute to acid rain, which you will read more about in later chapters. Acid rain can damage buildings, reduce forest growth, harm crops, and kill or injure plant and animal life in lakes and streams.

The increasing demand for fossil fuels threatens protected lands and wildlife. As fossil fuels have become harder to find, oil companies have sought permission to search for petroleum and natural gas in protected land such as wildlife refuges. Oil spills there or anywhere else on land can pollute soil and drinking water. Spills at sea can damage miles of coastline and kill thousands of organisms.

Renewable Energy

In general, renewable energy resources have a less damaging impact on the environment than nonrenewable energy resources. However, each type is limited in some way. None is usable everywhere. For example, water power can be used only in areas where dams can be built for water storage. In the United States, only about 10 percent of the electricity used is hydroelectric.

Wind power can be used only in areas with strong, steady winds. Windmills can interfere with television and radio reception. Windmill farms need a lot of land and can interfere with bird migration. There is also the problem of energy storage. Scientists have not found a reliable and

What Happens When an Oil Spill Occurs? Examine the affects of an oil spill on the surrounding environment.
Keycode: ES0703

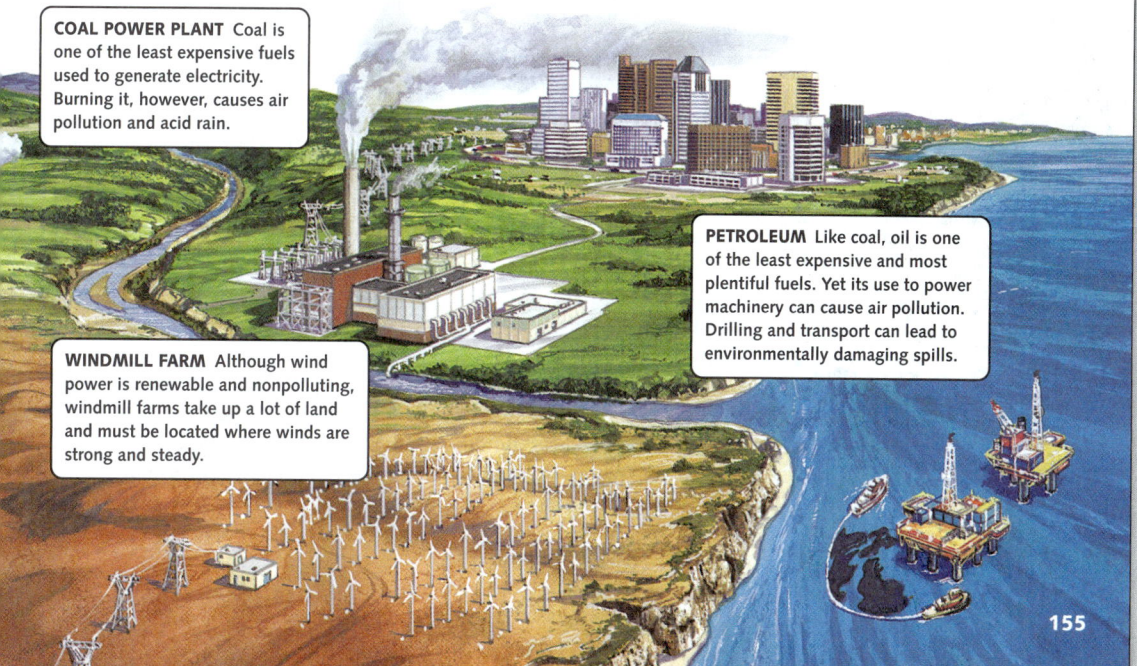

COAL POWER PLANT Coal is one of the least expensive fuels used to generate electricity. Burning it, however, causes air pollution and acid rain.

WINDMILL FARM Although wind power is renewable and nonpolluting, windmill farms take up a lot of land and must be located where winds are strong and steady.

PETROLEUM Like coal, oil is one of the least expensive and most plentiful fuels. Yet its use to power machinery can cause air pollution. Drilling and transport can lead to environmentally damaging spills.

CHAPTER 7 SECTION 3

What Happens When an Oil Spill Occurs?
Keycode: ES0703
Use a computer and projector to introduce the investigation in class. Then assign the investigation as homework.

📖 Use Internet Investigations Guide, pp. T25, 25.

More about...

Choosing an Electric Power Source
The production of electric power in the United States creates more pollution than any other single industry. However, within the industry, environmental impact differs greatly depending on the power source. As a result, an increasing number of state governments and private electric utilities are allowing customers to choose the power source that generates their electricity, including renewable energy resources, such as water or sunlight, if available. This policy is meant to support the development of renewable resources.

DIFFERENTIATING INSTRUCTION

Challenge Activity Have students read the key idea and look briefly at the head, subheads, photographs, and illustrations in this section to get a sense of what the section is about. Have students find articles in newspapers and magazines that relate to the section topics. Give students two or three days to bring articles to class as you teach this section. Have students summarize the articles they bring. Then discuss how they relate to students' lives both now and in the future.

Developing English Proficiency Students may have trouble comprehending the meanings of *alters* and *steady* as they are used in the captions on pages 154 and 155. Students might confuse *alter* with *altar*. Suggest students substitute *changes* for *alters* and *continuous* for *steady*.

CHAPTER 7 SECTION 3

CRITICAL THINKING

Ask students to identify two or more ways they could have used resources more wisely yesterday. **Answers might include examples of conservation, such as riding a bicycle to school or turning off lights when not in use, and examples of recycling, such as disposing of aluminum cans and newspapers in recycling bins.**

More about...

Recycling
There are two major types of recycling. In primary recycling, the recycled product is used to make products of the same type. For example, paper collected in primary-recycling programs is used to make newsprint, cardboard, and other paper products. In secondary recycling, collected products are used to make different products. For example, recycled glass bottles might be ground up to become part of road-paving material. Primary recycling is preferable because it reduces the amount of new resources used to make products by up to 90 percent. Also, the potential exists for all these products to be recycled again. In secondary recycling more new materials are needed. Also the materials can seldom be recycled a second time.

Why Is This Place Protected? Explore the unique features of National Parks and Monuments and decide if another place is worth protecting.
Keycode: ES0705

CONSERVATION By bicycling rather than driving cars, these commuters in Shanghai, China, conserve petroleum and reduce air pollution.

156 Unit 2 Earth's Matter

efficient method of storing the energy produced during strong winds for use during calms. Similar problems of storage and reliability affect solar power.

Geothermal energy provides little of the world's total energy supply. Few areas have hot bedrock near the surface. Those areas that do may not be near large population centers. The chemical-rich, superheated water can pollute lakes and streams. Cave-ins can occur when hot water drawn from the ground is not returned.

Using Resources Wisely

How can we slow the depletion of minerals, rock resources, and energy resources? Conservation and the development of efficient and reliable renewable energy resources are important. Likewise, recycling and legislation can help slow the rate at which resources are used and help protect the environment.

Conservation and Recycling

Think of **conservation** as the protection, restoration, and management of natural resources. Geological resources can be conserved, but because industrial societies are dependent on mined resources, it isn't likely that mining will stop completely. Conservation may include research that improves the efficiency of people's use of resources, so that reserves last longer. We can also find better ways of controlling the environmental impact of mining. For example, topsoil can be removed prior to mining and replaced and replanted after the mine is closed. Although the land will not be perfectly restored, the amount of damage can be kept to a minimum. Controlling erosion and preventing runoff that pollutes water can also limit environmental damage.

Reducing gasoline consumption can conserve oil resources. Fuel mixtures are one conservation measure. Gasohol is a mix of gasoline (usually 90 percent) and alcohol made from grain crops. Gasoline can also be conserved through the development of hybrid cars that run on both gasoline and electricity.

Individuals can practice energy conservation by reducing energy consumption and changing energy-wasting habits. They can, for example, walk or ride a bicycle for short trips, use car pools, adjust clothing instead of the thermostat, and turn off unneeded lights. Insulating a home means less energy is needed to cool and heat it. A well-maintained car pollutes less. Energy-efficient cars, houses, and appliances also help save energy. How can we conserve nonenergy resources? We can buy only the products we really need and plan on using. Avoiding waste is an important conservation method.

Reducing use and recycling can help make metal and nonmetal mineral resources last longer. To **recycle** is to collect and reuse materials from waste. Scrap iron, aluminum cans, and gold in computer chips are examples of metals that can be reused. Recycling glass bottles helps conserve sand and the other materials needed to make glass. Copper can be conserved by using glass-fiber cables rather than copper wires to carry

DIFFERENTIATING INSTRUCTION

Reading Support Help students understand the meanings of unfamiliar terms by relating them to the context of the sentences around them. For example, the word *depletion* occurs in the first sentence under the heading "Using Resources Wisely." Students may not know exactly what this word means. However, within the context of its sentence, they can figure out that it means "a using up" because from the sentence, students know that depletion of resources is something we want to slow. Other possible terms on this page that students can understand by their context include *legislation, efficiency, hybrid cars,* and *energy-efficient.*

156 Unit 2 Earth's Matter

SCIENCE & Technology

Harnessing the Power of the Sun

Is active solar power the only "real" solar power? Can passive solar technology also help to conserve Earth's fossil fuels and protect our environment? Researchers are investigating many ways of making use of the sun's energy.

How can a simple change to a common building material help to reduce the demand for electricity?

DEBBY TEWA is a solar energy expert. She tours throughout the United States, lecturing about the advantages of solar energy.

Engineers are working hard to perfect central solar power plants. Ideally, this type of solar power would replace coal or nuclear power. Yet the development of such complex systems can be expensive, and it may be decades before they are widely used. A low-tech, less expensive approach to solar energy offers improved prospects for its widespread and immediate use.

Traditionally, a passive solar heating system includes a thermal mass that stores heat—a thick concrete wall, for example, or barrels of water. Such materials can take up space and limit the design and location of a building, but these bulky systems may soon be obsolete. Researchers have been testing thermal-mass materials that are up to 14 times more efficient at storing and transferring heat. They've used these materials to make a new product, phase-change wallboard.

A phase change occurs when the state, or phase, of matter—solid, liquid, or gas—changes. Researchers have incorporated materials such as paraffin wax into normal gypsum drywall. When the outside temperature rises, the material in the wallboard slowly melts but remains at a constant temperature until it is entirely melted. This way it absorbs the sun's energy without getting hotter. At night, when the outside air cools, the material radiates heat while returning to its solid state. Phase-change wallboard can keep inside temperatures comfortable in summer and in winter.

Widespread use of this product could significantly reduce the demand for coal and nuclear power plants. Simulations indicate that its use could shift 90 percent of the air conditioning in Dallas to off-peak hours. Parts of California might even be able to eliminate air conditioning altogether.

Extension

SCIENCE NOTEBOOK

Consider the many ways fossil fuels are used. What other simple changes to commonly-used materials or products might help reduce the demand for fossil fuels?

Learn more about solar power.
Keycode: ES0704

PASSIVE SOLAR ENERGY systems, such as the one in this Chicago, Illinois, home, use the renewable energy of the sun, helping to conserve nonrenewable natural resources.

CHAPTER 7 SECTION 3

Learn more about solar power.
Keycode: ES0704

Why Is This Place Protected?
Keycode: ES0705

📖 Use Internet Investigations Guide, pp. T26, 26.

Science and Technology

Passive Solar Energy
All passive solar systems share certain design features. For example, in the Northern Hemisphere, the south wall contains large areas of glass. This allows the sun's rays to penetrate the building directly all year long. In winter, when the sun is low in the sky, sunlight passes directly through the windows, striking walls and floors of stone, concrete, adobe, or brick. These materials store heat during hours of sunlight and slowly release it at other times. When the sun is higher in the sky in summer, roof overhangs and shades prevent some sunlight from entering windows.

Science Notebook
Students might suggest using better insulation, gasohol, and other materials mixed with plastics to use less petroleum.

CHAPTER 7 SECTION 3

CRITICAL THINKING

Lead a discussion in which students consider whether people should be required by law to conserve and recycle or whether such actions should be voluntary. You might ask the following questions to stimulate discussion: Would you be more likely to recycle if you knew you would be fined if you did not? Does the government have the right to force people to recycle products and to use recycled products, or would this infringe on people's freedom? If you had the power, how would you encourage people to recycle and practice conservation?

RECYCLING Manufacturers used recycled plastics to make this playground equipment in Laguna Niguel, California.

telephone conversations and digital data. Many communities now have pilot recovery programs for recycling (or properly disposing of) the materials in old televisions and computers.

Legislation

Since the 1870s, federal and state laws have been passed that enabled government agencies to protect the environment and conserve resources. More recently, federal laws have allowed the Environmental Protection Agency (EPA) to monitor and set standards for drinking-water and air quality. Current laws control the production and disposal of toxic industrial chemicals and hazardous waste. The Pollution Prevention Act of 1990 was

Recycling Technician

When you sort paper, aluminum cans, and glass to be recycled, do you ever wonder what happens after the materials are taken to the recycling center? Recycling technicians work to ensure that the recycled materials are properly processed so they can be sold and reused. These materials include not only the consumer waste that you are familiar with but also industrial waste, such as scrap metal. In order to carry out their work, recycling technicians operate trucks, loaders, and other heavy equipment. Recycling technicians also identify, report, and remove hazardous waste from the rest of the materials to be recycled. Periodically, they sample groundwater and monitor chemical levels to ensure that the environment is not being harmed by the recycling process.

Since most recycling technicians receive on-the-job training, a high school degree and a valid driver's license are usually sufficient. Recycling technicians often find that organizational skills come in handy when collecting sampling data. Physical fitness and good coordination are also necessary, since the job requires heavy lifting and the operation of industrial equipment. Recycling technicians are dedicated to their jobs, since they must sometimes work in severe weather conditions. Although their career is challenging, recycling technicians take pride in knowing that community recycling and conservation efforts would not be possible without their work. ■

RECYCLING TECHNICIANS at the Alcoa Recycling Center in Tennessee are organizing thousands of crushed aluminum cans.

Find out more about a career in recycling.
Keycode: ES0706

158 Unit 2 Earth's Matter

DIFFERENTIATING INSTRUCTION

Reading Support After students have read the section, have them write two or three brief paragraphs that summarize what they have learned. Remind students that the summaries should contain the main facts and ideas presented in the section and should be written in their own words.

designed to control pollution and to encourage the conservation of energy, water, and other natural resources. The act promotes recycling, reduction of resource use, and sustainable agriculture. Many states have laws requiring manufacturers to use a minimum amount of recycled material in their products. Several states require lower levels of certain automobile exhaust gases than federal regulations demand.

Concerned individuals have joined or formed environmental and conservation organizations. These have influenced state, federal, and corporate agencies to adopt policies that help conserve resources and protect the environment. Locally, residents may work to make their communities more conservation minded. Many communities rely on landfills to get rid of solid wastes. Because landfills can pollute soil and water and take up valuable land, some communities now encourage conservation to reduce the need for landfills. They require recycling, thereby conserving aluminum, trees, and other resources. They fine those who throw out recyclable materials and charge residents for landfill-bound garbage by weight. Some communities have cut their waste by 50 percent by recycling a variety of materials and composting food waste and yard debris.

LEGISLATION Voicing your opinion about environmental issues can influence legislation that affects Earth's environment and its resources.

7.3 Section Review

1. What are the advantages and disadvantages of using nuclear energy?
2. Although renewable energy resources cause few environmental problems, why is the use of nonrenewable energy resources more common?
3. Name three ways of conserving nonrenewable energy resources. Be specific.
4. **CRITICAL THINKING** The United States currently recycles between 10 percent and 25 percent of its garbage. Propose two or three ideas for getting people in this country to recycle more.

Chapter 7 Resources and the Environment   159

CHAPTER 7 SECTION 3

Find out more about a career in recycling.
Keycode: ES0707

ASSESS

1. **Advantage:** The use of nuclear energy does not cause air pollution as does the burning of fossil fuels. **Disadvantages:** The use of nuclear energy produces toxic radioactive wastes that are difficult to store and cannot be safely disposed of. Accidents at nuclear power plants can release harmful radiation over large areas.

2. The use of renewable energy resources has more limitations than the use of nonrenewable energy resources. For example, water power can be used only where dams can be built. Wind power works only in areas with strong, steady winds. Solar power requires a mechanism for storing energy when sunlight isn't available. Geothermal energy works only where there is hot bedrock near the surface.

3. **Sample Response:** Nonrenewable energy resources can be conserved by carpooling, turning off lights when not in use, and using energy-efficient appliances.

4. **Sample Response:** Pass laws that make recycling mandatory; give financial incentives, such as higher trash collection fees or tax rebates that depend on the amount of material each household recycles; and institute fines for not recycling.

MONITOR AND RETEACH

If students miss . . .

Question 1 Refer students to the diagram on pages 154 and 155. Have them read the captions in the diagram that relate to nuclear power and fossil fuels.

Question 2 Refer students to the diagram on pages 154 and 155. Have them read the captions in the diagram that relate to water, wind, and solar power.

Question 3 Reteach "Conservation and Recycling." (pp. 156–158) Compile a list of ways to conserve resources, including information in the text as well as students' suggestions.

Question 4 Encourage students to think of possible incentives for recycling and punishments for not recycling.

CHAPTER 7 ACTIVITY

PURPOSE
To compare the operation and effectiveness of passive and active solar heaters

MATERIALS
Provide students with a copy of Lab Sheet 7. A copymaster is provided on page 139 of the teacher's laboratory manual.

Have students use water-based latex paints to paint their collectors.

Supply non-mercury thermometers for the lab.

PROCEDURE
❶ Be sure students follow the precautions on the label when using rubber cement.

❷, ❹ The holes in the shoeboxes could be made in advance if students are likely to have difficulty cutting them.

❽ If the class must do the lab on a cloudy day, students can use incandescent lamps instead of full sunlight. Be sure to connect the lamps to circuits with ground-fault interrupters.

CHAPTER 7 LAB Activity

Passive and Active Solar Heating

SKILLS AND OBJECTIVES
- **Construct models** of passive and active solar collectors.
- **Compare** the operation and effectiveness of solar collectors.

MATERIALS
- 2 shoeboxes of the same size
- 1 brick
- 2 thermometers
- 1 sheet of aluminum foil
- 2 paper cups
- 2 pieces of plastic wrap
- masking tape
- black paint
- rubber cement
- 1 sheet of graph paper
- scissors
- hole punch
- colored pencils

STEP 3

Two kinds of solar collectors are used for heating buildings. One is a *passive solar collector,* which uses the building itself to collect and store heat from the sun. The other is an *active solar collector,* which uses a system to collect, store, and transfer heat. The two collectors absorb and release heat at different rates. In this experiment, you will construct models of passive and active solar collectors and observe the differences between them.

Procedure

❶ Place the brick inside a shoebox 1 cm from the end as shown in the diagram below. Use rubber cement to glue the brick to the bottom of the box. Paint the inside of the box and brick black.

❷ Make a hole in the side of the shoebox. Position the hole so that you can center the bulb of the thermometer in the 1-cm space between the box and the brick, as shown in the diagram to the right. Use masking tape to secure the thermometer at the hole and to prevent air from escaping. You now have a passive solar collector.

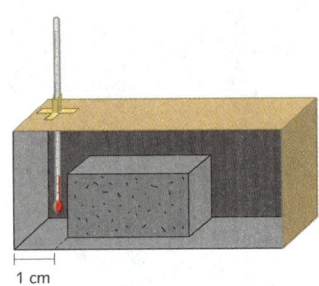

❸ On the outside of the second shoebox, use the open end of a paper cup to trace two circles diagonally opposite each other. The circles should be in the bottom corners of the short sides of the box, as in the diagram to the left. Cut out the two circles with scissors.

❹ Use the bottom of a paper cup to make flaps that can be opened and closed. Cut around the bottoms of the cups, but leave one eighth of the circumference still attached.

❺ From the inside of the box, insert the narrow end of a cup into one of the holes. Pull gently until the lip of the cup is even with the side of the box. Tape the cup in place. Repeat the procedure for the second cup.

❻ Punch a hole in the side of one of the paper cups at the point at which it joins the box. Insert a thermometer in the hole and position the bulb of the thermometer in the center of the cup. Tape the thermometer to the outside of the box, as shown in the diagram on the next page. Make sure that you can read the entire thermometer scale.

7. Line the inner sides of the box with aluminum foil and glue the foil in place. Cut the foil from the holes in the box. Paint the bottom of the box black. Cover the top of the box with plastic wrap and tape to the outside of the box. You now have an active solar collector.

8. Place the collectors in full sunlight. Tilt each collector so it faces the sun. The collectors will be at the correct angle when a pencil held perpendicular to the plastic cover has no shadow. Prop up the collectors with books and allow them to heat for 3 minutes.

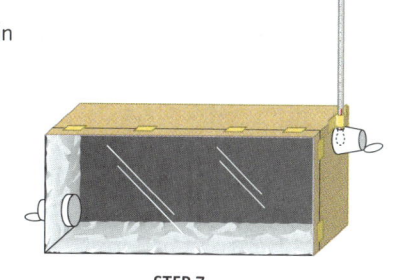
STEP 7

9. Read the temperatures on both thermometers. Copy the table below to record the data. Open the flaps on the active solar collector. For the next 16 minutes, read and record the temperatures on both thermometers every 2 minutes.

	Heating Phase									Cooling Phase							
Minutes	0	2	4	6	8	10	12	14	16	18	20	22	24	26	28	30	32
Temperature (°C), passive collector																	
Temperature (°C), active collector																	

10. Remove the collectors from the sunlight. As the collectors cool, read and record the temperatures every 2 minutes for 16 minutes.

11. Make a graph of your data by plotting time on the x-axis and temperature on the y-axis. Begin numbering the y-axis with the lowest temperature you recorded. Use two different colors to plot the data for the two collectors.

Analysis and Conclusions

1. Which collector reached the higher temperature? Which collector stayed warm longer?

2. If you were building a passive solar-heated house, what features would you include in the design to help the house hold heat?

3. There are three parts to an active solar collector. Which one is missing from the model you built? Explain why the missing part is needed.

4. Your collectors were painted black. Design an experiment that determines whether black is the most efficient color.

5. Do you think it would be wise to invest in research to improve the efficiency and reliability of solar energy? Explain your answer.

CHAPTER 7 ACTIVITY

ANALYSIS AND CONCLUSIONS

1. The active collector reached a higher temperature, but the passive collector stayed warm longer.

2. The walls would be made of brick and stone and would be well insulated to hold in heat.

3. The heat storage area is missing. It is needed because heat that is absorbed during times when the sun shines must be stored for release during nights or cloudy periods when there is no sunlight.

4. Students would paint several passive collectors different colors, one of which would be black. The collectors would be allowed to absorb solar energy for the same period of time. The one that registers the highest temperature is painted the color most efficient at absorbing solar energy.

5. Accept all reasonable answers, such as investment in solar energy would be wise because fossil fuel energy sources, which we depend on now, are nonrenewable and are being used up. Solar energy is also an inexhaustible resource that does not cause pollution.

Refer students to Safety in the Earth Science Laboratory on pages xxii–xxiii.

CHAPTER 7 REVIEW

VOCABULARY REVIEW

1. fossil fuels
2. renewable resource
3. environment
4. conservation
5. reserves

CONCEPT REVIEW

6. Oxygen is a renewable resource. It is replaced in nature as fast as it is used. The others are minerals, which are nonrenewable resources.

7. Students can list any metallic mineral, such as gold, silver, iron, copper, tin, or chromium. The metallic mineral is taken from the ground through an open pit or mine shaft. When necessary, ore is separated from gangue. The metallic mineral is then processed before use.

8. Uranium is a metallic mineral. Petroleum, a fossil fuel, is a liquid of organic origin. Both can be used as fuels to provide energy, and both are nonrenewable resources.

9. Both produce electricity. Coal is an inexpensive energy resource, but it is nonrenewable and produces a great deal of pollution that can cause health problems. Hydroelectric dams use water as their energy source, which is renewable and does not produce air pollution. However, dams can alter the environment around them and disturb ecosystems.

10. Students can list any type of conservation measure, recycling, or passing legislation that protects the environment.

11. peat; lignite; bituminous coal; anthracite

CRITICAL THINKING

12. Wood, trees, or forests can become nonrenewable resources if used faster than they are replaced in nature. Managing forests wisely—replacing harvested trees with the same or a greater number of new trees, for example—can help ensure that forests remain a renewable resource. Students can also list putting limits on logging, or finding substitutes for wood used as fuel, building materials, or other products.

13. The amount of air pollution and associated health problems could increase; coal resources would be exhausted sooner.

CHAPTER 7 REVIEW

Summary of Key Ideas

7.1 Earth's environment includes all the renewable and nonrenewable resources, influences, and conditions at Earth's surface. Renewable resources are replaced by nature at a rate close to their rate of use. Nonrenewable resources, which include minerals, exist in fixed amounts. A resource can be exhausted if the demand for and use of it is greater than the amount that exists in nature.

7.2 Humans use a variety of renewable and nonrenewable energy resources to meet their energy needs. Nonrenewable energy resources include uranium and fossil fuels such as coal, petroleum, and natural gas. Water, wind, solar, and geothermal are types of renewable energy sources.

7.3 The recovery and use of resources affects both the living and nonliving parts of the environment. Mining minerals and rocks can harm soil, water, and living organisms. Nonrenewable resources are limited and their use can cause pollution. Renewable energy sources use no fuel and do not pollute but can be more expensive to use than fossil fuels. Conservation, recycling, and environmental legislation can help protect resources and the environment.

KEY VOCABULARY

conservation (p. 156)
environment (p. 144)
fossil fuel (p. 148)
nonrenewable resource (p. 144)
ore mineral (p. 145)
recycling (p. 156)
renewable resource (p. 144)
reserve (p. 146)

Vocabulary Review

Write the term from the key vocabulary list that best completes the sentence.

1. Nonrenewable ___?___ form from the remains of plants and animals.
2. A(n) ___?___, such as water power, wind power, solar energy, and geothermal energy, uses no fuel and does not pollute.
3. All the living and nonliving things at Earth's surface are part of the ___?___.
4. Natural resources can be protected and managed through ___?___.
5. Knowing Earth's ___?___ allows scientists to predict how long resources will last.

Concept Review

6. Of gold, oxygen, iron, and sulfur, which is a renewable resource? Explain.
7. Identify a metallic mineral resource. Summarize how it might be recovered, processed, and used.
8. Compare and contrast uranium and petroleum. How do they differ? How are they similar?
9. Compare and contrast a coal-burning power plant and a hydroelectric dam. List advantages and disadvantages of each.
10. List three ways an individual can help protect the environment.
11. **Graphic Organizer** Copy and complete the flow chart below.

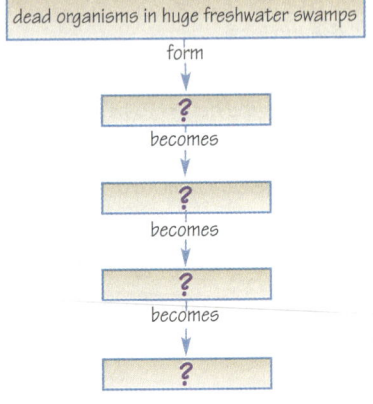

14. The reserves of a mineral are its known deposits that are economically workable. Estimates of reserves may change when new deposits are discovered, when poorer deposits become profitable to remove because of higher prices, or when new technology allows hard-to-reach deposits to be removed at a profit. Resources

Critical Thinking

12. **Evaluate** Forests have been called America's renewable resource. Under what circumstances might wood, trees, or forests be a nonrenewable resource? Describe a process for preserving forests as renewable resources.

13. **Predict** What types of problems would result if everyone heated their homes with coal? Explain your reasoning.

14. **Communicate** Make a presentation describing ways in which technology could affect mineral reserves. Explain why it is possible to increase reserves of a mineral but not resources of a mineral.

15. **Draw Conclusions** Which produces more energy, burning peat or burning an equal mass of anthracite? Explain. Keeping in mind your answer to the first question, explain why peat is still a popular fuel in some countries.

Interpreting Graphs

The graph shows the hourly averages of fine particulates in the air at an Austin, Texas, collection station. These pollutants, measured in micrograms per cubic meter of air, generally come from vehicle exhaust and industrial and residential activities.

16. How does the concentration of pollutants at 6 A.M. Tuesday compare with the concentration of pollutants at 6 A.M. Wednesday?

17. Air pollution can worsen the health of people with lung disease. According to the graph, when is the best time for such a person to run errands?

18. At what time does the concentration of pollutants peak on both days?

19. What could be the cause of the peaks in the amounts of pollution on the two days?

20. Falling rain tends to clean pollutants from the air. What evidence is there that no rain fell Tuesday morning before 6 A.M.?

Internet Extension

What Environmental Changes Can We See with Satellites? Use images taken from space to investigate cases of environmental change through time.
Keycode: ES0707

Writing About the Earth System

SCIENCE NOTEBOOK Your home state probably has its own interesting rock formations or geologic history. Use a variety of resources to find information about the rocks in your region. Try to identify which particular type of rock—igneous, sedimentary, or metamorphic—is more common than others. Describe any environmental changes—biological, water, or atmospheric events—that have helped shape the geologic features you see today.

21. Assume these trends continue into Thursday, June 21. Predict the time on that day when the amount of pollutants is most likely to be greatest.

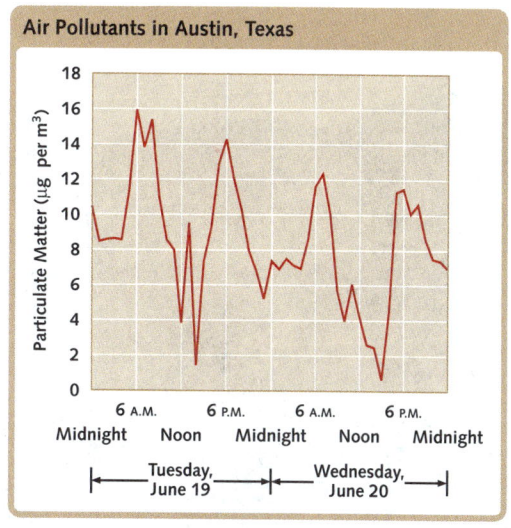

CHAPTER 7 REVIEW

What Environmental Changes Can We See with Satellites?
Keycode: ES0707

📖 Use Internet Investigations Guide, pp. T27, 27

(continued)

are estimates of the total amount of a mineral in Earth's crust and should not change unless new deposits are discovered.

15. Anthracite produces more energy because it has a greater percentage of carbon. Peat is still widely used in some areas because it is relatively cheap and available.

INTERPRETING GRAPHS

16. The concentration is higher on Tuesday, at about 16 micrograms/m^3, compared with a little more than 12 micrograms/m^3 on Wednesday.

17. mid-afternoon around 3:00 P.M.

18. about 6:00 A.M.

19. **Possible Response:** particulate pollution from vehicle traffic during rush hour

20. The particulate levels rose before 6:00 A.M. and peaked at 6:00 A.M.

21. 6:00 A.M.

WRITING ABOUT SCIENCE

Writing About the Earth System

Suggest students contact their state's geological survey or department of natural resources for information about the local geologic history.

UNIT 2 FEATURE

Focus

Objective
Describe the benefits of recycling in the Earth system and in modern human society.

Set a Purpose
Have students read to find answers to the question "In what ways does recycling help sustain Earth's limited resources?" **Recycling materials helps to cut back on the amount of raw material taken from Earth. This is especially important for nonrenewable resources. Recycling of plastics helps to conserve petroleum; recycling of paper saves trees. Recycling also reduces the need for landfill space, which can be considered a limited resource.**

UNIT 2 — The EARTH SYSTEM

Recycling Cents and Sensibility

When you think of the word recycling, *what comes to mind? While recycling has become a widespread community effort in the past few decades, the concept of recycling is nothing new to Earth's environment.*

Recycling is a natural interaction between Earth's spheres. For example, plants and animals die and decompose, becoming nutrients for living organisms. The water cycle replenishes the freshwater supply. Although recycling in nature existed long before humans, we now play an essential role in the sustainability of nature's limited resources.

In general, as technology improves production techniques, the amount of manufactured goods available has increased while their cost has decreased. As a result, we consumers are simply buying more than ever before. How do our buying habits impact the environment? Take the case of cars. The number of cars driven along global roads reached 532 million in 2000. More cars mean an increase in the use of fossil fuels, greater use of scarce metal resources, more crowding in junkyards, and a greater potential for pollution.

As we buy more, we also produce more waste. Think of the packaging that comes with food. While it may help prevent contamination or damage, it also results in more garbage. On average, each person in the United States throws away almost 1,500 pounds of trash annually. Most of this waste ends up in landfills or incinerators.

Many communities have found that recycling not only makes environmental sense, but economic "cents" as well.

RECYCLED PLASTIC can be made into Rhovyl'Eco, a fiber used in this fleece jacket.

CREATIVE RECYCLING This wall is made of old tires and aluminum cans, bound with adobe.

DIFFERENTIATING INSTRUCTION

Reading Support Have students read the title of this article and view the photographs and captions to predict what the focus will be. Tell students to write their predictions in their science notebooks. Have some students read their predictions to the class and discuss them. After they have read the article, have students check their predictions and describe in their notebooks the ways in which their predictions were correct or incorrect. Remind students to be as specific as possible.

and the ENVIRONMENT

By cutting down on garbage, you save your community money by reducing the cost of waste collection and disposal. Recycled materials are also valuable commodities for manufacturers. For example, the demand for recycled paper is growing as the anticipated needs of paper mills exceed the current supply of timber. Researchers estimate that by 2010, at least 47 percent of all paper will be made of post-consumer fiber.

Thanks to recent technological innovations and creative thinking, the possibilities of recycling are broadening as never before. Would you ever have thought that a plastic bottle would be reincarnated as next season's must-have fleece-wear? Engineering has enabled used consumer goods to be turned into synthetic materials that can be used in industries such as textiles. For example, *Rhovyl'Eco* is a synthetic yarn comprised of a blend of plastic bottles and natural fibers. Ingenious products such as this are helping to curb the 600 million plastic bottles thrown away each day.

Novel methods of recycling don't always involve sophisticated technology. For example, some communities use recovered glass to make *glassphalt,* an alternative to conventional asphalt used as road pavement that contains ground glass. Creative ways to recycle paper include uses as animal bedding and insulation. Can you think of unconventional uses for materials you would ordinarily discard?

For some innovative thinkers, recycling literally becomes an art form. Some notable sculptors use what would otherwise be considered garbage—such as discarded toys, kitchenware, and plastics—to create their masterpieces. Common everyday trash can take on priceless value and aesthetic meaning when redefined through an artist's eyes.

COWBOY SCULPTURES made from recycled trash.

UNIT INVESTIGATION
CLASSZONE.COM

Paper or Plastic—Which Type of Bag is Better for the Environment? Paper is manufactured from wood, a renewable resource, while plastic comes from petroleum, a nonrenewable resource. Both types of bags can be reused or recycled, but many end up decomposing in a landfill. Evaluate which type of bag is better for the environment by considering the effects that production and disposal of the bags have on the Earth system.
Keycode: ESU201

UNIT 2 FEATURE

INVESTIGATIONS
CLASSZONE.COM

Paper or Plastic—Which Type of Bag Is Better for the Environment?
Keycode: ESU201

📖 Use Internet Investigations Guide, pp. T28–T29, 28–29.

Investigation Photograph: A bulldozer gathers garbage at a landfill site in Cheltenham, England.

Unit 2 The Earth System and the Environment

UNIT 2 ASSESSMENT

ANSWERS

1. b, Some minerals consist of single elements.
2. d, This is the definition of a reserve.
3. b, This is the definition of a rock.
4. b, Minerals are inorganic.
5. a, Recycling saves nonrenewable resources so they can be used again.
6. d, The other terms describe rock properties besides mineral composition.
7. d, This is the definition of an element.
8. d, Uranium is an element—inorganic and not derived from fossils.
9. d, The other terms are not repeated processes.
10. a, The Northern Hemisphere is tilted away from the sun.
11. d, The rays strike Earth most directly at the Tropic of Capricorn; this is the winter solstice.
12. b, Rotation is the movement of Earth about its axis, as shown by the arrow.
13. sedimentary and metamorphic rocks
14. Dolomite is the parent rock. It was metamorphosed by high pressures and temperatures to marble.
15. The deposition of shells and skeletons from sea organisms is followed by cementation to form limestone.
16. Dolomite does not react to hydrochloric acid unless powdered, because dolomite is altered from the related mineral calcite.

UNIT 2 STANDARDIZED TEST Practice

Directions (1–12): For *each* statement or question, select the word or expression that, of those given, best completes the statement or answers the question. Record your answer on a separate answer sheet.

1. Which is not true of crystals?
   (a) Crystals have regular shapes.
   (b) Crystals are always formed from ions.
   (c) Each mineral has a crystal shape.
   (d) Atoms are arranged in a pattern.

2. The known amount of mineral that can be profitably mined at the present time is called
   (a) an ore
   (b) the gangue
   (c) the resource
   (d) a reserve

3. One or more minerals bound together make up
   (a) a rock-forming mineral
   (b) a rock
   (c) an element
   (d) an ore

4. Which of the following is not a characteristic of a mineral?
   (a) solid
   (b) organic
   (c) orderly atomic structure
   (d) definite chemical composition

5. Recycling is one way of
   (a) conserving nonrenewable resources
   (b) wasting renewable resources
   (c) creating nonrenewable energy resources
   (d) using renewable energy resources

6. The mineral composition of igneous rocks may be
   (a) pumice or granite
   (b) plutonic or extrusive
   (c) clastic, chemical, or organic
   (d) felsic, mafic, or intermediate

7. A substance that cannot be further broken down by ordinary chemical means is
   (a) a mixture
   (b) a metal
   (c) a mass
   (d) an element

8. Which is not a fossil fuel?
   (a) coal
   (b) oil
   (c) natural gas
   (d) uranium

9. The repeated series of events by which rock changes form is
   (a) metamorphism
   (b) deformation
   (c) the sorting process
   (d) the rock cycle

Base your answers to questions 10 through 12 on the diagram of Earth below.

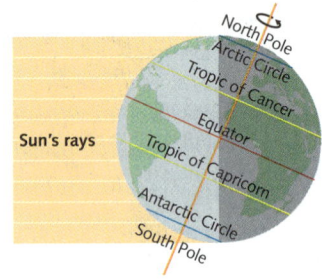

10. According to this diagram, the season in the Northern Hemisphere is
    (a) winter
    (b) summer
    (c) spring
    (d) autumn

11. When the Northern Hemisphere has its shortest day, the sun will be directly overhead at the
    (a) Tropic of Cancer
    (b) South Pole
    (c) equator
    (d) Tropic of Capricorn

12. The circular arrow at the North Pole indicates that Earth is
    (a) orbiting
    (b) rotating
    (c) revolving
    (d) at equinox

UNIT 2 ASSESSMENT

Directions (13–22): Record your answers on a separate answer sheet. Some questions may require the use of material in the Appendix.

Base your answers to questions 13 through 16 on the reading passage below.

Mosaic Canyon Field Trip

The entrance to Mosaic Canyon looks ordinary, but after just a short walk the canyon narrows dramatically to a deep slot cut into the face of Tucki Mountain. Smooth, polished marble walls enclose the trail as it follows the canyon's snakelike curves. The relatively recent uplift and erosion of Death Valley's mountain ranges exposed these rocks.

Mosaic Canyon's polished marble walls were carved out of the Noonday Dolomite and other Precambrian carbonate rocks. These rock formations began as limestone deposited about 850–700 million years ago when a warm sea covered the area. A later addition of magnesium changed the calcium carbonate limestone to dolomite, a calcium-magnesium carbonate.

Later, younger sediment accumulated, deeply burying the dolomite far below the surface, where high pressures and high temperatures altered the dolomite, thus changing it to marble.

13. What types of rocks—metamorphic, sedimentary, igneous—are discussed in the passage above?

14. What parent rock is named? Explain.

15. Describe the process that can form a rock of organic origins such as limestone.

16. Would you expect that dolomite does or does not react to hydrochloric acid? Explain.

Base your answers to questions 17 through 19 on Mohs Scale of Mineral Hardness on page 106.

17. Of the minerals listed on page 106, which is the softest? What is the rating of the hardest mineral?

18. Compare the hardness of a diamond to that of a steel file. Why is steel used more often than diamond on abrasive tools?

19. An unidentified mineral scratches apatite but not a steel file. Can you determine its exact hardness on Mohs scale? Explain. Name three other means by which the mineral might be identified.

Base your answers to questions 20 through 22 on the schemes for rock identification on pages 702–703 of the Appendix.

20. How might grain size help you identify whether an igneous rock was intrusive or extrusive? What is the relationship between grain size and texture?

21. Stone inside the Commerce Department Building in Washington, D.C., is coarse-grained and white to light gray with a bluish-gray tint. It has wavy, thin, dark lines formed when minerals in the rock dissolved under pressure. If the stone reacts positively to the acid test, which type of rock is it likely to be? Explain.

22. The stone used to build the steps to the west front of the Capitol in Washington, D.C., is felsic, gray to white, and has a grain size between one and ten millimeters. Its composition is about 65% feldspar, 27% quartz, and 8% biotite mica. Which type of rock is it likely to be? Explain.

(continued)

17. Talc is the softest mineral listed. The hardest mineral, diamond, has a rating of 10 on the Mohs Scale of Hardness.

18. Diamond, which has a hardness of 10, is harder than a steel file, which has a hardness of 6.5. Steel is made from the mineral iron, which is far less expensive than diamond.

19. The mineral's hardness is between 5 and 6.5, but more testing is necessary to determine its exact hardness. Few minerals can be identified by hardness alone. Tests of streak, cleavage, fracture, specific gravity, acid reaction, or other special properties—as well as inspection of color, luster, and crystal shape—could be used to identify the mineral.

20. Intrusive rocks have a grain size of 1 mm or larger with a coarse or very coarse texture. Extrusive rocks have a grain size less than 1 mm or are non-crystalline and have a fine or glassy texture. The scheme indicates that the larger the grain size, the coarser the texture.

21. The dark lines are a result of pressure on dissolved minerals and indicate that this is a metamorphic rock. The acid test indicates the rock contains calcite, and therefore is probably a type of marble.

22. The word *felsic* indicates that this must be an igneous rock. Its grain size and composition indicate that it is probably granite.

UNIT 3

UNIT OVERVIEW

Most students are acquainted with the view of Earth from space shown here. But why is the unit titled "Dynamic Earth"? As students work to define the word *dynamic* (characterized by continuous movement, change, or activity), have them examine the chapter images and ask students to suggest what motions might occur within Earth itself. In this unit students will learn that the movement of Earth's crust—in response to intense heat and pressure from deep within Earth—shapes the oceans, lands, and atmosphere around them.

VISUAL TEACHING

Chapter 8 Plate Tectonics

The sonar image—which is color-coded for depth in meters below sea level—shows a section of the mid-ocean ridge off the coast of Central America. The data used to make the image were gathered by a sonar device towed just above the sea floor by a ship. Computers processed the sonar data to produce a three-dimensional image. Students will learn how moving plates impact features on the ocean floor.

Answer: Sections of Earth's crust rise and sink as the crust changes.

Chapter 9 Volcanoes

Paricutin, a volcano in Mexico, began as an eruption in a cornfield in 1943. The eruption continued until 1952. During the first year, the volcano's cone grew to 335 meters (1,100 feet). For eight more years, eruptions continued

3 UNIT Dynamic Earth

The Earth System Even if you've never felt an earthquake, seen a volcano, or climbed a mountain, dynamic forces on and under Earth's surface affect you. Earth's oceans, atmosphere, weather, and soil are shaped by the events—both gradual and catastrophic—that we explore in this unit.

CHAPTER 8
Plate Tectonics
What can images like this show you about the movement of Earth's crust?

CHAPTER 9
Volcanoes
What processes lead to dramatic eruptions, such as this one in Central Mexico?

168 Unit 3 Dynamic Earth

CONNECTIONS TO . . .

Prior Knowledge In Unit 2, students learned about Earth's interior and the temperature and pressure changes below Earth's surface that form and alter rocks. This prepares them to study the forces originating in Earth's interior that affect Earth's surface. Unit 3 begins with a chapter on plate tectonics, providing students with the fundamental knowledge they use to understand volcanoes, earthquakes, and mountain building. Unit 4 builds on earlier units and proceeds with the study of the forces that shape Earth's surface.

Unifying Themes *Energy* is released during processes and events such as plate movements, volcanoes, earthquakes, and mountain building. As students study these chapters in Unit 3, point out how dynamic forces affect Earth's geosphere as well as the atmosphere, hydrosphere, and biosphere.

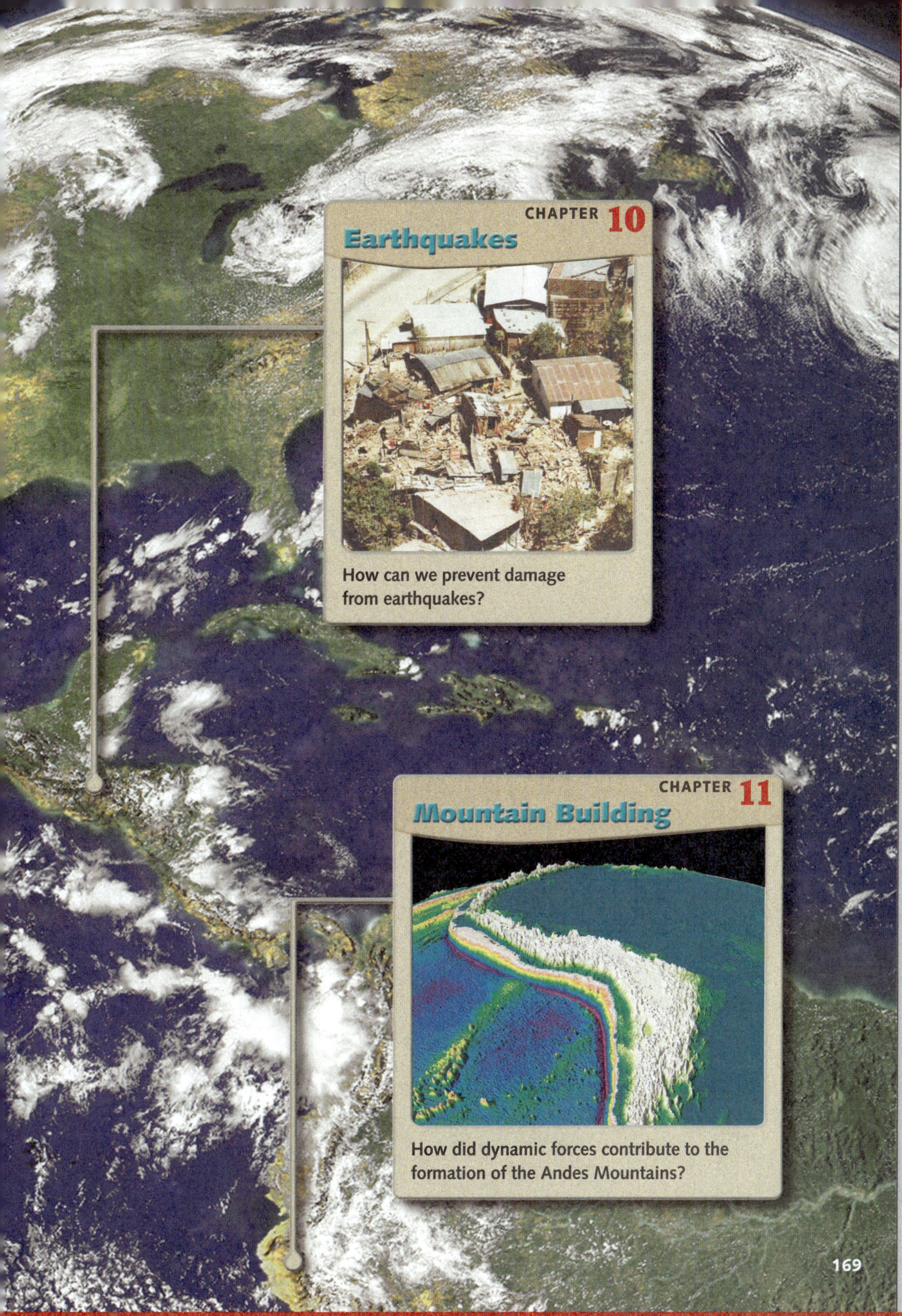

UNIT 3

CHAPTER 10 Earthquakes

How can we prevent damage from earthquakes?

CHAPTER 11 Mountain Building

How did dynamic forces contribute to the formation of the Andes Mountains?

(continued)

but added only another 88 meters (290 feet) to the cone's height. No one was killed directly by lava or ash, but lightning associated with the eruption killed three people. Students will learn why volcanoes like this one erupt.

Answer: Hot molten rock escapes to the surface in volcanic eruptions, as an oceanic plate is subducted under a continental plate.

Chapter 10   Earthquakes
The photograph shows damage caused by an earthquake in El Salvador. Students will learn how earthquakes such as this one occur and why they cause such devastation.

Answer: Prevention of damage is aided by new codes for buildings and advances in technology used to predict earthquakes.

Chapter 11   Mountain Building
The three-dimensional elevation model shows cordilleras, or parallel mountain ranges, of the Andes in South America. Color was added to emphasize the cordilleras. Students will learn how land crumples like this where two of Earth's continental plates collide.

Answer: As two plates collide, land is pushed together and thrust upward.

CONNECTIONS TO . . .

National Science Education Standards
- Students will develop an understanding of energy in the Earth system. Specifically, students will understand that the outward transfer of Earth's internal heat drives plate movements.
- Students will develop an understanding of Earth's dynamic geology.

Unit Features
The Earth System and the Environment
When Volcanoes Breathe Life pp. 250–251

How Can One Volcano Change the World?
Investigate images of Mt. Pinatubo before and after an eruption.
Keycode: ESU301

Standardized Test Practice, pp. 252–253

CHAPTER 8 PLANNING GUIDE: PLATE TECTONICS

	Section Objectives	Print Resources
Chapter-Level Resources		Laboratory Manual, pp. 35–38 Internet Investigations Guide, pp. T32–T34, 32–34 Guide to Earth Science in Urban Environments, pp. 5–12, 29–36 Spanish Vocabulary and Summaries, pp. 15–16 Formal Assessment, pp. 22–24 Alternative Assessment, pp. 15–16
Section 8.1 What Is Plate Tectonics?, pp. 172–175	1. Discuss some of the evidence that Alfred Wegener used to support his idea of continental drift. 2. Explain how the theory of plate tectonics supports the occurrences of earthquakes and volcanoes.	Lesson Plans, p. 24 Reading Study Guide, p. 24 Teaching Transparency 8
Section 8.2 Types of Plate Boundaries, pp. 176–179	1. Discuss the differences among the three types of plate boundaries. 2. Contrast the three different types of convergent boundaries.	Lesson Plans, p. 25 Reading Study Guide, p. 25 Teaching Transparencies 40, 41
Section 8.3 Causes of Plate Movement, pp. 180–181	1. Discuss mantle convection as a possible cause of plate movements. 2. Compare and contrast ridge push and slab pull.	Lesson Plans, p. 26 Reading Study Guide, p. 26 Teaching Transparencies 4, 9, 10
Section 8.4 Plate Movements and Continental Growth, pp. 182–187	1. Explain how Earth's landmasses have changed positions over the past 200 million years. 2. Discuss the roles of plate tectonics, igneous activity, and deposition in the formation of continental landmasses.	Lesson Plans, p. 27 Reading Study Guide, p. 27 Teaching Transparency 41

INVESTIGATIONS
CLASSZONE.COM

Section 8.1 INVESTIGATION
What Is the Earth's Crust Like?
Computer time: 45 minutes / Additional time: 45 minutes
Students examine volcanoes, earthquakes, and mountains to understand Earth's lithosphere.

Section 8.1 INVESTIGATION
How Old Is the Atlantic Ocean?
Computer time: 45 minutes / Additional time: 45 minutes
Students determine how long ago Africa and South America separated.

Technology/Online Resources

Electronic Teacher Tools
Visualizations CD-ROM
Classzone.com
Online Lesson Planner
Test Generator

Visualizations CD-ROM
Classzone.com
   ES0801 Investigation, ES0802 Investigation,
   ES0803 Visualization

Visualizations CD-ROM
Classzone.com
   ES0804 Visualization

Visualizations CD-ROM
Classzone.com
   ES0805 Visualization

Visualizations CD-ROM
Classzone.com
   ES0806 Visualization, ES0807 Visualization,
   ES0808 Visualization, ES0809 Career

Classroom Management

Gather these materials for minilab and map activities.

Minilab, p. 175
$8\frac{1}{2}$ in. × 11 in. sheet of paper
scissors
shoebox lid
2 different-colored markers

Map Activity, pp. 188–189
Appendix B Plate Boundaries Map, pages 712–713
Teaching Transparency 41

Chapter Review INVESTIGATION

How Fast Do Plates Move?
Computer time: 45 minutes / Additional time: 45 minutes
Students use maps and ages of volcanic rocks to calculate the rate of plate motion.

CHAPTER 8

Plate Tectonics

Iceland, an island country in the North Atlantic Ocean, lies on a boundary between two plates, or pieces of lithosphere.

Why does stream rise from immense breaks in the surface of Iceland?

INTRODUCE

Build Interest

Before sharing background information about the photograph with students, allow time for their questions and comments.

Ask questions like the following:

- Describe the landscape in the photograph. **rock that seems to have split apart**
- What do you see in the background of the photograph? **a mountain** What type of mountain might that be? **a volcano**
- What dangers might be associated with this place? **falling, sudden slippage of rock, release of hot steam or other material** Would you like to be where the person in the photograph is? Why or why not? **Responses vary. Curiosity might overcome any sense of danger.**

Respond

- Why does steam rise from immense breaks in the surface of Iceland? **Iceland sits atop a mid-ocean ridge, where an immense crack in Earth's crust connects with heat from deep within Earth. The heat sends water upward as steam.**

About the Photograph

The photograph shows a divergent plate boundary. The westward-moving North American plate (at the left) and the eastward-moving Eurasian plate (at the right) are moving away from each other at a rate of about 1.8 centimeters per year.

BEYOND THE TEXTBOOK

Print Resources 📖

- Reading Study Guide, pp. 24–27
- Transparencies 4, 8, 9, 10, 40, 41
- Formal Assessment, pp. 22–24
- Laboratory Manual, pp. 35–38
- Alternative Assessment, pp. 15–16
- Spanish Vocabulary and Summaries, pp. 15–16
- Internet Investigations Guide, pp. T32–T34, 32–34
- Guide to Earth Science in Urban Environments, pp. 5–12, 29–36

Technology Resources

- Visualizations CD-ROM
- Classzone.com
- Online Lesson Planner
- Electronic Teacher Tools
- Test Generator

170 Unit 3 Dynamic Earth

CHAPTER 8

PREVIEW

▶ **FOCUS QUESTIONS** In this chapter you will investigate plate tectonics and learn more about the key questions below.

Section 1 What evidence have scientists found to support the theory of plate tectonics?

Section 2 What are important features of different types of plate boundaries?

Section 3 What are some of the hypotheses scientists have about the causes of plate movements?

Section 4 How have plate movements caused changes in the positions and shapes of Earth's landmasses?

▶ **REVIEW TOPICS** As you investigate plate tectonics, you will need to use information from earlier chapters.

- mantle (p. 72)
- lithosphere (p. 73)
- asthenosphere (p. 73)
- magnetic field (p. 74)
- magma (p. 118)

▶ **READING STRATEGY**
USE TEST AIDS

Notice that, throughout the chapter, key terms appear in boldface type. The definition of each term is given where the term is boldfaced.

At our Web site, you will find the following Internet support:

DATA CENTER **LOCAL RESOURCES**
EARTH NEWS **CAREERS**
VISUALIZATIONS **INVESTIGATIONS**

- Polarity at Mid-Ocean Ridges
- Processes at Plate Boundaries
- Convection in the Mantle
- Breakup of Pangaea
- Predicted Plate Movement
- Continental Growth
- What Is the Earth's Crust Like?
- How Old is the Atlantic Ocean?
- How Fast Do Plates Move?

Chapter 8 Plate Tectonics 171

CHAPTER 8

PREPARE
Focus Questions
Find out what knowledge students have about plate tectonics. For each focus question, have students consider the section title and then develop two or three additional questions for each section. For example, Section 1: What is plate tectonics? Section 2: Are there any plate boundaries near where I live? Section 3: How fast do the plates move?

Review Topics
In addition to the earth science topics listed, students may need to review this concept:

Physical Science
- Convection is the transfer of heat energy in a liquid or gas through the circulation of currents of particles within the substance in response to differences in density.

Reading Strategy
Model the Strategy: Have students turn to page 172. Ask them to identify the second "key term" on that page. continental drift Have them write the term and its definition. a hypothesis that states the continents have moved or drifted from one location to another over time Tell students to skim the chapter, locating and writing the definitions of five other key terms. Review these terms and definitions. Point out that students can help study for a test by reviewing the key terms where they are boldfaced and by writing the terms and their definitions.

BEYOND THE TEXTBOOK

INVESTIGATIONS
- What Is the Earth's Crust Like?
- How Old is the Atlantic Ocean?
- How Fast Do Plates Move?

VISUALIZATIONS
- Animation of sea-floor spreading that shows alternating polarity of magnetic minerals
- Animations of processes that occur along plate boundaries
- Animation of convection in the mantle
- Animation of the breakup of Pangaea
- Animation showing growth of a continent

DATA CENTER
EARTH NEWS
LOCAL RESOURCES
CAREERS

Online Lesson Planner

Chapter 8 Plate Tectonics 171

CHAPTER 8 SECTION 1

FOCUS

Objectives
1. Discuss some of the evidence that Alfred Wegener used to support his idea of continental drift.
2. Explain how the theory of plate tectonics supports the occurrences of earthquakes and volcanoes.

Set a Purpose
Have students use the red and blue heads on pages 172–174 to propose an answer to the question "What is the theory of plate tectonics, and what types of evidence support this theory?"

INSTRUCT

More about...

Mesosaurus
Paleoecological evidence shows that *Mesosaurus* was a small, freshwater reptile. Thus, it is unlikely that it would have crossed the present-day Atlantic. If it *had* swum across the Atlantic, then it can be argued that it should have swum across other oceans and ended up as fossilized remains in rocks on other continents. These two arguments are strong evidence that South America and Africa were once joined.

8.1

KEY IDEA
The lithosphere is broken into rigid plates that move in relationship to one another on the asthenosphere.

KEY VOCABULARY
- plate tectonics
- continental drift
- mid-ocean ridge

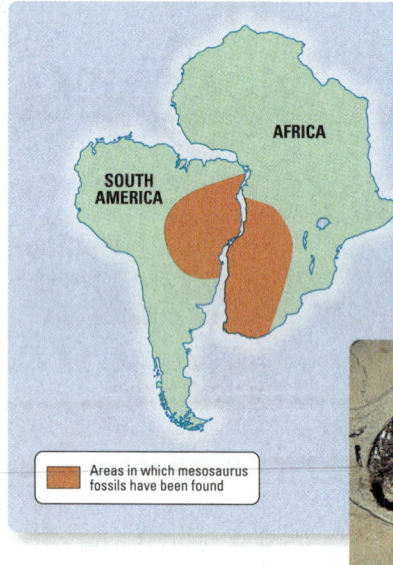

MAP *Mesosaurus* fossils have been found in South America and Africa, lending support to the hypothesis that the continents were once joined together.

What Is Plate Tectonics?

Earth's lithosphere is broken into plates that move on the asthenosphere. In some places, the plates are moving toward each other. In other places, they are moving apart, and in others, they are sliding past each other. **Plate tectonics** (tehk-TAHN-ihks) is a theory that describes the formation, movements, and interactions of these plates.

Early Ideas About Plate Movements

The idea that Earth's surface might be moving is not new. The theory of plate tectonics developed from early observations made about the shapes of the continents and from fossil and climate evidence.

In the early 1500s, explorers using maps noted the remarkable fit of the shape of the west coast of Africa and the shape of the east coast of South America. In 1596, a Dutch mapmaker suggested that the two continents may have been part of a larger continent that had broken apart.

In 1912, a German scientist named Alfred Wegener (VAY-guh-nuhr) proposed a hypothesis called **continental drift.** According to this hypothesis, the continents have moved, or drifted, from one location to another over time. Wegener used many observations to support his hypothesis. In addition to the similarities in the shapes of the continents, he noted that the fossil remains of *Mesosaurus*, a reptile that lived about 270 million years ago, are found only in parts of South America and Africa. This strange distribution is easily explained if the two continents were once joined, as suggested by the map below. Distinctive rock formations found on both continents would have matched up with each other if the continents had been joined in the past. Climate change evidence further supports the continental drift hypothesis.

One of the strongest objections to Wegener's hypothesis was that it did not explain *how* the continents moved. Wegener suggested that the continents might float on deeper, more fluid layers, and that Earth's internal heat could provide the energy needed to move the continents through these layers. He had no evidence to support that explanation, however. Scientists continued to debate Wegener's ideas about continental drift for a number of years. During his lifetime, Wegener continued his efforts to defend the continental drift hypothesis, but he was not successful.

FOSSIL EVIDENCE This fossil *Mesosaurus* was found in Brazil. Similar fossils have been found in Africa.

172 Unit 3 Dynamic Earth

DIFFERENTIATING INSTRUCTION

Challenge Activity Have students work in groups to brainstorm the type of evidence that would indicate a climate change and how this evidence supports the theory of continental drift. Fossils of tropical plants and animals found in a polar location such as Antarctica suggest that the climate of Antarctica was once tropical. It must have been located near the equator in the past. Evidence of past glaciation in a tropical climate such as Africa suggests that it was once located close to one of the poles.

Hands-On Demonstration Inform students that an integral part of Wegener's hypothesis was that Earth's continents were once joined as a single landmass he named Pangaea. Provide thin cardboard templates of Earth's continental landmasses. Challenge pairs of students to join the present-day continents to form a single landmass. Some overlapping of continents is acceptable to account for erosion, uplift, and subsidence. Have students compare their landmasses with Pangaea, which is shown on page 182 (Transparency 10).

CHAPTER 8 SECTION 1

The Theory of Plate Tectonics

In the 1950s and 1960s, discoveries about earthquakes, magnetism, and the age of rocks on the ocean floor added support to some, though not all, of Wegener's ideas. Evidence was strong that Earth's landmasses had moved over time; however, they did not move in the way Wegener had proposed. Instead, scientists proposed a new theory—the theory of plate tectonics.

According to the theory of plate tectonics, the continents are embedded in lithospheric plates. As these plates move, they carry the continents with them. The ocean basins are part of lithospheric plates as well.

The theory of plate tectonics is supported by a wealth of evidence and explains many important geological processes. The theory explains why earthquakes and volcanoes are likely to occur in particular locations and how new crust forms along the ocean floor.

What Is the Earth's Crust Like?
Examine the locations of volcanoes, earthquakes, and folded mountains to develop an understanding of Earth's lithosphere.
Keycode: ES0801

Locations of Earthquakes and Volcanoes

Data indicate that earthquakes and volcanic activity do not occur randomly throughout the world. Instead, they occur primarily in concentrated belts, as shown on the map below.

The theory of plate tectonics helps explain this pattern because the earthquake and volcano belts mark the locations of plate boundaries. These boundaries are places where two plates are pushing toward, pulling away, or sliding past each other. Strain builds up along plate boundaries, and when the strain becomes too great, fractures form and earthquakes occur. The boundaries are also areas of high heat flow, where molten rock moves upward to Earth's surface, causing volcanic activity.

MAP The map shows earthquakes with magnitudes ranging from 5 to 9, from 1996 through 2000.

Locations of Earthquakes and Volcanoes

- Earthquakes
- △ Volcanoes

What Is Earth's Crust Like?
Keycode: ES0801
Use a computer and projector to introduce this investigation in class.
📖 Use Online Investigation Guide, pp. T32, 32.

Visual Teaching

Interpret Diagram
Refer students to the map. Ask: Where in North America do most earthquakes occur? **the western coast and Alaska** Where do most volcanoes form? **in and around the Pacific Ocean** Ask students to think about why there are fewer volcanoes in the Atlantic Ocean than in the Pacific. **Some students may know that the plate boundaries in these two oceans are different: a divergent boundary exists in the mid-Atlantic; convergent boundaries are more prevalent in the Pacific.** Why is the area around the Pacific Ocean called the Ring of Fire? **the abundance of volcanoes around the rim of the Pacific, outlining plate boundaries**

Extend
Ask: How can you explain the occurrences of "isolated" earthquakes such as those in the eastern United States or mid-Pacific Ocean? **ancient plate boundaries or hot spots (Chapter 9, page 197)**
Use Transparency 8.

Chapter 8 Plate Tectonics 173

DIFFERENTIATING INSTRUCTION

Reading Support Have students make a table to distinguish between the hypothesis of continental drift and the theory of plate tectonics. Have them begin with the column heads *Continental Drift* and *Plate Tectonics*. Then state a fact about either idea. Have students reread the text, if necessary, in order to place the fact into the correct column. Repeat this activity until all pertinent information has been placed correctly in the table. Have students keep their tables for future reference.
Use Reading Study Guide, p. 24.

Challenge Activity Have students use the Internet to identify about ten locations each of recent volcanic eruptions and earthquakes. Students then plot the events on a world map and compare the map with the maps on pages 173 and 712–713. Ask: Did any of the events occur other than where you expected? **If so, have students confirm the accuracy of their work. If their plotting is correct, have students suggest an explanation for the anomalies. Points that fall along plate boundaries or at hot spots support the theory of plate tectonics.**

CHAPTER 8 SECTION 1

Visual Teaching

Interpret Diagram
Refer students to the diagram. Ask: What is the source of the rock? **the magma** What happens along a mid-ocean ridge? **Magma wells up and fills the crack that forms as two plates separate. Earth's magnetic polarity is preserved in the magma as it hardens.** What is the polarity of rocks closest to the ridge? **normal polarity** Where are the youngest ocean-floor rocks? **closest to the ridge** How old are the oldest rocks shown? **over 4 million years old** In this example, which are generally longer—periods of normal polarity or periods of reversed polarity? **periods of reversed polarity**

Extend
Describe the pattern of the bands. **The alternating bands of rock in this diagram are mirror images.** Suppose you observed correlative bands that weren't the same width. How could you explain your observation? **Asymmetrical bands would indicate slightly different rates of spreading on either side of the ridge.**

How Old Is the Atlantic Ocean?
Explore a model showing the shape and age of the Atlantic seafloor. Determine how long ago Africa and South America separated.
Keycode: ES0802

VISUALIZATIONS CLASSZONE.COM
Observe how alternating magnetic polarity is recorded in rocks at mid-ocean ridges.
Keycode: ES0803

Magnetism and the Age of the Ocean Floor

Important evidence that supports the theory of plate tectonics comes from studies of the magnetic properties and the ages of igneous rocks on the ocean floor. Plate tectonics explains patterns observed in data from these studies.

Some igneous rocks contain minerals that are magnetic. These minerals provide a record of the direction of Earth's magnetic field at the time when the molten matter that formed the rock cooled. One important discovery that geologists made when studying such igneous rocks was that some of the rocks recorded reversals in the direction of Earth's magnetic field. At certain times in the past, the present north magnetic pole became the south magnetic pole, and the present south magnetic pole became the north magnetic pole. A number of magnetic reversals have taken place at different times over a period of millions of years.

Geologists also studied the magnetic record of the rocks of the ocean floor, particularly the igneous rocks on the sides of mid-ocean ridges. A **mid-ocean ridge** is a long chain of volcanic mountains on the ocean floor with a deep central valley. It was discovered that the magnetic reversals are recorded in bands of rock on either side of a mid-ocean ridge. As shown in the diagram below, the pattern of the bands was mirrored on either side of the ridge, with the center of the ridge always showing the current orientation of the north and south poles.

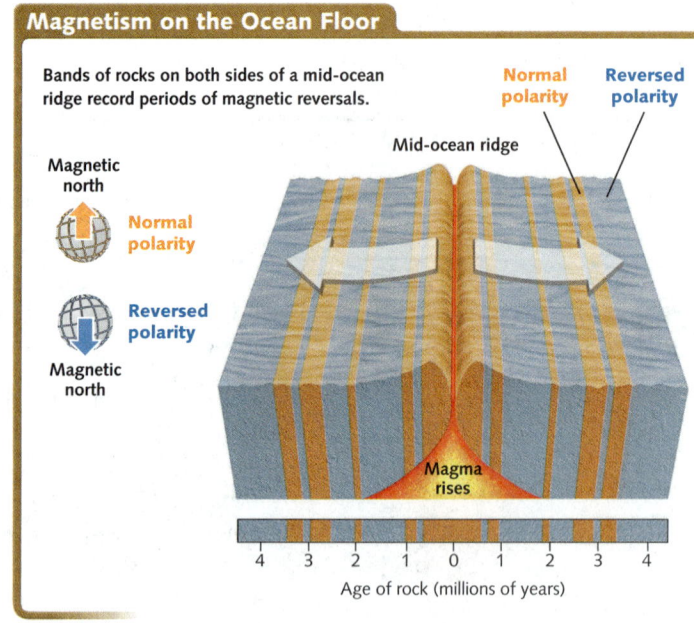

Magnetism on the Ocean Floor
Bands of rocks on both sides of a mid-ocean ridge record periods of magnetic reversals.

When geologists determined the ages of the rocks in the magnetic reversal bands on either side of mid-ocean ridges, they found that the rocks at the center of a mid-ocean ridge are the youngest. The rocks are older the farther out from the ridge on either side they are.

174 Unit 3 Dynamic Earth

DIFFERENTIATING INSTRUCTION

Hands-On Demonstration If your students did not do the Minilab in Chapter 4, you can model Earth's magnetic field here. Place a bar magnet on an overhead projector and then lay a transparency over the bar magnet (the stiffer the transparency, the better). Place about a half cup of iron filings in a large self-sealing plastic bag, and place the bag flat on top of the transparency. Gently tap or jostle the transparency to distribute the filings. The pattern created by the magnetic filings is indicative of Earth's magnetic field.

⚠ *Exercise caution when working with iron filings. Remind students that safety is everyone's responsibility. See Safety in the Earth Science Laboratory at the front of this book, pages xxii–xxiii.*

Based on studies of the ages and the magnetic properties of the rocks along the mid-ocean ridges, geologists have concluded that these ridges represent boundaries where lithospheric plates are moving apart.

The newer rocks along a mid-ocean ridge are formed by hot, molten rock rising between the spreading plates. As these new rocks form, the older rocks spread away from the ridge on either side.

In the computer-generated images below, colors indicate the ages of the rocks on the sides of mid-ocean ridges in the Atlantic and Pacific oceans. The red color represents the youngest rocks along a mid-ocean ridge. Moving away from the ridge on either side, the colors change gradually to indicate that the rocks of the ocean floor become increasingly older.

AGE OF THE OCEAN FLOOR The left-hand image shows the Mid-Atlantic Ridge, a mid-ocean ridge in the Atlantic Ocean. The right-hand image shows the East Pacific Rise, a mid-ocean ridge in the Pacific Ocean.

Measurements have shown that the flow of heat leaving the rocks along the mid-ocean ridges is unusually high. The amount of heat leaving the rocks decreases gradually farther away from the mid-ocean ridge. These observations support the conclusion that mid-ocean ridges are regions where new rock is being formed as Earth's plates move apart.

20-Minute Mini LAB

Ocean Floor Magnetism

Materials
- 8½ in. by 11 in. sheet of paper
- scissors
- shoebox lid
- 2 different-colored markers

Procedure

1. Cut the paper in half lengthwise, making two long strips.
2. Cut a slit in the shoebox lid. Insert each strip of paper into the slit, leaving about 3 centimeters of each strip showing. Fold the strips back as shown below. Color the part of each strip that is showing.

3. Pull out each strip another 3 centimeters. Mark new stripes in a different color.
4. Repeat Step 3 three times.

Analysis

What landform does the slit represent? What do the stripes represent? How might you make this model more realistic?

8.1 Section Review

1. What observations support the continental drift hypothesis?
2. How do observations of earthquake and volcanic activity support the theory of plate tectonics?
3. What evidence in support of plate tectonics is provided by studies of the ocean floor?
4. **CRITICAL THINKING** Identify the weaknesses in Wegener's continental drift hypothesis. Explain how the theory of plate tectonics differs from the continental drift hypothesis.
5. **PHYSICS** What would you expect to find if you measured the temperatures of rocks on the ocean floor at various distances from a mid-ocean ridge? Explain your thinking.

Chapter 8 Plate Tectonics 175

CHAPTER 8 SECTION 1

INVESTIGATIONS
CLASSZONE.COM

How Old Is the Atlantic Ocean?
Keycode: ES0802

📖 Use Internet Investigations Guide, pp. T33, 33.

VISUALIZATIONS
CLASSZONE.COM

Observe how alternating magnetic polarity is recorded in rocks at mid-ocean ridges.
Keycode: ES0803

💿 Visualizations CD-ROM

MINILAB
Ocean Floor Magnetism
Analysis Answers
a mid-ocean ridge; rocks formed as the result of spreading along the ridge; vary the width of the stripes to correspond with actual magnetic periods and add the orientation of the magnetic field to each set of stripes

ASSESS

1. shapes of continental coastlines, fossil distribution, climate change evidence, distinctive rock formations
2. Earthquakes and volcanoes occur in concentrated belts that correspond to present tectonic plate boundaries.
3. magnetic reversals in rocks on either side of the Mid-Atlantic Ridge; high heat flow along ridges; youngest crust is closest to ridge axis
4. Wegener could not explain *how* continents moved. The theory of plate tectonics states that continents and ocean basins move as the lithospheric plates on which they ride move. Continental drift stated that continents had changed their positions over geologic time.
5. Temperature would decrease with distance from the ridge axis because as the newly formed oceanic crust moves farther from the ridge, it cools.

Chapter 8 Plate Tectonics 175

MONITOR AND RETEACH

If students miss . . .

Question 1 Review the tables students made in the Reading Support strategy at the bottom of page 173.
Question 2 Compare the map on page 173 to the plate boundary map. (pp. 712–713)
Question 3 Review the diagram on page 174 and the Minilab on page 175.
Question 4 Reteach pages 172 and 173. Ask: What data would have allowed Wegener to complete his hypothesis? **ocean floor magnetic data**
Question 5 Review the diagram on page 174 and the text below that diagram. Help students conclude that the rising magma is the source of the high heat flow associated with ocean ridges. Newly formed crust cools as it moves away from the ridge.

CHAPTER 8 SECTION 2

▶ Focus

Objectives
1. Discuss the differences among the three types of plate boundaries.
2. Contrast the three different types of convergent boundaries.

Set a Purpose
Have students read this section to answer the question "What are the three types of tectonic plate boundaries, and how do they differ?"

▶ Instruct

Visual Teaching
Discussion

Have students locate the divergent boundary in the diagram on pages 176 and 177. Then ask: What type of landform is found at a divergent boundary on the ocean floor? **a mid-ocean ridge** Where is the rift valley in relation to the ridge? **The rift valley runs down the center of the ridge.** Name a divergent boundary. **the Mid-Atlantic Ridge** Have students use their fingers to trace where the fracture zones associated with mid-ocean ridges would be found. **perpendicular to the ridge axis**

8.2

KEY IDEA
Boundaries between plates are described generally as divergent, convergent, or transform, depending on how the plates move relative to each other.

KEY VOCABULARY
- divergent boundary
- rift valley
- rift
- convergent boundary
- subduction boundary
- deep-sea trench
- collision boundary
- transform boundary

Types of Plate Boundaries

Scientists classify boundaries between two plates according to plate movement. There are three main types of plate boundaries: divergent, convergent, and transform.

Divergent Boundaries

A **divergent boundary** is a boundary between two lithospheric plates that are moving apart. Divergent boundaries are sometimes called spreading centers. Most divergent boundaries lie along the ocean floor and have **rift valleys,** which are deep valleys at the center of a mid-ocean ridge. A rift valley runs along the entire length of the mid-ocean ridge.

The rift valley along a mid-ocean ridge forms the boundary between two diverging lithospheric plates. In a process sometimes called sea-floor spreading, molten rock forces its way upward through cracks, or **rifts,** along the valley. The molten rock cools, hardening into new oceanic crust. The older oceanic crust on either side of the valley moves away from the mid-ocean ridge.

A rift valley at a mid-ocean ridge is typically broken into segments that are offset from each other by breaks in the lithosphere called fracture zones. These fracture zones tend to lie perpendicular to the ridge. Movements along fracture zones have been found to be a source of the earthquakes that occur along mid-ocean ridges.

As shown on the plate boundary map on pages 712–713, two examples of mid-ocean ridges are the Mid-Atlantic Ridge and the East Pacific Rise. Along the rift valleys of these ridge systems, hot springs rise up from hydrothermal vents on the ocean floor. Scientists have found colonies of previously unknown organisms living around the hydrothermal vents.

SUBDUCTION BOUNDARY The Indonesian island of Krakatau has formed near the subduction boundary between two oceanic plates.

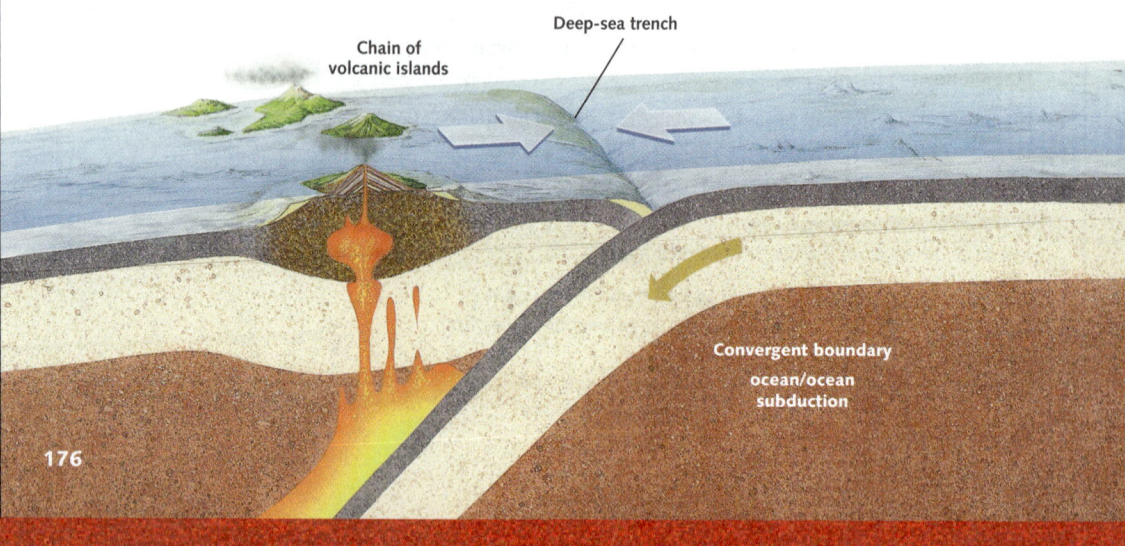

DIFFERENTIATING INSTRUCTION

Developing English Proficiency Have students find all forms of *convergent* and *divergent* used on these two pages: *divergent* and *diverging; convergent, converging, convergence,* and *converge.* Have students copy the sentences in which these words appear. Challenge them to write new sentences that correctly use another form of each term. **Sample sentences: When plates diverge, they move apart from each other. Mount Shasta formed when an oceanic plate converged with a continental plate.**
Use Spanish Vocabulary and Summaries, pp. 15–16.

176 Unit 3 Dynamic Earth

Convergent Boundaries

A **convergent boundary** is a boundary between two plates that are moving toward each other, or converging. Two broad classifications for convergent boundaries are subduction boundaries and collision boundaries.

Subduction Boundaries

When an oceanic plate plunges beneath another plate, the oceanic plate is said to be *subducting* (suhb-DUHK-tihng) beneath the overriding plate. The boundary between the plates is called a **subduction boundary.** One important feature of a subduction boundary is a long, deep trench called a **deep-sea trench** that forms along the boundary. Such trenches are the deepest parts of the ocean floor.

Subduction boundaries can occur at the convergence of two oceanic plates or at the convergence of an oceanic plate with a continental plate.

When two oceanic plates converge, the deep-sea trench that forms is accompanied by the formation of a chain of volcanic islands called a volcanic island arc on the overriding plate. For example, as the Pacific Plate subducts under the Philippine Plate, the Pacific Plate is pulled down to form the Mariana Trench. The leading edge of the overriding Philippine Plate is marked by a chain of volcanic islands, the Mariana Islands.

When an oceanic plate converges with a continental plate, the denser oceanic plate subducts beneath the less-dense continental plate. A deep-sea trench forms, as shown below. A mountain chain and volcanoes form inland on the overriding continental plate.

VOCABULARY STRATEGY
The prefix *sub-* in the words *subducting* and *subduction* means "below." *Subduction* comes from a Latin verb meaning "to draw away from below."

DIVERGENT BOUNDARY These hydrothermal vents are on the Mid-Atlantic Ridge.

SUBDUCTION BOUNDARY Mount Shasta is a volcano in California that has formed near a subduction boundary.

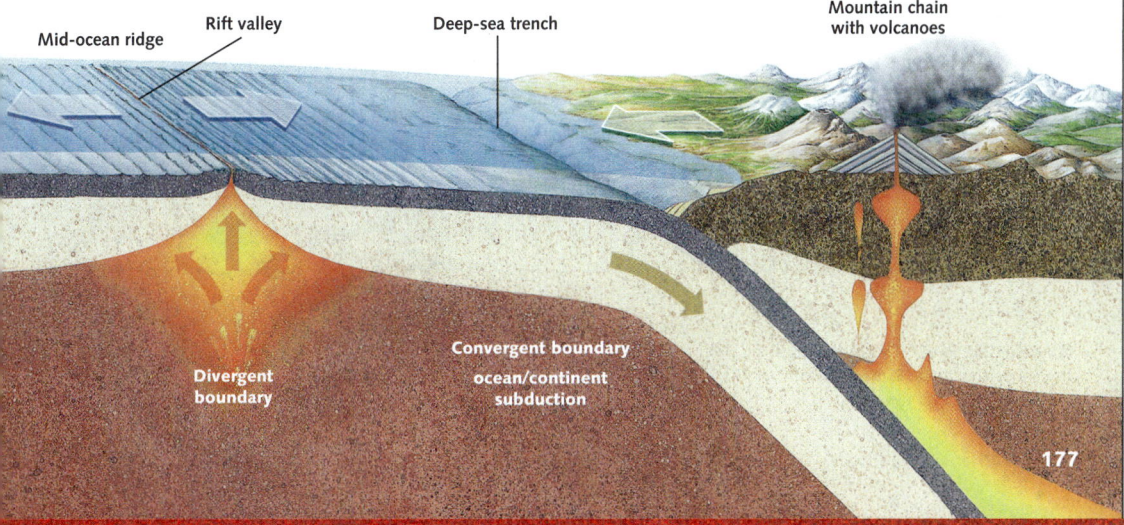

CHAPTER 8 SECTION 2

VISUAL TEACHING
Discussion
Refer students again to the diagram. Ask: Where are the two convergent boundaries? **on either end of the diagram** How are these boundaries the same? **Both are subduction boundaries that result in the formation of a deep-sea trench.** How do they differ? **The subduction boundary on page 176 involves two oceanic plates. The subduction boundary on page 177 involves an oceanic plate and a continental plate. During oceanic-oceanic convergence, an island arc forms. During oceanic-continental convergence, a mountain chain with volcanic peaks forms.**

Extend
Have a student read aloud the caption of the photograph of Mount Shasta. On a world map, locate Mount Shasta and other volcanoes in the Cascade Range. Ask: How did these volcanoes form? **These volcanoes formed when an oceanic plate converged with a continental plate in the geologic past. The subducted oceanic plate partly melted to form magma that fuels the volcanoes.**

DIFFERENTIATING INSTRUCTION

Reading Support Refer students to the Vocabulary Strategy on page 177. Have them make a list of other words that contain the prefix *sub-*. Lists might include *submarine, subway, submersible, subliminal,* and *subterranean.* Have students look up each word in the dictionary and discuss how its derivation and prefix relate to the meaning of the word.
Use Reading Study Guide, p. 25.

Reading Support Challenge students to create mnemonic devices that will help them distinguish between divergent and convergent boundaries. Samples: Plates <u>c</u>ome together at <u>c</u>onvergent boundaries. Plates <u>s</u>pread apart at <u>d</u>ivergent boundaries.

Chapter 8 Plate Tectonics 177

Discussion

Use the following questions to help students differentiate between subduction and collision boundaries: How are subduction boundaries like collision boundaries? Both are convergent plate boundaries. How do these two types of boundaries differ? Subduction boundaries involve at least one oceanic plate; collision boundaries involve two continental plates. Name one mountain range that formed as the result of convergence at each of these boundaries. The islands of Indonesia, the Mariana Islands, and the Andes all formed as the result of convergence involving an oceanic plate. The Himalayas formed and are still rising as the result of a collision between the Indian Plate and the Eurasian Plate.

Use Transparencies 40, 41.

More about...

Seismic Gaps
Portions of an active fault that have not experienced movement over a significant period of time are called seismic gaps. The probability of a major earthquake occurring along a seismic gap is often high.

Critical Thinking

Have students hypothesize why there is no subduction associated with collision boundaries. Continental crust is too light to be subducted into the mantle.

HIMALAYAS The world's tallest mountains, the Himalayas, have formed at a collision boundary.

VISUALIZATIONS CLASSZONE.COM
Observe animations of processes that occur along plate boundaries.
Keycode: ES0804

SAN ANDREAS FAULT This photograph shows a portion of the San Andreas Fault near Taft, California.

178 Unit 3 Dynamic Earth

For example, off the west coast of South America, the Nazca Plate is subducting under the South American Plate. The Peru-Chile Trench has formed between the plates. The Andes Mountains and active volcanoes have formed along the western edge of the South American continent.

Collision Boundaries

If two converging plates each carry continents, the two continents may become welded into a single, larger continent. The boundary that forms when two continents collide and are welded into a single, larger continent is called a **collision boundary.** The collision causes the crust at the boundary to be pushed upward into a mountain range.

The highest mountains in the world, the Himalayas, lie along a collision boundary where India is pushing northward into China at a rate of about 5 centimeters each year. The Indian subcontinent is now welded to the Eurasian continent. The Himalayas continue to grow even higher, and large numbers of earthquakes occur as the two plates push together. Mountain ranges have also formed at other collision boundaries in the past. The Ural Mountains in Russia formed about 250 million years ago when Europe collided with Siberia to form the Eurasian continent. The Appalachian Mountains formed by the same process during the collision of North America and northern Africa. At a much later time, the two continents moved apart, forming the Atlantic Ocean.

Transform Boundaries

A **transform boundary** is a boundary between two plates that are sliding past each other. The fracture zones that offset the segments of a mid-ocean ridge are transform boundaries. You will learn more about fracture zones in Chapter 23.

Another example of a transform boundary occurs in California, where the North American Plate and the Pacific Plate are sliding past each other along the San Andreas Fault. Southwestern California is part of the Pacific Plate, which is moving northwest. The rest of the United States is part of the North American Plate, which is moving southeast.

Movement along transform boundaries is not uniform. The rate of movement along the San Andreas Fault may be as high as 5 centimeters per year. However, some areas have not moved in over a century.

DIFFERENTIATING INSTRUCTION

Reading Support Emphasize the amount of information in the summary chart on page 179 as well as the usefulness of such charts as study tools by asking the following questions: Where do deep-sea trenches form? along convergent boundaries that involve an oceanic plate What type of geologic activity is associated with transform boundaries? earthquakes Where does sea-floor spreading create new oceanic crust? at divergent boundaries What type of boundary results in the formation of a volcanic mountain chain along a continental coastline? ocean-continent subduction Where can you find an example of this type of volcanic mountain chain? west coast of South America (the Andes)

178 Unit 3 Dynamic Earth

The chart below summarizes important information about each type of plate boundary.

SUMMARY: Types of Plate Boundaries

Type of boundary	Process involved	Characteristic features	Current examples
Divergent	sea-floor spreading	• mid-ocean ridges • rift valleys • earthquake activity at fracture zones along mid-ocean ridges • volcanic activity	• Mid-Atlantic Ridge • East Pacific Rise
Convergent	ocean-ocean subduction	• deep-sea trenches • volcanic island arcs • earthquake activity	• islands of Indonesia • Mariana Islands
	ocean-continent subduction	• deep-sea trench bordering continent • volcanoes along coast of continent • earthquake activity	• western coast of South America
	continent-continent collision	• high continental mountain chains • earthquake activity	• Himalayas
Transform	plates sliding past each other	• earthquake activity	• San Andreas Fault • North Anatolian Fault (Turkey) • fracture zones along mid-ocean ridges

8.2 Section Review

1. Explain how new oceanic crust is formed at a divergent boundary.
2. Describe two different types of subduction boundaries. Use the plate boundary map on pages 712–713 to identify an example of each type.
3. Describe what happens at a collision boundary. Identify a collision boundary on the map on pages 712–713.
4. Describe the movement of plates at a transform boundary, and give some examples.
5. What types of plate boundaries are *not* shown in the diagram on pages 174–175?
6. **CRITICAL THINKING** Explain how the densities of oceanic crust and continental crust influence what happens when an oceanic plate converges with a continental plate.
7. **GEOGRAPHY** In 2001, a large earthquake related to the movements of two plates occurred about 20 kilometers northeast of Olympia, Washington. Use the map on pages 712–713 to identify the plates and the type of boundary they share.

Chapter 8 Plate Tectonics 179

MONITOR AND RETEACH

If students miss . . .

Question 1 Review the Minilab on page 175.
Question 2 Use the diagram on pages 176 and 177 to contrast the two types of subduction boundaries.
Question 3 Use your hands (palms down) to simulate what happens when two continental plates collide.
Question 4 Again, use your hands to show the movement along a transform boundary.
Question 5 Review the chart on page 179.
Question 6 Review the concept of density and refer students to "Igneous Rock Descriptions" in Chapter 6, pp. 123–124.
Question 7 Use the table on page 179 to discuss how to use arrows to symbolize each type of plate boundary on a tectonic map.

CHAPTER 8 SECTION 2

VISUALIZATIONS
CLASSZONE.COM

Observe animations of processes that occur along plate boundaries.
Keycode: ES0804

Visualizations CD-ROM

ASSESS

1. Molten rock forces its way upward into the rifts that form when two lithospheric plates separate. As the molten rock cools, new oceanic crust forms. The older oceanic crust moves away from the mid-ocean ridge.

2. Oceanic-oceanic plate convergence: two features are a deep-sea trench, and a volcanic island arc on the overriding plate. Example: the Java Trench and the islands of Indonesia. Oceanic-continental plate convergence: A deep-sea trench and an inland mountain chain and volcanoes characterize this type of subduction boundary. Example: the Peru-Chile Trench and the Andes.

3. Two continental plates collide, forming a single, larger continent and pushing up the crust to form a mountain range. Examples: the Himalayas, Ural Mountains, and Appalachian Mountains.

4. Plates slide past each other at a transform boundary. The San Andreas, the North Anatolian, and the fracture zones along mid-ocean ridges are transform boundaries.

5. No collision boundaries or transform boundaries are shown. (Transform boundaries are generally found along ocean ridges, but they are not shown.)

6. When an oceanic plate converges with a continental plate, the oceanic crust subducts (sinks) under the continental crust. This is because oceanic crust is denser than continental crust.

7. The plates involved are the Pacific Plate (actually the Juan de Fuca Plate) and the North American Plate. They form a convergent boundary off the Washington coast.

CHAPTER 8 SECTION 3

▶ Focus

Objectives
1. Discuss mantle convection as a possible cause of plate movements.
2. Compare and contrast ridge push and slab pull.

Set a Purpose
Have students read this section to answer the question "What causes plate movements?"

▶ Instruct

Discussion
Refer students to the diagram of Earth's interior in Chapter 4, page 72. Remind them that the lithosphere includes the tectonic plates and both the crust and the upper portion of the mantle. Then ask: What causes convection in Earth's mantle? **heat is transferred to the mantle from Earth's inner and outer cores** What role might mantle convection play in plate motions? **The upwelling of magma at an ocean ridge occurs on one side of a convection cell. As the convection current moves away from the ridge, it drags the plate with it. The cooler, denser rock of the plate eventually sinks into the mantle at a subduction zone where downwelling occurs in the same convection cell.**

Use Transparencies 4, 9.

8.3

KEY IDEA
Three hypotheses describe how mantle convection, ridge push, and slab pull may cause plate movements.

KEY VOCABULARY
- mantle convection
- ridge push
- slab pull

VISUALIZATIONS
CLASSZONE.COM
Observe an animation of convection in the mantle.
Keycode: ES0805

Causes of Plate Movement

The fact that Earth's plates are moving is evident from the earthquakes and volcanic activity at plate boundaries. But what causes plate movements? Three major hypotheses describe how mantle convection, ridge push, and slab pull may each play a role in driving plate movements. All three hypotheses may be important in identifying the cause of plate movements.

To understand the hypotheses described below, it is important to remember that the asthenosphere—a layer in the upper mantle—provides the plates with a surface on which they can move. The asthenosphere has a composition similar to that of the lower lithosphere, but it has very different properties. Instead of being rigid, the asthenosphere is pliable because the rock materials there are hotter than those in the lithosphere.

Mantle Convection

Heat from Earth's inner and outer cores is transferred through the mantle by a process called **mantle convection.** According to the mantle convection hypothesis, the mantle may be moving the plates along with it as it convects.

The diagram below shows one model for how mantle convection may contribute to plate movement. In this model, magma that is hotter and less dense than its surroundings rises upward at a mid-ocean ridge. This upwelling occurs on one side of a convection cell. As the convection current moves away from the mid-ocean ridge, it drags the lithospheric plate with it. The cooler, denser rocks of the lithospheric plate sink down

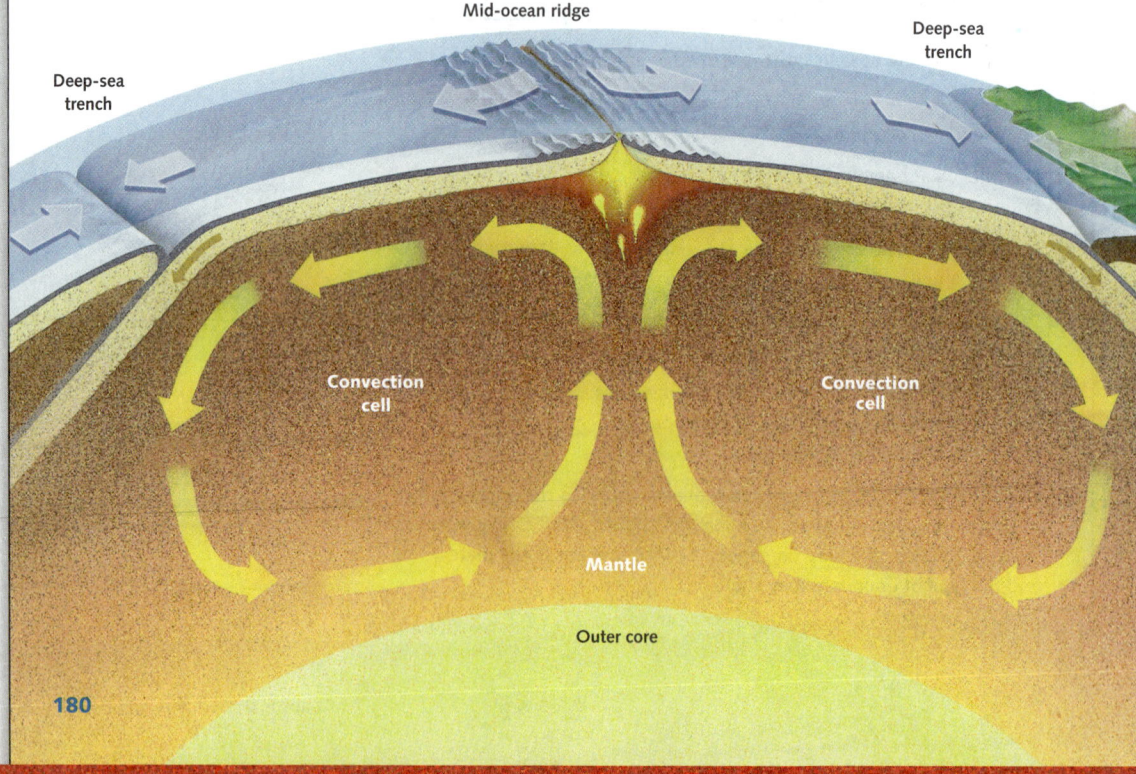

180

Unit 3 Dynamic Earth

DIFFERENTIATING INSTRUCTION

Hands-On Demonstration Use a small transparent coffeepot, a hot plate, water, and half a cup of rice to model mantle convection. Fill the pot three-quarters full with hot water and add the rice. As the mixture comes to a boil (5–10 minutes), have students observe how rice, as it is heated, rises from the bottom of the pot to the top. There it cools and sinks back to the bottom. These movements are the result of convection cells caused by unequal heating of the water.

Reading Support Tell students that descriptive vocabulary terms give clues to what those terms mean. Discuss the meanings of ridge push and slab pull as examples. As an extension, have students discuss with a partner why gravitational sliding might be a better name than ridge push for that process. **The ridge does not actually push the plates. Rather, gravity pulls (slides) the plates downhill and away from the ridge.**

Use Reading Study Guide, p. 26.

at a subduction boundary, along a downwelling zone in the mantle that compensates for the upwelling on the other side of the convection cell.

Many scientists think that mantle convection does not completely explain why plates move. In particular, the model does not account for the enormous force needed to move the lithospheric plates.

Ridge Push

The molten magma that rises at a mid-ocean ridge is very hot and heats the rocks around it. As the asthenosphere and lithosphere at the ridge are heated, they expand, and become elevated above the surrounding sea floor. This elevation produces a slope down and away from the ridge.

Because the rock that forms from the magma is very hot at first, it is less dense and more buoyant than the rocks farther away from the mid-ocean ridge. However, as the newly formed rock ages and cools, it becomes more dense. Gravity then causes this older, denser lithosphere to slide away from the ridge, down the sloping asthenosphere. As the older, denser lithosphere slides away, new molten magma wells up at the mid-ocean ridge, eventually becoming new lithosphere.

Scientists have used computer models to show that the cooling, subsiding rock exerts a force on spreading lithospheric plates that could help drive their movements. This force is called **ridge push,** though the phrase "ridge push" is somewhat misleading. It might be more accurate to refer to ridge push as gravitational sliding.

Slab Pull

At a subduction boundary, one plate is denser and heavier than the other plate. The denser, heavier plate begins to subduct beneath the plate that is less dense. The edge of the subducting plate is much colder and heavier than the mantle, so it continues to sink, pulling the rest of the plate along with it. The force that the sinking edge of the plate exerts on the rest of the plate is called **slab pull.**

Slab pull can be compared to the following situation: Suppose your jacket is resting on a table. You drop a heavy set of keys into a pocket that is dangling over the edge. The weight of the keys pulls downward on the rest of the jacket, causing it to slide toward the edge of the table.

Currently, many scientists consider slab pull to be a much stronger factor than ridge push or mantle convection in driving plate movements.

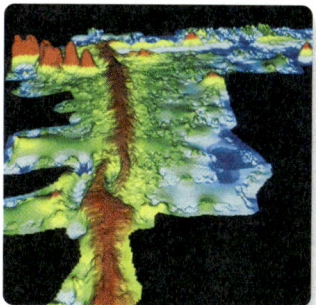

MID-OCEAN RIDGE In this computer-generated topographic map of a section of the East Pacific Rise, the highest elevations are shown in red. The sea floor slopes down from the mid-ocean ridge to lower elevations, which are shown in blue.

8.3 Section Review

1. How does the mantle convection hypothesis explain plate movements at divergent boundaries and subduction boundaries?
2. **CRITICAL THINKING** Describe the differences between the ridge push hypothesis and the slab pull hypothesis.
3. **PHYSICS** Construct and label a model that shows the processes of ridge push and slab pull.

CHAPTER 8 SECTION 4

FOCUS

Objectives
1. Explain how Earth's landmasses have changed positions over the past 200 million years.
2. Discuss the roles of plate tectonics, igneous activity, and deposition in the formation of continental landmasses.

Set a Purpose
Have students read this section to answer the question "How do continental landmasses change over geologic time?"

INSTRUCT

CRITICAL THINKING
Ask: On what present-day continent would you expect to find rocks similar in age to the rocks that make up the eastern coast of South America? **Africa** Coal forms in warm, swampy regions. Coal beds have been discovered in Antarctica. Explain. **This continent had a much warmer climate in the past.** How might you explain the existence of marine fossils high atop mountains? **The rocks that make up the mountains were once at or below sea level. Tectonic forces caused the rocks to become uplifted to form mountains.**

Use Transparency 10.

8.4

KEY IDEAS
Plate movements have caused Earth's continents to change their positions on the globe over time.

New material has been added to the continents as a result of plate tectonics.

KEY VOCABULARY
- Pangaea
- craton
- terrane

Plate Movements and Continental Growth

Understanding plate tectonics has enabled geologists to reconstruct what Earth's surface may have looked like millions of years ago.

Reconstructing the Past

Geologists rely on many different kinds of evidence to reconstruct Earth's past. For example, consider the Ural Mountains in Russia and the Appalachian Mountains in the eastern United States. The rocks in these mountains show evidence of past subduction, suggesting that they formed at a convergent boundary. Yet neither of the mountain belts is located near a plate boundary today. Geologic evidence indicates that each range formed in the distant past at a plate boundary that no longer exists. The Appalachian Mountains formed when North America collided with Africa hundreds of millions of years ago. The Ural Mountains formed when separate lithospheric plates carrying Europe and Asia converged in a process similar to the one that is forming the Himalaya Mountains today.

Geologists can also use data about the ages of the rocks that form the ocean basins to reconstruct Earth's past. In addition, the magnetic record of igneous rocks on the continents can reveal the latitude at which an igneous rock formed, even if the rock has since moved from its original position on the globe.

Fossils also provide clues about the past. Fossils of organisms that once lived in shallow seas have been found on high mountaintops.

Other information about Earth's past comes from rocks that show evidence of having been covered by glaciers in the past. Land areas that now lie in tropical regions of the Southern Hemisphere show evidence of having been covered by moving ice sheets at some time in the past, suggesting that the land areas were once located in a colder part of the globe.

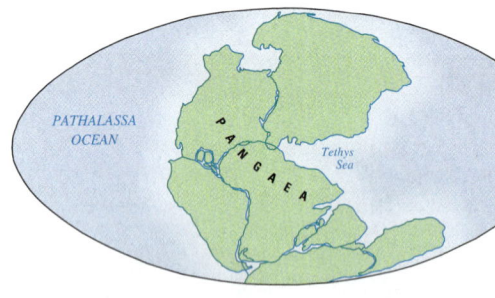

BREAKUP OF PANGAEA The maps below illustrate the movements of Earth's landmasses over the last 200 million years.

200 MILLION YEARS AGO Earth's landmasses were welded together. An area of water that geologists call the Tethys Sea would eventually become the Mediterranean Sea.

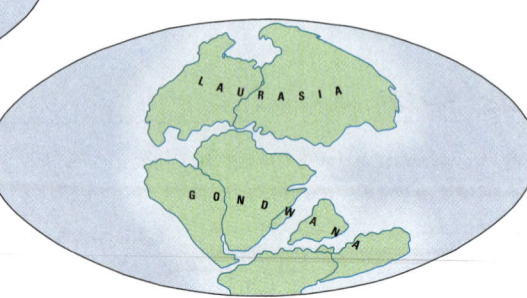

180 MILLION YEARS AGO Pangaea broke apart into two separate landmasses called Laurasia and Gondwana.

DIFFERENTIATING INSTRUCTION

Reading Support Have students change the first sentence under "Reconstructing the Past" into a question they can use to search for answers in the paragraphs that follow. **What are some different types of evidence geologists use to reconstruct Earth's past?** Next, have students write this question in their notebooks and list and describe five pieces of evidence explained in that section. **evidence of past subduction, data about the ages of rocks, magnetic record of igneous rocks, fossils, evidence of past glaciation**

Use Reading Study Guide, p. 27.

Plate Tectonics and Pangaea

Evidence indicates that around 250 million years ago, all the continents were welded together into one landmass. Geologists use the name **Pangaea** (pan-JEE-uh) to refer to this giant landmass.

Formation of Pangaea

How did Pangaea form? Geologists must study data from the continents to make models of what Earth may have looked like before the formation of Pangaea. They cannot use data from the ocean floor, because the oldest oceanic crust is less than 200 million years old. Subduction has destroyed older oceanic crust.

One proposal is that, before the formation of Pangaea, a large continental mass stretched between the south pole and the equator. Geologists use the name Gondwana (gahnd-WAH-nuh) for this landmass. Gondwana was made up of smaller landmasses that would eventually become South America, southern Europe, Africa, the Near East, India, Australia, New Zealand, and Antarctica. Other, smaller landmasses ranged over the rest of the globe. Eventually Gondwana moved northward and converged with other landmasses to form Pangaea, as shown in the map at the far left on page 182.

Breakup of Pangaea

Over time, the landmasses that had formed Pangaea began to break apart. As shown in the diagrams on page 182 and below, they broke into two separate landmasses, Gondwana and Laurasia (law-RAY-zhuh). Over time, Gondwana and Laurasia broke into smaller landmasses whose shapes began to resemble the shapes of the continents today.

Pangaea is still breaking up, and the process by which landmasses have broken apart and converged may have happened many times before the formation of Pangaea.

VOCABULARY STRATEGY
The word *Pangaea* comes from the Greek words *pan* and *gaia*. *Pan* means "all," and *gaia* means "Earth."

VISUALIZATIONS CLASSZONE.COM
Observe an animation of the breakup of Pangaea.
Keycode: ES0806
Visualizations CD-ROM

More about...

Ancient Tectonism
Geologic evidence suggests that several subduction zones existed around the periphery of the supercontinent Pangaea. These convergent boundaries ran along the northern edge of Eurasia, the western edges of North and South America, the present-day northern border of India, the eastern edge of Australia, and parts of Antarctica.

VISUALIZATIONS CLASSZONE.COM
Observe an animation of the breakup of Pangaea.
Keycode: ES0806

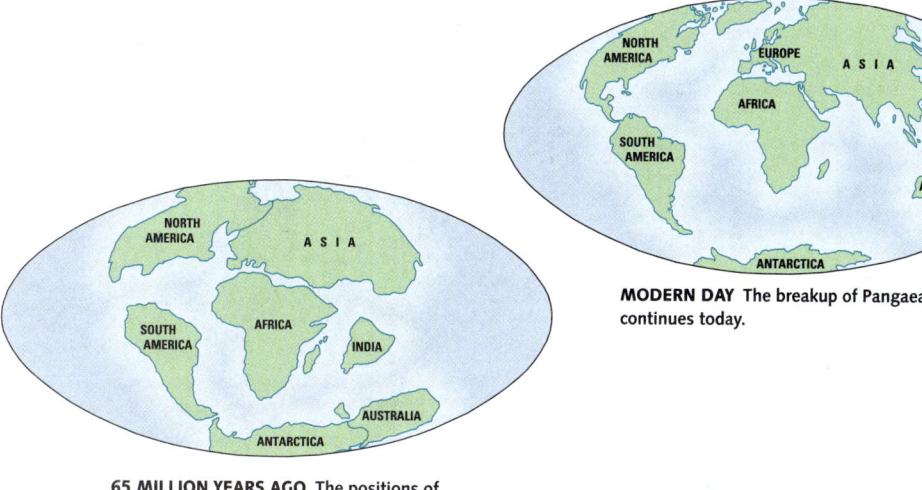

MODERN DAY The breakup of Pangaea continues today.

65 MILLION YEARS AGO The positions of the continents were beginning to resemble their modern-day positions.

Critical Thinking

Challenge students to use present-day rates and directions of continental movements to predict where Earth's landmasses might be 50 million years from now. Have students refer to the map on pages 712 and 713 to help them with this exercise.

Use Transparency 41.

DIFFERENTIATING INSTRUCTION

Support for Visually Impaired Students Make enlarged (full-page) color photocopies of the maps on these two pages. Have students arrange the maps chronologically. Challenge them to draw arrows on each continental landmass to show the general direction of movement over time.

CHAPTER 8 SECTION 4

VISUALIZATIONS CLASSZONE.COM

Examine an animation of plate movement predicted for the future.
Keycode: ES0807

Visualizations CD-ROM

Science and Technology

A Supercontinent Cycle
Many scientists believe Earth's continents have alternated between the supercontinent/superocean configuration and that of continent fragments scattered around the globe. Prior to Pangea, another supercontinent, Rodinia, existed around 800 million years ago. And once the ongoing subduction of the Pacific Ocean floor is completed, yet another supercontinent will form. The cycle seems to take around half a billion years to complete. Because of the constant generation of new continental crust, the continents get bigger with each cycle.

Geography
Maps should show continent positions intermediate between their present positions and those depicted in the diagram, with the Mediterranean Sea getting smaller and Australia and Antarctica getting closer together.

SCIENCE & Technology

Shifting Ground: Tracking Plate Movements

Don't panic, but the ground beneath your feet is moving!

Except during earthquakes, the movement of Earth's lithospheric plates is too slight for you to detect. Nevertheless, as a result of these plate movements, Earth's continents are slowly moving across the face of the globe.

How do scientists measure plate movement? How can they use technology to predict the positions of Earth's landmasses in the distant future?

SATELLITE LASER RANGING
A satellite covered with reflectors is used to measure plate movement.

Since Pangaea broke apart about 180 million years ago, North America and Europe have been moving apart at a rate of about 2.5 centimeters per year. Now the two continents are nearly 5000 kilometers apart.

Scientists can track this slow movement of lithospheric plates very precisely using a technique called satellite laser ranging (SLR). A laser pulse from an Earth-based station is directed toward a satellite covered with reflectors. These reflectors bounce the light directly back to the ground station. By timing how long it takes the light to travel to and from the satellite and multiplying this time by the speed of light, scientists can calculate the distance to the satellite.

A comparison of measurements taken over several years shows how the location of the ground station has changed, as well as the rate of movement for the plate on which the station is located.

For example, SLR measurements show that the Hawaiian island of Maui, which is part of the Pacific Plate, is moving northwest toward Japan at a rate of about 7 centimeters per year. Even such gradual movement of the Pacific Plate will cause substantial change over time.

Patterns of plate movement can be used to make predictions about how Earth's landmasses may shift in the future. One prediction forecasts that in 100 million to 150 million years, Australia and Antarctica will have become one landmass. Africa may have collided with Europe, forming a new mountain range where the Mediterranean Sea is today. ■

Extension

GEOGRAPHY Based on the map below, predict the positions of Earth's landmasses 75 million years from now and 100 million years from now. Sketch maps showing your predictions.

VISUALIZATIONS CLASSZONE.COM

Examine an animation of plate movement predicted for the future.
Keycode: ES0807

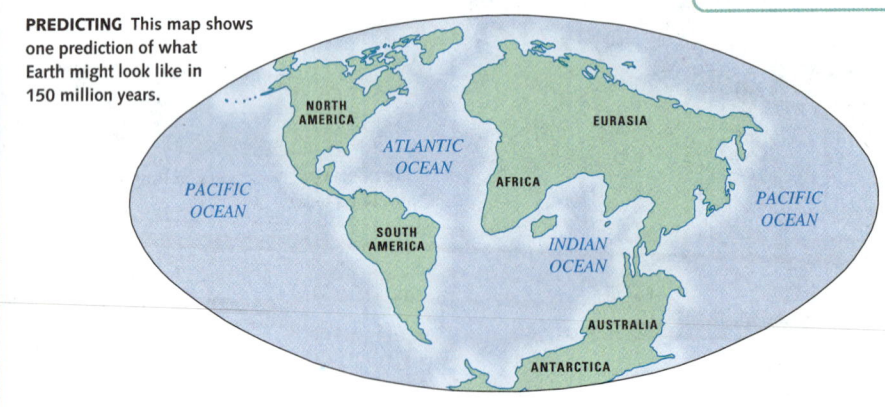
PREDICTING This map shows one prediction of what Earth might look like in 150 million years.

184 Unit 3 Dynamic Earth

Plate Tectonics and Continental Growth

Plate tectonics has affected the shapes of the continents as well as their positions. The ancestors of most modern continents were smaller. Processes associated with plate movements have added rock materials to the margins of the ancient continent cores. The shapes that are familiar today have formed gradually.

At the core of each continent is an expanse of ancient rock called the **craton** (KRAY-TAHN). Around 2.5 billion years ago, this core continental material stabilized. Before then, Earth's surface was probably too hot and unstable for continents to form. One example of a craton is the North American craton, which is exposed at the surface in most of eastern Canada. Geologists call this part of the craton the Canadian Shield. Some of the rock materials found there are among the oldest ever dated and are estimated to have formed about 3.96 billion years ago. The remainder of the North American craton lies buried under a platform of sediments.

The North American craton shows the approximate shape of the continent one billion years ago. The remainder of North America has been added to the craton as the continent developed into its present dimensions.

VOCABULARY STRATEGY
The word *craton* comes from a Greek word meaning "power."

North American Craton
- Canadian Shield (exposed ancient rock)
- Platform (buried ancient rock)

NORTH AMERICAN CRATON The Canadian Shield and the platform make up the North American craton. Newer materials, shown in green, have been added to the continent over time.

Sources of Growth Material
Material that is added to continents can come from a number of sources, including deep-sea sediments, igneous rock, river sediments, and terranes.

Deep-Sea Sediments
Deep-sea sediments can be added to the edges of a continent when an oceanic plate plunges under a continental plate at a subduction boundary. Some of the sediments from the ocean floor may be scraped off and left behind as growth material on the edge of the continent.

Visual Teaching

Interpret Diagram
Ask: Where are the oldest rocks in North America? *in the north-northeasternmost part of the continent (the dark orange area)* What are these rocks called? *the Canadian Shield* About how old are these rocks? *nearly 4.0 billion years old* What type of rocks do you think make up the Canadian Shield? *Most of the shield rocks are igneous in origin. Some are metamorphic rocks.*

More about...

Cratons
Cratons are extensive, flat, tectonically stable regions of continental landmasses. These ancient rocks underwent intense deformation billions of years ago; since then they have remained relatively stable. Most cratons are fringed by relatively young orogenic belts—regions that have undergone folding, faulting, metamorphism, and intrusive igneous activity. Orogenic belts that surround the Canadian Shield and the interior platform include the Appalachian fold belt (eastern United States) and the Cordillera orogenic belt (west coast of North and South America, including the Rockies).

DIFFERENTIATING INSTRUCTION

Developing English Proficiency Several terms on this page are used in science as well as in everyday language. These terms include *plate, ancestors, shield, core,* and *platform*. Have students work in pairs to explain why these terms are appropriate to describe geologic features.

CHAPTER 8 SECTION 4

DISCUSSION

Pose the following questions to aid students in understanding continental accretion (growth): **How do deep-sea sediments become a part of a continent?** *When an oceanic plate collides with a continental plate, ocean-floor sediments are scraped off the descending oceanic plate and become growth material on the edge of the continent.* **How does igneous activity contribute to continental growth?** *Intrusive igneous activity and volcanism add new material to continents.* **What role do rivers play in the growth of a continental landmass?** *Rivers deposit sediments that build up the edges of continents.* **What are terranes?** *Terranes are large blocks of rock that form in one place, are carried thousands of kilometers away by plate movements, and become welded to the edge of a different continent.*

More about...

Terranes
The Appalachian Mountains, which extend along the southeastern edge of much of North America, include numerous terranes from ancient Europe, Africa, oceanic islands, and crust from various episodes of convergence. Most of Florida was once a part of Africa. It was left behind when Pangaea began to fragment about 200 million years ago, sending North America and Africa adrift.

VISUALIZATIONS
CLASSZONE.COM
Observe an animation showing growth of a continent.
Keycode: ES0808

VOCABULARY STRATEGY
Do not confuse the geologic term *terrane* with the common word *terrain*. Both words come from the Latin word *terra*, meaning "earth" or "land."

Igneous Rock

A second source of growth material is igneous rock. Plutons formed from magma that rises beneath the surface and cools are one example. In addition, volcanoes at subduction boundaries eject lava, ash, and rock materials. These materials are added to the edges of continents. Chains of volcanic islands are also characteristic of many subduction zones. The volcanic rock that builds up on such islands may be added eventually to a continent by plate movements. Geologists have even found that volcanic rock from mid-ocean ridges has been added to land areas.

River Sediments

Another source of growth material are the sediments deposited by rivers that flow across continents. These sediments, which consist of weathered and eroded rock materials and soils, build up at the edges of continents. The Mississippi River delta is a region where sediments are building up at the edge of the North American continent.

VOLCANIC ROCK Rock materials ejected by Mount Katmai, a volcano in Alaska, have been deposited in this valley.

RIVER SEDIMENTS The Mississippi River deposits sediments at its mouth.

Terranes

A **terrane** (tuh-RAYN) is a large block of lithospheric plate that has been moved, often over a distance of thousands of kilometers, and attached to the edge of a continent. Terranes are found on all continents and range greatly in age. The attachment of terranes may have been the primary process contributing to continental growth in western North America.

Geologists use three characteristics to identify a terrane. First, each terrane block is bounded on all sides by major faults. Second, the rocks and fossils found in the terrane do not match those of neighboring terranes. And third, the magnetic record found in the terrane does not match that of neighboring terranes. All of these characteristics are strong evidence that a terrane formed in another place and was transported to its present location.

One example of a terrane is the Cache Creek terrane of British Columbia in Canada. The rocks of this terrane include shallow-water limestones deposited on oceanic crust. The limestones contain fossil shells of tiny ocean animals called fusulinids (FYOO-zuh-LY-nihdz). These shells are totally unlike fusulinid fossils found in rocks of the same age in other

186 Unit 3 Dynamic Earth

DIFFERENTIATING INSTRUCTION

Reading Support Students will have greater success understanding sources of continental growth material if they can connect their reading to previous knowledge from Chapter 6. Help them make connections by asking the following questions: What type of rock is added to continents by volcanoes? *igneous* Name one rock that could form by this process. *granite, gabbro, basalt, or any igneous rock* What type of rock is added to continents by rivers? *sedimentary* Name one rock that could form by this process. *sandstone, conglomerate, limestone, or any sedimentary rock* What type of rock could be included in a terrane? *igneous, sedimentary, or metamorphic* Name one rock that could be included in a terrane. *limestone, basalt, gneiss, or any other rock*

CAREER

Electromagnetic Geophysicist

Electromagnetic (EM) geophysicists travel to all corners of the globe. Using an assortment of sophisticated instruments, they measure electric and magnetic fields and gather data for a variety of purposes: mapping tectonic-plate boundaries, locating natural resources (especially oil and gas), and conducting environmental analyses. EM geophysicists' assignments range from locating plate boundaries in the Himalayas to finding diamond deposits in Canada's Northwest Territories.

Like all geophysicists, EM specialists are well-grounded in the principles of plate tectonics and related Earth processes. EM specialists usually have a bachelor's degree in geophysics and an advanced degree. Students take math, physics, electronics, chemistry, and computer-science classes. They often gain hands-on experience of EM methods by working on special projects. EM specialists must enjoy gathering, interpreting, and synthesizing data and be able to communicate their findings clearly and accurately. Because EM specialists often work in far-flung places—sometimes commuting to work via helicopter or camel—a sense of adventure and willingness to "rough it" also come in handy. ∎

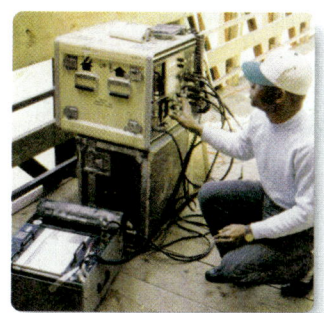

THIS ELECTROMAGNETIC GEOPHYSICIST uses ground-penetrating radar to analyze the earth's composition and stability underneath a bridge.

Find out more about a career in geophysics.
Keycode: ES0809

---

parts of North America. Instead, the Cache Creek fusulinids resemble shells found in Japan and large parts of Asia. From this evidence, scientists have hypothesized that plate movements carried the Cache Creek terrane thousands of kilometers across the Pacific Ocean before subduction processes welded it to the North American continent.

8.4 Section Review

1. Describe what scientists hypothesize Earth's surface looked like 200 million years ago and 180 million years ago.
2. What is a craton?
3. Describe at least two processes that contribute to the growth of continents over time.
4. **CRITICAL THINKING** Some regions in South America show evidence of having been covered by moving ice sheets. How might plate movements have made it possible for ice to form on land that now lies in tropical regions of the world?
5. **MATH** Due to plate movements, the Hawaiian island of Maui is moving toward Japan at a rate of about 7 centimeters per year. At this rate, about how long would it take Maui to move 7 meters? 7 kilometers?

Chapter 8 Plate Tectonics **187**

CHAPTER 8 SECTION 4

Observe the growth of a continent.
Keycode: ES0808

● Visualizations CD-ROM

Find out more about careers in geophysics.
Keycode: ES0809

ASSESS

1. About 200 million years ago, all of Earth's continents were joined as a single supercontinent named Pangaea. Geologists call the single ocean of that time the Pathalassa Ocean. About 180 million years ago, Pangaea began to fragment to form two large landmasses—Gondwana and Laurasia.
2. an expanse of ancient rock that forms the core of a continent
3. The addition of deep-sea sediments and terranes as the result of tectonic interactions are two processes involved in continental growth. Igneous activity also adds rocks to continental landmasses. Deposition of sediments by rivers is another process that plays a role in continental growth.
4. South America was once much closer to the South Pole. Thus, ice sheets (glaciers) could have once covered this continent.
5. 1 year/7 cm × 100 cm/1 m × 7 m = 100 years; 1 year/7 cm × 100 cm/1 m × 1000 m/km × 7 km = 100,000 years

---

MONITOR AND RETEACH

If students miss . . .

Question 1 Reteach "Plate Tectonics and Pangaea." (p. 183) Photocopy and cut out the maps from Transparency 10, removing the labels. Have students arrange the maps in chronological order.

Question 2 Use the map on page 185 to explain that a craton is the tectonically stable, interior portion of a continent.

Question 3 Have students paraphrase the information on continental growth under each of the blue heads on pages 185–186.

Question 4 Use the maps at the bottom of page 182 to point out the location of South America between 200 and 180 million years ago. Ask: What other continent might have experienced glaciation during this time? **Africa**

Question 5 Reteach the concept of plate motion using the tectonic map on pages 712 and 713 (Transparency 41). Review how to convert between SI units of distance (Appendix A, p. 696).

CHAPTER 8 ACTIVITY

PURPOSE
To use a GIS-derived map showing the locations of nuclear facilities, volcanoes, and earthquakes to evaluate the safety of those locations

MATERIALS
If necessary, provide each student with a transparent map—at the same scale as the map on page 189—that identifies each state by name. Students can use this overlay to assist them with answering Procedure Question 4.

PROCEDURE

1. Earthquake data points are the most numerous because, in general, seismic activity is more frequent than volcanism. Also, seismic activity tends to occur along belts. Nuclear facilities and volcanoes have singular locations.

2. Most, but not all, volcanoes occur along the plate boundary that runs along the western coast of the United States; earthquakes are common along this boundary as well but also occur in other locations in the United States.

3. Students might question those nuclear facilities located near earthquake cluster regions on the West Coast, as well as those along the Mississippi River and in the New England area, where seismic activity is also common.

CHAPTER 8 MAP Activity

Nuclear Sites and Plate Boundaries

SKILLS AND OBJECTIVES
- **Compare** locations of earthquakes and volcanoes with plate boundaries.
- **Assess** the value of GIS technology in decision-making.
- **Determine** categories of information pertinent to nuclear decision-makers.

MATERIALS
- Appendix B *Plate Boundaries Map*, pages 712–713

Nuclear energy is an important source of electrical power in the United States. The radioactive elements contained inside a nuclear reactor are very dangerous, and special precautions must be taken in order to prevent these elements from leaking into the biosphere. Most nuclear sites in the United States are situated near the population and industrial centers of the eastern half of the United States or along the west coast.

The map on the following page was produced by combining three map layers derived from Geographic Information Systems (GIS) software. GIS "smart maps" display many varieties of data. They can be customized for use by scientists, regulators, and others to combine data for answering any number of questions. These maps help people to visualize how environments, places, and people interact. In this case, the map shows locations of major earthquakes, volcanoes, and nuclear facilities in the continental United States.

Places where there are multiple blue marks on the map indicate that several earthquakes have occured in approximately the same location. Some nuclear sites, such as ones in California and New Hampshire, are obscured on the map because of a large amount of earthquake activity in the area. To whom would a map like this be of interest? Explain.

Procedure

1. Refer to the map comparing nuclear sites with the locations of known volcanoes and earthquake activity. Which of these three types of data is most numerous? Explain.

2. Compare the locations of volcanic and earthquake activity with the Plate Boundaries Map on pages 712 and 713. Write a sentence summarizing the pattern you observe between earthquake/volcanic activity and plate boundaries.

3. Suppose you are a regulator with the Nuclear Regulatory Commission, assigned to reviewing past decisions made on siting nuclear facilities. Use the map to help decide which facilities you would immediately call into question. Describe your reasoning.

4. Which facilities seem to be best sited? Explain.

Analysis and Conclusions

1. Does this map help predict future events that might occur in the United States? Explain.

CHAPTER 8 ACTIVITY

Learn more about GIS mapping.
Keycode: ES0811

(continued)

4. Nuclear facilities in Michigan, Wisconsin, Minnesota, Iowa, Florida, and Texas are situated fairly far from current seismic activity.

ANALYSIS AND CONCLUSIONS

1. The map can be used to help predict future earthquakes and volcanic eruptions because these types of activities tend to recur in tectonically active areas. However, earthquakes can occur anywhere without warning.

2. Data on subsurface faults might aid in the prediction of future earthquakes and volcanic eruptions. In addition, geologic data might show areas of possible volcanic activity. Data on geologic terranes might indicate areas of tectonism.

3. Regulators must consider places where demand for energy is high. Population density, industrialization, and climate might all contribute to the need for a nuclear facility in a particular region.

4. Some students might state that the map is too simplistic to be of value and that the density of some of the data makes careful evaluation difficult. Other students might argue that the map gives a general idea of places where nuclear facilities might not be plausible due to present or past tectonism.

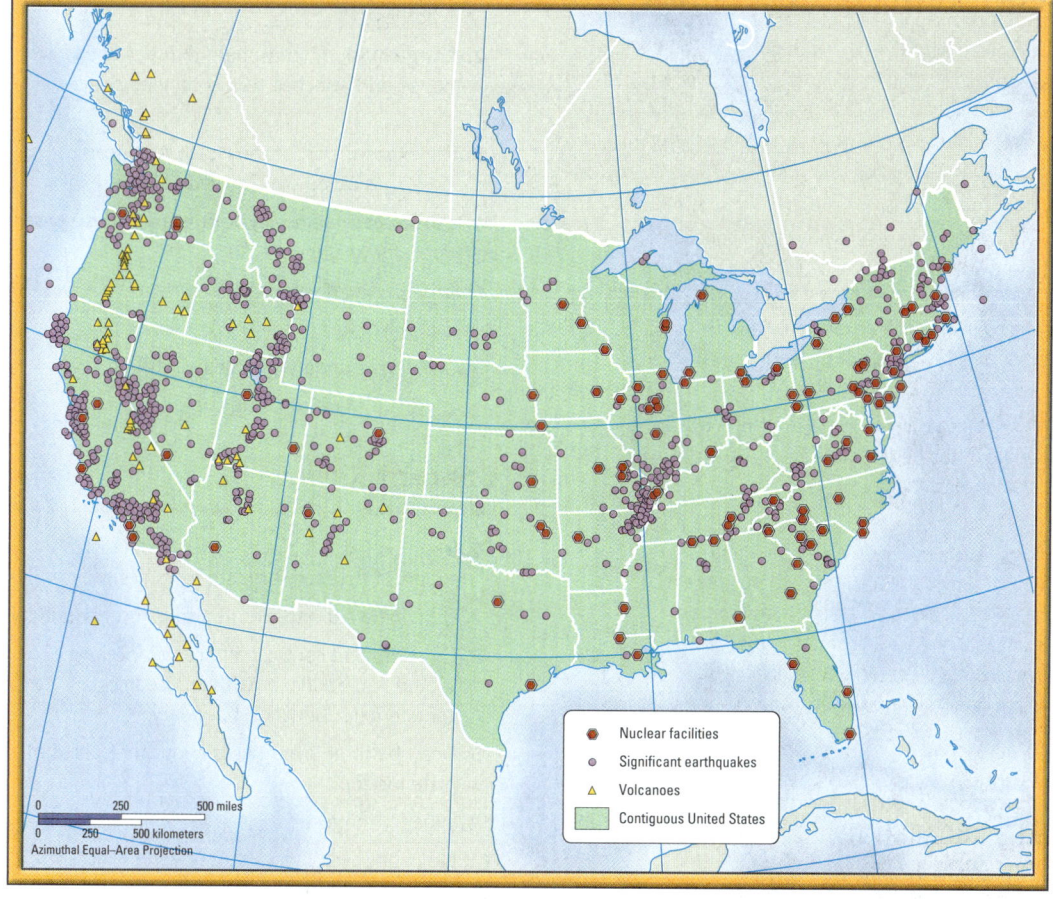

2. What additional data would you seek to help explain any earthquake or volcanic activity not located on plate boundaries? How might a GIS-derived map help you determine what factors contribute to earthquake or volcanic activity?

3. Certainly safety is a prime consideration when siting a nuclear facility, but what other factors must play into the decision-making process? List categories of data you would request in your role as a regulator reviewing past siting decisions.

4. Assess the value of this map for your review of nuclear siting decisions.

Learn more about GIS mapping.
Keycode: ES0811

Chapter 8 Plate Tectonics **189**

CHAPTER 8 REVIEW

Vocabulary Review

1. continental drift
2. terrane
3. craton
4. convergent boundary
5. subduction boundary

Concept Review

6. climate and fossil evidence, the puzzlelike fit of some continental margins, volcanic activity and frequency of earthquakes along plate boundaries, patterns in magnetic-reversal bands along the ocean floor, and heat flow patterns in the ocean crust

7. At a transform boundary, one plate slides horizontally past another plate.

8. With oceanic-oceanic subduction, a chain of volcanic islands forms on the overriding plate. At an oceanic-continental boundary, a chain of volcanic mountains forms as the oceanic plate subducts beneath the continental plate. At both types of boundaries, a deep-sea trench forms.

9. Mantle convection: Plates are moved along by convection within the mantle. Ridge push: Along the mid-ocean ridge, the older, denser plates move down and away from newly forming crust due to gravity. Slab pull: The descending plate at a subduction boundary "pulls" the rest of the plate down into the mantle.

10. Pangaea is a hypothetical landmass, thought to have formed about 250 million years ago, the precursor to modern continents.

11. through igneous intrusions and volcanic eruptions, by depositing of deep-sea sediments onto the continental plate at a subduciton boundary, by rivers depositing sediments along the edges of continents, and terranes

12. A terrane is bounded on all sides by major faults. The fossils and rocks that make up the terrane do not match those of the surrounding terranes. The magnetic polaritiy of the rocks in the terrane does not match that of the rocks that make up the surrounding terranes.

CHAPTER 8 REVIEW

Summary of Key Ideas

8.1 According to the theory of plate tectonics, the lithosphere is broken into plates that move on the asthenosphere. The theory of plate tectonics explains why volcanoes and earthquakes tend to occur in concentrated belts and why the ages of rocks on the ocean floor show a distinctive pattern.

8.2 Plates move apart at divergent boundaries, toward each other at convergent boundaries, and past each other at transform boundaries. Convergent boundaries can be classified as subduction boundaries or collision boundaries.

8.3 Mantle convection, ridge push, and slab pull are hypothetical models for the causes of plate movements.

8.4 Plate movements have caused the positions of Earth's landmasses to shift over time. The shapes of the landmasses have also changed. The North American craton contains the oldest rocks of the continent. Other materials have been added to the continent over time.

Key Vocabulary

collision boundary (p. 178)
continental drift (p. 170)
convergent boundary (p. 177)
craton (p. 185)
deep-sea trench (p. 177)
divergent boundary (p. 176)
mantle convection (p. 180)
mid-ocean ridge (p. 172)
Pangaea (p. 183)
plate tectonics (p. 172)
ridge push (p. 181)
rift (p. 176)
rift valley (p. 176)
slab pull (p. 181)
subduction boundary (p. 177)
terrane (p. 186)
transform boundary (p. 178)

13. Students might add the following: Convergent boundaries can involve two oceanic plates, two continental plates, or one of each. Divergent boundaries are where two plates move apart, either two oceanic plates or two continental plates. Transform boundares are where two plates move horizontally past each other. These boundaries can occur on land or on the ocean floor.

Vocabulary Review

Write the term from the key vocabulary list that best completes the sentence.

1. Wegener relied on fossil evidence, climate evidence, and continent shapes to support his hypothesis of ___?___ .

2. A large block of lithosphere that has been moved and attached to a continent is called a ___?___ .

3. The exposed part of the North American ___?___ is called the Canadian Shield.

Write the term from the key vocabulary list that best completes the analogy.

4. Ridge push : divergent boundary as slab pull : ___?___ .

5. Collision boundary : Himalayas as ___?___ : Mariana Islands

Concept Review

6. Describe some different types of data that support the theory of plate tectonics.

7. Describe what happens at a transform boundary.

8. Compare and contrast an oceanic-oceanic subduction boundary with an oceanic-continental subduction boundary.

9. Describe three hypotheses about the causes of plate motion.

10. What is Pangaea?

11. Describe several ways that new rock materials can be added to a continent.

12. Describe how geologists can identify a terrane.

13. **Graphic Organizer** Copy the concept map below. Then add to the concept map to include additional information about different types of convergent boundaries.

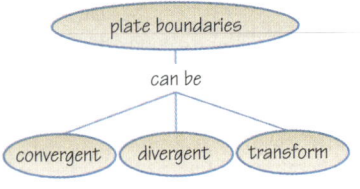

Critical Thinking

14. Coal is made of the remains of trees, ferns, and other plants. Such plants do not grow in Antarctica today because of the cold climate. Therefore, Antarctica must have been closer to the equator, with a warmer climate

CHAPTER 8 REVIEW

Critical Thinking

14. **Infer** How might coal deposits found in Antarctica provide evidence for plate tectonics?

15. **Analyze** The Great Pyramid of Giza in Egypt was built more than 4000 years ago. The structure faces slightly east of true north. Assuming they wanted the structure to face true north, did the Egyptian surveyors make a mistake in laying the pyramid's foundation or could there be some other explanation for the pyramid's position?

16. **Communicate** Draw a diagram to explain how the age and the elevation of the sea floor are related to distance from the rift valley along a mid-ocean ridge.

17. **Draw Conclusions** The oldest rocks of the continents are almost 4 billion years old, while the oldest rocks of the ocean basin are not even 200 million years old. Explain why this difference in age occurs and how it supports the theory of plate tectonics.

Interpreting Graphs

The graph shows a computer model of a slab of a lithospheric plate subducting into the asthenosphere. The subducting plate is shown in red, and the overriding plate is shown in blue. The dots represent earthquakes. The vertical axis of the graph shows the depth inside Earth. The horizontal axis of the graph shows the distance from the point where the plate starts to subduct. The graph also shows temperatures inside Earth.

18. Determine the temperature of the asthenosphere at a distance of 600 kilometers and a depth of 100 kilometers.

19. What is the approximate depth of earthquakes that occur at a distance of 200 kilometers from the point where the plate starts to subduct?

20. Describe the relationship between the distance from the point at which the plate begins to subduct and the depth of an earthquake.

21. How does the subducting plate appear to affect temperatures in the asthenosphere?

Internet Extension

How Fast Do Plates Move? Examine maps of Hawaii and the Pacific seafloor. Use the age of volcanic rocks to calculate the rate of plate motion.
Keycode: ES0810

Writing About the Earth System

SCIENCE NOTEBOOK It may seem that plate movements affect only the geosphere, by creating mountains, causing earthquakes, and forming volcanoes. Do research on how the movement of Earth's plates affects the other spheres of Earth, and write about what you discover in your Science Notebook.

22. What is the approximate depth at which earthquake activity in the subducting plate stops?

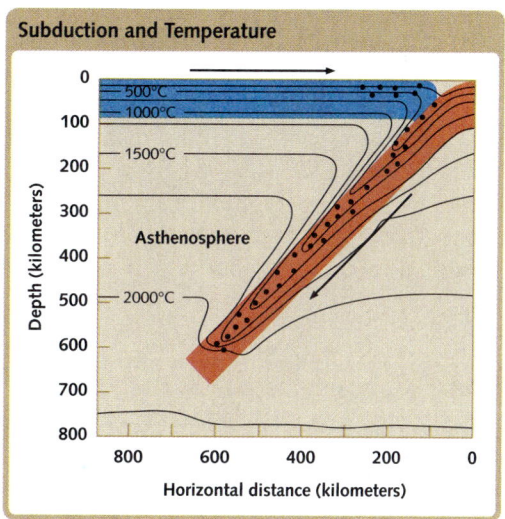

Chapter 8 Plate Tectonics **191**

How Fast Do Plates Move?
Keycode: ES0810

Use Internet Investigations Guide, pp. T34, 34.

(continued)

15. An error on the part of the ancient surveyors is unlikely. A more likely explanation is that the African plate shifted enough to rotate the pyramid slightly out of its original alignment.

16. Students' diagrams should show that as distance from an oceanic divergent boundary increases, the age of the sea floor increases and its elevation decreases. The new (younger) rocks forming along the ridge are elevated because they are hotter and less dense than rocks farther from the ridge.

17. Because continental crust is less dense than oceanic crust, it does not subduct; it remains at Earth's surface. Oceanic crust, however, is subducted, or recycled, along subduction margins.

INTERPRETING GRAPHS

18. The point is on an isotherm, a line of equal temperature, falling between 1000°C and 1500°C.

19. between about 150 and 200 kilometers

20. As distance increases, the depth of the earthquake increases.

21. The subducting plate lowers the temperature in the surrounding asthenosphere.

22. 600 kilometers

WRITING ABOUT SCIENCE

Writing About the Earth System

Students' findings should include how plate tectonics affects the biosphere, the atmosphere, and the hydrosphere. Possible answers include that the effects of tectonism on the biosphere can result in speciation by separating populations of a given species, allowing natural selection to follow two distinct paths. Effects of tectonism on the atmosphere include the addition of volcanic pollutants high in the atmosphere. These pollutants can be carried laterally for hundreds or even thousands of kilometers and reduce the amount of sunlight that reaches Earth's surface, and potentially lead to climate change. Effects on the hydrosphere include that earthquakes can disrupt river channels and generate tidal waves. Tectonic movements also affect the shapes and features of ocean basins.

CHAPTER 9 PLANNING GUIDE: VOLCANOES

	Section Objectives	Print Resources
Chapter–Level Resources		Laboratory Manual, pp. 39–44 Internet Investigations Guide, pp. T35–T37, 35–37 Guide to Earth Science in Urban Environments, pp. 29–36 Spanish Vocabulary and Summaries, pp. 17–18 Formal Assessment, pp. 25–27 Alternative Assessment, pp. 17–18
Section 9.1 How and Where Volcanoes Occur, pp. 194–198	1. Analyze how magma forms as a result of plate motion and interaction. 2. Explain why plate boundaries are the sites of most volcanic activity.	Lesson Plans, p. 28 Reading Study Guide, p. 28 Teaching Transparencies 8, 9, 11, 41
Section 9.2 Magma and Erupted Materials, pp. 199–201	1. Identify three types of magma and locate where each one forms. 2. Compare characteristics of magma and lava. 3. Describe pyroclastic materials.	Lesson Plans, p. 29 Reading Study Guide, p. 29
Section 9.3 Volcanic Landforms, pp. 202–204	1. Compare and contrast landforms that result from volcanic activity. 2. Hypothesize how predicting eruptions and emergency preparedness can reduce loss of life and loss of property.	Lesson Plans, p. 30 Reading Study Guide, p. 30
Section 9.4 Extraterrestrial Volcanism, pp. 206–207	1. Classify volcanic activity on other planets and moons. 2. Analyze the tectonics behind extraterrestrial volcanic activity.	Lesson Plans, p. 31 Reading Study Guide, p. 31

INVESTIGATIONS
CLASSZONE.COM

Section 9.1 INVESTIGATION
How Are Volcanoes Related to Plate Tectonics?
Computer time: 45 minutes / Additional time: 45 minutes
Students use 3-D maps to plot areas of volcanic activity.

Section 9.3 INVESTIGATION
How Fast Do Gases from Volcanic Eruptions Travel?
Computer time: 45 minutes / Additional time: 45 minutes
Students use satellite images of Mount Pinatubo to calculate speed of ash and gases moving through atmosphere.

Technology/Online Resources

Electronic Teacher Tools
Visualizations CD-ROM
Classzone.com
Online Lesson Planner
Test Generator

Visualizations CD-ROM
Classzone.com
  ES0901 Investigation, ES0902 Visualization,
  ES0903 Visualization, ES0904 Visualization

Visualizations CD-ROM
Classzone.com
  ES0905 Visualization

Classzone.com
  ES0906 Investigation

Classroom Management

Gather these materials for minilab and map activities.

Minilab, p. 200
  3 disposable cups
  water
  vegetable oil
  stop watch
  drinking straws (clear plastic)
  corn syrup or maple syrup
  eyedropper

Map Activity, pp. 208–209
  Map of plate boundaries, pp. 712–713
  Colored pencils
  Teaching Transparency 41
  Laboratory Manual, Teacher's Edition
    Lab Sheet 9, p. 140

Chapter Review INVESTIGATION
Is It Safe to Live Near Volcanoes?
Computer time: 45 minutes / Additional time: 45 minutes
Students analyze data from Mount St. Helens and develop an evacuation plan for the area around Mt. Rainier.

CHAPTER 9

Volcanoes

Molten lava at Hawaii's Mauna Loa volcano tends to keep flowing until it cools down to 770°C (1418°F). Scientists need to wear protective clothing when collecting lava samples.

Why would scientists risk their lives to study volcanoes?

INTRODUCE
Build Interest
Before sharing background information about the photograph with students, allow time for their questions and comments.

Ask questions like the following:
- Where is this person walking? near a volcano
- What is the red material? the black? hot lava, cooled lava
- What is the special clothing for? protection from heat and poisonous gases
- What dangers do you see in the photograph? erupting lava, unstable ground, heat

Respond
- Why might scientists risk their lives to gather data about active volcanoes? Such data may help predict future eruptions, thus making it possible to evacuate people in time to save lives.
- How close to a volcano would you like to go? Responses vary.

About the Photograph
In June of 1991, while photographing an eruption on Japan's Mount Unzen, two of the world's leading vulcanologists died from exposure to lethal gas. They were the husband and wife team of Maurice and Katia Krafft. This photograph, taken years earlier by Maurice, shows Katia in a heat protection suit.

BEYOND THE TEXTBOOK

Print Resources
- Reading Study Guide pp. 28–31
- Transparencies 8, 9, 11, 41
- Formal Assessment, pp. 25–27
- Laboratory Manual, pp. 38–44
- Alternative Assessment, pp. 17–18
- Spanish Vocabulary and Summaries, pp. 17–18
- Internet Investigations Guide, pp. T35–T37, 35–37
- Guide to Earth Science in Urban Environments, pp. 29–36

Technology Resources
- Visualizations CD-ROM
- Classzone.com
- Online Lesson Planner
- Electronic Teacher Tools
- Test Generator

192 Unit 3 Dynamic Earth

CHAPTER 9

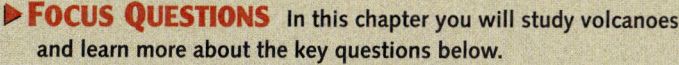

PREVIEW

▶ **FOCUS QUESTIONS** In this chapter you will study volcanoes and learn more about the key questions below.

Section 1 How and where do volcanoes form?
Section 2 Why are some volcanoes explosive?
Section 3 What kind of landforms are produced by volcanic eruptions?
Section 4 Where can volcanoes be found in the solar system?

▶ **REVIEW TOPICS** As you investigate volcanoes, you will need to use information from earlier chapters.
- asthenosphere (p. 73)
- magma (p. 118)
- plate tectonics (p. 170)
- subduction boundary (p. 177)

▶ **READING STRATEGY**
PREVIEW
Before you read, look through Chapter 9, noting the key ideas, key vocabulary, headings, images, and captions.

CLASSZONE.COM

At our Web site, you will find the following Internet support for this chapter.

DATA CENTER

EARTH NEWS

VISUALIZATIONS
- Volcanism at a Subduction Zone
- Volcanism at a Rift Zone
- Volcanic Island Formation over a Hot Spot
- Erupted Materials

LOCAL RESOURCES

INVESTIGATIONS
- How Are Volcanoes Related to Plate Tectonics?
- How Fast Do Gases from Volcanic Eruptions Travel?
- Is It Safe To Live Near a Volcano?

Chapter 9 Volcanoes 193

CHAPTER 9

PREPARE

Focus Questions
Find out what knowledge students have about volcanoes. For each focus question, have students consider the section title and then develop two or three additional questions for each section. For example, Section 2: What material is erupted during an explosion? Do people get hurt when a volcano explodes?

Review Topics
In addition to the earth science topics listed, students may need to review these concepts:

Physical Science
- Melting point temperatures vary depending on materials.
- Substances that are less dense rise above substances that are more dense. A familiar example is oil in water.
- Application of pressure to solid material increases its density.

Reading Strategy
Model the Strategy: To help students preview each section, use a version of the following: "I see that the title is 'How and Where Volcanoes Form.' This must relate to the main idea of the section. The Key Idea talks about magma reaching Earth's surface, and the first head in red is 'Magma Formation.' So, this text addresses the question '*How*.' In fact, the diagrams on pages 194 and 195 show hot magma rising. The next head is 'At Subduction Boundaries.' This must be a clue to '*Where*.' The diagram helps me to see the details." Have students continue to use this strategy for the remainder of the chapter.

BEYOND THE TEXTBOOK

INTERNET RESOURCES
CLASSZONE.COM

INVESTIGATIONS
- How Are Volcanoes Related to Plate Tectonics?
- How Fast Do Gases from Volcanic Eruptions Travel?
- Is It Safe To Live Near a Volcano?

VISUALIZATIONS
- Animation of volcanism at a rift zone and at a subduction boundary
- Animation of islands forming over a hot spot
- Video of erupted materials

DATA CENTER

EARTH NEWS

LOCAL RESOURCES

CAREERS

Online Lesson Planner

Chapter 9 Volcanoes 193

CHAPTER 9 SECTION 1

FOCUS

Objectives
1. Analyze how magma forms as a result of plate motion and interaction.
2. Explain why plate boundaries are the sites of most volcanic activity.

Set a Purpose
Have students read this section to find the answer to the question "How and where do volcanoes form?"

Use Transparencies 8 and 41 to relate volcano locations to plate boundaries.

INSTRUCT

More about...

Magma Formation
As temperatures within the upper mantle increase, the rock begins to melt, forming small, isolated pockets of magma. These pockets rise through fractures or natural conduits. The magma may then collect in weakened patches of the crust, called magma chambers. These act as holding tanks and may be just a few miles beneath Earth's surface. With pressure rising in the magma chamber, magma may be forced up through weak spots in the lithosphere to the surface and a volcano forms.

9.1

KEY IDEA
Volcanoes form where magma reaches Earth's surface.

KEY VOCABULARY
- volcano
- hot spot

INVESTIGATIONS
CLASSZONE.COM

How Are Volcanoes Related to Plate Tectonics? Plot volcano locations. Examine maps and 3D diagrams to explore volcanism at plate boundaries.
Keycode: ES0901

VOLCANO The diagram (inset) shows magma rising inside a volcano. Rising magma can lead to an explosive eruption like this one at Alaska's Mt. Augustine in 1986.

How and Where Volcanoes Form

One of the most dramatic activities associated with plate tectonics is the eruption of a volcano. The term **volcano** refers both to the opening in Earth's crust through which molten rock, gases, and ash erupt and to the landform that develops around this opening.

Magma Formation

A volcanic eruption occurs when magma—molten rock that has formed deep within Earth—rises to the surface. Most of the asthenosphere is solid because of the pressure exerted by the lithosphere above it. This pressure raises the melting temperatures of materials in the asthenosphere. Yet for magma to form, some of these materials must melt. The following three conditions allow magma to form:

- A decrease in pressure can lower the melting temperatures of materials in the asthenosphere. Such a decrease takes place along the rift valley at a mid-ocean ridge, where the lithosphere is thinner and exerts less pressure.
- An increase in temperature can cause materials in the asthenosphere to melt. Such an increase occurs at a hot spot.
- An increase in the amount of water in the asthenosphere can lower the melting temperatures of materials there. Such an increase occurs at subduction boundaries.

Conditions at both divergent and subduction boundaries are ideal for magma formation. Most volcanoes are found along mid-ocean ridges, where plates are moving apart, and at subduction boundaries, where plates are being forced under other plates.

Once magma forms, it tends to rise to the surface because its density is lower than that of the solid materials surrounding it. The characteristics of a magma and the rates at which it rises depend upon the amount of silica it contains.

194 Unit 3 Dynamic Earth

DIFFERENTIATING INSTRUCTION

Hands-On Demonstration Why does magma rise? A substance that is less dense than the material around it rises to the top. Show students three materials: a cork, a glass marble, and vegetable oil. Have students predict what will happen when all three materials are placed in a container. Place the cork and the marble in the bottom of a clear plastic container. Fill the container with oil. As the cork rises, ask students which material is less dense. **the cork** Explain that magma is less dense, thus it rises to the surface.

At Subduction Boundaries

As you may recall from Chapter 8, subduction boundaries are places where lithospheric plates collide. The diagrams below show the two processes by which volcanoes form: first, when an oceanic plate is forced beneath a continental plate, and second, when one oceanic plate is forced beneath another oceanic plate.

At a subduction boundary, volcanoes always form on the overriding plate. Where an oceanic plate collides with a continental plate, the volcanoes form on the overriding continental plate. The Cascade Range, which extends along the Pacific coast from northern California to British Columbia, Canada, is made up of volcanoes that have formed at an oceanic-continental subduction boundary.

Where two oceanic plates collide, the volcanoes create a chain of volcanic islands, called a volcanic island arc, on the overriding plate. The Mariana Islands in the Pacific Ocean are an example of a volcanic island arc formed at such a subduction boundary.

Observe an animation of volcanism at a subduction zone.
Keycode: ES0902

Volcanic Activity at a Subduction Boundary

BETWEEN AN OCEANIC PLATE AND A CONTINENTAL PLATE

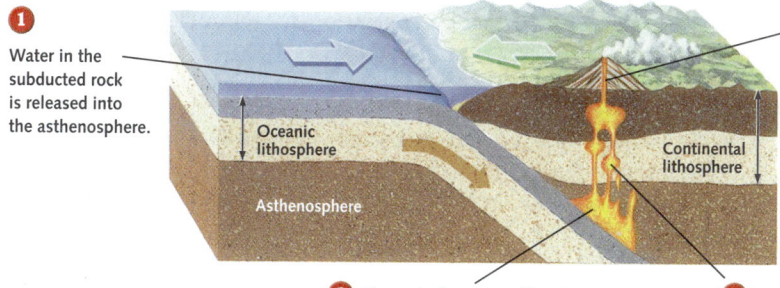

1. Water in the subducted rock is released into the asthenosphere.
2. The water lowers melting temperatures of materials in the asthenosphere, leading to the formation of magma.
3. The magma is less dense than its surroundings, so it rises.
4. Some of the magma reaches Earth's surface, and volcanoes form on the overriding continental plate.

BETWEEN OCEANIC PLATES

The process by which magma forms at an oceanic-oceanic subduction boundary is similar to the process at an oceanic-continental boundary. Notice that the difference between the two processes occurs at step 4.

4. Magma that reaches Earth's surface is underwater. Thus, an arc of volcanic islands forms underwater on the overriding oceanic plate.

Chapter 9 Volcanoes 195

CHAPTER 9 SECTION 1

> **More about...**
>
> **Magma Formation**
> The combination of high temperature and low pressure causes large amounts of magma to form at a divergent boundary. Rifts form over diverging limbs of adjacent convection cells, which bring mantle material from deeper, hotter regions within Earth. This heat flow brings the upper mantle close to its melting point. The decrease in pressure at rift zones lowers the melting point of the mantle material enough to form magma.
>
> *Use Transparency 9.*

> **Scientific Thinking**
>
> **ANALYZE**
> A risks and benefits chart may be helpful to students as they research and prepare their summaries. For example, under a column for *Risks*: erupting lava and ash, poisonous gases, damage to property, threats to humans and other life. Under *Benefits*: hydrothermal heated pools for recreation, economical source of renewable energy.

VISUALIZATIONS CLASSZONE.COM
Observe an animation of volcanism along a rift zone.
Keycode: ES0903

> **Scientific Thinking**
>
> **ANALYZE**
> Volcanic activity in Iceland can cause problems, but it also provides a nearly inexhaustible source of energy, such as that produced by the geothermal plant below. Find information about the uses of Iceland's geothermal resources.
>
> **COMMUNICATE**
> Use the information to present a summary of the risks and benefits associated with living in an area with Iceland's level of volcanic activity.

At Divergent Boundaries

Below a rift, mantle material rises from deeper, hotter regions within Earth. Also, because of the rift, the pressure is lower than it is elsewhere in the mantle. This decrease in pressure lowers melting temperatures. The combination of high temperature and low pressure causes large amounts of magma to form. Like the magma formed at subduction boundaries, the magma formed at divergent boundaries is less dense than the materials around it. The magma therefore rises through the rift to the surface.

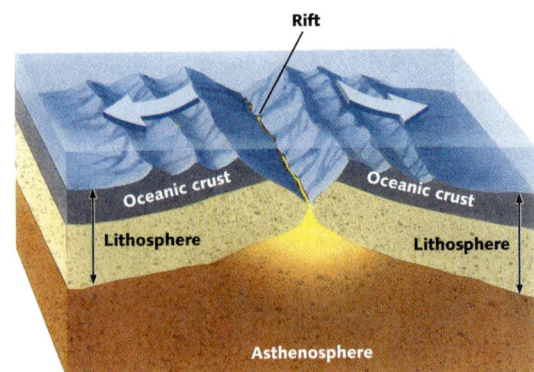

MAGMA FORMATION AT A MID-OCEAN RIDGE

Most of the magma that reaches Earth's surface does so at divergent boundaries, along the mid-ocean ridges. Because most mid-ocean ridges are underwater, you can understand why most of Earth's volcanic activity takes place beneath its oceans.

One place where a mid-ocean ridge rises above sea level is Iceland. Here the rifts associated with the Mid-Atlantic Ridge mark the surface of the land. Much of Iceland is volcanically active, and sometimes this volcanic activity is harmful to living things. In 1783 the Icelandic volcano Laki erupted, releasing lava, ash, and poisonous gases including sulfur dioxide. This eruption killed over three-fourths of Iceland's livestock. It also caused flooding. The destruction of crops and livestock resulted in the deaths of nearly 10,000 people from starvation or disease.

DIFFERENTIATING INSTRUCTION

Developing English Proficiency Some of the vocabulary in this section may present difficulty in reading comprehension. Have students list any words that are new to them, such as *exerted, geothermal, isolated, mid-ocean,* and *island chain*. Students may work in small groups to learn the meanings of these words. Encourage student to reread the sentence where each word first appears and then restate the meaning of each sentence in their own words.

Use Spanish Vocabulary and Summaries, pp. 17–18.

Unit 3 Dynamic Earth

Over Hot Spots

Not all volcanic activity occurs at plate boundaries. **Hot spots** refer to areas of volcanic activity that result from plumes of hot solid material that have risen from deep within Earth's mantle. As the material rises, it melts at areas of lower pressure. Some scientists estimate that the sources of magma producing such hot spots are anywhere from 670 kilometers to nearly 2900 kilometers beneath the surface.

A hot spot remains in the same place even as a lithospheric plate moves across it. As the diagram below shows, the Hawaiian Islands have formed as the Pacific plate has moved northwest over a hot spot. The largest and youngest island, Hawaii, currently lies above the hot spot. As you might expect, this island is volcanically active. Extending northwest is a chain of extinct volcanoes, many of which have sunk below sea level as erosion has worn them away. Each volcano in that chain formed over the hot spot that now lies below Hawaii. The chain also includes underwater volcanoes called seamounts which are not shown in the illustration below.

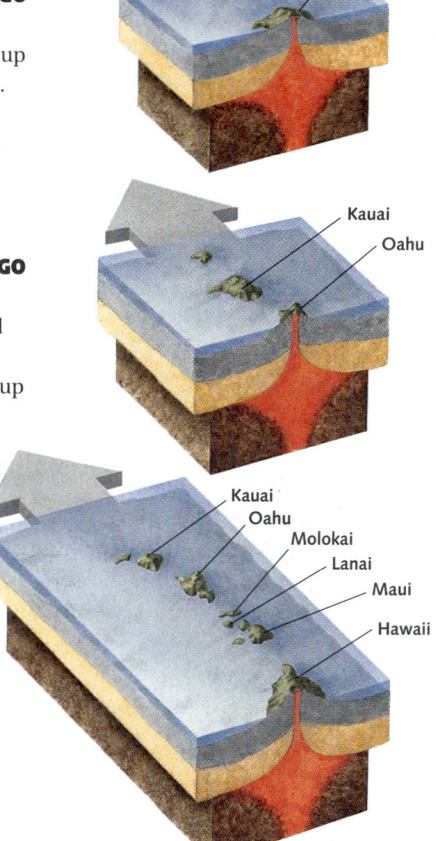

5.3 TO 4.9 MILLION YEARS AGO
The island of Kauai formed as molten rock hardened and built up on the seafloor over the hot spot. As the plate moved northwest, it carried the island away from the hot spot.

3.8 TO 2.5 MILLION YEARS AGO
The island of Oahu formed after the volcano on Kauai had moved away from the hot spot. Again, molten rock hardened and built up until Oahu rose above sea level.

TODAY The island of Hawaii now sits over the hot spot. The hot spot fuels three active volcanoes, including one underwater. (not shown here) This seamount is located to the east of the island.

VISUALIZATIONS
CLASSZONE.COM

Observe an animation of volcanic islands forming over a hot spot.
Keycode: ES0904

CHAPTER 9 SECTION 1

VISUALIZATIONS
CLASSZONE.COM

Observe an animation of volcanism along a rift zone.
Keycode: ES0903

Observe an animation of volcanic islands forming over a hot spot.
Keycode: ES0904

Visualizations CD-ROM

More about...

Hot Spots
The concept of hot spots was introduced by Jason Morgan of Princeton and J. Tuzo Wilson of Toronto to explain the occurrence of centers of volcanic activity within plates rather than only at plate boundaries. They proposed that such volcanic centers are the surface effects of jets or plumes of hot material that rise from deep within the mantle through the lithosphere to emerge on Earth's surface or on the ocean floor. Students need to understand that "Earth's surface" and the "ocean floor" are both part of the lithosphere, or Earth's crust. In other words, a lithospheric plate may be either an oceanic or a continental plate.

DIFFERENTIATING INSTRUCTION

Support for Visually Impaired Students Clarification of the diagrams on page 197 may be necessary. Students may need help differentiating the top two layers in each diagram. Close examination shows there is a water level and a layer that represents the oceanic plate above the tan layer of the lithosphere. The volcanic islands are formed on and above the oceanic crust. Try enlarging the diagrams on a photocopier so students have a larger illustration of the process. Using a large monitor to view the animation on the Web site may be especially helpful to visually impaired students.

CHAPTER 9 SECTION 1

VISUAL TEACHING
Discussion
The island of Hawaii will continue to change. Discuss what kinds of changes may take place over the next 100,000 years? **Hawaii will likely move northwest as the plate moves.** How might the land change as the island moves away from the hot spot? **Volcanic eruptions there will become less frequent or stop. Its mountains will begin to erode.**

ASSESS

1. Water from the subducting rock lowers the melting temperatures of materials in the asthenosphere, leading to magma formation.

2. Volcanic activity at a hot spot does not occur at a plate boundary, but rather at a point in the lithosphere where magma rising from the mantle reaches Earth's surface. Volcanic activity at the boundary between an oceanic and a continental plate occurs because of the collision of the two plates, with magma rising up through the continental crust.

3. Magma rises and moves through rifts to the surface at mid-ocean ridges.

4. Graphic organizers will vary but should indicate that volcanoes form on overriding plates at subduction or convergent boundaries. They should also indicate that volcanoes also form at divergent boundaries and at hot spots.

5. Cascade Range: Juan de Fuca plate and North American plate; Mariana Islands: Pacific plate and Philippines plate

---

Island formation continues about 30 kilometers off the coast of the island of Hawaii, where scientists are studying a young, very active underwater volcano called *Loihi Seamount*. These scientists predict that tens of thousands of years from now, the peak of Loihi will reach sea level and become a new Hawaiian island. In the meantime, studies of Loihi are revealing important information about hot spots, deep-sea geology, and the effects of seamounts on marine life.

LOIHI SEAMOUNT
This computer-generated image shows Loihi Seamount and the island of Hawaii.

Monitoring an underwater volcano involves unique difficulties. Although Loihi is taller than Mount St. Helens was before its eruption in 1980, its peak is still about 900 meters below sea level. Since 1996, scientists have been using submersibles to study the site. In 1997, an undersea observatory was deployed at Loihi's summit. This observatory allows scientists to monitor eruptions and earthquakes with an underwater microphone, a seismometer, and a pressure sensor.

9.1 Section Review

1. Describe how magma forms at a subduction boundary.

2. Explain the difference between the volcanic activity that occurs at a hot spot and the volcanic activity that occurs at a subduction boundary between an oceanic plate and a continental plate.

3. Describe the volcanic activity that occurs at a divergent boundary.

4. **CRITICAL THINKING** Make a concept map or other graphic organizer that shows how the following geologic features are related: subduction boundary, divergent boundary, hot spot, overriding plate, subducting plate, volcano. You may add other words to your concept map if you wish.

5. **GEOGRAPHY** Use the map of plate boundaries on pages 712–713. Identify the plates associated with the development of the Cascade Range and those associated with the development of the Mariana Islands.

MONITOR AND RETEACH

If students miss . . .
Question 1 Reteach "Magma Formation." (p.194) Then refer to the diagram of Volcanic Activity at a Subduction Boundary. (p.195) Have students explain the diagrams in their own words.

Question 2 Reteach "At Subduction Boundaries" (p. 195) and "Over Hot Spots." (p. 197) Have students make a list of the causes of volcanic activity in each case.

Question 3 Reteach magma formation at divergent boundaries. (p.196) Be sure students understand the concept that pressure is lower at the rifts, lowering temperatures, allowing magma to form.

Question 4 Set up a web-type organizer. Place *Volcano* in the center; *subduction*, *divergent*, and *hot spot* in surrounding circles. Leave remaining circles and details for students to fill in.

Question 5 Review the map. Identify plate boundaries.

Magma and Erupted Materials

Kilauea, a volcano on the island of Hawaii, has spewed molten rock for decades. In contrast, Mount St. Helens exploded violently in 1980 after more than a century of being a quiet, snow-capped mountain. Differences in the volcanic activity at these two sites result partly from differences in the magmas that rise to the surface there.

Types of Magma

Silica, a principal ingredient in all magmas, determines a magma's **viscosity,** or resistance to flow. Magmas high in silica resist flow, whereas magmas with lower silica content flow more easily. Basaltic magmas contain the least silica; as a result, they flow most easily. Andesitic and rhyolitic magmas contain more silica and are more resistant to flow.

Magmas also contain gases—mainly water vapor and carbon dioxide. The gases, dissolved in the magmas at the depths where they form, may bubble out of solution as the magmas rise. Most gases escape easily from basaltic magmas. When these magmas reach the surface, any remaining gases usually produce relatively harmless fountains and floods. Gases in more viscous andesitic and rhyolitic magmas, however, cannot escape as easily. As such magmas rise, the gases expand and propel the magma rapidly to the surface. The result can be an explosive eruption of gas and debris, such as the one at Mount St. Helens.

Each type of magma tends to form at specific locations. Basaltic magmas form at rifts and at oceanic hot spots. Andesitic magmas tend to form at subduction boundaries. Rhyolitic magmas generally form where hot spots underlie continental plates.

9.2

KEY IDEA
The composition of magma largely determines how explosive a volcanic eruption will be.

KEY VOCABULARY
- viscosity
- lava
- pahoehoe
- aa
- pillow lava
- pyroclastic material
- pyroclastic flow

SUMMARY: Characteristics of Magma

	Basaltic Magma	Andesitic Magma	Rhyolitic Magma
Silica content	Least (about 50%)	Intermediate (about 60%)	Most (about 70%)
Gas content	Least	Intermediate	Most
Viscosity	Least viscous	Intermediate	Most viscous
Type of eruption	Rarely explosive	Sometimes explosive	Usually explosive
Melting temperature	Highest	Intermediate	Lowest
Location	Rifts, oceanic hot spots	Subduction boundaries	Continental hot spots
	Kilauea	Mount St. Helens	Yellowstone caldera

CHAPTER 9 SECTION 2

FOCUS

Objectives
1. Identify three types of magma and locate where each one forms.
2. Compare characteristics of magma and lava.
3. Describe pyroclastic materials.

Set a Purpose
Have students read this section to find the answer to "What materials erupt from volcanoes?"

INSTRUCT

More about...

Silica in Magma and Lava
Silica, silicon dioxide (SiO_2), accounts for 40–70% of the weight in most igneous rocks. Magmas with high silica content are dense, light-colored, and slow moving. These are called *fel*sic magmas (from *fel*dspar and *si*lica). Magmas with low silica content are more fluid, darker in color and flow more easily. These are *mafic* magmas (from *ma*gnesium and *f*erric). Like magma, lava is classified by its silica content as either felsic or mafic. Mafic lavas pour out smoothly onto the surface. Felsic lavas are thick and stiff and do not move easily.

DIFFERENTIATING INSTRUCTION

Reading Support Show students how they can become better readers by reading actively. State the first subhead, "Types of Magma." Read aloud the first two sentences in the first paragraph. Model for students how to separate ideas so that they make sense: silica is the main ingredient in all magma; if *viscosity* means "resistance to flow," then magma with high viscosity must move very slowly, and magma with low viscosity must flow easily. The more silica there is in the magma, the more it resists flowing.
Use Reading Study Guide, p. 29.

CHAPTER 9 SECTION 2

MINILAB
Modeling Viscosity
Analysis Answers
- Water is most like basaltic magma, since its consistency is the thinnest and most fluid of the liquids.
- Vegetable oil is most like andesitic magma, the intermediate magma in terms of viscosity and thickness.
- The thick, highly viscous syrup is most like rhyolitic magma, the most viscous of all magmas.

For proper behavior in a laboratory setting, refer students to Safety in the Earth Science Laboratory on pages xxii–xxiii.

CRITICAL THINKING

Is it possible for magma to become so thick as it approaches the surface that it would stop flowing? Have students hypothesize what might happen if magma greatly increased its viscosity near the surface. Remind students that trapped gases are held within the magma. Suggest they take one of the following positions and describe what might happen:

- Magma hardens and totally blocks the opening. Flow may stop at that point, but magma may force its way out through other openings.
- Magma flow decreases but gases continue to build and escape. Explosions may occur as gases expand.
- Magma flow continues with slow seepage. Slow buildups of gases and magma result in intermittent explosions or flows.

25-Minute Mini LAB
Modeling Viscosity

Materials
- 2 disposable cups
- water
- stopwatch
- vegetable oil
- 3 straws
- eyedropper
- syrup

Procedure
1. Poke a hole in a cup and position a straw through the hole and down to the second cup as shown. The straw should be tilted about 50°.

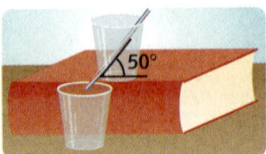

2. Fill an eyedropper with water. Squeeze the water into the straw. Time how long it takes for all of the liquid to exit the straw into the cup.
3. Repeat step 2, using vegetable oil. Then use syrup. Use a new straw for each liquid.

Analysis
Compare the three liquids to the types of magma described in the chart on page 199.
Which of the liquids is most like basaltic magma? Most like andesitic magma? Most like rhyolitic magma? Explain.

VOCABULARY STRATEGY
In the word *pyroclastic*, *pyro-* comes from a Greek word meaning "fire", and *-clastic* from a Greek word meaning "broken".

Lava Flows
Magma that reaches Earth's surface is called **lava**. Like magma, lava is primarily molten rock. The composition of lava may differ from that of magma, however, because materials may be added to or removed from the magma as it rises to the surface.

Lava Flows on Land
Basaltic lava flows are usually associated with less-explosive eruptions. As these flows cool, they form basaltic rock. The temperature and speed of a basaltic flow affect the appearance of its hardened surface. Volcanologists use two Hawaiian terms to describe solidified lava flows on land, **pahoehoe** (puh-HOH-eh-HOH) and **aa** (AH-ah).

Basaltic lava at a high temperature flows quickly out of vents, forming pahoehoe, lava with smooth, ropelike surfaces. Cooler basaltic lava moves more slowly. It cools quickly into aa, with rough, jagged surfaces.

PAHOEHOE Smooth, ropelike surfaces characterize this cooling pahoehoe.

AA Sharp, jagged surfaces characterize this cooling aa.

Underwater Lava Flows
Whether it comes from an underwater eruption or flows from land into the sea, lava that cools underwater has a distinctive shape—a rounded, pillowlike form with a hard crust. Pressure builds up inside the lava until its crust cracks, and more lava pours out, forming yet another pillow shape. The resulting mass of rounded lumps is called **pillow lava**.

Ash and Rock Fragments
More explosive eruptions usually involve magmas which contain trapped gases. When these gases are released, solid fragments called **pyroclastic** (PY-roh-KLAS-tihk) **material** may be ejected. A combination of pyroclastic material and fluid lava is common in most volcanic eruptions.

200 Unit 3 Dynamic Earth

DIFFERENTIATING INSTRUCTION

Developing English Proficiency First have students read through the entire section independently; then have them work in small groups to develop explanations of important concepts or terms. Have each group do a presentation on one of the following: types of magma, lava flows, ash and rock fragments.

Pyroclastic materials are classified by size. The smallest pieces are called *ash*, pieces of intermediate size are called *lapilli*, and the largest fragments are called *blocks* and *bombs*. The classification chart below compares pyroclastic materials.

Classification of Pyroclastic Materials

Ash	Lapilli	Block or Bomb
Diameter less than 2 mm	Diameter from 2 to 64 mm	Diameter greater than 64 mm

In some violent eruptions, pyroclastic material combines with hot gases to form a **pyroclastic flow**—a dense, superheated cloud that travels downhill with amazing speed. The cloud may follow the course of a valley, moving faster than 100 kilometers per hour. The eruption of Mount Vesuvius in A.D. 79 produced a pyroclastic flow that buried the Roman city of Pompeii under pumice and ash.

Examine video clips of erupted materials.
Keycode: ES0905

9.2 Section Review

1. How do basaltic, rhyolitic, and andesitic magmas differ?
2. Describe pahoehoe, aa, and pillow lava.
3. Explain why rhyolitic and andesitic magmas are associated with more-explosive eruptions, whereas basaltic magma is associated with less-explosive eruptions.
4. **CRITICAL THINKING** Use the information in this section to infer whether a volcano that erupts explosively would be more likely to form at a subduction boundary or at a divergent boundary.
5. **MATHEMATICS** One of the fastest recorded lava flows in Hawaii traveled over land at an average speed of 2.7 meters per second. Work with a partner to estimate your running speed. Predict how successful you would be if you tried to outrun such a lava flow.

CHAPTER 9 SECTION 2

Examine video clips of erupted materials.
Keycode: ES0905

Have students compare the rates of lava flows to speeds with which students are familiar: car on highway 55–65 mph, walking 2–3 mph, bicycle 10–15 mph.

 Visualizations CD-ROM

ASSESS

1. Basaltic magma; lower silica content; very hot, quick flowing; has low gas content, so is less explosive. Andesitic magma; greater silica content, greater viscosity; higher gas content can be more explosive. Rhyolitic magma; highest silica content, greatest viscosity; lowest melting temperature; associated with explosive eruptions.

2. Pahoehoe and aa lava flows occur on land. Hardened pahoehoe has a smooth, ropy surface, and aa lava has a rough, jagged surface. Pillow lava forms at underwater volcanic vents and has small, rounded lumps.

3. Andesitic and rhyolitic magmas are more likely to lead to explosive eruptions because of their high gas content.

4. A subduction boundary is the more likely site of an explosive eruption. The magma is more viscous and contains more gases.

5. **Sample Response:** Walking speed of 2 m/s would allow the walker to outpace volcanic lava flow.

MONITOR AND RETEACH

If students miss . . .

Question 1 Reteach "Types of Magma." (p. 199) Have students use the chart to summarize each magma type.

Question 2 Have students look at the photographs for pahoehoe and aa lava. (p. 200) Then have them describe or draw how pillow lava might look.

Question 3 Reteach characteristics of gases within magmas. (p. 199) Ask: Why do some gases create explosions? **Gases within viscous magmas cannot escape easily. The gases expand and cause magma to erupt rapidly.**

Question 4 Reteach subduction and divergent boundaries (Chapter 8, pp. 176–179) as well as the Characteristics of Magma chart. (p. 199) Refer students to the location category on the chart as well as the viscosity and gas content categories.

Question 5 Review measuring speed as meters per second. With a stopwatch, measure walking speed over 5 meters or running speed over 50 meters.

CHAPTER 9 SECTION 3

FOCUS

Objectives
1. Compare and contrast landforms that result from volcanic activity.
2. Hypothesize how predicting eruptions and emergency preparedness can reduce loss of life and loss of property.

Set a Purpose
Read aloud the key idea. Then have students read this section to find the answer to "What sort of landforms result from erupted volcanic materials?"

INSTRUCT

VISUAL TEACHING
Discussion
Have students identify various parts of the diagram such as the asthenosphere, the oceanic and continental lithosphere, and the oceanic and continental crust. Use the diagram to review the processes occurring at subduction boundaries and hot spots. Have students compare this diagram to those on pages 195 and 197 as they discuss these processes.

Stress that this landscape summarizes processes and landforms but does not represent a real area. Ask: If the scale were different, what area might it represent? *If the ocean were much wider, it could represent the Pacific Ocean from Hawaii to the west coast of the Americas.*

9.3

KEY IDEA
The shape of a volcanic landform is determined by the materials produced during an eruption.

KEY VOCABULARY
- shield volcano
- cinder cone
- composite volcano
- lahar
- caldera
- lava plateau

Volcanic Landforms

The term *volcano* refers not only to a volcanic vent, but also to the landform that develops as the materials from a volcanic eruption harden. The shape and structure of a volcano are determined by the nature of its eruptions and the materials it ejects.

Shield Volcanoes

Because of its low viscosity, basaltic lava tends to flow long distances before hardening. In some cases, the lava builds up in layers, forming **shield volcanoes** with broad bases and gently sloping sides. The broad base of a shield volcano can support a mountain of enormous height. For example, Mauna Loa, a volcano on the island of Hawaii, rises 4170 meters above sea level and its base is 5000 meters below sea level; thus, its total height is 9170 meters.

Because shield volcanoes discharge basaltic lavas, they tend to be less explosive than other types of volcanoes. Basaltic lava flows, however, may be frequent and copious, causing damage to homes, highways, and other property.

Cinder Cones

A **cinder cone,** perhaps the simplest type of volcano, forms when molten lava is thrown into the air from a vent. As it falls, the lava breaks into fragments that harden before hitting the ground. These fragments accumulate, forming a cone-shaped mound with an oval base. Cinder cones, which tend to be smaller than other types of volcanoes, typically form in groups and on the sides of larger volcanoes.

MAUNA LOA is a **shield volcano** on the island of Hawaii.

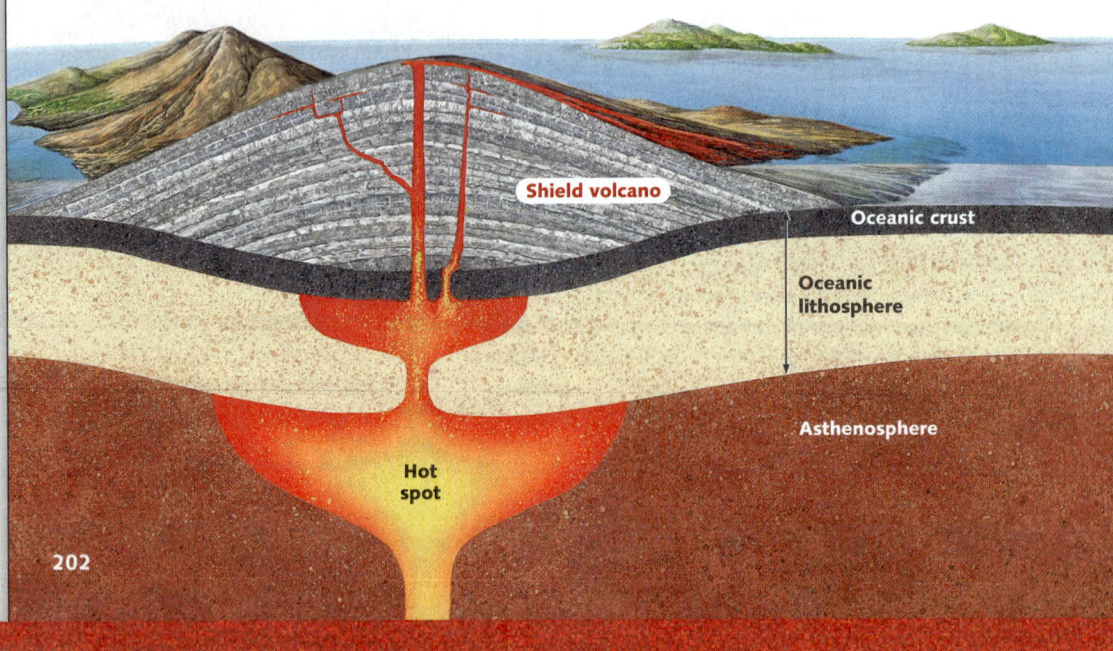

DIFFERENTIATING INSTRUCTION

Reading Support A concept web can clarify students' understanding as they read. Begin with Volcanic Landforms as the topic. The next level would indicate the types of landform, followed by details and the names and locations of various volcanic formations. Students may fill in the organizer as they read or at the end of the section.
Use Reading Study Guide, p. 30.

Hands-On Demonstration Use cardboard with a small hole cut in the center and a squeeze bottle filled with plaster to demonstrate how a lava cone forms. From below, place the neck of the squeeze bottle through the hole. Keeping the cardboard level, squeeze and release the tube to indicate a series of eruptions. Have students describe the shape of the resulting form. *cone-shaped, with an oval base*

202 Unit 3 Dynamic Earth

Composite Volcanoes

Composite volcanoes develop when layers of materials from successive explosive eruptions accumulate around a vent. The materials include hardened lava flows and other pyroclastic material.

The 1980 eruption of Mount St. Helens illustrates some of the forces and events involved in shaping a composite volcano. Except for a few minor eruptions, Mount St. Helens had been quiet since 1857. In the spring of 1980, however, earthquake activity increased, a bulge in the north face of the volcano's peak grew larger, and small eruptions of steam and ash occurred.

Eventually, an earthquake burst the bulge that had been forming on the volcano. Magma, water, and gases exploded in a massive cloud of superheated ash and stones. Some of the hot ash mixed with the snow and ice on the mountain to form a fast-moving mudflow called a **lahar** (LAH-hahr). In some areas, the debris from the lahar and landslides reached a depth of about 46 meters.

After a violent eruption, a composite volcano may remain relatively quiet for a long period of time. Beneath the surface, however, gas-rich magma may again be building up pressure, eventually leading to another explosive eruption.

The illustrations on these pages show three volcanic landforms and the processes which formed each type.

MOUNT SHASTA is a **composite volcano** located in California.

CAPULIN is a **cinder cone** located in New Mexico.

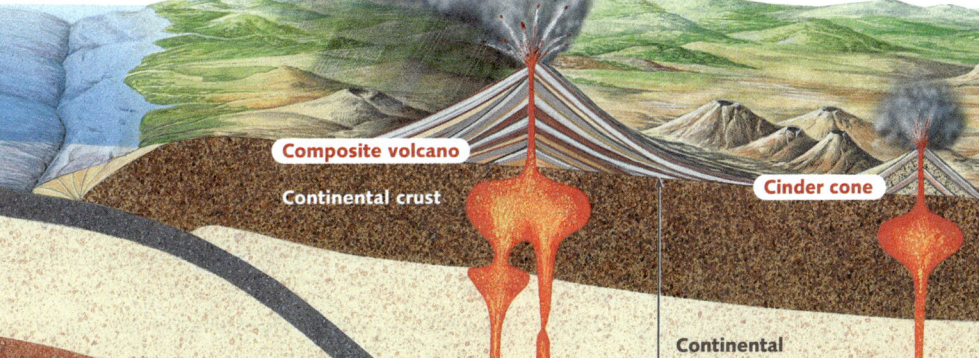

CHAPTER 9 SECTION 3

VISUAL TEACHING
Discussion
The photographs show examples of landforms, but the illustration shows the geologic processes that create the landforms. Ask: Which volcanic landforms are formed at subduction boundaries? **composite volcano, cinder cone** Which is formed at a hot spot? **shield volcano**

Extend
Compare and contrast the three types of volcanic landforms in the diagram. **All are results of magma forced to the surface, but shield volcanoes form over hot spots above the oceanic lithosphere. Composite volcanoes and cinder cones form above continental lithospheres at subduction boundaries.**

CRITICAL THINKING
Have students apply what they already know about magma. Ask: Do you think the magma in cinder cones is more viscous or less viscous than in shield volcanoes? Explain. **more viscous; basaltic lavas flow more easily and contain less gas; the lava in cinder cones contains more gases and is likely to be thicker and somewhat explosive**

DIFFERENTIATING INSTRUCTION

Developing English Proficiency Check for vocabulary that may hinder comprehension. Have small groups identify words or phrases that describe the volcanoes, such as *gently sloping sides*, *cone-shaped*, *oval base*, *bulge*, or *crater*. Model how to use context clues in visuals and in text to help students make meaning of unfamiliar words. For example, point out the "broad base" and "gently sloping sides" of Mauna Loa in the photograph. Students could recognize a *bulge* as a bulbous growth on a volcano because it "grew larger" on one side of the peak and it "burst."

Chapter 9 Volcanoes 203

CHAPTER 9 SECTION 3

Visual Teaching

Interpret Diagrams
Read the caption for the first illustration. Ask: When the volcano erupted, what happened to the magma? **It was released to the surface.** What happened when the magma chamber emptied? **It created a gap, or space, and the top of the cone collapsed to fill the space.** How did it become a lake? **Rainwater filled the caldera, held in by the remaining sides of the volcanic cone.**

ASSESS

1 Shield volcanoes are formed by fast-moving basaltic lava that travels long distances before hardening. Over time, layers of basaltic lava build up to form a large, broad, shield-shaped volcano. Cinder cones tend to be much smaller than shield volcanoes. Cinder cones form when droplets of lava harden into cinders as they are ejected from a vent. The cinders build up in a cone shape around the vent.

2 Composite volcanoes form at subduction boundaries. Their eruptions are very dangerous because the magma that forms there contains trapped gases, including steam. During an eruption, these hot gases may be explosively released.

3 Sample Response: People living near Mount Rainier may not realize that this beautiful mountain is also a very dangerous volcano. They may be endangered by huge clouds of hot gas and ash. They also may be threatened by lahars, the huge fast-flowing mudflows.

Calderas

Sometimes magma beneath a volcano is released after the top of the volcano collapses, forming a large crater-shaped basin called a **caldera** (kal-DAIR-uh). Many active volcanoes have calderas at their summits. In some cases, a caldera fills with water to form a lake.

CRATER LAKE is a caldera that has filled with water.

Crater Lake in Oregon is an example of a caldera. About 7700 years ago, a volcano exploded in a cataclysmic eruption. The collapse of the cone's top created a caldera. Gradually, the caldera filled with water to form Crater Lake, shown at left.

Some calderas, such as the one in Yellowstone National Park in the western United States, are still active. Beneath Earth's surface, magma superheats the water which feeds the hot springs and geysers at Yellowstone. Geologists estimate that the eruption that produced the Yellowstone caldera may have been 1000 times more powerful than the 1980 Mount St. Helens blast.

Formation of Crater Lake

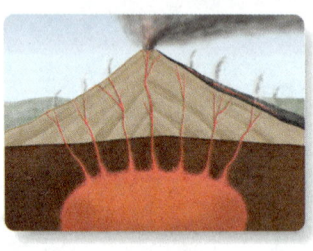

1 About 7700 years ago, a volcanic eruption partially emptied an underground magma chamber.

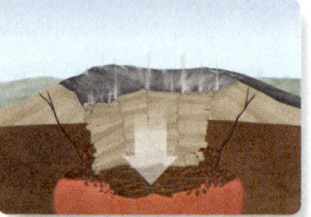

2 The top of the cone collapsed inward, forming a basin known as a caldera.

3 Over time, rainwater filled the caldera, forming Crater Lake. Additional volcanic activity formed the small cone in the center.

Lava Plateaus

Sometimes plate tectonics results in the formation of a long, narrow crack or fissure in Earth's surface. Basaltic lava pouring from the fissure spreads across the land, forming a **lava plateau.** The basaltic lava that formed the Columbia Plateau in the northwestern United States is over one kilometer thick in some places and covers an area of about 164,000 square kilometers.

9.3 Section Review

1 Compare and contrast the ways in which shield volcanoes and cinder cones are formed.

2 CRITICAL THINKING Describe the formation of a composite volcano.

3 WRITING The eruption of Mount Rainier, a composite volcano, could pose a serious threat to local residents. Write a description of the potential hazards that people living near Mount Rainier might face.

MONITOR AND RETEACH

If students miss . . .

Question 1 Reteach "Shield Volcanoes" and "Cinder Cones." (p. 202) Have students examine the diagram and describe the shapes of the two volcanic forms. Then have them scan the text to tell the type of material in each form.

Question 2 Reteach "Composite Volcanoes." (p. 203) As students examine the diagram, ask: Where have you seen this form before? **Students may remember a similar diagram in section 9.1, Volcanic Activity at a Subduction Boundary.** (p. 195)

SCIENCE & Society

Predicting Eruptions, Saving Lives

Consider two of the deadliest eruptions of the past century:

- **1902** An eruption of Mount Pelée on Martinique kills nearly 30,000 people.
- **1985** Lahars triggered by an eruption of Colombia's Nevado del Ruiz volcano sweep through the town of Armero, killing over 20,000 people.

Can such tragedies be prevented? Is it possible to predict a volcanic eruption in time to save lives?

DEVASTATION The 1985 eruption of Nevado del Ruiz, in Colombia, caused extensive damage and casualties.

When Mount Pinatubo, a volcano in the Philippines, rumbled to life in April 1991, scientists quickly set up a station to monitor its activity. They also collected samples of the gas escaping from the summit of the mountain. Toward the end of May, they detected an increase in sulfur dioxide emissions, which led them to conclude that magma was rising. In early June, they observed that the side of the volcano was swelling. Meanwhile, seismic activity was increasing. All evidence suggested that a large eruption was about to occur.

From June 9 through June 15, Mount Pinatubo erupted in a series of increasingly violent explosions. Fortunately, scientists and public officials had worked together to set up a plan for educating and alerting the population. Some 58,000 people had been evacuated from the areas predicted to be in the greatest danger. Eventually, as many as 200,000 people were forced to evacuate their homes. The evacuees were spared from the dangerous lahars the eruption produced. Thousands of lives had been saved from what proved to be one of the most destructive eruptions of the 20th century.

Extension

SCIENCE NOTEBOOK
Consider ways in which predictions made by scientists promote safety. Explain how scientific predictions might affect your life.

How Fast Do Gases from Volcanic Eruptions Travel?
Use satellite images of Mount Pinatubo to calculate the speed of volcanic ash and gases moving through the atmosphere.
Keycode: ES0906

EVACUATING People carry their belongings away from a dangerous area near Mount Pinatubo.

CHAPTER 9 SECTION 3

How Fast Do Gases from Volcanic Eruptions Travel?
Keycode: ES0906

Have students predict the speed of volcanic gases before doing their calculations.

📖 *Use Internet Investigations Guide, pp. T36, 36.*

Science and Society

Early Detection and Evacuation
Before a volcano erupts, there are measurable changes in seismicity, ground deformation, and other physical and chemical processes. Scientists monitoring Mount Rainier in Washington work with local emergency response teams. Orting, Washington, is the town most at risk for lahar activity from Mount Rainier. Emergency planners have developed a public education program, and there is a yearly volcano drill. School-children quickly board buses and are rushed out of the valley. Orting maintains a siren system. If sirens go off, people are expected to leave their homes and follow planned evacuation routes.

Science Notebook
These signs would help students predict when the next eruption might occur: seismic activity, release of sulfur dioxide gas, any swelling or bulges in the mountain.

MONITOR AND RETEACH

Question 3 Reteach pyroclastic flow (p. 201) and characteristics of composite volcanoes. (p. 203) Ask: What is most dangerous about volcanic eruptions? *clouds of superheated hot gases and pyroclastic materials exploding, explosions of steam and ash, and lahars*

CHAPTER 9 SECTION 4

Focus

Objectives
1. Classify volcanic activity on other planets and moons.
2. Analyze the tectonics behind extraterrestrial volcanic activity.

Set a Purpose
Read the title and allow students to discuss their understanding of *extraterrestrial*. Explain that *extraterrestrial* means "originating, located, or occurring outside Earth or its atmosphere." Have students read to find the answer to "Where do extraterrestrial volcanoes occur?"

Instruct

Critical Thinking
Discuss how scientists get information about extraterrestrial volcanoes. Ask: How do scientists know the age and type of lava on the moon? **from rock samples collected** Without traveling to Mars, Venus, or Io, how do scientists know there are volcanoes there? **Photographs taken from space show volcanic landforms on Mars and Io. Radar revealed volcanic features on Venus where clouds cover the planet.**

9.4

Key Idea
Extraterrestrial volcanoes have been active and are currently active in the solar system.

OLYMPUS MONS, MARS Images taken by the Viking I Orbiter were used to create this view of Olympus Mons.

Extraterrestrial Volcanoes

Earthbound scientists knew, even before probes landed on Mars, that volcanic activity existed elsewhere in our solar system. As scientists gather and interpret data from telescopes and orbiting probes, their understanding of volcanic activity elsewhere in the solar system grows.

The Moon
About 15 percent of the moon's surface is covered by dark areas. Early astronomers, looking at the moon without telescopes, originally thought that these dark areas might be seas. For this reason, the areas came to be known as *maria* (from the Latin word for seas). Scientist have determined the maria consist mostly of basaltic lava flows, the result of volcanic activity that began between 3 and 4 billion years ago. Radioactive elements beneath the moon's surface probably provided the energy necessary for the volcanic activity to begin.

Scientists rely, in part, on photographs to understand the history of volcanic activity on the moon. For example, if a photograph shows an impact crater filled with lava, scientists know that the lava flow is younger than the crater. They also use rock samples. The oldest samples of lunar lava that have been found are about 4.2 billion years old. The youngest are about 3.1 billion years old.

Mars
Mars is home to a number of shield volcanoes. One of these, Olympus Mons, is the largest known volcano in the solar system. It measures 600 kilometers across its base and towers almost 26 kilometers above the surrounding terrain. Its huge size suggests that Mars does not have moving plates. If it did, Olympus Mons would have moved away from the hot spot that formed it, and we would see a chain of smaller volcanoes rather than one volcano.

Are volcanoes on Mars still active? For many years, the answer to that question seemed to be no. But in 1999 the Mars Global Surveyor began mapping the Martian surface. Scientists studying the new data found lava

---

DIFFERENTIATING INSTRUCTION

Reading Support Model how to look for main ideas when taking notes: "The first paragraph is about what ancient astronomers thought about the moon—it even mentions seas. The important point about volcanoes here is that 'dark areas on the moon show basaltic lava flows.'"
Use Reading Study Guide, p. 31.

flows that appeared to be 20 to 60 million years old. Though this sounds like a great age, it really amounts to the blink of an eye in the long history of Mars. For this reason, some scientists wonder whether volcanoes on Mars could erupt in the future.

Venus

Orbiting spacecraft have used radar to penetrate the thick clouds around Venus and to map its surface, revealing more than 1600 large volcanoes, as well as volcanic features such as lava flows and calderas. Countless smaller volcanoes also cover the surface of Venus. Most of the volcanoes are shield volcanoes. The largest are as broad as those on Mars but much flatter.

Although most volcanoes on Venus are probably inactive, some may still be active. Computer models have suggested that volcanic activity could account for the scorching 460°C (860°F) surface temperatures on Venus. Gases escaping from volcanoes could be interacting with the planet's atmosphere to continually trap heat at Venus's surface.

Io

One of the most volcanically active places in the solar system is Io (EYE-oh), Jupiter's third largest moon. Io is caught in a gravitational tug of war between Jupiter and two other moons. As a result, some parts of Io's surface regularly move up and down by as much as 100 meters. The resulting friction is the source of heat that powers Io's volcanic activity, which continually changes this moon's surface. In the photograph to the right, the small white patch extending into space is a 140-kilometer-high plume erupting from a volcano called Pillan Patera.

Lava temperatures at Pillan Patera may be as high as 1720°C. Volcanic activity on Earth has not produced such high temperatures for billions of years, and some scientists think that lavas on Io may resemble those of early Earth. It is possible that studies of Io will provide a window into Earth's own history.

Io The volcanic plume of Pillan Patera is 140 kilometers high.

9.4 Section Review

1. What type of lava flows are found on the moon?
2. Why is Olympus Mons much larger than Earth's shield volcanoes?
3. How do scientists know that there are volcanoes on Venus?
4. What is the source of heat responsible for the volcanic activity on Io?
5. **CRITICAL THINKING** Would you expect to find composite volcanoes on Mars? Explain your reasoning.
6. **MATHEMATICS** Look back at the description of Mauna Loa on page 202. Use a ratio to compare its height with the height of Olympus Mons.

More about...

Io
In 1979 NASA's *Voyager* took the first pictures of Io, the closest Galilean moon of Jupiter. Observers were surprised to discover that Io was covered with volcanoes. Later flights from the *Galileo* program found hundreds to thousands of active volcanoes on Io. *Galileo's* instruments were able to measure volcanic temperatures. From this information, scientists were able to speculate that the lava is silicate material rich in magnesium.

Assess

1. basaltic lava flows
2. Olympus Mons has not moved away from the hot spot that formed it.
3. Radar-produced maps of Venus have shown that its surface is covered with shield volcanoes, calderas, and hardened lava flows.
4. Friction from movements of Io's surface is the heat source. These movements are caused by the gravitational pull of Jupiter and two other moons.
5. **Sample Response:** No; these types of volcanoes form at subduction boundaries and Mars does not have moving plates.
6. Accept $\frac{26{,}000}{9170}$ meters or any equivalent ratio.

CHAPTER 9 ACTIVITY

PURPOSE
To locate volcanic activity worldwide and analyze the resulting patterns.

MATERIALS
Provide students with a copy of Lab Sheet 9 (World Map with Latitude and Longitude Grid). A copymaster is provided on page 140 of the teacher's laboratory manual.

PROCEDURE
- ❷ Plot a few of the locations with the students first.
- ❸ Students should notice that a large number of volcanoes correspond with plate boundaries.

Use Transparency 41.

ANALYSIS AND CONCLUSIONS

1. North American/Caribbean, Caribbean/Cocos, South American/Nazca, Eurasian/North American, Eurasian/African, African/Australian, Australian/Eurasian, Philippine/Eurasian, Pacific/Australian, Eurasian/Pacific.

2. Many of the volcanoes occur because they are near plate boundaries, where the crust is under stress. Volcanoes may also occur above hot spots.

CHAPTER 9 MAP Activity

Where Are Active Volcanoes?

At this very moment, somewhere in the world a volcano is erupting or is about to erupt. In this map activity, you will use your knowledge of latitude and longitude to locate volcanic activity around the world. What patterns do you see?

SKILLS AND OBJECTIVES
- *Plot* the locations of active volcanoes on a world map.
- *Hypothesize* why volcanic activity occurs in certain locations.
- *Predict* where volcanic activity is likely to occur in the future.

MATERIALS
- **Lab Sheet 9** World Map with Latitude and Longitude Grid
- **Appendix B** World Map with Plate Boundaries, pages 712–713
- colored pencils

Procedure

❶ Refer to the table on page 209 showing approximate locations of volcanic activity. The table lists locations on land where volcanic activity was recorded during a five-month period.

❷ Use the latitude and longitude coordinates in the table to identify and mark each location on the map on Lab Sheet 9. (For help with plotting latitude and longitude, look back to the Chapter 7 Map Activity.)

❸ Compare the patterns of volcanic activity as shown on your map with the map of plate boundaries on pages 712–713.

❹ Record your answers to the Analysis and Conclusions questions in your science notebook.

Analysis and Conclusions

1. Identify the plate boundaries that are close to sites of volcanic activity.

2. Hypothesize why volcanic activity occurs in the locations you have marked on Lab Sheet 9.

3. Which volcanoes occur at subduction boundaries? at diverging plate boundaries? at hot spots?

4. From the information you have on your lab sheet, describe where you see the most volcanic activity. For example, is volcanic activity more prevalent at a particular longitude?

5. Of the volcanoes listed in the table, find the ratio of the number of active volcanoes in the Pacific Ring of Fire to the total number of active volcanoes. How can you use this ratio to predict the likelihood that an eruption occurring on Earth will occur in the Pacific Ring of Fire?

6. Based on your analysis, where would you expect most future volcanic activity on Earth to take place?

CHAPTER 9 ACTIVITY

DATA CENTER
CLASSZONE.COM

Find out more about current volcanic activity.
Keycode: ES0908

3. Subduction: Soufrière Hills, Mount Oyama, White Island, Krakatau, Etna, Popocatépetl, Kavachi, San Cristóbal, Tungurahua, Shishaldin, Usu, Guagua Pichincha, Mayon, Bezymianny, Pacaya, Karymsky.
Diverging plate boundaries: Hekla.
Hot spots: Piton de la Fournaise and Kilauea. Mount Cameroon's origins are unclear; rifts and/or hot spots are among the possibilities.

4. Most volcanic activity appears to be near subduction zones and/or at the edges of the Pacific Ocean. Some students may mention that there seem to be more volcanoes at the lower latitudes, as over half of the volcanoes listed occur within 30 degrees North and 30 degrees South latitude.

5. The ratio is 14/20. You may wish to indicate to your students that the area around Indonesia and the Philippines is included in the Pacific Ring of Fire, despite not being directly on the Pacific Plate. Using the ratio, students should predict that there is a 70 percent likelihood that an eruption occurring on Earth will occur in the Pacific Ring of Fire, as 14/20 = 0.7.

6. Student answers will vary, but they should realize that most future volcanic activity on Earth is likely to take place at subduction boundaries along mid-ocean ridges, and/or in the Pacific Ring of Fire.

Locations of Volcanic Activity

Volcano	Approximate Location
Soufrière Hills, Montserrat	17°N, 62°W
Mount Oyama, Japan	34°N, 140°E
White Island, New Zealand	38°S, 178°E
Piton de la Fournaise, Réunion	21°S, 56°E
Krakatau, Indonesia	6°S, 105°E
Etna, Italy	38°N, 15°E
Mount Cameroon, Cameroon	4°N, 9°E
Popocatépetl, Mexico	19°N, 99°W
Kavachi, Solomon Islands	9°S, 158°E
San Cristóbal, Nicaragua	13°N, 87°W
Tungurahua, Ecuador	1°S, 78°W
Shishaldin, USA	55°N, 164°W
Usu, Japan	43°N, 141°E
Guagua Pichincha, Ecuador	0°S, 79°W
Mayon, Philippines	13°N, 124°E
Bezymianny, Russia	56°N, 161°E
Pacaya, Guatemala	14°N, 91°W
Hekla, Iceland	64°N, 20°W
Kilauea, USA	19°N, 155°W
Karymsky, Russia	54°N, 159°E

7. Find the data on volcanic activity from the past 12 months. Plot the locations of this volcanic activity using a different colored pencil. Use the completed lab sheet to test the predictions you made in Question 6.

DATA CENTER
CLASSZONE.COM

Find out more about current volcanic activity.
Keycode: ES0908

Chapter 9 Volcanoes

CHAPTER 9 REVIEW

VOCABULARY REVIEW

1. hot spot
2. pillow lava
3. pyroclastic materials
4. viscosity

CONCEPT REVIEW

5. along mid-ocean ridges

6. The Hawaiian islands formed over a hot spot. Each island has formed as the Pacific plate has moved northwest over this hot spot. The island of Hawaii now lies over the hot spot and is volcanically active. Older islands such as Oahu and Kauai have moved away from the hot spot and are no longer volcanically active.

7. Some risks include the death of people and livestock, the destruction of property, and the loss of land used for living and farming. Benefits of volcanic activity may include the availability of geothermal energy, formation of land for living and farming (the Hawaiian Islands, for example), lakes (Crater Lake), and tourist attractions such as Yellowstone National Park.

8. Andesitic and rhyolitic; these types of magma are resistant to flow and contain trapped gases. The escape of these gases during an eruption cause the eruption to be highly explosive.

9. A **shield volcano** forms from basaltic lava flowing long distances before cooling. The lava builds up in layers, forming a large, broad-based volcano with gently sloping sides. Mauna Loa is a shield volcano. A **cinder cone** forms when fragments of molten lava expelled from a vent cool and harden in the air, then build up from an oval base to form a cone around a vent. Cinder cones tend to form in groups and on the sides of larger volcanoes. Capulin is a cinder cone. A **composite volcano** is formed from layers of lava and pyroclastic materials that build up around a vent. They form on continental crust near subduction boundaries. Mount Shasta is a composite volcano.

10. Lava plateaus form when basaltic lava pours out of fissures in Earth's surface. The lava spreads across great distances before cooling, forming a large plateau. The Columbia Plateau in the north-western United States is a lava plateau.

11. Heat from radioactive elements inside the Moon probably caused its volcanic activity. Io is regularly pulled and stretched by gravitational forces exerted by Jupiter and two other moons of Jupiter. The frictional heat from this pulling and stretching causes Io's volcanic activity.

CHAPTER 9 REVIEW

Summary of Key Ideas

9.1 Volcanic activity takes place primarily at subduction boundaries and at divergent boundaries, where the combination of temperature, pressure, and water content are right for the formation of magma. Magma and volcanoes also form at hot spots.

9.2 The amount of silica in magma affects its viscosity. Basaltic magma has the least silica, so it flows easily. Andesitic and rhyolitic magmas have more silica, so they are more resistant to flow. Gases escape easily from basaltic magma, generating relatively quiet eruptions. But gases tend to be trapped in andesitic and rhyolitic magmas, leading to explosive eruptions.

9.3 A volcano's shape and structure depend on how it erupts and what materials are released. Shield volcanoes are formed by basaltic lava that flows long distances before hardening. Cinder cones are formed when molten lava is thrown into the air from a vent and breaks into drops. These drops harden into cinders that form a steep cone around the vent. Composite volcanoes are formed by layers of pyroclastic materials and lava that have erupted in the past.

9.4 Volcanic activity has occurred in the past on the moon, Mars, and Venus. Jupiter's moon Io is currently volcanically active.

KEY VOCABULARY

aa (p. 200)
caldera (p. 204)
cinder cone (p. 202)
composite volcano (p. 203)
hot spot (p. 197)
lahar (p. 203)
lava (p. 200)
lava plateau (p. 204)
pahoehoe (p. 200)
pillow lava (p. 200)
pyroclastic flow (p. 201)
pyroclastic material (p. 200)
shield volcano (p. 202)
viscosity (p. 199)
volcano (p. 194)

210 Unit 3 Dynamic Earth

Vocabulary Review

Write the term from the key vocabulary list that best completes the sentence.

1. An area of volcanic activity in the middle of a lithospheric plate is called a ___?___.
2. When lava cools underwater, ___?___ often forms.
3. In an explosive eruption, a composite volcano may erupt ash and other solid fragments called ___?___.
4. Because it has greater resistance to flow, andesitic magma is said to have greater ___?___ than basaltic magma.

Concept Review

5. Where does most of Earth's volcanic activity occur?
6. How did the Hawaiian Islands form?
7. Describe some risks and benefits of living near a volcano.
8. Which types of magma are most likely to result in an explosive eruption? Why?
9. Describe three types of volcanoes and give an example of each.
10. Explain how lava plateaus form, and give an example of one.
11. What do scientists think has caused the volcanic activity on the moon? On Io?
12. **Graphic Organizer** Copy and complete the concept map below. There may be more than one possible answer. Justify your answer.

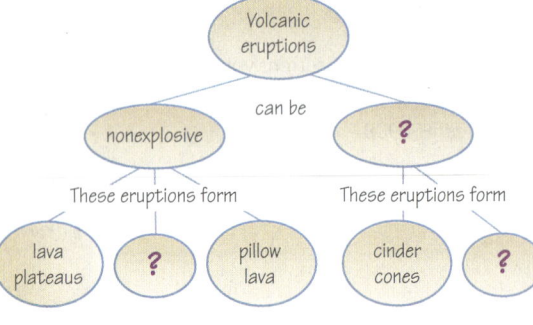

12. Volcanic eruptions can be explosive and form cinder cones as well as composite volcanoes, lahars, calderas, pyroclastic material, and pyroclastic flows. Nonexplosive eruptions form shield volcanoes and pahoehoe as well as lava plateaus and pillow lava.

Critical Thinking

13. Infer The rate of movement of subducting plates differs from one subduction boundary to the next.

  a. How might this fact affect the rate at which magma is produced in the asthenosphere? Explain your thinking.

  b. How might the rate of magma production affect the number of volcanic eruptions at a subduction boundary?

  c. How might your answers to parts (a) and (b) explain why Indonesia usually has one major volcanic eruption each year while the Cascade Range in Washington has only occasional eruptions? Explain.

14. Predict What relationship would you expect to find between the sizes of particles in a pyroclastic deposit and the distance from the volcanic source of the material? Explain your thinking.

Internet Extension

Is It Safe to Live Near a Volcano? Analyze what happened at Mount St. Helens to understand volcanoes in the Northwest United States. Develop an evacuation plan for the area surrounding Mount Rainier.

Keycode: ES0907

Writing About the Earth System

SCIENCE NOTEBOOK One of the most catastrophic volcanic events in history occurred when the Indonesian volcano Krakatau erupted in 1883. Do some research to find information about the 1883 eruption of Krakatau. Using the information, explain the impact of this eruption on the geosphere, the hydrosphere, the atmosphere, and the biosphere.

Interpreting Graphs

The graph shows some of the extinct volcanoes of the Hawaiian Island chain. The vertical axis of the graph shows the approximate age of each volcano, and the horizontal axis shows the distance of each volcano from Kilauea, the youngest volcano on the Big Island of Hawaii. Kilauea is still an active volcano.

15. Which volcano was over the hot spot about 7 million years ago?

16. What is the age difference between Haleakala, which is on the island of Maui, and Waianae, which is on the island of Oahu?

17. If a volcano of the Hawaiian Island chain were located 600 kilometers from Kilauea, what would you expect its age to be?

18. About how far away from Kilauea would an island be if its rocks were 4 million years old?

19. Which volcano in the graph is oldest? About how old is it, and about how far is it from Kilauea?

20. About how fast, in centimeters per year, has the Pacific Plate been moving over the hot spot? Explain your thinking.

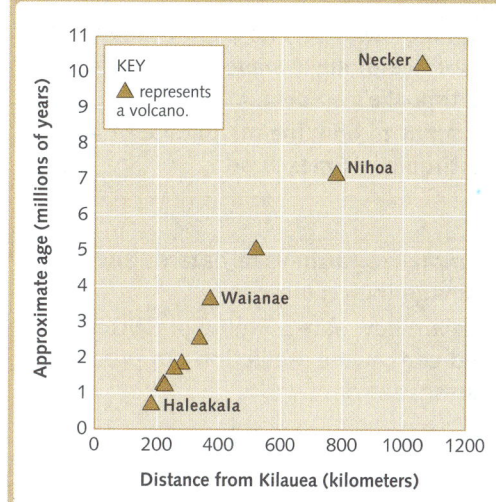

Note: In the graph above, only the volcanoes mentioned in the exercises are labeled.

CHAPTER 9 REVIEW

Is It Safe to Live Near a Volcano?
Keycode: ES0907

Use Internet Investigations Guide, pp. T37, 37.

CRITICAL THINKING

13. a. A more rapidly moving plate may tend to create more friction and therefore more melting. A rapidly moving plate would also have more volume subducted each year. As a result, a greater volume of rock could be melted.

  b. If more magma is produced, there would be more volcanic eruptions at the subduction boundary.

  c. The plate under Indonesia must be moving more rapidly than the plate under the Cascades.

14. When particles are suspended in water, the larger particles settle first. Volcanic deposits behave in the same way. Larger particles (blocks or bombs) are deposited nearer the volcano. Ash is carried farther away, in some cases completely around the world.

INTERPRETING GRAPHS

15. Nihoa

16. about 3 million years

17. about 6 million years

18. about 400 kilometers

19. Necker; over 10 million years (approximately 10.5 million years); about 1050 kilometers from Kilauea

20. According to the graph, Nihoa is about 800 kilometers from the hot spot. 800 km/8 million years = 80,000,000 m/8,000,000 years = 10 cm/year

WRITING ABOUT SCIENCE

Writing About the Earth System

The 1883 eruption of Krakatau generated tsunamis that killed an estimated 36,000 people. The eruption affected the geosphere as it destroyed much of the island of Krakatau. The biosphere was affected, as organisms were killed and their habitats destroyed. The atmosphere was affected because the eruption released enormous quantities of dust and gases. As the dust and gases dispersed through the atmosphere, they affected global temperatures and altered the ozone layer. The eruption also released steam, which added water vapor to the atmosphere, thus affecting the hydrosphere.

CHAPTER 10 PLANNING GUIDE EARTHQUAKES

	Section Objectives	Print Resources
Chapter-Level Resources		Laboratory Manual, pp. 45–46 Internet Investigations Guide, pp. T38–T40, 38–40 Guide to Earth Science in Urban Environments, pp. 29–36 Spanish Vocabulary and Summaries, pp. 19–20 Formal Assessment, pp. 28–30 Alternative Assessment, pp. 19–20
Section 10.1 How and Where Earthquakes Occur, pp. 214–216	1. Explain how most earthquakes result from strain that builds up along faults or near plate boundaries. 2. Describe how the energy released in an earthquake travels as waves.	Lesson Plans, p. 32 Reading Study Guide, p. 32 Teaching Transparency 12
Section 10.2 Locating and Measuring Earthquakes, pp. 217–221	1. Explain how a seismograph is used to record earthquake waves and locate an earthquake's epicenter. 2. Summarize how the magnitude of an earthquake is measured.	Lesson Plans, p. 33 Reading Study Guide, p. 33 Teaching Transparency 12
Section 10.3 Earthquake Hazards, pp. 222–227	1. Summarize earthquake hazards and the damage they can cause. 2. Explain how safe building practices and earthquake prediction can prevent earthquake damage.	Lesson Plans, p. 34 Reading Study Guide, p. 34
Section 10.4 Studying Earth's Interior, pp. 228–229	1. Describe how data from seismic waves are used to infer the structure of Earth's interior.	Lesson Plans, p. 35 Reading Study Guide, p. 35

CLASSZONE.COM

Section 10.1 INVESTIGATION
How Are Earthquakes Related to Plate Tectonics?
Computer time: 45 minutes / Additional time: 45 minutes
Students plot earthquakes on a map and analyze their location and depth at plate boundaries.

Section 10.2 INVESTIGATION
Where Was That Earthquake?
Computer time: 45 minutes / Additional time: 45 minutes
Students read seismograms and locate the epicenter of an earthquake.

Technology/Online Resources

Electronic Teacher Tools
Visualizations CD-ROM
Classzone.com
Online Lesson Planner
Test Generator

Visualizations CD-ROM
Classzone.com
  ES1001 Investigation, ES1002 Visualization

Classzone.com
  ES1003 Investigation, ES1004 Career

Visualizations CD-ROM
Classzone.com
  ES1005 Visualization, ES1006 Safety Tip,
  ES1007 Visualization, ES1008 Data Center

Visualizations CD-ROM
Classzone.com
  ES1009 Visualization

Classroom Management

Gather these materials for the laboratory activity.

Lab Activity, pp. 230–231
40 coffee stirrers (per model)
40 mini-marshmallows (per model)
metric ruler
2 identical shallow cardboard boxes
scissors
10–20 marbles
4 small rubber bands
stapler

Chapter Review INVESTIGATION
Which Fault Moved in the Northridge Earthquake?
Computer time: 45 minutes / Additional time: 45 minutes
Students use data and maps to pinpoint the fault that moved in that California earthquake.

CHAPTER 10

Earthquakes

Powerful earthquakes, such as the 1999 earthquake that hit Adapazarı, Turkey, can level buildings in seconds.

How can studying the causes and effects of earthquakes save lives?

INTRODUCE

Build Interest

Before sharing background information about the photograph with students, allow time for their questions and comments.

Ask questions like the following:

- What type of damage do you see in this photograph. *toppled buildings; damaged buildings with smashed windows, caved in roofs and walls; rubble on the ground*
- What damage might there be that you cannot see? *broken water pipes, gas pipes, and sewage systems*

Respond

- How can studying the causes and effects of earthquakes save lives? *The data scientists collect might help predict earthquakes, so that warnings could be issued. Data could help architects design buildings that resist earthquake damage.*
- If you lived in this area, and your home was rebuilt, would you return here to live? *Have students discuss reasons for their answers.*

About the Photograph

At 3:00 A.M. on August 17, 1999, an earthquake (magnitude 7.4) near Adaparzi, Turkey, caused the city to shake for 45 seconds. Shoddy construction and loose soil under the structures led to extensive damage. The earthquake killed tens of thousands of people in Turkey and left hundreds of thousands homeless.

BEYOND THE TEXTBOOK

Print Resources
- Reading Study Guide, pp. 32–35
- Transparency 12
- Formal Assessment, pp. 28–30
- Laboratory Manual, pp. 45–46
- Alternative Assessment, pp. 19–20
- Spanish Vocabulary and Summaries, pp. 19–20
- Internet Investigations Guide, pp. T38–T40, 38–40
- Guide to Earth Science in Urban Environments, pp. 29–36

Technology Resources
- Visualizations CD-ROM
- Classzone.com
- Online Lesson Planner
- Electronic Teacher Tools
- Test Generator

212 Unit 3 Dynamic Earth

CHAPTER 10

PREVIEW

▶ **FOCUS QUESTIONS** In this chapter you will study earthquakes and learn more about the key questions listed below.

Section 1 How and where do earthquakes occur?

Section 2 How do scientists locate and measure earthquakes?

Section 3 What types of damage do earthquakes cause, and how can people prepare for earthquakes?.

▶ **REVIEW TOPICS** As you investigate earthquakes, you will need to use information from earlier chapters.

- lithosphere (p. 73)
- mantle (p. 72)
- plate tectonics (p. 172)
- transform boundary (p. 178)

▶ **READING STRATEGY**

CONNECT

As you learned in Chapter 8, most earthquake activity takes place at or near plate boundaries. Before you read Chapter 10, review chapter 8 and write down key information about each type of plate boundary in your science notebook.

INTERNET RESOURCES
CLASSZONE.COM

At our Web site, you will find the following Internet support for this chapter.

DATA CENTER
EARTH NEWS
VISUALIZATIONS
- Earthquake Waves
- Effects of Earthquakes
- Earthquake Risk Map
- Movement of P and S waves

LOCAL RESOURCES
CAREERS
INVESTIGATIONS
- How Are Earthquakes Related to Plate Tectonics?
- Where Was That Earthquake?
- Which Fault Moved in the Northridge Earthquake?

Chapter 10 Earthquakes 213

CHAPTER 10

PREPARE

Focus Questions
Find out what knowledge students have about earthquakes. For each focus question, have students consider the section title and then develop two or three additional questions for each section. For example, Section 1: Why do earthquakes occur in some places and not in others? Section 2: What tools do scientists use to study earthquakes?

Review Topics
In addition to the earth science topics listed, students may need to review these concepts:

Physical Science
- Waves travel at different speeds through different materials, resulting in refraction. Refraction is the bending of waves as they pass from one material to another, due to a change in their speed.

Reading Strategy
Model the Strategy: Give students the framework for a chart on which they can organize the information on plate boundaries from Chapter 8, Section 2. Make a chart that compares the way the plates move and the effects of movement for the three types of boundaries: convergent, divergent, and transform. After students have filled in the chart, review the different types of plate boundaries with them.

BEYOND THE TEXTBOOK

INTERNET RESOURCES
CLASSZONE.COM

INVESTIGATIONS
- How Are Earthquakes Related to Plate Tectonics?
- Where Was That Earthquake?
- Which Fault Moved in the Northridge Earthquake?

VISUALIZATIONS
- Animations of Love and Raleigh waves in motion
- Video clips of earthquake destruction as it occurs
- Map of faults in the United States
- Animation of P and S waves moving through the Earth's interior

DATA CENTER
EARTH NEWS
LOCAL RESOURCES
CAREERS

Online Lesson Planner

Chapter 10 Earthquakes 213

CHAPTER 10 SECTION 1

FOCUS

Objectives
1. Explain how most earthquakes result from strain that builds up along faults or near plate boundaries.
2. Describe how the energy released in an earthquake travels as waves.

Set a Purpose
Read aloud the titles in blue and red. Rephrase these titles as questions students can answer as they read. Examples include "What causes an earthquake?" and "What is a body wave?"

INSTRUCT

Visual Teaching

Interpret Diagram
Have students locate and describe where the earthquake shown originated. below Earth's surface along a fault

Extend
Ask: What do the circles in the diagram represent? movement of seismic waves Is the focus really a point? No; it can be a large area along a fault but it can be treated as a point for the purpose of determining the epicenter.
Use Transparency 12.

10.1

KEY IDEA
Most earthquakes result from the release of stress that has built up at plate boundaries.

KEY VOCABULARY
- earthquake
- fault
- focus
- epicenter
- body waves
- P waves
- S waves
- surface waves

EARTHQUAKE These rows of lettuce were displaced by an earthquake in California in 1979.

How and Where Earthquakes Occur

More than 3 million earthquakes occur each year, or about one earthquake every ten seconds. Most of these are too small to be noticeable. Each year, however, a number of powerful earthquakes occur. Because such earthquakes are among the most destructive of natural disasters, it is important to understand how and where earthquakes occur in order to prevent the loss of lives and property.

Causes of Earthquakes

An **earthquake** is a shaking of Earth's crust caused by a release of energy. Earthquakes can occur for many reasons. The ground may shake as a result of the eruption of a volcano, the collapse of a cavern, or even the impact of a meteor. The cause of most major earthquakes is the strain that builds up along faults at or near boundaries between lithospheric plates. A **fault** is a break in the lithosphere along which movement has occurred.

Most of the time, friction prevents the plates from moving, so strain builds up, causing the plates to deform, or change shape. Eventually, the strain becomes great enough to overcome the friction, and the plates move suddenly, causing an earthquake. The plates then snap back to the shapes they had before they were deformed, but at new locations relative to each other. This model of an earthquake is called the *elastic-rebound theory*.

The point at which the first movement occurs during an earthquake is called the **focus** of the earthquake. The focus is the point at which rock begins to move or break. It is where the earthquake originates and is usually many kilometers beneath the surface. The point on Earth's surface directly above the focus is the **epicenter** of the earthquake. News reports about earthquakes usually give the location of the epicenters.

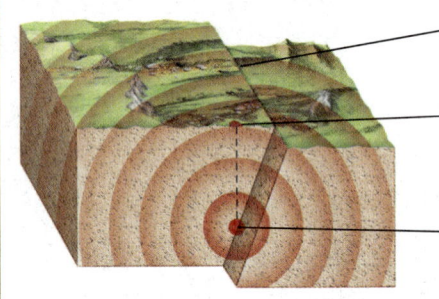

Focus and Epicenter of an Earthquake

FAULT Most earthquakes originate at faults along plate boundaries.

EPICENTER The epicenter is the point on Earth's surface directly above the focus.

FOCUS Energy is released at the focus and travels away from it in all directions.

The depth at which an earthquake originates depends upon the type of plate boundary involved. At divergent boundaries, such as the Mid-Atlantic Ridge, earthquakes tend to occur within 30 kilometers of the surface. Earthquakes also tend to occur at shallow depths along transform boundaries. At subduction boundaries, however, where plates plunge beneath other plates, the focus of an earthquake can be located as far as

---

DIFFERENTIATING INSTRUCTION

Developing English Proficiency Have students work together to research the meanings of non-vocabulary terms in this section that might be difficult to understand. Examples include *friction, deform, snap back, originates, stress, plunge, perpendicular, elliptical,* and *perceptible*. Students should record the meanings in their science notebooks.
Use Spanish Vocabulary and Summaries, pp. 19–20.

700 kilometers beneath the surface. The depth of an earthquake's focus can affect the amount of damage the earthquake causes.

Body Waves

The energy released in an earthquake travels in waves. Waves that travel from the focus of an earthquake through Earth are called **body waves** because they travel through the material of Earth's body. Every earthquake produces two different types of body waves, called P waves and S waves.

The body waves known as compressional waves, primary waves, or **P waves** squeeze and stretch rock materials as they pass through Earth. P waves can travel through any material—solid rock, magma, ocean water, even air.

The body waves called shear waves, secondary waves, or **S waves** cause particles of rock material to move at right angles to the direction in which the waves are traveling. S waves can travel through solid material, but not through liquids or gases.

How Are Earthquakes Related to Plate Tectonics?
Plot earthquakes on a world map. Analyze their location and depths at various plate boundaries.
Keycode: ES1001

Modeling P Waves and S Waves

P WAVE As a P wave travels through rock, the rock particles are (1) compressed and (2) expanded before returning to their (3) original positions.

S WAVE As an S wave travels through rock, the rock particles move at right angles to the direction in which the wave is traveling.

The rate at which P waves and S waves move depends upon the type and density of the material through which they travel. The velocity of these waves is greater through material that is rigid and dense than it is through material that is less rigid and less dense. Through all types of solid material, S waves usually travel at a little more than half the speed of P waves.

Chapter 10 Earthquakes 215

CHAPTER 10 SECTION 1

 Observe animations of earthquake waves.
Keycode: ES1002
Visualizations CD-ROM

ASSESS

1. Earthquakes are most likely to occur at boundaries between lithospheric plates because of stress that builds up there.

2. The focus is where an earthquake originates, the point at which rock begins to move or break. The focus is usually many kilometers below the surface. The epicenter is the point on Earth's surface directly above the focus.

3. Love waves cause particles of material to move from side to side, perpendicular to the waves' direction of travel. Rayleigh waves cause particles of material to move in elliptical patterns; they travel more slowly than Love waves.

4. Both body waves and surface waves are earthquake waves. Body waves travel from the focus of an earthquake through Earth, whereas surface waves travel along Earth's surface. Generally, the closer the focus is to the surface, the more intense the damage and the more concentrated it is in a given area.

5. The P waves will take 10 seconds to reach the recording station and the S waves will arrive about 17.6 seconds after that.

Observe animations of earthquake waves.
Keycode: ES1002

Surface Waves

As their name implies, **surface waves** are earthquake waves that travel along Earth's surface. When P waves and S waves reach Earth's surface, they produce surface waves. The two types of surface waves are Love waves and Rayleigh waves.

Love waves cause particles of material to move from side to side, in a direction perpendicular to the waves' direction of travel. Rayleigh waves travel more slowly than Love waves and cause particles of material to move in elliptical patterns. The Rayleigh wave pattern is similar to the movement of particles in the ripples that appear on the surface of a lake into which a pebble has been tossed.

Even though surface waves travel more slowly than either P waves or S waves, they are often perceptible far from the epicenter of the earthquake and can cause considerable damage.

EARTHQUAKE DAMAGE Surface waves can cause damage far from an earthquake's epicenter, as demonstrated by an earthquake that struck India in 2001. The photograph at the left is from the town of Bhuj, about 42 kilometers from the epicenter. The photograph at the right is from the city of Ahmedabad, about 253 kilometers from the epicenter.

10.1 Section Review

1. Explain where earthquakes are most likely to originate and why they originate in these places.

2. Describe the difference between the focus of an earthquake and the epicenter of an earthquake.

3. Draw and label a diagram illustrating two types of surface waves.

4. **CRITICAL THINKING** Compare and contrast body waves and surface waves. Explain how the depth of an earthquake's focus might determine the extent of the damage it causes.

5. **MATHEMATICS** Suppose an earthquake's P waves travel at an average speed of 6 kilometers per second, and its S waves travel at an average speed of 3.4 kilometers per second. How long will it take the P waves to reach a recording station that is 60 kilometers from the focus? How long after the P waves will the S waves reach the same station?

MONITOR AND RETEACH

If students miss . . .

Question 1 Reteach "Causes of Earthquakes." (p. 214) Ask: What is the cause of most major earthquakes?

Question 2 Reteach from the diagram of the Focus and Epicenter of an Earthquake. (p. 214) Have students describe the features shown in the diagram.

Question 3 Reteach the two types of surface waves. (p. 216) Emphasize how they affect particles of material through which they move.

Question 4 Reteach "Body Waves" (p. 215) and "Surface Waves." (p. 216) Suggest that students list ways in which the waves are alike and different.

Question 5 Provide students with the following equation:
time = speed/distance. Demonstrate how to use the equation.

Locating and Measuring Earthquakes

An instrument called a **seismograph** detects and records waves produced by earthquakes that may have originated hundreds, even thousands, of kilometers away. Scientists use data from seismographs to locate earthquakes' epicenters and to measure their magnitudes. There are more than 10,000 seismograph stations around the world.

Seismographs

Because earthquakes produce different types of wave motions, there are different types of seismographs. Some record side-to-side motions. Others record up-and-down motions. A modern seismograph station usually has three seismographs. One records up-and-down motions, another records side-to-side motions in a north-south direction, and a third records side-to-side motions in an east-west direction.

In one common type of seismograph, a heavy weight is attached to a base anchored in bedrock. The weight is attached in such a way that it stays almost perfectly still, even when the bedrock and the base are being shaken by an earthquake.

A record sheet called a **seismogram** is placed on a drum attached to the base. The drum is turned slowly by a clock. Attached to the heavy weight is a pen, with its point resting on the drum. As long as the bedrock is not moving, the pen makes a straight line on the seismogram. When the bedrock shakes, the drum shakes as well. The pen does not shake because it is attached to the heavy weight. As you might expect, when the pen is stationary and the drum moves, the pen produces a zigzag trace. The distance the pen departs from the center line is related to the amount of energy released in the earthquake. To assess how powerful an earthquake is, scientists must determine the distance of its epicenter from the seismograph station.

10.2

KEY IDEA
A seismograph is used to determine the magnitude of an earthquake and the location of its epicenter.

KEY VOCABULARY
- seismograph
- seismogram
- magnitude

SEISMOGRAM A seismologist at the National Earthquake Information Center studies a seismogram. Note that a *seismogram* refers to the record of an earthquake, while a *seismograph* refers to the instrument that records it.

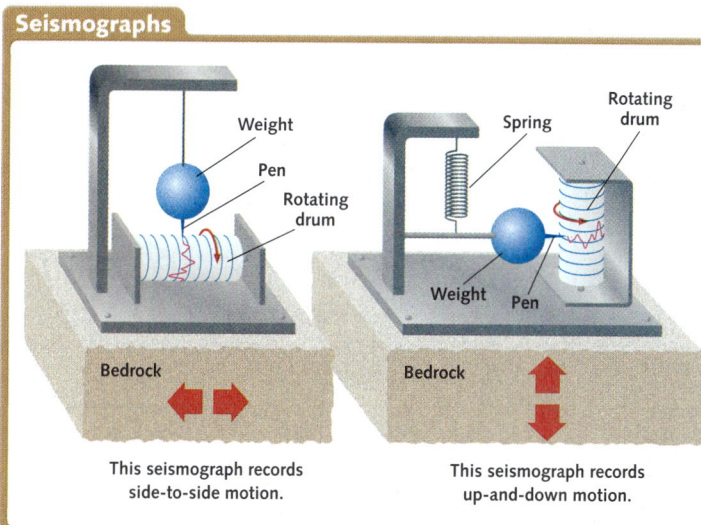

Seismographs

This seismograph records side-to-side motion.

This seismograph records up-and-down motion.

CHAPTER 10 SECTION 2

FOCUS

Objectives
1. Explain how a seismograph is used to record earthquake waves and locate an earthquake's epicenter.
2. Summarize how the magnitude of an earthquake is measured.

Set a Purpose
Have students read this section to find the answer to the question "How are scientists able to locate and measure earthquakes?"

INSTRUCT

Visual Teaching

Interpret Diagrams
Point out how the different orientation of the seismographs' drums allows them to record ground movements in different directions.

Extend
Ask: During an earthquake, how would the trace appear if the pen was not attached to the weight? Why? It would likely be a fairly straight line because it would move with the drum. Ask: If the drums are moving in response to motion in bedrock, what force is acting on the suspended pens? gravity

---

DIFFERENTIATING INSTRUCTION

Challenge Activity Devices used to detect earthquakes date back to the first century A.D. Chinese mathematician, astronomer, and geographer Chang Heng (A.D. 78–139) designed an "earthquake weathercock" that used carefully balanced and arranged balls to detect the motion and direction of even distant earthquakes. Have one group of students research and report upon Chang Heng's invention, presenting a diagram or model of the device. Have other groups of students research other earthquake detectors. These could include an 18th-century seismometer that used a pendulum and pointer balanced over a floating tray of sand and a 19th-century seismoscope that consisted of a bowl of mercury that would overflow into one of eight smaller bowls. Discuss what features these devices have in common. Ask: What do these devices detect or measure? usually direction, possibly relative strength

CHAPTER 10 SECTION 2

Visual Teaching

Interpret Diagram
Point out that this graph labels the horizontal axis at the top. Ask: What does the separation between the red curve and the blue curve represent? *the difference in arrival times of P and S waves* Be sure students understand how to use the horizontal axis of the travel-time graph to determine the arrival time difference for the two stations shown on the graph. For example, point to the light blue vertical line between 6 and 8. Explain that you understand that this line represents travel time at 7 minutes. Now trace the path of the S-wave curve. Tell the students that you want to know how far it has traveled in 7 minutes. Follow the curve until it hits the 7-minute line. Then, trace your finger from that point, along the horizontal line to the y-axis to find the distance. Demonstrate how to utilize the graph grid to help estimate time in minutes and seconds.
Use Transparency 12.

Interpreting a Seismogram

Because P waves travel faster than S waves, the P waves produced by an earthquake always arrive at a seismograph station before the S waves. The first major zigzag on the seismogram marks the arrival of the P waves at the station. As the seismogram below shows, the slower S waves arrive next, producing a different pattern. The seismogram indicates the arrival time of each type of wave.

As the P waves and S waves travel through the ground, the slower S waves lag progressively farther behind. Thus, the farther a seismograph station is from the epicenter, the greater the difference in the arrival times of the P and S waves. Suppose, for example, that a seismograph station in Berkeley, California, records an arrival-time difference of 1 minute 40 seconds between the P waves and the S waves from a given earthquake. (See the seismogram below.) For the same earthquake, a station in St. Louis, Missouri, records a difference of 2 minutes 45 seconds. From this information, you can determine that the St. Louis station is farther from the earthquake's epicenter.

A travel-time graph, such as the one below, shows the relationship between P-wave and S-wave arrival times and the distance from an earthquake's epicenter. Given the difference in arrival times between the first P waves and the first S waves, you can determine the distance to the epicenter. The graph shows that the distance from the Berkeley station to the epicenter is 1212 kilometers; from the St. Louis station, the distance is 2003 kilometers.

Interpreting a Travel-Time Graph

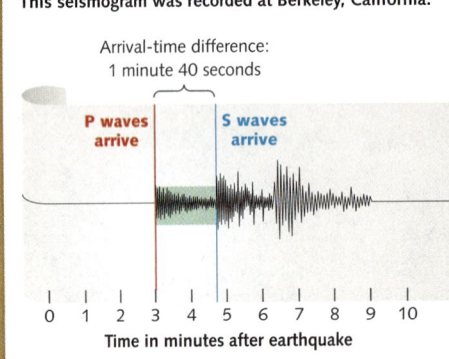
This seismogram was recorded at Berkeley, California.

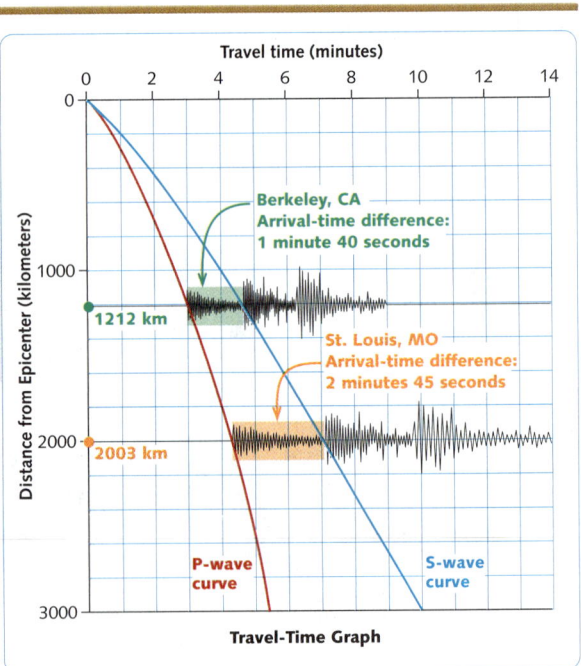

The red curve on the graph at the right shows the times at which an earthquake's first **P wave** reaches locations at various distances from the epicenter. The blue curve gives the same information about the first **S wave**. Two seismograms, each showing data from a different seismograph, are superimposed on the graph.

218   Unit 3 Dynamic Earth

DIFFERENTIATING INSTRUCTION

Support for Visually Impaired Students The numbers and details on graphs may appear too small for some students to view. Provide students with an enlarged, color photocopy of the seismogram and travel-time graph on page 218.

Locating the Epicenter

Although a seismogram tells scientists the distance between the seismograph station and the earthquake's epicenter, it does not actually give scientists enough information to locate the epicenter. For example, the information from the St. Louis station indicates that an earthquake is located somewhere within a 2003 kilometers radius of St. Louis. (See map below. That makes a very large circle of territory, with St. Louis at its center.) Because the direction from any single station to the epicenter is unknown, scientists need to know the distances from at least three different stations in order to plot an epicenter's location.

Think again about the earthquake with an epicenter 2003 kilometers from St. Louis and 1212 kilometers from Berkeley. Suppose a third station in Seattle, Washington, finds that the distance to the same earthquake's epicenter is only 770 kilometers. To locate the epicenter, one can draw three circles on a map. The center of the first circle is St. Louis; the radius of that circle is 2003 kilometers. Berkeley is at the center of a second circle, with a radius of 1212 kilometers. Seattle is at the center of the third circle, with a radius of 770 kilometers. The point where all three circles meet is the location of the earthquake's epicenter.

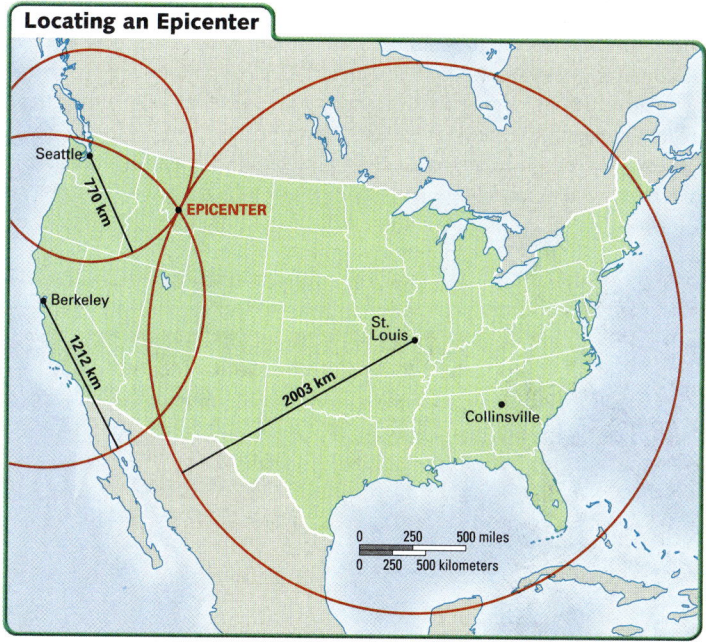

Locating an Epicenter

In the United States, some 2500 seismograph stations work together to collect earthquake data. Computer technology enables scientists at these stations to analyze and share data about recent earthquake activity quickly. In some areas, such as central California, it takes scientists only about a minute to determine the location of an earthquake and to relay that information to the public.

10-Minute Mini LAB

Interpreting a Travel-Time Graph

Procedure

1. Use the travel-time graph on page 218. Suppose a station in Collinsville, Georgia, records data from the same earthquake. The difference between the P-wave and S-wave arrival times at this station is 3 minutes 45 seconds. Estimate the distance of the Collinsville station from the epicenter.

2. Use the map below to check your answer.

Analysis

Suppose a seismograph station in Chapel Hill, North Carolina, records data from the same earthquake. Chapel Hill is 3006 kilometers from the epicenter. Predict the difference in P and S wave arrival times at this station. Explain how you determined these arrival times.

Where Was That Earthquake?
Read seismograms and use the information to locate the epicenter of an earthquake.
Keycode: ES1003

CHAPTER 10 SECTION 2

Where Was That Earthquake?
Keycode: ES1003

Use a computer and projector to introduce this investigation in class and to answer students' questions.

Use Internet Investigations Guide, pp. T39, 39.

MINILAB
Interpreting a Travel-Time Graph

Management
Be sure students understand how to interpret the travel-time graph. If necessary, review how to use the graph with students who are struggling.

Outcome
The distance of the Collinsville station from the epicenter is approximately 2600 km.

Analysis Answers
Responses vary. Since the graph shows a difference in arrival times of about 4 minutes 30 seconds for a distance of 3000 km, students may infer that for a distance of 3006 km, the difference in arrival times would be approximately equal to or slightly greater than 4 minutes 30 seconds.

For proper behavior in a laboratory setting, refer students to Safety in the Earth Science Laboratory on pages xxii–xxiii.

DIFFERENTIATING INSTRUCTION

Reading Support Show students how to check their understanding by asking themselves questions. After reading the first two paragraphs on page 219, they might ask themselves: Why can't the epicenter of an earthquake be located using the distance to the epicenter from only one seismograph station? If you use only one station's data to draw a circle, the epicenter could occur anywhere on that circle. Why can't the epicenter of an earthquake be located using the distance to the epicenter from only two seismograph stations? If you use only the data from two stations to draw two circles, the epicenter could occur at either of the two points where the circles intersect.

Use Reading Study Guide, p. 33.

CHAPTER 10 SECTION 2

More about...

The Richter Scale
In 1935, Charles Richter of the California Institute of Technology developed a system for determining the magnitude of earthquake motions recorded by seismic instruments. The modern Richter scale is a refined version of that early system now used worldwide. On this scale, earthquakes with a magnitude of less than 3.5 are not generally felt by humans; those of magnitude 5.5–6.0 cause slight damage to structures; those of magnitude 7.0–7.9 inflict serious damage; and those of magnitude 8.0 or greater produce total destruction of nearby communities. The strength of an earthquake increases exponentially with each whole number increase on the Richter scale. For example, an earthquake of magnitude 5 is 31 times more powerful than an earthquake of magnitude 4, and 31×31 times more powerful than an earthquake of magnitude 3.

Measuring an Earthquake's Magnitude

In addition to using seismograms to locate the epicenters of earthquakes, scientists use them to collect other data. For example, seismograms can be used to determine the strength, or **magnitude** (MAG-nih-TOOD), of an earthquake. Magnitude is a measure of the amount of energy released in an earthquake.

One widely used scale of earthquake magnitude was developed by Charles F. Richter (RIHK-tuhr) in 1935. Each increase of one whole number in Richter magnitude represents a 31-fold increase in energy. For example, a magnitude-6 earthquake is about 31 times more powerful than a magnitude-5 earthquake. A magnitude-7 earthquake is about 31 times more powerful than a magnitude-6 earthquake, or about 31×31 times more powerful than a magnitude-5 earthquake.

In the years since its introduction, the Richter scale has been shown to have limitations. For example, it does not indicate accurately the amounts of energy released in very large earthquakes. Another measure that scientists now use to describe earthquakes' power is called moment magnitude. Whereas Richter magnitude measures the intensity of ground movements, moment magnitude indicates the energy released at an earthquake's source.

Although not as easy to measure as Richter magnitude, moment magnitude more accurately indicates the total energy involved in an earthquake. For example, the 1906 San Francisco earthquake had an estimated Richter magnitude of 8.3, and the 1964 Alaskan earthquake had a Richter magnitude of 8.5. The Alaskan earthquake, however, released at least twice as much energy as the San Francisco earthquake because it involved greater movements along a much larger fault plane—moment magnitude reflects this difference in energy. The moment magnitude of San Francisco's 1906 earthquake was 7.9; Alaska's 1964 earthquake, 9.2!

ALASKA, 1964 The earthquake that struck along the coast of Prince William Sound in 1964 caused massive displacements along faults.

Major 20th-Century Earthquakes

Location	Date	Moment Magnitude
Chile	1960	9.5
Alaska	1964	9.2
Russia	1952	9.0
Ecuador	1906	8.8
Alaska	1957	8.8
Kuril Islands	1958	8.7
Alaska	1965	8.7
India	1950	8.6
Chile	1922	8.5
Indonesia	1938	8.5

DIFFERENTIATING INSTRUCTION

Developing English Proficiency Be sure students recognize the difference between a siesmograph and a seismogram. Both share a common root, the Greek word *seismos*, meaning "vibration." Have students look at the ending of each word. The ending *-graph* comes from the Greek verb *graphein*, meaning "to write." The ending *-gram* comes from the Greek noun *gramma*, meaning "letter" or "writing." To distinguish between the two, have students think of the seismograph as the instrument that does the writing and the seismogram as the physical record of that writing. Refer them to the photograph and diagram on page 217. Have them consider two similar words: telegraph and telegram.

CAREER

Earthquake Engineer

When an earthquake strikes, earthquake engineers find out how well they have done their jobs. Together with teams of architects and other engineers, earthquake engineers design buildings that can withstand earthquakes. They use 3-D computer modeling to evaluate how a building and the ground beneath it will act during a quake. Based on their findings they recommend suitable building materials and construction techniques. Earthquake engineers also evaluate older structures and find ways to reinforce them so that the structures will meet current building standards in earthquake-prone areas.

Most earthquake engineers have advanced degrees in civil or structural engineering. Core studies include math, physics, geology, and chemistry. On the job, a disposition for teamwork and a knack for designing creative, cost-effective solutions are important. Earthquake engineers often divide their time between analyzing data at the office and solving problems at building sites. The job is challenging but rewarding, especially when a building stands up to the ultimate test—an earthquake. ■

SCALE MODEL Earthquake engineers use a scale model to test the effects of earthquake tremors on a bracing system.

Find out more about a career in structural engineering.
Keycode: ES1004

10.2 Section Review

1. Explain how a seismograph records earthquake data.
2. How is the fact that P waves travel more quickly than S waves used to determine the distance of an earthquake's epicenter from a seismograph station?
3. Explain what is meant by *magnitude* with regard to an earthquake, and name two different scales that are used to measure an earthquake's magnitude.
4. **CRITICAL THINKING** Suppose one earthquake has a Richter magnitude of 3. A second earthquake has a Richter magnitude of 7. How many times more powerful is the magnitude-7 earthquake than the magnitude-3 earthquake? Explain how you got your answer.
5. **PAIRED ACTIVITY** Work with a partner to prepare a presentation on how to locate the epicenter of an earthquake.

CHAPTER 10 SECTION 3

FOCUS

Objectives
1. Summarize earthquake hazards and the damage they can cause.
2. Explain how safe building practices and earthquake prediction can prevent earthquake damage.

Set a Purpose
Have students read this section to relate earthquake hazards and risks to the area where they live. Have them answer the question "How would you describe the earthquake risk in this area?"

INSTRUCT

More about...

The Great Alaskan Earthquake One of the most violent earthquakes to jar North America occurred on March 27, 1964 in Prince William Sound, Alaska. This great earthquake and the tsunami that ensued took 125 lives. Anchorage, located about 120 kilometers northwest of the epicenter, sustained the most severe property damage, much of which landslides caused. In the downtown area alone, about 30 blocks of buildings were damaged or destroyed. A tsunami having a maximum wave height of 67 meters devastated many towns along the Gulf of Alaska.

10.3

KEY IDEAS
The amount of damage an earthquake causes depends on its magnitude and where it occurs. Safe building practices can limit loss of life and damage to property.

KEY TERMS
- liquefaction
- aftershock
- tsunami
- seismic gap

VOCABULARY STRATEGY
The words *liquid*, *liquefy*, and *liquefaction* are closely related. *Liquefaction* comes from a Latin word that means "to make liquid."

Earthquake Hazards

In January 1995, the city of Kobe, Japan, was rocked by an earthquake with a moment magnitude of 6.9. More than 5000 people were killed, and over 180,000 buildings were badly damaged or destroyed. The Kobe earthquake highlights the vulnerability of people living near Earth's major plate boundaries. By studying the causes and effects of such devastating earthquakes, scientists hope to minimize loss of life in the future.

Damage from Earthquakes

In an earthquake, ground shaking and foundation failure may cause buildings to collapse. Aftershocks, fire, and tsunamis may cause additional damage.

Ground Shaking and Foundation Failure

Ground shaking is produced by the waves set in motion by an earthquake's sudden release of energy. The ground vibrates in much the same way that a bell vibrates when struck. Some of these vibrations are up-and-down, but the largest vibrations at Earth's surface are side-to-side motions. Most buildings can withstand fairly violent up-and-down shaking; however, few buildings can survive side-to-side shaking, and as a result many buildings collapse.

As a result of severe ground shaking, soils under buildings may settle or even liquefy. **Liquefaction** occurs when loose soil temporarily takes on some of the properties of a liquid. A building located on soil that settles is no longer safely supported and may collapse. During the 1906 San Francisco earthquake, buildings located on solid rock experienced little damage, while buildings located on bog muds or soft landfill suffered severe damage or collapsed because of foundation failure.

Other earthquakes have caused foundation failure. San Francisco's Marina District sits on landfill that was used to extend the city into San Francisco Bay. In the 1989 Loma Prieta earthquake, this landfill liquefied,

EARTHQUAKE DAMAGE This building in San Francisco's Marina District collapsed during the 1989 Loma Prieta earthquake.

DIFFERENTIATING INSTRUCTION

Reading Support As students read the section, have them look for cause-and-effect relationships. Explain that one event or circumstance is often brought about by a second event or circumstance. The first event is the cause, and the second is the effect. To help them identify cause-and-effect relationships, students should look for phrases such as *because of*, *causes*, *as a result of*, and *due to*. Provide an example by asking: What causes liquefaction? **severe ground shaking** Be sure students understand that, in this example, severe ground shaking is the cause and liquefaction is the effect.
Use Reading Study Guide, p. 34.

---

222 Unit 3 Dynamic Earth

shaking buildings off their foundations. In Mexico City, buildings located on unstable lake sediments collapsed during a 1985 earthquake. Liquefaction of sediments also caused great destruction in the 1964 Alaskan earthquake and in a 1999 earthquake that devastated many areas in Turkey.

Aftershocks and Fire

Even after earthquake waves subside, danger remains. Aftershocks and fires can cause damage to buildings and other structures and can harm people.

A large earthquake may be followed by a series of smaller ones originating close to the focus of the large earthquake. These smaller earthquakes are called **aftershocks.** After a large earthquake the number of aftershocks can be as great as 1000 per day, though the frequency usually diminishes quickly over time. Aftershocks can cause damage to buildings and other structures, especially ones already weakened by ground shaking.

Whenever a large earthquake affects a heavily populated area, there is a danger of fire. In January 1994, an earthquake with a moment magnitude of 6.7 struck Northridge, California, just outside Los Angeles. One of the most damaging effects of this earthquake was the rupturing of gas lines by ground movement. When gas from these lines ignited, towering flames shot up through cracks in the roadways kindling many large fires.

Fires that broke out after the 1906 San Francisco earthquake raged uncontrolled, largely because the city was not adequately prepared to handle such a calamity. These fires destroyed about 3000 buildings and burned about 2800 acres of the city.

FIRE Many fires broke out after the Northridge, California earthquake in 1994.

Tsunamis

Underwater earthquakes and landslides sometimes cause huge ocean waves, called **tsunamis** (tsu-NAH-meez). A tsunami can travel very quickly across large expanses of water. Its speed depends on the depth of the water. The average depth of the ocean is about 4500 meters. In water that deep, a tsunami's speed is about 750 kilometers per hour. When a tsunami reaches shallower water near a shoreline, however, it slows down and increases dramatically in height.

In 1946, a large earthquake in the Aleutian Islands of Alaska produced one of the largest tsunamis ever recorded. Almost five hours after the earthquake occurred, the tsunami reached the Hawaiian Islands—about 4000 kilometers from the earthquake's epicenter. When it crashed into shore, it caused tremendous damage and destruction, killing 159 people.

In response to this devastating tsunami, officials created a tsunami warning system for regions in the Pacific Ocean. Scientists now carefully monitor seismic activity to identify earthquakes large enough to cause tsunamis. They try to detect tsunamis by monitoring wave activity at various stations throughout the Pacific.

Observe video taken during an earthquake.
Keycode: ES1005

CHAPTER 10 SECTION 3

Observe video taken during an earthquake.
Keycode: ES1005

Visualizations CD-ROM

More about...

Tsunamis
The U.S. West Coast/Alaska Tsunami Warning Center (WC/ATWC) was established in Palmer, Alaska, in 1967 after the great Alaskan earthquake that occurred in Prince William Sound on March 27, 1964. This earthquake compelled state and federal officials to find a way of providing timely and effective tsunami warnings and information for the coastal areas of Alaska. Since 1967, the WC/ATWC's area of responsibility has enlarged to include the issuing of tsunami warnings for potential tsunami-producing earthquakes that could affect the coasts of California, Oregon, Washington, British Columbia, and Alaska.

Chapter 10 Earthquakes 223

DIFFERENTIATING INSTRUCTION

Support for Hearing-Impaired Students As you introduce key terms relating to earthquake hazards, write out the terms and their definitions for students. Instead of writing on the board with your back to the class, write on a transparency using an overhead projector so that you can maintain eye contact with students while you write.

CHAPTER 10 SECTION 3

More about...

Predicting Earthquakes
Earthquakes have been accurately predicted over the short term in a few instances. An earthquake in northeast China was predicted in February 1975, only hours before it occurred. Rather large foreshocks preceded the earthquake, aiding the prediction that is believed to have saved tens of thousands of lives. The 1966 earthquake in Tashkent, Uzbekistan, was predicted by monitoring the level of radon gas in wells. Radon, which is produced by radioactive decay of radium, is normally locked within rocks. During the buildup of stress in rocks, however, newly formed cracks in the rocks facilitate the release of this gas.

DISCUSSION

Ask students to infer why they should move into the open if they are outdoors during an earthquake. *to avoid trees or heavy structures falling; to avoid current from broken electrical wires*

SAFETY TIPS
EARTHQUAKES

Before an earthquake, check your house for potential hazards, such as large, heavy objects that could fall and injure people. Keep disaster supplies readily available.

During an earthquake, take cover under a piece of heavy furniture if you are inside. If you are outside, move into the open, away from buildings, lights, and electrical wires.

After an earthquake, stay out of damaged buildings and away from downed electrical wires. If you are inside and hear a hissing noise or smell gas, get out of the building immediately, then call for help.

Learn more about earthquake safety
Keycode: ES1006

Preventing Earthquake Damage

Most large cities in earthquake-prone areas have building codes intended to prevent structural collapse during an earthquake. These codes are usually developed by scientists and engineers who study how buildings and other structures respond to earthquakes.

For example, in 1981, Japan adopted a revised building code for concrete-frame buildings. After the 1995 Kobe earthquake, engineers discovered that most of the concrete-frame buildings constructed after the new code was adopted had only light damage. Older concrete buildings had suffered far greater damage.

KOBE EARTHQUAKE, 1995 The photograph on the left shows the undamaged roof of a house. The photograph to the right, taken in the same neighborhood, shows that houses with heavy tile roofs collapsed.

Engineers may try to determine why some structures successfully withstand an earthquake while others collapse. Monitoring equipment installed in and around structures can provide precise records of how different parts of the structures have responded to ground shaking. Over time, the knowledge gained by analyzing such records can help engineers develop better designs for buildings, bridges, and other structures.

Earthquake Risk

Where in the United States is the risk of earthquakes the greatest? Almost everyone thinks regions lying near plate boundaries, such as California or Alaska, are at greatest risk. But many other areas of the United States are at risk as well. Three of the most powerful earthquakes in U.S. history did not occur anywhere near a plate boundary. They originated near the Mississippi River town of New Madrid, Missouri.

The three earthquakes that struck New Madrid during the winter of 1811 and 1812 were felt throughout most of the northeastern United States. The earthquake activity at New Madrid did not end in 1812. Today the New Madrid region is the most seismically active region east of the Rocky Mountains in the United States.

If New Madrid is not on a plate boundary, why is there so much seismic activity there? The activity is caused by many faults buried deep beneath sediments deposited by the Mississippi, Missouri, and Ohio rivers and beneath the thick layers of sedimentary rock that underlie these sediments.

DIFFERENTIATING INSTRUCTION

Challenge Activity Have students brainstorm a list of disaster supplies they should have available in case of an earthquake. *Important items include water, canned foods, a radio, batteries, a flashlight, clothing, and a first-aid kit.*

The New Madrid faults are significant because of the size of the area that could be affected by a major earthquake there and because that area is relatively unprepared for such a large earthquake.

The map below shows that some regions in the United States have a much higher risk of experiencing a damaging earthquake than others.

Examine a map showing earthquake risks.
Keycode: ES1007

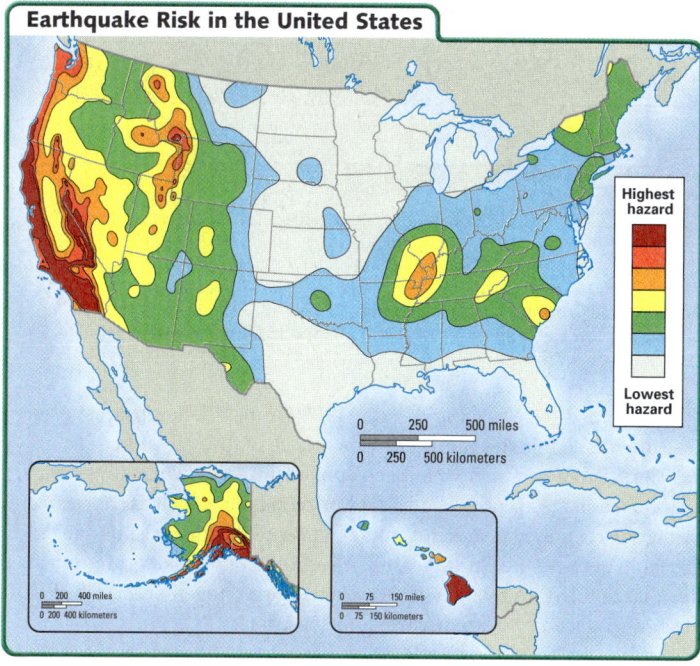

Examine a map showing earthquake risks.
Keycode: ES1007

Visualizations CD-ROM

VISUAL TEACHING

Discussion
Ask: What color is used to show areas of highest earthquake risk? **red** lowest earthquake risk? **white** Where is the risk of an earthquake higher, in Florida or Maine? **Maine**

Extend
Have students compare the earthquake risk in northern and southern Alaska and offer a possible explanation for any difference. **Earthquake risk is higher in southern Alaska, which is closer to a plate boundary or fault.** Ask: Why does the earthquake risk in Hawaii decrease from southeast to northwest? **The southeastern portion of Hawaii, including the Big Island, currently lies above a hot spot, which results in volcanic activity and the accompanying seismic activity. As the Pacific plate has moved northwest, the other islands have moved farther from the volcanic and seismic activity.**

Predicting Earthquakes

To be most effective, an earthquake prediction must correctly forecast three things—where an earthquake will occur, when it will occur, and what its magnitude will be.

Many prediction efforts are based on an assumption that earthquakes are periodic events. By studying past earthquake activity, scientists may be able to predict the probability that an earthquake will affect a particular region over the next few years or even decades. Such long-term predictions are useful. They can, for example, help city officials assess the costs and benefits of adopting stricter building codes. Officials in an area where there is a high probability of an earthquake occurring in the next decade might want to make sure that appropriate emergency services are in place.

Geologists use data from past earthquakes to evaluate activity along faults. By plotting earthquake foci along a fault, they can identify areas where the fault has not moved over a period. Such areas are called **seismic gaps.** Seismic gaps may be places where stress is building up. Scientists may hypothesize that these locations will be sites of earthquake activity in the future.

The word *focus* comes from the Latin word for "hearth." In many Latin-derived words, including *focus*, *-us* indicates the singular form, and *-i* indicates the plural form, as in *foci*.

DIFFERENTIATING INSTRUCTION

Challenge Activity Challenge students to design and create a display showing earthquake risk in the United States. Suggest they build their design around an enlarged version of the map on page 225. They can locate their home town on the map and compare its relative risk to that of major cities or locations they may be interested in, such as a favorite vacation spot or the home of friends or relatives.

CHAPTER 10 SECTION 3

Visual Teaching

Interpret Diagram

Be sure students understand that the black dots on the diagram represent earthquake foci and a scarcity of foci represents a seismic gap. Then have them identify any areas besides the labeled seismic gap where foci are scarce. **between San Francisco and Portola Valley, and outside of Parkfield**

ASSESS

1. Ground shaking can cause buildings to shake, which can lead to collapse. Ground shaking also causes liquefaction—loose soil beneath a building liquefies; this causes foundation failure and leads to building collapse.

2. By studying how buildings and other structures respond to earthquakes, engineers can develop building codes to help prevent structural collapse during an earthquake.

3. A seismic gap is an area along a fault where the fault has not moved over a period of time. Because it has not moved, stress is building up.

4. Areas of loose soil would not be suitable for development, since loose soil is likely to liquefy during an earthquake.

5. Long-term predictions help people make decisions about building codes and emergency services. Short-term predictions make it possible to warn and evacuate people before an earthquake.

The Loma Prieta Seismic Gap

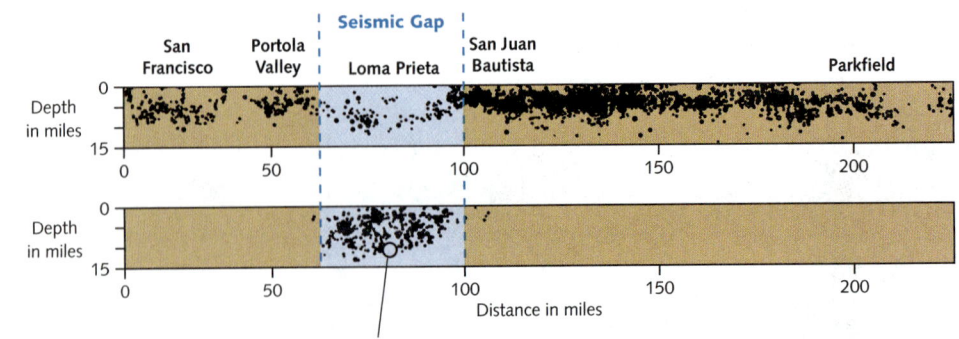

The top band in the diagram shows the depths of earthquake foci along a cross section of the San Andreas fault from January 1969 through July 1989. Notice that there are fewer data points in the seismic gap—the blue area.

The lower band in the diagram shows how the October 17, 1989, Loma Prieta earthquake (shown by the open circle) and its aftershocks filled in a seismic gap.

Some scientists have suggested that the 1989 Loma Prieta earthquake may have originated in a seismic gap on one branch of the San Andreas Fault.

The seismic-gap theory has been applied in Turkey, where numerous earthquakes have occurred along the North Anatolian Fault. In 1999, an earthquake centered near Izmit, Turkey, appeared to fill a seismic gap on this fault. Because the North Anatolian and San Andreas faults are very similar, scientists from both countries have worked together to understand the hazards faced by people living along these faults.

10.3 Section Review

1. Describe how ground shaking and liquefaction can cause buildings to collapse.

2. How can studying the effects of earthquakes help engineers improve the safety of buildings and other structures?

3. Describe a seismic gap and explain why it may be a site of future earthquake activity.

4. **CRITICAL THINKING** Local planning organizations often decide whether residential developments can be built within their cities or towns. How would the type of soil in an area near a fault affect the area's suitability for development of high-rise residences?

5. **SOCIAL STUDIES** Most earthquake predictions are long-term. They may state the probability that an earthquake will happen in the next few years or the next few decades. Work with others to develop a list of the benefits of long-term earthquake predictions. Explain why it would be valuable to find ways to improve short-term predictions.

MONITOR AND RETEACH

If students miss . . .

Question 1 Reteach "Ground Shaking and Foundation Failure." (p. 222) Have students describe ground shaking and liquefaction, and their results.

Question 2 Reteach "Preventing Earthquake Damage." (p. 224) Ask: What are building codes? How are they developed? What is their intention?

Question 3 Reteach "Predicting Earthquakes." (p. 225) Be sure that students understand that stress builds up when motion does not occur along a fault.

Question 4 Reteach "Ground Shaking and Foundation Failure." (p. 222) Emphasize how solid rock and loose soil are affected differently by ground shaking.

Question 5 Encourage small groups of students to brainstorm lists of benefits for both long- and short-term predictions. Suggest they spark ideas through library or Internet research.

SCIENCE & Technology

Earthquakes: Technology to the Rescue

On October 17, 1989, millions of TV viewers waited eagerly for a World Series game to begin in San Francisco. Suddenly, the announcer said, "We're having an earth . . . ," and the picture and sound from the ballpark were cut off. Few people had trouble guessing how the announcer had finished his sentence.

What are some of the emergency situations people face after an earthquake? How can technology help communities respond quickly to these emergencies?

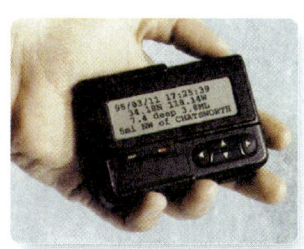

PAGERS Soon after an earthquake occurs, scientists and others receive critical data on pagers.

Earthquakes as strong as the 1989 Loma Prieta quake can cripple a city. Buildings crumble, trapping people inside. Phone and electric lines are downed. Natural gas escaping from broken lines may ignite and cause fires.

One technology that helps people respond quickly to these emergencies is the seismographic network. During an earthquake, remote seismographs send information to a central computer, which calculates the earthquake's epicenter and magnitude. Within minutes, scientists and emergency managers get this information on pagers. The scientists feed the information into a computer that relates it to local geological conditions. The computer produces a color-coded map showing where shaking is likely to have been strongest. (A map created after the 1994 Northridge earthquake is shown below.) Emergency managers use the map to determine where building damage and human injuries may be the most serious. Rescue crews are then sent to those areas.

Emergency managers also rely on simple radio technology to protect rescue workers from dangerous aftershocks. Because radio waves travel more quickly than seismic waves, radios can be used to give rescue workers advance warning of potential shaking. The benefits of such an emergency warning system were apparent after the Loma Prieta earthquake. The warning came just in time when workers clearing wreckage from a collapsed freeway were told that they had 25 seconds to move to safety before an aftershock would rumble through the area. ■

Extension

RESEARCH ACTIVITY
Research the supplies that should be included in an earthquake disaster kit. Create an emergency supply kit.

Learn more about the impact of earthquakes in populated areas.
Keycode: ES1008

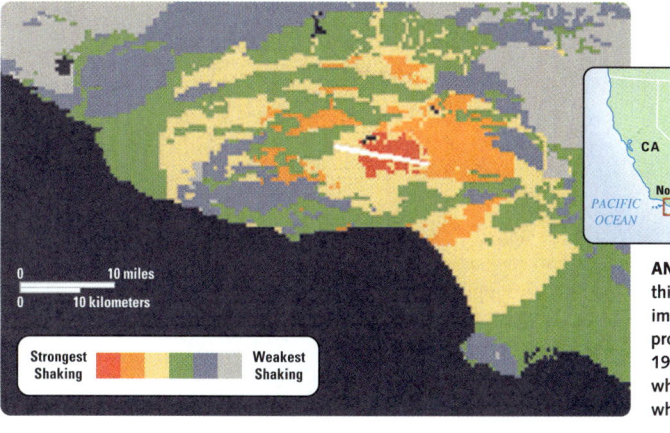

ANALYZING Emergency officials used this computer-generated map to make important decisions about how to provide emergency relief after the 1994 Northridge earthquake. The white line represents the fault along which the earthquake occurred.

CHAPTER 10 SECTION 3

Learn more about the impact of earthquakes in populated areas.
Keycode: ES1008

Science and Technology

Critical Thinking
The Loma Prieta quake occurred in an area that is renowned for earthquake activity. Discuss the choices people make about where to live. Ask: Why would people choose to live in areas, such as along the San Andreas Fault, where earthquake risk is high? *family ties, job opportunities, pleasant climate, and so on* How much weight would you place on natural disaster risk if you were choosing a new place to live? Are any regions in the U.S. safe from natural disasters? *Every region has some risks from hurricanes, tornadoes, volcanoes, blizzards, floods, droughts, forest fires, lightning strikes, and so on.*

Research Activity
A three-day survival pack can be assembled in a plastic trash barrel. Its contents should include: bedding, can opener, disposable dishes and utensils, clothing and toiletries, shovel, bucket, eyedropper, chlorine bleach, food, water, water purification tablets, money, flashlight, first aid kit, radio, batteries, prescription medicines.

CHAPTER 10 SECTION 4

FOCUS

Objective
Describe how data from seismic waves are used to infer the structure of Earth's interior.

Set a Purpose
Have students read this section to find the answer to the question "How do seismic waves help scientists learn about Earth's interior?"

INSTRUCT

Visual Teaching

Interpret Diagram
Make sure students understand what happens to the P and S waves as they move through different parts of Earth. Ask: What happens to P waves when they encounter the outer core? Why? **They bend because the material's density changes.** What happens to S waves when they encounter the outer core? **They stop because they cannot travel through liquids.** What happens to P waves when they encounter the inner core and then the outer core again? Why? **They bend because the material's density changes.** Point out the shadow zones on the diagram. Ask: Identify the continent in part of the shadow zone in the lower left. Which one is it? **South America**

10.4

KEY IDEA
Scientists use data from seismic waves to learn about the structure of Earth's interior.

Studying Earth's Interior

By studying earthquakes, scientists have learned about the structure of Earth's interior. Seismologists have observed that the velocities of both P waves and S waves increase when the waves travel through more dense material, and decrease when the waves travel through material that is less dense. In addition, evidence shows that S waves cannot travel through liquids. Knowledge about P-wave and S-wave velocities through materials of different densities has made it possible for geologists to infer the depths and characteristics of Earth's layers.

For example, scientists observe a sharp change in velocities at a depth of 2900 kilometers, at the core-mantle boundary. P waves are greatly slowed at this boundary, and S waves are stopped. Because S waves cannot travel through liquids, scientists conclude that the material directly below 2900 kilometers, the outer core, must be liquid. P-wave velocity increases again at a depth of about 5200 kilometers. This increase in speed suggests that the inner core, like the mantle, is solid.

The Shadow Zone

Even though an earthquake's waves travel in all directions from its focus, not all seismograph stations can receive information from the earthquake. Seismograph stations in what is called the earthquake's shadow zone cannot detect P waves or S waves from the earthquake. The shadow zone is a wide belt around the side of Earth opposite the focus of the earthquake.

Why does the shadow zone exist? The diagram below shows that P waves passing through the mantle are refracted (bent) in smooth arcs back to the surface. When P waves travel deep enough to enter the outer core, however, they are refracted again as they enter the outer core and yet again when they re-enter the mantle. As a result, seismograph stations that are between the earthquake's epicenter and the shadow zone receive both

VISUALIZATIONS
CLASSZONE.COM
Examine P and S waves moving through Earth's interior.
Keycode: ES1009

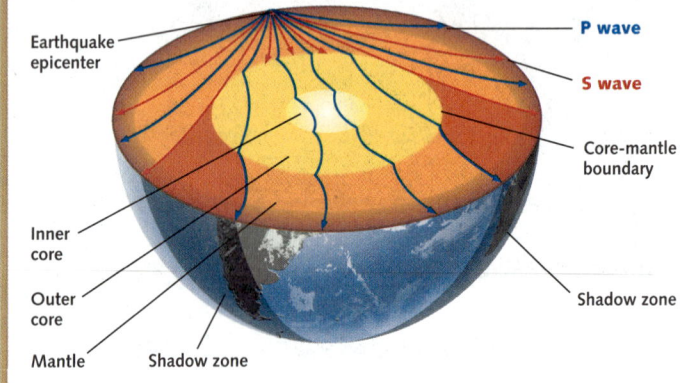

P and S Waves Inside Earth
Because P waves are refracted and S waves cannot travel through the liquid outer core, seismograph stations in the shadow zone do not receive any waves from the earthquake.

228 Unit 3 Dynamic Earth

MONITOR AND RETEACH

Reading Support Preview with students the headings within the section and demonstrate how they can use headings to predict section content. Ask what a *shadow* is **dark area where light is blocked** and what a *zone* is. **a region or an area** Then ask them to predict what the *shadow zone* might be in relation to Earth's interior. **a dark area in Earth or an area where something is blocked** Then ask what a *transition* is **a change** and have them predict what the *transition zone* might be. **an area where Earth's interior changes**

Use Reading Study Guide, p. 35.

P waves and S waves. Stations within the shadow zone receive neither P waves nor S waves. This is because P waves have been refracted away and S waves cannot pass through the liquid outer core. Stations that are beyond the shadow zone on the opposite side of Earth receive only P waves because the liquid outer core stops the S waves.

The Moho

An abrupt change in the velocities of P waves and S waves occurs at the boundary between the crust and the mantle. In 1909, Croatian seismologist Andrija Mohorovičić (ahn-DREE-yah maw-HAWR-aw-VEE- cheech) discovered this fact while studying seismograms of minor earthquakes. Several seismograms showed two distinct groups of P waves and S waves. One of the groups had traveled at an average velocity of 6.75 kilometers per second, but the other had sped up to 8 kilometers per second. He reasoned that the second group had gone through denser material below the crust. He calculated the depth of the denser material to be about 50 kilometers. The boundary he discovered is named the *Mohorovičić discontinuity*, or *Moho* for short. This boundary is where the dense rock of the mantle meets the less dense rock of the crust. Additional study has revealed that the Moho is located about 32 kilometers under the continents and between 5 and 10 kilometers under the oceans.

Scientific Thinking

COMMUNICATE

In 1691, astronomer Sir Edmund Halley proposed that Earth was hollow, with another sphere inside it that was illuminated and inhabited. Authors such as Jules Verne and Edgar Rice Burroughs later wrote fiction that was inspired by Halley's idea. Draw a diagram or write a description to explain how data from seismic waves disproves Halley's hypothesis.

The Transition Zone

Abrupt changes in P-wave and S-wave velocities occur when the waves pass across boundaries between rocks of different compositions. Yet experiments have shown that between depths of 400 and 670 kilometers, seismic-wave velocities increase more quickly than expected. These depths mark a region in the middle of the mantle called the transition zone. The transition zone separates the upper mantle from the denser lower mantle.

If the mantle is made up of the same kind of material throughout, what could cause the P-wave and S-wave velocities to change in the middle of the mantle? The answer to this question has to do with the fact that the material deeper within the mantle is under greater pressure from the overlying material. This increase in pressure compresses the material in the lower mantle. As a result, the material is more dense, and P waves and S waves travel more quickly through it.

10.4 Section Review

1. How was the Moho discovered?
2. **CRITICAL THINKING** Explain why the shadow zone of an earthquake in Japan would be different from the shadow zone of an earthquake in California.
3. **APPLICATION** Petroleum geologists can use seismic waves to locate oil beneath Earth's surface. Oil is usually trapped in porous rock that lies beneath a layer of denser rock. Predict how the speed of a seismic wave would change as it traveled through these layers.

CHAPTER 10 SECTION 4

CLASSZONE.COM

Examine P and S waves moving through Earth's interior.
Keycode: ES1009
 Visualizations CD-ROM

More about...

Earth's Interior
The size of the inner core was accurately calculated in the early 1960s as a result of underground nuclear testing conducted in Nevada. Because the precise time and location of the nuclear explosions were known, echoes from seismic waves generated by the explosions that bounced off the inner core provided an accurate means of determining the inner core's dimensions.

Scientific Thinking

COMMUNICATE
Diagrams and descriptions should include conditions of increasing pressure and heat that make habitation impossible.

ASSESS

1. Seismogram data showed two distinct groups of P and S waves traveling at different velocities in Earth, from which a seismologist inferred that a boundary occurred between rocks of different densities.
2. Since the shadow zone is a wide belt on the opposite side of Earth to the earthquake's focus, the position of a shadow zone depends on and varies with the position of the earthquake.
3. The speed of seismic waves would decrease as they move from the more dense to the less dense layer.

MONITOR AND RETEACH

If students miss . . .

Question 1 Reteach "The Moho." (p. 229) Review the seismic data that led to the discovery of rock layers of different density.

Question 2 Reteach "The Shadow Zone. (p. 228) Have students use the diagram to describe where the shadow zone is located relative to the earthquake.

Question 3 Reteach how P and S waves move through different materials. (p. 228) Be sure students understand that the velocity of the waves increases as the waves travel through more dense material.

CHAPTER 10 ACTIVITY

PURPOSE
To design, build, and test a model of an earthquake-resistant building

MATERIALS
Shallow cardboard boxes that work well in this activity include boxes that hold multiple soft-drink cans or other canned goods, the lids for photocopy-paper boxes, and pizza boxes.

PROCEDURE
6. Challenge students by holding a class design contest in which groups compete to build the most earthquake-resistant building.
11. Extend the activity by having students design and conduct a test to determine the effect of duration on structural damage. (Students can vary the shaking time on the platform.)

ANALYSIS AND CONCLUSIONS
1. Answers will vary regarding which model held up to the "earthquake" test the best. Aspects that help a structure resist earthquake stress include the building material's ability to absorb shock, stability with regard to the location of the center of gravity, and flexibility to move during an earthquake without buckling or breaking.

CHAPTER 10 LAB Activity
"Earthquake Engineering"

SKILLS AND OBJECTIVES
- **Design** a model of an earthquake resistant building.
- **Record** a procedure for repeatability.
- **Build** the design.
- **Test** the model using an earthquake simulation device.
- **Improve** the design based on initial trials.

MATERIALS
- 40 coffee stirrers (per model)
- 40 mini-marshmallows (per model)
- metric ruler
- 2 identical shallow cardboard boxes
- scissors
- 10–20 marbles
- 4 small rubber bands
- stapler

Earthquakes can occur even in areas where they are not fairly common. Sometimes the loss of lives and homes in such areas is greater than in high-risk areas due to a lack of earthquake preparedness. A building's design determines how well the structure withstands earthquake stress. In this activity, you will design, build and test an earthquake resistant structure.

Procedure

Part A: Build an Earthquake Platform

1. Using the scissors, cut the bottom of one of the boxes so that the piece will fit inside the other box with a 2-cm clearance on all sides.
2. Staple a rubber band to each inside corner of the intact box.
3. Fill the intact box with the marbles. Place the cardboard base cut-out on top of the marbles.

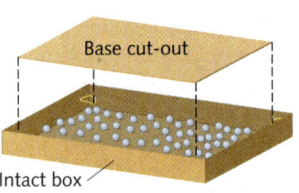
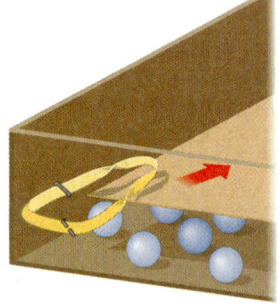

4. Staple the free end of each corner rubber band to the corresponding corner edge of the base cut-out, making sure that each rubber band connection is taut.
5. To simulate a model "earthquake," gently pull one side of the cut-out (the earthquake platform) to the edge of the outside box and let go.

Part B: Design and Build an Earthquake Resistant Structure

6. Design an earthquake resistant structure to be made of marshmallows and stirrers, that will most optimally resist collapse on the earthquake platform. The structure should be at least 50 cm tall. Record the procedure for assembling your design; make diagrams as necessary.

The building instructions should be detailed enough so that another person can accurately build your structure.

7 Build your structure using the available materials.

8 Design a procedure to test your model. Record the testing procedure, outlining the duration of shaking time and number of trials. Also detail the criteria by which you will judge the quality of your model.

9 Test the quality of your design by placing it on the earthquake platform. Be sure to follow your testing procedure. Use a chart such as the one shown below to record your results and observations.

Trial Number	Model Number	Shake Duration	Observations After the Simulation

10 Use what you have learned from your testing to design several more structures. Record or diagram the assembly instructions for each.

11 Build and test the additional structures.

Analysis and Conclusions

1. Which of your designs holds up to the "earthquake" test the best? What aspects of its structure do you think help it resist the earthquake stress? Do you know of any buildings that have a similar structure?

2. What type of earthquake motion does the "earthquake" tray simulate? How could you simulate other earthquake motions?

3. Based on your models, hypothesize how the duration or strength of the shaking affects structural damage.

4. When designing earthquake resistant buildings structural engineers focus on shock absorption. In what way, if any, does your building model provide shock absorption?

5. Besides the architectural design, what other factors do you think affect the ability of a building to resist damage during an earthquake?

Learn more about creating earthquake resistant structures.
Keycode: ES1011

Learn more about creating earthquake-resistant structures.
Keycode: ES1011

(continued)

2. The ground-shaking modeled by the "earthquake" tray is side-to-side motion. Up-and-down motion could be simulated by using a board attached to upright springs.

3. Students should hypothesize that the longer the duration or the greater the strength of the shaking, the greater is the structural damage.

4. The marshmallows in the building model help absorb shock.

5. Other factors that affect the ability of a building to resist damage during an earthquake include the composition of the ground on which the building stands (solid rock versus loose soil), the building materials used, and the proximity to other dangers that can damage the building, such as falling structures or fires.

Refer students to Safety in the Earth Science Laboratory on pages xxii–xxiii.

Courtesy of PBS series, *Newton's Apple*.
www.tpt.org/newtons

CHAPTER 10 REVIEW

Vocabulary Review

1. aftershocks
2. magnitude
3. tsunami
4. An earthquake's focus is the point at which the first movement occurs during the earthquake, whereas an earthquake's epicenter is the point on Earth's surface directly above the focus.
5. Body waves are earthquake waves that travel from the focus through Earth, whereas surface waves are earthquake waves that travel along Earth's surface.
6. A seismograph is an instrument that detects and records earthquake waves, whereas a seismogram is a record sheet on a seismograph.

Concept Review

7. Along plate boundaries, friction prevents plates from moving past each other, so strain builds up and the plates deform. When the strain overcomes the friction, the plates move suddenly, causing an earthquake. Then the plates snap back to their original shapes.
8. subduction boundaries
9. the magnitude of the earthquake and the distance to the epicenter
10. A magnitude-7 earthquake is 31 × 31, or 961, times more powerful than a magnitude-5 earthquake.
11. Liquefaction occurs when, as a result of severe ground shaking, loose soils move, taking on some of the properties of a liquid.
12. data about the location of earthquake foci along a fault
13. The shadow zone is a wide belt around the side of Earth opposite the focus of an earthquake in which seismograph stations cannot receive P or S waves.
14. surface waves → liquefaction → foundation failure (As a result of surface waves and the ground shaking they produce, liquefaction occurs, which can result in foundation failure.)

CHAPTER REVIEW 10

Summary of Key Ideas

10.1 An earthquake is a shaking of Earth's crust caused by a release of energy. Most earthquakes are caused by strain that builds up along faults at or near boundaries between lithospheric plates. Earthquakes produce body waves and surface waves.

10.2 A seismograph detects and records earthquake waves. Scientists use seismograph data to locate an earthquake's epicenter and determine its magnitude. The Richter scale and moment magnitude are two measures of magnitude.

10.3 Earthquake hazards include ground shaking and liquefaction, aftershocks, fires, and tsunamis. Damage and loss of life can be minimized through use of building codes. Long-term predictions about earthquakes help people decide how and where to implement building codes and emergency services.

10.4 Scientists use earthquake data and knowledge about earthquake waves to make inferences about Earth's interior.

Key Vocabulary

aftershock (p. 223)
body waves (p. 215)
earthquake (p. 214)
epicenter (p. 214)
fault (p. 214)
focus (p. 214)
liquefaction (p. 222)
Love waves (p. 216)
magnitude (p. 220)

P waves (p. 215)
Rayleigh waves (p. 216)
S waves (p. 215)
seismic gap (p. 225)
seismogram (p. 217)
seismograph (p. 217)
surface waves (p. 216)
tsunami (p. 223)

Vocabulary Review

Write the term from the key vocabulary list that best completes the sentence.

1. Smaller earthquakes called ___?___ may follow a large earthquake.
2. The Richter scale is used to describe the strength, or ___?___, of an earthquake.
3. An underwater earthquake may generate a large ocean wave called a ___?___.

Explain the difference between the terms in each pair.

4. epicenter, focus
5. body waves, surface waves
6. seismograph, seismogram

Concept Review

7. According to the elastic-rebound theory, what is a cause for earthquakes along plate boundaries?
8. At what type of plate boundary do very deep earthquakes tend to occur?
9. What information about an earthquake can scientists obtain from a seismogram?
10. Compare the strengths of an earthquake whose Richter magnitude is 5 and an earthquake whose Richter magnitude is 7.
11. What is liquefaction and how does it occur?
12. What data do scientists need to determine whether a seismic gap may exist along a fault?
13. How is the shadow zone of an earthquake related to the focus of the earthquake?
14. **Graphic Organizer** Use the following terms to complete the flow chart: *surface waves, foundation failure, liquefaction.*

Critical Thinking

15. The farther a seismograph station is from the epicenter, the greater the difference in arrival times of P and S waves. Similarly, the farther a lightning flash occurs from your location, the greater the difference in time between when you see the lightning and hear its thunder.
16. Other seismic gaps appear to exist outside of Parkfield and between San Francisco and Portola Valley. These regions may be sites of earthquake activity in the future.
17. 1800 km

Critical Thinking

15. Compare In what way is the time difference between the arrival of P waves and S waves at a seismograph station like the time difference between a lightning flash and thunder?

16. Predict The diagram on page 226 shows the Loma Prieta seismic gap. Identify another seismic gap on the diagram. What predictions might scientists make for the region where this seismic gap is located?

17. Analyze Use the travel-time graph on page 218. Determine the distance to the epicenter if difference in the arrival times of the P waves and the S waves is 2 minutes and 30 seconds.

18. Infer Rocks of the eastern United States are cooler and more dense than rocks near the West Coast. The cooler rocks transmit seismic waves better. How might these facts explain why the New Madrid earthquakes affected such a large region?

Internet Extension

Which Fault Moved in the Northridge Earthquake? Use epicenter location information and a shaking intensity map to pinpoint the fault that moved in this 1994 California earthquake.

Keycode: ES1010

Writing About the Earth System

SCIENCE NOTEBOOK During the last ice age, massive ice sheets covered the northern regions of North America. Some scientists have hypothesized that the New Madrid earthquakes are related to the melting of those ice sheets. When the ice melted, a great weight was removed from the continent. As a result, the continent may still be rebounding. Earthquakes occur from time to time as the land adjusts to the removal of the ice. According to this hypothesis, which of Earth's spheres are interacting? What other spheres are affected by their interaction?

Interpreting Data

The drawings at the right are seismogram tracings made at three seismograph stations for the same earthquake. The arrival times of the P waves and S waves are indicated on each tracing.

19. Which wave arrived at each seismograph station first? Why?

20. After how much time did the P wave arrive at station A? Give your answer to the nearest minute.

21. After how much time did the S wave arrive at station B? Give your answer to the nearest minute.

22. What is the difference between the arrival times of the P waves and the S waves at station C?

23. Station B was located nearest to the earthquake's epicenter, while station A was farthest away. Cite three types of evidence from the tracings to support this statement.

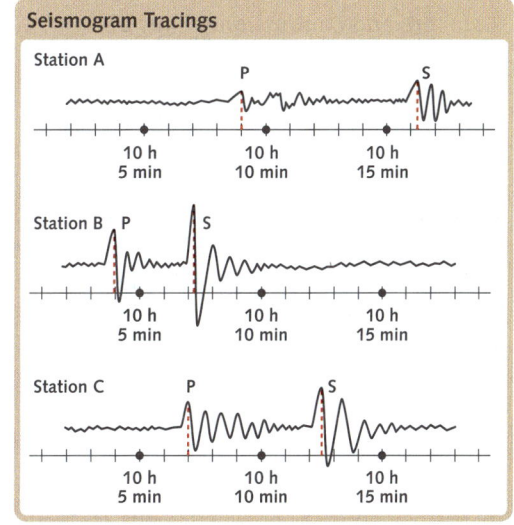

CHAPTER 10 REVIEW

Which Fault Moved in the Northridge Earthquake?
Keycode: ES1010

📖 *Use Internet Investigations Guide, pp. T40, 40.*

(continued)

18. The cooler, more dense rocks of the eastern United States transmitted the seismic waves from the New Madrid earthquake more effectively and thus farther than rocks near the West Coast would have. Therefore, this earthquake affected a large region.

INTERPRETING DATA

19. P waves; P waves travel faster than S waves.

20. 10 h 9 min

21. 10 h 7 min

22. 5.5 min

23. Sample Responses: P waves arrived the earliest at station B and the latest at station A. S waves arrived the earliest at station B and the latest at station A. The difference in P and S wave arrival times is least at station B and greatest at station A. The lengths of the zigzags of the tracing are longest at station B and shortest at station A.

WRITING ABOUT SCIENCE

Writing About The Earth System

The melting of ice sheet involves the hydrosphere. The rebounding of the land involves the geosphere. The interaction of these two spheres might also affect the biosphere, as organisms are impacted by changes in land and water, which in turn might also impact the atmosphere because of the exchange of gases between organisms and the atmosphere.

CHAPTER 11 PLANNING GUIDE MOUNTAIN BUILDING

	Section Objectives	Print Resources
Chapter-Level Resources		Laboratory Manual, pp. 47–50 Internet Investigations Guide, pp. T41–T43, 41–43 Guide to Earth Science in Urban Environments, pp. 21–28 Spanish Vocabulary and Summaries, pp. 21–22 Formal Assessment, pp. 31–33 Alternative Assessment, pp. 21–22
Section 11.1 Where Mountains Form, pp. 236–237	1. Explain how some of Earth's major mountain belts formed. 2. Compare and contrast active and passive continental margins.	Lesson Plans, p. 36 Reading Study Guide, p. 36 Teaching Transparencies 9, 40, 41
Section 11.2 How Mountains Form, pp. 238–242	1. Explain how compression, tension, and shear stress deform rocks. 2. Compare and contrast anticlines and synclines. 3. Distinguish among the three major types of faults—normal, reverse, and strike-slip.	Lesson Plans, p. 37 Reading Study Guide, p. 37 Teaching Transparency 13
Section 11.3 Types of Mountains, pp. 243–245	1. Classify mountain ranges by their most prominent features. 2. Compare and contrast folded mountains, dome mountains, and fault-block mountains.	Lesson Plans, p. 38 Reading Study Guide, p. 38 Teaching Transparency 13

INVESTIGATIONS
CLASSZONE.COM

Section 11.1 INVESTIGATION
How Are Mountains Related to Plate Tectonics?
Computer time: 45 minutes / Additional time: 45 minutes
Students examine mountain belts to determine their relationship to plate boundaries.

Section 11.2 INVESTIGATION
How Do Rocks Respond to Stress?
Computer time: 45 minutes / Additional time: 45 minutes
Students apply a force to rocks under different conditions.

Technology/Online Resources

Electronic Teacher Tools
Visualizations CD-ROM
Classzone.com
Online Lesson Planner
Test Generator

Classzone.com
  ES1101 Investigation

Visualizations CD-ROM
Classzone.com
  ES1102 Investigation, ES1103 Visualization,
  ES1104 Data Center

Visualizations CD-ROM
Classzone.com
  ES1105 Visualization

Classroom Management

Gather these materials for minilab and map activities.

Minilab, p. 240
  2 fist-sized pieces of different-colored clay
  wire

Map Activity, pp. 246–247
  Appendix B Physical World Map, pages 710–711
  Teaching Transparency 40
  Appendix B Plate Boundaries Map, pages 712–713
  Teaching Transparency 41
  Appendix B Physical United States Map, pages 708–709
  Teaching Transparency 39
  Appendix B Topographic Map: Harrisburg, Pennsylvania, page 718
  Teaching Transparency 38
  tracing paper, 10 cm $\times$ 15 cm

Chapter Review INVESTIGATION

What Forces Created These Geologic Features?
Computer time: 45 minutes / Additional time: 45 minutes
Students examine faults and folds, then illustrate the forces and conditions that created them.

CHAPTER 11

INTRODUCE
Build Interest
Before sharing background information about the photograph with students, allow time for their questions and comments.

Ask questions like the following:
- Describe the mountains in this photograph. *jagged, covered with ice and snow*
- What do you think the climate is like here? Explain your response. *very cold; Snow and ice cover the mountains, and the climber is wearing protective clothing.*
- What would you like and not like about going on an expedition to this place? *The adventure and firsthand learning opportunities of such an expedition might appeal to some students, but the dangers might not.*

Respond
- What causes mountains to grow? *Several processes cause mountain building. In the case of the Himalayas two continental plates are colliding, causing the rock layers to buckle (fold).*

About the Photograph
The climber is in the Himalayas on Mt. Lhotse, the fourth highest peak in the world (elevation: about 8516 meters). The Himalayas formed over a period of about 40 million years as two plates converged. The Himalayas continue to rise 5 to 8 millimeters per year.

Mountain Building

Geologists have gathered evidence showing that the Himalayas, the highest mountains in the world, are growing.

What causes mountains to grow?

BEYOND THE TEXTBOOK

Print Resources
- Reading Study Guide, pp. 36–38
- Transparencies 9, 13, 38, 39, 40, 41
- Formal Assessment, pp. 31–33
- Laboratory Manual, pp. 47–50
- Alternative Assessment, pp. 21–22
- Spanish Vocabulary and Summaries, pp. 21–22
- Internet Investigations Guide, pp. T41–T43, 41–43
- Guide to Earth Science in Urban Environments, pp. 21–28

Technology Resources
- Visualizations CD-ROM
- Classzone.com
- Online Lesson Planner
- Electronic Teacher Tools
- Test Generator

234 Unit 3 Dynamic Earth

CHAPTER 11

PREVIEW

▶ **FOCUS QUESTIONS** In this chapter you will study mountains and learn more about the key questions below.

- Section 1 Where do mountains form?
- Section 2 How do mountains form?
- Section 3 How can mountains be classified?

▶ **REVIEW TOPICS** As you investigate mountain building, you will need to use information from earlier chapters.

- pluton (p. 125)
- collision boundary (p. 178)
- fault (p. 214)
- volcano (p. 194)

▶ **READING STRATEGY**

QUESTION

Before reading the chapter, skim its contents and develop questions you would like to explore. Write your questions in your science notebook. As you read, record information.

At our Web site, you will find the following Internet support for this chapter.

DATA CENTER
EARTH NEWS
VISUALIZATIONS
- Fault Motion
- Formation of the Himalayas

LOCAL RESOURCES
INVESTIGATIONS
- How Are Mountains Related to Plate Tectonics?
- How Do Rocks Respond to Stress?
- What Forces Created These Geologic Features?

CHAPTER 11

PREPARE

Focus Questions
Find out what knowledge students have about mountain building. For each focus question, have students consider the section title and then develop two or three additional questions for each section. For example, Section 1: Can mountains form anywhere on Earth? Section 2: How long does it take a mountain to form?

Review Topics
In addition to the earth science topics listed, students may need to review these concepts:

Physical Science
- Stress is the strength of an applied force that tends to change the shape or volume of a body. It is expressed in force per unit area.
- Strain is a change in shape or volume caused by external forces.

Reading Strategy
Model the Strategy: Refer to students' focus questions from the exercise above. Direct students to use all parts of Section 1 including key ideas, photographs, diagrams, captions, key words, and the Internet Investigation to develop other questions for the section. For example, How did the Appalachians form? What is the difference between an active and a passive continental margin? Do all mountain ranges indicate a current or former plate boundary? Have students write down their questions and attempt to answer to them as they read the section. Have students repeat this activity for the other two sections in the chapter.

BEYOND THE TEXTBOOK

INVESTIGATIONS
- How Are Mountains Related to Plate Tectonics?
- How Do Rocks Respond to Stress?
- What Forces Created These Geologic Features?

VISUALIZATIONS
- Animations of fault motion
- Animation of the Himalayas forming.

DATA CENTER
EARTH NEWS
LOCAL RESOURCES
CAREERS

Online Lesson Planner

CHAPTER 11 SECTION 1

▶ FOCUS

Objectives
1. Explain how some of Earth's major mountain belts formed.
2. Compare and contrast active and passive continental margins.

Set a Purpose
Have students read this section to find the answer to "Where do most mountain belts form?"

▶ INSTRUCT

DISCUSSION

Ask: What is a mountain? **a large mass of rock that rises a great distance above its base** Where do most mountains form? **along convergent boundaries** What happens during convergence involving an oceanic plate? **From Chapter 8, students should recall that when two oceanic plates collide, one is forced into the mantle. It partly melts and the resulting magma rises to form a volcanic island arc. During convergence between an oceanic plate and a continental plate, the oceanic plate is forced into the mantle and partly melts. A mountain chain that contains volcanic peaks forms on the continental plate.** What is the major difference between an active continental margin and a passive margin? **Tectonism occurs along active margins creating young mountains.**

Use Transparencies 9, 40, and 41.

11.1

KEY IDEA
Events at plate boundaries and at continental margins result in the formation of mountains.

KEY VOCABULARY
- mountain
- continental margin

INVESTIGATIONS
CLASSZONE.COM
How Are Mountains Related to Plate Tectonics? Examine the locations and ages of mountain belts to determine their relationship to plate boundaries.
Keycode: ES1101

THE APPALACHIANS The Blue Ridge mountains are a mountain range in the Appalachians. This view is from an overlook in Shenandoah National Park, Virginia.

Where Mountains Form

A **mountain** is a large mass of rock that rises a great distance above its base. Most mountains result from forces associated with processes that occur at converging plate boundaries.

Mountain Belts

Most of the world's mountains occur in long belts that tend to follow converging plate boundaries. These mountain belts are regions where mountains are forming or have formed in the past. The North American Cordillera is a mountain belt that runs down the western side of North America from Alaska to Mexico. This mountain belt is made up of smaller groups of mountains called mountain ranges. For example, the Cascade Range is part of the North American Cordillera.

Some mountains, such as the Himalayas, lie along current plate boundaries. Other mountains, such as the Appalachians, do not. By analyzing how the rocks in the Appalachians are arrayed and by studying the ages of the rocks and the materials in them, geologists have concluded that the Appalachians formed at a plate boundary that existed millions of years ago.

Continental Margins

In order to understand how processes at convergent plate boundaries contribute to mountain building, geologists study what happens along Earth's continental margins. A **continental margin** is a boundary between continental crust and oceanic crust.

There are two types of continental margins. Active continental margins occur along plate boundaries. Passive continental margins do not occur at plate boundaries. For example, there is an underwater continental margin along the east coast of North America that marks the boundary between the North American continent and the oceanic crust. Both the continental crust and the oceanic crust along this continental margin are part of the North American Plate.

Mountain building takes place near active continental margins. One place where an active continental margin exists is along the west coast of South America. Here, the Nazca [NAHZ-kuh] Plate, which is carrying

DIFFERENTIATING INSTRUCTION

Reading Support Students can become active readers by trying to figure out the meaning of an unfamiliar word by context or by looking up the word in a glossary or dictionary. Often students will find that the word comes from a language other than English. This knowledge can enrich their reading. Model this process with the word *cordillera*. Say: *Cordillera* doesn't sound like an English word. What language do you think it is from? **Spanish** How can we find out what it means? **Spanish-speaking students may know, but a dictionary shows that it is derived from *cuerda*, which means "rope" or "cord."** Explain that only the mountain belts of western North and South America are called cordilleras, which makes sense given the influence of the Spanish in the early European exploration and settlement of these areas.

Use Reading Study Guide, p. 36.

oceanic crust, is subducting beneath the South American Plate, which is carrying continental crust. The west coast of South America is very mountainous and prone to earthquakes and volcanic activity. The slow process of mountain building is taking place today in this region.

Passive continental margins are stable areas because they do not occur at plate boundaries. These areas accumulate large amounts of sediment. Some of the sediment comes from skeletons and shells of marine organisms. But much of it consists of sediments that have been weathered away from the continent and transported to the continental margin by rivers. One example of a passive continental margin lies off the east coast of North America. A wedge of sediment 250 kilometers wide and as much as 10 kilometers thick has accumulated there. Eroded rock materials from the Appalachians now lie along this passive continental margin.

Sediment Buildup at a Passive Continental Margin

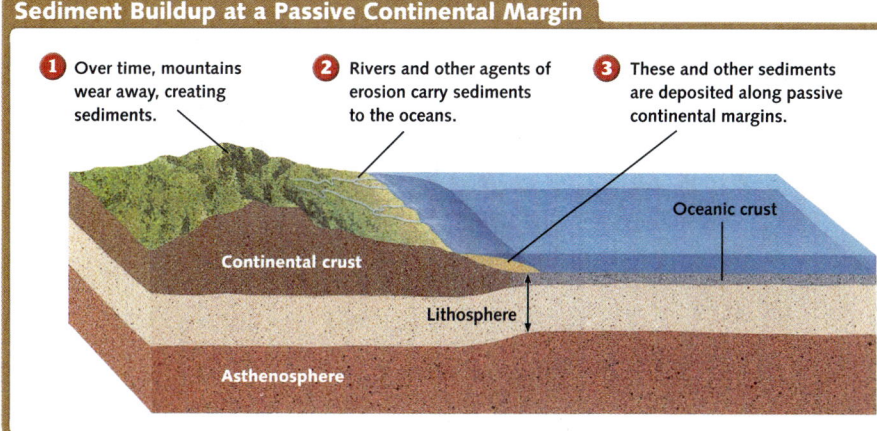

1. Over time, mountains wear away, creating sediments.
2. Rivers and other agents of erosion carry sediments to the oceans.
3. These and other sediments are deposited along passive continental margins.

How are passive continental margins related to mountain formation? These passive continental margins provide the materials that form mountains. The mountains of today were passive continental margins in the past. The active continental margin on the west coast of South America was a passive continental margin until about 200 million years ago. The Andes Mountains contain the sediments that were deposited on that passive continental margin.

11.1 Section Review

1. Where do most of the world's mountains form?
2. **CRITICAL THINKING** Compare and contrast passive and active continental margins.
3. **GEOGRAPHY** Look at the plate boundary map on pages 712–713. Identify at least two active continental margins and two passive continental margins. What mountain ranges appear to be associated with the active continental margins you identified?

Chapter 11 Mountain Building 237

CHAPTER 11 SECTION 2

FOCUS

Objectives
1. Explain how compression, tension, and shear stress deform rocks.
2. Compare and contrast anticlines and synclines.
3. Distinguish among the three major types of faults—normal, reverse, and strike-slip.

Set a Purpose
Apply enough force to gently bend a plastic ruler. Ask students what would happen if you were to increase the force. Have them propose an answer to the question "How do forces affect rocks?"

INSTRUCT

Visual Teaching

Interpret Diagrams
Compare and contrast each type of stress and its effect. **compression: squeezing makes rock shorter and thicker; tension: rock is stretched or torn; sheer stress: rock is deformed where one body of rock moves against another**

Extend
Which type of stress is associated with divergent boundaries? **tension** convergent? **compression** transform? **shear**
Use Transparency 13.

11.2

KEY IDEA
Rocks at converging plate boundaries are under stress, and this stress may lead to deformation.

KEY VOCABULARY
- anticline
- syncline
- normal fault
- reverse fault
- thrust fault
- strike-slip fault
- joint

INVESTIGATIONS CLASSZONE.COM
How Do Rocks Respond to Stress?
Apply a force to model rocks under different conditions and view the results.
Keycode: ES1102

How Mountains Form

A sculptor uses a hammer to hit a chisel, and the point of the chisel hits a rock. The rock breaks because the sculptor has concentrated the force of the hammer blow onto the relatively small area where the chisel meets the rock. The sculptor has increased the amount of stress being applied to the rock. *Stress* is a measure of the amount of force applied over a given area.

Stress is applied to rocks at converging plate boundaries. Rocks at Earth's surface are rigid, and usually respond to stress by fracturing. Deeper in Earth, high pressures and temperatures make rocks less rigid than those at the surface. For this reason, stress applied to rocks deep within the lithosphere may cause the rocks to fold or stretch without fracturing. Mountain belts are made of rocks that have been permanently deformed under stress.

Types of Stress

The illustration at the right models a series of rock layers. The bottom rock layers are under stress from the weight of layers above. Overall, however, the rock layers are relatively stable. The illustrations below show three different types of stress that might affect the rock layers.

COMPRESSION The rock layers are being squeezed inward. Stress that involves squeezing is called *compression*. Compression tends to make rock layers thicker and shorter.

TENSION The rock layers are being stretched. Stress that involves pulling and stretching is called *tension*. Tension tends to make rocks thinner and longer. In this illustration, some of the rock layers have fractured.

SHEAR STRESS The rock layers are being pushed in two different, opposite directions. Stress that involves forces moving in opposite directions is called *shear stress*. Shear stress tends to distort the shape of the rocks, especially along the plane between the opposing forces.

238 Unit 3 Dynamic Earth

DIFFERENTIATING INSTRUCTION

Reading Support Have students locate the italicized words on this page: *stress, compression, tension, shear stress.* Help students make a word web that defines each of the terms and relates them to one another. One type of web could have *Stress* at the top. Indicate that stress is a force that changes rocks. The next level would show that three types of stress are compression, tension, and shear. The next levels would indicate the kind of action involved (squeezing, twisting, stretching) and the effect that results (makes rocks shorter, longer, distorted).
Use Reading Study Guide, p. 37.

Compression, tension, and shear stress all contribute to the process of mountain building at plate boundaries. These three types of stress usually occur together. Different types of rocks may respond differently to the same stress. For example, a layer of sandstone may respond to stress by folding. A thin layer of shale lying next to the sandstone may fracture rather than fold.

Folds

During plate collisions, stress can cause rock layers along continental margins to crumple into folds. Although this folding occurs deep beneath Earth's surface, folded rock layers may be exposed after long periods of time by uplift, weathering, and erosion.

Geologists use a number of terms to describe folds in rocks. As shown in the diagram below, an **anticline** is an upfold in the rock layers. A **syncline** is a downfold. The two sides of a fold are called *limbs*. The steepness, or dip, of the limbs reflects the intensity of folding. Limbs may be gently dipping, steeply dipping, straight up and down, or even overturned. The compass direction of the fold or of the rock layers exposed at the surface along the fold is called the strike. Sometimes folding can cause extreme deformation. In other cases, rock layers have been pushed into gentle anticlines and synclines.

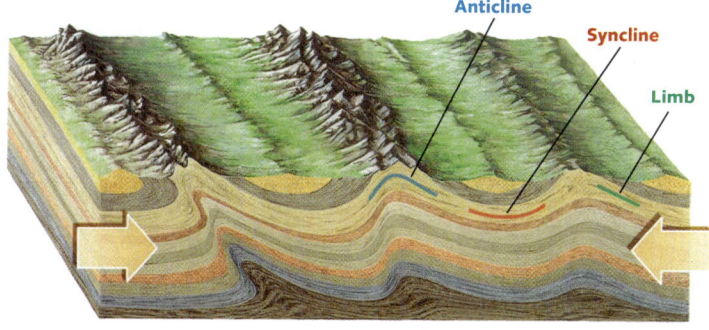

FOLDED ROCK LAYERS

A well-known example of folded mountains is the Valley and Ridge Province of the Appalachian Mountains. In this region, which extends from New York State to Alabama, the rock layers have not been badly crumpled. Instead, the stress of collision has formed the layers into long, straight folds. Interestingly, the valleys between the mountains do not correspond to fold synclines nor the ridges to fold anticlines, as one might expect. Instead, the locations of the valleys and ridges are controlled primarily by the weathering rates of the rocks in different areas of the folds.

Scientific Thinking

INTERPRET
The photograph to the left shows folded rock layers in New Zealand. Identify the types of stress that may have caused the folding of these rock layers. Explain your thinking.

CHAPTER 11 SECTION 2

INVESTIGATIONS
CLASSZONE.COM

How Do Rocks Respond to Stress?
Keycode: ES1102

Students should predict results based on conditions they choose.

📖 *Use Online Investigations Guide, pp. T42, 42.*

Scientific Thinking

INTERPRET
The rock layers are thicker and shorter than they were prior to deformation, which is indicative of compressive forces. Also, folds are usually the result of compression.

VISUAL TEACHING
Discussion
Point out the steep dipping of the folds in the photograph. Ask: Which part of the diagram most resembles the folds in the photograph? the sharp folds at the far left

CRITICAL THINKING
Have students imagine that the folded rocks in the diagram have been eroded to produce a flat surface. Ask: How could you tell which rocks belonged to the anticline and which rocks belonged to the syncline? In an eroded anticline, the oldest rocks are surrounded by younger rocks. In an eroded syncline, the youngest rocks are surrounded by older rocks.

DIFFERENTIATING INSTRUCTION

Reading Support Students who have difficulty visualizing may not be able to comprehend the critical thinking activity above. Have these students draw a steep, two-layer set of anticlines and synclines, labeling the top layer young (Y) and the bottom layer old (O). Also label the anticline and syncline below each feature. Next, erase the top half of the sketch to model erosion. Direct students to observe the patterns of layers when read left to right across the sketch: anticlines show young-old-old-young; synclines show old-young-young-old.

CHAPTER 11 SECTION 2

MINILAB
Modeling a Fault

Management
Have students work in groups of four. Each student in a group should model one of the faults shown on pages 240 and 241.

Analysis Answers
normal faults, tensional stress; reverse and thrust faults, compressive stresses; and strike-slip, shear stress

For proper behavior in a laboratory setting, refer students to Safety in the Earth Science Laboratory on pages xxii–xxiii.

VISUAL TEACHING
Discussion

Ask: **What is a fault?** a break in the lithosphere along which movement has occurred **What is a fault plane?** the surface between two moving pieces of crust **Where is the hanging wall relative to the footwall in a normal fault?** below **in a reverse fault?** above **Compare and contrast reverse and thrust faults.** Both are caused by compression, with the hanging wall moving upward relative to the footwall. The angle of the fault plane in a thrust fault is 45° or less from the horizontal; in a reverse fault the angle is greater than 45°. **What is a strike-slip fault?** a fault where movement is horizontal

Use Transparency 13.

Mini LAB — 20 Minutes
Modeling a Fault

Materials
- Two fist-sized pieces of different-colored clay
- wire

Procedure
1. Flatten each piece of clay until it is about 1 cm thick. Cut each piece into two congruent rectangles.
2. Stack all four rectangles, forming a block with layers of alternating colors.
3. Using the wire, cut through the block diagonally, as shown.
4. Move one side of the block up and the other down. Move the sides in the opposite directions.

Analysis
What types of faults have you modeled? For each fault, what type of stress did you exert? What would you change to model other types of faults?

Faults

As you learned in Chapter 10, a fault is a break in the lithosphere along which movement has occurred. Not only are movements along fault planes the primary cause of earthquakes, they are also an important part of the process of mountain formation.

The illustration at the right shows a fault plane, which is the surface between the two pieces of moving crust at a fault. The part of the fault above the fault plane is called the *hanging wall*. The part of the fault below the fault plane is the *footwall*.

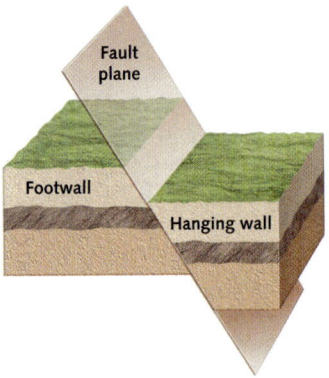

Geologists classify faults by the way in which the rocks on either side of the fault plane move with respect to each other. In each of the fault illustrations below, the large arrows indicate stress being applied to the rocks. The small black arrows show movement along the fault plane.

A **normal fault** occurs when the hanging wall moves down with respect to the footwall. Normal faults occur in areas where tension is pulling the crust apart.

A **reverse fault** occurs when the hanging wall moves up with respect to the foot wall. Reverse faults are caused by compression. Such compression may occur, for example, when two plates collide. The compression at a reverse fault thickens and shortens rocks.

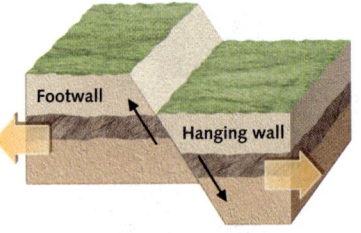
NORMAL FAULT

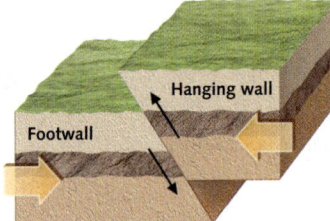
REVERSE FAULT

A **thrust fault** is a reverse fault in which the fault plane dips 45° or less from the horizontal. Thrust faults are common in many mountain belts, including the Appalachians. From studies of thrust faults, geologists have seen that compressive stresses can move sheets of rock over great distances.

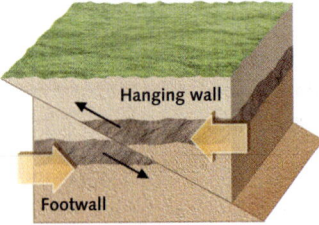
THRUST FAULT

240 Unit 3 Dynamic Earth

DIFFERENTIATING INSTRUCTION

Developing English Proficiency Help students find words on this page that are used both in science and in everyday language. Such words include *fault, crust, wall, stress, normal, reverse, plates, belts,* and *sheets*. Pair students acquiring English with English-proficient students. Have the English-proficient student use the word in a sentence that shows its common usage. Have students acquiring English read aloud a sentence from the text that includes the word. Let student pairs discuss why these terms may have been chosen to describe these natural phenomena.

Use Spanish Vocabulary and Summaries, pp. 21–22.

At a **strike-slip fault,** the rocks on opposite sides of the fault plane move horizontally past each other. The San Andreas Fault is a well-known example.

Extremely long strike-slip faults are found in the Himalayas. These faults result as the Indian Plate pushes rock materials aside on its drive into the Eurasian Plate. Some of the faults are thousands of kilometers from the plate boundary.

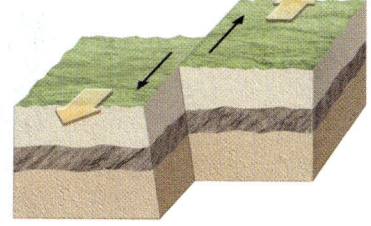

STRIKE-SLIP FAULT

Examine animations of fault motion.
Keycode: ES1103

Joints

Joints, like faults, are breaks in bedrock. Unlike faults, joints are breaks along which no apparent movement has occurred. Joints can be the result of the same stresses that lift, tilt, and fold rock layers into mountains. They are one of the most common rock structures.

The surface of a joint is usually a plane, although the surfaces of some joints are curved. The joint plane appears on the surface of a rock outcrop as a line. These lines can appear in parallel groups called joint sets. Sometimes one joint set is crossed at an angle by another joint set.

Joints provide channels through which fluids enter and move through bedrock. Hot fluids rising through the crust may fill a joint with quartz, calcite, or some other mineral to form a vein deposit. Caverns form when groundwater flows through and dissolves limestone along joint planes. In some areas, surface features are controlled by joint patterns. The spires at Bryce Canyon are the result of weathering along joints.

11.2 Section Review

1. Describe different ways that rocks can respond to stress.
2. Compare and contrast an anticline and a syncline.
3. Explain the difference between a normal fault and a reverse fault.
4. What is a thrust fault?
5. Describe the movement in a strike-slip fault.
6. **CRITICAL THINKING** What visual clues could a geologist use to decide whether a large crack in a rock cliff is a joint or a fault?
7. **MATHEMATICS** Geologists use a device called a clinometer to measure the dip of a limb in an anticline or syncline. The clinometer measures the angle that the limb makes with a horizontal line. Sketch and label an anticline with a limb whose dip is about 45°. Then sketch and label an anticline with a limb whose dip is about 30°.

CHAPTER 11 SECTION 2

Examine animations of fault motion.
Keycode: ES1103
Visualizations CD-ROM

ASSESS

1. Rocks at Earth's surface generally fracture in response to stress. Deep beneath Earth's surface, rocks fold or stretch when subjected to stress.

2. Both are folds in rocks that form as the result of compression. An anticline is an upward fold (hill); a syncline is a downward fold (valley).

3. In a normal fault, tension causes the hanging wall to move down relative to the footwall. In a reverse fault, the opposite is true—the hanging wall moves up relative to the footwall.

4. A thrust fault is a reverse fault in which the fault plane dips 45° or less from the horizontal.

5. At a strike-slip fault, plates move horizontally past each other.

6. Offset layers, like those shown in the faults on page 240, would indicate a fault rather than a joint. Also, deformation (metamorphism) of rocks is more common along fault planes than along joint planes.

7. Students' drawing should be similar to two of the anticlines shown in the figure on page 239. The limbs of the far right anticline are dipping about 30° with respect to the horizontal. The left limb of the middle anticline, which is overprinted by the blue line, is dipping at about 45° with respect to the horizontal.

MONITOR AND RETEACH

If students miss . . .

Question 1 Using silicon putty, slowly stretch the putty to demonstrate tensile deformation at depth. Quickly pull at the putty and cause it to snap to model how surface rock generally responds to tension.

Question 2 Use a telephone book to demonstrate how anticlines and synclines form in response to compression.

Question 3 Use Transparency 13.

Question 4 Use Transparency 13 and a protractor to measure the angle of the fault plane in the diagram of the thrust fault.

Question 5 Use Transparency 13.

Question 6 Ask: How do faults and joints differ? **Movement occurs along faults but not along joints.**

Question 7 Reteach anticlines and synclines on page 239.

CHAPTER 11 SECTION 2

Learn more about the Global Positioning System (GPS).
Keycode: ES1104

Science and Technology

GPS and Himalayan Measurements

Any measurement of Everest's elevation actually measures the height of the snowpack on top of Everest, not the rocky peak itself. The depth of snow atop Mt. Everest is unknown, but it is thought to vary by at least a meter throughout the year because summers in the region bring heavy snowfall, and winters bring strong winds to scour away the snowpack. Despite the increasing technology, it seems obtaining the actual height of the mountain will always be somewhat elusive. Perhaps a more important role of GPS in the Himalayan region is to monitor the relative motions of 26 places where survey points have been installed. These measurements reveal the rate of deformation in the Himalayas, and might help predict earthquakes.

Science Notebook

A sudden change could indicate that an earthquake is imminent. Any such insight gained might help scientists predict earthquakes.

SCIENCE & Technology

Measuring Mount Everest

How tall is the world's tallest mountain? And is it getting taller over time? These questions have not been easy to answer. First comes the challenge of climbing Mount Everest. Then comes the challenge of getting an accurate measurement.

How has technology helped scientists face these challenges?

In 1954, Everest's height was measured at 8848 meters (nearly 6 miles) tall. However, none of the surveying instruments were placed any closer than 48 kilometers from Everest's summit, a fact that increased the uncertainty of the measurement.

In the mid-1990s, new technology, the Global Positioning System (GPS), made it possible to get very accurate elevation measurements for any point on Earth. By processing signals from 24 satellites that orbit Earth, a GPS receiver can give information about the receiver's exact location. Scientists knew that GPS could be used to find Everest's elevation. The tricky part would be getting a receiver to Everest's frozen and forbidding summit.

In spring 1999, a climbing team began the trek to Everest's peak. The team carried lightweight GPS units with lithium batteries guaranteed to work at temperatures as cold as −40°C. Readings from the GPS receiver that the team placed at Everest's summit and readings from a second receiver placed at a lower elevation were used to determine the mountain's height. This new measurement, 8850 meters, was 2 meters higher than the 1954 measurement. Because GPS technology is unaffected by atmospheric or gravitational effects (which may have skewed the 1954 measurement), the new measurement is highly reliable. It is now accepted as Everest's official height.

THE GLOBAL POSITIONING SYSTEM includes hand-held receivers—such as this one—which pick up satellite signals.

By taking new GPS measurements year after year, scientists can track Mount Everest's height and position over time. As India continues to thrust under Nepal and China, Mount Everest apparently grows higher by about 4 millimeters each year. It also moves about 6 centimeters to the northeast each year.

Extension

SCIENCE NOTEBOOK

Consider the advantages and disadvantages of monitoring mountain formation using GPS technology. What might a sudden change in the pattern of mountain building indicate?

Learn more about the Global Positioning System (GPS).
Keycode: ES1104

MEASURING Scientists used surveying instruments that incorporated GPS technology to measure Mount Everest.

Types of Mountains

The forces involved in plate interactions produce features such as folds and faults. Mountains can be classified by these and other features.

Folded Mountains

When two plates carrying continental crust collide, rocks can fold and crumple into **folded mountains.** The Appalachians, the Alps, the northern Rocky Mountains, the Urals, and the Himalayas are examples of folded mountains. Although geologists studying these mountains may also find faults and evidence of igneous activity, folded rocks are an especially prominent feature.

Before two continents can collide, the ocean basin between them must close. This change occurs as the oceanic crust subducts beneath one of the continents. Subduction stops, however, once the ocean is gone and the continents are in contact because continental crust is not dense enough to subduct. Continued plate movements cause the rocks of the continental margins to crumple into mountains.

As shown by the diagram below, the formation of the Himalayas was preceded by the closing of an ocean between India and Tibet, a part of Eurasia. The oceanic crust disappeared into a subduction zone, although some of the crust was caught up in the collision, along with sediments that had accumulated on the edges of the colliding continents. Once the continents collided, subduction stopped. India continued to move north, pushing some rocks aside and crumpling others into the Himalayas. Mountain building is still going on today, as the Indian Plate continues to push into the Eurasian Plate.

11.3

KEY IDEA

Mountains can be classified by features that result from forces involved in plate interactions.

KEY VOCABULARY
- folded mountains
- dome mountain
- fault-block mountains

Observe an animation of the Himalayas forming.
Keycode: ES1105

Formation of Himalayas

 As India moved toward Tibet, the oceanic crust between the two continents subducted beneath the Eurasian Plate.

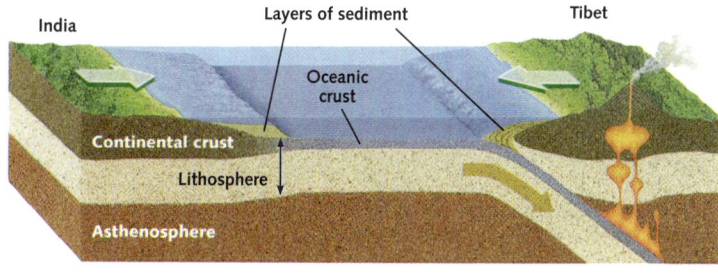

 Once the ocean had closed, subduction stopped. India and Eurasia collided, crumpling and folding rocks into the Himalayas.

CHAPTER 11 SECTION 3

Visual Teaching

Interpret Diagrams
Ask: How are these two mountains similar? **Both are dome mountains, which are nearly circular folded mountains.** How do their modes of formation differ? **The plutonic dome forms as the result of igneous activity, which causes the overlying rocks to bulge upward. The tectonic dome forms as the result of uplift.**

Extend
Suppose you observed a dome mountain in the field. How could you determine whether it was a tectonic dome or a plutonic dome? **An igneous core would be indicative of a plutonic dome. If all of the observed rocks were igneous, however, you could compare the ages of the core and the surrounding rock layers. A dome with an old core is indicative of a tectonic dome.**

CRITICAL THINKING

Ask: How do dome mountains differ from large anticlines? **Dome mountains form as the result of essentially vertical forces. Anticlines form when rocks are subjected to intense horizontal forces (compression).**

Dome Mountains

A **dome mountain** is a nearly circular folded mountain. Dome mountains are not found in mountain belts such as the Himalayas or the Appalachians. Instead, they are individual, isolated structures that tend to occur in areas of essentially flat-lying sedimentary rocks. These layers are bent upward in a dome shape as a result of uplifting forces. If the rocks above the dome mountain's center have eroded away, the layers of rock may stand out as sharp ridges around the edge of the mountain.

There are two basic types of dome mountains. One type is called a plutonic dome mountain. The other type is referred to as a tectonic dome mountain. Plutonic dome mountains form when overlying crustal rocks are pushed upward by an igneous intrusion, such as a laccolith. Because the intrusion occurs after the overlying crustal rocks have been formed, the igneous rock at the core of the mountain is younger than the sedimentary rocks around the core. Many examples of plutonic dome mountains are found on the border of the Colorado Plateau and the Rocky Mountains.

Tectonic dome mountains result from uplifting forces that arch rock layers upward. All the rocks in the dome were present before the uplift occurred. The rocks at the core extend under the rocks around the dome and, therefore, must be older. Two examples of tectonic dome mountains are the Adirondack Mountains of New York State and the Black Hills of South Dakota.

Formation of Dome Mountains

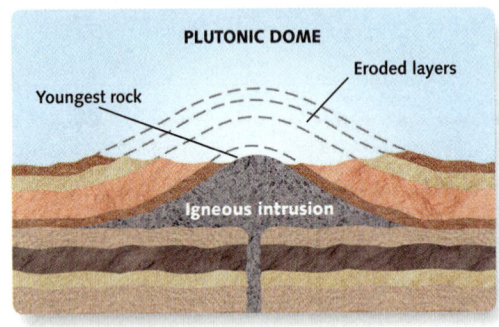

An igneous intrusion pushes up existing rock layers. When these rock layers wear away, the igneous rock is exposed at the center of the dome. The igneous rock is younger than all of the rock layers in the dome.

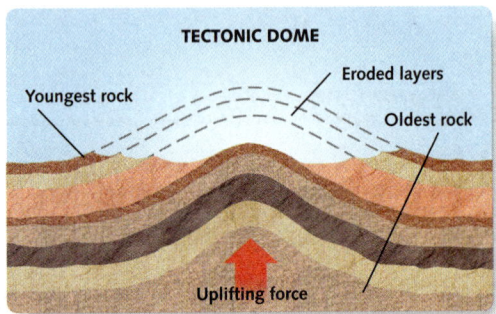

Uplifting forces arch rock layers into a dome. As the top layers wear away, older rock layers are exposed. The exposed layer at the center of the dome is older than the exposed layers around the edges of the dome.

Volcanic Mountains

Volcanic activity is a contributor to mountain formation. As you learned in Chapter 9, volcanic mountains such as the Cascades tend to form on continental crust near a subduction boundary.

Volcanic rocks are not common in the Himalayas, but some do occur on the northern edge of the range. These rocks may have formed when the

244 Unit 3 Dynamic Earth

DIFFERENTIATING INSTRUCTION

Reading Support As they read, have students create a three-column chart that lists the four main types of mountains, how each type forms, and an example of each type. Suggest that students used the red topic headings and boldface terms to help them complete the chart correctly. Developing the chart will not only set a purpose for reading, but it can also be used as a study aid.
Use Reading Study Guide, p. 38.

Indian Plate was still subducting under the Eurasian Plate. Once the plates collided, subduction stopped, and volcanic activity ceased.

Fault-Block Mountains

Although many mountain belts form as a result of compression, tension plays a large role in the formation of **fault-block mountains.** In some parts of the western United States, Earth's crust is slowly being uplifted. This uplift has caused the crust to stretch and crack, forming normal faults along the surface. As uplift continues, whole blocks of crust have been pushed up into fault-block mountains. Examples of such mountains are the Sierra Nevada of California, the Wasatch Range of Utah, and the Teton Range of Wyoming.

Horsts and Grabens

Tensional stress and normal faulting also result in features called horsts (hawrsts) and grabens (GRAH-buhns). Along the Great Rift Valley in East Africa, rising magma is forcing the crust upward. Tensional stress causes the crust to stretch. On either side of the wide rift valley, normal faults have formed. Between these normal faults, large blocks of crust have dropped. These blocks of crust are called grabens.

Sometimes a block of crust is thrust upward between two normal faults. This block of crust is called a horst. Horsts and grabens are found in the Basin and Range Province of Nevada.

GREAT RIFT VALLEY Tensional stress has resulted in normal faults along the sides of the Great Rift Valley in Africa.

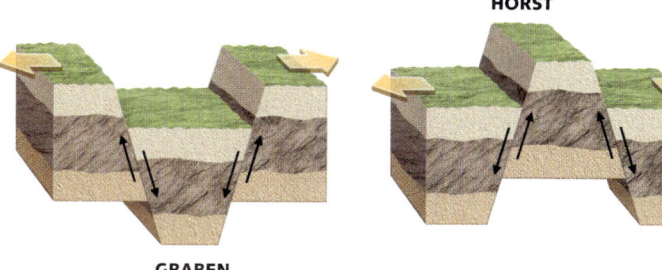

HORST

GRABEN

More about...

The Great Rift Valley
The rift valleys of Africa are a complex system of grabens, horsts, and tilted fault blocks that stretch 5600 km from Ethiopia in the north to Mozambique in the south. The rift, which has been forming over the last 30 million years, averages 30–40 km wide.

Use Transparency 13.

▶ Assess

1. Folded mountains tend to form when landmasses collide.
2. A dome mountain is a nearly circular folded mountain.
3. Volcanic mountains tend to form on continental crust near a subduction zone.
4. Fault-block mountains form when large blocks of rock are uplifted between normal faults.
5. Compression is a squeezing force. Most folded mountains form as two plates converge and collide, squeezing the rocks in between.
6. Student pairs should divide the work equally. Have students use the information presented in the text as well as information from other sources.

11.3 Section Review

1. What type of mountains tend to form when two landmasses collide?
2. What is a dome mountain?
3. Where do volcanic mountains tend to form?
4. How do fault-block mountains form?
5. **CRITICAL THINKING** How is compression involved in forming folded mountains?
6. **PAIRED ACTIVITY** Work with a partner to develop a presentation that illustrates one of the processes by which mountains are formed.

MONITOR AND RETEACH

If students miss . . .

Question 1 List on the board the steps involved in the formation of folded mountains and challenge students to sequence the steps correctly.

Question 2 Have students study the diagrams on page 244 as you read each caption aloud. Pause between sentences so that students can relate the text to the diagrams.

Question 3 Reteach "Volcanic Mountains." (pp. 244–245) Also review the material on subducting boundaries on page 177, Chapter 8.

Question 4 Reteach "Fault-Block Mountains." (p. 245) Refer students back to page 240. Ask: What kind of stress causes normal faults to form? **tension**

Question 5 Demonstrate how compression can form folds by exerting compressive forces on a stack of several sheets of paper or a paperback book.

Question 6 Have students review the chart they created for the Reading Support on page 244.

CHAPTER 11 ACTIVITY

PURPOSE
To explore the locations and features of folded mountains

PROCEDURE

2 The Rockies and the Appalachians are on the North American Plate. The Andes run along the western edge of the South American Plate. The Atlas Mountains are on the African Plate, and the Alps and the Himalayas are on the Eurasian Plate.

3 The Andes are at the boundary between the Nazca Plate and the South American Plate. The Atlas Mountains and the Alps are along the boundary between the African and Eurasian Plates. The Himalayas are at the Indian/Eurasian plate boundary.

5 The mountain ranges identified in Step 3 are found along convergent boundaries. Mid-ocean ridges are divergent plate boundaries. Mountain chains on continents form as the result of convergence, which causes the folding and faulting of the rocks involved. Mountain chains on the seafloor form as the result of divergence and are associated with volcanism.

6 Harrisburg is located in the Appalachian Mountains.

7 Peter's, Third, Second, and Blue mountains appear both on the map and the cross section, which was drawn in a predominantly N-S orientation.

8 Peter's and Second Mountains are both made of rocks that comprise the Pocono Sandstone.

CHAPTER 11 MAP Activity

Folded Mountain Range

SKILLS AND OBJECTIVES
- **Observe** the locations of the major mountain ranges of the world.
- **Correlate** mountain system locations with types of plate boundaries.
- **Compare** and **contrast** continental and oceanic mountain ranges.
- **Interpret** the structure and geology of an area from map and cross section data.
- **Infer** the relative resistance to weathering of various rock types.

MATERIALS
- Appendix B *Physical World Map*, pages 710–711
- Appendix B *Plate Boundaries Map*, pages 712–713
- Appendix B *Physical United States Map*, pages 708–709
- Appendix B *Topographic Map: Harrisburg, Pennsylvania*, page 718
- Tracing paper, 10 cm × 15 cm

Folded mountain ranges are the largest and most complex type of mountains found on continents. Most of these mountains consist of roughly parallel ridges of sedimentary rock. The energy needed to shape thousands of meters of sedimentary rock layers into folded mountains comes from the movements of Earth's lithospheric plates. The major mountain chains of today were formed, or are forming, at plate boundaries. The region around Harrisburg, Pennsylvania, displays some of the classic features of folded mountain ranges. The cross section on the next page shows the structure of the rock layers found in the Harrisburg area.

In this activity, you will explore the locations and features of folded mountains. In Part A, you will compare the locations of major mountain ranges with plate boundaries. In Part B, you will use the cross section on the next page and a topographic map of Harrisburg to study the physical features of folded mountains.

Procedure

Part A: Folded Mountains and Plate Boundaries

1 Use the Physical World Map on pages 710–711 to locate the Rocky, Appalachian, Andes, Atlas, Alps, and Himalayan mountains.

2 Use the Plate Boundaries Map on pages 712–713 to identify and list the plate (North American, Eurasian, African, etc.) on which each mountain range listed in Step 1 is located.

3 How many of the mountain ranges named in Step 1 are directly on or beside a plate boundary? Name the mountain range and identify the two plates that meet at that plate boundary.

4 Use the Physical World Map to locate the Mid-Atlantic Ridge, the East Pacific Rise, the Southeast Indian Ocean Ridge, and the Southwest Indian Ocean Ridge. Each of these mid-ocean ridges is an underwater mountain range.

5 Review the definitions of *divergent boundary* and *convergent boundary* on pages 176 and 177. Which term applies to the mountains identified in your answer to Step 3? Which term applies to mid-ocean ridge systems? In terms of plate motions, explain the basic difference between the formation of mountain chains on continents and the formation of mountain chains on the ocean floor.

Part B: Features of Folded Mountains

6. Using the Physical Map of the United States on pages 708–709, find the location of Harrisburg. In what mountain range is Harrisburg located?

7. Compare the topographic map of Harrisburg on page 718 with the geologic cross section shown at right. List the four mountains that occur on both the map and the cross section. In what general map or compass direction was the cross section drawn?

8. Two of the mountains shown on the cross section are from the same rock formation. Name these mountains and the rock formation.

9. Using Section 11.2 as a reference, what vocabulary term best describes the structure shown in the cross section above?

10. Compare the composition of the ridges to that of the valleys in the Harrisburg area. Which rock types form ridges? Which rock types form valleys?

11. Use tracing paper to outline the cross section above. Label Third Mountain. Locate Stone Glen on the topographic map. Where would it be located on the cross section? Label its location on your tracing paper. Repeat this procedure for Lucknow, the airway beacon, and the WHP TV-tower. Which of these features is located on or formed by the oldest rocks? Which is located on or formed by the youngest rocks?

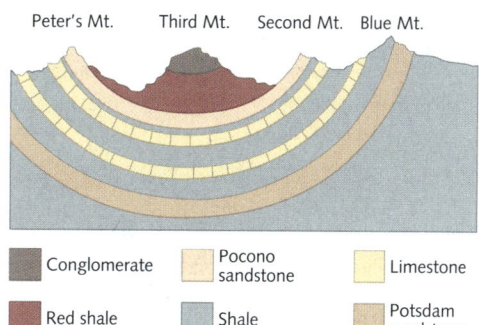

Conglomerate | Pocono sandstone | Limestone
Red shale | Shale | Potsdam sandstone

Analysis and Conclusions

1. The Andes Mountains are younger than the Appalachian Mountains. What evidence on the Physical World Map supports this statement?

2. Would you expect to find extensive mountain building taking place in Australia? Explain your answer.

3. Why is it impossible to determine if sedimentary rock layers have been overturned using only a topographic map?

4. Using the Physical Map of the United States, identify at least three eastern states (not including Pennsylvania) where folded rock layers would be expected to occur at Earth's surface. Why would folded rocks not be expected to occur in the Atlantic Coastal Plain?

CHAPTER 11 ACTIVITY

9. The overall structure is a syncline.

10. Sandstones and conglomerates make up the ridges, whereas limestones and shales make up the valleys.

11. Lucknow is located on the oldest rocks. Third Mountain is made of the youngest rocks.

ANALYSIS AND CONCLUSIONS

1. The Andes are located on an active continental margin and therefore are still forming. The Appalachians are on a passive continental margin. The Appalachians formed millions of years ago when the eastern coast of North America was also an active continental margin. Another indication of age is the overall elevation of each range—the Appalachians have been worn down over time by agents of weathering and erosion.

2. No; Australia is in the center of a tectonic plate and is surrounded by passive continental margins.

3. To determine structural deformation, it is necessary to examine the rocks themselves for features such as ripple marks, cross beds, and mud cracks, which are not shown on topographic maps.

4. Any state in which the Appalachian Mountains are found would be expected to have folded rock layers at Earth's surface. Virginia, West Virginia, Tennessee, and North Carolina are four examples of such places. Folded rock layers would not be expected to occur on the Atlantic Coastal Plain because it is a passive continental margin.

CHAPTER 11 REVIEW

Vocabulary Review

1. anticline, syncline
2. dome mountain
3. continental margin
4. Both joints and faults are breaks in rocks. However, only along faults has movement occurred.
5. Active continental margins occur along plate boundaries. Passive continental margins do not and thus are stable margins.
6. A normal fault occurs as a result of tension; the hanging wall moves down with respect to the footwall. A reverse fault occurs as a result of compression; the hanging wall moves up with respect to the footwall.

Concept Review

7. Most mountain building occurs near active continental margins where two plates converge.
8. Students' sketches should resemble the illustration on page 239.
9. A thrust fault forms when compressive stresses act over great distances. A thrust fault is a reverse fault in which the fault plane dips 45° or less from the horizontal.
10. The Himalayas are folded mountains that are forming as the Eurasian Plate collides with the Indian Plate. The collision is causing rocks to crumple and fold to form mountains.
11. Students who answer "no" should state that volcanism is common along subduction boundaries, which involve at least one oceanic plate. The convergent boundary involved in the formation of the Himalayas is between two continental plates. Students who answer "yes" might argue that slices or blocks of volcanic rock that formed during the subduction between India and Tibet might be found in the Himalayas.
12. Horsts and grabens form when tensional stress produces normal faults. Horsts are large blocks of crust that have been thrust upward between the faults. Grabens are large blocks of crust that have dropped down between the faults.

CHAPTER REVIEW 11

Summary of Key Ideas

11.1 In general, the world's mountains lie in belts that follow Earth's plate boundaries. Mountain building occurs along active continental margins. Large amounts of sediment accumulate in the relatively shallow water of passive continental margins.

11.2 Mountains are made of rocks that have been deformed under stress. Compression, tension, and shear stress are different types of stress. Stresses applied to rocks form folds, faults, and joints. A fault is a break in the lithosphere along which movement has occurred. Normal faults occur where tension pulls the crust apart. Reverse faults and thrust faults are caused by compression. Strike-slip faults are associated with shear stress. A joint is a break in the bedrock along which no apparent movement has occurred.

11.3 Mountains can be classified by features that result from forces associated with plate tectonics. Folded mountains can form when two plates carrying continents collide. Dome mountains are nearly circular folded mountains that result from uplifting. Fault-block mountains, horsts, and grabens are associated with normal faults.

Key Vocabulary

anticline (p. 239)
continental margin (p. 236)
dome mountain (p. 244)
fault-block mountains (p. 245)
folded mountains (p. 243)
joint (p. 241)
mountain (p. 236)
normal fault (p. 240)
reverse fault (p. 240)
strike-slip fault (p. 241)
syncline (p. 239)
thrust fault (p. 240)

248 Unit 3 Dynamic Earth

13. Types of stress are compression, *tension*, and *shear*. Compression is associated with reverse and *thrust* faults. Tension is associated with *normal* faults. Shear is associated with strike-slip faults.

Vocabulary Review

Write the vocabulary term or terms that best complete the sentence.

1. An upfold in rock layers is called a(n) ___?___ , and a downfold in a rock layer is called a(n) ___?___ .
2. Depending on how it formed, a(n) ___?___ can be classified as plutonic or tectonic.
3. The boundary between continental crust and oceanic crust is called a(n) ___?___ .

Explain the difference between the terms in each pair.

4. fault; joint
5. active continental margin; passive continental margin
6. normal fault; reverse fault

Concept Review

7. How are continental margins related to mountain building?
8. Sketch some folded rock layers and label the following features: anticline, syncline, limb.
9. Explain how a thrust fault is formed. How is a thrust fault related to a reverse fault?
10. What type of mountains are the Himalayas? How did the Himalayas form?
11. Would you expect to find some volcanic rock in the Himalayas? Why?
12. How are horsts and grabens similar? How are they different?
13. **Graphic Organizer** Complete the concept map below.

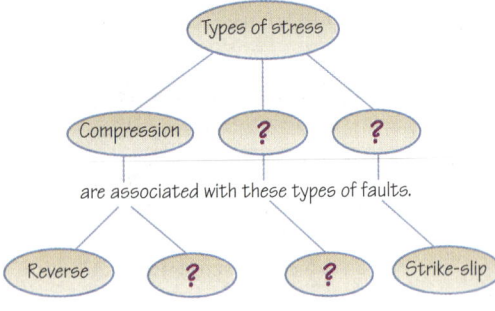

Critical Thinking

14. The west coast of Africa is a passive continental margin. Active continental margins occur along plate boundaries.
15. probably an eroded anticline because relatively old rocks are flanked by younger rocks

Critical Thinking

14. **Draw Conclusions** Use the Physical World Map on pages 710–711. Would you expect the west coast of Africa to be an active or a passive continental margin? Explain.

15. **Apply** A geologist standing on a low hill notes that rock layers stand out in sharp ridges all around the edge of the hill and dip gently away in all directions. The oldest rocks are at the center of the hill, and the rock layers away from the center are younger. What kind of structure could the geologist be standing on? Explain.

16. **Analyze** Plutonic and tectonic dome mountains can have an igneous rock core. How could contact metamorphism, which was discussed in Section 6.4, be used to determine whether a dome mountain is plutonic or tectonic?

Internet Extension

What Forces Created These Geologic Features?
Examine images of faults and folds. Annotate the images to show the forces and conditions that created them.
Keycode: ES1106

Writing About the Earth System

SCIENCE NOTEBOOK Explain how the volcanic mountains in the Cascade Range were formed. Identify the stages at which the hydrosphere, the geosphere, and the atmosphere were involved in the formation of volcanic mountains like those in the Cascade Range.

Interpreting Diagrams

The diagram below shows a cross section of the sedimentary rock layers in a folded area. The cross section was drawn from west to east along a line about 50 kilometers long. Points *A* through *F* are locations along the ground surface. The rock layers have not been overturned.

17. On which side of the anticline do the rock layers have the greater dip? On which side of the syncline do the rock layers have the greater dip?

18. How do the ages of the rocks change from point *B* to point *D*? From point *E* to point *D*?

19. On the basis of your answer to question 18, how does the age of rocks at the center of an anticline compare with the age of rocks at the center of a syncline?

20. If a normal fault occurs at line *XY*, what will be the effect on the distance between points *A* and *F*? What will be the effect on the distance if the fault is a reverse fault? Explain.

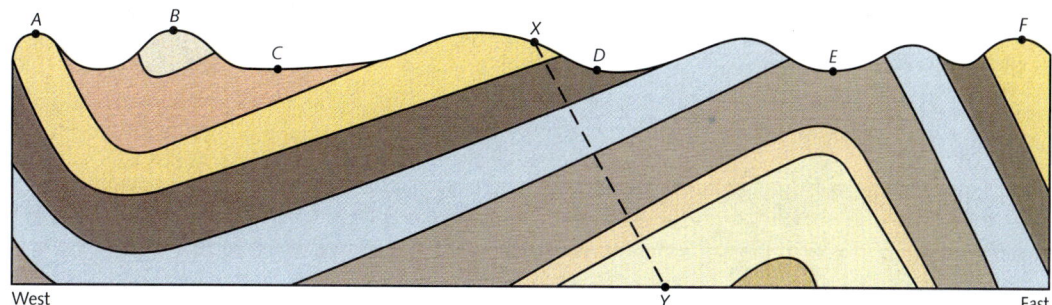

CHAPTER 11 REVIEW

What Forces Created These Geologic Features?
Keycode: ES1106

Use Internet Investigations Guide, pp. T43, 43.

(continued)

16. As an igneous plutonic body intrudes into existing rock, contact metamorphism occurs and would indicate that the mountain was plutonic. A tectonic dome mountain forms when internal forces uplift rock layers. There is generally no metamorphism involved.

INTERPRETING DIAGRAMS

17. The rocks that make up the eastern limb of the anticline dip more steeply than the rocks that make up the western limb. The western limb of the syncline dips more steeply than the eastern limb.

18. From point *B* to point *D*, the rocks get older. From point *E* to point *D*, the rocks get younger.

19. The rocks at the center of an anticline are older than the rocks at the center of a syncline.

20. The distance between points *A* and *F* will increase with normal faulting, as tensional forces pull the rock apart. The distance between the points will decrease as the result of reverse faulting, which is caused by compressive forces.

WRITING ABOUT SCIENCE

Writing About the Earth System

The volcanic mountains in the Cascade Range formed as the result of oceanic-continental convergence between the Juan de Fuca and North American plates. The small Juan de Fuca Plate was subducted beneath the North America Plate and partly melted. The magma was forced upward and gave rise to the volcanic peaks in the Cascades, which include Mount St. Helens, Mt. Shasta, Mt. Rainier, and Lassen Peak, among others. The geosphere, or solid Earth, was involved during convergence. The hydrosphere—Earth's water—was also involved during convergence. The atmosphere was and is still affected each time a volcano erupts.

UNIT 3 FEATURE

Focus

Objective
Identify positive as well as negative effects of volcanoes on the Earth system, especially the biosphere.

Set a Purpose
Have students read to find answers to the question "What are some ways in which volcanic eruptions can actually benefit living things?" Volcanic eruptions can provide new habitats for wildlife. Also, they allow some species to blossom as part of the natural disturbance/renewal cycles of local ecosystems.

The EARTH SYSTEM

UNIT 3

When Volcanoes Breathe Life

People can only imagine the very early features of Earth. It probably took many millions of years for a barren surface, shaken by earthquakes and scorched by volcanoes, to become hospitable to living things.

Today Earth is home to countless species living in a variety of unique ecosystems. Earth experiences regular episodes of volcanic activity, some of which permanently alter ecosystems. How do volcanic events, originating under Earth's crust, affect the biosphere?

In reading about volcanoes, you have learned about some of the destructive results of such eruptions. Volcanoes can also have positive effects on life and the environment.

Consider volcanic islands. Hawaii's islands, for example, formed and continue to form as the Pacific Plate moves slowly over a volcanic hot spot. When these volcanic islands first formed, they were completely barren. Over time, seeds and spores carried by wind from the distant mainland sometimes fell on the stark terrain. Some of these bits of life sprouted and took root, thus transforming the island environment. Eventually, marine life, birds, and other "accidental travelers" arrived on the island. Some thrived in the isolated habitat, because it was free of competitors. One such accidental traveler was a bird known today as the Hawaiian honeycreeper.

A few million years ago, the birdless shores of Hawaii offered a lush collection of foods. So when one finch-like species accidentally found its way to Hawaii, it had little competition and a breathtaking array of food choices. As a result, the original finch ancestor gave rise to many separate but closely related species, each of which developed to eat a particular type of food. Some are able to feed on the nectar of a variety of Hawaiian flowers, because they have developed long, curved bills which precisely fit those flowers. Other finches developed needle-like beaks adapted for catching and eating insects. Still others have short, squat bills that efficiently crack seeds and nuts.

The eruption of Mount St. Helens in the early 1980s was enormously destructive. In the rush of burning ash and lava, thousands of flattened trees blazed, while tons of ash piled up on what had been a lush forest floor.

Scientists who ventured into the area shortly after the blast first saw a desolate landscape. On closer inspection they were surprised to find that some plants and animals

HAWAIIAN HONEYCREEPER
This bird has a specialized bill for feeding on Hawaiian flowers.

DIFFERENTIATING INSTRUCTION

Reading Support Creating a graphic summary can help students comprehend and retain key information from their reading. Have students summarize the main idea and supporting details of "When Volcanoes Breathe Life" in a chart like this one.

Main Idea	Volcanoes can benefit living things.
Supporting Details	Example 1: Hawaii
	Example 2: Mount St. Helens

and the ENVIRONMENT

NEW GROWTH is shown colonizing the Spirit Lake area of Mount St. Helens.

had survived. Pockets of snow had shielded some areas of vegetation from volcanic debris. Following the eruption, these areas attracted new organisms to the region.

Some mammals, such as the northern pocket gopher and the Pacific jumping mouse, survived the eruption because they had burrowed beneath the ground. Underground ant colonies also escaped destruction.

Surviving plants such as prairie lupine and red alder were quick to take root in the ash-strewn soil of the area. Unlike most other plants, these plants have special bacteria in their roots that fix nitrogen into a biologically usable form. When lupines and alders decompose, they enrich the ashy soil with usable nitrogen. This makes the soil more fertile for plants requiring nitrogen.

Ecologists at Mount St. Helens have even discovered that in years following the blast some kinds of amphibians thrived in the area. Protected by snowpack and frozen ponds, the amphibians woke after the snowmelt to more sunlight and far fewer predators. Their populations boomed.

Scientists now hypothesize that events such as the blast that shook Mount St. Helens are a natural part of some ecosystems' disturbance routines. These events display the interaction between the lithosphere and the biosphere.

UNIT INVESTIGATION
CLASSZONE.COM

How Can One Volcano Change the World? Mount Pinatubo, a volcano on the island of Luzon in the Philippines, erupted explosively in June 1991. This geosphere event let to changes in all of Earth's spheres. Investigate before and after images from space, as well as photos and animations of damage. Check fact sheets about the eruption and about evacuation efforts. Discover how the eruption affected the local environment as well as the global one.
Keycode: ESU301

PRAIRIE LUPINES thrive in volcanic ash and enrich soil with nitrogen when they die and decompose.

UNIT 3 FEATURE

INVESTIGATIONS
CLASSZONE.COM

How Can One Volcano Change the World?
Keycode: ESU301

Use Internet Investigations Guide, pp. T44–T45, 44–45.

Investigation Photograph: Mount Pinatubo produced smoke and ash plumes 18–40 kilometers high.

UNIT 3 ASSESSMENT

ANSWERS

1. c, both are divergent boundaries
2. c, the islands are located within the Pacific plate, not at its boundaries
3. d, that feature located closest to the surface is most recent
4. c, loose soil takes on the properties of a liquid with the violent shaking that accompanies an earthquake
5. a, S waves will not travel through liquid: S waves stop at the outer core; changes in P-wave velocity suggest only the outer core is liquid, the inner core is solid
6. a, the plates are moving apart, not colliding, at a deep-sea trench
7. d, tension pulls the crust apart, so these are normal faults
8. a, outer core is liquid
9. a, uplift causes crust to stretch so normal faults result
10. c, high silica content means the magma is resistant to flow, so gases build up
11. b, lower silica content means the magma flows more easily, less buildup of gases
12. Sedimentary layers are deposited horizontally. Therefore, if a layer is tilted, then a force associated with mountain building was involved in tilting it.
13. The side of the rock layer that was originally on the bottom is now on the top
14. studying ripple marks, studying mud cracks

UNIT 3 STANDARDIZED TEST Practice

Directions (1–11) For *each* statement or question, select the word or expression that, of those given, best completes the statement or answers the question. Record your answers on a separate answer sheet.

1. At the Mid-Atlantic Ridge and the East Pacific Rise, lithospheric plates are generally
   (a) colliding
   (b) moving over a hot spot
   (c) moving apart
   (d) subducting

2. The Hawaiian Islands formed
   (a) at a subduction boundary between two oceanic plates
   (b) at a mid-ocean ridge
   (c) over a hot spot
   (d) at a subduction boundary between an oceanic plate and a continental plate

3. If a lava flow covers an impact crater on the moon, scientists may conclude that
   (a) the lava flow caused the crater to form
   (b) the lava flow is older than the crater
   (c) the lava flow is the same age as the crater
   (d) the lava flow is younger than the crater

4. Liquefaction is most likely to affect buildings when
   (a) the buildings are located on granite bedrock
   (b) the buildings are located near a lake
   (c) the buildings are located on loose sediments
   (d) the buildings are located on limestone bedrock

5. Scientists infer that Earth's outer core is liquid based primarily on their studies of
   (a) earthquake waves
   (b) magnetic polarity reversals
   (c) lithospheric heat flow
   (d) magma

6. Which of the following would NOT be found at or near a subduction boundary?
   (a) deep-sea trench
   (b) mountains and volcanoes
   (c) subducting oceanic crust
   (d) subducting continental crust

7. A graben is bounded on each side by
   (a) strike-slip faults   (c) reverse faults
   (b) thrust faults   (d) normal fault

8. Which of the following cannot transmit S waves?
   (a) outer core   (c) mantle
   (b) asthenosphere   (d) lithosphere

9. Fault-block mountains are associated with
   (a) normal faults   (c) thrust faults
   (b) reverse faults   (d) strike-slip faults

Base your answers to questions 10 and 11 on the table below.

Characteristics of Four Magma Samples				
	Magma A	Magma B	Magma C	Magma D
Silica content	60%	52%	69%	62%
Gas content	4%	1%	4.5%	3.5%
Melting point	Medium	High	Low	Medium

10. The magma most likely to produce an explosive eruption is
    (a) magma A   (c) magma C
    (b) magma B   (d) magma D

11. The magma most likely to produce a nonexplosive eruption is
    (a) magma A   (c) magma C
    (b) magma B   (d) magma D

252 Unit 3 Dynamic Earth

UNIT 3 ASSESSMENT

Directions (12–21) Record your answers on a separate answer sheet. Base your answers to questions 12 through 17 on the reading passage below. The reading passage discusses some methods that geologists use to determine whether rock layers have been overturned.

Studying Tilted Rock Layers

Most sedimentary rocks are formed in level layers. Therefore, the occurrence of tilted rock layers is evidence of mountain building.

Rock layers can become tilted in a number of ways. The folding of sedimentary rocks into anticlines and synclines is one way. Tilting can also result when rocks are pushed upward, or uplifted.

In some areas of the world, rock layers are so severely tilted that they may be bottom side up. Geologists use a number of methods to determine whether rock layers have been overturned.

One method involves by studying ripple marks. These features form on the floor of a quiet body of water when waves are moving gently across the surface. Ripple marks consist of miniature valleys between sharply pointed tiny hills. A rock layer containing ripple marks is right side up if the sharp hills point up.

Geologists can also study mud cracks to determine whether rock layers have been overturned. These features develop on the surfaces of areas such as mud flats when the mud and ooze dry. Individual cracks are wider at the top than at the bottom. When the cracks are preserved in rocks, they are still wider at the top if the layer is right side up.

12. Why is a tilted sedimentary rock layer evidence of mountain building?
13. What does it mean to say that a rock layer has been overturned?
14. List two different methods a geologist can use to determine whether a rock layer has been overturned.
15. Under what circumstances might ripple marks form in a rock layer?
16. Suppose a geologist finds vertical mud cracks in a layer of mudstone. The cracks are narrower at the top than at the bottom. What can the geologist conclude?
17. Could a geologist use the methods described in the passage to determine whether an igneous rock layer has been overturned? Explain your thinking.

Base your answers to questions 18 through 21 on the fault diagram below.

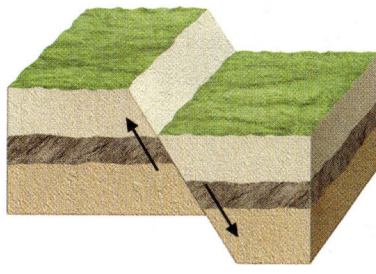

18. Explain how to identify whether the fault is a normal fault, a thrust fault, or a reverse fault.
19. What type of stress is associated with the fault shown above?
20. Explain how the fault shown is different from a strike-slip fault.
21. According to the elastic-rebound theory, why do earthquakes tend to occur along faults?

(continued)

15. If the sediments in a rock layer were originally on the bottom of a quiet body of water, ripple marks may have been left on top of the sediments. These marks remain when the sediments become sedimentary rock.
16. the rock layer has been overturned
17. No; ripple marks and mud cracks occur in sediments that later may become sedimentary rock. Igneous rock layers would not have these features.
18. The fault is a normal fault because the hanging wall has moved down relative to the footwall.
19. tension, the crust is pulling apart
20. At a normal fault, the hanging wall moves down with respect to the footwall. At a strike-slip fault, the rocks on either side of the fault plane move horizontally in opposite directions. The type of stress associated with a normal fault is tension, while the type of stress associated with a strike-slip fault is shearing.
21. Along a fault, friction prevents blocks of crust from moving past each other. Strain builds up at these points and the rocks along the fault deform. Eventually, the strain becomes great enough to overcome friction, and an earthquake occurs. That is, the blocks of crust snap back to the shape they had before they were deformed at new locations relative to each other.

UNIT 4

UNIT OVERVIEW

When introducing this unit, ask students for specific examples of natural features of Earth's surface that they are familiar with. Use the chapter images as springboards and ask if any of the features shown are common to the local area. Then have students work in pairs to predict the answers to the caption questions. With the exception of the action of wind and waves, most Earth-shaping forces discussed in this unit work over very long periods of time, far longer than a human life span.

VISUAL TEACHING

Chapter 12 Weathering, Soil, and Erosion

Many hikers visiting Bryce Canyon National Park choose to walk the Navajo Loop Trail that makes its way up and down Bryce Canyon. The Navajo Loop Trail is lined with rock pillars called hoodoos. Students will learn about processes that can produce such rock formations.

Answer: Over time, erosion by wind and water have carved away the sedimentary rock layers, leaving behind these distinctive rock formations.

Chapter 13 Surface Water

The Colorado River has cut through about 1.6 kilometers of the Colorado Plateau as the plateau has slowly uplifted. Students will learn how surface water can change the shape of the land.

Answer: about 5.5 million years

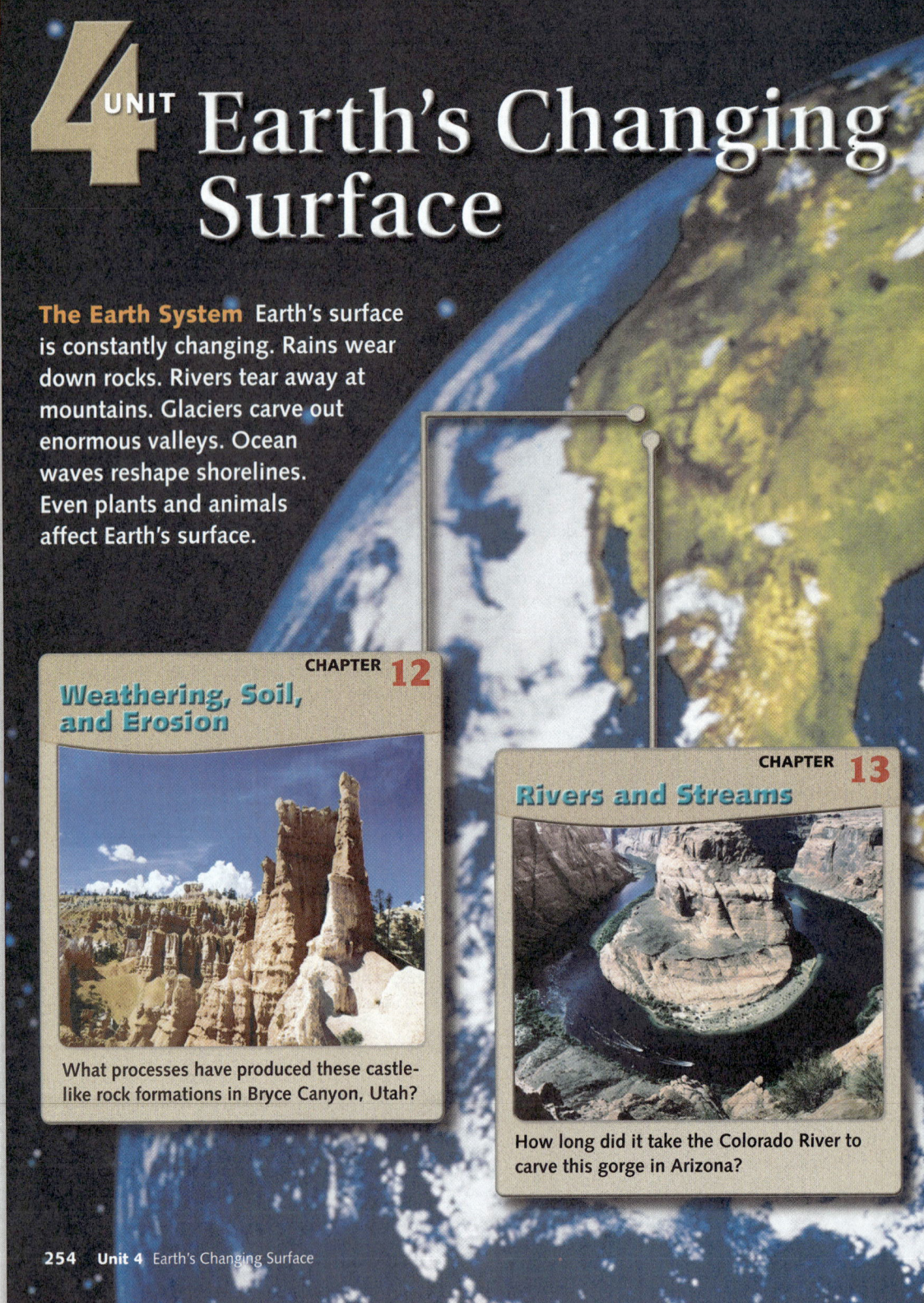

UNIT 4 Earth's Changing Surface

The Earth System Earth's surface is constantly changing. Rains wear down rocks. Rivers tear away at mountains. Glaciers carve out enormous valleys. Ocean waves reshape shorelines. Even plants and animals affect Earth's surface.

CHAPTER 12 Weathering, Soil, and Erosion
What processes have produced these castle-like rock formations in Bryce Canyon, Utah?

CHAPTER 13 Rivers and Streams
How long did it take the Colorado River to carve this gorge in Arizona?

CONNECTIONS TO...

Prior Knowledge In Unit 3, students learned how processes inside Earth form and change rocks and minerals. In Unit 4, they will learn how surface processes—weathering and erosion by wind and water—change rocks and minerals over time. Students will build upon this knowledge in Units 5 and 6 as they study the atmosphere and oceans in greater depth.

Unifying Themes The *constancy* of interactions among Earth's spheres, in the form of weathering and erosion, are the primary cause of *change* on Earth's surface. As students study the effects of weathering and erosion in Unit 4, reinforce how interactions among the hydrosphere, geosphere, biosphere, and atmosphere change the shape of the land.

UNIT 4

Chapter 15 Glaciers
How quickly does this glacier on Baffin Island move?

Chapter 14 Groundwater
What role did groundwater play in the formation of Mammoth Cave in Kentucky?

Chapter 16 Wind, Waves, and Erosion
How does the wind shape this Georgia sand dune?

Chapter 14 Groundwater
Mammoth Cave is the longest recorded cave system in the world. Over 540 kilometers of caves spanning multiple levels have been explored and mapped. Students will learn how groundwater can slowly produce these and other features near Earth's surface.

Answer: The cave formed as carbonic acid in groundwater slowly dissolved portions of the limestone bedrock.

Chapter 15 Glaciers
Canada's Baffin Island, the fifth largest island in the world, is covered with over 10,000 glaciers. Highway Glacier, shown in the photograph, is the largest of these glaciers. Students will learn how glaciers produce different landforms as they erode and deposit materials.

Answer: about 56 meters each year

Chapter 16 Wind, Waves, and Currents
Wild-turkey tracks cut across ripples on a sand dune on Cumberland Island along the coast of Georgia. In deserts and on beaches, wind blows loose sand into dunes that can reach more than 500 meters in height and migrate up to 25 meters per year. Students will learn about the role of wind and waves as agents of change.

Answer: The wind piles loose sand up to form dunes and then blows across the dune, causing ripples to form.

CONNECTIONS TO . . .

National Science Education Standards
- Students will come to understand that Earth's surface changes over long periods of time as a result of erosion and deposition of rock material.
- By studying interactions among Earth's hydrosphere, geosphere, biosphere, and atmosphere, students will develop an understanding of the ongoing evolution of the Earth system.

Unit Features
The Earth System and the Environment
Harnessing Earth's Water, pp. 358–359

What Are the Costs and Benefits of Damming a River? Analyze the costs and benefits of Glen Canyon Dam. *Keycode:* ESU401

Standardized Test Practice, pp. 360–361

CHAPTER **12** PLANNING GUIDE **WEATHERING, SOIL, AND EROSION**

	Section Objectives	Print Resources
Chapter-Level Resources		Laboratory Manual, pp. 51–54 Internet Investigations Guide, pp. T48–T49, 48–49 Guide to Earth Science in Urban Environments, pp. 5–12, 45–52 Spanish Vocabulary and Summaries, pp. 23–24 Formal Assessment, pp. 34–36 Alternative Assessment, pp. 23–24
Section 12.1 Weathering, pp. 258–263	1. Describe how mechanical weathering breaks down rocks. 2. Describe how chemical weathering breaks down rocks. 3. Name three factors that affect weathering rates.	Lesson Plans, p. 39 Reading Study Guide, p. 39
Section 12.2 Soil, pp. 264–267	1. Explain how soil forms. 2. Describe soil composition and the factors that affect it.	Lesson Plans, p. 40 Reading Study Guide, p. 40
Section 12.3 Mass Movements and Erosion, pp. 268–270	1. Give examples of mass movement. 2. Explain how erosion reshapes Earth's surface.	Lesson Plans, p. 41 Reading Study Guide, p. 41
Section 12.4 Soil as a Resource, pp. 271–273	1. Describe the ways in which human activity threatens soil fertility. 2. Summarize soil conservation methods that reduce soil erosion.	Lesson Plans, p. 42 Reading Study Guide, p. 42

INVESTIGATIONS
CLASSZONE.COM

Section 12.3 INVESTIGATION
When Is Mud Dangerous?
Computer time: 45 minutes / Additional time: 45 minutes
Students analyze the conditions that create potential mudflows.

Chapter Review INVESTIGATION
How Does Soil Vary from Place to Place?
Computer time: 45 minutes / Additional time: 45 minutes
Students compare and contrast soils from different regions.

Technology/Online Resources

Electronic Teacher Tools
Visualizations CD-ROM
Classzone.com
Online Lesson Planner
Test Generator

Visualizations CD-ROM
Classzone.com
  ES1201 Visualization, ES1202 Visualization

Classzone.com
  ES1203 Data Center

Visualizations CD-ROM
Classzone.com
  ES1204 Investigation, ES1205 Visualization

Classroom Management

Gather these materials for minilab and laboratory activities.

Minilab, p. 263
  two clear containers
  water
  two steel wool pads
  scissors

Lab Activity, pp. 274–275
  lab apron
  safety goggles
  5 250-mL beakers
  5 thermometers
  hot water (40–50°C)
  ice water
  5 effervescent antacid tablets
  stopwatch
  graph paper
  Map of Earth's Climates, pp. 470–471
  Teaching Transparency 25
  Laboratory Manual, Teacher's Edition
    Lab Sheet 12, p. 141

CHAPTER 12
Weathering, Soil, and Erosion

INTRODUCE

Build Interest
Before sharing background information about the photograph with students, allow time for their questions and comments.

Ask questions like the following:
- How would you describe the landforms you see here? **tall, jagged columns and cliffs of reddish brown rock**
- How do you think this place looked a million years ago? **The main column shown here was less fragile because it was not as eroded. Other formations existed that have since completely eroded.**
- Where else have you seen rocks like these? **Discuss local and distant places that show erosion of rock.**

Respond
- Why do some rock layers in the canyon wear away more quickly than others? **Some layers are composed of harder rocks and minerals, making them more resistant to weathering and erosion.**

About the Photograph
The tall pillars of rock in Bryce Canyon, called hoodoos, are the result of weathering by water and ice of sedimentary rock deposited about 60 million years ago. The reddish colors of the hoodoos are due to iron compounds in the rock. The hammer-shaped hoodoo in the photograph is called Thor's Hammer.

Over time, rainwater has worn down and eroded the rocks in Bryce Canyon, Utah.

Why do some rock layers in the canyon wear away more quickly than others?

BEYOND THE TEXTBOOK

Print Resources
- Reading Study Guide, pp. 39–42
- Transparency 25
- Formal Assessment, pp. 34–36
- Laboratory Manual, pp. 51–54
- Alternative Assessment, pp. 23–24
- Spanish Vocabulary and Summaries, pp. 23–24
- Internet Investigations Guide, pp. T48–T49, 48–49
- Guide to Earth Science in Urban Environments, pp. 5–12, 45–52

Technology Resources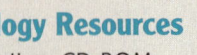
- Visualizations CD-ROM
- Classzone.com
- Online Lesson Planner
- Electronic Teacher Tools
- Test Generator

256 Unit 4 Earth's Changing Surface

CHAPTER 12

PREVIEW

▶ **FOCUS QUESTIONS** In this chapter you will study weathering, soil, and erosion and learn more about the key questions below.

Section 1 How does weathering break down rock materials?

Section 2 How does soil form?

Section 3 How does erosion affect Earth's surface?

Section 4 What measures can be taken to protect soil as a resource?

▶ **REVIEW TOPICS** As you investigate weathering, soil, and erosion, you will need to use information from earlier chapters.

- ions (p. 93)
- mineral (p. 96)
- rock cycle (p. 119)

▶ **READING STRATEGY**

SET A PURPOSE

The focus questions at the top of this page can help you set a purpose in your reading. As you read each section of Chapter 12, keep the focus question for that section in mind. In your notebook, write down ideas and information from the section that answer the focus question.

At our Web site, you will find the following Internet support for this chapter.

DATA CENTER
EARTH NEWS
VISUALIZATIONS
- Mechanical Weathering Effects
- Chemical Weathering
- Landscape Formation by Erosion

LOCAL RESOURCES
INVESTIGATIONS
- When Is Mud Dangerous?
- How Does Soil Vary from Place to Place?

CHAPTER 12

PREPARE

Focus Questions
Find out what knowledge students have about weathering, soil, and erosion. For each focus question, have students consider the section title and then develop two or three additional questions for each section. For example, Section 1: How long does it take weathering to break down rock? Section 2: What materials are in soil? Section 3: How have the landforms where I live been affected by erosion?

Review Topics
In addition to the earth science topics listed, students may need to review this concept:

Mathematics
- Consider one cubic meter of rock with a surface area of 6 square meters. Breaking that rock into 8 smaller cubes of 1/2 cubic meter each increases the surface area to 24 square meters.

Reading Strategy
Model the Strategy: Have students write the focus question for Section 1 at the top of a page in their notebook. Then have them find the two types of weathering (mechanical and chemical) in the first paragraph on page 258 and list them in two columns beneath the focus question in their notebook. As students read the section, tell them to make a bulleted list of the different ways that mechanical and chemical weathering each breaks down rock. Direct students to use the focus question for each section the same way as they work through the chapter.

BEYOND THE TEXTBOOK

INVESTIGATIONS
- When Is Mud Dangerous?
- How Does Soil Vary from Place to Place?

VISUALIZATIONS
- Images showing the effects of mechanical weathering
- Animation of chemical weathering
- Animation of erosion forming a landscape

DATA CENTER
EARTH NEWS
LOCAL RESOURCES
CAREERS

Online Lesson Planner

CHAPTER 12 SECTION 1

FOCUS

Objectives
1. Describe how mechanical weathering breaks down rocks.
2. Describe how chemical weathering breaks down rocks.
3. Name three factors that affect weathering rates.

Set a Purpose
Have students use the section title to form a question, such as "What is weathering and how does it occur?" Students should then read this section to answer their question.

INSTRUCT

Visual Teaching

Interpret Diagrams
Have students refer to the diagrams as they read about ice wedging. Then ask: What conditions exist in the first diagram that can lead to ice wedging? *Cracks have formed in the rock and rain enters them.* What change occurs that causes water to freeze and expand in the second diagram? *The temperature drops below the freezing point of water.* What might eventually happen to the rock in the second diagram? *Pieces of the rock might break off.*

12.1

KEY IDEA
Over time, rocks are broken down by mechanical and chemical weathering.

KEY VOCABULARY
- weathering
- mechanical weathering
- chemical weathering
- frost wedging
- abrasion
- exfoliation
- hydrolysis
- acid rain
- oxidation

Weathering

Granite forms deep underground, where pressures and temperatures are high, and water and oxygen are lacking. When granite is pushed up to Earth's surface, pressures and temperatures decrease dramatically. The granite is exposed to water and oxygen. As a result of these changes, the granite begins to weather. **Weathering** is the breakup of rock due to exposure to processes that occur at Earth's surface. Geologists group weathering processes under two headings—mechanical weathering and chemical weathering. **Mechanical weathering,** or disintegration, takes place when rock is split or broken into smaller pieces of the same material without changing its composition. Mechanical weathering is also called physical weathering. **Chemical weathering,** or decomposition, takes place when the rock's minerals are changed into different substances. Although mechanical and chemical weathering are often studied separately, they almost always act together.

Mechanical Weathering

Mechanical weathering processes include frost wedging, wetting and drying, abrasion by rock materials, plant and animal activity, and exfoliation that occurs as a result of upward expansion.

Ice and Water
Water occupies about 10 percent more space when it freezes. This expansion puts great pressure on the walls of a container. Consider a pail of water left outdoors in freezing weather. The pail may split open when the water freezes. In the same way, water held in the cracks of a rock wedges the rock apart when it freezes. This process is called **frost wedging,** or ice wedging, and is common in places where the temperature varies from below the freezing point of water (0°C) to above the freezing point. It occurs mostly in porous rocks and in rocks with many cracks.

FROST WEDGING The process of frost wedging widens cracks in rocks.

Frost Wedging

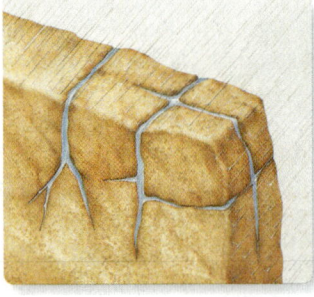

① Rainwater enters existing cracks in a rock.

② The water expands as it freezes, wedging the rock apart.

DIFFERENTIATING INSTRUCTION

Reading Support Teach students to look for cause-and-effect relationships. First, explain that a cause is an event or a circumstance that brings about another event or circumstance. The resulting event or circumstance is the effect. On page 259, point out the sentence "Repeated swelling and shrinking cause the rocks to break apart." In this example, repeated swelling and shrinking is the cause, and the rocks breaking apart is the effect. Have students identify the cause-and-effect relationship in the last sentence on page 259. *Upward expansion is the cause, and granite breaking along curved joints is the effect.*
Use Reading Study Guide, p. 39.

258 Unit 4 Earth's Changing Surface

Frost wedging also occurs in places where there are frequent freezes and thaws. Bare mountaintops are especially susceptible to frost wedging. The vast fields of large, sharp-cornered boulders that are often found on mountaintops are pieces of the mountain that have been broken off by frost wedging.

Frost wedging also causes potholes to form on paved streets and highways. Here the process is helped by ice heaving, which happens when water in the ground freezes and lifts the pavement above it. When the ice thaws, the pavement collapses, leaving a pothole.

In addition, liquid water plays a role in mechanical weathering. Repeated wetting and drying can break up shale and other rocks that contain clay. These clays, which absorb water easily, swell up when they are wet and shrink when they are dry. Repeated swelling and shrinking cause the rocks to break apart.

Abrasion

Water, wind, and ice are capable of moving rocks. Water can tumble sand, pebbles, and even boulders along streambeds. Wind can blow sand and small pebbles across a rocky plain. Glaciers, which are moving masses of ice, carry rocks over great distances. Even gravity moves rocks.

As moving sand, pebbles, and larger rocks grind and scrape against one another, these rock materials are worn away. This type of mechanical weathering is called **abrasion.** The sand that you walk on at the beach is a product of abrasion. Rocks and pebbles get ground down into fine particles of sand as they are carried by rivers, streams, and ocean waves.

Plants and Animals

The growth of plants and the activities of animals contribute to the mechanical weathering of rock. When lichens (LY-kuhnz), mosses, and other small plants grow on rocks, they wedge their tiny roots into pores and crevices. As the roots grow, the rock splits. Larger shrubs and trees may grow through cracks in boulders, causing the cracks to widen.

Ants, earthworms, rabbits, woodchucks, and other animals dig holes in the soil. These holes allow air and water to reach the bedrock and weather it. Burrowing animals also bring rock fragments to the surface, where the fragments weather more rapidly.

Upward Expansion

The upward expansion of rocks that are formed deep underground may result in a mechanical weathering process called **exfoliation** (ehks-FOH-lee-AY-shuhn). For example, granite becomes exposed when it is lifted up and the rocks above it are worn away. The removal of these overlying rocks reduces the pressure along the surface of the granite. Upward expansion causes the granite to break along curved joints that are parallel to the surface. Such joints can be seen in exposed

Observe the effects of mechanical weathering.
Keycode: ES1201

TREE ROOTS This tree's roots are growing between cracks in the rock below. As the tree grows, the roots widen the cracks.

CHAPTER 12 SECTION 1

Observe the effects of mechanical weathering.
Keycode: ES1201
Have students describe the cause of the mechanical weathering in each image.

 Visualizations CD-ROM

DISCUSSION

If necessary, distinguish between exfoliation and upward expansion and clarify their relationship. Explain that exfoliation is the breaking away of loosened sheets of rock. Upward expansion involves uplifting and subsequent expansion of rock, which causes the rock to break along curved joints, possibly leading to exfoliation.

DIFFERENTIATING INSTRUCTION

Hands-On Demonstration Use a rock sample with a relatively smooth surface, such as a piece of shale, and a piece of coarse sandpaper to demonstrate abrasion. Over a piece of newspaper, rub the sandpaper several times over the surface of the rock. Tell students that the sandpaper is like the tiny sand particles carried by wind, water, or ice. These transported particles scrape against rocks, causing abrasion. Allow students to examine the surfaces of the rock and sandpaper, as well as any abraded material that fell on the newspaper. Relate the abrading of the rock to the mechanical weathering process of abrasion.

CHAPTER 12 SECTION 1

Visual Teaching

Interpret Diagrams
Use the diagram to explain how the exfoliation dome in the photograph formed. Ask: How does granite become exposed at the surface? **Tectonic forces uplift the granite, and the overlying rock wears away.** How might the overlying rock wear away? **Sample Response: Weathering breaks the rock apart. Then wind or water carries the pieces away.** In what other forms of mechanical weathering does expansion play a role? **frost wedging and ice heaving** At what step in the diagram does exfoliation occur? **step 4**

Extend
Have students review Chapter 11 and then describe the tectonic forces that can uplift granite. **Students should describe the uplift that occurs when plates converge and rocks become folded or faulted. They should also describe the uplift that occurs as a result of an intrusion of magma.**

peaks or outcrops. Over time, large sheets of loosened rock break away from an exposed outcrop, as shown in the diagram below.

Exfoliation forms rounded mountain peaks called exfoliation domes. Spectacular exfoliation domes occur in Yosemite National Park, California. Other exfoliation domes include Stone Mountain in Georgia and Sugarloaf Mountain near Rio de Janeiro, Brazil.

EXFOLIATION Sheets of granite break away from Half Dome, an exfoliation dome in Yosemite National Park, California.

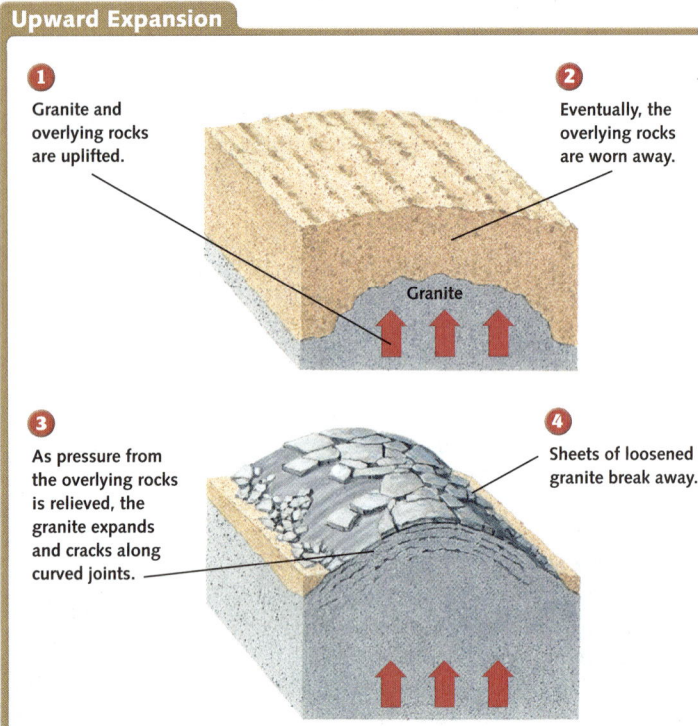

Upward Expansion
1. Granite and overlying rocks are uplifted.
2. Eventually, the overlying rocks are worn away.
3. As pressure from the overlying rocks is relieved, the granite expands and cracks along curved joints.
4. Sheets of loosened granite break away.

Chemical Weathering

Chemical weathering occurs almost everywhere because water or water vapor is found almost everywhere. All chemical weathering involves water or water vapor. Other agents of weathering include acids and oxygen.

Water and Chemical Weathering

The chemical weathering by reaction of water with other substances is called **hydrolysis** (hy-DRAHL-ih-sihs). Common minerals that undergo hydrolysis include feldspar, hornblende, and augite. When these minerals are exposed to water, they dissolve into ions. These ions slowly react with the water and form clay minerals such as kaolinite.

Water's chemical effect on minerals is increased by the presence of acids that are dissolved in the water. A common acid found in water is carbonic acid. This weak acid is formed when carbon dioxide, one of the

VISUALIZATIONS
CLASSZONE.COM
Observe the chemical weathering of feldspar to clay.
Keycode: ES1202

DIFFERENTIATING INSTRUCTION

Developing English Proficiency Help students learning English to understand and remember the vocabulary of chemical weathering. Write the terms *hydrolysis, carbonic acid,* and *oxidation* on the board, and underline the prefix *hydro-* and the roots *carbon* and *oxid.* Explain that *hydro-* means "water," *carbon* refers to the element carbon, and *oxid* comes from *oxide,* meaning "a chemical compound of oxygen with one or more other elements." Point out that these word parts provide clues to the definitions of the terms, namely a reaction with water (hydrolysis), an acid formed from carbon dioxide (carbonic acid), and a reaction with oxygen (oxidation).

Use Spanish Vocabulary and Summaries, pp. 23–24.

gases present in Earth's atmosphere, dissolves in rainwater falling through the air. When rainwater containing carbonic acid seeps into the ground, it reacts chemically with many common minerals, including feldspar, hornblende, augite, and biotite mica. Elements such as potassium, sodium, magnesium, and calcium are dissolved, and the original minerals are turned into clay minerals.

Carbonic acid has an even greater effect on calcite. When carbonic acid reacts with calcite, the calcite dissolves completely. Unless the calcite is impure, no clay minerals remain after the reaction. The dissolving action of carbonic acid has hollowed out great underground caverns in limestone bedrock. You will learn more about caverns in Chapter 14.

BARBADOS, WEST INDIES Rainwater containing carbonic acid has dissolved limestone bedrock, forming Harrison's Cave.

Sulfur dioxide, nitrogen compounds, and carbon dioxide released by industries react with water in the atmosphere to form **acid rain.** Acid rain is rainwater that contains unusually high amounts of acids that can be traced to these pollutants. In addition to the damage that acid rain inflicts on the environment, acid rain also increases the rate of chemical weathering. Acid rain can cause structures made of concrete, stone, and metal to wear out more quickly.

Acids that are formed by the decay of dead plants and animals are dissolved by rainwater and carried through the ground to the bedrock. These acids react with minerals and contribute to chemical weathering.

Oxygen and Chemical Weathering

The brown or red color of some exposed rocks may be the result of a process called **oxidation** (AHK-sih-DAY-shuhn). Oxidation, which is the chemical reaction of oxygen with other substances, is very effective at weathering minerals that have iron in their chemical formulas. These minerals include magnetite, pyrite, and the dark-colored ferromagnesian silicates—hornblende, augite, and biotite. Oxidation of these minerals results in the formation of different types of rust, or iron oxides. Sometimes the rust that forms is a red iron oxide called hematite. Sometimes a yellowish-brown rust called limonite is formed.

CHAPTER 12 SECTION 1

Observe the chemical weathering of feldspar to clay.
Keycode: ES1202

Have students use their own words to describe the chemical weathering of feldspar to a classmate.

 Visualizations CD-ROM

Scientific Thinking

INVESTIGATE
Have students work in small groups to make their predictions. First, have them use their own experiences to describe local climatic conditions. Students can then find out about the geology of the region by consulting a state geologic map available in libraries or on the Internet. Have groups share their predictions with the class.

Scientific Thinking

INVESTIGATE
Consider the climate and physical features of the region in which you live. Does it rain frequently? Is it arid or windy? How does the temperature vary during the year? What types of rocks are common in your region? Based on your knowledge of the climate and physical features of your area, predict what types of weathering are likely to occur.

DIFFERENTIATING INSTRUCTION

Challenge Activity Write the following chemical reactions on the board.

$4Fe + 3O_2 \rightarrow 2Fe_2O_3$

$H_2O + CO_2 \rightarrow H_2CO_3$

Challenge students to identify the three substances involved in each reaction and describe how each reaction relates to chemical weathering. Iron (Fe) combines with oxygen (O_2) during oxidation to form iron oxide (Fe_2O_3), or rust. Water (H_2O) reacts with carbon dioxide (CO_2) to form carbonic acid (H_2CO_3), which can weather minerals to form clay.

CHAPTER 12 SECTION 1

Visual Teaching

Interpret Diagrams
Review the meaning of surface area. Then have students compare the surface area of the rock sample in steps 1 and 2 of the diagram. **The rock sample's surface area is greater in step 2.** Ask: What caused the sample's surface area to increase? **Mechanical weathering broke the rock sample into smaller pieces.** How is this change likely to affect the rate at which the rock sample weathers? **It will increase the weathering rate.**

Extend
Have students predict how rain falling on the rock sample in step 2 might affect it. **Rainwater might enter cracks in the rock, freeze, expand, and cause mechanical weathering. If the rain contains acid, it might chemically weather the rock.**

VISUAL TEACHING
Discussion
Refer students to the photograph. Ask them to predict what will happen when the cap of more-resistant rock on the taller formation weathers. **The less-resistant rock below the cap will weather quickly until it wears away completely and a new layer of more-resistant rock is exposed, forming a new cap.**

Rates of Weathering

Under average conditions, weathering is a slow process. Several factors affect the rate of weathering, including the amount of a rock's surface area that is exposed to weathering influences. The composition of rocks and climate also affect rates of weathering.

Surface Exposure

The rate at which a rock weathers is affected by the amount of the rock's surface that is exposed to chemical weathering processes. As shown below, when a rock is broken into smaller pieces by mechanical weathering, more of its surfaces are exposed. Thus, breaking a rock into smaller pieces causes the rock to weather more quickly.

Surface Exposure and Weathering

1 Mechanical weathering breaks a large rock into smaller pieces.

2 More rock surfaces are exposed to weathering, contributing to the breakdown of the rock.

Composition of Rock

Depending on their composition, various rocks respond differently to the same weathering processes. The rock layers in the photograph at the left have different compositions, so, even though the rock layers have been exposed to the same weathering processes, the layers have weathered at different rates. In some places, remnants of rock that are more resistant to weathering have protected the layers of less-resistant rock beneath. The taller formation in the photograph has a cap of more-resistant rock.

The rate at which a given rock weathers also depends on the type of weathering to which it is exposed. Because quartz does not react much with water, oxygen, or acids, it is almost unchanged by chemical weathering. Because quartz is hard and does not have cleavage, it also resists mechanical weathering. Gradually, however, quartz is broken into smaller particles.

In contrast, feldspar, hornblende, biotite mica, augite, calcite, and gypsum are affected by both mechanical and chemical weathering. Mechanical weathering breaks these minerals into small fragments. Chemical weathering turns these fragments into clay minerals.

ROCK LAYERS This rock formation is in Palo Duro Canyon State Park, Texas. The sedimentary rock layers in the formation are primarily claystone and shale.

MONITOR AND RETEACH

If students miss . . .

Question 1 Have students reread the definitions of mechanical weathering and chemical weathering on page 258.

Question 2 Reteach "Mechanical Weathering." (pp. 258–259) Have students describe the processes of frost wedging, wetting and drying, and abrasion.

Question 3 Use the diagram on page 260 to explain the relationship between upward expansion and mechanical weathering.

Question 4 Reteach "Water and Chemical Weathering." (pp. 260–261) Have students define *hydrolysis* and describe the hydrolysis of minerals.

Some minerals, such as calcite, gypsum, and halite, are also dissolved and carried off in solution.

Sandstones, quartzites, and quartz-pebble conglomerates are only as durable as the cements that hold them together. When the cement gives out, these rocks fall apart into the grains that compose them. Rocks that are cemented with calcite are subject to more rapid weathering. Rocks cemented with silica are more resistant to weathering. Quartzites and silica-cemented sandstones and conglomerates are among the most durable of all sedimentary rocks.

Shales are the most easily weathered of the sedimentary rocks. They split easily between layers and, in time, crumble into the clay minerals from which they were formed.

Marbles and limestones are fairly resistant to mechanical weathering. However, the calcite that constitutes marble and limestone undergoes slow attack by acids in water. In moist climates, where rainwater contains dissolved acid, rocks made of calcite are less durable than quartzites or sandstones. In dry climates there is little dissolved acid, and limestones are among the most durable rocks.

Climate

Climate is another factor that affects weathering processes. In general, warm, wet climates are conducive to both chemical and mechanical weathering processes. For example, most igneous rocks and many metamorphic rocks weather rapidly in wet climates. When these rocks are exposed at the surface, they often become cracked because of differences in temperature and pressure. Mechanical weathering widens the cracks in the rocks, exposing minerals to chemical weathering processes.

In regions with cold or dry climates, mechanical weathering plays a greater role than chemical weathering in breaking down rocks. For example, windblown rock materials can wear away rock surfaces in dry desert climates.

20-Minute Mini LAB

Surface Area and Chemical Weathering

Materials
- two clear containers
- water
- two steel wool pads
- scissors

Procedure
1. Fill the two containers with water. Place one steel wool pad in one of the containers.
2. Cut up the other pad in eighths and put the pieces in the second container.
3. After three days, observe the contents of the two containers.

Analysis
Which steel wool pad rusted more quickly? Estimate how many times faster that pad rusted. Explain how you made your estimate. How could you make the steel wool pad rust even faster?

12.1 Section Review

1. Compare mechanical and chemical weathering.
2. How are ice and water involved in mechanical weathering?
3. How is upward expansion related to mechanical weathering?
4. What is hydrolysis, and how is it involved in chemical weathering?
5. Explain how oxygen is connected to chemical weathering.
6. Identify three factors that control the rate at which a rock weathers.
7. **CRITICAL THINKING** Compare the weathering processes that take place in a dry desert region with those that occur in a humid tropical region.
8. **SOCIAL STUDIES** Describe some ways that human activities affect weathering processes.

Chapter 12 Weathering, Soil, and Erosion 263

CHAPTER 12 SECTION 1

MINILAB
Surface Area and Chemical Weathering

Management
- Students should wash their hands thoroughly after handling steel wool. Provide gloves if possible.
- Have students observe their containers periodically.

Materials
Use sharp scissors or wire cutters.

Analysis Answers
- The cut pad rusted twice as fast.
- **Sample Response:** by comparing the total areas of rust in the two containers
- Cut the pad into more pieces.

ASSESS

1. Mechanical weathering physically breaks rock into smaller pieces. Chemical weathering breaks rock by causing its minerals to decompose.
2. frost wedging, repeated wetting and drying, abrasion
3. The upward expansion of rocks can result in exfoliation.
4. Hydrolysis is the reaction of substances with water. Chemical weathering occurs when minerals exposed to water dissolve into ions that react with water to form clays.
5. Oxygen is effective in chemically weathering minerals that contain iron by reacting with them.
6. the rock's surface area, its composition, and climate
7. In a desert, mechanical weathering plays a greater role than chemical weathering, whereas in a tropical region, both types of weathering tend to take place.
8. **Sample Response:** The release of chemicals into the air by industries results in acid rain, increasing chemical weathering. Ground breaking in construction causes mechanical weathering, which can increase rock surface area and thus chemical weathering.

MONITOR AND RETEACH

Question 5 Have students reread "Oxygen and Chemical Weathering." (p. 261) Ask: How does rust form?

Question 6 Have students reread the first paragraph on page 262 and list three factors that affect weathering rate.

Question 7 Reteach "Climate." (p. 263)

Question 8 As a class, brainstorm ways in which people change the environment.

CHAPTER 12 SECTION 2

FOCUS

Objectives
1. Explain how soil forms.
2. Describe soil composition and the factors that affect it.

Set a Purpose
Have students read this section to find the answer to the question "How does soil form?"

INSTRUCT

More about...

Soils
Scientists classify soil based on a number of criteria, including color, texture, and structure. Texture refers to the proportion of sand, silt, and clay contained in the soil. Soils that contain approximately equal amounts of sand, silt, and clay are called loam soils. Structure refers to the bonding of soil particles into aggregates called peds.

12.2

KEY IDEAS
Soil is made of weathered rock and organic material.

Climate and other factors affect the composition of soil.

KEY VOCABULARY
- soil
- parent material
- residual soil
- transported soil
- soil profile
- soil horizon
- topsoil
- subsoil

Soil

Weathering has attacked the rocks of Earth's surface since the beginning of geologic time. It has helped to wear down mountains and to shape countless landforms. Weathering has led to valuable mineral deposits and has provided materials for sedimentary rocks. Most important, weathering has helped form a priceless resource—Earth's life-supporting soil. **Soil** is made of loose, weathered rock and organic material in which plants with roots can grow.

How Soil Forms

The material from which a soil is formed is called its **parent material.** Based on a soil's parent material, soil can be classified as either a residual or a transported soil.

A soil whose parent material is the bedrock beneath the soil is called a **residual soil.** The soil of the Bluegrass region of Kentucky is an example of a residual soil. The parent material is the underlying limestone bedrock.

In some parts of North America, deposits left by winds, rivers, and glaciers have covered the bedrock. Soils formed from transported materials are called **transported soils.** The soils of New England and much of the Midwestern United States are transported soils. Their parent material is loose soil, boulders, sands, and gravels deposited by glaciers during the last ice age.

A soil forms as its parent material is weathered away. The rate of weathering differs from location to location, depending on the type of rocks that make up the parent material. Climate is also an important factor. Over time, a layer of soil covers the parent material. Organic material—decaying plant and animal remains—are mixed with materials that have weathered away from the bedrock. Soil scientists use the term *mature* to refer to soils that have had a long time to form.

Soil scientists can study soils by digging until they reach the parent material. The cross section of earth exposed by the digging is called the **soil profile.** In mature soils, three distinct zones, or **soil horizons,** can be seen in the soil profile. These are called the A-, B-, and C-horizons.

SOIL HORIZONS The A-, B-, and C-horizons can be seen in this profile of soil near Mobile, Alabama.

264 Unit 4 Earth's Changing Surface

DIFFERENTIATING INSTRUCTION

Reading Support Explain that every paragraph has two basic parts: the main idea and the supporting details. Identifying the main idea provides a key to understanding the paragraph because the main idea is the primary point being made by the writer. Consequently, it is usually located in the first sentence of the paragraph. Supporting details are evidence or information provided to support the main idea. Direct students' attention to the second to last paragraph on page 265. Read the first sentence, and identify it as the main idea. Then point out how the other sentences in that paragraph provide details that support the main idea.
Use Reading Study Guide, p. 40.

The soil of the A-horizon is called **topsoil.** Topsoil is generally gray to black in color. It tends to be darker than soil in other horizons because it usually contains humus. Humus is organic material that forms from decayed plant and animal materials. Although topsoil may contain both sand and clay, the clay tends to wash down to the B-horizon over time. Leftover sand makes the topsoil sandy.

The B-horizon begins with the **subsoil.** The subsoil is usually red or brown from iron oxides that formed in the A-horizon and have been washed down into the B-horizon. The subsoil also contains clay that has washed down from the topsoil. Soluble minerals such as calcium and magnesium may also have washed into soil of the B-horizon.

The C-horizon is made of slightly weathered parent material, such as rock fragments. The unweathered bedrock of the parent material lies directly beneath the C-horizon.

Characteristics of Soil Horizons

	Color	Composition
A-Horizon	Usually dark to light gray	Fine particles of weathered rock materials mixed with humus
B-Horizon	Often red or brown	Clay, iron oxides, and dissolved minerals washed down from the A-horizon
C-Horizon	Dependent on parent material	Partially weathered parent material (rock fragments)

Soil Composition

The rock material in soil contains three noticeable parts: sand, silt, and clay. The amounts of these materials in a particular soil can affect the soil's ability to hold water and air, both of which are essential for plant growth. For example, soil that is very sandy holds water easily but dries out quickly.

What factors affect soil composition? One factor is time. Another is the parent material, though this factor becomes less important as a soil matures. Plants and animals affect soil composition by increasing the amount of organic material in the soil. Topography can also affect soil composition. Water flows down a sloping surface quickly, so the soil that is not eroded away by the water is often dry and resistant to plant growth. Because plant growth is limited, less organic material builds up in the soil.

One of the most important factors affecting soil composition is climate. For example, scientists have found that mature soils in a wet tropical climate strongly resemble each other no matter what the parent material is. Heavy rains wash nutrients from these soils. Rainfall and high temperatures lead to the weathering of clay minerals. Much of the silica that results from this weathering is also washed out of tropical soils.

Scientists classify soils based on their composition and have identified thousands of different types of soil. The map on page 266 shows some broad classifications for soils in North America. These classifications show a strong link between soil composition and climate.

TROPICAL SOIL The red coloring in this tropical soil comes from iron oxides left behind after rainwater washes away other dissolved materials.

CHAPTER 12 SECTION 2

Visual Teaching

Interpret Diagram
Have students study the map and read the key. Then ask: Where do desert soils occur? *in the middle of the western half of North America* Which soils have shallow profiles, and where are they found? *Arctic soils; in the northernmost part of North America* Which soils are often fertile for farming? *temperate grassland and forest soils*

ASSESS

1. Residual soil forms from the bedrock beneath it. Transported soil forms from parent material that originated in a different location.

2. Sketches should show the A-horizon on top, the B-horizon in the middle, and the C-horizon on the bottom. The A-horizon is likely to contain humus and sand. The B-horizon is likely to contain iron oxides, clay, and soluble minerals such as calcium and magnesium. The C-horizon is likely to contain slightly weathered parent material.

3. sand, silt, and clay

4. It takes more rain to support a forest than it does to support grassland, thus, forest soil exists where there is more rain.

5. **Sample Response:** The type of soil affects how easily crops can be grown and thus how large a population can be supported by local food growth.

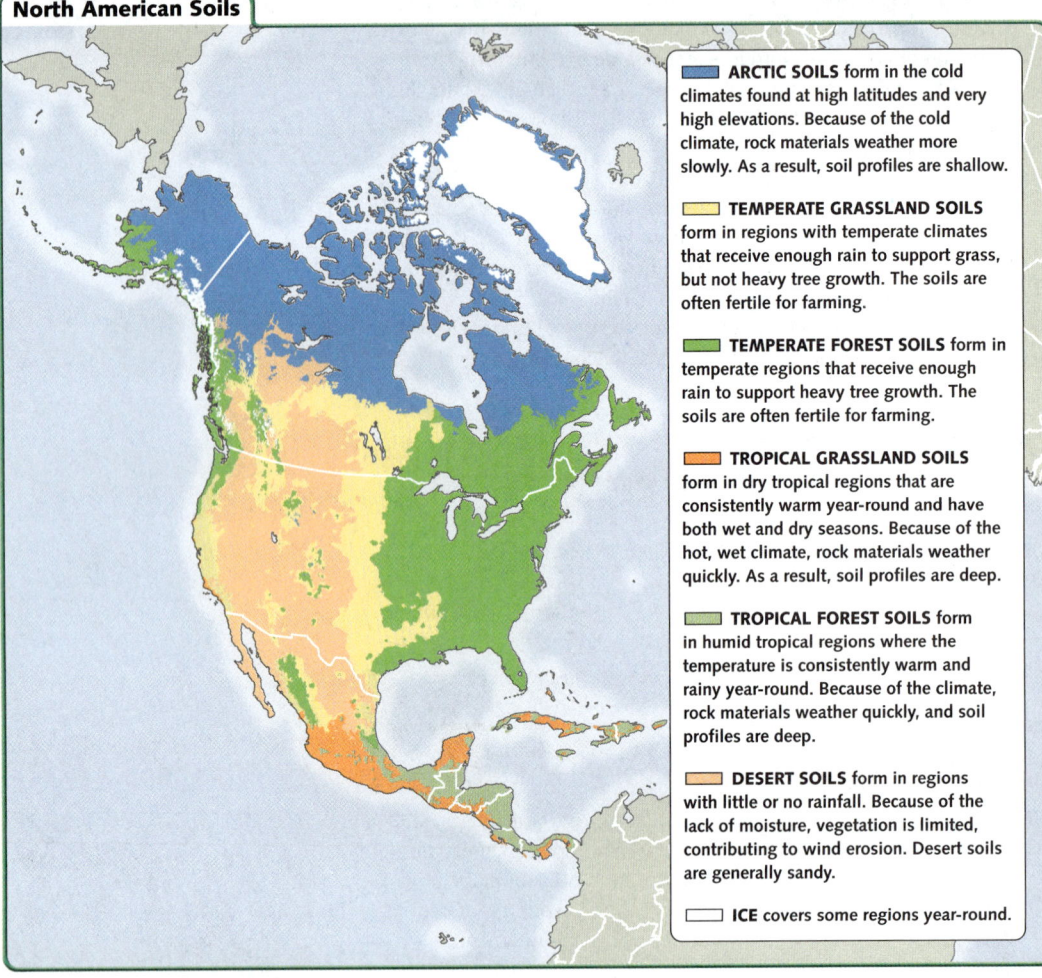

North American Soils

- **ARCTIC SOILS** form in the cold climates found at high latitudes and very high elevations. Because of the cold climate, rock materials weather more slowly. As a result, soil profiles are shallow.
- **TEMPERATE GRASSLAND SOILS** form in regions with temperate climates that receive enough rain to support grass, but not heavy tree growth. The soils are often fertile for farming.
- **TEMPERATE FOREST SOILS** form in temperate regions that receive enough rain to support heavy tree growth. The soils are often fertile for farming.
- **TROPICAL GRASSLAND SOILS** form in dry tropical regions that are consistently warm year-round and have both wet and dry seasons. Because of the hot, wet climate, rock materials weather quickly. As a result, soil profiles are deep.
- **TROPICAL FOREST SOILS** form in humid tropical regions where the temperature is consistently warm and rainy year-round. Because of the climate, rock materials weather quickly, and soil profiles are deep.
- **DESERT SOILS** form in regions with little or no rainfall. Because of the lack of moisture, vegetation is limited, contributing to wind erosion. Desert soils are generally sandy.
- **ICE** covers some regions year-round.

12.2 Section Review

1. Describe the difference between a residual soil and a transported soil.
2. Sketch a soil profile and label the A-, B-, and C-horizons. Describe what you are likely to find in each soil horizon.
3. What rock materials are found in soil?
4. **CRITICAL THINKING** Use the information in the map above. How does the amount of rainfall appear to affect whether a soil is a grassland or a forest soil? Explain your thinking.
5. **GEOGRAPHY** How do you think the type of soil found in a particular region might affect the lives of people living in the region?

266 Unit 4 Earth's Changing Surface

MONITOR AND RETEACH

If students miss . . .

Question 1 Have students read the definitions of residual soil and transported soil on page 264.
Question 2 Have students review the chart on page 265.
Question 3 Reteach "Soil Composition." (pp. 265–266)
Question 4 Have students review the map key on page 266.
Question 5 Have students reread "Soil Composition." (pp. 265–266) Ask them to focus on how soil composition and topography affect a soil's ability to hold water. Review the map key with students, focusing on how different soil types can support different kinds of vegetation.

266 Unit 4 Earth's Changing Surface

SCIENCE & Society

Preserving Rainforest Topsoil: The Environment

Looking at photographs of the lush tropical rainforests of Asia, Africa, and Latin America, you might assume that the soil is rich in nutrients. In fact, the topsoil there is thin and low in nutrients.

What happens to the nutrients stored in the plant and animal life of a rainforest? How can farmers tend their land to keep the topsoil productive?

SLASH-AND-BURN farming in Thailand is shown here as an area of rainforest is burned to expand agricultural land.

In the humid rainforest environment, dead organisms decompose rapidly, and the forest's living organisms directly absorb most of the nutrients released as a result. The soil is enriched only temporarily, however.

Forest-dwelling peoples have traditionally cleared small sections of forest by chopping down and burning trees, a technique know as slash-and-burn. Burning releases nutrients from the trees into the soil to provide good cropland for several years. Once the nutrients are used up, the farmers leave the land to lie fallow—they stop farming the land to allow natural growth to return—for decades before planting there again.

Today, however, urban sprawl, poverty, and other pressures are limiting the time left for the soil to recover. Overuse and single-crop farming soon strip the topsoil of nutrients so that the land ceases to produce plentiful harvests. Farmers are then forced to clear more land.

Such unsustainable slash-and-burn farming is now a major cause of deforestation in the rainforests. Addressing the problem is difficult because many slash-and-burn farmers have no other way to support themselves. One possible solution is for farmers to adopt the traditional farming methods still practiced by the Lacandon Maya of the Selva Lacandona rainforest in Mexico. The Lacandons slash and burn a small plot in the forest. Over several years there, they plant more than 70 different trees and plant crops that help to restore nutrients to the soil. Eventually, they let the plot lie fallow, much as their ancestors have done. This system, known as agroforestry, is recognized as an excellent way to farm the forest without damaging the soil. ∎

Extension

SCIENCE NOTEBOOK

Large-scale commercial logging is another major cause of deforestation and is also indirectly responsible for loss of topsoil. How do you think logging contributes to soil loss?

DATA CENTER CLASSZONE.COM

Learn more about rainforests and conservation programs.
Keycode: ES1203

SUSTAINING the land in Costa Rica, this Arbofilia environmental group member plants trees in the Puriscal region.

CHAPTER 12 SECTION 2

DATA CENTER CLASSZONE.COM

Learn more about rainforests and conservation programs.
Keycode: ES1203

Have students list some of the goals of rainforest conservation programs.

Science and Society

Why Rainforest Soils Are Poor
Laterites are the reddish soils in which most tropical rainforests grow. They are among the world's poorest soils because of the environmental conditions of the wet tropics and subtropics. Heavy rainfall constantly pouring downward through the soil quickly leaches any organic material that is decomposing in the upper layer. Because the nutrients of the rainforest ecosystem are contained in the vegetation, removing the forest leaves only sterile soil behind. Exposed to the harsh tropical sun, the soil bakes into a brick-like surface that can become totally unusable as farmland within a few years.

Science Notebook
The roots of trees help hold soil in place. When trees are removed, the loss of root systems makes it easier for heavy rainfall to wash away loose soil.

CHAPTER 12 SECTION 3

► FOCUS

Objectives
1. Give examples of mass movement.
2. Explain how erosion reshapes Earth's surface.

Set a Purpose
Ask: What do you think mass movement and erosion are? Have students read this section to find out if their predictions were correct.

► INSTRUCT

More about...

Mass Movement
Water plays an important role in the movement of rock material downslope. When the spaces between sediment particles become filled with water, the force of cohesion holding the particles to one another is destroyed and they easily slide past each other. This lubricating effect, as well as the additional weight added to the sediment by water, makes rock material more susceptible to being set in motion under the force of gravity.

12.3

KEY IDEA
Mass movements and erosion carry away weathered materials and reshape Earth's surface.

KEY VOCABULARY
- mass movement
- erosion
- talus
- landslide
- creep
- slump
- earthflow
- mudflow
- volcanic neck

TALUS has collected at the bottom of this slope in Idaho.

Mass Movements and Erosion

Typically, weathered rock materials do not stay in place. Wherever the ground slopes, gravity causes soil and rock fragments to fall, slide, or move at slow speeds to lower levels. Rain or wind may remove sand and dust from the side of a hill. A river transports weathered material downstream.

Mass movement refers to the downward transportation of weathered materials by gravity. **Erosion** is the removal and transport of materials by natural agents such as wind and running water.

Mass Movements

Soil partially protects the bedrock beneath it from weathering. By removing soil and loose rock materials, mass movements continually expose fresh bedrock to weathering, thus speeding up weathering processes. You can see the results of mass movement at the base of many steep slopes. Geologists use the term **talus** (TAY-luhs) to refer to rock fragments that have been weathered from a cliff and pulled down by gravity. Talus piles rest against the cliff at angles as great as nearly 40 degrees.

Landslide is a term commonly used for the movement of a mass of bedrock or loose soil and rock down the slope of a hill, mountain, or cliff. Landslides are most likely to occur on steep slopes. Roads or houses built on or into the sides of steep slopes are at risk of being damaged by a landslide. The risk is even greater in regions near volcanoes and in earthquake-prone regions, where eruptions and seismic tremors can trigger landslides.

People can avoid injury from landslides by knowing when the danger is greatest. Landslides tend to occur after heavy rains or during the spring, when large amounts of snow are melting. Rain and water from melted snow add weight to soil, making it easier for gravity to pull soil down a slope. Rain and snowmelt also may create a layer of water between the soil and the underlying bedrock. This layer may make it easier for gravity to pull soil down a slope.

Geologists use various terms—*creep, slump, earthflow,* and *mudflow,* for example—to describe different types of landslides.

Creep

Creep is a slow, imperceptible movement of soil down a slope. Its effects are noticeable, however, because creep causes fence posts, poles, and other objects fixed in the soil to lean downhill. The presence of water in the soil contributes to creep.

Slump

Sometimes blocks of land tilt and move downhill along a surface that curves into the slope. This type of movement is called **slump.** Slump tends to occur because a slope has become too steep for the bottom of the slope to support the soil at the top of the slope. For example, if the rock and soil at the bottom of a slope become worn away, the top of the slope becomes unstable and slumps downward.

268   Unit 4 Earth's Changing Surface

DIFFERENTIATING INSTRUCTION

Developing English Proficiency On the board, write the words *creep, slump,* and *neck.* Discuss the more common usage of these words, and compare them with their scientific usage in this section. For example, students may have heard *creep* used as a noun to refer to a disagreeable person or as a verb to describe a person moving slowly. Have a volunteer demonstrate a creeping motion. Then explain that the scientific usage of this word—the slow, downslope movement of soil—is likely derived from its meaning of moving slowly.

268   Unit 4   Earth's Changing Surface

Earthflows

During an **earthflow,** a mass of weathered material that has been saturated with water flows downhill. The downhill movement is slower and less fluid than a mudflow. Factors affecting the velocity of an earthflow include the amount of water present, the composition of the soil, and the steepness of the slope.

Some earthflows take place relatively quickly, perhaps over a period of days. Other earthflows may last for a period of years. Geologists have found evidence that part of the Slumgullion earthflow in Colorado has been moving for about 300 years. The active section of this massive earthflow is about 3.9 kilometers long.

Mudflows

A **mudflow** is the rapid movement of water that contains large amounts of suspended clay and silt. Mudflows contain more water than earthflows and have been known to travel at up to 100 kilometers per hour down steep mountains. Mudflows are capable of moving rocks, boulders, trees, and houses. They tend to occur in drier regions that experience infrequent but heavy rainfall. Although mudflows occur most often on steep, barren slopes that erode easily, they can also occur on gentle slopes that are prone to erosion.

Lahars, which you learned about in Chapter 9, are mudflows that accompany volcanic eruptions. Heat from erupted materials melts snow and ice on top of a volcano. Water cascades down the slopes of the volcano, carrying mud along with it. The photograph below shows a lahar that followed the 1985 eruption of Nevado del Ruiz, a volcano in Colombia. The devastating lahars from this eruption traveled about 100 kilometers. In the town of Armero alone, over 20,000 people were killed.

When Is Mud Dangerous? Analyze the conditions that create potential mudflows.
Keycode: ES1204

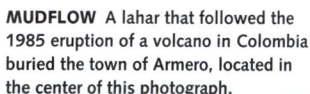

MUDFLOW A lahar that followed the 1985 eruption of a volcano in Colombia buried the town of Armero, located in the center of this photograph.

CHAPTER 12 SECTION 3

Examine a landscape formed by erosion.
Keycode: ES1205

Have students write a brief summary of events leading to the formation of the final landscape.

 Visualizations CD-ROM

VISUAL TEACHING
Discussion

Help students understand that the volcanic neck in the photograph was once surrounded by rock, making up the body of a volcano. Refer students to the cross sections of volcanoes on pages 202 and 203 of Chapter 9. Point out the column of magma that, after hardening and erosion, could become a volcanic neck like the one on this page.

ASSESS

1. Mass movement is the downward transportation of weathered materials by gravity. Examples include landslides, creep, slump, earthflows, and mudflows.

2. Climate determines the types of erosion that dominate in a region, which determine the topography. In humid climates, water is an effective agent of erosion. As a result, the topography tends to be rounded. In dry regions, the topography tends to be sharp and jagged.

3. Where lahars have occurred in the past, they may be likely to occur in the future. Thus, studying lahars may help prevent deaths and future damage to communities.

4. Gravity causes mass movement of materials down a mountain slope. Friction opposes gravity and thus acts to hold rocks on a mountain slope.

270 **Unit 4**

VISUALIZATIONS
CLASSZONE.COM

Examine a landscape formed by erosion.
Keycode: ES1205

Erosion and Landforms

Rivers and streams, glaciers, wind, and ocean waves and currents are all agents of erosion. By removing and transporting earth materials, these agents play as important a part in shaping a landscape as the forces associated with plate tectonics. Climate and the composition of rock are two other important factors that affect erosion in a given region.

The topography of a region depends on the balance at any given time between forces that uplift the land and agents of erosion that wear down the land. Even as rocks are uplifted, weathering and erosion are acting on the rocks. Sometimes uplift is more apparent than erosion. The Himalayas are being uplifted more quickly than they are being eroded. As a result, these mountains are growing higher over time. Their topography is rugged and sharp. In contrast, erosion is now the dominant process in the Appalachians. The effects of erosion are apparent in these mountains. Their topography is more smooth and rounded than the topography of the Himalayas.

The effect of erosion on topography is also influenced by climate. In regions with humid climates, water is the primary agent of erosion. Because water is such an effective agent of erosion, humid regions tend to have more rounded topography. The topography in regions with dry climates tends to be sharp and jagged.

The composition of rock also affects rates of erosion. Some types of rock are more resistant to erosion than other types of rock. For example, the rock structure in the photograph at the left is all that remains of an extinct volcano. The outer layers of the volcano have been eroded away, leaving behind a more resistant plug of igneous rock called a **volcanic neck.** The volcanic neck is made of hardened magma—the molten rock that originally rose upward in the volcano when it was active. In later chapters of this unit, you will read about how weathering and erosion are involved in shaping other landforms.

VOLCANIC NECK Agathla Peak, Arizona, is a core of hardened magma, the remains of an eroded volcano.

12.3 Section Review

1. Explain what the term *mass movement* means, and give some examples of mass movements.

2. Explain how the climate of a region affects erosion and topography.

3. **CRITICAL THINKING** Volcanologists often study the area around a volcano to identify places where lahars have occurred in the past. How could such studies benefit communities near the volcano?

4. **PHYSICS** Gravity is a force that pulls objects downward. Friction is a force that resists the sliding motion of one surface against another surface. Explain how these forces affect rocks and earth on a mountain slope.

270 Unit 4 Earth's Changing Surface

MONITOR AND RETEACH

If students miss . . .

Question 1 Have students define the boldface words on pages 268 and 269.

Question 2 Reteach "Erosion and Landforms." (p. 270) Ask: What is the relationship between climate, erosion, and topography?

Question 3 Review the dangers posed by lahars (p. 269), and discuss the benefits of predicting future geologic events.

Question 4 Review the definition of mass movement. (p. 268) Discuss the forces acting on a rock that sits on a steep slope.

Soil as a Resource

Because soil supports plant life, which in turn supports animal and human life, it is an important renewable resource. Yet less than 25 percent of Earth's land can be used to grow crops. Rough mountain regions often have no soil cover. Polar regions are too cold to grow crop plants. Desert regions are too hot and dry.

As human populations continue to climb, the amount of land available for farming is becoming even more limited. Areas with the most fertile soil are often the same places where people want to build houses. Thus, it is important to conserve and protect the soil that is available as a resource.

Soil Fertility

Soil fertility is the ability of soil to grow plants. The proportions of mineral matter, water, and organic matter in soil determine the types of plants that will grow in the soil. Soil that is fertile for potatoes may be less fertile for wheat. In part this is because potatoes require a different quantity of soil nutrients and water than wheat does. A number of problems, which include soil depletion and salinization, threaten soil fertility.

Soil Depletion

Crop plants and natural vegetation use up nutrients in soil. When the plants die, they decompose in the ground, and the nutrients are returned to the soil. When crops are harvested, however, the nutrients are removed from the soil. **Soil depletion** occurs when the soil gradually becomes so lacking or depleted in nutrients that it can no longer grow a usable crop.

Good farming practices can prevent soil depletion. Farmers, for example, can allow fields to lay fallow for a period, or they can rotate the types of crops grown on each field from year to year. Farmers do not always follow such practices, however, because it can be expensive in the short term to do so. Instead, farmers often add artificial fertilizers to soil. These fertilizers increase crop yields, but pose a serious threat to the environment. Over time, runoff from fields causes nutrients to build up in lakes and rivers, and in ocean environments where the nutrients stimulate plant growth. The plants, in turn, use up oxygen needed by other organisms. Eventually, the ecological balance of these water environments is destroyed.

Salinization

Irrigation can make desert soils very fertile. The problem, however, is that the water brought in to irrigate a desert contains dissolved minerals. As the desert's dry air rapidly evaporates the irrigation water, the minerals that had been dissolved in the water are left behind, deposited on the soil surface. In time, the soil contains so much mineral matter from the evaporated irrigation water that the soil can no longer sustain crop growth—a process called **salinization**. Soil affected by salinization is difficult to reclaim.

12.4

KEY IDEA
Soil is an important resource that can be conserved and protected.

KEY VOCABULARY
- soil fertility
- soil depletion
- salinization

SALINIZATION Evaporated irrigation water has caused salt deposits to accumulate in this soil, severely diminishing the soil's fertility.

CHAPTER 12 SECTION 4

More about...

Soil Erosion

Vegetation helps hold soil in place by slowing down wind and surface runoff. The slower that wind and water move, the less soil they are able to pick up and carry away. Also, when runoff slows down, it is more likely to collect and seep into the ground, helping to anchor the soil further. Plant roots also help to secure soil in place.

VISUAL TEACHING

Discussion

Point out that in both contour farming and terrace farming, crops are planted in rows perpendicular to the slope of the hill. Have students predict the effect that planting crop rows parallel to the slope would have. **Runoff would be concentrated between rows and channeled downslope at greater speeds, thereby increasing soil erosion.**

Erosion and Soil Conservation

Although erosion by wind and water occurs naturally, it is accelerated by human activity. In many regions, farming, construction, mining, and other activities have left soil relatively unprotected from erosion. As a result, erosion is removing topsoil more quickly than natural processes can restore it. As of 2001, one estimate indicated that soil in the United States was being eroded at a rate 17 times higher than the rate at which it was formed. About 90 percent of available cropland was losing soil faster than it could be replaced.

For farms to remain productive, soil erosion must be controlled using soil conservation methods. These methods include planting windbreaks, constructing terraces, and implementing erosion-reducing farming practices such as contour farming, strip-cropping, and no-till.

Windbreaks, also known as shelterbelts, are belts of trees planted along the edges of fields. These trees slow the wind and reduce wind erosion. Windbreaks are important on level plains where strong winds may blow nearly all the time. Windbreaks are common in the Great Plains region of the United States, where they were planted after the Dust Bowl disaster of the 1930s proved that protection against wind erosion was necessary. Other benefits of windbreaks include protection against water erosion and frost.

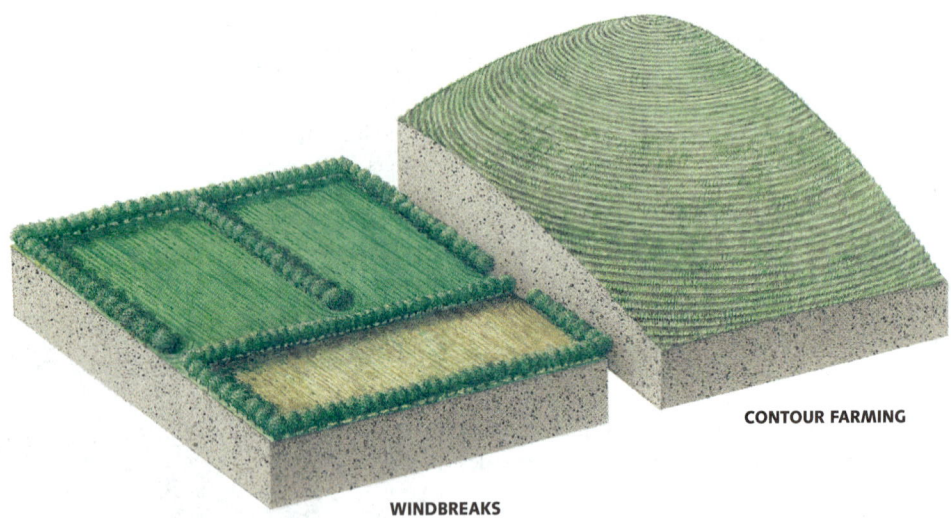

WINDBREAKS

CONTOUR FARMING

Contour farming is a method that inhibits water from flowing rapidly downhill and carrying soil with it. Instead of plowing up and down a slope, farmers plant crops in rows parallel to land contours. Contour farming is utilized in irrigation-dependent regions where slopes are moderately steep. It has been practiced in various parts of the world for centuries.

Flattening a slope into terraces slows the speed of runoff. As a result, the water can transport less topsoil. In some regions, terraces have been shown to reduce soil erosion by 50 percent or more. Terrace farming is used

DIFFERENTIATING INSTRUCTION

Challenge Activity Challenge pairs or small groups of students to plan and conduct an experiment to test the effect of salt content on soil fertility. Instruct students to state a hypothesis, devise a plan to test their hypothesis, and present you with their plan for approval. Remind them to plan an experiment in which there will be only one variable. Once the plan is approved, have students conduct their experiment and present the results to the class. Make sure that students explain how their experimental results support or refute their hypothesis.

in the cultivation of rice and other crops, and has been practiced for many centuries in places ranging from China to the Andes.

Another method of reducing soil erosion is strip cropping. Farmers alternate a crop that leaves bare ground between rows with a crop that completely covers the ground. For example, the ground between rows of corn plants is bare. Alfalfa is a crop that covers the ground. By planting alternating strips of a field with corn and alfalfa, farmers can protect the soil from erosion. To reduce water erosion, the strips are planted on the contour, perpendicular to the slope. To reduce wind erosion, the strips are planted perpendicular to the prevailing wind direction. Regions in which strip cropping is practiced include much of the Eastern and Central United States.

Still another method of reducing soil erosion is a technique called no-till. In this method, plowing, planting, fertilizing, and weed control are all done at the same time. Once the field is planted, the ground does not need to be disturbed again until harvest. Because the soil is left alone, there is less chance that it will be eroded. Regions where no-till has been used include the Southern and Midwestern states.

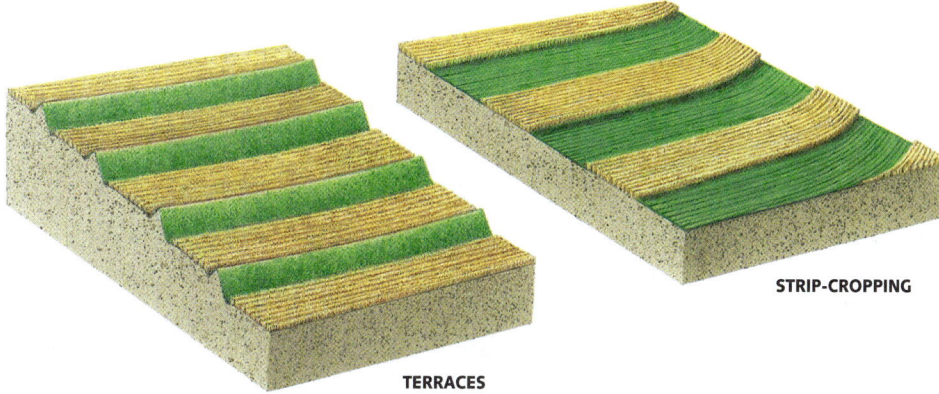

TERRACES   **STRIP-CROPPING**

12.4 Section Review

1. Describe some ways to prevent soil depletion.
2. How and where does salinization occur?
3. Describe at least three methods that can be used to protect a field from wind or water erosion.
4. **CRITICAL THINKING** Terrace farming is used to reduce erosion. Explain how terrace farming could also increase the amount of land available for farming.
5. **MATHEMATICS** One estimate for the rate at which natural processes can replace topsoil that has been eroded is about 2 millimeters per year. At this rate, how long would it take to form 10 centimeters of new topsoil? Show how you got your answer.

CHAPTER 12 SECTION 4

CRITICAL THINKING
Ask students to describe how strip cropping can incorporate other methods of soil conservation.
Sample Response: As in contour farming, strips are planted perpendicular to the slope to reduce water erosion. Like windbreaks, strips can be planted perpendicular to the prevailing wind direction to prevent wind erosion. Also, tall crops slow down the wind, protecting crops low to the ground.

ASSESS

1. **Sample Response:** allowing fields to rest for periods of time, allowing a crop to decompose on the soil, and changing the kind of crop grown on a field from year to year
2. Salinization occurs in desert areas when irrigation water evaporates rapidly and minerals dissolved in the water are left behind on the soil surface.
3. windbreaks: planting belts of trees along field edges; contour farming: planting crops in rows parallel to land contours; constructing terraces: flattening sloping areas by forming terraces; strip cropping: alternating a crop that leaves bare ground between rows with a crop that covers the ground; and no-till farming: doing planting, fertilizing, and weed control at the same time
4. Some slopes may be too steep for use in farming. By creating terraces on these slopes, unusable land becomes farmable land.
5. 2 mm/yr × 1 cm/10 mm = 0.2 cm/yr; 10 cm ÷ 0.2 cm/yr = 50 yr

---

MONITOR AND RETEACH

If students miss . . .
Question 1 Reteach "Soil Depletion." (p. 271) Discuss methods that farmers use to protect the soil.
Question 2 Have students study the photograph on page 271 and read the caption.
Question 3 Use the diagrams on pages 272 and 273 to explain methods that farmers use to protect a field from erosion.
Question 4 Have students examine the diagram of terraces on page 273 and imagine a much steeper slope prior to terracing. Discuss the effect that terracing would have on such a slope. If students have difficulty visualizing a terraced slope, sketch a side view of a terraced hill on the board.
Question 5 Remind students that there are 10 millimeters in 1 centimeter. Demonstrate an algebraic solution using ratios.

CHAPTER 12 ACTIVITY

PURPOSE
To determine how temperature affects the rate of chemical weathering

MATERIALS
Provide students with a copy of Lab Sheet 12. A copymaster is provided on page 141 of the teacher's laboratory manual.

You may want to use an insulated container, such as a large thermos, to provide a supply of hot water that does not cool down quickly.

PROCEDURE
❸ Students may have difficulty maintaining the correct temperature range of water within the beakers. To help solve this problem, you might maintain large sources of water at the correct temperatures and have students take water from this supply for each beaker just before using it in the experiment. Alternatively, you can divide the class into groups and assign each group one beaker to time. Then each student can use the data from all groups to plot a graph.

ANALYSIS AND CONCLUSIONS
1. most slowly: Beaker 1; most rapidly: Beaker 5

CHAPTER 12 LAB Activity

Chemical Weathering and Temperature

SKILLS AND OBJECTIVES
- **Model** a chemical weathering process.
- **Graph** the data from the model and **interpret** the graph.
- **Predict** what will happen when the model is modified.

MATERIALS
- lab apron
- safety goggles
- 5 250-mL beakers
- 5 thermometers
- hot water (40–50°C)
- ice water
- 5 effervescent antacid tablets
- stopwatch
- graph paper
- map of Earth's Climates, page 470

Whether it's the granite of a New Hampshire mountain breaking down into sand and clay or the limestone of Kentucky decomposing to form rich soil, all chemical weathering processes involve water. What effect does the temperature of the water have on the rate of chemical weathering?

Carbonic acid is a weak acid that forms when carbon dioxide dissolves naturally in water. A common chemical weathering process is the reaction of carbonate rocks, such as limestone and marble, with carbonic acid. In this lab activity, you will observe the dissolution of effervescent tablets in water of varying temperatures. These tablets contain sodium bicarbonate, which dissolves in water in much the same way that carbonate rocks dissolve in carbonic acid.

Procedure

❶ **CAUTION: Put on your lab apron and safety goggles.**

❷ Arrange five beakers in a row, numbered from 1–5. Place a thermometer in each beaker.

❸ Add combinations of hot and ice water to each beaker so that the temperature of the water matches the following: Beaker 1, 0–10°C; Beaker 2, 10–20°C; Beaker 3, 20–30°C; Beaker 4, 30–40°C; Beaker 5, 40–50°C. Each beaker should contain about 200 mL of water.

❹ Remove any ice from Beaker 1. Make sure that the water is within the correct temperature range. When the temperature reading stabilizes, record the start temperature of the water in Beaker 1 on a copy of the data table. Remove the thermometer from the beaker and set it aside.

❺ Drop an antacid tablet into Beaker 1. Start the stopwatch at the instant the tablet enters the water. Stop the stopwatch when the last piece of the tablet dissolves. (Do not wait for the bubbling to stop; wait only for all pieces of the tablet to dissolve completely.) Record the time on the stopwatch to the nearest whole second.

❻ Place the thermometer back in Beaker 1 and wait until the temperature stabilizes. Record the end temperature of the water in Beaker 1. Calculate the average temperature of Beaker 1 by adding the start and end temperatures and dividing by 2. Record the average temperature of the water in Beaker 1.

❼ Repeat Steps 4, 5, and 6 for each of the remaining beakers.

8 Plot a graph of the data for the five trials. Place "Average Temperature (°C)" on the x-axis and "Time (seconds)" on the y-axis. Connect the five points with a smooth curve.

Beaker No.	Start Temp. (°C)	Time (sec.)	End Temp. (°C)	Average Temp. (°C)
1				
2				
3				
4				
5				

Analysis and Conclusions

1. In which beaker did the reaction occur most slowly? In which beaker did the reaction occur most rapidly?

2. Based on your observations, hypothesize the relationship between temperature and the rate of chemical weathering. What are some possible reasons for this relationship? Explain.

3. Look at the temperatures you recorded. Are all of these temperatures likely to occur on Earth's surface? If so, where? Which of the beakers corresponds with the water temperature of your local area?

4. Turn to the map of Earth's Climates on page 470. Locate Rio de Janeiro in South America and Seattle in North America. The map key indicates that both cities have climates with abundant moisture.

    a. Compare the likely weathering rate of a limestone in Rio de Janeiro with that of a limestone in Seattle. Is there a difference?

    b. Which of the two locations is likely to have thicker soil?

5. Now locate Barrow, Alaska, on the map. Why is a limestone in Barrow likely to weather very slowly?

6. How would the rate of the reaction have been different if the tablets had been ground into a powder before they were dropped into the water? Would a graph for such a reaction result in a curve above or below the line of your actual data? Do you think the shape of the curves would be the same? Explain your answers.

Chapter 12 Weathering, Soil, and Erosion   275

CHAPTER 12 ACTIVITY

(continued)

2. As temperature increases, the rate of chemical weathering also increases. At higher temperatures, the molecules/ions in substances move more quickly, so chemical reactions, such as those that produce weathering, occur more quickly.

3. No; the highest temperature is unlikely to occur under normal circumstances on Earth's surface. The next highest temperature might occur near the equator, the middle temperature in temperate areas, the second to lowest temperature in fairly cold climates, and the lowest temperature in polar areas. The temperature that corresponds to your local area will depend on your location and the time of year.

4. a. Limestone in Rio de Janeiro should weather more quickly because the climate there is warmer.
    b. Rio de Janeiro is likely to have thicker soil because more weathering should produce more particles that form soil.

5. The climate there is cold and dry, which slows chemical reactions such as weathering.

6. The reaction rate would have been faster because grinding would increase the surface area exposed to the water. The curve would be below the actual curve because the ground-up tablets would dissolve in less time. The shape of the curves would be about the same because the effect of temperature would be essentially the same.

Refer students to Safety in the Earth Science Laboratory on pages xxii–xxiii.

Chapter 12 Weathering, Soil, and Erosion   275

CHAPTER 12 REVIEW

Summary of Key Ideas

12.1 Mechanical weathering breaks down rocks without changing their composition. Chemical weathering changes the composition of rocks. Some factors that affect the rate at which rock weathers are surface exposure, composition, and climate.

12.2 Soil is loose, weathered material capable of supporting rooted plants. A soil may be residual or transported, depending on its parent material. A soil's composition depends on many factors, one of the most important of which is climate.

12.3 Mass movements involve the downward transport of rock materials by gravity. Erosion is the removal or transport of rock materials by natural agents such as water and wind.

12.4 Appropriate farming practices can help protect soil fertility. Windbreaks, contour plowing, terracing, strip cropping, and no-till are methods used to prevent soil erosion.

Key Vocabulary

abrasion (p. 259)
acid rain (p. 261)
chemical weathering (p. 258)
creep (p. 268)
earthflow (p. 269)
erosion (p. 268)
exfoliation (p. 259)
frost wedging (p. 258)
hydrolysis (p. 260)
landslide (p. 268)
mass movement (p. 268)
mechanical weathering (p. 258)
mudflow (p. 269)
oxidation (p. 261)
parent material (p. 264)
residual soil (p. 264)
salinization (p. 271)
slump (p. 268)
soil (p. 264)
soil depletion (p. 271)
soil fertility (p. 271)
soil horizon (p. 264)
soil profile (p. 264)
subsoil (p. 265)
talus (p. 268)
topsoil (p. 265)
transported soil (p. 264)
volcanic neck (p. 270)
weathering (p. 258)

Vocabulary Review

1. Mechanical weathering breaks rock but does not change its chemical composition. Chemical weathering changes the minerals in rock into different substances.
2. Oxidation is the chemical reaction of oxygen with other substances. Hydrolysis is the chemical reaction of water with other substances.
3. Residual soil forms from the parent bedrock beneath it. Transported soil forms from materials that have been carried to an area.
4. Mudflow is the rapid movement of water containing large amounts of clay and silt. Creep is the slow movement of soil down a slope.
5. Topsoil is the top layer of soil. Subsoil is just beneath the topsoil.
6. Salinization is the accumulation of salt deposits in soil as the result of irrigation. Soil depletion occurs when soil becomes so lacking in nutrients that it can no longer grow a usable crop.

Concept Review

7. Ice wedging is the splitting apart of rock that occurs when water in cracks freezes and expands. It most often occurs where the temperature varies above and below the freezing point of water.
8. Mechanical weathering: Roots and other plant parts grow in the cracks of rocks, splitting the rocks. Some animals dig holes in soil that allow air and water to reach bedrock and weather it. Chemical weathering: Decaying plants and animals produce acids that dissolve in rainwater and are carried through the soil to bedrock.
9. Carbonic acid forms when carbon dioxide in the air dissolves in rainwater. When rainwater containing carbonic acid seeps into the ground, the acid reacts chemically with substances in rock, causing them to dissolve.
10. Factors include the amount of rock surface that is exposed to weathering, the composition of the rock, the climate, and the presence of pollutants.
11. in the A-horizon, or topsoil
12. partially weathered rock fragments
13. East: primarily fertile temperate forest soils; West: a mixture of fertile temperate grassland soils, sandy desert soils with little vegetation, and arctic soils (Alaska) with shallow soil profiles
14. rock fragments that have been weathered from a cliff and fall to form a pile at the cliff's base
15. Both flows are downhill movements of water and solid material, but mudflows contain more water and move downhill more rapidly than earthflows.
16. rain; wind; and flowing water such as rivers, glaciers, ocean waves, and ocean currents

Vocabulary Review

Explain the difference between the terms in each pair.

1. mechanical weathering, chemical weathering
2. oxidation, hydrolysis
3. residual soil, transported soil
4. mudflow, creep
5. subsoil, topsoil
6. salinization, soil depletion

Concept Review

7. What is ice wedging, and where is it likely to occur?
8. Explain how plants and animals contribute to mechanical weathering, and how they contribute to chemical weathering.
9. What is carbonic acid, and how does it affect chemical weathering processes?
10. Describe some factors that influence the rate at which a rock weathers.
11. Where are you likely to find dark, humus-rich soil in a soil profile?
12. Describe what you would expect to see in the C-horizon of a soil.
13. Use the map on page 266. Compare the soil types in the eastern United States with those in the western United States. What type of soil is found where you live?
14. What is talus?
15. Compare and contrast an earthflow and a mudflow.
16. Give some examples of agents of erosion.
17. Describe some farming methods that can be used to prevent soil erosion.
18. **Graphic Organizer** Copy and complete the cause-and-effect diagram below by giving the cause of soil depletion.

Critical Thinking

19. Predict How would the weathering of a bare mountain peak differ from the weathering of bedrock under a forest?

20. Compare Why are streets and highways damaged so much more in the winter months than in the summer months in much of the United States? Compare the processes of weathering in the two seasons.

21. Analyze Sandstones cemented by calcite usually weather much more rapidly than those cemented by silica. Why?

22. Predict What effect would a long, dry period have on the frequency of earthflows and mudflows?

Interpreting Graphs

The type of weathering that dominates in an area depends upon the climate in that area. Two key factors that affect climate are precipitation (rain and snow) and temperature. The graph shows the relationship between precipitation, temperature, and weathering. For example, a climate with an average yearly temperature (AYT) of 5°C and average yearly precipitation (AYP) of 75 centimeters would have moderate chemical weathering with frost action. Use the graph to answer the questions.

23. Determine the major type of weathering that occurs in Washington, D.C., where the AYT is 23°C and the AYP is 104 centimeters.

24. If the AYT in Washington, D.C., dropped 26°C but the AYP stayed the same, what kind of weathering would dominate?

25. Phoenix, Arizona, has an AYT of 20°C and an AYP of 20 centimeters. How would the climate in Phoenix have to change for moderate chemical weathering to become dominant?

26. According to the graph, no frost action occurs at a mean annual temperature above 13°C. What is a possible reason?

27. In general, how does a climate with strong chemical weathering differ from a climate with strong mechanical weathering?

Internet Extension

How Does Soil Vary from Place to Place? Compare and contrast soils from different regions.
Keycode: ES1206

Writing About the Earth System

SCIENCE NOTEBOOK Describe some ways that Earth's hydrosphere, atmosphere, biosphere, and geosphere are involved in the weathering and erosion of rock materials. Then look back at the diagram of the rock cycle in Section 6.1. Does the rock cycle involve all of these spheres? Explain.

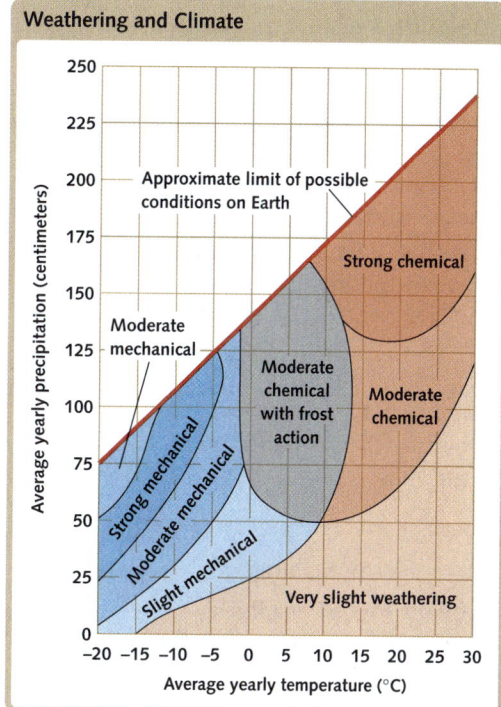

Chapter 12 Weathering, Soil, and Erosion 277

WRITING ABOUT SCIENCE

Writing About the Earth System

Hydrosphere: Frost wedging, hydrolysis, acid rain, and carbonic acid cause weathering. Flowing water causes erosion. Atmosphere: Oxidation and climate influences cause weathering. Wind causes erosion. Biosphere: Plant roots, animals that burrow, and acids formed by decay cause weathering. Geosphere: Geological uplift can break rock and expose it further to weathering and erosion.
Yes, Geosphere: Magma cooling, uplift, and crust movements form rock. Hydrosphere, Atmosphere: Water and wind contribute to weathering, erosion, and deposition that forms sedimentary rock. Water aids metamorphism. Biosphere: Animal and plant deposits can form sedimentary rock.

CHAPTER 12 REVIEW

How Does Soil Vary from Place to Place?
Keycode: ES1206

📖 Use Internet Investigations Guide, pp. T49, 49.

(continued)

17. planting windbreaks, constructing terraces, practicing contour farming, using strip cropping, and making use of no-till farming

18. loss of nutrients in soil

CRITICAL THINKING

19. The weathering of a bare mountain peak would probably occur faster because the rock would be directly exposed to elements that cause weathering. A layer of forest soil would likely protect bedrock from agents of weathering, making the process take longer.

20. Streets are damaged in winter through frost wedging and ice heaving. In cold winter weather, mechanical weathering processes such as frost wedging are most common. During summer, when temperatures are warmer, chemical weathering through processes such as hydrolysis occurs most often.

21. Silica is very resistant to weathering, whereas calcite is susceptible to chemical weathering, especially in moist climates where rainwater contains dissolved acid.

22. It would decrease the frequency of earthflows and mudflows.

INTERPRETING GRAPHS

23. moderate chemical weathering

24. moderate mechanical weathering

25. precipitation would have to increase

26. At this average temperature, it probably never gets cold enough for water to freeze.

27. A climate with strong chemical weathering has high AYT and high AYP. A climate with strong mechanical weathering has very low AYT and low to moderate AYP.

Chapter 12 277

CHAPTER 13 PLANNING GUIDE: SURFACE WATER

	Section Objectives	Print Resources
Chapter-Level Resources		Laboratory Manual, pp. 55–60 Internet Investigations Guide, pp. T50–T52, 50–52 Guide to Earth Science in Urban Environments, pp. 13–20, 21–28, 29–36 Spanish Vocabulary and Summaries, pp. 25–26 Formal Assessment, pp. 37–39 Alternative Assessment, pp. 25–26
Section 13.1 Streams and Rivers, pp. 280–282	1. Define river system. 2. Describe the characteristics of a stream or river that affect its ability to erode sediment.	Lesson Plans, p. 43 Reading Study Guide, p. 43
Section 13.2 Stream Erosion and Deposition, pp. 283–286	1. Describe how streams weather and erode Earth's surface. 2. Describe how streams transport and deposit sediments. 3. Explain how deltas and alluvial fans form.	Lesson Plans, p. 44 Reading Study Guide, p. 44
Section 13.3 River Valleys, pp. 287–289	1. Explain how river valleys form. 2. Describe the formation of rapids and waterfalls.	Lesson Plans, p. 45 Reading Study Guide, p. 45
Section 13.4 Flood Plains and Floods, pp. 290–293	1. Describe the features of a floodplain. 2. Explain why floods occur. 3. Summarize natural and artificial methods of flood prevention and control.	Lesson Plans, p. 46 Reading Study Guide, p. 46 Teaching Transparency 14

INVESTIGATIONS
CLASSZONE.COM

Section 13.1 INVESTIGATION
How Does Stream Flow Change over Time?
Computer time: 45 minutes / Additional time: 45 minutes
Students examine stream discharge and variables that influence it.

Section 13.2 INVESTIGATION
What Controls the Shape of a Delta?
Computer time: 45 minutes / Additional time: 45 minutes
Students compare deltas and analyze the processes that shape them.

Technology/Online Resources

Electronic Teacher Tools
Visualizations CD-ROM
Classzone.com
Online Lesson Planner
Test Generator

Classzone.com
ES1301 Investigation, ES1302 Career

Visualizations CD-ROM
Classzone.com
ES1303 Visualization, ES1304 Investigation

Visualizations CD-ROM
Classzone.com
ES1305 Visualization

Visualizations CD-ROM
Classzone.com
ES1306 Visualization, ES1301 Local Resources

Classroom Management

Gather these materials for minilab and map activities.

Minilab, p. 285
tall clear container
sand
water
clay
gravel
spoon

Map Activity, pp. 294–295
tracing paper
colored pencils
paper clips
Appendix B Physical United States Map, pages 708–709
Teaching Transparency 39

Chapter Review INVESTIGATION
Have Flood Controls on the Mississippi River Been Successful?
Computer time: 45 minutes / Additional time: 45 minutes
Students compare floods that occurred before and after construction of flood control structures.

CHAPTER 13

Surface Water

As it flows to the ocean, Virginia's James River forms curving meanders through its tidal marsh.

How and why does a river change over time?

INTRODUCE
Build Interest
Before sharing background information about the photograph with students, allow time for their questions and comments.

Ask questions like the following:
- How would you describe the course of the river in this photograph? **zigzag curves**
- Does the river appear to be flowing swiftly or slowly in this photograph? **slowly**
- Describe the relief of this landscape. **flat**
- How would the course of this river be different if this area had a steep slope? **The river would be straighter and flow much faster.**

Respond
- How and why does a river change over time? **Over time a river can change in its velocity, course, channel shape, and sediment load. These changes occur in response to changes in gradient, local topography, seasonal variations in precipitation, and runoff.**

> **About the Photograph**
>
> The James River is the largest river in Virginia. It rises in the Allegheny Mountains in western Virginia, where the Jackson and Cowpasture Rivers meet, and flows southeastward for about 547 kilometers, to empty into Chesapeake Bay.

BEYOND THE TEXTBOOK

Print Resources
- Reading Study Guide, pp. 43–46
- Transparencies 14, 39
- Formal Assessment, pp. 37–39
- Laboratory Manual, pp. 55–60
- Alternative Assessment, pp. 25–26
- Spanish Vocabulary and Summaries, pp. 25–26
- Internet Investigations Guide, pp. T50–T52, 50–52
- Guide to Earth Science in Urban Environments, pp. 13–20, 21–28, 29–36

Technology Resources
- Visualizations CD-ROM
- Classzone.com
- Online Lesson Planner
- Electronic Teacher Tools
- Test Generator

278 Unit 4 Earth's Changing Surface

CHAPTER 13

PREVIEW

▶ **FOCUS QUESTIONS** In this chapter you will investigate rivers and learn more about the key questions below.

Section 1 What are some important features of rivers?
Section 2 How do streams erode and deposit rock materials?
Section 3 How do river valleys form?
Section 4 How and why do river floods occur?

▶ **REVIEW TOPICS** As you investigate rivers, you will need to use information from earlier chapters.

- water cycle (p. 13)
- erosion (p. 258)
- weathering (p. 258)
- abrasion (p. 259)

▶ **READING STRATEGY:**
USE TEXT AIDS

At the beginning of each section in Chapter 13 there is a statement of the section's key ideas and a list of key vocabulary terms. Before you read a section, write the key terms in your notebook. As you read, take notes on how each key term relates to the key ideas of the section.

INTERNET RESOURCES
CLASSZONE.COM

At our Web site, you will find the following Internet support:

DATA CENTER
EARTH NEWS
VISUALIZATIONS
- Sediment Transportation
- Waterfall and Chasm Creation
- Change in a Meandering River

LOCAL RESOURCES
CAREERS
INVESTIGATIONS
- How Does Stream Flow Change over Time?
- What Controls the Shape of a Delta?
- Have Flood Controls on the Mississippi River Been Successful?

Chapter 13 Surface Water 279

CHAPTER 13

PREPARE
Focus Questions
Find out what knowledge students have about streams and rivers. For each focus question, have students consider the section title and then develop two or three additional questions for each section. For example, Section 2: How can a stream erode solid rock? Section 4: What type of damage do river floods cause?

Review Topics
In addition to the earth science topics listed, students may need to review this concept:

Physical Science
- Friction, which affects the flow of water through a channel, is the force that opposes motion.

Reading Strategy
Model the Strategy: Tell students to read the key idea at the top of page 280. Have students list each of the key vocabulary terms under that sentence. Then say, "The first key vocabulary term is *tributary*. Read the definition in the text. *a stream that runs into another stream or river* How does *tributary* relate to the key idea of this section?" *A tributary is one of the streams that carries precipitation from land back to the oceans.* Tell students to write a sentence that explains how each vocabulary term relates to a key idea as they read each section of the chapter.

BEYOND THE TEXTBOOK

INVESTIGATIONS
- How Does Stream Flow Change Over Time?
- What Controls the Shape of a Delta?
- Have Flood Controls on the Mississippi River Been Successful?

VISUALIZATIONS
- Animation of how sediment is transported by flowing water
- Animation of how river erosion creates waterfalls and chasms
- Images and animation showing changes in the channel of a meandering river

DATA CENTER
EARTH NEWS
LOCAL RESOURCES
CAREERS

Online Lesson Planner

Chapter 13 Surface Water 279

CHAPTER 13 SECTION 1

FOCUS

Objectives
1. Define *river system*.
2. Describe the characteristics of a stream or river that affect its ability to erode sediment.

Set a Purpose
Read aloud the section title and heads. Then have students read the section to find the answer to "What is a river system and what are the characteristics of streams and rivers?"

INSTRUCT

VISUAL TEACHING
Discussion
Ask students: What makes a river a tributary? *A river is a tributary if it flows into another river.* Then have students identify how many tributaries flow into each of the two main rivers shown in the diagram. *Two tributaries flow into the river on the left, and one flows into the river on the right.* Ask students how the two river systems differ. *The river system on the left levels out onto a plain leading to a delta (covered in the next section). The river on the right continues on its path through the mountains.*

13.1

KEY IDEA
Streams and rivers carry a portion of the precipitation that falls on land back to the oceans.

KEY VOCABULARY
- tributary
- river system
- drainage basin
- watershed
- divide
- gradient
- discharge

RIVER SYSTEM All the land whose runoff and tributaries feed into a river are part of that river's drainage basin. This diagram shows drainage basins for two different river systems. The drainage basins are separated by a divide.

Streams and Rivers

Almost half of the water that falls to Earth's surface eventually ends up in a stream or river, where it travels overland to the oceans. In this way, streams and rivers are an essential part of the water cycle. Streams and rivers also account for most of the erosion of Earth's surface.

River Systems

Rain running down a slope eventually reaches a permanent body of running water such as a stream. This stream may run into a larger stream. A stream that runs into another stream or river is called a **tributary.** The running water carries bits of eroded rock called *sediment*. Larger tributaries carry this sediment until it reaches a main river. The main river carries the sediment to a lake or an ocean and deposits it there. A river and all of its tributaries is called a **river system.**

The **drainage basin,** or **watershed,** of a river includes all the land that drains into the river either directly or through its tributaries. The high land that separates one drainage basin from another is called a **divide.** The major divide of the United States, called the Continental Divide, is located in the Rocky Mountains. Rain falling east of the Continental Divide eventually flows into the Atlantic Ocean. Rain falling west of it flows into the Pacific Ocean.

The largest single drainage basin in the United States is the Mississippi River system. Its western divide is the Continental Divide. Its eastern divide is in the Appalachian Mountains. The drainage basin between these two divides covers a land area of about 4.76 million square kilometers.

DIFFERENTIATING INSTRUCTION

Challenge Activity Have students trace a topographic map of North America, including major mountain chains, lakes, and rivers. Have them draw in red a rough outline of the borders of major drainage basins in North America (Pacific Ocean, Arctic Ocean, Hudson Bay, Gulf of Mexico, and Atlantic Ocean). Students should be prepared to describe for the class the different divides that define the borders of each drainage basin.

Developing English Proficiency *Tributary,* from the Latin word for tribe (*tribus*), can be used to describe a form of payment given to another, often to gain protection or to show appreciation. In a river system, a tributary might be thought of as paying into a larger river in the system. A watershed in a river system is defined by high points (divides) that determine to which basin the water will drain. In a more general way, *watershed* can be used to describe an event that marks a critical turning point for an individual or group.
Use Spanish Vocabulary and Summaries, pp. 25–26.

Characteristics of Streams and Rivers

The ability of a stream or river to erode and transport sediment is affected by many factors. These factors, which are interconnected, include the velocity of the water, the stream's gradient, its discharge, and the shape of its channel.

Velocity

The velocity of water in a stream or river is the distance that water travels in a given amount of time. The velocity of the water in a river is related to the amount of energy that the water has. A fast-moving river can erode materials more quickly and can carry larger particles than a slow-moving river. Many factors affect a river's velocity, including the steepness of the slope, the amount of water traveling downstream, and the shape of the path through which the water travels.

Gradient

The steepness of the slope of a stream or river is called its **gradient.** A river's gradient varies along its course. A river may plunge down steep hills or mountains near its source. There its gradient is very large. By the time a river approaches sea level, it may be traveling across a plain that slopes very gradually, so its gradient is very small.

GRADIENT Cullasaja Falls in North Carolina tumble down a steep slope, as shown in the photograph at the left. In contrast, the Manistee River in Michigan, in the photograph at the right, is flowing over a more level area.

Discharge

The **discharge** of a stream or river is the amount, or volume, of water that passes a certain point in a given amount of time. Discharge is not constant over the length of a river. In many rivers, discharge increases downstream because tributaries continually add more water. In rivers that flow into deserts, discharge may decrease downstream.

Discharge is not constant year-round. During times of increased precipitation or at times when snow is melting, more water runs into rivers. The velocity of the water also increases. Rivers become wider and deeper and may even flood their banks.

How Does Stream Flow Change over Time? Examine stream discharge and variables that influence it.
Keycode: ES1301

CHAPTER 13 SECTION 1

Learn more about a career as a hydrologist.
Keycode: ES1302

ASSESS

1. Rivers and streams transport almost half of the water that falls to Earth's surface, carrying it overland to the oceans.

2. A river system is a river and all of its tributaries.

3. Discharge is the amount of water in a stream or river that passes a certain point in a given amount of time. Velocity is the distance that water in a stream or river travels in a given amount of time. Gradient is the steepness of the slope of a stream or river. A channel is the physical depression through which the water flows.

4. Students should infer that water flows faster over steeper ground, and thus as a river's gradient increases, water velocity also increases.

5. The gradient of stream A is greater (or steeper) than that of stream B.

Hydrologist

Have you ever wondered whether the water you drink is safe? Thanks to the work of hydrologists, the source of the water most likely has been tested to ensure its purity.

While their specific duties may vary, all hydrologists work to solve problems involving water. Some hydrologists may study how water flows in rivers and streams. Other hydrologists specialize in studying erosion or the effects of pollution. Some hydrologists deal with weather-related issues such as flooding and droughts. All hydrologists use their creativity and research skills to help protect the quality and the availability of water for all uses.

Most hydrologists have at least a bachelor's degree in hydrology or a related field, such as environmental science, civil engineering, geology, or forestry. Also, subjects such as math and atmospheric science are important. Besides a solid educational background, an ability to work in teams and a willingness to travel and work outdoors are also useful. Some of the rewards of being a hydrologist include educating the public about water conservation and getting to share research with other scientists. Most importantly, hydrologists derive great satisfaction from knowing that they are helping to preserve a vital natural resource.

A HYDROLOGIST examines the water quality of a marsh.

Learn more about a career as a hydrologist.
Keycode: ES1302

Channel

The path through which the water flows in a stream or river is called its *channel*. The size and shape of a channel affect the velocity of water. The channel of a shallow, winding stream with many boulders has a great deal of surface area in contact with water flowing through it. Friction slows the water down. By contrast, a straight channel that is wide and deep has less surface area in contact with the water so the water's velocity is greater.

13.1 Section Review

1. Describe the role that rivers and streams play in the water cycle.
2. What is a river system?
3. Define these terms: discharge, velocity, gradient, and channel.
4. **CRITICAL THINKING** Explain how a river's gradient could affect its velocity.
5. **MATHEMATICS** Stream A runs down a slope that drops 3 meters in elevation for every 10 meters in horizontal distance. Stream B runs down a slope that drops 5 meters in elevation for every 20 meters in horizontal distance. Compare the gradients of the two streams.

282   Unit 4   Earth's Changing Surface

MONITOR AND RETEACH

If students miss . . .

Question 1 Reteach the introductory description of streams and rivers. (p. 280) If necessary, review the water cycle with students. (Chapter 1, p. 13)

Question 2 Reteach "River Systems." (p. 280) Ask students to describe the river system on the left side of the diagram on page 280.

Question 3 Reteach characteristics of streams and rivers. (pp. 281–282) Be sure students can explain the four terms in their own words.

Question 4 Reteach "Characteristics of Streams and Rivers." (pp. 281–282) Encourage students to infer the relative velocities of the rivers shown on page 281 and relate these to the rivers' relative gradients.

Question 5 Explain that gradient is equal to change in vertical distance (or elevation) divided by change in horizontal distance. Then have students draw a diagram of the gradient described and calculate the slope of each.

Stream Erosion and Deposition

Of all the agents of erosion, running water is the most effective. Gravity draws surface water downhill, eroding soil and rock materials along the way. These materials are deposited when a river or stream no longer has enough energy to transport them. The process by which materials are deposited is called **deposition**.

How Streams Weather and Erode Material

Running water in streams and rivers wears down Earth's surface by breaking up bedrock and by removing eroded rock and soil materials.

Running water breaks up the bedrock over which it flows primarily by mechanical means. Rapidly flowing water has a lifting effect that can split off and move rock fragments. However, most erosion occurs when running water abrades and hammers away at its bed by using sand, pebbles, and even boulders as cutting tools. Abrasion also wears down the cutting tools themselves, especially at their edges. In time, abrasion produces the rounded boulders, pebbles, and sand grains that are commonly found in the beds of streams and rivers.

Cutting tools are involved in the formation of **potholes** in the bedrock of a river. These deep oval or circular basins are formed when water in a river develops small whirlpools. As sand, pebbles, and small boulders swirl around in the whirlpools, they grind potholes in the bedrock. The cutting tools that have formed a pothole are often found at its bottom.

Potholes occur in many different types of rock and range in size. Potholes in the James River at Richmond, Virginia, have been ground out of hard granite. Most of these potholes are relatively shallow because granite is resistant to weathering. Potholes in the Mohawk River valley near Little Falls, New York, have been ground out of limestone. Some of these potholes are very deep because limestone is more easily weathered.

A plunge pool is a basin that has been worn away at the base of a waterfall by the action of falling water. Abrasion by churning particles also plays a part in forming a plunge pool. Many waterfalls have plunge pools. One example is Taughannock Falls in New York.

Running water's chemical weathering of bedrock consists of dissolving soluble minerals. Limestone and marble, as well as sandstones that are held together with calcite cements, are affected when rivers flowing over them form pits and holes in the riverbed. The running water also widens existing cracks and holes. In the sandstones, the water dissolves the cement and leaves the sand to be picked up mechanically by the running water.

13.2

KEY IDEA
As streams flow, they erode rocks and soil, transport these sediments over long distances, and deposit them in areas of quieter water.

KEY VOCABULARY
- deposition
- pothole
- load
- suspension
- bed load
- competence
- capacity
- delta

POTHOLES A pothole may form in the bedrock of a river when rock particles are churned about by whirlpool action.

Chapter 13 Surface Water **283**

CHAPTER 13 SECTION 2

FOCUS

Objectives
1. Describe how streams weather and erode Earth's surface.
2. Describe how streams transport and deposit sediments.
3. Explain how deltas and alluvial fans form.

Set a Purpose
Read the section title aloud and allow students to discuss their prior knowledge of erosion and deposition—where they have seen it and how it may have affected them. Then have students read to answer the question "How do streams erode and deposit sediment?"

INSTRUCT

More about...

Stream Erosion and Transport
Not all sediment carried by a stream is weathered and eroded by the waters in its channel. Tremendous amounts of weathered material are transported to streams by runoff, mass wasting (landslides), and groundwater to become part of a stream's load. Groundwater supplies the greatest proportion of dissolved material carried by streams.

DIFFERENTIATING INSTRUCTION

Reading Support Relate stream erosion and deposition to students' lives by discussing their experiences with streams. Focus on local streams and those that students might have seen, boated on, fished in, or lived near while on vacation or living in another region. Ask: What features of streams have you seen? meanders, sand bars, steep banks, clear or muddy water, rapids, slow-moving water, deltas, and so on Why do you think those features exist?
Use Reading Study Guide, p. 44.

CHAPTER 13 SECTION 2

Visual Teaching

Interpret Diagram
Direct students to the caption in the Stream Load diagram and ask: Why can't you see the material carried in solution? **Since this material is dissolved, its particles are too tiny for the unaided eye to observe.** Then have students rank the three types of stream load represented in the diagram from largest to smallest. **bed load, suspended material, dissolved material**

Extend
Ask students to imagine that they have just taken a water sample from a local river. The sample came from a depth of about halfway between the river's surface and bed. Have students use the diagram to infer the type of load and sediment their sample is likely to contain. **The sample would likely contain dissolved and suspended load, consisting of microscopic particles in solution as well as suspended particles of clay, silt, and fine sand.**

SUSPENSION The Colorado River carries much of its load in suspension.

VISUALIZATIONS CLASSZONE.COM
Observe how sediment is transported by flowing water.
Keycode: ES1303

How Streams Transport Material

The eroded rock and soil materials that are transported downstream by a river are called its **load.** A river transports, or carries, its load in three different ways: in solution, in suspension, and in its bedrock.

Mineral matter that has been dissolved from bedrock is carried in solution. Common minerals carried in solution by rivers include dissolved calcium, magnesium, and bicarbonate. Most of a river's solution load comes from groundwater seeping into the river. Before it reaches the stream, the groundwater has traveled through fractures in the bedrock, chemically eroding rock along the way.

When river water looks muddy, it is carrying rock material in **suspension.** Suspended material includes clay, silt, and fine sand. Although these suspended materials are heavier than water, the turbulence of the stream flow stirs them up and keep them from sinking. Turbulence includes swirls and eddies that form in water as a result of friction between the stream and its channel. The faster a stream flows, the more turbulent and muddy it becomes. A rough or irregular channel also increases turbulence.

A river may also transport rock materials in its **bed load.** The bed load consists of sand, pebbles, and boulders that are too heavy to be carried in suspension. These heavier materials are moved along the streambed, especially during floods. Boulders and pebbles roll or slide along the river bed. Large sand grains are pushed along the bottom in a series of jumps and bounces.

The relative amounts of a river's load that are carried in solution, in suspension, and in the bed load depend on the nature of the river, the climate, the type of bedrock, and the season of the year. As a general rule, most of the load carried by the world's streams and rivers is carried in suspension. The size of a river's suspended load increases with human land use. Road and building construction and removal of vegetation make it easier for rain to wash sediment into streams and rivers.

Stream Load

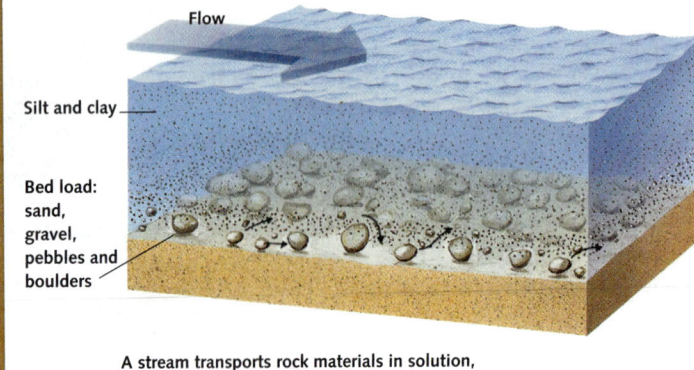

A stream transports rock materials in solution, in suspension, and in its bed load. Materials carried in solution cannot be seen.

284   Unit 4   Earth's Changing Surface

DIFFERENTIATING INSTRUCTION

Hands-On Demonstration Use a clear glass jar filled with warm water, salt, fine sand, and gravel to model the three types of stream load. First, drop a pinch of salt into the water, secure the jar lid on tightly, and shake the jar until the salt dissolves. Ask: If the water in the jar represents stream flow, what type of load am I modeling? **dissolved material** Next, add fine sand to the water in the jar, shaking the jar to suspend the sand. Ask: What type of load is the stream carrying now? **dissolved and suspended material** Finally, add gravel to the water in the jar. Ask: What type of load is the stream carrying now? **dissolved material, suspended material, and bed load** Be sure to identify for students each material as you add it to the water in the jar.

⚠ Exercise caution when handling the glass jar. Remind students that safety is everyone's responsibility.

284   Unit 4   Earth's Changing Surface

Two measures are used to describe the ability of a stream to transport materials. **Competence** is a measure that describes the maximum size of the particles a stream can carry. **Capacity** is a measure of the total amount of sediment a stream can carry. The competence and capacity of a stream depend on the stream's velocity and discharge. Because the velocity and discharge of a given stream are not constant, the competence and capacity of a stream are not constant. Competence and capacity vary along a stream and change throughout the year.

Streams moving at high velocity with high discharge can carry a larger amount of sediment and larger sizes of sediment particles than slow-moving streams with small discharge. When the velocity and discharge of a stream increase, the ability of the stream to carry larger particles in suspension increases as well. For example, during normal times, the lower Mississippi River may carry nothing larger than silt in suspension. During a flood, the river carries sand and pebbles as well.

Because the competence and capacity of a stream increase during a flood, much of the erosion caused by moving water occurs during floods. At such times, a stream can erode more deeply into its channel.

Stream Deposition

Rock materials and sediments that are transported by rivers and streams are eventually deposited. A river will deposit a part of its load of sediment when either its velocity or its discharge decreases.

A river's velocity may decrease if its channel widens or the river meets an obstruction in the form of a curving bank or a rock outcrop. For example, when a river rounds a bend, its velocity increases on the outside of the curve. Its velocity slows on the inside part of the curve, so sediments tend to be deposited there. However, the greatest decrease in a river's velocity occurs when it empties into a sea or a lake. At this point most of the river's remaining sediment is deposited.

A river's discharge may decrease when people divert water for irrigation or for city water supplies. Discharge may also decrease if the river passes through an arid region where it loses water by evaporation into the air and seepage into the ground. In humid regions, however, rivers usually grow larger as they approach a sea and are fed by new tributaries.

Discharge and velocity increase during a flood, then decrease as flood waters subside. As the river's discharge and velocity decrease, the river deposits its load of sediment.

Depositional Features

A **delta** is a fan-shaped deposit that forms when a river flows into a quiet or large body of water, such as a lake, an ocean, or an inland sea. The word *delta* comes from the Greek letter Δ, whose shape resembles the shape of a river delta. The Mississippi, the Nile, the Ganges, and other large rivers have deltas where they enter larger bodies of water.

25-Minute Mini LAB

Modeling River Sediments

Materials
- tall clear container
- sand
- water
- clay
- gravel
- spoon

Procedure

1. Pour the clay into the container to form a layer 2 cm thick. Cover the clay with a similar layer of sand and a layer of gravel. Pour 15 cm of water on top.
2. Stir the sediments until the layers have completely disappeared. Stop stirring and watch the sediments settle.
3. Stir the sediments again just fast enough to keep the sand suspended.

Analysis

In what order do the sediments settle during Step 2? Why do they settle in this order? In Step 3, what happens to the gravel? To the clay? What would happen if you slowed your stirring during Step 3?

KLAMATH RIVER, CALIFORNIA
Sediment is deposited on the inside of a river bend as water velocity decreases.

CHAPTER 13 SECTION 2

Observe how sediment is transported by flowing water.
Keycode: ES1303

Visualizations CD-ROM

MINILAB
Modeling River Sediments

Management
- Have paper towels available so students can immediately clean up spills.
- Some water will soak into the layers. Have students keep pouring water into the container until the water line is 15 cm above the gravel line.

Materials
For the clay, use dry clay-sized particles, not wet clay or modeling clay. You can do the Minilab using sets of other materials of different-sized particles, such as flour, rice, or beans.

Analysis Answers
The gravel settles first, followed by the sand, and then the clay. Heavier particles settle first because they require the greatest water velocity to be moved. In step 3, the gravel does not move, but the clay becomes suspended. If the stirring is slowed, the sand may start to settle, but the clay will probably continue to be suspended.

For proper behavior in a laboratory setting, refer students to Safety in the Earth Science Laboratory on pages xxii–xxiii.

DIFFERENTIATING INSTRUCTION

Developing English Proficiency As you discuss stream transport and deposition, consider substituting simpler or more familiar terms for those used in the text. For example, you might substitute the word *largest* for *maximum* when defining competence, *dry* for *arid*, or *looks like* for *resembles*.

CHAPTER 13 SECTION 2

What Controls the Shape of a Delta?
Keycode: ES1304

Use Internet Investigations Guide, pp. T51, 51.

ASSESS

 Fast-moving water lifts and moves apart rock fragments; abrasive particles carried by the water cut away at bedrock. Soluble minerals in bedrock are dissolved.

❷ Materials are carried in solution, in suspension (by turbulence), and as bed load (rolling or sliding along the river bed).

❸ When the velocity of a river increases, its ability to carry more and larger sediment increases, so its capacity and competence increase.

❹ when their velocity or discharge decreases

❺ Unlike an alluvial fan, a delta is formed in water, is made of finer sediments including silt and clay, and has a flat surface.

❻ Deposits would be minimal. The winding channel increases turbulence, thereby increasing the amount of suspended material and decreasing the likelihood of deposition.

❼ Alluvial fans are likely to be found in desert regions at the base of the Rocky Mountains in the Midwest and West and at the base of the Sierra Nevada in California.

River water comes almost to a standstill at a delta where most of a river's sediment is dropped. As long as the amount of sediment supplied by the river is larger than the amount that can be taken away by currents, the deposit grows. A river flowing over its delta splits into branches called distributaries. The distributaries are responsible for the delta's shape. Their smaller channels bring sediment to the front of the delta.

Delta formation is a delicate balance between deposition and erosion. Human use of rivers and deltas can cause a decrease in deposition, which leaves a delta more susceptible to wave and current erosion.

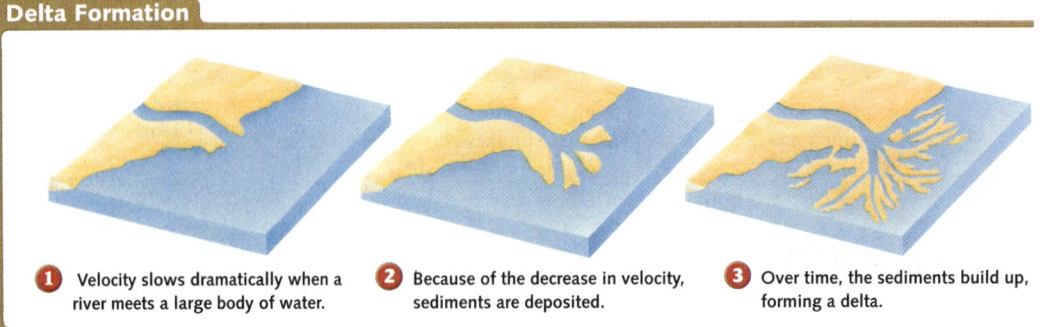

Delta Formation

❶ Velocity slows dramatically when a river meets a large body of water.

❷ Because of the decrease in velocity, sediments are deposited.

❸ Over time, the sediments build up, forming a delta.

What Controls the Shape of a Delta?
Compare deltas and analyze the processes that shape them.
Keycode: ES1304

A fan-shaped deposit called an *alluvial fan* may form when a steep mountain stream meets dry, level land at the base of the mountain. When it reaches the land, the stream's velocity decreases greatly. As a result, the stream drops a large part of its sediment load. An alluvial fan differs from a delta in several ways. First, the deposit is formed on land, not in water. Second, the sediments are coarse sands and gravels rather than fine silt and clay. Third, its surface is sloping, not flat like that of a delta.

13.2 Section Review

❶ Describe how rivers wear down rock materials.
❷ Describe how rivers transport materials.
❸ How does a river's velocity affect its competence and capacity?
❹ Under what circumstances do rivers and streams deposit sediment?
❺ How is a delta different from an alluvial fan?
❻ **CRITICAL THINKING** A river carrying sediments in suspension curves back and forth sharply. Predict what you might find if you studied deposits on the stream bed. Explain your prediction.
❼ **GEOGRAPHY** Alluvial fans are common in desert and semidesert regions. Use the map on pages 706–707 to identify a region in the United States where you might find an alluvial fan. Explain.

MONITOR AND RETEACH

If students miss . . .

Question 1 Reteach "How Streams Weather and Erode Material." (p. 283) Contrast mechanical weathering and chemical weathering by river water.

Question 2 Reteach "How Streams Transport Material." (p. 284) Use the diagram to explain materials in solution, in suspension, and in bed load.

Question 3 Reteach competence and capacity. (p. 285) Be sure students can define both measures and associate them with water velocity.

Question 4 Reteach "Stream Deposition." (p. 285) Discuss the effects of a river's velocity and discharge on the deposition of its sediment load.

Question 5 Reteach "Depositional Features." (pp. 285–286) Have students list similarities and differences between deltas and alluvial fans.

Question 6 Review water velocity and sediment deposition. (p. 285)

Question 7 On the map, help students find mountains that are located in desert or semidesert regions.

River Valleys

The formation of a river valley begins on a small scale. A single heavy rain may form a small valley in loose soil along a hill slope. When the rain ends, the stream may disappear, but the small valley remains. Such a feature is called a *gully*. Gullies grow in length, width, and depth every time it rains. The process by which land is worn away at the head of stream or gully is **headward erosion.** Headward erosion lengthens streams or gullies.

As a valley grows in length and depth, the stream may eventually cut down far enough to become permanent. When the tributary gullies also become permanent, a river system is born. Most of the world's river systems probably began in this way.

Canyons and V-Shaped Valleys

The streams of mountain regions and high plateaus are likely to have formed relatively recently. The amount of precipitation in a given area is one of the most important factors affecting the formation of the young or youthful river valleys, because more surface water means greater erosion.

Some river valleys have very steep, almost vertical sides. Such valleys, which are often called canyons, gorges, or chasms, tend to form in regions with less rainfall. They can also form when a river cuts into its bed rapidly or when the rock materials on the sides of the valley are resistant to erosion.

Factors that affect the length of time it takes a river to form a canyon include the type of rock the river must erode, the amount of water and sediment in the river, and the climate of the region. The Colorado River is thought to have taken millions of years to cut the mile-deep Grand Canyon into the rocks of the Colorado Plateau.

Most youthful river valleys are V-shaped. Such valleys are found in regions where there is enough rain to erode the sides of the valley. As the river cuts its way down into its channel, the upper valley walls are widened into a V-shape by erosion. The Yellowstone River, shown in the photograph below, is an example of a river that has eroded its valley into a V-shape.

V-SHAPED VALLEY
The Yellowstone River has eroded a V-shaped canyon in Yellowstone National Park.

13.3

KEY IDEA

The continuous erosion caused by running water tends to form V-shaped valleys that grow longer and wider over time.

KEY VOCABULARY
- headward erosion
- base level
- stream piracy

CHAPTER 13 SECTION 3

FOCUS

Objectives
1. Explain how river valleys form.
2. Describe the formation of rapids and waterfalls.

Set a Purpose
Have students read the section to find the answer to "How does erosion by running water form V-shaped valleys?"

INSTRUCT

More about...

Headward Erosion
The source of a river, or where it begins, is called the head or headwaters. A river can lengthen itself upstream by headward erosion. A river valley collects water at its head where sheets of surface runoff flow together. This concentration of water increases the water's velocity, resulting in increased erosion at the head. Through this headward erosion, the river valley lengthens, extending itself beyond its former head position.

DIFFERENTIATING INSTRUCTION

Support for Visually Impaired Students Use folded sheets of cardboard to help students compare the shapes and gradients of the sides of a canyon and V-shaped valley. To form a canyon, make two parallel folds in a large square of cardboard and arrange the sides adjacent to the folds at 90° angles. While you hold the cardboard canyon, have students run their hands inside it so they can note its steep, near vertical sides. Then form a V-shaped valley by making a single fold in a second square of cardboard. While you hold the cardboard to form a V shape, allow students to run their hands over it and note its shape and less-steep sides.

CHAPTER 13 SECTION 3

Visual Teaching

Interpret Diagrams
Have students describe in their own words the events shown in the diagram. Then ask: What can you infer about the amount of rainfall that occurs in the region where this valley formed? **The region must get relatively abundant rain, since enough rain falls here to erode the sides of the valley to form a V shape.**

Extend
Have students describe how the diagram would be the same and different if it showed canyon formation instead of V-shaped valley formation. **Diagram 1 would be the same, but diagrams 2 and 3 would show a valley with steep, almost vertical sides being cut by a river rapidly cutting its bed. Also, the surface terrain might not be green, due to less rainfall.**

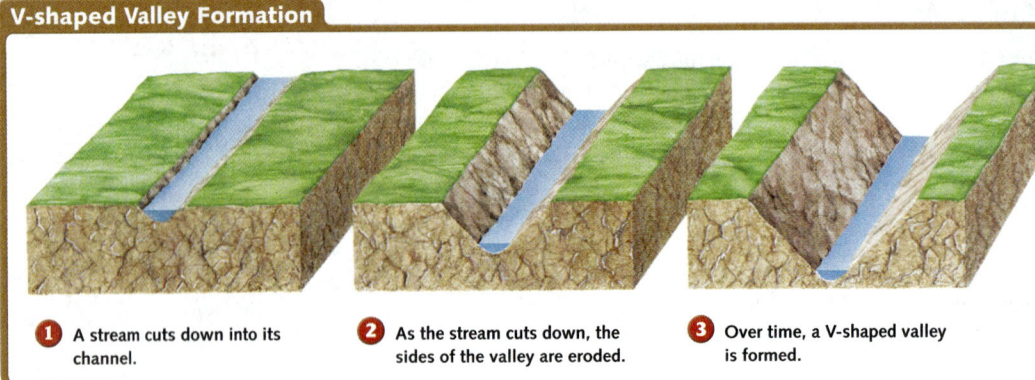

V-shaped Valley Formation
1. A stream cuts down into its channel.
2. As the stream cuts down, the sides of the valley are eroded.
3. Over time, a V-shaped valley is formed.

VOCABULARY STRATEGY
Sea level is called the ultimate base level for all rivers. In everyday usage, *ultimate* means "last in a series or progression."

VISUALIZATIONS CLASSZONE.COM
Observe river erosion creating waterfalls and chasms.
Keycode: ES1305

A stream cannot erode its bed to a level any lower than the level of the largest body of water into which it flows. This level is called the **base level** of the stream. For streams that flow into an ocean, the base level is sea level. Sea level is also the ultimate base level because all rivers eventually empty there. Lakes and rivers are local base levels for the streams that run into them.

As a stream approaches its base level, the gradient of the streambed and the velocity of the stream decrease. Thus, there is less erosion of the bed and more erosion on the sides of the channel. The resulting valley is wider with a broad floor and gently sloping walls.

The headward erosion that brings about the formation of a river valley is an important factor in a process called **stream piracy** or stream capture. Picture two rivers separated by a divide. Through headward erosion, one of the rivers wears through the divide and captures the headwaters of the other river. The first river grows larger and extends its drainage basin at the expense of the captured river. Geologists think that stream piracy has been an important factor in the early growth of many great river systems.

Rapids and Waterfalls

A stream running through a steep mountain region flows over ever-changing slopes. The riverbed may be steep enough to form white-water rapids. It may level out into a lake or a pond, or the stream may plunge over a cliff to form a waterfall.

Many processes have led to the development of the steep slopes and cliffs of rapids and waterfalls. The Great Falls of the Potomac have formed where the Potomac River flows over hard igneous rocks onto soft sedimentary rocks. The two great falls on the Yellowstone River in Yellowstone National Park flow over hard igneous rock and down onto more easily weathered rock below. The falls in Yosemite Valley in California were formed when glaciers eroded one valley more deeply than others.

DIFFERENTIATING INSTRUCTION

Challenge Activity Have pairs of students choose a famous or local waterfall and research to find out how it formed. Students can present their findings by using a multistage model or a computer animation.

Reading Support Visualizing helps students comprehend their reading, especially when reading about a process. Help students visualize stream piracy by diagramming it on the board or overhead projector. Add to the diagram a little at a time as a student reads each sentence aloud in the paragraph on stream piracy. When the diagram is complete, go through it again as a student reads the paragraph in its entirety.
Use Reading Study Guide, p. 45.

NIAGARA FALLS Goat Island divides Niagara falls into the American falls (left) and the Canadian falls (right). The Canadian falls are also called the Horseshoe Falls.

Because stream erosion is greatest at rapids and waterfalls, they are temporary features of streams. One way in which streams erode at waterfalls is by undermining. The water falling into the plunge pool at the base of the waterfall erodes the rock there, leaving the rocks at the top of the falls to overhang. From time to time, pieces break off the top. Each time this breakage occurs, the waterfall recedes farther upstream.

A famous example of a waterfall that recedes by undermining is Niagara Falls. Here the falls are the result of a nearly flat-lying, 20-meter-thick layer of tough dolostone. Dolostone is a rock similar to limestone, but it is made of dolomite rather than calcite. The rocks underneath the dolostone are almost all thin beds of shale. The rapidly eroding shale undermines the dolostone. Consequently, the base of the American Falls is littered with huge blocks of fallen dolostone.

The Niagara Gorge marks the path of recession of Niagara Falls. The gorge extends from the base of the falls to Lake Ontario, a distance of about 11 kilometers. At the end of the Ice Age 12,000 years ago, Niagara Falls was located on or near Lake Ontario. The recession of the falls since that time has carved the gorge. Horseshoe Falls, shown in the photograph above, erodes its ledge at a rate that can vary from eight centimeters to two meters per year.

NIAGARA FALLS Water action erodes weak shales at the base of the falls. This erosion undermines the tough dolostone layer above. From time to time, the dolostone breaks off, and the waterfall recedes.

13.3 Section Review

1. Explain how rivers form V-shaped valleys and canyons.
2. How does stream piracy happen?
3. **CRITICAL THINKING** Identify a river in your area. Identify a body of water that represents the local base level of the river. Is the local base level also the ultimate base level?
4. **APPLICATION** When a dam is built across a river, it raises the local base level of the river to the level of the water in the reservoir that forms behind the dam. How might changing the base level affect the river's ability to transport eroded materials? Explain your thinking.

CHAPTER 13 SECTION 4

FOCUS

Objectives
1. Describe the features of a floodplain.
2. Explain why floods occur.
3. Summarize natural and artificial methods of flood prevention and control.

Set a Purpose
Have students set a purpose for reading by restating each red heading as a question such as: What are the features of a floodplain? How do floods occur? How can floods be prevented and controlled?

INSTRUCT

VISUAL TEACHING
Discussion
Tell students that at an earlier time, the river valley shown in the diagram on page 290 looked like the river valley in the diagram on page 288. Have students summarize how the change in the river valley came about. Be sure they understand that the river on page 288 is primarily eroding downward, deepening the valley, whereas erosion in the valley on page 290 is primarily horizontal, widening the valley.
Use Transparency 14.

13.4

KEY IDEA
River floods are naturally occurring events that sometimes threaten human populations.

KEY VOCABULARY
- flood
- floodplain
- meanders
- oxbow lake
- natural levees
- flash flood

Observe changes in the channel of a meandering river.
Keycode: ES1306

FLOODPLAIN As a river approaches its base level, its valley is widened by erosion into a floodplain.

Floodplains and Floods
At times, a river may overflow its banks as a **flood.** The floodwater may cover part or all of the valley floor where the river runs. This part of the valley floor is the **floodplain.**

Features of a Floodplain
As a river cuts its bed lower and approaches its base level, its gradient and velocity decrease. The river current is more easily deflected sideward, and its course becomes more winding. The banks and valley walls are slowly eroded by the river, and the valley floor widens into a floodplain.

A river flowing through a floodplain typically winds back and forth in broad curves called **meanders** (mee-AN-ders). Meanders form because erosion is most rapid on the outside of a river bend. In time, the channel is deepened there. Swift currents send the water across the channel and downstream into the opposite bank. The process repeats itself, forming a series of meanders.

Meandering rivers constantly change. As a meander swings wider and wider, it becomes a loop that the river can break through. The river usually breaks through during a flood, forming a cutoff. The river then drops mud and silt at the ends of the abandoned meander. A curved body of water formed when deposits separate a meander from its river is an **oxbow lake.** (See diagram below.)

During a flood, a river carries large amounts of sediment. When a river overflows onto its floodplain, the velocity of the river slows and its sediment load is immediately deposited. Thick deposits build up alongside the stream banks. These deposits form elevated ridges called **natural levees** (LEHV-eez).

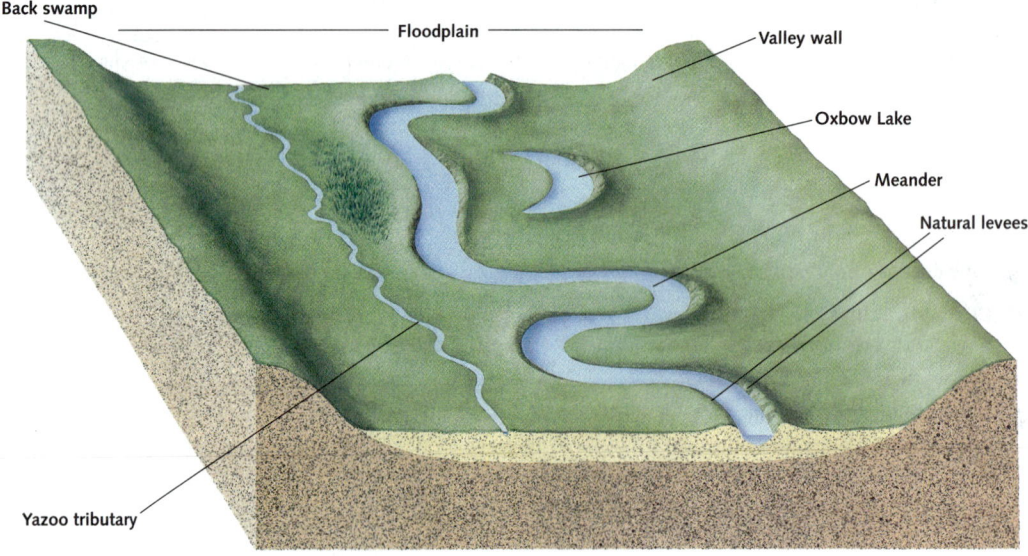

DIFFERENTIATING INSTRUCTION

Hands-On Demonstration Use a large cookie sheet, dampened sand, and water to model the formation of an oxbow lake from a meander. First, spread the sand evenly on the cookie sheet. Use a craft stick to make the channel of a meandering river, with one meander wider than the others. Have a student slowly but continuously add water to the channel using a single-spout watering can, not a sprinkling can. The "river" that forms should not overflow its banks. Make a couple of carvings with the craft stick, modeling the migration of the meanders toward one another. Make a final carving that connects two meanders. Form an oxbow lake by dropping sand at the ends of the abandoned meander. Ask: What factor in this model has been greatly exaggerated? *time*

Beyond the levees, the floodplain slopes away from the river. Swamps called *back swamps* may form in the lowest areas. New tributaries called *yazoo tributaries* may also form and flow through the back swamps. Some of these tributaries may flow for many kilometers parallel to the main stream before breaking through the levee to join the main stream.

Floods

River floods are naturally occurring events that can be constructive as well as destructive. Floods temporarily relieve the water and sediment overload of a river channel. Floods deposit minerals and nutrients on floodplains, making these areas fertile for agriculture. Unfortunately, floods are also tremendously destructive for the people living near rivers. For these reasons, the causes of floods are of great importance to people.

Most river floods result from heavy or long-lasting rains, the rapid melting of winter snows, or both. A single cloudburst can cause a **flash flood,** especially if the cloudburst occurs over the narrow valley of a young mountain stream. Towns at the bases of such valleys can suffer severe damage when hit by flash floods.

Large rivers, such as the Ohio, Missouri, and Mississippi, do not have flash floods. Their floods result from many days of steady rainfall over large parts of their vast drainage basins. Large rivers such as these also may flood after the thaws that occur in winter and early spring. These warm spells increase runoff, especially when the ground is still frozen, because frozen ground cannot absorb water from melting snow.

When a dam forms across a river, it floods the valley above the river up to the height of the dam. Dams are built to create reservoirs. They also form naturally. An ice jam, which is a common type of natural dam, can form when a frozen river breaks up during winter or spring thaws. Other natural dams can result when a volcano erupts and deposits ash, cinders, or lava across streams. Dams caused by landslides are also common.

Many bad floods have been caused by the failure of reservoir dams. The Johnstown, Pennsylvania, flood of 1889 happened when a reservoir dam made of earth collapsed after days of heavy rain. The dam had been built about 22.5 kilometers upstream from Johnstown. When the dam broke, the water in the reservoir burst down on the city, killing more than 2200 people.

Although most causes of floods are natural, human activities can worsen and even cause floods. Land covered with buildings and pavement does not absorb water. Thus, people increase the amount of runoff entering streams when they develop urban areas. Removal of vegetation from steep slopes can also increase runoff because vegetation helps the soil absorb water. Agricultural activities and urban development can increase runoff too by displacing wetlands, which act as natural sponges.

OXBOW LAKE An oxbow lake has formed along the meandering Congaree River in South Carolina.

MISSISSIPPI RIVER Heavy rains can cause large rivers to overflow their banks.

CHAPTER 13 SECTION 4

VISUALIZATIONS
CLASSZONE.COM

Observe changes in the channel of a meandering river.
Keycode: ES1306

Visualizations CD-ROM

CRITICAL THINKING

Tell students that the surface of the Huang He River in China is as much as 10 meters (33 feet) above the level of its floodplain. Ask students to explain how this might have come about. With each flood, levees built up higher, forming higher riverbanks. Sediments also accumulated on the floor of the river, raising its bed. This raised the surface of the river water to a level higher than the floodplain.

More about...

Dams and Flooding
To prevent dam failure, engineers study a variety of data to estimate likely local water volumes, including data from ancient flood deposits. Such a study of Australia's arid interior has shown that extremely destructive floods, called superfloods, that had been thought to occur once in several thousand years actually occur about once every 200 years. Engineers have found that a superflood would be beyond the engineering limits of the Warragamba Dam, near Sydney, Australia, putting it at risk of failure.

DIFFERENTIATING INSTRUCTION

Reading Support Skimming, or selective reading, is a skill that allows students to read quickly through a passage, looking for and noting important ideas but skipping secondary concepts. Skimming can be helpful when it is not necessary to read every word in the section, such as when looking for the answer to a question, finding a main idea, or getting an overview. Demonstrate the skill by skimming the section to find out how human activities cause floods. First, quickly read the heads, looking for the appropriate subsection. Then, read quickly to find key words or phrases such as *human activity* or *cause floods.*
Use Reading Study Guide, p. 46.

CHAPTER 13 SECTION 4

Scientific Thinking

ANALYZE
Have students imagine themselves as environmental scientists concerned with the preservation of river ecosystems and wetlands. Then, to get a different perspective, have students imagine that they are potential property owners living in an overcrowded community where only floodplain land remains for purchase and development.

ASSESS

1 A floodplain forms when a river approaches its base level and its gradient and velocity decrease. The river current is more easily deflected sideward, and the banks and valley walls become eroded, widening to form a floodplain. When a river overflows onto its floodplain, the river slows and its sediment load is deposited along the banks, forming levees. Meanders are broad curves of a river that form because erosion is most rapid on the outside of a river bend. In time, the channel is deepened there by erosion. Swift currents send the water across the channel and downstream into the opposite bank. The process repeats itself, forming a series of meanders. An oxbow lake forms when a meander is cut off from the river. During flooding, the river can cut a straight channel, leaving behind an abandoned meander, which in time becomes an oxbow lake.

2 heavy or long-lasting rain, rapid melting of snow, the formation of dams across a river, the failure of reservoir dams, agricultural activities, or urban development

3 replanting vegetation to control runoff, building dams and storing runoff in reservoirs, building artificial levees, and constructing spillways

MISSISSIPPI RIVER The photograph above shows a natural levee along the Mississippi River.

Scientific Thinking

ANALYZE
There are many benefits to controlling floods by allowing rivers to flood naturally. However, allowing a river to overflow onto its floodplain has costs as well. For example, areas preserved as wetlands cannot be used for farming or development. Describe the benefits and costs of controlling river floods. What issues must be considered in deciding how to control flooding?

Flood Prevention and Control

People rely on many different measures to control and prevent floods. Most of these measures have both advantages and disadvantages.

One way that people can lessen the impact of flooding is to restore natural flood protections that human activity has altered. In areas where vegetation has been removed, replanting helps to control runoff. Replanting is most important at the headwater regions of a river's drainage basin. Such measures cannot prevent floods, however.

Dams are often used to control and prevent floods. When dams are built across rivers, reservoirs can store excess runoff. One example is the Tennessee River system, where about 50 dams have been built. Some of the risks associated with dams are that they can break, causing floods that may be more destructive than natural flooding. Dams also trap sediment, which upsets the balance between erosion and deposition downstream. Less deposition means more erosion of a floodplain, which worsens flooding. Dams become less effective at controlling floods once they fill up with sediment. They can also damage river ecosystems.

Artificial levees, such as sandbags stacked on top of a river's natural levees, are another method of flood control. The principle behind using artificial levees is that a deeper river will hold more water. A problem occurs, however, when the water level rises within this deeper channel. As the river's height increases, so does its velocity, creating a powerful erosive force that can break through levees downstream, sometimes causing more damage than might have occurred if the river had been allowed to overflow the natural levee onto the floodplain.

On the lower Mississippi, spillways control flooding. Spillways on this river system are channels running parallel to the main river and through back swamps that then flow into the Gulf of Mexico. At critical points on the floodplain, water is guided into the spillways to relieve flooding.

Once a river has overflowed its natural levees, it spreads out across the floodplain, thus preventing a powerful torrent of flood waters. Flood problems have become more severe as floodplains are altered for human use. Natural floodplain management attempts to lessen flood damage by preserving sections of floodplains for overflow, restoring wetlands, and discouraging development of flood-prone areas.

13.4 Section Review

1 Describe how floodplains, levees, meanders, and oxbow lakes form.

2 What are some of the causes of floods?

3 Describe some of the artificial means used to control flooding.

4 **CRITICAL THINKING** Compare and contrast natural levees with artificial levees. Explain how each type of levee controls flooding. Describe some advantages and disadvantages of each type of levee.

5 **GEOGRAPHY** What are some of the benefits of living on or near a floodplain? What are some risks?

MONITOR AND RETEACH

If students miss . . .

Question 1 Reteach "Features of a Floodplain." (p. 290) Have students record and define all the boldface terms on page 290.

Question 2 Reteach "Floods." (p. 291) Be sure students can describe causes of floods that are both natural and related to human activities.

Question 3 Reteach "Flood Prevention and Control." (p. 292) Have students list on the board artificial and natural flood control methods.

SCIENCE & Society

Removing the Edwards Dam

When the Edwards Dam was built on the Kennebec River at Augusta, Maine, in 1837, it promised to bring prosperity to the area. By the 1990s, however, many citizens were calling for the removal of the dam.

What made people change their minds about the dam?

The Edwards Dam was built to harness the energy produced by the action of falling water. Factories powered by the dam sprang up along the riverbank and brought jobs to the area. One huge textile factory in the area employed more than 700 people. As some people had feared, however, the dam prevented certain types of fish that migrate from the ocean to the river from reaching their upstream spawning grounds. As a result, the river's fish population fell dramatically.

In 1913, generators were installed at the dam to produce electricity. Industry continued to thrive during the first half of the century but

THE EDWARDS DAM blocked the Kennebec River in Maine from 1837 to 1999.

declined during the 1970s and 1980s. The large textile factory closed in the early 1980s. The dam owners, however, continued to sell electricity even though the generators were capable of producing only a tiny fraction of Maine's electricity.

People began suggesting that the dam had outlived its usefulness. In 1997, various regional conservation groups helped convince the government that the dam should be dismantled—a process completed in 1999. Changes appeared almost immediately. Fish such as shad, alewives, and striped bass headed toward their spawning grounds for the first time in more than 160 years. Canoeists and kayakers no longer had to turn around at the dam. Improved recreational opportunities boosted the tourist economy as people traveled to the area to fish, canoe, camp, and hike. ■

REMOVAL of the Edwards Dam, which let the waters of the Kennebec River flow freely, helped to restore the fish population and improved recreational opportunities.

Extension

SCIENCE NOTEBOOK

As fish populations are restored, scientists also expect to see increased populations of wildlife such as bald eagles, kingfishers, and ospreys. Why do you think they make this prediction?

LOCAL RESOURCES CLASSZONE.COM

Find out more about dam removal projects in your area.
Keycode: ES1307

Chapter 13 Surface Water 293

CHAPTER 13 SECTION 4

LOCAL RESOURCES CLASSZONE.COM

Find out more about dam removal projects in your area.
Keycode: ES1307

(continued)

❹ Natural levees are created by the deposition of sediment by rivers during floods. Artificial levees are constructed by people stacking sandbags on or adding earth to natural levees. Both act as barriers to rising river water and serve to control flooding. Artificial levees can cause problems for areas downstream since the river continues to increase in velocity and discharge.

❺ Benefits include fertile land, a nearby irrigation source, water transportation, and water recreation. Risks include damage to property and threats to lives from floods.

Science and Society

Rethinking Dams
There are about 75,000 dams over six feet tall on American rivers. They provide 10 percent of the electric power in the United States, but dams also disrupt entire river ecosystems. As a result, Americans have been rethinking the value of dams, leading to a steep decrease in new dam construction.

Science Notebook
Eagles, kingfishers, and ospreys would become more plentiful because they eat fish.

MONITOR AND RETEACH

Question 4 Reteach "Features of a Floodplain" and "Flood Prevention and Control." (pp. 290, 292) To help students organize their responses, provide them with copies of a compare/contrast table with the column heads *Natural Levees* and *Artificial Levees* for them to fill out.

Question 5 Reteach floods and flood prevention and control. (pp. 291–292) Have students form small groups to brainstorm and list benefits and risks of living on or near a floodplain.

Chapter 13 Surface Water 293

CHAPTER 13 ACTIVITY

PURPOSE
To examine and identify the major river systems in the United States

PROCEDURE

4 For each river system, suggest that students first locate the main river of that system, such as the Colorado River, on the physical map and then on the map on page 295. Then they can follow the river upstream (away from the ocean) to locate all the tributaries in the river system.

ANALYSIS AND CONCLUSIONS

1. Students should name and label the river and all of its tributaries. If the river has no major tributaries, then students might identify other sources, such as lakes or groundwater.

2. The Continental Divide should bound the Columbia and Colorado River systems to the west and the Rio Grande and Mississippi River systems to the east. It represents the high land separating water that will eventually flow into the Atlantic Ocean from water that will eventually flow into the Pacific Ocean.

3. The source of the St. Lawrence River is Lake Ontario, which is fed by the other Great Lakes. The river flows northeast.

4. They are located in a basin, therefore they do not drain into the ocean.

CHAPTER 13 MAP Activity

Stream Divides and River Systems

SKILLS AND OBJECTIVES
- *Identify* the major river systems of the United States.
- *Deduce* the divides that separate these river systems.

MATERIALS
- tracing paper
- colored pencils
- paper clips
- **Appendix B** *Physical United States Map,* pages 708–709

Every second, over one-half of a million cubic feet of water pours from the Mississippi River into the Gulf of Mexico. This water originally fell as precipitation in places as far away as Montana and Pennsylvania. Over time, the water flowed through the various tributaries of the Mississippi River system, working its way to the Gulf.

In this exercise, you will examine the drainage basins for the Mississippi and other major river systems of the United States.

Procedure

1. Use paper clips to fasten tracing paper over the map of major rivers of the United States on the opposite page. Trace the outline of the United States coastline with a black colored pencil.

2. Locate the Mississippi River on the opposite page using the Physical United States Map as a reference. Trace the Mississippi River and all of its tributaries with a dark blue colored pencil.

3. With a lighter colored pencil, encircle the rivers outlined in Step 2. The circled region should include the entire Mississippi River and its tributaries, but should not include any of the other rivers. Label the enclosed area the *Mississippi River System* and lightly shade the area.

4. Repeat Steps 2 and 3 for the *Colorado River System,* the *Columbia River System,* the *Rio Grande System,* and the *St. Lawrence River System.* Use a different color to outline and shade each river system.

Analysis and Conclusions

1. Label the river on the tracing paper that is closest to your school. Name all of the sources of the water for this river.

2. Draw and label the Continental Divide on your map. What does the Continental Divide represent?

3. What is the source of the water in the St. Lawrence River? In what general direction does the river flow?

4. Locate the rivers in central Nevada. Why are these rivers not part of a major river system?

5. The headwaters of three rivers systems are located in Colorado along the Continental Divide. Identify the three river systems.

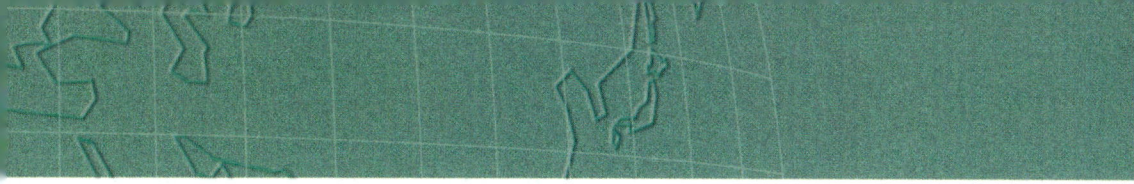

(continued)

5. the Rio Grande, Colorado, and Mississippi River systems

6. the Columbia, Colorado, and Mississippi River systems

7. The Continental Divide is associated with the Rocky Mountains, which determine the divide's position and form a high barrier preventing water west of the divide from flowing to the Atlantic Ocean.

8. The Erie Canal should run roughly east-west between Lake Erie and the Hudson River. The St. Lawrence River System now has two outlets to the ocean, the St. Lawrence and Hudson Rivers, which enlarges the outline of the river system.

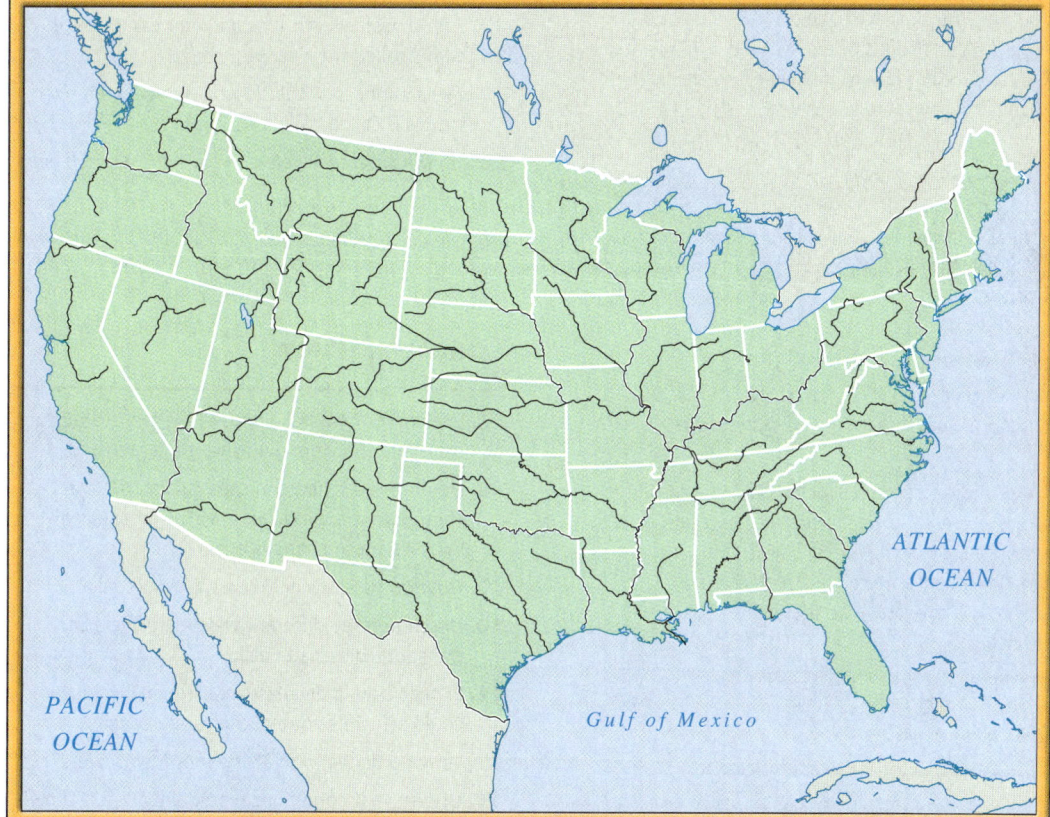

6. The headwaters of three different river systems are also found in Wyoming. Identify these three river systems.

7. Why is the Continental Divide located in the western United States and not halfway between the Atlantic and Pacific Oceans? With what topographic feature is the Continental Divide associated?

8. In 1825 engineers completed construction on the Erie Canal, which connects the Hudson River and Lake Erie. Draw and label the Erie Canal on your map. How does the Erie Canal affect the outline of the St. Lawrence River System?

CHAPTER 13 REVIEW

VOCABULARY REVIEW

1. suspension
2. base level
3. gradient
4. Competence is a measure of the maximum size of the particles a stream can carry, whereas capacity is a measure of the total amount of sediment a stream can carry.
5. An alluvial fan has a sloping surface and forms from coarse sands and gravels deposited at the base of a mountain by mountain streams. A delta is flat and forms from fine silt and clay deposited in water where a river flows into a large body of water.
6. A tributary is a single stream that runs into a larger stream or river, whereas a river system is a river and all of its tributaries.

CONCEPT REVIEW

7. A river and all its tributaries make up a river system, the watershed is the land that feeds water into that system, and a divide separates one watershed from another.
8. Discharge increases as the Mississippi River approaches the Gulf of Mexico because tributaries continually add more water to the river.
9. Running water breaks up rock materials with a lifting effect and through abrasion by the rock particles it carries.
10. in solution, in suspension, and as bed load
11. A decrease in velocity reduces the amount and sizes of sediment a river can carry, and thus decreases its competence and capacity.
12. At its mouth, a river flows into a large body of water and comes almost to a standstill, so the river drops its sediment, forming a delta.
13. A river forms a V-shaped valley in a region where rain is plentiful. It forms a steep-sided canyon in a region with little rainfall, when it cuts into its bed rapidly, or when the rock material on the valley sides is resistant to erosion.
14. Waterfalls are temporary features because stream erosion is greatest at these features.

CHAPTER 13 REVIEW

Summary of Key Ideas

13.1 A river system consists of a river and all of its tributaries. The drainage basin of a river system is all the land that is drained by the river and its tributaries. A river's velocity, gradient, discharge, and channel shape affect how it erodes materials.

13.2 Rivers wear down Earth's surface and erode and deposit materials. A river may carry materials in solution, in suspension, and in its bed load. Velocity and discharge affect how much material a river can transport.

13.3 Youthful rivers form steep-sided canyons and V-shaped valleys. The lowest level to which a river can erode its bed is called its base level.

13.4 A river that has cut down close to its base level tends to erode the sides of its valley, forming a meandering river in a wide flood plain. River floods are natural events that can have constructive as well as destructive effects.

KEY VOCABULARY

base level (p. 288)
bed load (p. 284)
capacity (p. 285)
competence (p. 285)
delta (p. 285)
deposition (p. 283)
discharge (p. 281)
divide (p. 280)
drainage basin (p. 280)
flash flood (p. 291)
flood (p. 290)
flood plain (p. 290)
gradient (p. 281)
headward erosion (p. 287)
load (p. 284)
meanders (p. 290)
natural levees (p. 290)
oxbow lake (p. 290)
pothole (p. 283)
river system (p. 280)
stream piracy (p. 288)
suspension (p. 284)
tributary (p. 280)
watershed (p. 280)

15. An oxbow lake forms at a river meander when the river breaks through the wide loop, forming a cutoff. The river drops sediment at the ends of the abandoned meander, forming a body of water separate from the river.

Vocabulary Review

Write the term from the key vocabulary list that best completes the sentence.

1. When a river is carrying small particles of clay and silt in ___?___, the river may appear muddy.
2. The lowest level to which a river can erode its bed is called the river's ___?___.
3. The steepness of the slope down which a river travels is called its ___?___.

Explain the difference between the terms in each pair.

4. competence, capacity
5. alluvial fan, delta
6. river system, tributary

Concept Review

7. Explain how the following terms are related to each other: divide, river system, watershed.
8. How does discharge in the Mississippi River change as it approaches the Gulf of Mexico? Why do these changes occur?
9. How do rivers wear down rock materials?
10. Describe three different ways that rivers can transport rock materials.
11. Explain how a decrease in a stream's velocity could affect its competence and capacity.
12. Why do deltas form at the mouths of rivers?
13. Under what conditions would you expect a river to form a V-shaped valley? a steep-sided canyon?
14. Why are waterfalls temporary features of rivers?
15. Describe how and where an oxbow lake is formed.
16. **Graphic Organizer** Where would you put the phrases *decrease in discharge* and *deposition* in the cause-and-effect diagram below? Explain.

16. A *decrease in discharge* (left box) causes *deposition* (right box). When a river carries less water, its ability to carry sediment can also decrease, and it deposits some of its load.

Critical Thinking

17. **Compare** How do the competence and capacity of a small, fast-flowing stream in the Rocky Mountains compare with the competence and capacity of the lower part of the Mississippi River?

18. **Predict** Suppose a region near a large river experiences a drought for a long period of time. Predict the effect of the drought on the amount of material that the river carries in suspension.

19. **Infer** Would you expect a tributary to form a delta at the point it enters the main stream of a river system?

20. **Analyze** On May 31, 2000, thunderstorms began dropping rain over parts of Minnesota, Wisconsin, Iowa and Illinois. Up to 7 inches of rain fell in about 12 hours. The rain ended by June 1, but flooding caused by the storm lasted through June 3. Why did flooding continue after the rain ended?

Internet Extension

Have Flood Controls on the Mississippi River Been Successful? Compare floods that occurred before and after construction of flood control structures.
Keycode: ES1308

Writing About the Earth System

SCIENCE NOTEBOOK Rivers are important agents of change that affect your local area's geology and water supply. Identify a river in your area and some of its tributary rivers. Explain how the rivers affect rocks and other parts of the geosphere. Describe how the rivers affect the water supply in your area.

Interpreting Graphs

The graph shows the relationship between particle diameter, in centimeters, and the speed of stream flow, in centimeters per second, needed to keep the particle in suspension. The graph also shows size ranges for clay, silt, sand, pebbles, cobbles, and boulders.

21. What is the range of diameters for particles that are classified as pebbles?

22. Suppose a particle has a diameter of 0.05 centimeters. What is the particle called?

23. What is the name of the particle that stays in suspension at the slowest stream speed?

24. What is the minimum stream speed needed to carry a boulder in suspension?

25. Name the particles that would be in suspension in a stream moving at 100 centimeters per second.

26. Describe in words the relationship between particle diameter and the velocity of stream flow needed to keep the particle in suspension.

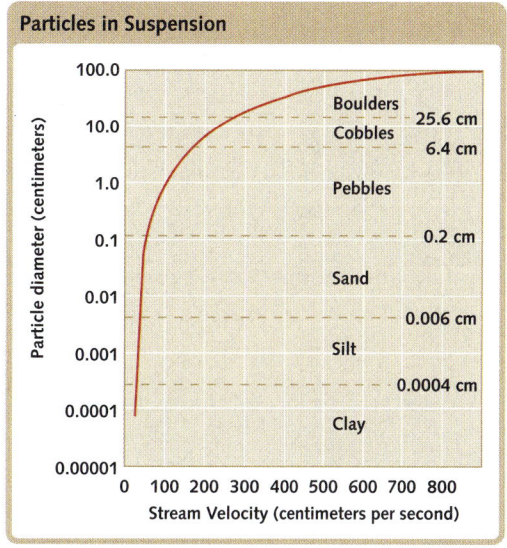

Chapter 13 Weathering, Soil, and Erosion 297

CHAPTER 13 REVIEW

Have Flood Controls on the Mississippi River Been Successful?
Keycode: ES1308
Assign this investigation for homework.

📖 *Use Internet Investigations Guide, pp. T52, 52.*

CRITICAL THINKING

17. Most likely, the competence of the fast-flowing small stream will be greater than that of the lower Mississippi River. However, the capacity of the wider Mississippi will be greater than that of the small stream.

18. During a drought, less water is flowing in the river, so it is likely to be less turbulent than normal, and thus carry less material in suspension.

19. No; deltas form when the water comes almost to a standstill, which would not be the case where a tributary enters a main stream.

20. Runoff resulting from the rain would continue to flow downslope after the rain ended and would take additional time to sink into the already saturated ground.

INTERPRETING GRAPHS

21. 0.2–6.4 cm
22. sand
23. clay
24. about 275 cm per second
25. clay, silt, sand, and pebbles
26. As particle diameter increases, the velocity of stream flow needed to keep the particle in suspension also increases.

WRITING ABOUT SCIENCE

Writing About the Earth System

Students might suggest that rivers erode, weather, and deposit rock materials and contribute water to the ground and to the atmosphere. Locally, rivers may contribute to the water supply, providing water for drinking, washing, and recreation.

Chapter 13 Surface Water 297

CHAPTER 14 PLANNING GUIDE: GROUNDWATER

	Section Objectives	Print Resources
Chapter-Level Resources		Laboratory Manual, pp. 61–66 Internet Investigations Guide, pp. T53–T54, 53–54 Guide to Earth Science in Urban Environments, pp. 13–20, 45–52 Spanish Vocabulary and Summaries, pp. 27–28 Formal Assessment, pp. 40–42 Alternative Assessment, pp. 27–28
Section 14.1 Water in the Ground, pp. 300–305	1. Explain how porosity and permeability affect the storage and movement of groundwater. 2. Describe the water table and features associated with it. 3. Explain how artesian formations affect groundwater. 4. Distinguish among hot springs, geysers, and fumaroles.	Lesson Plans, p. 47 Reading Study Guide, p. 47 Teaching Transparency 15
Section 14.2 Conserving Groundwater, pp. 306–308	1. List factors that affect a water budget. 2. Describe the results of overuse of groundwater. 3. Explain how groundwater becomes polluted.	Lesson Plans, p. 48 Reading Study Guide, p. 48
Section 14.3 Groundwater and Geology, pp. 309–312	1. Explain the presence of minerals in groundwater. 2. Describe how groundwater deposits minerals. 3. List three factors that can cause a spring to have a high mineral content. 4. Describe the role that groundwater plays in the creation of caverns and karst topography.	Lesson Plans, p. 49 Reading Study Guide, p. 49

INVESTIGATIONS
CLASSZONE.COM

Section 14.1 INVESTIGATION
How Does Water Move through the Ground?
Computer time: 45 minutes / Additional time: 45 minutes
Students compare and contrast the flow of groundwater through various materials.

Chapter Review INVESTIGATION
How Many People Can an Aquifer Support?
Computer time: 45 minutes / Additional time: 45 minutes
Students examine the ability of an aquifer to supply an increasing population.

Technology/Online Resources

Electronic Teacher Tools
Visualizations CD-ROM
Classzone.com
Online Lesson Planner
Test Generator

Visualizations CD-ROM
Classzone.com
  ES01401 Investigation, ES1402 Career,
  ES1403 Visualization

Classzone.com
  ES1404 Data Center

Visualizations CD-ROM
Classzone.com
  ES1405 Visualization

Classroom Management

Gather these materials for minilab and laboratory activities.

Minilab, p. 302
  coarse gravel
  graduated cylinder
  water
  2 beakers

Lab Activity, pp. 312–313
  calculator
  notebook paper

Section 17.4 INVESTIGATION
How Does the Ozone Layer Change over Time?
Computer time: 45 minutes / Additional time: 45 minutes
Students compare images of the amount of ozone over the Southern Hemisphere.

CHAPTER 14

Groundwater

These limestone formations in Australia's Northern Territory look as though they might have been weathered and eroded by surface water or wind. In fact, these formations are chiefly the result of the action of water beneath Earth's surface.

How does groundwater shape Earth's surface?

INTRODUCE

Build Interest
Before sharing background information about the photograph with students, allow time for their questions and comments.

Ask questions like the following:

- Imagine this landscape overlain by a "roof" of rock, soil, and vegetation. Then, what would the landscape you see in the photograph be called? **a system of caverns**
- Why is it significant that these rocks are limestone? **Limestone dissolves easily in groundwater, resulting in formations such as those shown.**
- Have you ever seen a landscape like this in person? Where? **Discuss students' experiences.**

Respond
- How does groundwater shape Earth's surface? **Water underground weathers and erodes rock, especially soft rock such as limestone. This process can form caverns and lead to the formation of sinkholes and other surface features of karst topography.**

About the Photograph
The limestone formations, also called karst formations, are in Australia's Gregory National Park. The area shows karst topography, which students will learn about on page 311.

BEYOND THE TEXTBOOK

Print Resources
- Reading Study Guide, pp. 47–49
- Transparency 15
- Formal Assessment, pp. 40–42
- Laboratory Manual, pp. 61–66
- Alternative Assessment, pp. 27–28
- Spanish Vocabulary and Summaries, pp. 27–28
- Internet Investigations Guide, pp. T53–T54, 53–54
- Guide to Earth Science in Urban Environments, pp. 13–20, 45–52

Technology Resources
- Visualizations CD-ROM
- Classzone.com
- Online Lesson Planner
- Electronic Teacher Tools
- Test Generator

298 Unit 4 Earth's Changing Surface

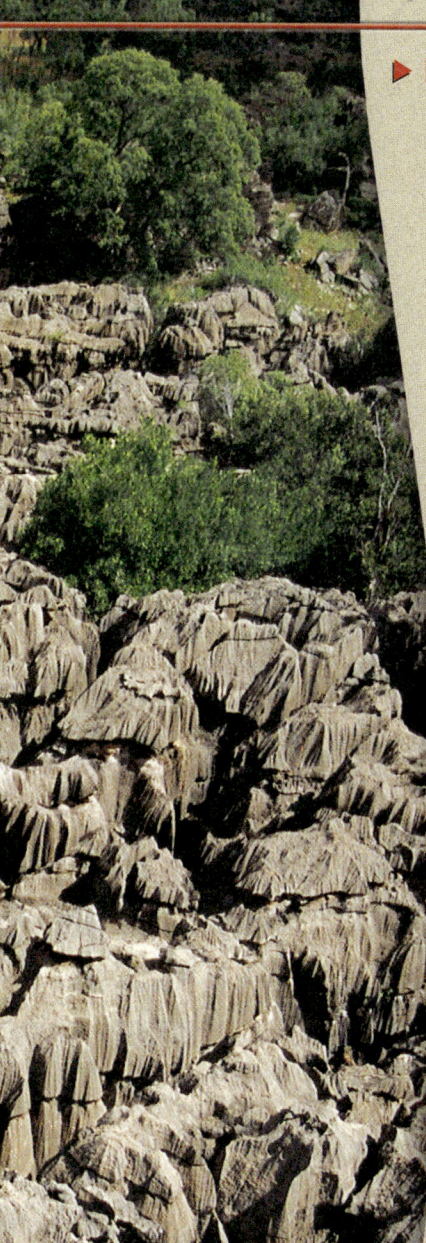

CHAPTER 14

PREVIEW

▶ **FOCUS QUESTIONS** In this chapter you will study water in the ground and learn more about the key questions below.

Section 1 What factors affect the storage and movement of water in the ground?

Section 2 What measures can be taken to conserve and protect groundwater supplies?

Section 3 How does groundwater erode and deposit rock materials?

▶ **REVIEW TOPICS** As you investigate groundwater, you will need to use information from earlier chapters.

- evapotranspiration (p. 13)
- mineral (p. 96)
- ions (p. 93)

▶ **READING STRATEGY**

IDENTIFY MAIN IDEA

At end of each section in Chapter 14, write down what you think are the main ideas in the section. Then write down important details that support these main ideas.

INTERNET RESOURCES
CLASSZONE.COM

At our Web site, you will find the following Internet support for this chapter.

DATA CENTER
EARTH NEWS
VISUALIZATIONS
- Geyser Eruptions
- Cave Formation

LOCAL RESOURCES
CAREERS
INVESTIGATIONS
- How Does Water Move through the Ground?
- How Many People Can an Aquifer Support?

Chapter 14 Groundwater 299

CHAPTER 14

PREPARE

Focus Questions
Find out what knowledge students have about groundwater and how it shapes the landscape. For each focus question, have students consider the section title and then develop two or three additional questions for each section. For example, Section 1: How does water move underground? Section 2: How does groundwater become polluted? Section 3: Which landscape features indicate groundwater erosion?

Reading Strategy
Model the Strategy: Help students identify main ideas even before they begin their reading. Use a version of the following: "The main ideas of the section are probably included in, or at least related to, the Key Idea for the section. By reading the Key Idea for Section 1, I can tell that the main ideas are going to include factors that affect the storage and movement of water in the ground. As I read, I can probably find a main idea for each factor." Have students use the Key Idea of each section to find main ideas.

BEYOND THE TEXTBOOK

INVESTIGATIONS
- How Does Water Move through the Ground?
- How Many People Can an Aquifer Support?

VISUALIZATIONS
- Animation showing how geysers erupt
- Animation of cave formation

DATA CENTER
EARTH NEWS
LOCAL RESOURCES
CAREERS

Online Lesson Planner

Chapter 14 Groundwater 299

CHAPTER 14 SECTION 1

FOCUS

Objectives
1. Explain how porosity and permeability affect the storage and movement of groundwater.
2. Describe the water table and features associated with it.
3. Explain how artesian formations affect groundwater.
4. Distinguish among hot springs, geysers, and fumaroles.

Set a Purpose
Have students read the section to answer the question "What factors affect the movement and storage of water in the ground?"

INSTRUCT

VISUAL TEACHING
Discussion

As students examine the diagrams, be sure they understand that materials with low porosity (at the left side of the diagrams) have less pore space and thus can hold less water than materials with high porosity (at the right side of the diagrams). Ask: Which particles in the diagram result in high porosity? *the spherical particles and the well-sorted particles* Explain that sorting is the process by which particles of similar size are deposited together. Sorting can occur naturally as sediments are deposited by water or wind.

Use Transparency 15.

14.1

KEY IDEA
Factors such as the porosity and permeability of soil and rock materials affect the storage and movement of water in the ground.

KEY VOCABULARY
- groundwater
- porosity
- permeability
- water table
- capillary action
- ordinary well
- spring
- aquifer
- artesian formation
- geyser

How Does Water Move through the Ground? Compare and contrast the flow of groundwater through various rocks and materials.
Keycode: ES1401

Water in the Ground

Although some precipitation becomes runoff, water also enters the ground, where it is stored as **groundwater**. One factor that affects the amount of water that seeps into the ground is the type of rock or soil on which the water falls. Other factors are climate, topography, vegetation, and land use. For example, in a hot, dry climate, water evaporates before it enters the ground. In mountainous regions, water tends to run off rocky slopes. In populated regions, paved areas prevent water from entering the ground.

Porosity

The amount of water that soil or rock can hold depends upon the amount of space, or pore space, between the grains of the material. **Porosity** (puh-RAHS-ih-tee) is the percent of a material's volume that is pore space.

Porosity depends upon a number of factors. One factor is particle shape. Materials consisting of rounded particles have a lot of pore space—picture the spaces between marbles in a jar. Materials consisting of flat or angular particles have less pore space because the particles fit together more tightly.

Another factor affecting porosity is sorting. The amount of pore space is greatest in materials made up of particles that are all the same size. Such materials are said to be well-sorted. For example, the porosity of deposits of well-sorted, rounded particles of sand with little or no cement between the grains may be more than 40 percent. When a material is poorly sorted, as in a mixture of gravel and sand, small particles fill the spaces between the large particles, reducing the total porosity.

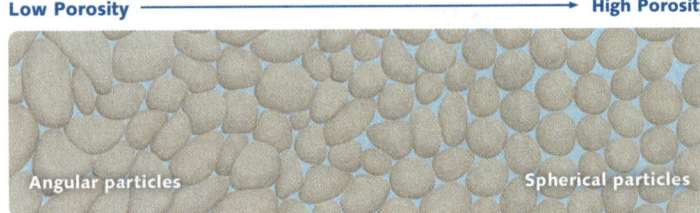

PARTICLE SHAPE Materials made of angular particles have less pore space than materials made of spherical particles.

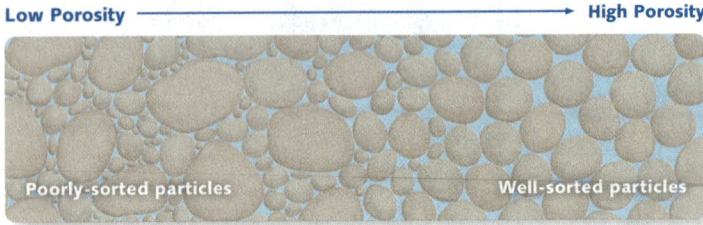

SORTING Poorly-sorted materials have less pore space than well-sorted materials.

300 Unit 4 Earth's Changing Surface

DIFFERENTIATING INSTRUCTION

Reading Support Have students preview the boldface vocabulary words. Before reading the section, ask students to write a sentence for each vocabulary word that demonstrates what they think the word means. Then, as students read the section, have them change their definitions as needed.
Use Reading Study Guide, p. 47.

Permeability

Porosity does not describe how easily water can pass through a material. **Permeability** (PUR-mee-uh-BIHL-ih-tee) is the rate at which water or other liquids pass through the pore spaces of a rock. In general, permeability increases with grain size because large-grained materials have larger pore spaces. Water passes easily through materials with large pore spaces, such as sand and gravel. It passes slowly through finer materials, such as silt. A material that water cannot pass through is called impermeable (ihm-PUR-mee-uh-buhl). Clays and shales, which are very fine-grained, are usually impermeable.

It is possible for a material to be highly porous but not at all permeable. For example, pumice has many pores, but the pores are not connected. Thus, water cannot pass through pumice. On the other hand, a nonporous rock such as granite may become permeable if cracks develop in the rock, because the cracks transmit the water.

Some of the water that passes through a sediment or rock sticks to particles, forming a film of water. This film of water is called capillary water, and can be removed only by evaporation and transpiration. For materials made of small particles, such as shale, capillary water may fill the pore space. Because the pores are so small, the water cannot drain, and the material is impermeable.

CAREER

Wastewater Treatment Plant Operator

If wastewater were not recycled, people would quickly run out of clean, fresh water. At wastewater treatment plants, operators closely monitor the purification of sewage and wastewater to ensure that clean, safe water is being returned to the environment. Their job includes making sure that harmful bacteria, toxins, and solid waste are removed. Plant operators keep a close eye on gauge measurements, test the water at different stages in the purification process, and make minor repairs to machinery. They are also knowledgeable about laws regulating water quality, and they work to comply with such laws.

Plant operators usually need a high school diploma with coursework in math, chemistry, and biology. Many operators also have a certificate in water quality or wastewater treatment. Wastewater treatment plants operate 24 hours a day, 365 days a year. With such a demanding schedule, plant operators work in one of three 8-hour shifts and rotate working on weekends and holidays. During emergencies such as floods, they often work overtime to keep the situation under control. Besides being reliable, plant operators must be willing to work both indoors and outdoors and not mind getting a bit dirty. Although their work can be challenging, wastewater plant operators find reward in knowing that they are responsible for renewing a vital resource for the public and the environment.

THESE WASTEWATER PLANT OPERATORS are sampling water from a clarifier tank to ensure that clean water is being returned to the environment.

Learn more about careers in wastewater treatment.
Keycode: ES1402

INVESTIGATIONS
CLASSZONE.COM

How Does Water Move Through the Ground?
Keycode: ES1401

Discuss the variables that students will change and have them predict how each variable will affect the flow of groundwater.

 Use Internet Investigations Guide, pp. T53, 53.

CAREERS
CLASSZONE.COM

Learn more about careers in wastewater treatment.
Keycode: ES1402

CRITICAL THINKING

Ask students to suppose that their community wants to build a small pond in a local park by digging a depression in the ground. There are no water sources near the proposed pond site for filling the pond, so it must be able to trap and hold rainwater. The soil at the site consists primarily of sand. Have students predict any problems with the plan and suggest a solution.

Sand is a permeable material, so rain will likely sink into the ground rather than stay in the depression. Lining the depression with an impermeable material such as clay will help trap water in the pond.

DIFFERENTIATING INSTRUCTION

Developing English Proficiency Write the words *permeable* and *impermeable* on the board. Explain that the prefix *im-* means "not." Thus, *impermeable* means "not permeable," or "not allowing to pass through." Encourage students to use the dictionary to list and define other words beginning with the prefix *im-* (such as *immature, impure, imbalance, immeasurable, impassible*).
Use Spanish Vocabulary and Summaries, pp. 27–28.

CHAPTER 14 SECTION 1

MINILAB
Measuring Porosity

Management
- Have paper towels available so students can wipe up spills immediately.
- Have students use the graduated cylinder to measure both the gravel and the water.
- In Step 3, students should measure the water poured from the gravel-filled beaker combined with the leftover water they measured in Step 2.

Materials
- Measurements and percentage calculations will be easiest if each group uses 50 or 100 mL of water and gravel.
- Students should use dry gravel to get the most accurate porosity measurements.

Analysis Answers
- Percent of the water that was poured into the gravel = [(total amount of water − amount of water left after pouring into gravel) ÷ total amount of water] × 100%
- Most students probably underestimated the percent of the gravel's volume that is air space, and they will be surprised by the large percent of water that was poured into the gravel.
- By observing that the amount of water measured in Step 3 was less than the amount of water measured in Step 1, students should conclude that they were not able to remove all of the water from the gravel.

For proper behavior in a laboratory setting, refer students to Safety in the Earth Science Laboratory on pages xxii–xxiii.

10-Minute Mini LAB
Measuring Porosity

Materials
- coarse gravel
- graduated cylinder
- water
- 2 beakers

Procedure
1. Measure equal amounts of gravel and water into two beakers. Estimate the percent of the gravel's volume that is air space.
2. Pour the beaker full of water into the gravel until it just covers the gravel. Measure the amount of water left over.
3. Carefully pour the water out of the gravel-filled beaker into the water beaker. Measure the amount of water.

Analysis
What percent of the water was poured into the gravel? How good was your estimate from Step 1? Were you able to remove all of the water from the gravel?

The Water Table

When rain falls to the ground, it enters the pores in the soil and sticks to the particles. If enough rain falls, the upper layers of soil will not be able to absorb all the water. When this happens, the water continues downward until it reaches an impermeable material. The water then begins to fill the pore spaces above the impermeable material. As rain continues, the water level in the ground rises higher as more pore spaces are filled. The part of the ground where all pore spaces are filled is called the zone of saturation. The upper surface of the zone of saturation is called the **water table.**

Between the water table and the surface, there is a section of the ground that can still hold some water. This section is called the zone of aeration because air can enter this region. It includes three parts. Just above the water table is the capillary fringe, where water rises because the water molecules are attracted to the soil particles. This interaction between the water and the soil is called **capillary action.** A familiar example of capillary action occurs when a paper towel is dipped in water. Just as water rises into the towel, water rises into the soil just above the water table. Above the capillary fringe is a section of soil where rainwater drains downward to the capillary fringe and to the zone of saturation below. The soil here is generally drier than the soil below and the soil above except during and shortly after rainfall. The third section of soil lies just below the soil surface. The soil here contains more organic material, which helps it hold water. A film of water may stick to the grains of topsoil in this upper section unless it evaporates or is absorbed by plant roots.

The water table's distance from the surface depends upon many factors. These include the amount of rainfall, the amount of time that passes between rains, the season, the slope of the ground surface, the thickness of the soil, and the climate. Humans can also affect the water table by drilling wells.

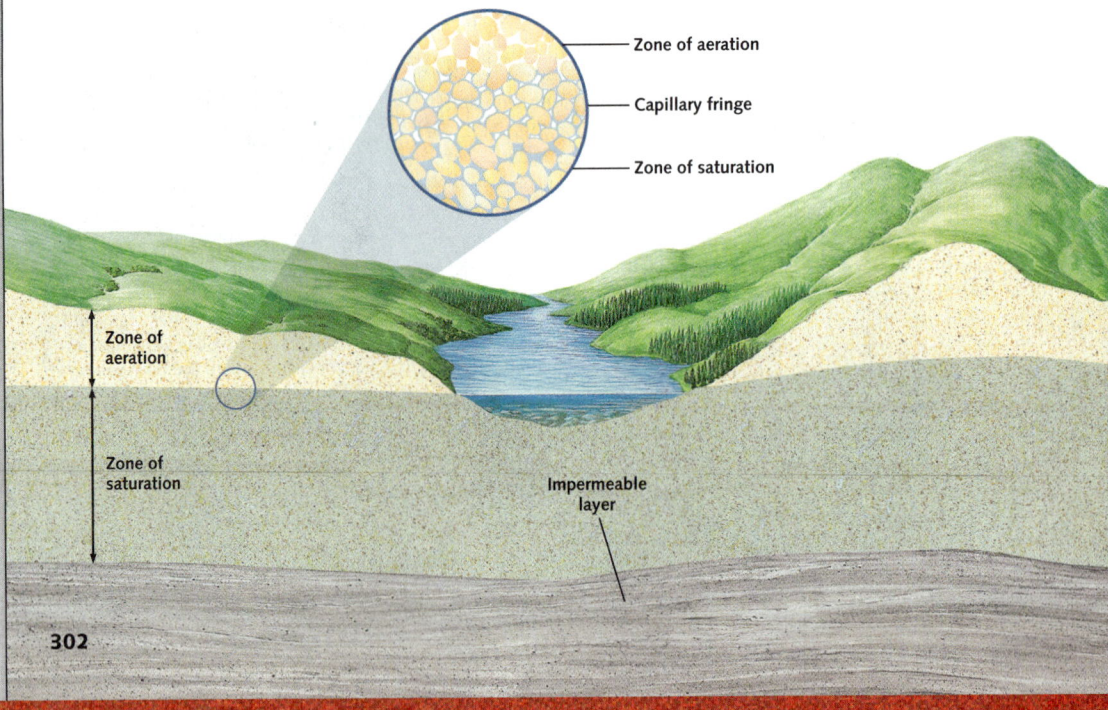

302 Unit 4 Earth's Changing Surface

DIFFERENTIATING INSTRUCTION

Hands-On Demonstration Hold a sheet of paper towel vertically over a pan of water dyed with food coloring. Dip the bottom edge of the towel in the water and allow students to observe the movement of water up the towel. Ask: In this model, what represents the water table? *the surface of the water in the pan* Where is the zone of saturation? *below the water surface* Where is the zone of aeration? *above the surface of the water* What does the paper towel represent? *soil particles in the capillary fringe* Have students relate capillary action to what they observed and to what occurs in the ground.

In humid climates, the water table is at the surface in places such as swamps, lakes, and rivers. In desert regions, the water table may be hundreds of meters below the surface. In woods, fields, and farmlands, the water table is likely to be much closer to the surface. In hilly country, it is generally nearer the surface in valleys than it is in hills. Notice in the illustration of the water table below that the water table has its own hills much like the overlying surface.

The water table is important in several ways. Seepage of water from the water table keeps streams flowing between rains and maintains the water levels of swamps and lakes. The water table also supplies drinking water to springs and human-made wells.

Ordinary Wells and Springs

In places where the water table does not reach the surface, humans can reach the groundwater by digging or driving wells into the ground. A well of this type, known as an **ordinary well,** contains water up to the level of the water table.

Recall that the depth of the water table depends in part on the season. A well must reach below the lowest level to which the water table is likely to fall in dry weather. If it does not, the well will not provide water all year. As the water table rises and falls, so does the level of the water in the well.

On a hillside where the water table meets the surface, groundwater may flow out as a **spring.** The diagram below shows a spring flowing from a perched aquifer. A perched aquifer forms on top of an impermeable layer that lies above the water table. Springs are more common in mountainous areas.

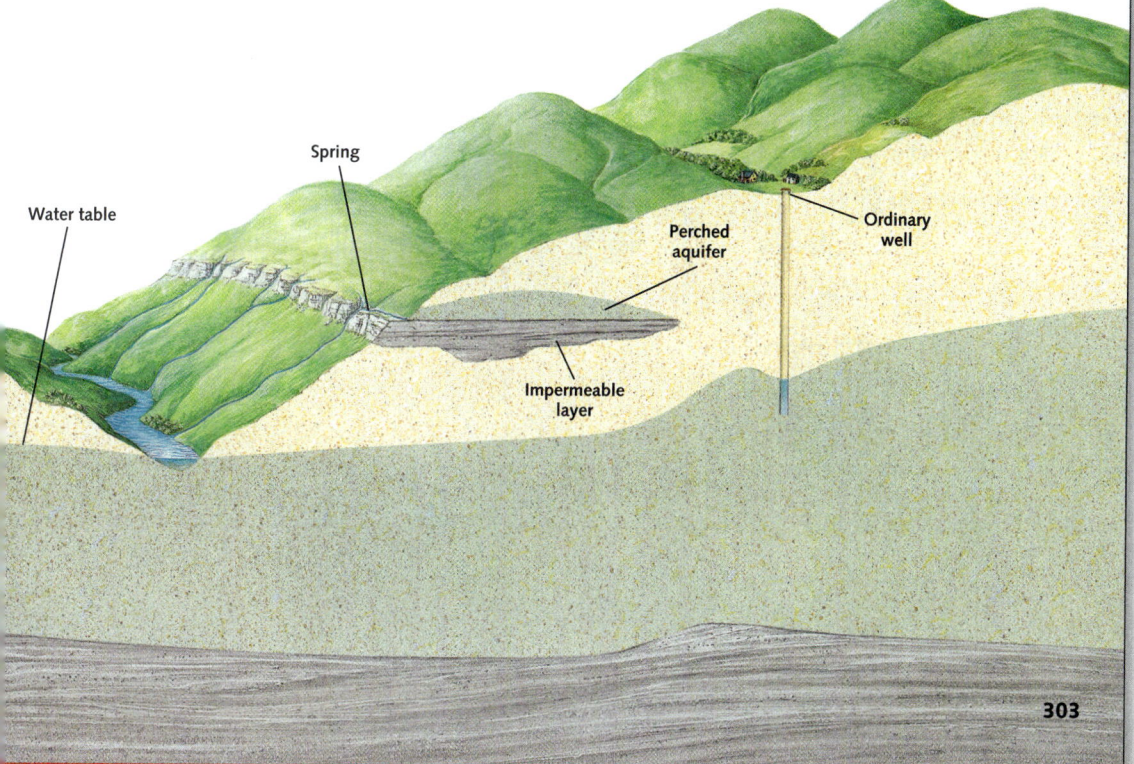

This illustration of a water table does not represent any specific geographic area.

CHAPTER 14 SECTION 1

Visual Teaching

Interpret Diagram
Have students explain why the aquifer in the diagram is an artesian aquifer. *The aquifer is sandwiched between two impermeable rock layers.* Ask: How does water enter the aquifer? *Rainwater enters from the surface, seeping into the permeable layer.* Will rain falling at the base of the mountains enter the aquifer? Why or why not? *No; in this area, water entering the ground becomes trapped above the impermeable cap rock.* How does water get from the surface at the top of the mountains to a point below the well? *by gravity and by being pushed by the weight of water above it*

Extend
Have students compare and contrast the aquifer of the artesian formation with the perched aquifer on page 303. *Both are bounded on the bottom by an impermeable layer. In the artesian formation, the top boundary of the aquifer is an impermeable cap rock, whereas the top boundary of the perched aquifer is a permeable layer. As a result, the water in the artesian formation is under greater pressure than the water in the perched aquifer.*

Artesian Formations

Permeable layers of rock and sediment that store and carry groundwater in enough quantity to supply wells are called **aquifers** (AK-wuh-fuhrz). The upper level of an aquifer that supplies an ordinary well is the water table. The best aquifers are uncemented sands and gravel, followed by porous sandstones.

Sometimes an aquifer dips underground between impermeable rock layers. A sandwich of permeable and impermeable rocks is formed. This arrangement is called an **artesian** (ahr-TEE-zhuhn) **formation.** The upper impermeable layer of an artesian formation, usually shale, is called the cap rock. Rain that enters the aquifer is trapped between the cap rock and the impermeable rock layer below. Gravity moves the water downward in the dipping aquifer. The water is pushed along by the weight of all the water above and behind it.

Great quantities of water may enter the aquifers of artesian formations. Like the water in a great sloping pipe, the water in the aquifer is under pressure. When a well is drilled, the pressure causes the water to rise up in the well. The water may even spout into the air if the water pressure is great enough. Wells drilled into artesian formations are called artesian wells.

Artesian wells vary greatly in depth. Generally, as the distance from the source of water increases, the depth of the aquifer increases. On the Great Plains, wells that are hundreds of kilometers from the mountains may need to be drilled down hundreds of meters to reach the aquifer.

Artesian formations may be broken by cracks in the cap rock called fissures. Artesian springs, or fissure springs, rise through these cracks. Such a spring may form a desert oasis.

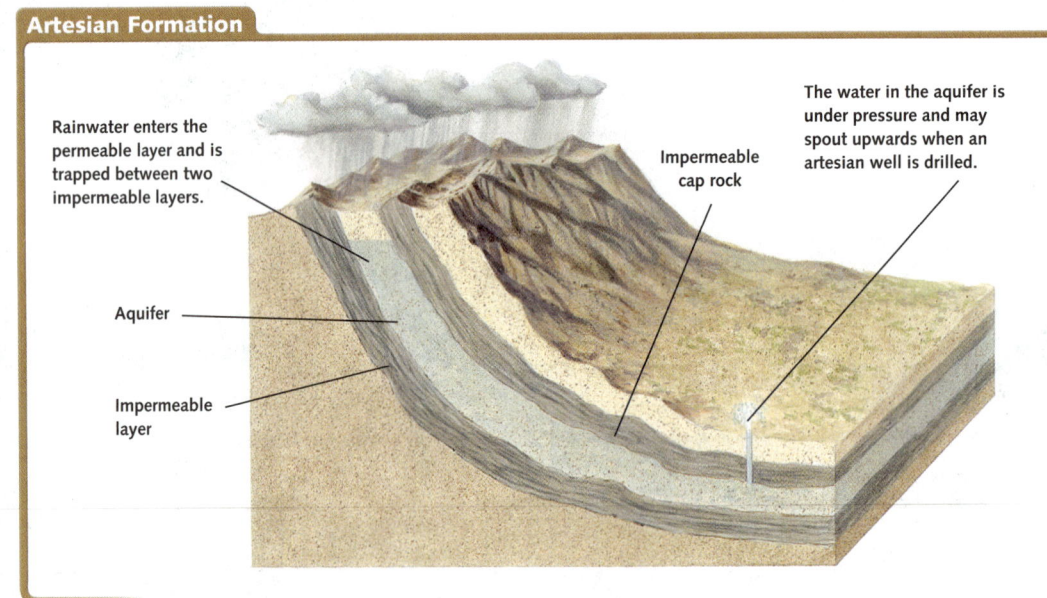

Artesian Formation

DIFFERENTIATING INSTRUCTION

Challenge Activity Challenge students to design a working model of an artesian well. Suggest they begin by studying the diagram on page 304 and listing materials they can use to model the diagram. *Possible materials include a plastic shoe box, sand, clay, a plastic straw, scissors, a watering can, and water.* Students should plan how they will build their model and write a brief summary of their plan. Once you have approved their plan, they can build their model and demonstrate for the class how it works.

Hot Springs, Geysers, and Fumaroles

Hot springs and other surface features are evidence that groundwater is sometimes heated beneath Earth's surface.

One way that groundwater can become heated is if it comes from a great depth. Because subsurface temperatures increase with depth, water from deep artesian wells or springs may be much warmer than water from ordinary wells or springs. At Hot Springs National Park in Arkansas, groundwater collects at depths of 1800 to 2400 meters. The water heats up and rises quickly to the surface through cracks and fissures. The average temperature of water reaching the surface is about 60°C (143°F).

Groundwater may be hot even without coming from great depths. In many regions of recent volcanic activity, igneous rocks near the surface are hot enough to boil water. The groundwater may come to the surface as boiling hot springs. Sometimes volcanic gases make hot groundwater acidic. The acidic water reacts with minerals from surrounding rocks, forming sticky clay minerals. As a result, sputtering springs called mud pots and mud volcanoes may appear at the surface. Some mud pots form when steam rises through silt.

A **geyser** (GY-zuhr) is a hot spring that intermittently shoots columns of hot water and steam into the air. It consists of a long, vertical, irregularly shaped tube that may extend hundreds of meters into the ground. When the tube is full of water, the water at the bottom is under so much pressure that its boiling point rises far above the boiling point at Earth's surface. The pressure prevents the water at the bottom of the tube from turning into steam, even though nearby igneous rocks have heated it to extremely high temperatures. Eventually, some of the water at the top of the tube boils out, or is pushed out because the superheated water below has expanded. The pressure is relieved, and the superheated water at the bottom of the tube boils instantly, producing an explosive burst of steam. The steam blows out the water above it, and the geyser erupts. The average time between eruptions at Old Faithful, a geyser in Yellowstone National Park, is about 89 minutes.

Where fairly recent volcanic eruptions have occurred, groundwater is released as steam, along with other gases, from fissures in the ground called fumaroles (FYOO-muh-ROHLZ). Fumarole fields in Iceland, Japan, and other countries are often a source of geothermal energy.

MUD POT Hot, highly acidic groundwater in Yellowstone National Park leads to the formation of mudpots.

VOCABULARY STRATEGY

The word *geyser* comes from the name of a bubbling hot spring in Iceland, Geysir. In Old Norse, a language in which modern Icelandic is rooted, *geysa* means "to gush."

14.1 Section Review

1. Explain how a rock can be porous but not permeable.
3. Compare and contrast an ordinary well and an artesian well.
2. Explain why the water at geysers and some springs is hot.
4. **CRITICAL THINKING** Describe where and how groundwater is stored in the region above the water table.
5. **PHYSICS** Explain the roles that temperature and pressure play in the eruption of a geyser.

CHAPTER 14 SECTION 1

VISUALIZATIONS
CLASSZONE.COM

Observe an animation showing how geysers erupt.
Keycode: ES1403

Have students make a diagram that explains how geysers erupt.

Visualizations CD-ROM

ASSESS

1. A rock may have many pores, making it porous, but if the pores are not connected, the rock will not be permeable.

2. Both ordinary wells and artesian wells are water sources that are drilled or driven into the ground. An ordinary well contains water up to the level of the water table, whereas water pressure in an artesian well may cause the water to rise into the well above the water table or even spout into the air.

3. The water at geysers and some springs comes from deep below Earth's surface where temperatures are high or from regions of volcanic activity where hot igneous rocks near the surface heat the water.

4. Water may rise upward into the soil from the water table by capillary action, it can drain downward from the soil's surface after a rainfall, and it can be held as a film sticking to grains of topsoil.

5. An increase in water pressure from the weight of the filled geyser tube causes an increase in boiling temperature. At such high pressure, the water becomes superheated but does not boil. When the pressure is eventually relieved, the water boils, leading to the geyser's eruption.

MONITOR AND RETEACH

If students miss . . .

Question 1 Reteach "Porosity" (p. 300) and "Permeability." (p. 301) Ask: Why is pumice porous but not permeable?

Question 2 Use the diagrams on pages 303 and 304 to discuss the geologic settings of ordinary wells and artesian wells.

Question 3 Have students reread "Hot Springs, Geysers, and Fumaroles." (p. 305) Ask them to explain in their own words how groundwater can become heated.

Question 4 Compare and contrast the three parts of the zone of aeration. (p. 302)

Question 5 Have students make a diagram that summarizes the process of geyser eruption, underlining any words related to temperature and pressure.

CHAPTER 14 SECTION 2

14.2

FOCUS

Objectives
1. List factors that affect a water budget.
2. Describe the results of overuse of groundwater.
3. Explain how groundwater becomes polluted.

Set a Purpose
Ask a student to read aloud the key idea. Then have students read to find the answer to the question "How is groundwater availability threatened by overuse and pollution?"

INSTRUCT

CRITICAL THINKING
Have students classify each of the following situations as a time of recharge, surplus, usage, or deficit.
- During a dry, hot summer, plants remove water from the soil. **usage**
- After weeks without rain, plants in a forest are unable to get the moisture they need from the soil and begin to wilt. **deficit**
- After weeks of heavy rains, the soil in a valley becomes extremely wet, and the level of water in local wells rises. **surplus**
- On a cool spring day, rain seeps into pore spaces in the soil. **recharge**

KEY IDEA
Groundwater is an important resource whose availability is threatened by overuse and by pollution.

KEY VOCABULARY
- water budget
- recharge
- surplus
- usage
- deficit

SAN JOAQUIN VALLEY The signs on the pole in this photograph mark the ground level in 1925, 1955, and 1977. Removal of groundwater caused the ground to subside over time.

Conserving Groundwater

About 50 percent of drinking water in the United States comes from groundwater. Because humans rely on groundwater, they must conserve and protect groundwater supplies.

Water Budgets

A budget is a statement of expected income versus expected spending or expenses. In a balanced budget, income and spending are equal. A **water budget** describes the income and spending of water for a region. In a water budget, the income is rain or snow. The spending includes water lost by use, by runoff, and by evapotranspiration.

Weather is a controlling factor in the evapotranspiration in a given region. For example, when air temperature is high, plants growing in the ground may use more moisture. At such times, evapotranspiration is high. When air temperature is low, plants may not use as much moisture; thus, evapotranspiration is low.

If it rains during a time when the plants need little moisture, the extra moisture soaks into the soil, where it is stored between the grains of soil. This is a time of soil water **recharge**. During recharge, the soil water storage is filling. If the rain continues so that the soil becomes saturated, the surplus water raises the water table or becomes part of stream runoff. Thus, a moisture **surplus** occurs when two conditions are true: the rainfall is greater than the need for moisture and the soil water storage is filled.

If the need for moisture is greater than the rainfall, the plants can draw water from the soil water supply. This is a time of soil water **usage**. If the need for moisture continues to be greater than the rainfall, all of the water available in the soil may be used up. A water **deficit** occurs when the need for moisture is greater than the rainfall and the soil water storage is gone.

Groundwater Conservation

In many regions, humans are using groundwater faster than natural processes can replenish supplies. In addition, pollution threatens groundwater supplies.

Overuse of Groundwater

When groundwater supplies are depleted, the water table drops, thus lowering the water level in wells and springs. Such depletion can cause them to go dry. In coastal areas, where fresh groundwater rests on top of salt water, salt water tends to seep upward into overused aquifers. Wells and springs become salty and unusable. This problem has damaged water supplies in many parts of the United States.

Another problem caused by overuse of groundwater is subsidence, which occurs when groundwater removal causes the ground to become compacted so much that the ground level drops, or subsides. From 1925 to 1975, California's San Joaquin Valley subsided by up to 9 meters in places because groundwater had been removed for irrigation. Removal of natural gas and petroleum also causes land to subside.

DIFFERENTIATING INSTRUCTION

Reading Support Have students make a graphic organizer, such as a concept map or chart, that clarifies the concept of a water budget. Tell them to make sure the words *recharge, surplus, usage,* and *deficit* are included in the graphic. Encourage students to share their graphic organizer with the class, using it to explain a water budget. *Use Reading Study Guide, p. 48.*

Subsidence can cause damage to structures such as foundations and underground pipes. In some coastal regions, subsidence has caused the land to drop below sea level.

What can be done to protect and replenish groundwater supplies? In regions where heavy groundwater use has lowered the water table, artificial methods of groundwater recharge are used. Instead of pouring used water into sewers where it becomes runoff, water can be pumped back underground through wells. Water is also pumped into ponds and allowed to seep back into the groundwater naturally.

Groundwater Pollution

Groundwater is recharged by rain seeping down through the soil. Thus, any polluting agent in the soil becomes part of the groundwater. Pollutants include oil washed from roads, nitrates from soil fertilizers, pesticides applied to plants, farm wastes, and sewage from septic tanks and sewers. Even salt, which is used to melt ice on roads in winter, is carried into groundwater supplies by melting snow and spring rains.

Hazardous wastes are poisonous byproducts of some industrial processes. Toxic chemicals from accidental spills, careless disposal, or rotting underground storage containers pollute the soil around them and the groundwater with which they come into contact. For many years, toxic wastes were dumped with little care, and their locations were not recorded. In some cases, houses were built on old toxic waste disposal sites. When the toxic waste was later discovered, families had to be moved.

As yet, no simple or inexpensive way to purify polluted groundwater is known. However, further pollution can be reduced or prevented. One way to reduce pollution is to restrict the use of pesticides and fertilizers. Another is to make sure that toxic wastes are disposed of in such a way that they cannot enter the environment.

Scientific Thinking

COMMUNICATE
Many household products, such as some paints, pesticides, and solvents, can pollute groundwater if not disposed of properly. Some toxic chemicals are illegally poured down storm drains into sewers. How could this practice lead to groundwater pollution?

SORTING HOUSEHOLD WASTES
In Los Angeles, California, workers collect hazardous household wastes for safe disposal.

14.2 Section Review

1. Explain what a water budget is, and how a water surplus and a water deficit can occur.
2. Describe how overuse of groundwater can affect groundwater supplies and how it can affect soil.
3. How can pollutants enter groundwater?
4. **CRITICAL THINKING** Suppose a large housing development is built in a region. Discuss the potential impact on the region's water budget.
5. **GEOGRAPHY** Suppose the population of a city is expanding rapidly. What are some precautions the city could take to protect its groundwater supply?

CHAPTER 14 SECTION 2

Learn more about regional and local aquifers.
Keycode: ES1404

Science and Society

High Plains Aquifer
The High Plains Aquifer, also known as the Ogallala Aquifer, is Earth's largest known aquifer. Most of the water in the aquifer was stored underground 15,000 to 30,000 years ago, toward the end of the last ice age. The aquifer's recharge area—where water seeps into the ground from surface sources—is along the western edge of the aquifer in the High Plains and the foothills of the Rocky Mountains. The aquifer's size has not protected it from depletion because the rate at which water is pumped out is far greater than the aquifer's recharge rate. The recharge rate is slow because the area of recharge is small, precipitation in the area is low, and the degree of evaporation is high.

Research Activity
Students might contact the local water utility, the state department of environmental protection, or the EPA's Office of Water Web site.

SCIENCE & Society

The Shrinking of the High Plains Aquifer

The High Plains aquifer lies east of the Rocky Mountains beneath parts of eight states. It serves as the water source for nearly 30 percent of the irrigated cropland in the United States. People once believed that the aquifer's water supply was limitless, but in recent years it has declined at an alarming rate.

Why is the High Plains aquifer being depleted? What measures have been taken to conserve this water supply?

THE HIGH PLAINS AQUIFER extends about 174,000 square miles beneath parts of eight states and is a critical water source for the region's agriculture.

Most of the water being removed from the High Plains aquifer is used for irrigation. Use of the aquifer for irrigation began in the 1930s and expanded rapidly in the 1940s. With irrigation, land that once had been unsuitable as cropland supported crops such as cotton, corn, and wheat. However, groundwater was being pumped out much faster than the aquifer could recharge. In the Texas Panhandle, which lies over the shallowest part of the aquifer, the problem was at its worst. By 1970, the level of the water table had dropped throughout much of this region. In one county, it dropped by 24 meters. Policymakers discovered that many irrigation systems were extremely wasteful. For example, up to 40 percent of the water in irrigation ditches either evaporated or was absorbed into ditch soil before it reached crops.

During the 1980s and 1990s, new irrigation technologies and practices were developed. In Texas, one of the most promising practices relies on a network of weather stations known as the North Plains Evapotranspiration Network. The system analyzes water-use data for various crops and relates the information to real-time weather conditions. The network then sends farmers information by fax about how much water they should deliver and when they should start and stop watering their crops. In 1999 farmers saved 76 billion liters of water from the High Plains aquifer. Better irrigation technologies have been a factor in slowing the emptying of the aquifer, but discharge still outpaces recharge, so further conservation measures are essential. ∎

Extension

RESEARCH ACTIVITY
Where does the water you use come from? What water conservation efforts are in place in your region? Do some research and present your findings in a written report.

Learn more about regional and local aquifers.
Keycode: ES1404

IRRIGATION of this cornfield in Nebraska via a sprinkler relies on water from the High Plains aquifer.

Groundwater and Geology

Some of the most beautiful places on Earth are underground. Majestic limestone caverns such as Mammoth Cave in Kentucky have been formed by the action of groundwater. By eroding and depositing rock materials, groundwater forms many distinctive geologic formations.

Minerals in Groundwater

When water evaporates to become water vapor, it leaves impurities behind. As a result, rainwater that forms when this water vapor condenses contains almost no dissolved mineral matter, although it may contain other liquids and dissolved gases such as carbon dioxide. When rainwater seeps into the ground, however, the situation changes. As groundwater passes through the lower soil layers or bedrock, it dissolves minerals. Much of this dissolved mineral matter remains in the groundwater. The type of rock through which groundwater travels, the distance the water travels underground, and the water's temperature all affect the type and amount of mineral matter dissolved in it.

When groundwater contains large amounts of ions from dissolved minerals, it is called hard water. For example, almost all the groundwater in regions that have limestone bedrock is hard water because the groundwater dissolves calcium ions out of the limestone. Calcium ions are the most commonly found ions in hard water.

The ions in hard water interfere with use of the water. Calcium, magnesium, and iron ions found in hard water react with soap to form scum instead of suds. In hot-water pipes, groundwater containing dissolved minerals leaves behind deposits called boiler scale.

Artesian water is usually harder than ordinary groundwater. Artesian water travels farther and may be warmer than ordinary groundwater, so it can dissolve more mineral matter. By contrast, ordinary groundwater is almost always harder than river water.

Mineral Deposits by Groundwater

When groundwater that contains dissolved minerals cools or evaporates, **mineral deposits** are left behind. For example, geyser waters dissolve silica from hot igneous rock beneath the surface. When the water reaches the surface and cools, the silica is deposited around a geyser opening as a white, porous substance called geyserite. Hot groundwater often leaves deposits of minerals in bedrock cracks and fissures. Such mineral veins may contain quartz, calcite, gold, and silver.

Petrified wood is formed when minerals dissolved in groundwater replace the decaying wood of buried trees. As each microscopic particle of wood is replaced by a grain of mineral matter, many details of the wood structure are reproduced. The petrified trees of Arizona, which formed in this way, consist of silica.

Perhaps the most important groundwater deposit is the cement that binds together sand grains and pebbles to form sedimentary rocks. Calcite is the most common cementing mineral, but silica and iron oxides also serve as natural cements.

GEYSERITE A mound of geyserite has been deposited around Lone Star Geyser in Yellowstone National Park.

14.3

KEY IDEA
Groundwater erodes and deposits rock materials, creating distinctive geologic formations.

KEY VOCABULARY
- mineral deposit
- cavern
- karst topography

CHAPTER 14 SECTION 3

Focus

Objectives
1. Explain the presence of minerals in groundwater.
2. Describe how groundwater deposits minerals.
3. List three factors that can cause a spring to have a high mineral content.
4. Describe the role that groundwater plays in the creation of caverns and karst topography.

Set a Purpose
Have students read the section to answer the question "What geologic formations does groundwater form?"

Instruct

More about...

Limestone
Though limestone is nearly insoluble in pure water, water containing even a small amount of carbonic acid easily dissolves it. Rainwater forms a weak solution of carbonic acid when carbon dioxide from the air and from decaying vegetation in the soil dissolves in it. When groundwater containing carbonic acid comes in contact with limestone, the acid reacts with the calcite in the limestone and forms calcium bicarbonate, which gets carried away by the water.

DIFFERENTIATING INSTRUCTION

Reading Support Explain to students that writers use comparison words to signal a similarity between two things or processes. Examples of comparison words include *like, likewise, just as, similarly,* and *in the same way.* Contrast words signal a difference between two things or processes. Examples of contrast words include *but, however, differ, more than, in contrast,* and *conversely.* Point out the use of *by contrast* in the fifth paragraph on page 309. Ask what is being discussed. the difference between ordinary groundwater and river water Then lead students to classify *more easily than* as contrast words in the first paragraph of "Caverns" on page 310. Ask what is being contrasted. limestone and other rock types

Use Reading Study Guide, p. 49.

CHAPTER 14 SECTION 3

Visual Teaching

Interpret Diagram

Have students who have been inside caves relate what they observed to what is shown in the diagram. Then ask students: In the diagram, what effect does the carbonic acid in the rainwater have? *It dissolves minerals in the rocks with which it comes in contact.* What is this process called? *chemical weathering*

Extend

Ask students to predict what will happen if the water table in the diagram drops. *The caves at the base of the diagram that are filled with water will drain and grow larger through erosion. In these caves, dripstone will be deposited, eventually forming stalactites, stalagmites, and columns.*

MAMMOTH HOT SPRINGS When water pouring from Mammoth Hot Springs cools, it leaves behind calcite deposits called travertine.

VISUALIZATIONS
CLASSZONE.COM

Observe an animation of cave formation.
Keycode: ES1405

Mineral Springs

A spring with a high concentration of mineral matter is called a mineral spring. The high mineral content of the water may be due to any of the factors below:

- The water passes through rock containing easily-dissolved minerals. The salt beds in Michigan are a region where this occurs.
- The water contains large quantities of gases, such as carbon dioxide and hydrogen sulfide, that form acids when mixed with water. The water at Saratoga Springs in New York contains carbon dioxide. The water at White Sulphur Springs in West Virginia contains sulfur dioxide.
- The water dissolves minerals more easily because it is very hot. The water at Hot Springs, Arkansas, is an example.

Some mineral spring areas have become health resorts. In desert regions, however, alkali mineral springs may be poisonous. For example, alkali springs in the southwestern United States may carry borax, sodium carbonate, and sodium sulfate in solution.

Sometimes mineral springs leave mineral deposits behind. At Mammoth Hot Springs in Yellowstone National Park, hot mineral water pours out of long hillside fissures in limestone bedrock. The water deposits some of its dissolved calcite as it cools. These calcite deposits are called travertine (TRAV-uhr-TEEN).

Caverns

Limestone is a common bedrock that dissolves more easily than some other types of rock. Limestone is dissolved by the carbonic acid found in groundwater. Dissolving the carbonate minerals in limestone either creates porosity in limestone or increases the porosity that is already present.

Frequently, limestone formations are split by fissures that run down from the surface and by cracks that run horizontally between the beds. As groundwater flows through the limestone, the carbonic acid slowly

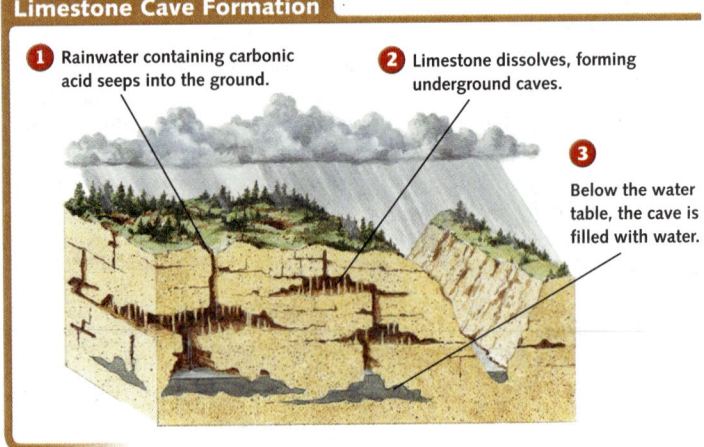

Limestone Cave Formation

1. Rainwater containing carbonic acid seeps into the ground.
2. Limestone dissolves, forming underground caves.
3. Below the water table, the cave is filled with water.

310 Unit 4 Earth's Changing Surface

DIFFERENTIATING INSTRUCTION

Challenge Activity Have students do library or Internet research to find out about Carlsbad Caverns in New Mexico or Mammoth Cave in Kentucky. They should research the geology of the area, find out the characteristics of erosional and depositional features, and be able to explain how the caves and other features formed. Students can share their findings with the class in the form of a poster, bulletin-board display, or oral presentation.

dissolves the limestone and carries ions away in solution. After thousands of years, the cracks between beds become so large that they form networks of underground tunnels, sometimes many kilometers long and hundreds of meters deep. These tunnels are called **caverns,** or caves.

Dripstone is a calcite deposit formed from dripping water in caverns. Dripstone can form only when a cave is at or above the water table, where water can evaporate. When groundwater drips from the roof of a limestone cave, it slowly deposits calcite. Slender deposits called stalactites hang like icicles from the roof along the routes of the dripping water. On the cave floor beneath the stalactites, blunt, rounded masses called stalagmites form. When stalactites and stalagmites meet, columns, or pillars, are formed.

Karst Topography

In some regions, rainwater enters the ground through sinkholes and fissures. Sinkholes are depressions that form in the surface because carbon dioxide in the water has dissolved some of the rock beneath the soil. Many shallow sinkholes form slowly. But some sinkholes form more quickly when part of a cave roof collapses. The resulting depression may be deep and have steep sides. A pond or lake may form when a sinkhole is deep enough to meet the water table.

Because rainwater drains through sinkholes and fissures, there are few surface rivers in regions marked by sinkholes. Lost rivers may form when surface streams disappear underground and flow out of caves many kilometers away. Regions characterized by sinkholes, sinkhole ponds, lost rivers, and underground drainage are said to have **karst topography.**

Karst topography forms in areas with bedrock made of calcite, dolomite, or other minerals that dissolve easily. The Mammoth Cave region of Kentucky has karst topography. Other states having karst topography include Alabama, Florida, Tennessee, and Indiana.

NEW ZEALAND Sinkholes have formed in the ground above the limestone cliff in this photograph.

14.3 Section Review

1. What factors determine the amount and type of mineral matter dissolved in groundwater?
2. What factors can cause a spring to have a high mineral content?
3. Why are caves often found in regions that have limestone as a bedrock?
4. Identify some features found in regions with karst topography.
5. **CRITICAL THINKING** Groundwater dissolves and deposits rock materials. Describe how these processes are involved in forming a cave.
6. **CHEMISTRY** Rainwater seeping into the ground is naturally acidic enough to dissolve limestone. Acid rain caused by pollution is even more acidic. What do you think happens when acid rain seeps into the ground?

CHAPTER 14 SECTION 3

Observe an animation of cave formation.
Keycode: ES1405

Have students list features that may form as a result of water flowing underground.

 Visualizations CD-ROM

ASSESS

1. the type of rock through which the water travels, the distance the water travels underground, and the water's temperature
2. The water passes through rock that contains easily dissolved minerals, the water is acidic due to a large amount of dissolved gases, and the water is very hot.
3. Limestone dissolves more easily than some other types of rock dissolve.
4. sinkholes, sinkhole ponds, fissures, lost rivers, caves, and underground drainage
5. Groundwater dissolves limestone as it flows through fissures and cracks in the limestone. By dissolving the limestone, the groundwater enlarges these cracks and creates networks of tunnels that form caves. Deposition of calcite by groundwater in caves creates structures such as stalactites, stalagmites, and columns.
6. The acid rain is more effective than rainwater in dissolving limestone and, consequently, in forming caverns.

MONITOR AND RETEACH

If students miss . . .

Question 1 Reteach "Minerals in Groundwater." (p. 309)
Question 2 Have students restate the three bulleted factors on page 310 in their own words.
Question 3 Reteach "Caverns." (pp. 310–311) Ask: What property does limestone have that makes it readily form caves?
Question 4 Have students reread "Karst Topography." (p. 311)

Question 5 Have students make a flowchart that shows the sequence of events that leads to the formation of caverns with stalactites, stalagmites, and columns.
Question 6 Be sure students understand the corrosive effects of acid on limestone. Lead them to infer that the more acidic a substance is, the more corrosive it is.

CHAPTER 14 LAB Activity

Water Budgets

SKILLS AND OBJECTIVES

- **Analyze** water usage, deficit, recharge, and surplus shown in water budget data and graphs.
- **Decide** water allocation for a community with limited water resources.

MATERIALS

- calculator
- notebook paper

The distribution of limited resources, such as water, is an important task for all communities. Groundwater supplies dictate whether a period is one of usage, deficit, recharge, or surplus. Water budgets must take these periods into account in determining how water is to be allocated.

The water budget graph on the next page shows moisture need and supply for Little Rock, Arkansas, over a hypothetical one-year period. The graph also shows incoming supply versus expected spending of groundwater, assuming the ground holds 100 millimeters of water. The table below shows the corresponding data. In this activity, you will determine the status of water availability in Little Rock for all the months of the year. You will then make decisions on how and when to allocate water to various users.

Procedure

1. Copy the table below. Locate the "Supply minus need" row and start with the value for June. Because the previous month's value is positive, June's negative value shows that more water will be drawn from the ground than will have entered it. June, therefore, is a month of water usage. On the line labeled "Water budget section," write "U" for usage in June.

2. Usage continues until the negative values total 100. When you add the −56 for June and the −87 for July, the answer exceeds 100. Therefore, July marks the transition from usage to deficit. Write "U/D" for July.

3. Continue to mark deficits with a "D" in the "Water budget section" until the values turn positive again. Positive values indicate recharge, which you should mark with "R."

4. When the total of consecutive positive numbers exceeds 100, that month defines the time when recharge changes to surplus. Write "R/S" for that month. Any subsequent positive value is considered a surplus and can be marked "S." Continue filling in the "Water budget section" until you have completed the entire row.

Water Budget Data Table for Little Rock, Arkansas (mm moisture)												
Month:	Jan	Feb	Mar	Apr	May	Jun	Jul	Aug	Sep	Oct	Nov	Dec
Supply	127	102	115	128	127	95	85	81	79	71	104	103
Need	7	11	35	65	110	151	172	157	110	65	21	10
Supply minus need	+120	+91	+80	+63	+17	−56	−87	−76	−31	+6	+83	+93
Water budget section												

CHAPTER 14 ACTIVITY

PURPOSE

To interpret water-budget data and use the data to think critically about water-use issues

PROCEDURE

3. August and September should be marked "D" because their supply-minus-need values are negative. October should be marked "R" because it is the first month when supply exceeds need.

4. November should be marked "R" because supply still exceeds need, and the sum of the recharges in October and November, (+6) + (+83) = (+89), is less than 100. As stated in the introduction, one hundred millimeters of groundwater is the assumed benchmark for the normal level of groundwater supply. December should be marked "R/S" because the sum of the recharges in October through December, (+6) + (+83) + (+93) = (+182), is greater than 100. The water budget has therefore shifted from recharge to surplus in December. January through May all have a greater supply of water than need for water, so all five months should be marked "S" for surplus.

ANALYSIS AND CONCLUSIONS

1. Answers may vary. The graph offers a clearer overview of the water budget and an easier means to understanding the relationships among need, supply, surplus, recharge, deficit, and usage. The table provides an easier means to making specific calculations.

2. A desert community will have a larger need for water, especially in the summer, and a much smaller supply year-round. This will result in long periods of deficit, with very short recharge periods in the winter and no surplus at all.

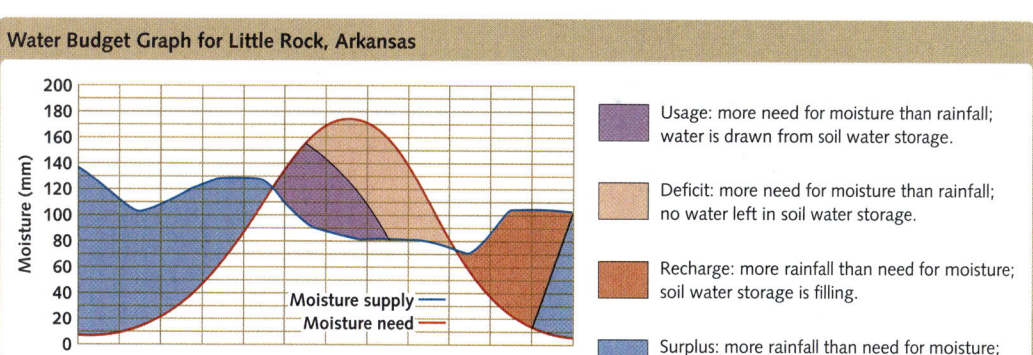

Analysis and Conclusions

1. Compare your assessment of the Water Budget Data Table with the Water Budget Graph for Little Rock. Which display of the data makes more sense to you, the table or the graph? Explain why.

2. Describe how you think the data would differ for a desert community.

3. Suppose you are a city administrator in Little Rock, charged with allocating groundwater use for the following year. A successful local business bottles and sells the region's spring water. During which months will you allow the business to bottle water? Explain why.

4. The owner of a large local farm has determined that people are interested in buying locally grown late-summer corn. Growing such a crop would require drawing from groundwater supplies to irrigate. Will you approve the farmer's application for this allocation of groundwater? Explain.

5. A developer has applied to construct 500 new homes. Is there enough water available to support this development? The Chamber of Commerce supports the development; local farmers oppose it. State your opinion, and explain your reasoning.

6. Meteorologists are predicting a summer drought. Discuss what temporary water conservation regulations you could instate to protect the local brown trout fishing industry. Enough groundwater must be conserved to keep the streams full during the drought. What other steps could you consider? Does this forecast affect the decision you made in Question 4?

Learn more about the water in your area.
Keycode: ES1407

CHAPTER 14 ACTIVITY

LOCAL RESOURCES
CLASSZONE.COM

Learn more about the water in your area.
Keycode: ES1407

(continued)

3. A water surplus is available from December through May. Bottling of water might be appropriate for any or all of these months.

4. The late summer months in Little Rock (July–September) are all months of water deficit, which means that no groundwater is available at that time for irrigation of crops. Students might decide that the project is important enough to store water from the surplus period in winter and spring to be used by the farmer for his project.

5. Students should agree that over the course of the entire year there is more water available for use than is being used. Students may disagree on how that extra water should be allocated. The water might be used for development or farming, depending on the economic and social benefits the community wishes to encourage. Any development project would require the construction of reservoirs to store surplus water from the winter and spring.

6. Water conservation steps could include the following: limiting the watering of gardens and lawns to nighttime or to odd/even numbered days; installing water-saving devices on showers and toilets; banning car washes and recreational uses of water; offering incentives for homes that reduce water usage. In addition, water could be piped into the community from other sources, such as lakes, rivers, and other communities. Students will probably say that the allocation of surplus water for irrigation in question 4 should be suspended until it is clear that there is enough water available.

Refer students to Safety in the Earth Science Laboratory on pages xxii–xxiii.

CHAPTER 14 REVIEW

Summary of Key Ideas

14.1 The amount of water that rocks or soil can hold depends on their porosity and permeability. The water table is the top of the water-saturated region of the ground. The depth of the water table depends on climate, season, and location. Groundwater reaches the surface through natural springs and through wells. In areas associated with volcanic activity, groundwater may be very hot, resulting in the formation of springs and geysers.

14.2 A water budget relates the recharge, surplus, usage, and deficit of soil water to the moisture needs and the moisture supply of an area. Overuse of groundwater leads to problems such as subsidence. Groundwater pollution is a serious threat to supplies of usable water.

14.3 Dissolved minerals in groundwater are often left behind as deposits such as travertine, geyserite, petrified wood, stalactites, stalagmites, and the cement that binds sedimentary rocks. Groundwater containing carbonic acid dissolves limestone, forming caverns and features of karst topography.

Key Vocabulary

aquifer (p. 304)
artesian formation (p. 304)
capillary action (p. 302)
cavern (p. 311)
deficit (p. 306)
geyser (p. 305)
groundwater (p. 300)
karst topography (p. 311)
mineral deposit (p. 309)
ordinary well (p. 303)
permeability (p. 301)
porosity (p. 300)
recharge (p. 306)
spring (p. 303)
surplus (p. 306)
usage (p. 306)
water budget (p. 306)
water table (p. 302)

Vocabulary Review

314 Unit 4 Earth's Changing Surface

Vocabulary Review

1. cavern
2. capillary action
3. karst topography
4. An ordinary well is driven into the ground and contains water up to the level of the water table. An artesian well is drilled into an artesian formation and contains water that rises up due to underlying pressure.
5. Permeability is the rate at which water or other liquids pass through the pore spaces of a rock. Porosity is the percent of a material's volume that is pore space.

Concept Review

6. Material consisting of well-sorted, rounded grains will have more pore space than material consisting of well-sorted angular grains. Angular particles can fit together more tightly than rounded particles.
7. Sketches should show the water table above the zone of saturation. The capillary fringe should extend upward a short distance from the water table, with the zone of aeration extending upward from the capillary fringe to the ground surface.
8. Increased rainfall will raise the water level in an ordinary well; lack of rainfall will cause the water level to drop. A population increase might lower water levels as people draw more water from the same aquifer.
9. The sandstone, which is permeable, would be sandwiched between layers of shale, which is impermeable. The impermeable shale layers trap water in the sandstone layer.
10. A geyser forms when water is trapped at the bottom of a long, vertical tube extending hundreds of meters into the ground where rocks are hot. The deep water becomes superheated and pressurized. When water at the top of the column boils out, it relieves pressure below. This allows the superheated water to instantly boil and thus produce an explosive burst of steam that blows out the water above.
11. Hard water is groundwater containing large amounts of ions from dissolved minerals. The calcium, magnesium, and iron ions often found in hard water react with soap to form scum instead of suds and can leave deposits in hot-water pipes.
12. Geyserite forms where geysers that have dissolved silica from hot igneous rocks below the surface deposit the silica around the geyser opening. Travertine is left behind at mineral springs in areas where the predominant dissolved mineral is calcite from limestone.
13. Limestone bedrock would be expected in a region with caves and sinkholes because limestone dissolves easily.
14. recharge

Vocabulary Review

Write the term from the key vocabulary list that best completes the sentence.

1. Stalactites and stalagmites are calcite deposits found in a(n) ___?___ .
2. Water moves from the zone of saturation to the zone of aeration by means of ___?___ .
3. Sinkholes and lost rivers may form in regions that have ___?___ because groundwater has dissolved the underlying bedrock.

Explain the difference between the terms in each pair.

4. ordinary well, artesian well
5. permeability, porosity

Concept Review

6. Is material with high porosity more likely to consist of well-sorted, round grains or well-sorted, angular grains? Explain your thinking.
7. Make a quick sketch that shows the location of the water table in relation to the zone of aeration, the zone of saturation, and the capillary fringe.
8. Explain how the amount of rainfall in a region could affect the water level in an ordinary well. How might an increase in the population of an area affect water levels in ordinary wells?
9. The rock layers in an artesian formation consist of shale and sandstone. Sketch how the rock layers could be arranged. Explain your thinking.
10. Why does water erupt from a geyser?
11. What is hard water? How does the hardness of water affect human activities?
12. Where are you likely to find geyserite? Travertine?
13. What type of bedrock would you expect to find in a region with caves and sinkholes? Why?
14. **Graphic Organizer** Use a term from the key vocabulary list to complete the flow chart.

usage → deficit → ? → surplus

Critical Thinking

15. **Analyze** In cold climates a smaller amount of water enters the ground during winter than in any other season. Why?

16. **Compare** Describe how the porosity and the permeability of well-sorted sand compares with the porosity and the permeability of a mixture of sand and silt.

17. **Predict** What is likely to happen to the height of the water table if rainfall increases but evapotranspiration is unchanged?

18. **Infer** Explain why dripstone forms only when a cave is at or above the water table.

19. **Analyze** In some regions, petrified wood is composed of silica. In other regions, it is composed of calcite. Why?

Internet Extension

How Many People Can an Aquifer Support?
Examine the ability of an aquifer to supply water for an increasing population.
Keycode: ES1406

Writing About the Earth System

SCIENCE NOTEBOOK Some sinkholes fill with water to form lakes or ponds. Explain how the geosphere and the hydrosphere are involved in the formation of a sinkhole pond or lake. How might the biosphere and the atmosphere be affected by the formation of a sinkhole pond or lake?

Interpreting Graphs

A water-budget graph summarizes the water budget of a given region. It shows two kinds of information: moisture need and moisture supply. Periods of water usage, deficit, recharge, and surplus can be determined from data on the graph. The graph displays information about Springfield, Illinois, a region that has four seasons.

20. During which months is the moisture need the lowest? During which month is it the highest? Give the moisture need values for these months.

21. During which months is there a water surplus for all or part of the month? Compare moisture need to moisture supply during these months.

22. During which months is there a moisture deficit for all or part of the month? Compare moisture need to moisture supply during these months.

23. During which month does the water storage finish refilling?

24. Use the words *supply* and *need* to complete these sentences.

    Usage and deficit occur when moisture ___?___ exceeds moisture ___?___ . Recharge and surplus occur when moisture ___?___ exceeds moisture ___?___ .

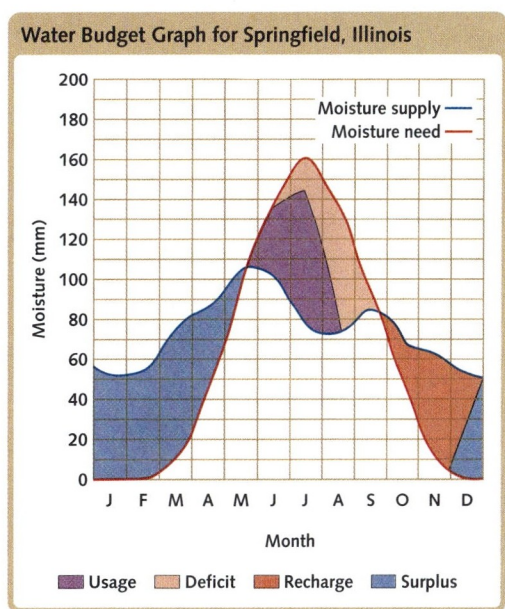

Writing About the Earth System

The geosphere contributes to the formation of a sinkhole lake by the limestone composition of the bedrock in the area. The hydrosphere provides the rainfall and resultant groundwater that fills the sinkhole, thus forming the lake. The lake becomes part of the biosphere when inhabited by plants, animals, and other organisms. Evaporation from the lake adds water vapor to the atmosphere, which can lead to the formation of clouds and fog.

CHAPTER 15 PLANNING GUIDE: GLACIERS

	Section Objectives	Print Resources
Chapter-Level Resources		Laboratory Manual, pp. 67–70 Internet Investigations Guide, pp. T55–T56, 55–56 Guide to Earth Science in Urban Environments, pp. 5–12, 13–20 Spanish Vocabulary and Summaries, pp. 29–30 Formal Assessment, pp. 43–45 Alternative Assessment, pp. 29–30
Section 15.1 What is a Glacier?, pp. 318–320	1. Explain what a glacier is. 2. Describe where glaciers form. 3. Explain how glaciers form. 4. Describe two types of glaciers.	Lesson Plans, p. 50 Reading Study Guide, p. 50
Section 15.2 Glacial Movement and Erosion, pp. 321–325	1. Describe how glaciers move. 2. Explain how glaciers cause erosion. 3. Compare and contrast the effects of erosion by valley and continental glaciers.	Lesson Plans, p. 51 Reading Study Guide, p. 51 Teaching Transparency 16
Section 15.3 Glacial Deposits, pp. 326–329	1. Describe two types of glacial deposits. 2. Describe landscape features characteristic of glacial deposits.	Lesson Plans, p. 52 Reading Study Guide, p. 52
Section 15.4 Ice Ages, pp. 330–333	1. Describe the ice ages that Earth has experienced and the evidence they left behind. 2. Summarize several hypotheses for the causes of the ice ages. 3. Explain how glaciation contributed to the formation of the Great Lakes.	Lesson Plans, p. 53 Reading Study Guide, p. 53

INVESTIGATIONS
CLASSZONE.COM

Section 15.4 INVESTIGATION
How Does Land Cover Affect Global Temperature?
Computer time: 45 minutes / Additional time: 45 minutes
Students predict how changing land cover might affect Earth's temperature balance.

Chapter Review INVESTIGATION
How Fast Do Glaciers Flow?
Computer time: 45 minutes / Additional time: 45 minutes
Students use the change in location of stakes on glaciers to calculate glaciers' speeds.

Technology/Online Resources

Electronic Teacher Tools
Visualizations CD-ROM
Classzone.com
Online Lesson Planner
Test Generator

Visualizations CD-ROM
Classzone.com
  ES1501 Visualization

Visualizations CD-ROM
Classzone.com
  ES1502 Visualization

Classzone.com
  ES1503 Data Center

Visualizations CD-ROM
Classzone.com
  ES1504 Investigation, ES1505 Visualization,
  ES1506 Visualization

Classroom Management

Gather these materials for minilab and laboratory activities.

Minilab, p. 320
  snow or crushed ice

Lab Activity, pp. 334–335
  pencil
  plastic box
  ice cube
  towel
  160 oz. box of cornstarch
  2-qt. mixing bowl
  1 to 2 cups of water
  spoon
  5 × 7 in. index cards
  5 wooden toothpicks
  5 or 6 large pebbles

CHAPTER 15

INTRODUCE

Build Interest

Before sharing background information about the photograph with students, allow time for their questions and comments.

Ask questions like the following:

- How could chunks of ice break off from the glacier? **Changing temperatures could cause chunks to break off along cracks in the glacier.**
- What do you think the people in the boat are doing? **The people might be scientists observing the glacier and making measurements.**
- Are the people in any danger? **They do not appear to be in any danger, but they could be harmed if they are too close to falling ice.**

Respond

- How and why do scientists study glacial ice? **They track the movement of glaciers to discover climate trends. They also analyze the composition of the ice to find out about Earth's climate and atmosphere in the past.**

About the Photograph

Glaciologists, such as those shown here, study various aspects of glaciers, including the separation of icebergs from glaciers, a process called calving. In March 2000, glaciologists observed a massive iceberg calving off Antarctica's Ross Ice Shelf. At 295 kilometers long and 37 kilometers wide, the iceberg is about twice the size of Delaware.

Glaciers

Massive icebergs break off into the sea from the massive ice sheet that covers Antarctica. One of the largest icebergs ever recorded was about 295 kilometers long.

How and why do scientists study glacial ice?

BEYOND THE TEXTBOOK

Print Resources

- Reading Study Guide, pp. 50–53
- Transparency 16
- Formal Assessment, pp. 43–45
- Laboratory Manual, pp. 67–70
- Alternative Assessment, pp. 29–30
- Spanish Vocabulary and Summaries, pp. 29–30
- Internet Investigations Guide, pp. T55–T56. 55–56
- Guide to Earth Science in Urban Environments, pp. 5–12, 13–20

Technology Resources

- Visualizations CD-ROM
- Classzone.com
- Online Lesson Planner
- Electronic Teacher Tools
- Test Generator

316 Unit 4 Earth's Changing Surface

CHAPTER 15

PREVIEW

▶ **FOCUS QUESTIONS** In this chapter you will study glaciers and learn more about the key questions below.

Section 1 What are glaciers and how do they form?

Section 2 How do glaciers move and reshape Earth's surface through erosion?

Section 3 How do glaciers deposit eroded rock materials?

Section 4 How have ice ages affected Earth's surface?

▶ **REVIEW TOPICS** As you investigate glaciers, you will need to use information from earlier chapters.
- carbon cycle (p. 14)
- weathering (p. 258)
- erosion (p. 268)

▶ **READING STRATEGY**

PREVIEW

Before you read this chapter, look through the pages at the key vocabulary lists at the beginning of each section. Look at the photographs, illustrations, and maps. Based on what you see, make a list of some topics you expect to learn about in this chapter.

At our Web site, you will find the following Internet support for this chapter.

DATA CENTER
EARTH NEWS
VISUALIZATIONS
- Seasonal Snowline Migration
- Glacial Erosion of Bedrock Surfaces
- Ice Sheet Retreat from North America
- Earth Orbit's Effect on Global Climate

LOCAL RESOURCES
INVESTIGATIONS
- How Does Land Cover Affect Global Temperature?
- How Fast Do Glaciers Flow?

Chapter 15 Glaciers **317**

CHAPTER 15

PREPARE

Focus Questions
Find out what knowledge students have about glaciers. For each focus question, have students consider the section title and then develop two or three additional questions for each section. For example, Section 1: How long does it take a glacier to form? Section 2: How fast does a glacier move?

Review Topics
In addition to the earth science topics listed, students may need to review this concept:

Physical Science
- Friction is a force that opposes the relative motion of two objects in contact with each other. Sliding friction occurs when one object, such as a glacier, slides over another object, such as a valley wall.

Reading Strategy
Model the Strategy: Have students read the list of Key Vocabulary terms on page 318. Ask students what they expect to read about based on these terms. *what a glacier is, different types of glaciers* Have them list other expected topics based on the graph on page 318 and the photograph on page 319. *definition of a snow line and how it relates to glaciers, valley glaciers* Have students do the same for the rest of the section. Were all of the topics they expected to be covered actually covered? What topics might they research themselves?

BEYOND THE TEXTBOOK

INVESTIGATIONS
- How Does Land Cover Affect Global Temperature?
- How Fast Do Glaciers Flow?

VISUALIZATIONS
- Animation of seasonal migration of the snow line
- Animation of how glaciers erode bedrock surfaces
- Animation of the retreat of ice sheets from North America
- Animation of changes in Earth's orbit that contribute to climate change

DATA CENTER
EARTH NEWS
LOCAL RESOURCES
CAREERS

Online Lesson Planner

Chapter 15 Glaciers **317**

CHAPTER 15 SECTION 1

15.1

FOCUS

Objectives
1. Explain what a glacier is.
2. Describe where glaciers form.
3. Explain how glaciers form.
4. Describe two types of glaciers.

Set a Purpose
Have students read the section to answer the question "What is a glacier?"

KEY IDEAS
Glaciers are huge ice masses that move under the influence of gravity.

Glaciers form from compacted and recrystallized snow.

KEY VOCABULARY
- glacier
- snow line
- firn
- valley glacier
- continental glacier
- ice cap

INSTRUCT

VISUAL TEACHING

Discussion
Point out that the graph shows how snow line elevation in meters (rather than feet) changes north of the equator. Ask: What should the two vertical lines between 0° and 15°N be numbered? **5°N, 10°N** What elevation does the horizontal line between 2000 and 3000 represent? **2500 m** Now have students compare the elevations of the snow line at 10° north and 60° north of the equator. **At 10° N, the snow line is 4000 meters higher than at 60° N.**

VISUALIZATIONS
CLASSZONE.COM
Examine seasonal migration of the snow line.
Keycode: ES1501

VOCABULARY STRATEGY
The word *firn* comes from a German word meaning "last year's snow."
The word *névé* is related to a Latin word meaning "cooled by snow."

What Is a Glacier?

About 75 percent of Earth's fresh water is frozen in glaciers. A **glacier** is a large mass of compacted snow and ice that moves under the force of gravity. A glacier changes Earth's surface as it erodes geological features in one place and then redeposits the material elsewhere thus altering the landscape.

Where Glaciers Form

Glaciers form in areas that are always covered by snow. In such areas, more snow falls than melts each year, as a result layers of snow build up from previous years. Climates cold enough to cause such conditions may be found in any part of the world. Air temperatures drop as you climb high above sea level and as you travel further from the equator.

Even in equatorial areas, however, a layer of permanent snow may exist on high mountains at high elevation. Further from the equator, the elevation need not be so high for a layer of permanent snow to exist. In the polar areas, permanent snow may be found even at sea level. The lowest elevation at which the layer of permanent snow occurs in summer is called the **snow line.** If a mountain is completely covered with snow in winter but without snow in summer, it has no snow line.

In general, the snow line occurs at lower and lower elevations as the latitudes approach the poles. The snow line also changes according to total yearly snowfall and the amount of solar exposure. Thus, the elevation of the snow line is not the same for all places at a given latitude.

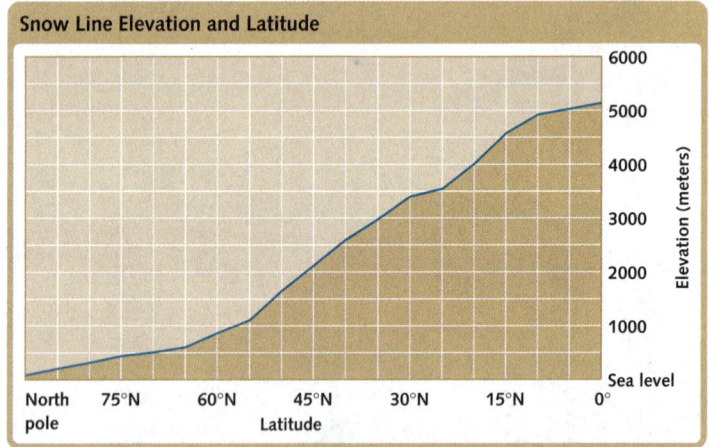

Snow Line Elevation and Latitude

How Glaciers Form

Except for bare rock cliffs, a mountain above the snow line is always buried in snow. Great basins below the highest peaks are filled with snow that can be hundreds of meters thick. In these huge snowfields, buried snow becomes compressed and recrystallizes into a rough, granular ice material called **firn** (feern) or *névé* (nay-VAY).

318 Unit 4 Earth's Changing Surface

DIFFERENTIATING INSTRUCTION

Reading Support Explain and model how to construct a concept map. If necessary, explain that a concept map is a tool that can help students organize broad concepts and their subconcepts. You might build a concept map by placing concepts (usually nouns) into ovals and connecting them with words that describe their relationship (usually verbs, verb phrases, prepositions, or adverbs). The broadest concept might be placed at the top. Two concepts and their connecting words could form a complete thought or sentence. Then help students construct a concept map about the types of glaciers. *Use Reading Study Guide, p. 50.*

Firn resembles the ice of a packed snowball. It is not fluffy, such as new-fallen snow, nor is it as hard as solid ice. The granules of firn start out no larger than grains of sand. As the layer of firn thickens, the firn's crystals may grow as large as kernels of corn. Within a layer of firn, the weight of the material at the top compresses the firn below turning that firn into solid ice. Under the weight of the overlying snow and firn, the ice begins to flow downward or outward. This moving mass of snow and ice is a glacier.

Types of Glaciers

There are two main types of glaciers, valley glaciers and continental glaciers. A **valley glacier** is a glacier that moves within valley walls. A **continental glacier** is a glacier that covers a large part of a continent.

Valley Glaciers

Many mountain ranges in the world have peaks and valleys high enough so that snowfall there exceeds snowmelt. The snow builds up and changes to ice as it accumulates in the valleys of such mountain ranges. The ice stays within valley walls, forming a large river of ice and snow, which moves slowly downhill under the influence of gravity. This long, slow-moving, wedge-shaped stream of ice is a valley glacier. Valley glaciers are also known as *alpine glaciers,* after the Alps in south-central Europe.

Valley glaciers form in regions where mountains are high enough to be in the colder part of Earth's atmosphere. Valley glaciers even form in equatorial regions where mountains are located at high elevations. Valley glaciers exist on all continents except Australia.

Valley glaciers vary in size. Small valley glaciers may be less than 2 kilometers long. Large valley glaciers may be over 100 kilometers long and hundreds of meters thick. Some of the world's largest valley glaciers are in southern Alaska. The world's tallest mountains, the Himalayas, also have very large valley glaciers.

CHAPTER 15 SECTION 1

VISUALIZATIONS
CLASSZONE.COM

Examine seasonal migration of the snow line.
Keycode: ES1501
Visualizations CD-ROM

More about...

How Glaciers Form
Glacier formation occurs when more snow falls during winter than melts during summer. As layers of new snow accumulate on top of old snow, a snowfield forms. The conversion of snow to firn begins when pressure from overlying snow causes the buried snow to compact and recrystallize, allowing snowflakes to change from six-sided crystals into small grains of ice. Firn has a gray or milky-white color that results from abundant bubbles of air trapped during compaction.

DENALI NATIONAL PARK Muldrow Glacier, a valley glacier in Alaska, is about 56 kilometers long.

DIFFERENTIATING INSTRUCTION

Hands-On Demonstration If students are not going to do the Minilab on page 320, you can do this demonstration instead, to provide students with an opportunity to compare snow and glacial ice up close. Use real snow if available or create some by scraping ice crystals off the sides of a freezer. Allow students to examine individual snowflakes under a microscope or powerful hand lens. Hold the snowflakes or crystals in one hand and press your hands together to compact them into a snowball. Break open the ball and have students use the microscope to examine a sample of the ice resulting from compaction. If necessary, point out the loss or deformation of the original snowflakes' six points, or sides, which have become crushed and melted under the pressure of your hands. Ask students to compare the formation of the snowball to that of glacial ice.

Chapter 15 Glaciers

CHAPTER 15 SECTION 1

MINILAB
Modeling Glacier Formation

Management
Provide extra paper towels with which to mop up melted snow.

Analysis Answers
- Squeezing causes the particles to partially melt, stick together, and form ice. When you stop applying pressure, the snow melts.
- Glacial ice forms when snow accumulates, becomes compressed, recrystallizes, and turns to ice, just as the snow in the snowball formed ice.

ASSESS

1. The snow line is the lowest elevation at which snow is present year-round.

2. In snowfields, buried snow becomes compressed and recrystallized to form firn. As firn becomes thicker, its crystals grow larger and its lower layers become compressed into solid ice. Under the weight of overlying layers, the ice begins to flow, becoming a glacier.

3. A valley glacier forms in the valley of a high mountain range and flows downhill. It varies in size. A continental glacier forms inland, in a polar region, moving outward in all directions to lower elevations. It is circular or oval in shape and generally larger and thicker than a valley glacier.

4. Snow line elevations decrease as latitude increases.

5. Oulu: at about 600 meters; Tucson: no, the snowline is over 3000 meters there

---

10-Minute Mini LAB
Modeling Glacier Formation

Materials
- snow or crushed ice

Procedure
1. Take a handful of snow or crushed ice and squeeze it between your hands.
2. Keep applying pressure until you notice a change in the snow or ice particles.

Analysis
What does squeezing seem to do to the particles? How does your applying pressure to the snow relate to the formation of a glacier?

Continental Glaciers

Over the inland areas of polar regions, the climate is so cold that all precipitation falls as snow. Even in the summer, the air is never warm enough to melt all the snow. In places such as Greenland and Antarctica, snow has been falling, for thousands of years, building up and changing to ice. Landmasses such as these are almost entirely covered by thick masses of ice that leave only the highest mountain peaks uncovered. The ice is thousands of meters thick and moves outward from its center in all directions to lower elevations. In some places the ice reaches the sea, where great chunks break off and float away as icebergs. This moving mass of ice is a continental glacier.

Continental glaciers are roughly circular or oval in shape. The Greenland glacier is about 1.7 million square kilometers in area (almost the size of Mexico) and up to 3 kilometers thick in places. The Antarctic glacier covers an even larger landmass, with an area of about 13.7 million square kilometers. This land area is greater than the combined areas of the United States and Mexico. In some places, the Antarctic glacier is over 4 kilometers thick. Along the coast the ice may descend more than 2 kilometers below sea level. Inland, mountain peaks called *nunataks* project through the ice.

A glacier that is less than 50,000 square kilometers in area is sometimes called an **ice cap.** Iceland, Baffin Island, Spitsbergen, and other large islands of the Arctic Ocean have ice caps.

ANTARCTIC PENINSULA The world's largest continental glacier is in Antarctica.

15.1 Section Review

1. What is the snow line?
2. Describe how a glacier forms.
3. Compare and contrast a valley glacier and a continental glacier.
4. **CRITICAL THINKING** The graph on page 318 shows how snow-line elevations change north of the equator. Predict how snow-line elevations change as latitude increases south of the equator.
5. **GEOGRAPHY** Use the graph on page 318. If you lived in Oulu, Finland (latitude 65° N), at what elevation would you expect to find a glacier? Suppose you lived in Tucson, Arizona (latitude 32° N). Would you expect to find a glacier at an elevation of 2000 meters? Why or why not?

MONITOR AND RETEACH

If students miss . . .

Question 1 Reteach "Where Glaciers Form." (p. 318) Discuss snow lines and what affects them.

Question 2 Reteach "How Glaciers Form." (pp. 318–319) Have students number and list stages in the formation of a glacier.

Question 3 Reteach "Types of Glaciers." (pp. 319–320) On the board, prepare a class compare/contrast table of the two types of glaciers.

Question 4 Review "Where Glaciers Form." (p. 318) Review the graph on page 318. Then ask students to make a general statement about how snow line elevation changes with distance from the equator.

Question 5 Reteach how to read the graph on page 318. Verify that students understand what the axes and graph line represent.

Glacial Movement and Erosion

How can something as heavy as a glacier move? In fact, because a glacier moves to lower elevations under the influence of gravity, it is the glacier's great weight that enables it to move. As it moves, it erodes rock materials.

How Glaciers Move

Some glaciers may move only a few centimeters a day, while others may move several meters a day for weeks or months. Occasionally, glaciers surge forward at even higher rates. Glaciers move more rapidly after winters of heavy snowfall, on steep slopes, and in summer. Friction between glacial ice and the valley floor and walls slows a glacier's movement along its base and sides. For this reason, a typical glacier moves more rapidly at its surface and at its center.

Basal Slip and Plastic Flow

At the base of some glaciers, the overlying weight causes grains of ice to partially melt and refreeze. The melting produces a great deal of water, which collects underneath the glacier. This water can form a thin layer of its own, or it can mix with sand and gravel, creating a slushy layer. Either type of layer reduces friction thus allowing the huge mass of ice to slide easily under the influence of gravity. This movement at the glacier's base is called basal slip (BAY-suhl slihp).

While basal slip generally accounts for the movement at the base of a glacier, it cannot propel the rest of the glacier. In fact, in extremely cold regions, basal slip does not even occur. Most ice in the interior of a glacier moves downhill through a process called plastic flow. In this process, the grains of ice actually deform, or change shape, continuously and permanently under the pressure of the overlying snow and ice. Near the bottom of the glacier, the ice grains are almost flat. Aided by the change in shape, the grains of ice slip past each other to create forward movement.

Glacial Movement

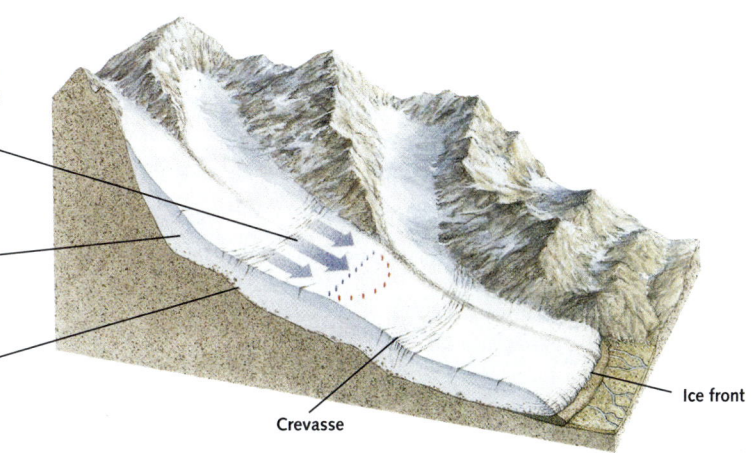

A glacier moves fastest at the surface and in the center, as shown by a line of stakes driven into the ice. The positions of the stakes after movement has occurred are shown in red.

Most of the ice in the interior of the glacier moves by a process called plastic flow.

Along the base of the glacier, the ice moves by a process called basal slip.

Crevasse

Ice front

15.2

KEY IDEAS

Glaciers move by the processes of basal slip and plastic flow.

As glaciers move, they erode rock materials in ways that alter a landscape.

KEY VOCABULARY
- crevasse
- ice front
- calving
- till
- moraine
- striations
- glacial valley
- cirque

CHAPTER 15 SECTION 2

> **More about...**
>
> **Crevasses**
> The upper rigid layer of glacial ice where crevasses form is called the fracture zone. Where ice pulls away from the head of a mountain valley, a crevasse called a bergschrund forms. When a glacier bends around a curve and falls sharply over a bedrock step, a mass of crevasses, called an ice fall, can result. The surface is reduced to cracks and ridges, and the ridges are known as seracs, or ice pinnacles.

ALASKA Crevasses have formed in the rigid ice at the top of this glacier in Glacier Bay National Park.

Observe how glaciers erode bedrock surfaces.
Keycode: ES1502

Like river valleys, glacial valleys have both steep and gentle slopes. When a valley glacier comes to a steep downward slope, great fissures, or cracks, called **crevasses** (krih-VAS-ihz) form across the width of the glacier. These cracks form because the ice near the surface of the glacier is rigid. This ice responds to the movement of the ice underneath it by breaking. Crevasses rarely go deeper than about 50 meters, since below this depth there is enough pressure for plastic flow to occur.

How do scientists study glacial movement? One way is by driving rows of stakes into the ice across a valley and observing the positions of the stakes regularly over time. The illustration on page 321 shows how such markers can change position. Geologists also use orbiting satellites to track the movements of glaciers.

Movement at the Ice Front

Most glaciers extend below the snow line. As a valley glacier moves downhill into lower, warmer elevations, the snow and ice melt away, causing the glacier to thin steadily. A glacier is thinnest at the elevation where the ice melts as fast as it moves. A glacier that loses most of its ice through melting ends here, at its **ice front.**

A glacier always moves forward. As long as the rates of movement and melting are equal, the ice front is stationary. After a series of winters with heavy snows, which add pressure to the bottom ice, a glacier may move faster than normal. Its ice front then advances. After several warm summers, the snow and ice near the ice front may melt more quickly than usual. If the glacier's rate of melting is greater than its rate of movement, its ice front will recede.

Icebergs

Where the snow line is close to sea level, as in Alaska and Greenland, many glaciers reach the sea. Even here they do not melt as fast as they move. As they extend into the sea, great blocks break off to become icebergs. This process is called **calving.**

Calving also occurs in Antarctica, where the snow line is at sea level. Here the ice sheet reaches the coastline almost everywhere. In a number of places, the ice sheet extends beyond the coast far into the sea to form huge ice shelves. The largest of these, the Ross Ice Shelf, is hundreds of kilometers wide. One massive iceberg that broke off the Ross Ice Shelf in 2000 was about 295 kilometers long and 37 kilometers wide. Its area was about that of Connecticut. After the iceberg broke off, it may have knocked loose additional large icebergs as it floated in the frigid waters off Antarctica.

How Glaciers Cause Erosion

Glaciers are powerful agents of erosion. Like rivers, they remove loose rock from the valleys through which they move. Glaciers can pick up and move particles ranging in size from fine powder to house-size boulders. Often rocks fall onto a glacier from the valley walls. Tributary glaciers may transport more material to a main glacier.

DIFFERENTIATING INSTRUCTION

Hands-On Demonstration Contrast for students what occurs when a rigid material versus a plastic material moves over a steep downward slope. Use stacked books to create a gently sloping surface that meets a steeper sloping surface. Use a slab of clay to represent the ice at the base of a glacier that experiences plastic flow. A rigid piece of toast stuck to the top of the clay slab will represent the rigid ice at the top of a glacier. Move the "glacier" down the slope until half of the stack is hanging over the steeper segment. Demonstrate how the clay bends to follow the slope surface, but the toast must break in order to stay on top of the clay. Relate the demonstration to the formation of a crevasse.

Rock material eroded by glacial ice can ride on top of the glacier, become incorporated into the glacier, or be dragged beneath the glacier. The unsorted and unstratified rock material that is deposited by a retreating, or melting, glacier is called **till**. The accumulation of glacial till is called a **moraine** (muh-RAYN).

The photograph to the right shows two long lines of rock pieces that have piled up along the sides of the glacier are called *lateral (side) moraines.* Sometimes two glaciers come together to form a larger, single glacier. When their lateral moraines are joined, the result is a single *medial (middle) moraine* appearing within the glacier.

LAMPLUGH GLACIER, ALASKA Medial and lateral moraines are seen as dark parallel bands in this photograph.

The mixture of clay- and silt-sized particles that is formed by the crushing of rock beneath a glacier is called *rock flour*. The meltwater that pours off a glacier is likely to contain suspended rock flour. Because the rock flour has a milky white color, such meltwater is called *glacial milk*.

Glaciers weather and erode bedrock through the cutting action of rock that have pieces frozen into the ice. These pieces are dragged over the bedrock by the forward movement of the glacier. Particles of silt and fine sand, acting like sandpaper, smooth and polish the bedrock. Coarse sand, pebbles, and sharp boulders leave long parallel scratches called **striations** (stry-AY-shuhnz). Striations show the general direction of ice movement. If the bedrock is soft, pebbles and small boulders may dig in so deeply as to leave long parallel grooves. The pebbles and boulders carried by the glacier also show signs of wear, becoming flattened and scratched.

A glacier may also erode materials through a process called plucking. When glacial ice is squeezed up against a rocky surface, the ice may melt from the pressure. The meltwater fills in cracks in the rock, then refreezes, enlarging the cracks through ice wedging. Blocks of rock that have been separated from the rocky surface are then carried off, or plucked, by the glacier.

SEQUOIA NATIONAL PARK The parallel scratches on this granite in California are striations carved by a glacier.

CHAPTER 15 SECTION 2

VISUALIZATIONS
CLASSZONE.COM

Observe how glaciers erode bedrock surfaces.
Keycode: ES1502

Have students predict the effects of the glacial erosion on the landscape.

Visualizations CD-ROM

CRITICAL THINKING

Have students define *striations* and explain how they form. Striations are long, parallel scratches in bedrock carved by rocks carried by glaciers. State that abundant striations have been found on bedrock in a part of Wisconsin that was once covered by glacial ice. Ask: If the striations are oriented roughly northeast-southwest, what can you infer about the direction of glacial movement? Explain your answer. Since striations show the general direction of ice movement, the glacier must have moved parallel to the striation direction, either southwest or northeast.

VISUAL TEACHING
Discussion

Have students find the dark medial moraine in the picture of the Lamplugh Glacier. Then have them find at least two other lighter-colored medial moraines. Ask: What does multiple medial moraines indicate? that several glaciers have joined the larger glacier at various places up slope

---

DIFFERENTIATING INSTRUCTION

Reading Support Help students understand and differentiate the definitions of a lateral moraine and a medial moraine by associating the terms *lateral* and *medial* with familiar concepts. Students might think of a *lateral* pass in football (short pass to the side) or the *median* of grass that runs along the middle of a highway, separating opposing traffic. Distinctions between different types of moraines will become even more important when students learn about ground moraines and terminal moraines in Section 15.3.

Developing English Proficiency Students may be confused by non-English terminology in this section. Explain that many French words have been adopted for use in the English language for both common and scientific usage. Tell students that *roches moutonnées*, *cirque*, and *arête* are French words. Review their pronunciation and check students' comprehension of these terms.
Use Spanish Vocabulary and Summaries, pp. 29–30.

CHAPTER 15 SECTION 2

Visual Teaching

Interpret Diagrams

Ask: **What type of glacier caused the erosion in stage 3 of Glacial Erosion shown on page 325?** valley glaciers **Contrast the landscape before and after glacial erosion occurred.** Before glacial erosion, the mountains appear smooth and rounded. After glacial erosion, they are steep and jagged with deep, U-shaped valleys. **How does a horn differ from an arête?** An arête is a narrow, sharp divide between two cirques; a horn is a pyramid-shaped peak between three or more cirques. **What is a cirque and where does it form?** a semicircular basin that forms at the head of a glacial valley

Extend

Have students describe how the landscape in stage 3 would appear if this diagram showed erosion by a continental glacier. The mountain peaks would be smoother and rounded.

Use Transparency 16.

Effects of Erosion

Valley glaciers and continental glaciers erode Earth's surface in characteristic ways. Glacial erosion leaves behind formations which can be seen when the glacial ice retreats. One such formation is an outcropping of rock termed *roches moutonnées* (RAWSH MOOT-uhn-AY), which means *"sheep rocks"* in French. From a distance, such a formation may look like a flock of sheep.

ROCHES MOUTONNÉES, HUDSON BAY, CANADA One side of the outcropping is smooth, polished in the direction of the advancing glacier. The other side has loose blocks of steep and rough rocks plucked away and then deposited by the retreating glacier.

Erosion by Valley Glaciers

A river touches only a small part of its valley floor. A valley glacier, on the other hand, touches the entire valley floor and a large part of the valley walls as well. As the valley glacier moves, it smooths the valley floor and carves the valley walls nearly vertical. The resulting formation is a roughly U-shaped valley, called a **glacial valley.**

Main valley glaciers are usually much more massive than their tributary glaciers. Thus, a main valley glacier erodes its valley more than its tributary glaciers erode theirs. Proof of this can be seen in regions where a change in climate has caused glaciers to disappear. The U-shaped valley carved out by a main glacier is deeper than the valleys carved out by the tributary glaciers. The tributary valleys are called hanging valleys. A river that flows through a hanging valley plunges to the main valley, forming a hanging-valley waterfall. Glacial valleys and hanging valleys are common in all glaciated mountain ranges. Yosemite Falls in California is a famous hanging-valley waterfall.

A semicircular basin called a **cirque** (seerk) can be found at the head of a glacial valley formed by a valley glacier. The cirque is where the valley glacier began. When a change in climate causes a valley glacier to disappear, the snowfields that once filled the cirque may melt. The cirque may thus fill with water, becoming a *cirque lake*. One example of a cirque lake is Lake Louise in the Canadian Rockies.

When two cirques are formed next to each other on a peak, the divide between them may be narrow and sharp. Such a divide is called an *arête* (uh-RAYT). When three or more cirques cut into the same peak, they may cut away so much that they leave behind a spectacular, pyramid-shaped peak, which is called a *horn*. The Matterhorn, a mountain on the border of Switzerland and Italy, is one such example.

YOSEMITE NATIONAL PARK Bridalveil Fall in California is an example of a hanging-valley waterfall.

DIFFERENTIATING INSTRUCTION

Hands-On Demonstration Demonstrate the formation of some or all of the features described on page 324. For example, demonstrate the transformation of a river valley system into a glacial valley system using salt clay or damp sand and an ice cube with gravel frozen into the bottom. Use the salt clay or sand to create a river valley landscape with several tributary valleys. Slowly rub the gravelly ice cube down the main valley several times to widen and deepen it. Use smaller ice cubes or pieces of the original cube to "erode" the tributary valleys less than the main valley, creating hanging valleys. Use the ice to gouge out cirques and form arêtes and horns.

In the United States, valley glaciers have left their mark in high mountain ranges such as the Rocky Mountains and the Sierra Nevada of California. These mountain ranges were glaciated in the recent past.

Glacial Erosion

1. Before glaciation, mountains have a relatively smooth, rounded relief.
2. Glaciers advance down the valleys, eroding rock materials. (Cirque, Glacier)
3. Glacial erosion results in U-shaped valleys and steep, jagged formations. (Horn, Arête, Cirque lake, U-shaped valley, Hanging valley waterfall)

Erosion by Continental Glaciers

Like valley glaciers, continental glaciers remove loose rock and soil. They smooth, striate, and groove bedrock. Erosion by continental glaciers, however, differs in a key way from erosion by valley glaciers. Because a continental glacier covers most mountaintops, it grinds down the peaks and leaves them polished and rounded. This type of landscape is familiar in the northeastern United States and in Canada. Valley glaciers, by contrast, sharpen mountain peaks by grinding away at their sides.

15.2 Section Review

1. What are the two main processes by which glaciers move?
2. What is a crevasse, and how does it form?
3. What causes a glacier's ice front to recede? to advance?
4. How does erosion by continental glaciers differ from erosion by valley glaciers?
5. **CRITICAL THINKING** What evidence would you look for to tell whether valley glaciers had once existed in a mountainous region?
6. **HISTORY** In the early 1800s, geologists noted that many outcrops in northern regions of Europe featured polished and scratched surfaces. Boulders and sediments in these regions differed in composition from the underlying bedrock. Some boulders could be traced to outcrops far to the north. Some scientists hypothesized that a flood had scoured the outcrops and transported the boulders. Give another hypothesis. What additional evidence would support your hypothesis?

CHAPTER 15 SECTION 2

ASSESS

1. basal slip and plastic flow
2. A crevasse is a great fissure or crack that forms across the width of a glacier. It forms when a valley glacier comes to a steep downward slope and the rigid ice near the glacier's surface breaks.
3. A glacier recedes when more ice melts in summer than accumulates in winter. A glacier advances when more ice accumulates in winter than melts in summer.
4. Valley glaciers erode valley walls and floors to create U-shaped valleys. They sharpen mountain peaks by grinding away at their sides. Continental glaciers grind down mountain peaks and leave them polished and rounded.
5. sharpened mountain peaks, U-shaped valleys, hanging valleys, cirques, arêtes, or horns
6. A glacier probably picked up and transported the boulders. Glaciers also polish surfaces and scratch them to form striations.

MONITOR AND RETEACH

If students miss . . .

Question 1 Reteach "Basal Slip and Plastic Flow." (pp. 321–322) Have students contrast the two types of movement.

Question 2 Discuss crevasse formation in valley glaciers.

Question 3 Reteach "Movement at the Ice Front." (p. 322) Have students take turns describing what causes glacial advance and retreat, and the conditions that result in a stationary ice front.

Question 4 Reteach erosion by valley glaciers and continental glaciers. (pp. 324–325) In a class discussion, compare and contrast erosion by each type of glacier.

Question 5 Have students review page 324 and list landscape features characteristic of erosion by valley glaciers.

Question 6 Review how glaciers move loose rock and create striations. (p. 323)

CHAPTER 15 SECTION 3

▶ FOCUS

Objectives
1. Describe two types of glacial deposits.
2. Describe landscape features characteristic of glacial deposits.

Set a Purpose
Read aloud the key idea. Then have students read the section to answer the question "How does glacial deposition alter the landscape?"

▶ INSTRUCT

More about...

Drumlins
The word *drumlin* is derived from the Irish Gaelic word *druim*, which means "ridge." Drumlins commonly occur in groups called swarms. They are particularly common in the Northeastern United States, Wisconsin, Nova Scotia, and Ireland. Partly submerged drumlins form islands in Boston Harbor. Perhaps the most famous drumlin is Bunker Hill in Massachusetts, although the revolutionary war Battle of Bunker Hill was actually fought on the adjacent drumlin called Breed's Hill.

15.3

KEY IDEA
Glaciers deposit materials in characteristic features that alter the landscape.

KEY VOCABULARY
- outwash
- erratic
- drumlin
- outwash plain
- esker
- kame
- kettle

Glacial Deposits

When a glacier melts, it leaves behind deposits of eroded rock materials. These deposits are called drift. Recall that unsorted and unstratified rock material deposited directly by glacial ice is called *till*. Materials deposited by glacial meltwater are called **outwash**. Like all stream deposits, outwash is sorted and stratified.

Glacial deposits are left behind as features that alter a landscape. These features include moraines, drumlins, outwash plains, eskers, kames, and kettles. Glacial deposits are also associated with certain types of lakes.

Moraines and Drumlins

When a glacier melts, its rock load remains in nearly the same places as in the glacier. When the glacier retreats, some of the rock material is left behind in deposits of till called *moraines*.

A large, fairly even blanket of till is called *ground moraine*. Materials in a ground moraine were deposited when the ice at the bottom of a glacier melted. Lateral and medial moraines form ridges running in almost the same direction as the glacier's advance.

At a glacier's ice front, rock materials brought forward by the glacier pile up as the ice melts, creating an end moraine. The longer the ice front stays in one place, the larger the end moraine there becomes. When a receding ice front stops in new places for any length of time, new end moraines are formed behind the first one. These moraines are called recessional moraines. The end moraine marking a glacier's farthest advance is called its terminal moraine.

Even a stationary ice front moves back and forth. End moraine deposits are spread over a broad belt in front of a glacier. Furthermore, no two parts of the ice front deposit exactly the same amount of material.

ESKER The long ridge in this photograph is an esker near Dahlen, North Dakota.

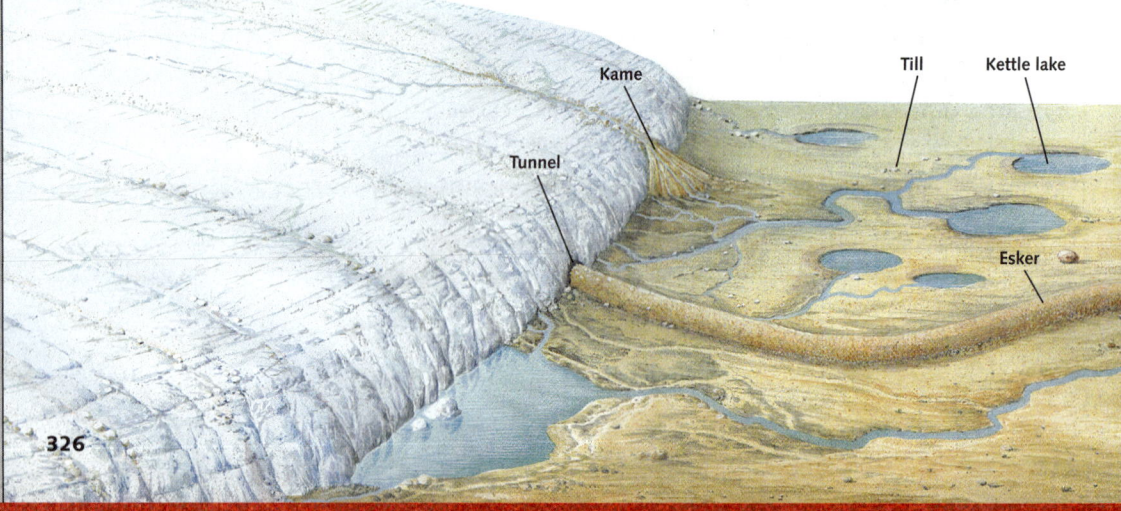

This illustration of glacial deposits does not represent any specific geographic area.

326

DIFFERENTIATING INSTRUCTION

Reading Support Remind students of the importance of connecting what they are reading with what they have read before. Explain that scientific topics are interconnected, and what was read previously often provides a foundation to help understand new concepts. As an example, tell students that in order to understand how outwash features form, they must understand stream transport and deposition, which they read about in Chapter 13, Section 2. Have students reread this section. Discuss how connecting to prior knowledge helps them understand topics such as sorting and stratification of outwash deposits, and the formation of eskers and kames.

Use Reading Study Guide, p. 52.

326 Unit 4 Earth's Changing Surface

For these reasons, terminal and recessional moraines are likely to have irregular hills and hollows, rather than being straight ridges. End moraines of continental glaciers may be hundreds of kilometers long, tens of kilometers wide, and hundreds of meters high.

The materials of a moraine range from boulders to clays. Large boulders that have been transported into an area by a glacier are called **erratics** (ih-RAT-iks). The composition of erratics differs from that of the surrounding bedrock.

Drumlins (DRUHM-lihnz) are long, smooth, canoe-shaped hills that are usually found in groups. Drumlins point in the direction of glacial movement. They are thought to form when an advancing glacier runs over an earlier glacial moraine, sweeping it into long strips. Notable groupings of drumlins are found in eastern Wisconsin and western and central New York State.

Outwash Plains and Eskers

Glacial meltwater pours out at the ice front in streams filled with rock flour, sand, and gravel. These streams deposit outwash that may extend for many kilometers beyond the terminal moraine. The deposits look like alluvial fans. In front of large glaciers, they overlap and form broad, flat areas called **outwash plains.**

Much of the water of a melting glacier falls to the base of the glacier through crevasses. Subglacial streams are formed that run in tunnels beneath the ice and come out at the ice front. The winding tunnels of these streams become partly filled with roughly stratified sands and gravel. When the glacier melts, these deposits slump down at the sides and form long, winding ridges called **eskers.** Eskers can be found in the Mississippi Valley, the north-central states, New York, and Maine.

DRUMLINS Glaciers formed these drumlins in Yorkshire, United Kingdom.

ERRATICS These large boulders were deposited by glaciers in what is now part of Yellowstone National Park.

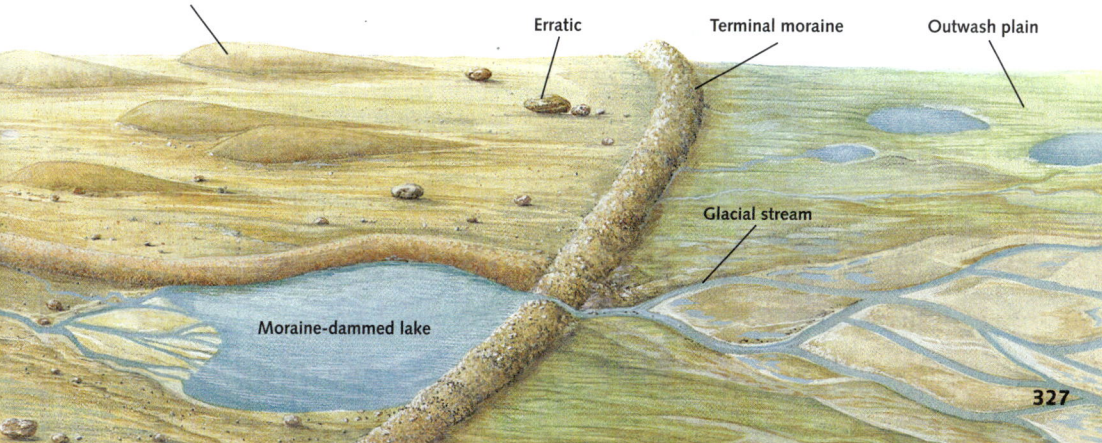

CHAPTER 15 SECTION 3

VISUAL TEACHING
Discussion
Have students compare the kettle lakes in the aerial photograph with those in the diagram on pages 326 and 327. Ask: **How could you distinguish kettle lakes from other lakes, such as those associated with karst topography?** look for other nearby outwash features such as kames, eskers, and erratics

ASSESS

1. Till deposits are unsorted, unstratified, and deposited directly by glacial ice, whereas outwash deposits are sorted, stratified, and deposited by glacial meltwater.

2. A terminal moraine represents a glacier's farthest advance, whereas a recessional moraine represents where a receding ice front stopped for a length of time.

3. Drumlins are thought to form when an advancing glacier runs over an earlier glacial moraine, sweeping it into long strips.

4. An outwash plain is a broad, flat area of glacial meltwater deposits formed in front of large glaciers.

5. Drumlins are made of till, so a geologist should look for unsorted, unstratified deposits. If the deposits are sorted and stratified, they are more likely kame deposits.

6. Kettle lakes form when kettle holes of moraines and outwash plains fill with water, whereas cirque lakes form when bedrock cirques fill with water. Kettle lakes occur on outwash plains, and cirque lakes occur in glaciated mountain valleys or peaks.

Kames and Kettles

Kames (KAYmz) are small, cone-shaped hills of stratified sand and gravel. They form when streams flowing across the top of a glacier deposit their sediments at the ice front or into lakes on top of the ice. The sediments pile up as the ice thins. When the ice melts completely, sediments that had been heaped up against it or resting on top of it slump to the ground. Like the sediments in eskers, deltas, and outwash plains, those in kames are deposited by water, not ice.

Meltwater can also collect at the sides of a retreating valley glacier. The sediment deposited there may form long hills attached to the valley walls. These hills are called kame terraces. Kame terraces may look like lateral moraines. However, the sediments in a kame terrace are composed of outwash, while those in a lateral moraine are composed of till.

Kettles are bowl–shaped hollows found in moraines and outwash plains. Kettles form where buried blocks of ice have been left behind by a retreating glacier. When the ice melts, the water drains, leaving the kettles.

Glacial Deposits and Lakes

Two types of lakes that are associated with glacial deposits are moraine-dammed lakes and kettle lakes.

Moraine-dammed lakes form where river valleys are blocked by glacial moraines. The river rises and floods its valley, creating a long, usually narrow lake. Many of the larger lakes of the northern United States were formed in this way. Lake George in New York is an example.

Some lakes were formed both by glacial erosion, which scoured out river valleys, and by the accumulation of glacial deposits, which dammed the rivers. The Finger Lakes of New York State are an example. Many of their former tributary valleys were left as hanging valleys above the main glacial valleys that became the lakes.

Kettle lakes form in large numbers in the kettle holes of moraines and outwash plains. They are filled with either rainwater or surface runoff. Kettle lakes and kettle ponds have no natural inlet or outlet, thus many of them fill up with vegetation and become bogs. Kettle lakes are common in Minnesota, Michigan, Wisconsin, New York, and New England.

ALASKA Large blocks of glacial ice embedded in till melted, leaving holes called kettles. Later, the kettles filled with water, as shown in this aerial photograph.

15.3 Section Review

1. How do till deposits differ from outwash deposits?
2. How is a terminal moraine different from a recessional moraine?
3. Give one explanation of how a drumlin might form.
4. Describe an outwash plane.
5. **CRITICAL THINKING** Suppose a geologist is trying to decide whether a glacial deposit is a kame or a drumlin. What evidence should the geologist look for?
6. **GEOGRAPHY** Compare how kettle lakes form with how cirque lakes form. Where are you likely to find each type of lake?

MONITOR AND RETEACH

If students miss . . .

Question 1 Reteach the characteristics and formation of till and outwash. (p. 326)

Question 2 Review the different types of moraines and how they form. (pp. 326–327)

Question 3 Reteach "Moraines and Drumlins." (pp. 326–327) Ask: What are drumlins and how do they form?

Question 4 Have students describe in their own words the outwash plain pictured on pages 326 and 327.

SCIENCE & Technology

Rivers of Ice

Antarctica, which has a surface that is 97 percent ice, covers an area larger than the United States and Mexico combined. Until recently, large swaths of the continent remained unmapped. New technology has made it finally possible to produce a complete, highly detailed map.

What does the new map reveal about the world's iciest continent?

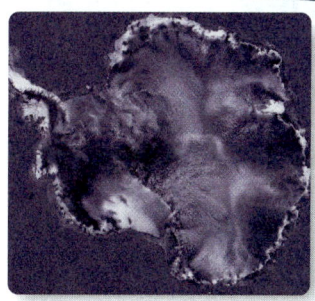

ANTACTICA is shown in a radar image obtained from the RADARSAT satellite.

In 1995, NASA launched the Canadian satellite RADARSAT. One of many goals set for RADARSAT was the collecting of images for a new map of Antarctica. Unlike those satellites that depend on sunlight to illuminate the surface they are imaging, RADARSAT provides its own microwave illumination, and its radar signals can penetrate the dense cloud cover that often shrouds parts of Antarctica. RADARSAT captured a complete set of images of Antarctica in just 18 days. A previous, less accurate satellite-image map took years to complete.

Some of the most intriguing features shown on the map are the ice streams that slice through the Antarctic ice sheet. The map reveals just how immense some of these riverlike glaciers are—up to 800 kilometers in length and up to 50 kilometers wide. Ice streams move along at the very speedy clip of about 900 meters per year. The ice streams are 100 times faster than the ice sheet they cut through. Like conveyor belts, they transport snow and ice from the continent's interior to the sea. Scientists estimate that one ice stream moves about 80 cubic kilometers of ice each year—enough to bury the city of Manhattan under nearly a kilometer of ice.

Scientists are currently trying to determine what role ice streams play in the advance and retreat of the West Antarctic ice sheet and why they sometimes slow down and even flow backward. One theory suggests they play a regulatory role—keeping the ice sheet from growing so large that it would collapse under its own weight. ■

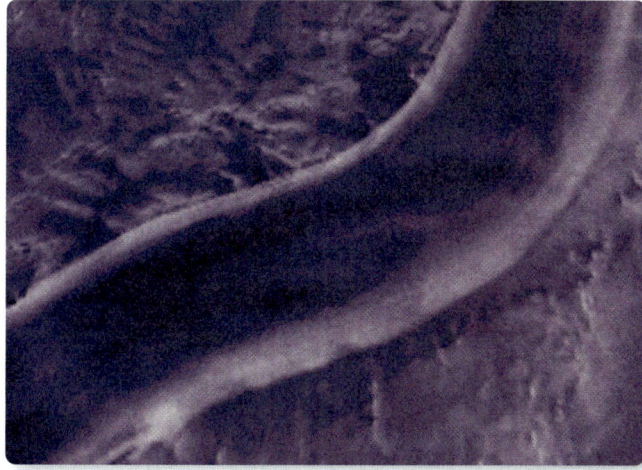

THE RECOVERY GLACIER, one of the principal channels composing the East Antarctic ice stream shown below, reaches more than 800 kilometers into the continent's interior.

Extension

RESEARCH ACTIVITY

RADARSAT's technology has allowed researchers to gather more information about Antarctica than ever before possible. What are some other ways RADARSAT could be used to study Antarctica? What would you like to know about Antarctica?

Learn more about Antarctic research.
Keycode: ES1503

CHAPTER 15 SECTION 3

Learn more about Antarctic research.
Keycode: ES1503

Science and Technology

Research in Antarctica
The world's nations have set up 30 scientific research stations in Antarctica. The United States has three sites: Amundsen-Scott South Pole Station, as well as McMurdo Station and Palmer Station on the coast. McMurdo is the largest facility, with a population of about 1100 scientists, technicians, and other specialists in summer. In winter, the population falls to fewer than 200. A variety of scientific research occurs in Antarctica. Geologists collect rock samples to piece together Earth's early history. Biologists observe wildlife along the coasts. Research on the effects of air pollutants on Earth's ozone layer has been conducted for several years.

Research Activity
RADARSAT might be used to make maps that compare the sizes of Antarctic ice sheets over time. Students might want to know about the work being done at the Antarctic research stations, the types of organisms that live in Antarctica, or how the nations of the world share the continent.

MONITOR AND RETEACH

Question 5 Have students reread "Moraines and Drumlins" (pp. 326–327) and "Kames and Kettles." (p. 328) Ask: What types of deposits make up drumlins and kames?

Question 6 Distinguish between erosional features of valley glaciers (p. 324) and depositional features of outwash plains. (p. 328)

CHAPTER 15 SECTION 4

▶ FOCUS

Objectives
1. Describe the ice ages that Earth has experienced and the evidence they left behind.
2. Summarize several hypotheses for the causes of the ice ages.
3. Explain how glaciation contributed to the formation of the Great Lakes.

Set a Purpose
Read the section title aloud. Ask students, "What is an ice age and what causes one?" Have students read the section to find out if their answers to the question was correct.

▶ INSTRUCT

Visual Teaching

Interpret Diagram
Have students estimate the southernmost latitude to which glacial ice extended during Earth's most recent ice age as indicated on the map. **approximately 37°N**

15.4

KEY IDEAS
Earth has experienced ice ages in its past. Glaciation in the most recent ice age led to the formation of the Great Lakes.

KEY VOCABULARY
- ice age

INVESTIGATIONS
CLASSZONE.COM
How Does Land Cover Affect Global Temperature? Predict how changing land cover might affect the planet's temperature balance.
Keycode: ES1504

Ice Ages

Glaciers leave clues about the past in the form of erosional and depositional landforms. By studying these clues, geologists have seen that Earth has gone through periods of extensive glaciation (GLAY-shee-AY-shun) in the past. These periods of glaciation are called **ice ages.** Such ice ages are associated with periods of global cooling.

Periods of Glaciation

Evidence shows that Earth has experienced a number of ice ages, some of which occurred hundreds of millions of years ago. Some have occurred more recently.

Beginning almost 2 million years ago and lasting until about 11,000 years ago, Earth underwent extensive periods of glaciation. Huge ice sheets advanced and retreated many times as Earth's climate alternately cooled and then warmed. As shown below, the ice sheets reached their greatest size about 20,000 years ago, when they extended through North America as far south as southern Illinois and eastward to central Long Island. Ice covered much of what is now the north-central and northeastern parts of the United States. In Europe, ice sheets covered most of Scandinavia, the British Isles, Belgium, and the Baltic countries, and reached far into Germany and Russia.

Could Earth experience another ice age? The last time the glaciers retreated was about 11,000 years ago. It is possible that Earth is now in a warm, interglacial period that will be followed by a return of the ice sheets in perhaps 10,000 years. Some scientists have hypothesized that the current warm period will last as long as 60,000 years.

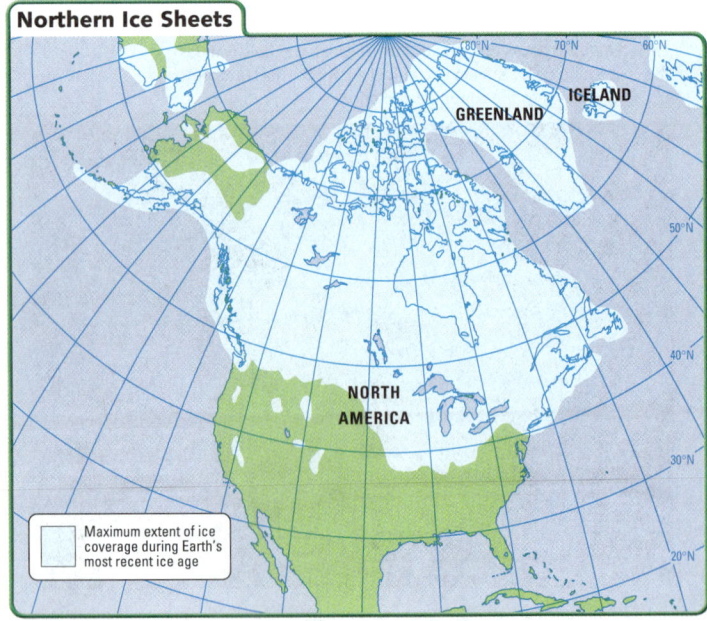

Northern Ice Sheets

Maximum extent of ice coverage during Earth's most recent ice age

330 Unit 4 Earth's Changing Surface

DIFFERENTIATING INSTRUCTION

Reading Support Remind students that they can often infer the meaning of unfamiliar science terms by analyzing word parts. Point out the term *glaciated* in the first paragraph on page 331. Write the word on the board, circle the root word *glaciate*, and point out its similarity to the word *glacier*. Explain that *glaciate* is a verb that means "to cover with ice or a glacier." Thus, a *glaciated area* is one that is covered by ice or a glacier.
Use Reading Study Guide, p. 53.

Evidence for Ice Ages

Evidence for relatively recent ice ages is given by the many erosional and depositional features. Drift, for example, covers much of Canada and the northern part of the United States. Erratics are found in glacial-till soils. Exposed bedrock is striated and polished, even on mountaintops. Mountain valleys have been eroded into glacial valleys. Deposits of till— unsorted, unstratified rock material—are found in many places. Lakes and swamps are far more common in glaciated areas. In the Rocky Mountains and the Sierra Nevada, markings caused by glacial erosion occur high up on valley walls. Geologists infer from such markings that the glaciers which occupied these valleys during past ice ages were much larger than the glaciers found in the region today.

Terminal moraines mark the southern limit of ice sheet advancement in North America. One of these terminal moraines is the Long Island, New York, moraine that extends almost 225 kilometers from Brooklyn to Montauk Point. Other terminal moraines stretch from New Jersey through Pennsylvania, Ohio, and Indiana and westward to Puget Sound, in the state of Washington. These terminal moraines are not all connected, nor can they be traced continuously across the continent. Outwash plains are found in many places south of the terminal moraines.

Evidence for ancient ice ages is found in ancient glacial sediments that have hardened into rock. When a deposit of till cements together it forms a rock is called *tillite*. Tillites, some as old as 2.3 billion years, have been found all over Earth. Studies of tillites indicate that a number of ice ages took place hundreds of millions of years ago.

Observe the retreat of ice sheets from North America.
Keycode: ES1505

Causes of Ice Ages

Geologists have proposed many hypotheses to account for ice ages. Most hypotheses identify factors that may have caused Earth's surface to cool enough for glaciation to occur. Such factors include:

- changes in the position of Earth relative to the sun
- changes in atmospheric carbon dioxide levels
- changes in the positions of Earth's landmasses
- changes in the amount of solar energy reaching Earth's surface.

It is likely that more than one of these factors contributes to global cooling. Thus, a combination of the factors may cause an ice age.

Earth's Position

Changes in Earth's orbit and in the tilt of its axis also affect global temperatures. Earth's elliptical orbit around the sun varies slightly over time, so that every 100,000 years Earth is at its farthest point from the sun. The tilt of Earth's axis also changes over time, causing seasonal differences to intensify every 41,000 years. Finally, because Earth's axis wobbles, every 23,000 years, Earth's axis points in such a way that the poles experience extremes in solar radiation. The result is that the polar regions and areas at higher latitudes receive even less solar energy than usual in the winter, and far more solar energy than usual in the summer.

Observe changes in Earth's orbit that contribute to climate change.
Keycode: ES1506

CHAPTER 15 SECTION 4

How Does Land Cover Affect Global Temperature?
Keycode: ES1504

Use Internet Investigations Guide, pp. T55, 55.

Observe the retreat of ice sheets from North America.
Keycode: ES1505

Have students work in pairs to view the visualization and summarize what they observe.

Observe changes in Earth's orbit that contribute to climate change.
Keycode: ES1506

Have students explain how the orbital cycles and changes shown are related to climatic changes.

 Visualizations CD-ROM

More about...

Earth's Orbit
The shape of Earth's orbit around the sun has a degree of eccentricity. That is, over a period of about 100,000 years, the shape varies from nearly circular, as it is currently, to more elliptical. As the orbit becomes more elliptical, the distance to the sun increases for most of the year, resulting in less solar radiation reaching Earth, possibly cooling global temperatures.

DIFFERENTIATING INSTRUCTION

Challenge Activity Have students do library or Internet research to find out about the ice ages that occurred on Earth hundreds of millions of years ago. Students can present their findings to the class in an oral report. Their reports should include a description of the evidence scientists use to infer the occurrence of ancient ice ages and an estimate of where and when continental glaciers covered Earth.

CHAPTER 15 SECTION 4

Scientific Thinking

COMMUNICATE
Students' posters should indicate that driving gas-powered cars, flying airplanes, and producing electricity at coal-burning plants all increase the amount of carbon dioxide in the atmosphere because the burning of fossil fuels releases carbon dioxide. Clear-cutting forests increases carbon dioxide levels because plants take in carbon dioxide from the atmosphere as they undergo photosynthesis.

More about...

Causes of Ice Ages
Some scientists think that the occurrence of volcanic eruptions over several centuries could trigger the advance of ice sheets. While volcanic eruptions increase the amount of carbon dioxide in the atmosphere, which could have a warming effect on Earth, the dust released during volcanic eruptions reflects solar energy and thus could contribute to the lowering of global temperatures.

Scientific Thinking

COMMUNICATE
What is the best way to list and evaluate the ways in which human activities influence the amount of carbon dioxide in the atmosphere?

Work with several research partners to design a poster that explains how the following activities affect carbon dioxide levels in the atmosphere:

- driving gas-powered cars; compare an SUV to a compact
- flying airplanes; find out how much CO_2 a jet engine emits
- producing electricity at coal-burning plants
- clear-cutting forests.

In the 1920s, a scientist named Milutin Milankovitch noticed that ice ages tended to occur in cycles that could be linked to these changes in Earth's tilt and orbital position. The idea that these combined cyclic changes in Earth's position are the underlying cause of ice ages is now called the *Milankovitch Theory*.

Carbon Dioxide

Carbon dioxide in the atmosphere keeps Earth warm by absorbing the heat that Earth radiates and releasing it in random directions. Some of the heat radiates from Earth directly to space, but much of the heat stays in the atmosphere or radiates back to Earth. Carbon dioxide levels have dropped in the past, however, and this drop could have led to global cooling, as more heat would have been able to radiate from Earth's surface to space.

Mountain building is one geologic process that may affect carbon dioxide levels. During periods of extensive mountain building, granite and other rocks are uplifted and exposed. Carbon dioxide (CO_2) gas from the atmosphere combines with rainwater to form *carbonic acid* (H_2CO_3). This weak acid weathers feldspar, for example, to form potassium ions (K^+), bicarbonate ions (HCO_3^-), and silica. Rivers and streams transport these new minerals until they are eventually deposited on the ocean floor. In this way, the carbon dioxide gas that reacted with exposed rocks is removed from the atmosphere, as long as the sediments remain on the ocean floor.

Positions of Earth's Landmasses

Changes in the positions of Earth's landmasses may account for the occurrence of some ice ages. The movement of landmasses to higher latitudes, for example, is associated with ice age events.

The positions of Earth's landmasses also affects changes in ocean current patterns. Because they mix warm and cold waters, ocean currents play an important role in regulating global temperatures. A change in the positions of Earth's continents could have inhibited the mixing of cold and warm water. Without this mixing, some regions might have become cold enough for glaciers to form.

Solar Energy

The amount of solar energy absorbed by Earth's surface affects global temperatures. The amount of energy that is absorbed depends, among other things, on snow coverage, because snow reflects solar energy back to space. During periods of mountain building, more land area rises above the snow line. Increased snow coverage could have meant that more solar energy was reflected back into space, leading to global cooling.

The Great Lakes

Continental glaciers cause enormous changes to Earth's surface as they advance and retreat. In North America, glaciation during the most recent ice age contributed to the formation of the Great Lakes. More than 1 million years ago, the region where the Great Lakes now lie was marked by river valleys which flowed through the easily eroded sandstones and

---

DIFFERENTIATING INSTRUCTION

Hands-On Demonstration Demonstrate that white objects, such as snow, reflect more light energy than do dark objects, such as soil. (Students will do a similar activity in Chapter 17.) Obtain two thermometers that have been kept at room temperature, and record the temperature indicated on each. Then place one thermometer under black paper and the other under white paper. Use a desk lamp to shine light evenly over both pieces of paper. Record the temperature on each thermometer every 10 minutes for 30 minutes. Ask students: Under which paper did the temperature increase less? **under the white paper** What caused this difference? **The white paper reflected more of the light energy than did the black paper.**

⚠ *Do not use mercury thermometers. Remind students that safety is everyone's responsibility. See Safety in the Earth Science Laboratory at the front of this book, pages xxii–xxiii.*

shales. The advancing ice sheet widened and deepened the valleys many times, thus depositing moraines to the south.

Around 14,000 years ago, the ice sheet started to melt. The weight of a glacier that was three kilometers thick had depressed the Earth's crust by as much as one kilometer, tilting up the land surface toward the north. Meltwater pooled in front of the ice sheet, thus forming the early Great Lakes to the south. Since ice blocked river outlets to the north, the rivers drained the lakes to the south, into what is now the Mississippi River.

As the ice sheet continued to melt, more and more water accumulated in the lakes. The lake rose and connected with the Mohawk, Hudson, and Susquehanna rivers. These rivers drained the lakes into the Atlantic Ocean.

By about 7000 years ago, the ice sheet had melted past Lake Erie, Lake Ontario, and Lake Michigan. The removal of the weight of the ice caused the land in the southern part of the region to rebound. That is, the land rose closer to its pre-glaciation elevation. The St. Lawrence River Valley in the northeast was freed of ice. The land surface still tilted north, so the flow of water through the lakes reversed direction, creating Niagara Falls between Lakes Erie and Ontario. Since then, all of the water from the Great Lakes has drained through the St. Lawrence River into the Atlantic Ocean.

Formation of the Great Lakes

14,000 YEARS AGO Meltwater pools in front of the melting ice sheets. Rivers form, draining meltwater to the south into what will become the Mississippi.

7000 YEARS AGO As the ice sheet melts and recedes, meltwater fills the lakes and drains westward to the Atlantic through the St. Lawrence River valley.

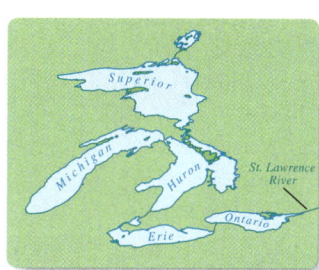

TODAY The Great Lakes drain into the Atlantic Ocean through the St. Lawrence River.

15.4 Section Review

1. What evidence is there for recent and long-past ice ages?
2. Describe at least four factors that could contribute to global cooling.
3. What role did glaciation play in the formation of the Great Lakes?
4. **CRITICAL THINKING** How was gravity important in the formation of the Great Lakes?
5. **BIOLOGY** Plants absorb carbon dioxide from the atmosphere. How might a significant increase or decrease in the amount of Earth's surface covered by vegetation affect global temperatures? Explain.

CHAPTER 15 SECTION 4

Visual Teaching

Interpret Diagrams
Ask: What was the original source of water that formed the Great Lakes? **glacial meltwater** How did lake drainage change? **Originally the lakes drained to the south, and later, as the ice melted more, they drained to the north.**

ASSESS

1. Evidence for recent ice ages includes erosional and depositional features, such as drift, erractics, striations, U-shaped valleys, terminal moraines, and outwash plains. Evidence for long-past ice ages is found in ancient glacial sediments that have hardened into rock to form tillites.

2. Factors include changes in the position of Earth relative to the sun, changes in atmospheric carbon dioxide levels, changes in the positions of Earth's landmasses, and changes in the amount of solar energy reaching Earth's surface.

3. An ice sheet advancing in the area widened and deepened existing river valleys and depressed the crust. As the ice melted, water pooled in front of it, forming the Great Lakes. With continued glacial melting, the lake levels rose and the land rebounded, reversing lake drainage toward the north.

4. Gravity causes water to flow downslope, so it affected the direction in which the Great Lakes drained.

5. Since plants absorb carbon dioxide, a significant increase in vegetation would increase absorption of carbon dioxide and decrease atmospheric carbon dioxide, leading to global cooling. A significant decrease in vegetation would decrease absorption of carbon dioxide and increase atmospheric carbon dioxide, leading to global warming.

MONITOR AND RETEACH

If students miss . . .

Question 1 Reteach "Evidence for Ice Ages." (p. 331) Have students list different kinds of evidence that suggests ice ages have occurred in the past.

Question 2 Have students review the bulleted list on page 331.

Question 3 Use the diagram on page 333 to discuss the formation of the Great Lakes.

Question 4 Reteach "The Great Lakes." (pp. 332–333) Discuss the role gravity plays in the movement of water.

Question 5 Reteach "Causes of Ice Ages." (pp. 331–332) Discuss how carbon dioxide levels in the atmosphere impact Earth's absorption of heat.

CHAPTER 15 ACTIVITY

PURPOSE
To model the movement of a glacier and its impact on the landforms over which it moves

MATERIALS
The plastic box should be approximately 12" × 6" × 4".

If the cornstarch mixture that students make in Step 5 is too runny, the index cards will not hold back the mixture's flow in Step 7. To better ensure the correct consistency of the cornstarch mixture, prepare a large batch of it just before the lab and have students take handfuls.

PROCEDURE
8, **10** Tell students to be careful when piling the mixture and removing the index card "dam." They must do so evenly to prevent the mixture from being pulled to one side or the other.

Do not dispose of the cornstarch mixture in sinks. Scoop it into garbage bags.

ANALYSIS AND CONCLUSIONS
1. basal slip; The cube slides down the incline on a thin layer of meltwater.

2. The ice cube needs a thin layer of water beneath it to slide. Without this layer, the ice cube would not overcome friction and slide down the incline.

3. The larger the magnitude of the box's tilt, the faster the ice cube should slide down the incline. As a test, students might gradually increase the box's tilt and time the movement of the ice cube down the incline.

4. plastic flow; because there is no thin layer of water at the base of the cornstarch mixture but the entire mass still moves downhill

5. The cornstarch mixture is easier to study because: it is much smaller, it exists at room temperature, it flows

334 Unit 4

CHAPTER 15 LAB Activity

Modeling Glacial Movement

SKILLS AND OBJECTIVES
- **Model** two types of glacial movement.
- **Interpret** the results.
- **Modify** the model and predict the outcome.
- **Design** alternative experiments.

MATERIALS
- pencil
- plastic box
- ice cube
- towel
- 16-oz. box of cornstarch
- 2-qt. mixing bowl
- 1 to 2 cups of water
- spoon
- 5 × 7 in. index cards
- 5 wooden toothpicks
- 5 or 6 large pebbles

Basal slip and plastic flow are the two main types of glacial movement. In this activity, you will model both of these types. You will also consider how to model other aspects of glacial movement.

Procedure

Part A: Glacial Movement Model A

1. Lay the pencil flat on a table. Place one end of the plastic box on top of the pencil to give the box a slight tilt.

2. Wrap the ice cube in the towel and rub it between your hands until it is slightly melted. Unwrap the ice cube, place it at the raised end of the box, and observe what happens. Record your observations.

3. Remove the ice cube from the box and make sure the box is completely dry.

4. Answer Questions 1–3 under Analysis and Conclusions.

Part B: Glacial Movement Model B

5. Pour the box of cornstarch into the mixing bowl. Slowly add water while stirring with a spoon until the mixture has the consistency of toothpaste. It should not be runny or wet.

6. Pour the cornstarch mixture into the raised end of the plastic box and observe what happens. Record your observations.

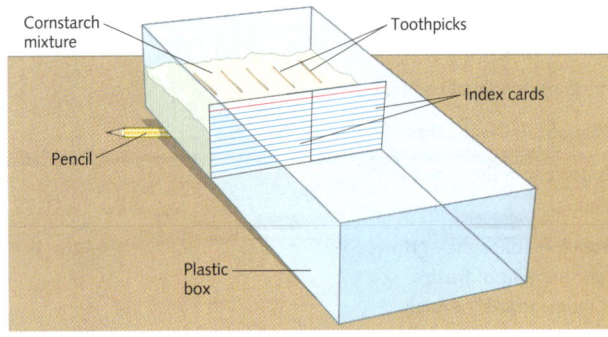

7. After the mixture has flowed through the entire box, scrape it up with your hands and pile it in the raised end of the box. Use the index cards to function as a dam across the valley to hold the mixture back, as shown in the diagram.

8. Lay the five toothpicks across the front of the mixture so that they are one inch apart and parallel to each other, as shown. Remove the dam and observe the way the toothpicks move as the glacier flows. Record your observations.

9. Repeat Step 7.

334 Unit 4 Earth's Changing Surface

10 Place a few pebbles in the path of the cornstarch mixture to create obstructions. Remove the dam and observe how the glacier interacts with the pebbles. Record your observations.

11 Answer Questions 4–9 under Analysis and Conclusions.

Analysis and Conclusions

Part A: Glacial Movement Model A

1. What type of glacial movement does this model represent? Explain.

2. Why do you think you are asked in Step 2 to slightly melt the ice cube? Hypothesize the results of Step 2 if you do not melt the ice cube before placing it in the raised end of the plastic box.

3. How might the magnitude of tilt of the box affect the results of Step 2? Design an experiment to test your hypothesis.

Part B: Glacial Movement Model B

4. What type of glacial movement does this model represent? Explain.

5. Compare the behavior of the cornstarch mixture to that of glaciers. List three reasons why the cornstarch mixture model is easier to study than an actual glacier is. List three ways in which the cornstarch mixture is unlike glacial material.

6. Compare and contrast the results of Steps 6, 8, and 10.

7. What do you think would happen if you tilted the shoebox more? In particular, how would the movement of the toothpicks in Step 8 change? Would the interaction of the cornstarch mixture with the pebbles be any different?

8. What does the movement of the toothpicks tell you about the speed of flow at different parts of a glacier? Design an experiment that would allow you to measure the speed of flow at different depths.

9. As glaciers move, they pick up material. The embedded material then scours the landscape, eroding Earth's surface. How could this activity be modified to allow the investigation of glacial scouring?

CHAPTER 15 ACTIVITY

Learn more about glaciers and glacier movement.
Keycode: ES1508

(continued)

more quickly. The cornstarch mixture and glacial material differ in that: they are made of different substances; the glacier has more water; chemical bonds hold the cornstarch mixture together, whereas the particles of ice in the glacier form a mass because of extreme pressure.

6. The mixture flow during Step 6 is unimpeded. The flow during Step 8 is similar, though the removal of the index cards may pull at the mixture. The toothpicks help show how the mixture is flowing at various points along its front. In Step 10, the mixture at first tries to ooze around the pebbles, but then flows over them. Once the mixture has flowed over the pebbles, its movement is basically the same as under the other conditions tested in this lab.

7. The mixture would flow more quickly. The toothpicks would move farther and be more spread out. The pebbles might be carried along with the cornstarch mixture.

8. The toothpicks indicate that, at the surface, the center of a glacier moves faster than the sides. To measure the speed of flow at different depths, toothpicks can be placed in a horizontal line up the center from near the bottom of the box to near the surface of the mixture.

9. **Sample Response:** Make ice cubes with particles of sand or gravel frozen in them and line the shoebox with a material, such as clay or soap, that can be scratched or scoured by something flowing over it.

Refer students to Safety in the Earth Science Laboratory on pages xxii–xxiii.

Courtesy of PBS series, *Newton's Apple*.
www.tpt.org/newtons

CHAPTER 15 REVIEW

Summary of Key Ideas

15.1 Much of Earth's fresh water is frozen in valley glaciers or continental glaciers. Glaciers form above the snow line from compressed snow called firn. The snow line elevation is high in areas close to the equator and gets lower in areas farther from the equator.

15.2 Gravity causes glaciers to move. The ice front of a glacier may advance or retreat depending on conditions. If a glacier's ice front is near the sea, icebergs may break off. As glaciers move, they erode the underlying surface, leaving behind characteristic erosional features.

15.3 When glaciers retreat, they leave behind deposits. Till is unsorted, unstratified material deposited directly by the ice. Outwash is sorted, stratified material deposited by meltwater from the ice.

15.4 Earth has experienced a number of periods of widespread glaciation. Evidence for such ice ages comes from erosional and depositional features. Scientists have identified a number of possible causes for glaciation.

Key Vocabulary

calving (p. 322)
cirque (p. 324)
continental glacier (p. 319)
crevasse (p. 322)
drumlin (p. 327)
erratic (p. 327)
esker (p. 327)
firn (p. 318)
glacial valley (p. 324)
glacier (p. 318)
ice age (p. 330)
ice cap (p. 320)
ice front (p. 322)
kame (p. 328)
kettle (p. 328)
moraine (p. 323)
outwash (p. 326)
outwash plain (p. 327)
snow line (p. 318)
striations (p. 323)
till (p. 323)
valley glacier (p. 319)

336 Unit 4 Earth's Changing Surface

Vocabulary Review

1. calving
2. snow line
3. terminal moraine
4. firn
5. Till is unsorted and unstratified rock material carried and then deposited directly by a glacier. Outwash is material deposited by meltwater; it is sorted and stratified.
6. A valley glacier moves within the walls of a valley. A continental glacier covers a large part of a continent.
7. A kame is a small, cone-shaped hill of sand or gravel deposited at an ice front by streams flowing over a glacier. A drumlin is a canoe-shaped hill formed by a glacier moving over a moraine.

Concept Review

8. Glaciers form in cold areas that are always snow-covered. Buried snow becomes compressed and recrystallizes into firn, which eventually becomes solid ice. The ice mass begins to flow under the weight of overlying snow and firn.
9. Ice at the base moves by basal slip, caused when melting ice reduces friction and allows the mass of ice to slide. Ice at the center moves by plastic flow, in which ice grains slip past each other creating forward movement.
10. An ice front moves if the glacier's rates of movement and melting are unequal. A glacier may move forward faster than normal after a series of winters with heavy snowfalls that add pressure to the bottom of the glacier. The ice front may move back after several warm summers.
11. Valley glaciers scour underlying rock and soil along valley floors and walls. Continental glaciers also scour the rock and soil under them. But unlike valley glaciers that sharpen mountain peaks by grinding away at their edges, continental glaciers cover and grind down mountain peaks, giving them a rounded shape.
12. As glaciers retreated at the end of the ice age, blocks of ice were left behind, buried in moraines and outwash plains. When the ice melted and drained, it left bowl-shaped hollows behind. Some of these filled with water to form kettle lakes.
13. A recessional moraine forms when the front of a receding glacier stops at a new spot and deposits till. A terminal moraine marks the glacier's point of farthest advance before it begins to recede.
14. Causes include: changes in Earth's position that cause polar regions to receive less solar energy than usual in winter; a drop in atmospheric carbon dioxide levels that decreases Earth's ability to hold in heat; cooling caused by movement of landmasses to higher latitudes and resulting changes in ocean currents; increased snow cover due to more land above the snow line that reflects solar energy.
15. from left to right: till, drumlin, esker

Vocabulary Review

Write the term from the key vocabulary list that best completes each sentence.

1. ___?___ produces icebergs.
2. As distance from the equator increases, the elevation of the ___?___ decreases.
3. The farthest advance of a glacier, is marked by the ___?___
4. The rough, granular ice that becomes a glacier is called ___?___ .

Explain the difference between each pair of terms.

5. till, outwash
6. valley glacier, continental glacier
7. kame, drumlin

Concept Review

8. How and where do glaciers form?
9. Compare how a glacier moves near its base with how it moves at its center.
10. Why is the ice front of a glacier not necessarily stationary?
11. Compare how valley glaciers and continental glaciers erode materials.
12. In Minnesota, there are many kettle lakes. Explain how these lakes formed.
13. How is a recessional moraine different from a terminal moraine?
14. Describe several possible causes of ice ages.
15. **Graphic Organizer** Copy and complete the concept map below using appropriate terms from the key vocabulary list.

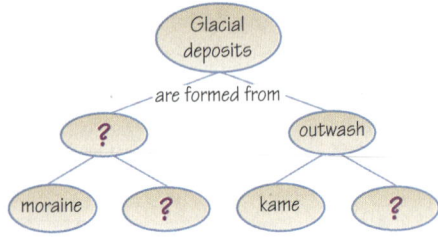

Critical Thinking

16. **Analyze** Typically, the greater a mountain's distance from the equator, the lower the elevation of the mountain's snow line. Yet, the elevation of the snow line is not the same for all mountains at the same latitude. Why?

17. **Predict** After the ice sheet retreated from New England, valley glaciers existed for some time in the White Mountains of New Hampshire. What evidence of these valley glaciers would you expect to find in the White Mountains today?

18. **Infer** How would the amount of sediment carried by the Mississippi River system today compare with the amount that was probably carried 11,000 years ago. Explain your reasoning.

19. **Draw Conclusions** In Wisconsin, fossil evidence of a forest has been preserved between two till layers. What can you conclude about the length of the interglacial period between the formation of the two till layers? Explain your thinking.

Interpreting Data

About 10 percent of the surface of Iceland is covered by glacial ice. The table at the right gives the areas (in square kilometers) of five of Iceland's glaciers for two different years.

20. For each glacier, find the decrease in area from 1958 to 1973. Which glacier lost the greatest amount of ice? Which glacier lost the least amount of ice?

21. Find the percent change in the area of each glacier from 1958 to 1973. You can find the percent change by using the formula below.

    Percent change = $\frac{1973 \text{ area} - 1958 \text{ area}}{1958 \text{ area}} \times 100$

22. Are the percentages of change you calculated in question 21 positive or negative? Why?

23. Which glacier had the greatest percent change in area? Which glacier had the smallest percent change in area?

24. Which set of calculations, those from Question 20 or those from Question 21, do you think are most helpful for comparing how the areas of the five glaciers changed? Why?

Areas of Icelandic Glaciers

Glacier	1958	1973
Glacier 1 (Vatnajökull)	8475 km²	8300 km²
Glacier 2 (Langjökull)	1022 km²	953 km²
Glacier 3 (Hofsjökull)	996 km²	925 km²
Glacier 4 (Myrdalsjökull)	701 km²	596 km²
Glacier 5 (Eyjafjallajökull)	107 km²	77.5 km²

Internet Extension

How Fast Do Glaciers Flow? Use the change in location of survey stakes on glaciers to calculate glaciers' speeds.
Keycode: ES1507

Writing About the Earth System

 SCIENCE NOTEBOOK During periods of glaciation, water that would ordinarily return to the oceans as precipitation or runoff instead builds up as snow and ice on land. Sea levels drop. Scientists note that the numbers of certain marine organisms tend to increase when sea levels decrease. Atmospheric carbon dioxide that is chemically modified to carbonic acid in ocean water is used by these organisms to make their shells, which accumulate in ocean sediment as the organisms die. Describe how an increase in the numbers of these organisms might affect global temperatures.

WRITING ABOUT SCIENCE

Writing About the Earth System

If an increasing amount of carbon dioxide was removed from the air and (after chemical changes) it was used to produce shells for a growing population of marine organisms, the level of carbon dioxide in the atmosphere would decrease. This could decrease the atmosphere's capacity to absorb and hold in heat, resulting in global cooling.

CHAPTER 15 REVIEW

How Fast Do Glaciers Flow?
Keycode: ES1507

Use Internet Investigations Guide, pp. T56, 56.

CRITICAL THINKING

16. Other factors, such as prevailing winds, nearby ocean currents, precipitation differences, nearness of large bodies of water, and local vegetation, can affect the climate and thus the position of the snow line.

17. Evidence includes: U-shaped valleys, hanging valley waterfalls, cirques, cirque lakes, arêtes, and horns.

18. Much liquid water that might have flowed through the Mississippi River system would have been bound up in the glacial ice sheet. With less water flow, the river would have carried less sediment. The ice sheet also covered some of the Mississippi River system.

19. The interglacial period would have been fairly long because the forest had to have sufficient time to grow and develop between periods of glaciation.

INTERPRETING GRAPHS

20. Glacier 1: 175 km²; Glacier 2: 69 km²; Glacier 3: 71 km²; Glacier 4: 105 km²; Glacier 5: 29.5 km²; Glacier 1 lost the most ice; Glacier 5 lost the least ice.

21. Glacier 1: −2%; Glacier 2: −6.8%; Glacier 3: −7.1%; Glacier 4: −15%; Glacier 5: −27.6%

22. They are negative because they represent a loss in glacial ice.

23. Glacier 5 had the greatest percent change; Glacier 1 had the smallest percent change.

24. Calculations from Question 21 are most helpful because they compare the amount of change to the original size of the glacier. Calculations from Question 20 only give the amount of change.

CHAPTER 16 PLANNING GUIDE: WIND, WAVES, AND CURRENTS

	Section Objectives	Print Resources
Chapter-Level Resources		Laboratory Manual, pp. 71–74 Internet Investigations Guide, pp. T57–T59, 57–59 Guide to Earth Science in Urban Environments, pp. 5–12, 21–28 Spanish Vocabulary and Summaries, pp. 31–32 Formal Assessment, pp. 46–48 Alternative Assessment, pp. 31–32
Section 16.1 Wind as an Agent of Change, pp. 340–343	1. Describe how wind shapes Earth's surface. 2. Identify landforms formed by wind erosion.	Lesson Plans, p. 54 Reading Study Guide, p. 54
Section 16.2 Waves in the Sea, pp. 344–348	1. Explain what causes waves. 2. Identify the features of waves. 3. Describe wave motion. 4. Identify how waves affect coastal landforms.	Lesson Plans, p. 55 Reading Study Guide, p. 55 Teaching Transparency 17
Section 16.3 Shoreline Features, pp. 349–353	1. Explain how waves cause coastal erosion and deposition. 2. Describe how coastal erosion and deposition shape shorelines.	Lesson Plans, p. 56 Reading Study Guide, p. 56

INVESTIGATIONS
CLASSZONE.COM

Section 16.1 INVESTIGATION
What Controls the Shape and Motion of Sand Dunes?
Computer time: 45 minutes / Additional time: 45 minutes
Students analyze variables that affect how dunes change over time.

Section 16.3 INVESTIGATION
Where Did This Sand Come From?
Computer time: 45 minutes / Additional time: 45 minutes
Students investigate the clues sand provides about its origins.

Technology/Online Resources

Electronic Teacher Tools
Visualizations CD-ROM
Classzone.com
Online Lesson Planner
Test Generator

Visualizations CD-ROM
Classzone.com
ES1601 Visualization, ES1602 Visualization, ES1603 Investigation

Visualizations CD-ROM
Classzone.com
ES1604 Visualization, ES1605 Visualization, ES1606 Visualization

Classzone.com
ES1607 Investigation

Classroom Management

Gather these materials for minilab and laboratory activities.

Minilab, p. 345
clear, rectangular baking pan
water

Lab Activity, pp. 354–355
200 mL Mixture A (75% sand, 25% gravel)
200 mL Mixture B (25% sand, 75% gravel)
clear rectangular container
masking tape
thin permanent marker
metric ruler
100-mL beaker
water
small wooden block (5 cm × 10 cm)
stopwatch

Chapter Review INVESTIGATION

How Do Storms Affect Coastlines?
Computer time: 45 minutes / Additional time: 45 minutes
Students analyze images of shorelines before and after storms.

CHAPTER 16

Wind, Waves, and Currents

Winds have formed sand dunes in the Namib Desert on the coast of Namibia. Winds also produce the ocean waves that crash against the shore.

What other effects do wind and waves have on Earth's surface?

INTRODUCE

Build Interest
Before sharing background information about the photograph with students, allow time for their questions and comments.

Ask questions like the following:
- Does this look like a typical beach to you? **no** Why? **It is more barren than most coastal areas.**
- What do you think forms the thin, dark green line along the shore? **probably seaweed washed onshore by the breaking waves**
- What evidence shows that wind is an important factor in shaping this landscape? **Sand dunes are prominent, showing that wind constantly reshapes the landscape.**

Respond
- What other effects do wind and waves have on Earth's surface? **They weather and erode rock, shaping coastal areas especially. Wind causes much soil erosion.**

About the Photograph
Namib Desert has the tallest sand dunes in the world. The highest dunes, such as the Sossus Vlei, are over 390 meters high and can be seen from space. The photograph shows the Skeleton Coast in the northern part of the desert, so-called because many shipwrecked sailors have died there. The coast receives less than 2 centimeters of rainfall annually.

BEYOND THE TEXTBOOK

Print Resources
- Reading Study Guide, pp. 54–56
- Transparency 17
- Formal Assessment, pp. 46–48
- Laboratory Manual, pp. 71–74
- Alternative Assessment, pp. 31–32
- Spanish Vocabulary and Summaries, pp. 31–32
- Internet Investigations Guide, pp. T57–T59, 57–59
- Guide to Earth Science in Urban Environments, pp. 5–12, 21–28

Technology Resources
- Visualizations CD-ROM
- Classzone.com
- Online Lesson Planner
- Electronic Teacher Tools
- Test Generator

338 Unit 4 Earth's Changing Surface

CHAPTER 16

PREVIEW

▶ **FOCUS QUESTIONS** In this chapter you will study wind, waves, and erosion and learn more about the key questions below.

Section 1 What role does wind play in weathering, eroding, and depositing rock materials?

Section 2 How do ocean waves form?

Section 3 How do ocean waves affect a shoreline?

▶ **REVIEW TOPICS** As you investigate wind, waves, and erosion, you will need to use information from earlier chapters.

- weathering (p. 258)
- glacier (p. 318)
- erosion (p. 268)

▶ **READING STRATEGY**

CONNECT

In this unit, you have learned how running water, groundwater, and glaciers change Earth's surface by weathering, eroding, and depositing rock materials. As you read Chapter 16, take notes on how wind and waves change Earth's surface. What similarities do you notice between these agents of change and those that you have studied in other chapters?

At our Web site, you will find the following Internet support for this chapter.

DATA CENTER
EARTH NEWS
VISUALIZATIONS
- Arch Formation
- Loess Deposits Formation
- Wave Motion
- Breaking Waves
- Wave Erosion

LOCAL RESOURCES
INVESTIGATIONS
- What Controls the Shape and Motion of Sand Dunes?
- Where Did This Sand Come From?
- How Do Storms Affect Coastlines?

Chapter 16 Wind, Waves, and Erosion 339

CHAPTER 16

PREPARE

Focus Questions
Find out what knowledge students have about wind, waves, and currents, and their effects on landforms. For each focus question, have students consider the section title and then develop two or three additional questions for each section. For example, Section 3: What shoreline features are formed by ocean waves?

Review Topics
In addition to the earth science topics listed, students may need to review this concept:

Physical Science
- Refraction is the bending of waves because of a change in their speed.

Reading Strategy
Model the Strategy: Have students make a chart on which they can compare and contrast the way various agents of weathering, erosion, and deposition work. The chart should have five columns with the headings *Running Water*, *Groundwater*, *Glaciers*, *Wind*, and *Waves* and three rows with the headings *Weathering*, *Erosion*, and *Deposition*. Instruct students to fill in the chart as they read the chapter with the way wind and waves weather, erode, and deposit rock and soil. Have students refer to previous chapters to fill in the information on running water, groundwater, and glaciers. Use the charts to discuss the similarities in the agents of change.

BEYOND THE TEXTBOOK

INVESTIGATIONS
- What Controls the Shape and Motion of Sand Dunes?
- Where Did This Sand Come From?
- How Do Storms Affect Coastlines?

VISUALIZATIONS
- Animation of arch formation
- Animation of the formation of loess deposits
- Animation of wave motion
- Animation of waves breaking on shore
- Animation of wave erosion

DATA CENTER
EARTH NEWS
LOCAL RESOURCES
CAREERS

Online Lesson Planner

Chapter 16 Wind, Waves, and Currents 339

CHAPTER 16 SECTION 1

FOCUS

Objectives
1. Describe how wind shapes Earth's surface.
2. Identify landforms formed by wind erosion.

Set a Purpose
Read the section title aloud. Then have students read the section to answer the question "How does wind change Earth's surface?"

INSTRUCT

More about...
The Dust Bowl
During the dust storms over the Great Plains in the 1930s, wind-borne dust and soil was so thick at times that the storms were called black blizzards. One storm lasted for 1½ days and produced a dust cloud 2000 kilometers (1200 miles) across. As the storm moved eastward, the airborne sediment created muddy rains in New York and black snows in Vermont.

Scientific Thinking
APPLY
no-till farming, terracing, contour farming, strip cropping, and planting windbreaks

16.1

KEY IDEAS
Winds transport silt, clay, and sand.

Windblown sand weathers rock surfaces.

Winds deposit particles in distinctive formations.

KEY VOCABULARY
- dust storm
- deflation
- desert pavement
- ventifact
- loess
- sand dune

Scientific Thinking
APPLY
Could dust storms like those that affected the southern Great Plains during the 1930s occur again? Farmers today use soil-conservation methods to protect against wind erosion. However, under drought conditions, wind erosion still affects unprotected areas. Describe some conservation methods that protect against wind erosion. You may want to look back at Chapter 12 for ideas.

Wind as an Agent of Change

Like rivers and glaciers, winds cause changes to Earth's surface. Winds pick up and move sediment. They also drive sediment against rocks and other materials, causing weathering of these surfaces. Winds also play a role in the formation of ocean waves, which erode shorelines.

Windblown Rock Materials

Wind erosion and weathering are most pronounced where sands, silts, and clays lie loose and dry. Optimum conditions for wind erosion are found in great deserts, such as the Sahara in Africa and the Mohave in the southwestern United States. On a smaller scale, these conditions also exist on beaches and in semiarid regions.

When strong, steady winds lift great amounts of silt and clay from the topsoil, a **dust storm** results. During the 1930s, destructive dust storms struck the southern Great Plains region of the United States. Most of the hardy, natural grasses in this region had been cleared to make way for farms, leaving the soil vulnerable to wind erosion. In addition, a drought had dried out the soil. Wind storms that struck carried dust high into the atmosphere.

DUST BOWL A drought that affected the southern Great Plains during the 1930s resulted in frequent dust storms. Thousands of people were forced to move away from hard-hit areas. In this 1935 photograph, wind carries dust over houses in Springfield, Colorado.

Dust in the atmosphere can be transported over great distances. Winds have carried dust from the Gobi Desert in Asia across the Pacific Ocean to the western United States. Likewise, winds have transported dust from the Sahara in Africa across the Atlantic Ocean to the Caribbean islands and the southeastern United States.

Sand grains are much larger and heavier than clay and silt particles. Experiments have shown that winds at least as strong as about 18 kilometers per hour are needed to move them. Sand grains carried by the wind do not move in a steady stream. Like sand grains in a riverbed, they move in short hops and bounces. Even in strong winds, most sand is carried within one meter of the ground.

DIFFERENTIATING INSTRUCTION

Reading Support Model for students how to ask questions that will increase understanding. Have students skim the section, paying close attention to headings, boldface vocabulary words, photographs, and diagrams. Ask: What would you like to learn about any of these as you read? Suggest questions such as "What is deflation?" or "How does wind form the landscape in this photograph?" or "How does wind form the sand dunes in the diagrams?" Have students write their questions on a sheet of paper and try to answer them as they read. Help students clarify their answers after they finish reading the section.

Use Reading Study Guide, p. 54.

Deflation

Deflation (dih-FLAY-shuhn) is the removal of loose rock particles by the wind. In many desert areas the sands and clays formed by weathering are blown away by winds, leaving pebbles and boulders. The resulting surface is called **desert pavement.** The surface shown in the photograph below is desert pavement. Desert pavement protects the materials beneath from further deflation. Stony surfaces of this type are common in the deserts of the southwestern United States and in the Sahara in Africa.

In semiarid regions, such as the Great Plains, deflation has formed thousands of hollows called blowouts. Most of these are shallow and small, but some are many thousand meters long and about a hundred meters deep. Blowouts form in dry regions with patchy vegetation. A small irregularity on the ground causes winds to focus on a small area. Over time, these winds remove more and more sands or clays from the area. If the bottom of the blowout reaches the water table, the wet ground stops deflation. The growth of vegetation also halts deflation. In desert regions, oases sometimes are found in deep blowouts that were formed by deflation.

BLOWOUT Deflation caused by wind erosion has formed a blowout in the Sand Hills of Nebraska.

Abrasion

Windblown silt and clay particles are too small, and often too soft, to wear away most rocks. Sand grains, however, are larger and tend to be made of more abrasive materials. Sand grains driven by winds grind and scour most surfaces they hit. Quartz sand grains, in particular, can wear away many materials. In some desert areas, rocks and telephone poles may be undercut at the base by windblown sand.

Blasts of desert sand grind boulders and small rocks into shapes called **ventifacts** (VEHN-tuh-FAKTS). The shape of a ventifact gives a clue about the direction of the wind that formed it. As shown in the photograph below, the side of a ventifact that faces the prevailing wind is worn into a smooth, flat surface. A second facet may form if the wind blows from different directions at different times of the year or if the boulder is turned so that a new side faces the wind.

DEATH VALLEY, CALIFORNIA
Wind-driven sands have weathered the windward sides of these ventifacts. The wind has also blown away sands and clays from the ground, leaving desert pavement around the ventifacts. Sand has piled up behind the ventifacts.

VISUALIZATIONS
CLASSZONE.COM

Observe an animation showing the formation of an arch.
Keycode: ES1601

VOCABULARY STRATEGY

The word *ventifact* comes from the Latin word for wind, *ventus,* and the Latin word *factum,* which refers to something that has been made.

Chapter 16 Wind, Waves, and Currents **341**

CHAPTER 16 SECTION 1

> **More about...**
>
> **Loess**
> Loess particles are so fine they feel like talcum powder. Winds can carry these fine-grained particles higher and farther than larger particles such as sand. Therefore, loess deposits often are found far away from their source. The origin of most loess in the United States is sediment left behind by retreating glaciers during the last ice age. Prevailing westerly winds blew the sediment across floodplains, depositing them in thick layers on the eastern sides of river valleys in parts of the central United States. Major deposits are found in South Dakota, Nebraska, Iowa, Missouri, Illinois, and Washington State. Because soils produced from loess are very fertile, these areas have very productive farmland.

WESTERN IOWA Loess erodes easily, leaving behind vertical cliffs like those shown here.

Observe the formation of loess deposits.
Keycode: ES1602

Loess

Wind can deposit sediment as well as remove it. Large areas in China, northern Europe, and the north-central United States are covered by deposits of material called **loess** (LOH-uhs). Loess is made of unlayered, silt-sized particles that are generally yellowish in color. Loess particles are angular in shape and have been weathered from many different minerals and rocks. Loess particles hold together so well that when loess erodes, pieces of it split off vertically to form clifflike slopes, as shown in the photograph at left. Loose particles of loess are light and small enough to be carried by winds in dust storms.

Loess deposits range in thickness from about 1 meter to as much as 300 meters, the depth of some loess deposits in China. The main deposits of loess in the United States are in the upper Mississippi and Missouri river valleys. The particles that make up these loess deposits were probably picked up by winds from the outwash left by glaciers. The loess found in northern China was probably blown from the great deserts of Mongolia.

Sand Dunes

Sand dunes are hills of sand deposited by winds. They form when sand piles up against shrubs, boulders, or other obstructions. Sand dunes are found wherever there are strong winds and enough loose sand, such as in the Sahara. Sand dunes also form on sandy river floodplains in semiarid climates and on sandy beaches.

Dunes range in size from a few meters high to more than a hundred meters high and many thousand meters long. Dunes often occur in groups, which may cover a large area. Over one million square kilometers of the Sahara is covered by dunes.

Most sand dunes are made of quartz sands, but there are exceptions. The dunes of White Sands National Monument in New Mexico are made of gypsum sands. Where limestone or coral is common, dunes are made of calcite sands. Dune sand may contain grains of additional minerals, such as feldspar, mica, and magnetite.

Where winds blow steadily from one direction, dunes tend to have a long, gentle slope on the windward side and a shorter, steep slope on the sheltered, or leeward, side. For example, winds blowing steadily from the west will generally form dunes with gentle slopes on their west sides and steep slopes on their east sides. Tiny sand ripples are likely to form on the windward slopes.

The shape of a dune seems to depend on the supply of sand, the strength and steadiness of the winds, and the amount of vegetation present. Strong, steady winds blowing over a limited supply of sand usually produce a crescent-shaped dune called a barchan (bahr-KAHN). The ends of a barchan point downwind. Where sand is more abundant, long, continuous ridges called transverse dunes form at right angles to the wind. U-shaped parabolic (PAR-uh-BAHL-ihk) dunes often form around blowouts. The open ends of parabolic dunes face upwind, and their long "arms" tend to become stabilized by vegetation. Transverse and parabolic dunes are commonly seen on beaches. Longitudinal dunes form in desert

---

DIFFERENTIATING INSTRUCTION

Hands-On Demonstration Model sand dune formation. Pour a layer of flour into a rectangular pan. Blow over the flour through a straw from one end of the pan. Ask: What do you observe? *Shapes that resemble sand dunes form in the flour.* What does my breath represent? *wind blowing over sand* Refer students to the diagram on page 343. Ask: What type of sand dune do the shapes in the flour most closely resemble? *barchans or parabolic dunes*

Types of Sand Dunes

BARCHANS form when winds blow over limited amounts of sand.

TRANSVERSE DUNES form when winds blow over abundant amounts of sand.

PARABOLIC DUNES tend to form around a blowout, with open ends facing upwind.

LONGITUDINAL DUNES form in desert regions having moderate amounts of sand.

regions with a moderate supply of sand. The winds that produce longitudinal dunes tend to come from the same general direction, though they may shift slightly. The winds blow the sand into long, straight ridges that are parallel to the prevailing wind direction. Dunes in areas of widely varying wind directions are less regular than the shapes described above.

Over time, some sand dunes migrate. Each time the wind blows against the windward side of such sand dunes, some of the surface sand is carried over the top. Then it falls down on the leeward side. In time, the whole dune is moved in the leeward direction. The movement of a dune in this manner is called dune migration. The rate of migration may be as much as 30 meters per year. Migrating dunes can bury towns, farms, and forests. Not all dunes migrate, however. In humid beach areas, grasses and shrubs that grow on the dunes keep them from moving.

INVESTIGATIONS
CLASSZONE.COM

What Controls the Shape and Motion of Sand Dunes? Examine images of dunes. Analyze variables that affect how they change over time.
Keycode: ES1603

16.1 Section Review

1. Is sand or silt more likely to be carried by wind? Explain.
2. How does a ventifact form?
3. How is desert pavement formed?
4. What are loess deposits?
5. What are three factors that determine the shape of a sand dune?
6. **CRITICAL THINKING** Draw a sketch that shows what happens to the sand in a migrating dune.
7. **PAIRED ACTIVITY** Work with a partner to create a display about sand dunes. Be prepared to give an oral report about your display.

Chapter 16 Wind, Waves, and Currents

MONITOR AND RETEACH

If students miss . . .

Question 1 Reteach "Windblown Rock Materials." (p. 340)

Question 2 Have students study the photograph on the bottom of page 341 and read the caption.

Question 3 Review the formation of desert pavement and the photograph on page 341.

Question 4 Reteach "Loess." (p. 342)

Question 5 Reteach "Sand Dunes." (pp. 342–343)

Question 6 Have students reread the last paragraph on page 343. Ask them to describe the process of dune migration in three steps.

Question 7 Have students use pages 342 and 343 as a starting point for their display.

CHAPTER 16 SECTION 1

Observe the formation of loess deposits.
Keycode: ES1602

 Visualizations CD-ROM

What Controls the Shape and Motion of Sand Dunes?
Keycode: ES1603

Use a computer and projector to introduce the investigation in class.

📖 Use Internet Investigations Guide, pp. T57, 57.

Visual Teaching

Interpret Diagrams
Have students compare and contrast the different types of sand dunes. Ask: Which type of dune is most likely to form in an area of deep sand deposits in the middle of the Sahara? **a transverse dune**

ASSESS

1. Silt; its particle size is smaller and therefore more easily moved by wind.
2. Wind-driven desert sand shapes and grinds rocks into ventifacts.
3. Wind removes loose rock from the ground, leaving behind pebbles and boulders.
4. surface material made of unlayered, silt-sized particles
5. wind strength and steadiness, sand supply, and amount of vegetation
6. Sketches should show wind hitting the dune, sand on the windward side blowing up and over the dune, and deposition of sand on the leeward side.
7. Displays might deal with dune variety, migration, extremes, and geographic locations.

Chapter 16

CHAPTER 16 SECTION 2

16.2

Focus

Objectives
1. Explain what causes waves.
2. Identify the features of waves.
3. Describe wave motion.
4. Identify how waves affect coastal landforms.

Set a Purpose
Read aloud the key ideas. Have students change each idea into a question to set a purpose for reading. For example, "What causes waves?"

KEY IDEAS
Most waves are caused by winds. Important features of waves are height, wavelength, period, and speed.

Wave motion changes as waves approach shore.

KEY VOCABULARY
- fetch
- wave height
- wavelength
- period
- refraction

Instruct

More about...

Ocean Waves
On a day when a steady, strong wind blows over a long stretch of open ocean, waves 4.2 meters (14 feet) high can result. However, the size of waves has limits. Even in stormy weather, ocean waves rarely top 12 meters (39 feet). On one rare occasion in 1933, a strong Pacific storm is said to have generated waves 34 meters (112 feet) high. A wave that tall would be higher than a 10-story building.

Use Transparency 17.

Waves in the Sea

An ocean wave is a rhythmic rise and fall of the water's surface. When ocean waves reach the shore, they erode and shape the shoreline.

Ocean waves are generally produced in one of three ways. Undersea earthquakes can cause waves. The gravitational pull of the moon produces waves by causing tides. Most commonly, winds cause waves.

Winds and Waves

When a gusty storm wind blows over open water in an ocean or a lake, ripples form. If the wind continues to blow, or if it blows over a long distance, the ripples grow larger until they become waves. The height of a wind-formed wave depends on three factors: the wind speed, the length of time that the wind blows, and the fetch. The **fetch** is the length of open water over which the wind blows. The fetch of a lake is shorter than that of an ocean. Thus, lake waves are not as high as ocean waves. Normal and even gale force winds rarely produce ocean waves higher than 10 meters. Hurricane winds are capable of producing waves that are even higher.

Strong wind gusts that change speed and direction cause choppy seas. The waves produced by such gusts have different heights and may come from different directions. As the waves clash and the winds tear off the wave crests, foamy whitecaps form.

Even on calm days the sea is likely to have steady, smooth waves, called swells. Swells come at regular intervals of up to ten seconds and are caused by winds and storms far off at sea.

Features of Water Waves

Wave height is the difference between a wave's high point, or crest, and its low point, or trough. **Wavelength** is the distance from one crest to the next. Strong winds produce waves with long wavelengths. Waves produced during a storm often have wavelengths of more than 150 meters. Typically, the wavelength of an ocean wave is 20 to 30 times its height. A wave 2 meters high would thus have a wavelength of between 40 and 60 meters.

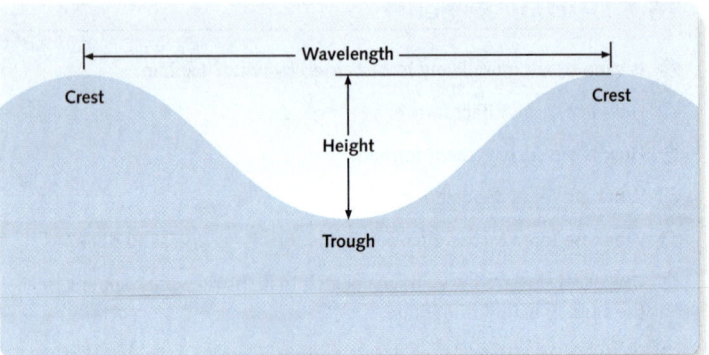

WATER WAVE FEATURES

344 Unit 4 Earth's Changing Surface

DIFFERENTIATING INSTRUCTION

Reading Support In order to focus on and retain the most important material in their reading, help students recognize main ideas. As they read the section, have students write one sentence that contains the main idea in each red heading and at least one supporting detail that supports the main idea. When students finish, they will have a simple outline of the most important material in the section. Encourage students to use their outlines as study aids.
Use Reading Study Guide, p. 55.

344 Unit 4 Earth's Changing Surface

The **period** of a wave is the time it takes one wavelength to pass a given point. Most ocean waves have periods ranging from two seconds to ten seconds. To find the speed of a wave, divide its wavelength by its period.

$$\text{Speed} = \frac{\text{wavelength}}{\text{period}}$$

An example of a wave with a long period is a tsunami (su-NAH-mee). As you learned in Chapter 10, a tsunami is typically caused by an underwater earthquake or landslide. A typical tsunami has a wavelength of 150 kilometers and a period of 12 minutes. Dividing 150 kilometers by 12 minutes gives a speed of 12.5 kilometers per minute.

Wave Motion

You may have seen objects bobbing in the water as waves passed beneath them. Waterborne objects are not carried forward with a wave; instead they move in place in a circular motion. The water in the wave moves in the same way. Furthermore, this same circular motion takes place below the surface as well. As depth increases, the water moves in smaller and smaller circles. As each water particle moves, it bumps into another particle, passing its energy along. In this way, the wave's energy is transferred through the water.

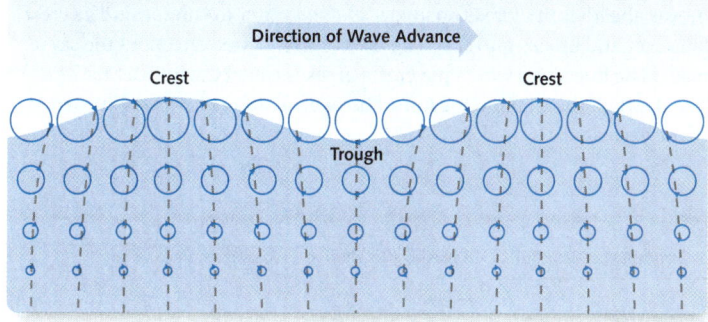

WAVE MOTION

Wave Refraction

Most waves approach the shoreline at an angle. Yet when a wave reaches shallow water, it tends to swing around until it approaches the shoreline more or less head on. This swinging or bending is called **refraction.** Refraction occurs because the end of the wave closest to shore scrapes bottom first and slows down. The end that is still in deeper water continues at its normal speed and catches up. Thus, the wave ends up nearly parallel to the shore.

Wave refraction helps explain why an uneven shoreline with shallow water is eventually worn away to a more even shoreline. As shown on page 346, incoming waves reach the shallow water at a headland first. Each wave slows down in front of the headland and refracts, striking the headland on all sides. Because the wave energy is concentrated at the headland, the headland wears away more quickly than the shoreline in the bay.

10-Minute Mini LAB

Making Waves

Materials
- clear, rectangular baking pan
- water

Procedure

1. Fill the pan about one-quarter full with water.
2. From one of the short sides of the pan, blow air gently over the water. Observe the resulting patterns. Blow harder and observe.
3. Add more water to the pan and repeat step 2.

Analysis

Identify the wave crests and the wave troughs. How does blowing harder affect the waves? How does adding water affect them? Describe other ways to generate waves in the pan.

VISUALIZATIONS
CLASSZONE.COM

Observe an animation of wave motion.
Keycode: ES1604

VISUALIZATIONS
CLASSZONE.COM

Observe an animation of wave motion.
Keycode: ES1604

Have students describe the circular motion of the molecules that make up waves.

 Visualizations CD-ROM

MINILAB
Making Waves

Management
- Have plenty of paper towels on hand for spills.
- The Minilab can also be done as a demonstration by placing a clear rectangular pan on an overhead projector. The waves will show up as alternating light and dark stripes on the projection screen.

Analysis Answers
- The wave crests are the tops of the ripples; the wave troughs are the bottoms of the valleys between adjacent wave tops.
- Blowing harder produces ripples that travel more quickly and farther. It also produces more interference from waves bouncing off the sides and bottom of the pan. Wave heights are increased, which means that the wave crests are farther apart.
- Adding water allows the waves to increase in size. The waves are smoother, due to less interference from waves hitting the bottom of the pan.
- Waves can also be generated by dropping items into the water or by shaking the pan.

CHAPTER 16 SECTION 2

VISUAL TEACHING
Discussion
Lead students to understand that although waves generally approach a shoreline at an angle, the waves bend, or refract, at the shoreline, causing them to move nearly parallel to it. Point out the parallel waves approaching the uneven shoreline in the diagram at the top of the page. Ask: Why is wave energy concentrated at the headlands? *The waves reach the headlands first, where they slow down and refract toward the headland, hitting it head-on from all sides.* How will the waves eventually change the shape of the shoreline? *They will gradually wear away the headlands, making the shoreline straighter.*

Visualizations
CLASSZONE.COM
Observe waves as they break on the shore.
Keycode: ES1605

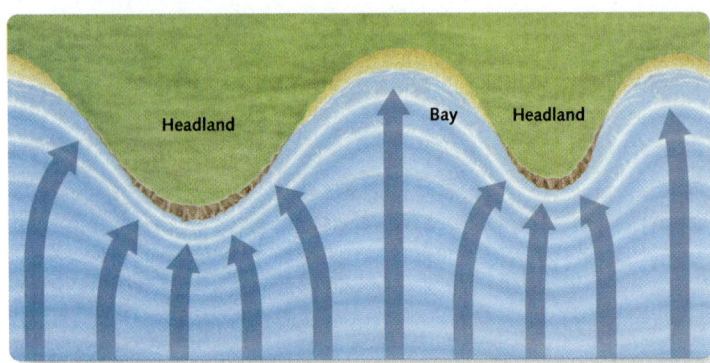

WAVE REFRACTION Because of refraction, ocean waves approach the shoreline more or less head on. As shown, wave energy tends to concentrate at the headlands and spread out in the bays.

Breakers

Waves usually approach the shoreline smoothly until they reach water so shallow that they touch the bottom. This occurs where the water depth is about half the wavelength.

As a wave scrapes bottom, the circular motion of the wave is distorted, and the lower part of the wave slows down. The upper part of the wave moves ahead until there is no longer enough water to support it. The crest falls over and breaks into surf. The next crest will break in about the same place. The line along which the crests break is called the line of breakers. The depth of the water at the line of breakers is about 1.3 times the height of the waves.

Visual Teaching

Interpreting Diagram
Have students contrast the water's depth and wavelength in different parts of the diagram. Ask: What geographical feature causes the water depth to change? *The ocean bottom slopes upward toward the shoreline.* Where would breakers form if the water depth remained constant from the outer edge of the diagram across to the shore? *Breakers would not form.*

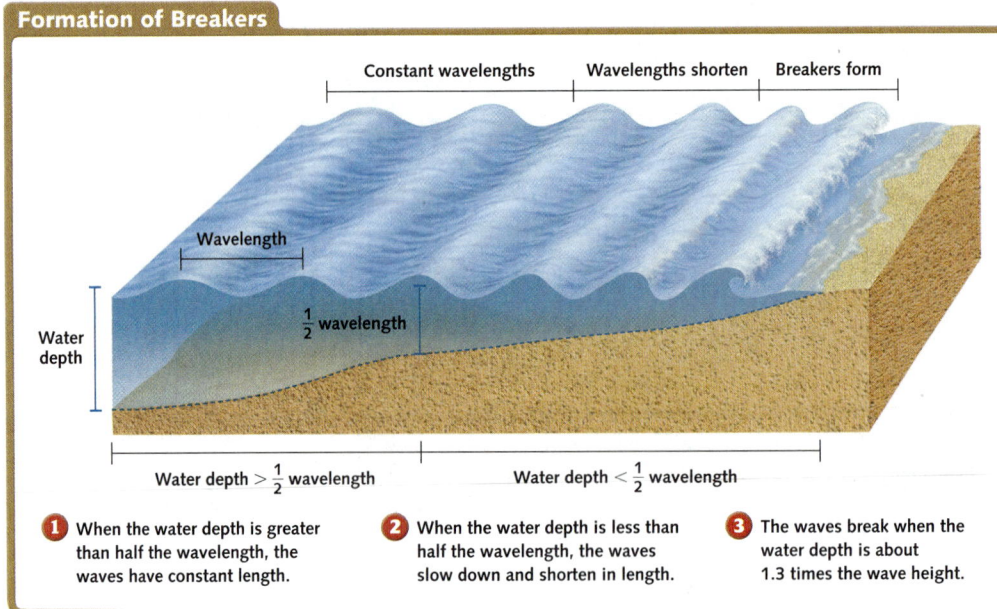

Formation of Breakers

① When the water depth is greater than half the wavelength, the waves have constant length.

② When the water depth is less than half the wavelength, the waves slow down and shorten in length.

③ The waves break when the water depth is about 1.3 times the wave height.

DIFFERENTIATING INSTRUCTION

Developing English Proficiency To help students understand the diagram at the bottom of the page, review these terms: *wave, wavelength, water depth, breaker, shore, ocean bottom*. Have students redraw the diagram, using the terms to label the parts. Ask: What happens to ocean water as it gets closer to the shore? *It becomes less deep and forms breakers.* Take this opportunity to introduce a new word. Tell students that the opposite of *deep* is *shallow*. Use Spanish Vocabulary and Summaries, pp. 31–32.

The surf formed by breaking waves is a powerful agent of erosion. On rocky shorelines it pounds the rocks and cliffs and wears them down. On beaches it scours the bottom and moves sediments along the shoreline. The surf grinds pebbles against each other, wearing them down into sand.

Shoreline Currents

Waves, like the winds that form them, may come from any direction. Thus, many waves approach the shoreline at an angle. When such waves break, large amounts of water and sand are pushed up the beach at an angle. The motion of water up the beach is called swash. Most of the water runs back down the beach under the next wave in a gentle current called backwash. (A very strong backwash is sometimes called undertow.) The sand that was pushed up at an angle is pulled straight back by the backwash. Each successive wave pushes and pulls the sand in this way. As a result, the sand drifts down the beach in a zigzag path. This process is called beach drift.

The water behind the line of breakers is also pushed toward shore by waves and pulled back by backwash. This movement forms a so-called longshore current, which runs almost parallel to shore. Swimmers often discover that they have been moved along the beach by a longshore current. Longshore currents are important in the movement of sand and in the formation of sandbars.

Rip currents are strong surface currents that flow away from the beach. They may form where too much water builds up in the area where surf forms. Excess water can build up where two longshore currents meet head-on, where breakers bring in more water than backwash can return, or where water is held back by a barrier such as a headland, a breakwater, or an underwater sandbar. Water held back by a barrier flows rapidly (often at speeds of up to five kilometers per hour) back to the sea through a gap in the barrier.

Rip currents are dangerous to swimmers. Fortunately, a swimmer who knows what to look for can avoid them. A rip current may carry a lot of sand, which makes it visible. Waves around the rip current may be steeper than the surrounding waves.

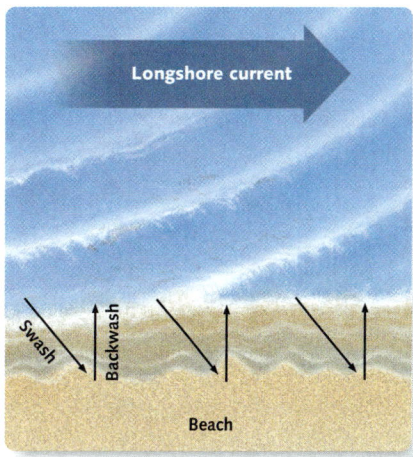

LONGSHORE CURRENT The movement of waves toward the shore combined with backwash causes a longshore current to form behind the line of breakers. Swash and backwash move sand down the beach in a zigzag path.

16.2 Section Review

1. What factors can affect the height of a wind-formed wave?
2. Explain what a wavelength, crest, trough, and period are.
3. Describe the motion of water in a wave.
4. What is wave refraction and how is it caused?
5. **CRITICAL THINKING** Would the line of breakers at a particular beach always form at the same distance from shore? Explain.
6. **MATHEMATICS** Find the speed of a wave that has a wavelength of 12 meters and a period of three seconds.

CHAPTER 16 SECTION 2

Examine an example of wave erosion.
Keycode: ES1606

 Visualizations CD-ROM

Science and Society

Saving a Lighthouse
Erosion often affects lighthouses perched on sandy beaches. Some lighthouses, undermined by shore erosion, have fallen into the ocean. Others, like Cape Hatteras Lighthouse in North Carolina, have been saved by moving them back from the sea. When Hatteras Light began to operate shortly after the Civil War, it stood about 460 meters from the sea. Over more than a century, most of the beach between Hatteras Light and the ocean was chewed up by erosion. By the late 1990s, the nation's tallest lighthouse stood just 36 meters from the surf. To save the lighthouse, engineers moved it 480 meters from the Atlantic.

Science Notebook
Some students may think it is necessary to replenish beach sand to protect structures from erosion; others may think that people who build structures at the shore are aware of the risks of erosion and should bear the cost without help from taxpayers.

SCIENCE & Society

Disappearing Beaches

Beach erosion is a serious problem along the coasts of the United States. As much as a meter of shoreline disappears from some areas each year; however, a large storm at high tide can remove ten times that much beach area in a single night. The rising sea level is also a factor in erosion.

Is it possible to prevent beach erosion? Can people prevent shoreline homes and other structures from collapsing into the sea?

In an effort to protect property and lives, authorities have spent millions of dollars attempting to save beaches from erosion. Some of these measures have involved the construction of seawalls. Seawalls temporarily stop storm waves from reaching homes but do not save beaches. In fact, the findings suggest that seawalls may speed erosion. Waves crashing against a seawall carry away more sand than they would if they spent their energy gently rolling onto a beach. Some towns have built jetties, structures in the sea that trap sand carried by longshore currents. However, jetties often trap the sand needed at other beaches down the coast, causing those beaches to erode more swiftly.

Jetties and seawalls do not seem to be the answer. In fact, in Maine, Oregon, Texas, and North and South Carolina, they have been banned. Recent "beach nourishment" efforts, in which offshore sand is dredged up and redeposited on a beach, have been somewhat successful. A drawback of beach nourishment is that eventually the sand washes back out to sea. For this reason, nourishment, which is costly, must be ongoing. New solutions being tested include a portable barrier structure that decreases wave energy by up to 30 percent. Because the movable barrier is in place only during the storm season, when severe erosion is most likely, it presents fewer problems than jetties or other permanent barriers. ■

A PORTABLE BARRIER is one new solution for preventing shoreline erosion. These structures can decrease wave energy by up to 30 percent.

Extension

 **SCIENCE NOTEBOOK**
Some people in favor of beach nourishment defend its high cost as a necessary expense, like that of maintaining roads or bridges. Explain whether you agree or disagree with this opinion.

Examine an example of wave erosion.
Keycode: ES1606

WAVES driven by hurricane winds in Virginia Beach, Virginia, destroy a seawall built to protect these beachfront homes.

Shoreline Features

Ocean waves change the shape of a shoreline by eroding rock materials and by depositing sediments.

Waves and Erosion

Breaking storm waves may strike rock cliffs with a force of thousands of kilograms per square meter. Such breakers easily remove large masses of loose sand and clay. Air and water driven into cracks and fissures may split bedrock apart. Sand and pebbles carried by the water abrade the bedrock. Waves pound loose rock and boulders into pebbles and sand. In addition, seawater dissolves minerals from rocks such as limestone.

When waves strike the headlands of a deep-water shoreline, they may cut away the rock up to the high-tide level, forming a notch. If the materials overhanging the notch collapse, a sea cliff results.

Cliffs made of soft materials such as soil and sand wear away very quickly. For example, waves washing up on Cape Cod in Massachusetts are carrying away materials from sand cliffs there so rapidly that the cliffs are receding at a rate of about one meter every year.

In cliffs made of harder rock materials, a notch may deepen until it becomes a sea cave. Waves may cut through the walls of sea caves to form sea arches. Arches may also form when waves cut through vertical cracks in narrow headlands. If the roof of a sea arch falls in, what remains is a tall, narrow rock island called a sea stack.

Sea caves, sea arches, and sea stacks can be seen on the coasts of California, Oregon, Washington, and Maine, on the Gaspé Peninsula of Canada, and in many parts of the Mediterranean Sea.

16.3

KEY IDEA
Waves erode shorelines and deposit sediments in characteristic formations.

KEY VOCABULARY
- beach
- sandbar
- fjord

BAJA PENINSULA Ocean waves have formed this sea stack and sea arch in Mexico.

Sea stack
Sea arch

CHAPTER 16 SECTION 3

FOCUS

Objective
1. Explain how waves cause coastal erosion and deposition.
2. Describe how coastal erosion and deposition shape shorelines.

Set a Purpose
Have students view the illustration on pages 350 and 351. Then have students read the section to answer the question "How do these shoreline features form?"

INSTRUCT

> **More about...**
>
> **Coastal Erosion**
> Many people think water by itself erodes shoreline rock. Actually, the particles of sand, pebbles, and other debris carried by waves cause much of the erosion. The erosive force depends on slope. Gently sloping shores dissipate the energy carried by waves over long distances, reducing a wave's force.

CRITICAL THINKING

Ask: Would sea stacks and sea arches form from cliffs on the sheltered bay side of a rocky island? **not as readily** Explain. **Waves pounding against the shore are needed. Wave action would not be as powerful in a sheltered bay.**

DIFFERENTIATING INSTRUCTION

Reading Support As students read this section, have them note cause-and-effect relationships. Tell them to restate each relationship in a sentence that follows this pattern: [cause] causes [effect]. Model this pattern, using the first two sentences under "Waves and Erosion." Read the sentences aloud and then write this cause-and-effect statement on the board: *Storm waves striking rock cliffs cause the removal of large masses of loose sand and clay.* Tell students that restating relationships this way will help them remember important information from the text.
Use Reading Study Guide, p. 56.

Chapter 16 Wind, Waves, and Currents

CHAPTER 16 SECTION 3

VISUAL TEACHING
Discussion
Have students describe how the landforms in the illustration at the bottom of pages 350 and 351 might change over time through natural processes. **The lagoon behind the baymouth bar might become a marsh; the spit might grow into a baymouth bar enclosing the bay behind it and forming a new lagoon; sand on the beach at the center might migrate eastward with the longshore current, building up the beach to the west of the headland; the sea cliff might continue to erode and perhaps form a sea cave; the sea stack might become smaller or might disappear altogether because of wave erosion; wave erosion might also cause the sea arch to collapse, forming a sea stack.** Ask: Where is the greatest force of wave erosion concentrated along this coast? **It is concentrated on the headland.** Explain your answer. **The waves on both sides of the headland hit it basically head on. Along much of the rest of this coast, a longshore current turns the waves slightly away from striking it head on.**

Where Did This Sand Come From?
Investigate the clues sand provides about its origin.
Keycode: ES1607

Beaches

Geologists define a **beach** as the area of shore between the high-tide level and the low-tide level. The gentler the slope of the sea floor, the wider the beach. Beaches may be sandy, pebbly, or even rocky. The makeup of a beach depends on both the available material and the slope of the shoreline.

Most of the sand on beaches has been deposited by rivers. Materials may also come from areas near the beach or from other beaches. Not all materials that are deposited on a beach remain on the beach. If the sea floor is steep, sands and clays are washed out to sea by backwash, leaving a pebble beach. If the sea floor slopes gently and a constant supply of sand replenishes the beach area, only clay is washed out, and the beach is composed of sand.

Most beach sands are grains of durable minerals, such as quartz and some feldspars. Other materials commonly found in beach sands are grains of magnetite and flakes of muscovite mica. The makeup of beach sand, however, depends not only on which minerals are durable but also on which minerals are common in the area where the beach sand originated. Beach sands that have been weathered from granite are mainly grains of quartz and feldspar. Beach sands on coral islands, such as Bermuda, are mainly coral (calcite) fragments.

BAYMOUTH BAR Waves have deposited sand across the mouth of a bay in South Africa.

SPIT Ocean waves off the coast of New Zealand have formed a spit.

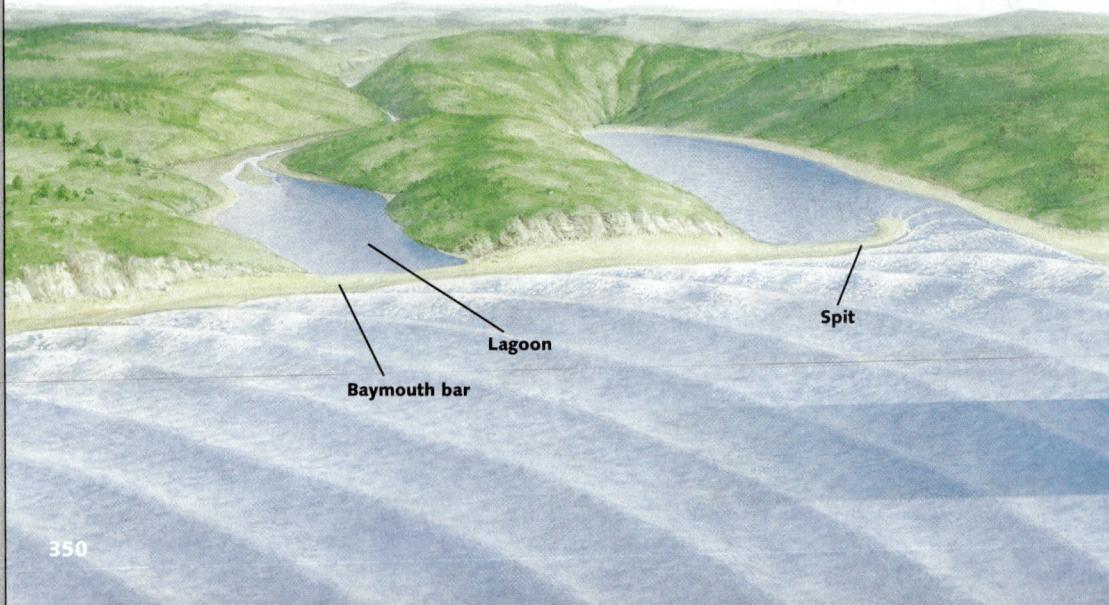

DIFFERENTIATING INSTRUCTION

Developing English Proficiency Have students make a series of drawings that shows the sequence of events in the formation of a sea arch or a baymouth bar. Take this opportunity to review vocabulary connected with sequencing such as *first, second, next, finally, then,* and *last.* When students finish this exercise, have them practice by using the sequence vocabulary to write simple sentences explaining the steps in another process.

Unit 4 Earth's Changing Surface

Sandbars

Along irregular shorelines, longshore currents carry away most of the sand and pebbles eroded from the headlands. Where a longshore current passes across the mouth of a bay or cove, some of the sediment is carried inland by waves. There it may form a sand or pebble beach.

Often, however, the current carries enough sand to form a **sandbar** across the mouth of the bay. The sandbar seems to grow right out of the end of the headland. Such a bar, attached at one end, is called a spit. In time it may grow completely across the bay to become a baymouth bar. In some places bars form between islands and the mainland.

Waves and crosscurrents may curve the end of a spit toward the shore. A spit with a curved end is called a hook. Sandy Hook in New Jersey is a well-known hook. Other famous hooks are Rockaway Beach on Long Island, New York, and the tip of Cape Cod in Massachusetts.

Sandbars usually protect the water behind them from strong winds and waves. The protected areas are called lagoons. As time passes, lagoons may fill with sediment and become salt marshes. Jamaica Bay is the lagoon behind the Rockaway Beach hook. It is an important shelter for shore birds, as are many other lagoons.

SEA CLIFF These cliffs on the coast of England are made of limestone.

This illustration of possible features associated with a shoreline does not represent the landscape of any specific geographic area.

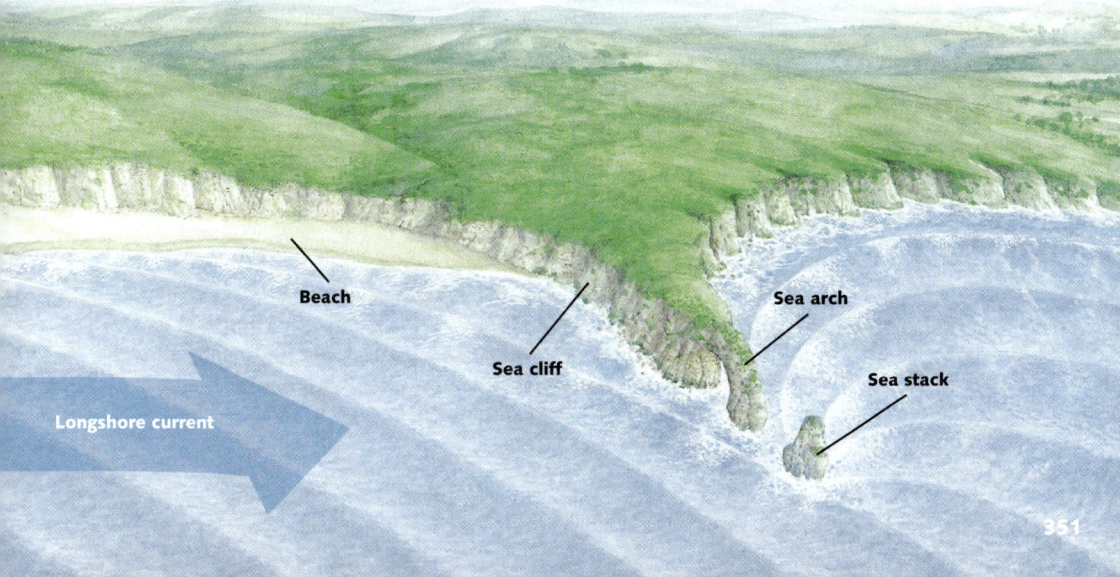

CHAPTER 16 SECTION 3

INVESTIGATIONS
CLASSZONE.COM

Where Did This Sand Come From?
Keycode: ES1607
Have students work in pairs.
📖 *Use Internet Investigations Guide, pp. T58, 58.*

More about...

Beach Erosion
Almost one-third of the U.S. shoreline is undergoing serious erosion, much of it along sandy beaches. A major cause is that the sea level has been rising over the past 20,000 years. This rise is due to the melting of polar ice since the last ice age and to rising global temperatures, which have expanded the volume of ocean water. Because beach property is very valuable, engineers have developed methods to stop or decrease beach erosion. These methods include placing groins—walls that extend from the beach into the water—on coasts to trap sand, building sea walls along the shore to protect property, and dredging sand from the ocean bottom offshore to replenish beach sand that has washed away. None of these methods stops beach erosion permanently. In many cases, beach protection methods stop erosion and replenish a beach in one area—only to have beach sand erode in another area.

DIFFERENTIATING INSTRUCTION

Support for Visually Impaired Students Students may have difficulty reading labels in the illustration at the bottom of the page. Reproduce the illustration and circle the labels. Review the illustration with students to make sure they understand the structure of these landforms. If possible, project slides of the landforms discussed in the section onto a large screen. All students would benefit from seeing different and larger images of these landforms.

CHAPTER 16 SECTION 3

CRITICAL THINKING

The sandy shores of barrier islands are favorite spots for building beach houses and hotels. Because the sands are always shifting and the low-lying islands are vulnerable during storms, the structures built on the shores often are destroyed or damaged by floods, hurricanes, or shifting sand. People often rebuild in the same locations, using insurance that is subsidized by taxpayers. Some experts in beach systems say that beach structures should be abandoned to allow beach sands to shift naturally. Some say that anyone who insists on building on beaches should not get government funds to rebuild if their home or business is destroyed. Others think property owners have the right to build on beaches and should get the same assistance as people who build in other areas prone to natural hazards, such as earthquakes and tornadoes. Present both sides of this argument to students and moderate a debate on the issue.

BARRIER ISLAND Fire Island, a barrier island off the coast of Long Island, New York, is about 51.5 kilometers long.

Sandbars may also form on coasts with straight shorelines. These bars are not attached to the shore. Instead, they run parallel to it at some distance offshore. They are called barrier islands. Barrier islands are found along the eastern coast of the United States from New York to Texas. Some well-known examples include Fire Island in New York, Atlantic City Beach in New Jersey, and Hatteras in North Carolina. Galveston, Texas, and the Florida cities of Palm Beach and Miami Beach are also located on barrier islands. Padre Island, a barrier island in Texas, is notably long, measuring about 160 kilometers from end to end.

Barrier islands, sandbars, and beaches are not permanent features. The sand that composes them is constantly being removed by waves, storms, and longshore drift. As long as the sand is replenished from rivers or from other beaches, the feature persists. However, if the amount of sand removed exceeds the amount replenished, then the feature erodes.

Types of Shorelines

Geologists have many ways to describe shorelines. One way to classify shorelines is by whether they are regular or irregular. An irregular shoreline is one with many inlets and bays. A regular shoreline is one that is relatively straight. Whether a shoreline is irregular or regular depends in part on how it was formed.

Irregular Shorelines

The coast of Maine, with its many inlets and bays, has an irregular shoreline. Maine's zigzagging shoreline, including the many offshore islands, measures about 5565 kilometers. The actual coast of Maine, however, extends over a distance of only 367 kilometers.

Most irregular shorelines appear to have formed when coastal areas were flooded by the sea. This flooding occurred either because the land sank or the sea rose, which the sea did at the end of the Ice Age. The coast of Maine is thought to have formed when rising seas flooded hilly regions cut by river valleys. As shown in the diagram at the top of page 353, the drowned main valleys became short, deep, narrow bays. The divides between the valleys became headlands. The drowned tributary valleys became branches of the bays. Many hills were partly drowned and became islands. Some were completely drowned and became shallow areas called shoals.

The land along the Atlantic coast from New York to Florida is a coastal plain. Much of this area also has an irregular shoreline formed by drowning of the land. However, unlike the river valleys of Maine, those of the coastal plain were wide and gently sloping. Thus, the bays of the coastal plain are wider and longer than those in Maine. The water close to shore is not as deep as in Maine, and there are fewer islands. Chesapeake Bay is the drowned valley of the lower Susquehanna River and its tributaries. New York Bay is the drowned valley of the lower Hudson River. Some irregular shorelines were formed by glaciers. During the last ice age, glaciers in Norway, Alaska, and other near-polar regions reached the

---

DIFFERENTIATING INSTRUCTION

Developing English Proficiency To help students learn the names of different landforms created by wave erosion, have them make picture dictionaries. Students should draw a simple sketch of each landform, followed by its name, a pronunciation guide, and a definition. Review the dictionaries with students to make sure their comprehension is correct. Students can add to their dictionaries as they read other sections in the text.

Formation of an Irregular Shoreline

1. Maine's irregular shoreline formed in a hilly region cut by river valleys.
2. A rise in sea level drowned the valleys, turning them into bays. Ridges became headlands, and hills became islands.
3. Erosion by waves and currents formed sea cliffs and spits.

oceans. Glacial valleys were formed below the present-day sea level. When the glaciers retreated, the sea flooded parts of the valleys, forming long, deep, steep-sided bays called **fjords** (fyawrdz). The tributary valleys of fjords are hanging valleys. Spectacular waterfalls drop down the walls of the fjords from these hanging valleys. Fjord shorelines are found in British Columbia, Greenland, Labrador, Chile, New Zealand, Norway, and Alaska.

Regular Shorelines

The shorelines of the west coasts of North and South America are comparatively regular. For the most part, they do not have many deeply indented inlets or drowned valleys, although they do have a number of shallow coves. The reason for their regularity is that these shorelines are, for the most part, on the boundary line between two sets of plates. As the denser Pacific Ocean plates meet the continental plates and subduct beneath them, mountains rise and deep undersea troughs form. Long lines of mountains run parallel to the coasts, and the ocean floor slopes steeply to great depths. These coasts are bordered in many places by sea cliffs and sea stacks.

16.3 Section Review

1. How are a sea arch and a sea stack related?
2. What is a beach?
3. Explain how a fjord is formed.
4. **CRITICAL THINKING** Create a chart that summarizes the ways that waves affect shorelines.
5. **GEOGRAPHY** Padre Island is a barrier island off the coast of Texas. Find Padre Island on the physical map on pages 708–709. Explain why its land area might change over time.

Chapter 16 Wind, Waves, and Currents 353

CHAPTER 16 SECTION 3

Visual Teaching

Interpreting Diagrams
Ask students to predict how the landscape in step 3 would have looked after the sea-level rise if the original landform had been a rolling plain with one high mountain rather than a hilly region cut by river valleys. The sea would have mostly covered the landscape, with the high mountain becoming an isolated island.

ASSESS

1. A sea stack forms when the roof of a sea arch falls into the sea.
2. the area of shore between the high-tide level and the low-tide level
3. During the last ice age, the sea level was lower. Glaciers carved river valleys, and when the glaciers retreated and the sea level rose, the sea flooded parts of the valleys, forming long, steep-sided bays called fjords.
4. Charts should include: erode sand from beaches and other landforms; deposit sand on beaches and other landforms; erode rocky cliffs; create sea caves, sea arches, and sea stacks.
5. Waves and storms constantly remove the sand that forms barrier islands. The islands remain as long as the sand is replenished from rivers or other beaches. If more of Padre Island's sand is removed than is replenished, Padre Island could decrease in size.

MONITOR AND RETEACH

If students miss . . .

Question 1 Use the photograph on page 349 to explain how a collapsed sea arch can become a sea stack.

Question 2 Have students reread "Beaches." (p. 350) If necessary, review the meanings of high tide and low tide.

Question 3 Reteach "Irregular Shorelines." (p. 352) To help students understand how a fjord forms, ask them to visualize a deep, steep-sided canyon that suddenly becomes partially filled with water.

Question 4 Have students reread the entire section. (pp. 349–353) Tell them to list ways in which waves act on shorelines.

Question 5 Have students reread the first two paragraphs on page 352. Ask: What makes up a barrier island? sand Where does the sand come from? rivers or other beaches What effect do waves have on these islands? Waves erode the sand.

CHAPTER 16 ACTIVITY

PURPOSE
To infer how wave frequency and beach slope affect erosion and deposition on shorelines

MATERIALS
The clear container should be approximately 12" × 6" × 4".

PROCEDURE
❷ When adding water, students should hold the beaker as close to the water's surface as possible to prevent splashing that might disturb the sand.

❻,❽ Students may have to replace sand-gravel mixture that was poured off with the water during the previous step.

CHAPTER 16 LAB Activity

Beach Erosion and Deposition

SKILLS AND OBJECTIVES
- **Compare** the effects of wave frequency on beach erosion and on deposition.
- **Describe** how shoreline slope affects the rate of beach erosion.
- **Determine** how beach composition affects erosion.

MATERIALS
- 200 mL Mixture A (75% sand, 25% gravel)
- 200 mL Mixture B (25% sand, 75% gravel)
- clear rectangular container
- masking tape
- thin permanent marker
- metric ruler
- 100-mL beaker
- water
- small wooden block (5 cm × 10 cm)
- stopwatch

If you visit a beach often, you will become familiar with the many features that make a beach unique. For example, you will begin to notice how the beach slope changes as you approach and enter the water. However, if you return to the beach after a major storm, you will see how quickly the beach can change as wind and waves remove or deposit large amounts of sand and other material. In this activity, you will explore how waves influence the development of shorelines through erosion and deposition.

Procedure

❶ Use Mixture A to construct a model beach at one end of the plastic container. The model beach should extend across the entire end of the container and slope gradually toward the opposite end. The beach should be 3 cm deep at its highest point and extend 5 cm into the container, as shown in the diagram below. (You may not need to use the entire 200 mL of mixture.)

❷ Tape a strip of masking tape to a vertical edge of the container opposite the model beach. Use the marker to mark the tape 2 cm from the bottom. Use the beaker to fill the container with water to the 2-cm mark. Add the water slowly at the opposite end of the beach so you do not disturb the setup.

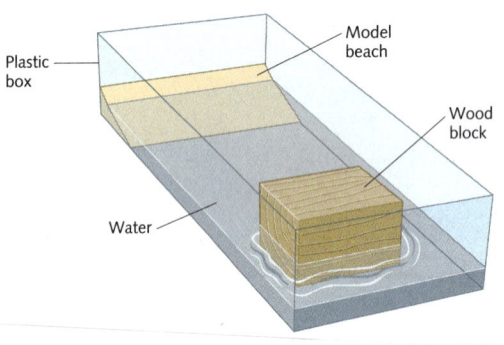

❸ Place the wooden block in the water at the end of the container opposite the shoreline. Place the long side of the block parallel to the shoreline (see diagram to the left). Without removing the block from the water, generate waves by moving the block up and down for two minutes at a rate of one wave per second. Make sure the rate and size of the waves remain constant.

❹ After two minutes, sketch a cross-section of the slope of the model. In your sketch, record the former slope and position of the model beach and show the new distribution of the sand and the gravel.

354 Unit 4 Earth's Changing Surface

5. Carefully pour the water from the container into a container provided by your teacher. Do not pour the water down the sink.

6. Reconstruct the model beach and refill the container with water as before.

7. Repeat Step 3, this time creating waves for two minutes at a rate of two waves per second. Sketch a cross-section of the beach, noting the former shoreline position and the new distribution of sand and gravel.

8. Repeat Step 6, making a steeper slope than in the first two trials. The sand-gravel mixture should extend 2 cm into the box.

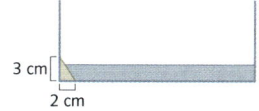

9. Repeat Step 3, using a rate of one wave per second. After two minutes, sketch a cross-section of the shoreline. Record the former shoreline position and the new distribution of sand and gravel.

10. Empty the water and the sand-gravel mixture into the container provided by your teacher.

11. Use Mixture B to construct a beach with the same slope as in Step 8. Repeat Steps 3–5.

Analysis and Conclusions

1. Based on your models, describe how the rate at which waves strike a shoreline affects the erosion of a gradually sloping model beach.

2. How does the initial slope of the beach affect erosion?

3. Describe the distribution of the sand compared to the gravel in each of your four trials. Why is there a difference?

4. Explain how the composition of a beach affects beach erosion. Give evidence for your answer.

5. Describe any depositional features formed due to wave action on your model beaches. In which trial was deposited material most noticeable?

6. Describe how a storm may vastly change the features of a beach.

7. Beach erosion is a significant problem for many seaside communities. Suggest ways to stabilize shorelines in order to prevent erosion.

CHAPTER 16 ACTIVITY

ANALYSIS AND CONCLUSIONS

1. The rate of beach erosion increases as the frequency of waves increases. Erosion may also occur higher up on the shore as the rate increases.

2. A steeply sloped beach erodes more rapidly than one with a gentler slope.

3. The smallest particles are eroded from the beach and transported just offshore, where they are deposited. The gravel is left exposed on the shore. Because gravel is heavier than sand, waves do not move gravel as far as sand particles.

4. The more gravel in the beach, the less erosion. Students can use one of their observations from the lab as evidence.

5. Students should notice the formation of a low mound of sand just offshore. The deposited sand may be more noticeable in trials with a more steeply sloped beach.

6. A storm produces large, powerful waves that hit the beach with great frequency. The waves also reach high up on the beach. This combination of factors makes storm waves a powerful erosion force—washing away beach sand, changing the slopes of beaches, and depositing eroded beach sand in new areas onshore and offshore.

7. **Sample Response:** building sea walls to prevent large waves from striking the beach or replenishing sand eroded by storm waves

Refer students to Safety in the Earth Science Laboratory on pages xxii–xxiii.

CHAPTER 16 REVIEW

Summary of Key Ideas

16.1 Wind changes Earth's surface by moving and depositing loose sediment. Desert pavement and blowouts result from wind erosion. Loess is an example of a wind-deposited sediment. Winds also deposit sand into formations called sand dunes. Sand supply, wind strength, and vegetation affect a sand dune's shape.

16.2 Most waves result from winds. A wave is described by its height, wavelength, period, and speed. As a wave passes through water, the water particles move in a circular motion. This motion extends downward to a depth of about one half the wavelength. As the water depth becomes more shallow, wave motion is distorted. Waves approaching a shoreline at an angle refract and approach the shore at less of an angle. Swash, backwash, longshore currents, and rip currents move water and sediments on beaches.

16.3 Waves affect the shoreline by eroding and depositing materials. Sea cliffs, sea caves, sea arches, and sea stacks result from shoreline erosion. Sandbars, spits, baymouth bars, hooks, lagoons, and barrier islands result from shoreline deposition. Irregular shorelines form when hilly, uneven coastal areas are flooded by the sea. The regular shorelines of western North and South America are the result of plate interactions.

Key Vocabulary

beach (p. 350)
deflation (p. 341)
desert pavement (p. 341)
dust storm (p. 340)
fetch (p. 344)
fjord (p. 353)
loess (p. 342)
period (p. 345)
refraction (p. 345)
sandbar (p. 351)
sand dune (p. 342)
ventifact (p. 341)
wave height (p. 344)
wavelength (p. 344)

356 Unit 4 Earth's Changing Surface

Vocabulary Review

1. sand dune
2. deflation
3. refraction
4. wavelength
5. fetch

Concept Review

6. Because sand particles are larger and heavier than clay and silt particles, stronger winds are necessary to move sand. Clay and silt particles are lifted higher into the air and are carried farther than sand.

7. Verifacts and desert pavement both result from wind erosion in a dry (desert) environment.

8. The shape of a dune depends on the supply of sand, the strength and steadiness of winds, and the amount of vegetation present—conditions that differ from place to place.

9. A long fetch (the length of open water over which the wind blows) causes bigger waves. Wave height is the distance between the wave crest and trough. Wavelength is the distance from one wave crest to the next, or the wave's linear size. The period of a wave is the time it takes one wavelength to pass a certain point.

10. Refraction causes waves to swing around to hit the shore more head on. Refraction occurs because the end of the wave closest to shore scrapes the bottom first and slows down. The end of the wave in deeper water continues at the faster speed. As that end catches up, the wave straightens out and ends up nearly parallel to the shoreline.

11. A longshore current forms as a result of the movement of waves toward shore behind the breaker line combined with backwash away from the shore.

12. The most important factors are the materials available in the area and the slope of the shoreline. For example, backwash removes sand and clay particles along shorelines where the nearby sea floor is steep. This leaves a beach of mostly pebbles. Where the sea floor slopes gently and a constant supply of sand is available to replenish the beach, only clay particles are washed away, leaving mostly sand.

13. The sand that forms beaches formed from weathered and eroded inland rock that is carried to the sea by rivers and also from the weathering and erosion of rocky headlands. Ocean waves and currents deposit sand in some places, forming beaches. Waves and currents also remove sand and destroy beaches in other areas. If a current carries enough sand, a sandbar can form across the mouth of the bay. Sandbars that form some distance from and parallel to the shore, but are unconnected to the mainland, are barrier islands.

14. The area near shore was once a hilly region cut by several river valleys. At the end of the last ice age, the sea level rose, causing the ocean to move farther inland. The river valleys were flooded, forming bays.

Vocabulary Review

Write the vocabulary term from the key vocabulary list that best completes the analogy.

1. ___?___ : barchan as sandbar : spit.
2. abrasion : ventifact as ___?___ : desert pavement.

Write the term from the key vocabulary list that best completes the sentence.

3. Waves that approach a shore at an angle are turned to be more parallel to the shore through a process called ___?___.
4. The distance from the crest of one wave to the crest of the next wave is the ___?___.
5. The height of a wind-driven wave is affected by ___?___, which is the distance of open water over which the wind blows.

Concept Review

6. Compare how clay, silt, and sand particles are carried by the wind. Which particles are likely to be carried the farthest by the wind? Why?
7. Why might you find ventifacts and desert pavement in the same region?
8. Why do sand dunes have different shapes?
9. Explain how the following terms relate to an ocean wave: fetch, wave height, wavelength, period.
10. Explain how wave refraction occurs at a shoreline.
11. Explain how a longshore current forms.
12. Identify some factors that help determine whether a beach is sandy or pebbly.
13. Explain how erosion and deposition are involved in forming a beach and the features found near a beach, such as sandbars and barrier islands.
14. Explain how Maine's irregular shoreline formed.
15. **Graphic Organizer** Complete the cause-and-effect diagram below.

Critical Thinking

16. Analyze Why is quartz sand a better tool of wind erosion than gypsum sand or calcite sand? (The table "Properties of Some Common Minerals" on pages 700–701 may be helpful.)

17. Communicate Make and label sketches showing the windward and leeward sides of sand dunes for winds blowing steadily from the north, the south, the east, and the west.

18. Hypothesize How might the sand grains in a sand dune be different from the sand grains in a river delta? Explain your thinking.

19. Calculate Find the speed of a wave that has a wavelength of 300 meters and a period of 25 seconds.

20. Analyze At what depth would a 300-meter-long wave first be slowed by the sea floor? At what depth would a 10-meter wave be slowed?

21. Predict Would you expect to find a longshore current in a lagoon behind a long sandbar? Explain your answer.

Interpreting Graphs

Particle sizes in sand dunes and loess vary, as do the amounts of various sizes of particles. The graph at the right shows the percent of each particle in a sand dune sample and the percent of each particle in a loess sample.

22. What is the size range for medium sand?

23. Give the name, size range, and percentage of the most abundant particle in the sand dune sample.

24. Give the name, size range, and percentage of the most abundant particle in the loess sample.

25. How does the percentage of coarse silt in the sand dune compare with the percentage of coarse silt of the loess?

26. How does the range of sizes for the sand dune compare with those of the loess?

27. In which sample are the particles more sorted? Explain. You may want to refer to Chapter 14 for information about sorting.

Internet Extension

How Do Storms Affect Coastlines? Analyze images of shorelines before and after storms.

Keycode: ES1608

Writing About the Earth System

SCIENCE NOTEBOOK The gray sand found on many beaches in Hawaii consists of small grains of weathered and unweathered basaltic lava, broken-down coral, and broken-down shells of various marine animals. Describe how the geosphere, the atmosphere, the hydrosphere, and the biosphere are involved in the formation of these beaches.

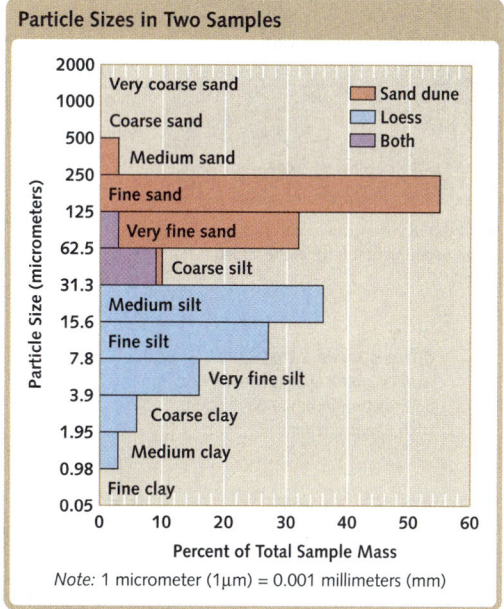

Note: 1 micrometer (1μm) = 0.001 millimeters (mm)

WRITING ABOUT SCIENCE

Writing About the Earth System

Geosphere: The basaltic lava grains are an igneous rock that reached Earth's surface through volcanic activity and was then weathered and eroded. **Atmosphere:** Rain and wind help weather and erode rocks and other particles that form the beach sand.

Hydrosphere: Rain, flowing water, and ocean waves help weather and erode rock that becomes beach sand. **Biosphere:** corals and other marine animals die and leave behind bone or shell that is broken down and transported to beaches by weathering and erosion.

CHAPTER 16 REVIEW

How Do Storms Affect Coastlines?
Keycode: ES1608

Use Internet Investigations Guide, pp. T59, 59.

(continued)
Ridges between the valleys became headlands. Some of the hills became islands.

15. underwater earthquake or landslide

CRITICAL THINKING

16. Quartz is harder than gypsum or calcite.

17. Students' sketches should show sand dunes with a long, gentle slope on the windward side and a short, steep slope on the leeward side. Dunes might be barchans or transverse.

18. A delta would probably have more small particles than a sand dune would. Rivers carry sand as well as silt and clay, all of which settle out as sediment in a delta. Dunes are composed primarily of sand particles.

19. 12 meters/second

20. less than 150 m; less than 5 m

21. No; A lagoon behind a sand bar is an area of sheltered, calm water where there would be no waves coming from the open ocean or backwash coming from the shore.

INTERPRETING GRAPHS

22. 250–500 micrometers

23. fine sand; 125–250 micrometers; 55 percent

24. medium silt; 15.6–31.3 micrometers; 36 percent

25. The sand dune contains a slightly larger percentage of coarse silt.

26. The size ranges of loess particles are smaller than those of sand dune particles.

27. The closer in size the particles in a sample are, the more sorted the sample. Because the sand dune sample has fewer particle-size ranges, it is more sorted than the loess sample.

UNIT 4 FEATURE

▶ Focus

Objective
Recognize the importance of the hydrosphere, particularly the supply of fresh water, to living things.

Set a Purpose
Have students read to find answers to the question "What are some successful techniques people have used to manage Earth's supply of fresh water?" Among the techniques used to manage water resources better are strict measurement and monitoring of water use, fixed allocation of water, and tiered pricing structure. Farmers have been given low-interest loans so that they can implement more efficient irrigation methods, such as drip irrigation, spray irrigation, surge flooding, and controlled runoff.

UNIT 4 The EARTH SYSTEM

Harnessing Earth's Water

Water is one of Earth's most precious resources. Civilizations have risen and fallen because of faulty water–management practices. Current demands on water are greater than ever before. Our future depends on how we use and preserve water.

Have you ever wondered where your drinking water comes from? Water is a finite resource. While the water supply may seem to be nearly infinite, this is possible only if the supply is carefully managed. Sustainable water management means meeting current demands on the water supply without jeopardizing the future supply.

Water covers more than 70 percent of Earth's surface. Nearly all of that is ocean water, which is salty. Most of Earth's fresh water is frozen in glaciers and icebergs or trapped as moisture in the soil and in the atmosphere. In fact, only 0.01 percent of Earth's fresh water is available or suitable for human and animal use.

SALT WATER IRRIGATION fed this cabbage field in Ashalim, Israel.

Overuse and poor control of water has caused water quality and quantity to decrease. In 2000, the World Health Organization estimated that 1.1 billion people did not have access to safe drinking water—one out of every six people on Earth. In the United States, the need for fresh water has led to massive projects such as the desalination plant in Yuma, Arizona. Desalting is an excellent way to cleanse and reuse water, but it involves an enormous cost that the government is currently subsidizing.

For the future, consider how Israel manages its water supply. Because their country is situated in a desert where water is extremely scarce, Israelis have had to figure out how to use their meager water resources effectively.

TOMATOES were grown with salt water on an experimental farm in the Israeli desert.

GRAIN SUPPLY A vital portion of the world's grain supply depends heavily on irrigation from the diminishing High Plains aquifer.

358

DIFFERENTIATING INSTRUCTION

Reading Support Have students create a graphic organizer. This will help them to organize the information they read in this article and to retain the important facts. Students can answer the question in Set a Purpose by filling in a chart like this one.

Location	Techniques for Managing Water
Arizona	Example 1: desalination plant
Israel	Example 2: strict water policies; farming methods that use less water
U.S. High Plains	
NW Texas	

358 Unit 4 Earth's Changing Surface

and the ENVIRONMENT

Israel's Water Commission developed strict water policies, including fixed allocation of water based on the crops and the area being cultivated, a tiered pricing structure so that using greater amounts of water involved paying more per unit, and strict measurement and monitoring of water use.

Israel has also developed farming methods that use less water, and has provided low interest loans to help farmers improve their irrigation systems. As a result, between 1951 and 1985, water use per hectare dropped 37 percent, while output per cubic meter of water nearly tripled. Long-term projects such as Israel's water-conservation policies prove that sustainable water management can be fair, successful, and practical.

The High Plains aquifer, a huge underground storehouse of water, has been collecting for several ice ages. Far more water is now being used than is seeping back in each year. In 1982, a federal study predicted that the High Plains aquifer (then known as the Ogallala aquifer) would be nearly used up by the year 2020.

Although the region irrigated by the High Plains aquifer represents a significant portion of the world's grain supply, the question of what to do when the aquifer runs dry has not yet been resolved. Plans proposed to divert other water sources—such as the rivers of British Columbia in Canada—carry with them huge price tags, as well as environmental and political problems.

Just finding more water from another source is not enough—at best, it is a temporary solution. Conservation is necessary. We need to take better care of the water we have and work to waste less of it. Many farmers in the United States have been working on water conservation. For example, the traditional flood irrigation methods waste half the water, as much of the water is lost to the atmosphere. New methods, such as drip irrigation, spray irrigation, surge flooding, and control runoff, greatly reduce the amount of wasted water and have been extremely successful in places like northwest Texas.

What does it take for people to appreciate the value of water? Do you believe that people would be more likely to appreciate the worth of their water if they knew what it costs to supply it? Water is necessary for life and must be distributed fairly and in a sustainable fashion. If we are to survive—as individuals, as cities, as countries, as a global community—we must develop ways to guarantee safe and accessible water to all of Earth's citizens.

UNIT INVESTIGATION
CLASSZONE.COM

What Are the Costs and Benefits of Damming a River?
Dams are constructed to provide hydroelectricity, flood control, and water for public use. However, damming a river dramatically changes the area's ecosystem. Large areas of land are submerged, and plants and animals adapted to shallow, moving water are replaced by species adapted to deep, still water. Analyze the costs and benefits of Glen Canyon Dam on the Colorado River, and evaluate proposed or existing dams in your own state or region.
Keycode: ESU401

UNIT 4 FEATURE

INVESTIGATIONS
CLASSZONE.COM

What Are the Costs and Benefits of Damming a River?
Keycode: ESU401

Use Internet Investigations Guide, pp. T60–T61, 60–61.

Investigation Photograph: The dam in the photograph is the Hoover Dam, located on the Arizona-Nevada border. The reservoir behind the dam forms Lake Mead. The dam provides hydroelectric power and water for irrigation and public use for much of Arizona, Nevada, and southern California.

UNIT 4 ASSESSMENT

UNIT 4 STANDARDIZED TEST Practice

ANSWERS

1. d, Backwash will carry away smaller particles.

2. c, The features form when carbon dioxide in the water reacts with the limestone bedrock.

3. c, The river's velocity decreases and particles settle out.

4. c, The pumping of water from the ground decreases the amount of groundwater.

5. a, Upward expansion of granite leads to cracks along curved joints and the breaking away of loosened granite sheets.

6. b, Wind transports the soil particles from one place to another.

7. c, Kames form when a stream flowing across a glacier deposits sediments at the ice front.

8. d, An oxbow lake forms from a meander, a river feature found only on flood plains.

9. d, Desert pavement forms under conditions of little or no soil movement.

10. b, The longer the fetch is, the higher the waves.

11. d, The graph on page 318 shows that Mount Mitchell is below the snow line.

Directions (1–12): For *each* statement or question, select the word or expression that, of those given, best completes the statement. Record your answer on a separate answer sheet.

1. A beach near a steeply sloping sea floor is likely to be covered with
   - (a) mud
   - (b) silt
   - (c) sand
   - (d) pebbles

2. Features of karst topography tend to form in areas where the bedrock is
   - (a) granite
   - (b) shale
   - (c) limestone
   - (d) sandstone

3. A river deposits sediments when
   - (a) its discharge increases
   - (b) its velocity increases
   - (c) it empties into a large body of water
   - (d) its channel narrows

4. The most likely cause for the lowering of a water table is
   - (a) salt water seeping into the ground
   - (b) rainwater seeping into the ground
   - (c) pumping water from the ground
   - (d) pumping water into the ground

5. Exfoliation of granite at Earth's surface occurs because
   - (a) the granite has undergone upward expansion
   - (b) the granite contains quartz
   - (c) the granite is not very porous
   - (d) the granite is impermeable

6. Loess is an example of
   - (a) residual soil
   - (b) transported soil
   - (c) depleted soil
   - (d) river sediment

7. A kame is composed of
   - (a) till
   - (b) ice
   - (c) outwash
   - (d) talus

8. An oxbow lake typically forms in a
   - (a) steep-sided canyon
   - (b) U-shaped valley
   - (c) V-shaped valley
   - (d) flood plain

9. Which of the following is not an example of a mass movement?
   - (a) talus
   - (b) lahar
   - (c) earthflow
   - (d) desert pavement

10. Ocean waves are often higher than lake waves because ocean waves
    - (a) form in deeper water
    - (b) have a greater fetch
    - (c) are more easily refracted
    - (d) are affected by backwash

Base your answers to questions 11 and 12 on the table below and the snow line elevations graph on page 318.

Elevations and Latitudes

Mountain	Elevation (meters)	Latitude (degrees)
Pikes Peak, Colorado	4301	38° N
Popocatépetl, Mexico	5465	19° N
Mount Mitchell, North Carolina	2037	36° N
Mount Whitney, California	4418	37° N
Mount McKinley, Alaska	6194	63° N

11. Which mountain is not likely to have permanent snow year-round?
    - (a) Pikes Peak
    - (b) Popocatépetl
    - (c) Mount McKinley
    - (d) Mount Mitchell

12. At approximately what elevation would you expect to find the snow line on Mount Whitney?
    - (a) 2000 meters
    - (b) 3000 meters
    - (c) 4000 meters
    - (d) 4400 meters

Directions (13–19): Record your answers on a separate answer sheet.
Base your answers to questions 13 through 15 on the reading passage below.

Stream Drainage Patterns

The pattern that the streams in a river system form on Earth's surface is influenced by the bedrock beneath. Geologists have identified different types of drainage patterns.

A dendritic drainage pattern is the most common pattern. This drainage pattern looks like the branches of a deciduous tree. It is common in areas in which the bedrock does not have any large-scale structure, or is fairly uniform. The Mississippi River system has a dendritic drainage pattern.

A rectangular drainage pattern occurs in areas where fractures in the bedrock affect the direction of stream flow.

In such regions, streams typically make right-angle bends to follow the fractures.

A trellis drainage pattern tends to form in areas where the terrain consists of alternating bands of rock that vary in their resistance to erosion. The streams typically run through valleys between rocky ridges that are more resistant to weathering. Hence, streams travel more or less parallel to each other.

A radial stream drainage pattern occurs when a number of streams flow away from a single point, such as from a mountain peak.

13. Identify each of the stream drainage patterns below.

A. 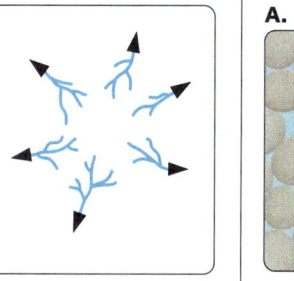 B.

14. Suppose a river and its tributaries flow across a terrain that has many straight joints in the bedrock. What type of stream drainage pattern would you expect to see?

15. In the Valley and Ridge Province of the Appalachians, rivers flow through valleys between raised ridges of rock. What kind of stream drainage pattern would you expect to find there?

Base your answers to questions 16 through 19 on the diagrams below. Each diagram shows a sample of rock particles.

A.  B.  C.

16. Explain the difference between the porosity and the permeability of a material.

17. Compare samples A and B. Which sample has greater porosity? Why?

18. Compare samples A and C. Which sample has greater porosity? Why?

19. Sample A is from a layer of coarse-grained sand. Is this sand layer likely to be permeable? Explain.

UNIT 4 ASSESSMENT

(continued)

12. b, This elevation is indicated above 37° N latitude on the graph.

13. a. dendritic; b. radial

14. rectangular

15. trellis

16. The porosity of a material is the percent of the material's volume that is pore space. The permeability of a material is the rate at which water or other liquids pass through the pore spaces in the material.

17. Sample A has greater porosity because its particles, compared to those in Sample B, are round and well sorted, leaving more space between the particles. In Sample B, the particles are more angular and fit together more closely.

18. Sample A has greater porosity because its particles, compared to those in Sample C, are well sorted. In Sample C, smaller particles fill up the pore spaces between the larger particles.

19. Yes; sand has high permeability because the pore spaces between the grains of sand are connected, so water can flow through.

UNIT 5

UNIT OVERVIEW
Discuss some decisions that students have made within the past week that were at least partly based on the weather and climate. Examples might include what clothes to wear, how to spend leisure time, or whether to take medication for asthma. Then help students identify the locations of the chapter images as pinpointed on the image of Earth. It might be helpful to use a classroom globe for this. Have students form small groups to discuss the question under each chapter image.

VISUAL TEACHING

Chapter 17 Atmosphere
The satellite image shows a dust cloud in Oman blowing out over the Arabian Sea in March 2000. Students will learn about the effects of dust and other air pollutants on the atmosphere.

Answer: Dust can be carried great distances by global winds. The dust particles act like tiny mirrors, reflecting sunlight back to space, lowering temperatures on Earth.

Chapter 18 Water in the Atmosphere
Cloud type is a good predictor of weather. Small, white, fluffy clouds, like those shown, indicate fair weather. Students will learn about different types of clouds and how they form.

Answer: The cumulus clouds in the photograph show very limited vertical growth. The cumulonimbus clouds associated with heavy rainfall show significant vertical growth.

UNIT 5 Atmosphere and Weather

The Earth System Day-to-day weather and regional climates influence many of the decisions you make in your life. This unit explores the processes that shape the atmosphere and how these processes affect past, present, and future climates.

CHAPTER 18 Water in the Atmosphere
How do these clouds differ from clouds that are associated with heavy rainfalls?

CHAPTER 17 Atmosphere
How could dust particles blown into the atmosphere affect temperatures throughout the world?

CONNECTIONS TO . . .

Prior Knowledge In Unit 4, students learned that moving water and wind erode Earth's surface. In Unit 5, they will study the atmosphere to learn what causes the rains, snows, and winds that cause erosion. Unit 6 proceeds with the study of the properties of water and features of oceans.

Unifying Themes Seasonal changes, differences in climate, and movement of storm systems across the planet allow Earth to remain in *equilibrium* with respect to *energy*. As students work through the chapters in Unit 5, emphasize the dynamic processes that help maintain equilibrium on our planet.

UNIT 5

CHAPTER 20 Weather

In what ways do tropical storms and hurricanes influence coastal communities?

CHAPTER 19 The Atmosphere in Motion

What causes the wind and heavy rain that strike India several months each year?

CHAPTER 21 Climate and Climate Change

What natural and artificial processes could modify the mild climate of Kyoto, Japan?

Chapter 19 The Atmosphere in Motion

Monsoons, which are seasonally changing winds, bring heavy rain and flooding to southern Asia every summer. Students will learn about different types of winds and factors that affect them.

Answer: In the summer, air rising over the warm land becomes part of a large convection cell that causes winds to blow from the oceans onto the land. This moist air rises over the mountains and produces heavy rains.

Chapter 20 Weather

The satellite photograph shows Typhoon Jelawat, which formed over the Pacific Ocean in August 2000. (A typhoon is a hurricane that forms in the western Pacific.) While over the ocean, Jelawat's maximum winds reached 200 kilometers per hour. Students will learn how different types of storms form.

Answer: These storms can cause massive flooding and severe wind damage.

Chapter 21 Climate and Climate Change

Kyoto is located on Honshu Island and is one of Japan's largest cities. Its average temperature in winter is 4°C (39.2°F) and in summer 26.5°C (79.7°F). Students will learn about climate zones and factors that affect climate.

Answer: natural processes: increase in sunspots, extensive volcanic activity; artificial processes: burning of fossil fuels, deforestation

363

CONNECTIONS TO . . .

National Science Education Standards
- Students will develop the understanding that the sun's unequal heating of Earth's surface causes convection currents to form in the atmosphere, producing wind.
- Students will understand that energy transfer at and near Earth's surface determines global climate and that climate is affected by various factors, including cloud cover, Earth's rotation, and the positions of the continents.

Unit Features

The Earth System and the Environment
Climate and Civilization, pp. 482–483

How Might Global Climate Change Affect Life on Earth? Investigate evidence of global climate change, its causes, and implications for Earth's future. *Keycode:* ESU501

Standardized Test Practice, pp. 484–485

363

CHAPTER 17 PLANNING GUIDE ATMOSPHERE

	Section Objectives	Print Resources
Chapter-Level Resources		Laboratory Manual, pp. 75–80 Internet Investigations Guide, pp. T64–T66, 64–66 Guide to Earth Science in Urban Environments, pp. 13–20, 37–44 Spanish Vocabulary and Summaries, pp. 33–34 Formal Assessment, pp. 49–51 Alternative Assessment, pp. 33–34
Section 17.1 The Atmosphere in Balance, pp. 366–368	1. Describe the formation of Earth's early atmosphere and the composition of the lower atmosphere. 2. Demonstrate how the Earth system continually recycles gases such as oxygen, carbon dioxide, and water vapor, and how natural events and human activities disturb an atmosphere in balance.	Lesson Plans, p. 57 Reading Study Guide, p. 57
Section 17.2 Heat and the Atmosphere, pp. 369–363	1. Describe how energy from the sun moves through the atmosphere by radiation, conduction, and convection. 2. Identify the characteristics of each atmospheric layer. 3. Analyze Earth's heat budget.	Lesson Plans, p. 58 Reading Study Guide, p. 58 Teaching Transparency 18
Section 17.3 Local Temperature Variations, pp. 374–377	1. Identify the factors that cause the intensity of insolation to vary from place to place. 2. Describe how the characteristics of a material affect its rate of solar absorption. 3. Analyze a temperature map.	Lesson Plans, p. 59 Reading Study Guide, p. 59 Teaching Transparency 19
Section 17.4 Human Impact on the Atmosphere, pp. 378–383	1. Discuss how human activities can affect the atmosphere. 2. Compare and contrast acid rain, smog, ozone depletion, and global warming.	Lesson Plans, p. 60 Reading Study Guide, p. 60

INVESTIGATIONS
CLASSZONE.COM

Section 17.2 INVESTIGATION
What Can You Learn from a Thermometer on a Rising Balloon?
Computer time: 45 minutes / Additional time: 45 minutes
Students analyze changes in temperatures to find the boundaries of atmospheric layers.

Section 17.3 INVESTIGATION
How Does the Temperature at One Location Change over a Year?
Computer time: 45 minutes / Additional time: 45 minutes
Students compare animations of insolation and temperature maps then list factors that affect variations.

Technology/Online Resources

Electronic Teacher Tools
Visualizations CD-ROM
Classzone.com
Online Lesson Planner
Test Generator

Classzone.com
  ES1701 Career

Visualizations CD-ROM
Classzone.com
  ES1702 Investigation, ES1703 Visualization

Visualizations CD-ROM
Classzone.com
  ES1704 Visualization, ES1705 Visualization, ES1706 Investigation

Visualizations CD-ROM
Classzone.com
  ES1707 Visualization, ES1708 Investigation, ES1709 Data Center

Classroom Management

Gather these materials for minilab and laboratory activities.

Minilab, p. 370
  small Styrofoam cup
  short thermometer
  water
  shallow pan

Lab Activity, pp. 384–385
  3 empty soup cans (1 painted dull white, 1 painted dull black, and 1 shiny silver, each with a hole punched in the unopened end for inserting a thermometer)
  3 250-mL beakers
  3 thermometers with ranges of 0°C to at least 50°C
  desk lamp or clip-on lamp with incandescent bulb of at least 100 watts
  ring stand
  3 ring clamps
  stopwatch
  topsoil
  water
  coarsely ground concrete
  2 sheets of graph paper
  Laboratory Manual, Teacher's Edition
    Lab Sheet 17, p. 142

Section 17.4 INVESTIGATION
How Does the Ozone Layer Change over Time?
Computer time: 45 minutes / Additional time: 45 minutes
Students compare images of the amount of ozone over the Southern Hemisphere.

CHAPTER 17

Atmosphere

The air you breathe is part of Earth's atmosphere.

What makes up the atmosphere, and how do human activities affect it?

INTRODUCE

Build Interest

Before sharing background information about the photograph with students, allow time for questions and comments.

Ask questions like the following:

- What does the photograph show? *an aerial view of a large coastal city*
- How does the quality of the air in the photograph compare to that of your community today? *Responses will vary depending on location, but may include mention of any visible air pollution such as smog.*
- How might the photograph be different if it was taken on the following day? *Cloud conditions and pollution levels might be different.*

Respond

- What makes up the atmosphere? *mainly gases—such as nitrogen, oxygen, and carbon dioxide—and water vapor; The atmosphere also contains tiny particles of dust and salt. Clouds form when droplets form around these particles.*
- How do human activities affect the atmosphere? *Students may be familiar with global warming, smog, acid rain, and ozone depletion.*

About the Photograph

The photograph shows Miami, Florida. Biscayne Bay is at the lower right. For a metropolitan area, Miami air pollution is relatively low, failing to meet air-quality standards on only a few days per year.

BEYOND THE TEXTBOOK

Print Resources

- Reading Study Guide, pp. 57–60
- Transparencies 18, 19
- Formal Assessment, pp. 49–51
- Laboratory Manual, pp. 75–80
- Alternative Assessment, pp. 33–34
- Spanish Vocabulary and Summaries, pp. 33–34
- Internet Investigations Guide, pp. T64–T66, 64–66
- Guide to Earth Science in Urban Environments, pp. 13–20, 37–44

Technology Resources

- Visualizations CD-ROM
- Classzone.com
- Online Lesson Planner
- Electronic Teacher Tools
- Test Generator

364 Unit 5 Atmosphere and Weather

CHAPTER 17

PREVIEW

▶ **FOCUS QUESTIONS** In this chapter you will study the atmosphere and learn more about the key questions below.

Section 1 How do other parts of the Earth system affect the composition of the atmosphere?

Section 2 How does heat move through and affect the atmosphere?

Section 3 Why does temperature vary?

Section 4 How does human activity affect the atmosphere?

▶ **REVIEW TOPICS** As you investigate the atmosphere, you will need to use information from earlier chapters.

- geosphere, hydrosphere, biosphere (pp. 8–12)
- atoms and molecules (pp. 88, 91)
- gases in magma (p. 197)

▶ **READING STRATEGY**

PREVIEW AND QUESTION

Start by skimming through Chapter 17. Notice the key ideas, the vocabulary, and the titles for each section and subsection. Look over any images or diagrams and read their captions. Record in your notebook any questions that come to mind. As you read, seek answers to your questions.

INTERNET RESOURCES
CLASSZONE.COM

At our Web site, you will find the following Internet support for this chapter.

DATA CENTER
EARTH NEWS
VISUALIZATIONS
- Auroras
- Seasonal Changes in Sunlight
- Infrared Images of Surface Temperature
- Forest Fires as Seen from Space

LOCAL RESOURCES
CAREERS
INVESTIGATIONS
- What Can You Learn from a Thermometer on a Rising Balloon?
- How Does the Temperature at One Location Change over a Year?
- How Does the Ozone Layer Change over Time?

CHAPTER 17

PREPARE

Focus Questions
Find out what knowledge students have about the atmosphere. For each focus question, have students consider the section title and then develop two or three additional questions for each section. For example, Section 3: How does temperature vary throughout the year? Which places are warmest? Which are coldest?

Review Topics
In addition to the earth science topics listed, students may need to review these concepts:

Physical Science
- Temperature and heat are not synonymous.
- Specific heat is the amount of heat needed to raise the temperature of 1 gram of a substance by 1°C.
- Different types of materials absorb and radiate heat at different rates.

Reading Strategy
Model the Strategy: To help students preview the chapter and develop questions, point to the headings and graphics in the first section. Make these observations: "The table on page 366 details the composition of dry air, and I can see from the headings and diagram on page 367 that maintaining the balance of gases in the atmosphere is a complex process. This suggests the following questions: 'How is this balance maintained?' and 'What happens when the balance is upset?'" Have students continue to use this strategy for the remainder of the chapter.

BEYOND THE TEXTBOOK

INTERNET RESOURCES
CLASSZONE.COM

INVESTIGATIONS
- What Can You Learn from a Thermometer on a Rising Balloon?
- How Does the Temperature of One Location Change Over a Year?
- How Does the Ozone Layer Change Over Time?

VISUALIZATIONS
- Videos and images of auroras as seen from the ground and from space
- Animation of seasonal changes in the amount of sunlight reaching locations on Earth

- Infrared images that show variation in surface temperature
- Images of forest fires as seen from space

DATA CENTER
EARTH NEWS
LOCAL RESOURCES
CAREERS

Online Lesson Planner

CHAPTER 17 SECTION 1

FOCUS

Objectives
1. Describe the formation of Earth's early atmosphere and the composition of the lower atmosphere.
2. Demonstrate how the Earth system continually recycles gases and how certain activities disturb an atmosphere in balance.

Set a Purpose
Have students read to find the answer to "How does the atmosphere remain in balance?"

INSTRUCT

More about...

Oxygen in the Atmosphere
Scientists use fossils and the presence of iron oxide in ancient rocks to trace the origin of oxygen in the atmosphere. Fossilized remains of cyanobacteria indicate that photosynthesis was occurring—and thus oxygen was in the atmosphere—roughly 3.5 billion years ago. In addition, Precambrian sedimentary beds contain banded iron formations, produced when iron in rocks reacted with free oxygen. Analyses of the rocks show that the amount of atmospheric oxygen increased significantly between 2.1 and 2.5 billion years ago.

17.1

KEY IDEAS

The composition of Earth's atmosphere remains fairly constant.

Gases move continually between the atmosphere and other parts of the Earth system.

The recycling of atmospheric materials maintains a delicate balance and local events may have global consequences.

PHOTOSYNTHESIS Most of the oxygen in the atmosphere is a byproduct of photosynthesis. Today photosynthesis occurs both in the oceans and on land.

The Atmosphere in Balance

In 1815, the violent eruption of the Tambora volcano in Indonesia propelled huge quantities of dust high into the atmosphere. This volcanic cloud blocked some incoming sunlight, and temperatures fell worldwide. In fact, the next year was known as the "year without a summer" in parts of Europe and North America. The global impact of this local event was dramatic but brief, and temperatures soon returned to normal levels.

The Composition of the Atmosphere

Scientists hypothesize that volcanic eruptions played the main role in forming Earth's early atmosphere. Gases released from volcanic eruptions—primarily carbon dioxide, sulfur dioxide, water vapor, and nitrogen—probably made up nearly all of this early atmosphere. Oxygen may have first entered the atmosphere as a result of sunlight splitting water vapor molecules into oxygen and hydrogen. However, the amount of oxygen in the atmosphere increased significantly as early life forms trapped the energy of sunlight through photosynthesis. The process of photosynthesis releases oxygen as a byproduct.

Today Earth's lower atmosphere is a mixture of many gases called air. The main gases in air are nitrogen and oxygen, which together form about 99 percent of dry air by volume. The remaining 1 percent is mostly argon and carbon dioxide. The atmosphere also contains tiny amounts of trace gases, such as helium, hydrogen, and neon. The table below shows the approximate percentages by volume of gases in dry air.

The percentages of nitrogen and oxygen are fairly constant throughout the atmosphere up to an altitude of about 80 kilometers. However, the amounts of some gases in the atmosphere vary from place to place and from time to time. For example, the amount of water vapor varies with location, season, and time of day. The water-vapor concentration is highest near the surface and decreases rapidly with altitude. Similarly, the amount of carbon dioxide in the air varies with the seasons. It is lowest during the periods of greatest photosynthesis (summer) and highest in winter.

Principal Gases of Dry Air	
Nitrogen	78.08%
Oxygen	20.95%
Argon	0.934%
Carbon Dioxide	0.036%

DIFFERENTIATING INSTRUCTION

Developing English Proficiency Students may not be familiar with the meaning of *trace* as used in the third paragraph on page 366, which is "a tiny or barely detectable amount." Help students arrive at this meaning from the word's context. Discuss with students other meanings of *trace*. You might have them look up the word in a dictionary.
Use Spanish Vocabulary and Summaries, pp. 33–34.

366 Unit 5 Atmosphere and Weather

In addition to gases, the atmosphere also contains a wide variety of tiny dust particles. Dust includes tiny grains of rock, dirt, pollen, salt crystals from sea spray, and soot from fires.

Recycling of Atmospheric Materials

The composition of the atmosphere has remained stable throughout Earth's recent history because our planet is an efficient recycling system. Elements and compounds are constantly moving between the atmosphere and the other parts of the Earth system—the geosphere, hydrosphere, and biosphere. An overall balance is maintained, however, because the amount of a given substance leaving the atmosphere equals the amount of that same substance entering the atmosphere over the same period of time.

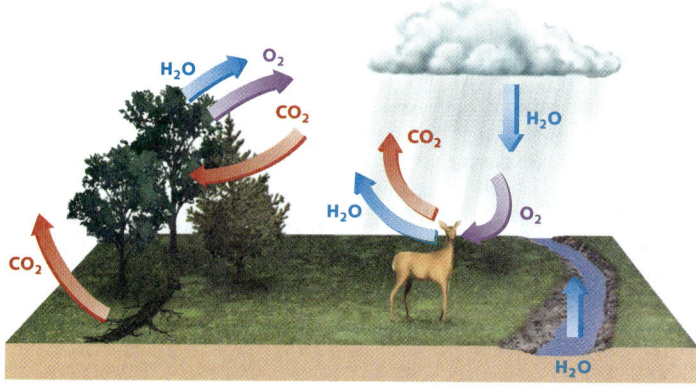

A COMPLEX BALANCE This illustration shows some of the key pathways by which substances important to living things cycle into and out of the atmosphere.

The illustration above provides a simplified view of the ways in which oxygen (O_2), carbon dioxide (CO_2), and water (H_2O) cycle in and out of the atmosphere. For example, plants take carbon dioxide from the air for photosynthesis and release oxygen as a byproduct of that same process. In contrast, animals inhale oxgyen and exhale carbon dioxide. Carbon dioxide is also returned to the atmosphere through the decomposition of organic materials. Finally, water vapor enters the atmosphere through evaporation, transpiration from plants, and the exhaled breath of animals. Water leaves the atmosphere in the form of precipitation.

A Delicate Balance

The balance maintained by recycling in the atmosphere may be disturbed by a variety of factors, both natural and of human origin. For example, a large body of evidence indicates that the amount of carbon dioxide in the atmosphere has steadily increased in recent years. Data collected at the Mauna Loa Observatory in Hawaii show that average CO_2 levels rose more than 16 percent between 1959 and 1999. This increase is due primarily to human activities, especially the burning of fossil fuels, such as coal,

---

CHAPTER 17 SECTION 1

DISCUSSION

Water is the only substance in the atmosphere found as a solid, liquid, and gas. Have students offer examples of water in all three states of matter. **solid: snow or hail; liquid: rain; gas: water vapor** Students may think that clouds are water vapor. Tell them that clouds consist of tiny water droplets that form when water vapor in the air condenses on small particles of dust, smoke, or salt crystals.

More about...

Trees in the Balance
Deforestation is another human activity that contributes to increasing levels of CO_2, and, therefore, global warming. Like all green plants, trees remove carbon dioxide from the air during photosynthesis. When large tracts of forests are cut down, overall rates of photosynthesis decrease and less CO_2 is removed from the atmosphere. The combined effect of deforestation and the burning of fossil fuels may be responsible for the 0.6°C increase in global temperatures over the past 100 years. If levels of atmospheric CO_2 continue to increase at current rates, average global temperatures may rise by as much as 4.5°C by the year 2060.

DIFFERENTIATING INSTRUCTION

Reading Support Books often include diagrams to clarify, reinforce, summarize, or extend concepts explained in the text. The diagram on page 367 summarizes the information discussed in the second paragraph under the head "Recycling of Atmospheric Materials." Ask students to correlate each arrow in the diagram with one of the sentences in the paragraph. For example, the arrow labeled H_2O that points downward from the cloud relates to the sentence "Water leaves the atmosphere in the form of precipitation." The arrow labeled H_2O that points upward from the stream relates to the phrase "through evaporation" in the sentence "Finally, water enters the atmosphere through evaporation, transpiration from plants, and the exhaled breath of animals."
Use Reading Study Guide, p. 57.

CHAPTER 17 SECTION 1

ASSESS

1. Earth's early atmosphere was made up primarily of carbon dioxide, sulfur dioxide, water vapor, and nitrogen. Earth's present atmosphere contains mainly nitrogen and oxygen—the oxygen is a byproduct of photosynthesis.

2. nitrogen and oxygen

3. Diagrams should show carbon dioxide entering the atmosphere through respiration and the decomposition of organic materials. Carbon dioxide leaves the atmosphere through the process of photosynthesis.

4. Less carbon dioxide would be removed from the air.

5. Microorganisms decompose organic materials and return carbon dioxide to the air. Certain bacteria also convert nitrogen in soil into a form that can be used by other living things.

CAREER

Environmental Consultant

Environmental consultants are able to combine their interest in the environment with their talent for creative problem solving. One aspect of the environment in which many consultants specialize is air quality. While some consultants focus their attention on outdoor air quality, others work to improve the indoor air quality of buildings and homes. By keeping up-to-date on government regulations and advances in industrial technology, these consultants help companies monitor and minimize air pollution in a cost-effective way. An environmental consultant may work with a company's employees to design a scrubber for a smokestack or a more effective air-ventilation system for the inside of a building.

Most environmental consultants hold a bachelor's or an advanced degree in environmental, chemical, or civil engineering. In order to specialize in air quality, a strong background in atmospheric science is especially helpful. Computer knowledge is also essential for developing the sophisticated computer models that are used to assess air-quality trends and potential problems. A willingness to travel is important, since environmental consultants must visit sites to perform air-quality tests. One reward of being an environmental consultant is knowing that you are helping to promote a cleaner environment.

ENVIRONMENTAL CONSULTANTS study both outdoor and indoor environments. This scientist is in Antarctica.

CAREERS CLASSZONE.COM
Learn more about a career as an environmental consultant.
Keycode: ES1701

gasoline, and natural gas. These rising CO_2 levels are suspected of contributing to global warming, which you will read about in Section 17.4.

The atmosphere's sensitive balance involves not only matter like gases but also energy. Energy from the sun is constantly entering the atmosphere. The movement of energy through the atmosphere and other parts of the Earth system plays a critical role in keeping Earth habitable. You will learn more about Earth's energy balance in the next section.

17.1 Section Review

1. Describe how the composition of the atmosphere has changed over time.

2. Which two gases make up most of the atmosphere?

3. Draw a diagram showing some of the ways in which carbon dioxide enters and leaves the atmosphere.

4. **CRITICAL THINKING** In what ways might the loss of vegetation as a result of deforestation affect the atmosphere's balance?

5. **BIOLOGY** Use print or Internet resources to research the role of microorganisms in returning nitrogen or carbon dioxide to the atmosphere.

MONITOR AND RETEACH

If students miss . . .

Question 1 Reteach "The Composition of the Atmosphere." (p. 366) Ask: What gases are released from volcanic eruptions? What gas is released during photosynthesis?

Question 2 Reteach "The Composition of the Atmosphere." (p. 366) Make a bar graph of the table showing the principal gases of dry air to help students appreciate the differences in the amounts of the gases.

Question 3 Review "Recycling of Atmospheric Materials." (p. 367) Have students describe each cycle.

Question 4 Reteach "A Delicate Balance." (pp. 367–368) Ask: What material is removed from the air during photosynthesis? If rates of photosynthesis decrease, would more or less of the material be present in the air?

Question 5 Review "Recycling of Atmospheric Materials." (p. 367) Have students reread the text, concentrating on the cycling of carbon dioxide.

Heat and the Atmosphere

Energy from the sun drives the weather and is essential to almost all life on Earth. How does energy from the sun reach Earth across nearly 150 million kilometers of space? What happens to the energy after it reaches Earth?

How Heat Energy Moves

Heat energy enters and moves through the atmosphere in three different ways—radiation, conduction, and convection.

Energy from the sun reaches Earth through radiation. **Radiation** is the transfer of energy through space in the form of visible light, ultraviolet rays, and other types of electromagnetic waves. All objects warmer than absolute zero emit some form of radiation. Unlike conduction and convection, radiation does not require a medium. Thus radiation is the only way that energy can travel through outer space.

Conduction is the transfer of heat energy through collisions of the atoms or molecules of a substance. When you walk barefoot on hot ground, for example, heat moves by conduction from the ground to the soles of your feet. Air touching warm ground is also heated by conduction.

Convection is the transfer of heat energy in a liquid or gas through the motion of the liquid or gas caused by differences in density. For example, in a pot of simmering water, the water at the bottom of the pot is heated by conduction and becomes less dense. Because it is less dense, it rises and is replaced by downward-flowing water that is colder and denser. Similarly, warm air in the atmosphere rises, transferring heat upward by convection.

Heating Water

- The heated water is less dense and so it rises, moving heat upward by **convection**.
- Objects near the burner are heated by **radiation**.
- The water touching the bottom of the pot heats by **conduction**.
- The bottom of the pot touches the burner and is heated by **conduction**.

Heat and Temperature

Although heat and temperature are related, they are not the same. The atoms or molecules that make up any substance, even a solid, are in constant motion. The faster the atoms or molecules are moving, the greater their kinetic energy, or energy of motion. The **temperature** of a substance is a measure of the average kinetic energy of the atoms or molecules in that substance. For example, molecules of water that is almost boiling move

17.2

KEY IDEA
Energy from the sun heats the atmosphere and Earth's surface. This heat spreads throughout the atmosphere and is also radiated back into space.

KEY VOCABULARY
- radiation
- conduction
- convection
- temperature
- heat
- troposphere
- stratosphere
- ozone
- mesosphere
- thermosphere
- ionosphere
- insolation

VOCABULARY STRATEGY
In the words *conduction* and *convection*, the prefix *con-* means "with" or "together"; *-duction* is from a Latin word that means "to lead" and *-vection* is from a Latin word that means "to carry."

CHAPTER 17 SECTION 2

FOCUS

Objectives
1. Describe how energy from the sun moves through the atmosphere by radiation, conduction, and convection.
2. Identify the characteristics of each atmospheric layer.
3. Analyze Earth's heat budget.

Set a Purpose
Have students study the illustrations and photographs in this section. Then have them read to find the answer to the question "How does heat energy move through the atmosphere?"

INSTRUCT

Visual Teaching

Interpret Diagram
Help students visualize the continuous nature of heat transfer, shown here in the interaction of the processes of conduction and convection. Water is heated at the bottom of the pot, becomes less dense, and rises. Ask: What happens to the heated water when it reaches the top of the pot? It cools, becomes more dense, and sinks.

DIFFERENTIATING INSTRUCTION

Hands-On Demonstration Place a beaker of water on a hot plate or over a Bunsen burner. Add a few drops of food coloring to the surface of the water, then heat the water. As the water heats, have students observe the swirling of the food coloring in the water. Ask: What causes the food coloring to swirl? the movements of the upward-flowing warm water and downward-flowing cold water In what direction is heat being transferred in the water? upward What is this kind of heat transfer called? convection

⚠ *Exercise caution when working with a hot plate or bunsen burner. Remind students that safety is everyone's responsibility.*

17 SECTION 2

DISCUSSION
Have students refer to page 696 to compare and contrast the Celsius and Fahrenheit scales. Have students convert 70°F to degrees Celsius. **21.1°C** Ask: When do you often see the Fahrenheit scale used? **in weather forecasts, on oven dials, on household thermometers** the Celsius scale? **in science textbooks, on laboratory equipment** Which do you prefer and why? **Answers will vary.**

MINILAB
Changes of State
Management
Have plenty of paper towels on hand to wipe up spills.

Materials
Do not use mercury thermometers.

Analysis Answer
Students should observe that when they add ice to the cup of liquid water, the temperature of the liquid decreases to 0°C and stays at that temperature as ice continues to melt in the cup. The explanation is that heat energy moves from the liquid water to the ice, causing it to melt, because the temperature of the water is higher than the temperature of the ice. As heat leaves the liquid water, the average kinetic energy of the water molecules decreases, thus the temperature of the water decreases.

For proper behavior in a laboratory setting, refer students to Safety in the Earth Science Laboratory on pages xxii–xxiii.

30-Minute Mini LAB
Changes of State

Materials
- small Styrofoam cup
- short thermometer
- water
- shallow pan

Procedure
1. Fill the cup half full of water. Place the thermometer in the cup. Freeze overnight.
2. Fill the pan with lukewarm water. Place the cup and thermometer in the pan.
3. Record the initial temperature. Record the temperature every minute until the ice has melted. Graph your data.

Analysis
Describe the trend in temperature as the ice melts. Why does the temperature stop rising as water changes phases? What is the heat source melting the ice?

VOCABULARY STRATEGY
The names of the layers of the atmosphere contain Greek or Latin roots that give clues to the properties of the layers: *tropo* ("change" or "turning"), *strato* ("spreading out"), *meso* ("middle"), and *thermo* ("heat").

370   Unit 5   Atmosphere and Weather

---

much more rapidly (and thus have more kinetic energy) than molecules of water that is nearly freezing. For this reason, the temperature of boiling water is greater than the temperature of freezing water.

Temperature is a measure of the average kinetic energy of the particles of a substance, but **heat** can be thought of as the total kinetic energy of all of the particles of the substance. For example, a large cup of tea has more heat than a smaller cup of tea at the same temperature. Heat always flows from a substance at a higher temperature to a substance at a lower temperature. For instance, when ice melts, it absorbs heat energy from its surroundings. The heat melts the ice instead of raising its temperature, so the melted water is at the same temperature as the ice even though the water has more heat energy.

A thermometer measures temperature, not heat. Scientists measure temperature using the Celsius scale, which is based on the properties of water. At sea level, ice melts at 0°C, and water boils at 100°C. For a comparison of Celsius and other temperature scales, see page 695.

Structure of the Atmosphere
The temperature of the atmosphere changes dramatically at varying altitudes. Scientists use these temperature differences to divide the atmosphere into four layers: (1) troposphere, (2) stratosphere, (3) mesosphere, and (4) thermosphere.

The Troposphere
The lowest layer of Earth's atmosphere is called the **troposphere** (TROH-puh-SFEER). In the troposphere, temperature decreases with altitude, as shown in the graph on page 371. Most of the sun's radiation is absorbed at Earth's surface, which in turn transfers heat to the atmosphere through conduction and radiation. Thus the air at the surface is warmest, and temperature generally decreases with altitude, or the distance from the warming effect of Earth's surface. The rate of cooling with altitude is highly variable, but on average the temperature of the air in the troposphere decreases about 6.5°C for each kilometer of altitude gain.

The temperature stops decreasing at the tropopause, the area between the troposphere and the stratosphere. The altitude of the tropopause varies according to latitude. At the equator, the tropopause is at an altitude of about 16 kilometers, but at the poles, it is at about 9 kilometers. (The jet stream is located just below the tropopause.)

Because the density of the atmosphere also decreases with altitude, the troposphere contains about 80 percent of the total mass of the atmosphere. It also contains most of the water vapor present in the atmosphere. Partly for this reason, almost all of Earth's weather occurs in the troposphere.

The Stratosphere
Above the tropopause lies a clear, dry layer of the atmosphere called the **stratosphere** (STRAT-uh-SFEER). As you can see in the graph on page 371, the lower part of the stratosphere is about as cold as the tropopause. The upper part of the stratosphere warms steadily up to its top, the stratopause, which is about 50 kilometers above Earth's surface.

Reading Support Discuss the importance of paying close attention to the headings within a section. Tell students that these elements are clues to the content. Ask: What would you expect to learn about under the heading "Structure of the Atmosphere"? **the different parts of the atmosphere** Point out that the headings vary in appearance. The larger red headings indicate a main topic. **for example, "Structure of the Atmosphere"** The smaller blue headings indicate subtopics that are related to the main topic. **for example, "The Stratosphere"** Use the Reading Study Guide, p. 58.

370   Unit 5   Atmosphere and Weather

The temperature increase in the stratosphere is caused by the presence of **ozone,** a form of oxygen gas. (An ozone molecule consists of three oxygen atoms; an oxygen molecule contains two oxygen atoms.) Ozone absorbs ultraviolet rays from the sun and then releases some of this energy in the form of heat. You will learn more about ozone in Section 17.4.

The Mesosphere and Thermosphere

The atmosphere's third layer, the **mesosphere** (MEHZ-uh-SFEER), extends between about 50 and 90 kilometers above Earth's surface. Because very little ozone is present in the mesosphere, temperatures again drop with increasing altitude.

Above 90 kilometers lies the atmosphere's fourth layer, the **thermosphere.** The atmosphere at this great altitude is extremely thin, but the few molecules and atoms present receive such intense solar radiation that temperatures can rise above 1000°C. The thermosphere is separated into layers of different gases, with heavier gases in the lower levels and lighter gases in the higher layers. The lowest layer is composed primarily of nitrogen molecules. Next, a layer of oxygen reaches to about 1000 kilometers. Above that layer is one of helium that extends to about 2400 kilometers. Above this helium layer, a layer of hydrogen thins out into space.

STRUCTURE OF THE ATMOSPHERE The graph below shows the approximate temperature at each altitude. The exact altitude and temperature of the pauses vary.

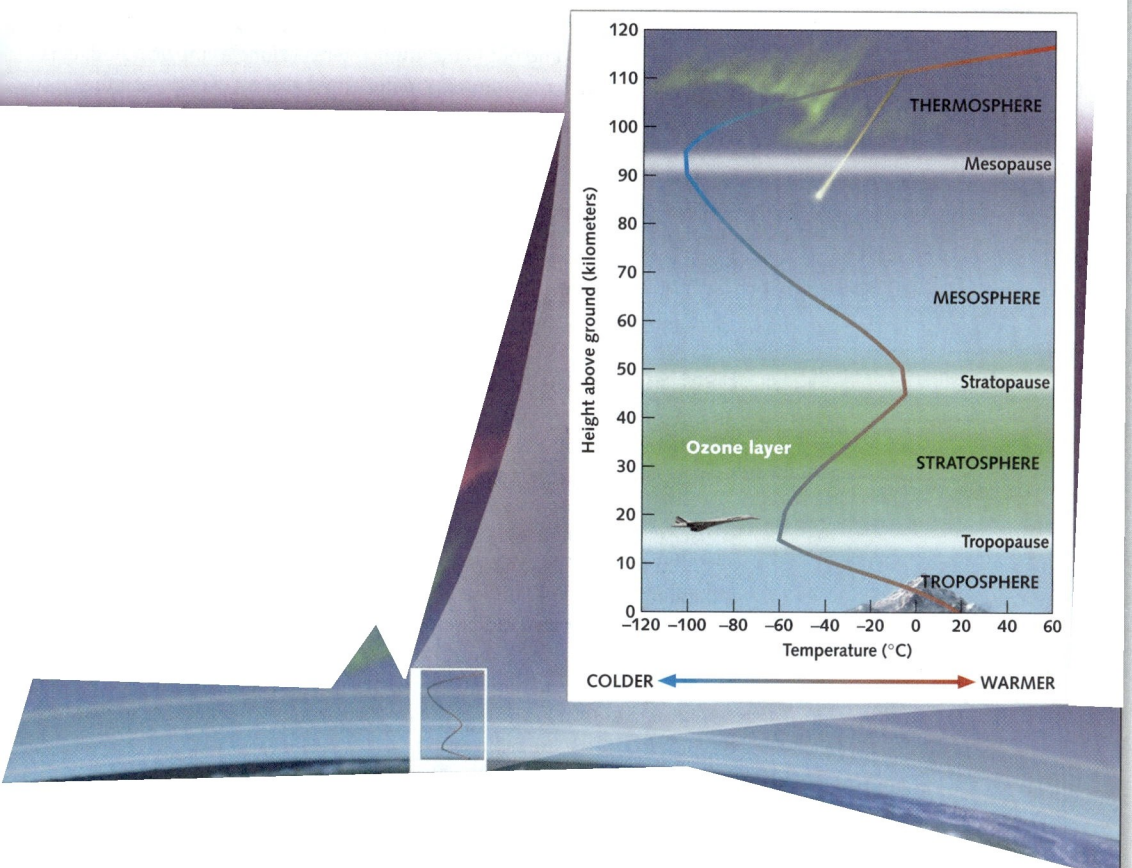

INVESTIGATIONS
CLASSZONE.COM

What Can You Learn from a Thermometer on a Rising Balloon?
Analyze changes in temperatures to find the boundaries of the atmospheric layers.
Keycode: ES1702

INVESTIGATIONS
CLASSZONE.COM

What Can You Learn from a Thermometer on a Rising Balloon?
Keycode: ES1702

If you have a projector, introduce the investigation in class.

📖 Use Internet Investigations Guide, pp. T64, 64.

More about...

The Tropopause
The tropopause can be defined as the minimum temperature, or point, at which temperatures in the troposphere stop decreasing. This minimum temperature varies from the equator to the poles in a somewhat surprising way. Above the poles, the minimum temperature is about −60°C in January. During the same month, the minimum temperature is a frigid −80°C above the equator. The different values are largely caused by differences in height. In January, the altitude of the tropopause is 8.5 kilometers above polar regions and 17.7 kilometers above equatorial regions. Also, air temperature changes with altitude at different rates over the two areas because of variations in moisture content.

Use Transparency 18.

Support for Visually Impaired Students Copy and enlarge the graph on this page. Then pass out copies to students with visual impairments. Or, pair the students with classmates who do not have visual impairments and have the students work together to redraw the graph.

The Ionosphere

The portion of the thermosphere between about 90 and 500 kilometers above Earth is also called the **ionosphere,** because the air there is highly ionized. These ions are formed when ultraviolet rays from the sun knock electrons off oxygen and nitrogen molecules and oxygen atoms.

The ionosphere is affected by solar events. Huge eruptions associated with sunspots send out large amounts of a form of radiation that disrupts radio communications. These solar eruptions also send out ionized particles. Because the particles are electrically charged, they are deflected by Earth's magnetic field to the North and South poles. At the poles, the ionized particles interact with air molecules to form auroras, colorful displays of light in the nighttime sky.

AURORA BOREALIS is the name of the aurora in the Northern Hemisphere.

Observe auroras as seen from the ground and from space.
Keycode: ES1703

Insolation and the Atmosphere

As you have seen, the atmosphere is a complex system in which energy is always flowing. The sun radiates energy into space in all directions. Earth's system receives only about one two-billionth of the sun's rays. This *in*coming *sol*ar radi*ation* is called **insolation** (IHN-soh-LAY-shuhn). Some insolation is absorbed by gases in the atmosphere, and some reaches Earth's surface without interference. However, much of the insolation is scattered by collisions with gas molecules and dust in the atmosphere. Some of the scattered insolation returns to space, some is absorbed by the air, and some makes it to Earth's surface.

Scientists use the model of a global heat budget to represent the overall flow of energy into and out of the atmosphere. As long as the heat budget is in balance, global temperatures remain fairly constant over time. If the budget should become out of balance, the Earth system would heat up or cool down.

The diagram on page 373 illustrates the heat budget in balance. Suppose the Earth system receives 100 units of energy from the sun. As the diagram shows, 30 units of this insolation are reflected back to space and the remaining 70 units are absorbed by the atmosphere or Earth's surface. When the budget is in balance, 70 units of energy are also radiated out to space by the atmosphere and by Earth's surface.

Heat Budget of Earth and the Atmosphere

As you can see in the diagram, only a small percentage of insolation is absorbed by the atmosphere. In fact, most of the atmosphere's energy is transferred from the surface by radiation, conduction, and the evaporation and subsequent condensation of water. This energy transfer from Earth's surface is one of the major causes of weather.

Although Earth's atmosphere allows half of the incoming solar radiation to reach the surface, much of that energy is radiated back into the atmosphere as infrared radiation. The accumulation of carbon dioxide and water vapor in the atmosphere, however, absorbs most of the infrared radiation, thus preventing it from radiating directly back to space. A condition known as the *greenhouse effect* results when infrared radiation remains in Earth's atmosphere.

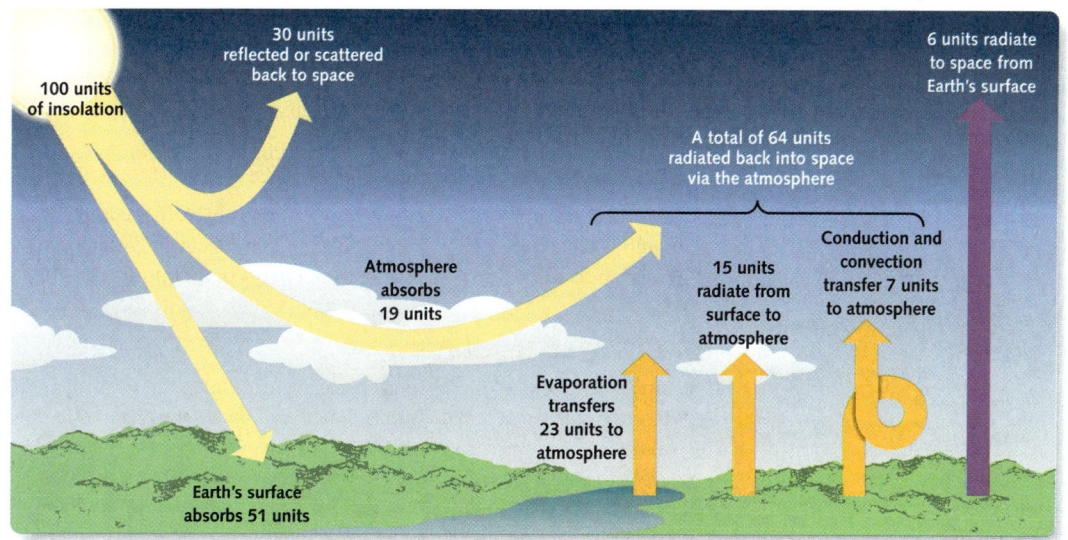

GLOBAL HEAT BUDGET This illustration, based on average data, shows what happens to the energy in 100 units of insolation. If the sky were clear and snow covered the ground, then more energy would reflect directly to space from Earth's surface. On an overcast day, the atmosphere either reflects or absorbs more insolation, so Earth's surface receives less.

Without the greenhouse effect, much of Earth's heat energy would be lost to outer space. In fact, Earth's average surface temperature would be about 33°C cooler than it is now—freezing! The greenhouse effect has actually helped Earth thrive as a planet. Recently, however, there has been such a significant increase in the levels of carbon dioxide in the atmosphere that Earth's heat budget may be out of balance. Many scientists warn of the possibilities of global warming. (See Section 17.4.)

17.2 Section Review

1. Describe three ways heat is transferred through the atmosphere.
2. Why does temperature generally decrease with altitude in the troposphere?
3. Compare the temperature changes in the stratosphere with those in the thermosphere. Include the role of ozone in your explanation.
4. Describe at least two paths that a unit of energy could take from its arrival at Earth's atmosphere until it is reradiated out to space.
5. **CRITICAL THINKING** Based on what you have learned about the layers of the atmosphere, explain why jets generally fly at or above the tropopause.
6. **PHYSICAL SCIENCE** Draw and label a diagram explaining why the water at the bottom of a pot of simmering water becomes less dense as it is heated.

 VISUALIZATIONS
CLASSZONE.COM

Observe auroras as seen from the ground and from space.
Keycode: ES1703

Visualizations CD-ROM

ASSESS

1. radiation (the transfer of energy through space in the form of electromagnetic waves), conduction (the transfer of energy through collisions of atoms or molecules), and convection (the transfer of energy through motion caused by differences in density in a liquid or gas)
2. Most insolation is absorbed by Earth's surface. Thus, temperatures are warmest near the surface and decrease with altitude.
3. Temperatures increase in the stratosphere because this layer contains ozone, which absorbs ultraviolet rays from the sun. Temperatures also increase in the thermosphere because this layer receives intense solar radiation.
4. **Sample Response:** A unit of energy could be reflected back into space or absorbed by the atmosphere.
5. to avoid inclement weather or to take advantage of jet streams
6. Diagrams should show that the molecules of water move apart as they are heated, making the water less dense.

If students miss . . .

Question 1 Review "How Heat Energy Moves." (p. 369) Use diagrams and gestures to demonstrate radiation, convection, and conduction.

Question 2 Review the diagram on page 371. Ask: Is more solar radiation absorbed high in the troposphere or near the surface? **the surface**

Question 3 Review the diagram on page 371. Ask: Which layer receives the most intense solar radiation? **the thermosphere** Which layer contains a gas that absorbs ultraviolet radiation from the sun? **the stratosphere**

Question 4 Review the diagram on p. 373. Have students trace each arrow on the diagram and describe the way heat is moving.

Question 5 Ask: Where is the jet stream located? **just below the tropopause** Where does most weather occur? **in the troposphere**

Question 6 Ask: What happens to the molecules of a liquid or gas when it is heated? **they move faster and farther apart**

FOCUS

Objectives
1. Identify the factors that cause the intensity of insolation to vary from place to place.
2. Describe how the characteristics of a material affect its rate of solar absorption.
3. Analyze a temperature map.

Set a Purpose
Have students relate examples of temperature variations that they've experienced. Then have them read the section to find the answer to "What causes temperatures to vary from one place to the next?"

INSTRUCT

VISUAL TEACHING
Discussion
Have students read the caption to the diagram showing angles of insolation. Ask: At what angle is insolation most intense? 90° Why? because sunlight is concentrated on a relatively small area What happens when the angle of insolation decreases? Sunlight is spread over a larger area, and the intensity of insolation decreases.

17.3

KEY IDEA
Many factors affect how much solar energy is absorbed by Earth's surface at any given time and place.

KEY VOCABULARY
- isotherm

Local Temperature Variations

Why does the temperature vary from place to place? The basic reason is that insolation heats Earth's surface and atmosphere unequally.

- The intensity of insolation varies with the time of day, the latitude, and the time of year.
- The characteristics of a material affect both how much insolation the material absorbs and how the absorbed energy affects the temperature. For example, on a sunny day dark pavement becomes hotter than grass.

Intensity of Insolation

The intensity of insolation depends on the angle at which the sun's rays strike Earth's surface. When the sun is directly overhead, the angle of insolation is 90°, and Earth's surface receives the maximum amount of energy. As the angle of insolation decreases, the energy of the rays is spread out over a larger area, so the energy per unit area decreases. The sunlight must also travel farther through the atmosphere, so that more insolation is absorbed or reflected before it reaches Earth's surface. Both factors reduce the amount of solar energy that reaches Earth's surface.

ANGLE OF INSOLATION When sunlight hits Earth's surface at an angle, its energy is spread over a greater area, so the insolation is less intense. The angle of sunlight varies with time of day, latitude, and time of year.

Time of Day

Because the sun's rays are closest to vertical at noon, the intensity of insolation is greatest then. However, the highest temperature is generally not at noon. Instead, the warmest hour of the day is usually in the afternoon. For several hours after noon, the lower atmosphere receives more heat from the ground than it loses. Thus, its temperature keeps rising until well into the afternoon. The coldest hour usually comes just before sunrise because the ground and the lower air lose heat all through the night.

Hands-On Demonstration Use a small flashlight and a globe to model the diagram on this page. Darken the room and stand about 1 meter from the globe. Shine the flashlight at a 90° angle on the equator. Have students describe how the intensity of light varies from the equator to the poles. The intensity of light is greatest over the equator and decreases toward the poles. Use this demonstration as a lead-in to the next topic in the text—the effects of latitude on the intensity of insolation.

Latitude

Because Earth is spherical, the sun's rays strike the surface at angles ranging from 0° to 90°. Near the equator, almost-vertical rays fall all through the year. Thus, these areas have hot climates. In high latitudes (near the poles, for example), the sun's rays generally strike the surface at low angles. Such areas may even have no sunlight at all for part of the year and are cold year-round.

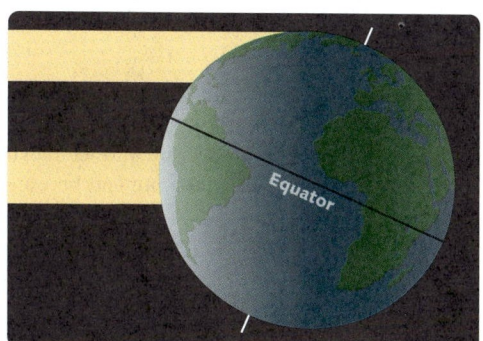

LATITUDE The angle of sunlight varies with latitude.

Time of Year

Locations in middle latitudes (like most of the United States) get near-vertical rays in summer, so their summers are hot. The angle of the rays is less vertical in winter, so winters in middle latitudes are cold.

Like the day's highest temperature, the year's highest temperature occurs after the time of maximum insolation. In middle latitudes of the Northern Hemisphere, the time of strongest sunlight occurs around June 21. However, July is usually the warmest month. Similarly, the year's weakest sunlight occurs around December 21, but January is usually the coldest month. In the Southern Hemisphere, the warmest and coldest months are reversed.

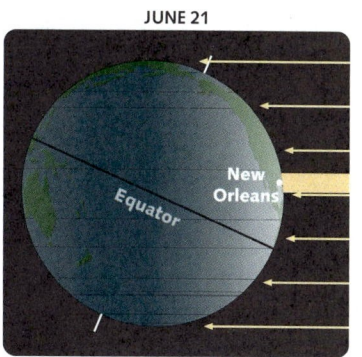 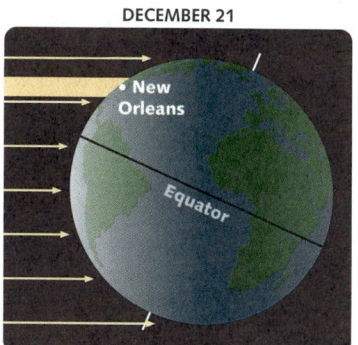

TIME OF YEAR The angle of sunlight varies with time of year.

CRITICAL THINKING

Assess students' understanding of rates of heat absorption by different land surfaces. Ask: Which surface is likely to be hotter on a sunny day—a parking lot surfaced with blacktop or one surfaced with crushed white gravel? Explain. **The blacktop parking lot absorbs more solar radiation so it would be the hotter surface; the white gravel, in fact, would reflect some of the insolation.**

More about...

Isotherms

The use of isolines, or lines that connect points of equal value, has a long tradition in cartography. A marine chart from the late 1500s contained isobaths, or lines of equal depth. Isotherms, however, were not introduced until 1817, when they appeared on a map by noted German scientist Alexander von Humboldt. Humboldt's map focused on average global temperatures, not daily variations in temperature. During this same period, Heinrich Wilhelm Brandes, a physics professor at the University of Breslau, proposed using isobars (lines of equal pressure) on daily weather maps and analyzing the resulting patterns. Thus, although Humboldt is considered the father of isotherms, Brandes is credited with introducing the concept of isolines on daily weather maps.

Cloud Cover

The intensity of the insolation that reaches Earth's surface depends not only on the angle of the sun's rays but also on how much energy makes it through the atmosphere without being absorbed, reflected, or scattered back to space. Because clouds reflect a significant amount of insolation back into space, more solar energy reaches Earth's surface on clear days than on cloudy days. Similarly, more radiation travels from Earth's surface out to space on clear nights than on cloudy nights.

Heating of Water and Land

On a sunny day in summer, dry beach sand is much warmer than the nearby water. At night the same sand cools faster than the water and becomes colder. On a larger scale, continents are warmer than nearby ocean waters in summer. In winter the same continents become colder than the nearby waters.

Why does the temperature of land vary more than the temperature of water? The reason is that water and land warm up and cool off at different rates. Water warms much more slowly than land for many reasons:

- Heat energy from insolation spreads through a greater depth in water than on land. On land, insolation warms only the top few centimeters of soil, but the sun's rays penetrate to a depth of many meters in water. In addition, water spreads heat easily by convection because it is a fluid.

- In water, some solar energy is used in the process of evaporation. Therefore, less solar energy is available to raise the temperature of the water.

- Water needs more energy than land to raise its temperature the same amount. The amount of heat needed to raise 1 gram of a substance by 1°C is called the substance's specific heat. The specific heat of water is almost three times the specific heat of land. The higher the specific heat, the slower the temperature change.

Just as water and land absorb heat differently, various types of land surfaces absorb heat differently too. For example, dark surfaces absorb more energy than light surfaces. Rough surfaces absorb more energy than smooth ones. Wet ground warms more slowly than dry ground because wet ground contains water. Meadows warm more quickly than forests, and pavements get warm long before grassy lawns. Snow and ice reflect sunlight and remain cold.

VISUALIZATIONS
CLASSZONE.COM
Examine infrared images that show variation in surface temperature.
Keycode: ES1705

Temperature Maps

The maps on the following page show mean, or average, temperatures in Asia and Australia for January and July. Notice that the warmest temperatures are to the south of the equator in January and to the north of the equator in July. The sun's rays are more vertical north of the equator in July. Notice also that the warmest and coldest temperatures are over land.

Reading Support Help students become more active readers. Point out that the bulleted items on this page summarize the main reasons why water heats up and cools down more slowly than land. Tell students that bullets are often used to list or summarize important concepts.

Land heats and cools more readily than water, so the continents are warmer than nearby oceans in the summer and colder than the nearby oceans in the winter.

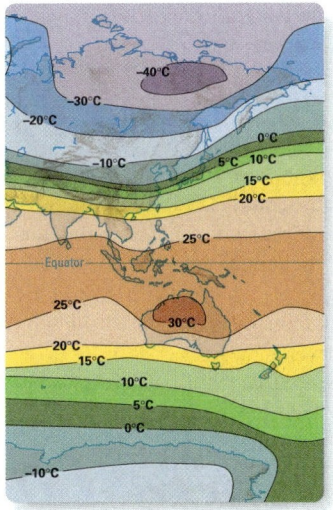

MEAN SEA-LEVEL TEMPERATURES IN JANUARY

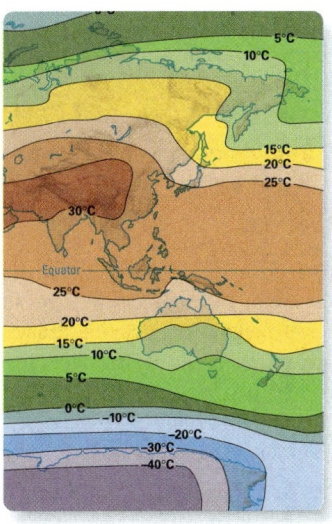

MEAN SEA-LEVEL TEMPERATURES IN JULY

The lines separating the temperature zones on the map are isotherms. **Isotherms** are lines that connect places with the same temperature. Because temperature decreases with altitude, the temperatures on the map are adjusted to eliminate the effect of altitude.

How Does the Temperature at One Location Change over a Year?
Compare animations of global insolation and temperature maps. List factors that affect global variations.
Keycode: ES1706

VOCABULARY STRATEGY

In the word *isotherm*, *iso-* means "equal" and *-therm* means "heat."

17.3 Section Review

1. How does the angle at which sunlight strikes Earth's surface affect the intensity of the sunlight?
2. When does the highest temperature of the day usually occur? Why?
3. Why is it warmer near the equator than near the poles?
4. In the United States, why is it colder in the winter and warmer in the summer?
5. Explain why water doesn't get as hot as land on a clear summer day.
6. **CRITICAL THINKING** The daily temperature range is the difference between the day's maximum and minimum temperatures. Why is the daily temperature range greater on clear days than on cloudy ones?
7. **PHYSICS** Work with a classmate to design an experiment that tests whether a black substance or a white substance would be better for collecting solar heat.

VISUALIZATIONS
CLASSZONE.COM

Examine infrared images that show variation in surface temperature.
Keycode: ES1705
Visualizations CD-ROM

How Does the Temperature of One Location Change Over a Year?
Keycode: ES1706
Use Internet Investigations Guide, pp. T65, 65.

ASSESS

1. As the angle of insolation decreases, the intensity of sunlight decreases.
2. afternoon; The lower atmosphere receives more heat from the ground than it loses, so its temperature continues to rise.
3. The sun's rays strike the equator at an almost-vertical angle, but strike the poles at a low angle.
4. The U.S. is located in the middle latitudes and receives less intense sunlight in winter than in summer.
5. **Sample Response:** Insolation spreads through a greater depth in water than on land.
6. More insolation reaches Earth's surface on clear days. On clear nights, this energy radiates back to space.
7. **Sample Response:** Place one thermometer under a black cloth and another under a white cloth. Position a lamp above them and periodically record temperatures to see which material absorbs heat more efficiently.

If students miss . . .

Question 1 Reteach "Intensity of Insolation." (p. 374) Have students review and describe the diagram on that page.
Question 2 Review "Time of Day." (p. 374) Remind students that the surface releases heat to the lower atmosphere throughout sunlit hours.
Question 3 Use a globe to review latitude. (p. 375) Have students model the angle of the sun's rays striking the equator and the poles.
Question 4 Reteach "Time of Year." (p. 375) Have students examine the diagram illustrating the tilt of Earth's axis on June 21 and December 21.
Questions 5 and 7 Review "Heating of Water and Land." (p. 376) Have students reread the text and summarize the bulleted points in their own words. Have students work with a classmate to perform item 7 for homework.
Question 6 Ask: How do clouds affect the amount of solar radiation that reaches the surface? Clouds reflect insolation, so less solar radiation reaches the surface.

FOCUS

Objectives
1. Discuss how human activities can affect the atmosphere.
2. Compare and contrast acid rain, smog, ozone depletion, and global warming.

Set a Purpose
Have students read to find the answer to "In what ways is human activity having a negative effect on the atmosphere?"

INSTRUCT

VISUAL TEACHING
Discussion
Have students read about the effects of common air pollutants in the chart. Then have students describe the photographs on pages 378 and 379. Point out that some materials that occur naturally, such as carbon dioxide, can become harmful. Ask: Does air pollution affect only humans? **No, air pollution affects all living things.** How might air pollution affect living and nonliving parts of the environment? **Air pollution can kill living things and can destroy wildlife habitats. It can harm the atmosphere and bodies of water. It also damages buildings and structures made of limestone and marble.**

17.4

KEY IDEA
Human activities affect the atmosphere by producing air pollutants and other substances which contribute to problems such as acid rain and ozone depletion.

KEY VOCABULARY
- air pollutant
- temperature inversion

SMOKE from forest fires contains carbon monoxide, carbon dioxide, and particulate matter. These substances contribute to global warming and smog.

Human Impact on the Atmosphere

The modern transportation and manufacturing technologies that provide many benefits for everyday life also have some negative effects on the environment. These effects include local air pollution as well as worldwide problems such as ozone depletion and global warming.

Common Air Pollutants

An **air pollutant** is any airborne gas or particle that occurs at a concentration capable of harming humans or the environment. Although some air pollution is caused by natural sources such as volcanoes and forest fires, human activity produces a significant amount of the pollutants that are of greatest concern today. Locally, pollution can affect individuals by limiting their outdoor activity and causing health problems. On a global level, pollution contributes to several environmental problems, including acid rain and the depletion of the ozone layer.

In the United States, the Clean Air Act of 1970 identified six key pollutants as indicators of air quality. The table below lists these pollutants, their major sources, and their environmental effects.

Common Air Pollutants

Air Pollutant	Major Sources	Effects
Carbon monoxide (CO)	automobile exhaust	Reduces delivery of oxygen to body tissues; impairs vision and reflexes
Nitrogen dioxide (NO_2)	burning of fossil fuels in power plants and automobiles	Irritates lungs and lowers resistance to respiratory infections; contributes to acid rain and smog
Sulfur dioxide (SO_2)	burning of fossil fuels in power plants, oil refineries, paper mills, volcanoes	Irritates respiratory system; contributes to acid rain
Particulate matter (dust, smoke, soot, ash)	factories, power plants, oil refineries, paper mills, volcanoes	Contributes to respiratory problems; linked to some cancers
Lead (Pb)	smelters, battery plants	Damages nervous and digestive systems
Ozone (O_3)	reactions of nitrogen oxides and hydrocarbons in the presence of sunlight	Reduces lung function and causes inflammation

As people have become more aware of the dangers of air pollution, efforts to prevent it have increased. The Environmental Protection Agency monitors levels of air pollution and establishes regulations aimed at reducing pollution. Scientists and environmental engineers are identifying

---

Reading Support Have students make a concept web based on the data in the chart. The first level should be titled *Common Air Pollutants*. The next level should list the pollutants, followed by sources and effects. This exercise will encourage students to organize their thoughts as they read.
Use Reading Study Guide, p. 60.
Use the Concept Web Transparency to help students.

Challenge Activity The Clean Air Act was originally passed in 1963. It was amended in 1965, 1966, 1967, 1969, 1970, 1977, and 1990. Have students research the Clean Air Act and its amendments. How did the act change over time? What standards are in place today for reducing air-pollution emissions? Have students evaluate the effects of the Clean Air Act. What recommendations, if any, would they make for further amendments to the act?

sources of pollution, tracking pollution trends, developing alternative energy sources, and refining devices and methods to limit industrial pollution. Private citizens are becoming more knowledgeable consumers and purchasing products that don't harm the atmosphere. People have also become active in conservation efforts, such as recycling and limiting their use of fossil fuels.

Acid Rain

Acid precipitation, which is commonly called acid rain, forms when pollutants such as sulfur dioxide and nitrogen oxides react with water vapor in the air. The resulting acid precipitation can fall to the ground as rain or snow.

Acidity is measured using pH. The scale at the right shows how the pH of acid rain compares with the pH of some common substances. Notice that rain is naturally slightly acidic with a pH just under 6, but pollution has significantly increased the acidity of rain in some areas. On the pH scale, reading from 7 to 0, each whole number represents a tenfold change in acidity. In other words, pH 5 is ten times more acidic than pH 6, and pH 4 is 100 times more acidic than pH 6.

Acid precipitation can harm both plant and animal life. Most life forms can survive only within a limited range of pH, and acid rain can lower the pH beyond that range. This effect is noticeable in thousands of lakes and streams throughout the United States and Canada, where many fish populations have died out due to the increased acidity. Acid precipitation also harms forests by stripping away vital nutrients from the soil. This loss of nutrients limits the growth of trees and makes them vulnerable to damage and parasites. These harmful effects have damaged regional tree populations, such as the spruce trees found in the Appalachian Mountain range of the eastern United States.

Acid precipitation also damages structures built of limestone or marble. Reactions between acid rain and the calcite in limestone and marble dissolve the calcite, resulting in roughened surfaces and loss of carved details. Cultural monuments that have been damaged by acid rain, like the ancient Parthenon in Greece, are expensive to restore and are sometimes permanently damaged.

Observe forest fires as seen from space.
Keycode: ES1707

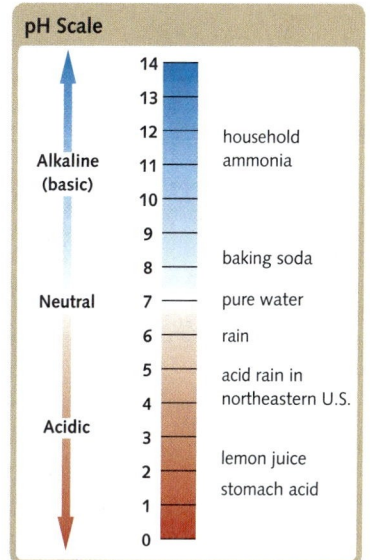

PH SCALE The pH scale ranges from 0 to 14, with a value of 7 considered neutral.

ACID RAIN These scientists are sampling the acidity of a lake where the fish population has died out due to acid rain.

Chapter 17 Atmosphere **379**

VISUALIZATIONS
CLASSZONE.COM

Observe forest fires as seen from space.
Keycode: ES1707

Have students discuss how smoke from a forest fire disperses throughout the atmosphere. Lead them to understand that the fire ultimately impacts areas that are not burned.

Visualizations CD-ROM

VISUAL TEACHING

Discussion

Make sure students understand that, because pH represents an exponent, each whole number on the pH scale represents a tenfold change in acidity. Vinegar has a pH of 3, or 1×10^3. Pure water has a pH of 7, or 1×10^7. Ask: How many more times acidic is vinegar than pure water? 1×10^4, or 10,000 times Students can refer to the Math Skills Handbook to review scientific notation.

Extend

Have students research the pH levels of other substances, such as orange juice (pH: 3.5), milk (pH: 6.5), and tears (pH: 7.4), and place the substances in their proper position on the pH scale.

Hands-On Demonstration Use litmus paper to demonstrate pH levels of common substances, such as lemon juice, distilled water, and household ammonia. Blue litmus paper will turn red in acidic solutions (lemon juice), and red litmus paper will turn blue in alkaline solutions (household ammonia). If time permits, gather samples of rainwater or pond water from your area. Have students analyze the results of a litmus test on the local water samples.

⚠ *Exercise caution when working with household ammonia. Remind students that safety is everyone's responsibility.*

Chapter 17 Atmosphere **379**

CHAPTER 17 SECTION 4

DISCUSSION

Ask students to explain why the phrase *temperature inversion* aptly describes a situation where warm air overlies cold air. In the troposphere, air temperature generally decreases with altitude. In the case of a temperature inversion, this relationship is inverted. Air temperature increases with altitude.

More about...

Ozone Depletion
CFCs are stable at low altitudes. When they enter the stratosphere, however, they react with ultraviolet radiation and release chlorine, which in turn reacts with and destroys ozone molecules. During the long polar winter, CFCs accumulate over polar regions. Little to no ultraviolet radiation reaches the poles, so the compounds remain stable. In spring, however, ultraviolet radiation strikes the poles and CFCs vigorously "attack" ozone molecules. This is largely why the thinning of the ozone layer is most pronounced over Antarctica in spring.

Smog

The term *smog* was coined in the early 1900s to refer to the smoky fog in London, which resulted from emissions of particulate matter and other pollutants from factories. Today, smog is primarily used to refer to photochemical smog, a brownish haze that forms in air polluted with nitrogen oxides and hydrocarbons that come mainly from automobile exhaust. When these pollutants are present, solar radiation can trigger a chain of chemical reactions that form ozone and other harmful substances. Although ozone in the stratosphere protects Earth from ultraviolet radiation, ground-level ozone is a powerful lung irritant that can cause respiratory problems and illness. Ozone also stunts the growth of plants by interfering with photosynthesis. As a result, the ozone in smog can reduce crop yields and hurt the agricultural industry.

The severity of smog depends on atmospheric conditions. Usually convection mixes warm air from near Earth's surface with the cooler air above, thus diluting any pollutants. Sometimes, however, the air at Earth's surface is colder than the air above, so convection does not occur. This situation is called a **temperature inversion.** The warm air lying above the cooler air acts as a lid on the underlying air pollutants, trapping them close to the ground and allowing smog to build to dangerous levels.

NO INVERSION NEAR SURFACE

INVERSION NEAR SURFACE

Ozone Depletion

In Section 17.2, you read about a layer of ozone in the stratosphere that absorbs harmful ultraviolet radiation. In the 1970s, scientists began to suspect that the ozone layer was being harmed by chlorofluorocarbons (KLAWR-oh-FLUR-oh-KAHR-buhnz), or CFCs, manufactured compounds that were widely used in products such as aerosol sprays, air conditioners, and solvents. Since the 1980s, scientists have documented a thinning of the global ozone layer. An extremely thin area of the ozone layer, popularly called the ozone hole, forms over Antarctica each spring.

Increased exposure to ultraviolet radiation results from ozone depletion and is dangerous to humans and the environment. Risks to

---

DIFFERENTIATING INSTRUCTION

Hands-On Demonstration Gather the following materials to demonstrate the effect of a temperature inversion on pollutants: clear wide-mouthed jar (about 1 gallon, available at bulk food stores) with lid, frozen ice pack or sealed bag of ice to fit on the bottom of the jar, two sealed sandwich bags filled with hot tap water, duct tape, sheet of black paper, drinking straw, and stick matches.

Tape the black paper to the outside of the jar, covering about half the jar. Place the ice on the bottom of the jar. Tape the bags of hot water to the mouth of the jar, keeping the bags inside the jar. Cover the jar with the lid and allow layers of cool and warm air to form inside the jar, for about two minutes. Insert the straw into the jar, keeping one end of the straw outside the jar and the lid closed as much as possible. Quickly light two matches, extinguish them, and drop them one at a time into the jar through the straw. Observe the smoke from the matches against the jar's black background. Ask: Why does the smoke stay in the bottom half of the jar? There are no convection currents in the jar because warm air is layered above cool air. Thus, the smoke remains trapped in the lower layer of cool air.

humans include certain types of skin cancer, cataracts, premature skin aging, and weakening of the immune system. Higher levels of radiation also harm crops and destroy sensitive marine life such as phytoplankton.

Concerns about ozone depletion have led more than 170 nations to sign the 1987 Montreal Protocol on Depletion of the Ozone Layer, a treaty to reduce and eventually eliminate the production of all CFCs and other ozone-depleting substances by 2006. Based on the success of this agreement, stratospheric ozone levels are expected to return to normal by about 2050.

OZONE HOLE
In this computer-generated image, the dark blue area indicates a region with very low ozone concentration in the Southern Hemisphere, October 1999.

Global Warming

Available data indicate that the average global temperatures have increased by about 1°C since the late 1800s. It is possible that this global warming is part of the natural cycle of temperature changes that has occurred throughout Earth's history. However, many scientists are concerned that human activities are most likely contributing to global warming.

Recall from Section 17.2 that carbon dioxide, water vapor, and other naturally occurring molecules in the atmosphere keep Earth's surface warm through the greenhouse effect. The level of greenhouse gases in the atmosphere has risen significantly in the past two centuries. This increase is due primarily to human activities such as the burning of fossil fuels and global deforestation, both of which release CO_2. Global deforestation also contributes to higher CO_2 levels both because burning trees releases CO_2 and because trees are no longer present to absorb CO_2 from the atmosphere after they have been cut down or burned.

The connections among human activity, greenhouse gases, and a rise in global temperatures are complex and difficult to predict. Atmospheric scientists are currently using computer models to investigate the potential impact of increased levels of greenhouse gases on Earth's temperatures. Most models predict that warming will continue, but there is still uncertainty about the extent of the warming and its likely effects.

INVESTIGATIONS CLASSZONE.COM

How Does the Ozone Layer Change over Time?
Use an interactive animation to compare images of the amount of ozone over the Southern Hemisphere by month or by year.
Keycode: ES1708

CHAPTER 17 SECTION 4

INVESTIGATIONS CLASSZONE.COM

How Does the Ozone Layer Change Over Time?
Keycode: ES1708
You may want to assign the investigation as homework.

📖 *Use Internet Investigations Guide, pp. T66, 66.*

More about...

Computer-Generated Images
Computer-generated images are often based on data gathered by satellites, such as NASA's Nimbus 7, which had on board an instrument called the Total Ozone Mapping Spectrometer (TOMS). TOMS took daily readings of levels of ozone concentration in the atmosphere. These levels are usually given in Dobson Units (DU), which indicate the thickness of a given volume of ozone in hundredths of millimeters. The colors in the image on this page correspond to Dobson Units, with dark blue representing the lowest concentration of ozone.

DIFFERENTIATING INSTRUCTION

Challenge Activity After students have completed the Investigation on this page, have them use the Internet to explore how levels of atmospheric carbon dioxide have changed over time. Students can use their research to make a graph showing CO_2 levels over the past 100 years. Have students compare and contrast this graph with the one completed during the Investigation. Ask: What patterns do you see? **Levels of atmospheric carbon dioxide have increased, while the concentration of stratospheric ozone has decreased.**

> **Scientific Thinking**
>
> **ANALYZE**
> Given the uncertainty about the precise causes and effects of global warming, some governments are choosing to reduce emissions of greenhouse gases, while others are resisting such actions. What are the potential benefits of reducing greenhouse gas emissions? What might be the risks?

Preliminary evidence indicates that possible effects include

- rising sea levels due to melting polar ice caps;
- increasing frequency and severity of storms and hurricanes;
- more frequent heat waves and droughts; and
- relocation of major crop-growing areas.

Given the potential impact of these effects, studies of greenhouse gases and climate will remain a critical area of research for years to come. Meanwhile, governments have begun to take action. Some nations have passed laws to limit their own production of greenhouse gases. The Kyoto Protocol, an international agreement established in 1997, proposed a dramatic reduction in greenhouse gas emissions worldwide. However, controversy about the terms of this agreement and its implementation continued into the early years of the twenty-first century.

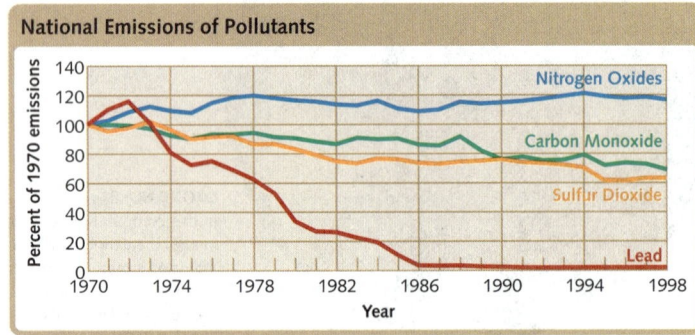

EMISSIONS OF POLLUTANTS What does the increase in the emissions of nitrogen oxides since 1970 tell you?

17.4 Section Review

1. List some examples of pollutants created by the burning of fossil fuels. What are the effects of these pollutants?
2. What are some of the damaging effects of acid rain?
3. Explain how a temperature inversion affects smog.
4. What are some human activities suspected of contributing to global warming?
5. **CRITICAL THINKING** A new car today produces 60 to 80 percent less pollution than a car in the 1960s, yet air quality has improved less than expected. What are some possible reasons why cleaner exhaust has not significantly reduced total emissions from cars?
6. **PAIRED ACTIVITY** Work with a partner to create a poster or other visual display contrasting surface-level ozone and ozone in the stratosphere. Be sure to include information about the formation and effects of ozone in these two different layers of the atmosphere.

SCIENCE & Technology

New Ideas Help Fight Air Pollution

Is air pollution an inevitable effect of modern society? Is it possible to run cars and generate electricity without polluting the air? As pollution issues become more critical, more of the world's best thinkers are tackling these problems.

What can be done to meet society's energy needs without degrading the environment?

Several promising developments will help reduce air pollution. Hybrid electric cars hit the market in 2000, sending the auto industry in a new direction. A hybrid car burns gasoline in a small combustion engine. Leftover kinetic energy recovered during braking keeps a charge in the battery that accelerates the car. This combination of electricity and gasoline creates the highest mileage efficiency of any car ever sold in the mass market, without restrictive range limitations.

Even these cars might be old news before the decade is out. Several automakers plan to sell cars powered by fuel cells that run on liquid hydrogen. Fuel cells make electricity from chemical reactions. With hydrogen as the fuel, the only waste products are water and heat. Fuel cells offer potential for powering homes and businesses, too.

In California, wind power produces electricity at a cost-competitive price. As the cost of generating wind power continues to fall, wind farms featuring hundreds of tower-mounted rotors are expected to crop up around the country.

Perhaps the most futuristic idea comes from NASA. The space agency is investigating Space Solar Power, which involves placing large solar collectors into orbit. These satellites would beam vast amounts of microwave energy to power stations on Earth, where it would enter local electrical power grids.

Advances in technology will lead to new solutions to today's daunting pollution problems and secure a sustainable environment for the future. ■

DEVELOPING TECHNOLOGY This engineer is removing a fuel cell from an electric generator. These small generators may change the way electricity is delivered.

REDUCING POLLUTION Fuel-cell–powered cars, such as this one, emit no harmful exhausts.

Extension

SCIENCE NOTEBOOK
There are tradeoffs to everything. Consider the possible issues with alternative energy solutions and cleaner energy options. Are the costs worth the benefits? Explain.

DATA CENTER CLASSZONE.COM
Learn how air pollution can be reduced.
Keycode: ES1709

DATA CENTER CLASSZONE.COM
Learn how air pollution can be reduced.
Keycode: ES1709

Science and Technology

Air Pollution in the Past
Contrary to popular belief, air pollution is not a phenomenon strictly limited to modern times. As legend has it, centuries ago the tribal leaders of the Hopi forbade the use of foul-smelling coal indoors—the Hopi used the coal to make pottery. In the 1200s, the wife of King Henry III of England complained about smoke emanating from coal fires burning in the town below Nottingham Castle. The sheer quantity of global emissions has increased in modern times, but air pollution itself has plagued humans since the first fires were burned in caves.

Science Notebook
Students should consider long-term health and environmental benefits versus short-term costs to the economy.

Question 4 Review "Global Warming." (p. 381) Ask: Which human activities release carbon dioxide into the atmosphere? **the burning of fossil fuels and deforestation**

Question 5 Review "Common Air Pollutants." (p. 378) Have students brainstorm human activities that both improve and worsen air quality.

Question 6 Reteach "Smog" (p. 380) and "Ozone Depletion." (pp. 380–381) Have students read the text. Ask: When is ozone helpful to living things? **when it occurs in the stratosphere** When is ozone harmful to living things? **when it occurs in the lower atmosphere**

PURPOSE

To investigate and compare rates of absorption and radiation by different materials.

MATERIALS

Provide students with a copy of Lab Sheet 17. A copymaster is provided on page 142 of the teacher's laboratory manual.

Gravel and broken pieces of plaster of paris or asphalt can be used instead of concrete.

PROCEDURE

❶ Before beginning the lab, make sure the water, concrete, and soil have been sitting at room temperature for several hours.

❺ Wrap the concrete in a cloth, then use a rock hammer to break the concrete into small pieces. Complete this step yourself or closely supervise students.

SKILLS AND OBJECTIVES

- **Record** temperature changes produced by heat absorption and radiation.
- **Compare** light and dark materials and dull and shiny materials in terms of the amount of heat they absorb and radiate.
- **Compare** the "land" and "ocean" in terms of the heat they absorb and radiate.

MATERIALS

- 3 empty soup cans (1 painted dull white, 1 painted dull black, and 1 shiny silver, each with a hole punched in the unopened end for inserting a thermometer)
- 3 250-mL beakers
- 3 thermometers with ranges of 0°C to at least 50°C
- desk lamp or clip-on lamp with incandescent bulb of at least 100 watts
- ring stand
- 3 ring clamps
- stopwatch
- topsoil
- water
- coarsely ground concrete
- 2 sheets of graph paper

Absorption and Radiation of Heat

The effects of the sun's radiation vary depending on the type of surface. In this investigation, you will compare the effect of heat absorption and radiation on objects of different colors and reflectivity. You will then investigate absorption and radiation on several types of surfaces: land, represented by soil; ocean, represented by water; and urban areas, represented by concrete.

Procedure

Part A: Light and Dark Materials

❶ Set up the apparatus and place the cans with their open ends down. Put thermometers through the holes in the closed ends of the cans. The thermometers should be at the same angle, slanting in the same direction.

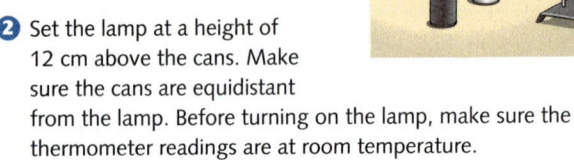

❷ Set the lamp at a height of 12 cm above the cans. Make sure the cans are equidistant from the lamp. Before turning on the lamp, make sure the thermometer readings are at room temperature.

❸ Copy the data table and record the room temperature for all 3 cans under 0 minutes. Then switch on the lamp. Record the temperatures of the cans every 2 minutes for the next 10 minutes. At the end of 10 minutes, turn off the lamp and continue to record the temperatures every 2 minutes for an additional 10 minutes.

❹ On a sheet of graph paper, plot the times and temperatures for the black can and connect the points with a smooth curve. Repeat for the white can and the shiny can. Label each curve on the graph.

	Heating and Cooling Temperatures (°C)										
		Heating					Cooling				
	TIME (minutes):	0	2	4	6	8	10	12	14	16	18	20
Part A	Black can temperature											
	White can temperature											
	Shiny can temperature											
Part B	Water temperature											
	Soil temperature											
	Concrete temperature											

384 Unit 5 Atmosphere and Weather

Part B: Land and Ocean

5 Fill half of one beaker with water at room temperature. Fill half of the second beaker with soil. Fill half of the third beaker with the concrete.

6 Arrange the apparatus as shown to the right. Set up one thermometer in the soil with its bulb just below the surface. Repeat the setup for the beakers filled with concrete.

7 Set the lamp at a height of 30 cm above the water, concrete, and soil surfaces. Make sure the beakers are equidistant from the lamp.

8 Repeat Step 3 for the beakers filled with water, soil, and concrete.

9 Repeat Step 4 for the data collected for the water, soil and concrete.

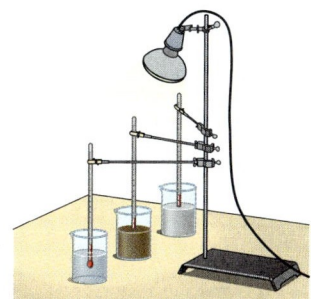

Analysis and Conclusions

1. Study the three curves on your graph from Part A. Which can warmed up most in the first 10 minutes? Which can warmed up the least?

2. The greater the amount of heat absorbed or radiated, the greater the change in temperature. Which can absorbed the most heat? Which can radiated the most heat? Explain your answer.

3. How do color and shininess affect absorption and radiation? How might the choice of color or of metallic paint affect the temperature inside a car?

4. Which material in Part B absorbed more heat in the first 10 minutes?

5. Compare the graphs from Parts A and B. Which surfaces absorbed and radiated the most heat? How are they similar?

6. The sun shines with equal intensity on a stretch of tilled land, a concrete road, and the ocean. Which area will heat up most during the day? Which area will cool the slowest at night? Explain.

7. Portland, Maine, and Pierre, South Dakotah, are located at approximately 44°N latitude. Portland is close to the Atlantic coast and Pierre is in the middle of North America. Which city is warmer on average in the summer? In the winter, Earth's surface tends to lose heat. On average, which city do you predict will be colder in the winter? Explain your answers.

ANALYSIS AND CONCLUSIONS

1. The black can warmed the most; the shiny can warmed the least.

2. Because its temperature changed the most during the experiment, the black can both absorbed and radiated the most heat.

3. Dark, dull objects absorb and radiate more heat than light or shiny objects do. A car with a light-colored, metallic paint will have lower inside temperatures than a car with a dark-colored, dull paint.

4. the soil

5. The black can and the soil absorbed and radiated the most heat. They are the darkest and roughest (least reflective) of the six surfaces.

6. The tilled land will heat up more during the day because its surface is darker and rougher. The ocean will cool more slowly because water has a higher heat capacity than land does and conducts heat more slowly.

7. Pierre is warmer because it is inland where more heat is absorbed. Pierre is also colder during the winter for the same reason—it is inland and the land radiates more heat than the ocean.

Refer students to Safety in the Earth Science Laboratory on pages xxii–xxiii.

CHAPTER 17 REVIEW

Summary of Key Ideas

17.1 Earth's atmosphere contains nitrogen and oxygen, with small amounts of other gases. Gases move between the atmosphere and other parts of the Earth system, yet the composition of the atmosphere remains fairly constant. Local events can change the composition of the atmosphere, with global consequences.

17.2 Heat moves by radiation, convection, and conduction. The atmosphere is divided into layers based on temperature. An imbalance in Earth's heat budget changes Earth's mean temperature.

17.3 The intensity of insolation depends upon the angle at which sunlight strikes Earth's surface. The intensity is greatest at low latitudes, during the summer, and around noon. Land heats and cools more readily than water. Isotherms shift with the seasons, more dramatically over land than over water.

17.4 Pollutants can react with water vapor to form acid precipitation, be trapped by temperature inversions to cause thick smog, reduce the amount of ozone in the ozone layer, and contribute to global warming.

Key Vocabulary

air pollutant (p. 378)
conduction (p. 369)
convection (p. 369)
heat (p. 370)
insolation (p. 372)
ionosphere (p. 372)
isotherm (p. 377)
mesosphere (p. 371)
ozone (p. 371)
radiation (p. 369)
stratosphere (p. 370)
temperature (p. 369)
temperature inversion (p. 380)
thermosphere (p. 371)
troposphere (p. 370)

Vocabulary Review

Explain the difference between the terms in each pair.

1. conduction, convection
2. convection, radiation
3. troposphere, stratosphere

Write the term from the key vocabulary list that best completes the sentence.

4. The bottom layer in a(n) ___?___ of air is colder than the air above it.
5. On a map, all of the points on a(n) ___?___ are the same temperature.
6. Although in the stratosphere it is beneficial to humans, ___?___ near Earth's surface is harmful to humans.

Concept Review

7. Which gas makes up most of the atmosphere?
8. Describe several ways in which gases enter and leave the atmosphere.
9. Why is air warmer at the top of the stratosphere than at the top of the troposphere?
10. Describe three ways energy is transferred from Earth's surface to the atmosphere.
11. If there were no greenhouse gases in the atmosphere, how would Earth's temperature be different than it is today? Explain your thinking.
12. Why are the poles colder than the equator?
13. Describe at least two ways humans increase the amount of carbon dioxide in the atmosphere.
14. **Graphic Organizer** Copy and complete the table below to summarize how insolation depends on location and time.

SUMMARY	Intensity of Insolation	
	Greatest	**Least**
Time of day	around noon	nighttime
Time of year	?	?
Latitude	?	?

Critical Thinking

15. A tropical rainforest increases oxygen and decreases carbon dioxide due to photosynthesis; it increases water vapor due to transpiration. Wheat fields increase the dust and pollen content of the atmosphere.

16. The diagram will show a convection current that has surface water cooling, then sinking as it becomes more dense. This water is replaced by warmer water from the bottom, which in turn cools and sinks. Convection gradually cools the whole lake.

Vocabulary Review

1. Conduction is the transfer of energy through collisions. Convection is the transfer of energy through currents.
2. Convection can transfer energy only within a liquid or gas; it does not require a medium.
3. troposphere: lowest layer of atmosphere, temperature decreases with altitude; stratosphere: above troposphere, temperature increases with altitude
4. temperature inversion
5. isotherm
6. ozone

Concept Review

7. nitrogen
8. Plants remove carbon dioxide from and add oxygen to the air during photosynthesis. Animals remove oxygen from the air when they breathe and add carbon dioxide when they exhale. Carbon dioxide is also released as matter decomposes. Water vapor enters the atmosphere through evaporation, plant transpiration, and animal exhalations and leaves through precipitation.
9. Ozone near the top of the stratosphere absorbs ultraviolet rays from the sun.
10. Radiation transfers energy through space in the form of electromagnetic waves. Conduction transfers energy through collisions of atoms or molecules. Convection transfers energy through currents caused by differences in density in a fluid.
11. Earth would be colder because more of the heat radiated by its surface would escape directly into space.
12. The angle of insolation at the poles is less than at the equator.
13. The burning of fossil fuels releases carbon dioxide into the air. Deforestation reduces the number of trees available to remove carbon dioxide from the air.
14. Greatest: around noon, summer, near equator
    Least: nighttime, winter, near poles

Critical Thinking

15. Contrast Explain how the composition of the atmosphere differs from the atmosphere over a field of wheat and from the atmosphere over a tropical forest.

16. Communicate Draw and explain a diagram to describe how convection helps to cool the waters of a lake in autumn.

17. Hypothesize The atmosphere of Venus is about 97 percent carbon dioxide. How would Venus's heat budget differ from Earth's? Explain.

18. Analyze Describe at least two ways in which Earth's heat budget could become out of balance.

19. Hypothesize Refer to the section about albedo in Unit 1 (pages 17–18). Based on what you know, write a hypothesis about whether buildings in warm areas should be painted with light colors or dark colors.

20. Infer Why are temperature inversions near Earth's surface unlikely on cloudy nights?

Internet Extension

Learn more about instruments used to study the atmosphere.
Keycode: ES1710

Writing About the Earth System

SCIENCE NOTEBOOK Because of plate tectonics, the continents were once closer to the equator than they are today. How would the temperature of the continents at that time have been different from their temperatures today? Currently, much of the land at high latitudes is ice covered. Was there more or less ice on Earth when the continents were closer to the equator? How might this difference in ice cover have affected the heat budget? Explain your thinking.

Learn more about instruments used to study the atmosphere.
Keycode: ES1710

(continued)

17. Carbon dioxide is a greenhouse gas. Venus's atmosphere thus traps heat much more efficiently than Earth's. Venus's surface temperatures are higher than Earth's because less heat escapes.

18. If stratospheric ozone were to decrease, then less of the sun's radiation would be reflected and more insolation would reach Earth's surface. If the amount of greenhouse gases were to increase, Earth would radiate less energy into space than it received.

19. light colors, because they absorb less heat than dark colors do

20. On cloudy nights, less radiation escapes from Earth's surface into space. Thus, the air near the surface is warmer than on clear nights, so conditions are not right for a temperature inversion.

Interpreting Graphs

A student performed an experiment to compare the heating and cooling rates of soil and water. She placed identical containers, one filled with soil and the other filled with water, at equal distances from a light source. A thermometer was placed in each container to record the temperatures. She turned the light on and recorded the temperatures each minute for 10 minutes. Then she turned the light off and again recorded the temperatures each minute for 10 minutes. A graph of her data is shown at right.

21. In which container did the material warm faster?

22. In which container did the material cool faster?

23. Which container held the water?

24. Did the soil and water both warm the same amount? Why or why not?

25. The smallest daily temperature ranges at Earth's surface are less than 1°C. Where are they and why? What sort of location would usually have the largest temperature range? Why?

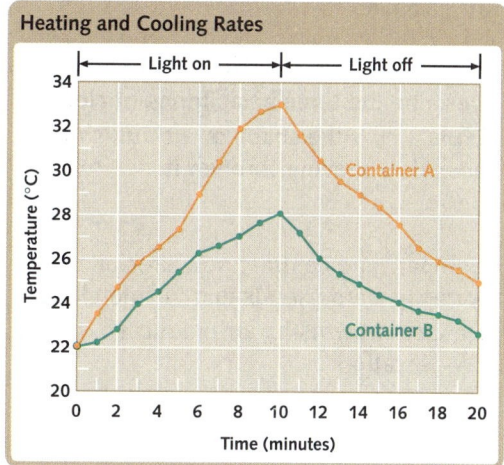

INTERPRETING GRAPHS

21. Container A
22. Container A
23. Container B, water absorbs heat more slowly than soil
24. No; water needs more energy than does soil to raise its temperature the same amount.
25. smallest: over oceans because they heat up and cool down more slowly than land; largest: in a city because its surfaces absorb a great amount of heat during the day, then cool quickly at night

Writing About the Earth System

The continents were warmer because the angle of insolation is most intense near the equator. The amount of ice cover was less because of the warmer temperatures. In addition, there was less ice near the poles because polar areas were surrounded by ocean rather than land, and water does not get as cold as land. Because there was less ice cover, less insolation was reflected back into space.

Chapter-Level Resources		Laboratory Manual, pp. 81–82 Internet Investigations Guide, pp. T67–T68, 67–68 Guide to Earth Science in Urban Environments, pp. 13–20, 29–36, 37–44 Spanish Vocabulary and Summaries, pp. 35–36 Formal Assessment, pp. 52–54 Alternative Assessment, pp. 35–36
Section 18.1 Humidity and Condensation, pp. 390–395	1. Describe the three states in which water vapor can exist in the atmosphere. 2. Explain how air temperature affects the amount of water vapor air can contain. 3. Analyze how condensation occurs in the atmosphere.	Lesson Plans, p. 61 Reading Study Guide, p. 61
Section 18.2 Clouds, pp. 396–401	1. Describe the three basic forms of clouds. 2. Explain how the shape of a cloud shows how air is moving through it.	Lesson Plans, p. 62 Reading Study Guide, p. 62
Section 18.3 Precipitation, pp. 402–407	1. Compare and contrast how precipitation forms in warm clouds and in cold clouds. 2. Describe how rising air produces condensation.	Lesson Plans, p. 63 Reading Study Guide, p. 63 Teaching Transparency 20

INVESTIGATIONS
CLASSZONE.COM

Section 18.3 INVESTIGATION
Which Way Does the Wind Blow?
Computer time: 45 minutes / Additional time: 45 minutes
Students determine the direction of prevailing winds by examining precipitation and vegetation.

Section 18.3 (Science and Society) INVESTIGATION
How Acidic Is Your Rain?
Computer time: 45 minutes / Additional time: 45 minutes
Students examine local precipitation data and determine local trends in acid rain.

388A   Unit 5 Atmosphere and Weather

Electronic Teacher Tools
Visualizations CD-ROM
Classzone.com
Online Lesson Planner
Test Generator

Visualizations CD-ROM
Classzone.com
  ES1801 Visualization, ES1802 Visualization

Visualizations CD-ROM
Classzone.com
  ES1803 Visualization, ES1804 Career

Visualizations CD-ROM
Classzone.com
  ES1805 Visualization, ES1806 Investigation

Classroom Management

Gather these materials for minilab and laboratory activities.

Minilab, p. 393
  thermometer
  cotton gauze
  rubber band
  paper fan
  room-temperature water

Lab Activity, pp. 408–409
  lab apron
  latex gloves
  safety goggles
  225 mL sulfuric acid
  water
  pH probe or pH indicator strips
  12 250-mL beakers
  50-mL graduated cylinder
  3 cups potting soil
  alfalfa sprouts
  metric ruler
  medicine dropper
  balance
  4 pieces of concrete
  8 steel nails
  steel wool
  masking tape
  small scrubbing brush
  tongs

CHAPTER 18

Water in the Atmosphere

Water in the atmosphere makes air feel humid and produces clouds.

How do clouds form, and why do they sometimes produce rain?

INTRODUCE

Build Interest
Before sharing background information about the photograph with students, allow time for their questions and comments.

Ask questions like the following:
- How would you describe the cloud in this photograph? **thick, tall, white cloud with a fluffy or cottony appearance**
- If you saw this cloud in the sky, what type of weather would you expect? **stormy because the cloud appears to be building and darkening (assume the cloud is approaching the viewer)**

Respond
- How do clouds form? **They form as moist air rises in the atmosphere, cools, and then condenses into droplets around tiny particles in the atmosphere. These droplets form clouds.**
- Why do clouds sometimes produce rain? **Rain often occurs when small cloud droplets coalesce and become large enough and heavy enough to fall from the sky as precipitation.**

About the Photograph
The photograph shows an anvil-shaped cumulonimbus cloud. Cumulonimbus clouds, also called thunderclouds, can extend as much as 18,000 meters from their base near the ground. They usually produce heavy rains and lightning.

BEYOND THE TEXTBOOK

Print Resources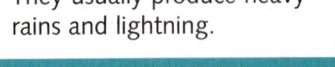
- Reading Study Guide, pp. 61–63
- Transparency 20
- Formal Assessment, pp. 52–54
- Laboratory Manual, pp. 81–82
- Alternative Assessment, pp. 35–36
- Spanish Vocabulary and Summaries, pp. 35–36
- Internet Investigations Guide, pp. T67–T68, 67–68
- Guide to Earth Science in Urban Environments, pp. 13–20, 29–36, 37–44

Technology Resources
- Visualizations CD-ROM
- Classzone.com
- Online Lesson Planner
- Electronic Teacher Tools
- Test Generator

CHAPTER 18

PREVIEW

▶ **FOCUS QUESTIONS** In this chapter you will study water in the atmosphere and learn more about the key questions below.

Section 1 In which states of matter does water exist in the atmosphere? Why does condensation occur?

Section 2 How do clouds form?

Section 3 What causes precipitation?

▶ **REVIEW TOPICS** As you investigate water in the atmosphere, you will need to use information from earlier chapters.

- radiation, conduction, convection (p. 369)
- temperature and heat (pp. 369–370)
- heating of land and water (p. 376)

▶ **READING STRATEGY**
CONNECT

As you read Chapter 18, notice the clouds in the sky each day. Try to connect what you see with the information you are reading. Record in your notebook any observations or questions that come to mind as you observe the sky.

INTERNET RESOURCES
CLASSZONE.COM

At our Web site, you will find the following Internet support for this chapter.

DATA CENTER
EARTH NEWS
VISUALIZATIONS
- Water Vapor
- Advection Fog
- Cloud Formation and Dissipation
- Hail Formation

LOCAL RESOURCES
CAREERS
INVESTIGATIONS
- What Way Does the Wind Blow?
- How Acidic Is Your Rain?

PREPARE

Focus Questions
Find out what knowledge students have about clouds and precipitation. For each focus question, have students consider the section title and then develop two or three additional questions for each section. For example, Section 2: What are the types of clouds and what kind of weather does each bring? Section 3: What are the different types of precipitation?

Review Topics
In addition to the earth science topics listed, students may need to review these concepts:

Physical Science
- Matter exists in three states: solid, liquid, and gas.
- Matter changes from one state to another, depending on temperature.
- Solids that gain heat and pass directly into the gaseous state, without becoming liquid first, are said to undergo sublimation.

Reading Strategy
Model the Strategy: Before starting the chapter, have students brainstorm a list of observations they might make each day. Possibilities include: Are there clouds in the sky? What shape and color are the clouds? Are they high or low in the sky? Do certain types of clouds accompany specific types of weather? Tell students to keep their notebooks handy and determine a specific time or times each day to observe the sky. Have students share their observations in class.

INTERNET RESOURCES
CLASSZONE.COM

INVESTIGATIONS
- Which Way Does the Wind Blow?
- How Acidic Is Your Rain?

VISUALIZATIONS
- Images and animations of water vapor
- Images and animations of advection fog forming
- Images of clouds forming and dissipating
- Animation of hail forming

DATA CENTER
EARTH NEWS
LOCAL RESOURCES
CAREERS

Online Lesson Planner

CHAPTER 18 SECTION 1

◗ FOCUS

Objectives
1. Describe the three states in which water vapor can exist in the atmosphere.
2. Explain how air temperature affects the amount of water vapor air can contain.
3. Analyze how condensation occurs in the atmosphere.

Set a Purpose
Ask students what they think of when they think of water in the atmosphere. Have students read this section to answer the question "What is humidity?"

◗ INSTRUCT

More about...

Indoor Humidity
Indoor air can become uncomfortably dry, especially in winter when furnaces heat and dry the air. A humidifier can raise the indoor relative humidity by vaporizing water and forcing it into the air. By contrast, in hot, humid weather, a dehumidifier can lower the relative humidity. A fan draws in humid air, passing it over cold refrigerator coils. The water vapor condenses on the coils and drips into a collection area.

18.1

KEY IDEAS
Water exists in the atmosphere in each of the three states of matter.

The amount of water vapor that air can contain depends on the air temperature.

If air cools to its dew point, condensation occurs.

KEY VOCABULARY
- water vapor
- condensation
- specific humidity
- saturated
- relative humidity
- dew point
- condensation nuclei

Humidity and Condensation

The molecules of liquid water are always in motion. Even in a glass of water, though the water appears to be still, millions of water molecules are moving about in random directions. Water molecules are also moving through the air around you. This water vapor, while invisible, strongly affects the weather.

Characteristics of Water

Water is unique because it is the only substance that commonly exists in all three states of matter. Depending on its temperature, water can be either a solid, a liquid, or a gas.

- Water is in a solid state at temperatures of 0°C or below, appearing as ice, snow, hail, and ice crystals.
- Water is in a liquid state between 0°C and 100°C, present as rain and cloud droplets.
- At 100°C or above, water evaporates and enters the atmosphere as **water vapor,** an invisible gas. The bubbles in boiling water are an example of water vapor. Clouds and steam are liquid droplets, not gas.

Although you can't see water vapor, sometimes you can feel it. The more water vapor the air contains, the more humid the air feels.

Water often changes state in the atmosphere. When water changes from one state to another, energy is either absorbed or given off. (See the diagram on page 391.) The change from water vapor to liquid water is called **condensation.** Products of condensation include dew, fog, and clouds. The change from liquid water to water vapor is called evaporation. While condensation releases heat, evaporation absorbs heat. The process of condensation slows down the rate at which air cools. Evaporation, by contrast, is a cooling process: after you get out of a swimming pool, for example, you may feel chilly, because the water molecules on your skin are absorbing your body heat as they evaporate.

Frost forms by *deposition* when water vapor condenses as a solid. Snowbanks can become smaller not only through melting but also through *sublimation,* when water changes directly from a solid to a gas, without first becoming a liquid.

FROST The formation of frost is an example of deposition of water vapor.

390 Unit 5 Atmosphere and Weather

DIFFERENTIATING INSTRUCTION

Reading Support Remind students that the design of the book can be used as an aid in organizing ideas. Red headings are used for major topics within the section. The blue headings subdivide each topic into important aspects of that topic. Model this process for the students. For example, "Humidity [red heading] is one of three major topics covered in this section. In relating humidity to atmosphere and weather, you will also need to know about relative humidity and measuring humidity [blue headings]." Have students read the headings before reading this section.

Use Reading Study Guide, p. 61.

States of Water

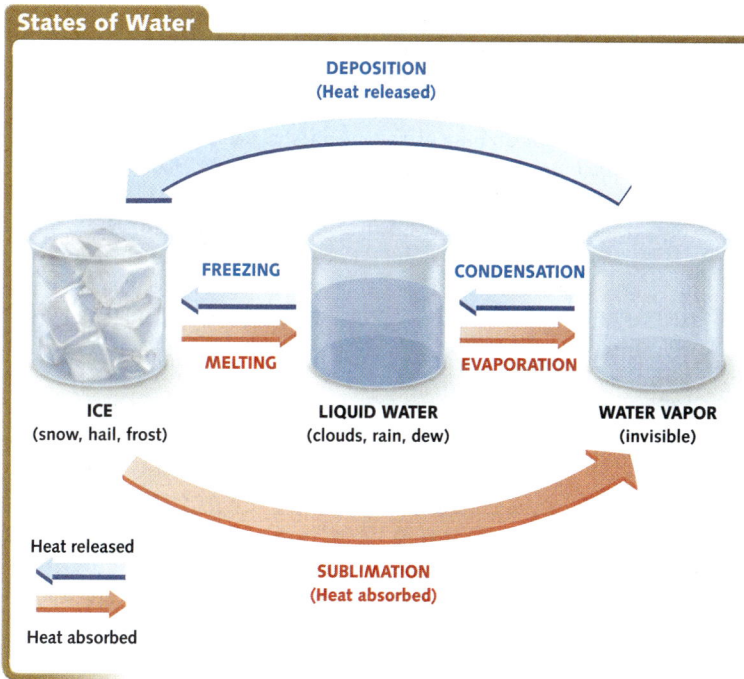

Humidity

The amount of water vapor present in the air varies widely. The actual amount of water vapor in the air at a given time and place is called the **specific humidity.** It is expressed as the number of grams of water vapor per kilogram of air. On a humid summer day, for example, the specific humidity may be about 20 grams per kilogram.

There is a limit to the amount of water vapor that can be present in the air. To understand how this works, consider the water and air inside a fish tank with a glass lid. Some water molecules have enough energy to escape from the surface and enter the air as water vapor. In other words, they evaporate. At the same time, other water vapor molecules lose energy and return to the liquid state through condensation. As temperatures hold steady, the rate of condensation increases as the amount of water vapor in the air increases. When there is so much water vapor in the air that the rate of condensation equals the rate of evaporation, the air is **saturated.** If any additional water evaporates into saturated air, an equal amount will condense. This fact explains why water droplets may form on the lid of the fish tank. These drops confirm that as water continues to evaporate, an equal amount of water condenses from the saturated air.

The amount of water vapor present in saturated air depends on the temperature of the air. The warmer the air, the more water vapor it can contain. The water vapor capacity of air roughly doubles for every rise in temperature of about 11°C.

Visual Teaching

Interpret Diagram
Make sure students understand that in the diagram the full column represents the maximum amount of water that air can hold and the blue shaded part represents the actual amount that is held. Help students relate these amounts to the formula. Then ask students to find relative humidity if the air holds 15 grams per kilogram of water vapor, and it can contain at most 60 grams per kilogram. 15 g/kg ÷ 60 g/kg × 100% = 25%.

Extend
Have students work in pairs to make up a similar problem for which their partner must find the relative humidity. Students can check each other's answers.

CRITICAL THINKING
Ask: If the relative humidity of the air is 25 percent, and the temperature outside rises while the amount of humidity in the air stays the same, what happens to the relative humidity? It decreases. Why? If the temperature of the air rises, the amount of water vapor the air can hold increases. Because relative humidity is the ratio of the water vapor in the air compared to the amount it can hold, raising the amount the air can hold without increasing the amount of water vapor in the air lowers the relative humidity.

VISUALIZATIONS CLASSZONE.COM
Observe animated satellite images of water vapor.
Keycode: ES1801

Relative Humidity
When meteorologists refer to the **relative humidity**, they are telling us how near the air is to its maximum capacity for holding water vapor. Relative humidity compares the actual amount of water vapor in the air with the maximum amount of water vapor that can be present in air at a given temperature and pressure. Relative humidity is usually stated as a percentage. Saturated air has a relative humidity of 100 percent; air that contains no water vapor has a relative humidity of 0 percent.

To calculate the relative humidity of a kilogram of air, divide its specific humidity by its maximum capacity. The example below calculates the relative humidity for air at 26.5°C with a specific humidity of 11 grams per kilogram.

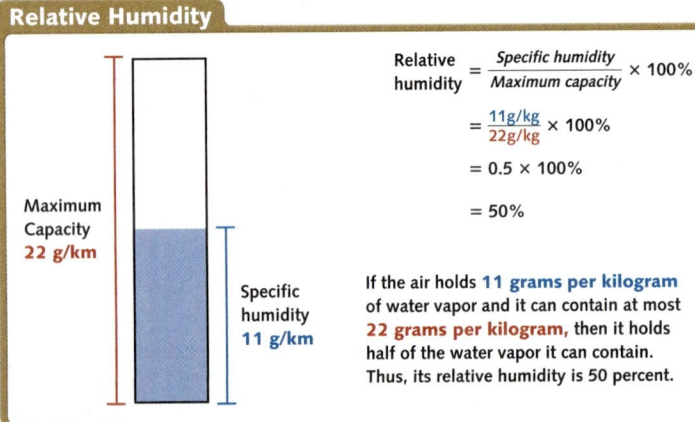

Relative Humidity

$$\text{Relative humidity} = \frac{\text{Specific humidity}}{\text{Maximum capacity}} \times 100\%$$

$$= \frac{11\text{g/kg}}{22\text{g/kg}} \times 100\%$$

$$= 0.5 \times 100\%$$

$$= 50\%$$

Maximum Capacity 22 g/km
Specific humidity 11 g/km

If the air holds **11 grams per kilogram** of water vapor and it can contain at most **22 grams per kilogram**, then it holds half of the water vapor it can contain. Thus, its relative humidity is 50 percent.

Measuring Humidity
Humidity is typically measured with a psychrometer (sy-KRAHM-ih-tuhr), an instrument that works on the principle that evaporation causes cooling. A psychrometer consists of two thermometers. One is a dry-bulb thermometer that shows the air temperature. The other is a wet-bulb thermometer that has a water-soaked wick wrapped around its bulb. Fans are set to circulate air past the two thermometers. The wet-bulb thermometer usually shows a lower temperature than the dry-bulb thermometer, because water evaporating from the wick cools the wet bulb. The drier the air, the greater the cooling from evaporation, and the greater the difference in the readings. If both the wet-bulb and dry-bulb thermometers read the same, this shows that no water is evaporating from the wet bulb, and the air is saturated.

Relative humidity can be determined by using a psychrometer along with a table like the one on page 393. For example, the table shows that at a dry-bulb temperature of 16°C, a difference of 3°C between the readings indicates a relative humidity of 71 percent.

Reading Support Point out the pronunciation of *psychrometer* given in parentheses in the text. Suggest students use this tool to correctly pronounce the term.

Support for Visually Impaired Students The formula for relative humidity in the diagram may be difficult for students to see. Write the formula and the solution to the problem on the board, saying aloud the information as you write.

Relative Humidity

Dry-Bulb Temperature (°C)	Difference Between Wet-Bulb and Dry-Bulb Temperatures (°C)						
	0	3	6	9	12	15	18
0	100	46					
8	100	63	29				
16	100	71	46	23			
24	100	77	56	37	20	5	
32	100	80	62	46	32	20	9

Condensation

How does condensation, the change from vapor to liquid, usually happen in the atmosphere? Consider this example. On a sunny spring day the air is not saturated. At night, however, as the air cools rapidly, its capacity to contain water vapor diminishes. The air becomes saturated. If the air continues to cool past the point of saturation, condensation occurs.

The water vapor may condense into droplets, forming a cloud or fog. If the water vapor condenses on a surface, such as grass, it is called dew. The temperature at which saturation occurs and condensation begins is called the **dew point.**

The dew point is a measure of the amount of water vapor in the air. The more water vapor the air contains, the less the air has to cool in order for condensation to start, so the higher the dew point.

DEW ON SPIDER WEB The beads of dew on this spider web formed from droplets of condensed water vapor.

15-Minute Mini LAB

Measuring Humidity

Materials
- a thermometer
- cotton gauze
- a rubber band
- a paper fan
- room-temperature water

Procedure

1. Measure and record the room temperature.
2. Wrap gauze around the bulb of the thermometer and fasten it with the rubber band. Dip the gauze-wrapped bulb in room-temperature water.

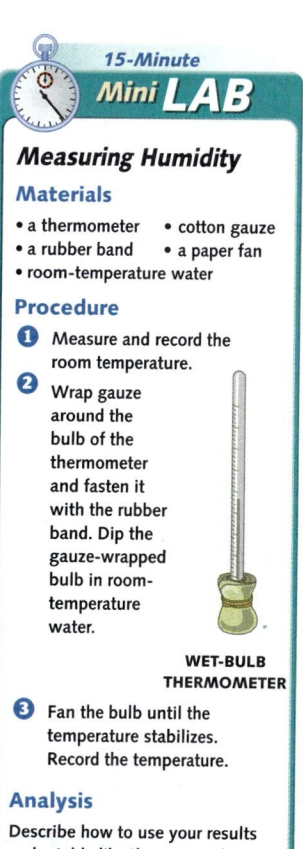

WET-BULB THERMOMETER

3. Fan the bulb until the temperature stabilizes. Record the temperature.

Analysis

Describe how to use your results and a table like the one on this page to estimate the relative humidity. How could you make a more accurate estimate?

VISUALIZATIONS
CLASSZONE.COM

Observe satellite images of water vapor.
Keycode: ES1801

Visualizations CD-ROM

MINILAB
Measuring Humidity

Analysis Answer

Record both the initial temperature reading (dry-bulb) and the wet-bulb temperature. Determine the difference between the two temperatures. On the chart, find the relative humidity in the cell where the row containing the dry-bulb reading and the column containing the wet-bulb difference intersect. If the readings obtained do not match those on the chart, use the more detailed chart provided in the Appendix, page 704. A more accurate reading might come by fanning the wet bulb longer to see if the temperature continues to drop.

For proper behavior in a laboratory setting, refer students to Safety in the Earth Science Laboratory on pages xxii–xxiii.

CRITICAL THINKING

Present this scenario to students: On a warm summer day, the relative humidity is about 60 percent. That night, the temperature does not drop very much at all. Predict if you will find dew on car windshields the next morning. Probably not. If the temperature doesn't drop very much, the air probably will not reach full saturation, and hence will not reach its dew point.

Developing English Proficiency Phrases that contain more than one new word may present difficulty in reading comprehension for students. For example, in the phrase "its capacity to contain water vapor diminishes" on page 393, students may need help with the words *capacity* and *diminishes*. Substitute appropriate terms such as *ability* and *becomes less*, respectively, to help students comprehend meaning. Have them restate the meaning of the sentence in their own words.

Scientific Thinking

INFER

The frozen water on the surface of the oranges is 0°C. But heat released when the water freezes tends to stabilize the temperature of the fruit underneath the ice layer. The heat prevents the fruit from cooling to the point where the liquid in the plant cells would freeze and destroy plant tissues.

Extend

Ask: Why would the melting of the ice on the fruit in the morning be a concern? If spraying stops too soon, the fruit can freeze due to cooling from evaporation. Fruit growers continue to spray the fruit during the day until all of the ice on the fruit has melted.

More about...

Condensation Nuclei

A cubic centimeter of air can contain as many as 10,000 condensation nuclei, derived from sources such as soot produced by forest fires, wind erosion of soil, particles ejected from erupting volcanoes, salt from seawater spray, and emissions from industrial smokestacks and home chimneys. Condensation occurs most readily on nuclei that are fairly large or those likely to absorb water.

Scientific Thinking

INFER

Some fruit growers spray their trees with water on cold nights. As the water cools and freezes, it releases heat. While frozen, the water maintains a temperature of 0°C. Explain why spraying water on fruit keeps the fruit from freezing.

Observe images of advection fog.
Keycode: ES1802

Cooling and Condensation

Two conditions are necessary for water vapor to condense: (1) there must material for water vapor to condense onto and (2) air must cool below its dew point. When fog or clouds form, the water vapor is condensing on tiny particles called **condensation nuclei** (KAHN-dehn-SAY-shuhn NOO-klee-EYE). Even when air is cooled below its dew point, condensation to fog or clouds may not occur if there are no condensation nuclei available. Air cooled below its dew point is said to be supersaturated.

Condensation nuclei are usually substances such as salt, sulfate particles, or nitrate particles. Salt enters the air when fine sea spray evaporates. The sulfates and nitrates come from natural sources and from the burning of fuels. Condensation nuclei are so tiny that one puff of smoke contains millions of them. Just as these particles are necessary for the formation of fog or clouds, ice nuclei are required for the formation of ice crystals. Some types of bacteria and clay particles contaminated with organic material make good ice nuclei.

Air may cool, or lose heat, through the following ways:

- contact with a colder surface
- radiation of heat
- mixing with colder air
- expansion as it rises

Dew and frost form when moist air contacts a colder surface. Fog forms when air cools through contact and mixing. In Section 18.2, you will learn how air expands and cools as it rises.

Formation of Dew and Frost

When air cools to its dew point through contact with a colder surface, water vapor condenses directly onto that surface. If the air temperature is above 0°C, dew forms. Dew may form on the ground, on leaves and grass, and on other surfaces. At night these surfaces become cooler than the air, because they lose heat more rapidly than air does.

If the air temperature is below 0°C, the water vapor becomes frost through deposition. This kind of frost, resulting from atmospheric moisture, is not responsible for so-called killing frosts. A killing frost occurs when the temperature near the ground drops below −2°C for several hours, causing liquid in the cells of some plants to freeze. As the liquid freezes it expands, bursting the cell walls and killing the plants.

Formation of Fog

Fog forms when a cold surface cools the warmer moist air above it. As water vapor condenses in the air, tiny droplets fill the air and form fog. Each droplet is centered around a condensation nucleus. The droplets are so tiny that they fall slowly. The slightest air movement keeps them suspended in the air. At very cold temperatures, the fog may consist of tiny ice crystals.

Hands-On Demonstration How does a cold surface cause water vapor in the air to condense? Heat a pot of water on a burner or hot plate, and then set the pot on a table. Have ready a metal tray that has been chilled for a half hour. Make sure there is no condensation on its surface. Ask: Is any water vapor in the air above the pot? **yes** Have students predict what will happen to this invisible water vapor when you hold the cold tray above the pot. **Water droplets form on the metal surface above the pot.** Explain that the cold temperature of the surface causes the temperature of the air in contact with it to drop below the dew point—the point where condensation occurs.

⚠ *Exercise caution when handling the pot of hot water. Take care to protect the table surface. Remind students that safety is everyone's responsibility.*

FOG This fog is an example of a radiation fog.

One kind of fog, radiation fog, forms when the night sky is clear and the ground loses heat rapidly through radiation. As the ground cools, light winds mix the cooled bottom air with the warmer air a short distance above it. Eventually, the whole layer of air cools to its dew point. The resulting fog at ground level is colder than the layer of air above it. This arrangement of warmer air above cooler air is called a temperature inversion.

Radiation fogs are common in humid valleys near rivers or lakes. They are most frequent in the late fall and in winter. These fogs are thickest in the early morning, becoming "burned away" by the later morning sunshine.

Another kind of fog, an advection fog, may form when warm, moist air blows over a cool surface. In the northern United States and southern Canada, advection fogs form when warm, moist southerly winds blow over snow-covered ground. The famous fogs of Newfoundland form when warm, moist air from the Gulf Stream blows over the cold Labrador Current. Summer fogs in coastal California form when warm ocean air moves over cold coastal waters. Winter fogs form along parts of the Gulf Coast when cold Mississippi River waters chill the warm gulf air at the river's mouth.

VOCABULARY STRATEGY

The prefix *ad-* means "toward," and *vection* means "to bring." *Advection* describes horizontal air movement.

18.1 Section Review

1. Identify the three states of water that exist in the atmosphere and give examples.
2. Use a Venn diagram to compare and contrast specific humidity and relative humidity.
3. Describe what happens when air is cooled to its dew point.
4. Compare and contrast the formation of radiation fog with the formation of advection fog.
5. **CRITICAL THINKING** What do you think is the source of condensation nuclei for the advection fogs of Newfoundland?
6. **MATHEMATICS** At 15.5°C, the maximum amount of water vapor that can be present in the air is 11 grams per kilogram. If the specific humidity is 4 grams per kilogram, what is the relative humidity?

ASSESS

1. Water exists in the solid state as ice, in the liquid state as liquid water, and in the gaseous state as water vapor.
2. Diagrams will vary but might include a circle labeled Specific Humidity within a larger circle labeled Saturation Vapor Pressure, and the entire diagram labeled Relative Humidity.
3. Saturation occurs in the air, and water vapor condenses into droplets of liquid water.
4. Both types of fog occur when moist air near the surface is cooled, causing moisture to condense into water droplets. Radiation fogs form when the ground loses warmth rapidly, and the air cools below the dew point. They often occur in early morning, in valleys or near rivers and lakes. Advection fogs form when warm, moist air blows into an area where the surface is cooler, causing condensation. They often form where warm ocean air blows over cooler offshore waters.
5. Salt from the seawater most likely would be a source of condensation nuclei.
6. Students should set up this equation to find relative humidity: 4 g/kg × 11 g/kg × 100% = 36%.

If students miss . . .

Question 1 Review the three states of matter. (p. 390) Give everyday examples as they apply to water.
Question 2 Review the relationship between specific humidity and relative humidity. (pp. 391–392) Review how to draw a Venn diagram.
Question 3 Review dew point. (p. 393) Remind students that water vapor condenses at the dew point when full saturation is reached.
Question 4 Review the two types of fog: how they differ in formation and where they are most likely to form. (p. 395)
Question 5 Review condensation nuclei. (p. 394) Ask: What kinds of particles act as condensation nuclei?
Question 6 Review the equation for determining relative humidity. (p. 392) Identify the specific humidity in the question as 4 g/kg and the maximum capacity (saturation specific humidity) as 11 g/kg.

FOCUS

Objectives
1. Describe the three basic forms of clouds.
2. Explain how the shape of a cloud shows how air is moving through it.

Set a Purpose
Have students read this section to find the answer to the questions "What are the major types of clouds and how do they form?"

INSTRUCT

More about...

Unusual Clouds
The clouds shown at right are those commonly seen in the sky. There are several other types of clouds that are only rarely seen. For example, lenticular (lens-shaped) clouds are middle-level clouds that form over mountains as steady winds blow at a right angle to the mountain range. They have an unusual shape that often makes them look like thick, stacked disks. The clouds remain stationary for extended periods of time and, because of their shape, have on occasion been reported to authorities as UFOs hovering in the sky.

18.2

KEY IDEAS

There are three basic forms of cloud: cumulus, stratus, and cirrus.

The shape of a cloud shows how air is moving through it.

As water vapor condenses, it releases heat.

KEY VOCABULARY
- stratus
- cumulus
- cirrus
- condensation level
- dry-adiabatic lapse rate
- moist-adiabatic lapse rate

Clouds

Like fog, clouds form when the air cools below its dew point. Clouds can form at any altitude in the troposphere. At temperatures above freezing, clouds are made of water droplets. Below freezing, clouds are usually mixtures of snow crystals and supercooled water, which is water that has cooled below 0°C without freezing. When supercooled droplets contact ice nuclei, they form snow and ice crystals. At temperatures below −20°C, clouds are made up mostly of snow and ice crystals.

Types of Clouds

There are four types of clouds: low clouds, middle clouds, high clouds, and clouds of vertical development. Clouds are classified according to their height, or altitude—low, middle, or high—and their shape. If air movement is mainly horizontal, clouds form in layers; these are *stratiform* clouds. If air movement is mainly vertical, clouds grow upward in great puffs; these are *cumuliform* clouds.

When stratiform and cumuliform clouds appear at altitudes between 2000 and 7000 meters, they are called altostratus and altocumulus clouds. When they appear above 7000 meters, they are called cirrostratus and cirrocumulus clouds. Cirrostratus clouds are thin, smooth or fibrous sheets that sometimes cause halos around the sun or moon and may indicate snow or rain. Stratocumulus clouds are layers of puffy clouds; they often

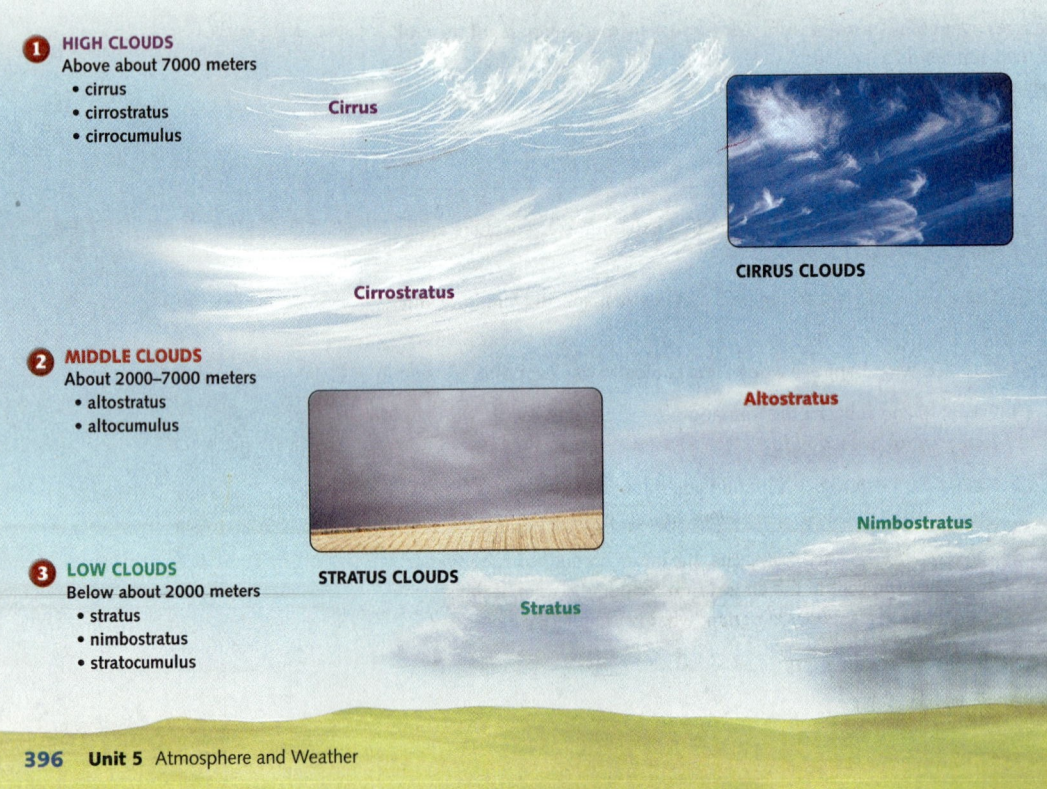

① HIGH CLOUDS
Above about 7000 meters
- cirrus
- cirrostratus
- cirrocumulus

② MIDDLE CLOUDS
About 2000–7000 meters
- altostratus
- altocumulus

③ LOW CLOUDS
Below about 2000 meters
- stratus
- nimbostratus
- stratocumulus

Reading Support Have students use the illustration at the bottom of pages 396 and 397 to enhance their understanding of cloud types as they read. Read aloud the subhead "Types of Clouds." Then read the first sentence of the first paragraph. Have students look at the illustration to find each of the four groups of clouds. Tell students to refer to each section in the illustration as they read about that group of clouds in the text. Have students try to match the description of the cloud in the text with its shape and location in the illustration.
Use Reading Study Guide, p. 62.

cover the whole sky, especially in winter. Among clouds found below 2000 meters are stratocumulus and nimbostratus clouds. Nimbostratus clouds are dark gray layers of cloud that produce steady rain. Cumulonimbus clouds, another kind of rain cloud, grow to great heights and produce heavy rain with thunder, lightning, and occasionally hail.

Cloud heights are measured as distance above the ground, not above sea level. The height ranges mentioned are average heights in the troposphere for the middle latitudes. Clouds reach higher altitudes along the equator and lower altitudes in the polar regions.

Note that cloud names are formed from one or more of the same five words or word parts.

- *Stratus* and *strato-* describe clouds that form in layers.
  Stratus clouds are layered, low clouds.
- *Cumulus* and *cumulo-* describe clouds that grow upward. *Cumulus* is the Latin word for a heap.
  Cumulus clouds are fluffy clouds with flat bases.
- *Cirrus* and *cirro-* describe feathery clouds. *Cirrus* is the Latin word for a curl of hair.
  Cirrus clouds are high, feathery ice clouds.
- *Alto-* is used in describing clouds located between 2000 and 7000 meters.
- *Nimbus* and *nimbo-* refer to dark rain clouds.

VOCABULARY STRATEGY

Alto- (AL-toh) comes from the Latin word meaning "high."

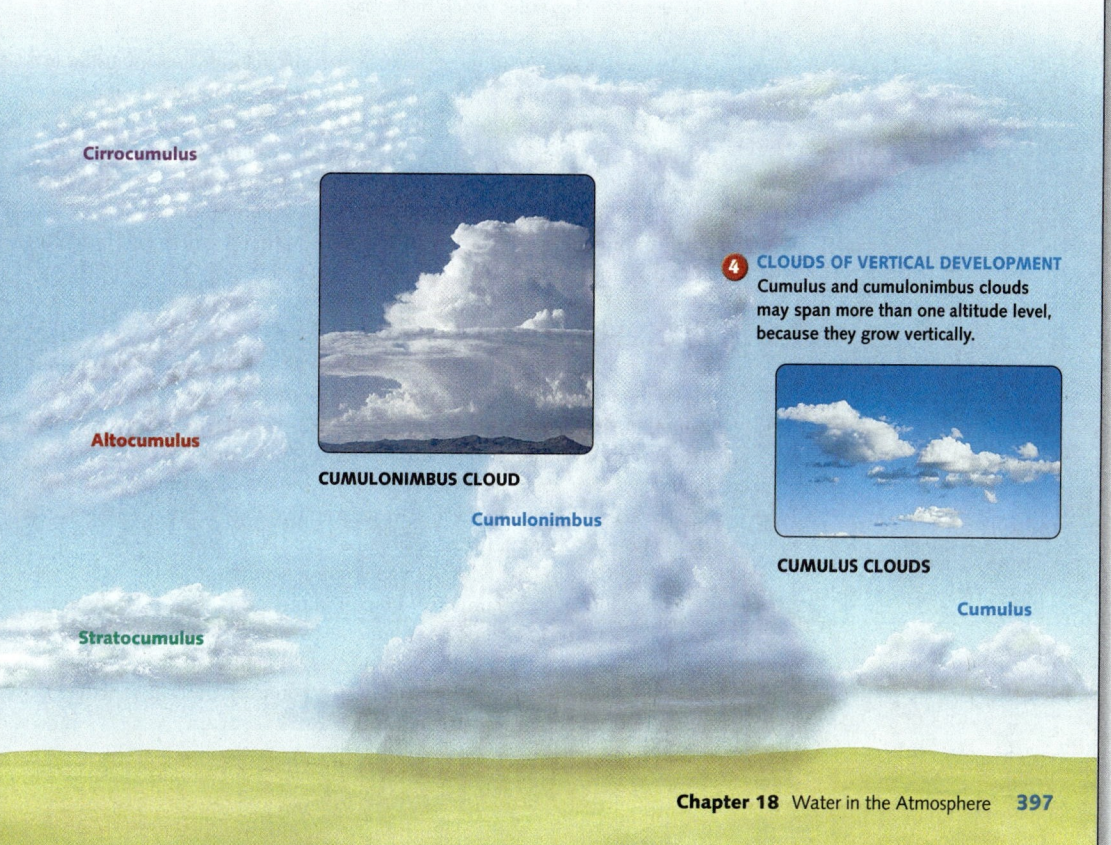

Cirrocumulus

Altocumulus

CUMULONIMBUS CLOUD

Cumulonimbus

④ CLOUDS OF VERTICAL DEVELOPMENT
Cumulus and cumulonimbus clouds may span more than one altitude level, because they grow vertically.

CUMULUS CLOUDS

Cumulus

Stratocumulus

Chapter 18 Water in the Atmosphere 397

More about...

Naming Clouds

Luke Howard was a pharmacist who spent most of his spare time looking at the sky and forming theories about what he saw. Clouds fascinated him, and in 1803 Howard wrote a paper in which he proposed a system for naming them. Howard's simple system used Latin names to classify clouds by shape. It was Howard who first used the classification *cumulus*, the Latin word for "heap." Strangely enough, before Howard's system, clouds had no classification system to differentiate them. Although meteorologists have added to and modified the system since Howard proposed it 200 years ago, the system we use today for naming and classifying clouds is largely the one he devised.

Observe clouds form and dissipate.
Keycode: ES1803

Cloud Formation

The shape of a cloud shows how the air moves through it. Cumulus clouds and other clouds with vertical development form when rising air currents are buoyant, or lighter than the surrounding air.

On a sunny day, darker areas of ground absorb more of the sun's heat. The warmer ground heats the air above it, making that air less dense than the surrounding air. As the air rises, it cools. If the air cools to its dew point, a cumulus cloud forms. The atmospheric level at which condensation occurs is called the **condensation level.**

CUMULUS CLOUD FORMATION Air in a growing cumulus cloud moves upward because it is buoyant and its temperature is warmer than that of the surrounding air. The flat cloud base shows where the water vapor began to condense. Below the condensation level, the air is unsaturated; no water vapor is condensing.

Without a steady influx of moist, warm air, a cloud will soon evaporate. If there is a continuous supply of moist air, however, and if the air in the cloud remains warmer and therefore less dense than the surrounding air, the air in the cloud will continue to rise, and the cloud will grow in height. A cumulus cloud ceases to grow when it loses its buoyancy, which happens when the air in the cloud reaches the same temperature as the air surrounding the cloud at the same altitude. At that point, the air inside the cloud attains the same density as the surrounding air.

Dry- and Moist-Adiabatic Lapse Rates

The adiabatic lapse rate is the rate at which air cools as it rises. Unsaturated air cools at a rate of about 10°C for every kilometer it rises. This rate is called the **dry-adiabatic** (dry-AD-ee-uh-BAT-ihk laps rayt) **lapse rate.** The cooling is caused only by the air's expanding. The air expands as it rises because it is surrounded by areas of lower pressure. The **moist-adiabatic lapse rate** is the rate at which saturated air cools as it rises; this rate varies from about 5°C per kilometer to about 9°C per kilometer.

Remember that when water vapor condenses it releases heat. This release of heat is a very important factor in the formation of clouds and storms. Air rising in a cloud does not cool as fast as rising dry air does, because the condensing water vapor releases heat into the air, causing it to cool more slowly. The heat released during condensation can fuel the growth of huge cumulonimbus clouds.

Support for Visually Impaired Students The illustrations and photographs of clouds on several pages in this chapter may be difficult for some students to see in detail. Collect large illustrations or photographs that show some of the major types of clouds. Label them, and review the cloud types with students.

Cumulonimbus Clouds

Like cumulus clouds, cumulonimbus clouds begin forming when moist air rises and cools to its dew point. Cumulonimbus clouds also have a cloud base at condensation level. The tallest cumulonimbus clouds, however, can grow to more than twice the height of Mount Everest. The heat from condensation keeps the air rising inside the clouds warmer and less dense than the air outside.

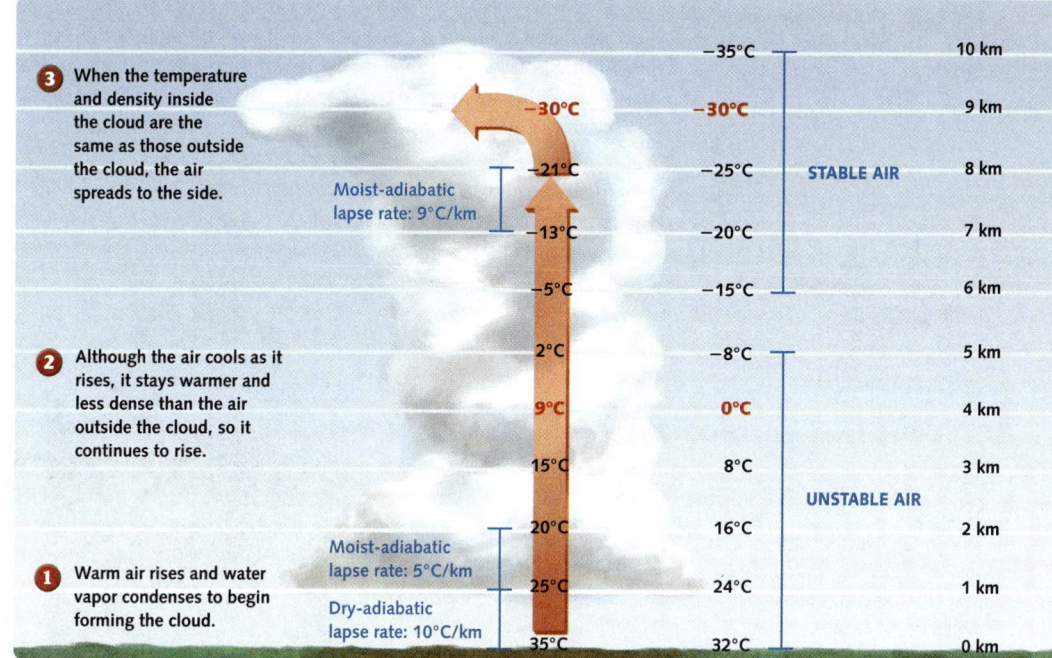

CUMULONIMBUS CLOUD FORMATION A cumulonimbus cloud can grow to great heights.

When meteorologists know the air temperatures outside a cloud and the rate at which the rising air cools, they can predict how high the cloud will develop vertically. Instruments attached to a high-altitude balloon measure temperatures outside the cloud. Actual temperatures can be more erratic than those shown above.

In the illustration, notice that below 5 kilometers the temperature outside the cloud decreases more quickly with altitude than the temperature inside the cloud. As a result of this faster decrease, the rising air inside the cloud becomes increasingly less dense than the air outside the cloud. Under these conditions, meteorologists say that the air is unstable. If the layer of unstable air is shallow, cumulus clouds can form. If the layer of unstable air is deep, cumulonimbus clouds may form.

If a cumuliform cloud receives a continuous supply of moist air, it will grow until it enters a layer of stable air. Air is said to be stable if the temperature of air rising inside a cloud decreases more quickly than the temperature outside the cloud.

Observe clouds form and dissipate.
Keycode: ES1803
Have students discuss what they observe and think about how long it actually takes clouds to form and dissipate. Ask these questions to get students thinking: Have you ever seen clouds form and dissipate as quickly as on the Web site? No, the action was sped up considerably. Do clouds usually look the same way all day long? No, cloud patterns change throughout the day. Students should come to the conclusion that clouds change over a period of time that ranges from a few minutes to a few hours.

Visualizations CD-ROM

More about...

Cumulonimbus Clouds
Some cumulonimbus clouds continue to grow beyond the point at which the temperature inside the cloud is the same as the temperature outside the cloud. The cause of this continued growth is the currents inside the cloud, which are so strong that the momentum keeps the air moving upward.

Reading Support Remind students that illustrations, especially diagrams, often summarize major concepts presented in the text. Such is the case with the diagrams on pages 398 and 399. If they understand the diagrams, including the captions and labels, they understand the major concepts. Suggest that students not only study the diagrams during reading but also review them after reading. Then they should explain in their own words what the diagrams show.

More about...

Formation of Layer Clouds
Students learn that clouds can form when moist air rises over mountains. Sheets of stratiform clouds also can form as air rises at the boundary of a frontal system. As a warm air mass moves into an area of cold air, the lighter warm air gently rises up over the denser cold air at the front. As it does, the air cools, and condensation occurs. Thick layers of stratus clouds form near the front, causing a wide area of precipitation that often lasts for several hours while the front passes.

DISCUSSION

After students have read the sections on Layer Clouds and Lifting Condensation Levels ask: Suppose there is a significant drop in temperature between 11:00 A.M. and 3:00 P.M, how does that affect the formation of stratus clouds? Remind students that the air cools more quickly with altitude and that the saturation point is reached more quickly in colder air. The lifting condensation level is likely to drop to a lower level in the late afternoon, after the drop in temperature. Stratus clouds forming later in the afternoon will form at a lower altitude.

Layer Clouds

Stratiform, or layer, clouds form in stable air. In stable air, air cannot easily move up or down, so it tends to spread out horizontally in layers instead. If the air is forced to rise to its condensation level, layer clouds form. For example, suppose the temperature of the air is uniform throughout a thick layer. If warmer, moist air is present, it can rise only a short distance before cooling to the temperature of the surrounding air, at which point it stops rising. Clouds can form in stable air in two ways:

- As described above, air can be forced slowly upward to its condensation level. This happens with air moving up a mountainside or over a layer of colder, denser air. (Air can also be forced to rise when it is part of a low-pressure system. You will learn more about low-pressure systems in Chapters 19 and 20.)
- A layer of air can cool to its dew point by radiating heat or mixing with cooler air. The water vapor in the air will then condense to form clouds.

CAREER

Air Force Meteorologist

When you think of the United States Air Force, fighter pilots and ground support crew may come to mind. Yet meteorologists are also essential to the workings of the Air Force. Air Force meteorologists apply their weather expertise to support flight operations, missile launches, the army's ground operations, and NASA's space-shuttle launches. A few meteorologists are even members of highly classified special-operations teams that deal with missions in hostile territories.

Given the costliness and sensitivity of these projects, accuracy in predicting weather is crucial. For example, during a space-shuttle launch, Air Force meteorologists must constantly monitor atmospheric conditions, noting cloud formation, storm development, and wind speeds. Based on their observations they advise the launch director either to continue or to abort the mission.

Like civilian meteorologists, those in the Air Force need at least a bachelor's degree in meteorology or atmospheric science. Since meteorologists have to analyze large amounts of data from satellite information and computer models, a successful career requires a solid foundation in subjects such as physical science, computer science, and statistics. Air Force meteorologists work long hours on assignment, knowing that their efforts are vital to the nation's security. ■

AIR FORCE meteorologists often work as members of an airplane crew, collecting and analyzing weather data.

Find out more about careers in meteorology.
Keycode: ES1804

If students miss . . .

Question 1 Review cumuliform and stratiform clouds. (p. 396) Then have students compare both types of clouds in the illustration. (pp. 396–397)

Question 2 Reteach dry-adiabatic lapse rate and moist-adiabatic lapse rate. (pp. 398–399) Emphasize that the condensation of water vapor in moist air adds heat as the air rises, slowing the air's rate of cooling.

Question 3 Review the formation of cumulonimbus and layer (stratus) clouds. (pp. 399–401) Ask students to tell in their own words what is happening in the "Cumulonimbus Cloud Formation" diagram. (p. 399)

Question 4 Review the definition of cirrus clouds. (p. 397) Then refer students to the diagram on pp. 396–397. Have them note that cirrus clouds form at higher, colder altitudes than other cloud types.

Predicting the Lifting Condensation Level

As warm air rises and expands not only does its temperature decrease, but its dew point also decreases. Given the temperature and the dew point at ground level, meteorologists can predict the condensation level. For dry air, the rate of cooling by expansion is about 10°C for every kilometer. As the air rises, its dew point falls at a rate of about 2°C for every kilometer. When the temperature and dew point are the same, condensation occurs.

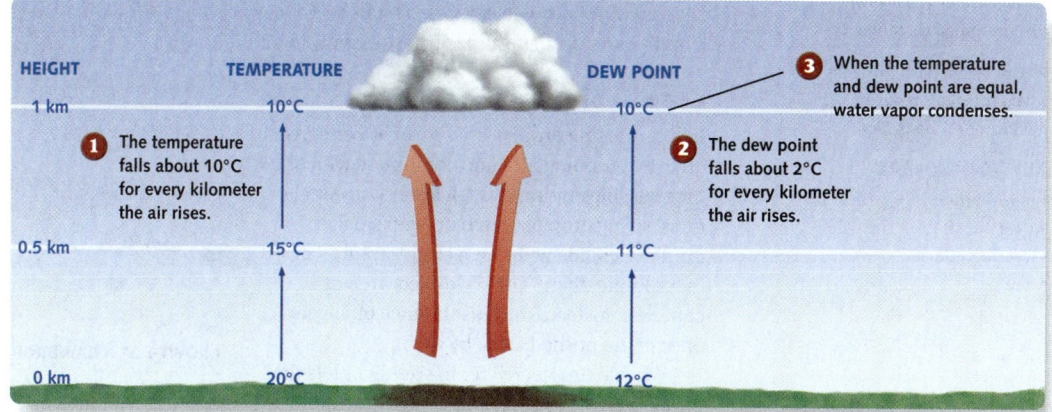

CONDENSATION LEVEL The cloud base forms at an altitude where the temperature of the rising air equals the dew point. Above the condensation level, the air in the cloud is saturated.

Knowing the condensation level is important for forecasting the weather. For example, by knowing the height at which clouds will form and how high they will grow, meteorologists can predict the severity of a resulting storm.

18.2 Section Review

1. Describe the difference between cumuliform clouds and stratiform clouds. Name some examples of each.
2. Why is the moist-adiabatic lapse rate lower than the dry-adiabatic lapse rate?
3. Compare and contrast the formation processes of cumulonimbus and stratus clouds.
4. Describe cirrus clouds and explain why they can form only at higher altitudes.
5. **CRITICAL THINKING** Compare the growth of a cumulonimbus cloud to the eruption of a volcano. What causes the vertical motion in each case? How are the causes similar? How are they different?
6. **MATHEMATICS** If the temperature at the surface is 20°C and the dew point at the surface is 8°C, what is the condensation level? Draw a diagram like the one above to justify your answer.

CHAPTER 18 SECTION 3

FOCUS

Objectives
1. Compare and contrast how precipitation forms in warm clouds and in cold clouds.
2. Describe how rising air produces condensation.

Set a Purpose
Have students read the section to find the answer to the question "What is precipitation and how does it form?"

INSTRUCT

More about...

Rain and Cloud Droplets
Students may think that raindrops falling through the air look like raindrops running down a window pane—a sphere with a tail. Raindrops in air look quite different. The shape of a raindrop depends on its size; small drops are spherical, but large drops are not. Rain droplets are gigantic when compared to the droplets that make up clouds. A typical rain droplet is about 2 millimeters in diameter. A cloud droplet might have a diameter of about 20 micrometers. If that width is hard to picture, consider the fact that it would take about 100 cloud droplets to make up one droplet of rain.

18.3

KEY IDEAS
In warm clouds, water droplets grow by colliding with each other; in cold clouds, ice crystals grow through collision and by gaining water vapor from evaporating supercooled water.

Precipitation falls in regions where moist air tends to rise, producing condensation.

KEY VOCABULARY
- precipitation
- sleet
- freezing rain
- hail

GROWTH OF AN ICE CRYSTAL
Water vapor evaporates from water droplets and is deposited on ice crystals.

402 Unit 5 Atmosphere and Weather

Precipitation

When the water droplets or ice crystals in clouds grow heavy enough to fall, precipitation occurs. **Precipitation** is any form of water that falls from a cloud to Earth's surface. Rain, snow, sleet, and hail are all examples of precipitation.

How Precipitation Forms

How do water droplets and ice crystals grow heavy enough to fall out of a cloud? Water droplets and ice crystals grow in different ways.

Growth of Water Droplets

In a cloud, tiny droplets that form by condensation grow by bumping into and combining with other droplets. Bigger droplets fall faster than smaller ones, so big droplets catch up with smaller droplets, collide with them, and "capture," or incorporate, them. Other droplets are not captured; instead, they just bounce off bigger ones or are pushed aside by them.

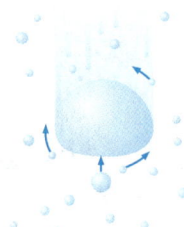
GROWTH OF A RAINDROP

Why do droplets differ in size? Droplets that have been in the cloud longer have had more time to grow. In addition, some cloud droplets start out larger because they formed around larger condensation nuclei than others did. Smaller droplets include those that were mixed into air that was not saturated, whereupon they shrank through evaporation. The mixing of air from different parts of the cloud and the falling of larger drops from higher up bring droplets of different sizes together.

Growth of Ice Crystals

Except in the shallowest clouds in the warm tropics, temperatures in the upper layers of clouds are usually below freezing. Both ice crystals and supercooled droplets are present in these clouds. Some of the supercooled water evaporates, and the resulting water vapor becomes deposited on the ice crystals. When the ice crystals get heavy enough, they start to fall. The falling crystals then grow by capturing both smaller ice crystals and water droplets in their path. If the temperature in the lower layers of the cloud is above freezing, the crystal may melt to form a water droplet, which grows through collision.

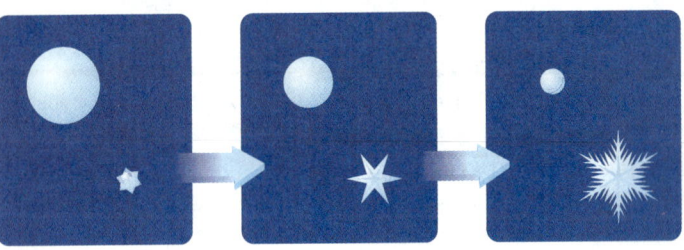

DIFFERENTIATING INSTRUCTION

Reading Support The strategy of asking questions helps students clarify and connect to the text. Sometimes questions are provided in the text for this purpose, such as the first sentence in the fourth paragraph on page 402, "Why do droplets differ in size?" Model for students how to form questions that deepen understanding. For example, while reading the sections "Growth of Water Droplets" and "Growth of Ice Crystals," students could ask themselves: How do water drops grow? How does the growth of ice crystals differ from the growth of water droplets? Review whether students have been able to answer their questions after reading the text.
Use Reading Study Guide, p. 63.

Kinds of Precipitation

Precipitation comes in many forms, including drizzle, rain, snow, sleet, freezing rain, and hail.

Drizzle consists of very fine water drops that fall slowly and close together. Raindrops are larger and fall faster and farther apart. Snow forms when ice crystals in a cloud collide and clump together. When snowflakes fall into warm air, they partially melt into sticky, wet clusters. If the snowflakes melt completely, they fall as rain.

Sometimes rain falls into a layer of cold air. There the raindrops become supercooled. If the raindrops freeze, they fall as **sleet**. However, if the layer of cold air is thin, the supercooled raindrops might not freeze until they hit solid surfaces. Then they freeze instantly. This kind of rain, called **freezing rain**, causes sheet ice, or glaze, on sidewalks, trees, roofs, and power lines. If the ice becomes heavy enough, trees and power lines may break under its weight.

In the summer, frozen precipitation usually melts before reaching the ground. An exception is **hail**, precipitation in the form of balls or irregular clumps of ice. A hailstone begins as a frozen raindrop or small, dense clump of ice crystals. It grows by collecting smaller ice particles, cloud droplets, and supercooled raindrops that freeze onto it. The growing hailstone is kept aloft by strong updrafts until it becomes too heavy and falls. The stronger the updrafts, the larger the hailstones can become.

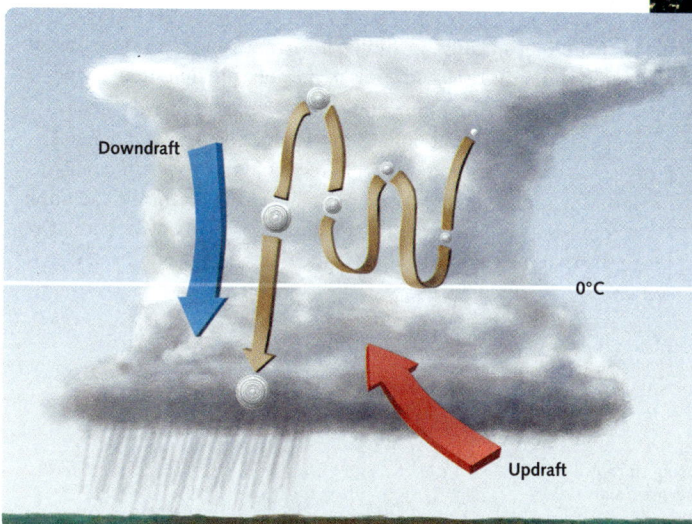

FORMATION OF HAIL

HAILSTONES can become quite large, as this photograph illustrates.

A hailstone has a layered structure like that of an onion. The layers form as the hailstone encounters different forms of moisture and different temperatures—which cause partial melting and refreezing—on its journey through the cloud. A hailstone may be as large as 14 centimeters across.

CHAPTER 18 SECTION 3

VISUALIZATIONS
CLASSZONE.COM

Examine an animation of hail forming.
Keycode: ES1805

After students observe the formation of hailstones, have them write a paragraph describing what they saw.

Visualizations CD-ROM

VISUALIZATIONS
CLASSZONE.COM

Examine an animation of hail forming.
Keycode: ES1805

More about...

Hail
Hailstones generally range from pea size to the size of a golf ball. Occasionally, larger hailstones form. The largest hailstone on record fell on Kansas in 1970. It weighed 758 grams and was 14.2 centimeters across. Hailstones cause almost $1 billion in damage in the United States each year. Because of the great damage that hail can cause, efforts are underway to find ways to suppress hail. One such attempt involved seeding thunderstorm clouds (cumulonimbus) with silver iodide particles that would act as condensation nuclei. These particles were meant to form many small hailstones that would melt before they hit the ground, preventing the formation of huge, damaging hailstones. Neither this method of hail suppression nor any other has yet been successful.

DIFFERENTIATING INSTRUCTION

Developing English Proficiency Help students differentiate the different types of precipitation by discussing the harmful effects that each type can cause for people and property.

- **drizzle:** no serious damage
- **rain:** flooding, especially when rain is heavy
- **snow:** blocked roads, avalanches, collapsed roofs
- **sleet:** motor vehicle accidents, people falling on icy roads and sidewalks
- **freezing rain:** motor vehicle accidents, down power lines and branches, people falling and injuring themselves
- **hail:** damage to cars, buildings, crops and livestock; injury to people

Measuring Precipitation

The National Weather Service reports rainfall in hundredths of an inch. Rainfall is measured by an instrument called a *rain gauge*. The recorded, and reported, measurement represents what the depth of water would be if the rain did not soak into the ground, flow away, or evaporate.

Snowfall is measured with a measuring stick. But because a dry snow is deeper than an equal weight of wet snow, the depth of snow is not an accurate measure of how much water it contains. The rain equivalent of the snowfall is determined by melting the snow.

Where Does Precipitation Occur?

Precipitation occurs in every part of the world. In some places there may be no precipitation for years at a time, while in other places it may rain almost every day. One of the main causes of precipitation is the rising and cooling of moist air. The warmer the air before it rises, the more moisture it can contain. Furthermore, the higher the air rises, the more precipitation it can release. Therefore, areas that receive the most precipitation are those where warm, moist air rises high in large quantities. Following are descriptions of the kinds of areas where these weather conditions exist.

- Heat from the sun produces high land temperatures near the equator, and these in turn cause the air near the surface to become very warm and to rise. The result is almost daily thunderstorms. Because of the heavy rain, the land around the equator is home to dense tropical forests.
- In storm areas of all kinds, including hurricanes and many low-pressure areas and fronts, air rises and cools to produce precipitation.
- In areas where moist air often blows across a mountain range, the windward side of the mountains—that is, the side the wind reaches first—may receive large amounts of precipitation. When wind reaches the mountain range, it must rise up the side. Because the air cools as it rises, some of its moisture condenses and may fall as rain or snow. One range with a rainy windward slope is the Cascade Mountains in the northwestern United States.

Which Way Does the Wind Blow?
Determine the direction of prevailing winds by examining precipitation and vegetation patterns on mountains.
Keycode: ES1806

MOUNTAIN RANGES Air is forced to rise on the windward side of mountain ranges. As the air rises, it cools. Clouds form, and precipitation falls.

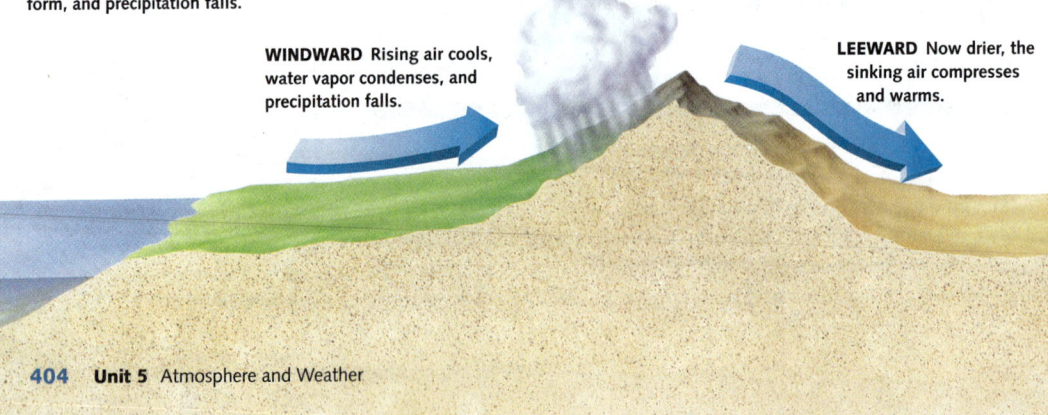

WINDWARD Rising air cools, water vapor condenses, and precipitation falls.

LEEWARD Now drier, the sinking air compresses and warms.

Global Precipitation

Cold air sinks near the North Pole. This area is dry.

Cascade Range

The windward side of the Andes Mountains is wet.

The leeward side of the Andes Mountains is dry.

Air near the equator rises. Equatorial areas are usually wet.

Cold air sinks near the South Pole. This area is dry.

Mean Annual Precipitation (centimeters): Less than 25 | 25 to 50 | 50 to 100 | 100 to 250 | More than 250

Where is there little or no precipitation? Since precipitation falls when moist air rises and cools, there is no precipitation in areas where air sinks and warms. Where sinking air persists, the dry conditions produce deserts.

As you will learn in Chapters 19 and 20, air generally sinks in high-pressure areas. Areas of persistent high pressure occur at about 30 degrees latitude north and south. Some of the great deserts of the world are located at these latitudes, including the Sahara in Africa. Polar regions also experience high pressures. The poles are dry not only because air sinks there but also because the air is so cold that it contains little water vapor.

Air also sinks on the leeward, or downwind, side of mountains. If precipitation removes moisture from the air on the windward side of the mountains, the leeward slopes can be very dry. They are also warmer than the windward slopes, because while the air traveling up the windward side cools at the moist-adiabatic lapse rate (about 6°C per kilometer), the air traveling down the leeward side warms at the dry-adiabatic lapse rate (10°C per kilometer).

Warm, dry winds formed on the leeward side of mountains have different names in different parts of the world. The winds that blow down the eastern slopes of the Rocky Mountains are called Chinook (shih-NUK) winds. The chinooks can raise temperatures by more than 20°C in an hour. Santa Ana winds blow toward the Pacific Ocean from mountains of southern California. If the Santa Ana winds blow too long, forests, fields, and wooden buildings become so dry that a careful watch must be kept against fire.

CHAPTER 18 SECTION 3

 INVESTIGATIONS
CLASSZONE.COM

Which Way Does the Wind Blow?
Keycode: ES1806

Have students work in pairs to complete the investigation.

Use Internet Investigations Guide, pp. T67, 67.

Visual Teaching

Interpret Diagram

The map shows the world-wide pattern of precipitation. Ask students to describe the precipitation pattern around the Cascade Range, which runs north-south near the western coast of North America. **Rain is more plentiful on the western side of the mountains than on the eastern side.** Then tell students to use this pattern to determine from which direction the wind generally blows. **It blows from west to east; generally more rain falls on the windward side of mountains.**

Extend

Have students use the map to answer these questions: Which continent is generally driest? **Antarctica** Why? **The combination of cold and sinking air over the continent prevents the formation of large amounts of precipitation.**

DIFFERENTIATING INSTRUCTION

Challenge Activity To reinforce the concept that rainfall amounts differ on the windward and leeward sides of mountains, have students look for photographs in magazines or on the Internet that show contrasting landscapes on the western and eastern flanks of the Cascades. Have students compare and contrast the flora and fauna native to the two areas. Students can put together a computer presentation for the class.

CHAPTER 18 SECTION 3

ASSESS

1. Both grow by capturing other droplets or ice crystals. Cloud droplets grow by combining with other drops. Ice crystals grow when supercooled water evaporates and the water vapor gets deposited on the ice crystals.

2. Students can list any five of the following: drizzle, rain, snow, sleet, freezing rain, hail.

3. rain—with a rain gauge; snow—with a measuring stick.

4. Air rising over the mountains cools; its water vapor condenses, causing precipitation to fall on the western flank. The air that flows down the leeward side is now dry. It is compressed and its temperature rises. Air is generally cooler on the windward side because rising air cools at the moist-adiabatic rate, but descending air warms at the dry-adiabatic rate.

5. Pellets of frozen carbon dioxide (dry ice) dropped into a cloud cool the cloud until ice crystals form and fall as precipitation. The cloud also can be seeded with condensation nuclei of silver iodide crystals.

6. Relative humidity increases as air travels up the windward side because its capacity to hold moisture decreases as it ascends and cools. It drops moisture as precipitation; relative humidity decreases. Then as the air travels down the leeward side, it warms, and its relative humidity continues to decrease.

7. Where air sinks, it becomes compressed and warms. Precipitation requires the opposite conditions.

Weather Modification

Often it does not rain when or where rain is needed most, so scientists have developed various methods of producing rain. Scientists are also trying to prevent hail and to eliminate fog at airports. Implementing methods of changing the weather is called weather modification.

One method of making rain is to "seed" a supercooled cloud by dropping pellets of frozen carbon dioxide, or "dry ice," into it. These pellets cool the cloud so much that tiny ice crystals form. The crystals grow by normal processes until they are heavy enough to fall. In another method of seeding, artificial ice nuclei are put into a cloud. Often these are tiny silver-iodide crystals, which are very much like ice crystals in shape. Once ice crystals form on the artificial nuclei, precipitation grows by normal processes.

Both of these methods of making rain require that clouds already be present. When rain follows seeding by one of these methods, it is hard to prove that seeding caused the rain, since the rain could also have fallen naturally.

RAINMAKING This airplane is dropping artificial ice nuclei into a cloud in an effort to produce rain.

18.3 Section Review

1. Compare and contrast the growth of cloud droplets and the growth of ice crystals.
2. Name and describe at least five kinds of precipitation.
3. How are rain and snow measured?
4. Describe what happens to moist air as it travels over the Cascade Mountains. Include an explanation of how and why the temperature changes.
5. Describe two methods used to make rain.
6. **CRITICAL THINKING** Describe what happens to the relative humidity of moist air as the air travels over the Cascade Mountains. Explain your reasoning.
7. **PHYSICAL SCIENCE** Explain why areas of sinking air are dry.

MONITOR AND RETEACH

If students miss . . .

Question 1 Review "Growth of Water Droplets" and the "Growth of Ice Crystals." (p. 402) Then have students explain what is happening in the diagrams on the page.

Question 2 Review "Kinds of Precipitation." (p. 403) Have students make a chart that lists the six types of precipitation, defines each, and explains how each forms.

Question 3 Reteach how precipitation is measured. (p. 404)

Question 4 Refer students to the diagram that explains how mountain ranges affect precipitation (p. 404). Ask students to explain the diagram in their own words.

Question 5 Reteach "Weather Modification." (p. 406) Then have students describe each process.

SCIENCE & Society

Who Will Stop the (Acid) Rain?

All rainwater is slightly acidic. Acid rain, however, is so acidic that while it might not cause a burning sensation, it is highly toxic. While there are natural causes of acid rain—including volcanic eruptions and lightning—the primary cause is the exhaust from power stations and factories.

What can we do now to reduce the environmental damage of acid rain?

Acid rain forms when the pollutants sulfur dioxide (SO_2) and various nitrogen oxides combine with water vapor in the atmosphere. It is not a new problem. During the 1800s, coal-powered factories belched tons of sulfur-containing smoke. In industrial cities a suffocating blanket of smog choked residents. As a solution, factory smokestacks were made taller. Smoke was then discharged higher up, leaving the air at ground level more breathable.

Unfortunately, taller smokestacks only made the problem worse. These smokestacks gave pollutants a head start toward the stratosphere, allowing for more pollutants to be transformed into acids. By the 1980s, acid precipitation and its causes had become a serious concern. In parts of the United States and Europe, precipitation was recorded with pH's between 2 and 3, making it about 1000 times more acidic than normal rain. Scientists discovered that coal- and oil-burning electricity plants were the major causes. Such plants are responsible for about 70 percent of the sulfur dioxide pumped into the atmosphere and for about 30 percent of the nitrogen oxides contributed. Exhaust from cars, trucks, trains, and planes is also a notable source of nitrogen oxides.

In the United States, the Clean Air Act has legislated strategies to cut back emissions of these pollutants. Cars sold in the United States are required to have catalytic converters, which break down nitrogen oxides. Many electricity plants have installed scrubbers to remove sulfur from smoke before the smoke exits the smokestack. These strategies and others have succeeded in reducing some of the pollutants that cause acid rain. Whether they will put an end to acid rain—and whether ecosystems already damaged will fully recover—remains uncertain. ■

GREAT SMOKY MOUNTAINS These trees show damage by acid rain.

BEFORE AND AFTER Acid-rain damage over the years is evident on this New York City statue of George Washington, shown in 1935 (left) and 1994 (right).

Extension

SCIENCE NOTEBOOK

Individuals can do their part to reduce acid rain by cutting down on fossil-fuel and electricity use. List five specific things you could do to help reduce the pollutants responsible for acid rain.

INVESTIGATIONS CLASSZONE.COM

How Acidic Is Your Rain?
Examine local precipitation data and determine local trends in acid rain.
Keycode: ES1807

CHAPTER 18 SECTION 3

INVESTIGATIONS CLASSZONE.COM

How Acidic Is Your Acid Rain?
Keycode: ES1807

📖 Use Internet Investigations Guide, pp. T68, 68.

Science and Society

Acid Rain's Effects on Trees
Acid rain damages trees by weakening them, thereby making them more prone to disease, insect infestation, drought, and cold. Acid rain can damage leaves directly. It also dissolves and flushes nutrients from the soil, while triggering the release of toxic substances into the soil. The buffering capacity of soil largely determines the amount of damage acid rain does to forests. Buffering is the ability of soil to resist acid-rain damage. The presence of substances, such as calcium and magnesium, can neutralize (buffer) the acid. The buffering effect depends on the thickness of the soil and the chemical composition of the bedrock.

Science Notebook
Take more trips on foot or on bikes instead of riding in cars; use buses or trains instead of individual cars; adjust thermostats to use less energy for heating and cooling; turn off lights when not in use; purchase energy-efficient appliances.

MONITOR AND RETEACH

Question 6 Review relative humidity (p. 392) and how mountains influence the movement of air and formation of precipitation. (pp. 404–405) Focus on how the air's temperature and capacity to hold moisture changes as it moves up one side of the mountain and down the other.

Question 7 Review weather and climate conditions caused by sinking air. (pp. 404–405) Stress that anything that prevents air from rising or makes it sink prevents precipitation from forming.

CHAPTER 18 ACTIVITY

PURPOSE
To examine how various concentrations of an acidic solution affect different materials

MATERIALS
If bare steel nails are not available, you can use metal washers or pieces of rebar. You also can use galvanized nails and rub off the coating with steel wool.

Have students look for and bring in small pieces of loose concrete around sidewalks, curbs, or driveways that are cracked or chipped.

PROCEDURE
① You may want to consult Material Safety Data Sheets (MSDS) for the information listed for sulfuric acid.

⑥ The concrete pieces need not all be the same mass because after students measure the mass of the pieces in step 11, they can find the percentage of change for each piece.

ANALYSIS AND CONCLUSIONS
1. Plants in Beaker 1 grow the most. Those in Beaker 4 grow the least. Some plants in Beakers 3 and 4 may die. Roots may be brown, but leaves should not be affected. Students should conclude that the damage to plants increases with an increase in acid concentration.

CHAPTER 18 LAB Activity

Effects of Acid Rain

SKILLS AND OBJECTIVES
- **Observe** and **measure** the effects of acid rain on plants, concrete, and steel.
- **Relate** this lab activity to the effects of acid rain in nature.

MATERIALS
- lab apron
- latex gloves
- safety goggles
- 225 mL sulfuric acid
- water
- pH probe or pH indicator strips
- 12 250-mL beakers
- 50-mL graduated cylinder
- 3 cups potting soil
- alfalfa sprouts
- metric ruler
- medicine dropper
- balance
- 4 pieces of concrete
- 8 steel nails
- steel wool
- masking tape
- small scrubbing brush
- tongs

The burning of fossil fuels releases many gaseous compounds, such as sulfur dioxide, into the air. These compounds, combine with moisture in the air, dissolving in water droplets that form. When this moisture falls as rain, its acidity affects both natural and manmade structures on the ground. In this experiment, you will compare the effects of water with differing concentrations of sulfuric acid on plants, concrete, and steel.

Procedure

①   **CAUTION: Wear your safety goggles, lab apron, and latex gloves throughout this experiment.**
Organize 12 beakers or cups into three groups of four. Use the masking tape to label the beakers in the three groups: Plants, Concrete, and Steel. Number each beaker within a group from 1–4. Write your name on each beaker.

② Gently pull out a plug of alfalfa sprouts about 1–2 cm in diameter. Measure in centimeters the length of the longest sprout from root tip to leaf tip. Copy the data table from the next page and record the length under the column "Beaker 1, Before".

③ Fill Plant beaker 1 halfway with potting soil. Hold the sprouts upright in the center of the beaker. Gently sift some potting soil around the sprouts to plant them. Cover the bottom halves of the stems and press down the soil gently.

④ Repeat Steps 2 and 3 for the remaining beakers in the Plant group.

⑤ Water the sprouts with the following liquids:
Beaker 1: 50 mL of tap water
Beaker 2: 10 drops of sulfuric acid plus water to make 50 mL of liquid
Beaker 3: 5 mL (100 drops) of sulfuric acid plus water to make 50 mL
Beaker 4: 50 mL of sulfuric acid

⑥ Fill the Concrete beakers with 100 mL of tap water. Dampen four pieces of concrete using a wet paper towel. Assign each piece to a beaker. Measure the mass of each and record it in the data table in the "Before" column under the appropriate beaker. Place each piece in its assigned beaker.

⑦ Fill the beakers labeled Steel with 100 mL of tap water. Use steel wool to rub off any plastic coating or corrosion on the eight steel nails. Measure the mass of the nails in sets of two. Place one set in each beaker and record the masses in the "Before" column as appropriate.

CHAPTER 18 ACTIVITY

Effects of Acid

	Beaker 1: Tap Water		Beaker 2: Tap Water + 10 Drops Acid		Beaker 3: Tap Water + 100 Drops Acid		Beaker 4: Acid	
	Before	After	Before	After	Before	After	Before	After
Plants, length (cm)								
Concrete, mass (g)								
Steel, mass (g)								
Water, pH								

8 Add 10 drops of sulfuric acid to Concrete and Steel beakers 2, 5 mL (100 drops) of acid to beakers 3, and 50 mL acid to beakers 4. Using pH indicator strips or an electronic pH probe, record the pH of the water in the four Concrete beakers in the appropriate "Before" column.

9 Allow the 12 beakers to sit undisturbed for four days.

10 After four days, use the tongs to gently lift the plants from the soil in each beaker. Rinse the sprouts and measure their lengths. Record the length of the longest sprout in the "After" column for each beaker.

11 Use the tongs to lift the concrete pieces and steel nails from their beakers. Rinse in water and carefully brush off any loose flakes or corrosion. Measure the mass of each piece of concrete and each set of nails and record these masses in the appropriate "After" column.

12 Measure the pH of the water in the Concrete beakers and record the pH level in the appropriate "After" column.

Analysis and Conclusions

1. Summarize your observations of each plant sample after four days. Describe the appearance of the roots and leaves. What conclusions can you draw?

2. Summarize the results obtained when you exposed the concrete and steel nails to different concentrations of acid. Describe the appearance of the two materials as the amount of acid increased.

3. Explain why your answers to Questions 1 and 2 are relevant to both urban areas where pollution is high, and to agricultural areas.

4. How does this lab activity effectively simulate the effects of acid rain? How is it different from acid rain? Which of the four acid concentrations would you expect to best simulate acid rain?

Learn more about acid rain and its effects on architectural structures.
Keycode: ES1809

DATA CENTER
CLASSZONE.COM

Learn more about acid rain and its effects on architectural structures.
Keycode: ES1809

(continued)

2. Corrosion of steel nails increased as the concentration of acid increased. The mass of the concrete samples decreased by a greater percentage as the acid concentration increased.

3. The air in many urban areas has high concentrations of air pollutants that can form acid rain. Acid rain can damage or weaken structures made of concrete and steel, such as buildings and bridges. In agricultural areas, acid rain can damage or kill plants.

4. Similar to acid rain: The mixture of water and sulfuric acid simulates acid rain. Adding the acid mixture to soil, steel, and concrete simulates the effects of acid rain as it falls on crops and forests, as well as on steel and concrete structures. Different from acid rain: Acid rain is composed of a number of chemicals in addition to sulfuric acid and water; acid rain acts on a wider variety of materials than represented in the experiment; some building materials have coatings that protect them from acid rain; some materials in the environment (such as limestone) are basic, which tends to neutralize the effects of acid rain. Acid rain has a pH of 5.7 or less, so solutions with a pH of 5.7 or less best simulate acid rain.

Refer students to Safety in the Earth Science Laboratory on pages xxii–xxiii.

CHAPTER 18 REVIEW

Vocabulary Review

1. water vapor
2. condensation
3. precipitation
4. condensation nuclei
5. cirrus

Concept Review

6. 18°C air because the higher the air temperature, the greater is its capacity to hold moisture
7. Specific humidity is the amount of water vapor actually present in the air. Relative humidity is a ratio that compares the amount of water vapor in the air with the maximum amount it can hold at a specific temperature.
8. A psychrometer consists of a wet-bulb thermometer and a dry-bulb thermometer. Depending on the degree of saturation in the air, air moving over the wet-bulb thermometer will cause evaporation, lowering the wet-bulb temperature. The less humid the air, the greater is the evaporation and temperature difference between the two thermometers. The resulting temperatures are used to determine relative humidity from a humidity table.
9. It means the air is holding the maximum amount of water vapor it can hold at a certain temperature.
10. Dew forms when water vapor condenses on an object. If the air temperature is below 0°C, the water vapor condenses as frost.
11. As moist air rises, it cools and has less of a capacity to hold moisture. The excess moisture condenses.
12. As air becomes warmer and less dense, it rises through denser, colder air around it. As the warm air rises, it cools and condenses. A cloud forms.
13. If raindrops fall through a layer of cold air, they become supercooled and freeze, forming sleet. If the layer of cold air is thin, the supercooled raindrops freeze only when they hit a solid surface, forming freezing rain.
14. Hail forms when frozen raindrops or small, dense clumps of ice crystals are kept aloft by updrafts in cumulonimbus clouds. The hailstone grows bigger within the cloud, as it collects ice particles, cloud droplets, and supercooled rain droplets. When the hailstone becomes heavy enough, it falls to the ground.
15. rises; windward

CHAPTER 18 REVIEW

Summary of Key Ideas

18.1 Water exists in the atmosphere as a solid, a liquid, and a gas. Evaporating water absorbs heat from the surroundings, which become cooler. Condensing water releases heat to the surroundings. Warmer air can contain more water vapor. Fog and clouds form when air is cooled to its dew point and water in the air condenses on condensation nuclei. Dew or frost forms if contact with cold ground cools air to the dew point.

18.2 Clouds are classified by their height above ground and their shape. Cumuliform clouds are formed by rising air, and stratiform clouds form in horizontal layers. Rising air cools at the dry-adiabatic lapse rate with no condensation, and at the moist-adiabatic lapse rate with condensation. Heat released through condensation within a cloud can cause air within the cloud to rise to great heights.

18.3 Water droplets in clouds grow by colliding with each other. Ice crystals grow from collisions and by using water vapor that evaporated from super-cooled drops. Precipitation's type depends on the conditions as it forms and falls.

Key Vocabulary

cirrus (p. 397)
condensation (p. 390)
condensation level (p. 398)
condensation nuclei (p. 394)
cumulus (p. 397)
dew point (p. 393)
dry adiabatic lapse rate (p. 398)
freezing rain (p. 403)
hail (p. 403)
moist-adiabatic lapse rate (p. 398)
precipitation (p. 402)
relative humidity (p. 392)
saturated (p. 391)
sleet (p. 403)
specific humidity (p. 391)
stratus (p. 397)
water vapor (p. 390)

Vocabulary Review

Write the term from the key vocabulary list that best completes the analogy.

1. ice : solid as ___?___ : gas.
2. freezing : melting as ___?___ : evaporation.
3. stratus : cloud as rain : ___?___ .

Write the term from the key vocabulary list that best completes the sentence.

4. The tiny particles on which water condenses in the air are called ___?___ .
5. High clouds made of ice crystals are known as ___?___ clouds.

Concept Review

6. Which contains more water, saturated 8°C air or saturated 18°C air? Explain your thinking.
7. Explain the difference between specific humidity and relative humidity.
8. Explain how a psychrometer works.
9. What does it mean for air to be *saturated*?
10. Compare and contrast the formation of dew and the formation of frost.
11. Explain why condensation may occur when moist air rises.
12. Explain how differences in density can cause cloud formation.
13. Compare and contrast sleet and freezing rain.
14. Describe how hail forms.
15. **Graphic Organizer** Complete the concept map.

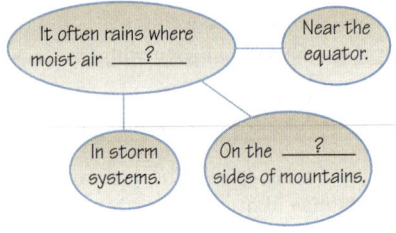

Critical Thinking

16. The relative humidity of the air is high, and it is holding almost all the water vapor it can hold.
17. Sublimation would make it colder because the resulting water vapor has absorbed heat.

Critical Thinking

16. Apply Why doesn't sweat evaporate on a humid day?

17. Infer Which process could make a snowbank colder, deposition or sublimation? Explain your thinking.

18. Communicate Write an explanation of why condensation occurs when air cools to its dew point. You may want to use a diagram.

19. Draw Conclusions Exhaust from jet engines contains water vapor and particles that can act as ice nuclei. What effect would you expect jet traffic to have on the cloud cover at altitudes of 10–13 kilometers?

20. Analyze Why do cumulus clouds form more often during the afternoon than in the early morning?

21. Predict Look at the physical map of the United States on pages 708–709. Locate Maine and Mississippi. In which state would the average annual precipitation be greater? Explain your thinking.

Problem Solving

Monitor local precipitation using current radar images.
Keycode: ES1808

Writing About the Earth System

SCIENCE NOTEBOOK As you will read in Chapter 20, the heat from condensation fuels hurricanes. In order for condensation to continue throughout the lifetime of the hurricane, the hurricane must be continually supplied with water vapor. Explain why hurricanes form over warm ocean water. Why do they weaken if they move over land? Over cold ocean water?

Interpreting Graphs

Copy and complete the table by calculating the height at which clouds will form (the lifting condensation level) for the temperatures and dew points shown. Then answer the following questions.

Condensation Level

Temperature	Dew Point	Condensation Level
16°C	12°C	?
20°C	12°C	1000 m
20°C	16°C	?
28°C	12°C	?
28°C	20°C	?

22. Describe a situation in which the temperature could rise while the dew point stays the same. Explain your thinking.

23. If the dew point stays the same, what effect does a rise in temperature have on the cloud base? Use examples from your table to support your answer.

24. Describe a situation in which the dew point would rise. What would happen to the temperature in this situation? Explain your thinking.

25. If the temperature stays the same, what effect does a rise in the dew point have on the cloud base? Use examples from your table.

Chapter 18 Water in the Atmosphere 411

WRITING ABOUT SCIENCE

Writing About the Earth System

Hurricanes form over warm ocean waters, in part because the heat and moisture from the ocean provides a continuous supply of warm, moist air that builds into huge hurricane cloud systems. Hurricanes weaken as they move over land and cold ocean water because the hurricane's source of heat and moisture needed to maintain itself is cut off.

CHAPTER 18 REVIEW

Monitor local precipitation using current radar images.
Keycode: ES1808

(continued)

18. When air cools to its dew point, it reaches its saturation point and can hold no more moisture. Water vapor condenses out of the air to form liquid water.

19. Ice droplets could form in the wake of the jet, helping to form cirrus clouds.

20. The ground is warmer and heats the air above it, causing patches of warm air to rise. As the rising air cools, water vapor condenses to form clouds.

21. Mississippi; The warmer climate contributes to more rising air, cloud formation, and precipitation.

INTERPRETING GRAPHS

The missing condensation levels are 500 m, 500 m, 2000 m, and 1000 m.

22. if the amount of moisture in the air also increases; Ordinarily, if the temperature rises and the amount of moisture stays the same, the air has to cool more for condensation to occur. So the dew point would decrease. If the dew point remains the same while the temperature increases, more moisture would have to be added for condensation to occur at the same temperature.

23. Cloud bases form at a higher altitude. If the dew point is 12°C, clouds form at 1000 m (1 km) when the temperature is 20°C. They form at 2000 m (2 km) when the temperature is 28°C.

24. if the amount of water vapor in the air increased but the temperature stayed the same; In that case, the air would have to cool less for condensation to occur.

25. Cloud bases form at a lower altitude. If the temperature is 28°C, clouds form at 2000 m (2 km) when the dew point is 12°C. They form at 1000 m (1 km) when the dew point is 20°C.

CHAPTER 19 PLANNING GUIDE: THE ATMOSPHERE IN MOTION

	Section Objectives	Print Resources
Chapter-Level Resources		Laboratory Manual, pp. 83–86 Internet Investigations Guide, pp. T69–T70, 69–70 Guide to Earth Science in Urban Environments, pp. 29–36, 37–44 Spanish Vocabulary and Summaries, pp. 37–38 Formal Assessment, pp. 55–57 Alternative Assessment, pp. 37–38
Section 19.1 Air Pressure and Wind, pp. 414–418	1. Define air pressure. 2. Describe how changes in elevation, temperature, and humidity affect air pressure. 3. Explain what makes the wind blow.	Lesson Plans, p. 64 Reading Study Guide, p. 64
Section 19.2 Factors Affecting Winds, pp. 419–421	1. Describe the Coriolis effect. 2. Explain how the Coriolis effect, friction, and pressure gradients affect wind direction.	Lesson Plans, p. 65 Reading Study Guide, p. 65 Teaching Transparency 21
Section 19.3 Global Wind Patterns, pp. 422–426	1. Identify factors that affect global wind patterns. 2. Describe the strengths and weaknesses of the three-celled circulation model, and use the model to explain prevailing winds and pressure regions.	Lesson Plans, p. 66 Reading Study Guide, p. 66
Section 19.4 Continental and Local Winds, pp. 427–429	1. Describe the effects of seasons and continents on wind patterns. 2. Explain the circulation of sea, land, mountain, and valley breezes.	Lesson Plans, p. 67 Reading Study Guide, p. 67

INVESTIGATIONS
CLASSZONE.COM

Section 19.2 INVESTIGATION
How Does the Jet Stream Change through the Year?
Computer time: 45 minutes / Additional time: 45 minutes
Students compare and contrast the jet stream during different seasons.

Chapter Review INVESTIGATION
Could You Break the Record for an Around-the-World Balloon Flight? Computer time: 45 minutes / Additional time: 45 minutes
Students apply knowledge of the atmosphere in motion to plan a balloon trip around the world.

Technology/Online Resources

Electronic Teacher Tools
Visualizations CD-ROM
Classzone.com
Online Lesson Planner
Test Generator

Visualizations CD-ROM
Classzone.com
  ES1901 Visualization, ES1902 Visualization, ES1903 Visualization

Visualizations CD-ROM
Classzone.com
  ES1904 Visualization, ES1905 Visualization, ES1906 Investigation

Classzone.com
  ES1907 Data Center

Classroom Management

Gather these materials for minilab and laboratory activities.

Minilab, p. 423
  4 paper drinking cups
  2 stiff cardboard strips
  pencil with eraser
  push pin
  stapler

Lab Activity, pp. 430–431
  thermometer
  barometer
  wind vane or flag
  anemometer (see page 423)
  wet-bulb thermometer (see page 393)
  rain gauge
  relative humidity chart (see page 704)
  cloud-type photographs (see pages 396–397)
  Laboratory Manual, Teacher's Edition
    Lab Sheet 19, p. 143

CHAPTER 19

The Atmosphere in Motion

Air is constantly moving across the face of Earth, swirling and eddying, blowing in strong currents and occasionally coming to a near standstill. Hang gliders can use air motion to stay aloft for hours.

What causes winds?

INTRODUCE

Build Interest

Before sharing background information about the photograph with students, allow time for their questions and comments.

Ask questions like the following:
- What special gear is the hang glider wearing? *a protective helmet, a harness*
- What force is keeping the hang glider aloft? *wind*
- What other kind of activities depend on this force? *kite flying, parasailing, parachuting*

Respond
- What causes winds? *Air moves in response to differences in air pressure. The resulting winds blow from areas of high pressure to areas of low pressure.*
- What effects let you know the wind is blowing? *leaves scattering, trees bending, and dust twirling in the air, whistling sounds of wind, the feel of wind on your body*

About the Photograph

The person in the photograph is hang gliding over Yosemite Valley in Yosemite National Park. The valley is located in California's Sierra Nevada mountains, at an elevation of about 1200 meters.

BEYOND THE TEXTBOOK

Print Resources
- Reading Study Guide, pp. 64–67
- Transparency 21
- Formal Assessment, pp. 55–57
- Laboratory Manual, pp. 83–86
- Alternative Assessment, pp. 37–38
- Spanish Vocabulary and Summaries, pp. 37–38
- Internet Investigations Guide, pp. T69–T70, 69–70
- Guide to Earth Science in Urban Environments, pp. 29–36, 37–44

Technology Resources
- Visualizations CD-ROM
- Classzone.com
- Online Lesson Planner
- Electronic Teacher Tools
- Test Generator

412 Unit 5 Atmosphere and Weather

CHAPTER 19

PREVIEW

▶ **FOCUS QUESTIONS** In this chapter you will study air pressure and winds and learn more about the key questions below.

Section 1 What is air pressure, and what causes winds?

Section 2 How do friction and Earth's rotation affect wind speed and direction?

Section 3 How does Earth's rotation affect global wind and pressure patterns?

Section 4 How do continents, seasons, and topography affect wind?

▶ **REVIEW TOPICS** As you investigate air pressure and winds, you will need to use information from earlier chapters.

- insolation (p. 375)
- heating of land and water (p. 376)
- seasonal temperature shifts (p. 377)
- precipitation (pp. 404–405)

▶ **READING STRATEGY**

PREVIEW

Before you read, look through Chapter 19, noting the key ideas, key vocabulary, heads, images, and captions. In your science notebook, make a list of things you expect to learn about in the chapter.

CLASSZONE.COM

At our Web site, you will find the following Internet support for this chapter.

DATA CENTER
EARTH NEWS
VISUALIZATIONS
- Air Pressure Affecting a Rising Balloon
- Barometric Pressure Changes
- Land and Sea Breezes

LOCAL RESOURCES
INVESTIGATIONS
- How Does the Jet Stream Change through the Year?
- Could You Break the Record for an Around-the-World Balloon Flight?

Chapter 19 The Atmosphere in Motion **413**

CHAPTER 19

PREPARE

Focus Questions

Find out what knowledge students have about wind. For each focus question, have students consider the section title and then develop two or three additional questions for each section. For example, Section 1: How fast does wind blow? How can you measure wind speed and direction?

Review Topics

In addition to the earth science topics listed, students may need to review these concepts:

Physical Science
- Pressure is a force exerted on an area. Air pressure is the force that air exerts per unit area.
- A gradient is the rate of change of a variable. For atmospheric studies, the pressure gradient is the rate of change in pressure between two points at the same elevation.

Reading Strategy

Model the Strategy: To help students preview each section, use a version of the following: "I see that key ideas are listed at the beginning of each section. Section 1 has two key ideas. The first one is, 'Air pressure is the force of air molecules exerted on a given area.' I can find statements in the text that support this key idea. One of them is: 'Air pressure is exerted in all directions because air molecules move in all directions.' For each key idea, I can find supporting statements in the text." Have students continue to use this strategy for the remainder of the chapter.

BEYOND THE TEXTBOOK

CLASSZONE.COM

INVESTIGATIONS
- How Does the Jet Stream Change through the Year?
- Could You Break the Record for an Around-the-World Balloon Flight?

VISUALIZATIONS
- Animation of how air pressure affects a rising balloon
- Animation and graph of how barometric pressure changes with weather conditions

- Animation of land and sea breezes
- Animation of the Coriolis effect
- Animation of how the Coriolis effect influences wind direction

DATA CENTER
EARTH NEWS
LOCAL RESOURCES
CAREERS

Online Lesson Planner

Chapter 19 The Atmosphere in Motion **413**

CHAPTER 19 SECTION 1

FOCUS

Objectives
1. Define *air pressure*.
2. Describe how changes in elevation, temperature, and humidity affect air pressure.
3. Explain what makes the wind blow.

Set a Purpose
Have students read this section to find the answer to the question "What is the relationship between air pressure and wind?"

INSTRUCT

Discussion
Ask: Why does a bicycle tire inflate when air is pumped into it? **Air molecules push against the inner surface of the tire, causing the tire to expand.** Tell students that the force exerted by the air per unit area is called air pressure. Offer this example to clarify the concept: If the air pressure on the inside of the tire is 60 pounds per square inch, that means that a force of 60 pounds is exerted on each square inch of the tire. Ask: How does the air inside the tire differ from that outside? **The air inside the tire has a greater density.** Does air pressure inside the tire increase or decrease when air is added? **increase**

19.1

KEY IDEAS
Air pressure is the force of air molecules exerted on a given area.

Changes in temperature and humidity change the air pressure.

KEY VOCABULARY
- air pressure
- isobar
- high-pressure area (high)
- low-pressure area (low)
- pressure gradient

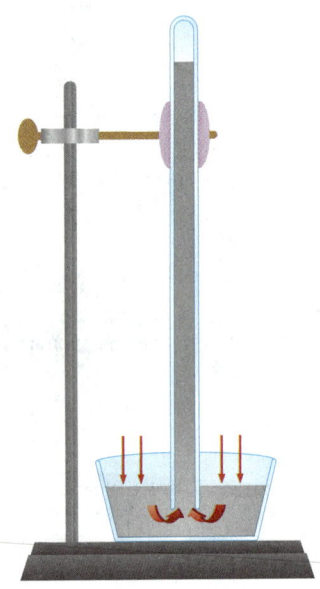

MERCURY BAROMETER

414 Unit 5 Atmosphere and Weather

Air Pressure and Wind

Swirling in complex patterns, air moves constantly across Earth's surface. Wind, which is the horizontal movement of air, helps to moderate surface temperatures, distribute moisture, and generally cleanse the atmosphere. Although several forces affect the direction in which the air moves, differences in air pressure are what set the air in motion.

What Is Air Pressure?
"Light as air" is a common enough expression, but a column of air that stretches from the top of the atmosphere to sea level, in fact, has considerable weight, equivalent to 14.7 pounds per square inch. The weight of the atmosphere as it pushes down upon Earth's surface exerts a force per unit of area called **air pressure.**

Air pressure is exerted in all directions because air molecules move in all directions. If you hold your hand out, palm up, the force of air pushing down on your hand is greater than your own weight. Yet you do not feel the air pushing down because there is also air under your hand, pushing up, and there are forces inside your body pushing out. All of these forces are in balance. So critical is the balance of these forces to the human body that astronauts in space must work in pressurized cabins or wear pressurized suits that re-create air pressure similar to that experienced on Earth.

Because air pressure at any point on Earth's surface depends on the weight of the air above, air pressure decreases as elevation increases. In general, air pressure decreases by about 50 percent for each five-kilometer increase in elevation. Mile-High Stadium in Denver, Colorado, is about 1.6 kilometers above sea level. The air exerts a pressure about 82 percent of that experienced at sea level (an 18 percent decrease). A jet aircraft flying over Denver at an altitude of 10 kilometers moves through air that exerts a pressure only 25 percent of that experienced at sea level.

Air pressure is probably most noticeable when you experience a change in air pressure, for instance when riding in an airplane or elevator. With a rapid ascent, the air pressure outside your body rapidly decreases, and your eardrums may hurt or pop because of greater internal pressure. With a rapid descent, the increase in pressure is external, creating greater pressure on the outside of your eardrums.

Measuring Air Pressure
The instrument used to measure air pressure is the barometer. There are two main types of barometers—mercury and aneroid.

The diagram to the left shows the basic setup of a mercury barometer. A glass tube filled with mercury is inverted and placed upright in a dish of mercury. The mercury flows from the tube into the dish until the weight of the column of mercury is equivalent to the air pressure on the mercury in the dish. (The space at the top of the tube is a vacuum and so exerts no air pressure.) At sea level, air pressure supports a column that is about 760 millimeters (about 30 inches) high. If the air pressure increases, the mercury column rises. If the air pressure drops, the mercury column falls.

DIFFERENTIATING INSTRUCTION

Reading Support Help students become active readers by having them identify unfamiliar words or phrases, then research their meanings. For example, students might research the meaning of *aneroid*. **not using liquid** Ask: Why is an aneroid barometer aptly named? **Unlike a mercury barometer, which contains liquid mercury, an aneroid barometer does not use liquid to measure changes in air pressure.**
Use Reading Study Guide, p. 64.

Hands-On Demonstration To show that air exerts pressure, try this demonstration. Place a large plastic sandwich bag inside a wide-mouth glass jar. Lap the top of the bag over the jar rim, and press the bag against the inside of the jar to expel air between the bag and the jar. Hold the bag tightly around the rim. Challenge a student to pull the bag out of the jar by tugging on the inside bottom of the bag. Ask: Why won't the bag come out? **A lot of air presses down on the bag but little or no air presses up under the bag.**
⚠ Exercise caution when handling the jar.

414 Unit 5 Atmosphere and Weather

An aneroid barometer uses a metal capsule to register changes in air pressure. Air is pumped from the capsule, and a spring or similar device is used to keep the capsule from collapsing. The capsule will expand if air pressure decreases and contract if air pressure increases. The movement is magnified by a series of levers attached to the capsule that cause a pointer to move over a scale indicating air pressure. A barograph is an aneroid barometer with a pen attached to the pointer. The pen records air pressure over time by tracing a line on a rotating chart.

Observe how air pressure affects a rising balloon.
Keycode: ES1901

ANEROID BAROGRAPH

Recording Air Pressure

Several units of measure are used to express air pressures. Inches and millimeters are used to record the height of the mercury column in a mercury barometer. While weather forecasters may often refer to the mercury rising or falling in their reports, they actually work with a different unit of measure: the millibar. Millibars are used on U.S. weather maps. At sea level, the average air pressure is about 1013 millibars; at Mile High Stadium in Denver, Colorado, the normal surface pressure decreases to about 835 millibars. To make use of air-pressure measurements from weather stations across United States, meteorologists adjust the measurements to eliminate the effect of elevation. The adjusted air pressure is referred to as standard sea-level air pressure.

AIR IN MOTION A satellite captures the image of a hurricane, with swirling winds that move in excess of 120 kilometers per hour.

CHAPTER 19 SECTION 1

Observe how air pressure affects a rising balloon.
Keycode: ES1901

 Visualizations CD-ROM

More about...

Air Molecules
The vast majority of the molecules that make up the atmosphere are found between Earth's surface and an altitude of 31 kilometers—a zone that encompasses the troposphere and part of the stratosphere. Air density in this region is high. Molecules are separated by an average of only 0.000008 centimeters and thus constantly collide. In contrast, molecules in the lower thermosphere (about 90 kilometers above Earth's surface) are separated by an average of nearly 0.8 kilometers and can travel relatively long distances without colliding with another molecule.

CRITICAL THINKING

In addition to adjustments for elevation, measurements from a mercury barometer must be corrected for temperature. Ask students to infer why. Mercury expands when heated and contracts when cooled. Thus, temperature affects the height of a column of mercury in a mercury barometer.

DIFFERENTIATING INSTRUCTION

Developing English Proficiency The discussion of air pressure in this section includes several related pairs of terms with opposite meanings, among them: *increase/decrease, ascent/descent, expand/collapse.* Encourage English-proficient students to help others by using hand gestures to clarify the meanings of these terms as they read the sentences aloud.
Use Spanish Vocabulary and Summaries, pp. 37–38.

Chapter 19 The Atmosphere in Motion

CHAPTER 19 SECTION 1

VISUAL TEACHING

Discussion

The caption of the diagrams indicates that the amount of water in the humid-air illustration is exaggerated. Ask: **How would the illustration change if it did not exaggerate the amount of water?** With the number of molecules shown (21), none of them would be water. With humid air containing 2 percent water vapor, the illustration would have to show 50 molecules to have just 1 water molecule. **How would you change the illustration on the left to show cool air versus warm air?** Cool air would have more molecules in the given space; warm air would have fewer molecules.

Scientific Thinking

COMPARE
Before drawing their diagrams, suggest students make a chart that compares and contrasts contour lines and isobars. For example, students may note that both contour lines and isobars connect points of equal value and indicate an extreme gradient when close together. Contour lines refer to elevation, whereas isobars refer to air pressure.

VISUALIZATIONS
CLASSZONE.COM
Examine how barometric pressure changes with weather conditions.
Keycode: ES1902

VOCABULARY STRATEGY
The prefix *iso-* in the term *isobar* is from a Greek word meaning "equal." All the points along an isobar have equal air pressure.

Why Does Air Pressure Change?

Changes in elevation are not the only causes of changes in air pressure. Both temperature and humidity affect air pressure as well.

In general, air pressure at sea level decreases as temperature increases. When air is warmed, its molecules mover farther apart. When air is cooled, its molecules move closer together. As a result, there are fewer molecules exerting pressure in a given volume of warm air than there are in an equal volume of cool air. If warm air replaces cool air over a region, the air pressure in that region will decrease. The arrival of cooler air in a region will be accompanied by an increase in air pressure.

Changes in humidity also cause changes in air pressure. The more water vapor air contains, the lighter the air is, because water molecules have less mass than oxygen or nitrogen molecules. Dry air is about 99 percent oxygen and nitrogen; air containing 2 percent water vapor—humid air—is only 97 percent oxygen and nitrogen. The water vapor has pushed out an equal volume of the more massive oxygen and nitrogen.

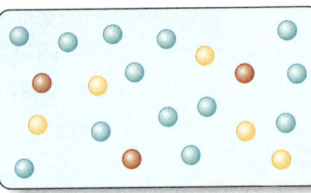

DRY AIR | **HUMID AIR**

 Nitrogen Oxygen Water

HUMID AIR is lighter than dry air because water molecules (H_2O) are less massive than oxygen (O_2) and nitrogen (N_2) molecules. (This illustration exaggerates the amount of water in humid air.)

Changes in air pressure give a simple, although not always accurate, way of forecasting the weather. A decrease in air pressure often signals the approach of warmer, more humid air, along with rain or snow. An increase in air pressure often signals the arrival of cooler, drier air and fair weather.

Meteorologists analyze air pressures by plotting isobars on weather maps. An **isobar** is a line that joins points having the same air pressure. A closed isobar is one that forms a closed loop on a weather map. If the air pressure steadily increases toward the center of a set of closed isobars, the area defined by the isobars is called a **high-pressure area** or **high**. If the air pressure steadily decreases toward the center of a set of closed isobars, the area defined by the isobars is called a **low-pressure area** or **low**. High- and low-pressure areas can be more than 1500 kilometers across. A large high-pressure area can cover most of the North American continent.

Closely spaced isobars on a weather map indicate an area where air pressure changes quickly. Dividing this pressure change by the distance over which the pressure changes yields the **pressure gradient.**

DIFFERENTIATING INSTRUCTION

Developing English Proficiency In most text discussions, certain non-scientific words and phrases may present difficulty in reading comprehension. When possible, suggest students substitute simpler language. Here are some substitutions for words and phrases on page 416: Instead of "exerting pressure," use "pushing on things." Instead of "be accompanied by" and "signals," use "means." Instead of "yields," use "gives."

Air Pressure on a Weather Map

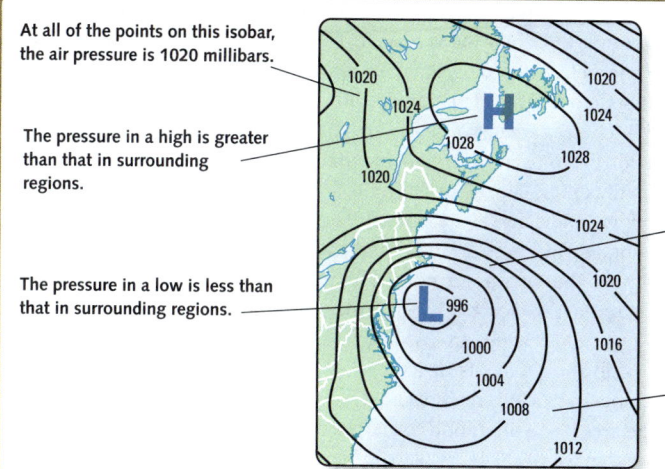

At all of the points on this isobar, the air pressure is 1020 millibars.

The pressure in a high is greater than that in surrounding regions.

The pressure in a low is less than that in surrounding regions.

Isobars that are close together indicate a strong pressure gradient.

Isobars that are far apart indicate a weak pressure gradient.

What Makes the Wind Blow?

If you puncture a tire, air will rush out. Air flows from the area of higher pressure inside the tire to the area of lower pressure outside. If no other forces act on it, air on Earth's surface will flow from high-pressure areas to low-pressure areas. The greater the difference in air pressure between two points, the stronger the wind that blows between them. Closely spaced isobars indicate a strong pressure gradient and strong winds; widely spaced isobars indicate a weaker pressure gradient and weaker winds.

Pressure differences and winds are ultimately caused by unequal heating of Earth's surface. The illustration below shows how a sea breeze forms. The island absorbs more of the sun's radiation than the surrounding ocean. As the air over the island is heated by that radiation, it becomes less dense than the ocean air surrounding it. The hot air rises and a sea breeze blows in, moving from an area of higher pressure to one of lower pressure. The speed of the breeze depends on the pressure gradient. The hotter the island, the lower the pressure over it, and the stronger the breeze.

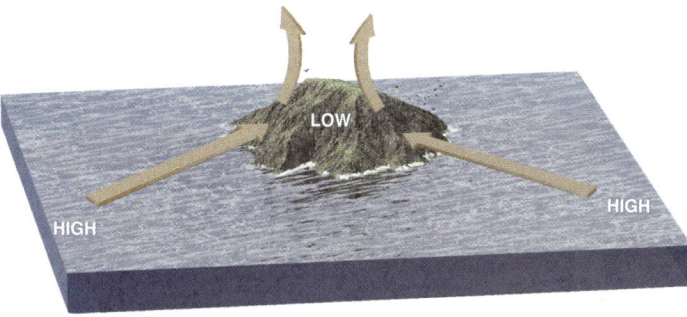

SEA BREEZES are winds that blow along the ocean surface onto land.

Scientific Thinking

COMPARE

Compare contour lines on a topographic map (see pages 54–55) with isobars on a weather map. Consider not only the appearance of each, but also how each is used. Draw and label two diagrams that demonstrate the similarities and differences between contours and isobars.

Observe an animation of land and sea breezes.
Keycode: ES1903

CHAPTER 19 SECTION 1

Examine how barometric pressure changes with weather conditions.
Keycode: ES1902

Have students infer whether current weather conditions in your area are related to high or low air pressure.

 Visualizations CD-ROM

Observe an animation of land and sea breezes.
Keycode: ES1903

Have students draw diagrams contrasting land and sea breezes.

 Visualizations CD-ROM

Visual Teaching

Interpret Diagram

Ask: What part of the world is shown on the weather map? **eastern United States and southeastern Canada** What is the name of the looping lines on the map? **isobars** What happens to the isobars that run off the map? **They eventually connect beyond the borders of the map.** Where on the map would you expect to find the strongest winds? **where isobars are close together** The weakest winds? **where isobars are far apart** Why? **Isobars that are close together indicate a large pressure gradient, whereas those that are far apart indicate a small pressure gradient. The larger the gradient is, the stronger the wind.**

DIFFERENTIATING INSTRUCTION

Hands-On Demonstration Use a balloon to demonstrate how air moves from an area of high pressure to an area of low pressure. Inflate the balloon and hold its neck to keep air from escaping. Ask: Where is air pressure highest—inside or outside the balloon? **inside** Release the neck of the balloon, then have students relate their observations to the movement of air on Earth. **Air rushed out of the balloon (an area of high pressure) and into the room (an area of low pressure). In a similar way, air on Earth moves from high-pressure areas to low-pressure areas.**

Challenge Activity Have students search the Internet for weather maps that include isobars, preferably for the area in which they live. They can print out the maps and draw arrows that show the directions of air flow. Also have students circle and label areas where winds are strongest and weakest.

CHAPTER 19 SECTION 1

More about...

Wind Speed
Wind speed can be estimated without instruments by using the Beaufort Wind Force Scale, developed in 1805 by Admiral Francis Beaufort of the British navy. Beaufort correlated the strength of wind to its effects on ships at sea. The scale was later modified to include phenomena on land. According to the scale, smoke rises vertically and the ocean's surface is smooth during a calm wind. During a light breeze, leaves rustle and small wavelets form. Roof and fence damage occurs during a strong gale, while high waves and streaks of foam form on the sea. The Beaufort Scale is a handy tool for amateur weather observers but is rarely used by meteorologists, who generally rely on more precise instruments to measure wind speed.

ASSESS

1. Air pressure is the force air exerts per unit area. Air pressure decreases as elevation increases.

2. Water vapor molecules are lighter than oxygen and nitrogen molecules.

3. The tail of a wind vane, which is larger than the arrowhead, offers greater resistance to the wind. It is pushed back, causing the arrowhead to rotate so that it points into the wind.

4. The net force of the air is north. Students' diagrams should show three arrows: one pointing northwest, a second pointing northeast, and a third pointing north.

Measuring Surface Wind Direction and Speed

Every day you make inferences about weather conditions. For example, if you observe that the wind is blowing hard outside, you might decide to wear a jacket. Meteorologists make more precise observations so that in addition to describing current weather, they can predict future weather.

Wind vanes are instruments meteorologists use to determine the direction of winds. A wind vane has an arrowhead on one end and a broad tail on the other. The tail resists the wind, so if a wind is blowing from the west, it will blow the tail to the east. The arrowhead then points west, into the wind. A wind vane always points to the direction from which the wind is coming. Thus, winds are named for their place of origin. A wind that blows from west to east is a west (or westerly) wind.

Meteorologists use an instrument called an *anemometer* (an-uh-MAHM-ih-tuhr) to measure wind speeds ten meters above the ground. Wind speeds can also be estimated based on wind's effects on water, smoke, trees, and other objects. For example, whitecaps on waves are fairly frequent in winds of about 20 to 30 kilometers per hour, umbrellas are difficult to use once winds reach 40 kilometers per hour, and trees are uprooted in winds of more than 90 kilometers per hour.

ESTIMATING WIND SPEED Umbrellas are hard to use in winds of more than 40 kilometers per hour.

19.1 Section Review

1. What is air pressure? How does air pressure vary with elevation?
2. Explain why humid air is lighter than dry air.
3. Explain why a wind vane always points into the wind.
4. **PHYSICS** Suppose there are two forces acting on the air at a given location. Suppose also that the pressure gradient causes a force that is directed northwest toward a low. If another equally strong force is directed northeast, what is the direction of the net force of the air? Draw a diagram to support your answer.

418 Unit 5 Atmosphere and Weather

MONITOR AND RETEACH

If students miss . . .

Question 1 Reteach air pressure. (p. 414) Ask: Why is air pressure greatest at Earth's surface? **overlying layers of air press down on the surface**

Question 2 Review the diagram on page 416 and read the caption.

Question 3 Reteach how to measure surface wind direction and speed. (p. 418) Have students use cardboard to construct simple wind vanes and observe the movements of the tail and arrowhead in front of a fan.

Question 4 Have students start with a point to represent the location. From it, they draw two arrows of equal length (equal force), at a 90° angle. One points NW, one NE. The arrow representing the wind is placed evenly between the two, pointing N.

Factors Affecting Winds

If Earth were perfectly smooth and did not rotate, air would flow straight from high-pressure areas to low-pressure areas. In reality, winds seldom move in a straight path. In this section you will read about how the Coriolis effect, friction, and pressure gradients affect the direction of the wind.

The Coriolis Effect

The **Coriolis** (KOR-ee-OH-lihs) **effect** is the tendency of an object moving freely over Earth's surface to curve away from its path of travel. The effect is created by Earth's rotation, as well as its curvature. In the Northern Hemisphere, the path of an object will curve to the right; in the Southern Hemisphere, the path will curve to the left. For example, a rocket launched from the North Pole toward a target located on the equator will veer to its right, causing it to land west of its target. The Coriolis effect can be described more fully as follows:

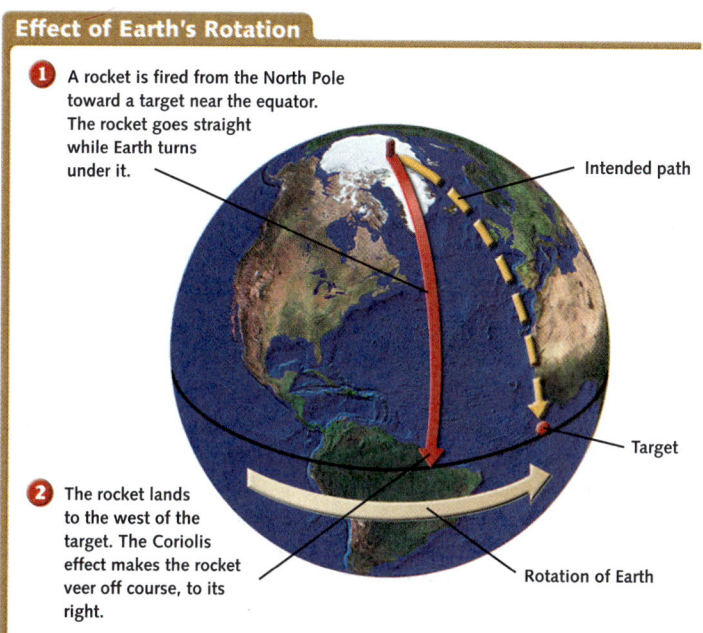

Effect of Earth's Rotation

1. A rocket is fired from the North Pole toward a target near the equator. The rocket goes straight while Earth turns under it.
2. The rocket lands to the west of the target. The Coriolis effect makes the rocket veer off course, to its right.

- Intended path
- Target
- Rotation of Earth

- In the Northern Hemisphere, the Coriolis effect deflects objects to their right; and in the Southern Hemisphere, to their left.
- The Coriolis effect is greatest near the poles and least near the equator.
- The effect is greater on faster objects than it is on slower objects.
- The effect does not depend on an object's direction of movement.
- The effect is generally noticeable only for objects traveling at great speeds over great distances, such as airplanes and rockets, or large masses traveling great distances, such as winds and ocean currents.

19.2

KEY IDEAS

Earth's rotation causes the Coriolis effect.

Wind direction depends upon the the Coriolis effect, the pressure gradient, and friction.

KEY VOCABULARY
- Coriolis effect
- jet stream

Observe an animation of the Coriolis effect over Earth's surface.
Keycode: ES1904

CHAPTER 19 SECTION 2

Visual Teaching

Interpret Diagrams
Have students study the two diagrams. Ask: How does the organization of the captions help you understand the diagrams? *For each diagram, the captions refer to similar features in the same order.* Make sure students understand that the red arrows represent air being deflected from its path and not a definition of the pathway itself. Because of the countervailing force of friction, the air will, once it's deflected, start to move along the isobars, creating a spiral of air flowing around the high and low.

Extend
Have students use tracing paper to recreate the diagrams, but this time without the gray arrows. Then have them take a red pencil and connect the tips of the arrow heads, creating a circle surrounding the high and a circle surrounding the low. Ask: What is the overall direction of the air flow around the area of high pressure? *clockwise* of low pressure? *counterclockwise*
Use Transparency 21.

Because of the Coriolis effect, winds in the Northern Hemisphere tend to blow clockwise out of areas of high pressure and counterclockwise into areas of low pressure, as shown below. In the Southern Hemisphere, the pattern is reversed: winds veer toward their left, so they flow counterclockwise out of highs and clockwise into lows.

Coriolis Effect on Wind in the Northern Hemisphere

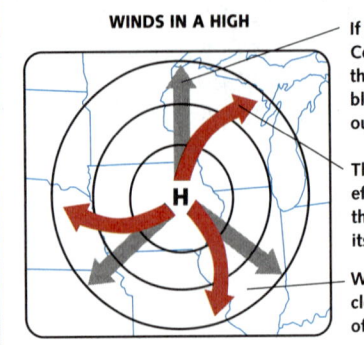

WINDS IN A HIGH
- If there were no Coriolis effect, the wind would blow straight out of a high.
- The Coriolis effect turns the wind to its right.
- Winds spiral clockwise out of the high.

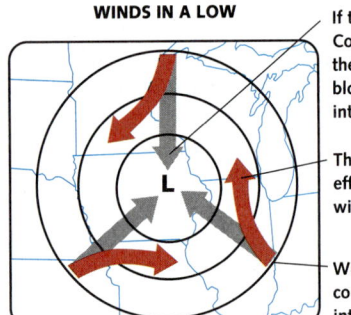

WINDS IN A LOW
- If there were no Coriolis effect, the wind would blow straight into a low.
- The Coriolis effect turns the wind to its right.
- Winds spiral counterclockwise into the low.

VISUALIZATIONS CLASSZONE.COM
Observe how the Coriolis effect influences wind direction.
Keycode: ES1905

Friction

As you might expect, friction between the air and the ground slows surface winds. Because the Coriolis effect is weaker for slower-moving objects, friction reduces the impact of the Coriolis effect on surface winds. The more friction there is, the less the wind will be deflected as it moves across isobars. Over rough land, surface winds blow at an average angle of about 30° to the isobars (as suggested by the red arrows in the diagrams above). The smoother the surface, the faster the wind, the stronger the Coriolis effect. Fast-moving winds blow more nearly parallel to the isobars because the deflection is greater. Over the ocean, which has a surface generally smoother than land, the angle between surface winds and isobars is usually about 10°.

Winds at greater altitudes above Earth's surface are less affected by friction. At heights greater than one or two kilometers, the Coriolis effect is as strong a factor in determining wind direction as is a pressure gradient. As a pressure gradient causes air to flow toward an area of low pressure, the Coriolis effect curves the wind clockwise or counterclockwise. A balance is reached with the wind flowing nearly parallel to the isobars.

Near the top of the troposphere, winds are very fast (120 to 240 kilometers per hour), and friction with the ground is negligible. Isobars at this level appear as wavy lines that stretch across weather maps in an east-west direction, as shown on the map on the next page. Bands of swiftly moving winds form, moving east to west, creating a **jet stream.** A jet stream is typically thousands of kilometers long, hundreds of kilometers wide, and about one kilometer from top to bottom. Sometimes a jet stream will split into two separate streams, which may or may not rejoin. Occasionally a jet stream will stretch around the entire Earth.

420 Unit 5 Atmosphere and Weather

DIFFERENTIATING INSTRUCTION

Reading Support Point out the value of the bulleted list on page 419 as a study tool and a way of summarizing a major concept. Have students make a similar bulleted list that summarizes important points about the effects of friction on wind speed and direction. Have students work in small groups to review one another's lists and keep them as study aids.
Use Reading Study Guide, p. 65.

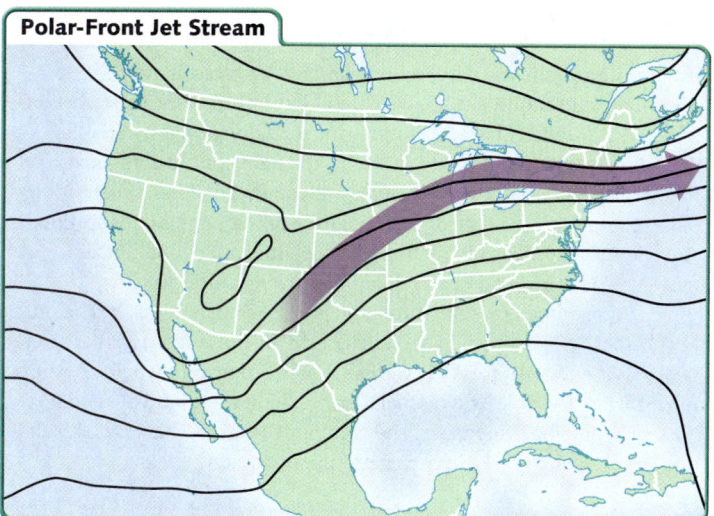

JET STREAM The polar-front jet stream (purple arrow) moves east to west, parallel to the isobars shown on this weather map of the upper troposphere.

Cool air from the northern polar regions forms the polar-front jet stream. This jet stream has a great effect on weather in the United States, supplying energy to storms and directing their paths. It can reach as far south as central Florida in winter, but generally moves over Canada and the northern states in summer. Its speed depends on the pressure gradient in the upper troposphere, which in turn depends on temperatures at Earth's surface. The polar-front jet stream is fastest in winter, when there are great temperature contrasts at the surface.

The warmer air in the tropics of the Northern Hemisphere also forms a jet stream at high altitude. The tropical-easterly jet stream is weaker than the polar-front jet stream.

19.2 Section Review

1. List three factors that affect wind direction.
2. Describe the path of air as it travels into a low-pressure area in the Northern Hemisphere.
3. **CRITICAL THINKING** Demonstrate the effect of friction by drawing diagrams of the winds around two Northern Hemisphere lows—one located above mountainous terrain and the other above an ocean.
4. **APPLICATION** What effect could the polar-front jet stream have on the speed of an airplane flying eastward? an airplane flying westward?

Chapter 19 The Atmosphere in Motion 421

CHAPTER 19 SECTION 2

VISUALIZATIONS
CLASSZONE.COM

Observe how the Coriolis effect influences wind direction.
Keycode: ES1905

Visualizations CD-ROM

INVESTIGATIONS
CLASSZONE.COM

How Does the Jet Stream Change Through the Year?
Keycode: ES1906

Use Internet Investigations Guide, pp. T69, 69.

Visual Teaching

Interpret Diagram

Ask: What does the path of the jet stream suggest about factors affecting wind direction and speed? For the high-altitude winds of the jet stream, a path running parallel to the isobars suggests that the Coriolis has a strong effect, while friction has little. The close spacing of the isobars suggests a strong pressure gradient to set the air in motion.

ASSESS

1. the pressure gradient, the Coriolis effect, and friction
2. spirals counterclockwise into the low
3. Both diagrams should show winds flowing counterclockwise as they move around the lows. Ocean wind will move nearly parallel to the isobars because there is less friction, so the Coriolis effect is stronger. Mountain winds will angle across the isobars.
4. increase the speed of eastward-flying airplane; decrease the speed of a westward-flying airplane

---

MONITOR AND RETEACH

If students miss . . .

Question 1 Have a student read aloud the first paragraph of this section. (p. 419)

Question 2 Reteach "The Coriolis Effect." (pp. 419–420) Have students sketch the diagram of winds in a low.

Question 3 Reteach "The Coriolis Effect" and "Friction." (pp. 419–420) Remind students that a rough surface **mountains** lessens the Coriolis effect; winds move across isobars. A smooth surface **ocean** means less friction; Coriolis effect deflects wind more nearly parallel to isobars.

Question 4 Review "Friction." (pp. 420–421) Ask: In which direction does the polar front jet stream flow? **west to east**

Chapter 19 421

CHAPTER 19 SECTION 3

19.3

FOCUS

Objectives
1. Identify factors that affect global wind patterns.
2. Describe the strengths and weaknesses of the three-celled circulation model, and use the model to explain prevailing winds and pressure regions.

Set a Purpose
Read aloud the key ideas. Point out that these ideas refer to general air movements around the globe. Then have students read the section to find the answer to "How do global wind patterns form?"

INSTRUCT

Visual Teaching

Interpret Diagram
Have students read the captions in the diagram. Ask: What happens to air at the equator? *It is warmed, becomes less dense, and then rises and flows toward the poles.* What happens to air at the poles? *It cools, becomes denser, and then sinks and flows toward the equator.*

Extend
Where in this simplified model would you expect to find cloudy conditions? *near the equator* Why? *Cloud formation is associated with areas of low pressure and rising humid air.*

KEY IDEAS
Both Earth's rotation and the uneven heating of Earth by the sun affect wind patterns.

The three-celled circulation model helps to explain prevailing winds and pressure regions.

KEY VOCABULARY
- polar front
- middle latitudes (mid-latitudes)
- intertropical convergence zone (ITCZ)
- trade winds
- prevailing winds

Global Wind Patterns

In some parts of the world, winds are fairly predictable. In northern Australia, for example, surface winds usually blow from the southeast. In much of the United States, however, wind direction is highly variable.

Many factors affect winds, including the temperature difference between equatorial and polar regions, the rotation of Earth, the locations of the continents, the time of year, and local topography. In this section, you will learn how Earth's rotation and the temperature differences between equatorial and polar regions affect wind patterns.

To understand how Earth's rotation affects wind patterns, it is helpful to consider first what would happen if Earth did not rotate and there were no Coriolis effect. Air from the warmer equatorial regions would rise and move toward the poles. Colder polar air would move toward the equator, where it would be heated, rise, and continue the cycle. This would result in one large circulation cell in each hemisphere, as shown below.

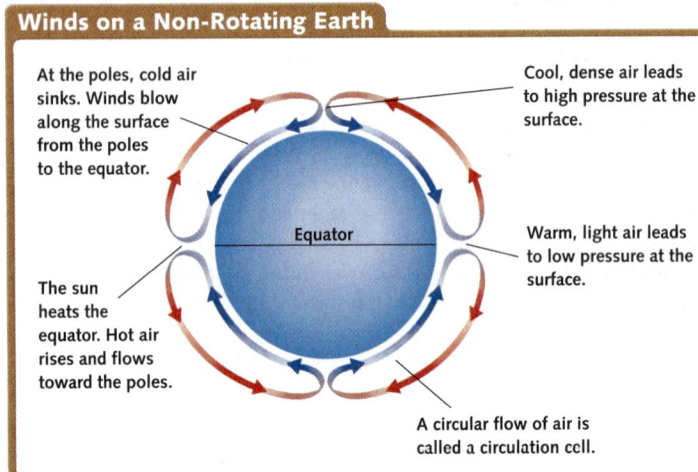

Winds on a Non-Rotating Earth

At the poles, cold air sinks. Winds blow along the surface from the poles to the equator.

Cool, dense air leads to high pressure at the surface.

The sun heats the equator. Hot air rises and flows toward the poles.

Warm, light air leads to low pressure at the surface.

A circular flow of air is called a circulation cell.

Effects of Earth's Rotation

Because Earth rotates, the Coriolis effect prevents air from flowing straight from the equator to the poles. Instead, air flowing northward from the equator is deflected to its right, and air flowing southward from the equator is deflected to its left. Also, the air cools and sinks long before it reaches the polar regions. According to one popular model of Earth's atmosphere, the air-circulation pattern is better represented using three circulation cells in each hemisphere. Scientists realize that this idealized model is not entirely accurate, yet it is still helpful for understanding Earth's wind patterns.

Three-celled Circulation Model

In the three-celled model of Earth's wind patterns, the Northern and Southern Hemispheres each have three circulation cells: one between the

422 Unit 5 Atmosphere and Weather

DIFFERENTIATING INSTRUCTION

Reading Support Students may be confused by the use of the word *cell* in the context of global wind patterns. Ask students what they think of when they think of a cell. *Possible answers may include a jail cell, a cell in the human body, a battery cell.* Point out that each of these is, in some way or another, a compartment. The same applies here. The atmosphere surrounding Earth can be modeled as consisting of six large doughnut-shaped cells (three in each hemisphere). Winds within a particular cell tend to flow in the same overall pattern. Refer students to the diagram on page 423. Make sure students understand that the cells depicted in cross section to the left of Earth show the overall pattern of up-down circulation, while the arrows across the face of Earth show overall crosswise movement. Each set of arrows defined by the same pair of latitude lines belongs to the same three-dimensional cell.

Use Reading Study Guide, p. 66.

422 Unit 5 Atmosphere and Weather

equator and 30° latitude, the second between 30° and 60° latitude, and a third between 60° latitude and the pole. The direction of air circulation changes from each cell to the next, as shown below.

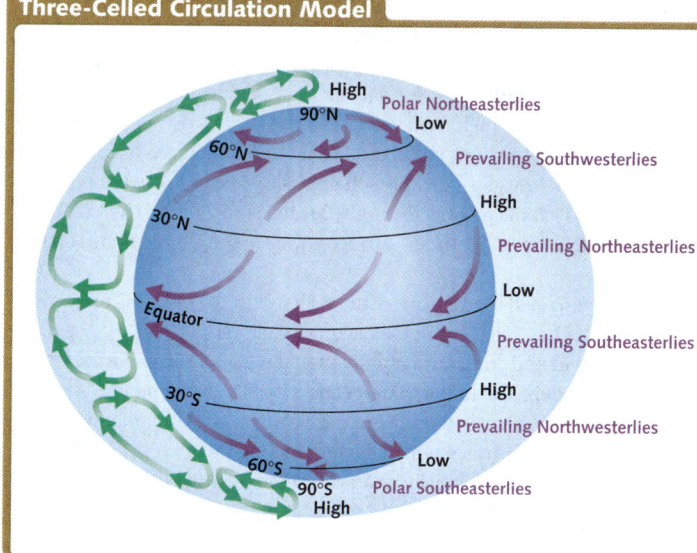

Three-Celled Circulation Model

The circulation cells are caused by alternating bands of high and low pressure at Earth's surface. At the polar region in each hemisphere, cold, dense air sinks, resulting in high pressure at the surface. At 60° latitude, warmer air rises resulting in low pressure at the surface. The boundary at 60° latitude, where air flowing away from the polar regions collides with warmer air moving up from the lower latitudes, is called the **polar front.** The air from the two adjacent cells is forced to rise. A similar boundary exists at 30° latitude. At 30° latitude, air that has risen from the equator cools and sinks. creating high pressure at the surface. This air flows back toward the equator where the air pressure is lower.

As winds blow from areas of high pressure to areas of low pressure, they are deflected by the Coriolis effect. In the Northern Hemisphere, the winds to shift to their right. This means that surface air flowing southward from the North Pole turns to its right, creating prevailing winds that blow east to west (easterlies). Between 30° and 60° N latitude, the winds blow west to east (westerlies), as northward-flowing air is shifted to its right. Between 30° N latitude and the equator, prevailing winds blow east to west, as southward-flowing air is deflected to its right. (See the diagram above.)

In the Southern Hemisphere, the pattern is reversed because the Coriolis effect deflects winds to their left. Easterly winds blow between the South Pole and 60° S latitude, and between 30° S latitude and the equator. Westerly winds blow between 30° and 60° S latitude.

VOCABULARY STRATEGY
Remember that winds are named for the directions from which they blow. Winds that blow from northeast to southwest are called *northeasterlies*.

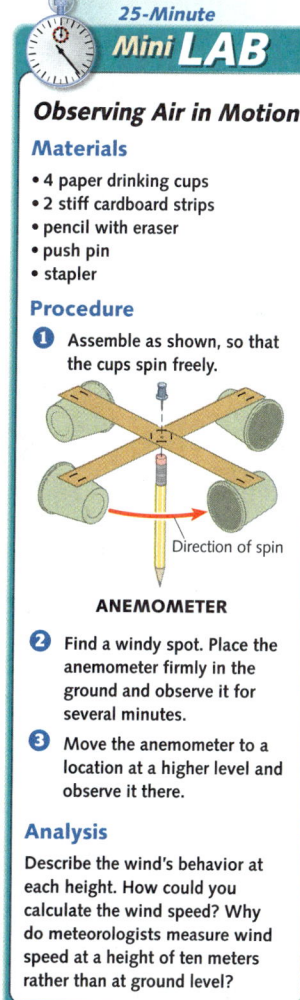

25-Minute Mini LAB

Observing Air in Motion

Materials
- 4 paper drinking cups
- 2 stiff cardboard strips
- pencil with eraser
- push pin
- stapler

Procedure

1. Assemble as shown, so that the cups spin freely.

ANEMOMETER

2. Find a windy spot. Place the anemometer firmly in the ground and observe it for several minutes.

3. Move the anemometer to a location at a higher level and observe it there.

Analysis
Describe the wind's behavior at each height. How could you calculate the wind speed? Why do meteorologists measure wind speed at a height of ten meters rather than at ground level?

Chapter 19 The Atmosphere in Motion 423

CHAPTER 19 SECTION 3

Visual Teaching

Interpret Diagram
Make sure students understand the wind-speed scale included with the image. Point out that the different colors represent different wind speed. Tell students that wind speed is generally measured in knots. One knot is equal to approximately 2,026 yd/h or 1,852 m/h. Have students use the scale to locate areas with the highest wind speeds. *yellow areas in the Southern Hemisphere* and those with the lowest wind speeds. *blue-green areas scattered around the globe*

Extend
In the scale shown, wind speed is also given in meters per second. Have students calculate the wind speed in meters per second for a wind blowing at 11 knots. *11 knots × 1,852 m/h × 1 h/60 min × 1 min/60 s = 5.66 m/s* Data for wind speed over the continents is not included. Ask: What effect would you expect the landmasses to have on wind speed? *lower speeds, because landmasses would interfere with wind movement*

Weaknesses of the Three-celled Model

Like all models, the three-celled model of Earth's wind patterns is accurate in some ways and inaccurate in others. The three main weaknesses of the model are that it gives a simplified view of circulation between 30° and 60° latitude, that it does not take into account the effects of the continents or seasons, and that it is based on a simplified view of upper-level winds.

Between latitudes of about 30° and 60°, often referred to as the **middle latitudes,** or **mid-latitudes,** surface winds are determined by the locations of transient high- and low-pressure systems. As you learned in Section 19.2, surface winds spiral around highs and lows. As a result, the surface winds at any given location change is response to the high- and low-pressure systems passing through. Because most of the United States is in the northern mid-latitudes, surface winds in the United States change often.

The three-celled model also assumes that Earth's surface is uniform, yet continents have significant effects on surface-wind patterns. The continents heat and cool more rapidly than the oceans, causing the pressure belts depicted in the three-celled model to break apart.

Finally, the three-celled model may give the impression that upper-level winds generally travel north or south, depending on the cell. Actually, upper-level winds are primarily westerly except near the equator, where the Coriolis effect is weak.

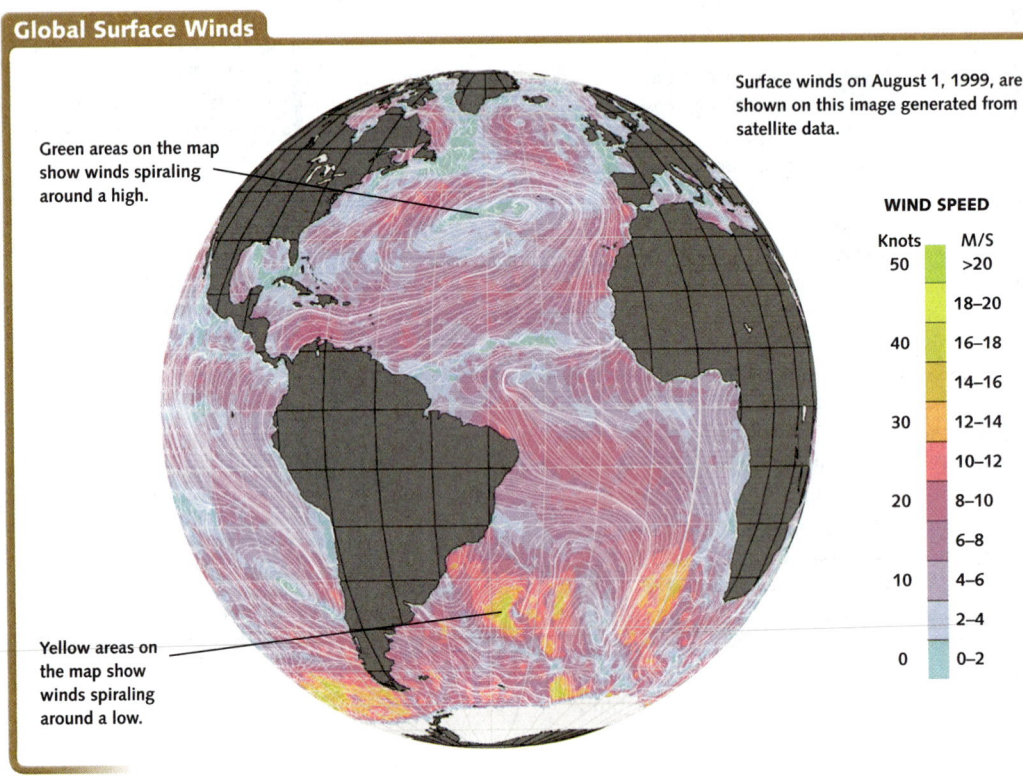

Global Surface Winds

Surface winds on August 1, 1999, are shown on this image generated from satellite data.

Green areas on the map show winds spiraling around a high.

Yellow areas on the map show winds spiraling around a low.

424 Unit 5 Atmosphere and Weather

DIFFERENTIATING INSTRUCTION

Developing English Proficiency Having discussed how the word *cell* is used in the context of global wind patterns (Reading Support, p. 422), ask students if they know another word that sounds like *cell* but is spelled differently. *sell* Tell students that *cell* and *sell* are homophones, which are words that are pronounced the same but are spelled differently and have different meanings. Encourage students to come up with a list of homophones and give the meanings of the words. *examples: to, two, too; hear, here; sun, son; bear, bare; hole, whole; deer, dear; ate, eight*

Strengths of the Three-celled Model

The three-celled model does give a fairly accurate image of the general patterns of surface winds and pressures outside the mid-latitudes. There actually is a region of low pressure near the equator, there are often high-pressure regions near 30° latitude, and there are often high-pressure regions near the poles. The three-celled model also gives a picture of wind patterns and pressure systems that is useful for climate studies, because understanding climates involves averaging patterns over long periods.

Description of Wind and Pressure Belts

Because the sun heats the tropics more than other parts of Earth, the warm, rising air creates a low-pressure zone at the surface in the tropics. This zone is called the **intertropical convergence zone (ITCZ)** because surface winds from the two hemispheres come together there. In the ITCZ, the air is hot and humid, there is little or no wind, and rain is common. This region has historically been called the doldrums. Due to the lack of wind, sailing vessels were often stuck in the doldrums for days at a time.

Between the latitudes of 20° and 35°, air generally sinks, forming the so-called subtropical highs. Because the sinking air yields very little precipitation, these regions contain many of Earth's deserts. They have historically been called the horse latitudes. The story goes that because sailing ships were likely to be becalmed in this area, with little or no wind, horses being transported would be thrown overboard to lighten the load.

Between the doldrums and the horse latitudes, the easterly **trade winds** blow. The trade winds are warm and relatively steady in both direction and speed. They are known as trade winds because merchants used them, whenever possible, as for trade routes for their sailing ships. The predictable southeast winds of northern Australia are trade winds.

The high-pressure regions where cold air sinks at the poles are called the polar highs. Surface winds in the polar regions are usually easterly. Winds that usually blow from the same direction, such as the trade winds and polar easterlies, are described as **prevailing winds.**

ITCZ The band of clouds near the equator marks the warm, moist, rising air of the ITCZ. Notice also the spiraling clouds in the mid-latitudes and the clear skies of the subtropical highs and polar highs.

19.3 Section Review

1. How is the direction of wind flow changed at the equator? at the poles? Why?
2. What determines the direction of surface winds in the mid-latitudes?
3. At which latitudes would you expect to find high surface pressures?
4. **CRITICAL THINKING** Why are the trade winds in the Southern Hemisphere southeasterly?
5. **APPLICATION** Modern sailboats have motors so that they do not get stranded when there is no wind. According to the three-celled model, would a sailor be more likely to need a motor when sailing from Georgia to southern Florida, or when sailing from Massachusetts to Maine? Explain.

Chapter 19 The Atmosphere in Motion 425

CHAPTER 19 SECTION 3

Explore how our understanding of global winds has improved over time.
Keycode: ES1907

Science and Society

Trade Winds

The trade winds blow steadily across the subtropical regions of the Pacific and Atlantic Oceans at average speeds of 17 to 21 kilometers per hour. Many people mistakenly believe that the trade winds got their name because they enhanced trade between Europe and the New World. Actually, the name comes from the old meaning of the word *trade*, which referred to a steady track. The trade winds provide a steady push westward. The winds are situated between a low-pressure band near the equator and high-pressure belts in the mid-latitudes.

Science Notebook

Even non-sailing ships are still affected by global winds. Students might say they would send back high-tech navigation tools, such as the Global Positioning System, or radio systems to improve communications.

SCIENCE & Society

Trade Winds

In the 15th century, Europeans wanted to find a way to sail to India, but the winds, which blew in the wrong direction, thwarted the sailors. Eventually, some clever Europeans found a way to use the wind belts to their advantage.

How can trade winds and westerlies be used to sail from Europe to India and back?

Planetary winds have long dictated the course of world exploration. In the 15th century, during the Age of Exploration, Europeans were desperate for a sea route to the riches of India. The challenges of sailing around Africa had long stymied the best seafarers.

Several ships sailed out of the Mediterranean and caught the Atlantic trade winds to the Canary Islands west of Africa. The trade winds always blow in the same direction, however, making it difficult to get home.

In the early 15th century, Portuguese sailors learned a trick called *volta do mar*, or "bend in the sea." Today it is known as tacking. Bartolomeu Dias used this technique in 1488 and successfully sailed around the southern tip of Africa. He might have reached India, but his crew revolted and forced him home.

Sailing south near the African coast, Dias encountered southerly trade winds. He decided to tack southwest into the wind, thereby gaining distance to the south as well as to the west. This westward sailing might have seemed counterproductive, as it took his ship farther from India, not closer. It did, however, take him to a wind belt starting at 30 degrees south latitude with prevailing westerly winds. These winds blew Dias east around the tip of Africa.

Vasco da Gama finally opened the ocean trade route to India in 1497. His four ships reached the Cape of Good Hope by sailing three times the distance Columbus sailed to reach the New World. Da Gama sailed southwest on the trade winds for 84 days and probably was very close to South America before he found the westerlies that carried him around Africa and into the Indian Ocean. The voyage was uncertain and tedious and included several sweltering weeks spent in the windless doldrums.

After years of searching, the Portuguese finally had found their searoute to the Asian markets. ■

TACKING, or sailing diagonally across the trade winds, executed by a Spanish caravel.

TRADE WINDS are shown on this 1736 copper engraving of a global map.

Extension

 SCIENCE NOTEBOOK
Consider the ways that modern technology makes ocean sailing easier today. Are today's sailors still affected by planetary winds? What equipment would you send back in time to Vasco da Gama if you could? Explain why.

Explore how our understanding of global winds has improved over time.
Keycode: ES1907

Continental and Local Winds

Because of the tilt of Earth, the relative position of the sun changes over the course of the year, creating seasons. As Earth's surface temperatures change with the seasons, so do the global winds. These winds are affected by large-scale shifts in temperature as well as by the positions of continents. In fact, because of such seasonal changes, the highest temperatures on Earth are not usually at the equator. For example, in the Northern Hemisphere summer, the hottest temperatures are north of the equator, where the sun causes air to heat, rise, and flow toward the poles.

Effects of Seasons and Continents

In the summer, continents become hotter than the surrounding oceans as they absorb more radiant heat. The hot land heats the air above it, causing the air to become less dense, creating an area of low air pressure. In general, you can expect the average air pressure over the continents in summer to be low. Because ocean waters do not absorb radiant heat as quickly as land, the ocean air will be cooler and denser in the summer, creating an area of high air pressure.

Because winds spiral out of high-pressure areas and into low-pressure areas, the resulting highs and lows determine the directions of the prevailing winds at various locations. For example, in the Northern Hemisphere summer an Atlantic high brings prevailing southerly winds to the eastern United States.

With the coming of winter, the air over the continents becomes cold and dense. The average air pressure is therefore high. In comparison, the average air pressure over the oceans is lower.

Because air pressures change seasonally, the directions of winds also change seasonally. Winds that change direction seasonally are called **monsoons** (mahn-SOONZ). The most dramatic monsoons occur in southern Asia, where seasonal changes in pressure caused by the heating and cooling of the Asian continent produce a complete wind reversal.

19.4

KEY IDEA

Seasons, land masses, and topography cause winds to vary from the global patterns depicted in the three-celled model.

KEY VOCABULARY
- monsoon

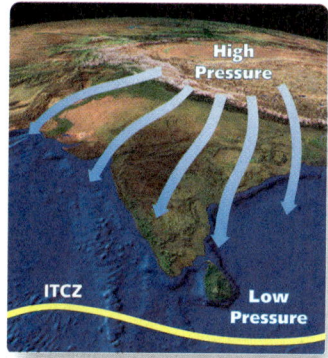

WINTER MONSOON Cold, dry air blows from the high-pressure area over land to the low pressure of the ITCZ (intertropical convergence zone).

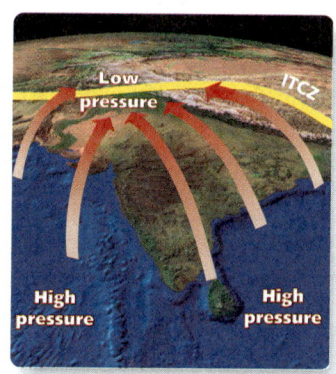

SUMMER MONSOON Warm, moist air blows from the high-pressure area over water to the low-pressure area over land. As the air rises, it releases rain.

CHAPTER 19 SECTION 4

Scientific Thinking

INTERPRET
A heavy rain is falling, so the photograph was taken during the summer monsoon. Have students draw upon their own personal experiences and knowledge of current affairs as they discuss how people may be affected by the seasonal cycle of monsoons. For example, ask: What are some problems associated with heavy rains? *floods, disruption of activities* What are some benefits? *ample moisture for crops and for drinking* Conversely, what are the advantages and disadvantages of an extended dry spell? *advantages: fair weather; disadvantages: no growing season, less drinking water*

CRITICAL THINKING

Show students a world map and point out the Himalayas. Tell students that during the summer monsoon, warm, moist air reaches these mountains and rises, causing heavy rains to fall. Ask: What sort of climate would you expect to find on the opposite side of the Himalayas? *dry* Why? *After rising and dropping its moisture, the air descending on the other side of the mountain range is dry.*

Scientific Thinking

INTERPRET
Explain whether the photograph above was taken during the winter monsoon or the summer monsoon.

Discuss how people's ways of life can be affected by a seasonal cycle of monsoons.

During the winter, the cold, sinking air over the continent flows toward the low pressure over the sea. The resulting cold, dry winds are the winter monsoons. In summer, the low-pressure areas over the continent cause humid air to flow onto the continent from the surrounding waters. These winds are the summer monsoons. When the warm, moist air reaches the highlands of India and Southeast Asia, heavy rains fall.

Another way continents and seasons affect winds is by shifting the location of the ITCZ, the zone where winds from the Northern and Southern Hemispheres converge. In the Northern Hemisphere summer, the highest temperatures, and therefore the lowest pressures, are as far north as 30°N in southwestern Asia and northern Africa. Air flows into the low-pressure belt from the north and south. The ITCZ moves north as tropical air moves toward the lower pressures. In the Southern Hemisphere summer, the ITCZ and the hottest temperatures move south. The ITCZ does not move as far over the oceans as over land because ocean temperatures do not vary as much as land temperatures do.

Local Winds

A local wind is a wind that extends for a distance of 100 kilometers or less. Like larger-scale winds, local winds are caused mainly by differences in temperature. Examples of local winds include land and sea breezes, as well as mountain and valley breezes.

In Section 19.1 you learned that sea breezes form when an island is heated more than the surrounding water. However, sea breezes can form along any coastline, not just over islands. During the day, coastal land is warmer than the nearby water. The pressure decreases over the land, and a pressure gradient develops between the ocean and the land. The pressure gradient causes a cool sea breeze to blow inland.

The sea breeze is part of a wind pattern called a *sea-breeze circulation*. The air rises over the land and flows out to sea; and as it cools, it sinks to replace the cool air that is flowing inland. As the diagram below shows, the airflow forms a complete circuit—from the sea to the land, up, back to the sea, and down again.

SEA-BREEZE CIRCULATION

A gentle sea breeze usually begins in the late morning along a seacoast. It increases in speed until midafternoon and dies down toward sunset. A sea breeze may be felt more than 20 kilometers inland from the shore.

MONITOR AND RETEACH

If students miss . . .

Question 1 Reteach "Effects of Seasons and Continents." (pp. 427–428) Ask: During which season is the air pressure over a continent low? *summer* Why? *The land is hot, heating the air above it. This causes the air to become less dense, resulting in low air pressure.*

Question 2 Reteach "Effects of Seasons and Continents," referring to the diagrams on the bottom of page 427. (pp. 427–428) Have students describe the general flow of surface air. *Air flows from areas of high pressure to areas of low pressure.* Ask: Where are the lowest pressures found during summer in the Northern Hemisphere? *as far north as 30°N*

Question 3 Review "Local Winds." (pp. 428–429) Have students trace the flow of air in the diagrams. Point out that the red arrows represent warm air, and the blue arrows represent cool air.

Question 4 Reteach the last paragraph under "Local Winds." (p. 429)

At night, the air over the land becomes colder than the air over the water. The resulting pressure gradient causes a cool land breeze to blow out to sea. There, the air rises and flows inland. A land breeze starts long before midnight and dies down after sunrise.

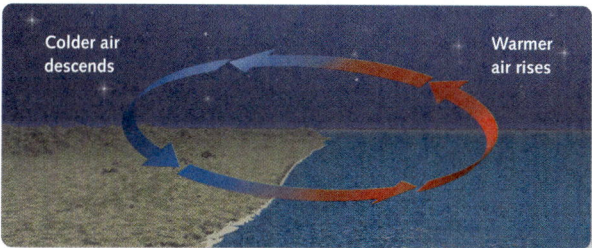

LAND-BREEZE CIRCULATION

Temperature differences also cause mountain and valley breezes. Air near a mountain can have a density different from that of the surrounding air. For example, at night a mountain will cool the air next to it, making the air more dense. The cool air sinks, flowing down into the valleys nearby. The narrower the valley, the stronger the breeze. Because the breeze comes from a mountain, it is called a mountain breeze. During the daytime, mountain air becomes warmer and less dense than the surrounding air. The mountain air rises, causing a valley breeze to blow up from nearby valleys. The speed of an uphill valley breeze is generally much lower than that of a downhill mountain breeze.

19.4 Section Review

1. How does average air pressure over the southwestern United States in the winter compare to the average air pressure in the summer?
2. Explain how and why the location of the ITCZ (intertropical convergence zone) varies from season to season.
3. Compare and contrast a sea-breeze circulation with a land-breeze circulation.
4. Explain why a cool breeze may blow down a valley at night.
5. **CRITICAL THINKING** Explain why Southeast Asia experiences heavy rains in the summertime. Why doesn't Southeast Asia experience heavy rains in the wintertime?
6. **MATHEMATICS** During the summer monsoon, parts of India may receive as much as 75 centimeters of rain in one day. If you stood outside in such a rain for 20 minutes, about how much water would fall on your head? Express your answer in centimeters and as an estimated volume.

CHAPTER 19 SECTION 4

More about...

The Santa Ana
A well-known example of a local wind is the Santa Ana in southern California. As this dry wind rushes down the mountains surrounding the Los Angeles area, it warms by compression and can reach a temperature of 38°C. The Santa Ana occurs during winter when a semipermanent high forms over the western United States. The dry conditions associated with the high combine with the strong wind to worsen the threat of brush fires.

ASSESS

1. a low in summer, a high in winter
2. The ITCZ moves as far north as 30°N during the summer in the Northern Hemisphere because the lowest pressures are found there. For the same reason, it moves south during the summer in the Southern Hemisphere.
3. During a sea breeze, cooler, denser air over the ocean moves inland, warms, rises, and then moves back out to sea. During a land breeze, cooler, denser air over the land moves toward the ocean, warms, rises, and then moves inland again.
4. At night, mountain air becomes cooler and denser than the surrounding air. The cool air sinks, flowing down the valley.
5. In the summertime, low-pressure areas over Southeast Asia cause moist air to flow from the ocean to the land. When the warm, moist air rises over highlands, heavy rains fall. In the wintertime, the air pressure over the land is high.
6. 60 minutes/hour × 24 hours/day = 1,440 minutes/day; 75 cm/day ÷ 1,440 minutes/day = 0.05 cm/minute; 0.05 cm/minute × 20 minutes = 1 cm; head = 20 cm in diameter; 300 cm³ (300 mL) of rain

MONITOR AND RETEACH

Have students make diagrams that describe the circulation of mountain and valley breezes.

Question 5 Reteach "Effects of Seasons and Continents." (pp. 427–428) Ask: What type of pressure system forms over Southeast Asia in the summertime? **low** What weather conditions are associated with this pressure system? **clouds and rain**

Question 6 Suggest that students think of the volume of rain falling on their head as a cylinder, the diameter of which is the diameter of their head (for example, 20 cm). The formula for volume is $\pi r^2 h$, where $r = 10$ cm. The height of the cylinder will be equal to the amount of rain that falls in 20 minutes (1 cm). From this they can compute the volume.

CHAPTER 19 ACTIVITY

PURPOSE
To measure and observe weather variables, then use collected data to predict the weather

MATERIALS
Provide students with a copy of Lab Sheet 19. A copymaster is provided on page 144 of the teacher's laboratory manual. If a rain gauge is unavailable, use a transparent container with a plastic metric ruler attached to the inside. If other equipment, such as a barometer, is difficult to obtain, supplement the data with measurements taken from local weather watchers, the Internet, or news organizations.

PROCEDURE
❸ If possible, set up a weather station with all the necessary equipment so that every measurement is taken at the same location.

ANALYSIS AND CONCLUSIONS
1. In general, high pressure is associated with clear skies and cool days. Low pressure is usually associated with stormy, warmer weather.

CHAPTER 19 LAB Activity

Correlating Weather Variables

SKILLS AND OBJECTIVES
- **Record** weather variables over a period of weeks.
- **Correlate** changes in different weather variables.
- **Predict** local weather.

MATERIALS
- thermometer
- barometer
- wind vane or flag
- anemometer (see page 423)
- wet-bulb thermometer (see page 393)
- rain gauge
- relative humidity chart (see page 393)
- cloud-type photographs (see pages 396–397)

Wind direction, temperature, clouds, and precipitation are examples of weather variables that can be observed and recorded for a specific location. Changes in one weather variable can be related to changes in other variables. Analyzing patterns in these weather variables can help you to accurately predict future weather conditions. This activity is designed to help you obtain a better understanding of the weather in your area. By observing the weather daily for three weeks, and organizing your data in a chart, you will begin to notice patterns and correlation among the variables.

Procedure

❶ Produce enough copies of the Daily Weather Chart such that you will be able to record three weeks of daily weather observations.

❷ Write the date and time in the row labeled "Date/Time".

❸ Use a thermometer to record the air temperature at a location just outside of your school. Record the temperature on your chart in °C.

❹ Use a wet-bulb thermometer to determine the relative humidity at the same locations in Step 3. (See the Minilab on page 393 on how to use a wet-bulb thermometer.) Record the wet-bulb temperature on the weather chart in °C. Use the relative humidity chart on page 393 to estimate the relative humidity. Record on the weather chart.

❺ Use a barometer to determine the air pressure at the same location. Record on the weather chart, noting the units of pressure.

❻ Use a wind vane or flag to determine the direction of the wind. Record on the weather chart the direction from which the wind is blowing: north, south, east, west, northeast, northwest, southeast, or southwest.

❼ Determine the wind speed using an anemometer. Record the speed on the chart, indicating the units of measurement. If you choose to use the anemometer from the Minilab on page 423, record the wind speed in revolutions per minute.

❽ Place a rain gauge or a can calibrated in centimeters in a spot open to the sky where the gauge or can will not be disturbed. On subsequent days, measure the amount of rain that has collected in the gauge or can. Record the amount on your chart, then empty the gauge or can for the next day's measurement.

9. Observe the sky and estimate the percentage of the sky covered in clouds. Record this number on your chart. Use the photographs on pages 396 and 397 to determine the types of clouds in the sky. Record your observations in the chart under "Cloud type(s)".

10. Briefly describe the present weather conditions (sunny, foggy, hazy, partly cloudy, rainy, cold, and so on) on the weather chart.

11. Repeat Steps 2–10 for three weeks. Try to have at least one person in your class obtain observations from this location on the weekends.

Analysis and Conclusions

1. Use your observations to describe the weather during periods of high air pressure. Describe the weather during periods of low air pressure.

2. Based on your observations, describe the relationship between wind direction and temperature.

3. How did rapid changes in air pressure affect the wind speed? How did the wind speed behave when the air pressure was steady?

4. Was there a relationship between the sky state (percentage of clouds) and the next day's weather in your observations? Explain.

5. Describe the relationship between relative humidity and precipitation. Describe the relationship between relative humidity and weather conditions.

6. Which variables on your weather chart are the best indicators of the next day's weather? Use your observations of today's weather to predict tomorrow's weather conditions, and explain your reasoning.

Daily Weather Chart					Week of: _____	
Date/Time						
Air temperature						
Wet-bulb temperature						
Relative humidity						
Air pressure						
Wind direction						
Wind speed						
Precipitation in last 24 hours						
Percent clouds						
Cloud type(s)						
Present weather conditions						

DATA CENTER CLASSZONE.COM

Find out more about past and current methods of forecasting weather.
Keycode: ES1909

CHAPTER 19 ACTIVITY

DATA CENTER CLASSZONE.COM

Find out more about past and current methods of forecasting weather.
Keycode: ES1909

(continued)

2. Winds from the north bring cooler air from higher latitudes, while winds from the south bring warmer air from lower latitudes. Depending upon local topography, there may also be local winds such as sea breezes and land breezes.

3. Wind speed changes as air pressure changes. Winds remain light when air pressure is steady.

4. The state of the sky is usually not a good indicator of future weather because the sky that is visible at any given spot usually represents weather that will pass through in a matter of minutes or hours.

5. Precipitation is associated with high humidity. Low humidity is associated with fair weather.

6. Wind direction and air pressure are the two most useful indicators of future weather. In addition, cloud type and humidity can indicate changes in precipitation and temperature. Answers will vary based on students' location and past experiences with the weather.

Refer students to Safety in the Earth Science Laboratory on pages xxii–xxiii.

CHAPTER 19 REVIEW

VOCABULARY REVIEW

1. isobars
2. low
3. Coriolis effect
4. polar front
5. intertropical convergence zone
6. monsoon

CONCEPT REVIEW

7. In the Southern Hemisphere, winds circulate around highs in a counterclockwise direction and around lows in a clockwise direction.

8. Diagrams should include the major wind belts shown in the Three-Celled Circulation Model on page 423. The winds would originate at the same latitudes but blow in opposite directions. Prevailing southwesterlies, for example, would blow from southeast to northwest.

9. Before sunrise, the sailboats could use a land breeze to sail out to sea. During midafternoon, the boats could use a sea breeze to sail back to land.

10. Atmospheric pressure is exerted on the human body in all directions. In a similar way, water exerts pressure on a scuba diver in all directions.

11. The low-pressure circle should include the phrases *humid air, warm air, stormy weather,* and *counterclockwise flow in Northern Hemisphere.* The high-pressure circle should include the phrases *dry air, cold air, fair weather,* and *clockwise flow in Northern Hemisphere.*

CRITICAL THINKING

12. Air rises, cools, and condenses during the formation of a cumulonimbus cloud. Thus, air pressure at the surface would lower.

13. A sea breeze develops because of sharp pressure gradients between land and water. These gradients are caused by the unequal heating of Earth's surface. If the land is very warm, then the pressure gradient between land and water is large and the resulting breeze is strong. So a sea breeze would extend farther inland in low latitudes.

CHAPTER REVIEW 19

Summary of Key Ideas

19.1 Air pressure is caused by the weight of the atmosphere over a given area. It is directed equally in all directions. Changes in temperature and humidity change the air pressure. Winds are caused by differences in air pressure.

19.2 In the Northern Hemisphere, the Coriolis effect turns winds to their right. Winds blow counterclockwise around lows and clockwise around highs. In the Southern Hemisphere, the Coriolis effect turns winds to their left.

19.3 Low pressure and rising air are common near the equator. Surface high pressure and sinking air are common at latitudes near 30 degrees and the poles. In the mid-latitudes, winds and pressure are determined by transient low-pressure systems. Except near the equator, upper-level winds are generally westerly.

19.4 In summer, air over continents has low air pressure and air over seas has high air pressure. Winds tend to blow more from the sea to the continents. The pattern reverses in winter. The reversing winds are called monsoons. Local winds along coasts and mountains reverse daily.

KEY VOCABULARY

air pressure (p. 414)
Coriolis effect (p. 419)
high-pressure area (high) (p. 416)
intertropical convergence zone (ITCZ) (p. 425)
isobar (p. 416)
jet stream (p. 420)
low-pressure area (low) (p. 416)
middle latitudes (mid-latitudes) (p. 424)
monsoon (p. 427)
polar front (p. 423)
pressure gradient (p. 416)
prevailing winds (p. 425)
trade winds (p. 425)

432 Unit 5 Atmosphere and Weather

14. No; the Coriolis effect is only noticeable for objects traveling at great speeds over great distances, such as airplanes, or large masses traveling great distances, such as winds and ocean currents.

15. (a) On the side of Earth facing the sun, air pressure and wind belts would be similar to the diagram of Winds on a Non-Rotating Earth shown on page 422.

Vocabulary Review

Write the vocabulary term or terms that best complete the sentence.

1. Lines on a weather map that connect points with the same air pressure are called __?__.

2. A set of closed isobars in which the pressures decrease toward the center represents a(n) __?__.

3. Winds in the Northern Hemisphere are turned to the right by the __?__.

4. The boundary between warm and cold air at about 60 degrees latitude is called the __?__.

5. The low-pressure zone at the equator is the __?__.

6. A wind that reverses direction with the seasons is a(n) __?__.

Concept Review

7. In what direction do winds circulate around highs in the Southern Hemisphere? In what direction do winds circulate around lows in the Southern Hemisphere?

8. What would Earth's major wind belts look like if Earth rotated from east to west? Draw a diagram.

9. During what time of day might a fleet of sailboats make the best use of land and sea breezes?

10. How is atmospheric pressure similar to the pressure felt by scuba divers?

11. **Graphic Organizer** Copy the circles below and write each of the following phrases in the appropriate circle: humid air, dry air, warm air, cold air, stormy weather, fair weather, counterclockwise flow in Northern Hemisphere, clockwise flow in Northern Hemisphere.

Low Pressure   **High Pressure**

Warm, less dense air from the equator would rise and move toward the poles. There, the air would cool, become more dense, and sink. The cool air would move toward the equator, become warm and less dense, and the cycle would continue. (b) At 10 kilometers above the surface, air pressure would decrease substantially to about 25 percent of that

Critical Thinking

12. **Predict** Explain what might happen to the air pressure at the surface when a deep cumulonimbus cloud is growing overhead.
13. **Infer** Sea breezes do not extend very far inland. Do sea breezes extend farther inland at low latitudes or at high latitudes? Explain your thinking.
14. **Apply** Would you expect the Coriolis effect to affect the movement of water going down a drain? Explain your thinking.
15. **Hypothesize** Suppose Earth did not rotate, and one side always faced the sun.
    a. Describe the winds and pressures at the surface.
    b. Describe the winds and pressures about 10 kilometers above the surface.
16. **Hypothesize** How would the atmospheric circulation be affected if Earth's speed of rotation were doubled? Explain your thinking.

Interpreting Maps

The map shows sea-level pressures over part of the eastern coast of North America.

17. What is the air pressure at point *E*?
18. Which point is in a low-pressure area?
19. At which point on the map are the winds strongest?
20. At point *B*, in what direction are the winds blowing? Explain your thinking.
21. At which point are winds approximately southerly?
22. At point *A*, what direction does the wind blow in the afternoon? Explain your thinking.
23. How is the air pressure at point *D* on the map different from the actual air pressure at point *D*? Explain your thinking.

Internet Extension

Could You Break the Record for an Around-the-World Balloon Flight? Apply your understanding of the atmosphere in motion to plan a trip around the world in a balloon.
Keycode: ES1908

Writing About the Earth System

SCIENCE NOTEBOOK The hydrosphere and geosphere both have a significant effect on winds. In what ways do the hydrosphere and geosphere affect air pressures and wind direction? In what ways do the geosphere, hydrosphere, and biosphere affect wind speed?

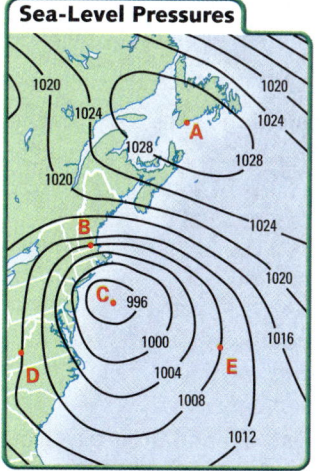

Sea-Level Pressures

Chapter 19 The Atmosphere in Motion 433

CHAPTER 19 REVIEW

Could You Break the Record for an Around-the-World Balloon Flight?
Keycode: ES1908

Use Internet Investigations Guide, pp. T70, 70.

(continued)

experienced at the surface. There would be no Coriolis effect or friction with surface features. So winds would blow fast from highs to lows.

16. The Coriolis effect would be stronger, so wind belts would be more strongly deflected to the right in the Northern Hemisphere and to the left in the Southern Hemisphere.

INTERPRETING GRAPHS

17. 1008 millibars
18. C, pressure decreases across the isobars
19. Winds are strongest at point B where the isobars are closest together.
20. Winds in the Northern Hemisphere blow in a counterclockwise direction into a low. The winds at point B are blowing from the northeast to the southwest.
21. E, winds are coming from the south, into a low-pressure area
22. The wind probably blows from the ocean to the land because a sea breeze forms during the day.
23. The map shows sea-level pressures. That means that the actual air pressure at D is lower than indicated on the map because D is located in the Appalachian Mountains—a higher elevation and therefore a lower air pressure.

WRITING ABOUT SCIENCE

Writing About the Earth System

Oceans cool down and heat up more slowly than land. Thus, air over an ocean will generally be cooler and denser in summer than air over land. The situation reverses in winter. The changes in air pressure between the hydrosphere and geosphere cause monsoons, which are winds that change direction seasonally, and local winds such as land breezes and sea breezes. The differences in air pressures between and within the hydrosphere and geosphere affect wind speed as well. The greater the pressure gradient, the greater the speed. Within the geosphere and biosphere, wind speed decreases because of friction with mountains, trees, and other features. Conversely, wind speed increases over oceans because friction is reduced.

Chapter 19 The Atmosphere in Motion 433

CHAPTER 20 PLANNING GUIDE WEATHER

	Section Objectives	Print Resources
Chapter–Level Resources		Laboratory Manual, pp. 87–94 Internet Investigations Guide, pp. T71–T72, 71–72 Guide to Earth Science in Urban Environments, pp. 13–20, 29–36, 37–44 Spanish Vocabulary and Summaries, pp. 39–40 Formal Assessment, pp. 58–60 Alternative Assessment, pp. 39–40
Section 20.1 Air Masses and Weather, pp. 436–438	1. Identify factors that determine the characteristics of an air mass. 2. Compare and contrast different types of air masses.	Lesson Plans, p. 68 Reading Study Guide, p. 68 Teaching Transparencies 22, 23
Section 20.2 Fronts and Lows, pp. 439–444	1. Describe the weather conditions associated with different types of fronts. 2. Describe the life cycle of a mid-latitude low.	Lesson Plans, p. 69 Reading Study Guide, p. 69 Teaching Transparency 24
Section 20.3 Thunderstorms and Tornadoes, pp. 445–449	1. Describe the conditions necessary for the formation of thunderstorms and tornadoes. 2. Describe the hazards of thunderstorms and tornadoes, and discuss related safety measures.	Lesson Plans, p. 70 Reading Study Guide, p. 70
Section 20.4 Hurricanes and Winter Storms, pp. 450–454	1. Describe the formation and effects of hurricanes, and the measures taken to mitigate their damage. 2. Describe the weather conditions associated with mid-latitude lows during winter.	Lesson Plans, p. 71 Reading Study Guide, p. 71
Section 20.5 Forecasting Weather, pp. 455–459	1. Compare and contrast the different technologies used to gather weather data. 2. Analyze weather symbols, models, and maps. 3. Describe how weather forecasts are made.	Lesson Plans, p. 72 Reading Study Guide, p. 72

INVESTIGATIONS

CLASSZONE.COM

Section 20.2 INVESTIGATION
How Does a Mid-Latitude Low Develop into a Storm System?
Computer time: 45 minutes / Additional time: 45 minutes
Students analyze the development of a storm as it moves across North America.

Chapter Review INVESTIGATION
How Well Can You Predict Tomorrow's Weather?
Computer time: 45 minutes / Additional time: 45 minutes
Students use current weather data to make a forecast for the local area.

Technology/Online Resources

Electronic Teacher Tools
Visualizations CD-ROM
Classzone.com
Online Lesson Planner
Test Generator

Visualizations CD-ROM
Classzone.com
ES2001 Visualization

Visualizations CD-ROM
Classzone.com
ES2002 Visualization, ES2003 Investigation

Visualizations CD-ROM
Classzone.com
ES2004 Visualization, ES2005 Safety Tips,
ES2006 Visualization, ES2007 Safety Tips

Visualizations CD-ROM
Classzone.com
ES2008 Visualization, ES2009 Safety Tips,
ES2010 Safety Tips, ES2011 Data Center

Classzone.com
ES2012 Career

Classroom Management

Gather these materials for minilab and map activities.

Minilab, p. 441
  5 sheets of graph paper
  tape
  straightedge
  pencil

Map Activity, pp. 460–461
  tracing paper
  colored pencils
  paper clips

CHAPTER 20

INTRODUCE

Build Interest

Before sharing background information about the photograph with students, allow time for their questions and comments.

Ask questions like the following:

- How do you think the waves caused damage to this house? **The waves probably undercut the soil or supports of the house, which then fell into the ocean, rather than smashing the house directly.**
- If this was your house, would you rebuild in the same place? **Some students might say that it is not worth the risk; others may disagree.**
- What can people do to avoid such damage? **Not build along the coast or protect the homes from pounding waves.**

Respond

- How are you able to predict the weather? **Students might suggest by observing the clouds, changes in temperature and air pressure, and considering incoming weather farther away.**

About the Photograph

During California's 1997–1998 winter season, strong winds and torrential rainstorms powered by El Niño caused widespread floods and mudslides. Waves 12 meters high crashed into beaches near Malibu. The storms in California resulted in hundreds of millions of dollars' worth of damage.

Weather

Severe weather can devastate populated areas, such as this coastal community in Malibu, California.

How are you able to predict the weather?

BEYOND THE TEXTBOOK

Print Resources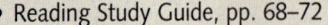

- Reading Study Guide, pp. 68–72
- Transparencies 22, 23, 24
- Formal Assessment, pp. 58–60
- Laboratory Manual, pp. 87–94
- Alternative Assessment, pp. 39–40
- Spanish Vocabulary and Summaries, pp. 39–40
- Internet Investigations Guide, pp. T71–T72, 71–72
- Guide to Earth Science in Urban Environments, pp. 13–20, 29–36, 37–44

Technology Resources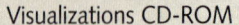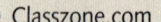

- Visualizations CD-ROM
- Classzone.com
- Online Lesson Planner
- Electronic Teacher Tools
- Test Generator

434 Unit 5 Atmosphere and Weather

CHAPTER 20

PREVIEW

▶ **FOCUS QUESTIONS** In this chapter, you will study weather and learn more about the key questions below.

Section 1 What are air masses? How and where do they originate?

Section 2 How do fronts and pressure systems affect weather patterns?

Section 3 What atmospheric conditions cause thunderstorms and tornadoes?

Section 4 How do scientists define hurricanes and winter storms? How do these phenomena start?

Section 5 By what means do meteorologist predict weather?

▶ **REVIEW TOPICS** As you investigate weather, you will need to use information from earlier chapters.

• Coriolis effect (p. 75)

▶ **READING STRATEGY**

PREVIEW

Before you read, look through Chapter 20, noting the key ideas, key vocabulary, images, and captions. In your science notebook, make a list of things you expect to learn about in the chapter.

INTERNET RESOURCES
CLASSZONE.COM

At our Web site, you will find the following Internet support for this chapter.

DATA CENTER
EARTH NEWS
VISUALIZATIONS
• Air Mass Movement across North America
• Warm and Cold Fronts
• Thunderstorm
• Tornado
• Hurricane

LOCAL RESOURCES
CAREERS
INVESTIGATIONS
• How Does a Mid-Latitude Low Develop into a Storm System?
• How Well Can You Predict Tomorrow's Weather?

Chapter 20 Weather **435**

CHAPTER 20

PREPARE

Focus Questions
Find out what knowledge students have about weather. For each focus question, have students consider the section title and then develop two or three additional questions for each section. For example, Section 5: What instruments are used to gather information about the weather? What do the symbols on weather maps mean?

Review Topics
In addition to the earth science topic listed, students may need to review this concept:

Physical Science
• Unlike electrical charges attract. Positive and negative charges flow toward each other to produce a lightning bolt.

Reading Strategy
Model the Strategy: To help students preview each section, use a version of the following: "I see that many of the topics in this section discuss weather events that I'm familiar with. I've seen lightning and heard thunder numerous times. I recently lost power during a heavy snowstorm, and a tree in my yard blew down years ago after a strong wind swept through the area. Do you have any similar experiences with weather? You can use these real-life experiences to better understand science topics." Have students continue to use this strategy for the remainder of the chapter.

BEYOND THE TEXTBOOK

INVESTIGATIONS
• How Does a Mid-Latitude Low Develop into a Storm System?
• How Well Can You Predict Tomorrow's Weather?

VISUALIZATIONS
• Animations of air masses moving across North America
• Animations of warm fronts and cold fronts

• Animation of a thunderstorm
• Animations of a tornado
• Animation of a hurricane

DATA CENTER
EARTH NEWS
LOCAL RESOURCES
CAREERS

Online Lesson Planner

Chapter 20 Weather **435**

CHAPTER 20 SECTION 1

FOCUS

Objectives
1. Identify factors that determine the characteristics of an air mass.
2. Compare and contrast different types of air masses.

Set a Purpose
Have students read the key ideas to find the definition of *air mass*. Then have them read this section to answer the question "How do air masses affect weather?"

INSTRUCT

More about...

Air-Mass Modification
Air-circulation patterns cause air masses to move from one region to another. In the process, the characteristics of air masses often are modified. Air-mass modification occurs when an air mass begins to take on the temperature and humidity characteristics of the new region over which it travels. In both the Northern and Southern Hemispheres, the modification of maritime tropical (mT) air, in particular, forms clouds and rain as the air moves toward the poles.

20.1

KEY IDEAS
An air mass is a large body of air that has similar characteristics throughout.

The temperature and humidity of an air mass depend on where the air mass originates.

KEY VOCABULARY
- meteorology
- air mass

DESERT AIR The hot, dry air over this desert used to be cool, moist air over the Pacific Ocean. The air was modified when it passed over mountains. As it rose, it lost moisture. When it sank on the other side of the mountains, it was compressed and warmed.

Air Masses and Weather

Weather influences our lives every day. It helps determine, for example, whether we carry an umbrella or put on sunblock, and whether our airplanes are delayed. Because of weather's influence, we depend on knowing ahead of time what weather to expect. **Meteorology,** the study of processes that govern Earth's atmosphere, helps make weather predictions possible. Meteorologists, the scientists who specialize in this study, perform tasks ranging from making weather forecasts to studying how tornadoes develop. To perform their work, they need to know how air masses form and also how the interactions of air masses generate weather.

Origin of an Air Mass

An **air mass** is a large body of air in the lower troposphere that has similar characteristics throughout. An air mass can be several thousand kilometers in diameter and several kilometers high. Two or three air masses can cover all of the continental United States. Throughout an air mass, temperature and humidity are nearly uniform.

The temperature and humidity of an air mass depend on where the air mass originates. For example, in polar regions, the lack of sunlight in winter causes the ground to be very cold. If an air mass stays in a polar region for days or weeks, it becomes cold as well. In the tropics, sunlight strikes directly and heats the ground. Thus an air mass staying in the tropics for an extended period becomes hot. The moisture content of an air mass also depends on the underlying surface. For example, an air mass that stays over land for a long time becomes dry, whereas an air mass over the ocean absorbs water vapor and becomes moist.

When an air mass travels from one area to another, it takes with it the temperature and humidity of its place of origin. For example, when the air in Chicago, Illinois, turns crisp and cool in the fall, it is because an air mass that originated in Canada has moved southward. When Chicago becomes hot and muggy in the summer, it is because an air mass from the Gulf of Mexico has moved northward.

As an air mass travels, its characteristics may change. For example, when a cold polar air mass moves south, it affects not only the weather of the area it enters but also gradually heats up as it moves over a warmer surface. Earth's topography can also contribute to changes in the temperature and humidity of an air mass as it travels.

DIFFERENTIATING INSTRUCTION

Developing English Proficiency Have students scan the text to identify unfamiliar words that may give them trouble, such as *crisp, topography, muggy,* or *maritime*. Students may find it helpful to work with a partner to learn the meanings of the words. Encourage them to focus on context. The meaning of *muggy*, for instance, can be inferred by rereading the sentence in the fourth paragraph on page 436 and zeroing in on clues, such as *hot* and *summer*. *Use Spanish Vocabulary and Summaries, pp. 39–40.*

436 Unit 5 Atmosphere and Weather

Types of Air Masses

Meteorologists classify air masses according to where they originate. The diagram below shows the principal North American air masses. The temperature of each type of air mass depends on whether the air mass originates in an arctic, polar, or tropical region. The humidity depends on whether the air mass comes from land (continental) or sea (maritime).

Continental Arctic

Continental arctic (cA) air masses originate in the arctic regions, where the air becomes extremely cold. Although these air masses may warm as they move southward, they are still capable of causing extreme cold waves in the regions they enter, particularly if they pass over snow-covered ground along the way. Because cold air is incapable of containing much moisture, cA air masses are also very dry.

Continental Polar

Continental polar (cP) air masses originate over the inland regions of Alaska and Canada. These air masses are somewhat warmer than cA air masses, but the difference in temperature, as well as in humidity, between cA and cP air can be slight. The brilliantly sunny, cold winter days in some regions of the United States are usually caused by cP air. Although this air is usually just cold and dry, in some regions it can create precipitation. For example, if cP air passes over the Great Lakes in the late fall when the water is still warm, the cold, dry air picks up moisture from the lakes, then deposits the moisture downwind from the lakes as heavy snow, called lake-effect snow.

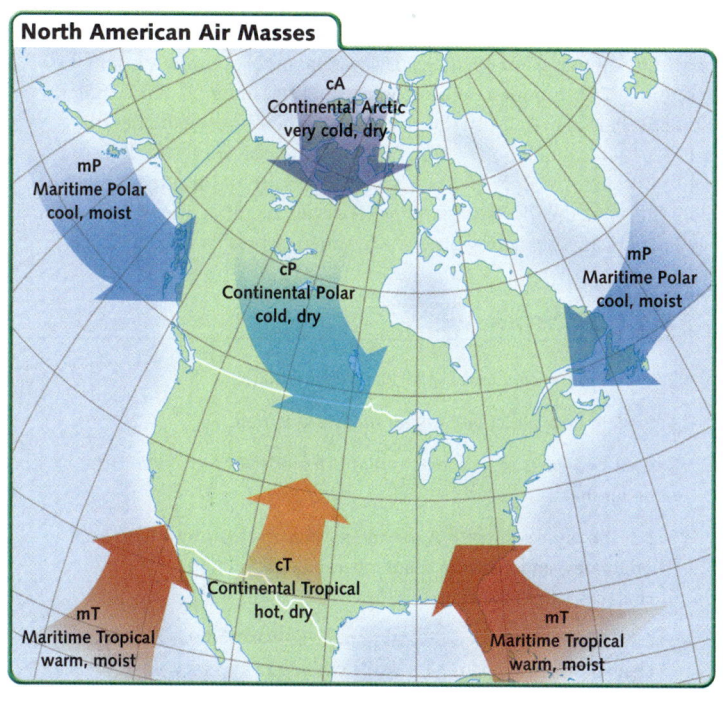

AIR MASSES The characteristics of an air mass depend on where it originates.

CHAPTER 20 SECTION 1

CRITICAL THINKING

Ask: Would snowfall in the Sierra Nevada be heaviest on the eastern or western slopes? **western** Use your knowledge of air masses to explain your answer. **Maritime polar air from the North Pacific Ocean would move inland, reach the mountains, and drop its moisture as it rose over the western slopes. The air would be much drier as it descended the eastern slopes.**

ASSESS

1. Air in a region takes on the characteristics of the region. Cold air masses originate in arctic and polar regions. Hot air masses originate in tropical regions.

2. Moist air masses originate over water, dry, over land.

3. **Sample Response:** the movement of an air mass into a new region with different temperature and humidity characteristics, and the movement of an air mass over a mountain

4. Continental arctic air originates in arctic regions and is very cold and dry. Continental polar air originates over polar regions and is cold and dry. Maritime polar air originates over oceans in high latitudes and is cool and moist. Maritime tropical air originates over warm tropical oceans and is warm and moist. Continental tropical air originates over deserts and is hot and dry.

5. Continental polar air is cold and dry and thus brings clear, cold weather to some regions. If cP air travels over water, however, its moisture increases and it can cause snow.

MARITIME POLAR AIR These low clouds over the coast of Lane County, Oregon, may have been caused by continental cooling of maritime polar air.

Maritime Polar

Maritime polar (mP) air masses originate over the ocean in high latitudes. These air masses are both cold and damp. However, mP air is not usually as cold as cP air because of the contrast in temperature between the land (colder) and the oceans (warmer). If mP air cools to its dew point, fog, clouds, or precipitation results. For example, in the Pacific Northwest, ocean air masses passing inland hit the Olympic and Cascade mountain ranges and rise abruptly, expanding and cooling; this sudden cooling brings the air to its dew point, resulting in the wet winter climate of the area. Occasionally, mP air from the North Atlantic brings heavy snowstorms, called nor'easters, to the East Coast in the winter and cool, clear weather to the area in the summer.

Maritime Tropical

When an air mass originates over a warm tropical ocean, it acquires both warmth and moisture. Such air masses are known as maritime tropical (mT) air masses. In the summer, mT air from the Bahamas and the Gulf of Mexico moves clockwise around the high pressure over the Atlantic Ocean bringing heat and humidity to the Midwestern and Eastern United States. Because the air contains a large amount of moisture, thunderstorms often develop during the heat of the day. As the summer sun heats the ground, the moist mT air is heated from below and rises, forming thunderclouds. When the sun sets, the clouds and thunderstorms dissipate as the surface cools. Thunderstorms will be covered in greater detail in Section 20.3.

Continental Tropical

Air masses that originate over deserts are hot and dry and are known as continental tropical (cT) air masses. Often, a cT air mass originates as a maritime air mass, but becomes dry as it passes over mountains. In the summer, cT air produce tremendous heat waves in much of the United States. Whereas mT air usually produces temperatures no higher than 100°F, cT air is much hotter, with temperatures exceeding 100°F. In addition, since cT air is dry, it doesn't bring clouds or thunderstorms to cool the air as the moist mT air does. If cT air advances into an agricultural region and stays for a long time, it can cause serious damage to crops.

20.1 Section Review

1. Explain how cold and hot air masses originate.
2. Explain how moist and dry air masses originate.
3. Give two examples of factors that can modify the characteristics of an air mass.
4. List the principal North American air masses, identify where each originates, and describe the temperature and humidity characteristics of each.
5. **CRITICAL THINKING** Explain how continental polar air can bring clear weather to one region while causing heavy snow in another.

MONITOR AND RETEACH

If students miss . . .

Question 1 Reteach "Origin of an Air Mass." (p. 436) Ask: How do meteorologists classify air masses? **according to their place of origin**

Question 2 Ask students: Which air mass would contain more moisture—one that formed over land or one that formed over the ocean? **over the ocean**

Question 3 Have students reread "Origin of an Air Mass" and the caption on page 436.

Question 4 Refer students to the map on page 437 to list and describe the different types of North American air masses.

Question 5 Have students reread the paragraph about continental polar air masses on page 437.

Fronts and Lows

Weather in the mid-latitudes can be quite changeable. One day might be mild and sunny, the next cold and clear, the next snowy and windy. These changes in weather result mostly from the movement of low-pressure systems and their associated frontal systems.

What Is a Front?

Air masses of different types don't easily mix. The boundary that separates opposing air masses is known as a **front.** The width of a front can range from 200 meters to 200 kilometers. It can be as high as 5 kilometers and as long as 2000 kilometers and can affect weather patterns in areas hundreds of kilometers wide. Fronts are most common at mid-latitudes, where southward-moving polar air masses and northward-moving tropical air masses often meet.

The air masses on either side of a front may differ in temperature, in humidity, or in both. At the front, the less-dense air mass is forced to rise over the denser air mass. As a result, the front is roughly wedge shaped, as shown in the illustration below, which depicts warm air being forced to rise over advancing denser and colder air. You will learn about the characteristics of different kinds of fronts later in this section.

Fronts can have steep slopes. The slope may range from about 1 over 50 to 1 over 300. A slope of 1 over 50 means that the frontal surface rises 1 kilometer for every 50 kilometers of horizontal distance. Such a slope is much steeper than one that rises 1 kilometer for every 300 horizontal kilometers.

Fronts usually bring precipitation. At the frontal surface, the less-dense air rises high into the troposphere. The air cools as it rises and, if the air is humid enough, clouds and precipitation form.

20.2

KEY IDEA

The movements of fronts and lows greatly influence the weather at mid-latitudes.

KEY VOCABULARY
- front
- cold front
- warm front
- occluded front
- stationary front

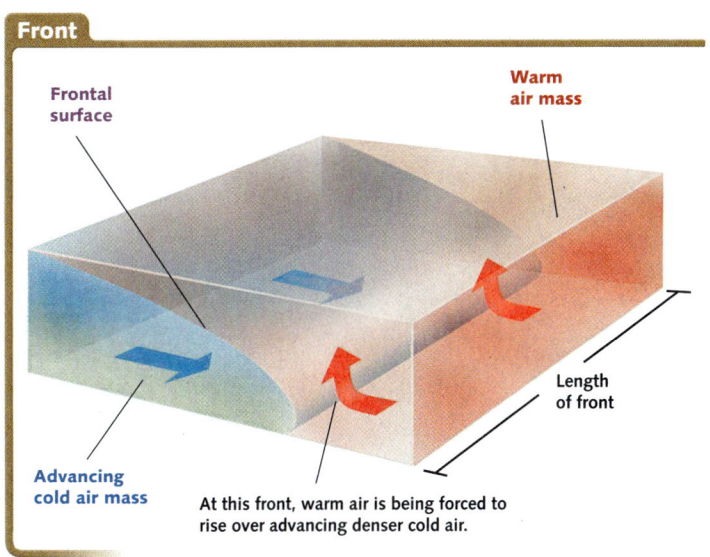

Front
- Frontal surface
- Warm air mass
- Advancing cold air mass
- Length of front

At this front, warm air is being forced to rise over advancing denser cold air.

CHAPTER 20 SECTION 2

FOCUS

Objectives
1. Describe the weather conditions associated with different types of fronts.
2. Describe the life cycle of a mid-latitude low.

Set a Purpose

Have students read this section to answer the question "How do fronts and lows influence weather at mid-latitudes?"

INSTRUCT

Visual Teaching

Interpret Diagram

Use the diagram to review the concept of cloud formation, which was discussed in Chapter 18. Ask: What happens to the warm air mass as it rises? **It cools.** Assume that the warm air contains moisture. Describe what happens next. **As the air rises and cools, the water vapor in the air condenses and clouds form.**

Extend

Ask: Would the steepness of the frontal slope affect the process of cloud formation? **yes** Explain. **Along a steep slope, warm air is forced sharply upward. Thus, clouds of vertical development (cumulonimbus) can form more easily than along a gradually sloping front.**

DIFFERENTIATING INSTRUCTION

Reading Support Help students distinguish between the scientific and common meanings of the word *front*. Have them brainstorm different meanings for *front* and use each meaning in a sentence.
Sample Responses: The house fronts the street. I'm standing near the front of the room. Tell students that the scientific meaning of a word sometimes has its basis in a common definition. One definition of *front*, for example, is "the area where two conflicting armies meet and clash." This definition is the basis for the meteorological meaning of front.
Use Reading Study Guide, p. 69.

CHAPTER 20 SECTION 2

Visual Teaching

Interpret Diagram

After students read about cold fronts and warm fronts, have them reexamine the diagram. Ask: How does the movement of warm air vary along a cold front and a warm front? *Along a cold front, warm air is forced sharply upward by the advancing cold air mass. Along a warm front, warm air slides gradually over the retreating cold air mass.* How does the area of precipitation vary along the two fronts? *Precipitation falls over a narrow band of ground along a cold front. Along a warm front, precipitation extends over a wide area.*

Use Transparency 23.

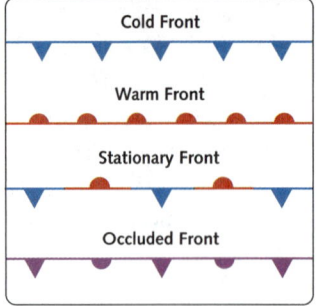

INTERNATIONAL WEATHER SYMBOLS

Compare and contrast warm and cold fronts.
Keycode: ES2002

Kinds of Fronts

The weather associated with a particular front depends on the types of air masses involved and the speed at which the front is moving. Although fronts occasionally occur between air masses that have the same temperature (but different humidity), scientists usually define fronts according to the temperature of the advancing air mass. There are four basic kinds of fronts: cold, warm, occluded, and stationary.

Cold Fronts

A **cold front** is the boundary between an advancing cold air mass and the warmer air mass it is displacing. Since cold air is denser than warm air, the cold air slides underneath the warm air and forces it upward. Friction from moving along the ground causes the lower part of the cold air mass to lag behind the upper part, as shown in the illustration below. Cold fronts have steep slopes, often as steep as 1 in 50.

The type of weather a cold front brings depends to a large degree on the type of air mass it is displacing. For example, when cold cP air displaces warm, humid mT air in the summer, thunderstorms often form along the front as the moist mT air is forced upward. On the other hand, if the cold air displaces hot, dry cT air, the rising air contains little moisture so there may be no precipitation. The passing front may cause no greater change than a shift in wind direction. In the summer, some cold fronts cause a

Movement of Cold and Warm Fronts

DIFFERENTIATING INSTRUCTION

Challenge Activity Have students examine the diagram on page 440 and then use the information in the text on page 441 to draw diagrams of a stationary front and an occluded front. Students should label the air masses on their diagrams and use arrows to show direction of movement. Tell them to use the chart of International Weather Symbols on page 440 for additional guidance.

decrease in humidity but little change in temperature. In the winter, a cold front may be marked by rain or snow showers.

Because a cold front has a steep slope, the precipitation associated with the front covers only a narrow band of ground. A cold front moves relatively quickly, and precipitation usually ends shortly after the front passes.

Warm Fronts

If warm air displaces cold air, the boundary between the air masses is known as a **warm front.** The advancing warm air rises above the denser cold air mass, which retreats slowly. The slope of a warm front—usually only about 1 in 150—is more gradual than that of a cold front, and the weather changes associated with the warm front are less dramatic.

The first signs of the approach of a warm front are high cirrus clouds, which are followed by cirrostratus and lower and thicker stratiform clouds. Such clouds form in the warm, stable air sliding up the frontal surface. The clouds may stretch 1500 kilometers ahead of the place where the warm front touches the ground; thus cirrus clouds may warn of approaching precipitation more than a day before it arrives.

Following the cirrus and cirrostratus are altostratus clouds, which almost screen out the sun and the moon. Finally, the heavy nimbostratus clouds arrive, and steady rain or snow begins. This area of rain and snow can stretch hundreds of kilometers ahead of where the front touches the ground, and the precipitation may last for a day or more. Although thunderstorms occasionally form, they are not typical of a warm front. After the front passes, the weather warms.

Occluded and Stationary Fronts

Cold fronts typically move about twice as fast as warm fronts. If a cold front "catches up" to a warm front, the result is an **occluded front.** The warm air that is caught between the two colder air masses is forced to rise. As this warm air rises, it cools, often causing cloudiness and precipitation.

If a front is not moving forward, it is called a **stationary front.** As with other fronts, the warmer air rises over the denser, colder air, and clouds and precipitation may result. If the front remains stationary for too long, flooding can occur.

Life Cycle of a Mid-Latitude Low

Fronts are usually connected to mid-latitude low-pressure systems. In fact, a low-pressure system often starts out as a small ripple on a stationary polar front where cold polar air meets warm tropical air. Under certain conditions, this small ripple can grow into an intense storm system.

Shortly after the end of World War I, Norwegian meteorologists developed a theory explaining the life cycle of mid-latitude low-pressure systems. This theory has been modified over time as scientists have gathered new information, but the basic aspects of the theory remain a cornerstone of modern meteorology.

25-Minute Mini LAB

Graphing a Front

Materials
- 5 sheets of graph paper
- tape
- straightedge
- pencil

Procedure
1. Cut the sheets of paper in half lengthwise.
2. Tape the sheets of paper side by side to form one piece that is more than 500 grid squares long.
3. Draw a front that is 5 kilometers high and has a slope of $\frac{1}{100}$. Use the scale 1 square grid length = 1 kilometer. Label the warm and cold fronts.

Analysis
Describe the slope of the front drawn on the graph. Using the straightedge, estimate the slope between the warm and cold fronts.

CHAPTER 20 SECTION 2

VISUALIZATIONS
CLASSZONE.COM

Compare and contrast warm and cold fronts.
Keycode: ES2002

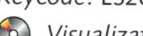 Visualizations CD-ROM

MINILAB
Graphing a Front

Management
- This activity will help students appreciate the size of weather fronts, as well as the gradual slope between warm and cold fronts.
- Tape the sheets of paper so that they are side by side in the landscape orientation.
- The number of sheets of graph paper needed may differ, depending on the size of the grids. The number given in the activity applies to paper with grids that are approximately 0.5 cm².

Materials
A meterstick may be used instead of a straightedge.

Analysis Answers
- The slope of the front is very gradual.
- To estimate slope, students should measure the length and rise for a given section of the intersection of the two fronts. For a length of 40 mm, the change in height for the warm front is 7 mm. Thus, the slope equals approximately 7/40.

For proper behavior in a laboratory setting, refer students to Safety in the Earth Science Laboratory on pages xxii–xxiii.

DIFFERENTIATING INSTRUCTION

Support for Physically Impaired Students The Minilab requires manual dexterity and thus may be difficult for some students to complete. Encourage these students to participate in the activity by pairing them with other students who can handle scissors, tape, pencils, and straightedges. Have students with physical impairments serve as the observer and recorder for the pair by keeping track of the results and later sharing them with the class.

CHAPTER 20 SECTION 2

Visual Teaching

Interpret Diagrams

Have students examine the map to infer why the low-pressure system weakens after it reaches the occluded stage. Ask: **What happens to the warm front as it moves northward?** *It moves over cooler ground and takes on the characteristics of the new surface. Thus, the temperature of the air decreases.* **How might this affect the storm's strength?** *There no longer are sharp temperature differences among the air masses, so the storm will eventually weaken.*

Extend

Ask students where they would expect to find the strongest winds and heaviest precipitation. *at or near the cold front* Where would they expect to find clear skies? *behind the cold front*

Use Transparency 24.

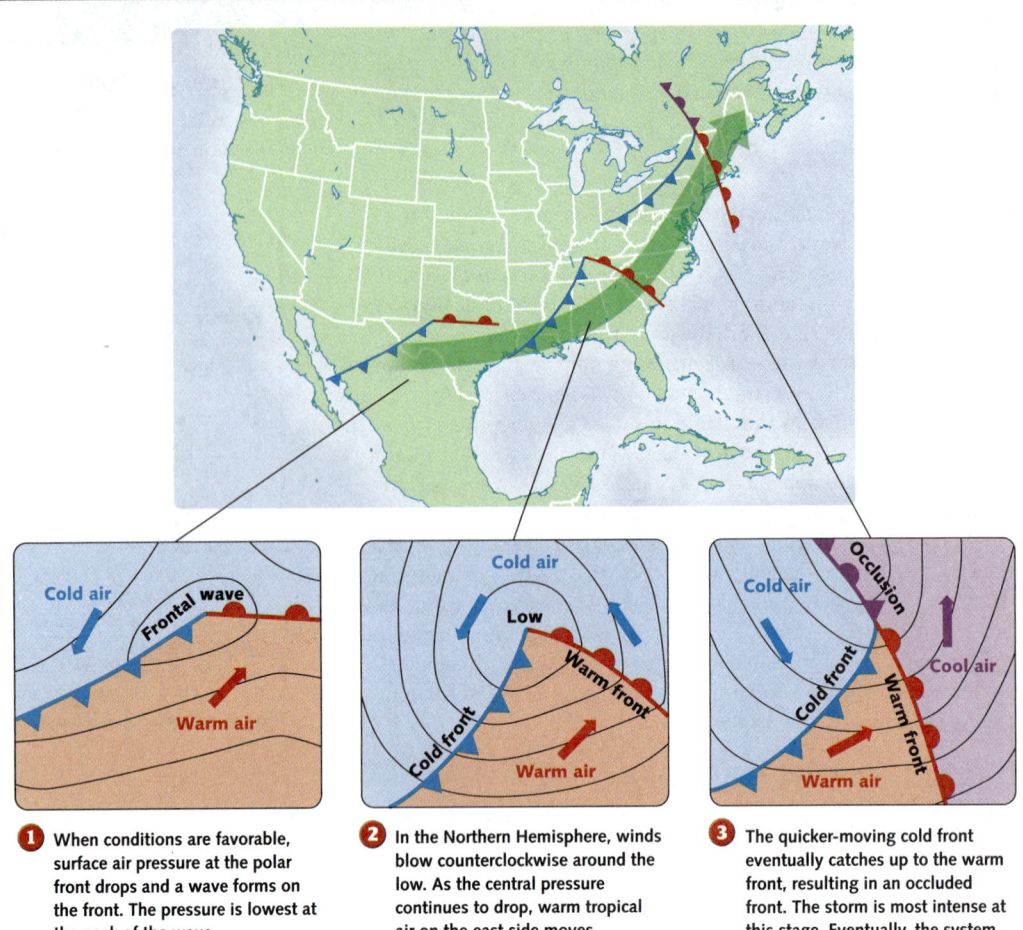

Norwegian Model of Low-Pressure System

① When conditions are favorable, surface air pressure at the polar front drops and a wave forms on the front. The pressure is lowest at the peak of the wave.

② In the Northern Hemisphere, winds blow counterclockwise around the low. As the central pressure continues to drop, warm tropical air on the east side moves northward, and cold polar air on the west side moves southward.

③ The quicker-moving cold front eventually catches up to the warm front, resulting in an occluded front. The storm is most intense at this stage. Eventually, the system loses energy and begins to weaken.

It generally takes 12 to 24 hours for a low-pressure system to pass through the first two stages shown above. After reaching the occluded stage, the low can last for three days or more. Although the original low usually weakens during the occluded stage, meteorologists must continue to monitor the system, since the low can still be strong, with high winds and rains. Sometimes a new low will develop where the cold, warm, and occluded fronts meet.

Not all lows are like the simple one shown in the diagram. Some are associated with more than two fronts, while others are associated with only one front. Yet if you read weather maps over a period of time, you will notice that many low-pressure systems follow a life cycle similar to the one described by the Norwegian model.

442 Unit 5 Atmosphere and Weather

DIFFERENTIATING INSTRUCTION

Support for Visually Impaired Students The diagrams on this page are fairly complex and some students may have difficulty making out the details. Enlarge the diagrams and the accompanying map on a photocopier so students can clearly see all labels and weather symbols. You might add crosshatching, stippling, and shading to the diagrams to distinguish among the three different kinds of air.

Upper-Air Flow

Throughout the life of a low-pressure system, upper-air flow is what controls the surface low's path and intensity. At Earth's surface, air is constantly spiraling into the low-pressure system. You might expect that this addition of air would increase the pressure in the center of the low-pressure system, making it weaker. Instead, though, lows often strengthen over time. The behavior of upper-air flow can help explain this unexpected fact.

If you look at a weather map, you will see that the upper-air flow is seldom straight, but curves back and forth like a winding road. Meteorologists call some bends *ridges* and others *troughs*, depending on the air pressure.

Troughs are like traffic jams in the upper-air flow. As upper-level winds approach a trough, the air crowds together like cars slowing down before a traffic jam. Some of the air sinks to Earth's surface, like cars taking a detour around the traffic jam. This sinking air increases the air pressure at the surface. If more air enters the high from above than spirals out of it at the ground, the high strengthens.

Eventually, the air spreading out from the high at the surface will spiral into a surface low. Here, the air rises and rejoins the upper-air flow beyond the trough. If more air rises out of a surface low than spirals into it at the ground, the low will strengthen.

The amount of air that sinks into surface highs and rises out of surface lows depends both on how sharp a bend the upper-air flow makes and on how fast the upper-air flow is moving. Meteorologists must monitor the upper-air flow to predict whether a surface high or low is likely to strengthen or weaken.

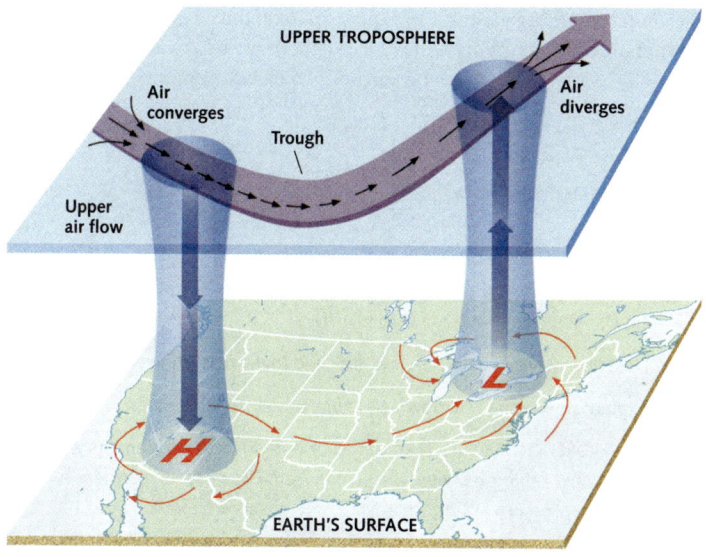

TROUGHS The upper-level flow is similar to cars slowing down and getting closer together (converging) as they approach a traffic jam (the trough), then speeding up and getting farther apart (diverging) after the traffic jam breaks up.

INVESTIGATIONS
CLASSZONE.COM

How Does a Mid-Latitude Low Develop into a Storm System?
Analyze the development of a storm as it moves across North America.
Keycode: ES2003

CHAPTER 20 SECTION 2

INVESTIGATIONS
CLASSZONE.COM

How Does a Mid-Latitude Low Develop into a Storm System?
Keycode: ES2003

Have students work in pairs to relate cloud cover and precipitation to the surface analysis map showing air masses, frontal boundaries, and low-pressure centers.

📖 *Use Internet Investigations Guide, pp. T71, 71.*

More about...

Ridges and Troughs
The isobars on a typical U.S. weather map are useful for identifying strong highs and lows and associated ridges and troughs. A strong pressure system is characterized by more closed isobars than a weak pressure system. When two strong lows are found on a surface weather map, they usually are separated by a ridge, which is a long, narrow region of high pressure. Conversely, when two strong highs are found on a surface weather map, they often are separated by a trough, or a long, narrow region of low pressure.

DIFFERENTIATING INSTRUCTION

Challenge Activity Have students research and write brief reports about upper-air charts and how they are used to monitor upper-air flow. In an upper-air chart, contour lines are used to indicate surfaces of constant pressure. The 500-millibar pressure surface, for example, corresponds to an altitude of approximately 5500 m. The charts allow meteorologists to track the movements of upper-air troughs and ridges and related wind patterns, all of which influence surface weather.

Developing English Proficiency Some students may be unfamiliar with the term *traffic jam* and thus not understand the analogy between a traffic jam and upper-air flow. Point out visual clues in the diagram. The trough, for example, resembles a road. Have students read the caption on page 443 for additional clarification.

CHAPTER 20 SECTION 2

VISUAL TEACHING
Discussion
The colors in the radar image represent different rates of precipitation. In the storm system shown here, the lightest rates of precipitation are shown in green. Yellow represents steady to relatively heavy rainfall or snow. Red corresponds to the heaviest rates of precipitation.

ASSESS

1 A front is the boundary that separates opposing air masses.

2 A warm front is associated with steady rain or snow over a large area. A cold front may be associated with thunderstorms over a narrow band of ground or with little change in weather. The conditions differ because of the air masses involved and the speed at which the fronts are moving.

3 First, air pressure at a polar front drops and a wave forms on the front. Next, warm tropical air on the east side of the front moves northward and cold polar air on the west side moves southward. The cold front catches up to the warm front, forming an occluded front. The resulting storm system eventually loses energy and weakens.

4 As upper air approaches a trough, some of the air sinks, which increases air pressure at the surface. If more air enters the high from above than spirals out from below, the high strengthens.

5 The low-pressure system is east relative to the observer because the winds are blowing counterclockwise around the low.

6 Cold temperatures are expected for Georgia because Georgia is north of Florida. The Florida low is spinning counterclockwise and therefore pulling cold air from the north.

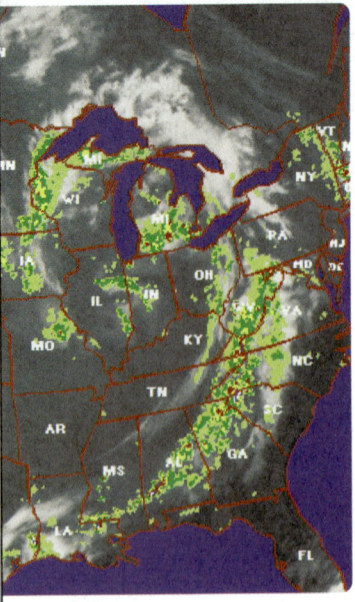

RADAR IMAGE This color radar image of the eastern coast of the United States shows the weather associated with a low-pressure system.

Weather Associated With Pressure Systems

When a low-pressure system passes, the weather you will experience depends on where the center of the low passes in relation to you. For example, in the Northern Hemisphere, if a low passes north of you, a warm front may move over you, followed by a cold front. If the low passes to the south, no fronts will move over you, but you may experience steady snow or rain.

The text below describes the weather that you would experience if a low heading east passes to the north of you.

- As the warm front approaches from the west, cirrus clouds lead to cirrostratus, altostratus, nimbostratus, and stratus clouds. Steady snow or rain, followed by drizzle, marks the front's approach.
- When the warm front passes, the temperature warms, winds shift, and the skies may slowly clear. If the air is humid, showery precipitation may occur, particularly nearer the center of the low.
- As the cold front approaches, it is preceded by scattered showers and possibly thunderstorms.
- As the cold front passes, the temperature drops, winds shift again, and the sky clears.

High-pressure areas are associated with fair weather. Because the air in a high is sinking, the skies are clear. Days may be hot, depending on the ground cover, and nights may be cold as heat radiates out to space. Inversions can form in the mornings, trapping pollution until the sun warms the ground.

Winds blow outward from a high. In the Northern Hemisphere, the winds spiral outward in a clockwise direction. There is little or no wind in the center of the high. The still air there takes on the characteristic temperature and humidity of the area, so highs are where air masses generally form. Whereas a low is surrounded by two or more air masses, a high represents a single air mass.

20.2 Section Review

1 What is a front?

2 Describe the weather conditions associated with warm and cold fronts. Why do these conditions differ?

3 Describe the life cycle of a mid-latitude low-pressure system.

4 Explain how upper-air flow can cause a high to strengthen.

5 **CRITICAL THINKING** If an observer in the Northern Hemisphere sees snow falling heavily with winds from the northeast, where is the low-pressure system relative to the observer? Explain your reasoning.

6 **APPLICATION** A low is predicted to pass through Florida in the winter. Do you forecast warm or cold temperatures for Georgia? Why?

MONITOR AND RETEACH

If students miss . . .
Question 1 Reteach the definition of *front* on page 439.

Question 2 Reteach "Kinds of Fronts." (pp. 440–441) Have students focus on the diagram on page 440.

Question 3 Have students reexamine the diagram on page 442.

Question 4 Reteach "Upper-Air Flow." (p. 443) Ask: Does sinking air increase or decrease air pressure at the surface? increase

Questions 5, 6 Use part 2 of the diagram on page 442 to discuss the effect of the counterclockwise air flow around the low.

Thunderstorms and Tornadoes

Storms come in various sizes. Mid-latitude storms, such as the low-pressure systems you studied in Section 20.2, can be 2000 kilometers across, and hurricanes, hundreds of kilometers across. In contrast, **thunderstorms**—a storm with lightning, thunder, rain, and sometimes hail—may be only a few kilometers across. A storm's size is not a good indicator of the damage it can cause. For example, a thunderstorm can produce a deadly tornado along a narrow path, whereas a typical mid-latitude low may produce only rain or snow over a large area. In this section, you will learn about some of the more violent types of storms.

Thunderstorms

Thunderstorms, and their cumulonimbus clouds, form in warm, moist, unstable air. The storm clouds may attain heights of up to 20 kilometers in the atmosphere, and the weather they bring may include torrential rain, damaging winds, lightning, thunder, hail, and tornadoes.

Although thunderstorms can occur at any hour, they often occur during the afternoon because surface warming throughout the day causes the air to become unstable. Within this unstable atmosphere, a thunderstorm develops like this: Some "trigger" forces part of the air to begin rising. The trigger can be a mountainside or a front standing in the way of the unstable air, or the air can collide with an opposing wind. As the rising air reaches the condensation level, the heat released during condensation makes this air warmer and less dense than the surrounding air. A cumulonimbus cloud quickly grows (as described in Chapter 18), and a thunderstorm begins.

20.3

KEY IDEA
Thunderstorms, which form in warm, moist, unstable air, can result in destructive weather, including tornadoes.

KEY VOCABULARY
- thunderstorm
- squall line
- supercell
- lightning
- tornado

Observe an animation of a thunderstorm.
Keycode: ES2004

Life Cycle of a Thunderstorm Cell

① CUMULUS STAGE Air rises and a cumulus cloud forms. The rising air is called an updraft. The updraft prevents precipitation from reaching the ground.

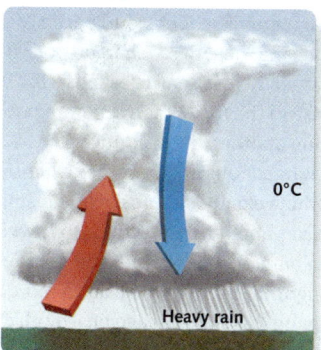

② MATURE STAGE The precipitation becomes heavy enough to fall through the updraft and reach the ground. The falling precipitation creates a downdraft.

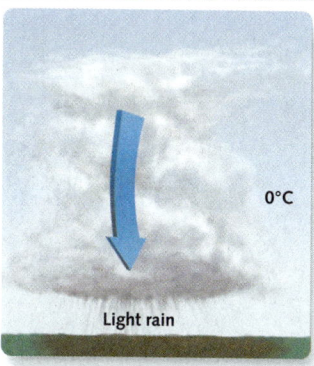

③ DISSIPATING STAGE The downdraft weakens the updraft, eventually cutting off the supply of moist air rising to the cloud. The cloud begins to evaporate.

CHAPTER 20 SECTION 3

More about...

Thunderstorms

Thunderstorms strike some areas far more frequently than others. In general, the regions between 60° latitude and the poles do not receive enough intense insolation to produce convection cells. Alaska, for example, experiences an average of only ten thunderstorms per year, and areas farther north may have only one thunderstorm every several years. Other places receive sufficient insolation but are too dry for a towering cumulonimbus cloud to develop. For this reason, low-latitude deserts may experience an average of only five thunderstorms annually. In the contiguous United States, the southeastern portion of the country receives the most thunderstorms each year, in part because of the movement of maritime tropical air masses.

SAFETY TIPS
LIGHTNING

Stay or go indoors, and stay away from ungrounded appliances, electrical cords, metal pipes, and running water. Do not use the phone.

If caught outside, get into a hardtop car, not a convertible. Do not touch the metal frame.

Avoid tall objects like lone trees, flagpoles, and telephone poles. Avoid metal objects like umbrellas, bicycles, and wire fences. If caught in the open, squat low to the ground. Do not lie down.

Keep away from bodies of water, and do not stand in water. Watch out for flooding in low-lying areas.

NEVER stay under a lone tree.

Learn more about lightning safety.
Keycode: ES2005

Thunderstorms consist of one or more convection cells, each of which usually has a lifetime of an hour or less. The life cycle of a single thunderstorm cell is shown in the illustration on page 445. The formation of new cells may extend the lifetime of a thunderstorm. Some multicell thunderstorms may cover a whole state and last as long as a day.

Thunderstorms often form along fronts because the frontal boundary forces air to rise. Frontal thunderstorms are associated with large-scale low pressure systems and may start and stop for days.

Frontal thunderstorms often occur in lines hundreds of kilometers long along the frontal surface. They also occur in lines ahead of the front, called **squall lines.** Strong winds often precede squall lines. This happens because the falling rain within a thunderstorm causes the air to cool. Since cold air is denser than warm air, the cold air sinks and spreads out ahead of the squall line. The sinking air within a thunderstorm is known as a downdraft. The downdraft of the thunderstorm's mature stage can transport upper-level winds to Earth's surface. Because frontal systems are often associated with strong winds aloft, frontal thunderstorms can cause severe wind damage. If a front or squall line moves slowly, its heavy rains can cause flooding.

Supercells are very large single-cell thunderstorms with particularly strong updrafts. The updraft of a supercell rotates, which can cause a tornado to develop. It is the supercell thunderstorm that produces the highest wind, the most damaging hail, and the most destructive tornadoes.

Lightning

All thunderstorms produce lightning. **Lightning** is a discharge of electricity from a thundercloud to the ground, to another cloud, or to another spot within the cloud itself. Lightning can also occur in the clouds of snowstorms, dust storms, or volcanic eruptions. No one is completely sure how clouds become electrically charged. Most scientists believe that when larger and smaller cloud particles collide within a cloud, the larger ones become negatively charged and the smaller ones positively charged. Most larger, heavier particles fall to the cloud's bottom, while most smaller ones rise to the top. When separate positive and negative charges flow toward one another, they produce a spark, which travels along the narrow path of charged ions and becomes a lightning bolt. The lightning heats the air, which expands explosively, causing the sound of thunder. Because sound takes about three seconds to travel a kilometer, and light travels at 300,000 kilometers per second, you see the lightning before you hear the thunder. Lightning can heat the air to more than 25,000°C. It can severely burn any living thing or damage nearby electrical equipment.

Lightning moving from a cloud to the ground often strikes the highest point available, and is more likely to strike protruding objects than flat ones. Once lightning strikes the ground, it may flow along the surface for a distance. Because lightning is an electrical charge, it flows best through materials that conduct electricity. These include metal, water, and wet ground.

LIGHTNING bolts like these seen in a time-lapse photograph over Tucson, Arizona, are formed when separate positive and negative charges in a thundercloud flow toward one another and create a spark.

446 Unit 5 Atmosphere and Weather

DIFFERENTIATING INSTRUCTION

Challenge Activity Have students search earth science Web sites that address commonly asked questions about thunderstorms. Have the class brainstorm some questions about thunderstorms. Then have them search the Web sites to find answers to their questions. For example, students may want to know how often lightning strikes or what the odds are of getting struck by lightning. Have students compile the facts into a booklet or, if your school has a Web site, post the answers there.

Tornadoes

A thunderstorm's most-destructive possible byproduct is a tornado. A **tornado** is a violently rotating column of air that usually touches the ground. Because tornadoes can cause severe property damage, personal injury, and loss of life, scientists are trying to determine how they form so that they will be easier to predict.

Tornado Formation

Because tornadoes are destructive and often unpredictable, they are difficult to study directly. Scientists have tried to place instruments in the paths of tornadoes, but it is difficult to do so without damaging the instruments and without putting people in danger. Predicting the path of a tornado also is difficult.

By observing from a distance, scientists have discovered that for a severe thunderstorm to produce a tornado, it must contain a rotating updraft called a mesocyclone. Such an updraft occurs when low-altitude winds are blowing at a different speed and in a different direction than winds higher up.

Before a tornado develops, the rotating clouds of the mesocyclone may become visible at the base of the storm and may lower to form a wall cloud. About 10 to 20 minutes later, a tornado may descend from the wall cloud. However, only a third of all mesocylones produce tornadoes, and a tornado may form without a visible mesocyclone or wall cloud.

TORNADO Sometimes referred to as a "twister," the rotating column of air at the base of a tornado is visible because it contains either moisture or dust and debris.

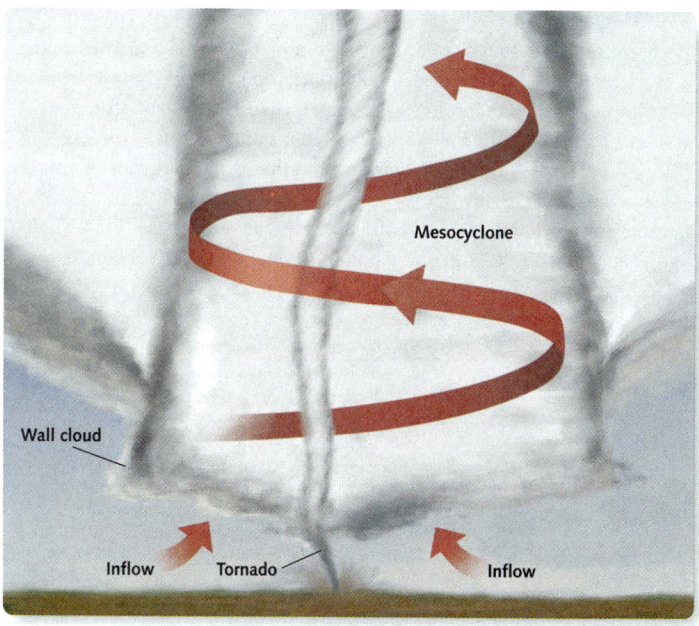

TORNADO FORMATION A rotating updraft of air may form a spinning column called a mesocyclone, which can eventually touch down on the ground as a tornado.

Examine an animation of a tornado.
Keycode: ES2006

CHAPTER 20 SECTION 3

DATA CENTER
CLASSZONE.COM

Learn more about lightning safety.
Keycode: ES2005

VISUALIZATIONS
CLASSZONE.COM

Examine an animation of a tornado.
Keycode: ES2006

Have students describe the stages of tornado formation in sequence.

 Visualizations CD-ROM

VISUAL TEACHING
Discussion

Use the diagram to discuss tornado formation. Ask: In the illustration, is air pressure at the surface increasing or decreasing? How do you know? **Air pressure is decreasing. Air is flowing upward into the storm system, creating lower pressure at the surface.** In which direction is the mesocyclone rotating? **counterclockwise** In which hemisphere is the storm occurring? **the Northern Hemisphere** Explain. **Lows rotate in a counterclockwise direction in the Northern Hemisphere.**

Extend

Tell students that highs are often referred to as anticyclones. Ask them to infer why. **Lows are called cyclones. Highs are the opposites of lows and thus are called anticyclones.**

DIFFERENTIATING INSTRUCTION

Support for Hearing-Impaired Students Each time you engage in a lengthy oral explanation or a classroom discussion, assign one student the task of taking notes. The notes should be dated and identified by topic, and a different student should take notes each time. Provide photocopies of the notes to students who are hearing impaired, as well as any other students who might benefit from reading the notes.

More about...

Tornado Damage
In addition to developing the tornado intensity scale, Professor Tetsuya "Ted" Fujita of the University of Chicago has extensively researched patterns of tornado damage. His studies indicate that some tornado funnels develop suction vortices, which can be thought of as smaller funnels that rotate around a common center. Suction vortices move with the storm, but they rotate at much higher speeds than the parent funnel and thus are far more destructive. Because they are only a few meters wide, suction vortices are likely responsible for the erratic damage associated with many tornadoes, wherein a funnel can pass over a building leaving some parts unscathed and other parts destroyed.

TORNADO DAMAGE Because a tornado can be only meters across, it can destroy some buildings completely while leaving structures next door virtually untouched.

A tornado often appears as a vortex or a funnel-shaped cloud of flying debris. Some tornadoes are rope shaped, and others wedge shaped with more than one vortex. A tornado's funnel cloud results when the air pressure at its center is very low and air sucked into the funnel expands and cools; water vapor in the air condenses, forming a funnel-shaped cloud. If the air is drier, or the pressure inside the tornado is higher, the tornado may consist of a cloud of dust and debris, creating a loud roaring sound. Tornadoes can be just a few meters or more than a kilometer wide.

A tornado usually appears at the back edge of its parent thunderstorm and travels with the thunderstorm. Because upper-level winds often guide thunderstorms to the northeast, tornadoes often appear at the southwest edges of storms. However, tornadoes can travel in any direction and often move erratically. If a tornado is approaching, it may be hidden behind rain falling from the main body of the thunderstorm.

Tornadoes can occur anywhere throughout the United States and at any time of year. However, they occur most often in an area called tornado alley in the spring and early summer. Tornado alley extends from Texas northward to South Dakota. The increased sunlight of spring and early summer warms the air near the ground while the air higher up stays quite cold. The temperature difference makes the air unstable, resulting in severe thunderstorms. Wind conditions in tornado alley can be ideal for the formation of mesocyclones within the storms because wind direction often changes with elevation. Warm, moist air may blow from the south near the ground, while cool, dry air may blow from the southwest or west higher up.

Effects of Tornadoes

The weakest tornadoes can damage chimneys, break tree branches, and blow over signs, while the most violent ones can rip sturdy houses from their foundations. Meteorologists classify tornadoes as weak, strong, or violent according to the Fujita scale, which is named for the tornado researcher Tetsuya "Ted" Fujita. Weak tornadoes, which are most common, are strong enough to down trees or shift mobile homes off their foundations. A violent tornado, which is rare, can destroy everything in its path, causing fatalities and doing millions of dollars' worth of damage.

Fujita Tornado Intensity Scale

Category	F-Scale	Estimated Wind Speed	Effects
Weak	F0	65–118 km/h	Minor damage, snaps small branches, breaks some windows
	F1	119–181 km/h	Downs trees, shifts mobile homes off foundations
Strong	F2	182–253 km/h	Rips roofs off houses, destroys mobile homes, uproots large trees
	F3	254–332 km/h	Partially destroys buildings, lifts cars
Violent	F4	333–419 km/h	Levels sturdy buildings, tosses cars
	F5	420–513 km/h	Lifts and transports sturdy buildings

DIFFERENTIATING INSTRUCTION

Hands-On Demonstration After students have read about Doppler radar on page 449, use a spring to demonstrate the Doppler effect. Hold one end of the spring and have a student, the receiver, hold the other so that the spring is stretched out. Walk toward the student while the class observes the compression of the spring. Tell students that the Doppler effect refers to the change in wave frequency that occurs when a source of energy is in motion relative to a receiver or an observer. Ask: When a storm is moving toward a radar receiver, would the reflected waves of energy be compressed or stretched out? **compressed** Does the frequency of the waves increase or decrease? **increase**

Predicting Tornadoes

Improvements in radar have helped meteorologists to spot and even predict tornadoes more reliably. They use conventional radar to create a map of precipitation in the surrounding area, and newer, Doppler radar can identify which way winds are moving within a storm.

Conventional radar works by emitting pulses of microwaves, some of which bounce back to the radar receiver when they hit drops of precipitation. The time it takes the microwaves to return indicates how far away the precipitation is.

A tornado often appears on conventional radar as a hook-shaped area of precipitation. However, sometimes the hook does not form until after the tornado touches ground; sometimes it does not form at all. Thus conventional radar is not always reliable for predicting tornadoes.

Doppler radar uses a principle called the Doppler effect to identify mesocyclones that may produce tornadoes. The Doppler effect is what causes the sound of a car to drop in pitch when the car drives by you quickly. When the car is approaching, its sound seems higher pitched than when it is going away from you. Meteorologists can use Doppler radar to tell whether a part of a storm cloud is moving toward or away from the radar receiver. By analyzing how the air in the cloud is moving, meteorologists can identify a rotating mesocyclone and give people in the area about 20 minutes advance warning of tornado formation.

Storm or Tornado Watches and Warnings

The National Weather Service issues watches and warnings to give people time to find shelter before a severe thunderstorm or tornado arrives. A severe thunderstorm is one that has wind gusts of at least 80 kilometers per hour, hail that is at least two centimeters in diameter, or a funnel cloud or tornado. If there is a chance that a severe thunderstorm will form, meteorologists issue a thunderstorm watch. If a severe thunderstorm approaches, they issue a thunderstorm warning for the areas in its path. If conditions are such that a tornado may form, they issue a tornado watch, and volunteer tornado watchers in the area begin looking for tornadoes. If they spot one, or if meteorologists detect one using radar, radio and television stations broadcast a tornado warning. In some communities, a siren sounds. Although the tornado warning system is not completely effective, deaths from tornadoes have decreased over the past 30 years despite a significant increase in population.

20.3 Section Review

1. What conditions are necessary for the formation of a thunderstorm?
2. What is a mesocyclone?
3. What should you do if a tornado is approaching?
4. **CRITICAL THINKING** Can thunder occur without lightning? Explain.

SAFETY TIPS
TORNADOES

Stay or go indoors.

If possible, go quickly to a basement or storm cellar. Otherwise, go to a small inner hallway or room without windows, such as a bathroom or closet, on the ground floor of a strong building. Avoid large spaces like auditoriums and malls. Stay away from windows and outside walls. Get under a mattress or sturdy piece of furniture and hold onto it. Protect your head and neck.

Avoid mobile homes and cars.

Do not try to outdrive a tornado.

If you are caught outside, lie in a ditch or low-lying area. Protect your head and neck. Watch for flooding.

Learn more about tornado safety.
Keycode: ES2007

CHAPTER 20 SECTION 4

▶ Focus

Objectives
1. Describe the formation and effects of hurricanes, and the measures taken to mitigate their damage.
2. Describe the weather conditions associated with mid-latitude lows during winter.

Set a Purpose
Read aloud the key ideas. Then have students read this section to answer these questions: "What is a hurricane? How does it differ from a winter storm?"

▶ Instruct

More about...

Hurricane Structure
The diameter of hurricanes varies from 150 to 1500 km. In comparison, the diameter of a hurricane's eye is small—usually within the range of 22–40 km. Some powerful hurricanes, however, have had very small eyes. For example, Hurricane Camille, classified as a category 5 on the Saffir-Simpson Hurricane Scale, had an eye that measured only 6.5 km across. Camille struck the Gulf Coast in 1969 and caused an estimated $5.2 billion in damages (1990 dollars).

20.4

Key Ideas
Hurricanes develop over warm water, fueled by the heat water releases during condensation.

Snow, wind, and freezing-cold temperatures can occur with mid-latitude lows in winter.

Key Vocabulary
- hurricane
- storm surge
- Saffir-Simpson scale
- blizzard

Hurricanes and Winter Storms

While thunderstorms and tornadoes are short-lived and cover a relatively small area, a hurricane or winter storm can last for a day or more and affect an area several hundred kilometers wide.

Hurricanes

A **hurricane** is a large rotating storm of tropical origin that has sustained winds of at least 119 kilometers per hour. The air pressure at the center of a hurricane is very low. Unlike a mid-latitude low, which gets its energy from the temperature contrast between air masses, a hurricane gets its energy from the heat of surface ocean water. Hurricanes have different names, depending on where they occur. For example, when they occur in the Pacific Ocean, they are called typhoons.

Hurricanes have a spiral structure, as seen in the satellite image below. At a hurricane's outer edge, rain and wind are comparatively mild. They increase as the winds spiral in toward the low-pressure area, or eye, at the hurricane's center. Winds and rain are strongest at the eye wall surrounding the eye, where the air rises to form giant thunderstorms. When the eye wall passes over an area, great damage can result. Inside the eye, winds are mild and there is no rain.

HURRICANE LINDA This computer-enhanced satellite image shows the structure of the hurricane as it approaches Baja, California, in 1997.

EYE The air inside a hurricane's eye is sinking, so the sky is almost cloudless.

FEEDER BANDS Often, bands of thunderstorms known as feeder bands spiral in toward the center of the hurricane.

EYE WALL A ring of violent thunderstorms surrounds the eye. The strongest winds and heaviest rain are found here.

DIFFERENTIATING INSTRUCTION

Developing English Proficiency Develop a simple language lab for students. Have a student read aloud the first page of this section into a tape recorder. Tell the student to pause for several seconds after each sentence. Lend the tape to students who are developing their English language skills. They can play the tape at home and practice proper inflection by repeating each sentence aloud.

Hurricane Formation

For a hurricane to form, there must be a supply of warm, moist air for a long period of time. Water that evaporates from the ocean later condenses within storm clouds, releasing large amounts of heat. This heat fuels the gradual development of a full-blown hurricane.

Hurricane formation begins in the tropics with a mild atmospheric disturbance that causes humid air to rise. As the air rises, it cools, and water condenses, releasing heat. Humid air flows in at the surface to replace the rising air, and as this humid air rises, more water condenses and releases heat. This process continues for as long as humid air is available at the surface and air is leaving the disturbance at higher levels.

For a mild atmospheric disturbance to develop into a hurricane, it must begin to rotate. The Coriolis effect causes air flowing into the disturbance at the surface to flow counterclockwise in the Northern Hemisphere and clockwise in the Southern Hemisphere. Near the equator, the Coriolis effect is so weak that hurricanes do not form. Consequently, hurricanes are born over warm water, generally between 5° and 20° latitude.

Scientists classify a tropical disturbance according to the speed of its strongest winds. A tropical depression has wind speeds of up to 61 kilometers per hour, a tropical storm's strongest winds are between 61 and 119 kilometers per hour, and a hurricane has wind speeds of 119 kilometers per hour or greater. Each year, only a few tropical depressions develop into hurricanes.

Because hurricanes rely on the transfer of heat from the ocean, they form only when surface ocean waters are sufficiently warm, and they weaken as soon as they make landfall or move over cold ocean water. Surface ocean water is warmest in summer and early fall, so the hurricane season in the United States is usually June through November.

Throughout their development, hurricanes are steered by the global wind patterns that you read about in Sections 19.3 and 19.4. Hurricanes that form in the Atlantic Ocean north of the equator initially move west or northwest, but then often curve toward the north and may eventually head east. However, the actual paths of hurricanes can vary considerably, and erratic changes in a path can take forecasters by surprise.

Effects of Hurricanes

Although a hurricane weakens rapidly once it makes landfall, it can still cause a tremendous amount of damage along the coast and inland. The damage can include winds, inland flooding, and a large wave of water called a storm surge.

A **storm surge** results, in part, from the strong winds of the eye wall, which blow water into a broad dome. When the hurricane makes landfall, this dome of water can raise the sea level several meters higher than it would be otherwise. When the storm surge coincides with high tide, the sea level can be dangerously high, swamping low-lying coastal areas. Hurricane winds also create huge waves, which batter the shore. One of the greatest hurricane disasters in United States history was the

Observe an animation of a hurricane.
Keycode: ES2008

SAFETY TIPS
HURRICANES

Listen to weather updates. Gather a portable radio, fresh batteries, flashlights, drinking water, food, and medicines. Set refrigerator to high and open only if necessary.

Outdoors, secure loose objects and moor boats. Shutter or board up windows. Secure doors. If in a mobile home, check tie-downs and go to a shelter.

Have a flood-free evacuation route planned. If ordered to evacuate, do so immediately. Shut off water and electricity.

During the hurricane, stay indoors, away from windows. After, beware of downed wires, unsafe roads, flooded areas, animals sheltering indoors, and broken gas lines.

Learn more about hurricane safety.
Keycode: ES2009

CHAPTER 20 SECTION 4

DISCUSSION

Tell students that in 1999, the National Weather Service issued an evacuation advisory in response to Hurricane Floyd, which was expected to strike the southeastern coast of the United States. More than 2 million people left their homes, and highways were snarled with traffic jams that stretched for miles. Hurricane Floyd did cause extensive damage in North Carolina, but it largely bypassed states such as Florida. As a result, many people questioned whether evacuation was necessary. Have students debate whether, given the unpredictable nature of a hurricane's path, they would heed an evacuation advisory. Lead students to understand that the inconvenience of an evacuation is a small price to pay to avoid loss of life.

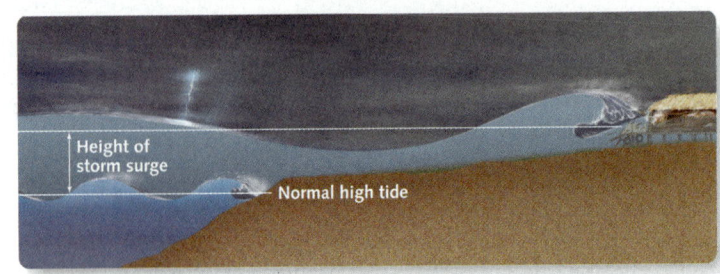

STORM SURGE When a storm surge coincides with high tide, it can raise the sea surface to dangerous levels.

great Galveston, Texas, hurricane of 1900, which produced a storm surge that killed 6000 people.

Hurricane Watches and Warnings

Meteorologists can track a hurricane throughout its life by satellite. When they expect a hurricane to arrive in a community within 24 to 36 hours, they issue a hurricane watch. When they expect the hurricane within 24 hours or less, they issue a hurricane warning. Because a hurricane's path is often erratic, meteorologists must issue the warning for a stretch of coastline that is much wider than the size of the hurricane path. Along the East Coast, meteorologists also announce the percent chance that the eye of the hurricane will pass within 105 kilometers of a particular community.

Because such a large area of the coast must receive each hurricane warning, often people feel they have unnecessarily spent time and money on preparation if the hurricane does not hit their community. However, loss of life due to hurricanes has decreased significantly over the last century. Property loss, on the other hand, has increased significantly, probably because of the population increase in coastal areas.

Saffir-Simpson Hurricane Scale

Category	Wind Speed (km/h)	Storm Surge (m)	Damage
1	119–153	1.0–1.7	Minimal. Trees and unanchored mobile homes damaged. Some coastal flooding.
2	154–177	1.8–2.6	Moderate. Minor damage to buildings. Some trees blown down.
3	178–209	2.7–3.8	Extensive. Some structural damage to small buildings. Mobile homes destroyed.
4	210–250	3.9–5.6	Extreme. Some roofs destroyed. Evacuations as far as 10 kilometers inland.
5	>250	>5.7	Catastrophic. Buildings destroyed. Evacuations as far as 16 kilometers inland.

DIFFERENTIATING INSTRUCTION

Reading Support Help students become more active readers by having them summarize the information in this section. Students can write brief paragraphs about hurricane formation, hurricane effects, hurricane safety, and winter storms. Afterward, review the paragraphs and choose several examples of particularly well-developed summaries. Look for those that zero in on key ideas. Read the examples aloud to the class to help students who have difficulty organizing their thoughts. *Use Reading Study Guide, p. 71.*

Meteorologists rate a hurricane's strength using the **Saffir-Simpson scale,** shown on page 452. This scale can help predict the damage that will occur when the hurricane makes landfall.

Winter Storms

A mid-latitude-low pressure system occurring in winter, known as a winter storm, can bring several types of weather. It can produce heavy snow, ice, or rain depending on the storm track, its intensity, and the temperature of the atmosphere. Often, a single region will have several zones containing different types of precipitation. The northern zone may produce only snow. Slightly to the south might be a zone of freezing rain and sleet. Farthest south and closest to the warm front, only rain may occur.

One special breed of winter storm is a blizzard. Although people often use the term to refer to heavy snowstorms, a true blizzard involves more than heavy snow. A **blizzard** is a winter storm characterized by high winds, low temperatures, and falling or blowing snow. To be classified as a blizzard, a storm must meet three criteria:

- The winds must exceed 56 kilometers per hour.
- The temperature must be −7°C or lower.
- The falling and/or blowing snow must reduce visibility.

A severe blizzard is one with winds greater than 72 kilometers per hour and temperatures of −12°C or lower.

In the United States, blizzards occur most often in the northern Great Plains states. However, they have also occurred, though rarely, as far south as Texas.

On the East Coast, storms known as nor'easters produce extremely heavy snow when a low-pressure system over the North Atlantic Ocean causes maritime polar air to blow over the land from the northeast. However, the temperatures are usually not cold enough for the storm to qualify as a blizzard. A major exception was the blizzard of 1888. Snowfall amounts from this nor'easter ranged from 50 to 100 centimeters, with drifts much higher. The blizzard brought life to a standstill from Philadelphia, Pennsylvania, to Albany, New York. Although reports vary, as many as 400 people were killed, half of the deaths occurring in New York City alone. The blizzard of 1998 has not been equaled since.

BLIZZARD When a blizzard strikes, it may last for days and dump several feet of snow, the removal of which takes several more days.

SAFETY TIPS
WINTER STORMS

Be sure you have food that doesn't need cooking, flashlights, batteries, medicines, fire extinguishers, and a way to safely make heat if electricity or gas is cut off. Check smoke detectors.

Stay indoors and dress warmly. Turn faucets on to drip. Conserve fuel.

If you are caught outside, find shelter from the wind and stay dry. Exercise to stay warm but avoid sweating. Avoid walking on ice. Cover your mouth. Watch for signs of frostbite and hypothermia.

If you are in a car, leave it running only if the exhaust pipe is clear of snow. Do not drive in icy conditions.

Learn more about winter storm safety.
Keycode: ES2010

20.4 Section Review

1. Describe how a hurricane forms.
2. What is the difference between a snowstorm and a blizzard?
3. **CRITICAL THINKING** Explain why heavy rain accompanies a hurricane.
4. **BIOLOGY** Describe some positive and negative ways in which hurricanes may affect plants.

Chapter 20 Weather 453

CHAPTER 20 SECTION 4

Learn more about winter storm safety.
Keycode: ES2010

More about...

The Blizzard of 1888
The nor'easter that occurred along the East Coast in March 1888 struck New York State particularly hard. Roughly half of the 400 recorded deaths during the four-day storm occurred in New York City. Gravesend, New York, recorded the highest drift at 15.5 m.

ASSESS

1. A mild atmospheric disturbance causes humid air in the tropics to rise, cool, and condense, releasing heat. The disturbance begins to rotate. If a steady supply of warm, moist air continues to flow in at the surface to replace the rising air, the atmospheric disturbance can develop into a full-blown hurricane.

2. A snowstorm involves a steady snowfall. A blizzard is a winter storm characterized by high winds, low temperatures, and falling or blowing snow.

3. Air rises to form a ring of violent thunderstorms around the eye of the hurricane.

4. **Sample Response:** The storm surge of a hurricane can carry sand and other materials inland, creating new habitats for plants. At the same time, the high winds of a hurricane can uproot and destroy existing plants.

MONITOR AND RETEACH

If students miss . . .

Question 1 Reteach "Hurricane Formation." (p. 451) Ask: What fuels the gradual development of a tropical disturbance into a hurricane? *heat released during condensation*

Question 2 Ask students to reread the bulleted points on page 453.

Question 3 Have students examine the image on page 450. Ask: What surrounds the eye of the hurricane? *a ring of violent thunderstorms that makes up the eye wall*

Question 4 Review "Effects of Hurricanes." (pp. 451–452) Have students work in small groups to make a chart that lists the positive and negative effects of hurricanes on plants.

CHAPTER 20 SECTION 4

Explore more about hurricanes.
Keycode: ES2011

Science and Technology

Hurricane Hunters

Hurricane hunting actually began on a dare during World War II when an Air Force officer successfully flew a training aircraft into the eye of a hurricane. One year later in 1945, the 53rd Weather Reconnaissance Squadron was given the mission of collecting weather data to aid the war effort; part of this mission included flying into hurricanes. The 53rd squadron has been known ever since as the Hurricane Hunters. Following the war, the squadron continued to gather hurricane data. Today, the 53rd Squadron shares its duties with the 815th Weather Reconnaissance Squadron, also known as the Storm Trackers.

Science Notebook

Safety measures take time to implement, and that these safety measures are based largely on the severity of the storm. In some cases, people can simply take shelter. In more severe cases, entire cities may have to be evacuated.

SCIENCE & Technology

Extreme Science: Flying into the Storm

Unlike a tornado, a hurricane can span hundreds of kilometers, and can damage thousands of square kilometers of land. Because of this potential for destruction and loss of life, meteorologists need every type of technology available to predict a hurricane's severity before it reaches the shoreline.

Since weather-model predictions can vary, how can a forecaster know for certain whether an approaching hurricane is extremely dangerous? Is there a way meteorologists can obtain absolutely accurate measurements?

Satellite photography can give a meteorologist some idea of a hurricane's strength as it approaches land. However, the only foolproof way to determine the severity of a hurricane is to fly into it. Pilots known as hurricane hunters fly directly into the storms, in specially designed aircraft. Instruments on the planes gather and feed out information on the wind strength in the eye wall of the hurricane and on the sea-level pressure in the eye.

The eye wall is the most intense part of the hurricane, consisting of violent winds and torrential rains. It surrounds the storm's eye, the central region where the storm is relatively mild, and has lower wind speed and less precipitation.

For a hurricane hunter to take measurements in the eye, the plane must "punch through" the eye wall. Once in the eye, the plane drops an instrument known as a dropsonde. A dropsonde measures the barometric pressure, which indicates the hurricane's strength. After obtaining this measurement, the pilot relays it to the National Hurricane Center, which then issues statements based on the measurement. ■

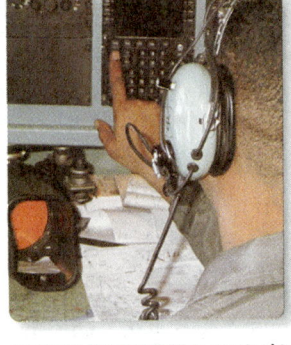

CAREFUL NAVIGATION is required to fly safely through a hurricane and to make the desired scientific observations.

Extension

SCIENCE NOTEBOOK

Since all hurricanes are dangerous, why is it important to know the exact severity of one that is approaching? Consider how you might respond differently to different hurricane forecasts.

Explore more about hurricanes.
Keycode: ES2011

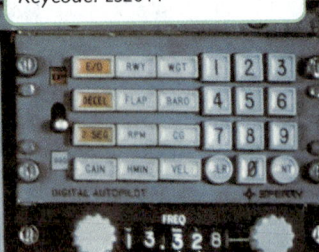

A HURRICANE HUNTER uses radar (bottom center) and other instrumentation to navigate near the eye of a hurricane (left).

454 Unit 5 Atmosphere and Weather

Forecasting Weather

Weather forecasting is a complex process that requires the cooperation of countries around the world. First, meteorologists worldwide gather weather information for their regions. Then they send this information to computer centers, where other meteorologists compile it and then use it to create weather maps and forecasts. Finally, the weather maps and computer-generated forecasts go out to local forecasters, who modify the forecasts to reflect local conditions.

Gathering Data

Huge amounts of data are necessary for weather forecasting. The sources of these data include satellites, instruments attached to balloons, weather stations, weather radar, airplanes, and ships.

Satellites

Satellite images provide weather information about every spot on Earth, even oceans and sparsely populated areas. Meteorologists use two basic types of satellite images: visible images and infrared images.

A visible satellite image is basically a black-and-white picture of Earth. The white color shows sunlight reflected off clouds or snow cover. The brighter the white, the thicker the clouds. Gray often represents land, and black often represents water. By tracking the movement of clouds in visible satellite images, meteorologists can estimate wind speed and direction or can track storms. They can also use visible satellite images to estimate the stage and severity of hurricanes. A disadvantage of visible satellite images, however, is that they aren't available at night.

In contrast infrared (IHN-fruh-REHD) satellite imagery uses temperature, rather than light, to create pictures. Therefore, infrared images can be taken day or night. On an infrared satellite image, bright areas represent cold temperatures and usually indicate the cold, high tops of clouds. Darker areas are warmer and represent lower clouds or a lack of clouds. Because temperature decreases with height, meteorologists can use the temperatures of cloud tops to determine how tall the clouds are. The taller the cumulus clouds, the more severe the thunderstorms they produce. By tracking clouds on infrared images, meteorologists can estimate wind speed and direction occurring at different altitudes. Infrared satellites can also provide images of water vapor in the air, which help to determine temperature and humidity at different altitudes.

Radiosondes

Meteorologists use devices called radiosondes to measure the temperature, pressure, and humidity of air at different altitudes. A radiosonde is an instrument package attached to a balloon, which carries it up into the atmosphere. The radiosonde transmits weather information to the ground, where computers track the instrument's position. Temperature and humidity data from radiosondes can help meteorologists estimate air stability and the likelihood that cumulus clouds and thunderstorms will

20.5

KEY IDEA

To prepare accurate weather forecasts, meteorologists must gather, distribute, and analyze huge amounts of atmospheric data.

KEY VOCABULARY
- station model

VISIBLE SATELLITE IMAGE This visible image shows thunderclouds over Ohio.

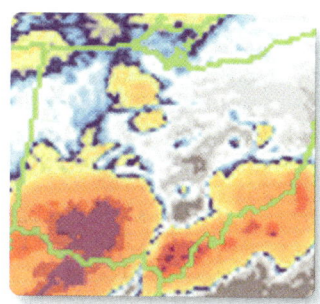

INFRARED SATELLITE IMAGE The various colors in this image indicate regions of differing temperatures.

Chapter 20 Weather 455

CHAPTER 20 SECTION 5

FOCUS

Objectives
1. Compare and contrast the different technologies used to gather weather data.
2. Analyze weather symbols, station models, and weather maps.
3. Describe how weather forecasts are made.

Set a Purpose
Have students preview this section, looking at the headings and the images. Then have them read to answer the question "What steps are involved in forecasting weather?"

INSTRUCT

VISUAL TEACHING
Discussion

Have students read the text and examine the two satellite images on this page. Ask: Where are the thickest clouds in the visible satellite image? *in the brightest areas* Where are warmer temperatures in the infrared satellite image? *in the darker areas* What is the main advantage of infrared imagery over visible light imagery? *Infrared imagery does not require light.*

DIFFERENTIATING INSTRUCTION

Reading Support Help students become more active readers by using visual aids to preview the material in this section. Supply surface weather maps from the local newspaper or download weather maps from the National Weather Service Web site. Point out the symbols on the maps. Remind students that they have already learned the symbols for fronts in Section 20.2. Tell them that in this section, they will learn about other weather symbols, such as those shown in the chart on page 457.
Use Reading Study Guide, p. 72.

Chapter 20 Weather 455

More about...

Surface Observations
Meteorologists often use shelters known as Stevenson screens to house instruments that record surface weather data. Stevenson screens minimize influences that can cause inaccurate readings of air temperature, such as direct sunlight or radiation reflected from hot surfaces. A Stevenson screen stands aboveground and usually has slatted sides to reduce the effects of reflected radiation and to promote a free flow of air throughout the shelter. It is placed in an open area away from buildings, which can influence readings. A typical Stevenson screen contains minimum and maximum thermometers, a barometer, and a psychrometer (used to measure the dryness of air).

develop. Wind speed and direction at different altitudes can help predict how air masses might move and how severe the weather might become. For example, if the wind speed and direction change rapidly with height, then the potential for tornadoes may exist. By piecing together the wind data from a network of weather stations, scientists can create weather maps for different vertical levels of air. These maps show the shape and intensity of the jet stream, which is important in determining the movement of air masses and storms.

Surface Observations

Even with the data made available by satellites and radiosondes, meteorologists still require more information in order to predict weather. Only by analyzing data from weather stations can meteorologists diagnose where in a storm it is raining and where it is snowing, or where fronts and lows are located.

On land, most weather stations are at airports, where they can easily retrieve data from commercial jets equipped with automatic weather recorders. Information on conditions over the oceans comes from ships and from automated stations on moored buoys. Weather stations typically report conditions every hour, though they may report more often if they detect significant weather. The information they provide includes temperature, dew point, barometric pressure, wind speed and

CAREER

Weather Observer

Serving as a weather observer on the top of Mount Washington is an interesting and challenging career. Mount Washington is a perfect location for observing all kinds of weather. It is the highest point in New England and receives adverse weather from three different storm tracks. The orientation of the mountain range also accelerates the wind; the strongest wind recorded there was measured at 371 kilometers per hour.

Duties of the Mount Washington staff include measuring the wind, temperature, precipitation, and humidity and analyzing cloud types. Data for various research projects must also be monitored. Other duties include disseminating weather information on the radio and making sure the equipment works properly. An important part of an observer's job is to educate the public on topics such as mountain weather and the fragile nature of the mountain environment.

Being an observer on Mount Washington isn't for everyone. However, seeing the beauty of a sunrise or a sunset from the top of the mountain is an unforgettable experience.

People interested in this career need a bachelor's degree in meteorology, as well as good communication skills. For people who have those interests,

MOUNT WASHINGTON OBSERVATORY WEATHER OBSERVERS are cleaning ice off instruments during a winter sunset in New Hampshire.

Mount Washington Observatory offers an internship program. ■

Learn more about becoming a weather observer.
Keycode: ES2012

DIFFERENTIATING INSTRUCTION

Challenge Activity Have students set up a weather station on school grounds. Students will need instruments to monitor temperature, pressure, relative humidity, and wind speed. They should also record changes in cloud cover and type. Based on their observations, have students predict the weather for the following day and then compare their predictions to actual weather conditions. Have them continue making observations and predictions for at least one week.

direction, visibility, precipitation, the height of clouds, and the amount of cloud cover. As you read in Section 20.3, information on areas of precipitation comes from radar. Doppler radar can be used to predict tornadoes.

Station Model

Using the data they receive from weather stations, meteorologists create weather maps. To help fit the huge amounts of data onto a compact map, scientists developed the **station model,** which includes information on temperature, dew point, weather conditions, wind speed and direction, barometric pressure, and cloud cover. Station models can be read by the meteorologists of any country.

The station model shown below indicates that when the station took its measurements, the wind was coming from the northwest at a speed of about 20 knots, the temperature was 31°F, the dew point was 30°F, the sky was completely covered with clouds, and it was snowing. The pressure, adjusted to sea level, was 1024.7 millibars, and it had risen 2.8 millibars in the past three hours.

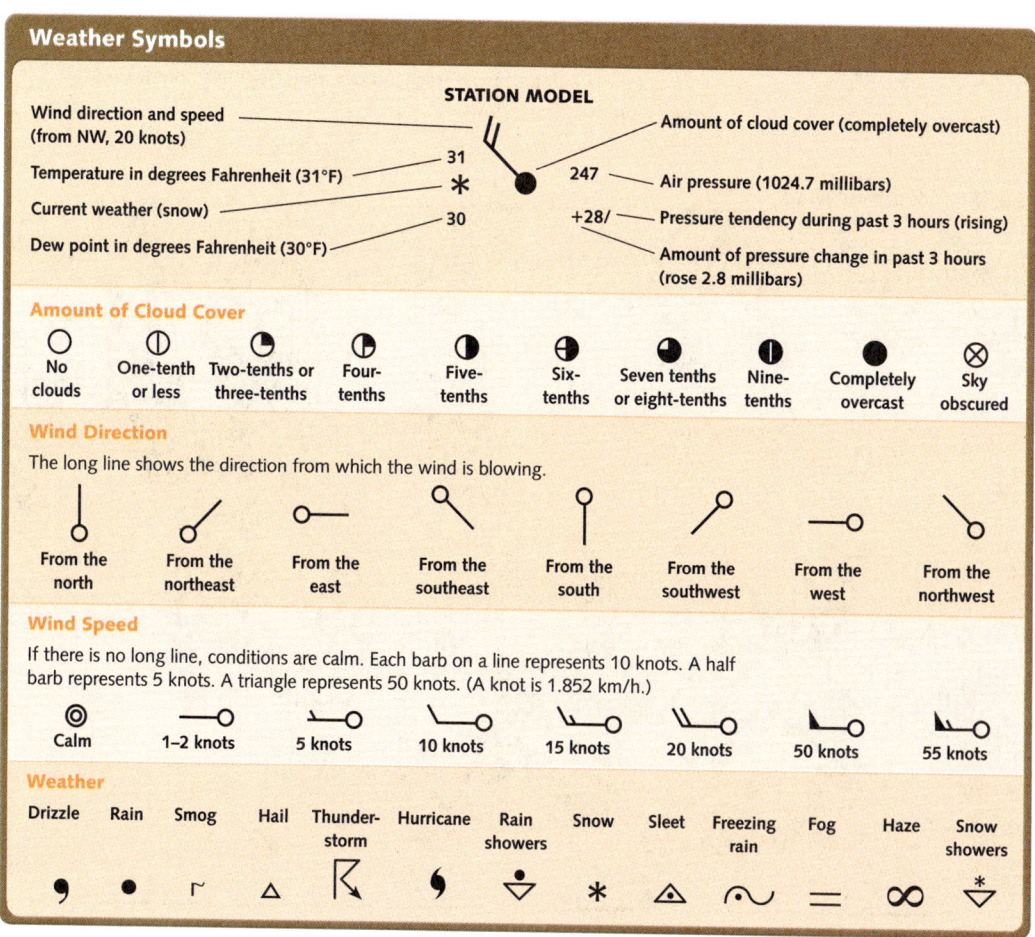

CHAPTER 20 SECTION 5

Learn more about becoming a weather observer.
Keycode: ES2012

VISUAL TEACHING
Discussion

Help students apply their knowledge of atmospheric conditions to station models. After students have examined the station model shown in the chart, ask: If a front passed through the area, what would you expect to happen to wind direction? *It would change.* How might current weather conditions change if the temperature increased by ten degrees? *The snow would likely turn to rain.* If a high-pressure system moved into the area, what might happen to the amount of cloud cover? *It would likely decrease.* Afterward, have students point out the symbols that represent the amount of cloud cover and current weather conditions for their area.

Chapter 20 Weather 457

DIFFERENTIATING INSTRUCTION

Developing English Proficiency Have students examine the chart of weather symbols for unfamiliar words, such as *overcast, drizzle,* or *haze.* Point out clues to help students make sense of the words. For example, the symbols for the amount of cloud cover are sequenced in a logical progression from no clouds to completely overcast. Thus, the word *overcast* must represent extensive cloud coverage. Another clue to a potentially unfamiliar word—*drizzle*—is the similarity between the symbol for drizzle and the symbol for rain.

Support for Visually Impaired Students Some of the Section Review questions will direct students to interpret the symbols on the Weather Symbols chart and the map on page 458. Enlarge both the chart and the map on a photocopier, and hand out copies to students with visual impairments. If students still have difficulty making out the details, draw the station models on the board.

CHAPTER 20 SECTION 5

Visual Teaching

Interpret Diagram
Have students interpret the weather map on this page. If necessary, have on hand a U.S. map that shows the names of the states. Ask: What common storm system do you see on this map? **a mid-latitude low** Describe the structure of this storm system. Identify the different types of fronts involved and their locations. **The mid-latitude low begins as a stationary polar front over New Mexico. The air moves in a counterclockwise direction, with the cold front moving southward and the warm front moving northward. The faster-moving cold front catches up with the warm front, forming an occluded front over Ohio.** Pinpoint an area that shows how wind direction, temperature, and air pressure change with the passage of a front. **Sample Response: southern Texas**

Extend
In which season do you think the data shown in the map were gathered? **during spring or fall** Explain. **Temperatures across the country are mild.**

Making a Surface Weather Map

Surface weather maps are essential tools that give weather forecasters the "big picture" of current weather conditions. Official U.S. weather maps are produced every three hours by the National Center for Environmental Prediction (NCEP), located in Silver Spring, Maryland.

It takes several steps to make a surface weather map. First, a computer draws a map showing station models. Using the sea-level air pressures reported by the stations, it then draws isobars every four millibars, and identifies highs and lows. After the computer draws the map showing station models, isobars, highs, and lows, forecasters draw in the fronts. Forecasters use three types of data from the station model to find fronts: temperature, wind direction, and dew point. In locating fronts, they apply the following rules:

- Wind direction changes behind fronts.
- Temperature changes sharply across fronts.
- Dew point changes sharply across fronts.

For example, if a station reports light southwest winds and high temperature and humidity, and a second station, 100 miles to the west, has strong, northwest winds, a temperature 10 degrees cooler, and a dew point 20 degrees cooler, then a forecaster can assume that a cold front lies between the two stations.

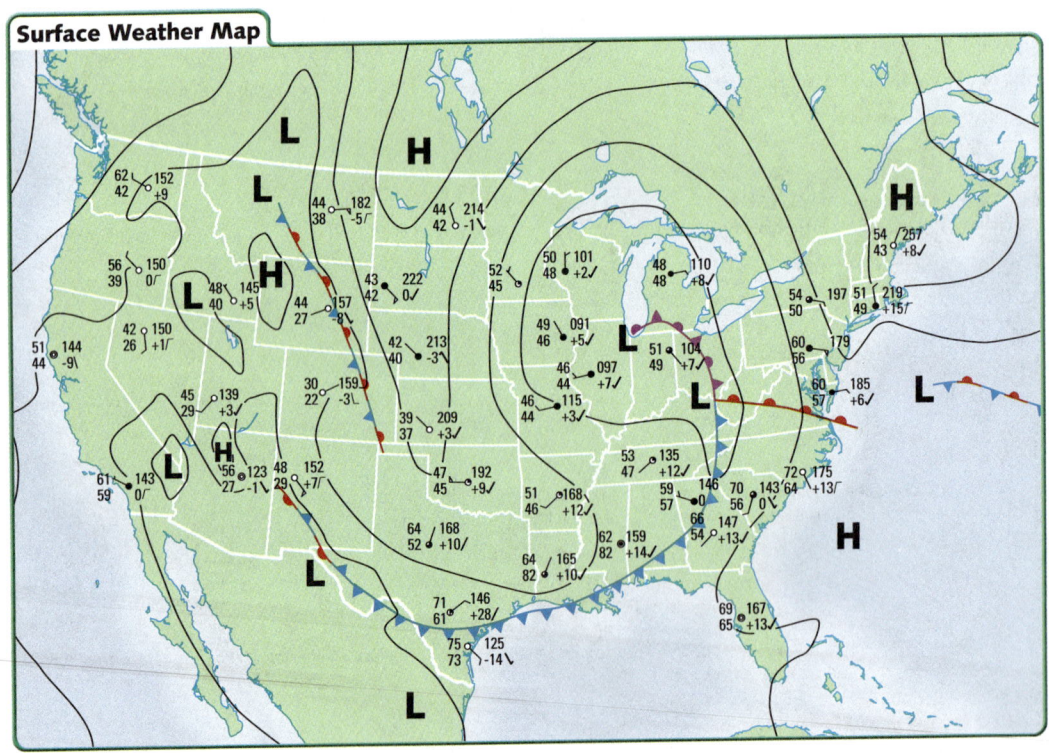

Surface Weather Map

458 Unit 5 Atmosphere and Weather

MONITOR AND RETEACH

If students miss . . .

Question 1 Read aloud the first paragraph under "Gathering Data" on page 455. Have students take notes as you read.

Question 2 Have students draw the station model and then refer to the drawing as they reexamine the chart of weather symbols on page 457.

Question 3 Have students reread "Making a Surface Weather Map" (p. 458) and then list in order the steps involved in making a surface weather map.

Question 4 Reteach "Forecasting." (p. 459) Ask: How long would it take to analyze all the data gathered by stations without using computers? **an exceptionally long time** Can computers account for all variations in weather? **no**

Forecasting

Modern-day weather forecasts are based on computer weather models. Weather models are large computer programs containing mathematical equations designed to simulate atmospheric processes. Many different models are used to make forecasts. Some models are used to predict the weather up to two days in advance, while others provide information up to ten days in advance. The models used for two-day forecasts run four times per day.

Models begin with observed current weather data, such as temperature, humidity, and wind at various levels of the atmosphere. The models transfer these data into their mathematical equations and predict the future state of the atmosphere. The predictions include possible temperatures, winds, sea-level pressure, precipitation, and configurations of the jet stream.

Since the models cannot predict the weather perfectly, different models may produce different predictions. A meteorologist must understand the advantages and disadvantages of each model in order to decide which model to use as a basis for a forecast.

Although models are useful tools in the forecasting process, subtle factors not included in the models can have a large impact on the weather. One of these factors is urbanization. In large cities, concrete covers most of the ground. Since concrete is more likely than vegetation to absorb and retain heat from the sun, cities are often warmer than the surrounding countryside. Urbanization plays a large role in the intensity of heat waves. Whereas the suburbs may get nightly relief from hot summer days, the cities can remain hot through the night. This continual heat not only affects air-conditioning costs but also can be dangerous to the elderly and the very young.

Urbanization can also play a subtle, yet important, role in winter storms when the temperature is at the critical point between rain and snow. For example, a storm that produces cold rain in New York City might cover the surrounding suburbs in several inches of snow. Such local effects are too subtle for models to predict, so forecasters must use their knowledge of local weather patterns to modify the predictions they receive from models.

Scientific Thinking

APPLY

Cities tend to generate and trap heat. These urban "heat islands" can result in unique weather patterns; for example, thunderstorms can pop up around cities when cool air replaces warm air rising above a city. Using what you know about how heat is absorbed and radiated, list several characteristics of cities that help generate or trap heat. Brainstorm ways this heat might be decreased. Include both individual and community actions.

20.5 Section Review

1. Describe several ways of gathering weather data for forecasts.
2. Interpret the station model over Florida in the map on page 458.
3. Describe how station-model data are used to make a weather map.
4. What are some advantages and disadvantages of computer models?
5. **CRITICAL THINKING** If a city is growing, what will happen to its nighttime temperatures? Explain.
6. **APPLICATION** Predict the weather for the next few hours for the station whose model is shown on page 457. Explain your reasoning.

CHAPTER 20 SECTION 5

Scientific Thinking

APPLY
Remind students that dark surfaces absorb heat more efficiently than light surfaces do. Ask them to describe the surfaces found in cities. streets, sidewalks, parking lots Have students work in small groups to brainstorm how these surfaces affect heat absorption and what can be done to lessen their effect. The surfaces found in cities increase rates of heat absorption and reradiate heat into the air. This phenomenon could be lessened if individuals planted more trees or grass, or if the community set aside more areas for parks.

ASSESS

1. satellites, instruments attached to balloons, weather stations, weather radar, airplanes, and ships
2. Winds are calm, temperature is 69°F, dew point is 65°F, air pressure is 1016.7 mb, pressure rose 1.3 mb in the past three hours.
3. First, a computer draws a map showing station models. Using the air pressures reported by the stations, the computer then draws isobars and identifies highs and lows. Forecasters draw in the fronts.
4. **Sample Response:** Computers can quickly analyze large amounts of data and simulate atmospheric processes, but they cannot account for subtle factors that can have a large impact on weather.
5. Its nighttime temperatures would likely increase because urbanization increases the intensity of heat waves.
6. Air pressure is rising, so a high may be entering the area. Cloud cover and precipitation may decrease.

MONITOR AND RETEACH

Question 5 Students may find it helpful to review their answers to the Scientific Thinking feature on page 459.

Question 6 Review "Station Model." (p. 457) Ask: Does a rising barometer indicate an approaching high or low? high In general, what sort of weather is associated with a high? fair weather

CHAPTER 20 ACTIVITY

PURPOSE
To identify the characteristics of a severe storm by analyzing weather maps

MATERIALS
As an alternative to having students trace the maps, make photocopies of the maps and pass them out to students.

PROCEDURE
❷ The center of the system can be located by finding the center of the pressure contours. Differences in air masses can be determined by comparing temperature and dew points. The boundary between air masses is marked by an abrupt change in temperature or humidity.

CHAPTER 20 MAP Activity

Severe Storms

Skills and Objectives
- **Identify** characteristics of severe weather on a weather map.
- **Analyze** the causes and effects of severe weather.

Materials
- tracing paper
- colored pencils
- paper clips

In March 1993, a severe winter storm hit the eastern United States. The storm, known as "Superstorm '93," brought record low temperatures, low pressures, and snowfalls. The next page contains two surface weather maps of conditions during the evening and early morning of March 12–13, 1993. Studying these maps can lead to insight on the characteristics and effects of severe weather associated with mid-latitude low-pressure systems.

Procedure

❶ Use paper clips to fasten tracing paper over the maps on the opposite page. Trace the outline of the United States coastline on both maps.

❷ The upper map shows surface conditions at 7 P.M. EST on March 12. Group the weather stations readings in the map into three different air masses: warm and moist, cold and moist, and cold and dry. Lightly shade each air mass a different color. Label the center of the low-pressure system with an "L." Label the cold front, extending out of the low, that divides the warm, moist air from the cool, moist air.

❸ Repeat Step 2 for the lower map showing conditions 12 hours later.

❹ Indicate precipitation falling in the form of snow with star symbols on both maps. Use raindrops where precipitation is falling as rain.

Analysis and Conclusions

1. Are temperatures generally higher to the east of the low-pressure system or to the west? How do these temperatures relate to the wind that is swirling into the low?

2. This low-pressure system originated over Mexico due to the heating of land and moved northwest toward the eastern United States. What was the source of the warm, moist air brought by the low? What caused the warm, moist air to become precipitation as rain or snow?

3. Record low temperatures hit the southeastern United States after the storm passed. (Birmingham, Alabama, reached 2°F.) What was the source of this cold, dry air?

4. Did the low strengthen (drop further in pressure) or weaken as it moved over land? What happens to a hurricane as it moves over land?

5. Give three reasons why this severe winter storm caused more damage to the eastern United States than typically caused by a major hurricane?

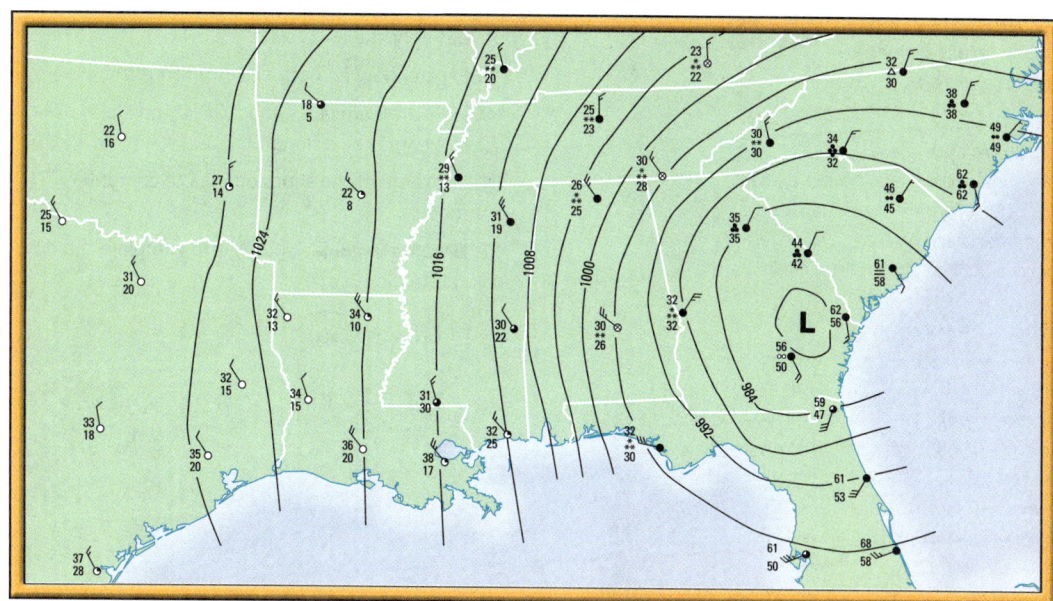

7 P.M. EST, March 12, 1993

7 A.M. EST, March 13, 1993

CHAPTER 20 ACTIVITY

ANALYSIS AND CONCLUSIONS

1. Temperatures are higher to the east of the low-pressure system. The winds swirl counterclockwise into the low, bringing warm air north to the eastern side of the system and cold air south to the western side of the system.

2. The source of the warm, moist air was the Gulf of Mexico and the Caribbean Sea. This air mass met cooler, drier air and, being less dense than the cool air, was forced to rise. As it rose, its temperature cooled. Cool air cannot hold as much moisture as warm air, so the water vapor in the air condensed and fell as precipitation.

3. The source of the cold, dry air was the northern part of the United States and Canada; air masses that form over land in winter are cooler and drier than those that form over water.

4. The winds are stronger on the second map than on the first map. Thus, the low-pressure center strengthened as it moved over land. Hurricanes typically weaken as they move over land.

5. Hurricanes generally weaken as they hit land. This winter storm became stronger. Also, the storm's extremely cold temperatures and large amounts of snowfall are potential dangers not associated with hurricanes. Lastly, this winter storm affected a larger area than most hurricanes do.

Chapter 20 Weather 461

CHAPTER 20 REVIEW

Vocabulary Review

1. occluded front
2. supercell
3. hurricane
4. winter storm
5. station model

Concept Review

6. polar: north; tropical: south
7. A warm front at first brings cloudy skies with rain or snow, shifting winds, and then warmer temperatures and gradually clearing skies. A cold front brings colder temperatures, shifting winds, and clear skies if the front replaces hot, dry air; however, if it replaces warm, humid air, thunderstorms may result.
8. The air mass is cold throughout and therefore resists rising.
9. A cold front catches up to a warm front. The warm air caught between the two cold air masses is forced upward. As the warm air rises, it cools. Condensation occurs, resulting in cloudiness and precipitation.
10. Hurricanes and mid-latitude lows both start out as small atmospheric disturbances. Hurricanes develop over tropical oceans where there is a steady supply of warm, moist air. Mid-latitude lows often develop along stationary fronts where cold polar air clashes with warm tropical air.
11. Computer models cannot predict the weather perfectly. Thus, different models may generate different predictions. Also, the models cannot account for subtle factors, such as urbanization, that can influence the weather.
12. top left oval: warm, moist, unstable air; top right oval: rising air; bottom oval: cumulonimbus cloud

Critical Thinking

13. Warm and cold fronts are usually weakest in the summer months over the continental United States because the temperature gradient between air masses is less pronounced at that time of year.
14. Computer-model maps can be used to simulate atmospheric processes, using current weather data gathered from hourly weather observations. Hourly weather observations can also be used to pinpoint where rain is occurring. Surface weather maps can be used to find out where fronts and lows are located. Temperature and humidity data from radiosondes can be used to estimate the likelihood that cumulonimbus clouds will form. Doppler radar can be used to identify rotating mesocyclones that develop into tornadoes. Satellite data can be used to track storms and predict their severity.

CHAPTER 20 REVIEW

Summary of Key Ideas

20.1 Scientists classify an air mass based on whether it originates in an arctic, in a polar, or in a tropical region and whether it forms over land (continental) or sea (maritime).

20.2 A front is the band of air between opposing air masses. Scientists classify a front based on the temperature of the advancing air mass. Fronts are usually connected to mid-latitude low-pressure systems. Upper-level air flow influences the convergence or divergence of air into and out of pressure systems.

20.3 Thunderstorms form in warm, moist, unstable air. They produce lightning, a discharge of electricity. Tornadoes can develop in thunderstorms containing rotating updrafts.

20.4 Hurricanes are large, rotating storms originating over tropical oceans. They are classified based on wind speed. Winter storms are mid-latitude low-pressure systems that occur over land in the winter.

20.5 Weather forecasters must gather huge amounts of data in order to make their predictions. They rely on sensing instruments and on computer models to provide the information they need.

Key Vocabulary

air mass (p. 436)
blizzard (p. 453)
cold front (p. 440)
front (p. 439)
hurricane (p. 450)
lightning (p. 446)
meteorology (p. 436)
occluded front (p. 441)
Saffir-Simpson scale (p. 453)
squall line (p. 446)
station model (p. 457)
stationary front (p. 441)
storm surge (p. 451)
supercell (p. 446)
thunderstorm (p. 445)
tornado (p. 447)
warm front (p. 441)

Vocabulary Review

Write the term from the key vocabulary list that best completes the sentence.

1. When a cold front "catches up" to a warm front, the result is a(n) ___?___ .
2. The type of thunderstorm that has the most damaging winds and precipitation is a(n) ___?___ .
3. A rotating storm from a warm ocean area, the ___?___ has sustained winds of at least 119 kilometers per hour.
4. To be classified as a blizzard, a(n) ___?___ must reduce visibility and fall into specific temperature and wind-speed ranges.
5. To fit huge amounts of information onto a compact map, forecasters use a(n) ___?___ .

Concept Review

6. In what directions would polar and tropical air masses move in the Southern Hemisphere?
7. What kind of air-mass weather is likely to follow the passing of a warm front? Of a cold front?
8. A wintertime continental Arctic air mass starts out very stable. Explain.
9. Explain how an occluded front forms.
10. How are hurricanes and mid-latitude lows alike? How are they different?
11. List some of the shortcomings of computer models.
12. **Graphic Organizer** Copy and complete the flow chart below.

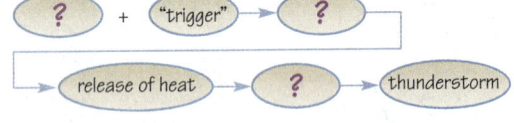

Critical Thinking

13. **Hypothesize** Write a hypothesis suggesting why warm and cold fronts are usually weakest in the summer months in the continental United States.

14. **Predict** You are a forecaster in Oklahoma. How will you use each of the following to foresee and then to track the occurrence of a tornado: computer-model maps, radiosonde maps, the surface-weather map, hourly weather observations, satellite data, and radar?

15. **Infer** If you hear thunder 15 seconds after seeing lightning, how far away is the storm?

16. **Infer** How can you use air masses to explain the fact that hurricanes have no fronts?

Internet Extension

How Well Can You Predict Tomorrow's Weather?
Use current weather data to make a forecast for your local area.

Keycode: ES2013

Writing About the Earth System

SCIENCE NOTEBOOK The deadliest hurricane in United States history occurred in Galveston, Texas, in September 1900. The coastal town had no seawall for protection from flooding, and scientists did not yet have the technology to foresee a hurricane far in advance. Six thousand people died in the hurricane, including 100 residents of an orphanage. Research the Galveston hurricane. Using the information you find, compare the weather information Galveston meteorologists had on that day with what would be available now. How might specific aspects of today's forecasting methods change the human outcome of such a hurricane if it were to happen today?

Interpreting Maps

The weather map at right shows two lows and two highs. The arrows indicate wind directions.

17. How many kinds of fronts appear on the map?

18. Which color represents areas of low pressure? Which color represents areas of high pressure? What evidence do you have to support your answers?

19. Is this most likely a winter map or a summer map? How can you tell?

20. What were the probable weather conditions along the northwestern coast at the time of this weather map?

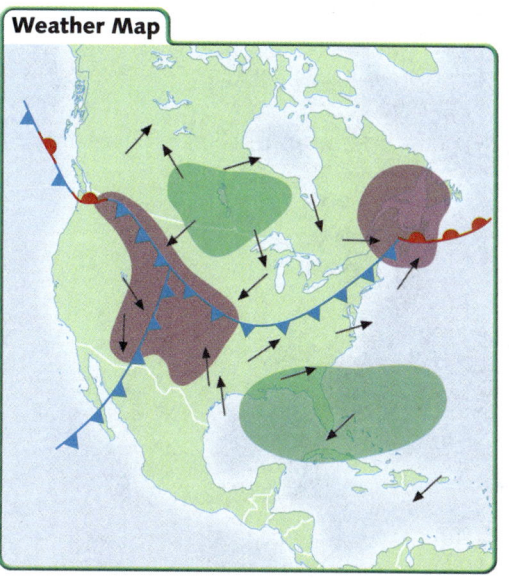

Weather Map

CHAPTER 20 REVIEW

How Well Can You Predict Tomorrow's Weather?
Keycode: ES2013

Use Internet Investigations Guide, pp. T72, 72.

(continued)

15. Sound travels at a speed of 3 seconds per kilometer: 15 seconds ÷ 3 seconds/kilometer = 5 kilometers.

16. Fronts are the boundaries that separate air masses that have different temperature and/or humidity characteristics. Hurricanes form from a single mass of rotating warm, moist air over the ocean.

INTERPRETING MAPS

17. 3: warm, cold, and stationary

18. Purple represents areas of low pressure, and green represents areas of high pressure. The direction of the arrows offers evidence. In the Northern Hemisphere, winds blow counterclockwise into a low and clockwise out of a high.

19. This is most likely a winter map. The cold front is fairly far to the south, indicating that the polar air masses are pushing southward. In the summer, the cold air would be farther to the north.

20. The stationary front in that area indicates that there was probably a long period of steady precipitation.

WRITING ABOUT SCIENCE

Writing About the Earth System

In 1900, the Galveston meteorologists did not have satellite images, computer models, radiosondes, or radar to help them make predictions. Although they did have surface observations, they did not have an adequate communication system with which to relay information. Had the Galveston meteorologists had modern technology at their disposal, many lives would likely have been saved.

CHAPTER 21

PLANNING GUIDE: CLIMATE AND CLIMATE CHANGE

	Section Objectives	Print Resources
Chapter-Level Resources		Laboratory Manual, pp. 95–96 Internet Investigations Guide, pp. T73–T74, 73–74 Guide to Earth Science in Urban Environments, pp. 13–20, 37–44 Spanish Vocabulary and Summaries, pp. 41–42 Formal Assessment, pp. 61–63 Alternative Assessment, pp. 41–42
Section 21.1 What is Climate?, pp. 466–468	1. Define climate. 2. Describe the factors that influence climate.	Lesson Plans, p. 73 Reading Study Guide, p. 73
Section 21.2 Climate Zones, pp. 469–473	1. Describe Earth's major climate zones. 2. Explain how climate zones are characterized.	Lesson Plans, p. 74 Reading Study Guide, p. 74 Teaching Transparency 25
Section 21.3 Climate Change, pp. 474–477	1. Describe the way Earth's climate changes over time. 2. Explain the causes of climate change.	Lesson Plans, p. 75 Reading Study Guide, p. 75

INVESTIGATIONS
CLASSZONE.COM

Section 21.1 INVESTIGATION
What Factors Control Your Local Climate?
Computer time: 45 minutes / Additional time: 45 minutes
Students use maps and local weather data to infer which factors most influence their climate.

Chapter Review INVESTIGATION
How Do Ice Cores of Glaciers Tell Us about Past Climates?
Computer time: 45 minutes / Additional time: 45 minutes
Students interpret the record of climate change written in ice cores.

Technology/Online Resources

Electronic Teacher Tools
Visualizations CD-ROM
Classzone.com
Online Lesson Planner
Test Generator

Classzone.com
  ES2101 Investigation, ES2102 Data Center

Visualizations CD-ROM
Classzone.com
  ES2103 Visualization

Visualizations CD-ROM
Classzone.com
  ES2104 Visualization

Classroom Management

Gather these materials for minilab and laboratory activities.

Minilab, p. 472
thermometer
graph paper

Lab Activity, pp. 478–479
3 1-qt. glass jars
3 Celsius thermometers
masking tape
marker
oven mitts
lab apron
dry ice
tongs
steam
sunlamp (optional)
colored pencils
graph paper
Laboratory Manual, Teacher's Edition
  Lab Sheet 21, p. 144

Chapter 21 Climate and Climate Change **464B**

CHAPTER 21

Climate and Climate Change

Climate is controlled by many factors, including elevation, latitude, topography, and vegetation.

Why is the climate in the plains of southern Nepal so different from the climate in the mountains of northern Nepal?

INTRODUCE

Build Interest

Before sharing background information about the photograph with students, allow time for their questions and comments.

Ask questions like the following:

- Describe the climate in the foreground of the photograph. *warm and moist, with a combination of trees and open grassland or cropland*
- Describe the climate in the background of the photograph. *cold and snowy*

Respond

- Why is the climate in the plains of southern Nepal so different from the climate in the mountains of northern Nepal? *The mountains rise to a very high altitude. Temperature decreases with elevation, so even in areas with a mild climate, high mountain peaks can have a cold, harsh climate.*

About the Photograph

The village in the photograph is Dhulikhel in Nepal's Kathmandu Valley. Dhulikhel has a population of about 10,000 and lies 5500 feet above sea level. The snow-capped Himalayas loom in the background. Terraced farm fields are visible in the valley.

BEYOND THE TEXTBOOK

Print Resources 📖

- Reading Study Guide, pp. 73–75
- Transparency 25
- Formal Assessment, pp. 61–63
- Laboratory Manual, pp. 95–96
- Alternative Assessment, pp. 41–42
- Spanish Vocabulary and Summaries, pp. 41–42
- Internet Investigations Guide, pp. T73–T74, 73–74
- Guide to Earth Science in Urban Environments, pp. 13–20, 37–44

Technology Resources

- Visualizations CD-ROM
- Classzone.com
- Online Lesson Planner
- Electronic Teacher Tools
- Test Generator

464 Unit 5 Atmosphere and Weather

CHAPTER 21

PREVIEW

▶ **FOCUS QUESTIONS** In this chapter you will study climate and learn more about the key questions below.

Section 1 What is climate, and what factors affect it?
Section 2 What are the characteristics of some of Earth's climates?
Section 3 What causes climate change?

▶ **REVIEW TOPICS** As you investigate Earth's climates, you will need to use information from earlier chapters.

- plate tectonics (pp. 172–173)
- heat budget (pp. 372–373)
- temperature variations (pp. 374–377)
- global warming (pp. 381–382)
- where it rains (pp. 404–405)

▶ **READING STRATEGY**

CONNECT

As you read Chapter 21, connect what you learn to the climate you live in and climates you have visited. If you know people who live in different climates, ask them about their climates.

INTERNET RESOURCES
CLASSZONE.COM

At our Web site, you will find the following Internet support for this chapter.

DATA CENTER
EARTH NEWS
VISUALIZATIONS
- Climate Zones
- Climate Change Recorded by Nature

LOCAL RESOURCES
INVESTIGATIONS
- What Factors Control Your Local Climate?
- How Do Ice Cores of Glaciers Tell Us about Past Climates?

Chapter 21 Climate and Climate Change **465**

CHAPTER 21

PREPARE
Focus Questions
Find out what knowledge students have about climate. For each focus question, have students consider the section title and then develop two or three additional questions for each section. For example, Section 1: What influences climate where I live? Section 3: How might Earth's climate change?

Review Topics
In addition to the earth science topics listed, students may need to review this concept:

Life Science
- A biome is a major community characterized chiefly by its climate and vegetation. World biomes include desert, tropical rain forest, semidesert, grassland, mountain, and tundra.

Reading Strategy
Model the Strategy: Refer students to the climate map on pages 470–471. Have students find their location, name its climate, and record this information in their notebook. Then have students write the location and climate of other places they have lived or visited. If possible, identify several places that have the same climate. Tell students to think about these places as they read about those climates in the chapter. Have them consider how the images and descriptions of the climates compare to their own experiences. Is there variation within a climate zone? What factors might account for this variation? *elevation differences within a climate zone, urban versus suburban or rural environment, proximity to bodies of water*

BEYOND THE TEXTBOOK

INTERNET RESOURCES
CLASSZONE.COM
INVESTIGATIONS
- What Factors Control Your Local Climate?
- How Do Ice Cores of Glaciers Tell Us about Past Climates?

VISUALIZATIONS
- Map and images of different climate zones
- Images of how nature records climate change

DATA CENTER
EARTH NEWS
LOCAL RESOURCES
CAREERS

Online Lesson Planner

Chapter 21 Climate and Climate Change **465**

CHAPTER 21 SECTION 1

FOCUS

Objectives
1. Define *climate*.
2. Describe the factors that influence climate.

Set a Purpose
Have students read this section to find the answer to "Why does climate differ in various places?"

INSTRUCT

> **More about...**
>
> **How Climate Controls Interact**
> In general, temperatures decrease from the equator to the poles. But other climate controls, such as ocean currents, can modify this pattern. The climate of Western Europe is warmer than other areas on Earth with the same high latitude because of warm ocean currents nearby. The Gulf Stream flows from the Gulf of Mexico northeast across the Atlantic Ocean. It then swings west of the British Isles and Scandinavia as the North Atlantic Current. Prevailing winds blowing over these warm waters bring warmth to Western Europe, moderating the climate. For example, although most of Norway's western coast lies in the Arctic, this coast is ice-free during winter because of the warm current nearby.

21.1

KEY IDEA
An area's climate is its long-term pattern of weather, which varies depending on several factors.

KEY VOCABULARY
- climate
- climate controls

What Is Climate?

Climate is an area's long-term pattern of weather. A description of an area's climate includes many characteristics, including how hot the summers are, how cold the winters are, and how much precipitation falls at different times of the year. The description should also include how the precipitation occurs—in thunderstorms, as gentle rain, or in the form of snow. Although temperature and precipitation are the two main characteristics of an area's climate, other characteristics include the number of days and hours of sunlight; the direction, speed, and steadiness of the wind; and the occurrence of severe weather conditions.

An area's climate is controlled by its latitude, elevation, nearby bodies of water, nearby ocean currents, topography, prevailing winds, and vegetation.

In this section and the next, you will read about local climates. Section 21.3 addresses the global climate and global climate changes.

Temperature and Precipitation

An average temperature helps to describe how hot or cold a climate is, but averages alone do not provide a complete picture of a climate. A description of how the weather varies is also important. For example, Beijing, China, and Valdivia, Chile, have almost identical yearly average temperatures. In Valdivia, the temperature ranges from an average of 7°C in the coldest month to an average of 16°C in the hottest month. Beijing, on the other hand, has much colder winters and much warmer summers. Beijing's average temperature in the coldest month is only −4°C, and in the hottest month it is 26°C. The climates of Beijing and Valdivia are not the same, even though they have the same average temperature.

Temperature ranges describe variations in temperature. The daily temperature range is the difference between the highest and lowest temperatures of the day. The annual temperature range is the difference between the average temperature of the warmest month and the average temperature of the coldest month. For example, Beijing's annual temperature range is 30°C and Valdivia's is only 9°C.

As is the case with average temperature, average precipitation can be misleading. For example, in Bombay, India, almost all of the rain falls in the four months of the summer monsoon. Mobile, Alabama, has almost as much rain each year as Bombay, but the rain is spread throughout the year.

Beijing, China

Valdivia, Chile

DIFFERENTIATING INSTRUCTION

Reading Support Have students create a table in which they can record the characteristics of climate in their own community. Students should use the categories of climate data as described in "What Is Climate?" Then students can make connections to their own observations by filling in climate data for their community. Have volunteers share information from their tables. Ask students to think of other characteristics that describe the climate where you live. As students read about other climates throughout the chapter, they can extend their chart with information for each climate zone.

Use Reading Study Guide, p. 73.

466 Unit 5 Atmosphere and Weather

Climate Controls

The climate of a location depends on a set of conditions called **climate controls**, most of which you have already studied.

SUMMARY: Climate Controls

	Temperature	Precipitation
Latitude	Temperatures are generally colder toward the poles.	The low-pressure areas of the ITCZ and the mid-latitudes have precipitation. The poles and horse latitudes usually have little precipitation.
Elevation	Temperatures are generally colder at higher elevations.	Air at higher elevations generally contains less moisture.
Nearby water	The temperature range of large bodies of water is small, so coastal areas often have mild climates.	Large bodies of water add water vapor to the air, so precipitation is more likely downwind of large bodies of water.
Ocean currents	Warm ocean currents warm nearby coasts. Cold currents, including upwelling currents, cool nearby coasts.	Some ocean currents cause fog.
Topography	The leeward side of a mountain range may be warmer than the windward side. Mountains can act as barriers to air masses.	The windward side of a mountain range may be wetter than the leeward side. Mountains can act as barriers to air masses.
Prevailing winds	Prevailing winds may determine whether air masses arrive from a hot region or from a cold region. See also note at topography.	Prevailing winds may determine whether air masses arrive from a wet region or from a dry region. See also note at topography.
Vegetation	Vegetation affects how much insolation Earth's surface absorbs and how quickly the surface heats or cools.	Vegetation releases water vapor into the air through transpiration.

At each location, some climate controls are more important than others. For example, London, England, is about 1100 kilometers closer to the North Pole than is Cleveland, Ohio, yet London's average temperature is about the same as Cleveland's because of nearby warm ocean currents. For London, ocean currents are a more important climate control than latitude.

INVESTIGATIONS CLASSZONE.COM

What Factors Control Your Local Climate? Use maps and local weather data to infer which factors most influence the climate where you live.
Keycode: ES2101

21.1 Section Review

1. Why doesn't average temperature give a complete picture of climate?
2. Describe several ways mountains can affect climate.
3. **CRITICAL THINKING** Choose three of the climate controls and describe how they affect the climate where you live.
4. **MATHEMATICS** In Plymouth, New Hampshire, the average temperature in January is −9°C and the average temperature in July is 19°C. What is Plymouth's annual temperature range?

Chapter 21 Climate and Climate Change   467

CHAPTER 21 SECTION 1

INVESTIGATIONS CLASSZONE.COM

What Factors Control Your Local Climate?
Keycode: ES2101

Have students work in small groups.

📖 *Use Internet Investigations Guide, pp. T73, 73.*

CRITICAL THINKING

Have students use the table to compare and contrast temperature and precipitation in two cities. City A is on a lowland upwind of a large lake. City B is 70 kilometers from City A at the same latitude, downwind of the same lake, and located at a slightly higher elevation on the windward side of a mountain range. *City B probably has a lower temperature due to its higher elevation. Precipitation probably is higher at City B because it is downwind of a lake and is on the windward slope of a mountain range.*

ASSESS

1. because locations with the same average temperature can have very different temperature ranges
2. The windward side of a mountain is cooler than the leeward side, and it can receive more precipitation; at higher elevations, air contains less moisture and temperatures are cooler.
3. Answers will depend on location.
4. 28°C

MONITOR AND RETEACH

If students miss . . .

Question 1 Review "Temperature and Precipitation." (p. 466) Have students explain why Beijing and Valdivia have the same average temperature but different climates.

Question 2 Direct students' attention to the Summary of Climate Controls. (p. 467) Have them summarize the topography information in the chart.

Question 3 Review the first three paragraphs of "What Is Climate?" (p. 466) Have them list the factors that affect climate. Help them relate these factors to their own community.

Question 4 Review "Temperature and Precipitation." (p. 466) Have students define "average temperature range."

Chapter 21 Climate and Climate Change   467

CHAPTER 21 SECTION 1

Learn more about the technologies used to monitor climate change.
Keycode: ES2102

Science and Society

Effects of El Niño
El Niño is the Spanish term for the Christ Child and also the name of a weather pattern that typically appears around Christmas. No one knows when the first El Niño occurred, but the El Niño of 1982–1983 was the most severe in about 100 years. That El Niño caused disasters such as severe drought in Australia and torrential rains in northwestern South America, and prompted intense research into the causes of the phenomenon. The destructive weather seemed linked to major changes in ocean currents, wind, and atmospheric pressure, so meteorologists searched for a relationship between specific ocean and temperature conditions and El Niño events. Their studies led to technology that now helps predict El Niños.

Science Notebook
Drawbacks might include causing panic and economic disruption. Repeated false alarms might lead people to ignore future warnings.

SCIENCE & Society

El Niño

El Niño is blamed for erratic weather around the world, but what is El Niño and how can it be predicted? Why does it have such farreaching effects? Scientists are monitoring temperatures in the Pacific to determine in advance the start of the next El Niño.

How does technology help scientists study El Niño?

In 1997 surprised fishermen were suddenly netting warm-water fish off the coasts of Washington and Alaska. Devastating floods hit Peru; Indonesia saw drought. Fires raged in Sumatra, Borneo, and Malaysia. Parts of the United States suffered torrential rains and mudslides. These were some of the consequences wrought by the 1997–1998 El Niño.

In normal years, the trade winds blowing westward along the equatorial Pacific drive surface ocean water to the west. Cold bottom water rises off South America to replace surface water blown west by the trade winds.

Every few years the trade winds weaken, and the westward currents and upwelling off South America cease. These changes trigger a chain reaction known as El Niño that disrupts the weather around the world.

Only recently has technology been developed to reduce the global consequences of El Niño—often thousands of lost lives and billions of dollars in property damage. In 1992 the United States and France jointly launched the TOPEX/Poseidon satellite, allowing scientists to monitor global sea levels in an effort to predict El Niño events.

In 1994 a consortium of countries finished installing about 70 buoys across the tropical Pacific. Known as the TAI/Triton Array, the buoys measure winds, water and air temperatures, currents, and humidity, then transmit the data via satellite to meteorologists.

Results came quickly. Scientists predicted the 1997–1998 El Niño months in advance. Governments stockpiled emergency supplies and readied international aid to prepare for the inevitable floods, fires, drought, and disease. No longer would El Niño take the world by surprise. ■

THE TAI/TRITON ARRAY consists of about 70 buoys. These buoys, shown being deployed, measure the sea and air conditions, such as temperature, currents, and humidity.

Extension

SCIENCE NOTEBOOK
Consider the advantages and disadvantages of predicting an El Niño event. Are there any drawbacks to telling people that extreme weather is on the way?

DATA CENTER
CLASSZONE.COM
Learn more about the technologies used to monitor climate change.
Keycode: ES2102

THE TOPEX/POSEIDON satellite recorded heights of the sea surface during the 1997–1998 El Niño to produce these images. White indicates greater sea-surface height and warmer water, purple indicates lower height and cooler water.

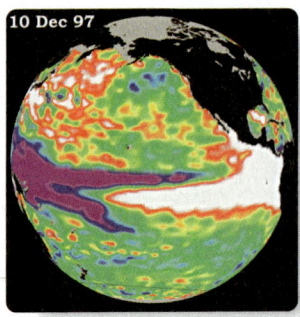
10 Dec 97

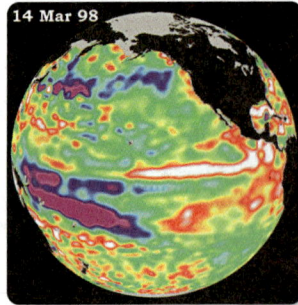
14 Mar 98

Climate Zones

Every place on Earth has a distinctive climate that depends on local factors such as latitude, nearby bodies of water, and elevation. To aid in understanding and studying the different climates around the world, scientists classify climates based on their similarities.

21.2

KEY IDEAS

Earth has a great diversity of climates.

Scientists categorize Earth's many climates into climate zones.

SUMMARY: Climate Zones

Climate Zone	Subclimate	Description
Polar	Tundra	Always cold and dry with short, cool summers
	Icecap	Freezing temperatures all year
Dry	Desert	Hot enough that evaporation exceeds precipitation
	Semiarid	Not as dry as desert, but evaporation still exceeds precipitation
Humid tropical	Tropical wet	Hot and very rainy most of the year
	Tropical wet and dry	Hot all year with wet and dry seasons
Moist mid-latitude with mild winters	Humid subtropical	Hot, humid summers and mild winters
	Marine west coast	Mild and rainy all year
	Mediterranean	Hot, dry summers and mild, rainy winters
Moist mid-latitude, with severe winters	Humid continental	Warm summers and cold, snowy winters
	Subarctic	Short summers and long, cold, snowy winters
Highland	None	Temperature and precipitation vary greatly with latitude and elevation

Polar Climates

Polar climates are very cold. In the winter, there is little or no daylight, and in the summer, the sun hits Earth's surface at a low angle, providing little solar energy. Much of the sunlight is reflected back out to space by snow or ice, and the little energy that is absorbed may melt snow and ice rather than raise the temperature. There is generally very little precipitation in polar climates.

The polar climate type has two subclimates, tundra and icecap. The vegetation of the treeless tundra consists of grasses, sedges, mosses, and lichens. Almost a tenth of Earth's land area is covered by icecaps, which are regions of permanent snow and ice.

POLAR CLIMATE This ice fjord near Ilulissat, Greenland, is an example of a polar climate zone.

CHAPTER 21 SECTION 2

FOCUS

Objectives
1. Describe Earth's major climate zones.
2. Explain how climate zones are characterized.

Set a Purpose
Have students read this section to find the answer to "What are the characteristics of Earth's major climate zones?"

INSTRUCT

More about...

Arctic Lichens
Lichens are organisms that thrive in polar climates. Actually, lichens are two organisms—a fungus and an alga. The alga lives inside the fungus. Some Arctic lichens have been found to be 4500 years old. Lichens easily grow on rocks because they absorb water and minerals from the air. For this reason, they are easily affected by air pollution. Scientist often use lichens as indicators of air quality.

DIFFERENTIATING INSTRUCTION

Hands-On Demonstration Tell students that in the tundra only the surface area of soil thaws during summer. Use a hammer and nail to punch four holes in the bottom of two coffee cans. Fill each can about one-third full of soil. Place one can aside. Slowly add water to the other can until it begins to drain from the bottom. Drain the excess water, then pack the soil firmly. Place the can in a freezer overnight. Remove the can from the freezer on the next day and pour a cup of water over each can. Ask students to observe what they see. The water poured over the frozen soil does not drain. Help students see that although little rain falls in the tundra, water is plentiful because the frozen soil does not allow the little rain that does fall to drain. As a result, lakes and ponds are common.

CHAPTER 21 SECTION 2

VISUAL TEACHING

Discussion
Write the names of the following cities on the board: London, Anchorage, Cairo, Manila. Tell students to find each of these cities on a world map. Then have them use the map and key to determine the climate zone in which each city is located. **London—marine west coast; Anchorage—subarctic; Cairo—desert; Manila—tropical wet**

Use Transparency 25.

More about...

Desert Climates
Possibly the world's driest climate is in the Atacama Desert of northern Chile, wedged between the Pacific Ocean and the Andes. The Andes act as a moisture barrier, blocking movement of moist air from the east. The Peru Current moving up the coast brings cold Antarctic waters north, causing a thermal inversion that produces fog and clouds, but no rain. The Atacama also lies in a global high-pressure area, further contributing to the lack of rain. This cold desert averages less than 1.5 cm (0.5 in.) of rain each year. Inhabitants place large "nets" in the hills and mountains. As fog passes through the nets, water condenses and collects. A large pipeline carries the water to local villages.

VISUALIZATIONS CLASSZONE.COM
Observe images of different climate zones.
Keycode: ES2103

DESERT CLIMATE The Southwest U.S., including parts of California, as shown here, are desert climate zones.

Dry Climates
Dry climates cover about 30 percent of Earth's land surface. Dry climates are regions that lose more water to evaporation than they receive from precipitation. By itself, the amount of precipitation a climate receives is not an accurate indicator of the dryness of the climate. Because evaporation increases as temperature increases, a hot dry climate may actually receive as much rainfall as a cool wet climate. Dry climates often exist on the leeward sides of mountain ranges and where air sinks in the horse latitudes. The dry climate type has two subclimates, desert and semiarid.

CLIMATE ZONES
- Polar
  - Tundra
  - Icecap
- Dry
  - Desert
  - Semiarid
- Humid Tropical
  - Tropical wet
  - Tropical wet and dry
- Moist Mid-Latitude, Mild Winters
  - Humid subtropical
  - Marine west coast
  - Mediterranean
- Moist Mid-Latitude, Severe Winters
  - Humid continental
  - Subarctic
- Highland
  - Highland

470 Unit 5 Atmosphere and Weather

DIFFERENTIATING INSTRUCTION

Developing English Proficiency Some students may confuse the words *desert* and *dessert*. Explain that a dessert is a sweet food, such as cake or ice cream, usually eaten after a meal. Explain that the dry climate zone students are learning about here is a desert. Help students differentiate the pronunciation of the two words as well.
Use Spanish Vocabulary and Summaries, pp. 41–42.

Reading Support Students can use the Summary chart on page 466 as a check on their reading comprehension throughout this section. As they read about each climate zone, they should go back to the chart and compare their understanding to the descriptions in the chart. If the descriptions do not match their understanding, have them skim or reread the text under that climate zone subheading. Stress that the chart lists only major points of each subclimate but that these are the points students should most remember.
Use Reading Study Guide, p. 74.

Although many people think of deserts as hot, dry, lifeless regions of endless sand dunes, this description fits very few deserts. Deserts can be cold, especially in the winter and at night. They are home to a variety of plants and animals that have adapted to live in a dry environment. For example, some plants are good at storing water, while others have very long roots that can reach water deep under the ground.

Semiarid climates are not as dry as deserts and are often home to dense grasses. The Great Plains of the midwestern United States are semiarid.

MOIST MID-LATITUDE Southern Japan, shown here, has a milder climate than the country's northern region.

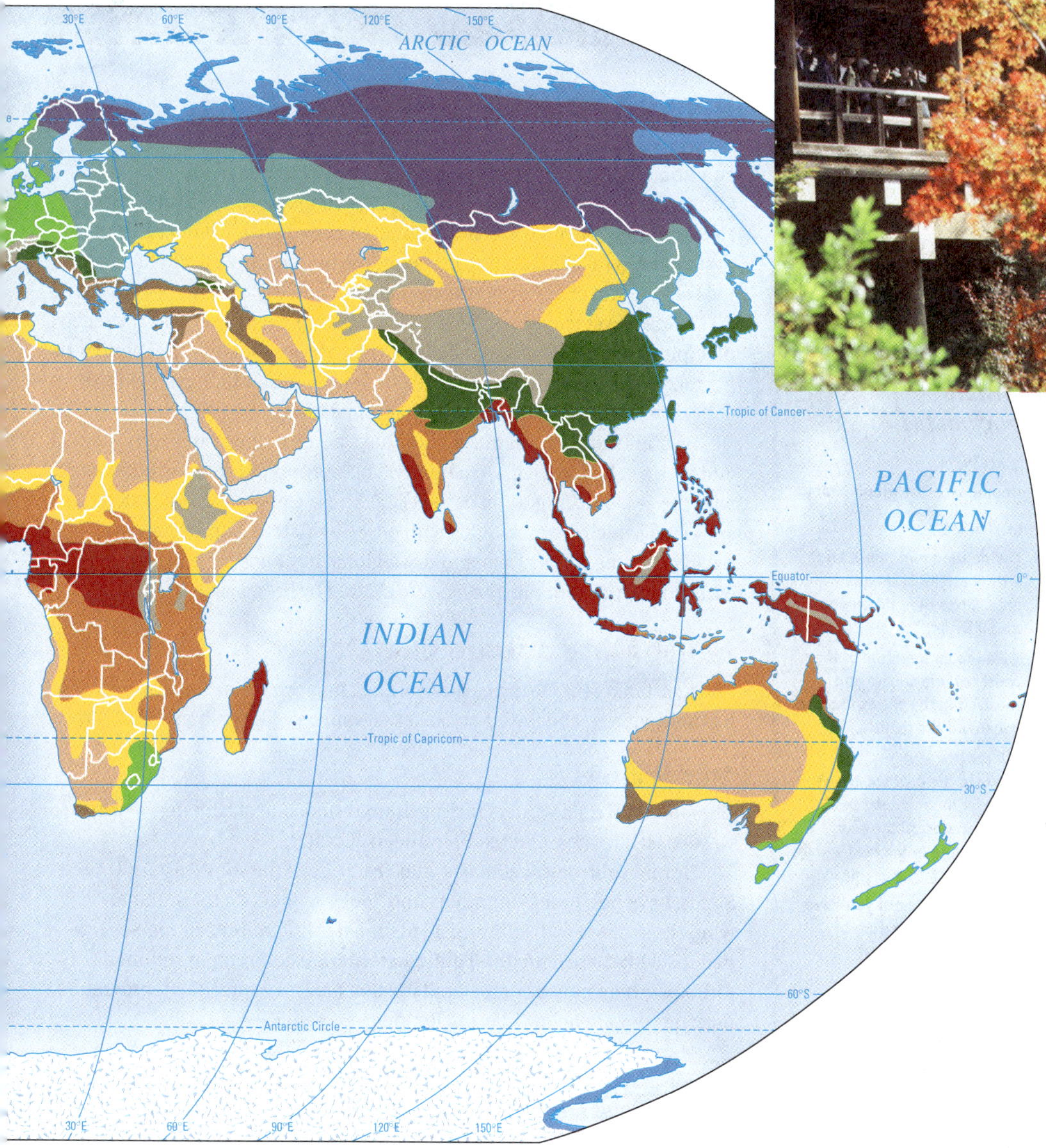

CHAPTER 21 SECTION 2

VISUALIZATIONS
CLASSZONE.COM

Observe images of different climate zones.
Keycode: ES2103

Visualizations CD-ROM

More about...

Desert Adaptations
Desert organisms are adapted to surviving long periods without water. After an infrequent but heavy rainfall, the flesh of a large saguaro cactus can soak up a ton of water, just about doubling its girth. Roots spread out in a shallow but wide area around the cactus, absorbing what little water falls on the ground before it evaporates or soaks deep into the soil. A waxy cuticle on the plant retards evaporation from the plant. During the hottest part of midday, the cactus also shuts its pores to stop water from escaping. Spines discourage desert animals from feeding on the plants, and in some species, point downward to direct scarce rainwater to trickle down the sides of the plant to its base, where the roots can soak it up.

Chapter 21 Climate and Climate Change **471**

DIFFERENTIATING INSTRUCTION

Challenge Activity Have small groups of students gather photographs from magazines or personal collections that show different landscapes. Have students identify important features, such as vegetation and soil, as well as temperature and humidity derived from evidence in the photographs. After students have discussed the content of each photograph, have them sort the photographs according to the climate zone in which they most likely belong. They can then create a visual to display the photographs.

Chapter 21 Climate and Climate Change **471**

CHAPTER 21 SECTION 2

MINILAB
Classroom Microclimates

Management
As an alternative, divide the classroom into a finer grid and assign each student a square or two.

Analysis Answers
Students will probably find that the temperature is not the same in every square owing to sunlight, shadows, heater, air conditioner, windows, doors, and distribution of students.

For proper behavior in a laboratory setting, refer students to Safety in the Earth Science Laboratory on pages xxii–xxiii.

More about...

Temperate Rain Forests
The world's largest remaining temperate rain forest is on the northwest coast of North America, straddling British Columbia and Alaska. Alaska's Tongass National Forest occupies a 500-mile-long coastal zone at the heart of this wilderness. The Tongass is a land of huge, ancient trees; glaciers; and clear, rushing streams. It is home to a large population of grizzly bears and bald eagles. In 2000, the federal government took steps to protect the Tongass by banning road building and logging in large, remote areas. But because the towering trees of the Tongass are a valuable resource, the safety of the forest is still in question.

RAIN FORESTS Tropical wet climates, which are hot and humid year-round, are home to lush rain forests, such as these in Indonesia.

MiniLAB
Classroom Microclimates

Materials
- thermometer
- graph paper

Procedure
1. Divide the room into a grid with 4 sections of equal area. Draw the grid on your graph paper.
2. Take the temperature at the center of one square and record it in the corresponding square on the graph-paper grid.
3. Move to the next square. Wait for your thermometer to readjust. Record the new temperature in the next square on the graph-paper grid.
4. Repeat step 3 until you have taken the temperature of every square.

Analysis
Was the temperature the same in every square? If so, why do you think that is? If not, hypothesize what is causing each change in temperature.

Humid Tropical Climates

Humid tropical climates are hot year-round because they are near the equator. These climates are generally in or near the Intertropical Convergence Zone (ITCZ), where winds from the Northern and Southern Hemispheres converge and hot, humid air rises, causing large amounts of precipitation to fall. The subclimates of the humid tropical climate include the tropical wet climate and the tropical wet and dry climate.

Tropical wet climates, which are closest to the equator, typically receive precipitation during most or all of the year. These climates cover almost 10 percent of Earth's land area and are home to tropical rain forests, which contain a great diversity of plant and animal life.

Regions farther from the equator have a wet summer and a dry winter, and so are called tropical wet and dry. The wet and dry seasons correspond with the seasonal migration of the ITCZ. These climates are home to savannas, which are grasslands with occasional trees. Some scientists believe that the tropical wet and dry regions may once have been covered with woodlands, but the forests were cleared by humans.

Moist Mid-Latitude Climates

The moist mid-latitude climates are divided into two main climate zones. One has mild winters, and the other has severe winters with persistent snow cover.

Mild Winters

Climates with mild winters include humid subtropical climates, marine west coast climates, and Mediterranean climates.

Humid subtropical climates, such as those of the southeastern United States, have hot, muggy summers and cooler winters. In the summer, winds associated with subtropical highs carry hot, humid air masses into regions with these climates. Thunderstorms are common in summer. Although there may be occasional frost or snow in winter, the temperature does not remain below freezing for long.

Marine west coast climates, such as those of the western coast of Canada and the Pacific Northwest of the United States, have cool summers and mild winters. The winters are milder than in other areas at similar latitudes because of the moderating effects of the nearby ocean.

DIFFERENTIATING INSTRUCTION

Support for Physically Impaired Students Allow students with limited mobility to work with a partner to complete the Minilab. The students with limited mobility can take charge of setting up the chart and recording the temperature data. The partner can move around the room, placing the thermometer and reading temperature data for recording. Have students work together to analyze the data they gather.

Mediterranean climates can be found along the western coast of the southwest United States and around the Mediterranean Sea. In the summer, subtropical highs cause these climates to be dry. In the winter, the subtropical highs move south, so mid-latitude storms bring rain. Although there may be occasional frost in winter, snow is rare.

Severe Winters

In moist mid-latitude regions with severe winters, snow often covers the ground in the winter, yet in the summer temperatures are still warm enough to support trees. Subclimates include the humid continental climate and the subarctic climate.

Humid continental climates are found in the interiors of continents and on eastern coasts, including in the northern United States east of the Great Plains. Temperatures can become very cold because there are no nearby bodies of water to moderate wintertime cooling of the land. Although winters may be very cold, summers can be hot and humid.

Subarctic climates are found closer to the poles and so have short summers. Temperatures in winter can be extremely cold. For example, in Verkhoyansk, Siberia, temperatures of –50°C are typical in January. Yet temperatures in the summer still exceed 10°C, and trees can grow. Although trees in this climate are short and scrawny because of the harsh conditions, they form the largest continuous forests in the world.

Highlands

Mountainous regions often have several different climates within a small area. For example, in the Sierra mountains in California, the bases of the mountains are in desert, while the tops of some of the mountains are covered with snow all year. Climate varies not only with elevation, but with the direction of the slope. For example, temperatures may be consistently colder on a north-facing slope than on a nearby south-facing slope. Because it is impractical to show all of the separate climates of mountainous regions on a global map, mountainous regions are labeled as highland climates. Highland climates indicate mountainous regions in which multiple climates exist within a small area.

HIGHLAND CLIMATE A mountainous region such as Patagonia in Chile can contain several different climates in a small area.

Scientific Thinking

HYPOTHESIZE
The map on pages 470–471 shows that there is a long band of highland climate in South America. Which of the five major climates could make up this highland climate zone south of 30 degrees south latitude? Explain your reasoning.

21.2 Section Review

1. Why are polar climates very cold?
2. What is a desert like?
3. Describe humid tropical climates.
4. Why do Mediterranean climates have dry summers and wet winters?
5. **CRITICAL THINKING** Explain why dry climates often exist on the leeward sides of mountain ranges and in the horse latitudes.
6. **GEOGRAPHY** Look at the map on pages 470–471. Compare the climate in central Greenland with the climate along the southern coast of Greenland.

▶ Focus

Objectives
1. Describe the way Earth's climate changes over time.
2. Explain the causes of climate change.

Set a Purpose
Have students read this section to find the answer to "What are the major causes of climate change?"

▶ Instruct

Visual Teaching
Discussion
Direct students' attention to the graph and have them analyze the pattern of temperature fluctuation in Antarctica. **Sample Response:** Temperatures near to today's levels appear about every 100,000 years.

More about...

Milutin Milankovitch
Periodic changes in Earth's motion are known as Milankovitch cycles. Milankovitch theorized that ice ages are triggered when the cycles cause minimum amounts of solar energy to be received at high latitudes in summer, causing winter snows to last through summer and accumulate over many years to form huge ice sheets.

21.3

KEY IDEA
Earth's climate is constantly changing, due to the effects of the energy budget and interactions among the four spheres of Earth.

Climate Change

Studies of Earth's climate show that it is constantly changing. Changes in Earth's heat budget apparently need not be very large to trigger climate changes. Global mean temperatures during the last Ice Age were only 5°C cooler than today's global mean temperatures. Local temperatures show greater variation. The graph below shows Antarctic temperatures over the past 420,000 years.

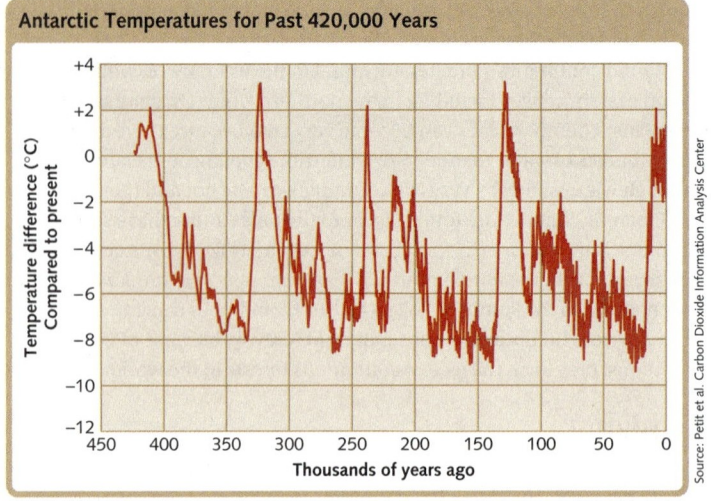

ANTARCTIC TEMPERATURES Scientists drilled deep into the Antarctic icecap and analyzed the ice at different levels to get these data.

Causes of Climate Change

Earth's climate depends on the heat budget. To see what can cause changes in the global climate, it is helpful to consider the processes involved in the heat budget. For example, light coming from the sun may be reflected back to space, absorbed by Earth's surface, or absorbed in the atmosphere. If an increase in cloud or ice cover causes more sunlight to be reflected out to space, Earth may cool.

Local climates are affected by the global climate as well as by local changes in the climate controls you read about in Section 21.1. The climate controls can change over geologic time. For example, a location's elevation can increase because of mountain building, causing the temperature to decrease. Similarly, if a continent moves closer to the poles because of plate tectonics, its temperature decreases.

The causes of climate change are not yet fully understood, but there are many hypotheses. Several of these hypotheses are described next.

Earth's Motions

Astronomer Milutin Milankovitch related climate changes over long time periods to periodic changes in Earth's motions.

474 Unit 5 Atmosphere and Weather

Reading Support Understanding causes and effects can help students understand and remember information in this section. As students read, have them organize information into a table. Tell them to list causes in the first column and the effect in the second column. For example, for "Earth's Motions," students might include "climate changes" in the *Cause* column and "major changes in ice cover during the last Ice Age" in the *Effect* column. Remind students that a cause can have more than one effect, and an effect can have more than one cause.

Use Reading Study Guide, p. 75.

- The shape of Earth's orbit varies with a period of about 100,000 years.
- The tilt of Earth's axis varies between 22.1° and 24.5° with a period of about 41,000 years.
- Earth's axis of rotation precesses (wobbles) with a period of about 23,000 years.

Evidence shows that these variations, which primarily affect the intensity of the seasons, may be responsible for major changes in ice cover during the last Ice Age.

Plate Tectonics

The continents have not always been where they are today. Scientists believe that Greenland once had a warmer climate because it was closer to the equator. Tropical plant fossils have been found on Greenland.

Changes in the positions of the continents may affect wind patterns, ocean currents, and the amount of solar radiation reflected back to space from land and water. When there are large landmasses near the poles, the temperature difference between the poles and the equator increases, and glaciation may occur on the continents near the poles. During glaciation, the increased reflection of sunlight back into space may contribute to global cooling. Similarly, when there are no large landmasses near the poles, the decrease in the amount of light reflected back to space may contribute to a warmer global climate.

Sunspots

Some scientist think that changes in Earth's temperature may be related to changes in the number of sunspots on the surface of the sun. The average amount of energy given off by the sun increases slightly when the number of sunspots increases. If the sun emits more energy, there is more energy available to reach Earth.

Volcanoes

Explosive volcanic eruptions inject dust and sulfur dioxide gas into the stratosphere. The sulfur dioxide reacts with water vapor to form tiny droplets of acid. The dust and the acid droplets are so small that they can stay suspended in the atmosphere for several years. During this time, the dust and droplets reflect sunlight back into space. Because less sunlight reaches Earth's surface, mean temperatures may decrease by a few tenths of a degree Celsius. Once the dust and droplets fall into the troposphere, they are washed out by rain.

Although a single volcanic eruption may affect weather for only a few years, extensive volcanic activity can have a longer-lasting effect on climate. The relative warmth during the Cretaceous period is thought to be related to extensive volcanic activity that injected large amounts of carbon dioxide into the atmosphere. As you read in Chapter 17, carbon dioxide helps keep Earth's surface and lower atmosphere warm by absorbing and remitting heat radiated from Earth. An increase in the amount of carbon dioxide in the atmosphere may cause an increase in the temperature of Earth's surface.

VOLCANO In 1994, the eruption plume from Rabaul Volcano in Papua New Guinea carried gas and ash more than 18,000 meters into the atmosphere.

More about...

Sunspots
A recent study suggests that the sun's energy cycle affects Earth's climate in more indirect ways than previously thought. The particles that comprise the solar wind act as a shield, deflecting many high-speed atomic particles (cosmic rays) from elsewhere in the galaxy. Thus, the size and deflecting strength of the shield rise and fall with that of the solar cycle. Those cosmic rays that reach Earth collide with atoms in the atmosphere. The collision rate is higher when the solar wind is weaker and the sunspot cycle is at its minimum. When a cosmic ray collides with an atom in the atmosphere, a positively charged ion forms. A reaction follows in which many ions form. These ions serve as condensation nuclei that attract water particles, which in turn form clouds.

More about...

Volcanoes
The June 1991 eruption of Mount Pinatubo in the Philippines sent an estimated 20–30 megatons of sulfurous aerosols into the air. Global temperatures fell about 0.5°C (0.9°F) the following year.

DIFFERENTIATING INSTRUCTION

Support for Visually Impaired Students Photocopy the graphs in this section, enlarging them as necessary so that students are better able to see the data. Or reproduce them on clear acetate film and project them with an overhead projector. If necessary, trace over lines on the graphs to make them easier to see. Alternatively, trace over the graph with a wide, smoother line, showing general trends rather than details.

CHAPTER 21 SECTION 3

VISUAL TEACHING
Discussion
Ask: Does the graph on this page indicate that global warming could be occurring? Explain your answer. Yes; Since about 1970, the temperature change generally has been trending upward. The greatest global-temperature change above the stated average occurred during the most recent time period on the graph—the late 1990s. This trend supports the idea that a gradual warming of Earth's atmosphere is occurring.

CRITICAL THINKING
Tell students that one of the effects of global warming could be a rise in sea level as the polar ice caps melt. Some melting of huge ice shelves at the edges of Antarctica is already occurring. Ask: What would be the effect if the ice caps began to melt and the level of the sea began to rise? Low-lying areas along coasts could be threatened with flooding. If the flooding was severe enough, cities, farms, beaches, housing developments, or islands along coastal lowlands could disappear, forcing many people to relocate.

Human Effect on Climate

As a result of deforestation and the burning of fossil fuels such as oil and coal, the amount of carbon dioxide in the atmosphere has increased since the late 1800s. Deforestation contributes to carbon dioxide levels in two ways. First, burning or decomposing trees release carbon dioxide into the air. Second, deforestation decreases the number of trees available to remove carbon dioxide from the air. Many scientists agree that the increase in carbon dioxide is contributing to global warming.

Earth is currently experiencing a warming trend. Because the consequences of global warming could be severe, scientists are trying to predict whether the warming will continue.

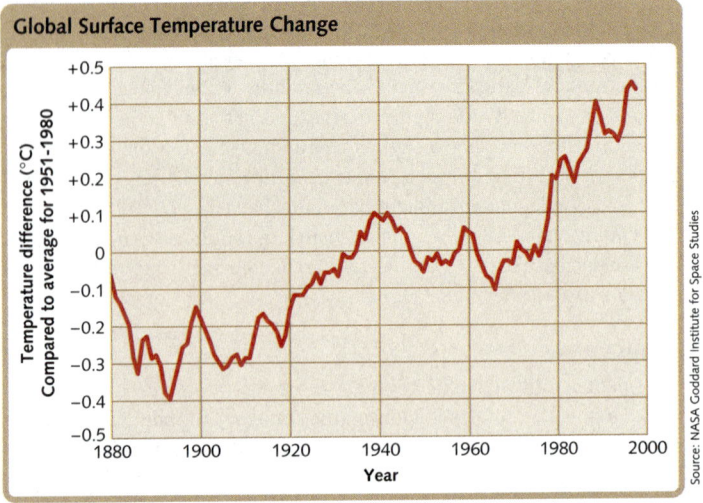

RECENT GLOBAL TEMPERATURE CHANGE This graph is based on adjusted data from weather stations.

Climatologists are attempting to use computers to model Earth's climate and the factors that affect it. The problem is that these factors are numerous and complex. Climate models must take into account not only the factors described in this section, but also a wide variety other factors, including the effects of oceans, changes in greenhouse gases, and the effects of different cloud types on sunlight at different elevations. Researchers are also trying to predict future climate change by studying past climate changes.

Measuring Climate Change

How do scientists know about past climate changes? Records of direct measurements of weather data go back only a few centuries, so researchers must use indirect evidence such as sea-floor sediments, glacier samples, and tree rings to study past climate changes.

Sea-floor sediments contain the shells of microorganisms that once lived at the surface of the ocean. The type and number of the microorganisms living at the surface depend on the surface water temperature. When organisms die, their shells sink to the ocean floor. By

---

DIFFERENTIATING INSTRUCTION

Challenge Activity Tell students that approximately 9 kg of carbon dioxide are produced for each gallon of gasoline that is burned in an automobile. Have students figure out how much carbon dioxide would be produced each year by driving a car to school from home. Tell students to assume the car averages 20 miles per gallon. Students should find the distance from their home to school and multiply it by 2 to determine mileage for a round-trip. They should then multiply this number by the number of school days per year. Next, they should divide the number by 20 to find how many gallons of gasoline they would use in a year. Finally they should multiply that number by 9 kg to determine the amount of carbon dioxide produced.

studying the distribution and type of shells in the sediments on the ocean floor, scientists can determine how the temperature of the surface water has changed over time.

Scientists also use the remains of shells in sea-floor sediments to study the ratio of different types of oxygen atoms in ocean water. Some oxygen atoms are heavier than others. When ocean water evaporates, the water molecules containing the heavier oxygen atoms tend to be left behind in the water. During periods of heavy glaciation, much of the water that evaporates from the oceans falls as snow and is stored in glaciers. As a result, there is less water in the oceans and the ratio of heavy oxygen atoms to light oxygen atoms in the ocean water is greater than at other times. Because the shells in sea-floor sediments contain oxygen from the ocean water, the ratio of heavy oxygen to light oxygen in these shells indicates how much water was held in glaciers when the shells formed. Studying the ratio of heavy oxygen to light oxygen is called oxygen isotope analysis.

Oxygen isotope analysis is also used to study glacial ice. More water that contains heavy oxygen evaporates into the atmosphere during warm periods than during cold periods. As a result, rain and snow during warm periods contain more heavy oxygen than they do during cold periods. By performing oxygen isotope analysis on glacier ice, scientists can tell whether the ice formed from snow that fell during a warm period or from snow that fell during a cold period. Scientists can also analyze bubbles of ancient air trapped in the glacier ice.

Old or fossilized trees can also provide evidence about past climates. Each year, a tree grows a new layer of wood under its bark. This layer is called a *growth ring*. The ring's thickness depends on temperature and precipitation during the year it formed. Scientists can study a tree's growth rings to determine climate changes that occurred during the tree's life.

Indirect evidence of climate change can be found in many other places as well, including soil layers, fossils, pollen, corals, fish bones, stalactites, historical documents, and various geological features such as hardened sand dunes and glacial sediments.

VISUALIZATIONS
CLASSZONE.COM

Observe how nature records climate change.
Keycode: ES2104

GLACIER SAMPLES may be more than 2000 meters long and may provide more than 200,000 years of data.

21.3 Section Review

1. In what ways do Earth's motions change over time?
2. Give an example of how plate tectonics can affect climate.
3. Explain how sunspots may affect climate.
4. Describe two ways in which volcanoes can affect climate.
5. Describe three ways scientists learn about past climate changes.
6. **CRITICAL THINKING** Look at the heat budget in Section 17.3. How can volcanoes alter the heat budget? How might sunspots alter the heat budget?
7. **WRITING** Why is it difficult to predict climate changes?

CHAPTER 21 ACTIVITY

PURPOSE
To model the greenhouse effect and compare the ability of greenhouse gases to hold in heat

MATERIALS
Provide students with a copy of Lab Sheet 21. A copymaster is provided on page 144 of the teacher's laboratory manual.

Make sure thermometers are small enough to fit inside the glass jars.

You can collect steam from boiling water, a facial steamer, or a vaporizer.

If you have access to additional gases, such as nitrogen and helium, you can have students include jars of these gases as part of the lab.

To save class time, you may want to provide students with jars that have already been labeled and filled with gases.

PROCEDURE
3. Make sure the glass jars have airtight lids that fit properly. Confirm that students understand the importance of screwing the lids on tightly.
4. Always use dry ice in a well-ventilated area.
8. Use the sunlamp only if sunlight is not available.

ANALYSIS AND CONCLUSIONS
1. Carbon dioxide heated up the most; air heated up the least.
2. If the gases start at the same temperature and each jar has the same number of moles of gas molecules, air will cool fastest because it has the lowest specific heat. However, answers will vary because the variables are not controlled well enough to accurately predict the answer. Given that the gases are likely to have different starting temperatures, the gas with the highest starting temperature is likely to have the fastest cooling rate.

3. An ideal greenhouse gas would have a high maximum temperature and a slow rate of cooling. Of the gases tested, carbon dioxide is the most effective greenhouse gas. An experiment might first attempt to determine the amount of dry ice needed to fill a jar with the maximum amount of carbon dioxide. Then four identical jars would be filled with different amounts of carbon dioxide: one full of CO_2, one $\frac{3}{4}$ full, one $\frac{1}{2}$ full, and one $\frac{1}{4}$ full. The jars would then be tested with the same procedure used in this lab. The data would be used to determine if the percentage of gas in the jar affects the rates of heating and cooling.

CHAPTER 21 LAB Activity

Observing Greenhouse Gasses

SKILLS AND OBJECTIVES
- **Model** the greenhouse effect.
- **Graph** and **compare** the effectiveness of different naturally occurring greenhouse gasses.

MATERIALS
- 3 1-qt. glass jars
- 3 Celsius thermometers
- masking tape
- marker
- oven mitts
- lab apron
- dry ice
- tongs
- steam
- sunlamp (optional)
- colored pencils
- graph paper

In a greenhouse, transparent glass windows allow light from the sun to radiate in, but do not allow infrared radiation to leave. "Trapped" infrared radiation causes the air in the building to heat up. The gases in our atmosphere act somewhat like windows in a greenhouse; they absorb infrared radiation instead of allowing it to escape into space. Some gases absorb radiation more than others. *Greenhouse gases* are gases that are particularly good at absorbing radiation. The related warming is called the *greenhouse effect*.

In this experiment, you will observe several different gases and compare their effectiveness as greenhouse gases. Dry ice is solid carbon dioxide, which converts directly to gas in a process called *sublimation*.

Procedure

1. Use the masking tape and a marker to label the three jars with the following: Air, Carbon dioxide, and Water vapor.

2. Tape a thermometer to each jar lid so that the thermometer's bulb hangs suspended in the jar, as shown on the right.

3. Leave the jar labeled Air open for a few seconds, and then seal the jar.

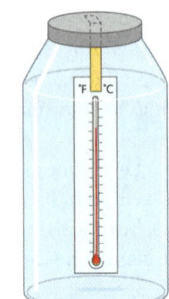

4. **CAUTION: Put on your oven mitts and lab apron. Do not touch dry ice with your bare hands.** Wear oven mitts when handling the jars labeled Carbon dioxide and Water vapor.

5. Drop a piece of dry ice into the jar labeled Carbon dioxide. As soon as all the dry ice has sublimated, seal the jar.

6. Hold the jar labeled Water vapor upside down over boiling water. Once steam has filled the jar, seal.

7. Once all the jars are sealed, use a copy of the data table shown on the next page to record the starting temperature of each jar.

8. Move the jars into direct sunlight or under a sunlamp. Record the temperature of each jar every 5 minutes for the next 30 minutes.

478 Unit 5 Atmosphere and Weather

CHAPTER 21 ACTIVITY

Learn more about the greenhouse effect and global warming.
Keycode: ES2106

Time (minutes)	Heating							Cooling					
	0	5	10	15	20	25	30	35	40	45	50	55	60
Air													
Carbon dioxide													
Water vapor													

9 Move the jars into shade away from any heat sources, and record the temperature of each jar every 5 minutes for an additional 30 minutes.

10 Graph your results with "Time (minutes)" on the *x*-axis and "Temperature (°C)" on the *y*-axis. Use a different colored pencil for each jar.

Analysis and Conclusions

1. Which jar heated up the most in the sunlight? The least?

2. To calculate the rate of cooling, subtract the final temperature from the temperature recorded for each jar just before it was moved into shade; then divide by the number of minutes the jar cooled (30). Calculate the rate of cooling for each jar. Which jar cooled most quickly? least quickly? How do these compare with your answers to Question 1?

3. Imagine an ideal greenhouse gas. How would it behave in terms of maximum temperature and rate of cooling? From your answers to Questions 1 and 2, identify the most effective greenhouse gas among the gases you tested. Design an experiment to understand how the percentage of that gas affects heat retention.

4. Brainstorm ways that humans affect the amount of each gas that you studied in the atmosphere. In particular, try to identify human activities that contribute to the increase of these gases in the atmosphere.

5. Considering your responses to Question 3, brainstorm ways humans can decrease the amount of this gas released into the atmosphere.

6. What other gases in the atmosphere can you think of that might influence the greenhouse effect? How common are these gases? How might you collect them in jars for experimentation?

7. Describe what Earth would be like if the greenhouse effect didn't exist. Now describe what Earth would be like if the greenhouse effect were much greater than it is today. Explain.

Learn more about the greenhouse effect and global warming.
Keycode: ES2106

Chapter 21 Climate and Climate Change 479

---

5. Answers may include the following: burn less fossil fuels by switching to alternative energy sources for generating electricity, developing alternative-fuel vehicles, and increasing the use of mass transit; better management of forests as a renewable resource; cut fewer trees by recycling paper and finding nonwood alternatives for some wood products

6. Answers may include oxygen, nitrogen, methane, and helium. Oxygen and nitrogen are very common in the atmosphere, while methane and helium are less so. All could be purchased. Oxygen would be available from tanks commonly used in health care, while helium tanks are common in stores where balloons are blown up. Some students may suggest recovering these gases from the atmosphere by chemical processing.

7. Without the greenhouse effect, Earth would be much colder than it is today. Most water would be frozen, and life as we know it would not exist. If it were greater, polar ice caps would melt and the sea level would rise, flooding many low-lying and coastal areas; climates would be drastically different from today's climates, with droughts in some places and excessive rainfall in others. Habitats suitable for people and wildlife can live would change, with the possibility that some species might die out.

Refer students to Safety in the Earth Science Laboratory on pages xxii–xxiii.

---

4. Air: Human activities don't affect the amount of air but do affect its composition. Manufacturing processes and activities that burn fossil fuels (such as driving cars and trucks) produce pollutants; cutting trees can increase the amount of atmospheric carbon and decrease the release of oxygen; Water vapor: anything that causes a rise of global temperature can increase evaporation from Earth's surface, causing the amount of water vapor in the air to increase; Carbon dioxide: combustion of fossil fuels in power plants and motor vehicles is a major source of this gas

CHAPTER 21 REVIEW

Vocabulary Review
1. climate
2. climate controls

Concept Review
3. seasonal temperatures and precipitation; days and hours of sunlight; wind speed, direction, and steadiness; occurrence of severe weather conditions
4. Temperatures are colder at higher elevations; the air is usually drier
5. Coastal areas have mild climates; precipitation is often higher downwind of an ocean. Warm ocean currents warm coastal areas; cold currents cool coastlines. Some currents cause fog.
6. They are located near the equator, in or near the Intertropical Convergence Zone. They are humid because winds from the Northern and Southern Hemispheres converge here and rise. The constantly rising supply of hot, humid air causes large amounts of precipitation.
7. Wet and dry seasons occur because the Intertropical Convergence Zone, where great amounts of rain fall, migrates north and south with the seasons. It brings rain when it is over the tropical wet and dry zone; a drier season prevails when it is not.
8. Mid-latitude climates with mild winters: hot, muggy summers with frequent thunderstorms and cool winters with little snow or frost; Mid-latitude climates with severe winters: hot and humid summers, cold winters with large amounts of snow
9. Polar climates: very cold, little precipitation, ground often covered with ice and snow; Dry climates: relatively low levels of precipitation; can be either hot or cold
10. It is likely that Denver's climate is cooler and drier than Kansas City's because of Denver's higher elevation. Other factors such as any nearby large bodies of water and the prevailing winds could also affect the climates of these two cities.
11. first column: dust, colder; second column: reflecting sunlight; third column: warmer, absorbing and reemitting heat radiated from Earth

CHAPTER REVIEW 21

Summary of Key Ideas

21.1 The climate of a region is its long-term weather. Temperature and precipitation can be described by averages and ranges. Many factors act together to determine Earth's climate. They are latitude, elevation, distance from large bodies of water, ocean currents, topography, prevailing winds, and vegetation. At each location, some climate controls have a greater effect than others.

21.2 Earth has a great diversity of climates. Scientists categorize Earth's many climates into zones: polar, dry, humid tropical, moist mid-latitude with mild winters, and moist mid-latitude with severe winters. Each climate zone is further divided into subclimates. In mountainous regions, climates can vary greatly within a small area.

21.3 Earth's climate is constantly changing. Global climate changes can be caused by changes in Earth's energy budget. Causes for climate change include variations in Earth's motions, the location of the continents, variations in sunspots, and gases and ash ejected into the atmosphere by volcanoes. Human activities are increasing the amount of greenhouse gases in the atmosphere, which may be contributing to current global warming. The causes of climate change are complex, so predicting climate change is difficult. Scientists study past climate change by analyzing sea-floor sediments, glacier samples, tree rings, and other indirect evidence.

Key Vocabulary

climate (p. 466)   climate controls (p. 467)

Critical Thinking
12. vegetation; climate might be warmer because the asphalt and cement surface of cities would absorb more heat than grass and trees; climate might also be drier because vegetation, which releases water vapor into the air through transpiration, would be largely absent.

Vocabulary Review

Write the term from the key vocabulary list that best completes the sentence or the analogy.
1. mood : personality as weather : ___?___.
2. The climate of a location is determined by ___?___.

Concept Review
3. What information should be included in a description of a location's climate?
4. How does a location's elevation affect its climate?
5. How can a nearby ocean affect a location's climate?
6. Where are humid tropical climates located? Why are they humid?
7. Why do tropical wet and dry climates have a wet season and a dry season?
8. Compare and contrast moist mid-latitude climates with mild winters and moist mid-latitude climates with severe winters.
9. Compare and contrast polar and dry climates.
10. Denver and Kansas City are at the same latitude, but Denver is at a higher altitude. How does Denver's climate compare to the climate in Kansas City?
11. **Graphic Organizer** Complete the concept map.

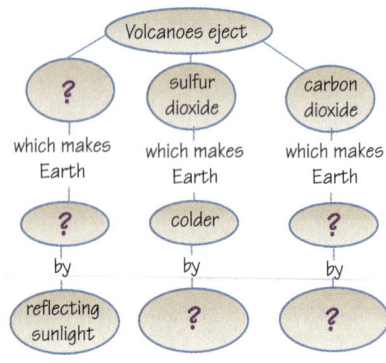

13. No; it depends on whether the location is upwind or downwind of the ocean. Areas downwind of large bodies of water are more likely to have heavier precipitation.
14. Because of their high latitude, polar regions receive indirect sunlight. There are relatively few hours of daylight during much of the year. These factors limit

Critical Thinking

12. **Predict** If the United States were completely covered by cities, which climate control(s) would be affected? How would the climate change? Explain your thinking.

13. **Infer** Do locations near oceans always have heavy precipitation? Why or why not?

14. **Communicate** Explain how and why the length of day varies in polar regions. How does the variation in length of day affect climate?

15. **Analyze** Compare and contrast marine climates and continental climates.

16. **Analyze** Explain how plate tectonics has helped to change climate over time.

17. **Hypothesize** Modern civilization is increasing the amounts of carbon dioxide and particulate matter (such as smoke particles) in the atmosphere. Which pollutant might have the greatest effect on the climate? Why?

18. **Hypothesize** If Earth continues warming, would you expect cloud cover to increase or decrease? How would the change in cloud cover affect Earth's heat budget? What effect would the change in cloud cover have on Earth's temperature? Explain your thinking.

Interpreting Maps

The map shows what North America would look like if it were moved so that the United States lay in the trade-wind zones and Canada was in the horse latitudes.

19. Describe the average annual temperature and precipitation (a) on the United States West Coast, (b) just east of the Rocky Mountains in Wyoming, and (c) in Philadelphia, Pennsylvania.

20. What would the climate be like in Canada? Why?

Internet Extension

How Do Ice Cores of Glaciers Tell Us about Past Climates? Interpret the record of climate change written in ice cores.
Keycode: ES2105

Writing About the Earth System

SCIENCE NOTEBOOK Describe how the geosphere affects climate. In your discussion, be sure to describe the effects of volcanism and plate tectonics. Which has the greater effect on climate? Explain your thinking. How do the hydrosphere and biosphere affect climate? How does the climate affect the biosphere? Use examples to support your argument.

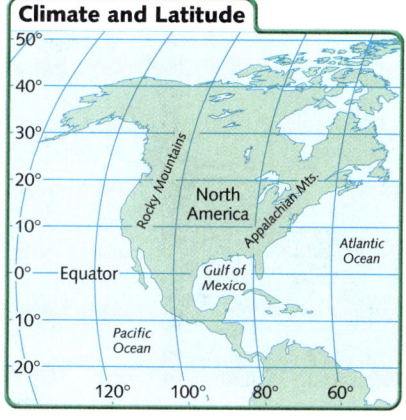

Climate and Latitude

CHAPTER 21 REVIEW

How Do Ice Cores of Glaciers Tell Us about Past Climates?
Keycode: ES2105

Use Internet Investigations Guide, pp. T74, 74.

(continued)
the amount of solar energy reaching polar regions each day.

15. Marine climates are moderated by nearby ocean waters and generally have cool summers and mild winters. Continental climates are more severe because they lack the moderating influence of large bodies of water. Their winters can be very cold and their summers very hot.

16. Areas that are now in the tropics might have once been in polar regions and vice versa. Changes in the positions of continents may have also changed wind patterns, ocean currents, and the amount of solar radiation reflected back to space from land and water.

17. Carbon dioxide would probably have more effect because it does not settle out or fall with precipitation, as particulates do.

18. increase; Some clouds might trap heat and increase global warming.

INTERPRETING MAPS

19. (a) the climate would be hot and dry, the West Coast being leeward to the trade winds; (b) drier than the East Coast, being inland; (c) very moist, being near the ocean and windward to the trade winds

20. desert because of descending air in the horse latitudes

WRITING ABOUT SCIENCE

Writing About the Earth System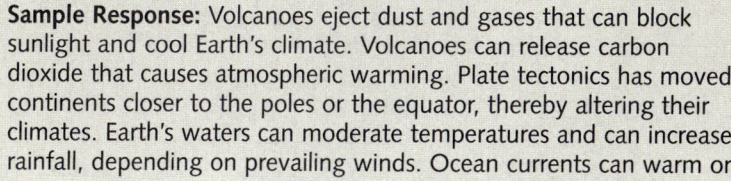

Sample Response: Volcanoes eject dust and gases that can block sunlight and cool Earth's climate. Volcanoes can release carbon dioxide that causes atmospheric warming. Plate tectonics has moved continents closer to the poles or the equator, thereby altering their climates. Earth's waters can moderate temperatures and can increase rainfall, depending on prevailing winds. Ocean currents can warm or cool coastal areas. Vegetation affects the amount of solar energy absorbed by Earth's surface, which can alter temperature and precipitation. Climate affects the biosphere, with different types of plants and animals living in different areas depending on the temperature and precipitation.

UNIT 5 FEATURE

Focus

Objective
Describe possible ways in which interactions between the hydrosphere, atmosphere, geosphere, and biosphere might affect Earth's climate.

Set a Purpose
Have students read to find answers to the question "What Earth-system interactions have been implicated as causes of the Little Ice Age and the more recent global warming?" **Little Ice Age:** decrease in solar radiation related to the sunspot cycle, ice reflecting solar radiation, increase in atmospheric particles, possibly caused by volcanic activity, a change in atmospheric wind patterns; **Global warming:** burning of fossil fuels. Students may also mention the increase of chlorofluorocarbons (CFCs) in the atmosphere and their effect on the ozone layer.

UNIT 5 The EARTH SYSTEM

Climate and Civilization

The end of the last major ice age spurred the development of human civilization. Interactions between human activity (biosphere), the weather and climate (atmosphere), and oceans and glaciers (hydrosphere) will shape Earth's future. Future climate changes once again could alter the course of human history.

Predicting how people's actions will affect Earth's climate is a topic which scientists and politicians earnestly debate. Because the interaction between Earth's atmosphere and the other spheres is so complex, it is very difficult to establish simple cause-and-effect relationships regarding Earth's climate.

Climate plays a large role in determining people's way of life and the availability of Earth's resources. The significance of this role is underscored by a recent climate change that has impacted life all over the planet.

The Little Ice Age was a period of cooler-than-normal global temperatures that lasted from the 13th century until the mid-19th century. The Little Ice Age was also marked by an increase in glaciation and ice formation near Earth's poles.

The cooler climate had a profound effect on lives around the world. Cooler temperatures led to shorter growing seasons and crop failures. The resulting decrease in food supplies was an important cause of famine and local population decreases during this period. Droughts in Africa were worse than the well-documented droughts of the 20th century. Avalanches and landslides caused by advancing glaciers plagued northern Europe.

Changes brought on by the Little Ice Age were not all disastrous, however. Native American migrations, forced by the cold, led to peace pacts such as the League of the Iroquois. Winters that were cold enough to freeze New York's harbor allowed people to walk from Manhattan to Staten Island.

Scientists point to historical and environmental records as evidence for the existence of the Little Ice Age, but they find it much more difficult to determine the causes of the Little Ice Age. One possible cause is a slight decrease in solar radiation related to the sun's sunspot cycle. Decreasing energy from the sun leads to cooler temperatures, which lead to increased ice formation at Earth's poles. When more of Earth's surface is covered with ice, more solar radiation is reflected from the surface, further decreasing temperatures.

Another possible cause of cooler temperatures is an increased amount of dust and ash in the atmosphere. An increase in atmospheric particles, which could be caused by more volcanic activity, would cut down the amount of solar radiation able to reach Earth's surface.

During the Little Ice Age, the 1815 eruption of the Tambora

NORSE COLONIES on Greenland were isolated from mainland Europe by sea ice and did not survive.

482 Unit 5 Atmosphere and Weather

DIFFERENTIATING INSTRUCTION

Reading Support Point out to students that the large blue letters indicate the various sections of this article. Have students write a summary of the key points made in each section and then read their section summaries to the class. Discuss how effectively the summaries relate the content of each section. Remind students that summarizing is a valuable technique for learning from assigned readings.

and the ENVIRONMENT

UNIT 5 FEATURE

How Might Global Climate Change Affect Life on Earth?
Keycode: ESU501

📖 *Use Internet Investigations Guide, pp. T75–T76, 75–76.*

Investigation Photograph: Landsat data show that this glacier in Iceland has receded 2 km since 1973. In this image, land appears red, water is dark blue, bare glacial ice is lighter blue, and snow-covered glacial ice is white.

FROZEN RIVER An artist's view of the frozen East River between Manhattan and Brooklyn.

volcano in Indonesia resulted in colder temperatures that led 1816 to be called the "year without a summer." That year, New England and northern Europe were afflicted by frost and snow in June and July. An increase in the number and the strength of volcanic eruptions may have contributed to the cooler temperatures of the Little Ice Age.

THE ATMOSPHERE shows that distinct bands appeared due to dust and ash from the 1991 Mount Pinatubo eruption.

Other causes for the Little Ice Age have also been proposed. For example, a change in atmospheric wind patterns could have affected the transfer of heat away from the equator. Changing winds would have affected ocean currents, such as the Gulf Stream, which transport heat from equatorial regions to the higher latitudes.

One lesson of the Little Ice Age is that the onset of climate change can be sudden. Evidence taken from ice cores in Wyoming suggest the Little Ice Age ended abruptly over a span of about ten years. This fact supports the concern that human-derived changes in our environment could trigger a sudden change in global climate.

The increase in greenhouse gases in the atmosphere due to the burning of fossil fuels would by itself cause an increase in global temperatures. However, the complex interaction of Earth's atmosphere with its hydrosphere, biosphere, and geosphere makes it difficult to determine exactly how Earth's systems could be impacted. Understanding how and why climate changes occur is a crucial step toward a more complete understanding of our planet.

UNIT INVESTIGATION
CLASSZONE.COM

How Might Global Climate Change Affect Life on Earth?
Scientific evidence indicates that global climate change is occurring, perhaps faster than ever before in Earth's history. Data also shows that human activity is contributing to these changes. A changing world climate will impact ecosystems and Earth processes. Investigate evidence for global climate change and its causes. Predict the potential environmental impacts that might result from a changing global climate.
Keycode: ESU501

UNIT 5 ASSESSMENT

ANSWERS

1. d, All objects warmer than absolute zero emit some form of radiation.

2. d, Lead has the lowest specific heat, 0.03 calories/g·C°. This means that less energy is required to raise lead's temperature compared to the other materials listed, so lead will show the greatest increase.

3. d, 540 calories/g released during condensation versus 80 calories/g released during freezing; melting and evaporation absorb energy

4. b, dew point found on the table, using 10°C as the dry bulb temperature and 3°C as the difference between the dry- and wet-bulb temperatures

5. b, Mid-latitude low-pressure systems are connected to precipitation-bearing fronts.

6. c, Winds blow from regions of high pressure to regions of low pressure.

7. b, Sinking air brings high air pressure.

8. c, Cold air is denser than warm air, so it slides beneath warm air and forces the warm air upward.

9. c, Areas near the equator receive relatively direct sun rays all year, thus remain warm.

10. The average surface temperature could rise.

11. Humans could work to minimize emissions of greenhouse gases.

12. A warm ocean current can bring droughts or increased rainfall and flooding, depending on location.

13. The graphs show a lack of water vapor at the tropopause and temperatures near −55°C.

UNIT 5 STANDARDIZED TEST Practice

Directions (1–9): For *each* statement or question, select the word or expression that, of those given, best completes the statement or answers the question. Record your answer on a separate answer sheet. Some questions may require the use of material in the Appendix.

1. Energy is radiated by
   (a) the sun.
   (b) Earth's surface.
   (c) the atmosphere.
   (d) all of the above.

2. Look at the table of specific heats on page 706. If 1 kilogram of each material listed absorbed the same amount of heat, which material would have the greatest temperature change?
   (a) liquid water
   (b) basalt
   (c) granite
   (d) lead

3. Look at the table of properties of water on page 705. Heat energy is measured in calories. Which releases the most heat energy?
   (a) 1 kilogram of water freezing
   (b) 1 kilogram of water melting
   (c) 1 kilogram of water evaporating
   (d) 1 kilogram of water condensing

4. Look at the tables of relative humidity and dew point temperatures on page 704. When the dry-bulb temperature is 10°C and the wet-bulb temperature is 7°C, then the
   (a) dew point is −9°C.
   (b) dew point is 4°C.
   (c) relative humidity is 65°C.
   (d) relative humidity is 24%.

5. Precipitation in the middle latitudes falls
   (a) during the summer rainy season.
   (b) in low-pressure areas.
   (c) where air diverges.
   (d) when warm air flows over cold water.

Base your answers to questions 6 and 7 on the simplified weather map below, which shows isobars and winds.

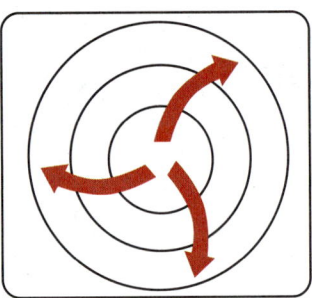

6. The region inside the isobars is a
   (a) storm system
   (b) low-pressure area
   (c) high-pressure area
   (d) front

7. Air above the region shown on the map is
   (a) rising
   (b) sinking
   (c) humid
   (d) motionless

8. A cold front
   (a) is the boundary between a retreating cold air mass and an advancing warm air mass
   (b) is associated with fair weather
   (c) forces warm air ahead of it to rise
   (d) is marked on a weather map by a slightly curving isobar

9. Areas near the equator have
   (a) high average temperatures and large yearly temperature ranges
   (b) long days and short nights
   (c) high average temperatures and small yearly temperature ranges
   (d) very long nights during the winter

Directions (10–19): Record your answers on a separate answer page. Some questions may require the use of material in the Appendix. Base your answers to questions 10 through 12 on the passage below.

El Niño

El Niño is a warm surface current in the Pacific Ocean. Climate researchers still do not understand exactly what causes El Niño in some years and not in others, or just how El Niño causes major changes in the world's weather. Evidence continues to mount that many regions of the world—some halfway around the globe from the waters where El Niño originates—can experience drastic changes in their normal weather patterns during El Niño events. Some of these regions are subject to droughts, while others experience increased rainfall and flooding.

El Niño events appear to be happening more often and with greater intensity than in the past. By studying records of temperature and rainfall around the world, climatologists have concluded that before 1975 El Niño events occurred on average about every seven years. But since the mid-1970s, El Niño events have been occurring more frequently and with greater intensity. The two strongest El Niño events ever recorded were in 1982–1983 and 1997–1998. In addition, according to some research, the period from 1992 to 1995 was an exceptionally long El Niño event.

Many climatologists think the increase in the frequency and strength of El Niños may well be the result of global warming.

10. State one effect global warming could have on surface ocean water.

11. Describe one thing humans could do to minimize negative effects of El Niño.

12. Explain one way a warm ocean current can affect weather.

13. State two specific reasons why rain is unlikely at the tropopause. Refer to the graph of selected properties of Earth's atmosphere on page 705.

Base your answers to questions 14 through 19 on the weather map below. The map shows isobars, fronts, and winds.

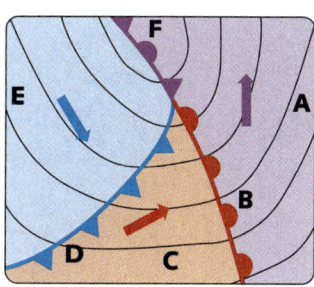

14. At which point is the air pressure lowest?

15. Where on Earth is the region shown on the weather map? In which hemisphere? At roughly what latitude?

16. What type of front is between points B and C? Describe what is happening at this front.

17. What type of front is between points D and E? Describe what is happening at this front.

18. What type of front is next to point F? Describe what is happening at this front.

19. Describe the weather you would expect to find at each of points A–F.

(continued)

14. point F, Winds spiral into areas of low pressure.

15. Fronts are usually connected to mid-latitude low-pressure systems, between 30 degrees N and 60 degrees N. The system shown here could be over North America. Because the winds are circulating counterclockwise around the low, we know it is in the Northern Hemisphere.

16. The symbols indicate a warm front between points B and C. Warm air at C is gradually displacing cold air at B, bringing steady rain or snow.

17. The symbols indicate a cold front between points D and E. Cold air at E sliding along the ground is forcing warm air at D upward. Depending on the season, the result could be thunderstorms, rain or snow showers, or simply a decrease in humidity.

18. Point F is next to an occluded front. Warm air is caught between two colder air masses and is forced to rise. Cloudiness and precipitation are the result.

19. Point A: fair weather with southerly winds and perhaps approaching high cirrus clouds. Point B: heavy nimbostratus clouds and steady rain or snow. Winds shift to westerly. Point C: warmer weather, continuing precipitation. Point D: strengthening wind shifting to the north; drop in temperature; heavy rains or snow, perhaps thunderstorms. Point E: fair weather, colder temperature.

UNIT 6

UNIT OVERVIEW

Oceans cover most of Earth's surface, and they look fairly uniform when observed from above. However, when the ocean floor is mapped with the help of satellites, submersibles, and sonar, a variety of features reveal themselves. Ask students to relate experiences they have had with the oceans. How are these experiences different than those with freshwater lakes and streams? Oceans affect all our lives. Oceans provide food and allow transport of goods. Ocean currents affect weather and climate. For some students, their family's livelihood may be linked to the ocean.

VISUAL TEACHING

Chapter 22 The Water Planet

The range of temperatures on Earth, combined with water's freezing and boiling points (0°C and 100°C, respectively), allow us to experience water in all three phases: solid, liquid, and gas. However, water is most commonly found as a liquid on Earth. Oceans contain 95.96% of Earth's water, glaciers and ice caps about 2.97%, groundwater about 1.05%, and rivers and lakes about 0.009%. The remaining 0.011% is contained in the atmosphere as water vapor and within living organisms. Students will learn about the general properties of water, as well as the specific properties of ocean water.

Answer: Since water heats up more slowly than land, a dip in the ocean can cool us off in summer. Salinity affects our ability to swim in the ocean.

UNIT 6 Earth's Oceans

The Earth System This view of Earth from above the Pacific Ocean emphasizes how Earth truly is a water planet. There is barely a hint of the continental land masses on which most of us live. This unit explores the processes that shape the oceans which occupy over two-thirds of the surface of Earth.

CHAPTER 22
The Water Planet

How do the properties of water affect our experience of the ocean?

CONNECTIONS TO . . .

Prior Knowledge In Unit 5, students learned about water in the atmosphere. Unit 6 focuses on the properties of water and its movement in the oceans. It also builds on what students learned about the features of the ocean floor in Unit 3. Unit 7 proceeds with the study of space where students can apply their knowledge of the dynamic processes of Earth to other planets and their moons.

Unifying Themes Physical *models* of the ocean floor and computer *models* of ocean currents and temperatures have allowed oceanographers to better understand and *explain the evidence* they have gathered of ocean processes. As students study the chapters in Unit 6, emphasize the methods oceanographers use to study the oceans.

UNIT 6

CHAPTER 23 The Ocean Floor

The Ocean Floor

What geological processes led to the formation of the Hawaiian Islands?

Sonar and satellite images of the ocean floor have taught scientists a great deal about its features. This image of Hawaii shows that it is a chain of volcanic islands. However, only the youngest islands of the chain (the southern or bottommost islands in the image) have active volcanoes. Students will learn about the features of the ocean floor and how they are mapped.

Answer: The Hawaiian Islands have formed as the Pacific plate has moved northwest over a hot spot.

CHAPTER 24 The Moving Ocean

The Moving Ocean

What forces cause water levels to rise and fall in this bay north of San Francisco?

Because water enters the San Francisco Bay through a narrow channel under the Golden Gate Bridge, the difference between high and low tides is especially great, and the change from high to low tides is rapid. This phenomenon is called tidal bore. Students will learn about tides and other motions of the ocean such as currents.

Answer: The gravitational forces of the moon and sun cause the water level in the bay to rise and fall each day.

487

CONNECTIONS TO . . .

National Science Education Standards
- Students will come to understand that the sun's heating of Earth creates ocean currents.
- Students will understand that the transfer of heat from inside Earth to the surface causes convection currents in the mantle that move Earth's tectonic plates, resulting in mid-ocean ridges and deep-sea trenches.

Unit Features
The Earth System and the Environment
Ocean Effects on Weather and Climate, pp. 548–549

Can We Blame El Niño for Wild Weather?
Analyze data to look for a relationship between extreme weather and ocean temperatures.
Keycode: ESU601

Standardized Test Practice, pp. 550–551

487

CHAPTER 22 PLANNING GUIDE: THE WATER PLANET

	Section Objectives	Print Resources

Chapter-Level Resources		Laboratory Manual, pp. 97–98
		Internet Investigations Guide,
		  pp. T79–T80, 79–80
		Spanish Vocabulary and Summaries,
		  pp. 43–44
		Formal Assessment, pp. 64–66
		Alternative Assessment, pp. 43–44

Section 22.1 Oceanography, pp. 490–491	1. Summarize the beginnings of oceanography. 2. Describe some methods of modern ocean research.	Lesson Plans, p. 76 Reading Study Guide, p. 76
Section 22.2 Properties of Water, pp. 492–494	1. Contrast the densities of liquid water and ice. 2. Explain how the polarity of water molecules affects the behavior of water. 3. Describe how the properties of water change when it combines with ocean salts.	Lesson Plans, p. 77 Reading Study Guide, p. 77
Section 22.3 Properties of Ocean Water, pp. 495–498	1. Identify factors that affect salinity levels in the oceans. 2. Describe the three temperature layers of the oceans.	Lesson Plans, p. 78 Reading Study Guide, p. 78
Section 22.4 Ocean Life, pp. 499–503	1. Explain the importance of phytoplankton to ocean life. 2. Identify important marine animals. 3. Describe how certain bacteria living near deep-sea vents produce food.	Lesson Plans, p. 79 Reading Study Guide, p. 79

INVESTIGATIONS
CLASSZONE.COM

Section 22.3 INVESTIGATION
How Do Temperature and Salinity Affect Mixing in the Oceans?
Computer time: 45 minutes / Additional time: 45 minutes
Students examine temperature and salinity to predict if water masses will sink or float.

Chapter Review INVESTIGATION
What Is Responsible for Smaller Shrimp Catches?
Computer time: 45 minutes / Additional time: 45 minutes
Students analyze seasonal changes in waters of the Gulf of Mexico.

Technology/Online Resources

Electronic Teacher Tools
Visualizations CD-ROM
Classzone.com
Online Lesson Planner
Test Generator

Visualizations CD-ROM
Classzone.com
  ES2201 Visualization

Classzone.com
  ES2202 Investigation

Visualizations CD-ROM
Classzone.com
  ES2203 Visualization, ES2204 Career,
  ES2205 Visualization

Classroom Management

Gather these materials for minilab and laboratory activities.

Minilab, p. 493
  rubber balloon
  piece of silk cloth or fake fur
  water faucet

Lab Activity, pp. 504–505
  3 500-mL beakers
  fresh water sample
  simulated ocean water sample
  simulated Great Salt Lake water sample
  scissors
  drinking straws
  modeling clay
  salinometer (salinity probe)
  rinse bottle
  metric ruler
  vegetable oil spray

CHAPTER 22

INTRODUCE
Build Interest
Before sharing background information about the photograph with students, allow time for their questions and comments.

Ask questions like the following:

- What instruments can you identify on the buoy? **A wind vane is visible. This implies the buoy carries other weather instruments.**
- Who operates the ship on the right? **The National Oceanic and Atmospheric Administration (NOAA).**

Respond
- How important is the ocean in the Earth system? **The ocean's full impact is not yet known, but we do know it holds the key to worldwide climate. It also supports organisms that contribute oxygen to the atmosphere and that are critical to the worldwide food web.**

About the Photograph
Workers from NOAA's research ship *KA'IMIMOANA* (Ocean Seeker) service a moored buoy, one of about 70 such buoys spanning the equatorial Pacific Ocean. They are equipped to measure surface winds, air temperature, relative humidity, water temperature, salinity, and pressure. The data are transmitted via satellites and are used to understand and predict El Niño and La Niña events, among other phenomena.

The Water Planet

The Ocean covers most of Earth, and is home to millions of plant and animal species.

How important is the Ocean in the Earth system?

BEYOND THE TEXTBOOK

Print Resources
- Reading Study Guide, pp. 76–79
- Formal Assessment, pp. 64–66
- Laboratory Manual, pp. 97–98
- Alternative Assessment, pp. 43–44
- Spanish Vocabulary and Summaries, pp. 43–44
- Internet Investigations Guide, pp. T79–T80, 79–80

Technology Resources
- Visualizations CD-ROM
- Classzone.com
- Online Lesson Planner
- Electronic Teacher Tools
- Test Generator

Unit 6 Earth's Oceans

CHAPTER 22

PREVIEW

▶ **FOCUS QUESTIONS** In this chapter you will study Earth's ocean environments, and learn more about the key questions below.

Section 1 What is oceanography, and why is it important?
Section 2 What are the unique properties of pure water molecules, and how can these properties help you understand ocean environments?
Section 3 How is ocean water different from pure water, and how does it change as it descends to greater depths?
Section 4 What plants and animals live in the ocean, and in which ocean layers do they thrive? Why?

▶ **REVIEW TOPICS** As you investigate oceans, you will need to use information from earlier chapters.

- hydrosphere (p. 9)
- ocean waves (p. 344)
- heating of water and land (p. 376)

▶ **REVIEW TOPICS**

CONNECT

As you read Chapter 22, think about what you have learned previously about the ocean's importance to the Earth's system.

At our Web site, you will find the following Internet support for this chapter.

DATA CENTER
EARTH NEWS
VISUALIZATIONS
- Water Molecules
- Areas of Highest Plant Productivity in the Oceans
- Coral Reefs

LOCAL RESOURCES
CAREERS
INVESTIGATIONS
- How Do Temperature and Salinity Affect Mixing in the Oceans?
- What's Responsible for Smaller Shrimp Catches?

Chapter 22 The Water Planet **489**

CHAPTER 22

PREPARE

Focus Questions
Find out what knowledge students have about the ocean. For each focus question, have students consider the section title and then develop two or three additional questions. For example, Section 2: Why are the various properties of ocean water important? How can we detect them?

Review Topics
In addition to the earth science topics listed, students may need to review these concepts:

Physical Science
- Density is the ratio of mass to volume in a substance.
- Several factors can affect density.

Life Science
- Photosynthetic organisms consume carbon dioxide and emit oxygen.
- Ecosystems depend on complex interactions between living and nonliving things.

Reading Strategy
Model the Strategy: Refer students to page 13 in Chapter 1. Use a version of the following: "Look at the diagram to review the water cycle. The ocean affects all four spheres of Earth—it affects the weather, shapes the landscape, and provides what is necessary for life. What else have we learned about oceans that shows their importance in the Earth system?" Have students skim through the book to review sections on weather, erosion, currents, tides, and so on. Encourage them to continue seeking connections between what they know and what they are learning.

BEYOND THE TEXTBOOK

INVESTIGATIONS
- How Do Temperature and Salinity Affect Mixing in the Oceans?
- What's Responsible for Smaller Shrimp Catches?

VISUALIZATIONS
- Animation of unique properties of water molecules
- Animations of changes in phytoplankton blooms—areas of highest plant productivity in the oceans
- Correlations of coral reefs with ocean data

DATA CENTER
EARTH NEWS
LOCAL RESOURCES
CAREERS

Online Lesson Planner

CHAPTER 22 SECTION 1

▶ Focus

Objectives
1. Summarize the beginnings of oceanography.
2. Describe some methods of modern ocean research.

Set a Purpose
Have students read the key idea. Then tell them to read this section to answer to the question "How do scientists study the ocean?"

▶ Instruct

More about...

The *Challenger* Expedition
The H.M.S. *Challenger*, a converted naval steamship, sailed from Portsmouth, England, and made stops in Bermuda, the South Atlantic island of Tristan da Cunha, Australia, New Zealand, the Philippines, Japan, China, Hawaii, and Tahiti. During its voyage, the *Challenger* logged over 120,000 km and became the first steamship to cross the Antarctic Circle. It took about 20 years to organize and report upon the volume of information they obtained during the expedition. Today, scientists consider the *Challenger* expedition the beginning of modern oceanography.

22.1

Key Idea
Oceanography utilizes several scientific fields, along with innovative technology, to study all aspects of Earth's oceans.

Key Vocabulary
• oceanography

Oceanography

Throughout history, poets and sailors alike have been awed by the size of the world's oceans. When viewed from space, Earth appears mostly blue, because oceans cover more than 70 percent of its surface. For this reason, Earth is sometimes called the "water planet." The immense volume of water on Earth is hard to imagine; the average depth of the ocean is more than four times the average elevation of the continents.

As vast and interesting as they are, the oceans remain largely a mystery. Aided by new technology, scientists are continually discovering and learning about interactions between the ocean and Earth's other environments.

The Beginnings of Oceanography

Oceanography is the scientific study of the ocean using chemistry, biology, physics, geology, and other sciences. In the mid-1800s, a U.S. Navy officer named Matthew F. Maury conducted one of the first modern ocean studies. Maury used the logbooks of U.S. Navy captains to compile charts of ocean currents and winds. In 1855, he published his findings in the book, *The Physical Geography of the Sea*.

The first large-scale ocean research project came in 1872, when scientists aboard the British ship, H.M.S. *Challenger*, used their onboard laboratory equipment to measure ocean depths, take water samples, record temperatures, and study currents. Oceanographers today still use data gathered during this expedition.

World War II brought about the next great advance in oceanography. The military's development of submarines and surface ships led to more accurate ocean charts and new instruments, such as sonar and magnetic recorders. These devices became the basic tools of modern ocean research.

H.M.S. *CHALLENGER* The research aboard the H.M.S. *Challenger* lasted nearly four years and produced over 50 volumes of reports.

490 Unit 6 Earth's Oceans

DIFFERENTIATING INSTRUCTION

Reading Support Have students write the key idea for each section of the chapter at the top of a sheet of paper. As they read, tell them to jot down the specifics they find in the text for each idea. As students complete each section, have them rewrite their notes into summary paragraphs. Ask volunteers to share their summaries with the class as a way to review the chapter.
Use Reading Study Guide, p. 76.

490 Unit 6 Earth's Oceans

Modern Ocean Research

Today, teams of explorers, engineers, and scientists from every continent collaborate to broaden their understanding of the oceans. Some scientists work at sea; some use robotic exploration vehicles. Others work in laboratories on land, using sophisticated computers and tools.

Research vessels are ships that have laboratories and scientific instruments onboard. Among the instruments are deep-sea corers, which collect sediment samples, and sonars, which measure ocean depth. One research vessel, the JOIDES *Resolution,* includes a four-story laboratory used to study samples ranging from rocks to bacteria. Another famous research vehicle, *Alvin,* is a three-person mini-submarine. *Alvin* has explored deep-sea trenches, hydrothermal vents, and the sunken ship *Titanic.* Investigators also use robotic mini-subs, such as *Argo.* Outfitted with cameras and lights, *Argo* can view areas of the deep ocean inaccessible to humans.

Oceanographers use innovative technology such as satellites and moored buoys to collect data. *Moored buoys* (BOO-eez) float on the ocean surface, take specific measurements, and relay data to scientists via satellites. Scientists use the data to make computer models of the ocean floor or to track global trends such as ocean temperatures.

An increasingly important task for oceanographers is the study of how humans affect the oceans. Pollution, oil spills, and deep-sea mining all threaten marine ecosystems. Researchers are also investigating effects of global warming, which may influence the stability of the polar ice caps, the salinity of the oceans, and the level of the sea worldwide.

Some of the exciting discoveries made in ocean research could directly benefit humans. For example, a chemical found in a rare coral species shows promise as a cancer-fighting drug. Mineral towers standing near ocean floor vents may offer valuable natural resources. Newly discovered hydrothermal vents and natural gas deposits are possible energy sources. The ocean plays a vital role in regulating Earth's weather and climate.

DSV *ALVIN* The Deep Submersible Vehicle (DSV) *Alvin* can take two scientists and a pilot to depths of over 4 kilometers.

More about...

Ocean Mineral Resources
Exploration of the oceans has led to the ocean-mining of many mineral resources. Currently, sediments along the continental shelf are mined for sand, gravel, tin, copper, iron, silver, titanium, and platinum. The as-yet-untapped resources of the ocean floor include hand-sized manganese nodules, which may cover 20 percent of the world's ocean floors. They are not only rich in manganese but also contain nickel, copper, and cobalt. There are also mineral towers near ocean floor vents, some 15 stories high, that contain gold, silver, copper, and zinc. However, mining of ocean minerals could cause pollution and disturb ocean life.

22.1 Section Review

1. What percentage of Earth's surface is covered by water?
2. List three scientific devices oceanographers use in their research.
3. **CRITICAL THINKING** Hypothesize how global warming might affect the polar ice caps and salinity of ocean water.
4. **HISTORY CONNECTION** Explain how World War II advanced exploration of the oceans.

ASSESS

1. more than 70%
2. sonar, satellites, moored buoys, deep-sea corers, magnetic recorders, robotic vehicles, and computers
3. Polar ice caps will melt; the ice caps are predominately fresh water, so their melting will reduce the salinity of ocean water.
4. Scientists first developed submarines and scientific devices, such as sonar and magnetic recorders, for military use. They later used this technology for oceanographic research.

MONITOR AND RETEACH

If students miss . . .

Question 1 Have students reread the introductory paragraph on page 490. Ask: Why is Earth called "the water planet"?

Question 2 Have students reread the section. As a class, make a list of the scientific devices mentioned.

Question 3 Remind students that 70 percent of Earth's fresh water is frozen, locked up in glacial ice (Chapter 1, p. 9). Ask: Are the ice caps made of fresh water or salt water? **fresh water** What do you think would happen to the ice caps if the climate warmed? **They would melt.** How would the addition of fresh water affect the ocean water? **The ocean water would become less salty.**

Question 4 Reteach "The Beginnings of Oceanography." (p. 490). Ask: What new oceanographic tools did scientists develop for military use during World War II? **sonar and magnetic recorders**

CHAPTER 22 SECTION 2

FOCUS

Objectives
1. Contrast the densities of liquid water and ice.
2. Explain how the polarity of water molecules affects the behavior of water.
3. Describe how the properties of water change when it combines with ocean salts.

Set a Purpose
Read aloud the section title and have students tell what they know about the properties of water. Then have them read the section to answer the question "How do the unique chemical properties of water explain the nature of Earth's oceans?"

INSTRUCT

Scientific Thinking

PREDICT
Ask these questions to help students make their predictions: Is the density of snow the same as the density of water? **no** How does it differ? **Snow is less dense.** Why do icebergs float? **They are less dense than water.** What would happen if icebergs were denser than water? **They would sink.**

22.2

KEY IDEA
Water has unique chemical properties that help explain the nature of Earth's oceans.

KEY VOCABULARY
- polarity
- hydrogen bonding
- aqueous solution
- boiling point
- freezing point
- buffer

Scientific Thinking

PREDICT
What would happen if water behaved like other substances and ice was denser than the liquid form? Would ponds freeze from the top to bottom? Would 6 inches of snow equal more or fewer inches of rain? What would happen to icebergs?

492 Unit 6 Earth's Oceans

Properties of Water

Water is a small molecule with a big influence. In the atmosphere, the soil, the rivers and seas, and even in the deep underground reaches of the planet—water, due to its unique chemical properties—plays a vital role in Earth's ecology. In this section, you will learn the ways in which some of these properties influence oceans and ocean life.

Density of Water

As you read in Chapter 4, the density of a substance is the ratio of its mass to its volume. The ratio varies depending on the matter's physical state and its temperature. Usually, the density of matter increases as it changes from a gas to a liquid, and again when it changes from a liquid to a solid. Density also increases as temperature decreases, because as the temperature lowers, so does the kinetic energy of the atoms. Thus, the atoms do not vibrate or move around as much, which allows them to be packed more densely together. Water, however, behaves differently from other compounds: The solid form of water (ice) is *less* dense than the liquid form of water. The density of pure water at 4°C is 1.0 g/cm^3, while the density of ice at 0°C is only 0.92 g/cm^3.

WATER   **ICE**

DENSITY OF WATER Patches of ice are less dense than liquid water, so they float.

Liquid water and ice have different densities because, in ice, the molecules are spread farther apart in a rigid framework, filling more space than the molecules in liquid water. This is why people commonly note that water expands as it freezes. You may be able to think of everyday examples illustrating this density difference. For instance, ice cubes float at the top of your drinking glass because they are less dense than the water. A full bottle of water placed in the freezer may pop its lid off as the liquid becomes solid and less dense. During cold winters, highway pavement buckles as the moisture underneath turns to ice and expands. A molecular property called *polarity* helps explain the low density of ice.

DIFFERENTIATING INSTRUCTION

Hands-On Demonstration Demonstrate that water expands as it freezes. Fill a small plastic bottle with water and place a cap on it. Place the bottle in a freezer until the water freezes. Remove the bottle from the freezer and allow students to observe that the sides of the bottle bulge outward. Ask: What caused the sides to bulge? **The water in the bottle expanded and took up more space as it froze.**

Polarity of Water

A water molecule is made up of one oxygen atom covalently bonded to two hydrogen atoms. While each hydrogen atom has only one electron to form a bond with the oxygen atom, the oxygen has eight, so six of them remain unbonded. These additional electrons create a slightly negative charge around the oxygen atom, which leaves a slightly positive charge around the hydrogen atoms. **Polarity** refers to the uneven distribution of charge between the molecule's two ends. Because a water molecule has two poles—one negative, one positive—scientists classify water as a *dipolar* (*di-* = "two") molecule.

When there are many water molecules, the slightly positive and negative regions behave like magnets: The positive end of one molecule attracts the negative end of another and so on. **Hydrogen bonding** is the attraction between oppositely charged regions of different water molecules. This is different from the ionic bond that holds together two atoms within the *same* molecule.

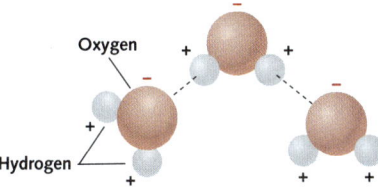

HYDROGEN BONDS Oppositely charged regions on neighboring water molecules (H_2O) are attracted to each other.

As a result of these hydrogen bonds, the attraction among water molecules is very strong. In order for water to change from liquid to solid, hydrogen bonds must break. Because the molecules tend to stick together, water has a high boiling point, meaning it requires a higher temperature than similarly weighted compounds before it turns to vapor. By contrast, some molecules of comparable size and mass exist as gases at room temperature. The strong links among water molecules also make pure water stable and less likely to change form over a range of temperatures. At sea level, water remains liquid in temperatures ranging from 0°C to 100°C.

Hydrogen bonds determine the structure of water in its solid state. As the water molecules freeze, they are rearranged in a rigid pattern with relatively large spaces between them. This results in an open, lattice-like crystal structure that is more spacious and less dense than the structure of liquid water. (See photograph on left with diagrams of water molecules.)

Solutions of Water

Until now, you have learned about the characteristics of pure water; many properties change when water is combined with other substances, such as ocean salts. An **aqueous solution** is a mixture that forms when a salt dissolves in water. Properties such as density, boiling point, and freezing point vary depending on the salt concentration in an aqueous solution.

25-Minute Mini LAB

Observing Water's Polarity in Action

Materials
- rubber balloon
- piece of silk cloth or fake fur
- water faucet

Procedure

1. Inflate and tie the balloon.
2. Rub the piece of silk or fake fur on the balloon to give the balloon a static charge.
3. Turn the faucet on low to form a thin, steady stream of water.
4. Hold the balloon near the stream of water without letting it touch the water.

Analysis

Describe what you observe. When the balloon is statically charged, electrons are transferred from the piece of silk or fake fur to the balloon's surface. If you could observe the molecules of water in the stream, which part of the water molecule, the oxygen atom or the two hydrogen atoms, would be attracted to the balloon? Explain your reasoning.

Explore a 3-D model of water molecules.
Keycode: ES2201

CHAPTER 22 SECTION 2

 VISUALIZATIONS
CLASSZONE.COM

Explore a 3-D model of water molecules.
Keycode: ES2201
Have students review covalent bonding in Section 5.1.

 Visualizations CD-ROM

MINILAB
Observing Water's Polarity in Action

Materials
If silk cloth and fake fur are unavailable, have students use their hair to statically charge the balloon.

Analysis Answers
- Students should observe the stream of water "bending" toward the charged balloon.
- The balloon is negatively charged due to the transfer of electrons to its surface. Since opposites attract, the water molecules' positively charged hydrogen atoms are attracted to the negatively charged balloon.

For proper behavior in a laboratory setting, refer students to Safety in the Earth Science Laboratory on pages xxii–xxiii.

VISUAL TEACHING
Discussion

Direct students' attention to the diagram. Ask: What do the dotted lines represent? **hydrogen bonds** What is a hydrogen bond? **the attraction between the negatively charged region around the oxygen atom of one water molecule and the positively charged region around the hydrogen atoms of another water molecule**

DIFFERENTIATING INSTRUCTION

Reading Support Have students write each boldface vocabulary word on one side of an index card and its definition on the opposite side. Tell students to write the definitions in their own words rather then copying them from their textbook. (You might have students do the same for the boldface words in the other sections of this chapter as well.) Then have small groups share their definitions. Allow time for students to correct the definitions on their cards if necessary.
Use Reading Study Guide, p. 77.

Support for Physically Impaired Students During the Minilab, some students may have difficulty inflating and tying the balloon. If necessary, inflate and tie the balloons ahead of time. Several volunteers could complete the task.

CHAPTER 22 SECTION 2

ASSESS

❶ The polarity of water molecules leads to hydrogen bonding between molecules. The strength of this bonding causes water to have a high boiling point and makes pure water stable and less likely to change form over a wide range of temperatures. As water freezes, the hydrogen bonds between the water molecules cause the molecules to form an open, lattice-like crystal structure that is more spacious and less dense than the structure of liquid water.

❷ Compared to pure water, salt water boils at a higher temperature and freezes at a lower temperature.

❸ The salt dissolved in ocean water lowers its freezing point; thus, ocean water remains liquid even when the air temperature is below freezing.

❹ Blood has dissolved salts that buffer the body's pH in the same way that dissolved salts in ocean water maintain a stable pH environment for aquatic life.

THE DEAD SEA Located between the countries of Israel and Jordan, the Dead Sea has the highest salt concentration of any sea in the world. The water is so saturated that the salt precipitates to the bottom of the sea. The high concentration of salt increases the water's density, allowing swimmers to float easily.

The density of a solution increases as salt concentrations rise, because dissolving more material in the same volume of water increases the mass of the solution. The levels of salt in seawater make it 2–3 percent more dense than pure water.

An aqueous solution's boiling and freezing points also change as salt concentrations increase. The **boiling point** of a solution, the temperature at which it turns from liquid to gas, increases as salt concentrations rise, while the **freezing point,** the temperature at which it turns from liquid to solid, decreases. This is why, after a snowfall, a sprinkling of salt prevents sidewalks from becoming slippery—the snow becomes a liquid, an aqueous solution, instead of a solid, ice, as it mixes with the salt. The raising of boiling point and lowering of freezing point mean that aqueous solutions such as ocean water are even more stable than pure water. The range of temperature over which seawater remains a liquid is approximately −2°C to 100.3°C, more than a two-degree increase over pure water.

Aqueous solutions can exhibit another property vital to oceans and to humans. A combination of salts in a solution can affect changes in its pH. An aqueous solution that acts to stabilize changes in pH is called a **buffer**. In the oceans, buffering is crucial in maintaining a stable pH environment suitable for aquatic life.

22.2 Section Review

❶ Describe three ways that polarity of water molecules affects the behavior of water.

❷ Compare the boiling and freezing points of pure water with the boiling and freezing points of salt water.

❸ **CRITICAL THINKING** Explain how the properties of ocean water make it possible for ships to sail into port even when the air temperature is below freezing.

❹ **BIOLOGY** The normal pH range of blood in the human body is between 7.35 and 7.45. A pH below 6.8 or above 7.8 is deadly. Form a hypothesis describing how the body maintains a steady pH despite environmental changes.

MONITOR AND RETEACH

If students miss . . .

Question 1 Use the diagrams on pages 492 and 493 to explain the effects of polarity on the behavior of water.

Question 2 Review the definitions of *boiling point* and *freezing point*. (p. 494) Ask: What would happen if you added salt to a pot of boiling water? Why do people sprinkle salt on icy sidewalks?

Question 3 Have students reread "Solutions of Water." (pp. 493–494) Ask students to create a graphic that summarizes the effects of salt concentrations on the density, boiling point, and freezing point of aqueous solutions.

Question 4 Review the importance of buffering in the oceans. (p. 494) Ask: How might blood be similar to ocean water?

Properties of Ocean Water

Recall that water is a dipolar molecule with unusual properties and behavior. The characteristics of water change further still when pure water is mixed with salts to form an aqueous solution. In this section, we'll examine some factors—salinity and temperature—that affect the behavior of the world's largest aqueous solution, the ocean.

Salinity

Salinity is a measure of the dissolved salts in water. You can measure the salinity of a sample of seawater in a beaker by evaporating the water and weighing the salt that remains at the bottom of the beaker. Finding the ratio of the mass of the salts to the initial mass of the water gives you the salinity of the water sample. We usually state percentages in parts per hundred, symbolized by %. Since the salt concentration in seawater is fairly low, however, we state salinity in parts per thousand, or ‰. On average, 1000 grams of seawater contain 35 grams of salts, or 35‰.

Oceanographers calculate salinity by measuring the electrical conductivity of seawater. Ions dissolved in seawater enable an electric current to pass through it. The greater the quantity of dissolved salts, the stronger the electric current. Salinity measured by conductivity is expressed in practical salinity units, abbreviated as psu. The average salinity of seawater is 35 psu, the same number as expressed in parts per thousand.

Salinity is an important factor in identifying water masses. A **water mass** is a body of water with distinct properties based on where it originates. When oceanographers know salinity and other properties of different water masses, they can trace a water mass as it moves through the ocean.

The Composition of Seawater

Ocean water contains salts in the form of dissolved ions. Sodium chloride, the most common salt in seawater, consists of a positive sodium ion and a negative chloride ion. Sodium chloride accounts for more than 85 percent of the ocean's salt content—just one cubic kilometer of seawater holds about 27 million tons of sodium chloride! Other ions in seawater include magnesium, sulfate, calcium, and potassium. The table lists percentages of the most common ions. Altogether, seawater contains more than 70 elements, but most, such as gold and uranium, occur only in trace amounts.

You might not think of the ocean as a mining site, but in fact miners extract two important minerals—common salt and magnesium—from seawater.

Other elements within seawater are essential for marine life. For example, calcium is required for the hard shells of some microscopic organisms and shellfish. Nitrogen and phosphorous must be extracted from the water for photosynthesis.

SALT MINING In Sousse, Tunisia, water evaporates from large pans leaving behind heaps of salt for industrial use.

Ions in Seawater

Chloride	55.04%
Sodium	30.61%
Sulfate	7.68%
Magnesium	3.69%
All others	2.98%

22.3

KEY IDEA

Salinity and temperature are two characteristics of water masses that help scientists study oceans and their behavior.

KEY VOCABULARY
- salinity
- water mass
- mixed layer
- thermocline

CHAPTER 22 SECTION 3

FOCUS

Objectives
1. Identify factors that affect salinity levels in the oceans.
2. Describe the three temperature layers of the oceans.

Set a Purpose
Have students preview the section by looking at the headings and illustrations. Then have them read this section to answer the question "How do salinity and temperature vary in parts of the ocean?"

INSTRUCT

VISUAL TEACHING

Discussion
Have students read the table listing salt concentrations in seawater and then view the photograph. Remind them that table salt (NaCl) is made of sodium and chlorine. Ask: Will the salt in the photograph be pure table salt? Explain. No; the salts left behind after evaporation will mirror the composition shown in the table. Further refining is necessary to produce table salt.

DIFFERENTIATING INSTRUCTION

Challenge Activity Nobel Prize winner Fritz Haber (1918) tried to extract gold from ocean water as a means of rescuing Germany from bankruptcy after World War I. Have students research his attempts and outcomes, and report their findings to the class.

More about...

Ocean Salinity
In 1691, the English astronomer Edmund Halley noticed that lakes with no outlets, such as the Caspian and Dead seas, were quite salty. Therefore, he reasoned that rivers carry salts into the oceans. He suggested that Earth's age could be determined by measuring the rate at which various salts are added to the oceans by weathering and then calculating the time it took for the oceans to arrive at their present salinity. While sophisticated for its time, this idea failed to take into account complicated salt recycling processes that occur in the ocean and the fact that much salt is delivered by underwater volcanic activity.

VOCABULARY STRATEGY
The word *salinity* comes from the Latin word *salinus*, which means "salt."

How Do Temperature and Salinity Affect Mixing in the Oceans?
Examine ocean maps of temperature and salinity to predict whether water masses will sink or float.
Keycode: ES2202

Variations in Salinity

Ocean water salinity varies. In deeper ocean waters, the salinity is close to the average of 35‰. Near the surface, it can vary between 33‰ and 37‰. Furthermore, water masses in different parts of the world have different salinities based on local conditions. Salinity is below average in places where large amounts of fresh water enter the ocean. The Baltic Sea, for example, has a salinity of only 30‰ because many rivers and glaciers drain into it. Ocean salinity is also lower in areas of heavy rainfall, such as those near the equator.

Salinity tends to be above average in areas that are extremely hot or cold. In hot, dry climates, the oceans lose water rapidly through evaporation, leaving the salts behind. These areas typically lie at 20° to 40° latitude north and south of the equator. The Mediterranean Sea and the Red Sea are located in these dry belts, and their salinities can be as high as 40‰. Salinity may also be above average in polar waters. When seawater freezes, only freshwater ice forms at first, leaving salts behind in the remaining water and thus increasing its salinity.

One amazing fact about seawater is that even though the salinity may vary, the relative percentages of the dissolved ions do not change. To illustrate, imagine making lemonade using premixed lemonade powder. You can make your drink more or less concentrated, depending on your taste, but no matter how much mix you stir into a glass of water, the ratio of sugar to lemon flavoring is constant. Likewise, the ratio between the salt ions within a water body is constant, regardless of its salinity.

Scientists believe that the salt composition of the oceans has not changed very much in the past 200 million years, even though salts naturally enter and exit ocean water all the time. Salts enter the ocean when underwater volcanoes release minerals and gases that react with seawater to form salts. Salts also enter the ocean when particles erode from mineral-rich rocks and are carried to the sea. Marine organisms contribute salts when they die, because their bodies decompose and return minerals to the water around them.

Salts leave the ocean as marine plants and animals take up dissolved minerals, which are essential for the structure and processes of living things. Salinity also decreases when minerals precipitate out and settle to the ocean bottom as sediment.

Temperature Profile of the Oceans

Almost all the energy that heats the oceans comes from the sun, yet solar light and heat do not penetrate very deeply into the oceans. In fact, most solar radiation is absorbed in the top few meters of seawater.

Because ocean water does not heat readily, ocean temperature decreases rapidly with depth. A typical set of readings for latitude 40° N, the approximate latitude of New York City, might be 20°C at the surface, 11°C at 500 meters, 5°C at 1000 meters, and 2°C at 4000 meters. You can make two observations from these data. First, the rate at which temperature decreases with depth is not constant. Second, with the exception of the surface water, most ocean water is just a few degrees above freezing.

DIFFERENTIATING INSTRUCTION

Support for Visually Impaired Students Redraw the graph on page 497 on white paper to eliminate the dark background. Use a black marker to draw the lines and add additional labels to show the three zones. You might also glue yarn to the paper to represent the curve and have students use their fingers to follow the curve.

Reading Support Have students make a concept web for each layer of the ocean: surface, middle zone, and deep water. They should write the name of the layer in a center circle and then surround it with additional information, such as salinity and temperature characteristics. As students proceed to the next section, they can add to their webs.
Use Reading Study Guide, p. 78.

Oceanographers conventionally divide the ocean into three temperature layers: a surface zone, which is warmed by sunlight; a middle zone penetrated by little light, in which temperatures change rapidly; and a very cold deep zone without light.

VOCABULARY STRATEGY
The word *thermocline* is made up of two parts. *Thermo-* means "heat," while *-cline* means "to slope."

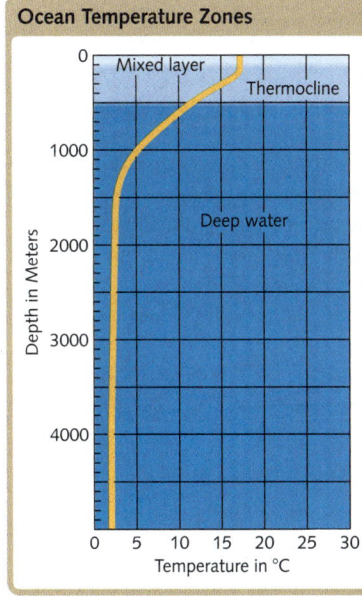

FISH POPULATIONS This school of surgeon fish swims through the warm mixed layer of the Pacific Ocean near the Philippines.

The Surface

The ocean's surface region is called the **mixed layer** because wind and waves mix heat evenly throughout this zone. The warm water of the mixed layer makes up only about 2 percent of the ocean's volume, yet it is very important to life in the ocean. The mixed layer absorbs almost all the sun's light, and therefore grows almost all the marine plants upon which most other marine organisms depend.

At high latitudes and near the equator, the mixed layer may extend to depths of about 50 meters. In middle latitudes, the layer may be as deep as 100 meters. In one or two rare spots, it can be 300 meters deep.

Seasonal changes also play a part in the temperature of the mixed layer. Near the equator, where air temperatures are always high, the mixed layer may be as warm as 30°C all year. Close to the poles, where air temperatures are always relatively cold, the temperature of the mixed layer may remain at about −2°C all year. The largest seasonal temperature changes in the mixed layer occur in the middle latitudes, because in these areas the air temperature changes most with the seasons. At latitude 40°N, for example, a difference of 10°C between summer and winter surface temperatures is not unusual.

CHAPTER 22 SECTION 3

Scientific Thinking

PREDICT
Deep water, open ocean, and polar zones would provide the coldest, least-disturbed ocean environments, thus the best-preserved shipwrecks.

ASSESS

1. Evaporate a measured amount of seawater, then weigh the salt left behind. The ratio of the mass of the salt to the initial mass of the seawater gives the salinity of the seawater. Electrical conductivity is also another measure of salinity. The higher the conductivity, the greater is the salinity.

2. Common table salt is sodium chloride. Salt extracted from seawater will also include sulfate, magnesium, and other dissolved ions.

3. Factors that cause salinity to increase include evaporation, freezing, runoff from land, underwater volcanic activity, and decay of marine organisms. Factors that cause salinity to decrease include precipitation and sedimentation of minerals to the ocean floor and use of minerals by marine organisms.

4. Oceanographers can measure salinity and other properties to identify ocean-floor water.

5. Students should hypothesize that the number of living creatures decreases with ocean depth because living conditions become less favorable.

Scientific Thinking

PREDICT
Shipwrecks are best preserved in cold, still seawater. The cold temperatures slow down bacterial processes that cause decay while the salt content in the ocean helps preserve materials. The lack of water movement prevents parts of the ship from being worn and washed away.

Where do you think shipwrecks would be best preserved? In shallow or deep water? Near coasts or out in open ocean? In tropical zones or polar zones? Explain your reasoning.

U.S.S. YORKTOWN This aircraft carrier sank on June 6, 1944, in the Pacific Ocean during the Battle of Midway. Deep-sea explorer Robert Ballard and a team from National Geographic found the *Yorktown* on May 9, 1998.

The ocean's mixed layer provides the friendliest habitat for marine animals, but some species are adapted to life at other levels. Most organisms require the mixed layer's warmer temperatures and richness of oxygen—which comes from the air overhead (in the atmosphere) and from the plant photosynthesis within. Warmth diminishes with depth, as does the oxygen supply, but animals such as some eels and fish are hardy enough to survive in the lower thermocline. Even at the deep-sea floor, where the scant oxygen and deep cold would make life seem impossible, a few species manage to thrive.

The Middle Zone

A thermometer sinking through the mixed layer registers almost the same temperature throughout the layer. However, below the mixed layer and to the depth of about 1000 meters temperatures drop rapidly. The layer of ocean directly beneath the mixed layer, in which the temperature changes rapidly is called the **thermocline.**

The water at the bottom of the thermocline is very cold. Even near the equator the temperature at the bottom of the thermocline may be only about 5°C. The temperature continues to drop, but more slowly, from the bottom of the thermocline to the ocean floor, where temperatures may be only about 2°C. The temperature of water near the ocean floor is fairly constant all over the globe.

Deep Water

Polar water masses receive the least amount of sunlight, and thus are cold at all depths, from top to bottom. Such cold water is denser than other ocean waters and tends to sink beneath the warmer masses. This cold water moves away from the polar regions along the ocean floor. As a result, polar water is found beneath other ocean water at almost all latitudes. The exceptions are inland seas with shallow, narrow openings to the ocean. For example, the Straits of Gibraltar prevent polar water from entering the Mediterranean Sea. The temperature at the bottom of the Mediterranean remains warm all year, in some places as high as 12°C.

22.3 Section Review

1. Describe two ways to measure salinity.
2. Explain how salt extracted from seawater might be different from common table salt.
3. a) Describe one factor that causes ocean salinity to increase.
   b) Describe one factor that causes salinity to decrease.
4. **CRITICAL THINKING** Explain how oceanographers can tell that most of the cold water on the ocean floor originates in polar regions.
5. **BIOLOGY** Work with a partner to hypothesize how the number of living creatures would change with the depth of the ocean.

MONITOR AND RETEACH

If students miss . . .

Question 1 Reteach "Salinity." (p. 495) Ask: How could you weigh the salts in seawater? What effect do ions dissolved in water have on the movement of electric current?
Question 2 Have students review the table on page 495.
Question 3 Refer students to "Variations in Salinity." (p. 496) Have them compare factors that increase salinity with those that decrease salinity.

Question 4 Review the definition of *water mass*. (p. 495) Have students discuss characteristics that distinguish polar ocean water.
Question 5 Ask students to identify factors that are necessary to support life. **oxygen, particular temperature range, source of energy** Have them evaluate how the availability of each of these factors changes as ocean depth changes and how these changes affect the number of organisms that can survive.

Ocean Life

The diversity of life found in the oceans is extraordinary. Scientists estimate that the oceans harbor as many as 10 million species, many of which have yet to be discovered. The life forms range from microscopic animals and plants, such as plankton, to the largest known animal, the blue whale.

Marine Plant Life

Sunlight is vital to ocean life. Like land plants, most sea plants need sunlight to grow. However, as you learned in Section 22.3, sunlight penetration decreases rapidly with depth. Only within the mixed layer is there enough sunlight for most plants to carry out photosynthesis.

While many types of plants live in the ocean's mixed layer, one of the most important groups is the microscopic phytoplankton. **Phytoplankton** are typically single-celled plants that float freely in the ocean's surface waters. One of the most abundant kind of phytoplankton is the **diatom**, a one-celled plant with a delicate, thin shell made of silica. Phytoplankton, including diatoms, make up the base of the ocean's food chain and are the primary energy source for the marine ecosystem.

When large numbers of phytoplankton concentrate in one area, they can change the color of the water. Such formations are called *blooms*. Large blooms are visible from space and can help scientists predict where to find groups of life forms.

The survival of phytoplankton populations depends on factors like ocean currents, temperature, and the amount of nutrients available. Areas with large phytoplankton populations can support large numbers of microscopic marine animals, which eat the phytoplankton. During photosynthesis, the phytoplankton consume the animals' carbon dioxide waste and then give off oxygen, which the animals then use to survive.

When diatoms die, their shells settle to the sea bottom and become part of the sediment. Marine geologists can use shells preserved in this way to trace changes in diatom populations, to determine the age of the sediment, and even to hypothesize the water temperature at the time the diatoms lived.

22.4

KEY IDEAS

Marine organisms are an important part of the ocean and provide clues to the ocean's history.

While most marine life need many of the same nutrients that land plants and animals do, some have adapted to use other resources.

KEY VOCABULARY

- phytoplankton
- diatom
- zooplankton
- coral
- nekton
- black smoker

Discover areas of highest plant productivity in the ocean.
Keycode: ES2203

DIATOMS This microscope photograph shows that one-celled diatoms are remarkably diverse. (Magnification is approximately 150X.)

Chapter 22 The Water Planet **499**

CHAPTER 22 SECTION 4

FOCUS

Objectives
1. Explain the importance of phytoplankton to ocean life.
2. Identify important marine animals.
3. Describe how certain bacteria living near deep-sea vents produce food.

Set a Purpose
Ask students what they know about the marine life shown in the diagram on page 501. Then have them read to answer the question "How do marine organisms interact?"

INSTRUCT

More about...

Blooms—Red Tides
Blooms of dinoflagellates, a type of phytoplankton, are more commonly called "red tides," given their color. The plankton produce a substance toxic to humans. This toxin can become concentrated in the tissue of shellfish, making the shellfish dangerous to eat.

 INVESTIGATIONS
CLASSZONE.COM

Discover areas of highest plant productivity in the ocean.
Keycode: ES2203

 Visualizations CD-ROM

DIFFERENTIATING INSTRUCTION

Developing English Proficiency Students should be familiar with many of the organisms shown and discussed in this section. However, the common names of the organisms probably differ in meaning in different languages. "Hatchet fish" translated into the Spanish equivalent may not be the name used for the fish. Duplicate and enlarge the illustration on page 501. Post the illustration and have students add the names of the organisms in their native languages.
Use Spanish Vocabulary and Summaries, pp. 43–44.

CHAPTER 22 SECTION 4

More about...

Zooplankton
Krill are an important form of zooplankton that range from 1 cm to 15 cm in length and occur in vast numbers in some oceanic regions—sometimes coloring the water pink. These small shrimplike animals are important food for baleen whales and other marine animals, especially in Antarctic waters. Recent evidence suggests that global warming might be contributing to their decline. Samplings in some oceans show decreased numbers of krill and corresponding population decreases of certain species that feed on them. Adélie penguins, for example, have suffered at least a 30-percent population decline on King George Island, which is close to Antarctica.

CRITICAL THINKING

Ask: What might be the effect on marine food chains if phytoplankton populations were to decrease? Decreases in phytoplankton populations would cause decreases in populations at all levels of marine food chains.

VOCABULARY STRATEGY
The words *phytoplankton* and *zooplankton* come from the ancient Greek word *planktos*, which means "wandering." The prefix *zoo-* comes from the ancient Greek word for "living being," while the prefix *phyto-* comes from the ancient Greek word for "plant."

Marine Animal Life

Many animals, rely on phytoplankton for nourishment. The first step in the food chain occurs when zooplankton eat phytoplankton. **Zooplankton** are microscopic animals that float in ocean water and eat phytoplankton or smaller zooplankton. Other animals, in turn, eat the zooplankton; species from small fish and birds to giant whales feed on these tiny animals.

Zooplankton come in many varieties. Some, such as radiolarians, are single-celled organisms. Others are very small crustaceans with hard outer shells; Daphnia is one example. Still other zooplankton are simply the larvae or immature forms of larger animals, such as squid, crab, or jellyfish.

Zooplankton help maintain balance in ocean environments. Besides serving as food for other animals in the food chain, they help moderate the ocean's salinity by absorbing mineral ions with which they build their hard body parts, and by returning them back to the water when they die and decay. Further, wastes from zooplankton fall, promoting the flow of nutrients and minerals from the surface to deeper parts of the ocean.

The **coral** is another important marine life form. This tiny sea animal lives in colonies fastened to rocky seafloors. The word *coral* can also refer to the animal's skeletal remains. Corals live in waters with temperatures between 18°C and 21°C and at depths no greater than 90 meters—where sunlight reaches. Since corals do not move, they depend on ocean currents to bring in part of their food supply. Protists—plant-like, single-celled organisms that live within the coral colonies—provide another food source.

CAREER

Marine Biologist

Some oceanographers focus their work on studying the exotic life forms found in the oceans. Marine biologists discover and study marine plants and animals and their interaction with their environments. Some may study the population growth of species in a given area, while other researchers focus on the effects of fishing or pollution. Marine biologists also can be involved in organizing ocean cleanup efforts or marine rescue operations for dolphins, whales, and manatees that may become stranded on land. One way marine biologists share their findings with their peers and the public is by publishing their research in scientific journals.

Marine biologists typically have a bachelor's degree or higher in marine biology or a related field such as biology, zoology, or ecology. Scuba diving and being comfortable in the water, on boats, and in submersible vessels are often necessary for those seeking a career in marine biology. Students exploring a career in marine biology may find it worthwhile to volunteer or intern at aquariums or laboratories or aboard research ships. ■

MARINE BIOLOGISTS study all kinds of marine life. This biologist, equipped with an underwater notepad and marker, is observing a manatee.

Learn more about a career as a marine biologist.
Keycode: ES2204

DIFFERENTIATING INSTRUCTION

Reading Support Have students read the Vocabulary Strategy about the origins of the words *phytoplankton* and *zooplankton*. Check their understanding of the root word by asking: What do you think plankton are? microscopic marine organisms that are either plants or animals What prefixes are added to this root word to distinguish between plants and animals? *phyto-* and *zoo-* Why is *plankton* an appropriate root word for these organisms? They float in the ocean, "wandering" with the ocean water. What is a good way to remember the difference between phytoplankton and zooplankton? One way is to associate a zoo with animals, thus remembering that zooplankton are microscopic animals, not plants.

Use Reading Study Guide, p. 77.

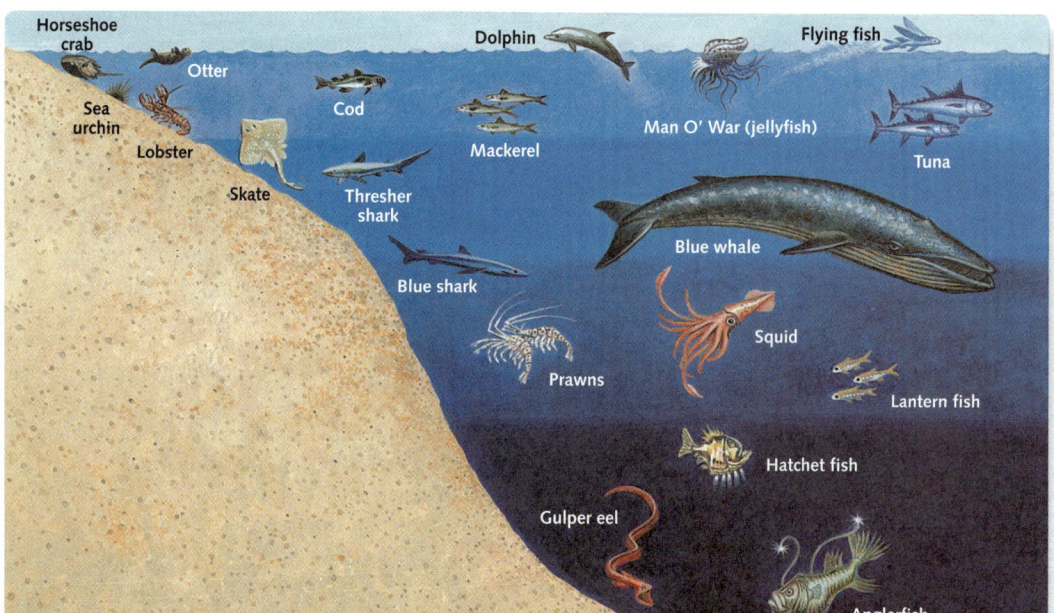

OCEAN LIFE Species have evolved to occupy many different ocean levels. For example, lobsters crawl along the continental shelf, blue whales may dive hundreds of meters but return to the surface to breathe, and anglerfish produce light chemically to see in water where sunlight does not penetrate.

Corals create an important ecosystem for fish and other aquatic life. The shells of corals form from the lime in seawater. When corals die, their shells remain, and new corals grow on the old, empty shells. Large buildups of coral shells are called coral reefs. The intricate canyons, caves, and surfaces of the reefs provide food and shelter to many plants and animals. Reefs also protect the coastlines of more than 100 countries by preventing coastal erosion.

The group of marine animals you are probably most familiar with are **nekton**—free-swimming organisms that include fish, reptiles, whales, squid, and jellyfish, among others. The areas in which each species of nekton can survive depend on temperature, salinity, and the availability of nutrients. For animals that live on the sea floor, the type of sea bottom is also a determining factor. Most nekton are found in the mixed layer.

Nekton are important food supplies both for each other and for humans. Nearly 1 billion people around the world rely on fish for a significant part of their diet. However, human activity is increasingly threatening the survival of nekton such as fish. Overfishing depletes their numbers, while trawling harms sea-floor ecosystems. Pollution that enters the oceans in the form of fertilizer runoff or dumped human waste also harms marine environments, affecting not only fish but all ocean plants and animals.

Deep Ocean Life

The small amount of oxygen in the deep ocean comes from the cold, dense water of the polar regions, which sinks and circulates into all parts of the

---

CHAPTER 22 SECTION 4

Learn more about a career as a marine biologist.
Keycode: ES2204

VISUAL TEACHING
Discussion
Have students identify the mixed-layer, thermocline, and deepwater zones in the illustration and the kinds of organisms that live in each zone. Lead students to recognize that the illustration is not drawn to scale. Ask: Which zone actually is largest? **the deepwater zone** What is the source of food for organisms in the deepwater zone? **The main food source is organic matter that settles from above. (On page 502, students will learn that near deep-sea vents, chemosynthetic bacteria use hydrogen sulfide to produce food. Other organisms around the vents feed on the chemosynthetic bacteria.)**

Extend
Ask: What challenges do whales face when they dive to depths of several hundred meters? **The whales must be able to deal with the enormous changes in pressure.**

DIFFERENTIATING INSTRUCTION

Challenge Activity Coral reefs are important marine ecosystems, so much so that scientists supplement naturally formed reefs by creating artificial reefs from sunken ships, oil rigs, and other scrap material. Have students research some of these artificial reefs to find out where they are located, what was used to create them, and how they have affected the area around them.

CHAPTER 22 SECTION 4

ASSESS

1. Phytoplankton use carbon dioxide that is given off by marine organisms as waste; they give off oxygen, which marine animals use to survive.

2. Answers may vary. Zooplankton help balance the salinity in the oceans by recycling minerals. They absorb mineral ions to build hard body parts. When zooplankton die and decay, their remains may dissolve. Zooplankton also help cycle nutrients from the surface to the deeper layers of the water when their wastes drop to the bottom of the water.

3. Chemosynthetic bacteria use hydrogen sulfide. This prevents the gas, which is poisonous to most other life-forms, from rising and harming organisms in the upper levels.

4. Answers may include individual actions and/or societal laws. For example, individuals can recycle so that less waste is dumped into seawater. They can decrease fertilizer use or develop alternatives so that chemical runoff does not end up in the oceans. Nations can pass international treaties and laws to stop overfishing, destructive methods of trawling, and the destruction of coral reefs.

---

sea. Because no light penetrates to such depths, plants cannot survive at deep levels. This means that plants neither provide oxygen nor consume carbon dioxide through photosynthesis. As a result, carbon dioxide accumulates in deep ocean currents.

Even under such harsh conditions, however, some animals do survive. Researchers aboard the mini-submarine *Alvin* discovered some marine animal communities living as far as two and a half kilometers below the ocean's surface. These animals do not need sunlight for energy or phytoplankton for food. They live near vents, found along mid-ocean ridges, that seep hydrogen sulfide from beneath the ocean floor.

Ocean-floor vents occur in two forms. In one form, cracks in the sea floor release gentle drifts of hot water at temperatures of about 16°C. Other vents, called **black smokers,** spew water of about 380°C from volcano-like chimneys. In both cases, the hot water begins as cold seawater seeping down through cracks in the ocean floor. The water sinks until it is heated by hot rocks deep in the basaltic crust. As it heats, the water dissolves minerals and gases from the basalt, then moves upward to the sea floor through either type of vent. If it erupts through a black smoker, the hot water reacts with surrounding seawater to form clouds of black iron sulfide particles.

Hydrogen sulfide gas is crucial to life near ocean-floor vents. Certain bacteria thrive on it. They use a process called *chemosynthesis* to produce food. Chemosynthetic bacteria then become food for larvae and other organisms, which then become food for barnacles, giant clams, white crabs, giant tube worms, and other unique animals that live near the vents.

For most life on Earth, hydrogen sulfide is a poisonous gas. Fossil fuel refineries generate hydrogen sulfide as a toxic and flammable by-product. Yet at deep-sea vents, hydrogen sulfide is the primary source of energy. The existence of organisms around these vents is a remarkable example of life's ability to adapt to severe environments.

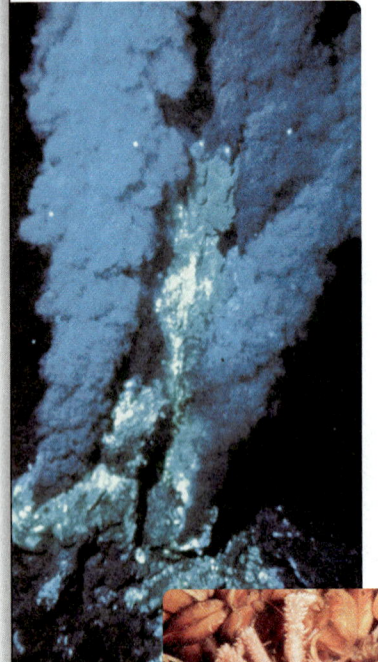

DEEP-SEA VENTS provide a hospitable environment for many unique sea animals.

22.4 Section Review

1. In addition to providing a food source, in what other ways do phytoplankton affect the lives of marine animals?

2. Describe one way that zooplankton help recycle ocean resources.

3. **CRITICAL THINKING** You've learned that organisms in the mixed layer depend on one another to survive. The same is true among life forms in deep ocean regions. How might the behaviors of deep-water species affect the balance of life higher up?

4. **ECOLOGY** Describe two ways in which people can protect the environments of marine plants and animals.

MONITOR AND RETEACH

If students miss . . .

Question 1 On the board, write the following equation for photosynthesis: sunlight + carbon dioxide + water → food + oxygen. Have students identify what phytoplankton use and produce as they undergo photosynthesis.

Question 2 Review with students the process by which zooplankton build their hard body parts. (p. 500) Ask: What happens to the minerals in the hard body parts when the animals die?

Question 3 Have students summarize the process by which bacteria living near deep-sea vents produce food. (p. 502)

Question 4 Have students read through the section and list threats to marine organisms. Then have them brainstorm solutions.

SCIENCE & Society

Troubled Waters: Coral Reefs

Coral reefs, which are among Earth's oldest and most important ecosystems, provide food and shelter for about one-quarter of all ocean species. Many coral reefs are being seriously damaged, however.

Why are coral reefs in danger? Can anything be done to save them?

Human activities as well as naturally occurring events threaten coral reefs. The exotic reef fish that lurk in the crannies of reefs fetch high prices from owners of expensive restaurants and saltwater aquarium keepers. To catch these fish, fishers in some parts of the world toss homemade bombs into the water or pump it full of poisonous cyanide. Millions of coral organisms, as a result, are poisoned or reduced to lifeless rubble.

Pollution is also a threat. Soil and fertilizer runoff from coastal farms enter the water; this adds nutrients that spur algae growth. Large algae blooms deprive coral of necessary oxygen and sunlight.

Damage also seems to result from changes in ocean temperature. In 1997–1998, a major El Niño climate event sent warm Pacific Ocean surface waters eastward. The warm water caused coral organisms to expel the algae that live within them. These algae help coral manufacture food and give them their vivid tints. Portions of reefs turned a deathly white and may eventually die. According to the Global Coral Reef Monitoring Network, 16 percent of the world's reefs experienced "bleaching" in 1997–1998.

Steps have been taken to stop coral reef destruction and to allow damaged reefs to recover. In some nations, including the United States, reefs are protected by law. Other nations are considering using tourism earnings to pay for reef conservation. Some nations have called for cutbacks in greenhouse-gas emissions, which may be linked to global warming and the increased severity of El Niños. ■

BLEACHED CORAL can be caused by warm water temperatures, which cause coral organisms to expel algae that live within them. This deprives the coral of nutrients.

Extension

SCIENCE NOTEBOOK
It is clear that eliminating blast and cyanide fishing and outlawing coastal farming would help to save reefs. What social and economic issues would be involved in stopping these practices?

Explore coral reefs around the world.
Keycode: ES2205

HEALTHY CORAL REEFS are important to sustain. They provide a home for a remarkable diversity of marine life, including this lionfish.

Chapter 22 The Water Planet 503

CHAPTER 22 SECTION 4

Explore coral reefs around the world.
Keycode: ES2205

 Visualizations CD-ROM

Science and Society

Further Risks to Coral Reefs
According to studies done by the World Resources Institute, more than a quarter of the coral reefs in the world are at high risk. Coastal development and overfishing pose enormous threats. Overfishing has upset the balance of the delicate reef ecosystems in many parts of the world. Coastal development has resulted in the dredging of harbors and shipping channels, with the dumping of dredge spoils on reefs. Airports and other construction projects have been built directly on reef communities. Excess sewage resulting from development has spurred deadly algae blooms that block sunlight necessary for photosynthesis by reef organisms.

Science Notebook
Students should understand that the outlawing of coastal farming and reef fishing might hurt some people's means of support and way of life. The subsequent unemployment and economic problems could be catastrophic in some areas.

CHAPTER 22 ACTIVITY

CHAPTER 22 LAB Activity

Ocean Water and Fresh Water

Purpose
To infer the relative densities of different water samples by measuring their salinity, both directly and indirectly

Materials
Using distilled water will give more accurate measurements.

To make the simulated ocean water sample, add $\frac{1}{8}$ cup of table salt for every liter of water and stir until dissolved. This will result in a salinity of 35–37 parts per thousand. (Alternatively, 35 g of salt for every 965 g of water will produce a salinity of exactly 35 parts per thousand.)

To make the simulated Great Salt Lake water sample, add $\frac{5}{8}$ cup of table salt for every liter of water and stir until dissolved. (Alternatively, 175 g of salt for every 825 g of water will produce a salinity of exactly 175 parts per thousand.)

A conductivity probe may be used instead of a salinity probe. However, the measurements must be converted to parts per thousand or psu.

Procedure

6 If the range of the salinometer does not extend to 175 parts per thousand, the saltwater samples must be diluted. To dilute a 200-mL sample to 25 percent of its original concentration, add 50 mL of the sample to 150 mL of fresh water. Students should multiply the reading by 4 to find the salinity of the undiluted sample.

It might be necessary to perform the above dilution twice for the Great Salt Lake water sample. (Then multiply by 8.)

Students can make and test their own saltwater samples or test other liquids, such as vegetable oil, vinegar, and rubbing alcohol, with the hydrometer.

Skills and Objectives
- **Measure** the density and salinity of salt and fresh water.
- **Evaluate** indirect versus direct measures of salinity.

Materials
- 3 500-mL beakers
- fresh water sample
- simulated ocean water sample
- simulated Great Salt Lake water sample
- scissors
- drinking straws
- modeling clay
- salinometer (salinity probe)
- rinse bottle
- metric ruler
- vegetable oil spray

A hydrometer is an instrument used to measure the density of liquids. The greater the density, the higher the hydrometer will float. Because increasing the salinity of water also increases the density, this instrument helps to determine salinity.

The electrical conductivity of water also increases with salinity. A salinometer detects salinity differences by measuring electrical conductivity.

In this lab you will explore these two methods of measuring salinity differences in water samples. The first method is an indirect salinity measure; the second is a direct measurement.

Procedure

Part A: Hydrometer

1 Make a hydrometer. Cut several straws into 10-cm lengths; coat the inside of one end of each straw with vegetable oil. Roll a ball of modeling clay approximately 2 cm in diameter. Starting with the coated end, push one straw completely through the clay. Observe how well the straw floats upright when placed in a beaker of water. Vary the length of straw and the amount of clay until you create a hydrometer that floats vertically.

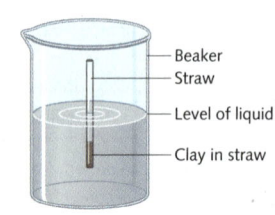

HYDROMETER

2 Measure the total length of your hydrometer to the nearest millimeter.

3 Your teacher will have room-temperature samples of fresh water, simulated ocean water, and simulated Great Salt Lake water. Fill a beaker with each sample and label the beakers.

4 Float your hydrometer in each sample. Use a ruler to measure to the nearest millimeter the length of straw extending from the surface of the water. Subtract this length from the total hydrometer length to determine the submerged length for each water sample. Copy the data table below to record all measurements.

Water Properties Data Table			
	Temperature (°C)	Length of Submerged Hydrometer (mm)	Salinity (psu)
Fresh water			
Ocean water			
Great Salt Lake water			

504 Unit 6 Earth's Oceans

5 For the ocean water and Great Salt Lake water, calculate the percent difference in the length of submerged hydrometer from that of the fresh water hydrometer. Use this formula:

$$\text{Percent difference} = \frac{\text{difference between the two values}}{\text{fresh water value}} \times 100$$

Part B: Salinometer

6 Obtain a salinometer from your teacher. This instrument determines salinity by measuring how easily the water conducts electricity. The more ions in the water, the more easily electricity passes through it. This conductivity is expressed in psu (practical salinity units). Place the salinometer in the fresh water sample; when the red light appears, record the temperature and the salinity from the readout. Record these values on your data chart. Rinse the probe with fresh water from the rinse bottle and take readings for the next two samples.

7 For the ocean water and Great Salt Lake water, calculate the percent difference in salinity from that of the fresh water.

Analysis and Conclusions

1. Compare the percent differences that you derived from your hydrometer with those you derived from the salinometer. Make a bar graph showing your results. What might account for the differences between the two methods? Which method is more reliable? Discuss the strengths and weaknesses of each type of measurement.

2. Approximately how much more saline is the Great Salt Lake water than the ocean water sample?

3. Use density to explain the differences in the length of straw that remain submerged in the water for each sample. How does this relate to salinity? Explain.

4. Explain why it was important that all three samples be the same temperature during this investigation. In what way would temperature differences affect your data?

5. If you were a shipping executive calculating the weight of cargo your tankers could safely carry from port to port, how would the observations you made in this lab affect your decision-making? Might the weight capacity of tankers differ depending on where they operate? Explain.

CHAPTER 22 ACTIVITY

ANALYSIS AND CONCLUSIONS

1. Results from both methods should be similar. Those students who made long hydrometers will have more accurate results than those who made short hydrometers, simply because the length of straw extending above the water will be greater for the longer hydrometer and therefore measurable with a smaller margin of error. Using a salinometer is more reliable because there is less chance for human error.

2. The Great Salt Lake water should be about five times more saline than the ocean water.

3. The greater the density of the liquid, the smaller is the length of submerged straw. This is because the volume of liquid equal in weight to the hydrometer is smaller when the density of the liquid is greater. This relates to salinity because dissolved salts in water increase the weight of any given volume. Increased weight per unit volume means increased density.

4. It was important that all three samples were at the same temperature to ensure that the lab activity had only one independent variable—salinity. Temperature affects density and conductivity.

5. The hydrometer floated lower in fresh water than in salt water because a given volume of fresh water weighs less than an equal volume of salt water. Thus, ships traveling through bodies of fresh water, such as the Great Lakes, have smaller weight capacities than those traveling though oceans.

Refer students to Safety in the Earth Science Laboratory on pages xxii–xxiii.

CHAPTER 22 REVIEW

VOCABULARY REVIEW

1. water mass
2. black smokers
3. coral
4. zooplankton
5. freezing point

CONCEPT REVIEW

6. chemistry: to study the ocean as an aqueous solution; biology: to study forms of ocean life; physics: to study circulation patterns; geology: to study ocean sediments and plate tectonics

7. In solid water, hydrogen bonds arrange the water molecules in a rigid pattern with relatively large spaces between them; thus, the molecules are packed less densely than in liquid water.

8. Salinity is a measure of the dissolved salts in water; 35‰ means 1000 grams of seawater contains 35 grams of salt. Salinity can fall in places where large amounts of fresh water enter the ocean. Salinity can rise in hot areas due to evaporation or in cold areas due to freezing.

9. Knowing these properties allows for identification of water masses, which can be traced as they move through the ocean.

10. At high latitudes and near the equator, the mixed layer can be as deep as 50 meters; in middle latitudes, the mixed layer can be as deep as 100 meters. Degree of mixing and season of the year determine the temperature of the mixed layer.

11. In one form, cracks in the seafloor release gentle drifts of hot water at temperatures of about 16°C. The other form, known as black smokers, spews water that is about 380°C.

12. Oxygen in the ocean's upper waters comes from the atmosphere and from plant photosynthesis within the water. Oxygen reaches the ocean's deep waters by the sinking and circulation of the cold, dense water of the polar regions. Carbon dioxide accumulation in the deep ocean is due to the lack of plants, which take in carbon dioxide during photosynthesis.

CHAPTER REVIEW 22

Summary of Key Ideas

22.1 Modern oceanographers work in land laboratories and on research vessels. They use satellites to collect data from moored buoys.

22.2 The water molecule is dipolar. Hydrogen bonds between water molecules make water stable over a wide temperature range. When water and salt are in solution, the temperature range widens.

22.3 Oceanographers measure salinity, the dissolved salts in seawater, to trace water masses. The relative proportions of ions is the same in all seawater, even when salinity is different. The mixed layer is the only zone with enough light to grow marine plants. Below it is the thermocline, a zone of rapid temperature drop. Deep water is very cold.

22.4 Phytoplankton are microscopic sea plants. Zooplankton, microscopic animals, eat phytoplankton. At hydrogen sulfide vents near mid-ocean ridges, hydrogen sulfide-eating bacteria, rather than sunlight-using phytoplankton, are the basis of the chain. Dissolved oxygen is most abundant near the ocean surface and decreases with depth. The concentration of carbon dioxide is high near the ocean floor.

KEY VOCABULARY

aqueous solution (p. 493)
black smoker (p. 502)
buffer (p. 494)
coral (p. 501)
diatom (p. 499)
hydrogen bonding (p. 493)
mixed layer (p. 497)
nekton (p. 501)
oceanography (p. 490)
phytoplankton (p. 499)
polarity (p. 493)
salinity (p. 495)
thermocline (p. 498)
water mass (p. 495)
zooplankton (p. 500)

13. Animals that live in the deep ocean depend on chemosynthetic bacteria that use hydrogen sulfide gas to produce food.

Vocabulary Review

Write the term from the key vocabulary list that best completes the sentence.

1. A region of ocean water having the same temperature and salinity throughout is a(n) ___?___ .

2. The hot springs of mid-ocean ridges are called ___?___

3. Although the individual animals are themselves very tiny, ___?___ lives in a large colony fastened to the sea floor.

4. Not only extremely small species of sea animals but also the the immature versions of larger animals such as crabs, can be found among ___?___ .

5. Ocean water has a lower ___?___ than pure water does.

Concept Review

6. Name some of the scientific fields involved in oceanography. Tell what role each might play in the study of Earth's oceans.

7. Explain how hydrogen bonding causes solid pure water to be less dense than liquid pure water.

8. What is salinity? What does 35‰ mean when applied to sea water? Identify some factors that cause ocean salinity to fall below 35‰ and to rise above 35‰.

9. Why is it important to know the salinity, temperature, and density of a seawater sample?

10. Describe the thickness of the mixed layer as it relates to latitude. Identify two factors that determine the temperature of the mixed layer.

11. Describe the two different types of ocean floor vents.

12. What are two sources of oxygen for the upper waters of the ocean? How does oxygen reach the deep waters of the ocean? Why does carbon dioxide accumulate in deep ocean waters?

13. How are the animals that live in the deep ocean different from other ocean life?

Critical Thinking

14. **Hypothesize** The concentration of silica in rivers flowing into the ocean is more than three times the concentration of silica in the ocean. Write a hypothesis explaining this difference in concentration.

15. **Predict** What relationship would you expect to find between the amount of dissolved oxygen in a water mass and the "age" of a water mass moving along the bottom of the ocean?

16. **Infer** Why is the bottom water of the Mediterranean Sea likely to be poor in oxygen?

17. **Analyze** The percentage of sodium and chloride ions in seawater is much greater than their percentages in Earth's crust. Propose a reason.

Interpreting Graphs

The graph at right shows variations in the density of water masses, in g/cm³, relative to their temperature and salinity. Salinity, in ‰, increases toward the right on the horizontal scale while temperature, in °C, increases upward on the vertical scale.

18. The average temperature and salinity of the Atlantic Ocean are 3.7°C and 34.8‰. Water flowing from the Mediterranean Sea into the Atlantic may have a temperature and salinity of 12°C and 37.2‰. Would this Mediterranean water be expected to rise or to sink when it enters the Atlantic? Why?

19. The ocean's densest water mass is the Antarctic Deep Water mass, which has a temperature of −1.9°C and a salinity of 34.6‰. Extrapolate (extend the graph's information) to approximate this water's density.

20. When two water masses with the same density but different temperature and salinity mix, the properties of the new water mass fall at some point on a straight line joining their original locations on the graph. In general, how does the density of a new water mass compare with the density of the original water masses? (Hint: Lay a straightedge between any two points on the same density line.)

Internet Extension

What Is Responsible for Smaller Shrimp Catches? Analyze seasonal changes in waters of the Gulf of Mexico.

Keycode: ES2206

Writing About the Earth System

SCIENCE NOTEBOOK Explain how a sample of ocean water is an example of the effects of the interactions among the geosphere, the biosphere, the atmosphere, and the hydrosphere.

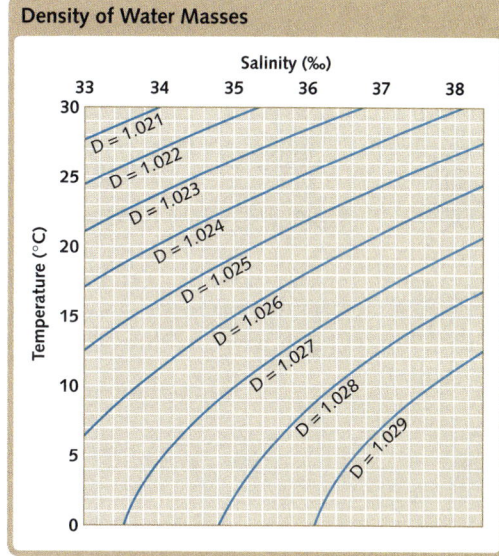

Density of Water Masses

Writing About Science

Writing About the Earth System

A sample of ocean water, which is part of the hydrosphere, has salt and possibly sediment that derives from the geosphere, phytoplankton and zooplankton that are part of the biosphere, and dissolved gases that result from wave action mixing the water with the atmosphere.

CHAPTER 23 PLANNING GUIDE THE OCEAN FLOOR

	Section Objectives	Print Resources
Chapter-Level Resources		Laboratory Manual, pp. 99–102 Internet Investigations Guide, pp. T81–T82, 81–82 Spanish Vocabulary and Summaries, pp. 45–46 Formal Assessment, pp. 67–69 Alternative Assessment, pp. 45–46
Section 23.1 Studying the Ocean Floor, pp. 510–513	1. Identify methods scientists use to map the ocean floor. 2. Describe the techniques of echo sounding and core sampling.	Lesson Plans, p. 80 Reading Study Guide, p. 80 Teaching Transparencies 26, 40, 41
Section 23.2 The Continental Margin, pp. 514–516	1. Describe the parts of the continental margin. 2. Compare and contrast active and passive continental margins. 3. Explain the origins of submarine canyons.	Lesson Plans, p. 81 Reading Study Guide, p. 81 Teaching Transparency 27
Section 23.3 The Ocean Basin, pp. 517–522	1. Describe the features of the ocean basin. 2. Explain how ocean basin features change over time.	Lesson Plans, p. 82 Reading Study Guide, p. 82
Section 23.4 Ocean Floor Sediments, pp. 523–525	1. Describe three sources of ocean sediments. 2. Explain why studying ocean sediment is important.	Lesson Plans, p. 83 Reading Study Guide, p. 83

INVESTIGATIONS
CLASSZONE.COM

Section 23.1 INVESTIGATION
What Does the Ocean Floor Look Like?
Computer time: 45 minutes / Additional time: 45 minutes
Students explore features of the seafloor and explain how they formed.

Chapter Review INVESTIGATION
When Were the Atlantic and Pacific Oceans Separated by Land?
Computer time: 45 minutes / Additional time: 45 minutes
Students date the formation of the Isthmus of Panama.

Technology/Online Resources

Electronic Teacher Tools
Visualizations CD-ROM
Classzone.com
Online Lesson Planner
Test Generator

Classzone.com
 ES2301 Investigation

Visualizations CD-ROM
Classzone.com
 ES2302 Visualization, ES2303 Visualization, ES2304 Data Center

Visualizations CD-ROM
Classzone.com
 ES2305 Visualization, ES2306 Career

Classroom Management

Gather these materials for minilab and laboratory activities.

Minilab, p. 516
 Ocean Floor Map, pp. 512–513
 Teaching Transparency 26
 World Map, pp. 710–711
 Teaching Transparency 40

Lab Activity, pp. 526–527
 shoebox
 modeling clay
 aluminum foil
 masking tape
 sharp pencil
 coffee stirrer or thin straw
 fine-tip marker or pen
 graph paper
 colored pencils
 metric ruler

CHAPTER 23

The Ocean Floor

This research image of Monterey Bay, California, shows some features of the ocean floor.

What other features do you expect to find on the ocean floor?

INTRODUCE

Build Interest

Before sharing background information about the image with students, allow time for their questions and comments.

Ask questions like the following:
- What predominant feature can you identify in the picture? a submarine canyon
- What do you think the different colors represent? various depths
- Who might find such a picture useful? oceanographers, ship captains, fishers

Respond
- What other features do you expect to find on the ocean floor? mid-ocean ridges, underwater peaks and plains, as well as deep trenches

About the Photograph

This image shows the submarine Monterey Canyon system. Orange represents the underwater Santa Cruz continental shelf and dark blue is the deepest water. The main canyon begins about 100 m offshore from Moss Landing, beyond the right border of this image. The canyon system extends about 80 kilometers and at its deepest part, the canyon walls are more than 1900 meters high, making Monterey Canyon system comparable in size to the Grand Canyon.

BEYOND THE TEXTBOOK

Print Resources
- Reading Study Guide, pp. 80–83
- Transparencies 26, 40, 41
- Formal Assessment, pp. 67–69
- Laboratory Manual, pp. 99–102
- Alternative Assessment, pp. 45–46
- Spanish Vocabulary and Summaries, pp. 45–46
- Internet Investigations Guide, pp. T81–T82, 81–82

Technology Resources
- Visualizations CD-ROM
- Classzone.com
- Online Lesson Planner
- Electronic Teacher Tools
- Test Generator

CHAPTER 23

PREVIEW

▶ **FOCUS QUESTIONS** In this chapter you will study the ocean floor and learn more about the key questions listed below.

Section 1 What tools and methods do scientists use to study the ocean floor?

Section 2 What are continental margins?

Section 3 What are the topographical features of the ocean basin?

Section 4 What are the different types of ocean sediments? How can they help us understand Earth's history?

▶ **REVIEW TOPICS** As you investigate the ocean floor, you will need to use information from earlier chapters.

- magma accumulation at mid-ocean ridges (p. 196).
- seafloor sediments as evidence of Earth's climate changes (p. 476).

▶ **READING STRATEGY**

SUMMARIZE

As you investigate the ocean floor, take down notes on the main ideas. Read each section of Chapter 23 thoroughly. Then write a brief description, in your own words, of each section's key idea.

INTERNET RESOURCES
CLASSZONE.COM

At our Web site, you will find the following Internet support for this chapter.

DATA CENTER **LOCAL RESOURCES**
EARTH NEWS **CAREERS**
VISUALIZATIONS **INVESTIGATIONS**

- Black Smoker
- Stages of Atoll Formation
- Origins of Ocean Floor Sediments

- What Does the Ocean Floor Look Like?
- When Were the Atlantic and Pacific Oceans Separated by Land?

Chapter 23 The Ocean Floor **509**

CHAPTER 23

PREPARE
Focus Questions
Find out what knowledge students have about the ocean floor. For each focus question, have students consider the section title and then develop two or three additional questions for each section. For example, Section 1: How is modern technology helpful for ocean floor studies? What have we learned by studying the ocean floor?

Review Topics
In addition to the earth science topics listed, students may need to review these concepts:

Mathematics
- Distance = Rate × Time

Reading Strategy
Model the Strategy: To help students summarize each section, use a version of the following: "Note that the authors have already provided a starting place for your section summaries. Look on page 510, which is the first page of section 1. Find the Key Idea written on the top left of the page, referring to the many methods and technologies used to study the ocean floor. Remind yourself to restrict the summary you write only to descriptions of oceanographic methods and technologies. Additional information, such as the history of these technologies, would not go in this summary." Encourage students to keep their summaries brief and concise as they continue through the chapter.

BEYOND THE TEXTBOOK

INVESTIGATIONS

- What Does the Ocean Floor Look Like?
- When Were the Atlantic and Pacific Oceans Separated by Land?

VISUALIZATIONS

- Interactive visualizations and photographs of a black smoker at a mid-ocean ridge
- Image of islands in various stages of atoll formation
- Video of the origins of ocean floor sediments

DATA CENTER
EARTH NEWS
LOCAL RESOURCES
CAREERS

Online Lesson Planner

Chapter 23 The Ocean Floor **509**

CHAPTER 23 SECTION 1

FOCUS

Objectives
1. Identify methods scientists use to map the ocean floor.
2. Describe the techniques of echo sounding and core sampling.

Set a Purpose
Have students read this section to find the answer to the question "How do scientists study the structure and composition of the ocean floor?"

INSTRUCT

VISUAL TEACHING
Discussion

Use the diagram to lead a discussion on how the technique of echo sounding works in practice. Tell students to suppose that after emitting a sound signal, the ship receives the echo in 6 seconds. Explain that the average speed of sound through seawater is 1500 meters per second. Ask: What is the water's depth? **4500 meters** (Some students may give an answer twice the depth. Lead them to understand that it took only 3 seconds for the sound signal to reach the ocean floor. Another 3 seconds were required for the echo to reach the ship.) Now tell students to suppose that the signal comes back loud and clear. Ask: What is the likely composition of the sea floor—rock and gravel, or mud? How do you know? **rock and gravel; The intensity of reflected sound waves is greater when the sea floor is rock and gravel than when it is mud.**

23.1

KEY IDEA
Scientists have developed many different technologies and methods for studying the ocean floor.

KEY VOCABULARY
- echo sounding
- core sampling

INVESTIGATIONS
CLASSZONE.COM

What Does the Ocean Floor Look Like? Drain the oceans and explore mountains, valleys, and volcanoes of the sea floor. Examine some features and explain how they formed.
Keycode: ES2301

Studying the Ocean Floor

The advent of submersibles, satellites, and other technology has enabled scientists to study better the structure and composition of the ocean floor.

Echo Sounding

In the days of the first oceanographic surveys, scientists measured the distance to the sea floor with a lead weight on a line. They lowered the weight until it touched bottom, measured how much line they had let out, and then hauled the weight back up. In deep seas, a single depth reading might take an entire day. The process was tiresome and resulted in very limited information.

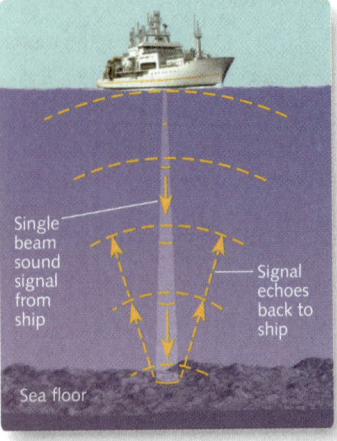
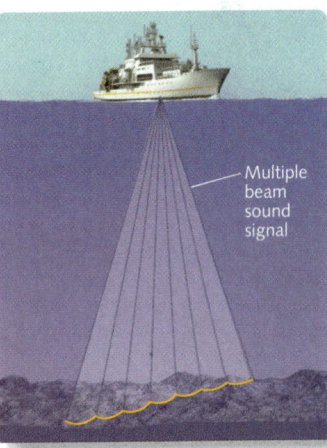

ECHO SOUNDING Single beam soundings provide the depth only along a ship's course, while multiple beams measure depth over a much larger area.

Today ships use **echo sounding,** or sonar, to find the distance to the ocean floor. A special device called a precision depth recorder sends a sound signal through the water to the sea floor. By tracking how long it takes for the signal to reach the bottom and echo back to the ship, scientists can measure the water's depth. The recorder traces a continuous profile of the area over which the ship is sailing. Such profiles help oceanographers make accurate and detailed maps of the sea floor.

Multi-beam echo sounding allows scientists to cover an area that is up to twice the depth of the water over which the research vessel is traveling. Accurate maps can be made more efficiently and thus be more readily compared with previous maps.

Scientists can also measure the intensity of reflected sound beams to determine sea floor composition. For example, rock and gravel reflect sound waves more strongly than mud. Then scientists may combine

510   Unit 6 Earth's Oceans

DIFFERENTIATING INSTRUCTION

Reading Support Have students outline the section in their science notebooks. Instruct them to use the blue heading as the overall title of their outlines. Then have them copy the red headings into their journals and number the headings, leaving plenty of space beneath each heading. Students should use this space to fill in the supporting details. After they have finished, ask students which supporting details they filled in, and list them on the board. Monitor students' work to be sure no key ideas were omitted.
Use Reading Study Guide, p. 80.

information about the shape and composition of ocean floor features. Scientists have used such data to pinpoint places along continental slopes that could experience landslides triggering dangerous tsunamis.

Sediment Sampling

Scientists gather samples of the ocean floor in order to study its composition. They have discovered that the ocean floor consists of layer upon layer of sediment. By studying these layers, scientists can learn a great deal about Earth, including how its atmosphere and climate have changed over millions of years. You will learn more about ocean sediments in section 23.4.

Scientists obtain some sediment samples through a process called **core sampling,** in which a hollow instrument removes a long cylinder of material—called a *core*—from the ocean floor. Depending on the type of coring device, the sediment layers in the core may remain intact so scientists can examine specific layers. Other sampling methods—such as dredging, in which a large scoop is dragged along the bottom—do not preserve the layers, although they are helpful for gathering rock samples.

Scientists use several kinds of coring devices. One is a gravity corer: a hollow, weighted tube with one open end, attached to a ship by a cable. The gravity corer drops through the water and plunges into the sea floor, driven by its own weight. Then crew members draw it back up and remove the core sample. In soft sediments, a gravity corer can obtain samples from one to ten meters long. Other types of corers can remove cores of up to 1500 meters in total length by retrieving them in shorter sections.

Satellite Observations

Echo sounding is useful for mapping the ocean floor, but it has limitations. Ships move slowly and can map only a fraction of the sea floor as they travel. When echo sounding was the best mapping method available, huge geographic features, such as the Louisville seamount chain in the Pacific, went undiscovered, despite their great size, found only by chance during random ship crossings.

Today, satellites provide greater range and speed in the mapping process. A satellite can gather far more data more quickly than a seagoing vessel. Although a signal sent from a satellite cannot reach the ocean floor, it can bounce off the ocean's surface. The ocean surface varies depending on what lies beneath it; it is slightly higher over undersea mountains and slightly lower over undersea trenches. Using a device called a radar altimeter, newer satellites detect these variations to within centimeters. Computers then process the data from many measurements to produce a high-resolution sea floor image. See pages 512 and 513 for a map of the ocean floor.

TOPEX POSEIDON SATELLITE
Launched in 1992, this satellite is a collaborative project between France and the United States. Dedicated to observing the Earth's oceans, the *Topex Poseidon* can monitor global ocean circulation, detect weather patterns, and measure global sea levels.

CHAPTER 23 SECTION 1

INVESTIGATIONS
CLASSZONE.COM

What Does the Ocean Floor Look Like?
Keycode: ES2301

Use a computer and projector to introduce the investigation in class.

Use Internet Investigations Guide, pp. T81, 81.

More about...

Mapping the Ocean
Variations in the height of the ocean surface are caused by variations in Earth's gravitational field. Massive undersea mountains have extra gravitational attraction that causes water to bulge over them. Deep trenches offer less gravitational attraction and account for minute dips in the ocean surface. The Geosat satellite launched by the U.S. Navy in 1985 had the ability to detect such sea surface variations with an accuracy of up to 0.03 m. Other satellites collecting similar data have led to the high-resolution images of the oceans available today. The maps are insufficient for assessing navigational hazards but can help scientists analyze ocean floor obstructions and constrictions for their effects on major ocean currents. The maps can also be used to locate shallow seamounts harboring important fisheries.

DIFFERENTIATING INSTRUCTION

Developing English Proficiency The word *sounding* might cause confusion because of the many meanings of the word *sound* in English. Point out that *sound*, used as a noun, can mean air vibrations that stimulate the ear or a body of water. *Sound*, used as an adjective, means being in good condition. In this section, *to take a sounding* means to measure the depth of water. The word *sounding* predates echo sounding technology. It is merely by coincidence that echo sounding involves measurements of actual sound waves. Have students write the various meanings of the word *sound* in their science notebooks.

Use Spanish Vocabulary and Summaries, pp. 45–46.

CHAPTER 23 SECTION 1

VISUAL TEACHING

Discussion
Have students study the ocean topography map. Ask: Where are the shallowest areas of the oceans located? *along the edges of continents* Where are the deepest areas? *in the mid-ocean basins and trenches, shown in purple* What feature of the crust does this map show? *plates, as indicated by some of the plate boundaries formed by the mid-ocean ridge system*

Extend
Have students use the information on the map to hypothesize the result of a major drop in sea level, as happens during ice ages. *Students should infer that the orange areas would become exposed, thus increasing the sizes of the landmasses.*

Use Transparencies 26, 40, 41.

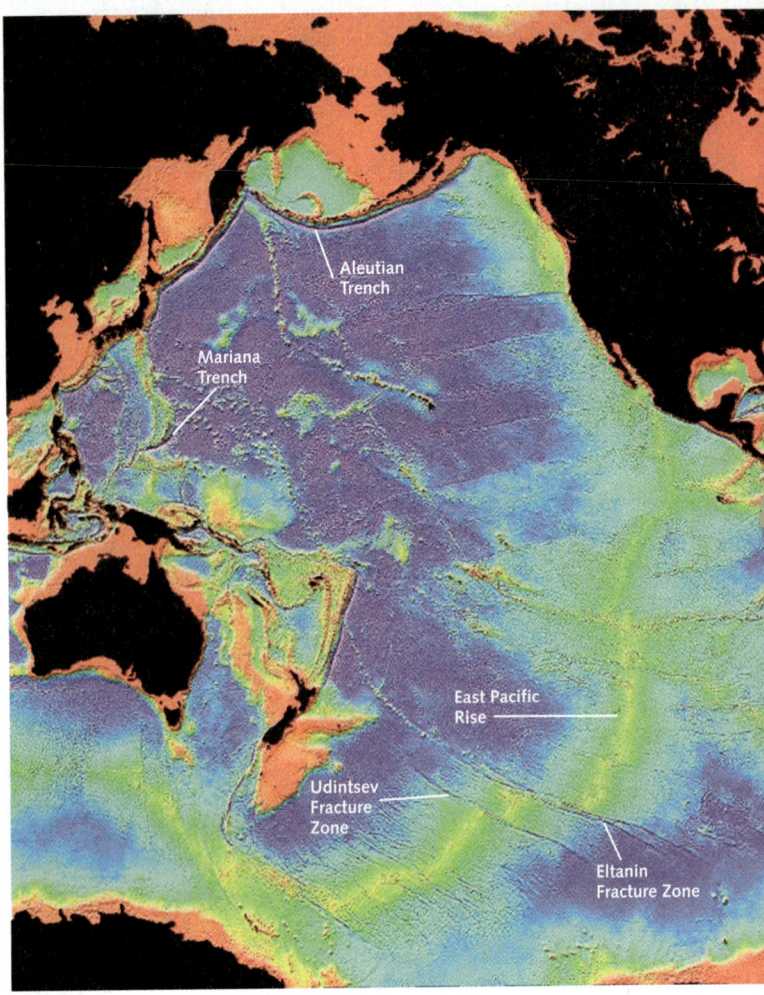

Map of the Ocean Floor
The above map of Earth's ocean topography was generated by combining data taken by radar altimeters and echo soundings. Land is shaded black, red represents the shallowest water and violet represents the deepest water. An 80,000 kilometer ridge system extending throughout the world's oceans shows up clearly as lines of green and yellow. Mid-ocean basins with depths over four kilometers are shown as large areas of dark blue and purple. The deep trenches lining the Western Pacific ocean are shown as thin lines of purple near Asia and Australia. You will learn more about all these features in section 23.3.

DIFFERENTIATING INSTRUCTION

Challenge Activity Have students use the map to draw a general profile of the Atlantic Ocean. They might choose to draw their profile along a line extending from North America to Africa. Profiles should show a shallow slope near the continents, a sharp drop-off toward the ocean basins, and a rise at the mid-ocean ridge. The profiles should be somewhat symmetrical on either side of the ridge.

Support for Visually Impaired Students It might be necessary to clarify the ocean floor map shown on these two pages. Try enlarging the diagram, then using a black felt-tip marker to make dotted lines around the violet areas of deepest water. Use a black outline with diagonal crosshatches to show the orange continental shelf regions. Mark trenches and mid-ocean ridges in black and label them. Explain the marks, and go over the enlargement with students, having them point out the features.

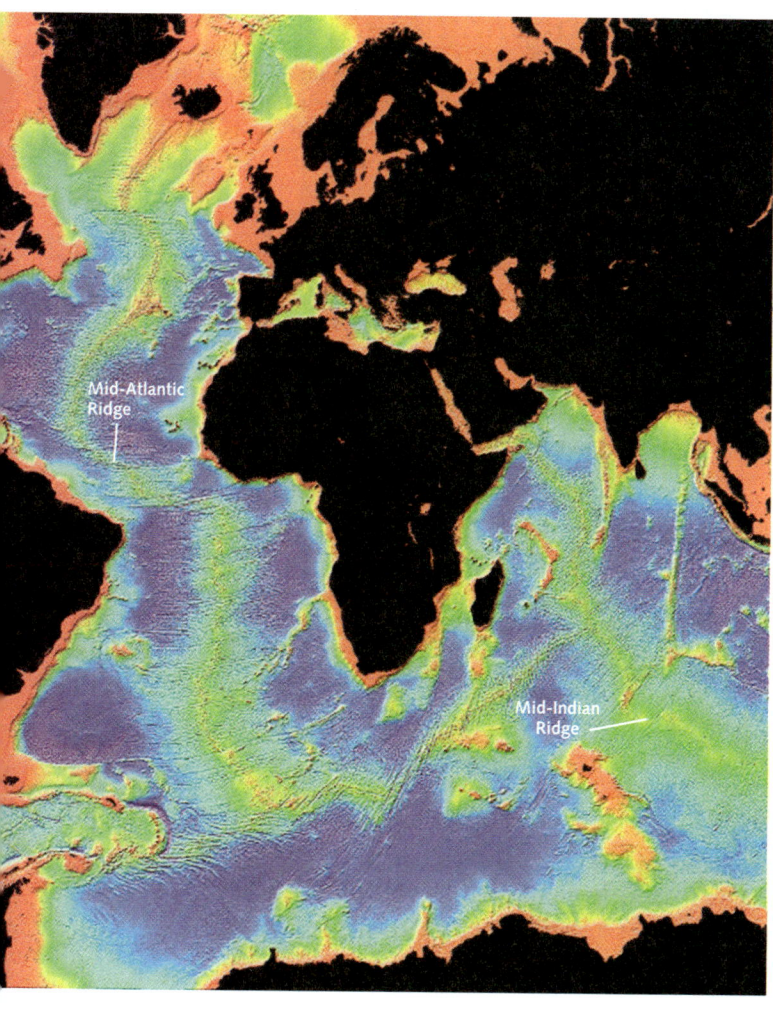

23.1 Section Review

1. Describe why echo sounding is a better method of mapping the ocean floor than the lead weight on a line.
2. What information can scientists learn from a sediment core?
3. How do satellites determine the terrain of the ocean floor?
4. **CRITICAL THINKING** What might be the benefits of understanding the topography of the ocean floor?

CHAPTER 23 SECTION 1

ASSESS

1. Taking soundings with a lead weight on a line is time-consuming and results in little information. Echo sounding traces a continuous profile of the area under the ship, allowing accurate maps to be made. Multiple-beam echo sounding measures depths over a large area. Echo sounding can also reveal sea floor composition.

2. Sediment cores reveal information about the composition of the ocean floor and how Earth's atmosphere and climate have changed over millions of years.

3. Satellites sense minute differences in the ocean surface that correspond to sea floor topography.

4. **Sample Response:** Knowing the topography of the ocean floor allows for safe and efficient submarine navigation, identification of prime fishing areas, and identification of potential landslide areas that could trigger tsunamis.

MONITOR AND RETEACH

If students miss . . .

Question 1 Use the diagram on page 510 to discuss the advantages of echo sounding.
Question 2 Have students reread "Sediment Sampling." (p. 511)
Question 3 Reteach "Satellite Observations." (p. 511) Ask: What does a radar altimeter do? **take readings off the ocean's surface** Ask: How does this compare to echo sounding? **readings are taken off ocean floor**

Question 4 Have students brainstorm jobs that might require the use of ocean floor maps.

CHAPTER 23 SECTION 2

FOCUS

Objectives
1. Describe the parts of the continental margin.
2. Compare and contrast active and passive continental margins.
3. Explain the origins of submarine canyons.

Set a Purpose
Have students read the key idea. Then have them read this section to find the answer to the question "What topographical features are located at the margins of continents?"

INSTRUCT

VISUAL TEACHING
Discussion
Use questions to lead a discussion on features of the continental margin shown on this page. Ask: What features show that this is a passive margin? *a broad continental shelf ending at a continental slope that descends toward a continental rise, and a lack of trenches and rugged mountains* Ask: Where does the ocean basin officially begin? *at the base of the continental slope*
Use Transparency 27.

23.2

KEY IDEA
The continental margins are the underwater edges of continents and include several types of topographical features.

KEY VOCABULARY
- continental shelf
- continental slope
- continental rise
- active continental margin
- passive continental margin
- submarine canyon
- turbidity current

The Continental Margin

The continental margin is the underwater part of the continental crust. It consists of the continental shelf and the continental slope, and, depending on whether it is an active or a passive margin, it may lie adjacent to a trench or a continental rise—both features of the ocean basin.

Parts of the Continental Margin

The **continental shelf** extends from the shoreline out toward the continental slope. Continental shelves are very flat, and their width varies. At some points, they are almost nonexistent; at others, they extend as far as 1280 kilometers (800 miles) out to sea.

The **continental slope** begins at the shelf edge, where water depth begins to increase rapidly. On average, a continental slope is 20 kilometers wide and descends to a depth of about 3.6 kilometers. It has about the same average slope as a movie-theater aisle, descending 70 meters over 1 kilometer. Sediments tend to build up temporarily on the slope; eventually they become unstable and tumble downward to form the **continental rise.**

The continental rise is generally several kilometers thick, and it lies on the ocean crust. It descends gradually from the continental slope to the ocean floor at between four and eight meters per kilometer. It is considered part of the ocean basin rather than part of the continent.

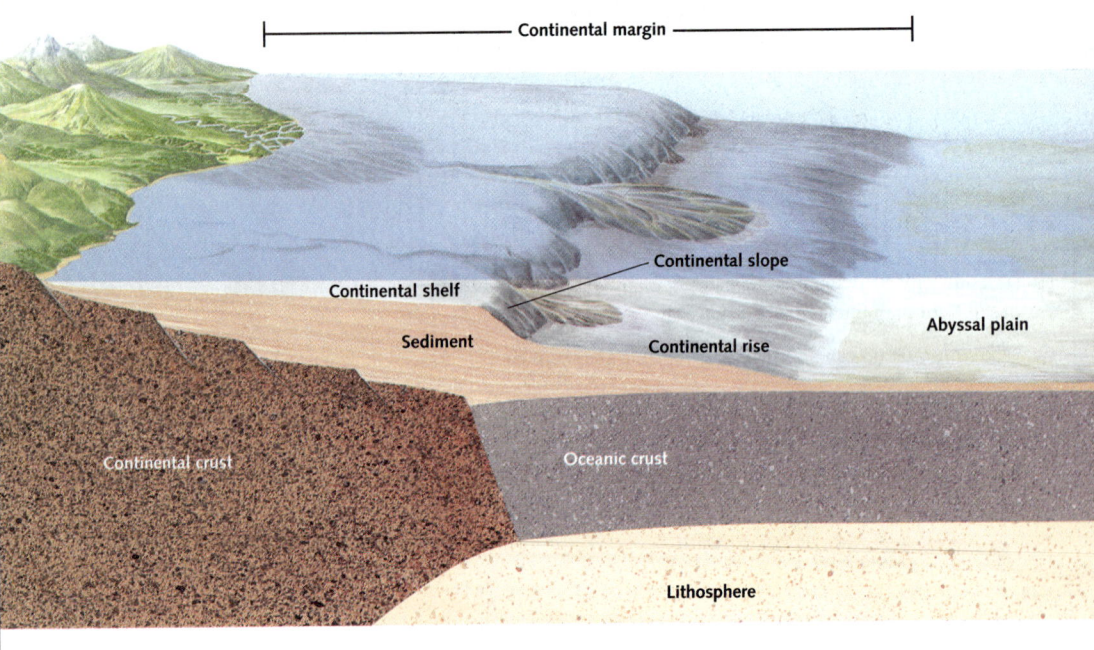

CONTINENTAL MARGIN This diagram shows a passive continental margin.

514 Unit 6 Earth's Oceans

DIFFERENTIATING INSTRUCTION

Reading Support Help students implement the strategy of restating facts and ideas. Have students read the section, then write in their science journals the key facts and ideas for each heading that the section contains. For example, under the first heading "The Continental Margin," they might write that the continental margin is the part of a continent that is under the ocean. Check to see that students are properly discriminating important ideas. With practice, students will improve their ability to do so.
Use Reading Study Guide, p. 81.

514 Unit 6 Earth's Oceans

Active and Passive Margins

Not all continental margins possess the same features. The features of a specific margin depend on where it lies in relation to a subduction zone or transform fault. Continental margins can be active or passive.

Where the oceanic plate is sliding beneath, or subducting under, the continental plate, there is an **active continental margin.** Here, the continental shelf is very narrow and the continental slope is steep; instead of moving down toward a broad continental rise, the continental slope falls away into a deep oceanic trench.

At an active margin where subduction is taking place, the continental rise is either small or nonexistent. Another type of active margin occurs where two plates are sliding past one another. These transform fault boundaries experience many earthquakes, as demonstrated by the San Andreas fault system in California.

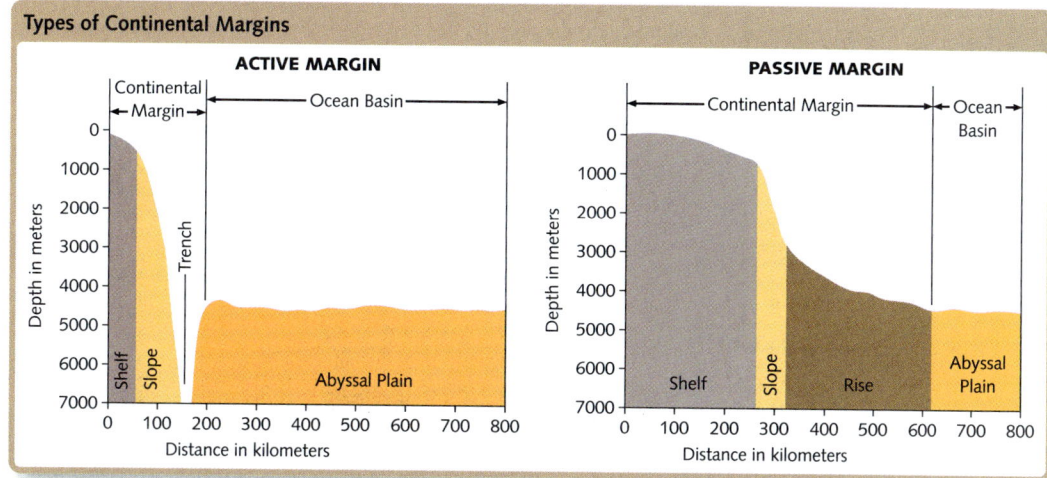

Active margins encircle the Pacific Ocean and include the Peru-Chile Trench, where the Nazca Plate is being subducted beneath the South American Plate. Parallel to the shoreline are rugged coastal mountains.

Unlike active continental margins, **passive continental margins** are not located at plate boundaries. At passive continental margins, there are no ocean trenches or rugged coastal mountains. The continental shelf is broad—sometimes more than 300 kilometers wide—and ends at the continental slope, which descends toward the continental rise. The Atlantic coast of North America is a good example of a passive continental margin.

Submarine Canyons

Most small valleys that cut through the continental shelf and the continental slope probably resulted from mudslides. Larger **submarine canyons,** however, have a much different origin. Some of these are bigger than the Grand Canyon and slice from the continental shelf clear to the ocean floor.

CHAPTER 23 SECTION 2

MINILAB
Continental Margin
Analysis Answers

- The margin east of Buenos Aires is a passive margin, whereas the margin east of Tokyo is an active margin.
- The Atlantic Ocean contains mostly passive margins, whereas the Pacific Ocean contains mostly active margins.
- The plate boundaries in the Atlantic Ocean run through the center of the ocean at the mid-ocean ridge. The plate boundaries in the Pacific Ocean are located at the continental margins.

ASSESS

1. Both active and passive continental margins occur at edges of continents and are the underwater part of the continental crust. Active margins occur where continental margins are along plate boundaries (subduction or transform fault). No plate boundaries occur at passive margins. Passive margins have a wide continental shelf and a broad continental rise. Active margins have a deep trench and may have no continental rise.

2. Answers will vary. Generally, places where continents meet the Pacific Ocean have active continental margins, and places where continents meet the Atlantic Ocean have passive continental margins. Active margins have trenches; passive margins have broad continental shelves.

3. When sea level was low, some rivers transported sediment all the way to the continental rise. After sea level rose, rivers transported sediment only to the continental shelf, where turbidity currents and wave action then moved the sediment.

Mini LAB
Continental Margin
Materials
- Ocean Floor Map, pp. 512–513
- World Map, pp. 710–711

Procedure
1. Find the east coast of South America, near Buenos Aires, Argentina, on the Ocean Floor Map.
2. Sketch a profile of the continental margin east of Buenos Aires. Label any shelves, slopes, rises, or trenches.
3. Repeat Step 2 for the continental margin east of Tokyo, Japan.

Analysis
Which profile shows an active margin and which shows a passive margin? Does the Atlantic Ocean contain mostly active or passive margins? What about the Pacific Ocean? Where are the plate boundaries located in the Atlantic Ocean? the Pacific Ocean?

516 Unit 6 Earth's Oceans

The Hudson Canyon and its associated channel systems extend more than 400 kilometers out to sea, from the mouth of New York's Hudson River out along the continental rise, where it reaches a depth of more than two kilometers. Geologists think Hudson Canyon, and others like it, began forming during the Ice Age. Sea level then was perhaps 100 meters lower than it is now, and broad areas of the continental shelves were above water. Rivers such as the Hudson cut valleys to the edge of the continental shelves. When the glaciers melted, the sea level rose, covering the valleys.

Some large submarine canyons, however, are too deep to have ever been above sea level, while others do not line up with existing rivers. Geologists think these canyons may be the result of powerful turbidity currents. Triggered by earthquakes or simply by gravity, **turbidity currents** are great landslides of mud and sand that speed down the continental slopes. The speed of turbidity currents makes them powerful agents of erosion. They may continue beyond the continental slope and onto the continental rise, carving out channels. Eventually, they slow and drop the sediments they carry: first the larger particles, such as coarse sand, and then the fine particles of clay.

Submarine Canyon

SUBMARINE CANYON A river formed the upper part of this canyon. Turbidity currents formed the deeper part, including the abyssal fan, or deep-sea fan. Turbidity currents deposit coarser sediments first, thus building thick abyssal fans on the continental rise.

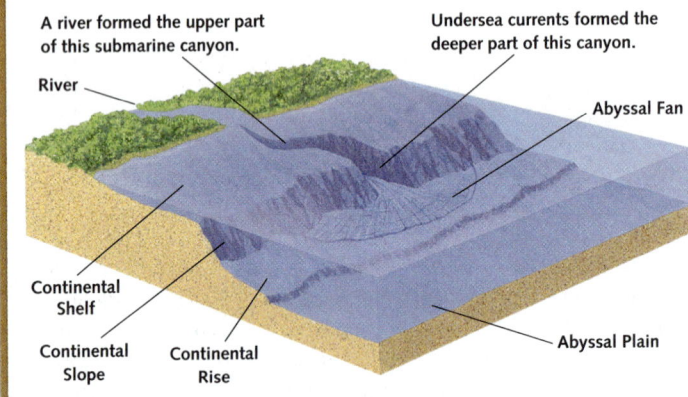

23.2 Section Review

1. Compare and contrast active and passive continental margins.
2. Working with a partner, use the map on pages 512–513 to find examples of active and passive continental margins. What geographic features does each display?
3. **CRITICAL THINKING** The way sediment was transported from the land during times of low sea level differs from how it is transported now. Explain how.

MONITOR AND RETEACH

If students miss . . .

Question 1 Have students use the graphs on page 515 to compare and contrast the features of active and passive continental margins.

Question 2 Have students correlate the depth information provided for the map on page 512 with the depths of continental-margin features shown in the graphs on page 515.

Question 3 Reteach "Submarine Canyons." (pp. 515–516) Have students generalize what they read about the formation of the Hudson Canyon during and after the Ice Age.

The Ocean Basin

The ocean basin is not the featureless, underwater expanse people once thought it was. It has a remarkable and interesting terrain that, like all of Earth's environments, changes over time.

Abyssal Plain and Abyssal Hills

One feature of the deep sea floor is the **abyssal plain,** the flattest of all Earth's surface areas. The abyssal plain off Argentina's coast, for example, rises fewer than 3 meters over 1300 kilometers (10 feet over 800 miles). Abyssal plains are composed of sediments, most of which came from continents. How could material from continents reach the deep sea floor? The answer lies in turbidity currents. During periods when the sea level is low, such as ice ages, rivers deposit sediment at the edges of continental shelves. Turbidity currents then carry the material down the continental slopes when the sea level rises again and spread it over the continental rise and abyssal plain. The sediments in an abyssal plain can be more than one kilometer thick.

Although they occur in all oceans, abyssal plains are especially well developed and widespread in the Atlantic Ocean, where the continental margins have few trenches that can trap continental sediment.

Abyssal hills are another part of the deep ocean basin. These small, rolling hills often occur in groups next to oceanic ridge systems. In the North Atlantic, abyssal hills form two strips that run parallel to the mid-Atlantic Ridge for almost its entire length.

Individual hills are typically several meters to several hundred kilometers across and rise no more than a few hundred meters above the abyssal plain. Interestingly, scientists have found entire systems of abyssal hills beneath the layers of sediment that blankets the abyssal plains. These sediment-covered hills represent the original sea-floor surface that formed at mid-ocean ridges.

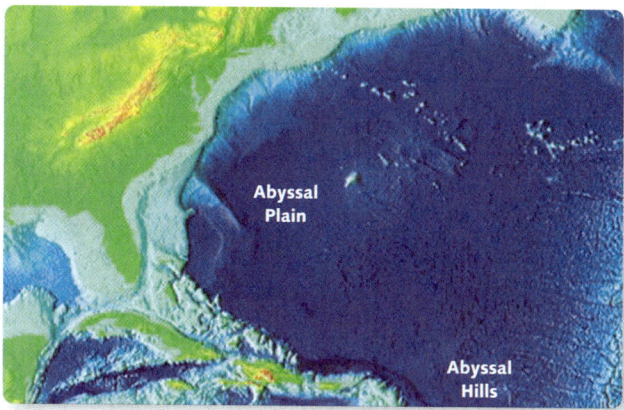

ABYSSAL PLAIN AND HILLS Generated from satellite and ship-board measurements of sea floor and land elevations, this map uses color variation to represent changes in elevation. The uniformity of the abyssal plain indicates little change in elevation while slight changes can be detected for the abyssal hills.

23.3

KEY IDEA

The ocean basin has a wide range of topographical features. Natural forces change these features over time.

KEY VOCABULARY
- abyssal plain
- abyssal hill
- island arc
- fracture zone
- seamount
- guyot
- coral atoll

VOCABULARY STRATEGY

"Abyss" comes from the ancient Greek word for "bottomless," *abussos.*

CHAPTER 23 SECTION 3

FOCUS

Objectives
1. Describe the features of the ocean basin.
2. Explain how ocean basin features change over time.

Set a Purpose
Have students read this section to find the answer to the question "How have natural forces formed and changed the topography of the ocean basin?"

INSTRUCT

VISUAL TEACHING
Discussion

Have students locate on the map the continental shelf, continental slope, continental rise, and abyssal plain and hills. Challenge them to also find a trench and mid-ocean ridge. Students can refer back to pages 512 and 513 for help. The lower right corner of the map shows a portion of the mid-Atlantic Ridge. The Puerto Rico Trench is clearly visible as a dark curved line to the left of the abyssal hills. This trench marks a subduction zone—a feature that students probably would not expect to see in the Atlantic region.

DIFFERENTIATING INSTRUCTION

Reading Support Have students make a concept web for this section by following the color scheme of the headings. Students should use the blue heading "The Ocean Basin" as the central theme in their web, branching out to the red headings—"Abyssal Plain and Abyssal Hills," "Deep-Sea Trenches," "Deep-Ocean Vents," "Mid-Ocean Ridges," "Seamounts and Guyots," and "Corals and Coral Atolls." Students can complete the final level of the concept map with the boldface vocabulary words and their definitions. *Use Reading Study Guide, p. 82.*

CHAPTER 23 SECTION 3

More about...

Deep-Sea Trenches
The deepest part of the world's oceans is located along the Marianas Trench, which lies in the Pacific Ocean just east of the Philippines, where the fast-moving Pacific Plate subducts beneath the slower Philippine Plate. The deepest spot along the trench was found in 1951 by the British navy ship *Challenger*. Thus came the name *Challenger Deep* for the deepest place on Earth. It is nearly 11 km to the bottom, a depth that would entirely submerge Mt. Everest and leave almost two kilometers of water above it. A U.S. Navy mini-submarine reached the bottom of the trench in 1960, and in 1995 a Japanese probe made the most accurate measurement of its depth. Adjacent to the trench lies a volcanic island arc known as the Mariana Islands.

Deep-Sea Trenches

Deep–sea trenches are long, narrow, steep-sided troughs that run parallel to continental margins or to volcanic island chains called **island arcs.** These trenches exist at subduction zones—convergent plate boundaries where subduction occurs. In the diagram below, for example, the oceanic plate sinks beneath the continental plate. Deep–sea trenches are common sites of earthquakes and volcanic activity.

Trenches are often thousands of kilometers long but usually less than 100 kilometers wide. They may extend as deeply as six kilometers below the neighboring ocean floor. The bottoms of the deep-ocean trenches are narrow, flat, and filled with sediment. On the descending plate, there is a bulge in the sea floor, which scientists think results from the bending of the plate as it is subducted beneath the overriding plate. The overriding plate may feature a submarine ridge made up of sediments that have been scraped off the descending plate.

If one plate is oceanic and the other is continental, a marginal trench forms. Here, where the oceanic plate is descending below the continental plate, a line of volcanoes stands on the overriding continental plate, generally at least 100 kilometers from the deep-ocean trench. Such lines of volcanoes form mountain chains similar to the Andes (see Chapter 11).

If the plates involved in the subduction are both oceanic plates, an arc of volcanic islands forms on the overriding oceanic plate. Trenches in the western Pacific are usually associated with island arcs. The Java Trench, in the Indian Ocean, is associated with the islands of Indonesia; the Philippine and Manila trenches are associated with the Philippines.

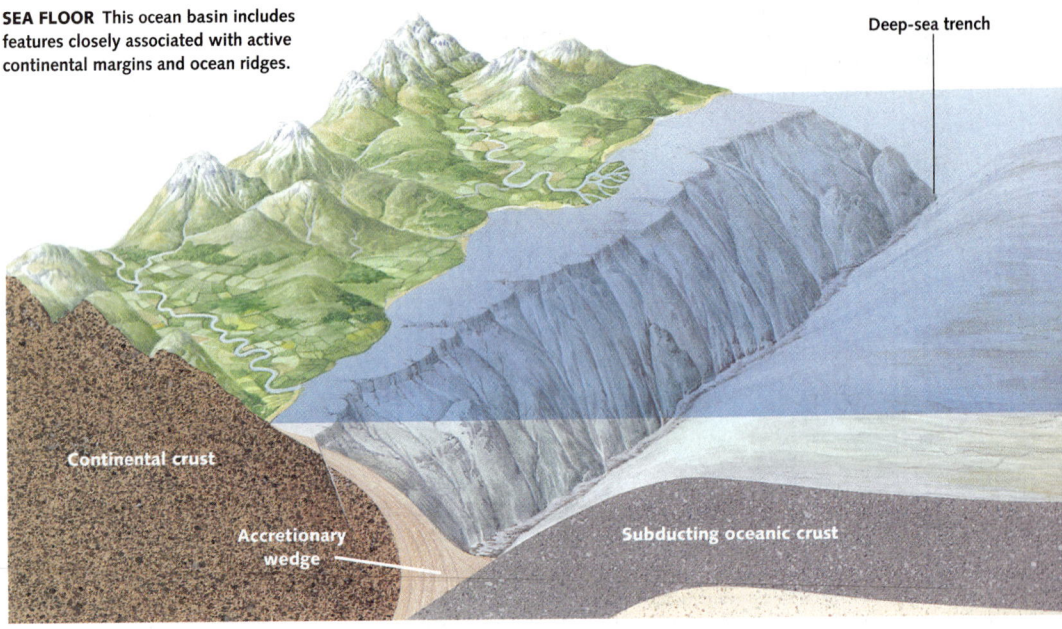

SEA FLOOR This ocean basin includes features closely associated with active continental margins and ocean ridges.

Deep-sea trench

Continental crust

Accretionary wedge

Subducting oceanic crust

DIFFERENTIATING INSTRUCTION

Challenge Activity Challenge students to investigate and report on the magnitude of water pressure in the deepest ocean trenches and how submersibles and probes have been engineered to withstand such extreme pressure.

Deep-Ocean Vents

A deep-ocean vent is a geyser that erupts underwater. Cold ocean water seeps through the sea floor and is heated by molten material. The hot water then gushes back into the ocean through sea floor cracks. When the hot vent water mixes with the cold ocean water, minerals carried from beneath the surface precipitate onto the surrounding sea floor, creating thick deposits rich in sulfides. Most deep-ocean vents are closely associated with mid-ocean ridges.

Mid-Ocean Ridges

The most obvious features of the ocean basin are mid–ocean ridges, great undersea mountain ranges. They form a chain about 70,000 kilometers long that wraps around the planet and crosses every ocean. They are usually more than 1000 kilometers wide, with a ridge crest that stands 1000 to 3000 meters above the neighboring sea floor. On average, mid-ocean ridges lie 2500 meters underwater, though in places their highest peaks rise above sea level as islands.

Mid-ocean ridges form at divergent plate boundaries where two lithospheric plates are moving apart. Magma rises between the plates and cools, forming a ridge of new sea floor. Ocean ridges spread at different rates depending on how quickly they add new material: slow (up to 50 mm/year), medium (up to 90 mm/year), and fast (up to 160 mm/year). Slow-spreading ridges, such as the mid-Atlantic Ridge, feature rugged topography and a rift valley at their crest. Fast-spreading ridges, such as the East Pacific Rise, have gentler topography and a minimal rift valley.

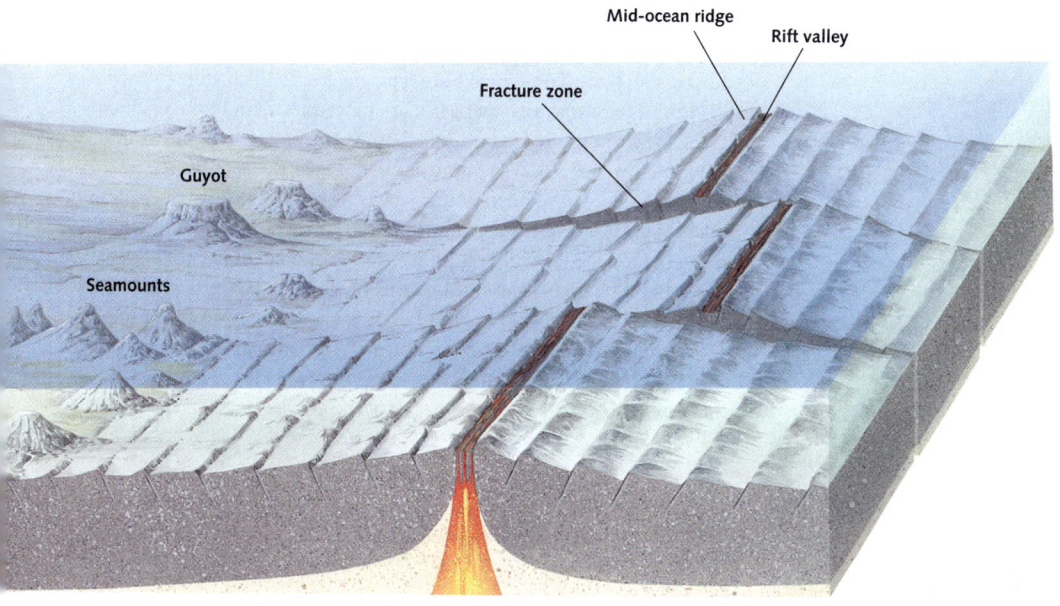

More about...

Mid-Ocean Ridges
The long chain of mountains that winds through each of Earth's ocean basins significantly isolates portions of the ocean basin, impeding the flow of deep water. As a result, water on either side can differ in temperature and salinity and can contain very different populations of organisms. Places along the mountain chain where slopes are gentle are referred to as rises (Melanesian Rise, East Pacific Rise); steeper areas are called ridges (mid-Atlantic Ridge, mid-Indian Ridge). However, some maps refer to all undersea mountains as ridges, regardless of their steepness.

Ocean ridges are not seamless. Hundreds of strike-slip faults, called transform faults, break them into separate pieces; some of the pieces are hundreds of kilometers long. Together with the rugged terrain around them, the transform faults make up **fracture zones.**

The pieces of ridge are offset relative to each other. Between them, the crustal plates are moving in opposite directions. The grinding and straining that results from this movement cause earthquakes along these sections of the faults. Beyond the offset ridge, however, the pieces of plate are moving in the same direction. Earthquakes do not occur in those areas. Some fracture zones may form high submarine cliffs, and others may extend across an entire ocean basin.

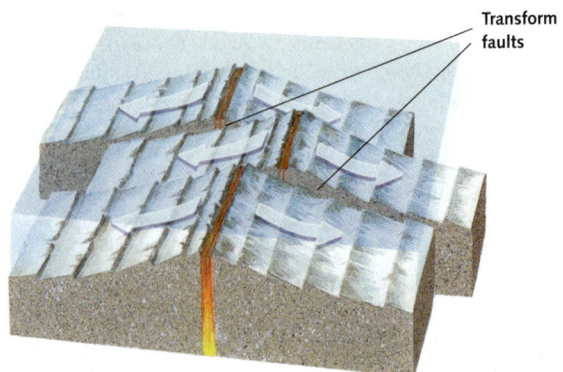

RIDGE WITH FRACTURE ZONES

Seamounts and Guyots

Seamounts are cone-shaped mountain peaks that rise high above the deep ocean floor. These peaks are typically found in clusters or rows. Seamounts occur in all oceans but are most abundant in the Pacific.

Volcanic in origin, seamounts seem to be related to plate boundary activity. However, some seamounts are not located near plate edges. These seamounts may have been pushed away from a boundary by sea floor spreading. Other isolated seamount groups probably originated over hot spots. In fact, the Hawaiian Islands, a famous chain of hot-spot volcanoes, were once a chain of seamounts, some of which have risen above sea level.

Some seamounts look as though their tops have been sliced off. These flat-topped seamounts are called **guyots** (GEE-oh, not JEE-ohz). Wave action removed their tops when they projected above sea level. Later, the sinking of the oceanic crust lowered the tops of the guyots, some as deep as two kilometers below sea level.

CORAL ATOLL This atoll is part of Maldives in the Indian Ocean. The ring around the island is made of coral, which will continue to grow upwards even as the volcanic island sinks and disappears under the sea.

DIFFERENTIATING INSTRUCTION

Challenge Activity Have students research the origin of the word *guyot*. They will learn that the flat-topped seamount was named for Arnold Henry Guyot, a prominent geologist of the mid-19th century. Have students research Guyot's contributions, as well as other places named for him.

Corals and Coral Atolls

Another product of crustal sinking is the **coral atoll**, a ring-shaped coral island. An atoll begins to form when a coral reef develops around a volcanic island. Corals are tiny sea animals that live in shallow, warm waters, and reefs form when new corals grow over the skeletons of the old. As the ocean crust beneath the volcano sinks, the corals sink with it, but new corals continue to form on top of the old. Eventually, the mountain is completely below sea level, leaving behind an atoll (circular reef) with a central lagoon where the mountaintop once was.

How a Coral Atoll Forms

❶ A fringing reef forms around a volcanic island. The island starts to sink.

❷ As the island sinks, the fringing reef grows and becomes a barrier reef, separated from the island by a lagoon.

❸ The island sinks completely below the sea, leaving the barrier reef behind to form an atoll reef surrounding a lagoon.

Coral reefs and atolls provide food and shelter for many marine organisms, and a natural sea wall to help dissipate ocean waves. Coral also acts as a filter on sea water. However, water pollution and rising water temperatures have led to the partial decay of coral reefs and atolls worldwide, endangering many coastal ecosystems and fisheries.

23.3 Section Review

❶ Why are abyssal plains so flat?

❷ How do seamounts differ from island arcs, both of which have their origins in volcanic activity?

❸ What actions occur along ocean ridges?

❹ **CRITICAL THINKING** The oceanic plate between the East Pacific Rise and South America is subducting beneath the western edge of the continent. What might happen to the East Pacific Rise as the plate continues to subduct?

CHAPTER 23 SECTION 3

Learn more about ocean research projects.
Keycode: ES2304

Science and Technology

NEPTUNE Beyond Earth
Some scientists affiliated with the NEPTUNE project seek information that might be pertinent to research into extraterrestrial life. Jupiter's moon Europa, for example, is thought to have moving water beneath its icy surface and a volcanic core similar to Earth's. Could Europa be harboring microbes similar to the chemosynthetic organisms near mid-ocean ridge geothermal vents? The idea is intriguing enough that detailed studies of vent communities are included in the wide scope of the NEPTUNE project. Information gleaned might help scientists identify ways to focus their data gathering in the search for extraterrestrial life.

Science Notebook
If the data result in breakthroughs in predicting earthquakes, tsunamis, or volcanic eruptions, property destruction and deaths may be avoided. Better understanding of factors affecting fish populations might affect fisheries, which might affect the economy.

SCIENCE & Technology

The NEPTUNE Project

A project, known as NEPTUNE, is taking shape and promises to give scientists the chance to take a long, hard look at what goes on beneath the ocean's surface.

What is NEPTUNE? How will it help us learn more about the undersea world?

The North East Pacific Time-series Undersea Networked Experiments, or NEPTUNE, will consist of a network of high-speed fiber-optic and power cables laid along the Pacific Ocean floor on the Juan de Fuca tectonic plate. Located off the coast of Washington, Oregon, and British Columbia, the network will link 20 to 40 seafloor observatories fitted with thousands of different sensing instruments and robotic devices. The instruments will capture many types of data and relay it to land-based laboratories.

The Juan de Fuca plate is an ideal site for the network because many major earth and ocean processes occur there. NEPTUNE is scheduled to be up and running by 2005 and will operate for about 30 years. Because the network will operate continuously over a long period, the data and observations should help scientists gain new insights into the complex processes responsible for producing earthquakes and volcanoes, forming minerals, and transporting sediments. Scientists also expect to learn more about factors that affect fish populations and to understand better the organisms that thrive in geothermal vents.

The network will be highly interactive. As scientists receive real-time data, they will be able to make adjustments when circumstances change. NEPTUNE's findings will not be the exclusive domain of scientists, however. One of the goals of the project is to make real-time data and images available via the Internet to classrooms so that students will be able to interact with the virtual ocean environment.

A DEEP-SEA ROVER is being deployed off of R/V Atlantis II for exploration of the ocean floor.

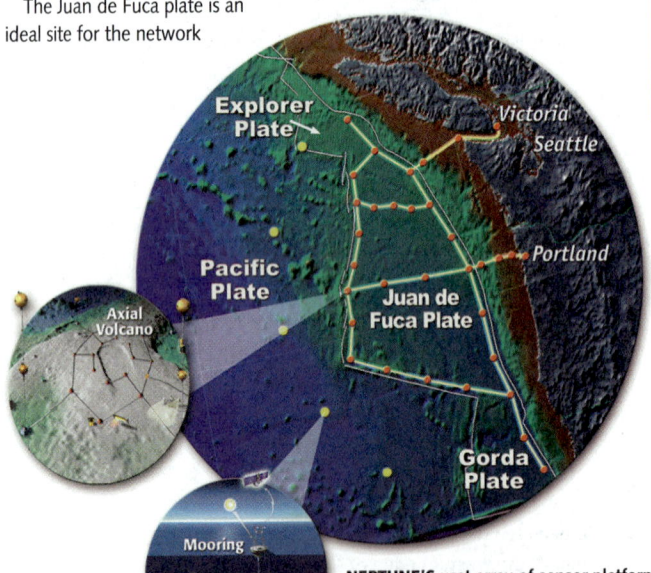
NEPTUNE'S vast array of sensor platforms and vehicles will be linked together by fiber-optic cable. Scientists at sea and on land will be able to access real time data through this network.

Extension

SCIENCE NOTEBOOK
How might constant access to ocean floor data benefit society?

Learn more about ocean research projects.
Keycode: ES2304

Ocean Floor Sediments

As you have read, sediments cover the deep ocean floor, sometimes in layers thick enough to bury abyssal hills. Sedimentary material reaches the sea floor in several ways. Some arrives in turbidity currents, and some falls down slowly from above. Some even settles to the ocean basin from melting icebergs far overhead.

There are four main classes of ocean floor sediments and they vary in size and composition. Some sediments consist of mostly sand and larger granules, while others—such as muds and clays—have finer particles. **Oozes** are sediments made from microscopic shells. These shells are the remains of tiny floating plants and animals found living in the mixed layer at the ocean surface. When these organisms die, their shells settle to the bottom.

Origins of Ocean Floor Sediments

Scientists classify sea floor sediments in a number of ways. One of the most useful systems is to describe sediments based on their origins. Sediments come from one of three sources: continental rocks and minerals, remains of marine plant- and animal-life, or chemical reactions in seawater.

Terrigenous (teh-RIDGE-uh-nuhs) **sediments** come originally from continental rocks and minerals broken down through weathering and erosion. The resulting fragments wash into rivers, whose currents carry the fragments out to sea. The largest fragments range in size from a grain of sand to a piece of gravel—these larger particles are deposited nearer the coastline, where layers of sediment build up quickly. Smaller particles can remain in a current much longer, as wind and water carry them thousands of kilometers out to sea before the particles settle to the ocean floor.

Sometimes it takes years for these tiny grains and flakes to reach their final destinations, so they may change chemically during their journey. For example, iron in a particle may react chemically with seawater to produce iron oxide (rust). A fine coating of rust gives sediment particles a reddish color.

Some terrigenous particles leave their continents as parts of glaciers. When glaciers calve, they crack and dropping off massive sections of the ice sheet into the ocean—the resulting icebergs carry dirt and rocks out to sea. When the icebergs melt, the dirt and rocks sink to the ocean floor.

23.4

KEY IDEA
The sediments covering the ocean floor have different origins and textures and vary by location. They can provide useful information about past changes in the ocean and in global climate.

KEY VOCABULARY
- terrigenous sediments
- biogenous sediments
- hydrogenous sediments

Observe the origins of some ocean floor sediments.
Keycode: ES2305

CALVING GLACIER The dark areas in this photograph of a calving glacier contain the fragments of minerals and rock which will eventually settle on the ocean floor as terrigenous sediments.

CHAPTER 23 SECTION 4

FOCUS

Objectives
1. Describe three sources of ocean sediments.
2. Explain why studying ocean sediment is important.

Set a Purpose
Have students read the section to find the answer to "Where do sediments in the ocean come from, and what can we learn from them?"

INSTRUCT

More about...

Origins of Sediments
Some sea floor sediments originated in outer space. They result from space particles that are constantly bombarding Earth's surface. Most space particles are very small and burn completely in the atmosphere. Some larger ones melt materials on Earth's surface upon impact. These particles are the source of splash-form tektites and microtektites that can be found both on land and on the sea floor.

Observe the origins of some ocean floor sediments.
Keycode: ES2305
 Visualizations CD-ROM

DIFFERENTIATING INSTRUCTION

Hands-On Demonstration Use a ball of steel wool to demonstrate that iron oxide forms in seawater. Submerge the steel wool in a beaker filled with salted water and keep it submerged overnight. Have students inspect the steel wool the following day. They should observe the beginnings of a layer of iron oxide. Ask students what must be present in the water for the iron oxide to form. **oxygen**

⚠ *Exercise caution when working with glass. Remind students that safety is everyone's responsibility.*

CHAPTER 23 SECTION 4

More about...

Siliceous Oozes

Certain microscopic marine organisms, such as diatoms and radiolarians, form shells composed of silica dioxide. The remains of these abundant marine organisms precipitate onto the ocean floor as siliceous ooze, which then forms the sedimentary rock known as chert. The process is similar to the one by which calcareous oozes lithify to chalk. Both chert and chalk are often found with marine limestone deposits. Chert was an important raw material for prehistoric humans. Its fine texture allowed it to be chipped to form sharp-edged tools and weapons.

VOCABULARY STRATEGY

The prefixes found in the names of the sediment types give you clues to their origins. *Terri-* comes from *terra*, the Latin word for "earth." *Bio-* comes from the Greek word for "life," and *hydro-* comes from the Greek word for "water."

Biogenous (by-AHJ-uh-nus) **sediments** come from living sources; they are oozes, composed mostly of shells and skeletons from tiny marine animals and decomposed plant-life. These sediments are most common on sea floors below surface waters with high plant and animal populations. There are two types of biogenous sediments: calcareous oozes and siliceous oozes.

Calcareous (kal-CAIR-ee-uhs) oozes contain at least 30 percent calcium carbonate ($CaCO_3$), in the form of calcite or aragonite. Calcareous oozes are the most common biogenous sediments, covering about half of the world's sea floor. When the surface layer organisms that make calcareous materials die, their shells and skeletons sink through the water column and begin to dissolve as they approach the *calcite compensation depth* (CCD), the depth below which sediments contain no calcium carbonate. Calcareous shells and skeletons will completely dissolve if they pass the CCD, which averages 4500 meters in depth.

Siliceous (sih-LISH-uhs) oozes contain silicon dioxide (SiO_2) rather than calcium carbonate. Because siliceous skeletons are tiny, they tend to drift rather than fall in large enough quantities to form an ooze; they reach the sea bottom more quickly when they form larger clumps (sometimes

CAREER

Research Vessel Officer

Research Vessel (R/V) officers get to combine the adventures of maritime travel with the gratification of promoting research. R/V captains and other officers operate ships owned by private companies, research institutions, or the government. Depending on the size of the research vessel, the crew can vary in number from two to many more. The officers and crew maintain the ship and provide support for the scientists who conduct research onboard. R/V officers supervise the crew, determine the course and speed of the vessel, and monitor the vessel's position. In addition, they perform maintenance and keep thorough logs of the ship's movements. A vital role of officers is to make sure proper safety practices are followed by everyone aboard the ship.

R/V officers must be prepared mentally and physically to meet the demands of working and living at sea. For example, they must be willing to work in all kinds of weather conditions and be able to endure being out on the ocean for months at a time. Captains and officers who work aboard commercial or private research vessels must be licensed by the United States Coast Guard. Other R/V officers are commissioned by the National Oceanic and Atmospheric Administration (NOAA) and serve aboard government research vessels. In addition to being physically fit, NOAA officers must hold a bachelor's degree in engineering, math, or science and complete a basic officer training course. ■

TWO RESEARCH VESSEL CREW MEMBERS participate in training for the National Oceanic and Atmospheric Administration Deep Worker project.

Learn more about becoming a research vessel crew member.
Keycode: ES2306

524 Unit 6 Earth's Oceans

DIFFERENTIATING INSTRUCTION

Reading Support Write the two objectives for this section on the board. (See page 523 of this Teachers Edition.) Then ask students to write a summary of the section that meets the two objectives. Read the best summaries aloud to the class and solicit comments.
Use Reading Study Guide, p. 83.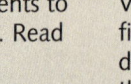

Developing English Proficiency Refer students to the Vocabulary Strategy on this page. Tell them to use the dictionary to find another word that begins with each prefix and write down the definitions of the words. Discuss how the definitions are related to the meanings of the prefixes.

called "marine snow"). Siliceous oozes are most common in the waters near the equator and around Antarctica, whose sediments contain 75 percent of the ocean's silica supply. In the deep sea, both calcareous and siliceous oozes tend to accumulate more than ten times faster than terrigenous sediments.

Hydrogenous (hy-DRAHJ-uh-nus) **sediments** form when chemical reactions cause minerals to crystallize from seawater. Manganese nodules are probably the best known of the hydrogenous sediments. Unlike other sediment materials, they form where they lie on the sea floor rather than settling from above. These lumps of minerals are rich in manganese and iron oxides and contain small amounts of nickel, cobalt, copper, and other metals. Manganese nodules appear in all oceans and in some lakes, in areas where sediments accumulate slowly. Manganese nodules form, layer by layer, at a rate of a few millimeters per 1 million years. Most of the material forming the nodule comes from seawater and the sediments around the nodule. Some of the metals in manganese nodules, including cobalt, are important in human industries, but they are extremely difficult and expensive to obtain.

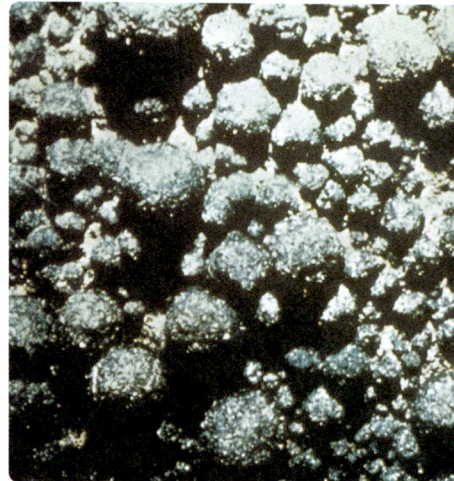

MANGANESE NODULES Although these nodules are a metal–rich resource, they lie kilometers below the ocean surface, making them difficult to mine.

The Importance of Sediments

Exploring the deep sea can be expensive and dangerous, and when the prize is a core sample of ooze, it may not at first glance seem worthwhile. But in the layers of sediment is a record of Earth's history going back millions of years. By studying the remains of sea creatures in the sediments, scientists can determine the extent of former polar ice sheets, the history of water temperatures on the sea floor, the past behaviors of prevailing winds, and the pattern of changes in Earth's climate. This prehistoric information can help scientists predict what changes might occur next and how we can prepare for them.

The sediments, unique organisms, magnetic records, and industrial resources available on the ocean floor represent a significant, if hidden, portion of Earth's hydrosphere, biosphere, and geosphere. As one of the least explored domains of our planet, there are many scientific discoveries waiting to be made about the deep ocean.

Scientific Thinking

COMMUNICATE
Explorers believe that the deep ocean holds vast amounts of valuable mineral resources. But scientists worry that deep-sea mining efforts may harm or even drive to extinction slow-growing life forms that live in these areas. What do you think?

23.4 Section Review

1. What is the source of terrigenous sediments? How do they move from their source to the deep sea?
2. What materials compose biogenous sediments?
3. Describe how hydrogenous sediments form.
4. **CRITICAL THINKING** What factors might account for the fact that manganese nodules rarely become covered by sediment, although they form far more slowly than sediment is deposited?

CHAPTER 23 SECTION 4

Learn more about becoming a research vessel crew member.
Keycode: ES2306

Scientific Thinking

COMMUNICATE
Have students work with a partner to make a chart that lists the pros and cons of deep-sea mining. Afterward, allow students to debate the issue in class.

ASSESS

1. Terrigenous sediments come from continents. After weathering from rocks, they are washed into rivers, which carry them to the sea. Currents carry small fragments to the deep sea.
2. Biogenous sediments come from living sources and are mainly composed of shells and skeletons from tiny marine animals and decomposed plants.
3. Hydrogenous sediments form when chemical reactions cause minerals to crystallize from seawater.
4. Manganese nodules do form in areas of slow sedimentation. Ocean current may keep the nodules free of sedimentation.

MONITOR AND RETEACH

If students miss . . .

Question 1 Refer students to the Vocabulary Strategy on page 524 to help them figure out the answers to the questions.

Question 2 Have students reread page 524. Ask: Where are biogenous sediments most commonly found? **parts of the ocean with a high concentration of life**

Question 3 Contrast the formation of hydrogenous sediments, as described on page 525, with the formation of other types of ocean floor sediments.

Question 4 Review how manganese nodules form, as described on page 525.

CHAPTER 23 ACTIVITY

PURPOSE
To make a contour map of an unseen model sea floor

MATERIALS
A bamboo skewer may be substituted for a coffee stirrer or straw. Cut off the sharp tip of the skewer and caution students to avoid sinking the skewer into the clay.

PROCEDURE
9 In choosing new locations to measure the depth, students should concentrate on areas where they might have unreliable data, areas with steep or rapidly changing slopes, or areas where there seems to be conflicting contour measurements.

CHAPTER 23 LAB Activity

Mapping an Unknown Surface

SKILLS AND OBJECTIVES
- **Develop** a plan to *examine* the shape of a unknown surface.
- **Create** a contour map of a surface based on measurements at different locations.
- **Compare** the actual surface with a contour map.

MATERIALS
- shoebox
- modeling clay
- aluminum foil
- masking tape
- sharp pencil
- coffee stirrer or thin straw
- fine-tip marker or pen
- graph paper
- colored pencils
- metric ruler

Sea-going vessels must carry detailed maps of the seafloor. Ship captains use these charts to pinpoint the ship's location and to avoid hitting the bottom. Because seafloor maps are so important, they require accurate information. In small groups, you will create your own seafloor topography. Then you will switch with another group to figure out what their seafloor looks like without actually seeing it.

Procedure

Part A: Create a Model Seafloor

1 Use clay to model a unique seafloor on the bottom of the shoebox. Vary the height from the bottom of the box to just below the top to represent both the deep ocean and islands or coastline. Include both smooth, flat areas and areas where the height changes rapidly.

2 Tear off a piece of aluminum foil that is larger than the shoebox. Tape the foil to the top of the box so that the surface is smooth and flat.

Part B: Map an Unknown Seafloor

3 Trade your box with a group whom you did not observe while they were creating their surface.

4 Use the ruler to measure the length and width of the shoebox. Draw a rectangle of the same size on a sheet of graph paper. On the graph paper, mark a maximum of 30 points with an "X" where you plan to measure the depth. You may choose points in any pattern you wish. The goal is to survey the model seafloor as thoroughly as possible.

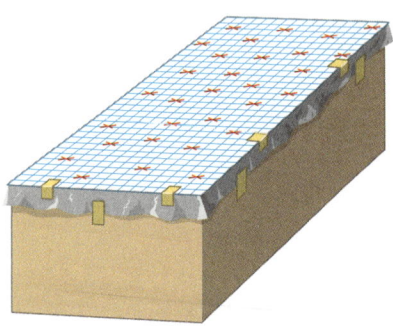

STEP 5

5 Each person in the group should copy the shoebox outline and the pattern of "X"s from the original graph paper onto their own sheet of graph paper in order to record the measurements of depth.

6 Secure the graph paper to the top of the shoebox with tape so that the outline of the box on the graph paper matches the top of the box. Use a sharp pencil to poke a hole through the graph paper and the aluminum foil at each location you have chosen to make a depth measurement.

7 Measure the height of the shoebox. Use your ruler to make marks on a coffee stirrer or thin straw every half a centimeter from one end so you can measure depths as large as the height of the shoebox.

526 Unit 6 Earth's Oceans

8. Gently poke the coffee stirrer through each hole and measure the depth the stirrer sinks before it contacts the clay inside the box to the nearest half-centimeter. Be careful not to dig into the clay. Hold the stirrer vertically so that you will make an accurate measurement. Record the depth on your graph paper next to the appropriate "X."

9. On your graph paper, note locations where you would like to have more information about the seafloor inside the box. As a group, pick up to 10 more locations in which to make measurements. Mark them with an "X" on your graph paper. Poke holes corresponding to these locations through the graph paper and foil on top of the shoebox.

10. Measure the depths at these new locations with your coffee stirrer and record them on your graph paper next to the appropriate "X."

11. Use a dark pencil to encircle all the locations that have a depth of 1 centimeter or less. Choose a lighter colored pencil and color in this area on your graph paper.

12. Now connect points representing 2 centimeters in depth with the dark pencil. Color in the space between the lines connecting 1 and 2 centimeter depths with a different color than you used in Step 11. Continue this process for 3 centimeter depths and larger until you have colored in the entire area of the box.

STEP 8

Analysis and Conclusions

1. Carefully remove the foil and graph paper from the shoebox. Compare your contour map to the surface inside. Evaluate how well your map represents the surface inside. Note specific characteristics of the surface. How well does your contour map show these features?

2. Evaluate your method for choosing locations at which to probe the depth. Where would it have been better to make more measurements? Is there any area where you needed fewer measurements than you took?

3. Which kinds of features were easiest to predict, and which were the most difficult? Describe those features that you found too difficult to map accurately.

4. When the group you traded with is also finished, place the two boxes side-by-side so that you may compare them. Which seafloor was more difficult to map? Do the two boxes have any common characteristics?

Learn more about mapping the ocean floors.
Keycode: ES2308

CHAPTER 23 ACTIVITY

Learn more about mapping the ocean floors.
Keycode: ES2308

ANALYSIS AND CONCLUSIONS

1. Answers will vary. Students should see widely spaced contours where the slope is gentle and narrowly spaced contours where the slope is steep. Peaks should be indicated by small, closed contours.

2. Answers will vary depending on the method chosen. Starting out with locations that are evenly spaced is probably the best method to use. Students might say that it would have been better to make more measurements where slopes changed rapidly, contour measurements seemed to conflict, or data seemed unreliable. Fewer measurements might have been needed in areas where the slope did not change or changed gradually.

3. Answers will vary. The most difficult features might be those located near the edges of the box and those with rapidly changing heights.

4. Answers will vary. Sea floors with a greater variety of features were probably more difficult to map. Common features might include continental shelf, continental slope, ridges, trenches, and plains.

Refer students to Safety in the Earth Science Laboratory on pages xxii–xxiii.

Chapter 23 The Ocean Floor 527

CHAPTER 23 REVIEW

VOCABULARY REVIEW

1. echo sounding
2. continental slope
3. an active continental margin
4. abyssal plain
5. biogenous sediments

CONCEPT REVIEW

6. From core sampling, scientists can learn how Earth's atmosphere and climate have changed over millions of years.

7. The continental rise is part of the ocean basin, whereas the continental shelf and continental slope are part of a continent.

8. During periods when the sea level is low, such as ice ages, rivers deposit sediment at the edges of continental shelves. When the sea level rises, turbidity currents carry the sediment down the continental slopes and spread it over the continental rise and abyssal plain.

9. Geologists think many submarine canyons formed during ice ages, when large rivers flowed across exposed continental shelves and cut deep canyons into them. The canyons were later flooded when sea levels rose. Other submarine canyons may be the result of turbidity currents that sped down the continental slopes and caused massive erosion.

10. Earthquakes occur due to friction between the overriding plate and the subducting plate. Volcanic eruptions occur as the subducted material is forced back to the surface by high pressures.

11. Mid-ocean ridges form where two lithospheric plates move apart. Magma rises between the plates and cools, forming a ridge of new sea floor.

12. The coral sinks as the ocean crust beneath the volcano that it surrounds sinks. New corals continue to form on top of the old.

13. Oozes are sediments made from microscopic shells. Other ocean sediments either derive from the breakdown of continental rocks and minerals or from chemical reactions that cause seawater minerals to crystallize.

14. Deep-ocean trenches exist where one lithospheric plate is being subducted beneath another, which occurs at active continental margins. Sketches should show both the overriding and subducting plates, with arrows indicating directions of plate movement. The trench should be labeled at the base of a steep continental slope, which also should be labeled.

CHAPTER REVIEW 23

Summary of Key Ideas

23.1 To determine the shape and composition of the ocean floor, scientists use techniques such as echo sounding, sediment sampling, and satellite observation.

23.2 A continental margin (the underwater edge of a continent) can be active or passive, depending where it lies in relation to a subduction zone or transform fault.

23.3 The ocean basin's topography varies widely and includes features such as abyssal plains and hills, deep-ocean trenches, and mid-ocean ridges.

23.4 Ocean-floor sediments vary in composition. Scientists classify each sediment based on where it originated.

KEY VOCABULARY

abyssal hill (p. 517)
abyssal plain (p. 517)
active continental margin (p. 515)
biogenous sediments (p. 524)
continental margin (p. 514)
continental rise (p. 514)
continental shelf (p. 514)
continental slope (p. 514)
coral atoll (p. 521)
core sampling (p. 511)
deep-ocean trench (p. 518)
echo sounding (p. 510)
fracture zone (p. 520)
guyot (p. 525)
hydrogenous sediments (p. 525)
island arc (p. 518)
mid-ocean ridge (p. 518)
passive continental margin (p. 515)
seamount (p. 520)
submarine canyon (p. 515)
terrigenous sediments (p. 515)
turbidity current (p. 516)

Vocabulary Review

Write the term from the key vocabulary list that best completes the sentence.

1. A precision depth recorder uses the technique of ___?___ to determine ocean depth.

2. The continental shelf extends from the shoreline to the ___?___ , where water depth begins to increase rapidly.

3. The margin where an oceanic plate is subducting under a continental plate is called ___?___ .

4. Composed of ocean-floor sediments, a(n) ___?___ stretches out from the continental rise and is the flattest of all Earth's surfaces.

5. Oozes are the decomposed remains of tiny plants and animals, also called ___?___ .

Concept Review

6. In what ways is core sampling useful to our understanding of Earth and its oceans?

7. Is the continental rise part of the ocean basin, or part of a continent? What about the continental shelf? the continental slope?

8. Most of the sediments that compose the abyssal plain originated on continents. Describe how the sedimentary materials could have moved from land to the deep-ocean floor.

9. Explain two possible origins of large submarine canyons.

10. Why are deep-ocean trenches such common sites for earthquakes and volcanic eruptions?

11. How do mid-ocean ridges form?

12. How can a coral atoll be attached to the ocean floor when corals cannot live (or, therefore, form reefs) in the deep ocean?

13. What is the difference between an ooze and other sediments?

14. Are you more likely to find a deep-ocean trench along an active continental margin along a passsive continental margin? Sketch and label a diagram to explain your answer.

CRITICAL THINKING

15. Since little deposition occurs at the active margins surrounding the Pacific Ocean, sediments that would form abyssal plains are unavailable. Also, trenches trap sediments that travel down the continental slopes.

Critical Thinking

15. Hypothesize Unlike the Atlantic Ocean, the Pacific Ocean has few abyssal plains. Propose a reason for this difference.

16. Infer Why are the mouths of rivers likely sources for turbidity currents?

17. Drawing Conclusions
  a. Sound travels at 1500 m/s in sea water. Find the depth of the ocean floor at a point where it takes a sound pulse 12 seconds to reach bottom and return.
  b. Which sea-floor features are most likely to be located beneath this point?

18. Hypothesize Slow-spreading mid-ocean ridges have rugged topography and a rift valley at their crest. Fast-spreading mid-ocean ridges have gentler topography and a minimal rift valley. What might account for this difference?

19. Hypothesize During the Ice Age, sediment deposition by turbidity currents dramatically increased on the abyssal plains of the North Atlantic Ocean. Propose at least one reason for this increase.

Interpreting Charts

The chart at right shows some of the sources of sediment for the abyssal plains. Use the chart to answer the questions below.

20. What is the process by which oozes settle from the ocean surface to the sea floor?

21. What natural event turns marine plants and animals into sedimentary material?

22. How do sediments deposited on the continental shelf reach the abyssal plain?

23. Why are shelf sediments depicted in a square rather than in an oval?

24. Would you expect to find hydrogenous sedimentary material drifting down from waters above the abyssal plain?

25. Where on the chart would you list wind-blown continental dust that settles onto the ocean surface?

Internet Extension

When Were the Atlantic and Pacific Oceans Separated by Land? Examine fossils and other data to date the formation of the Isthmus of Panama.

Keycode: ES2307

Writing About the Earth System

SCIENCE NOTEBOOK Scientists study ocean sediments hoping to understand how Earth's biological and chemical cycles have changed over time. Human activity influences these cycles. For example, as people cut down more forests, erosion rates increase; rivers carry more dirt and rocks out to sea, which results in a higher rate of sedimentation. Given what you know about the origins of sediments, how might global warming (another human-caused effect) influence sedimentation? How might toxic chemical dumping influence sedimentation? (See the chart below for ideas.)

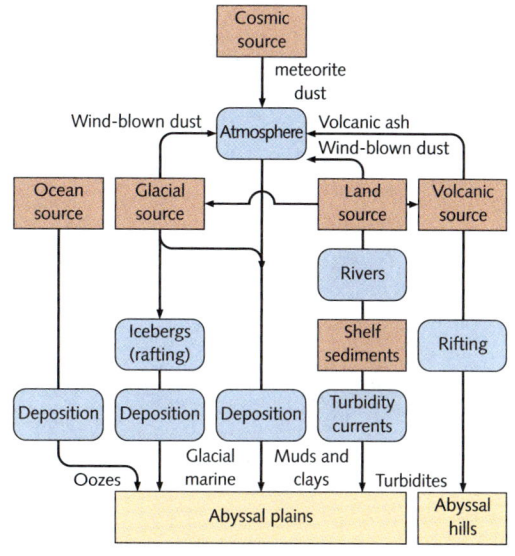

CHAPTER 23 REVIEW

When Were the Atlantic and Pacific Oceans Separated by Land?
Keycode: ES2307

📖 Use Internet Investigations Guide, pp. T82, 82.

(continued)

16. Mouths of rivers are places where sediments accumulate.

17. a. 9000 meters
  b. deep-ocean trenches

18. Unlike fast-spreading mid-ocean ridges, slow-spreading mid-ocean ridges might be subject to more faulting, which would create fracture zones and a more rugged topography.

19. The Ice Age might have been long enough to result in the deposition of large amounts of sediment at the edges of continental shelves. These sediments may have stimulated a larger than usual amount of turbidity currents, which carried the sediments to the abyssal plains.

INTERPRETING CHARTS

20. deposition
21. deposition
22. turbidity currents
23. because they are a source of sediments that make up abyssal plains
24. no, because many of them form where they lie on the sea floor
25. next to an arrow that extends from the "atmosphere" oval to the "ocean source" rectangle

WRITING ABOUT SCIENCE

Writing About the Earth System

Global warming and its associated climatic changes might result in more storms and thus more winds that carry dust. This would be a source of increased sedimentation. Storms would also increase river flow, which in turn would increase the deposition of shelf sediments by rivers. Toxic chemical dumping might lead to increased chemical reactions that could result in increased rates of hydrogenous sedimentation.

CHAPTER 24 PLANNING GUIDE THE MOVING OCEAN

	Section Objectives	Print Resources
Chapter-Level Resources		Laboratory Manual, pp. 103–108 Internet Investigations Guide, pp. T83–T85, 83–85 Guide to Earth Science in Urban Environments, pp. 13–20, 37–44 Spanish Vocabulary and Summaries, pp. 47–48 Formal Assessment, pp. 70–72 Alternative Assessment, pp. 47–48
Section 24.1 Surface Currents, pp. 532–535	1. Describe the relationship between winds and surface currents. 2. Describe patterns of different types of surface currents.	Lesson Plans, p. 84 Reading Study Guide, p. 84 Teaching Transparency 28
Section 24.2 Currents Under the Surface, pp. 536–540	1. Describe the flow of density currents. 2. Identify factors that affect the density of ocean water.	Lesson Plans, p. 85 Reading Study Guide, p. 85
Section 24.3 Tides, pp. 541–543	1. Explain how the moon and the sun affect tides. 2. Compare and contrast tidal ranges of different bodies of water.	Lesson Plans, p. 86 Reading Study Guide, p. 86 Teaching Transparency 29

INVESTIGATIONS
CLASSZONE.COM

Section 24.1 INVESTIGATION
How Can One Ocean Current Affect the Whole North Atlantic?
Computer time: 45 minutes / Additional time: 45 minutes
Students explore how the Gulf Stream affects ocean conditions, climate, and fishing.

Section 24.3 INVESTIGATION
How Do Tides Work?
Computer time: 45 minutes / Additional time: 45 minutes
Students explore how the sun and moon affect Earth's tides.

Technology/Online Resources

Electronic Teacher Tools
Visualizations CD-ROM
Classzone.com
Online Lesson Planner
Test Generator

Visualizations CD-ROM
Classzone.com
  ES2401 Visualization, ES2402 Visualization, ES2403 Investigation

Visualizations CD-ROM
Classzone.com
  ES2404 Visualization, ES2405 Visualization,

Classzone.com
  ES2406 Investigation

Classroom Management

Gather these materials for minilab and laboratory activities.

Minilab, p. 537
   round dish
   small paper circles
   drinking straw

Lab Activity, pp. 544–545
   aquarium
   cardboard
   scissors
   5 long thermometers
   metric ruler
   temperature probe
   magic marker
   4 small beakers
   4 different colors of dye
   4 Styrofoam cups
   masking tape
   clear tape
   ice cold water
   hot water
   salt
   spoon
   sharp pencil
   graph paper
   colored pencils
   Laboratory Manual, Teacher's Edition
     Lab Sheet 24, p. 145

Chapter Review INVESTIGATION

Could You Break the Record for an Ocean Sailboat Race?
Computer time: 45 minutes / Additional time: 45 minutes
Students apply knowledge of winds and ocean currents to plot a course for a sailboat race.

CHAPTER 24

INTRODUCE
Build Interest
Before sharing background information about the photograph with students, allow time for their questions and comments.

Ask questions like the following:
- How do tides affect this island? Building on the island is restricted to the area above the high-tide level.
- Does this photograph show high tide or low tide? low tide Why? because land around the island is exposed
- Why do you think people chose to build on this island in medieval times? High tide would have provided the town protection against invaders by land.

Respond
- What is the relationship between the land and the moving ocean? The ocean helps shape the land, especially along coasts. Currents affect climates along the coasts, and tides cover and expose land.

About the Photograph
Mont-Saint-Michel is located approximately 1 kilometer off the northwestern coast of France in Mont Saint Michel Bay. The large structure that rises above all others on the island is an abbey. The water surrounding the island rises as much as 14 meters between low and high tides. A paved road, built in 1879, connects Mont-Saint-Michel with the mainland.

The Moving Ocean

Due to its location in a funnel-shaped bay, the island of Mont-Saint-Michel, in France, experiences extreme high and low tides, as shown here.

What is the relationship between the land and the moving ocean?

BEYOND THE TEXTBOOK

Print Resources
- Reading Study Guide, pp. 84–86
- Transparencies 28, 29
- Formal Assessment, pp. 70–72
- Laboratory Manual, pp. 103–108
- Alternative Assessment, pp. 47–48
- Spanish Vocabulary and Summaries, pp. 47–48
- Internet Investigations Guide, pp. T83–T85, 83–85
- Guide to Earth Science in Urban Environments, pp. 13–20, 37–44

Technology Resources
- Visualizations CD-ROM
- Classzone.com
- Online Lesson Planner
- Electronic Teacher Tools
- Test Generator

530 Unit 6 Earth's Oceans

CHAPTER 24

PREVIEW

▶ **FOCUS QUESTIONS** In this chapter you will study ocean movement and learn more about the key questions below.

Section 1 What types of currents flow at the ocean's surface?
Section 2 What drives ocean currents below the surface?
Section 3 What factors influence the rising and falling of the tides?

▶ **REVIEW TOPICS** As you investigate the moving ocean, you will need to use information from earlier chapters.

- relationship between hurricanes and ocean surface temperature (p. 450)
- climate zones (p. 469)

▶ **READING STRATEGY**
OUTLINING

As you read the chapter, organize the main ideas by writing them in your notebook. Scan each section to identify its main topics. Write the main topics in list form, leaving space below each topic. Then, as you read the section fully, list subtopics and vocabulary definitions below each main topic.

INTERNET RESOURCES
CLASSZONE.COM

At our Web site, you will find the following Internet support for this chapter.

DATA CENTER
EARTH NEWS
VISUALIZATIONS
- Global Surface Currents
- Monsoon Direction Changes
- Deep and Surface Currents Circulation
- Upwelling

LOCAL RESOURCES
INVESTIGATIONS
- How Can One Current Affect the Whole North Atlantic?
- How Do Tides Work?
- Could You Break the Record for an Ocean Sailboat Race?

Chapter 24 The Moving Ocean   531

CHAPTER 24

PREPARE
Focus Questions
Find out what knowledge students have about movements of the ocean. For each focus question, have students consider the section title and then develop two or three additional questions for each section. For example, Section 1: How is the movement of the ocean's surface waters different from its deeper waters? How do currents differ?

Review Topics
In addition to the earth science topics listed, students may need to review these concepts:

Physical Science
- Usually, warm water or air rises above cool water or air.
- Water changes form by freezing, evaporation, and condensation.

Reading Strategy
Model the Strategy: To help students create outlines, use a version of the following: "I can use the headings in the section as most of my outline headings. The blue section heading 'Surface Currents' will be roman numeral I in my outline. The red headings will be my lettered outline headings. Then, as I read the chapter, I'll fill in numbered statements or phrases under each lettered heading. In the book, I see that there's a lot of information about surface currents before the first red heading. So I think my first lettered heading in my outline will be 'What are surface currents?'" Have students continue this strategy for each section in the chapter.

BEYOND THE TEXTBOOK

INVESTIGATIONS
- How Can One Current Affect the Whole North Atlantic?
- How Do Tides Work?
- Could You Break the Record for an Ocean Sailboat Race?

VISUALIZATIONS
- Animation of global surface currents
- Images showing the results of the monsoon changing direction

- Animation of how deep and surface currents circulate as a system
- Animation of how upwelling occurs

DATA CENTER
EARTH NEWS
LOCAL RESOURCES
CAREERS

Online Lesson Planner

Chapter 24 The Moving Ocean   531

CHAPTER 24 SECTION 1

▶ FOCUS

Objectives
1. Describe the relationship between winds and surface currents.
2. Describe patterns of different types of surface currents.

Set a Purpose
Read aloud the key idea. Then have students read to find the answer to the question "What are surface currents and how does wind affect them?"

▶ INSTRUCT

Visual Teaching

Interpret Diagram

Be sure that students understand the parts of the map on page 532. Ask: What do the red arrows represent? **warm currents** What do the blue arrows represent? **cool currents** In what general direction do cool currents flow? **away from the poles toward the equator** In what general direction do warm currents flow? **away from the equator toward the poles** In what direction do circular currents in the Northern Hemisphere turn? **clockwise** in the Southern Hemisphere? **counterclockwise**

Use Transparency 28.

24.1

KEY IDEAS
Surface currents are driven primarily by the wind.

There are several types of surface currents.

KEY VOCABULARY
- ocean current
- surface current
- cold-core ring
- warm-core ring
- countercurrent

VISUALIZATIONS
CLASSZONE.COM
Examine global surface currents.
Keycode: ES2401

Surface Currents

The water in the oceans is constantly on the move. Some motions, such as waves, are obvious. Other types of motion are so subtle or so deep that they are barely noticeable. These movements, called *ocean currents*, usually involve large water masses. An **ocean current** is any continuous flow of water along a broad path in the ocean. Ocean currents may flow at the surface or far below it.

A **surface current** is an ocean current that generally flows in the upper 1000 meters of the ocean. Surface currents are primarily driven by the wind. The Global Ocean Currents map below shows surface currents of the world. You can make several observations from the map. First, the Atlantic Ocean and the Pacific Ocean each have two circles of ocean currents, one in the Northern Hemisphere and another in the Southern Hemisphere. Consider as well that the direction in which each current circulates depends on the Coriolis Effect (Chapter 19): hence ocean currents in the Northern Hemisphere turns clockwise, and in the Southern Hemisphere, counterclockwise. As you see, the circular currents of the North Atlantic and North Pacific turn clockwise, while the currents of the South Atlantic and South Pacific turn counterclockwise. (Chapter 19)

You can also see from this map that the temperature of the currents follows a pattern. Currents flowing away from the equator carry warm water. Currents flowing toward the equator carry cold water. This occurs because areas near the equator have warmer temperatures and areas near the poles have colder temperatures.

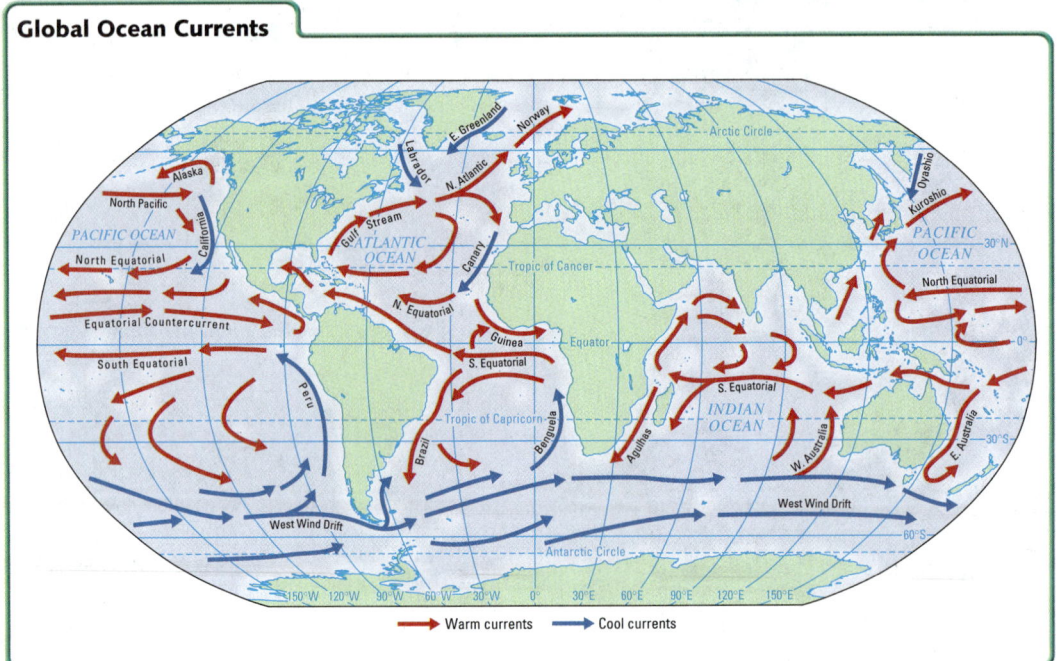

Global Ocean Currents

532 Unit 6 Earth's Oceans

DIFFERENTIATING INSTRUCTION

Reading Support Model for students how to preview a section. For example, you might read aloud the section title and the first sentence on page 532. Then say, "This section seems to be about how ocean water moves. The map shows the locations of different ocean currents and the directions in which they flow. This page must be about how currents move water to different parts of the oceans." Tell students that as they read the text, they will find that all the information on page 532 is about the movement of ocean currents. *Use Reading Study Guide, p. 84.*

532 Unit 6 Earth's Oceans

As a result of these patterns of circulation and temperature, the western sides of ocean basins have warm ocean currents moving away from the equator, while the eastern sides have cool ocean currents moving toward the equator. The western side of the North Atlantic Ocean has the warm, north-flowing Gulf Stream, while the eastern side has the cool, south-flowing Canary Current. Similarly, the western side of the North Pacific Ocean has the warm, north-flowing Kuroshio Current, while the eastern side has the cool, south-flowing California Current.

Currents and Winds

Although Earth's rotation and the presence of continents influence the paths of currents, surface currents are ultimately caused by the wind. As you read in Chapter 19, winds tend to blow in fairly constant directions at different latitudes on Earth's surface. Two sets of prevailing winds are involved in forming most ocean currents, the trade winds and the westerly winds. The trade winds, which are very steady, blow from the northeast in the Northern Hemisphere and from the southeast in the Southern Hemisphere. The westerlies blow from the southwest in the Northern Hemisphere and from the northwest in the Southern Hemisphere. In the North Atlantic, the trade winds push the North Equatorial Current westward, while the westerlies push the North Atlantic Drift eastward. The illustration below shows the interrelationship of winds and currents.

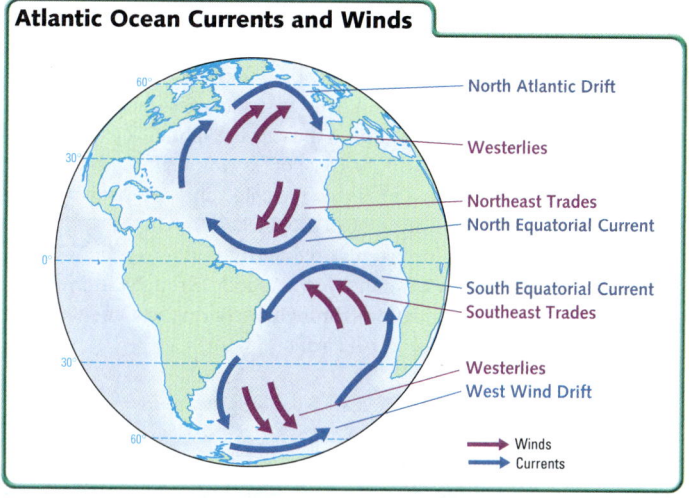

Atlantic Ocean Currents and Winds

Although seasonal changes in wind direction do not usually change the direction of ocean currents, there are exceptions. Seasonal winds called *monsoons* blow from one direction in summer and from the opposite direction in winter. When the winds reverse direction, the surface currents also reverse direction. The monsoon in the Arabian Sea off the west coast of India provides one well-known example of how changes in wind direction can cause a seasonal change in the direction of ocean currents.

Observe how the monsoon changes direction.
Keycode: ES2402

Chapter 24 The Moving Ocean **533**

CHAPTER 24 SECTION 1

VISUALIZATIONS
CLASSZONE.COM

Examine global surface currents.
Keycode: ES2401

Visualizations CD-ROM

VISUALIZATIONS
CLASSZONE.COM

Observe how the monsoon changes direction.
Keycode: ES2402

In addition to phytoplankton productivity, have students speculate about other changes in animal and plant life due to current reversals.

 Visualizations CD-ROM

Visual Teaching

Interpret Diagram
Explain that the purple arrows on the map show wind patterns and the blue arrows show current patterns. Have students locate the North Equatorial Current. Ask: Which winds create the current? **Northeast Trades** Which winds create the North Atlantic Drift? **the westerlies** What is the relationship between the surface currents and the winds that produce them? **They move in the same direction.**

Extend
Ask students to predict the direction of hurricane winds along the east coast of North America. **They would travel toward the northeast.**

DIFFERENTIATING INSTRUCTION

Developing English Proficiency Refer students to the visual information in the maps on pages 532 and 533. As they read, have them locate each current and wind on one of the maps. Assist with any difficult vocabulary such as *circulation*, *prevailing winds*, *clockwise*, and *counterclockwise*. Have students describe the patterns of current and wind circulation in their own words.
Use Spanish Vocabulary and Summaries, pp. 47–48.

CHAPTER 24 SECTION 1

More about...

The Gulf Stream
Scientists have estimated that the flow of cold water from the Arctic has decreased in the last 50 years. The decrease may be due to the melting of polar ice. The dense Arctic current flows in the ocean depths toward the equator where it causes warm water to flow back toward the Arctic. The Gulf Stream is part of this warm-water flow. Scientists are concerned that a reduced Gulf Stream could decrease the average winter temperature of the United Kingdom by an average of 11°C.

Scientific Thinking

EXPLAIN
Have students relate what they read about cold currents on page 534 to the location of the *Titanic* when it sank. Ask: What are two hazards that ships are likely to encounter where the Labrador Current meets the Gulf Stream? **icebergs and fog**

Warm Currents

Warm ocean currents flow away from the equatorial region on the western side of ocean basins. The Gulf Stream in the North Atlantic and the Kuroshio Current in the North Pacific are examples of warm currents. Of all the warm currents, the Gulf Stream has been studied most extensively.

A narrow, intense flow of warm water, the Gulf Stream begins in the Caribbean Sea and follows the east coast of the United States northward around Cape Hatteras, North Carolina. There the current veers northeastward across the Atlantic Ocean, where it is called the *North Atlantic Drift*. The current carries warm water to Iceland and the British Isles. As a result, these places have warmer climates than they would otherwise. (Hence there are palm trees along the coast of Ireland.)

The Gulf Stream forms the western and northern boundary of the Sargasso Sea, which is located in the middle of the North Atlantic Ocean. An area of warm water and light winds, the Sargasso Sea has relatively calm seas. Great amounts of floating brown seaweed called *sargassum* are typically found on the surface water there. Similar conditions exist in other oceans, but nowhere are they as well developed as in the North Atlantic.

Cold Currents

Cold currents flow toward the equator on the eastern side of ocean basins. Examples of cold ocean currents include the Canary Current in the North Atlantic, the California Current in the North Pacific, and the Benguela Current in the South Atlantic.

Cold currents can also flow out of far northern regions. The Labrador Current flows out of Baffin Bay and past Labrador, the coastal part of the Canadian province of Newfoundland. The current carries icebergs from Baffin Bay, creating a hazard for ships in the North Atlantic. The Labrador Current meets the Gulf Stream off the coast of Newfoundland. When warm, moist air from the Gulf Stream blows over the cold Labrador Current, water vapor condenses. This results in some of the thickest fogs in the world.

Two other important cold currents originate in northern regions. The East Greenland Current flows into the North Atlantic through the Strait of Denmark. The Oyashio Current flows through the Bering Strait between Siberia and Alaska and into the North Pacific.

Scientific Thinking

EXPLAIN
The *Titanic* sank where the Labrador Current meets the Gulf Stream in the Atlantic Ocean. (Review the Global Ocean Currents map on page 532.) Why is this area particularly dangerous for ships? What sort of hazards might the area pose?

ICEBERG As this computer-enhanced image shows, only a small portion of an iceberg extends above the surface of the water.

DIFFERENTIATING INSTRUCTION

Hands-On Demonstration As warm surface currents flow from the equator, some of the water's heat is transferred to the atmosphere. Demonstrate that the water temperature affects the temperature of the surrounding air. Obtain two identical jars with lids. Poke a hole in each lid and use string to suspend a thermometer in each jar. Monitor the jars until the air temperatures are the same. Then add cold water to one jar and an equal amount of warm water to the other jar. Leave space above the water for the thermometer. Cover the jars and record the air temperature every minute for 10 minutes. **The air temperature decreases in the jar with cold water and increases in the jar with warm water.** Refer students to the map on page 532. Ask: Would you expect average temperatures to be higher in cities along the coast of California or along the southeastern coast of the United States? Explain. **Cities along the southeastern coast have higher average temperatures partly because of the warming effect of the Gulf Stream.**

⚠ *Do not use mercury thermometers.*

Gulf Stream Rings

Ocean currents do not flow in the same channel all the time. For example, the Gulf Stream wanders and sends out offshoot streams. Occasionally, the Gulf Stream develops eddies or whirlpools that break away from the edge of the current. These eddies become structures called Gulf Stream rings. The satellite image at the right illustrates how Gulf Stream rings form from a bend in the Gulf Stream.

As each ring forms, it takes with it a column of water from the opposite side of the Gulf Stream. A ring that forms on the Sargasso Sea side of the Gulf Stream has a core of cold water from the continent side, which is called a **cold-core ring.** Conversely, a ring that forms on the continent side of the Gulf Stream has a core of warm water from the Sargasso Sea and is called a **warm-core ring.** While about 10 to 15 rings form each year, some rings last as long as two years.

Countercurrents

Countercurrents flow in the opposite direction of the wind-related currents. They can flow at or beneath the surface. For example, the Equatorial Countercurrents are surface currents. They flow eastward in a narrow belt of water between the westward-moving North and South Equatorial currents. The belt occurs because the trade winds and the North and South Equatorial currents pile water up on the western sides of the ocean basins. The countercurrents return some of that water to the eastern side of the basin. Equatorial Countercurrents are best developed in the Pacific Ocean but also occur in the Atlantic and Indian oceans.

The Cromwell Current is a subsurface countercurrent in the Pacific Ocean. Flowing eastward about 30 meters below the ocean surface, the Cromwell Current flows beneath the westward-flowing South Equatorial Current. The Cromwell Current is a major ocean current due to its size and volume. The Atlantic and Indian oceans also have countercurrents.

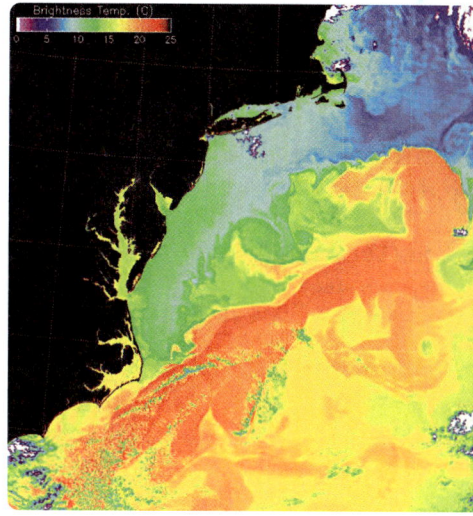

GULF STREAM This false-color satellite image represents values of heat radiation from the sea and overlying atmosphere. The red shows warmer areas, the green is intermediate, and the blue represents the coolest. The large yellow area is part of the Sargasso Sea.

How Can One Ocean Current Affect the Whole North Atlantic? Explore how the Gulf Stream affects ocean conditions, climate, and fishing. *Keycode:* ES2403

24.1 Section Review

1. What is the primary cause of surface currents?
2. What type of Gulf Stream ring is found between the Gulf Stream and the coast of the United States?
3. List two examples of countercurrents.
4. **CRITICAL THINKING** Would you expect the water off the Pacific coast of the United States to be warmer or colder than the water off the Atlantic coast at the same latitude? Explain your reasoning.
5. **GEOGRAPHY** Use the map on page 532 to compare and contrast the Brazil Current and the Gulf Stream.

CHAPTER 24 SECTION 2

FOCUS

Objectives
1. Describe the flow of density currents.
2. Identify factors that affect the density of ocean water.

Set a Purpose
Have students read the section to answer the question "Where do density currents flow and what causes them to form?"

INSTRUCT

Visual Teaching

Interpret Diagram
Have students follow the arrows in the map to observe that deep currents and surface currents together form a complete path of circulation. Ask: At what locations do deep currents rise to the surface? **in the Indian Ocean and north Pacific** Where do surface currents sink to become deep currents? **in the north Atlantic**

Extend
Have students compare the locations of the surface currents on page 532 and of the deep currents on this page. Ask: How are surface currents able to flow in the same areas as the deep currents? **The surface currents are much higher than the deep currents and flow over them.**

24.2

KEY IDEA
Deep ocean currents are driven by properties that determine the density of water, such as salinity and temperature.

KEY VOCABULARY
- density current
- upwelling

SURFACE AND DEEP CURRENTS
This simple model of ocean current circulation is called the *Global Conveyor Belt*. The model gives you a sense of how water moves from polar to equatorial regions.

Currents Under the Surface

Unlike surface currents, which are driven by wind, deep ocean currents are driven by gravity density and differences in density. A **density current** is heavier and denser than surrounding water; such dense water masses sink from the surface toward the bottom of the ocean where they circulate in the deep ocean for 500 to 2000 years before resurfacing.

Density Currents

Density currents can be found in all the world's oceans. As shown in the map below, the deep currents connect with global surface currents to form a complete path of circulation. This circulation of ocean water is an efficient heat-transport system. Similar to the conveyor belt at a grocery store check-out line, the circulating currents transport warm water to colder areas and cold water to warmer areas. For this reason, the global circulation of ocean currents is often modeled as a Global Conveyor Belt.

Density currents move very slowly compared to surface currents. Despite their slow movement, they are important to marine animals living in the deep ocean. Density currents retain the oxygen absorbed at the surface layer, as well as the temperature, salinity, and density. Such deep currents are the only source of oxygen for deep-sea life.

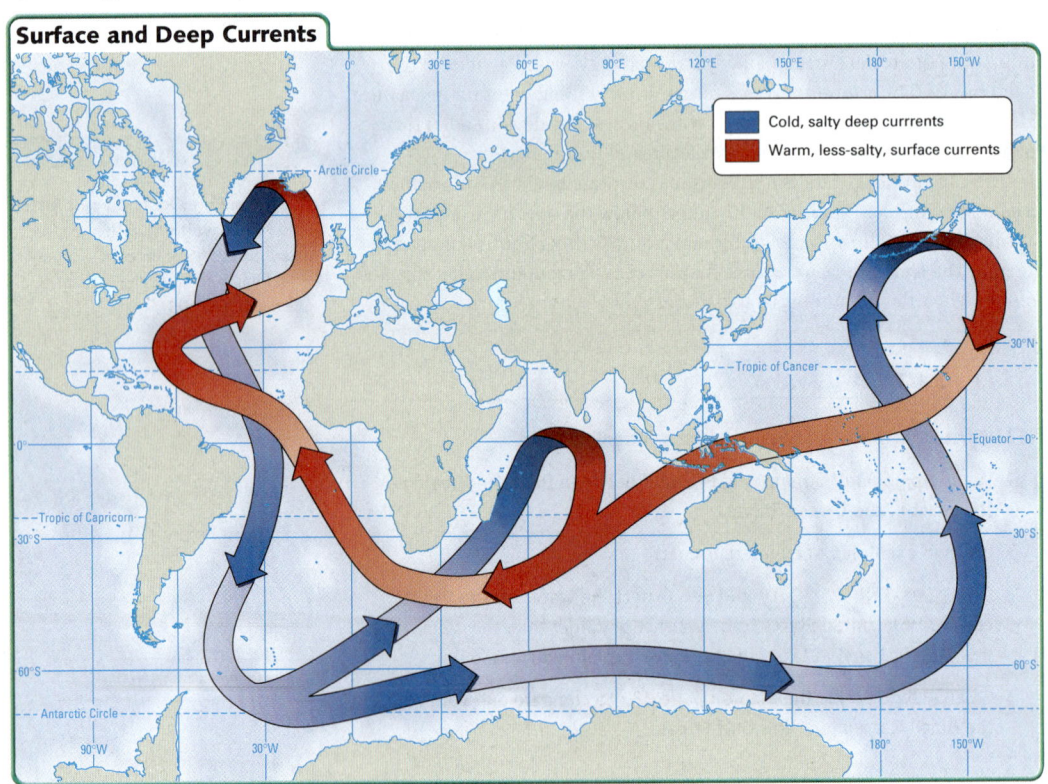

Surface and Deep Currents

536 Unit 6 Earth's Oceans

DIFFERENTIATING INSTRUCTION

Support for Visually Impaired Students In the diagram of Atlantic Ocean Currents on page 537, the different blue-colored layers may be difficult to distinguish, even with magnification. It may be necessary to photocopy the diagram for students and draw lines between the different layers.

Density currents flowing in the oceans can be vertical as well as horizontal. Examples of vertical currents are turbidity currents (discussed in Section 23.4). Sand, silt, and other particles mix with water to form dense water masses that sweep down continental slopes onto the abyssal plains. Turbidity currents are just one type of density currents. Cooling, freezing, and evaporation are other common causes of changes in density that can form deep ocean density currents.

Density Currents from Polar Water

The densest water in the oceans comes from polar regions, due to the intense cooling and freezing that occur there. Cooling causes sea water to contract, hence the water takes up less space. As water molecules become packed closely together, the water's density increases. Freezing also affects the density of polar water. When seawater freezes, most of the salt is left behind. As salinity increases, the density of the seawater also increases.

Three great water masses are formed as density currents in the polar regions. They are Antarctic Bottom Water, North Atlantic Deep Water, and Antarctic Intermediate Water. The coldest, densest water in the oceans is the Antarctic Bottom Water. This water is produced in large amounts in the Weddell and Ross seas off the coast of Antarctica. This dense water has an average temperature of −0.4°C and a salinity of 34.6 psu. The water sinks to the ocean floor and spreads northward across the equator to 40 degrees north latitude, which is off the northern coast of California.

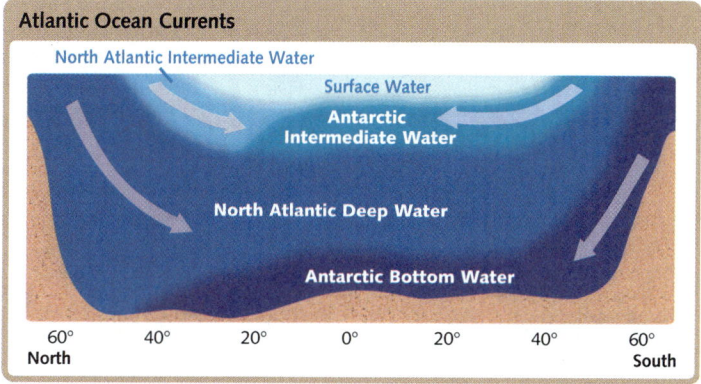

DEEP CURRENTS from polar areas flow throughout the Atlantic Ocean. The depth of each current depends on its density.

The second densest type of seawater is the North Atlantic Deep Water. Evidence suggests that this water forms near Greenland as higher-saline Gulf Stream water with cold water from the Arctic Ocean. The water has a temperature as low as 3°C and a salinity of about 34.5 psu. North Atlantic Deep Water is less dense than Antarctic Bottom Water and thus flows above it. It spreads through the Atlantic to about 60 degrees south latitude, which is around South America's Cape Horn. There, some of the water returns to the surface. North Atlantic Deep Water also flows west to form deep water in the Indian and Pacific oceans.

Observe how deep and surface currents circulate as a system.
Keycode: ES2404

Simulating Ocean Currents

Materials
- round dish
- drinking straw
- small paper circles

Procedure

1. Fill a dish halfway with water, let stand until the water is still. Sprinkle paper circles on the water's surface.
2. Blow through the straw at a low angle to the water. What is the best way to position the straw if you want to simulate a clockwise current? Draw a diagram showing straw's angle of position; use arrows to show air and water motion.
3. Repeat Step 2, making a counterclockwise current.
4. Repeat Step 2 again, making two currents.

Analysis
Relate this simulation to what actually happens with ocean currents. How could you make the simulation more realistic?

CHAPTER 24 SECTION 2

Observe how deep and surface currents circulate as a system.
Keycode: ES2404

 Visualizations CD-ROM

MINILAB
Simulating Ocean Currents

Management
- Punch out the paper circles before students do the Minilab.
- Students should sprinkle the paper circles evenly across the water's surface.

Materials
- An 8-inch glass pie plate would be ideal.
- Punch out the circles from stiff paper, such as a manila envelope.

Analysis Answers
- On Earth, winds blow from high pressure to low pressure, producing circular surface currents like those made in this simulation.
- **Possible Response:** Have more than one person blow through a straw to simulate multiple currents.

VISUAL TEACHING
Discussion

Refer to the diagram of Atlantic Ocean Currents. Explain that degrees of latitude are across the bottom of the diagram. Ask: How far does the North Atlantic Intermediate Water extend? from about 45° N to 10° N Which current is densest? Antarctic Bottom Water Which currents are closest to the surface water? North Atlantic Intermediate Water and Antarctic Intermediate Water

DIFFERENTIATING INSTRUCTION

Support for Physically Impaired Students The positioning of the drinking straw in the Minilab is crucial, and it must be held steadily in place. If necessary, have an assistant hold the straw in position as the student blows through it.

CHAPTER 24 SECTION 2

VISUAL TEACHING
Discussion
Have students analyze the salinity of the water in the diagram. Ask: Why is the salinity of the water in the Mediterranean higher than that of the water in the Atlantic? **Because the climate is hot and dry, water in the Mediterranean evaporates at a high rate, leaving salt behind.** How does salinity affect the density of water? **The higher the salinity of water, the greater is its density.** Why do surface currents from the Atlantic flow into the Mediterranean? **The surface currents bring in less-dense Atlantic water to take the place of the dense water that flows out of the Mediterranean.**

More about...

Upwellings
El Niño is an ocean current that brings unusually warm water to the west coast of South America. These currents occur on average about every four years. During El Niño years, the trade winds that drive surface water off the coast slacken. As a result, the upwelling that normally occurs near the Peruvian and Chilean coasts is reduced, greatly harming those countries' fishing industries.

THE STRAIT OF GIBRALTAR is the narrow passage connecting the Atlantic Ocean and the Mediterranean Sea. Taken from Spain (foreground), this photograph shows Morocco across the strait.

Around 55 degrees south latitude, Antarctic Intermediate Water forms. This water has a temperature of 2.2°C and a salinity of 33.8 psu. It is the least dense of the three polar currents. Antarctic Intermediate Water spreads northward at a depth of 700 to 800 meters to about 30 degrees north latitude, which is close to the latitude of Tallahassee, Florida.

Density Currents Caused by Evaporation

The density of seawater in warm, dry climates is often affected by evaporation. When seawater evaporates, the salt in the water is left behind. As with freezing, this increases the salinity, and thus the density, of the water that remains.

An example of a density current formed by evaporation is found in the Mediterranean Sea. The hot, dry climate causes much more water to evaporate than is received from rainfall or rivers. Hence, Mediterranean seawater has a higher salinity and density than average seawater.

The Mediterranean Sea and the Atlantic Ocean are connected by the narrow Strait of Gibraltar. On the Atlantic side of the strait, the salinity of the seawater is about 35 psu. On the Mediterranean side, the salinity is much higher—about 40 psu. A two-way flow of water results due to the difference in density. The heavy, dense Mediterranean water flows over the sill of the Strait of Gibraltar and into the Atlantic Ocean. Less-dense Atlantic Ocean water pours into the Mediterranean, over the denser water.

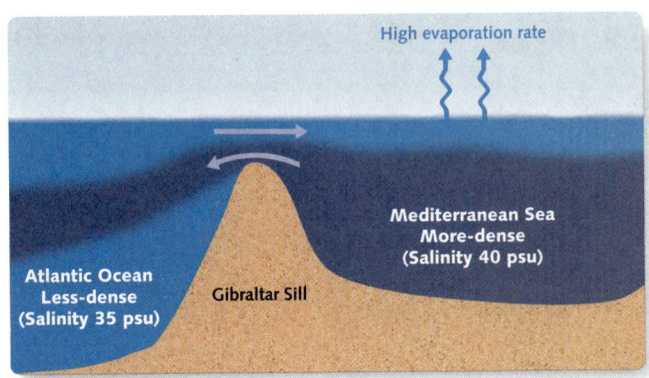

GIBRALTAR SILL Notice that there is a two-way flow of water crossing over the sill.

Outside the strait, the dense Mediterranean water sinks in the Atlantic to a depth of about 1000 meters. Dense Mediterranean water forms a stream many times larger than the Mississippi River. Branches of Mediterranean water have been traced to Greenland and Bermuda.

Upwelling

Upwelling is another kind of vertical current in the oceans. **Upwelling** occurs when cold deep water comes to the surface. Although upwelling can occur anywhere, it is most common on the western sides of continents where prevailing winds blow along the coastline toward the equator.

DIFFERENTIATING INSTRUCTION

Reading Support Have students monitor their reading by summarizing the section. For each red or blue heading, have them write the main idea. For example, for "Density Currents," students might write, "Density currents are deep currents that flow below the ocean's surface currents." Then have students use their main-idea sentences to write a paragraph that summarizes the section.
Use Reading Study Guide, p. 85.

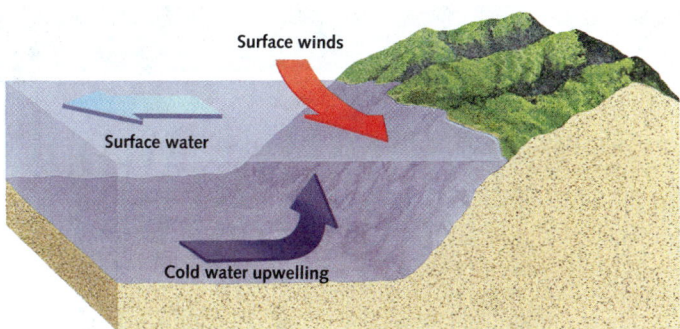

UPWELLING occurs when surface water is displaced by winds or density currents. The cold deep water rises to replace the surface water.

These winds push the coastal surface water, which is turned by the Coriolis Effect, and flows away from the coast. Cold water rises to replace the surface water that has been moved out to sea. Locations where this type of upwelling occurs include the coasts of California, Morocco, southwestern Africa, Peru, and western Australia.

Another kind of upwelling occurs in the Antarctic Ocean at about 60 degrees south latitude. Here the North Atlantic Deep Water returns to the surface. Upwelling in this area is caused both by winds and by density currents. Two major wind belts blow on either side of the upwelling region. One wind blows toward the east, while the other blows toward the west. Surface water is pushed by the winds and turned by the Coriolis Effect to flow away from the region, and deeper water rises to replace it. The other cause of upwelling in this area is due to density currents. Both Antarctic Bottom Water and Antarctic Intermediate Water sink nearby. The North Atlantic Deep Water upwells to replace water that has moved away.

Upwelled water contains large amounts of nutrients that phytoplankton need in order to grow. Phytoplankton, the microscopic organisms that are the basic food for other living things in the ocean, thrive in areas of upwelling. Animals higher in the food chain live with the phytoplankton in the areas of upwelling. As a result, these areas are usually major commercial fishing areas.

24.2 Section Review

1. Why are density currents so important to marine life in the deep ocean?
2. Where does the densest water in the ocean come from? Explain why.
3. Explain how evaporation, freezing, and cooling each cause seawater density to increase. If necessary, draw a concept make to help illustrate your explanation.
4. Describe two causes of upwelling.
5. **CRITICAL THINKING** Antarctic Bottom Water has a temperature of −0.4°C. Why does it not freeze?
6. **BIOLOGY** Why do areas of upwelling contain large amounts of fish?

CHAPTER 24 SECTION 2

ASSESS

1. Density currents carry oxygen absorbed from the surface.
2. The densest water comes from polar regions due to the intense cooling and freezing that occur there.
3. Evaporation and freezing leave salt behind in the seawater, thereby increasing the water's salinity and consequently its density. Cooling causes seawater to contract and take up less space, making it denser.
4. Upwelling is caused by winds that push away surface waters and by sinking density currents. In both cases, other waters upwell to replace water that has moved away.
5. Antarctic Bottom Water does not freeze because of its high salinity. The higher the salinity of a solution, the lower is its freezing point.
6. In areas of upwelling, nutrients are brought up close to the surface. There, phytoplankton populations thrive due to the availability of the nutrients. Fish feed on phytoplankton. Thus, the increased phytoplankton populations support larger fish populations.

MONITOR AND RETEACH

If students miss . . .

Question 1 Have students reread the second paragraph under "Density Currents" on page 536.
Question 2 Reteach "Density Currents from Polar Water." (pp. 537–538).
Question 3 If students did not draw a concept map, have them do so. The concept maps should show that both evaporation and freezing affect the water's salinity, whereas cooling affects the spacing of the water molecules.
Question 4 Have students reread "Upwelling." (pp. 538–539) Ask them to draw sketches that illustrate upwelling due to the movement of surface water and upwelling due to the sinking of density currents.
Question 5 Ask students to think about how the Antarctic Bottom Water differs from tap water, besides being much lower in temperature.
Question 6 Have students reread the last paragraph on page 536. Ask: What organisms feed on phytoplankton? fish and other marine animals

CHAPTER 24 SECTION 2

Observe how upwelling occurs.
Keycode: ES2405

Visualizations CD-ROM

Science and Society

Anchovies and the Food Chain
Anchovies are a critical link in the coastal marine food chain. When anchovies feed on phytoplankton, they help control the growth of algae in the coastal waters. In addition to clearing waters of phytoplankton, anchovies are also food for larger fish, birds, and mammals. Anchovies are part of the balance that keeps ocean waters suitable for life.

Science Notebook
Students might suggest management issues such as less fishing during El Niño years or seeding of additional fish spawn. Other suggestions might include technological improvements, such as ways to force cold-water currents to the top in fishery areas.

SCIENCE & Society

Upwelling and Anchovies off Peru

How does upwelling affect an important fish species? Is it possible to see a fishery decline when upwelling fails?

What lessons are learned when El Niño calls a halt to upwelling off Peru?

Off the coast of Peru, cold, deep water rises in the pattern typically seen off the western coasts of continents. The upwelling that delivers nutrients to the surface there has created one of the world's richest anchovy fisheries. Microscopic organisms known as phytoplankton bloom in the nutrients sent from below. Anchovies feed directly on the phytoplankton instead of eating smaller fish. This unusually efficient feeding method that shortcuts the food chain explains the abundance of anchovies off the coast of Peru.

In the 1950s, a huge commercial anchovy fishery was begun off the coast of Peru. Boats towing seine nets harvested the anchovies at sea. Anchovies processed into fishmeal were sold all around the world as feed for livestock and poultry. Millions of sea birds fed on the anchovies, too. Bird droppings on coastal rock outcroppings not only attested to the bounty of fish but were also harvested for use as fertilizer on Peruvian farms.

Trouble came in 1972, however, when El Niño struck. Diminishing trade winds halted the upwelling. The phytoplankton supporting the anchovies vanished, while the fishing fleet kept on fishing. The anchovy fishery collapsed, with global repercussions. Meat prices rose around the world; poultry prices in the United States jumped more than 40 percent. When the birds that fed on the anchovies died, farmers in Peru lost the source of fertilizer for their fields. Eventually, the trade winds resumed and restored the upwelling, and Peruvians began managing the fishing fleet to prevent overfishing.

Despite several more El Niño events, anchovy stocks have slowly recovered. Marine scientists hope that the hard lesson learned after the 1972 El Niño will help preserve the fish species that depends so heavily upon coastal upwelling. ■

Extension

SCIENCE NOTEBOOK
In the event of another El Niño, what do you think would be Peru's best strategy for protecting the anchovy fishery? Explain what you would suggest and why.

Observe how upwelling occurs.
Keycode: ES2405

FISHERMEN pull a load of anchovies onto their boat off the coast of Peru.

Tides

Tides are the periodic rise and fall of the ocean surface due to the gravitational pulls of the moon and sun. The positions of the moon and sun relative to Earth affect both the time and height of the tides. The heights of tides along the coast also depend on the shape of the shoreline. Because tides can have a significant effect on shorelines, knowledge of tides is important to people who live on the coast.

The Moon and Tides

Around the world, tides go in and out throughout the day. When the tide reaches its highest point on the shore, it is called high tide. When the tide reaches its lowest point, it is called low tide. This movement is influenced by the gravitational pulls of the moon and the sun.

When the moon is new or full, tides rise higher than normal. Tides are lower than normal during the time of the quarter moon. Because of these observations, people have known for many years that the moon and the tides are related. However, it took Sir Isaac Newton to explain that gravity was the cause of the tides.

Gravity is stronger when objects are closer together. Because of this, the moon pulls the water on the side of Earth nearest it more strongly than it pulls on Earth itself. This difference in force causes a bulge in the ocean on the side of Earth nearest the moon. This bulge is the direct high tide.

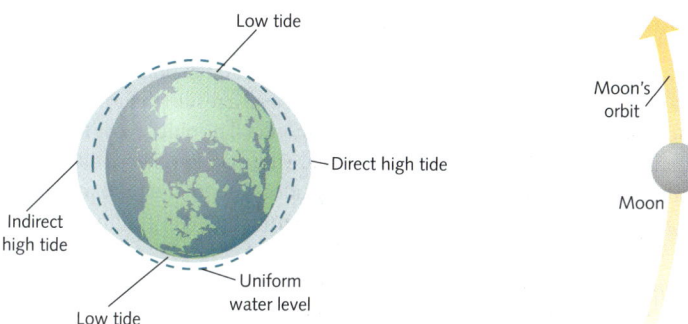

HIGH AND LOW TIDE

At the same time, another high tide, the indirect high tide, occurs on the side of Earth opposite the moon. This tide occurs because the moon pulls more strongly on Earth's center than it pulls on the water on the far side of the planet. Earth itself, therefore, is drawn away from the water on the far side, leaving a bulge of water behind. These bulges on opposite sides of Earth pull water away from the areas that lie between the two high tides. These areas experience low tides.

Because the moon moves around Earth, the bulges also move around Earth over the 29 days that make up the lunar month. The moon rises about 50 minutes later each day and so do the tides. The tides in a given area can vary somewhat because of the influence of the shapes of the ocean basins, ocean floors, and shorelines.

24.3

KEY IDEA
Tides, the periodic rise and fall of ocean waters, are affected by the moon, the sun, and the shape of the shoreline.

KEY VOCABULARY
- tides
- perigee
- apogee

INVESTIGATIONS
CLASSZONE.COM

How Do Tides Work? Explore how the sun and moon affect Earth's tides.
Keycode: ES2406

Chapter 24 The Moving Ocean **541**

CHAPTER 24 SECTION 3

FOCUS

Objectives
1. Explain how the moon and the sun affect tides.
2. Compare and contrast tidal ranges of different bodies of water.

Set a Purpose
Read aloud the section's title and headings. Then rephrase them as questions that students can answer as they read the section. For example, "What are tidal ranges and how do they vary?"

INSTRUCT

DISCUSSION

Have students describe situations where high or low tides can be dangerous or problematic. **Sample Responses:** Walking out to an island during low tide and being trapped when high tide comes in. Entering sea caves during low tide. Docking a ship during low tide. Ask: What force causes tides? gravity

INVESTIGATIONS
CLASSZONE.COM

How Do Tides Work?
Keycode: ES2406

Have students work in pairs to develop the model.

Use *Internet Investigations Guide,* pp. T84, 84.

DIFFERENTIATING INSTRUCTION

Reading Support Students may find the concept of tides difficult to understand. The diagrams on pages 541 and 542 can help them comprehend what they read. Have students preview the diagrams. Direct them to read the captions and labels carefully. Tell students to refer to the diagrams frequently as they read the section. At the end of the section, students can use the diagrams to write a summary explaining the effects of the moon and the sun on tides.
Use Transparency 29 and Reading Study Guide, p. 86.

Chapter 24 The Moving Ocean **541**

CHAPTER 24 SECTION 3

VISUAL TEACHING
Discussion

Have students examine the relative positions of the sun, moon, and Earth in the top diagram. Ask: Explain what happens to the tides when the sun, moon, and Earth are aligned. **High tide is especially high and low tide is especially low because the sun's gravitational pull adds to the moon's gravitational pull.** Why do spring tides occur twice a month? **because the sun, moon, and Earth are aligned when the moon is full and when it is new, each of which occurs once a month** Lead students to recognize that in the second diagram, both the sun and the moon pull on Earth but from different directions. Thus, the moon's tidemaking effect is lessened. Ask: During neap tide, what effect does the sun have on tides? **The gravitational pull of the sun causes high tide to be less high and low tide to be less low.**

Use Transparency 29.

> **VOCABULARY STRATEGY**
> Apogee has its roots in the ancient Greek word *apogaios*, which means "far from Earth" *(apo,* "away" and *gaia,* "earth"). Based on this and what you know about apogee and perigee, what do you think the prefix *peri–* means?

The Sun's Effect on Tides

The sun has the same kind of effect on Earth's waters as the moon does. However, because it is so much farther away, the sun's tidemaking effect is only about half that of the moon. The sun, however, can enhance or detract from the moon's effects.

Tides are always high in line with the moon and low midway between the high-tide points. When the sun is in line with the moon and Earth, as shown in the figure below, the sun's entire tidemaking effect is added to the moon's. This alignment of the moon, the sun, and Earth occurs at the times of the new and the full moons. During these times, high tides are especially high, and low tides are especially low. These tides occur twice a month and are called spring tides.

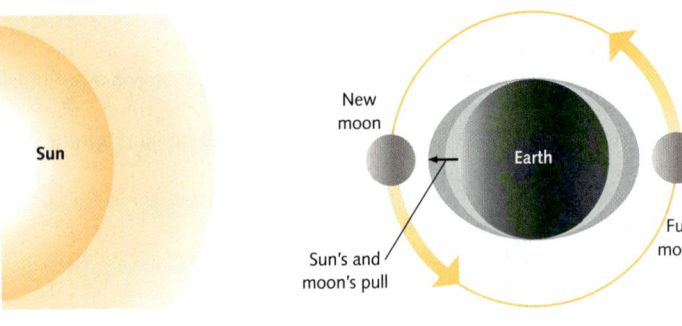

SPRING TIDE

At quarter phases, the moon and the sun are not in line (see the figure below). They are at right angles to each other. As a result, the sun's entire tidemaking effect is subtracted from the moon's. The outcome is high tides that are not very high and low tides that are not very low. These tides also occur twice a month and are called neap tides. Another factor that adds to the tidal effect is the moon's proximity to Earth. When the moon is at **perigee,** the closest point to Earth in its orbit, the tidal effect is greater, especially if perigee occurs during the new or full moon phases. If the moon is at **apogee,** its farthest point from Earth, the tidal effect is less.

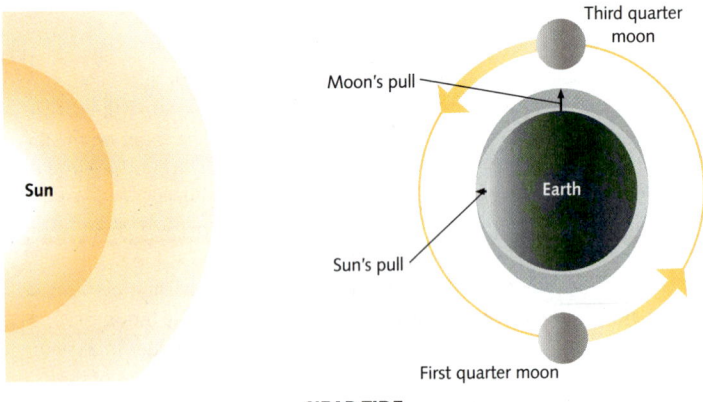

NEAP TIDE

542 Unit 6 Earth's Oceans

DIFFERENTIATING INSTRUCTION

Challenge Activity Provide students with the following problem: On Saturday, Audrey set sail in her boat at high tide, 8:20 A.M. She would like to set sail again at high tide on Tuesday. What time should she set out? **10:50 A.M.** If she wants to go again in a week, when would high tide be on the following Saturday? **2:10 P.M.** You may wish to provide students with a tide schedule and have them create their own word problems to solve.

HIGH TIDE

LOW TIDE

TIDES IN THE BAY OF FUNDY
These photographs clearly demonstrate significant tidal range from high to low tide in the Bay of Fundy, shown here on the coast of Nova Scotia, Canada.

Tidal Range

If you were sailing on the ocean, you would not notice the tides much, if at all. However, if you were standing on shore, the difference between high and low tide would be obvious. The difference in the water level between high tide and low tide is called the tidal range.

Tidal ranges vary from one body of water to the next and tend to be more noticeable on oceans than on lakes. Small lakes show no tides at all. The largest of the Great Lakes, Lake Superior, has a tidal range of only a few centimeters. The tidal range of the open ocean averages less than one meter.

Tidal ranges on ocean shores are most noticeable, but they also vary greatly. In the Gulf of Mexico, the tidal range may be only half a meter. In the Bay of Fundy on the coast of Nova Scotia, the range can be as great as 20 meters. What causes these differences? The answer lies in the shape of the shoreline. The Bay of Fundy is a long, V-shaped bay. Water from the ocean tide is funneled into the wide end of the V. When the water reaches the narrow end of the V, it piles up high. In the Gulf of Mexico, the opposite occurs. The Gulf has a shoreline much broader than its mouth. As the ocean tide enters the Gulf, the water spreads out over the long shoreline.

24.3 Section Review

1. What are tides?
2. Why do tides occur 50 minutes later each day?
3. Explain the differences between spring and neap tides.
4. How does the shape of an inlet affect its tidal range and tidal currents?
5. **CRITICAL THINKING** If the moon's mass was decreased by half, how would the tides be different? Explain your reasoning.
6. **PHYSICS** The gravitational force between two objects depends on the distance between them. As Earth revolves around the sun, the distance between the two bodies changes. Do Earth's tides vary with the seasons? Explain your reasoning.

CHAPTER 24 SECTION 3

CRITICAL THINKING

Ask: When the moon and sun are at right angles relative to Earth, why don't their tidemaking effects cancel each other out? The tidemaking effect of the sun is less than that of the moon. What would be necessary for them to cancel out? The sun would have to be larger and/or closer to Earth, or the moon would have to be smaller and/or farther from Earth.

ASSESS

1. Tides are the periodic rise and fall of the ocean surface due to the gravitational pulls of the moon and sun.
2. The moon rises 50 minutes later each day.
3. During spring tides, the sun, moon, and Earth align; the tidemaking effect of the sun is added to that of the moon. Thus, high tide is especially high, and low tide is especially low. During neap tides, the sun and moon are at right angles relative to Earth; the tidemaking effect of the sun is subtracted from the tidemaking effect of the moon. Thus, high tide is not very high, and low tide is not very low.
4. A wide mouth but narrow shoreline results in a wide tidal range. A shoreline broader than the mouth results in a narrow tidal range.
5. A smaller moon would have less gravitational pull. As a result, the tides would be less extreme.
6. Yes, when Earth is closest to the sun, the tidal range is greatest. When Earth is farthest from the sun, the tidal range is smallest.

MONITOR AND RETEACH

If students miss . . .

Question 1 Review the definition of *tides* on page 541.

Question 2 Use the diagram on page 541 to explain how the moon's orbit around Earth produces tides. Point out that as the moon revolves, its position relative to a particular location on Earth at a given time each day changes a little each day.

Question 3 Have students describe the diagrams on page 542 aloud, stating the relative positions of the sun, moon, and Earth in each.

Question 4 Reteach "Tidal Range." (p. 543) Have students examine maps that show the shapes of the Bay of Fundy and the Gulf of Mexico.

Question 5 Contrast the sizes and gravitational pulls of the moon and sun and their resulting tidemaking effects. (p. 542)

Question 6 Have students consider the effects of the sun during spring and neap tides at different times of the year.

CHAPTER 24 ACTIVITY

PURPOSE
To model the movement of oceanic water masses that differ in temperature and salinity

MATERIALS
Provide students with a copy of Lab Sheet 24. A copymaster is provided on page 145 of the teacher's laboratory manual.

An aquarium the size of a shoe box is ideal. Use the smallest Styrofoam cups available.

PROCEDURE

5 Make sure to tape the Styrofoam cups to the aquarium securely so that they do not tip over when the water is added.

6, 7 Do not add more than a couple of drops of dye to each water mass, otherwise the dyes will mix and make it difficult to distinguish water layers. Use the most extreme water masses that are safe and available; that is, make the water as hot and as cold as possible, and add as much salt as will dissolve. There should be no undissolved salt in the Styrofoam cups.

10 Try to poke holes in the Styrofoam cups as close to simultaneously as possible. Take care not to make the holes too large.

CHAPTER 24 LAB Activity

Oceanic Water Masses

SKILLS AND OBJECTIVES
- **Observe** the interaction of water with different properties.
- **Measure** the temperature of water at different depths.

MATERIALS
- aquarium
- cardboard
- scissors
- 5 long thermometers
- metric ruler
- temperature probe
- magic marker
- 4 small beakers
- 4 different colors of dye
- 4 Styrofoam cups
- masking tape
- clear tape
- ice cold water
- hot water
- salt
- spoon
- sharp pencil
- **Lab Sheet 24** Temperature vs Depth Data Table
- graph paper
- colored pencils

The world's oceans contain layers of water called water masses. Each water mass has a source at the surface where the water takes on specific properties, such as temperature and salinity. The water then sinks to a certain level based on its density. Denser water will sink below water that is less dense. In this activity you will observe the interaction of different sources of water in a simulated ocean basin.

Procedure

1 Cut out a rectangular piece of cardboard. Construct a bridge over the aquarium by cutting out the bottom center of the cardboard strip as shown in the diagram.

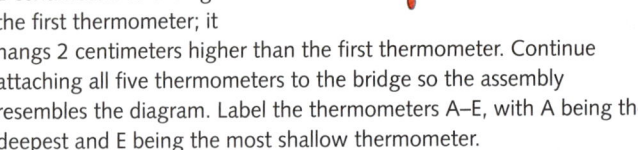

2 Tape one thermometer to the cardboard bridge so it hangs vertically with its bulb 1 centimeter off the bottom of the aquarium. Tape a second thermometer 2 centimeters to the right of the first thermometer; it hangs 2 centimeters higher than the first thermometer. Continue attaching all five thermometers to the bridge so the assembly resembles the diagram. Label the thermometers A–E, with A being the deepest and E being the most shallow thermometer.

3 Hang the thermometer bridge across the center of the aquarium. Tape the bridge to the aquarium so it remains in place when water is added.

4 Carefully add water at room temperature to the aquarium to a depth of 10 centimeters. The surface of the water should be about 1 centimeter above the bulb of thermometer E.

5 Use masking tape to attach four cups securely to the four corners of the aquarium. See the diagram on the next page. The bottoms of the cups should be a couple centimeters below the surface of the water. Label the cups 1–4.

6 Fill two beakers with ice cold water. Add green dye to one beaker. Add a heaping tablespoon of salt to the second beaker and stir until the salt is dissolved. Add blue dye to the beaker of cold salt water.

544 Unit 6 Earth's Oceans

⑦ Fill the other two beakers with hot water. Add red dye to one beaker. Add a heaping tablespoon of salt to the other beaker and stir until the salt is dissolved. Add yellow dye to the beaker of warm salt water.

⑧ Carefully fill Cup 1 with the blue water, Cup 2 with the green water, Cup 3 with the red water, and Cup 4 with the yellow water.

⑨ Read the temperatures on the five thermometers and record them on your data table for time = 0 minutes.

⑩ Carefully use a sharp pencil to poke a *small* hole in the bottom of each of the four cups so the dyed water flows out. Observe the movement of the dyed water out of each of the four cups.

⑪ Every 60 seconds after releasing the dye, record the temperature of the five thermometers on your table. Continue to observe the dyed water and record temperatures for 10 minutes.

⑫ Draw one graph that contains the data you took for all five thermometers. Place time on the x-axis and temperature on the y-axis. Connect the data for each thermometer with a smooth curve and label.

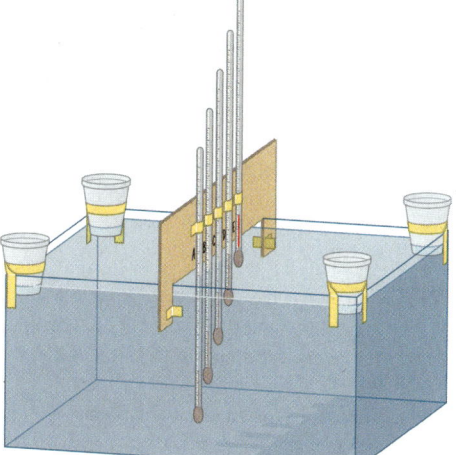

Analysis and Conclusions

1. Use your observations of the dye to rank the four sources of water in order from least dense to most dense.

2. Use your graph and your observations of the dye to determine the source of the water near the bulb of each thermometer at the end of 10 minutes. Without the dye, could you identify in which layer each thermometer was submerged from just the temperature readings?

3. Explain why the layer of water that formed second from the top is less dense than the water that was in the third layer from the top.

4. Ocean water contains water that originated in a variety of places where the water can be cold, warm, salty, or fresh. For each of the four cups, give an example of a water source that contains the properties you modeled in the experiment. (For example, the yellow-dyed warm and salty water could represent water from a subtropical sea such as the Mediterranean Sea.)

CHAPTER 24 ACTIVITY

ANALYSIS AND CONCLUSIONS

1. warm and fresh, cold and fresh, warm and salty, cold and salty

2. The layers are as described in the answer to Question 1. The thermometers should show a decrease in temperature from top to bottom, with a brief temperature rise indicating the warm and salty third layer. The dye was necessary to observe the different layers because the thermometers do not provide salinity information.

3. Although cold water is denser than warm water, the second layer is less dense than the third layer because it contains much less salt than the third layer.

4. Cold and fresh water could come from polar rivers or melting icebergs or glaciers. Warm and fresh water could come from rivers entering the ocean near the equator. Warm and salty water could come from salty, subtropical seas. Cold and salty water could come from water left behind as ice and glaciers form in polar regions.

Refer students to Safety in the Earth Science Laboratory on pages xxii–xxiii.

CHAPTER 24 REVIEW

VOCABULARY REVIEW

1. a, Trade winds affect surface currents.
2. c, These are two kinds of Gulf Stream rings that break away from the edge of the main current.
3. c, In upwelling, cold, deep water comes to the surface, bringing nutrients with it.
4. b, At perigee, the moon is at its closest point to Earth.

CONCEPT REVIEW

5. Monsoon winds blow in one direction in the summer and in the opposite direction in the winter. When the winds reverse direction, the ocean surface currents also reverse direction.
6. The Gulf Stream carries warm water currents to Iceland and the British Isles.
7. Density currents bring oxygen to the deeper parts of the ocean, as well as moderate temperatures and salinity.
8. The eastern side has cool ocean currents moving toward the equator. The western side has warm ocean currents moving away from the equator.
9. A high rate of evaporation over the Mediterranean Sea results in an increased salinity of the water, which causes a density current to flow into the Atlantic Ocean.
10. When surface winds push away warm ocean surface water, cold, deep water rises to replace it.
11. The moon's gravity pulls the water on the side of Earth nearest the moon more strongly than it pulls on Earth itself. This causes a bulge in the ocean on the side of Earth nearest the moon. At the same time, the moon's gravity pulls on Earth's center more strongly than it pulls on the water on the far side of Earth, thereby drawing Earth away form the water, producing a bulge in the ocean on the far side as well.
12. The gravity of the sun causes tides but, because the sun is so far away from Earth, the sun's tidal effect is less than that of the moon. When the sun aligns with the moon and Earth, the sun's tidal effect enhances that of

CHAPTER REVIEW 24

Summary of Key Ideas

24.1 Wind is the driving force of most ocean surface currents. Many surface currents come from either trade winds or westerly winds. Warm currents, including the Gulf Stream, flow away from the equator on the western side of ocean basins. Cold currents flow toward the equator on the eastern side. A countercurrent flows in the opposite direction of a wind-related current.

24.2 Density is what drives ocean currents under the surface. When an area of water becomes more dense than the waters around it, the denser water moves beneath the less dense water, forming a flowing current. Temperature and salinity are two factors that influence the density of an area of water. The densest seawater comes from the polar regions. Density currents can flow vertically or horizontally.

24.3 The twice-daily rise and fall of Earth's oceans—known as the tide—is a result of gravitational pulls from the moon and the sun. Tides reach different levels depending on the Earth's location in relation to the moon and sun. The shapes of individual shorelines influence how high tidewaters rise, and how low they fall.

KEY VOCABULARY

apogee (p. 542)
cold-core ring (p. 535)
countercurrent (p. 535)
density current (p. 536)
ocean current (p. 532)
perigee (p. 542)
surface current (p. 532)
tides (p. 541)
upwelling (p. 538)
warm-core ring (p. 532)

Vocabulary Review

Choose the best answer for each question.

1. Which would not cause a density current?
   (a) trade winds (b) evaporation (c) freezing
2. Eddies in the Gulf Stream may break off to form
   (a) countercurrents, (b) deep currents (c) cold- or warm-core rings
3. Which is true of upwelling?
   (a) It occurs only near Antarctica.
   (b) It is caused by evaporation.
   (c) It brings nutrients to the surface.
   (d) It forms cold-core rings.
4. Which term describes the moon's position when its gravitational pull most strongly affects Earth's tides?
   (a) apogee
   (b) perigee
   (c) countercurrent

Concept Review

5. Seasonal changes in wind direction do not usually change the direction of ocean currents. Give one example of an exception to this rule.
6. Account for the fact that Iceland and the British Isles have warmer climates than you might expect, given their high latitudes.
7. In what ways are density currents important to life in the deep sea?
8. Which side of an ocean basin has cold currents flowing toward the equator? Which has warm currents flowing away from the equator?
9. What causes the density current flowing from the Mediterranean Sea?
10. What causes upwelling?
11. Explain how the moon causes tides.
12. How does the sun affect tides? At what times does the sun have the greatest tidal effects? At what times does it have the smallest tidal effects?
13. What are spring tides? How and when do they occur? What are neap tides? How and when do they occur?

the moon. When the sun and moon are at right angles to each other, the sun's tidal effect detracts from that of the moon.

13. Spring tides are especially high and low tides that occur twice a month. They occur during the new and full moons, when the sun is aligned with the moon and Earth. Neap tides occur twice a month, during the first-quarter and third-quarter moons, when the sun and moon are at right angles to each other relative to Earth. The tide-making effects of the sun and moon detract from each other, so these tides are not especially high or low.

Critical Thinking

14. Apply If you were standing on the western coast of a continent in the Southern Hemisphere, what kind of ocean current would you expect off shore? In what direction would the current be flowing?

15. Analyze Explain how freezing and evaporation are similar processes when it comes to seawater.

16. Infer Mediterranean water flowing into the Atlantic Ocean sinks to a depth of 1000 meters. At that location, the Atlantic Ocean is 3000 meters deep. Why would Mediterranean water sink only to 1000 meters?

17. Hypothesize The tidal ranges of lakes and other minor bodies of water are so small that they are almost invisible. Why might this be so?

Internet Extension

Could You Break the Record for an Ocean Sailboat Race? Apply your understanding of winds and ocean currents to plot a course for a sailboat race.
Keycode: ES2407

Writing About the Earth System

SCIENCE NOTEBOOK Sports fishers know that the best time to fish from the shoreline is when the tide is rising. During those hours, the incoming tide carries fish closer to the shore. Obtain a tide chart for any area of your choice. Design a "best hours to fish" chart for someone going there for a one-week vacation (you choose the dates).

Interpreting Graphs

The graph at right shows how the average speed of a deep current in the western Atlantic Ocean changes with height above the sea floor. The horizontal axis of the graph shows the average current speed in centimeters per second. The vertical axis shows height above the sea floor in meters. Refer to the graph to answer the questions below.

18. Determine the thickness, in meters, of the mixing layer and of the layer of smooth flow. How many times thicker is the mixing layer?

19. What is the average current speed at 1 meter above the seafloor? At 0.01 meters above it?

20. In which layer is current speed constant with height above the seafloor?

21. Why is the current speed nearly zero at the bottom of the graph?

22. The mixing layer is where seawater properties (salinity, temperature, sediment content, etc.) become thoroughly mixed together. If a deep-sea storm causes the speed of the current to increase, what is likely to happen to the thickness of the mixing layer?

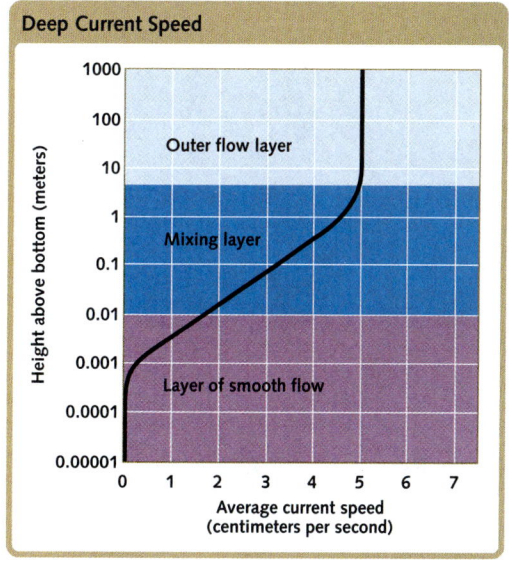

Chapter 24 The Moving Ocean 547

CHAPTER 24 REVIEW

Could You Break the Record for an Ocean Sailboat Race?
Keycode: ES2407

Use Internet Investigations Guide, pp. T85, 85.

CRITICAL THINKING

14. You would expect cold water currents flowing north from the polar region toward the equator.

15. Both freezing and evaporation of seawater leave behind seawater with increased salinity and therefore increased density.

16. Although Mediterranean water is higher in salinity than Atlantic water, it is also warmer. Thus, Mediterranean water's increased density due to salinity is counteracted by the decreased density due to temperature.

17. Compared to oceans, there is relatively little water in lakes and other bodies of water for the moon's gravity to pull on.

INTERPRETING GRAPHS

18. The mixing layer is about 7 meters. The layer of smooth flow is about 0.01 meter. The mixing layer is about 700 times thicker.

19. 1 meter: about 4.5 centimeters per second; 0.01 meter: about 1.75 centimeters per second

20. the outer flow layer

21. It is very close to the seafloor, and friction with the seafloor prevents movement.

22. The thickness of the mixing layer will likely increase.

WRITING ABOUT SCIENCE

Writing About the Earth System

Students should note that high tides occur twice a day. These times may occur at inconvenient times, such as in the middle of the night. Some students may choose not to list those hours, although zealous sports fishers will go out at any time. The best times for fishing will follow the tide chart and vary from day to day.

UNIT 6 FEATURE

FOCUS

Objective
Describe interactions between the hydrosphere and the atmosphere that might cause climate changes.

Set a Purpose
Have students read to find answers to the question "What are some ways in which the ocean affects climate?" There are negative effects. For example, El Niño, an ocean current, has caused monsoon seasons to fail; brought drought to southeast Asia; torrential rains to North America; and crop failure to South America. Potential disruption of deep-ocean currents due to atmospheric pollution could also seriously alter climates worldwide. In a positive way, the ocean acts as a thermal buffer, so that at the shore there are cool breezes in summer and slowly released stored heat in winter.

UNIT 6 The EARTH SYSTEM

Ocean Effects on Weather and Climate

The El Niño and La Niña phenomena are examples of the complex interactions between the world's oceans (hydrosphere) and global climatic conditions (atmosphere).

WILDFIRE This 1998 Mexican wildfire was brought on by months of El Niño-related drought.

Ocean currents influence the world's weather. El Niño—an ocean current that is warmer than usual—alters the patterns of atmospheric circulation around the world. The normal pattern is for easterly trade winds to blow Pacific Ocean water westward along the equator and for cold water to upwell in the eastern Pacific, replacing the water sent west. Every three to seven years, however, this pattern is reversed: The upwelling stops, ocean temperatures in the eastern Pacific rise, and—most importantly—weather changes around the world. The Indian monsoon fails and Southeast Asia sees drought and fires, while torrential rains hit parts of the United States and crops fail in South America.

How can a change in one ocean current cause so many drastic changes in weather? Scientists are only beginning to understand the complexity of the ocean-atmosphere interaction.

Head to the shore on a hot summer weekend, and whether you know it or not, you are benefiting from the ocean's well-known role as a thermal buffer. While continental interiors bake in the summer sun, the coasts enjoy cool ocean breezes. Water takes longer to heat and longer to cool than almost any other substance on Earth; therefore, it remains cool even while absorbing the most powerful solar rays of the year. The ocean heats slowly and continues to hold that heat long after winter arrives. During the colder seasons people living along the shore, for example, experience warmer weather than people living inland because the ocean moderates the air temperature. This effect is intensified for those living on the western coasts of continents, where

TROPICAL VEGETATION thrives far north of tropical latitudes in Cornwall, England due to warm ocean currents.

548   Unit 6   Earth's Oceans

DIFFERENTIATING INSTRUCTION

Reading Support Creating a graphic summary can help students comprehend and retain key information from their reading. Have students summarize the main idea and supporting details of "Ocean Effects on Weather and Climate" in a chart like this one.

Main Idea	El Niño changes the climate.	The ocean is a thermal buffer.	The ocean might regulate global warming.
Supporting Details	Example: Upwelling stops in the eastern Pacific. Ocean temperatures in the eastern Pacific rise. Drought occurs in southeast Asia.		

and the ENVIRONMENT

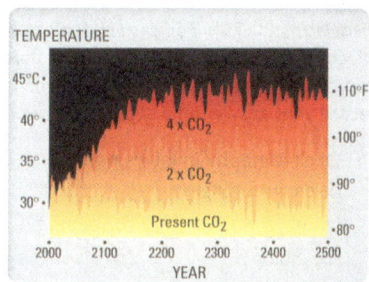

RISING TEMPERATURES are expected with an increase of the carbon dioxide levels in the atmosphere.

prevailing winds blow in from the ocean. Even communities as far north as Juneau, Alaska, enjoy moderate climates due to this buffering effect.

The ocean's role in regulating heat is pertinent today in the face of increases in atmospheric greenhouse gases and the prospect of global warming. If the ocean does have a deeper capacity for storing heat than scientists have realized, global warming could be delayed. On the other hand, an increase in atmospheric greenhouse gases could alter oceanic current systems and lead to effects more significant than currently projected.

Water moves around the ocean depths in a pattern known as thermohaline circulation. Extremely cold and salty surface waters in the polar regions sink to the bottom, then take thousands of years to creep across ocean basins. These cold waters reach the surface in upwelling areas, such as those off the coast of Peru where El Niños begin. Disruptions in this deep circulation are thought to have contributed to several big climate swings during the last ice age nearly 10,000 years ago. Some computer models show that increases in carbon dioxide will affect thermohaline circulation. It has also been suggested that the greater severity of recent El Niño events are due to global warming. Could we be witnessing the beginnings of changes in thermohaline circulation that ultimately could plunge us into another ice age? That's far from the scenario most people envision when greenhouse gases are the topic of discussion.

The complexities of these issues are mind-boggling. The National Oceanic and Atmospheric Administration is conducting long-term studies to analyze ocean water at various depths from locations around the world. The purpose is to find out which changes in the ocean result from human activities affecting the atmosphere and which changes are due to Earth's natural cycles.

Earth is a water planet—oceans cover most of its surface. When the temperature of ocean water rises or falls, it disturbs marine ecosystems and affects the global climate in ways that compel scientists to take notice.

UNIT INVESTIGATION
CLASSZONE.COM

Can We Blame El Niño for Wild Weather? Changes in water temperature in the Pacific Ocean occur each season. During some years, these changes are greater than normal, producing conditions called El Niño or La Niña. Explore images of sea surface temperatures as well as data on global climate cycles and weather patterns. Look for a relationship between extreme weather events and Pacific Ocean temperatures.
Keycode: ESU601

UNIT 6 FEATURE

Can We Blame El Niño for Wild Weather?
Keycode: ESU601

Use Internet Investigations Guide, pp. T86–T87, 86–87.

Investigation Photograph: A landslide, brought on by rains that were probably triggered by El Niño, buried this car in Boulder Creek, California.

Unit 6 The Earth System and the Environment 549

UNIT 6 ASSESSMENT

ANSWERS

1. d, Oceanography includes many branches of science.
2. b, Polarity is the uneven distribution of charge.
3. c, Hydrogen bonding brings together oppositely charged regions.
4. c, The mixed layer is the uppermost layer of the ocean.
5. a, Core sampling preserves the order and thickness of sediment layers.
6. b, The other choices characterize passive margins.
7. d, The other choices are mountain formations associated with volcanic activity at the ocean floor.
8. b, Submarine canyons are formed by turbidity currents, not volcanoes.
9. b, Oozes are formed from living sources.
10. d, Density currents are formed by water that becomes denser due to cooling, freezing, and evaporation.
11. a, Upwelling occurs when cold, deep water rises to replace surface water pushed out to sea by prevailing winds.
12. d, Shapes of shorelines determine tidal ranges.

UNIT 6 STANDARDIZED TEST Practice

Directions (1–13): For *each* statement or question, select the word or expression that, of those given, best completes the statement or answers the question. Record your answer on a separate answer sheet.

1. Oceanography involves numerous scientific fields, including
   - (a) physics
   - (b) biology
   - (c) geology
   - (d) all of the above

2. The uneven distribution of charge at opposite ends of a water molecule is known as
   - (a) density
   - (b) polarity
   - (c) hydrogen bonding
   - (d) salinity

3. The attraction between oppositely charged regions of two different water molecules is called
   - (a) density
   - (b) polarity
   - (c) hydrogen bonding
   - (d) salinity

4. The only ocean layer with enough sunlight to sustain marine plants is the
   - (a) deep ocean
   - (b) thermocline
   - (c) mixed layer
   - (d) Antarctic Intermediate Water

5. Core sampling is a useful form of sediment sampling because it
   - (a) preserves sediment layers
   - (b) helps determine water density
   - (c) records ocean depth
   - (d) allows retrieval of large rocks

6. An active continental margin is characterized by
   - (a) a large abyssal plain
   - (b) an oceanic plate subducting under a continental plate
   - (c) a long, gradual continental slope
   - (d) the absence of a continental shelf

7. The flattest surface on Earth is
   - (a) an island arc
   - (b) the top of a guyot
   - (c) a seamount
   - (d) the abyssal plain

8. Which of the following are not connected in some way to volcanoes?
   - (a) coral atolls
   - (b) submarine canyons
   - (c) seamounts
   - (d) guyots

9. Which sediment type is also called an ooze?
   - (a) terrigenous
   - (b) biogenous
   - (c) hydrogenous
   - (d) igneous

10. Two causes of density currents are
    - (a) polarity and evaporation
    - (b) tidal range and evaporation
    - (c) freezing and condensation
    - (d) evaporation and freezing

11. Upwelling is an example of
    - (a) a vertical density current
    - (b) a tide at the moon's apogee
    - (c) a tide at the moon's perigee
    - (d) a cold-core ring

12. Tidal ranges vary according to
    - (a) the temperature of local waters
    - (b) the salinity of local waters
    - (c) the density of local waters
    - (d) the shapes of individual shorelines

13. The sun has less effect on Earth's tides than does the moon because the sun
    - (a) is hotter than the moon
    - (b) is farther away than the moon
    - (c) has a lower gravitational pull than the moon
    - (d) is larger than the moon

UNIT 6 ASSESSMENT

Directions (14–19): Record your answers on a separate answer sheet.
Base your answers to questions 14 through 16 on the article below.

Bleaching, Coral's Way: Making the Best of a Bad Situation

For an organism that can't move, coral turns out to be pretty nimble.

Coral has a critical partnership with certain algae that absorb sunlight and convert it to the energy needed to feed a complex array of life found in the reef ecosystem. The loss of these algae—a common consequence of pollution or climate change—"bleaches" a reef, making it unable to produce energy from sunlight.

Coral bleaching has increased widely in recent decades. Because it often precedes coral death and the loss of the reef itself, conservationists are naturally concerned that many of the world's reefs are in trouble.

New findings, however, suggest that when coral is threatened, bleaching may be part of the solution.

It now appears that coral colonies, when confronted with dramatic environmental changes, may purge themselves of existing algae to make room for other algae more capable of thriving in the challenging conditions. Bleaching, then, may not signify coral's imminent demise, but its ability to tough out new conditions.

In one set of experiments, marine scientist Andrew C. Baker of the New York Aquarium found that corals which undergo bleaching after being exposed to sudden environmental change are more—not less—likely to survive in the long run.

"This counters conventional wisdom that bleaching is detrimental from all perspectives," Baker said.

14. State one form of pollution that could lead to coral-reef bleaching.

15. Explain how corals that undergo bleaching are more able to survive sudden environmental change than corals that do not undergo bleaching.

16. Explain why an increase in coral-reef bleaching is still harmful to the environment.

Base your answers to Questions 17 through 19 on the satellite image of the Gulf Stream, at the right.

17. Without referring to the colors in the image, explain why point A is colder than point B.

18. State whether ring C is a warm-core ring or a cold-core ring. Do the same for ring D.

19. Describe the process by which an eddy such as the one at position E forms a ring. Be sure to indicate whether the ring that forms at position E will be a warm-core ring or a cold-core ring.

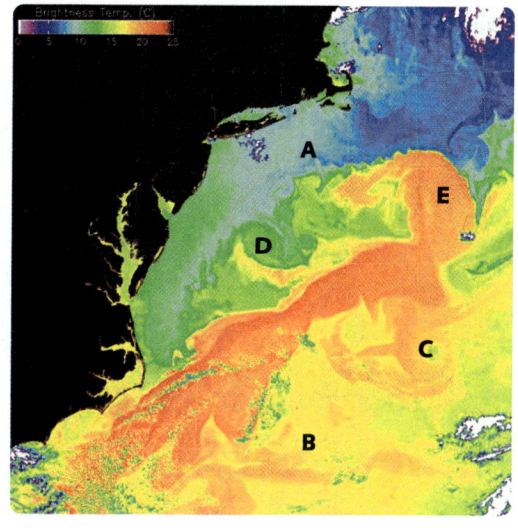

(continued)

13. b, The sun has a smaller gravitational pull on Earth because it is much farther from Earth than the moon is.

14. Forms of pollution that could lead to coral-reef bleaching include industrial waste, sewage, temperature pollution, and agricultural runoff.

15. Bleaching might allow algae that are better adapted to the new water conditions to move into the coral reef, replacing the old algae and making the coral more likely to survive under the new conditions.

16. Increased coral-reef bleaching is a sign of increased water pollution.

17. Water to the north of the Gulf Stream is colder than water to the south of it, so point A is colder than point B.

18. Ring C is a cold-core ring, and ring D is a warm-core ring.

19. As the eddy grows in size, it might eventually pinch itself off and form a ring. At position E, a warm-core ring would be formed as the warmer Sargasso Sea water is enclosed on the northern side of the Gulf Stream.

Standardized Test Practice 551

UNIT 7

UNIT OVERVIEW

The background image is a picture of deep space taken by the Hubble Space Telescope. Ask students what they think all the specks of light are. Some may think they are stars, but in fact they are distant galaxies made up of billions of stars. With advances in technology, including the development of the Hubble Space Telescope, we have learned much about the age and evolution of the universe—how our moon, the planets, stars, and galaxies change over time.

VISUAL TEACHING

Chapter 25 Earth's Moon

The lighter areas of the moon are rugged highlands. The dark patches are basins and plains made of dark basalts that formed from lava flows. These dark areas are called *maria* (seas) because early lunar observers thought they were areas of water. Students will learn how the moon and lunar features formed.

Answer: The dark patches indicate that the moon experienced volcanic activity.

Chapter 26 The Sun and the Solar System

The outer layer of the sun, the corona, is constantly expanding into space as a result of the intense heat inside the sun. This expansion causes a solar wind of charged particles that move through the solar system. Students will learn how solar wind affects Earth.

UNIT 7 Space

The Earth System As we have studied other planets, stars, and galaxies, we have come to appreciate the Earth system's complexity.

What have we learned about the universe that surrounds us?

Earth's Moon   CHAPTER 25

What do the dark patches on its surface tell us about the Moon's past?

552 Unit 7 Space

CONNECTIONS TO . . .

Prior Knowledge Previous units have focused on the Earth system. Unit 2 discussed the origin of Earth and the other planets orbiting the sun. Unit 7 continues the discussion of origins by discussing the origins of stars, galaxies, and the universe. The focus on Earth is resumed in Unit 8, which discusses Earth's history.

Unifying Themes The universe is an expanding *system* of galaxies. Galaxies are *systems* of millions or billions of stars. Some stars, such as the sun, are the center of planetary *systems*. As students study the chapters in Unit 7, point out the different levels of *systems* in the universe.

UNIT 7

CHAPTER 26 The Sun and the Solar System

What effect does the solar wind have on the Earth system?

CHAPTER 27 The Planets and the Solar System

What processes have formed the surface of Mars shown above?

CHAPTER 28 Stars and Galaxies

How does a star become a supernova?

(continued)
Answer: The solar wind causes auroras and can disrupt communications systems.

Chapter 27 The Planets and the Solar System

The image shows the great equatorial canyon of Mars—Valles Marineris. This gash is a system of canyons more than 5000 kilometers long (longer than the width of the United States) and 6 kilometers deep, on average. Students will learn about the features of Mars and other planets.

Answer: Scientists think this set of canyons was formed by the collapse of a nearby plateau as a result of tectonic activity. Stream erosion and landslides may have enlarged the canyons.

Chapter 28 Stars and Galaxies

A star exploded in February 1987, releasing three rings of glowing gas. This event is referred to as Supernova 1987A. The explosion occurred about 169,000 light years from Earth in a nearby galaxy called the Large Magellanic Cloud. The image of the gas rings, shown here, was taken by the Hubble Space Telescope. Students will learn about supernovas and other events in the life cycles of stars.

Answer: As a star ages, nuclear fusion produces an iron core. The core collapses, resulting in a tremendous rise in temperature. The star explodes violently. This event is a supernova and is visible initially as a very bright light.

CONNECTIONS TO . . .

National Science Education Standards
- Students will understand how the sun and other stars produce energy.
- Students will develop an understanding of how stars form and what galaxies are.
- Students will understand the "big bang" theory of the origin of the universe.

Unit Features
The Earth System and the Environment
Life in Extreme, pp. 640–641

Could Mars Support Life? Explore the possibility that Mars supported life in the past and the potential for Mars supporting life today or in the future.
Keycode: ESU701

Standardized Test Practice, pp. 642–643

CHAPTER 25

PLANNING GUIDE EARTH'S MOON

	Section Objectives	Print Resources
Chapter-Level Resources		Laboratory Manual, pp. 109–112 Internet Investigations Guide, pp. T90, 90 Spanish Vocabulary and Summaries, pp. 49–50 Formal Assessment, pp. 73–75 Alternative Assessment, pp. 49–50
Section 25.1 Origin and Properties of the Moon, pp. 556–561	1. Explain various hypotheses about how the moon formed. 2. Describe features and properties of the moon.	Lesson Plans, p. 87 Reading Study Guide, p. 87 Teaching Transparencies 4, 31
Section 25.2 The Moon's Motions, pp. 562–565	1. Describe the motions of the moon. 2. Explain the reason the moon goes through phases. 3. Analyze how the Earth-moon-sun geometry causes lunar and solar eclipses.	Lesson Plans, p. 88 Reading Study Guide, p. 88 Teaching Transparency 30

INVESTIGATIONS
CLASSZONE.COM

Chapter Review INVESTIGATION

What If Earth and the Moon Were Hit by Twin Asteroids?
Computer time: 45 minutes / Additional time: 45 minutes
Students examine images of impact craters to predict the results of each impact.

Technology/Online Resources

Electronic Teacher Tools
Visualizations CD-ROM
Classzone.com
Online Lesson Planner
Test Generator

Visualizations CD-ROM
Classzone.com
  ES2501 Visualization, ES2502 Data Center

Visualizations CD-ROM
Classzone.com
  ES2503 Visualization, ES2504 Visualization, ES2505 Visualization

Classroom Management

Gather these materials for minilab and laboratory activities.

Minilab, p. 557
  calculator

Lab Activity, pp. 566–567
  pan
  surface material
  sieve or sifter
  dry tempera paint
  balance
  safety goggles
  3 spheres of varying size and weight (impactors)
  meter stick
  metric ruler
  graph paper
  colored pencils
  Laboratory Manual, Teacher's Edition
    Lab Sheets 25a–c, pp. 146–148

CHAPTER 25

Earth's Moon

Cold and airless, the moon has long been the subject of scientific study.

What can our exploration of the moon tell us about Earth and the universe?

INTRODUCE

Build Interest

Before sharing background information about the photograph with students, allow time for their questions and comments.

Ask questions such as the following:
- What proportion of the moon is visible? **one quarter**
- What are the circular features? the dark areas? **craters; maria**
- Why is part of the moon darkened? **Only half of the moon is ever illuminated. Only part of that half is visible from Earth.**

Respond

- What can our exploration of the moon tell us about Earth and the universe? **Rocks from the moon can help us date Earth and the universe, and can yield information on how our planet formed. This localized information can help us understand the formation of other planets in our solar system.**

About the Photograph

The photograph shows clearly some of the topographic features of the moon, Earth's nearest neighbor. At center is a large impact crater with rays extending from it, the Crater Copernicus. To the right of the crater are the Carpathian Mountains. The dark area beyond the Carpathians is Mare Imbrium ("Sea of Rains"). These features suggest a geologically active past.

BEYOND THE TEXTBOOK

Print Resources
- Reading Study Guide, pp. 87–88
- Transparencies 4, 30, 31
- Formal Assessment, pp. 73–75
- Laboratory Manual, pp. 109–112
- Alternative Assessment, pp. 49–50
- Internet Investigations Guide, pp. T90, 90
- Spanish Vocabulary and Summaries, pp. 49–50

Technology Resources
- Visualizations CD-ROM
- Classzone.com
- Online Lesson Planner
- Electronic Teacher Tools
- Test Generator

554 Unit 7 Space

CHAPTER 25

PREVIEW

▶ **FOCUS QUESTIONS** In this chapter you will study the moon and learn more about the key questions listed below.

Section 1 How did the moon form? What are its properties?
Section 2 How does the moon move in relation to Earth?

▶ **REVIEW TOPICS** As you investigate the moon, you will need to use information from earlier chapters.
- density (p. 72).

▶ **READING STRATEGY**
PREVIEW
Before you read, look through the chapter and examine the headings, photographs, and other visuals. Does any visual surprise you? Write any questions you might have in your Science Notebook.

INTERNET RESOURCES
CLASSZONE.COM

At our Web site, you will find the following Internet support for this chapter.

DATA CENTER
EARTH NEWS
VISUALIZATIONS
- Moon Formation
- Moon Phases
- Lunar Eclipse
- Solar Eclipses

LOCAL RESOURCES
INVESTIGATIONS
- What if Earth and the Moon Were Hit by Twin Asteroids?

Chapter 25 Earth's Moon   555

CHAPTER 25

PREPARE
Focus Questions
Find out what knowledge students have about the moon. For each focus question, have students consider the section title and then develop two or three additional questions for each section. For example, Section 2: How do the moon's motions affect Earth? How can we tell the moon is moving?

Review Topics
In addition to the earth science topics listed, students may need to review these concepts:

Mathematics
- Angles are measured in degrees.
- Arcs are measured in degrees.
- Diameter is the distance across a circle.

Reading Strategy
Model the Strategy: To help students examine the visuals, use a version of the following: "Look at the photograph across the bottom of pages 558 and 559. What is it showing? It is reasonable to expect that in the photograph the features relate to the features listed in the blue headings: Lunar Maria, Lunar Highlands, Lunar Craters and Rays, and Lunar Soil. Can you identify these features in the photograph? Throughout the chapter, the photographs and diagrams help you visualize the content. Try to write your notebook questions in such a way as to connect the visuals to the content throughout the chapter." Encourage students to follow this strategy as they continue through the chapter.

BEYOND THE TEXTBOOK

INVESTIGATIONS
- What If Earth and the Moon Were Hit by Twin Asteroids?

VISUALIZATIONS
- Animation illustrating the impact theory of the moon's formation
- Animation of the phases of the moon from Earth and space
- Animations and video of lunar eclipses
- Animations and video of solar eclipses

DATA CENTER
EARTH NEWS
LOCAL RESOURCES
CAREERS

Online Lesson Planner

Chapter 25 Earth's Moon   555

CHAPTER 25 SECTION 1

FOCUS

Objectives
1. Explain various hypotheses about how the moon formed.
2. Describe features and properties of the moon.

Set a Purpose
Have students read this section to answer the question "How did the moon probably form and what is it like?"

INSTRUCT

Visual Teaching

Interpret Diagrams
Discuss the mechanisms behind how the debris ring might have coalesced to form the moon. *gravitational attraction of the particles, one molten blob bumping into another to become one*

Extend
Point out that the diagram shows the steps in just one of the four theories about how the moon formed. Direct students to Chapter 2, pages 32 and 33, where the other theories are illustrated, explained, and disproved. Discuss why each is no longer accepted.

25.1

KEY IDEAS
The moon was probably formed by a large impact between Earth and a planet-sized object. The moon has a heavily cratered surface that consists of older highlands and younger maria.

KEY VOCABULARY
- meteoroid
- crater
- micrometeoroid
- mascons
- maria
- rille
- ray
- regolith

VISUALIZATIONS
CLASSZONE.COM
Observe images illustrating the impact theory of the moon's formation.
Keycode: ES2501

Origin and Properties of the Moon

When probes from the Soviet *Luna* spacecraft first encountered the moon in 1959, the human perspectives on Earth and space changed dramatically. In 1969—a mere ten years later—*Apollo* astronauts from the United States landed on the moon. By the end of the final *Apollo* mission in 1972, about 400 kilograms of lunar rock had been brought back to Earth, most of it by *Apollo* spacecraft missions. Scientists are still using these rocks, along with thousands of lunar photographs and other data, to study the moon.

Origin of the Moon

There are several theories about the moon's origin. One suggests that Earth and the moon formed simultaneously, with the moon in orbit around Earth. Another suggests the early Earth was spinning so fast that a chunk of it spun off into orbit. Still another theory suggests that the moon formed elsewhere in the solar system and was later captured by Earth's gravitational field. These three theories are plausible; however, a fourth theory has become widely accepted by most planetary scientists. This theory proposes that the moon formed about 4.5 billion years ago as a result of a collision between Earth and a planet-sized object.

Development of the Moon

Approximately 4.5–4.0 billion years ago, Earth and the moon continued to grow from additional impacts. Most of these collisions involved **meteoroids.** A meteoroid is a celestial body that can range in size from a speck to an object weighing thousands of kilograms. For the first half billion or so years of the moon's life, frequent impacts melted its surface layers, forming a huge "magma ocean." Lighter materials, such as aluminum compounds, floated to the top of the ocean. When the crust cooled and hardened, additional impacts continued to gouge out **craters,**

Formation of the Moon—Impact Theory

① Earth is hit off-center by a planet-sized object.

② The impact heats and deforms both bodies. Some rocky debris remains in orbit around Earth.

③ The debris ring—comprised of rock from the lighter outer layers of Earth and the impactor—gradually coalesces, forming the moon.

DIFFERENTIATING INSTRUCTION

Reading Support Demonstrate outlining techniques that will help students summarize information in the text. Have students use their science notebooks to outline this section by rephrasing titles as questions, then writing answers to those questions as they read each section. Model the most effective way to answer these questions. For example, rephrase the first red title on this page as "How did the moon originate?" Point out that the best answer summarizes the information in the paragraph: "Four plausible theories have been proposed about the moon's origin, but the most accepted theory states that the moon formed 4.5 million years ago as a result of collision between Earth and a planet-sized object."

Use Reading Study Guide, p. 87.

or depressions in the moon's surface, and to raise up mountains. The largest impacts blasted out great basins, and formed cracks through which lava flowed from the interior. Tiny particles ground and pitted the surface. Rock fragments and dust spread over the landscape.

Approximately 4.0–3.0 billion years ago, the impacts slowed after Earth and the moon had swept up and absorbed most of the debris in their orbits. Over millions of years, magma—richer in iron than the original surface layers—rose to the surface, filling the largest impact basins. When the lava cooled, it formed large, flat plains that are darker than the older, lighter highlands. The interior of the moon gradually cooled and became geologically inactive.

The Moon Today

For approximately 3 billion years, the interior of the moon has been relatively quiet. However, impacts have continued to change the lunar surface. None of these impacts has been large enough to blow out new basins, but smaller ones have dug many new craters. The bombardment going on now is mainly by **micrometeoroids,** tiny objects no larger than sand grains. Micrometeoroids are the major cause of erosion on the moon today. The moon's lack of an atmosphere allows micrometeoroids, which would burn up in Earth's atmosphere, to reach the surface.

Properties and Features of the Moon

The moon turns once on its axis in the same time period in which it completes one orbit around Earth. Thus, the same side of the moon always faces Earth. We see only the "front side" of the moon.

The moon's diameter is 3476 kilometers, or more than one-fourth Earth's diameter. The moon's mass is only about one-eightieth Earth's mass, and the density of the moon, about 3.3 grams per cubic centimeter, is lower than Earth's (5.5 g/cm^3). The moon's lower density supports the impact theory whereby the moon would have formed from the less dense materials from the outer layers of early Earth and the object that had impacted it.

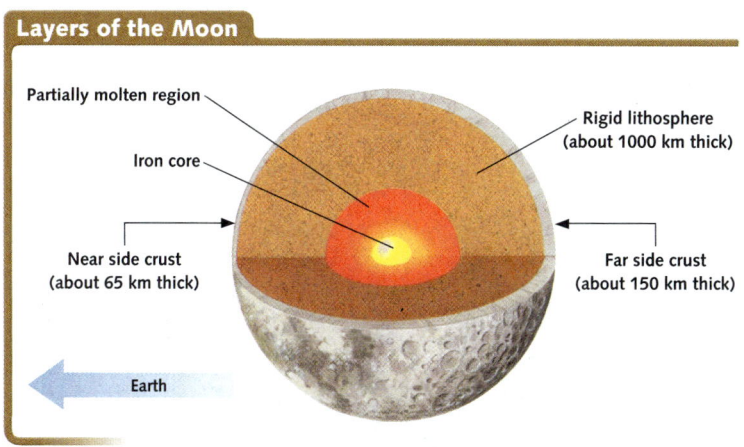

Layers of the Moon

- Partially molten region
- Iron core
- Near side crust (about 65 km thick)
- Rigid lithosphere (about 1000 km thick)
- Far side crust (about 150 km thick)
- Earth

25-Minute Mini LAB

Weights on the Moon and Earth

Materials
- calculator

Procedure

1. This table lists the approximate weight of objects on the moon.

Object	Weight (lb)
Blue whale	40,000
Mid-size car	500
Refrigerator	29
VCR	1.5
Loaf of bread	0.2

2. Estimate how much these objects weigh on Earth.
3. Using your estimates, determine the ratio between what objects weigh on Earth and what they weigh on the Moon.

Analysis

Based on your calculated ratio, how much would you weigh on the moon? How can you check to see if your answer is correct?

Chapter 25 Earth's Moon

CHAPTER 25 SECTION 1

 VISUALIZATIONS
CLASSZONE.COM

Observe images illustrating the impact theory of the moon's formation.
Keycode: ES2501

Visualizations CD-ROM

MINILAB

Weights on the Moon and Earth

Management
- Students can estimate the Earth-weight of each object in pounds, then convert to kilograms.
- As a hint, tell students that a typical loaf of bread weighs a little over a pound (0.45 kg).

Analysis Answers
The ratio is 6:1. A person who weighs 125 lbs. (57 kg) on Earth would weigh one-sixth of that, or 21 lbs. (9.5 kg), on the moon.

Visual Teaching

Interpret Diagram
Ask students if they notice anything unusual about the moon's crust. Crust is thinner on side near Earth.

Extend
Have students brainstorm why the crust would be thinner on the near side. gravitational attraction of Earth; The near side is where the maria are found.

Use Transparencies 31 and 4 to compare moon layers to Earth layers.

DIFFERENTIATING INSTRUCTION

Hands-On Demonstration Density differences between materials in the molten moon led to natural separation into layers. The same process formed Earth's layers, and it can be easily demonstrated. On a lab table, place a clear 1000-mL graduated cylinder next to containers of glycerin, vegetable oil, and water dyed with food coloring. Ask students to predict what will happen when you pour the three materials into the graduated cylinder. Pour the materials, stir them a bit for effect, and observe. Ask: Why do the materials separate into layers? They have different densities. Discuss why moon materials would have different densities. The minerals composing the moon vary in their densities. For example, the aluminum compounds found at the moon's surface are low in density, thus rose easily in the "magma ocean." Iron and other heavy minerals sunk deep into the core.

CHAPTER 25 SECTION 1

More about...

Mascons
Scientists involved in the Lunar Prospector mission have been able to map the moon's gravity pattern. Unlike Earth, the moon has a fairly irregular gravitational field, due not only to mascons but also to the variations in crustal thickness between the near and far sides. While LP (Lunar Prospector) orbited the moon, scientists used the Doppler effect to detect small changes in its velocity. Since lunar gravity was the primary force acting on LP, scientists knew that variations in the moon's gravity field caused changes in LP's speed. The mapping mission was completed by the time LP crashed to the moon's surface on July 31, 1999.

CRITICAL THINKING

Discuss the age of rocks in the lunar maria. Ask: Why would the maria rocks be the youngest found on the moon? As the moon solidified, only the lightest materials would have surfaced during this separation process. Yet the maria basalts are of denser material, indicating that a later process brought them to the surface. A solid crust must have been present for lava to have flowed onto a surface.

Lunar Maria

To the unaided eye, the moon appears as a pattern of light and dark areas. The light areas are lunar highlands, which are rugged mountains pockmarked with craters. The dark areas are great basins and level plains, formed when lava spewed up to the surface through the fractures made by earlier giant impacts. The first observers to look at the moon with telescopes thought the basins were filled with water and so named them **maria** (MAH-ree-uh), the Latin word for seas (singular *mare*, MAH-ray).

The first two *Apollo* missions to land on the moon explored the maria. Rock samples returned by these missions strongly resemble the basalts in lava flows from Hawaiian and Icelandic volcanoes. Like those basalts, the mare basalts are fine-grained crystalline rocks. They are dark gray or black and contain mostly plagioclase feldspar and pyroxene. Some mare basalts contain olivine and ilmenite (a mixture of oxygen, iron, and titanium). Scientists who have studied the moon's rock have determined that the mare basalts are the youngest lunar rocks. They range in age from 3.1 to 3.8 billion years.

In the 1960s, lunar orbiters found that the moon's gravity was greater over some of the more circular maria. Higher gravity readings indicate if the material beneath the surface has a different density from that of the surrounding rock. Some lunar geologists suspect that denser material from deep inside the moon—the same type of material that spewed up to create a mare basin—solidified within the fractures at a fairly shallow depth beneath the surface. These areas of higher gravity are called **mascons**, short for "mass concentrations."

A distinctive feature in maria bedrock is a **rille**. Rilles are trenchlike valleys running through maria bedrock. The best known is Hadley Rille on the floor of Mare Imbrium. This rille may have formed when a river of molten lava flowed along the surface. After a hard crust formed over the river, the molten lava drained away, leaving a hollow tunnel, the roof of which later caved in.

MARIA ROCK Lunar basalts, like this one, formed from molten lava flows.

DIFFERENTIATING INSTRUCTION

Support for Visually Impaired Students Students may have difficulty differentiating between the features seen in the photograph at the bottom of these pages. Try enlarging the image on a photocopier, then outlining the features with a bold marking pen. Ask students to identify specific features.

Lunar Highlands

The lunar highlands appear brighter than the maria because their rocks are lighter in color and thus reflect more sunlight. Within the lunar highlands are a few mountain ranges and many craters.

Most lunar mountain ranges lie at the edges of maria. One great range forms the western border of Mare Imbrium. This range includes the lunar Alps, Apennines, and Caucasus Mountains. These mountains tower as high as 5 kilometers—over 13 times the height of the Empire State Building—above the mare floor.

Lunar scientists think that the Apennines were thrust up by the impact that created Mare Imbrium. Perhaps all lunar mountains that border maria were formed in this way. There are two explanations for how such great masses of rock could be thrown so high. First, the moon has no atmosphere to slow down flying particles. Second, the moon's weak gravity does not exert the same downward force that Earth's gravity exerts.

Two types of rock have been retrieved from the lunar highlands. One is a light-colored, coarsely crystalline igneous rock. The composition of this rock is similar to gabbro and anorthosite. Scientists think that this anorthositic gabbro makes up all the moon's solid crust except in areas where mare basalts cover it.

The other specimens brought back from the lunar highlands are lunar breccias. Breccias are rocks made of angular fragments cemented together with fine material. Breccias on Earth are often caused by volcanic eruptions. On the moon, breccias were formed by meteoroid impacts that melted the rocks together.

Most highland rock specimens are between 4.0 and 4.3 billion years old. A few specimens collected by the *Apollo 17* mission, however, have been dated at between 4.2 and 4.5 billion years, nearly the age the moon itself is thought to be. This correspondence in age supports the hypothesis that the lunar highlands are the original lunar crust.

HIGHLAND ROCK Lunar breccias, such as this one from the highlands, formed after meteoroid impacts melted lunar rocks.

CHAPTER 25 SECTION 1

More about...

Moon Misconceptions
Misconceptions about the moon endured until the *Apollo* missions landed on the moon's surface between 1969 and 1972. Until then, some scientists still believed the craters were the result of live volcanic action, or that such a thick layer of dust covered the surface that a safe landing would be impossible. One of the biggest questions answered by the *Apollo* missions was the age of moon rocks. No one expected them to be as old as they were, and scientists learned the moon had been there for most of Earth's 4.5 billion-year existence.

CRITICAL THINKING

Certain recesses of the moon never receive direct light from the sun, but are dimly lit only by sunlight reflected from Earth called Earthshine. Ask students how Earth's seasons might affect this light. Scientists have measured this light to sense regular variations in Earth's reflectiveness of up to 20 percent, depending on the season. Variations outside this seasonal pattern have also been detected. Ask students to hypothesize reasons for these changes in Earth's reflectiveness. Ask students how solar activity might affect Earthshine. Scientists believe some variability is due to 11-year cycle in solar sunspot activity.

DIFFERENTIATING INSTRUCTION

Developing English Proficiency The names of minerals on pages 558 and 559, such as plagioclase feldspar and anorthosite, may be particularly difficult to read. List the mineral names on the board and pronounce them before students read these pages. Also, use hand motions to portray certain descriptive actions described in text to clarify their meanings. Focus on the following phrases: "pockmarked with craters," "lava spewed," "running through," "drained away," "caved in," "thrust up," "exert," and "retrieved."
Use Spanish Vocabulary and Summaries, pp. 49–50

CHAPTER 25 SECTION 1

ASSESS

1. the collision about 4.5 billion years ago between Earth and a planet-sized object

2. The angle and size of the impactor, as well as the depth of the lunar soil at the impact site, might affect the length of the rays.

3. The moon's iron core comprises only about 4 percent of the moon's total mass, compared with Earth's iron core, which comprises about 30 percent of Earth's mass. According to the theory, the moon was formed from the lighter, iron-poor outer layers of Earth and the impacting object, resulting in an iron core that comprises less of its total mass.

4. Volatiles such as water would have turned to vapor from the heat of the impact that created the moon. Volatiles would have escaped into space from the early molten moon.

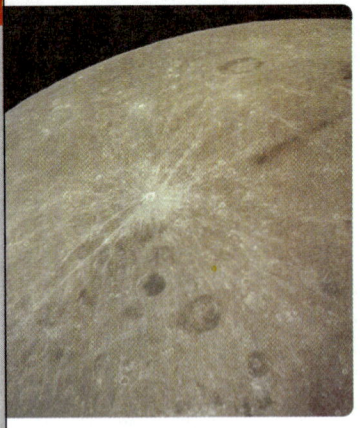

RAYS extending out from the crater Copernicus

VOCABULARY STRATEGY
In the word *regolith*, *rego-* is from a Greek word for "blanket," and *-lith* is from a Greek word for "stone."

Lunar Craters and Rays

Lunar craters are circular hollows on the moon's surface. Craters form after meteoroid strikes. The smallest craters are microscopic pits, but the largest, the Aitken Basin, is nearly 2100 kilometers across. Scientists think the Aitken Basin formed when a large object, such as an asteroid or meteoroid, collided with the moon.

Crater rims are rugged cliffs. In large craters, the rims may tower thousands of meters above the surrounding plains, and their floors may lie a thousand meters below the plains. Most crater floors are dotted with many small craters and include peaks that reflect light just as the lunar highlands do.

Most craters are named after people from around the world, including scientists (Einstein), inventors (Edison), explorers (Shackleton), mathematicians (Fermat), and astronauts (Scobee). The 93-kilometer-wide crater named Copernicus is fringed with bright streaks called rays.

Rays radiate from a number of craters, including Copernicus. Some may be hundreds of kilometers long. For a long time, the origin of the rays was a mystery. Recent evidence suggests that rays consist of shattered rock and dust that were splashed out by the meteoroids that formed the craters.

Lunar Soil

Lunar soil is not really soil. Scientists prefer to call it **regolith,** which means loose rock materials. Regolith is a grayish brown mixture of small rock pieces and fine particles that range in size from sand grains to fine dust. Unlike Earth soil, regolith contains no water or organic material. Regolith is formed by the smashing impact of meteoroids of all sizes. When large meteoroids explode, they mix rock fragments over broad areas. This stirring of the regolith is called *gardening*.

The regolith ranges in depth from approximately 2 to 20 meters. It is likely to consist of chips from many different kinds of rocks and minerals. Regolith also contains tiny beads of glassy material which form from rock melted by high-speed meteoroid impacts. Droplets of the melted rock solidify to form glassy beads. Some of the melted rock forms a glaze on other rocks.

25.1 Section Review

1. According to the most widely accepted theory, what event led to the formation of the moon?

2. What factors might affect the length of the rays formed after an impact?

3. **CRITICAL THINKING** How does the size of the moon's core support the impact theory of the moon's formation?

4. **CHEMISTRY** Materials, such as water, that become gaseous easily at room temperature are called *volatiles*. How does the complete lack of volatiles in lunar rocks support the impact theory of the moon's formation?

MONITOR AND RETEACH

If students miss . . .

Question 1 Reteach the moon origin theories. (p. 556) Ask: What are the four theories on the moon's origin? Which one has become widely accepted?

Question 2 To demonstrate how different variables affect the pattern of material ejected from an impact have students do the Lab Activity for this chapter, pp. 566–567.

Question 3 Review "Properties and Features of the Moon" and the illustration Layers of the Moon. (p. 557) Review the layers of Earth using the illustration Earth's Interior from Chapter 4. (p. 72)

Question 4 Reteach the impact theory, using the illustration Formation of the Moon. (p. 556)

SCIENCE & Technology

One Small Step into the Cosmos

The *Apollo* missions not only led to important discoveries about the moon but also opened the way for further explorations, including those of the space shuttle and the International Space Station.

Where will *Apollo*'s legacy lead us next?

Today, space missions take place so regularly that they may not attract much attention. But this wasn't always the case, especially in the late 1960s and early 1970s, when the name *Apollo* captured everyone's imagination.

The *Apollo* missions followed years of intense effort to find out whether humans could survive in space, and whether there was anywhere safe to land on the moon. The *Ranger* probes of the early 1960s sent back photographs of the moon's surface, and *Surveyor 3* landed softly on the moon in 1967 to study the crust. Manned missions *Mercury* and *Gemini* allowed scientists to determine that humans could indeed survive in outer space.

Using a multistage rocket, *Apollo* astronauts made several attempts to reach the moon. Tragedy struck the *Apollo* program when three astronauts died in a fire on the launch pad. But on July 20, 1969, Neil Armstrong broadcast to the world the immortal words, "Houston, Tranquility Base here. The Eagle has landed." *Apollo 11* had reached the moon safely, and the course of human history had been changed forever.

Apollo's success made possible other space-exploration projects. The *Voyager* probes have given us an up-close look at parts of the solar system, and *Mars Sojourner* has crawled about the surface of another planet. Probes, including the *Lunar Prospector* and *Galileo,* have further expanded human understanding. The International Space Station promises further discoveries. And again and again, the space shuttle flies home from space, borne on the wings of *Apollo.* ■

LUNAR PROSPECTOR, shown in an artist's rendering, was NASA's first lunar mission in 25 years.

ON THE MOON *Apollo 12* commander Charles Conrad, Jr., examines *Surveyor 3,* an unmanned probe that paved the way for *Apollo* missions.

Extension

SCIENCE NOTEBOOK
Consider the qualities astronauts must have, especially those who worked for the space program in its early years.

DATA CENTER
CLASSZONE.COM
Explore information on recent and historical missions to the moon.
Keycode: ES2502

CHAPTER 25 SECTION 1

DATA CENTER
CLASSZONE.COM

Explore information on recent and historical missions to the moon.
Keycode: ES2502

Science and Technology

Mars Ho?
Mars, the most hospitable body in our solar system other than Earth, seems the logical next step for exploration. In fact, on the 20th anniversary of the *Apollo 11* landing, in July 1989, President Bush directed NASA to place humans on Mars by 2019. This initiative was shelved because of projected costs: $450 billion. The high cost of such a mission is not the only serious challenge. The voyage would require astronauts to spend $2\frac{1}{2}$ years in space with unknown health consequences. Also, the amount of fuel and equipment necessary for a conventional mission to Mars is beyond the capacity of the most powerful rockets developed so far. Finally, too little is known about the Martian environment to risk placing astronauts there. The current focus for Mars exploration is the use of space probes at a far lower cost.

Science Notebook
top physical conditioning; courage and inventiveness to survive the rigorous conditions and uncertain outcomes

CHAPTER 25 SECTION 2

FOCUS

Objectives
1. Describe the motions of the moon.
2. Explain the reason the moon goes through phases.
3. Analyze how the Earth-moon-sun geometry causes lunar and solar eclipses.

Set a Purpose
Have students read this section to answer the question "What causes the moon phases and the eclipses?"

INSTRUCT

More about...

The Moon's Orbit
While the moon takes 27.3 days to complete one orbit around Earth, the lunar cycle takes 29.5 days to complete. The disparity arises from the combination of Earth's and the moon's movements around the sun. Remind students that Earth is not stationary. Point out that the moon circles Earth as it accompanies Earth on Earth's long orbit of the sun. The 2.2-day difference amounts to the extra time it takes the moon to keep up with Earth as Earth travels around the sun.

25.2

KEY IDEA

Moon phases, lunar eclipses, and solar eclipses are all caused by the moon's changing position relative to Earth.

KEY VOCABULARY
- phases
- waxing
- waning
- gibbous
- umbra
- penumbra
- lunar eclipse
- solar eclipse

The Moon's Motions

The moon travels in a regular and predictable motion. This motion explains why the moon sometimes appears as a thin crescent in the sky, and sometimes appears as a fully illuminated disk. By keeping track of the moon's motion, astronomers can predict exactly when the moon will pass between Earth and the sun, and exactly where the shadow of the moon will fall upon Earth.

The Moon's Orbit

The moon rises in the east and sets in the west. Like the sun's rising and setting, this motion is apparent—it's really a result of Earth's turning on its axis. But unlike the sun, the moon is in orbit around Earth, taking about $27\frac{1}{3}$ days to complete each orbit. When the moon is on the side of Earth opposite the sun, it is seen mostly in the night sky. When it is between Earth and the sun, it is seen mostly in the daytime sky.

The moon orbits in a different plane than Earth does. This difference is important in determining how often eclipses occur; you will read more about eclipses later in this section.

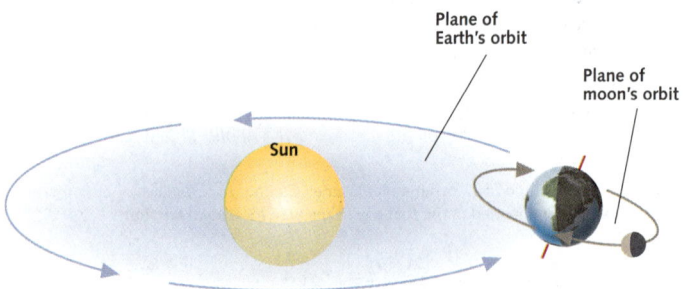

THE MOON'S ORBIT is not in exactly the same plane as Earth's orbit. The angle between the two orbits is about 5 degrees (exaggerated in the illustration for clarity).

The moon rises above the horizon at a different time each day (or night). This happens because every time Earth spins around once, the moon moves about 13° eastward along its orbit. Thus, Earth must rotate an extra 13° more each day for a point on its surface to be roughly under the moon again. Since Earth takes about 50 minutes to spin 13°, the moon rises about 50 minutes later each day and sets about 50 minutes later as well. So if you were to see the moon rise above the horizon at 9:00 tonight, you could expect to see the moon rise at about 9:50 tomorrow night.

The moon's orbit around Earth is elliptical. The moon's average distance from Earth is about 384,000 kilometers, about a hundred times the distance between New York and Los Angeles. When the moon is nearest Earth, it is said to be at perigee. When farthest from Earth, it is at apogee.

DIFFERENTIATING INSTRUCTION

Hands-On Demonstration Discuss the importance of the 5-degree angle between the planes of Earth's and the moon's orbits. Ask: Why doesn't the moon block sunlight from reaching Earth once each orbit? Use a lamp (sun), a globe (Earth), and a golf ball (moon) to show how the moon is usually higher or lower than Earth with respect to the sun. Ask: What needs to occur in order for the moon to block the sun's rays? Orbit the golfball around the globe to show how the inclined orbits can intersect, thus making eclipses possible.

The Moon's Phases

The **phases** of the moon are the daily changes in the moon's appearance as viewed from Earth. Moon phases occur for two reasons. One is that we see the moon only because it reflects sunlight. The other is that the moon is in orbit around Earth.

The sun lights the half of the moon that is facing it. However, we see changing amounts of the sunlit half. From Earth, the face of the moon changes from all dark to all light, or from new moon to full moon, in about two weeks. During this time, the moon is said to be **waxing**. During the next two weeks, the face of the moon gradually changes from all light back to all dark, or from full moon back to new moon. During this time, the moon is said to be **waning**.

The diagram below shows the moon at eight points in its orbit. Although the half of the moon facing the sun is always fully lit, a different portion of the bright half is visible from Earth during each phase.

When the moon is new, the bright half faces away from Earth and the moon cannot be seen. At the two crescent phases, only one edge of the bright half faces Earth. At the two quarter phases, the half of the moon facing Earth is half bright and half dark. At the two **gibbous** phases, almost all of the bright half of the moon faces Earth. When the moon is full, the entire bright half faces Earth. Each of these phases occurs each month.

Examine the phases of the moon from Earth and space.
Keycode: ES2503

CHAPTER 25 SECTION 2

Examine the phases of the moon from Earth and space.
Keycode: ES2503

 Visualizations CD-ROM

Visual Teaching

Interpret Diagram
Have students identify those phases in the diagram that are waxing phases and waning phases. 1 through 4 are waxing; 5 through 8 are waning

Extend
Have students imagine how Earth would appear if they were standing on the moon at positions 5 and 1. To an observer on the moon at position 5, the full-moon phase, Earth would be seen in the new phase; during position 1, the new-moon phase, the observer would see a full Earth.

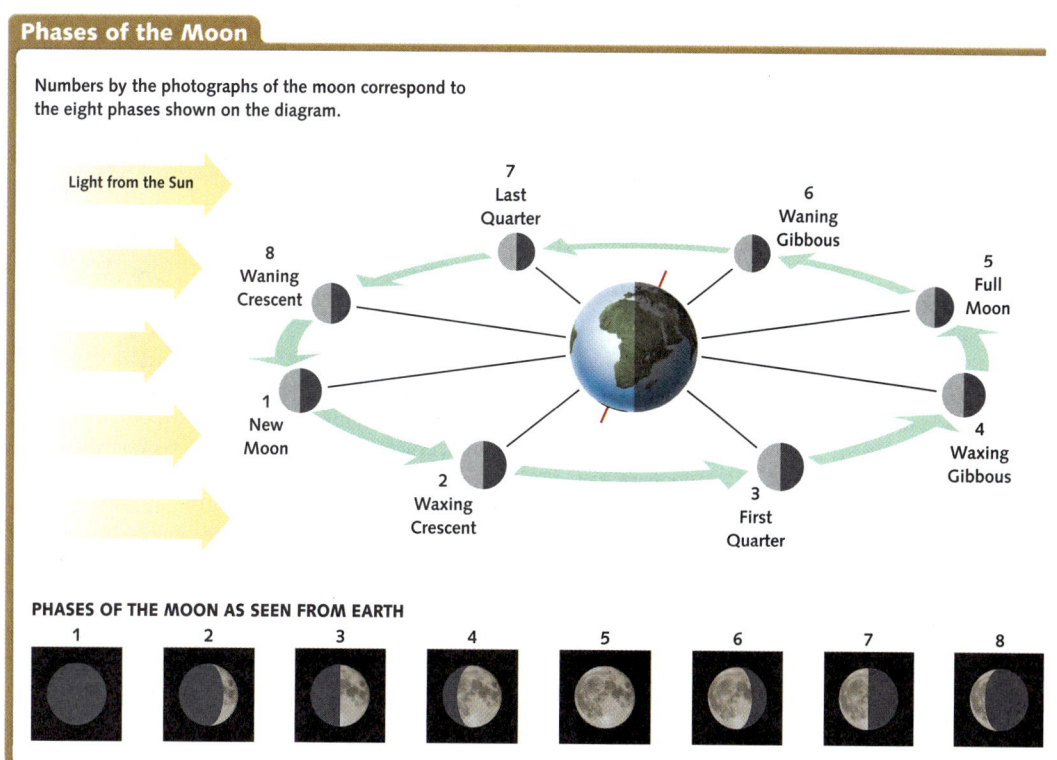

Phases of the Moon

CRITICAL THINKING

Discuss the names of moon phases. Ask: Why are the moon phases at positions 3 and 7 called quarter moons instead of half moons? Only one-half of the moon faces us. Half of that half, or one-quarter, is lighted during this phase. Ask: What, then, would be an appropriate name for the phase at position 5? If the same reasoning applied, it would be called a half moon.

Chapter 25 Earth's Moon **563**

DIFFERENTIATING INSTRUCTION

Reading Support This section includes several vocabulary words that can be grouped in related pairs: waxing/waning, umbra/penumbra, lunar eclipse/solar eclipse. Point out that recognizing this relationship helps in understanding the meanings of the terms—the meanings are similar in some ways and different in other ways.
Use Reading Study Guide, p. 88.

Chapter 25 Earth's Moon **563**

CHAPTER 25 SECTION 2

Observe a lunar eclipse.
Keycode: ES2504

Observe solar eclipses.
Keycode: ES2505

Visualizations CD-ROM

Visual Teaching

Interpret Diagrams

Have students compare the diagrams on these two pages. Ask: Why are you are much more likely to experience a lunar eclipse than a total solar eclipse? *The diagrams show how large a shadow Earth casts on the moon during a lunar eclipse, versus the small shadow the moon casts during a total solar eclipse. Anyone under clear skies on the nighttime side of Earth can witness the lunar eclipse. Only those in the narrow umbra will witness a total solar eclipse.*

Use Transparency 30.

Lunar Eclipses

The shadow cast by any opaque object has two parts: the **umbra** is the area of total shadow, and the **penumbra** is the area of partial shadow surrounding the umbra. Both Earth and the moon cast shadows into space. Earth's umbra is shaped like a long, narrow cone, with its tip stretching nearly 1,400,000 kilometers beyond Earth, well past where the moon orbits. The penumbra is also cone-shaped, but as it stretches out into space, it becomes wider and more faint. Because the moon is smaller than Earth, the moon's shadows are smaller and shorter.

A **lunar eclipse** is an event during which Earth's shadow prevents the sunlight from reaching the moon. A lunar eclipse can occur only at the full moon phase. Even though a full moon occurs every month, a lunar eclipse occurs less often, because of the 5° angle between the plane of Earth's orbit and the plane of the moon's orbit. The full moon is usually above or below Earth's umbra, and no eclipse occurs. When an eclipse does occur, the moon usually remains visible, but has a dusky red or coppery color. This color results when Earth's atmosphere bends some sunlight—mostly longer red wavelengths—into the umbra.

LUNAR ECLIPSE Light from the sun can refract through Earth's atmosphere and onto the moon, giving the moon a coppery color.

Lunar Eclipse

A lunar eclipse occurs when the moon passes into Earth's umbra.

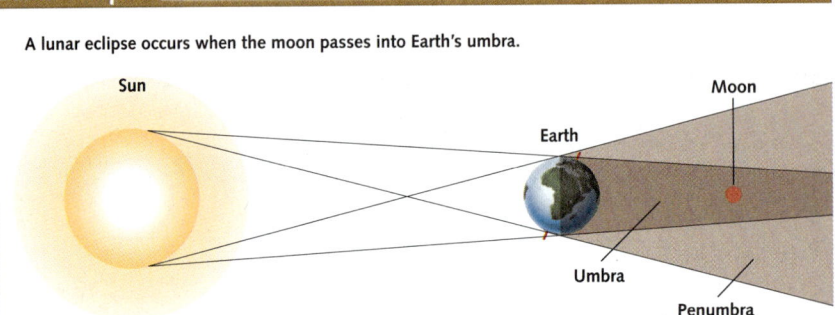

Observe a lunar eclipse.
Keycode: ES2504

A total lunar eclipse occurs when the moon is fully within Earth's umbra. A partial lunar eclipse occurs when only a portion of the moon is in Earth's umbra. On average, at least one total lunar eclipse occurs every year. In good weather it is visible from the entire nighttime half of Earth. If the moon travels through the center of the umbra, a total lunar eclipse may last as long as two hours.

Solar Eclipses

A **solar eclipse** occurs when the moon comes between the sun and Earth, and the moon's shadow hits Earth's surface. The entire shadow—the umbra and the penumbra—is about 7000 kilometers wide, wider than the continental United States. However, the diameter of the umbra, where the effects of the eclipse are most dramatic, never exceeds 270 kilometers, which is only about as wide as Indiana.

564 Unit 7 Space

DIFFERENTIATING INSTRUCTION

Developing English Proficiency Some of the vocabulary in this section may present difficulty in reading comprehension. Have students go through the section and list words that are new to them, such as *waxing, waning, crescent, gibbous, umbra, penumbra, lunar eclipse, solar eclipse*. Have them create flash cards with the word on one side and the definition on the other. Students can work in pairs to quiz each other.

Hands-On Demonstration With a golf ball, an overhead projector, and a screen, you can demonstrate how the moon or Earth casts a two-part shadow. Turn on the projector and hold the golf ball between it and the screen. Adjust the focus of the projector and the distance between the golf ball and the screen until you successfully cast a clear umbra and penumbra onto the screen.

Solar Eclipse

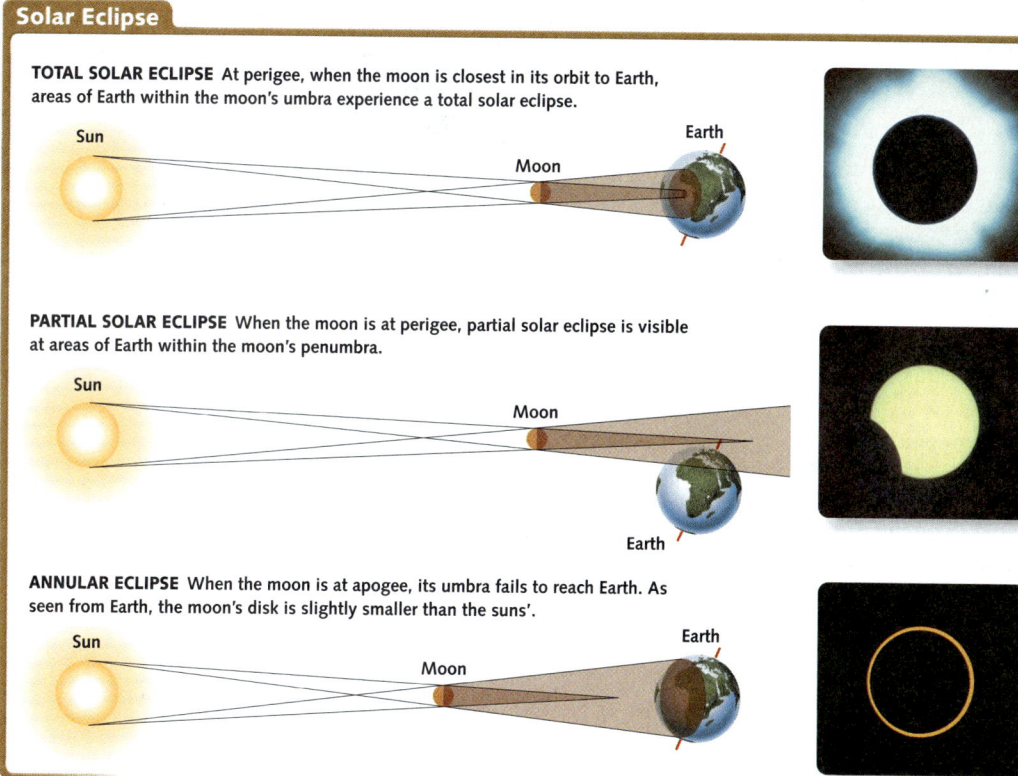

TOTAL SOLAR ECLIPSE At perigee, when the moon is closest in its orbit to Earth, areas of Earth within the moon's umbra experience a total solar eclipse.

PARTIAL SOLAR ECLIPSE When the moon is at perigee, partial solar eclipse is visible at areas of Earth within the moon's penumbra.

ANNULAR ECLIPSE When the moon is at apogee, its umbra fails to reach Earth. As seen from Earth, the moon's disk is slightly smaller than the suns'.

A solar eclipse occurs only at the new-moon phase. Because of the plane of moon's orbit, the moon's shadow usually falls above or below Earth. Therefore, like a lunar eclipse, a solar eclipse does not occur every month.

Although at least one solar eclipse occurs every year, a given area experiences a total solar eclipse only once every three or four centuries. This rarity is due to the small size of the moon's umbra. The moon's revolution makes the narrow shadow, called the eclipse path, race across Earth at over 1500 kilometers per hour. The eclipse path may be thousands of kilometers long, but at any one place a total solar eclipse can last at most only $7\frac{1}{2}$ minutes.

Observe solar eclipses.
Keycode: ES2505

25.2 Section Review

1. Why does the moon rise at a different time each day or night?
2. What is the difference between a partial and a total lunar eclipse?
3. What is the difference between an annular and a total solar eclipse?
4. **CRITICAL THINKING** How many days would you expect to fall between the new moon and the first quarter? The last quarter?
5. **PAIRED ACTIVITY** Use a light source to show how shadows from a small object and a large object change with distance.

CHAPTER 25 SECTION 2

Visual Teaching

Interpret Diagrams
Have students locate the umbra and the penumbra of the moon's shadow.

Extend
Ask students to imagine the effects if the moon were twice the size shown. The eclipse path would be wider, and a total solar eclipse would occur even at apogee.

Use Transparency 30.

ASSESS

1. Every time Earth rotates once, the moon has moved 13 degrees along its own orbit, and so Earth must rotate another 13 degrees, taking about 50 minutes to catch up.
2. A partial lunar eclipse occurs when the moon is only partly in Earth's umbra, whereas a total lunar eclipse occurs when the moon is fully within Earth's umbra.
3. During a total solar eclipse, the moon is at perigee, so the tip of its umbra reaches Earth, completely blocking out the sun. During an annular eclipse, which occurs at apogee, the moon's umbra falls short of Earth's surface. The moon blocks out most of the sun, but the outermost circle of the sun's surface remains visible.
4. about 7 days; about 21 days
5. Students should observe that the closer an object is to a light source, the larger a shadow it casts.

MONITOR AND RETEACH

If students miss . . .

Question 1 Reteach "The Moon's Orbit." (p. 562) Ask: How far in its orbit does the moon move each day? How long does Earth take to spin this far?

Question 2 Reteach "Lunar Eclipses." (p. 564) Ask: What is the difference between the umbra and the penumbra? Which one must the moon pass through to create a total lunar eclipse? a partial lunar eclipse?

Question 3 Reteach from the Solar Eclipse diagram. (p. 565) Have students contrast the total solar eclipse diagram with the annular eclipse diagram.

Question 4 Remind students how much time it takes for the moon to complete one orbit (27.3 days; p. 562), and a lunar cycle (29.5 days). Have them refer to the Phases of the Moon diagram on p. 563 and do simple division to calculate the amounts of time between phases.

Question 5 Have students diagram what they observed in this activity. Have them summarize their findings in one or two sentences.

CHAPTER 25 ACTIVITY

PURPOSE
To model the formation of impact craters and to relate those models to craters observed in the solar system.

MATERIALS
Provide students with a copy of Lab Sheets 25a, 25b, and 25c. Copymasters are provided on pages 146–148 of the teacher's laboratory manual.

Larger pans are best—at least 25 × 30 cm on the sides and 7.5 cm deep. Lids from copy-paper boxes work well, too, or pans made of plastic, aluminum, or cardboard. Do not use glass.

Consider the following for surface materials: all-purpose flour; baking soda mixed 1:1 with table salt; cornmeal; sand mixed 1:1 with cornstarch. The cornmeal and the sand/cornstarch mixture must be stored frozen in airtight containers if you want to reuse these materials.

You can substitute glitter or powdered drink mix for tempera paint.

PROCEDURE
1. Consider doing this activity outside, or spread newspapers under each container.
6. After each impact, have students mix in the layer of tempera paint, smooth the surface with a ruler, then sprinkle a new layer of paint on top (so craters and rays will be more visible).
11. You can demonstrate a more forceful impact using a slingshot after students have completed their experiments. Point out to students that the drop height-velocity chart applies only to stationary objects dropped to the ground on Earth.

566   Unit 7 Space

CHAPTER 25 LAB Activity

Making Impact Craters

SKILLS AND OBJECTIVES
- **Model** the creation of impact craters.
- **Compare** and **evaluate** factors involved in the making of impact craters.

MATERIALS
- pan
- surface material
- sieve or sifter
- dry tempera paint
- balance
- safety goggles
- 3 spheres of varying size and weight (impactors)
- meter stick
- metric ruler
- graph paper
- colored pencils

The surface of the moon is covered with impact craters—circular features caused by projectiles striking the moon with enough force to mark the surface. When a projectile, or impactor, hits, material flies in all directions, causing a circular dent, or crater. The impact may leave streaks of material, called rays, running outwards from the circle. Impact craters vary greatly in size, and can be found on all the terrestrial planets and on many moons. In this experiment, you will model the formation of impact craters. By altering variables, you will better understand some of the factors that affect the size and shape of impact craters.

Procedure

1. Fill a pan with surface material to a depth of about 7.5 cm. Smooth the surface, then tap the pan to make the material settle evenly.

2. Sift a fine layer of dry tempera paint evenly over the entire surface.

3. Using the balance, measure the mass of the first impactor in grams. Copy the data table below and record the mass.

4. **CAUTION: Put on a pair of safety goggles.**

5. Drop the first impactor into the pan from a height of 30 cm, using the meter stick to determine height. Record the height on the data table.

Impactor Mass (g) ____	Drop Height (m) ____			
	Trial 1	Trial 2	Trial 3	Average
Crater diameter (cm)				
Crater depth (cm)				
Average ray length (cm)				

6. Using the metric ruler, measure and record the diameter and depth of the resulting crater. Count and measure the length of any rays. Determine the average ray length and record it on the data table for that impactor. Smooth the surface of the material in the pan, and evenly sift a new layer of dry tempera paint over it.

7. Repeat Steps 5 and 6 twice more for a total of three trials. Compute the average value for each measurement taken during the trials.

8. Complete more trial drops for the first impactor, increasing the drop heights to 60 cm, 90 cm, and 2 meters. Use a separate copy of the data table for each height.

9. Repeat Steps 3 through 7 for the other two impactors.

566   Unit 7 Space

10. Draw a graph with "Impactor Velocity (cm/s)" on the x-axis and "Average Crater Diameter (cm)" on the y-axis. Use the chart on the right to determine the velocity based on height. Use different colors for each impactor. Connect each impactor's data points with a smooth curve.

11. Draw a second graph, this time with "Average ray length (cm)" on the y-axis. Again place "Impactor Velocity (cm/s)" on the x-axis, use different colors for each impactor, and connect with a smooth curve.

Drop height	Velocity
30 cm	242 cm/s
60 cm	343 cm/s
90 cm	420 cm/s
2 meters	626 cm/s

Analysis and Conclusions

1. Based on your observations, describe the appearance of an impact crater. Draw a picture of an impact crater. Label the rays and include an arrow to represent the path of the impactor.

2. According to your observations, what is the relationship between the velocity of the impactor and crater size? What about the relationship between the velocity of the impactor and ray length? Why is this?

3. If an impactor were dropped from 6 meters, would the crater be larger or smaller than the crater made by the same impactor dropped from 2 meters? How much larger or smaller? Note that the velocity of the impactor would be 1,084 cm/s. Explain your reasoning.

4. The size of a crater made during an impact depends on the amount of kinetic energy possessed by the impacting object. Kinetic energy, the energy of motion, is described using the equation $KE = 0.5\ mv^2$, in which KE stands for kinetic energy, m stands for mass, and v stands for velocity. During impact, the kinetic energy of the projectile is transferred to the target surface, breaking up material and moving rock particles. How would you expect kinetic energy to relate to crater diameter?

5. Use the equation in Question 4 to compute the kinetic energy for each object at each height. Write your answer underneath the table and record as "Kinetic Energy of Impactor (g-cm^2/s^2)".

6. Draw another graph, this time with "Average Crater Diameter (cm)" on the x-axis and "Kinetic Energy of Impactor (g-cm^2/s^2)" on the y-axis.

7. Which of the following was the most important factor controlling the kinetic energy of your impactors: the object's diameter, mass, or velocity? Explain. Compare your results to the kinetic energy equation.

Learn more about impact craters on planets and moons.
Keycode: ES2507

CHAPTER 25 ACTIVITY

Learn more about impact craters on planets and moons.
Keycode: ES2507

ANALYSIS AND CONCLUSIONS

1. An impact crater is a circular indentation, with a rim, and rays emanating from the center.

2. The greater the velocity is, the larger the crater, and the longer the rays. This relationship could be due to the greater force applied to the surface by an impactor traveling at greater velocity. A faster impactor displaces more material and throws it farther.

3. An impactor dropped from 6 m would make a larger crater than one dropped from 2 m. Students could attempt to extrapolate the crater size from their graphs.

4. The greater the kinetic energy is, the larger the crater.

5. Answers will vary depending on the mass of the objects used.

6. The graph is intended to show that kinetic energy increases dramatically with the size of the crater. Students may have difficulty choosing an appropriate scale for the x-axis. If you wish to skip question 6, students can use the values they calculated in question 5 to observe the relationship between kinetic energy and crater size.

7. A look at the equation in question 4 reveals that velocity is the most important factor. Because kinetic energy depends on the *square* of the velocity, velocity change will have a greater effect than mass. Object diameter is not in the equation at all and affects crater diameter only at small heights and velocities.

Refer students to Safety in the Earth Science Laboratory on pages xxii–xxiii.

CHAPTER 25 REVIEW

Vocabulary Review

1. phases
2. rilles
3. crater
4. meteoroids
5. umbra

Concept Review

6. They think Earth was struck by a planet-sized object early in its formation, and that rock ejected from both Earth and the impactor coalesced to form the moon.
7. Lunar maria are dark colored and mostly basalt; highlands are light colored and consist of gabbros and anorthosite.
8. Maria and rilles seem to have been formed by volcanic action.
9. Regolith contains no water or organic matter.
10. The moon "increases" in size from new moon, waxing crescent, first quarter, waxing gibbous, full moon, waning gibbous, last quarter, waning crescent, and back to new moon.
11. A lunar eclipse occurs when Earth is between the moon and the sun; a solar eclipse occurs when the moon comes between Earth and the sun.
12. craters and rays formed by meteoroids; maria made of basalts; highlands, which are light colored

CHAPTER REVIEW 25

Summary of Key Ideas

25.1 Scientists think the moon formed after a large object, about the size of a planet, hit Earth. The same side of the moon always faces Earth. Dark areas called maria are great basins and level plains on the moon. They are younger than the lunar highlands. Lunar rocks have textures similar to Earth rocks but differ in composition. Lunar highland rocks are older than mare rocks. Most lunar craters were caused by the impact of meteoroids; rays were splashed out by the impacts. Regolith is the loose rock material covering the moon's surface.

25.2 The moon's orbit is tilted 5 degrees relative to the plane of Earth's orbit. The moon's movement around Earth causes it to rise later each day and to go through phases. A lunar eclipse occurs when Earth passes between the sun and the moon, and the moon is within Earth's shadow. A solar eclipse occurs when the moon passes between the sun and Earth, and the moon's shadow falls on Earth.

Key Vocabulary

crater (p. 556)
gibbous (p. 563)
lunar eclipse (p. 564)
maria (p. 558)
mascon (p. 558)
meteoroids (p. 556)
micrometeoroids (p. 557)
penumbra (p. 564)
phases (p. 563)
ray (p. 560)
regolith (p. 560)
rilles (p. 558)
solar eclipse (p. 564)
umbra (p. 564)
waning (p. 563)
waxing (p. 563)

568 Unit 7 Space

Vocabulary Review

Write the term from the key vocabulary list that best completes the sentence.

1. The daily changes in the moon's appearance are called the ___?___ of the moon.
2. Long trench-like valleys running through the bedrock of lunar maria are called ___?___ .
3. The Aitken Basin on the moon is the largest known ___?___ in the solar system.
4. Craters form after ___?___ strike the surface of the moon.
5. A total solar eclipse occurs when the tip of the moon's ___?___ reaches Earth at perigee.

Concept Review

6. What theory do most scientists think best explains the formation of the moon?
7. How do lunar maria differ from lunar highlands?
8. What evidence do scientists have of past volcanic activity on the moon?
9. How does lunar soil, or regolith, differ from the soil found on Earth?
10. Describe how the moon's appearance to an observer on Earth changes during the course of a month.
11. Explain how a lunar eclipse differs from a solar eclipse.
12. **Graphic Organizer** Copy and complete the concept map below.

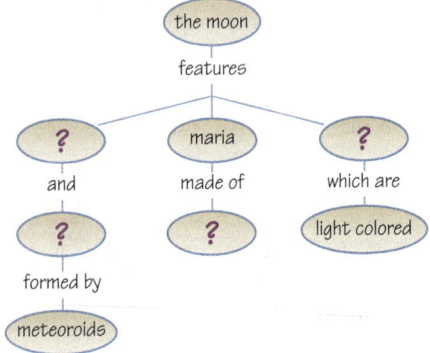

Critical Thinking

13. **Hypothesize** Write a statement suggesting how solar and lunar eclipses would be affected if the moon and Earth kept their present sizes and separation, but were five times farther from the sun (at about the orbit of Jupiter).

14. **Communicate** Write an explanation of the importance of the five-degree difference in the orbital planes of the moon and Earth.

15. **Infer** Could an astronomer stationed in a moon base observe a meteor shower? Explain your reasoning.

16. **Analyze** Evidence indicates that the moon and Earth formed at about the same time; however, rocks found on the moon are older than rocks found on Earth. How could you explain these two sets of evidence?

Interpreting Graphs

Each phase of the moon is visible only at a particular time of day or night. For example, a full moon cannot be seen at 12 noon because it is on the side of Earth opposite the sun. The approximate times that each phase is visible can be determined. The figure at right is similar to the one on page 563 except that times are shown—12 noon toward the sun, 12 midnight away from the sun, 6 A.M. at sunrise, and so on. Use a piece of paper to cover the daytime side of Earth and the moon phases on that side. The full moon is at its highest point in the sky at about midnight. It rises 6 hours earlier, at 6 P.M., and sets 6 hours later, at 6 A.M.

17. Cover the times when the waxing quarter phase CANNOT be seen. At what time is the waxing quarter at its highest point? What time does it rise? What time does it set?

18. Determine the time of moonrise and moonset for the waning gibbous phase.

19. When the waning crescent phase is at its highest point, what time is it?

20. Which phases could never be seen at 3 P.M.?

Internet Extension

What if Earth and the Moon Were Hit by Twin Asteroids? Examine images of impact craters to predict the results of each impact.
Keycode: ES2506

Writing About the Earth System

SCIENCE NOTEBOOK The moon is distant from Earth's atmosphere, hydrosphere, geosphere, and biosphere. However, the moon impacts the Earth system. Choose one way in which the moon affects Earth, such as its impact on tides. Write a description on how changing tides affect the life and the geology of an area.

21. When the waning quarter phase is midway between moonrise and its highest point, what time is it?

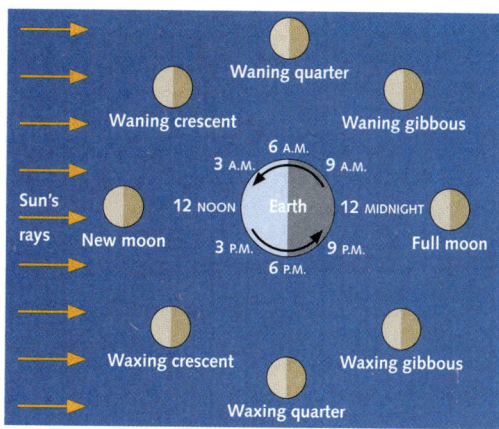

CHAPTER 25 REVIEW

What If Earth and the Moon Were Hit by Twin Asteroids?
Keycode: ES2506

Use Internet Investigations Guide, pp. T90, 90.

CRITICAL THINKING

13. Students should demonstrate understanding that the greater distance will result in smaller shadows.

14. If the angle were larger, say 25 degrees, eclipses would occur with the same frequency but would be of shorter duration. If the angle were 0 degrees, solar and lunar eclipses would occur once each lunar cycle.

15. No. A meteor shower results when meteoroids burn up in Earth's atmosphere. The moon has no atmosphere, thus meteoroids do not burn up.

16. Moon rocks have not been subjected to the continuous changing of rocks, such as erosion and volcanic changes, that Earth rocks have in the rock cycle.

INTERPRETING GRAPHS

17. 6 P.M.; 12 noon; 12 midnight
18. rises at 9 P.M.; sets at 9 A.M.
19. 3 A.M.
20. full moon, waning gibbous, and waning quarter; they are on the opposite side of Earth from 3 P.M. Waxing gibbous would just be setting, and waning crescent would just be rising.
21. 9 A.M.

WRITING ABOUT SCIENCE

Writing About the Earth System

Ocean tides resulting from the moon's gravitational pull on Earth wield a major influence on life at the ocean edges. Life forms have evolved that endure conditions in the intertidal zone, where they are both submerged in salt water and exposed to the air at different times of the day. In addition, tides help move nutrients on which many species rely. Therefore tides also affect the distribution of organisms near the ocean shore, as can be seen by how closely fishers consult tide tables when deciding when and where to fish. Tide-generated currents also affect geology through erosion and deposition. Storm-driven surf during a high tide can cause dramatic erosion in a single day.

CHAPTER 26

PLANNING GUIDE: THE SUN AND THE SOLAR SYSTEM

	Section Objectives	Print Resources
Chapter-Level Resources		Laboratory Manual, pp. 113–114 Internet Investigations Guide, pp. T91–T92, 91–92 Spanish Vocabulary and Summaries, pp. 51–52 Formal Assessment, pp. 76–78 Alternative Assessment, pp. 51–52
Section 26.1 The Sun's Size, Heat, and Structure, pp. 572–576	1. Explain the structure of the sun and its energy source. 2. Describe the effects of sunspots, solar wind, and magnetic storms on Earth and explain the role of Earth's magnetic field.	Lesson Plans, p. 89 Reading Study Guide, p. 89 Teaching Transparency 31
Section 26.2 Observing the Solar System: A History, pp. 577–581	1. Describe the early models of the movements of planets and stars. 2. Explain Newton's Law of Gravitation.	Lesson Plans, p. 90 Reading Study Guide, p. 90

INVESTIGATIONS
CLASSZONE.COM

Section 26.2 INVESTIGATION
Why Does the Size of the Sun Appear to Change?
Computer time: 45 minutes / Additional time: 45 minutes
Students view an animation of the sun throughout the year and make inferences about Earth's orbit.

Chapter Review INVESTIGATION
How Does the Sunspot Cycle Affect Earth?
Computer time: 45 minutes / Additional time: 45 minutes
Students compare sunspot activity with other data to find out how sunspots affect Earth.

Technology/Online Resources

Electronic Teacher Tools
Visualizations CD-ROM
Classzone.com
Online Lesson Planner
Test Generator

Visualizations CD-ROM
Classzone.com
ES2601 Visualization, ES2602 Career

Classzone.com
ES2603 Investigation

Classroom Management

Gather these materials for minilab and laboratory activities.

Minilab, p. 578
  tennis ball
  string

Lab Activity, pp. 582–583
  meter stick
  metric ruler
  calculator
  spherical objects of various sizes

CHAPTER 26

The Sun and the Solar System

In this x-ray image, our sun resembles a sphere-shaped volcanic inferno. Although the sun is millions of kilometers away, it is Earth's primary source of energy.

How do the reactions inside the sun affect our lives?

INTRODUCE

Build Interest

Before sharing background information about the photograph with students, allow time for their questions and comments.

Ask questions like the following:

- How does this image of the sun appear similar to volcanic activity? **looks very hot and appears to have eruptions**
- What do you think the bright, whitish areas are? **eruptions, flare-ups, solar flares**
- What patterns do you see in the eruptions? **Most are in the lower part of the sphere. There seem to be swirls around the eruptions.**

Respond

- How does the sun's energy affect our lives? **We get heat and light from the sun. Sun's energy can also damage people through burns and skin cancers.**
- How do you depend on the sun in your daily life? **daylight to see; light and warmth to be comfortable and provide food**

About the Photograph

The sun's outer atmosphere is so hot, reaching millions of degrees centigrade, that it emits x rays. The brightest areas in this false-colored x-ray image of the sun are areas of highest x-ray emission and therefore highest temperature.

BEYOND THE TEXTBOOK

Print Resources

- Reading Study Guide pp. 89–90
- Transparency 31
- Formal Assessment, pp. 76–78
- Laboratory Manual, pp. 113–114
- Alternative Assessment, pp. 51–54
- Spanish Vocabulary and Summaries, pp. 51–52
- Internet Investigations Guide, pp. T91–T92, 91–92

Technology Resources

- Visualizations CD-ROM
- Classzone.com
- Online Lesson Planner
- Electronic Teacher Tools
- Test Generator

CHAPTER 26

PREVIEW

▶ **FOCUS QUESTIONS** In this chapter you will study the sun and learn more about the key questions listed below.

Section 1 What is the sun's structure and source of energy?

Section 2 How have observations made by scientists in the past contributed to our understanding of the sun and the universe today?

▶ **REVIEW TOPICS** As you investigate the sun, you will need to use information from earlier chapters.
- scientific theories and laws (p. 32–33)
- Earth's revolution (p. 80)
- elements (p. 84)
- atoms and their structure (p. 85)

▶ **READING STRATEGY**

CONNECT

As you read Chapter 26, consider your everyday experiences involving the sun. Make connections between what you observe of the sun and what you are reading.

INTERNET RESOURCES
CLASSZONE.COM

At our Web site, you will find the following Internet support for this chapter.

DATA CENTER
EARTH NEWS
VISUALIZATIONS
- Sun at Different Wavelengths

LOCAL RESOURCES
CAREERS
INVESTIGATIONS
- Why Does the Sun Appear to Change Size?
- How Does the Sunspot Cycle Affect Earth?

Chapter 26 The Sun and the Solar System **571**

CHAPTER 26

PREPARE

Focus Questions
Find out what knowledge students have about the sun. For each focus question, have students consider the section title and then develop two or three additional questions for each section. For example, Section 1: How much larger is the sun than Earth? What is in the center of the sun? Is the sun's energy made up of light or heat?

Review Topics
In addition to the earth science topics listed, students may need to review these concepts.

Physical Science
- The four states of matter are solids, liquids, gases, and plasma.
- Throughout history, even cultures that were not technologically advanced left evidence of precise astronomical observations.

Reading Strategy
Model the Strategy: To help students connect with what they read, explain that since the sun is very much a part of their everyday lives, students will be able to use their own observations and experiences to relate to the reading. Provide examples similar to the following: "When I read the first sentence I realize that the sun is huge. But I notice that the sun in the sky seems small. That must be because it is so far away." "I've seen pictures of the light and colors of the auroras. I can really see how the sun's energy interacts with Earth's atmosphere."

BEYOND THE TEXTBOOK

CLASSZONE.COM

INVESTIGATIONS
- Why Does the Size of the Sun Appear to Change?
- How Does the Sunspot Cycle Affect Earth?

VISUALIZATIONS
- Images and animations of the sun at different wavelengths

DATA CENTER
EARTH NEWS
LOCAL RESOURCES
CAREERS

Online Lesson Planner

Chapter 26 The Sun and the Solar System **571**

CHAPTER 26 SECTION 1

FOCUS

Objectives
1. Explain the structure of the sun and its energy source.
2. Describe the effects of sunspots, solar wind, and magnetic storms on Earth and explain the role of Earth's magnetic field.

Set a Purpose
Have students read to answer the question, "How does the sun's size, heat, and structure affect Earth?"

INSTRUCT

Visual Teaching

Interpret Diagram

As students examine the diagram, ask: Which is the lighter element? **hydrogen** How many hydrogen nuclei are needed to fuse into one atom of helium? **4** How many protons are in one helium nucleus? **2** How many neutrons? **2** Where does the energy that is released come from? **Energy is released when the 4 protons fuse to become 2 protons and 2 neutrons, losing mass in the process.**

Extend

Have students use the information in the diagram to write in their own words an explanation of fusion.

26.1

KEY IDEA

The sun, which is vastly larger than the rest of the solar system combined, gets its energy from the fusion of light elements into heavier ones.

KEY VOCABULARY
- fusion
- plasma
- photosphere
- chromosphere
- corona
- sunspot
- solar wind
- aurora

The Sun's Size, Heat, and Structure

Compared to Earth, the sun is enormous. It has a diameter of about 1,400,000 kilometers, which is more than three times the distance from Earth to the moon, the longest distance humans have traveled in space. It would take a jet flying at three times the speed of sound more than two months to fly all the way around the sun. If multiple Earths could be placed inside the sun, more than a million would fit inside.

Although these examples give you an idea of how large the sun is compared to Earth, the sun is not a large star. If the sun's diameter were the size of a milk-bottle cap (about 1.3 inches), then the diameter of the largest star known, Epsilon Aurigae, would be the size of a football field.

The Sun's Energy

All stars get their energy from fusion. **Fusion** is the combining of the nuclei of lighter elements to form a heavier element. You may be familiar with the famous equation $E = mc^2$ (energy is equal to mass times the speed of light squared). This equation expresses that matter can be converted into energy, which is what happens during fusion.

A star is a place of intense heat and pressure—so intense that atoms are torn apart into their component nuclei and electrons. As a result, elements such as hydrogen and helium exist as plasma.

A **plasma** is a fourth state of matter consisting of charged particles—the nuclei, or ions, which have a positive electric charge, and electrons, which have a negative charge. The nuclei normally repel each other. Due to the speed at which they move, however, and the crowding and the heat, this normal repulsive force may be overcome, and the nuclei fuse.

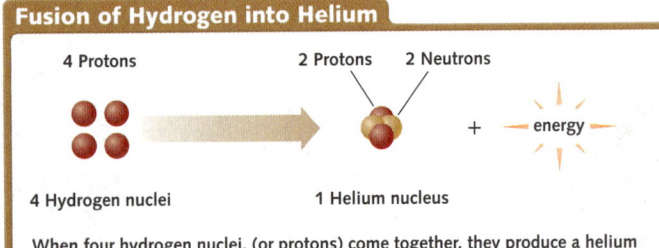

Fusion of Hydrogen into Helium

4 Protons → 2 Protons + 2 Neutrons + energy

4 Hydrogen nuclei → 1 Helium nucleus

When four hydrogen nuclei, (or protons) come together, they produce a helium nucleus of two protons and two neurons. Energy is released in this reaction.

When the nuclei fuse, some of their mass is converted into energy. This mass conversion is what the equation $E = mc^2$ predicts. The mass of the particles involved at the start of a fusion reaction is greater than the mass of the particles at the end; the missing mass has been converted to energy. The amount of energy produced varies depending on the kinds of elements involved in the fusion reaction.

DIFFERENTIATING INSTRUCTION

Reading Support A KWL chart can help students become active readers as they draw on what they already know as a base to expand learning. In the first column, students list facts that they **K**now, such as: the sun emits heat and energy, it is many times the size of Earth, and so on. Suggest that students skim the section heads as they fill in what they **W**ant to know. After students finish reading, they can then fill in the last column: what they have **L**earned.

Use Reading Study Guide, p. 89.

The Sun's Layers

Although astronomers have never actually observed the interior of the sun, they have developed models of the sun's structure. The energy produced inside the sun that pushes outward is balanced by the force of gravity drawing the outer layers inward.

The sun's core consists mostly of hydrogen and helium ions in a plasma state, which is more than 100 times as dense as water. Temperatures in the core reach about 15,600,000°C.

Around the core lies the radiative zone, another layer of plasma. It is cooler than the core; its temperature ranges from about 8,000,000°C near the core to about 2,000,000°C near the convection zone. In the convection zone, rising and falling currents of plasma carry energy to the sun's surface, where it is radiated out into space as sunlight.

At the **photosphere,** the visible surface of the sun, the tops of these currents form structures called granules. A granule may be about 1,000 km wide and last about 20 minutes. The photosphere is much cooler than the convection zone, with a temperature of about 6,000°C.

The sun has an atmosphere, although it is radically different from Earth's. The inner layer of the sun's atmosphere, the **chromosphere,** extends thousands of kilometers above the photosphere. Its 20,000°C temperature causes the hydrogen within it to emit light with a distinctive reddish color. Among the chromosphere's features are solar prominences, dense clouds of material suspended above the sun's surface by magnetic fields. They can erupt off the sun in just a few minutes or hours, extending thousands of kilometers into space before falling back to the sun's surface.

The sun's **corona** is its thin outer atmosphere, which is a million times less bright than the photosphere. Even so, the corona is surprisingly hot, with a temperature ranging from 1,000,000°C to 3,000,000°C.

Examine the sun at different wavelengths.
Keycode: ES2601

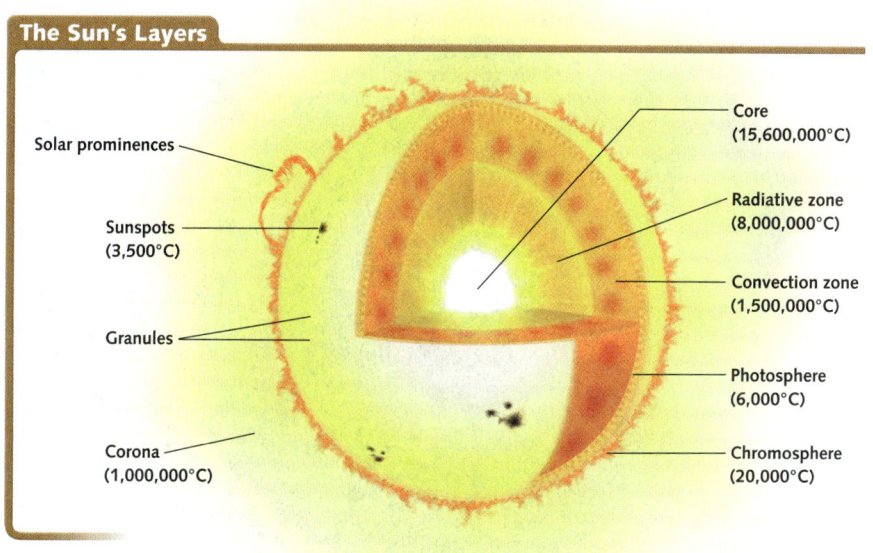

The Sun's Layers

- Solar prominences
- Sunspots (3,500°C)
- Granules
- Corona (1,000,000°C)
- Core (15,600,000°C)
- Radiative zone (8,000,000°C)
- Convection zone (1,500,000°C)
- Photosphere (6,000°C)
- Chromosphere (20,000°C)

Examine the sun at different wavelengths.
Keycode: ES2601

Visualizations CD-ROM

Visual Teaching

Interpret Diagram
Have students use the diagram to identify the sun's layers, going from the inside out. **labels on right side, going in descending order, plus corona** For each layer, have a student look in the text and give one fact about that layer. Ask students what is distinctive about the corona and sunspots. **corona has much higher temperature than either lower atmosphere (chromosphere) or surface (photosphere); sunspots have lower temperature than surface**

Extend
Have students convert the temperatures given to scientific notation. Ask: Do the exponents help you get a sense of how one layer compares to another? **the exponents can serve to categorize temperatures of similar scale**

Refer students to the section on Scientific Notation, Appendix C, page 731.

Use Transparency 31.

DIFFERENTIATING INSTRUCTION

Developing English Proficiency In addition to the key vocabulary, students may be unfamiliar with other words and concepts. Assign a small group of students to each of the red subheads in the section. Groups can work together to list and define additional vocabulary words in their subhead. Work with groups to define the words and explain concepts. Then have each small group explain what their section is about and present the new vocabulary definitions to the larger group.

Use Spanish Vocabulary and Summaries, pp. 51–52.

CHAPTER 26 SECTION 1

VISUAL TEACHING
Discussion
Have students use the illustration on pages 574 and 575 to follow the path of a solar wind. Be sure they understand that the blue lines indicate the solar wind. Have students follow a blue line as it approaches Earth. Ask: What do the bright blue lines around Earth represent? **Earth's magnetic field** For each caption on the diagram, have a student explain what happens to the particles as the solar wind approaches Earth. For example, some particles travel back along the magnetic field lines.

Features on the Sun

The sun is about 150 million kilometers away from Earth (a distance astronomers call an astronomical unit, or AU). Despite this distance, astronomers are able to observe changes on the sun's surface, such as the sunspot cycle, and recognize the effects the sun has on Earth.

Sunspots

Sunspots are dark spots on the photosphere. Some sunspots are barely visible, but others are four times larger than Earth's diameter. Smaller sunspots may last only a few hours, whereas larger ones may remain visible for a few months. Sunspots are actually very hot and bright; they look dark in photographs only because the surrounding photosphere is so much hotter and brighter. The magnetic field associated with a sunspot is about 1000 times stronger than the magnetic field of the surrounding photosphere.

As seen from Earth, sunspots move from left to right across the sun's surface. This motion was the first hint that the sun rotates on its axis. Because the sun is not solid, its rate of rotation varies from place to place. The rate of rotation at the sun's equator is a little over 25 days for one rotation. Near the poles, the rate is about 34 days for one rotation.

The number of sunspots visible on the photosphere changes from day to day. At times of peak sunspot activity, more than 100 can be counted on the sun's surface. During periods of low sunspot activity, several days may pass when no spots are visible. The sunspot cycle averages about 11 years from one period of peak activity to the next.

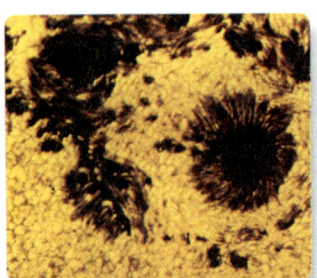

SUNSPOTS The frequency of sunspots is predictable, following an 11-year cycle.

Particles that make up the solar wind fly off into space from the sun's corona.

574 Unit 7 Space

DIFFERENTIATING INSTRUCTION

Challenge Activity Have students research to find out when the next period of peak sunspot activity is expected to occur. **about 2011** Suggest they make a graph of the sunspot cycle over the past half century. Discuss the pattern exhibited by this cycle and where we currently fall within the pattern.

The Solar Wind and Magnetic Storms

The corona gives off a constant stream of electrically charged particles called the **solar wind.** These particles—mostly protons and electrons—fly into space in all directions at a speed of about 450 kilometers per second, reaching Earth in a few days. There, they are deflected by the Earth's magnetic field, as shown below.

Some solar events produce huge gusts of solar wind. Large openings, called coronal holes, sometimes appear in the corona. Solar wind pours from coronal holes in a great torrent of particles.

Solar flares are another source of solar wind bursts. Solar flares are outbursts of light that rise up suddenly in areas of sunspot activity. Small solar flares last only minutes; large ones may last for hours. The number of solar flares increases as the number of sunspots increases.

Earth's magnetic field shields the planet's surface from the solar wind. Without the magnetic field, Earth's surface would be bombarded by particles that are very harmful to life.

As the solar wind blows past Earth, some particles interact with Earth's magnetic field and upper atmosphere, causing **auroras,** which are displays of color and light appearing in the upper atmosphere. Auroras, also called northern and southern lights, are common events in the regions near Earth's magnetic poles.

AURORA IN SPACE Auroras like this one result from the solar wind interacting with Earth's magnetosphere and atmosphere.

CHAPTER 26 SECTION 1

CRITICAL THINKING

Discuss how life on Earth would be different if its magnetic field were weaker or nonexistent. To promote discussion, have students consider the following:

- How might electrically charged particles be harmful to life? cause cancer and create other physical problems, kill vulnerable or fragile animals and plants, limit the variety of life on Earth.

- What would happen in between electrical surges? Life might grow and thrive only to be damaged or killed off when increased electrical surges happen. Alternatively, life might adjust itself and adapt to the increase in electrical surges.

- Could life survive with occasional bursts of solar particles? if the particles were steady? If there were only occasional bursts, life might have a chance of growing between bursts, but if the particle flow were steady, the constant bombardment might prohibit growth.

- How would the auroras change if the magnetic field were weaker or nonexistent? If Earth's magnetic field were weaker, the range in which the auroras could be seen would increase. Auroras would be seen farther away from the poles in the middle latitudes. If there were no magnetic field, there would be no auroras.

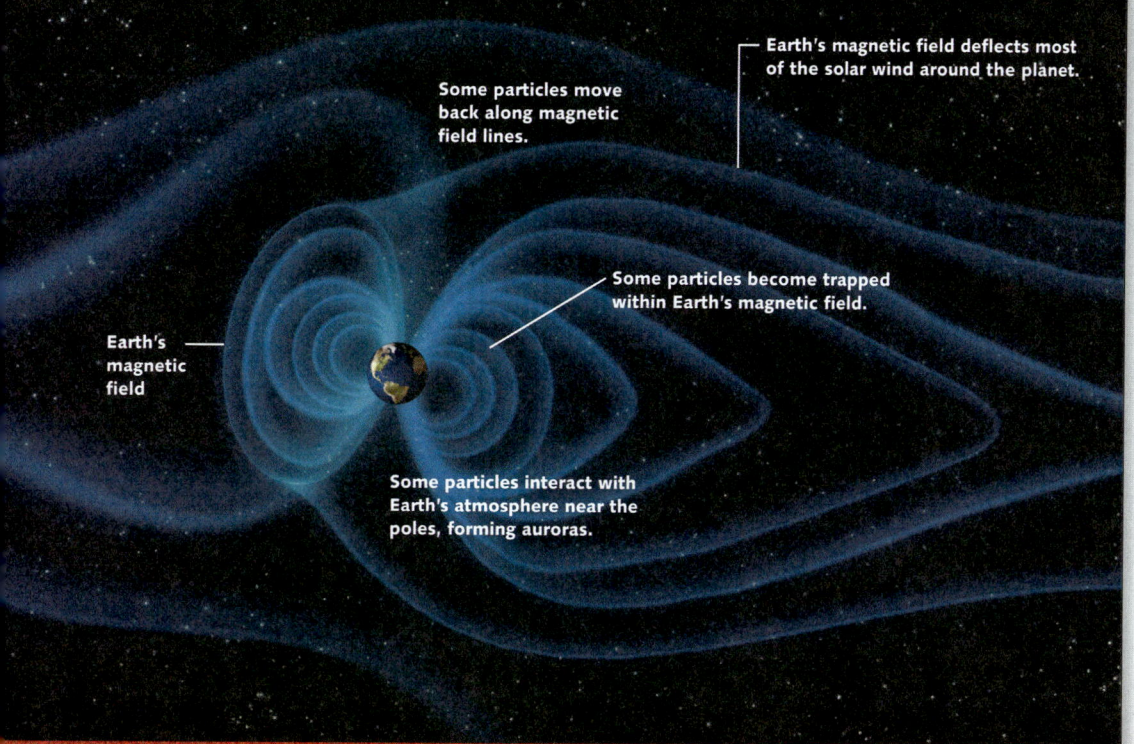

DIFFERENTIATING INSTRUCTION

Hands-On Demonstration Sunspots can be observed using a small telescope or binoculars, not by direct observation. Set up a stiff white paper or mat board in a position facing the sun. Line up the wide end of a telescope or binoculars with the sun so that light flows through it to appear on the paper screen. If the paper screen is firm and the scope is held securely, you may be able to sketch the position of the sunspots. Repeat the exercise over five days to observe the movement of sunspots across the face of the sun.

⚠️ Be sure to emphasize that students are not to use the telescope or binoculars for direct viewing of the sun.

CHAPTER 26 SECTION 1

Learn more about a career as a solar physicist.
Keycode: ES2602

More about...

Magnetic Storms
In March 1989, the sun emitted one of the largest solar flares recorded in the last 30 years. For days after, Earth's magnetosphere was hit by solar winds. In Quebec, Canada, TV reception faded; garage doors began to malfunction; and a power outage affected 6 million Canadians.

ASSESS

1. Plasma, unlike gas, consists of charged particles.
2. The inner core of the sun is very hot, but each layer is a little cooler as it nears the surface. In the atmosphere the chromosphere and corona become much hotter.
3. The solar wind blows past Earth due to its magnetic field. Solar wind particles are seen as auroras.
4. Solar particles move along the magnetic field lines that are strongest at the poles. The farther from the poles you are, the less you will see.
5. Solar particles will constantly bombard the surface of Mars. These particles are harmful to life.

CAREER

Solar Physicist

Solar physicists have the challenge of studying an object that is located almost 150 million kilometers away. Despite the sun's distance from Earth, solar physicists can use a variety of methods to study it. They study the sun's features by analyzing data from satellite instruments, making computer models, and applying the laws of physics. As a result, solar physicists are better able to understand how the sun works and to more accurately predict its effects on Earth. Many solar physicists enjoy the freedom to choose their topics of research. Some develop hypotheses based on theoretical physics. Some others study effects of the sun on our everyday life. For example, some solar physicists study how solar flares affect electronic devices such as cellular phones.

Most solar physicists find a background in math, physics, and computer science to be very useful. Solar physicists usually hold doctoral degrees in astrophysics or another related field. After receiving their degrees, most go on to postdoctoral positions at research institutions, where they gain research experience and refine their area of interest.

Many solar researchers are affiliated with the Solar and Heliospheric Observatory (SOHO), a satellite with instruments onboard that are devoted to studying different features of the sun.

SOLAR PHYSICISTS such as Theresa Kucera, a SOHO team scientist, use a variety of instruments to study the sun.

Learn more about a career as a solar physicist.
Keycode: ES2602

Magnetic storms occur on Earth when the particles thrown out by coronal holes and solar flares are added to the constant solar wind produced by the corona. At such times, auroras are visible in middle latitudes as well as in polar areas. Electrical surges following large solar flares may disrupt cellular telephone service and damage unprotected electrical appliances.

26.1 Section Review

1. How does a plasma differ from a gas?
2. Describe how the temperature of the sun changes as you move from the sun's core out to the corona.
3. How does the solar wind affect Earth?
4. **CRITICAL THINKING** Use your knowledge about Earth's magnetic field and solar winds to explain why auroras are not always visible from temperate regions.
5. **BIOLOGY** Mars has either no magnetic field or a very weak one. Why does this fact make it unlikely that life exists at the present time on the surface of Mars?

576 Unit 7 Space

MONITOR AND RETEACH

If students miss . . .

Question 1 Reteach the explanation of plasma. (p. 572) Point out that plasma is a fourth state of matter.

Question 2 Reteach from the diagram of the layers of the sun. (p. 573) Point out how the temperatures change, moving from the inside out.

Question 3 Reteach "The Solar Wind and Magnetic Storms." (pp. 575–576)

Question 4 Use the diagram showing the magnetic field. (p. 575) Ask students to point out the North and South Poles. Then have them show where auroras are likely to happen. Finally, have them determine where the temperate regions are. **closer to the equator, farther from the poles**

Question 5 Review solar particles. (p. 575) Emphasize that electrically charged particles are harmful to life. Determine that Earth's magnetic field protects it from solar winds. Mars lacks this magnetic field.

576 Unit 7 Space

Observing the Solar System: A History

When you look up and notice that the sun has moved from east to west over the course of a few hours, doesn't it seem as though Earth is standing still and the sun is moving across the sky? This perception—that Earth is stationary—was the basis of ancient Greek theories that prevailed for over two thousand years.

The Movements of Planets and Stars

For thousands of years, the predominant model of the universe stated that Earth stood still at the center of the universe. Such a model of the universe is called a **geocentric** (JEE-oh-SEN-trihk), or Earth-centered, model.

As long as 6000 years ago, astronomers were recording the movements of the stars. They noted that the stars appeared to move across the sky, but they did not move in relation to each other. To explain the apparent motions of the stars, early astronomers envisioned the stars as holes in a solid celestial sphere that surrounded Earth. Beyond the sphere, they imagined, was a source of intense light that shone through the holes. They concluded that the stars moved around Earth as the sphere rotated.

Early astronomers also noticed that the same constellations, or groups of stars, became visible at the same time every year. People in many different cultures noticed this phenomenon and used the changing constellations as a basis for a calendar.

Not all points of light in the sky are fixed in constellations. A few seem to wander across the sky, changing position over the course of days, weeks, and months. These wandering points of light are planets, and early astronomers inferred correctly that planets are closer to Earth than the stars are.

Early astronomers noticed that most of the time the planets moved eastward in front of the background of constellations. Periodically, however, the planets stopped moving eastward and moved westward for a few weeks, then resumed their eastward paths. This pattern of apparent backward motion is called retrograde motion.

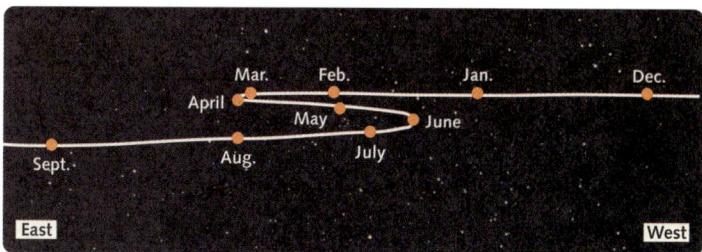

RETROGRADE MOTION Viewed from Earth, the planet Mars seems to go backward during its orbit. The effect is due to Earth catching up with and passing Mars.

26.2

KEY IDEA
Throughout history scientists have developed models to account for their observations of the stars and planets. The work of past scientists contributes to today's sun-centered model of the universe.

KEY VOCABULARY
- geocentric
- heliocentric
- gravitation

Scientific Thinking

DEVELOP MODELS
Design a way to demonstrate retrograde motion.

CHAPTER 26 SECTION 2

MINILAB
Orbital Forces

Management
Students should do this lab outdoors or in a gym. They should swing the ball in a circle over their head, not in front of them, and they should stand well away from other students.

Materials
To attach the ball securely to the string, first tie the string to the middle of a paper clip. Poke a hole in the ball, slip the paper clip through, and pull on the string so that the clip catches on the inside of the ball.

Expected Outcome
The ball should continue to go straight in the direction it was going when the student let go. So in the drawing, the arrow should be a tangent to the circle.

Analysis Answers
The ball represents Earth, while the student represents the sun. The string represents the gravitational force. Gravitational force between the sun and Earth causes Earth to revolve around the sun. The arrow indicates that if the gravitational force were ever removed, Earth would start to travel in a straight line from the point of release. This is due to inertia, the tendency of a body to resist any change in its state of motion.

⚠ Students need to take care when spinning the ball and string and when releasing it.

For proper behavior in a laboratory setting, refer students to Safety in the Earth Science Laboratory on pages xxii–xxiii.

25-Minute Mini LAB
Orbital Forces

Materials
- tennis ball
- string

Procedure
1. Attach the tennis ball to one end of the string.
2. Hold the other end of the string and swing the ball in a circle.
3. Let go of the string and observe the direction in which the ball travels.
4. On paper, draw a circle, representing the ball's movement before you released it. Label the center of the circle as you. Draw an arrow from the circle to represent the ball's movement after release

Analysis
What does the ball represent in the Earth-Sun system? What do you represent? What does the string represent? Explain the concepts behind your drawing, especially the arrow.

VOCABULARY STRATEGY
In the word *heliocentric*, *helio-* is from a Greek word for sun, and *centric* means "having as a center." Likewise, *geo-* is from a Greek word for Earth, so *geocentric* means "having Earth as a center."

Ptolemy's Geocentric Model

The Greek astronomer Ptolemy (TAHL-uh-mee) lived in Egypt in the second century A.D. Ptolemy was puzzled by retrograde motion, so he developed a system that allowed him to predict where planets would appear in the sky at a given time. To make his predictions, he needed to account for retrograde motion.

Ptolemy succeeded in developing the first model that could be used to predict the locations of the planets. He imagined the planets on small circular orbits, called epicycles. The center of each small orbit moved around Earth on a larger circular orbit called a deferent. Retrograde motion occurred when the planet moved along the part of the epicycle that an observer on Earth could see. Ptolemy's model didn't work perfectly—observations didn't always correlate with what the model predicted. However, Ptolemy's model was used by astronomers until the 16th century.

Copernicus's Heliocentric Model

The Polish astronomer Nicolaus Copernicus (koh-PUR-nuh-kuhs) (1473–1543) proposed the **heliocentric** (HEE-lee-oh-SEN-trihk), or sun-centered, model of the solar system. Copernicus suggested that Earth was a planet, that it rotated, and that Earth and the other planets revolved around the sun. The heliocentric model is the basis for our modern understanding of the universe.

Astronomers observe what appears to be retrograde motion because each planet orbits the sun from a different distance and at a different speed. For example, Earth moves faster in its orbit than Mars does. Whenever Earth overtakes and begins to pass Mars, Mars appears to move west, or backward, among the stars. The effect is similar to what you experience when the car you are riding in down the highway passes another car; both cars are still moving forward, but the slower car appears to move backward.

After Earth has fully passed Mars, Mars's eastward motion appears to resume. However, the planet has never really stopped moving eastward.

Tycho, Kepler, and Planetary Motion

Tycho Brahe (TEE-koh BRAH) was a 16th-century Danish nobleman and observational astronomer. Like Copernicus and Ptolemy, he studied the heavens without the aid of a telescope. Instead, he devised his own instruments to observe the moon, planets, and other celestial objects.

Unlike other observers, Tycho and his assistants studied the movement of the moon and planets throughout their orbits, rather than just at certain points. These more complete observations led Tycho to identify a number of unexpected occurrences, particularly in Mars's orbit. His records were the most precise made before the invention of the telescope.

Tycho died in 1601, before he was able to apply his data. Tycho's assistant Johannes Kepler, however, built on Tycho's lifetime of work. Through analyzing Tycho's data, Kepler discovered the unexpected

---

DIFFERENTIATING INSTRUCTION

Support for Physically Impaired Students Some students may have difficulty spinning the ball on the string in the Minilab. If possible, substitute a ball hung from a post, such as that used in a tether ball game. Have students hypothesize what would happen if the string broke or was released. They can then draw an arrow to represent that possible event.

occurrences could be explained if the planets' orbits were elliptical, rather than round. That discovery led him to develop three laws that have become the foundation of celestial mechanics.

Kepler's first law of planetary motion states that the planets travel in elliptical orbits with the sun at one focus. Instead of having only a center, as a circle does, an ellipse has two foci (foci = plural of focus) on opposite sides of its center. Because the sun is at one focus of the ellipse, a planet's distance from the sun will change throughout its orbit.

Kepler's second law of planetary motion, known as the equal area law, states that each planet moves around the sun in such a way that an imaginary line joining the planet to the sun sweeps over equal areas of space in equal periods of time. Because a planet's orbit is an ellipse with the sun at one focus, the equal area law means that the speed at which a planet travels around the sun is not constant. Kepler determined that planets travel faster when they are closer to the sun, although he was not able to explain why.

Kepler's third law of planetary motion is the harmonic law. The time it takes a planet to travel one orbit around the sun is its period. The third law of planetary motion states that the period (P) of a planet squared is equal to the cube of its mean distance (D) from the sun, or $P^2 = D^3$. The formula is used to find the mean distance between the sun and a planet if the period is known, or to find the period if the mean distance is known.

TYCHO'S OBSERVATORY Given an island by the Danish king, Tycho Brahe built Uraniborg, the best European observatory of the time.

Kepler's Equal Area Law

KEPLER'S EQUAL AREA LAW states that a line connecting Earth to the sun will pass over equal areas of space in equal times. Because Earth's orbit is elliptical, Earth moves faster when it is nearer to the sun.

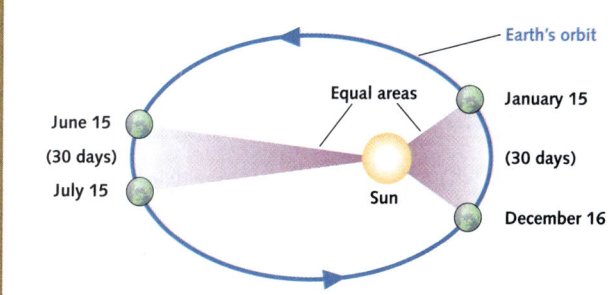

According to Kepler's third law, the farther a planet is from the sun, the longer its period of revolution. One reason is that its orbit is larger. Another is that it moves more slowly than planets closer to the sun. The average speed of Earth in its orbit is about 30 kilometers a second. Mercury, nearest to the sun, moves about 49 kilometers a second.

Why Does the Size of the Sun Appear to Change? Watch an animation of the sun throughout the year. Make inferences about Earth's orbit around the sun.
Keycode: ES2603

CHAPTER 26 SECTION 2

VISUAL TEACHING
Discussion
After students have read the caption under the portrait of Sir Isaac Newton, ask: Why does gravitation keep planets in orbit around the sun? *Gravitation is a force of attraction between two masses. The sun is larger, so its force affects the planets more.* What would happen to the orbits of the planets if there were no gravitation? *Their movement would keep them moving in a straight line instead of in an ellipse.*

ASSESS

1. Geocentric places Earth at the center; heliocentric places the sun at the center.
2. Ptolemy said that the planets followed small circular orbits called epicycles. When these epicycles were seen from Earth, they gave the appearance of retrograde motion.
3. Copernicus correctly suggested that the planets revolved around the sun and that Earth was but one of the planets. Each planet also rotated.
4. Models or diagrams should indicate that the different speeds of the orbits of Earth and Mars cause the apparent retrograde motion.
5. The formula is $P^2 = D^3$. Restated for the problem is $X^2 = 5.2^3$, or approximately 12 years.

Isaac Newton and the Law of Gravitation

Believing a force was required to keep the planets in motion around the sun, Kepler incorrectly connected that force with the sun and its rotation. Isaac Newton (1642–1727), an English scientist and mathematician, developed a very different explanation for what kept the planets in motion.

Newton's contribution to modern science is enormous. He invented calculus, studied the composition of light, and established classical physics. By synthesizing the work of his predecessors, including Kepler and Galileo, he discerned three laws of motion and the law of gravitation, which he showed mathematically to be a universal law.

Newton's first law of motion states than an object will move forever in a straight line at the same speed unless some external force changes its direction or speed. A baseball thrown into the sky would keep going into space, except for the gravitational force (and the friction of the air), which acts on the ball, bringing it back to Earth. **Gravitation,** as described by Newton, is a force of attraction between any two objects with mass; the strength of the attraction is proportional to their masses and inversely proportional to the distance between them. That means in the baseball example, both the ball and Earth exert a gravitational attraction on each other, but because Earth is more massive, its force is stronger.

Similarly, the sun is more massive than the planets, so its force affects the planets more than the planets' gravitational forces affect the sun. The sun's gravitational pull deflects the planets from the straight-line paths they would follow under Newton's first law of motion and keeps them in elliptical orbits around the sun.

Kepler had realized in his second law of motion that the planets did not move with constant speeds. With the law of gravitation, Newton was able to explain why they do not. Newton's laws still explain almost all the large-scale interactions we see in the universe.

SIR ISAAC NEWTON described gravitation, the force that keeps the planets in orbit around the sun.

26.2 Section Review

1. What is the main difference between geocentric and heliocentric models of the solar system?
2. How did Ptolemy account for retrograde motion in his model of the solar system?
3. Explain how Copernicus's model contributed to modern understanding of the solar system.
4. **CRITICAL THINKING** Develop a model or draw a diagram to demonstrate why Mars appears to travel from east to west during part of the year and from west to east during part of the year.
5. **MATHEMATICS** According to Kepler's third law, the period of a planet can be determined if the mean distance from the sun is known. Earth's distance from the sun is 1 AU (Astronomical Unit), and its period is 1 year. Jupiter's mean distance from the sun is 5.2 AU. Use Kepler's law to find Jupiter's period.

MONITOR AND RETEACH

If students miss . . .
Question 1 Reteach "The Movements of Planets and Stars." (pp. 577–578) Suggest students read the vocabulary strategy (p. 578) and apply that information to the solar system.
Question 2 Reteach "Ptolemy's Geocentric Model." (p. 578) Have students describe elements of the model.
Question 3 Reteach "Copernicus's Heliocentric Model." (p. 578) Explain that Copernicus's model was the first correct representation of the solar system.
Question 4 Reteach retrograde movement of planets (p. 577) and Copernicus's model (p. 578). Have students work together to brainstorm ways of making a diagrammatic or a moving model.
Question 5 Reteach Kepler's third law of planetary motion (p. 579) and review the math necessary to solve the formula.

SCIENCE & Society

Changing the Face of Science

Galileo Galilei transformed astronomy and the practice of science, and his influence lives on to this day.

How different would astronomy be if Galileo had never experimented with a telescope?

What do a swinging chandelier, a new Dutch invention for seeing across distances while at sea, and a love of mathematics have in common? All inspired the work of a man who revolutionized science: Galileo.

Born Galileo Galilei in Italy, in 1564, Galileo made his first important discovery as a teenager, when he observed that the swinging motion of a lamp took the same time no matter how large the swing. He measured the time by his own pulse—a sign of the belief in mathematical rationality that marked his entire career.

Experiments, mathematical quantification, and observation were all essential to Galileo's method. After hearing about a Dutch invention that made faraway objects look near, he experimented and created his own version of the device—which we know as the telescope. Galileo turned his telescope to the heavens and made revolutionary discoveries: that the sun had spots, the moon had craters and mountains, and the stars and planets looked different when viewed through the telescope, leading him to deduce that the stars are farther away from Earth than the planets.

Among the questions of physics Galileo studied was that of the behavior of falling bodies. Scientists since the Greek philosopher Aristotle had believed that heavy objects would fall faster than light ones. Galileo's studies showed this belief to be wrong. Although legend has it that to prove his point, Galileo dropped objects off the Leaning Tower of Pisa, the truth is less glamorous: he rolled objects down an inclined plane.

Galileo's work with the telescope changed astronomy forever. His work in physics showed that Aristotle had not been right about everything, as many scientists of Galileo's time had believed. Galileo tested his ideas with experiments and sought the mathematical evidence behind the results.

Through his discoveries and his practices, Galileo did much to change science from a philosophic pursuit to the evidence-driven discipline we know today. ■

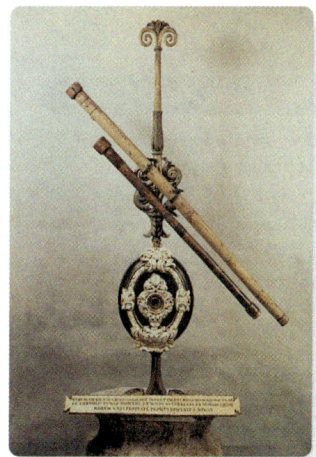

OBSERVATIONS with his telescopes, such as this one, led Galileo to new ideas about the universe.

GALILEO is best known for his astronomy studies but he also did groundbreaking work in other areas of science.

Extension

SCIENCE NOTEBOOK
Think about the work you have done throughout this course. Describe a hypothesis that you made recently.

Learn more about Galileo's telescopes.
Keycode: ES2604

CHAPTER 26 SECTION 2

Learn more about Galileo's telescopes.
Keycode: ES2604

Science and Society

Galileo's Telescopes
Galileo was one of the first to figure out how to use simple lenses to make distant objects seem closer. His first telescope was three-powered, made from lenses he bought from local spectacle makers. He taught himself how to grind lenses so that he could make stronger telescopes. An eight-powered telescope won him a doubling of his salary at the university. In 1609 he began observing space with a 20-powered telescope. His telescopes led to surprising discoveries (there were many more stars than once believed, the moon had craters, Jupiter had four moons) and marked Galileo as the grandfather of modern astronomy.

Science Notebook
Remind students of the laboratory work they have done in the course and hypotheses they have either written or thought through before doing a lab. Refer them also to critical thinking exercises in the Chapter Reviews they may have done. Have students state the hypothesis and then explain the steps they used to prove it.

CHAPTER 26 ACTIVITY

PURPOSE
To represent the dimensions of the solar system in a scale model.

MATERIALS
As an alternative to making spherical models, students might draw their models on paper. They could use adding machine tape and glue models of the sun and planets onto the tape, thereby allowing the sun and larger planets to be wider than the tape.

PROCEDURE

1. Students may need to use the length of the classroom or a larger space.
2. Students should choose a fairly large object, such as a beach ball, that allows the planets to be larger in scale and more easily visible. If necessary, they can use an object that is not spherical, such as a large box. To simplify the exercise, designate the object for the sun yourself or eliminate the sun from the model altogether.

ANALYSIS AND CONCLUSIONS

1. Two different scales are required because otherwise either the size of the planets or the distances to the planets would be visibly out of proportion.

CHAPTER 26 LAB Activity

Scale Model of the Solar System

SKILLS AND OBJECTIVES
- **Determine** appropriate scales for creating a model.
- **Construct** a scale model of the solar system.
- **Examine** the model to gain insight on the solar system's structure.

MATERIALS
- meter stick
- metric ruler
- calculator
- spherical objects of various sizes

The solar system occupies a region of space that is about ten billion kilometers across. Light from the sun reaches Earth after eight minutes, but takes over five hours to reach Pluto. The sun's volume is larger than that of all the planets combined. Jupiter's volume is larger than that of all of the other planets combined, while Pluto is smaller than Earth's moon. It is often difficult to imagine distances and size differences so great in magnitude. The dimensions of the solar system can be better understood by studying a scale model. A scale model shrinks an object proportionally so that the entire object can be viewed and studied at once. In this activity, you will have the opportunity to develop a scale model of the solar system.

Procedure

1. First choose appropriate scales for your model. Working in groups, determine how large an area your model will occupy. You may choose an area as large as you wish, but try not to allow your model to overlap other groups' models. Measure or estimate the extent of your model's area, and convert to centimeters. Divide this distance into the distance from the sun to Pluto. Round your answer to the nearest whole number. Copy the data table on the next page and record this number in the blank next to "Orbital distance scale."

2. Choose a large object to represent the sun. Describe this object in the data table at the end of the row labeled "Sun". Measure the diameter of this object in centimeters. Divide this number into the actual diameter of the sun. Round to the nearest whole number. Record your answer in the blank next to "Planetary size scale" in the data table.

3. Use the scales you have chosen in Steps 1 and 2 to fill in the columns of the table under "Diameter" and "Distance from Sun" for the model solar system. Round your answers to the nearest tenth of a centimeter for diameter and to the nearest centimeter for orbital distance. *Hint:* The diameter you calculate for the model sun should nearly equal the diameter of the object you have chosen to represent the sun.

4. Choose or make objects to represent the nine planets. The objects should be as close as possible to the diameters calculated in Step 3 for your model. Record your choices under the "Description of Object" column in the table.

5. Place the object representing the sun at the center of your model area. Use the meter stick or metric ruler to place the objects representing the planets at the appropriate distances away from the sun.

CHAPTER 26 ACTIVITY

Planetary Data for Scale Model						
Planet	Actual Solar System		Model Solar System		
	Diameter (10^3 km)	Distance from Sun (10^6 km)	Diameter (cm)	Distance from Sun (cm)	Description of Object
Sun	1392.0	0			
Mercury	4.9	58			
Venus	12.1	108			
Earth	12.8	150			
Mars	6.8	228			
Jupiter	142.8	778			
Saturn	120.0	1427			
Uranus	51.8	2869			
Neptune	49.5	4496			
Pluto	2.3	5900			

Planetary size scale: 1 cm = _____ × 10^3 km   Orbital distance scale: 1 cm = _____ × 10^6 km

Analysis and Conclusions

1. Why is it necessary to use two different scales for the size of the planets and the distance of the planets from the sun?

2. How many times larger is Earth than the object representing Earth in your model? How many times farther is Earth from the sun than the model Earth is from the model sun?

3. Did you consider anything other than size when choosing objects for your model? (For example, did you choose a blue Earth? Does your Saturn have rings?)

4. Describe two limitations of your model that prevent it from accurately representing the actual solar system.

5. Compare your model with the models of other groups. Compared to your model, which aspects of the other models more accurately depict the solar system? Which aspects of your model are more accurate?

6. The nearest star to our solar system is Proxima Centauri, which is about 38 trillion kilometers away from the sun. If you included Proxima Centauri in your model, what distance would it be from the sun?

(continued)

2. To answer this question correctly, students must be aware of the units within each scale. If the scale for orbital distance is 1 cm = 10×10^6 km, then the actual proportion is $1:10^{12}$, or 1 to 1 trillion. If the scale for planetary size is 1 cm = 30×10^3 km, then the actual proportion is $1:3 \times 10^9$, or 1 to 3 billion.

3. Responses will vary and indicate how creative the students were in constructing their model.

4. **Possible Responses:** using two different scales, not having objects that look like the planets, Pluto's orbit is not circular and is inclined, having a one- or two-dimensional model as opposed to the real three-dimensional solar system, not having an asteroid belt or moons.

5. Students may notice strengths such as having a two-dimensional layout as opposed to a linear one, constructing objects to be the exact size necessary instead of using approximations, or having a better relationship between distance and size.

6. Proxima Centauri is 3.8×10^{13} km away from the sun. Answers will vary depending on the model scale. To find the correct answer for an appropriate scale, use the following formula:

$$\frac{\text{Model Distance to Proxima Centauri (cm)}}{1 \text{ cm}} = \frac{3.8 \times 10^{13} \text{ km}}{\_\_\_ \times 10^6 \text{ km}}$$

Refer students to Safety in the Earth Science Laboratory on pages xxii–xxiii.

CHAPTER 26 REVIEW

Vocabulary Review

1. solar wind
2. heliocentric
3. fusion
4. gravitation
5. corona

Concept Review

6. Plasmas consist of charged ions; gases do not.
7. The motion of sunspots on the surface indicates rotation.
8. because Earth overtakes Mars during its orbit
9. granules
10. For traveling two periods of equal time, imaginary lines joining the planet to the sun sweep over equal areas of space.
11. Venus
12. The sun is the center of Earth's rotation, gets its energy from fusion, consists of several layers, produces the solar wind that on Earth causes auroras.

Critical Thinking

13. The following equation can be used in finding how long it takes a particle in the solar wind to reach Earth:

    $$\text{time} = \frac{150{,}000{,}0\cancel{00} \cancel{\text{km}}}{45\cancel{0} \cancel{\text{km}}} \times \frac{\cancel{s}}{1} \times \frac{1 \text{ day}}{86{,}4\cancel{00} \cancel{s}}$$

    Kilometers, seconds, and three zeros will cancel.

    $$\text{time} = \frac{150{,}000}{45} \times \frac{1 \text{ day}}{864}$$

    $$= \frac{150{,}000}{38{,}880}$$

    $$= 3.85, \text{ about 4 days}$$

14. According to Kepler's third law and Newton's law of gravitation, orbital speed increases as gravitational force increases. A mass coming closer to Earth's surface speeds up due to greater gravitational force.

15. Fusion occurs at very high temperatures and pressures, such as at the center of stars. These conditions are difficult to duplicate under controlled conditions on Earth.

16. Life might not exist at all because the magnetic field shields Earth's surface from the solar wind; a weaker field might not give life the protection it needs to develop.

17. **Sample Response:** Something happens in the sun every 11 years to cause sunspot changes; for example, the sun's magnetic field flips over.

CHAPTER REVIEW 26

Summary of Key Ideas

26.1 The sun is enormous compared to Earth. Its surface temperature is about 5500°C; its interior is even hotter. The photosphere, chromosphere, and corona are layers of the sun's atmosphere. Granules, solar prominences, sunspots, and solar flares appear on the sun's surface. The solar wind is a stream of charged particles from the sun's corona. Some solar events cause changes in the solar wind that can affect Earth. The sun's energy is the result of the conversion of hydrogen to helium in nuclear fusion. The mass that does not convert to helium is not lost, but becomes energy.

26.2 Ptolemy proposed a geocentric (earth-centered) solar system to explain planetary motion. Nicolaus Copernicus proposed a heliocentric system, in which the planets orbit the sun. Johannes Kepler used Tycho Brahe's data to develop three laws that explained the motions of the planets. Isaac Newton developed the universal law of gravitation, which helped explain the motions of planets in the solar system.

Key Vocabulary

aurora (p. 575)
chromosphere (p. 573)
corona (p. 573)
fusion (p. 572)
geocentric (p. 577)
gravitation (p. 580)
heliocentric (p. 578)
photosphere (p. 573)
plasma (p. 572)
solar wind (p. 575)
sunspot (p. 574)

Vocabulary Review

Write the term from the key vocabulary list that best completes the sentence.

1. The ___?___ is a stream of charged particles from the sun.
2. In a(n) ___?___ solar system, the planets travel around the sun.
3. A type of nuclear reaction called ___?___ provides the sun's energy.
4. The force of ___?___ between two objects depends on their mass and their distance apart.
5. The ___?___ is the outermost layer of the sun's atmosphere.

Concept Review

6. How does a plasma differ from a gas?
7. What evidence do scientists have for the sun's rotation?
8. Why does Mars appear to move backwards at certain times during its orbit?
9. What surface feature indicates that the sun's inner layers are in motion?
10. Describe Kepler's equal-area law.
11. According to Kepler's third law of planetary motion, which planet would take longer to orbit the sun, Mercury or Venus?
12. **Graphic Organizer** Copy and complete the concept map below.

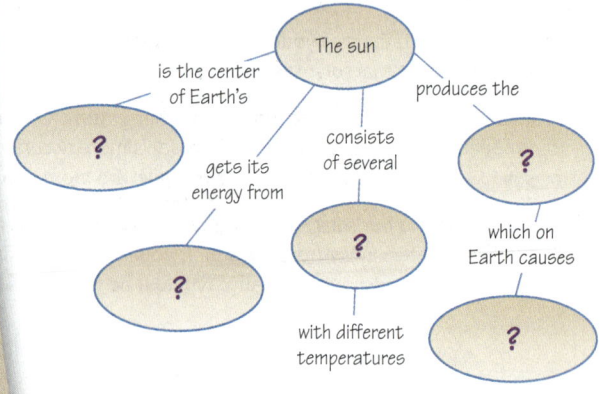

584 Unit 7 Space

Critical Thinking

13. Apply The average speed of the coronal solar wind is 450 km/s. How long does it take a particle in such a wind to reach Earth? (Assume that the distance from Earth to the sun is 150,000,000 km.)

14. Infer The Russian space station *Mir* entered Earth's atmosphere and burned up in 2001. As it spiraled toward Earth, its orbital speed increased. Why?

15. Communicate Explain why it is difficult to achieve nuclear fusion except near the center of stars.

16. Infer What would be some possible consequences for life on Earth if Earth had a weaker magnetic field?

17. Hypothesize Write a statement explaining why sunspots might vary on an 11-year cycle.

Internet Extension

How Does the Sunspot Cycle Affect Earth?
Compare yearly sunspot activity with other data to find out how sunspots affect Earth.
Keycode: ES2605

Writing About the Earth System

 SCIENCE NOTEBOOK The Earth system derives the major part of its incoming energy from the sun. Choose one event in Earth's atmosphere, hydrosphere, or biosphere and explain how the sun affects the transfer of materials in this event.

Interpreting Graphs

When Venus is observed from Earth, it is never more than 45 degrees of arc from the sun. Figure A shows the positions of Earth, the sun, and Venus in a geocentric system. Three *phases*, or lighted portions of Venus visible from Earth, are also shown. In the geocentric system, Venus and the sun are always on the same side of Earth. Figure B shows the positions of Venus, Earth, and the sun in a heliocentric system and shows eight phases of Venus. Study Figures A and B and answer Questions 18–21.

18. Which phases in Figure B match phases 1, 2, and 3 shown in Figure A?

19. When Galileo looked at Venus with his telescope, he saw Venus in phases that are impossible in the geocentric system shown in Figure A. What phases did Galileo see?

20. Why did Galileo's observations of Venus help to disprove the geocentric model of the solar system?

21. Are there other planets that would show phases, as Venus does? Explain.

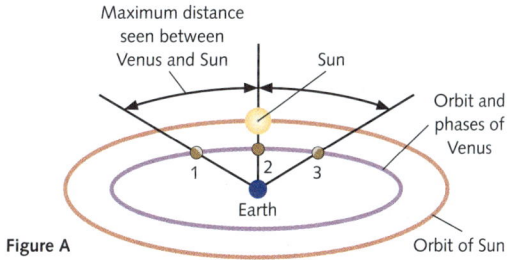

Figure A

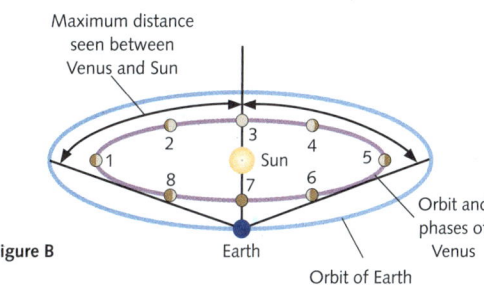

Figure B

CHAPTER 26 REVIEW

INVESTIGATIONS
CLASSZONE.COM

How Does the Sunspot Cycle Affect Earth?
Keycode: ES2605
Review this investigation in class before assigning for homework.

Use Internet Investigations Guide, pp. T92, 92.

INTERPRETING GRAPHS

18. Figure B, phases 8, 7, and 6 match Figure A, phases 1, 2, and 3, respectively.

19. Figure B, phases 1, 2, 4, and 5. He could not see phase 3 because it was blocked by the sun at the time; if students list phase 3, however, it should not be counted wrong. The point is to identify which phases are not possible in a geocentric system.

20. In a geocentric system, Venus would be a crescent of a totally dark disk; Galileo saw Venus as an almost full disk.

21. Yes; Mercury, the other planet with an orbit inside of Earth's orbit, would show phases.

WRITING ABOUT SCIENCE

Writing About the Earth System

Answers should recognize that the sun's energy moves through layers in Earth's atmosphere, providing light and heat, which allows for plant and animal growth within the biosphere. In the hydrosphere, the sun affects tides, the evaporation of bodies of water, cloud formation, and weather.

CHAPTER 27 PLANNING GUIDE: THE PLANETS AND THE SOLAR SYSTEM

	Section Objectives	Print Resources
Chapter-Level Resources		Laboratory Manual, pp. 115–118 Internet Investigations Guide, pp. T93–T94, 93–94 Spanish Vocabulary and Summaries, pp. 53–54 Formal Assessment, pp. 79–81 Alternative Assessment, pp. 53–54
Section 27.1 The Inner Planets, pp. 588–593	1. Describe characteristics of the four inner planets. 2. Compare the positions of the inner planets in orbit.	Lesson Plans, p. 91 Reading Study Guide, p. 91
Section 27.2 The Outer Planets, pp. 594–598	1. Describe characteristics of each of the outer planets. 2. Explain the orbiting patterns of Neptune, Pluto, and Charon.	Lesson Plans, p. 92 Reading Study Guide, p. 92
Section 27.3 Planetary Satellites, pp. 599–601	1. Describe the satellites of the planets. 2. Compare and contrast the Galilean satellites of Jupiter.	Lesson Plans, p. 93 Reading Study Guide, p. 93
Section 27.4 Solar-System Debris, pp. 602–605	1. Identify smaller components of the solar system. 2. Explain the effects of objects colliding with Earth.	Lesson Plans, p. 94 Reading Study Guide, p. 94

INVESTIGATIONS
CLASSZONE.COM

Section 27.2 INVESTIGATION
How Fast Does the Wind Blow on Jupiter?
Computer time: 45 minutes / Additional time: 45 minutes
Students analyze images of Jupiter's atmospheric features to calculate wind speeds.

Chapter Review INVESTIGATION
What Processes Shape Planetary Surfaces?
Computer time: 45 minutes / Additional time: 45 minutes
Students analyze images of planetary features to compare them with features on Earth.

Technology/Online Resources

Electronic Teacher Tools
Visualizations CD-ROM
Classzone.com
Online Lesson Planner
Test Generator

Visualizations CD-ROM
Classzone.com
 ES2701 Visualization, ES2702 Visualization, ES2703 Data Center

Classzone.com
 ES2704 Investigation, ES2705 Career

Visualizations CD-ROM
Classzone.com
 ES2706 Visualization, ES2707 Visualization

Classroom Management

Gather these materials for minilab and laboratory activities.

Minilab, p. 592
 graph paper
 colored pencils

Lab Activity, pp. 606–607
 colored pencils
 Laboratory Manual, Teacher's Edition
  Lab Sheet 27, p. 149

CHAPTER 27

The Planets and the Solar System

INTRODUCE

Build Interest

Before sharing background information with students about the photograph (a satellite image), allow time for their questions and comments.

Ask questions like the following:

- Do you think it is possible to count the rings of Saturn? *There are too many small rings to count by observing images like this one.*
- The rings look solid in the image. Do you think they are? What do you think the rings are made of? *They are not solid planes, rather, they are made of solid particles of various sizes.*

Respond

- What do we know about the planets? *They circle the sun in fixed orbits. They vary in size with Pluto the smallest and Jupiter the largest. They all are made up of different materials and gases.*
- Which planet might you like to visit if it were possible?

About the Photograph

This composite image was produced from images sent by the *Voyager* probes that flew near Saturn in the 1980s. Saturn has seven rings that consist of thousands of ringlets. Mimas, one of Saturn's many satellites, is visible as a black dot below the rings.

Surrounded by rings, Saturn is a spectacular planet. Each of our solar system's nine planets has features that make it unique.

What do we know about the planets?

BEYOND THE TEXTBOOK

Print Resources
- Reading Study Guide, pp. 91–94
- Formal Assessment, pp. 79–81
- Laboratory Manual, pp. 115–118
- Alternative Assessment, pp. 53–54
- Spanish Vocabulary and Summaries, pp. 53–54
- Internet Investigations Guide, pp. T93–T94, 93–94

Technology Resources
- Visualizations CD-ROM
- Classzone.com
- Online Lesson Planner
- Electronic Teacher Tools
- Test Generator

CHAPTER 27

PREVIEW

▶ **FOCUS QUESTIONS** In this chapter you will study the planets of Earth's solar system and learn more about the key questions below.

Section 1 How are the inner planets alike?
Section 2 How are the Jovian planets different from Earth?
Section 3 What are some of the characteristics of planetary moons?
Section 4 What other objects are part of the solar system?

▶ **REVIEW TOPICS** As you investigate the planets, you will need to use information from earlier chapters.

- the movement of planets (p. 569)
- the moon's origin (pp. 556–557)
- properties of the moon (p. 557)

▶ **READING STRATEGY**
SET A PURPOSE

Before each section, read the appropriate focus question as well as the questions at the end of the section. Use those questions to set a purpose for reading each section.

At our Web site, you will find the following Internet support for this chapter.

DATA CENTER
EARTH NEWS
VISUALIZATIONS
- Distances between the Planets
- Radar Mapping of Venus
- Comet's Passage through the Solar System
- Meteor Showers

LOCAL RESOURCES
CAREERS
INVESTIGATIONS
- How Fast Does the Wind Blow on Jupiter?
- What Processes Shape Planetary Surfaces?

Chapter 27 The Planets and the Solar System **587**

CHAPTER 27

PREPARE

Focus Questions
Find out what knowledge students have about the planets. For each focus question, have students consider the section title and then develop two or three additional questions for each section. For example, Section 1: Why are they called inner planets? Which planet is most like Earth? Could humans live on any of the other planets?

Review Topics
In addition to the earth science topics listed, students may need to review these concepts:

Life Science
- Primitive life needs the conditions of warmth and water to develop.
- Humans thrive in the exact mix of elements in the atmosphere present on Earth. They could not survive in the differing atmospheres of any of the other planets.

Reading Strategy
Model the Strategy: Model for students how to set a purpose for reading by using a version of the following: "The Section 1 focus question gives me an idea that the inner planets are more alike than different, but when I read the section review questions I can see that there are differences as well. So my purpose for reading this section will be to find out what characteristics of Mercury, Venus, Earth, and Mars make them alike and different." Have students continue to use this strategy for the remainder of the chapter.

BEYOND THE TEXTBOOK

INVESTIGATIONS
- How Fast Does the Wind Blow on Jupiter?
- What Processes Shape Planetary Surfaces

VISUALIZATIONS
- Animation showing the vast differences between planets in the solar system
- Animation of how radar was used to map Venus
- Animation of a comet's passage through the solar system
- Animation of a meteor shower

DATA CENTER
EARTH NEWS
LOCAL RESOURCES
CAREERS

Online Lesson Planner

Chapter 27 The Planets and the Solar System **587**

CHAPTER 27 SECTION 1

▶ FOCUS

Objectives
1. Describe characteristics of the four inner planets.
2. Compare the positions of the inner planets in orbit.

Set a Purpose
After students examine the illustration of the solar system, have them read the section to answer the question "What are the characteristics of the inner planets?"

▶ INSTRUCT

More about...

How Planets Were Named
All the planets except Earth are named after figures from Greek and Roman mythology. In the time of early astronomers, people were interested in Greek and Roman cultures. Naming the planets paid tribute to these mythological figures.
- Mercury—messenger of the gods; a god with wings on his ankles
- Venus—goddess of love and beauty
- Mars—god of war
- Jupiter—leader of the ancient Roman gods
- Saturn—god of agriculture
- Neptune—god of the sea
- Uranus—god of the sky
- Pluto—god of Hades, the underworld of the dead

27.1

KEY IDEA
The four planets closest to the sun all have a rocky crust, dense mantle layer, and very dense core.

KEY VOCABULARY
- inner planets
- outer planets

VISUALIZATIONS CLASSZONE.COM
Examine the vast distances between planets in the solar system.
Keycode: ES2701

The Inner Planets

The sun is at the center of the solar system. Around it orbit nine planets, each with unique characteristics. Although astronomers have known about the existence of other planets for centuries, it's only within the last few decades that technology has given scientists an up-close look at our nearest neighbors in space.

Two Planetary Neighborhoods

The planets in our solar system are divided into two groups. The four nearest the sun—Mercury, Venus, Earth, and Mars—are called the **inner planets**. All of the inner planets have rocky crusts, dense mantle layers, and very dense cores. Because of their earthlike characteristics, these planets are sometimes called the terrestrial (earthlike) planets.

Just beyond the orbit of Mars is a belt of small bodies called asteroids. This asteroid belt separates the inner planets from the **outer planets**, Jupiter, Saturn, Uranus, Neptune, and Pluto. The first four of these planets, called the Jovian, or Jupiter-like, planets, are considerably larger than Earth. They are gaseous, with an outer layer that is mostly hydrogen gas. Closer to the planet's center, the hydrogen is compressed to a hot liquid. The Jovian planets are much less dense than Earth, and all have ring systems.

Pluto is the oddity of the solar system. It is not dense enough to be considered a terrestrial planet, and it is too small to be a Jovian planet. You will read more about the outer planets in the next section.

THE SOLAR SYSTEM This illustration shows the relative sizes of the planets but does not represent the distance between them.

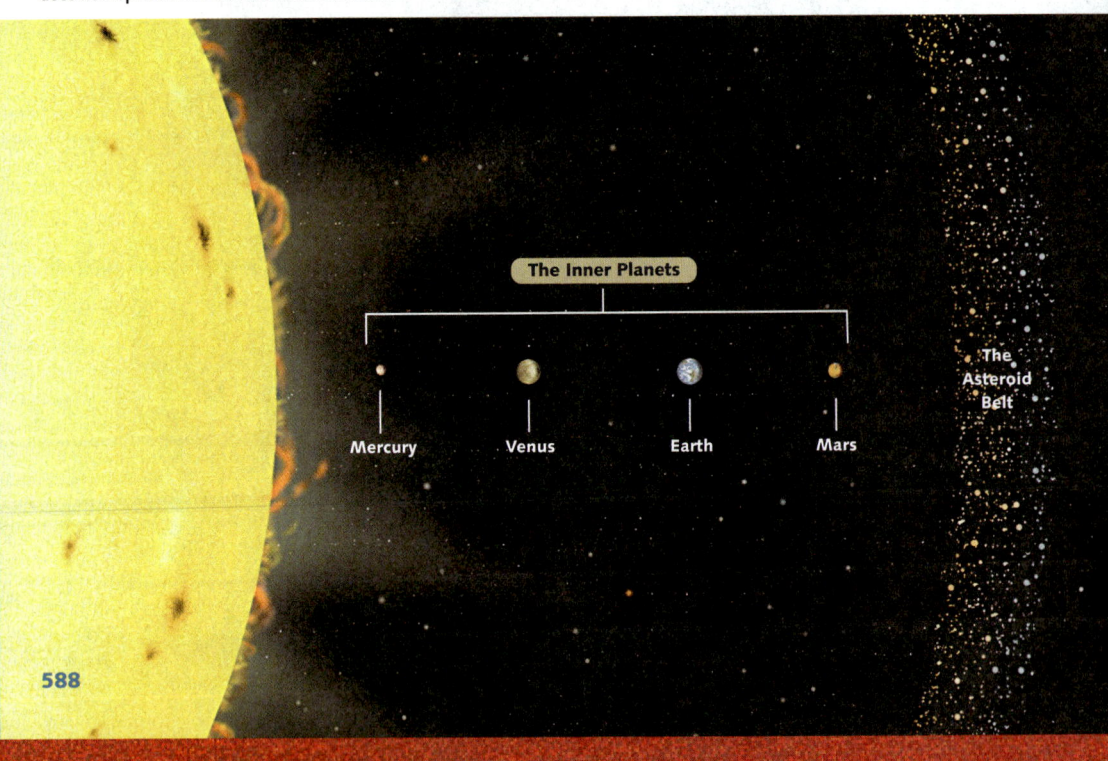

588

DIFFERENTIATING INSTRUCTION

Reading Support As students read about the four inner planets, have them construct a word web or chart to help organize the facts and increase comprehension. Students can set up their webs or charts showing the planets in order from the sun or they can group the planets as "inner planets," in preparation to add the outer planets later. Demonstrate how to record the important facts to represent each planet. For example: "Mercury—nearest the sun, shortest orbit," and so on. Caution students to provide enough room to include important facts about each planet.

Use Reading Study Guide, p. 91.

588   Unit 7   Space

Mercury

Mercury is the planet nearest the sun. Of all the planets, it orbits the sun in the shortest period of time—88 Earth days. Mercury is the smallest of the four terrestrial planets. Its diameter is about 38 percent that of Earth, as is its gravity. Mercury's magnetic field is about one percent as strong as Earth's.

Little was known about Mercury until *Mariner 10* photographed it in 1974 and 1975. These photographs show that Mercury's surface is heavily cratered, similar to Earth's moon. Like the craters on Earth's moon, these impact craters probably formed when huge rocks smashed into Mercury. Impact craters are Mercury's most abundant landform. The rest of the surface is smooth plains that may have been formed by lava flowing out of cracks in the surface.

Mercury turns on its axis once every 59 days. This slow rate, combined with Mercury's nearness to the sun, causes daytime temperatures of more than 400°C. In the nighttime, heat radiates away quickly and the temperature may be nearly −200°C.

Mercury has weak gravity, which prevents it from retaining an atmosphere. High daytime temperatures cause any particles above its surface to move at high speeds, allowing gases to escape into space.

MERCURY These images of Mercury are composed of many photos taken by *Mariner 10*, the only probe that has flown by Mercury.

The Outer Planets: Jupiter, Saturn, Uranus, Neptune, Pluto

CHAPTER 27 SECTION 1

VISUALIZATIONS
CLASSZONE.COM

Examine the vast distances between planets in the solar system.
Keycode: ES2701

Visualizations CD-ROM

VISUAL TEACHING
Discussion

Discuss why the planets are grouped as inner and outer planets. Ask: What is the asteroid belt? **the band of small bodies that separates the inner and outer planets** Which planets are inside the asteroid belt? **Mercury, Venus, Earth, Mars** What two reasons explain why the inner planets are grouped together? **They are inside the asteroid belt; they are terrestrial planets with rocky crusts, dense mantles, and iron cores.**

Discuss the attributes of the solar system model shown in the illustration. Ask: How is this model a true portrait of the solar system? **It shows the relative sizes of the sun, planets, and asteroids; the usual order of their orbits; and their colors.** How is this model *not* a true portrait of the solar system? **The distances between the bodies are not true (as pointed out in the caption).** How would the illustration have to change in order to show correct relative distance? **The sun and planets would have to be drawn as dots.**

DIFFERENTIATING INSTRUCTION

Developing English Proficiency As students read the text, have them pull out words or phrases that describe the planets, such as Mercury—*small, nearest the sun, heavily cratered, smooth plains, weak gravity,* and *high temperatures.* They should use these words and phrases to write a description of each planet. For example, "Mercury is a small planet nearest the sun. It is a heavily cratered planet with smooth plains and high temperatures."
Use Spanish Vocabulary and Summaries, pp. 53–54.

CHAPTER 27 SECTION 1

VISUAL TEACHING

Discussion

After students read about Venus, focus on the significance of its atmosphere. Point out the top photograph. Ask: **What prevents astronomers from seeing the surface of Venus?** *thick yellowish clouds* **How were scientists able to analyze the atmosphere of Venus?** *They sent weather balloons into the atmosphere.* **Describe the atmosphere.** *dense with mostly carbon dioxide and a little nitrogen; covered with clouds of sulfuric acid; atmospheric pressure is about 90 times greater than Earth's; very hot on the surface* **How did scientists examine Venus's surface?** *radar images and radar-mapping* **What did they discover?** *Venus has volcanic features, faulting, and impact craters.*

CRITICAL THINKING

Point out that the names of spacecraft are meaningful. *Mariner 10,* for example, is so named because mariners are sailors, and for hundreds of years most explorers were mariners. Ask: **Why do you think the *Magellan* spacecraft is so named?** *Magellan was a famous explorer in the early 1500s, and members of his crew were the first people to circumnavigate the globe. Also, mapping was a large part of Magellan's expedition.*

VENUS'S surface is hidden by thick clouds of sulfuric acid.

Observe how radar was used to map Venus.
Keycode: ES2702

MAAT MONS This view of the eight-kilometer-high volcano Maat Mons was synthesized from *Magellan* radar images.

Venus

Venus has been called Earth's sister planet because the two are near each other and are similar in diameter, mass, and gravity. Unlike Earth, however, Venus has a very weak or non-existent magnetic field. Unlike the other planets, Venus rotates from east to west. Thick, pale yellow clouds in Venus's atmosphere make its surface impossible to see from Earth. Most of our knowledge of Venus's surface comes from radar-mapping done by the *Magellan* spacecraft, beginning in 1990. Radar images show that the surface of Venus has some similarities to Earth. *Magellan* revealed a landscape dominated by volcanic features, faulting, and impact craters. About 80 percent of the surface is covered with lava, with more recent lava flows lying on top of previous ones. Venus has fault and fracture systems as well, indicating that tectonic activity has occurred on Venus in the past. Whether its volcanoes continue to erupt and faulting is still going on is uncertain. The oldest crust on Venus is estimated to be about 800 million years old; the oldest crust on Earth is about 4.3 billion years old.

In 1985, two balloons carrying weather instruments were placed in the atmosphere of Venus to take measurements. The data showed that the dense atmosphere is mostly carbon dioxide with about 3 percent nitrogen. Venus's yellow clouds are made of droplets of concentrated sulfuric acid. The surface atmospheric pressure is about 90 times greater than it is on Earth.

Despite Venus's thick clouds, its surface gets very hot. Carbon dioxide in the atmosphere acts like the glass roof of a greenhouse. The carbon dioxide atmosphere prevents much of the sun's heat that reaches the surface from escaping back into space. The result of this greenhouse effect is a surface temperature of about 475°C.

DIFFERENTIATING INSTRUCTION

Challenge Activity Students may have noticed that Earth was the only planet not named after Greek or Roman mythological figures. How did Earth get its name? Have students hypothesize why Earth was not named after figures from mythology. Then have them research the naming of Earth and report back on their findings.

Evening and Morning Stars

Venus is visible to observers on Earth at either evening or morning twilight almost all year.

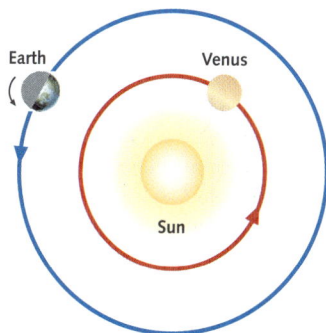
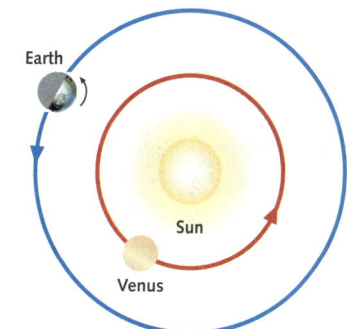

EVENING STAR When Venus is behind Earth in its orbit, the sun sets first and Venus is seen in the evening twilight of the western sky. At such times, Venus is called an evening star. It may remain visible as long as three hours after sunset.

MORNING STAR When Venus is ahead of Earth in its orbit, it rises before the sun and is seen in the eastern sky as a morning star.

Venus can be seen from Earth as a morning or evening star, as can Mercury. However, Mercury is much more difficult to see because its orbit is closer to the sun and it is smaller and less bright than Venus.

Mars

Mars is the fourth planet from the sun and the first planet outside Earth's orbit. Mars takes 687 days to orbit the sun. Its diameter is about one half that of Earth. The gravity of Mars is about two fifths of Earth's. Mars has a very weak or non-existent magnetic field.

Mars's axis is tilted at almost the same angle and in the same direction as Earth's. This tilt gives Mars four seasons similar to Earth's. However, because a Martian year is about twice as long as an Earth year, each Martian season is also twice as long. Because it is farther from the sun, Mars is colder than Earth. On a summer day, some areas may be as warm as 27°C, but on a winter night the temperature drops as low as −125°C. The thin Martian atmosphere is about 95 percent carbon dioxide and 5 percent nitrogen and argon, with traces of other gases. Because the atmosphere is so thin, atmospheric pressure is less than 1 percent that of Earth.

Like Earth, Mars has polar ice caps. The north cap is probably water ice, and the south cap may contain frozen CO_2. The caps increase in size during each Martian winter and shrink during each summer. The temperature difference between the polar caps and soil warmed by the spring sun leads to strong winds and great swirling dust storms that often cover the entire planet.

MARS Dust storms swirl across the Martian surface in this photo taken by the Hubble Space Telescope.

CHAPTER 27 SECTION 1

VISUALIZATIONS
CLASSZONE.COM

Observe how radar was used to map Venus.
Keycode: ES2702

Visualizations CD-ROM

Visual Teaching

Interpret Diagrams
Be sure students understand that the diagrams show views of the orbits as if looking down on the plane of the orbits. The viewer is looking down on the north polar region of Earth. In the left diagram, have students examine the illustration of Earth. Ask: In its rotation, which part of Earth is approaching morning? **bottom** approaching evening? **top** Which would have the better view of Venus? **the evening part of Earth** Repeat the process with the diagram on the right, emphasizing the morning view.

Extend
Have students infer why Venus is sometimes behind Earth in its orbit and sometimes before Earth. **Since Venus is closer to the sun, its orbit is shorter and its revolution is faster.** Ask: When would Venus not be visible from Earth? **when its orbit places it on the opposite side of the sun**

DIFFERENTIATING INSTRUCTION

Reading Support Help students compare and contrast to help them remember important facts. They can use their chart of the planets (see Reading Support, p. 588) to compare and contrast characteristics of the four inner planets, such as distance from the sun, length of orbit, age, temperature, and atmospheric composition. Get them thinking about comparisons by asking: Which planet is most similar to Earth in size? **Venus** in atmosphere? **Mars** in temperature? **Mars** What might a planet be like if it had the same atmosphere as Earth? **There might be life. We could visit the planet.** Then have students add to their charts as they compare and contrast other inner planets with Earth.

Chapter 27 The Planets and the Solar System 591

CHAPTER 27 SECTION 1

MINILAB
Design a Martian Calendar

Management
- Have students work with partners, pairing any students who need extra support with more able students.
- Provide visually impaired students with large grid paper with dark lines.

Analysis Answers
- Answers will vary, but months should add up to 668 or 669 days.
- Solutions include having a decimal day or a leap day three out of every five years. Students should at least be aware that the decimal has to be accounted for.

ASSESS

1. inner and outer planets; inner planets—rocky crusts, dense mantles, and iron cores; outer planets—gaseous, have rings

2. Mercury has very weak gravity. Its high daytime temperatures cause particles and gases above its surface to escape into space.

3. Thick clouds cover Venus, and the carbon dioxide atmosphere prevents heat from escaping out into space.

4. Mars's axis is tilted.

5. The orbits of Venus and Mercury are closer to the sun, so they are only seen near the sun.

6. Water may have been on Mars far in the past. But when the atmosphere changed, life could no longer develop.

MARS Canyons, mountains, and valleys mark the surface of Mars.

Cameras aboard several spacecraft have photographed the surface of Mars. Spacecraft have also visited the surface to take close-up photographs, to record quakes and weather, and to test soil samples. Photographs show that Mars's northern hemisphere is a smooth lowland plain of volcanic material, with few craters. Its southern hemisphere is a highland fractured by many large craters and cut by small channels.

Rising above the northern plains are several extinct volcanoes. The largest, also the largest known volcano in the solar system, is the shield volcano Olympus Mons (Mount Olympus). It is more than 500 kilometers across and about 26 kilometers high. In comparison, Earth's highest volcano, Mauna Loa, rises about 9 kilometers above the Pacific Ocean floor. Unlike many of Earth's volcanoes, the formation of Martian volcanoes does not seem to be related to plate motions, because the crust of Mars is one solid piece. Mars's crust varies in thickness from 80 kilometers in the southern hemisphere to about 35 kilometers in the north.

Cutting across the craters of the southern hemisphere is the Valles Marineris, a canyon system as long as the United States is wide. About 4 billion years ago, Mars may have had a thick atmosphere, blue skies, and abundant liquid water on its surface. The depths of Valles Marineris may have once held lakes of liquid water. Over the next billion years or so, however, most of Mars's atmosphere disappeared. At present, liquid water cannot exist on the surface of Mars, because it would quickly boil, evaporate, and freeze—all at the same time. However, much of the liquid water that once flowed across Mars may be trapped as ice beneath the surface.

Is there life on Mars? Since liquid water probably existed on Mars's surface in the past, it is possible that primitive life developed there before the atmosphere thinned and the surface water froze. If liquid water still exists in geothermal pools beneath the surface of Mars, then Martian life may still exist. Whether life existed or still exists on Mars is a question that will be answered only with further exploration.

Design a Martian Calendar

Materials
- graph paper
- colored pencils

Procedure

1. A Martian year includes 668.6 Martian days, and you're designing the first calendar. How many days are in your week? In your month?

2. On graph paper, outline a calendar. Make one square represent each day.

3. Color in the first month. Use different colors to fill in the other months, and label days and months.

Analysis

How many months are on your calendar? How did you compensate for the fact that the Martian year doesn't include a whole number of days?

27.1 Section Review

1. Into what two groups are the planets divided? What are the main characteristics of each group?

2. Why does Mercury have no atmosphere?

3. Why is the surface of Venus so hot?

4. Why does Mars have seasons?

5. **CRITICAL THINKING** Why does neither Venus nor Mercury appear high in the night sky on Earth?

6. **BIOLOGY** Why is it more likely that life existed on Mars billions of years ago rather than it does now?

MONITOR AND RETEACH

If students miss . . .

Question 1 Reteach from the illustration of the planets. (pp. 588–589) Have students read the text (p. 588) to find out what the planets are made of.
Question 2 Reteach "Mercury." (p. 589) Have students read about Mercury's gravity and compare it to that of Earth.
Question 3 Reteach "Venus." (p. 590) Ask students what they know about the greenhouse effect. Have them compare the carbon dioxide atmosphere of Venus to the atmosphere of Earth.
Question 4 Reteach "Mars." (p. 591) Have students describe how the axis tilt of Earth creates seasons.
Question 5 Review the Evening and Morning Stars diagram. (p. 591) Point out that the orbits of both Venus and Mercury are inside the orbit of Earth, thus they are usually seen in a position close to the sun.
Question 6 Emphasize that water must be present for life to exist. Ask students to read to find out about water on Mars. (p. 592)

SCIENCE & Technology

Seeking the Red Planet's Secrets

Three successful landings of probes on the Martian surface have told us a great deal about our planetary neighbor.

What other discoveries lie ahead?

Mars is enough like our own planet to be familiar, yet alien enough to inspire curiosity. Was there ever liquid water on Mars? Could life have existed there in the past—or could it exist now?

Earth scientists have been trying to reach the red planet for decades. Attempts to send a probe to Mars started in 1960, with the Soviet Union's Mars probe *1960A*, which failed to reach Earth orbit. Not until 1965 did a probe from the United States, *Mariner 4*, succeed in reaching Mars. It flew within 6,118 miles of the Martian surface and provided the first close-up photos of the planet. Successive *Mariner* missions came even closer to Mars, and one, *Mariner 9*, became the first U.S. spacecraft to enter orbit around another planet in November 1971.

Mars 3, a Russian probe, made the first successful Mars landing on December 2, 1971, but failed after relaying only 20 seconds of data to its orbiter.

The *Viking* missions, which took place in the late 1970s, gave American scientists a close look at the surface, providing more than 52,000 photos before they were shut down.

The *Mars Global Surveyor*, launched on November 7, 1996, and still in orbit around Mars, has made detailed maps of the planet's terrain. Also in 1996, *Mars Pathfinder*, consisting of a lander and a surface rover named *Sojourner*, sent back spectacular photos from the surface of Mars.

NASA plans future missions that will not only land spacecraft on the planet but also return them to Earth with soil samples. For now, scientists are still analyzing the evidence they have to try to unravel the secrets of Mars. ■

AIR-FILLED BALLOONS cushioned *Mars Pathfinder's* landing, providing an inexpensive solution to the problem of having the spacecraft land on the planet intact.

Extension

 **SCIENCE NOTEBOOK**

Using what you know about Mars's environment and geology, design an experiment to be conducted on a future Mars mission.

Find out about the latest missions to Mars.
Keycode: ES2703

VISITING MARS *Mars Pathfinder's* roving robot unit, *Sojourner*, was a remote-controlled "astronaut" exploring the area around *Pathfinder's* landing site.

CHAPTER 27 SECTION 1

Find out about the latest missions to Mars.
Keycode: ES2703

Science and Technology

Life on Mars in the Past
In order for life to begin on a planet, water, energy, and organic compounds must be present. Those conditions are not currently met on Mars as we know it. But what about Mars in the past? Recent discoveries have shown that volcanic activity of a colossal scale occurred on Mars. This volcanism changed the climate. Most likely water and carbon dioxide were released, possibly creating a greenhouse effect that warmed the surface of Mars above freezing. Perhaps these conditions led to running water that carved the valleys of Mars. Scientists have found some evidence that life existed on Mars more than 3.6 billion years ago, and the research indicates that conditions may once have been favorable to life.

Science Notebook
Students might design experiments to observe how water behaves on Mars, to locate permanent hot spots or other sources of the volcanoes on Mars, or to determine whether life exists on Mars.

CHAPTER 27 SECTION 2

FOCUS

Objectives
1. Describe characteristics of each of the outer planets.
2. Explain the orbiting patterns of Neptune, Pluto, and Charon.

Set a Purpose
After students read the first paragraph, have them read on to answer the question "How do the outer planets differ from the inner planets?"

INSTRUCT

Visual Teaching

Interpret Diagrams
Have students read the caption under the illustration of Saturn. Ask: Which layer is in a state similar to liquid metal? **metallic hydrogen** What causes the metallic state? **compression by the outer layer of molecular hydrogen** After students read the caption under the picture of Neptune, ask: Why don't Uranus and Neptune have a state of metallic hydrogen? **The planets are smaller, and the outer layer is not thick enough to compress the middle hydrogen layer enough to cause the metallic state to occur.**

27.2

KEY IDEAS

Four of the outer planets are gaseous and huge compared to Earth.

Pluto is the smallest and coldest planet in our solar system.

The Outer Planets

Beyond the orbit of Mars, the solar system becomes a very strange place of huge gaseous planets, peculiar ring systems, and many strange moons. The three outermost planets—Uranus, Neptune, and Pluto—were discovered only after the invention of the telescope.

The Jovian Planets

The Jovian planets—Jupiter, Saturn, Uranus, and Neptune—are unlike the terrestrial planets in several ways. First, Jovian planets are much larger. The smallest Jovian planet, Uranus, is nearly 15 times more massive than the largest terrestrial planet, Earth. Second, Jovian planets do not have solid surfaces; instead their "surfaces" comprise the uppermost gas layer. Third, Jovian planets are composed mainly of the light elements hydrogen and helium, while terrestrial planets are made of iron, silicon, oxygen, and other heavy elements.

All Jovian planets have a three-layered structure. The temperature and density of the planets' interiors increase with depth.

All the Jovian planets have ring systems. The ring systems have three common properties. First, they consist of many particles in independent orbits around the planet. Second, the rings are closer to the planet than its major moons. Third, the rings orbit over the planet's equator. Saturn's rings are highly visible; the faint rings of the other Jovian planets were discovered in the late 1970s.

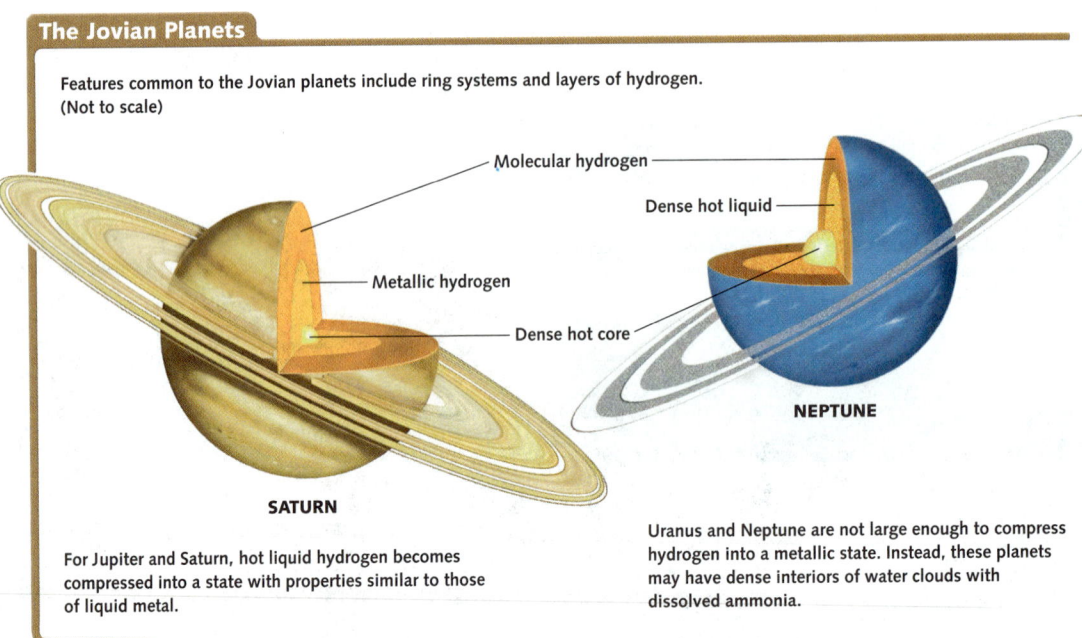

The Jovian Planets
Features common to the Jovian planets include ring systems and layers of hydrogen. (Not to scale)

SATURN
For Jupiter and Saturn, hot liquid hydrogen becomes compressed into a state with properties similar to those of liquid metal.

NEPTUNE
Uranus and Neptune are not large enough to compress hydrogen into a metallic state. Instead, these planets may have dense interiors of water clouds with dissolved ammonia.

594 Unit 7 Space

DIFFERENTIATING INSTRUCTION

Reading Support Have students add the outer planets to the web or chart they started in Section 27.1. Once the charts are complete with facts about each planet, students can organize the facts into a summary of the solar system.
Use Reading Study Guide, p. 92.

Developing English Proficiency Students may not understand why *surfaces* is within quotation marks in the second paragraph. Explain that the quotation marks mean the word is not being used in its typical sense. Say: The surfaces of these planets are not like the kinds of surfaces we usually think of. They are not solid surfaces.

Jupiter's surface

JUPITER is the largest planet in the solar system and has the most striking surface features of any planet.

- Dark belts are areas of sinking gases.
- Bright zones are areas of rising gases.
- Jupiter's largest moon, Ganymede.
- Between zones and belts, high-velocity winds blow parallel to the equator.
- Great Red Spot surrounded by turbulent atmosphere.

Jupiter

Jupiter, the fifth planet from the sun, takes 11.9 Earth years to complete one orbit. It rotates faster than any other planet—once in just under 10 hours. It is the largest planet in the solar system and has more than twice the total mass of all other planets combined.

Jupiter has the strongest known magnetic field of the solar system's planets. As on Earth, the interaction between the solar wind and the magnetic field causes brilliant auroras, colored displays of light.

Jupiter radiates about twice as much heat back to space as it receives from the sun. The extra heat is thought to come from Jupiter's original heat of formation and from contraction due to gravity.

The Great Red Spot is the most striking feature of Jupiter's surface. However, it is just one of several spots. Some spots appear and disappear quickly, while others remain for decades. Photographs indicate that the spots may be relatively calm areas that rotate slowly within the turbulent atmosphere.

On December 7, 1995, the *Galileo* probe successfully entered Jupiter's atmosphere, the first time an Earth spacecraft had entered the atmosphere of a giant planet. To scientists' surprise, the probe found no thick, dense clouds. The probe also revealed that the temperatures and pressures in the upper atmosphere were higher than scientists had expected, and the deep atmosphere is convective. These discoveries have raised intriguing questions for scientists to explore over the coming years.

INVESTIGATIONS
CLASSZONE.COM

How Fast Does the Wind Blow on Jupiter?
Analyze images of Jupiter's atmospheric features to calculate wind speeds.
Keycode: ES2704

CHAPTER 27 SECTION 2

INVESTIGATIONS
CLASSZONE.COM

How Fast Does the Wind Blow on Jupiter?
Keycode: ES2704

- Some student might need help using calculations.
- Provide calculators with large numbers for visually and physically impaired students.

📖 Use Internet Investigations Guide, pp. T93, 93.

Visual Teaching

Interpret Diagram
Have students view the image of Jupiter and note the different bands of color. Ask: How do the areas of light and dark differ? **dark—areas of sinking gases; light—areas of rising gases** Describe Jupiter's atmosphere. **turbulent**

Extend
After students read the page, have them find the Great Red Spot on Jupiter. Ask: What do you think is at the center of the Great Red Spot? What makes it red or reddish orange? **Possible Responses: It may be rotating winds. It may reflect energy or gases from the core. It may be a hot spot.**

CRITICAL THINKING

As students perform the Investigation, have them compare and contrast the atmospheric storms to weather phenomena on Earth, such as hurricanes and tornadoes.

DIFFERENTIATING INSTRUCTION

Reading Support Help students formulate questions to focus reading and pique interest. Tell them to look at the photograph of Jupiter and list the features they notice most readily. **probably various circular and elliptical features, swirls, and bands** List some questions students have about these features. **What causes the swirls? What makes some of the features orange? Why are some bands dark and some light?** Students should write the questions in their notebooks, then record the answers as they find them. Unanswered questions can prompt independent discussion and research.

Developing English Proficiency The following terms from this section may be difficult for students learning English: *massive, comprise, independent, interaction, radiates, contraction, turbulent, convective, internal, unaided eye, fainter, peculiarities, conductive,* and *mean temperature.* Have students write the terms on a sheet of paper and then work with partners to use context and dictionaries to find their meanings.

Saturn

Saturn, the sixth planet from the sun, takes nearly 30 Earth years to complete one orbit. Saturn turns on its axis about every 10 hours. Like Jupiter, the surface of Saturn has colored zones and belts, which are areas of rising and sinking gases. Saturn, however, has fewer zones and belts than Jupiter. Saturn has the lowest density of any planet. In fact, its density is lower than that of water. If there were an ocean of water big enough, Saturn would be able to float on it.

Saturn radiates more energy than it receives from the sun. Like Jupiter, it has sources of internal heat. Saturn's magnetic field is weaker than Jupiter's, but still much stronger than Earth's magnetic field.

SATURN has the most visible ring system of all the Jovian planets. The rings are believed to be made of billions of chunks of ice.

More about...

Saturn's Rings

Saturn's rings are made up of icy chunks with a little rock mixed in. The size of these chunks varies from as large as cars to as small as grains of sand. The rings are thought to be the remains of asteroids, comets, or possibly small moons that were crushed and held in by Saturn's gravity. The rings are relatively thin—*Voyager 2* data indicates they are no more than 200 meters thick. In fact, when the rings are seen edge-on from Earth, they are so thin that they seem to disappear. This view of the rings from Earth, called a ring plane crossing, happens every 15 years or so and gives astronomers a chance to look for small moons or faint objects near the planet that can't be seen in the glare of the rings.

CAREER

Mission Specialist

Astronauts come from all walks of life. However, despite their diverse backgrounds, all of them are accomplished professionals within their fields and are in excellent physical shape.

During the course of their training, they become accustomed to weightlessness, learn to work as a team, and acquire the skills needed to respond to emergencies in space. Some astronauts combine space exploration with scientific research. A mission specialist is an astronaut with expertise in fields such as astronomy, engineering, biology, meteorology, or medicine. Aboard the space shuttle, mission specialists perform scientific experiments and assist with other missions, such as the deployment of satellites.

All American astronauts are employees of the National Aeronautics and Space Administration (NASA). To become a mission specialist, candidates apply to NASA. Applicants must have at least a bachelor's degree in engineering, biological sciences, physical sciences, or math. Most candidates find that a solid background in math and science, starting in high school, is advantageous. Advanced degrees and work experience are also helpful. Chosen candidates must meet strict physical and medical requirements. They then complete a two-year training program before being assigned to space missions. Although the preparation is demanding and the job potentially dangerous, mission specialists are motivated by the knowledge that they may have the rare opportunity to travel in space.

MAE JEMISON became the first woman of color in space when she voyaged in the space shuttle *Endeavor* on September 12, 1992.

Learn more about becoming an astronaut.
Keycode: ES2705

DIFFERENTIATING INSTRUCTION

Challenge Activity Ask: How do the lengths of days (rotation) and years (revolution) of other planets compare to those of Earth? Have students research to create a table comparing the days and years of the planets measured by Earth time. For example, Saturn: 1 day equals 10 Earth hours, 1 year equals 30 Earth years. The accompanying table provides data for rotation in Earth time and revolution in Earth time for each planet:

Planet	Rotation	Revolution
Mercury	59 days	88 days
Venus	243 days	226 days
Earth	24 hours	365 days
Mars	25 hours	687 days
Jupiter	10 hours	11.9 years
Saturn	10 hours	30 years
Uranus	17.2 hours	84 years
Neptune	16.1 hours	165 years
Pluto	6.3 days	248 years

Uranus

Uranus (YUR-uh-nuhs), the seventh planet from the sun, takes 84 Earth years to complete one orbit. Because Uranus is not easily visible to the unaided eye from Earth, it was not discovered until 1781 when more powerful telescopes became available. Uranus is about 19 times farther from the sun than Earth is. Sunlight there is about 370 times fainter than on Earth, and the average surface temperature is only about −200°C.

Uranus turns on its axis once every 17.2 hours. More unusual is its axis of rotation—it is tipped almost completely over, so that Uranus orbits the sun on its side. Some scientists think that the planet was tipped by a collision with an Earth-sized mass early in the history of the solar system.

When *Voyager 2* flew past Uranus in 1986, it discovered something surprising about the planet's magnetic field. Even though the planet is tipped over, the magnetic field is not. For most planets, the axis of rotation and the magnetic field differ by only a few degrees. On Uranus the difference is 60 degrees. This difference causes the planet's magnetic field to trace a spiral pattern in the solar wind as the planet rotates.

URANUS has a turquoise color, due to the methane gas in its atmosphere.

Neptune, Pluto, and Charon

Neptune, the most distant of the Jovian planets, was discovered in 1846, after astronomers had predicted its existence mathematically. Its period of rotation is 16.1 Earth hours, and it takes about 165 years to orbit the sun.

Like Uranus, Neptune has some peculiarities. In 1989, *Voyager 2* discovered that the planet's magnetic axis is tipped 47 degrees in relation to its axis of rotation. In addition, it is offset from the planet's center by about 13,500 kilometers. Scientists think that motions of a conductive material (possibly water) in its middle layers generate the magnetic field.

Neptune is a harsh planet, where winds have been clocked at 2000 kilometers an hour, and the mean temperature is about −225°C. Neptune's atmosphere is mostly hydrogen (74 percent) with smaller amounts of helium (25 percent) and methane (1 percent).

Although it is usually the eighth planet from the sun, Neptune occasionally becomes the ninth planet from the sun. Every 248 years, Pluto's peculiar orbit brings it closer to the sun than Neptune. When this happens, Pluto becomes the eighth planet and Neptune the ninth for about 20 years. This switch happened most recently in 1979, and Pluto returned to its place as the outermost planet on February 11, 1999.

While not always the most distant planet, Pluto is always the smallest, with a diameter estimated to be only 2,274 kilometers (in comparison, the distance from New York to Oklahoma City is 2,140 kilometers). It is smaller than seven of the solar system's moons, including Earth's moon.

NEPTUNE The great dark spot has vanished since this photo was taken by *Voyager 2*.

CHAPTER 27 SECTION 2

Find out more about becoming an astronaut.
Keycode: ES2705

VISUAL TEACHING
Discussion
Ask: Why do you think the surfaces of the planets look different? **Possible Responses:** The colors and textures depend on what the planets are made of. Each planet has different substances. Some have clouds and gases.

CRITICAL THINKING
Ask: Do you think Saturn has auroras? Why or why not? **Saturn has a magnetic field that is stronger than Earth's. Therefore, it is likely to have auroras.** What do you think the auroras would look like on Uranus? **Answers should mention the spiral pattern that occurs.**

DIFFERENTIATING INSTRUCTION

Support for Visually Impaired Students The colors and patterns on each of the planets help communicate information and identify the planets. For those unable to distinguish this information, spend extra time on the descriptions of the planets' atmospheres or surfaces. Identify and describe verbally such unique features as the Great Red Spot on Jupiter.

CHAPTER 27 SECTION 2

Scientific Thinking

INFER

Explain that most comets are made up of frozen water, dust particles, and gases. They orbit out beyond the range of Neptune. Have students compare the elements that make up Pluto with those of the other outer planets. **Pluto—rock and frozen water; outer planets—gaseous** Then have students make up a list of questions they might have about Pluto, such as: Why is its orbit irregular? Why is it so different from the other planets? Is Pluto closer to the Kuiper Belt or to Neptune? After students have gathered the information they need, have them use the facts or data to explain the reasoning behind their inference.

ASSESS

1. Jovian planets—larger, don't have solid surfaces, composed mainly of hydrogen and helium, have ring systems.

2. Extra heat comes from the original heat of formation and from contraction due to gravity.

3. It is tipped over so that Uranus orbits the sun on its side.

4. Pluto is made up of rock and water. The dark areas are likely to be rock; the light areas are likely to be frozen water.

5. The last switch was in 1979. It will happen again 248 years from that date: 1979 + 248 = 2227. The year would be 2227.

598 Unit 7 Space

Scientific Thinking

INFER

The Edgeworth-Kuiper Belt, at the far edge of the solar system, is believed to be the source of short-period comets. Scientists think it is possible that the belt contains as-yet-unidentified objects that are as large as Charon or even Pluto. Do you think that Pluto originated in this belt? Explain your reasoning.

Pluto was thought to be larger than it is until 1978, when astronomers discovered that it had a moon about half its size. Charon, Pluto's moon, has a diameter of about 1172 kilometers. Given this similarity in size, some scientists consider Pluto and Charon to be a double planet, rather than a planet-moon system.

Pluto is so far away from Earth—an average of 39.5 AUs from the sun—that it was not discovered until 1930. Its surface temperature probably varies between −235°C and −210°C; at such temperatures, most of its atmosphere is frozen, thawing out slightly when the planet is nearest the sun. Although its density is not known, scientists believe it consists of about 70 percent rock and 30 percent water.

Because no spacecraft have yet visited it, Pluto remains largely a mystery. However, it shows us that we still have much to learn about our own solar system, even as we turn our attention to other, distant systems.

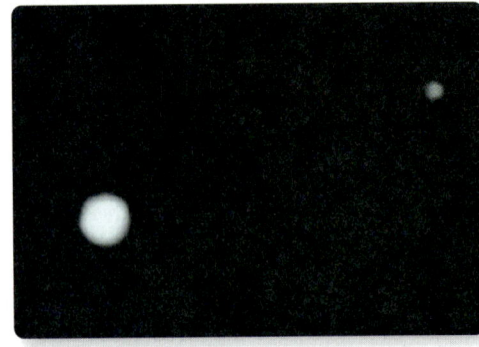

PLUTO AND CHARON have the smallest size difference of all the solar system's planet-moon systems.

27.2 Section Review

1. What are some main differences between the Jovian and terrestrial planets?

2. Why do Jupiter and Saturn give off more heat than they receive from the sun?

3. What is unusual about Uranus's axis of rotation?

4. **CRITICAL THINKING** Scientists have observed areas of light and dark on Pluto's surface. Given what you know about this planet, what do you think the areas might be?

5. **MATHEMATICS** In what year will Pluto once again be the eighth planet from the sun?

MONITOR AND RETEACH

If students miss . . .

Question 1 Reteach "The Jovian Planets." (p. 594)
Question 2 Help students find sentences on pages 595 and 596 that tell about the sources of internal heat for Jupiter and Saturn.
Question 3 Reteach the characteristics of Uranus. (p. 597) Emphasize the unusual discovery *Voyager 2* made as it flew past the planet in 1986.
Question 4 Reteach characteristics of Pluto and Charon (pp. 597–598) Ask students to read to find out what elements Pluto consists of.
Question 5 Reteach the orbits of Neptune and Pluto. (p. 597) Point out that the event happens every 248 years. To find the next event, take the year of the last switch and add 248 to that number to get the next event.

Planetary Satellites

Bodies that revolve around planets are called satellites, or moons. Except for Mercury and Venus, each planet has at least one natural satellite.

Satellites of Earth and Mars

The moon is Earth's only natural satellite. Mars has two tiny moons, Phobos (FOH-buhs) and Deimos (DEE-mohs). Both have irregular shapes and are marked with impact craters. Phobos, the larger of the two, is only 27 kilometers at its widest. It is closer to Mars than Deimos, and circles the planet more than three times a day.

27.3

KEY IDEA
Of the inner planets, only Earth and Mars have moons; all outer planets have moons, some of which are quite large.

DEIMOS

PHOBOS

DEIMOS AND PHOBOS These illustrations depict the Martian moons' irregular shapes and cratered surfaces.

Jupiter's Moons

Jupiter has at least 28 moons. The four largest—Io, Europa, Ganymede, and Callisto—are known as the Galilean satellites in honor of their discoverer, Galileo Galilei. In the late 1990s, a spacecraft called *Galileo* investigated these moons further and made some surprising discoveries.

Io, nearest of the Galilean satellites to Jupiter, is geologically active. At least nine active volcanoes have been observed, and some have been photographed in eruptions that reach more than 300 kilometers above the surface. The sulfur, sulfur dioxide, and other sulfur compounds from the volcanoes cause Io's distinctive surface color, which varies from yellow-orange to red to black. Unlike most solar-system objects, Io shows no signs of impact craters. If craters existed on Io, they have been erased by volcanic material.

Io's density is about 3.5 g/cm^3, which is close to that of Earth's moon. Io has a high-altitude ionosphere, detected by the *Galileo* spacecraft, and is thought to have a very thin sulfur-dioxide atmosphere. Its surface is covered with layers of sulfur and frozen sulfur dioxide.

The *Galileo* spacecraft also found that Io has an iron core that may take up half its diameter of over 3600 kilometers. A layer of molten silicate rock surrounds the core. Io gets its internal heat from friction due to tidal forces that change in strength continuously as Io moves along its elliptical orbit.

Europa, next out from Jupiter, was also found to have an atmosphere by the *Galileo* spacecraft. Europa's smooth and shiny white surface appears to

VOLCANOES ON IO Loki and Pele are two volcanoes on Io.

CHAPTER 27 SECTION 3

VISUAL TEACHING
Discussion
Lead a discussion to compare and contrast the Galilean satellites. Ask: Which moons have surfaces with relatively few impact craters? **Io and Europa** Which have frozen surfaces? **Europa, Ganymede, and Callisto** How are Ganymede and Callisto alike? **Both have a subsurface ocean.** How does Callisto differ from the other moons? **Callisto has traces of oxygen.**

More about...

Satellite Names
Moons and satellites get their names from mythology and literature.
- Deimos—son of the Roman god Mars
- Phobos—another son of the Roman god Mars
- Ganymede—Greek boy kidnapped to be a cup bearer to the gods
- Io, Europa, and Callisto—Greek female characters who attracted the attention of Zeus
- Titan—child of Uranus
- Titania, Oberon, Ariel, and Miranda—characters in Shakespeare's plays
- Umbriel—character in an Alexander Pope poem
- Triton—son of Neptune
- Charon—ferryman who conveyed the dead over the River Styx to Hades

EUROPA'S SURFACE This false-color, high-resolution image of Europa's surface shows what appear to be icebergs floating atop a liquid ocean. Notice that some of the structures fit together like puzzle pieces.

be a shell of water ice up to 100 kilometers thick. Europa has relatively few craters on its surface, suggesting that recent geologic activity—possibly within the last few million years—has erased the scars of meteorite bombardments. Europa's surface is marked by a crisscross pattern of bright and dark lines that resemble cracks in ice floes in Earth's polar oceans.

High-resolution photographs and magnetic-field data from the *Galileo* mission provide evidence that a liquid ocean 100 kilometers deep may exist below Europa's surface. If so, it is possible that life forms may have developed there, perhaps getting their energy from subsurface volcanic vents like those on Earth's ocean floor.

Ganymede, next in order from Jupiter, is the largest moon in the solar system. With a diameter of 5268 kilometers, it is larger than Pluto and Mercury, and about three-quarters the size of Mars. If Ganymede orbited the sun rather than Jupiter, it would no doubt be classified as a planet. Ganymede's density is less than 2 g/cm^3, which indicates that it is most likely composed of a large amount of ice, around a rocky core. The mantle is probably ice and silicates, and the crust is believed to be a layer of water ice.

Ganymede is the only one of Jupiter's moons known to have a magnetic field. This magnetosphere probably protects it from the magnetic influence of Jupiter. Callisto, the farthest out of the Galilean satellites, is the most heavily cratered object in the solar system and is thought to be the oldest landscape in the solar system as well. About 4800 kilometers in diameter, it is the least dense of the Galilean satellites (about 1.86 g/cm^3). The *Galileo* spacecraft detected oxygen on Callisto, probably released by sunlight striking its icy surface.

Evidence from the *Galileo* spacecraft has led scientists to believe that Callisto's internal structure does not consist of separate layers. Instead, scientists now think that the interior is a mixture of ice and rock, with a rocky core. This would make Callisto's structure different from the other Galilean satellites; scientists suspect that Callisto's distance from Jupiter has allowed it to avoid the gravitational stresses of Io, Europa, and Ganymede.

Magnetic-field data from the *Galileo* spacecraft suggest that a subsurface ocean of salty liquid water several kilometers deep may exist beneath the surfaces of Ganymede and Callisto.

The Galilean Satellites of Jupiter — Io, Europa, Ganymede, Callisto

600 Unit 7 Space

DIFFERENTIATING INSTRUCTION

Reading Support Have students identify cause-and-effect relationships to help them understand what they read. Draw a cause-and-effect diagram on the board consisting of two rectangular boxes with an arrow pointing from the left box to the right box. Write *Ganymede's magnetic field* in the left box. Tell students this is a cause. Have them read the information about Ganymede to find the effect. **protects it from magnetic influence of Jupiter** Write that information in the right box. Encourage students to find other cause-and-effect relationships in the chapter.

Saturn's Moons

Until the Space Age, Saturn was thought to have nine moons, all discovered before 1900. Recently, new moons have been discovered through telescopes and with spacecraft. The latest count is 24, although some have not yet been confirmed. The largest and most interesting is Titan.

Titan is the second-largest moon in the solar system. Its density is just under 2 g/cm^3, and it seems to be about half rock and half frozen water.

Titan is the only moon known to have a substantial atmosphere. Its atmospheric pressure is about 1.5 times Earth's. Like Earth, its principal gas is nitrogen, which is estimated to make up between 90 and 95 percent of the total atmosphere. Most of the remaining gas is methane, with traces of hydrogen cyanide and acetylene. Titan's surface temperature is about −180°C. This is cold enough to turn methane and other gases to liquid. The resulting droplets form a dense orange smog that hides Titan's surface.

The Moons of Uranus and Neptune

Uranus is known to have 21 moons. The five major moons are Titania, Oberon, Umbriel, Ariel, and Miranda. All are alike in that they lack atmospheres and have many impact craters on their surfaces. But differences between the moons are visible in *Voyager* photographs. Titania has huge, faulted valleys. Oberon's impact craters are partly flooded with dark material. Umbriel has an unusual dark surface, and Ariel's cratered surface is crisscrossed by valleys and faults. Miranda proved to be the most startling of all. Its surface is deeply scarred with V-shaped grooves and parallel ridges. Some scientists theorize that Miranda has been shattered as many as five times during its existence, and after each shattering reassembled with parts of the original surface buried and parts of the core exposed. Another theory suggests the surface features resulted from upwelling of partly melted ice. However, the real reason remains a mystery.

Neptune has eight moons. Triton, the largest of Neptune's moons, is about four-fifths the size of Earth's moon. Triton's southern ice cap is made of methane and ammonia. Ice volcanoes of nitrogen were erupting from the surface of Triton as *Voyager 2* passed by. Triton has a very thin atmosphere.

MIRANDA, one of Uranus's large moons, shows the scars that scientists think resulted from being shattered as many as five times.

27.3 Section Review

1. What are the moons of the inner planets?
2. What makes Titan an interesting moon?
3. Why are small moons, like Phobos, potato shaped, while larger moons are spherical?
4. **CRITICAL THINKING** Hypothesize why the Jovian planets have so many moons. Where do you think the moons came from?
5. **BIOLOGY** Which Galilean moon is most likely to support life? Explain your reasoning.

CHAPTER 27 SECTION 1

ASSESS

1. Earth's moon, Phobos, and Deimos
2. substantial atmosphere; heavier atmospheric pressure than Earth
3. Small moons, due to their smaller mass, do not have gravitational fields strong enough to pull their matter into a spherical shape.
4. The Jovian planets have great gravitational forces that hold the moons in orbit. The moons could have been pieces left over from the origin of the solar system, or they could have been comets or orbiting bodies that came close enough to get pulled in by the planet's gravity.
5. Europa: it has water and energy from subsurface volcanic vents.

MONITOR AND RETEACH

If students miss . . .

Question 1 Reteach "Satellites of Earth and Mars" (p. 599) and review "The Inner Planets." (p. 588)

Question 2 Reteach Saturn's moons. (p. 601) Have students explain how Titan is similar to Earth.

Question 3 Review the photographs of the various moons. (pp. 599–601) Have students compare the sizes of Deimos and Phobos to the other moons.

Question 4 Reteach "Jupiter's Moons." (pp. 599–600)

Question 5 Reteach "Jupiter's Moons." (pp. 599–600) Have students speculate on the conditions necessary for life to begin.

CHAPTER 27 SECTION 4

FOCUS

Objectives
1. Identify smaller components of the solar system.
2. Explain the effects of objects colliding with Earth.

Set a Purpose
Have students preview the key idea and read the section to answer the question "What smaller objects are found in the solar system?"

INSTRUCT

VISUAL TEACHING

Discussion
Focus students' attention on the photographs of the comets. Ask: Are comets solid or gaseous? **solid** What causes the "tails" of the comets? **The sun heats the comet's surface, forming a cloud of gas and dust.** What does the solar wind do to the coma? **pushes the material away from the sun back out into space, giving the appearance of a "tail"** Is the tail behind or in front of the comet? Explain. **As the comet approaches the sun, the tail is behind it; when the comet moves away from the sun, the tail precedes it.**

27.4

KEY IDEA
The solar system contains countless smaller objects—comets, asteroids, and meteoroids—which can and have collided with Earth.

KEY VOCABULARY
- comet
- asteroid
- meteor
- meteorite
- meteor shower

VISUALIZATIONS
CLASSZONE.COM
Observe an animation of a comet's passage through the solar system.
Keycode: ES2706

Solar-System Debris

The sun, planets, and many moons are the largest components of our solar system. However, our solar system is also made up of smaller objects. Some of these have collided with Earth in the past, leaving great scars on the surface. Collisions continue to occur today.

Comets and TNOs

Comets have been described as dirty snowballs. They are made of dust particles trapped in a mixture of water, carbon dioxide, methane, and ammonia. Comets spend most of their time far out beyond Neptune's orbit, where they consist only of a solid main body called a nucleus. Vast numbers of comets orbit in the cold region of our solar system beyond Neptune called the Edgeworth-Kuiper Belt and in the much more distant Oort Cloud. More than 375 large bodies up to several hundred kilometers in diameter have been detected in the Edgeworth-Kuiper Belt. They are known as Trans Neptunian Objects (TNOs).

A few comets, however, move in highly elliptical orbits that take them closer to the sun. When this happens, they can become visible in Earth's night sky. When a comet moves close enough to the sun (around the orbit of Jupiter), energy from the sun heats the comet's icy surface, causing it to form a coma, a cloud of gas and dust that expands into space. At a comet's closest point to the sun, its coma can be almost as wide as Jupiter. The solar wind pushes material from the coma far out into space, forming dramatic tails that always point more or less away from the sun and may extend for millions of kilometers. As the comet moves away from the sun back toward the outer solar system, its tails actually precede it.

The famous Halley's Comet returns to the inner solar system every 76 years. The comet is named for Edmund Halley, an 18th-century English astronomer. In studying records of comets, Halley noticed that bright comets had appeared in 1531, 1607, and 1682. He thought that these were all one comet with an orbital period of about 76 years. He correctly predicted its return sometime in 1758 or 1759. It returned again in 1835, 1910, and 1986.

HALLEY'S COMET last visited Earth in 1986. Its nucleus and tail show up clearly in this false-color photograph.

TWO TAILS The tail of Comet Hale-Bopp is split. The lower tail is made of ionized molecules. The upper tail is made of dust.

602 Unit 7 Space

DIFFERENTIATING INSTRUCTION

Developing English Proficiency Some words that are very close in spelling, such as *comet/coma* and *meteor/meteoroid/meteorite* may provide difficulty in this section. Work with students to separate words into word parts and define the meanings. Have students use each word in a sentence to establish comprehension.

Reading Support Model how to use the first paragraph under the blue heading as an introduction to the section and a preview to what's coming up. Say: It's clear that the section will discuss the smaller objects in the solar system and evidence that some of these objects struck Earth.

Use Reading Study Guide, p. 94.

602 Unit 7 Space

Comets

Name	Orbital Period (in years)	Perihelion Date	Perihelion Distance (AUs)	Absolute Magnitude
Halley	76.1	02-09-1986	0.587	5.5
Encke	3.30	12-28-2003	0.340	9.8
d'Arrest	6.51	08-01-2008	1.346	8.5
Tempel-Tuttle	32.92	02-28-1998	0.982	9.0
Kohoutek	6.24	12-28-1973	1.571	12.1
Wild2	6.39	09-25-2003	1.583	6.5
Wilson-Harrington	4.29	03-26-2001	1.000	9.0

Asteroids

Asteroids are solid, rocklike masses. Most seem to have irregular shapes, which explains why their brightness changes as they rotate. There are thousands of asteroids in the solar system, but only the two largest, Ceres and Pallas, are spherical. Ceres has a diameter of about 1000 kilometers. Most asteroids are less than 1 kilometer long. Scientists think that asteroids are leftover material from the solar system's formation.

THE ASTEROID 243 IDA, photographed by the *Galileo* probe, is about 55 kilometers across at its widest point. It has a tiny moon just 1.5 kilometers across, named Dactyl.

Asteroids revolve around the sun in the same direction as the planets. Most asteroid orbits are nearly circular and lie between Mars and Jupiter in the asteroid belt. A few, however, have long oval orbits. Some come close to Mercury at perihelion, when they are closest to the sun. Asteroids can collide, and have collided, with Earth. Many scientists believe that an asteroid or a comet collided with Earth 65 million years ago, leading to the extinction of the dinosaurs. In 1908 a much smaller object exploded with the force of a 10-megaton nuclear detonation over Siberia, leveling trees over a vast but remote area. A similar explosion over a populated area would cause massive loss of life, and the impact of an asteroid or comet could inflict terrible damage on Earth and its life systems. Although no known asteroids or comets are currently considered at risk of striking Earth, some scientists are looking into ways to prevent collisions by diverting objects before they reach Earth.

CHAPTER 27 SECTION 4

Observe an animation of a comet's passage through the solar system.
Keycode: ES2706

Visualizations CD-ROM

VISUAL TEACHING
Discussion

Refer students to the chart. Ask: What do the comets orbit? **sun** What does *perihelion* mean? **point closest to the sun** What does the perihelion date mean? **date the comet is closest to the sun** What is perihelion distance? **distance from the sun at perihelion** What does AU stand for? **astronomical unit** What does 1 AU represent? **distance between the sun and Earth** Which comet will be about the same distance from the sun as Earth is? **Wilson-Harrington** Tell students that absolute magnitude is a measurement that expresses the brightness of the objects. Ask: Which comet would be brightest if all the comets were all seen at the same distance? **Kahoutek**

DIFFERENTIATING INSTRUCTION

Challenge Activity How is it possible that an asteroid could lead to the extinction of dinosaurs? Have students research the theory that an asteroid or comet that collided with Earth about 65 million years ago caused the extinction. Suggest they consider evidence of how large it was, where it landed, what happened immediately afterwards, and so on. Students should report their findings to the class.

CHAPTER 27 SECTION 4

More about...

Meteors, Meteorites, and Impact Craters
Though many meteors blaze through Earth's atmosphere, very few actually make it to Earth's surface. Those that don't burn up immediately travel between 10–70 km/sec, slowing down considerably as they travel deeper into the atmosphere. Over 22,000 meteorites have been discovered on Earth. Of those, 18 originated from Mars and 23 from the moon. Most meteorites are small. Only meteorites of several hundred tons make craters. The largest meteorite ever found is in Namibia and weighs 60 tons. A large meteorite that fell to Earth in 1908 in Siberia was about 60 meters in diameter, but it broke up before hitting the surface and so it left no crater. According to the geological record, every million years Earth can expect about three meteorites large enough to make impact craters 10 km or larger. An impact the size that may have affected the dinosaurs may occur once every 100 million years.

VISUALIZATIONS CLASSZONE.COM
Observe an animation of meteor showers.
Keycode: ES2707

Meteors and Meteoroids

A **meteoroid** is a rock or an icy fragment traveling in space. Meteoroids are different from asteroids only in their smaller size—from less than 100 meters in diameter down to the size of a sand grain. A **meteor,** also called a shooting star, is the light made by a meteoroid as it passes through Earth's atmosphere. The light is caused by friction between the rapidly moving meteoroid and the atmosphere.

On a clear, dark night about 5 to 15 meteors can be seen every hour. However, this is a small portion of all meteors. Scientists estimate that anywhere from a million to a billion meteoroids enter the atmosphere daily. Most are tiny and burn or vaporize in the air.

Sometimes, large numbers of meteors streak across the night sky within a few hours of one another. Such an event is called a **meteor shower.** A meteor shower occurs when Earth passes through the tail of a comet, and particles from the tail plunge through the atmosphere as meteors.

Because Earth's orbit crosses the paths of comets around the same time each year, many meteor showers occur at predictable times. Meteor showers are named for the constellation from which they appear to originate; for example, the Perseid meteor shower, which happens in August, seems to come from Perseus.

Meteorites

A **meteorite** is part of a large meteoroid that survives its trip through the atmosphere and strikes Earth's surface. There are three basic types of meteorites. Most, about 94 percent, are stony meteorites, which resemble Earth's dark igneous rocks. They are composed primarily of silicates.

Iron meteorites make up about 5 percent of all meteorites. They consist of large crystals made mostly of iron with a small amount of nickel. The large crystals indicate that they cooled over millions of years, suggesting that they were formed inside large asteroids that later broke apart. About 1 percent of meteorites are called stony-iron meteorites. They appear to have formed when molten silicates came into contact with molten metal.

The most abundant source of meteorites is the Antarctic ice cap. Meteorites that have fallen there are exposed at the surface when wind erosion removes the ice around them. Thousands of meteorites have been recovered from Antarctica, providing an enormous increase in the supply of extraterrestrial material available for study.

STONY About 94 percent of meteorites that strike Earth are stony.

IRON meteorites account for 5 percent of meteorites that strike Earth.

STONY-IRON are the rarest meteorites, making up about 1 percent of the total.

DIFFERENTIATING INSTRUCTION

Hands-On Demonstration If students did not do the Lab Activity on Impact Craters in Chapter 25, you can demonstrate impact craters. Place 5–10 cm of sand in a box. Sift a thin layer of flour or cocoa over the top, for color contrast. The sides of the box should act as a backsplash for the powdered material. Use "meteorites" of three different sizes such as metal nuts, marbles, fishing weights, rocks, or similar items. Drop the three different objects from the same height. Ask the class to predict what the difference will be in the impact craters for the three different objects. Use a toothpick to measure depth and a ruler to measure diameter. Compare the size of the crater to the object. Craters are much larger. Discuss the results of the ejecta pattern, crater shape, rim, and any other details students observe.

Impact Craters

Impact craters are bowl-shaped depressions that remain after a meteor or other object strikes Earth, another planet, or a moon. Earth is not as heavily cratered as the moon, but it still bears scars of its encounters.

Impact craters are rare features on Earth; only about 150 are known to exist. One reason is that Earth's atmosphere burns up most meteoroids before they strike the surface. Another is that Earth is geologically active, and so it continually erases the marks of impacts. Earth's oldest crater, the Vredefort Crater in South Africa, is 2 billion years old. Few craters on Earth are more than half a billion years old.

One of the best-known craters is also one of the younger ones. Arizona's Barringer Meteor Crater is thought to have formed about 49,000 years ago when an iron meteorite about 45 meters in diameter struck Earth and exploded, leaving behind a crater about 1200 meters in diameter.

Impacts change Earth geologically. Very large impacts may leave rings in the surface like ripples in a pond. Other structures resulting from impacts may become reservoirs for oil and gas deposits. An impact near Sudbury, Ontario, Canada, about 1.85 billion years ago may have resulted in that area's large nickel and copper deposits.

BARRINGER METEOR CRATER, Arizona, is about 1200 meters in diameter and nearly 200 meters deep.

27.4 Section Review

1. What happens to comets as they approach the sun?
2. Where are most asteroids found?
3. Explain the difference between meteoroid, meteor, and meteorite.
4. **CRITICAL THINKING** Most meteorites formed between 4.55 billion and 4.65 billion years ago, making them a little older than the oldest moon rocks. Infer why moon rocks are younger than most meteorites.
5. **MATHEMATICS** When will Halley's Comet return to Earth? Comet Kohoutek?

CHAPTER 27 SECTION 4

VISUALIZATIONS
CLASSZONE.COM

Observe an animation of meteor showers.
Keycode: ES2707

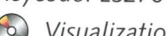 *Visualizations CD-ROM*

VISUAL TEACHING
Discussion

Be sure students understand the scale of the Barringer Crater. Point out that 1200 meters (crater diameter) is about 3/4 of a mile. Compare that size to a local feature such as a lake, park, or certain number of city blocks. Ask: What was the size of the meteor that created the crater in the photograph? **45 meters in diameter** Why didn't the meteor burn up in the atmosphere? **It was too large** How are very old craters "erased"? **Possible Responses:** *erosion, volcanic activity, plate movement*

ASSESS

1. Heat from the sun forms a coma that solar winds push back to create a tail.
2. in a belt in circular orbit around the sun, located between Mars and Jupiter
3. meteoroid—a rock or an icy fragment traveling in space; meteor—meteoroid as it passes through Earth's atmosphere; meteorite—part of the meteoroid that survives to strike Earth's surface
4. The meteoroids were formed before the moon was formed.
5. Halley's comet in 2062; Kohoutek in 2003

MONITOR AND RETEACH

If students miss . . .

Question 1 Reteach "Comets and NTOs." (p. 602) Have students choose one of the photographs and describe what is happening.

Question 2 Review the illustration of the Asteroid Belt. (pp. 588–589) Ask: Between which planets is the asteroid belt? **Mars and Jupiter**

Question 3 Have students review page 604 and describe how a meteoroid approaches, enters the atmosphere, and strikes Earth.

Question 4 Reteach meteoroids (p. 604) and review the origin of the moon. (pp. 32–33, 556)

Question 5 Reteach orbital periods of comets from the chart. (p. 603) 76 years from 1986 (1986 + 76 = 2062); 6.24 years in cycles from 1973 would be approximately 2003

CHAPTER 27 ACTIVITY

PURPOSE
To graph the positions of the Galilean moons and determine patterns in the their motions

MATERIALS
Provide students with a copy of Lab Sheet 27. A copymaster is provided on page 149 of the teacher's laboratory manual. You might use binoculars or a telescope to let students view the moons themselves. Use an almanac or a newspaper to find out when Jupiter is observable.

PROCEDURE

❸ You could connect D and R with a dashed line to represent the time when the moon passes behind Jupiter and is not visible.

❹ To view how the moons of Jupiter would appear in the sky at a given time, hold a straightedge across the completed graph and note the positions of the four colored lines when they intersect with the straightedge. The moons will appear at these intersections.

CHAPTER 27 LAB Activity

Galilean Moons of Jupiter

SKILLS AND OBJECTIVES

- **Graph** the positions of the Galilean moons relative to Jupiter.
- **Analyze** the graph to determine patterns in the moons' motion.
- **Compare** the orbits of the four moons.

MATERIALS
- colored pencils
- Lab Sheet 27 *Galilean Moon Orbit Graph*

Learn more about Jupiter's satellites.
Keycode: ES2709

First discovered by Galileo in 1610, the four largest moons of Jupiter can be observed with a small telescope or binoculars. Observations of the moons have played an important role in the history of science. For example, charts predicting eclipses of Jupiter's moons helped ships at sea to calibrate their clocks. In 1676 Ole Rømer found that the eclipse charts contained systematic errors, and concluded that the reason for the errors was due to the fact that the speed of light was not infinite. Rømer then became the first scientist who attempted to calculate the speed of light.

In this activity, you will graph the positions of the Galilean moons over a period of 18 days between December 19, 2001, and January 5, 2002.

Procedure

❶ Obtain a copy of the Galilean Moon Orbit Graph. Color in the space between the two central lines. This space represents the diameter of Jupiter. Fill in the key with the color you have used for Jupiter.

❷ The data table gives observational data for the positions of the Galilean moons relative to Jupiter as they appear from Earth. The data include the times each moon disappears and reappears after passing in front of or behind Jupiter. An example of a complete cycle of Io's orbit is shown on the Galilean Moon Orbit Graph.

❸ Use a colored pencil to plot the remaining data on the orbit graph for Io. As shown in the example, connect the events labeled E and D with a parabola to represent the portion of Io's orbit observed to the right of Jupiter. Connect the events from R to I with a parabola to represent Io's orbit to the left of Jupiter. Connect events I to E with a straight line to represent Io passing in front of Jupiter.

❹ Complete the curves for the other three Galilean moons. Fill in the key with the color you have chosen for each moon.

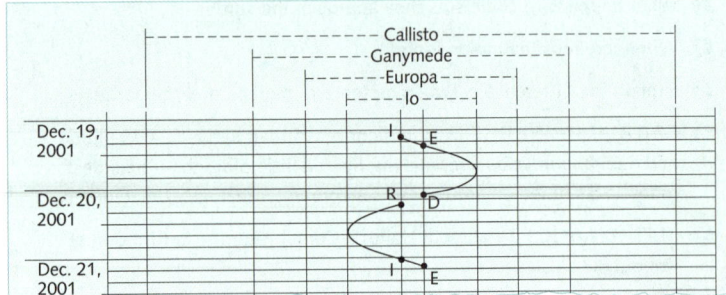

CHAPTER 27 ACTIVITY

Learn more about Jupiter's satellites.
Keycode: ES2709

Analysis and Conclusions

1. If you observed the moons of Jupiter with binoculars on 12/22/01 at 2 A.M. GMT, in what order would the moons and Jupiter appear from left to right? In what order would they appear on 12/30/01 at 2 A.M. GMT?

2. On 01/02/02 at 1 A.M. GMT, you observe the moons of Jupiter with a telescope, and all four moons appear to the left of Jupiter. Explain how this could happen.

3. How does the distance of a moon's orbit from Jupiter relate to the time it takes the moon to complete one orbit?

4. What is the approximate period of revolution for each moon?

5. To an observer situated above the north pole of Jupiter, do the moons revolve in a clockwise or counterclockwise direction?

6. If you wanted to observe the four moons when they are spread farthest apart and would be easily distinguishable from each other, what time would you choose from those shown on your graph? Be sure to choose a time when it is dark where you live. Note: Greenwich Mean Time (GMT) is five hours ahead of Eastern Standard Time and eight hours ahead of Pacific Standard Time.

GALILEAN MOON ORBITAL EVENTS
I – moon begins to pass in front of Jupiter on the left side
E – moon ends passage across Jupiter
D – moon disappears behind Jupiter on right side
R – moon reappears on the left side of Jupiter

Galilean Moon Orbital Data

Date	Time (GMT) – Event (I, E, D, R)			
	Io	Europa	Ganymede	Callisto
Dec. 19, 2001	6:35 – I 8:49 – E	11:56 – I 14:40 – E		3:02 – D 5:44 – R
Dec. 20, 2001	3:33 – D 6:08 – R			
Dec. 21, 2001	1:01 – I 3:15 – E 22:02 – D	5:35 – D 14:40 – E	16:21 – D 5:44 – R	
Dec. 22, 2001	0:33 – R 19:26 – I 21:41 – E			
Dec. 23, 2001	16:31 – D 18:59 – R	1:03 – I 3:52 – E		
Dec. 24, 2001	13:52 – I 15:55 – E	18:52 – D 22:02 – R		
Dec. 25, 2001	10:59 – D 13:25 – R		7:01 – I 10:03 – E	
Dec. 26, 2001	8:18 – I 10:33 – E	14:10 – I 16:58 – E		
Dec. 27, 2001	5:28 – D 7:51 – R			
Dec. 28, 2001	2:44 – I 4:58 – E 23:57 – D	8:10 – D 11:08 – R	20:20 – D 23:44 – R	11:52 – I 14:31 – E
Dec. 29, 2001	2:17 – R 21:10 – I 23:21 – E			
Dec. 30, 2001	18:25 – D 20:43 – R	3:18 – I 6:06 – E		
Dec. 31 2001	15:35 – I 17:50 – E	21:27 – D		
Jan. 1, 2002	12:53 – D 15:09 – R	0:27 – R	10:16 – I 13:18 – E	
Jan. 2, 2002	10:01 – I 12:16 – E	16:25 – I 19:13 – E		
Jan. 3, 2002	7:19 – D 9:38 – R			
Jan. 4, 2002	4:27 – I 6:42 – E	10:35 – D 13:32 – R	23:56 – D	
Jan. 5, 2002	1:45 – D 4:07 – R 22:53 – I		3:23 – R	17:06 – D 13:18 – E

Analysis and Conclusions

1. 12/22/2001 2 GMT: From left to right—Callisto, Europa, Ganymede, Io, Jupiter

   12/30/2001 2 GMT: From left to right—Ganymede, Europa, Jupiter, Io, Callisto

2. The chart indicates that all four moons are to the right of Jupiter at 1 A.M. GMT on 1/3/2002. If all four moons appear on the left, the mirrors inside the telescope have flipped the image. In fact, when viewing objects in space with a telescope, images can appear upside down and backwards.

3. The farther away a moon is from Jupiter, the longer it takes the moon to orbit Jupiter.

4. periods of revolution for the Galilean moons (rounded): Io 1.8 days (1 day, 19 hours); Europa 3.6 days (3 days, 14 hours); Ganymede 7.2 days (7 days, 5 hours); Callisto 16.7 days (16 days, 17 hours)

5. The moons orbit counterclockwise around Jupiter.

6. Students must pick a time when it is dark and when all four moons are visible. Two ideal times are between 0:00 and 6:00 GMT on either December 24, 2001, or December 31, 2001.

CHAPTER 27 REVIEW

Vocabulary Review

1. c, The inner planets are considered terrestrial—Saturn is one of the outer (gaseous) planets.
2. d, Mars is the smallest terrestrial planet. It is also the planet closest to the sun.
3. c, The diagram on page 591 shows why Venus appears as an evening and a morning star.
4. b, Mars, like Earth, has polar caps. The difference is in their makeup: frozen carbon dioxide instead of water.
5. a, The Jovian planets are characterized by their gaseous composition. It is the terrestrial planets that have rocky surfaces.

Concept Review

6. The inner planets have rocky crusts, dense mantle layers, and very dense cores.
7. Comets are frozen water and gases. Asteroids are solid, rock-like masses. Meteoroids are similar to asteroids but smaller in size.
8. Pluto consists of rock and frozen water rather than gas and light elements.
9. Neither of the two inner planets, Mercury and Venus, has a moon.
10. *Jovian planets* have *rings* and many moons and include *Jupiter, Saturn, Uranus*, and *Neptune*. Terrestrial planets have *rocky crusts*, dense mantles and *cores*, and include *Mercury, Venus, Mars*, and Earth.

CHAPTER 27 REVIEW

Summary of Key Ideas

27.1 The planets are grouped by position as inner or outer and by properties as terrestrial or Jovian. The inner planets are Mercury, Venus, Earth, and Mars. All have rocky crusts, dense mantle layers, and very dense cores.

27.2 The Jovian planets are large, consist mostly of gases, and are surrounded by ring systems and many moons. They are Jupiter, Saturn, Uranus, and Neptune. Pluto is too small to be a Jovian planet and not dense enough to be a terrestrial planet.

27.3 A planetary satellite or moon is a smaller body that revolves around a planet. Except for Venus and Mercury, each planet has at least one satellite.

27.4 The solar system also includes debris such as comets, asteroids, and meteoroids.

Key Vocabulary

asteroid (p. 603)
comet (p. 602)
inner planets (p. 588)
meteor (p. 604)
meteorite (p. 604)
meteor shower (p. 604)
outer planets (p. 588)

Vocabulary Review

Choose the best answer. Write the letter of the best answer on another sheet of paper.

1. Which is not a terrestrial planet? (a) Mars (b) Mercury (c) Saturn (d) Venus
2. A planet that has no atmosphere because of its high temperature and low gravity is (a) Mars, (b) Pluto, (c) Venus, (d) Mercury.
3. A planet that can be seen only as a morning star or as an evening star is (a) Jupiter, (b) Uranus, (c) Venus, (d) Mars.
4. Which planet has polar caps of frozen carbon dioxide? (a) Pluto (b) Mars (c) Venus (d) Jupiter
5. The Jovian planets do NOT have (a) rocky surfaces, (b) moons, (c) rings, (d) magnetic fields.

Concept Review

6. Describe the qualities the inner planets share.
7. How do comets, meteoroids, and asteroids differ?
8. Why is Pluto, one of the outer planets, not considered a Jovian planet?
9. What planet does not have a moon?
10. **Graphic Organizer** Copy and complete the concept map below.

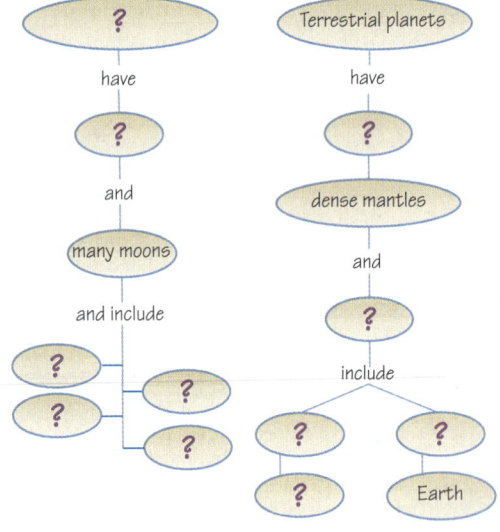

Critical Thinking

11. **Infer** Two astronomers with telescopes, one on Venus and one on Mars, are observing Earth. Who would have an easier time and why?

12. **Apply** How many complete orbits has Earth made in your lifetime? How many orbits has Mars made in your lifetime?

13. **Interpret** Neptune was discovered in 1846. Since then, a complete orbit of the sun by Neptune has not yet been observed. Why?

14. **Deduce** Phobos revolves around Mars from west to east faster than Mars rotates on its axis from west to east. Viewed from Mars, in what direction does Phobos rise and set?

15. **Hypothesize** The orbits of Neptune and Pluto cross periodically. Write an hypothesis describing whether or not you think it is likely that Neptune and Pluto could ever collide.

Interpreting Graphs

The straight lines on the graph show the speed (in km/s) needed by several gas molecules to escape a planet relative to the absolute, or kelvin (K), temperature of that planet's atmosphere. Points representing the planets are also on the graph.

16. If the escape speed of a gas from a planet's atmosphere is directly related to the mass of the planet, according to the graph, which planet has the greatest mass? Is this planet the largest?

17. According to the graph, which planet has the least mass? Is this planet the smallest?

18. Using the graph, identify two pairs of planets that must have nearly the same mass because gases can escape from their atmospheres at nearly the same speeds.

19. According to the graph, which two planets have no atmosphere? A gas is not held by a planet if the line for that gas is above the point for the planet.

20. Which planets have both hydrogen and helium in their atmospheres?

Internet Extension

INVESTIGATIONS CLASSZONE.COM

What Processes Shape Planetary Surfaces? Analyze images of planetary features to compare them with features on Earth.
Keycode: ES2708

Writing About the Earth System

SCIENCE NOTEBOOK On what planet do you think scientists should concentrate their research efforts? What do you think such exploration will mean for Earth? Formulate a persuasive argument for your choice and write it in your Science Notebook.

21. According to the graph, how does the atmosphere of Earth differ from that of Mars?

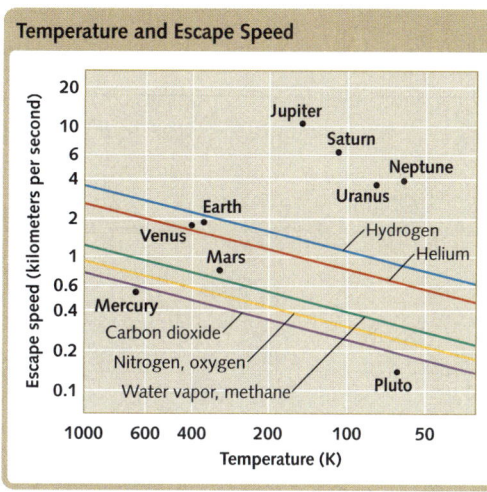

Temperature and Escape Speed

CHAPTER 27 REVIEW

INVESTIGATIONS CLASSZONE.COM

What Processes Shape Planetary Surfaces?
Keycode: ES2708

Use Internet Investigations Guide, pp. T94, 94.

CRITICAL THINKING

11. Although Venus is closer, it has a thick atmosphere and is covered in clouds that would be difficult to see through. It would be easier to view Earth from Mars.

12. Answers will vary. For a 14-year-old, it would be 14 Earth orbits and 6–7 Mars orbits.

13. Neptune's orbit is about 165 years. A full orbit from the time of discovery will be completed in 2011.

14. rises in the west, sets in the east

15. Since the orbit patterns cross, it is possible that at some point the two orbiting bodies will be in position to collide.

INTERPRETING GRAPHS

16. Jupiter has the greatest mass, and it is the largest planet.

17. Pluto has the least mass, and it is the smallest planet.

18. Venus and Earth must have nearly the same mass.

19. Mercury and Pluto have no atmosphere.

20. Venus, Earth, Jupiter, Saturn, Uranus, and Neptune

21. Earth has helium within its atmosphere; Mars does not.

WRITING ABOUT SCIENCE

Writing About the Earth System

Sample Responses: Scientists should concentrate research efforts on Mars. It is similar to Earth in many ways: its axis tilt is similar, it has seasons, it has polar caps. It is possible that there are signs of life on Mars. There may be traces of water or frozen ice. We may be able to study how life begins and evolves by studying Mars.

Venus would be a good planet on which to concentrate research efforts. It is close to Earth, and it has a similar diameter, mass, and gravity. It is like a sister planet. Its surface has some similarities to Earth as well. It has features that indicate tectonic activity. We could learn more about Earth by studying this planet.

CHAPTER 28 PLANNING GUIDE: STARS AND GALAXIES

	Section Objectives	Print Resources
Chapter-Level Resources		Laboratory Manual, pp. 119–126 Internet Investigations Guide, pp. T95–T96, 95–96 Spanish Vocabulary and Summaries, pp. 55–56 Formal Assessment, pp. 82–84 Alternative Assessment, pp. 55–56
Section 28.1 A Closer Look at Light, pp. 612–616	1. Describe the characteristics of electromagnetic radiation. 2. Explain techniques for analyzing light to obtain information about stars. 3. Explain the Doppler effect and how it gives information about star motions.	Lesson Plans, p. 95 Reading Study Guide, p. 95 Teaching Transparency 32
Section 28.2 Stars and Their Characteristics, pp. 617–625	1. Explain why the positions of constellations in the sky change with the seasons. 2. List three units astronomers use to measure distances to stars. 3. Describe characteristics of stars, including mass, size, temperature, color, and luminosity. 4. Describe variable stars.	Lesson Plans, p. 96 Reading Study Guide, p. 96 Teaching Transparency 33
Section 28.3 Life Cycles of Stars, pp. 626–630	1. Describe the birth of a star. 2. Compare and contrast the life cycle of a star like the sun and of a star more massive than the sun. 3. Describe the remnants of supernovas.	Lesson Plans, p. 97 Reading Study Guide, p. 97 Teaching Transparency 34
Section 28.4 Galaxies and the Universe, pp. 631–635	1. Tell what a galaxy is and describe the various types of galaxies. 2. Explain the origin of the universe according to the big bang model.	Lesson Plans, p. 98 Reading Study Guide, p. 98

INVESTIGATIONS
CLASSZONE.COM

Section 28.2 INVESTIGATION
What Does the Spectrum of a Star Tell Us about Its Temperature?
Computer time: 45 minutes / Additional time: 45 minutes
Students determine the temperature of a star by looking at the light it produces.

Chapter Review INVESTIGATION
What Happens as a Star Runs Out of Hydrogen?
Computer time: 45 minutes / Additional time: 45 minutes
Students use stellar properties to predict the stages of stars' lives.

Technology/Online Resources

Electronic Teacher Tools
Visualizations CD-ROM
Classzone.com
Online Lesson Planner
Test Generator

Visualizations CD-ROM
Classzone.com
  ES2801 Visualization, ES2802 Visualization

Visualizations CD-ROM
Classzone.com
  ES2803 Investigation, ES2804 Career,
  ES2805 Visualization, ES2806 Visualization

Visualizations CD-ROM
Classzone.com
  ES2807 Visualization

Visualizations CD-ROM
Classzone.com
  ES2808 Visualization, ES2809 Visualization

Classroom Management

Gather these materials for minilab and laboratory activities.

Minilab, p. 616
  buzzer
  tape
  ball

Lab Activity, pp. 636–637
  round balloon
  binder clip or paper clip
  fine-tip marker
  graph paper
  piece of string
  metric ruler
  Laboratory Manual, Teacher's Edition
    Lab Sheet 28, p. 150

Chapter 28 Stars and Galaxies **610B**

CHAPTER 28

Stars and Galaxies

Galaxies, including galaxy NGC 4414 shown here, are home to thousands of stars.

What does our knowledge about our solar system and galaxy suggest about the universe?

INTRODUCE
Build Interest
Before sharing background information about the image with students, allow time for their questions and comments.

Ask questions like the following:
- Why is this galaxy so bright? It is made of billions of stars.
- What shape is this galaxy? pinwheel-shaped
- How do we obtain images like this? through large telescopes on Earth and in space

Respond
- What does our knowledge about our solar system and galaxy suggest about the universe? that all galaxies are moving and that stars in those galaxies are in various stages of their life cycles

About the Photograph

The image of spiral galaxy NGC 4414 was produced by the Hubble Space Telescope using three different color filters. The center of the galaxy primarily contains older red and yellow stars. Young blue stars are found in the galaxy's outer spiral arms. The dark patches in the outer arms are clouds of interstellar dust. Based on brightness measurements of variable stars in NGC 4414, scientists have determined that the galaxy is about 60 million light-years from Earth.

BEYOND THE TEXTBOOK

Print Resources
- Reading Study Guide, pp. 95–98
- Transparencies 32, 33, 34
- Formal Assessment, pp. 82–84
- Laboratory Manual, pp. 119–126
- Alternative Assessment, pp. 55–56
- Spanish Vocabulary and Summaries, pp. 55–56
- Internet Investigations Guide, pp. T95–T96, 95–96

Technology Resources
- Visualizations CD-ROM
- Classzone.com
- Online Lesson Planner
- Electronic Teacher Tools
- Test Generator

610 Unit 7 Space

CHAPTER 28

PREVIEW

▶ **FOCUS QUESTIONS** In this chapter you will study stars, galaxies and other objects in the universe and learn more about the key questions listed below.

Section 1 What is the electromagnetic spectrum, and how does it help astronomers learn about stars?

Section 2 What are the characteristics of stars?

Section 3 What are the phases of a star's life cycle?

Section 4 What are galaxies? Where do scientists think the universe came from?

▶ **REVIEW TOPICS** As you investigate stars and galaxies you will need to use information from earlier chapters.

- structure of the sun (p. 565)
- gravitation (p. 572)

▶ **READING STRATEGY**

CONNECT

As you read this chapter, pay attention to the stars visible in the night sky. Try to connect what you observe with the information about the stars and galaxies described in this chapter.

At our Web site, you will find the following Internet support:

DATA CENTER
EARTH NEWS
VISUALIZATIONS
- Exploded Star at Different Wavelengths
- Doppler Effect
- Eclipsing Binary Stars
- Life Stages of Stars
- Milky Way Galaxy
- Regular, Irregular, and Very Peculiar Galaxies

LOCAL RESOURCES
CAREERS
INVESTIGATIONS
- What Does the Spectrum of a Star Tell Us about Its Temperature?
- What Happens as a Star Runs Out of Hydrogen?

Chapter 28 Stars and Galaxies **611**

CHAPTER 28

PREPARE

Focus Questions
Find out what knowledge students have about stars and galaxies. For each focus question, have students consider the section title and then develop two or three additional questions for each section. For example, Section 2: How do we categorize stars? What material makes up stars?

Review Topics
In addition to the earth science topics listed, students may need to review these concepts:

Physical Science
- Light intensity, or magnitude, follows the inverse-square law: magnitude varies in inverse proportion to the square of the distance.

Mathematics
- Large numbers are expressed by using bases and exponents. For example, $10^3 = 10 \times 10 \times 10 = 1,000$.

Reading Strategy
Model the Strategy: To help students connect the observable night sky to the information in this chapter, assign as homework the task of diagramming groups of stars they notice in the sky. Then have them compare their diagrams with those on pages 617 and 714–717. Have students continue to inspect the night sky, and to write connections they make with the content as you continue through the chapter. Depending on the amount of light pollution where they live, students might be able to observe subtle differences in star magnitudes and color, and to observe the Milky Way.

BEYOND THE TEXTBOOK

INVESTIGATIONS
- What Does the Spectrum of a Star Tell Us about Its Temperature?
- What Happens as a Star Runs Out of Hydrogen?

VISUALIZATIONS
- Images of an exploded star at different wavelengths
- Animation of the Doppler effect
- Animation of eclipsing binary stars
- Animation of the life stages of stars
- Views of the Milky Way galaxy
- Images of regular, irregular, and very peculiar galaxies

DATA CENTER
EARTH NEWS
LOCAL RESOURCES
CAREERS

Online Lesson Planner

Chapter 28 Stars and Galaxies **611**

CHAPTER 28 SECTION 1

▶ FOCUS

Objectives
1. Describe the characteristics of electromagnetic radiation.
2. Explain techniques for analyzing light to obtain information about stars.
3. Explain the Doppler effect and how it gives information about star motions.

Set a Purpose
Have students read the section to find the answer to "How do astronomers learn about stars by analyzing starlight?"

▶ INSTRUCT

More about...

Electromagnetic Radiation
Generally speaking, long-wavelength, low-frequency electromagnetic radiation carries less energy than short-wavelength, high-frequency electromagnetic radiation. However, the intensity of electromagnetic radiation is a function of wave amplitude. The larger the wave amplitude, the higher the intensity, and the more energy is carried by the waves. Thus, it is possible, for example, for high-intensity visible light to carry more energy than a low-intensity X-ray.

28.1

KEY IDEA
Astronomers analyze light from objects in space in order to learn about the composition and movement of the objects.

KEY VOCABULARY
- electromagnetic radiation
- electromagnetic spectrum
- continuous spectrum
- emission spectrum
- absorption spectrum

A Closer Look at Light

Light is more than what we get when we turn on a lamp. Light also refers to a form of radiation that stars and other celestial objects emit. Most of what we know about the universe we have learned from analyzing the light that reaches us from distant stars and galaxies.

What Is Light?

Light is a form of **electromagnetic radiation,** which is energy that travels in waves. Although you may not realize it, you are already familiar with some types of electromagnetic radiation. When you listen to music on the radio, the music is being broadcast by means of a type of electromagnetic radiation called radio waves. If you have an x-ray taken, the x-ray machine is using another type of electromagnetic radiation to produce an image of your bones. As you read this page, your eyes are gathering visible light to help you see.

All types of electromagnetic radiation travel in the form of waves, at a speed of about 300,000 kilometers per second—the speed of light. The lengths of these waves determine the characteristics of each form of electromagnetic radiation. The distance from one wave crest to the next is called the *wavelength*. Radio waves have the longest wavelengths—in some cases longer than a soccer field. Gamma rays have the shortest wavelengths.

The various types of electromagnetic radiation can be arranged in a continuum, with the longest wavelength at one end and the shortest at the other. This continuum is called the **electromagnetic spectrum.** Stars such as the sun emit a wide range of wavelengths.

A NATURAL SPECTRUM Visible light, one type of electromagnetic radiation, can be broken up into a spectrum of its component wavelengths. A rainbow is an example of a visible spectrum that occurs naturally when rain or mist refracts sunlight.

612   Unit 7 Space

DIFFERENTIATING INSTRUCTION

Reading Support Have students preview the illustration on page 613 showing the electromagnetic spectrum. Give them an opportunity to discuss what they know about the different types of electromagnetic radiation listed in the illustration, including their uses and any hazards associated with them. Then have students read the text under the heading "What Is Light?" Tell them to focus on the physical characteristics of electromagnetic radiation that are discussed, as they will be important in understanding the rest of the section.

Use Reading Study Guide, p. 95.

The Electromagnetic Spectrum

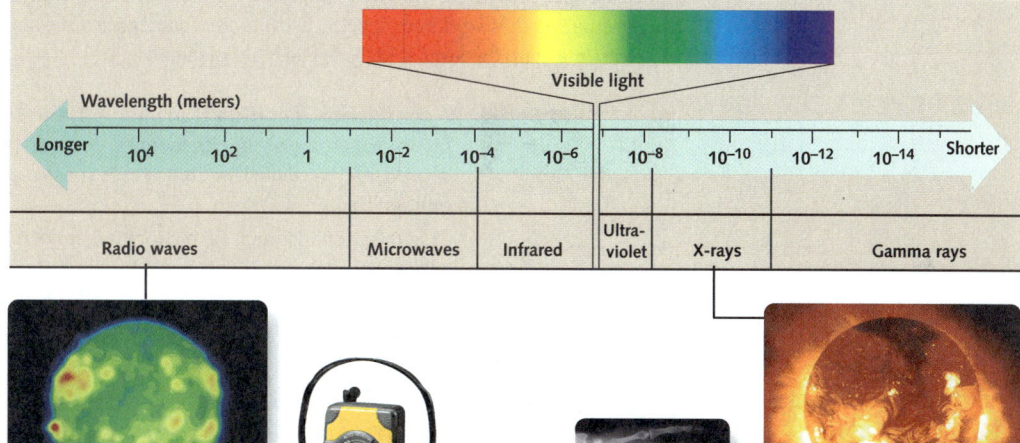

Visible light, represented by the rainbow-like spectrum at top, is just a small part of the electromagnetic spectrum. The sun emits electromagnetic radiation of all types, which can be used to produce different images. Human beings use electromagnetic waves for a variety of purposes such as radio and television broadcasting and x-ray imaging.

Unlike water waves or sound waves, which need a medium such as water or air in which to travel, electromagnetic waves can travel through empty space. Electromagnetic waves emitted by an object provide information about the elements present in the object and about the object's motion.

The Spectroscope

Visible white light is actually made up of light of various colors, each with a different wavelength. These are the colors seen in a rainbow or in the spectrum produced as sunlight passes through a triangular glass prism. Red light has the longest wavelength, and violet has the shortest. When light waves pass from air into a glass prism and out again, they are bent, or refracted, forming a band of colors called the *visible spectrum*. Longer wavelengths are refracted less than shorter wavelengths.

Astronomers use spectra of distant stars to learn more about the stars. To separate starlight into its colors, astronomers use a tool called a *spectroscope*, which uses a prism to split the light gathered by a telescope into a spectrum.

VISUALIZATIONS
CLASSZONE.COM

Observe an exploded star at different wavelengths.
Keycode: ES2801

VOCABULARY STRATEGY

Spectrum is the Latin word for "appearance." It comes from the Latin verb *specere*, "to look at." *Spectra* is the plural form of *spectrum*.

CHAPTER 28 SECTION 1

CLASSZONE.COM

Observe an exploded star at different wavelengths.
Keycode: ES2801

 Visualizations CD-ROM

Visual Teaching

Interpret Diagram
Review the electromagnetic spectrum. Ask: Which types of electromagnetic waves have longer wavelengths than visible light waves? **radio waves, microwaves, infrared waves** shorter? **ultraviolet waves, X-rays, gamma rays** According to the diagram, which color of visible light has the longest wavelength? **red light** shortest wavelength? **violet light**

Extend
Have students research the definition of *frequency* as it relates to waves. **Frequency is the number of crests that pass a given point in a given time.** Ask students to explain the relationship between wavelength and frequency. **As wavelength increases, frequency decreases.** Then refer students to the diagram of the electromagnetic spectrum and have them identify the electromagnetic radiation with the highest frequency. **gamma rays**

Use Transparency 32.

DIFFERENTIATING INSTRUCTION

Challenge Activity Have students investigate how astronomers use parts of the electromagnetic spectrum outside of the visible range to learn about the universe, for example, radio waves. Students should know that radio waves are different from sound waves. Although radio waves are coded to carry sound information for commercial broadcasting, they are, in fact, electromagnetic waves and do not, in themselves, transmit sound. Have students make informational posters and present their findings to the class.

More about...

Spectroscopy
As scientists have gained a better understanding of how atoms absorb and emit radiation, spectroscopy has become a more powerful tool for obtaining information about stars. The types of information scientists can get from analyzing starlight include detailed chemical composition of the star, temperature and pressure of the star's atmosphere, size and luminosity of the star, and radial motion and rotation measurements.

Types of Visible Spectra

Spectroscopes break light into three different types of spectra. By analyzing and comparing these different spectra, astronomers can figure out what elements make up the atmospheres of stars and planets.

A **continuous spectrum** is an unbroken band of colors, which shows that its source is emitting light of all visible wavelengths. Continuous spectra are emitted by
- glowing solids, such as the hot filament of an electric light
- glowing liquids, such as molten iron
- the hot, compressed gases inside stars

An **emission spectrum** is a series of unevenly spaced lines of different colors and brightnesses. The bright lines show that the source is emitting light of only certain wavelengths. Glowing thin gases produce emission spectra. Since every element has a unique emission spectrum, scientists are able to identify the elements in objects by analyzing the spectra of light emitted.

An **absorption spectrum** is a continuous spectrum crossed by dark lines. These lines form when light from a glowing object passes through a cooler gas, which absorbs some of the wavelengths. In fact, the elements in the gas absorb exactly the same wavelengths that they would emit if they were in the form of glowing gases. By comparing emission and absorption spectra, scientists can determine what elements are present in the cooler gas that is absorbing some of the light.

A star's absorption spectrum indicates the composition of the star's outer layer. The sun's spectrum is a good example. The hot, compressed gases of the sun's interior radiate a continuous spectrum. As these electromagnetic waves pass through the sun's cooler outer layers, the photosphere and chromosphere, some of the waves are absorbed. (Review chapter 26.1.) As a result, the sun's spectrum is crossed by many dark lines. By matching these lines with the emission spectra of gaseous elements heated in a laboratory, scientists have identified more than 67 elements in the sun's outer layer.

DIFFERENTIATING INSTRUCTION

Hands-On Demonstration Make a simple spectroscope, using a shoe box with lid and 3-cm-square piece of diffraction grating. (Diffraction grating can be purchased from a science supply store.) Cut out a 2-cm-square hole in the center of one end of the shoe box. Place the diffraction grating over the hole and secure it with tape. In the center of the opposite end of the box, cut out a vertical slit about 2 cm long and 0.5 cm wide. Place the lid over the box. Students can observe the spectra of various light sources by looking through the diffraction grating while pointing the slit at the light. If the spectrum appears to the side of the slit, untape the diffraction grating and rotate it until the spectrum extends from both sides of the slit. Students might view light from an incandescent bulb, fluorescent bulb, candle, neon sign, and street lamp. Tell students to draw the different spectra they observe.

⚠ *To view sunlight, students can point the spectroscope outside but not directly at the sun. Remind students that safety is everyone's responsibility.*

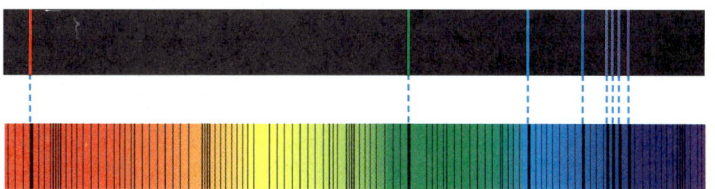

Emission spectrum of hydrogen produced in the laboratory.

Absorption spectrum of a star, showing absorption lines characteristic of hydrogen.

EVIDENCE OF HYDROGEN Scientists can determine what elements are present in a star by comparing the emission spectra of various elements with a star's absorption spectrum. The comparison shown here reveals that hydrogen (top spectrum) is present in the star.

Absorption spectra can also be used to determine the composition of a planet's atmosphere. A planet shines by reflecting sunlight. If the spectrum of a planet contains dark lines that are not found in the sun's spectrum, then these lines must be caused by substances in the planet's atmosphere.

The Doppler Effect

Spectral analysis tells scientists not only what stars are made of but also how they are moving in relation to Earth. Using spectral analysis to determine the movement of stars is possible because of a phenomenon known as the *Doppler effect*.

You may have already experienced the Doppler effect without realizing it. Perhaps one day you were standing on a sidewalk as a fire engine roared by with its siren blaring. Did you notice how the sound of the siren changed as the fire engine moved toward you and then again as it moved away from you. The siren's pitch grew higher as the fire engine approached and then the sound dropped suddenly as the fire engine passed you.

What causes the drop in pitch? As the fire engine approaches, the sound waves from the siren are compressed. When the wavelength decreases—that is, when the distance between wave crests becomes shorter—the sound's pitch becomes higher. As the fire engine moves away from you, the wavelength of the sound increases, so the pitch drops.

Because the Doppler effect applies to light as well as to sound, astronomers can use the Doppler effect to determine whether a star is moving toward Earth or away from it. The Doppler effect does not, however, reveal whether the star is moving across the line of sight.

When the spectra of stars are compared with the emission spectra of elements, the dark lines of the stars' spectra may be shifted to the left or to the right. These shifts are produced by the Doppler effect. They are evidence of the star's motion relative to Earth. If a star is approaching Earth, the wavelengths of light it emits become shorter—just as the sound waves from the fire engine's siren did. By becoming shorter, the light waves shift toward the blue end of the spectrum. Therefore, astronomers call this phenomenon a *blueshift*.

Observe the change in a star's spectrum as its motion changes.
Keycode: ES2802

CHAPTER 28 SECTION 1

MINILAB
Simulate the Doppler Effect

Management
Take the class outside or to the gym to perform this activity. Have students form groups of three, and spread the groups as far apart as possible.

Procedure
- Caution students to throw the ball high enough so as not to hit their partner.
- Tell students to pay attention to the pitch of the buzzer's sound as the ball is thrown back and forth.

Analysis Answers
- The pitch of the buzzer's sound increases as the ball moves toward the observer and decreases as the ball moves away from the observer.
- If the ball was thrown faster, the pitch would increase and decrease more quickly.
- Light moves too quickly to detect wavelength changes.
- The light would blueshift as the ball moved toward the observer and redshift as the ball moved away from the observer.

ASSESS

1. All are forms of electromagnetic radiation.
2. A spectroscope separates visible light into its component colors.
3. Scientists compare the dark lines in the star's absorption spectra with the colored lines in the emission spectra of various elements. If the lines match, those elements are found in the star's atmosphere.
4. Because the objects are moving, the wavelength of the light decreases if the objects are moving toward the observer (blueshift) or increases if the objects are moving away from the observer (redshift).
5. It suggests that the atmosphere of each star has a unique combination of elements.

616 Unit 7 Space

Mini LAB — 15-Minute
Simulate the Doppler Effect

Materials
- buzzer
- ball
- tape

Procedure
1. Tape the buzzer to the ball.
2. Have two people stand about 20 feet apart. Stand between them.
3. Turn the buzzer on, and have them toss the ball back and forth over your head.

Analysis
Describe what you hear. What would you expect to hear if the ball were thrown faster? slower? Why can't you do this experiment with light instead of sound? If you could somehow do this experiment with light, what would you expect to see?

Redshift and Blueshift

The motion of a star away from Earth causes the star's spectral lines to shift toward the red end of the spectrum. In the second spectrum below, the hydrogen lines have moved left, indicating a redshift. When a star is moving toward Earth, its spectral lines shift toward the blue end of the spectrum, as shown in the third spectrum below, where the lines of the hydrogen emission spectrum have moved right, indicating a blue shift.

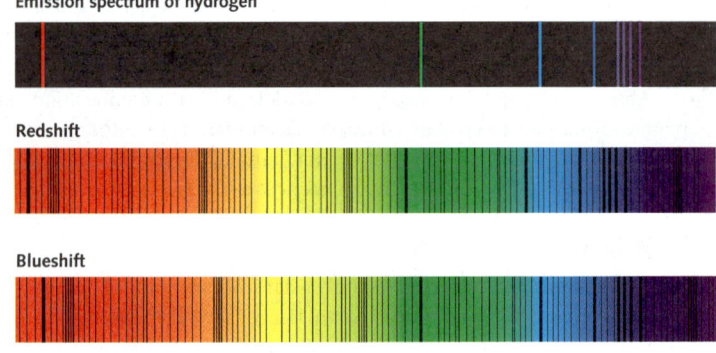

Emission spectrum of hydrogen

Redshift

Blueshift

Similarly, if a star is moving away from Earth, the wavelengths of light increase, just as the sound waves of the siren did after the fire engine passed you. All the star's spectral lines shift toward the red end of the spectrum, where wavelengths are longer. Astronomers call this phenomenon a *redshift*.

From the blueshift or the redshift of a star's spectrum, astronomers can also determine how fast a star is moving toward or away from Earth. Dividing a spectral lines' shift in wavelength by the wavelength determined in the laboratory, gives astronomers the ratio of the star's velocity to the speed of light. For example, if a star has a redshift of 0.001, it is moving away from Earth at one-thousandth the speed of light, or about 300 kilometers per second.

28.1 Section Review

1. What do radio waves, visible light, and gamma rays all have in common?
2. What does a spectroscope do?
3. How do scientists use different spectra to figure out the composition of a star's outer layer?
4. Why is the light reaching us from some celestial objects shifted toward the red or blue end of the spectrum?
5. **CRITICAL THINKING** Every star has a unique spectrum. What does this suggest about stars?

616 Unit 7 Space

MONITOR AND RETEACH

If students miss . . .

Question 1 Refer to the diagram on page 613.
Question 2 Demonstrate that a prism, which is found in some spectroscopes, separates white light into its component colors.
Question 3 Reteach "Types of Visible Spectra." (pp. 614–615) Have students explain how each type of spectrum is produced.
Question 4 Use a spring to model the compression and stretching of waves as the wave source moves toward and away from the observer.
Question 5 Refer students to the absorption spectrum on page 615. Explain that the black lines were produced because different elements in the star's atmosphere absorbed different wavelengths of light.

Stars and Their Characteristics

The sun is just one star among many. Depending on where you live, if you look up at the sky on a clear night, you may see hundreds of stars. The light from those stars has traveled great distances to get to Earth. For example, the light from Proxima Centauri, the star nearest the sun, left that star more than four years ago. Light from other stars and celestial objects may have taken thousands, millions, or even billions of years to get here.

Early Observations

Although some people consider astronomy to be among the most modern of sciences because astronomers use many high-tech instruments, watching the stars is one of the oldest of human pursuits. Some of the observations made by ancient astronomers are still used today.

Constellations

When ancient peoples looked at the night sky, they saw the stars in much the same way we do. The names they used to describe groups of stars referred to mythic heroes (*Hercules*), animals (*Leo*, the lion; *Taurus*, the bull), monsters (*Draco*, the dragon; *Centaurus*, the centaur), and familiar objects (*Crater*, the goblet; *Lyra*, the lyre or harp). From ancient tablets found in the Euphrates River valley, we know that some of these figures were familiar as long ago as 2450 B.C.E.

Some of the names ancient astronomers gave to groups of stars, called **constellations,** are still being used today. Other constellations, such as Telescopium (the telescope) and Microscopium (the microscope) were conceived only in the last few centuries.

28.2

KEY IDEA
Stars differ from one another in mass, size, surface temperature, and distance from Earth.

KEY VOCABULARY
- constellation
- apparent magnitude
- astronomical unit
- light-year
- parallax
- parsec
- luminosity
- absolute magnitude
- Cepheid variable

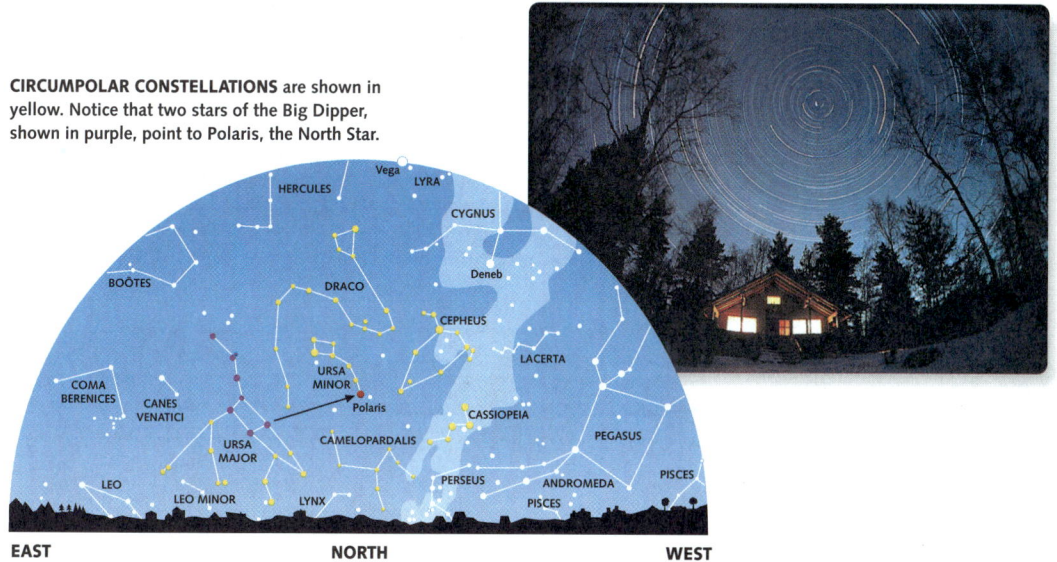

CIRCUMPOLAR CONSTELLATIONS are shown in yellow. Notice that two stars of the Big Dipper, shown in purple, point to Polaris, the North Star.

CIRCUMPOLAR STARS make circular trails around Polaris in this time-exposure photograph.

CHAPTER 28 SECTION 2

Visual Teaching

Interpret Diagram
The diagram might give students the impression that the sun is blocking or "eclipsing" Earth's view of certain constellations. Point out that the diagram is not drawn to scale and, in fact, the reason some constellations are not visible from Earth during certain seasons is that at those times the constellations rise in the sky during the daytime.

Extend
Tell students that during a total solar eclipse, the sky turns dark and stars appear. Ask: Which constellation in the diagram would appear during a total solar eclipse that occurred in June? **Orion**

Use Transparency 33.

Eighty-eight constellations can be seen from Earth's northern and southern hemispheres. Remember that constellations are not natural groupings, like solar systems and galaxies; they are human inventions. The stars in most constellations appear to be together only as they are viewed from Earth. They are actually at widely varying distances from Earth and are moving in relation to one another. However, because stars are so far away, it takes thousands of years before their motions alter the pattern of a constellation.

The Big Dipper—probably the best-known *asterism,* or small star grouping—is part of a large constellation known as *Ursa Major,* or the *Great Bear.* The dipper can be used to find other constellations. Imagine a line drawn through the two stars farthest from the dipper's hand. The line through these "pointer stars" leads to the last star in the handle of the Little Dipper (part of *Ursa Minor,* or the *Little Bear*). This is *Polaris,* or the North Star. At the same distance on the opposite side of the Little Dipper is a large, lopsided W—the constellation Cassiopeia (KAS-ee-uh-PEE-uh).

The apparent regular movement of the constellations across our sky is caused by Earth's motions—its rotation and revolution. Because Earth turns from west to east, the whole sky appears to turn from east to west. That is why the sun, moon, and stars are said to rise in the east and set in the west. The sections of sky directly above Earth's poles, however, seem stationary as Earth turns on its axis. Thus Polaris, the North Star, seems fixed in the sky while the stars nearby move counterclockwise around it. They are called *circumpolar stars.* Ursa Major, Ursa Minor, and Cassiopeia are examples of constellations comprising circumpolar stars. Viewed from some northern latitudes, these constellations never set below the horizon and can be seen all year long.

A constellation's position in the sky changes with the seasons. Viewed on a fall evening, the Big Dipper can be seen near the northern horizon. On a spring evening, it is high overhead. Cassiopeia is nearly straight overhead in the fall but is just above the northern horizon in the spring. These changes are caused by Earth's changes in position as it orbits the sun.

Seasonal Changes of Constellations

In the Northern Hemisphere's summer, the night side of Earth faces the part of the sky containing Lyra, while the daytime half of Earth faces Orion. As a result, Lyra can be seen at night in the summer, while Orion is lost in sunlight.

Six months later, Earth has moved about halfway around its orbit of the sun. So in the Northern Hemisphere's winter, the daytime side of Earth faces the part of the sky containing Lyra, while Orion is visible at night.

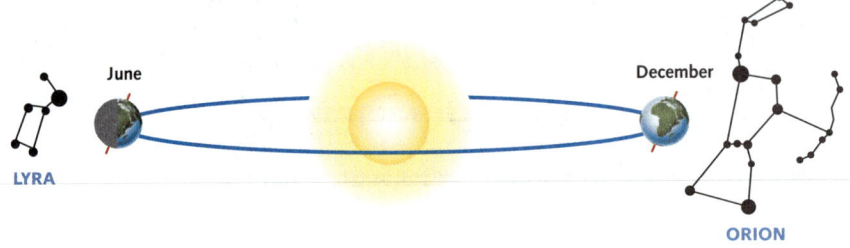

DIFFERENTIATING INSTRUCTION

Hands-On Demonstration Demonstrate the seasonality of constellations. Place a basketball on a table to represent the sun. Hold a golf ball representing Earth. Give one student a sign reading "LYRA" and another a sign reading "ORION." Have those students stand facing the basketball on opposite sides, holding their signs in front of them. Orbit the golf ball around the "sun" to show students how daytime light obliterates views of the out-of-season constellation. Have students determine where in the model the circumpolar constellations would be located. **above the basketball**

Some constellations can be seen only in certain seasons. For example, in the Northern Hemisphere, the best view of Lyra is in the summertime; Orion the Hunter, on the other hand, can be seen best in the winter. Such seasonal differences in viewable constellations are also due to Earth's movement around the sun. (See the star maps on pages 714–717.)

Apparent Magnitude

Besides noting the different patterns the stars seemed to form, ancient astronomers also noticed differences among the stars. Some stars seemed bright, others faint, and others fainter still.

Around 120 B.C.E., the Greek astronomer Hipparchus devised a system of classifying stars by how bright they looked. He rated the brightness of 850 stars on a scale of 1 to 6. Stars with a 1 rating were the brightest he could see, and those with a 6 rating were the faintest. Years later, the Alexandrian astronomer Ptolemy expanded Hipparchus's scale to include more than 1000 stars. Each value on the scale is the apparent magnitude.

The **apparent magnitude** of a star is a measure of how bright a star appears to be to an observer on Earth. The lower a star's magnitude number, the brighter the star is. Some of the brighter stars in the sky are classified as *first-magnitude stars*. The faintest stars that can be seen with the unaided eye are called *sixth-magnitude stars*. Some stars are even brighter than first-magnitude stars; they have magnitudes less than one. A few of the brightest stars even have negative magnitudes. For example, the apparent magnitude of Sirius is about -1.45; and, because it is so close to Earth, the sun's apparent magnitude is -26.7.

In the modern magnitude system, each magnitude differs from the next by a factor of about 2.5. A first-magnitude star is 2.5 times brighter than a second-magnitude star; a second-magnitude star is 2.5 times brighter than a third-magnitude star, and so on. A first-magnitude star is by definition 100 times brighter than a sixth-magnitude star. With telescopes, astronomers can see stars that are far dimmer than those Hipparchus observed, stars well beyond the twentieth magnitude.

Distances to Stars

What units of measure would you use to determine the length and width of your classroom? Feet, perhaps, or meters. You probably wouldn't use millimeters, because they are too small to be useful in this case. The problem astronomers face in expressing distance in space is that the units useful here on Earth, such as kilometers, are too small for measuring large distances; the numbers become so large they are unwieldy. As a result, astronomers have devised special units to use for measuring the vast distances through space.

The closest star to Earth is, of course, the sun. The average distance between Earth and the sun is about 150 million kilometers. This distance is called one **astronomical unit,** or AU. How far away is the next star? Suppose that Earth and the sun are dots one centimeter apart. On that scale, the next nearest star, Proxima Centauri, would be more than 2.5 kilometers away. In fact, Proxima Centauri is about 40 trillion kilometers from the sun, or more than 260,000 AU. Astronomers use AUs to express distances within the solar system. Jupiter, for example, is 5.2 AUs from the sun.

ORION Betelgeuse and Rigel, both in the constellation Orion, are among the stars with the lowest apparent magnitudes.

Scientific Thinking

USE MATHEMATICS

How far away in Astronomical Units (AU) would the following stars be?

Sirius (8.6 light-years)

Vega (26 light-years)

Aldebaran (65 light-years)

Regulus (77 light-years)

Deneb (1400 light-years)

CHAPTER 28 SECTION 2

VISUAL TEACHING
Discussion
Refer students to the diagram illustrating parallax. Have them explain why the diameter of Earth's orbit is 2 AUs. **Earth's distance from the sun, 1 AU, is the radius of Earth's near-circular orbit. The diameter of a circle is equal to two times its radius.** Ask: How would the angle in the diagram change if the distance of the star from Earth decreased? **The angle would increase.**

As you can see, neither kilometers nor astronomical units are satisfactory for expressing the great distances in space. Instead, astronomers use two other units: light-years and parsecs. A **light-year,** despite its name, is a unit that measures the distance that a ray of light travels in one year. Light travels about 300,000 kilometers per second. Thus, in one year, light travels about 9.5 trillion (9.5×10^{12}) kilometers. Proxima Centauri is about 4.2 light-years away from Earth.

One way of measuring the distance to the nearest stars is based on parallax. **Parallax** is a change in an object's direction due to a change in the observer's position. To experience parallax, hold your thumb out at arm's length. Close one eye, and note your thumb's position. Then open your eye, close the other, and look again; the position of your thumb will have shifted slightly in relation to background objects.

Because Earth orbits around the sun, astronomers experience parallax when they observe the stars. At different times of the year, for example, a nearby star does not seem to be in exactly the same position against the backdrop of distant stars. Astronomers can calculate the distance to a star, however, by knowing the angle between two observed positions and the distance between the observation points. (See the diagram below.) To express such measurements, astronomers use a special unit of distance, the **parsec**—short for "parallax second." A parsec is equal to 3.258 light-years, or 3.086×10^{13} kilometers.

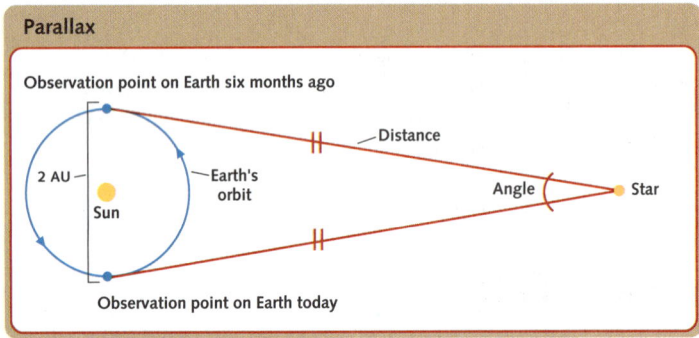

Elements in Stars

A star is a sphere of super-hot gases—mostly hydrogen and helium, although one or two percent of a star's mass may consist of heavier elements, such as oxygen, carbon, nitrogen, and sodium. At its surface, for example, the sun is about 69 percent hydrogen and 29 percent helium, while heavier elements make up the remaining 2 percent.

Although all stars consist mostly of hydrogen and helium, no two stars contain exactly the same elements in the same proportions. Astronomers use spectral analysis to determine a star's makeup. The wavelengths of light that a star radiates depend on both its composition and its temperature, two qualities that are different in every star. As a result, each star has a spectrum that is as unique as your fingerprints.

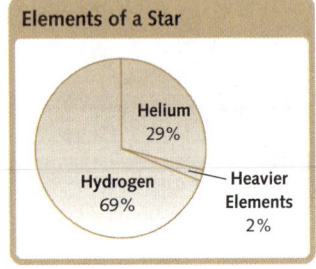

DIFFERENTIATING INSTRUCTION

Developing English Proficiency Review with students the meanings of the words *million, billion,* and *trillion.* Write the words on the board with the equivalent numeral next to each word: million = 1,000,000 or 1×10^6; billion = 1,000,000,000 or 1×10^9; trillion = 1,000,000,000,000 or 1×10^{12}. Challenge students to use each word in a sentence, using the information they read about distances to stars as well as other topics with which they are familiar. **Example sentences: Proxima Centauri is about 40 trillion kilometers from the sun. About 6 billion people live on Earth.**

Use Spanish Vocabulary and Summaries, pp. 55–56.

Mass, Size, and Temperature of Stars

Stars vary greatly in their masses, sizes, and densities. Unlike a star's position, a star's mass is something that we cannot observe directly. We can only calculate what its mass might be on the basis of other observations. Mass, you may recall, is a measure of the total amount of material in a body. It can be determined either by the inertial properties of the body or by its gravitational influence on other bodies. A star with a great mass will have a strong gravitational effect on the bodies around it. The gravitational effect of a star with lesser mass will be weaker.

Stellar masses are expressed as multiples of the mass of the sun, which is called *one solar mass*. Some stars are five, ten, or more times more massive than the sun. Other are less massive than the sun, perhaps with only one-fifth or one-tenth the sun's mass.

Stars vary more in size than they do in mass. The smallest stars are smaller than Earth. The largest star known has a diameter more than 2000 times that of the sun. Stars differ even more in density. Betelgeuse is about one ten-millionth as dense as the sun. One star near Sirius is so dense that one teaspoonful of it would weigh more than a ton on Earth.

Comparison of Star Sizes

Star	Type of Star	Mass (solar masses)	Radius (solar radii)
Sirius	main sequence	2.3	2.5
Rigel	blue supergiant	20	36
Betelgeuse	red supergiant	20	1000
Aldebaran	red giant	5	20
Deneb	yellow supergiant	14	60
Capella	red giant	3.5	13
Pollux	red giant	4	8
Altair	main sequence	2	1.5

1 solar mass ≈ 1×10^{30} kg ≈ 330,000 Earth masses

Temperature and Color of Stars

In addition to varying in mass, size, and density, stars vary in temperature. The range of colors a star emits depends on its surface temperature. The effects of temperature on color can be seen when iron is heated. As iron starts to heat up, it first glows red; as it gets hotter, its color changes from red to orange to yellow to white. Very hot objects glow with a bluish light.

As you can tell by looking at the night sky, stars are mostly whitish; yet they are tinged with faint colors. Cool stars are redder in color—for example, Betelgeuse, with a surface temperature of 3000°C, is slightly reddish white but astronomers call it red. The sun, with a surface

What Does the Spectrum of a Star Tell Us about Its Temperature?
Determine the temperature of a star by looking at the light it produces.
Keycode: ES2803

CHAPTER 28 SECTION 2

What Does the Spectrum of a Star Tell Us About Its Temperature?
Keycode: ES2803

Have students work in pairs.

 Use Internet Investigations Guide, pp. T95, 95.

More about...

Mass of Galaxies
While astronomers can calculate the masses of stars, measurements of the masses of galaxies have proved more difficult than expected. Measurements of redshift and blueshift, along with rotational curves in galaxies, have shown that galaxies rotate at speeds faster than their observable mass would indicate. Thus, astronomers speculate that up to 95 percent of the mass of a galaxy like the Milky Way is invisible to astronomers. Dubbed dark matter, this missing mass is so far unidentified and continues to perplex astronomers.

DIFFERENTIATING INSTRUCTION

Challenge Activity Give students construction paper, scissors, metersticks, compasses, string, and pushpins. Have them create scale models showing the relative radii of the stars listed in the chart on page 621. Caution students to consider carefully the size they choose to equal one solar radius, keeping in mind the size of Betelgeuse. As an extension, have students calculate the ratio of mass to radius for each star and to infer the stars' relative densities. Challenge them to explain density differences.

CHAPTER 28 SECTION 2

Stellar Color and Temperature

Surface Temperature (°C)	Color	Prominent Elements in Spectrum	Spectral Class
Above 30,000	bluish white	ionized helium	O
9500–30,000	bluish white	neutral helium	B
7000–9500	white	metals, hydrogen	A
6000–7000	yellow white	metals, hydrogen	F
5200–6000	yellow	metals, hydrogen	G
3900–5200	yellow to orange	metals, hydrogen	K
Below 3900	red	titanium oxide	M

More about...

Spectral Classes
Spectral classes were first established in 1863 by the Jesuit astronomer Angelo Secchi. Today, astronomers recognize the seven spectral classes shown in the table. They have learned that the heat in a star's outer layer has a greater effect on its spectrum than the elements it contains. The elements in stars have a variety of stages of ionization, and each element has a characteristic temperature at which it best produces absorption lines. Knowing these characteristic temperatures, astronomers can then analyze absorption spectra patterns to learn a star's temperature.

temperature of about 5500°C, is white with a tinge of yellow. The sun appears very yellow to us here on Earth because our atmosphere scatters blue light, thus causing the sun to appear more yellow than it is. Blue-hot stars, such as Sirius, are tinted slightly bluish white. Astronomers group stars by temperature and color into spectral classes. This system, called the *Harvard Spectral Classification Scheme*, was devised by Annie Jump Cannon in the 1920s. It has since been expanded to include other stars.

CAREER

Science Writer

The origin of the universe, medical marvels, deep-ocean wonders, and scientific discoveries believed impossible just a few years ago—all these topics and more are within the realm of science writers. Science writers bring news of the latest technologies and scientific discoveries to a mainstream audience in a clear and enjoyable nonfiction style.

There are no "typical" science writers. They may report for newspapers and magazines (both specialized and general-interest publications) or work for research institutions doing public relations work or writing grants. They may write books and educational materials. Some work in offices; others are freelancers working at home. Science writers find initiative, resourcefulness, and research skills to be their greatest assets.

Science writers have diverse backgrounds and qualifications, although most find an undergraduate or a graduate degree in a scientific field helpful. Regardless of their educational backgrounds, they find that effective writing skills are essential. Many writers hone their writing through experience. Some take part in special science-writing programs at prestigious academic research institutions to increase their knowledge of science. Through careful investigation and detailed crafting of words, science writers convey complex scientific topics and news in a way the general public can understand and appreciate. ∎

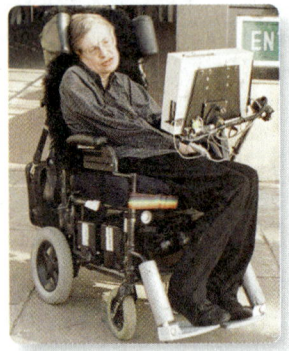

STEVEN HAWKING is a British theoretical physicist who actively teaches, researches, and writes. He has written volumes of research papers and has published best-selling books.

Learn more about becoming a science writer.
Keycode: ES2804

622 Unit 7 Space

DIFFERENTIATING INSTRUCTION

Challenge Activity Have students research the spectral class to which each of the following stars belongs: Arcturus, Aldebaran, Antares, Betelgeuse, Capella, Canopus, Procyon, Rigel, Spica, the sun, and 10 Lacertae. Arcturus, Aldebaran—Class K; Betelgeuse, Antares—Class M; the sun, Capella—Class G; Canopus, Procyon—Class F; Rigel, Spica—Class B; 10 Lacertae—Class O Have them find out more about several of these stars, such as the constellations to which they belong, distance from Earth, and their mass and size.

Luminosity and Absolute Magnitude

The actual brightness of a star is its **luminosity.** A star's luminosity depends only on its size and temperature; distance from Earth is not a factor. If two stars had the same surface temperature, the larger star would be more luminous. If two stars were the same size but had different temperatures, the hotter star would be more luminous.

Remember that a star's apparent magnitude is a measure of how bright the star appears to an observer on Earth. The measure does not indicate how bright the star actually is, since it depends on the star's distance from us as well as it luminosity. For example, a 100-watt light bulb has greater luminosity than a flashlight bulb. If both are viewed at the same distance, the greater brightness of the 100-watt bulb is obvious. But what if the flashlight was one foot away, and the 100-watt light bulb, one half a mile away? Under those conditions, the flashlight would look brighter than the 100-watt light bulb; that is, its apparent magnitude would be greater.

So how do astronomers express the true brightness, or luminosity, of a star? They use a scale of absolute magnitude. A star's **absolute magnitude** is a measure of how bright the star would be if all stars were at the same distance—ten parsecs—from Earth.

Apparent Magnitude vs. Absolute Magnitude

How do apparent magnitude and absolute magnitude compare? Here are the apparent magnitudes and absolute magnitudes of some of the brightest stars.

Star	Apparent Magnitude	Absolute Magnitude
Aldebaran	+0.87	−0.65
Algol	+2.09	−0.15
Antares	+1.06	−5.38
Betelgeuse	+0.45	−5.09
Capella	+0.08	−0.48
Polaris	+1.97	−3.59
Procyon	+0.41	+2.62
Rigel	+0.18	−6.75
Sirius	−1.44	+1.42
Spica	+0.98	−3.55

Variable Stars

Most stars shine with a steady brightness. Some, however, show a regular variation of brightness over cycles that last from days to years. These stars are called *variable stars*. There are several different kinds of variable stars.

Stars that change in brightness as they expand and contract are called *pulsating stars*. When they contract, they become hotter and brighter.

CHAPTER 28 SECTION 2

CLASSZONE.COM

Learn more about becoming a science writer.
Keycode: ES2804

VISUAL TEACHING
Discussion
Tell students that the brightest star in the sky as seen from Earth, besides the sun, is listed in the chart on page 623. Which is it? **Sirius with an apparent magnitude of −1.44** Remind students that the brighter the star is, the lower the apparent magnitude and that the apparent magnitude of the sun is −26.7. Ask: Why is the sun's apparent magnitude so different from that of other stars? **It is much closer to Earth.** Point out that the sun's absolute magnitude, however, is +4.8.

More about...
Comparisons of Magnitude
Astronomers use 10 parsecs as the standard distance to determine absolute magnitude. If a star farther than 10 parsecs away were moved to 10 parsecs, it would appear brighter than it really is. Its apparent magnitude would be greater than its absolute magnitude. A star closer than 10 parsecs, once "moved" to 10 parsecs, would be fainter. So its apparent magnitude would be lower than its absolute magnitude.

DIFFERENTIATING INSTRUCTION

Hands-On Demonstration Use a plain white bedsheet and a few flashlights of different brightnesses to demonstrate the difference between apparent magnitude and absolute magnitude. Have two students spread out the sheet by holding it up at two of its corners. Position the flashlights at various distances from behind the sheet, and darken the room. Ask students looking at the lights from the opposite side of the sheet to identify the brightest and dimmest lights. Then line up all the flashlights the same distance from the sheet, and ask students to identify once again the brightest and dimmest lights. Ask: When did you compare the apparent magnitudes of the lights? **when the flashlights were different distances from the sheet** When did you compare the absolute magnitudes of the lights? **when the flashlights were the same distance from the sheet**

CHAPTER 28 SECTION 2

Examine eclipsing binary stars from several perspectives.
Keycode: ES2805

 Visualizations CD-ROM

ASSESS

1. Galaxies are natural groupings of stars in space, whereas constellations are not. A constellation is a group of stars that appear to be together as viewed from Earth.

2. A light-year is the distance a ray of light travels in one year, equal to 9.5×10^{12} kilometers. A parsec equals 3.258 light-years, or 3.086×10^{13} kilometers.

3. second row: Temperature, Composition; third row: Luminosity

4. The coolest stars are red. As star surface temperature increases, star color changes to orange, then yellow, and then white. The hottest stars are bluish-white.

5. The sun would be more luminous in both cases because a star's luminosity depends on its temperature and size.

6. Stars with apparent magnitudes differing by 5 differ in brightness by a factor of 100. Thus, a star that has an apparent magnitude that is larger by 10 (11.5–1.5) is 10,000 times dimmer.

When they expand, they become cooler and dimmer. One important class of pulsating stars is known as Cepheid (SEE-fee-ihd) variables. **Cepheid variables** are yellow supergiants whose cycles of brightness range from about 1 day to 50 days. Most have cycles of about 5 days.

Astronomers have found that the absolute magnitude of a Cepheid is related to the length of time between its periods of maximum brightness. The slower the cycle, the greater the luminosity of the star. Astronomers have worked out the absolute magnitudes of Cepheids having cycles of many different lengths. By comparing a Cepheid's apparent and absolute magnitudes, astronomers can determine the distance from Earth to the star. In this way, astronomers can calculate the distances to galaxies in which they can identify Cepheid stars.

A nonpulsating "star" may change in brightness because it is, in fact, not one star but two or more stars. Unlike the sun, most stars are parts of systems in which two or more stars revolve around each other. Such star systems are called *eclipsing binaries*. If the orbits of stars in such a system are aligned so that, from our perspective, one star eclipses (passes in front of) another star, then, as seen from Earth, the star system dims at regular intervals. One eclipsing binary is the second-magnitude star Algol, in which a bright star is eclipsed by a dim companion. Algol's brightness appears to decrease by about one-third every 2.9 days.

Examine eclipsing binary stars from several perspectives.
Keycode: ES2805

28.2 Section Review

1. How do constellations differ from other groupings of stars, such as galaxies?

2. What is a light-year? a parsec?

3. Copy and complete the concept map below:

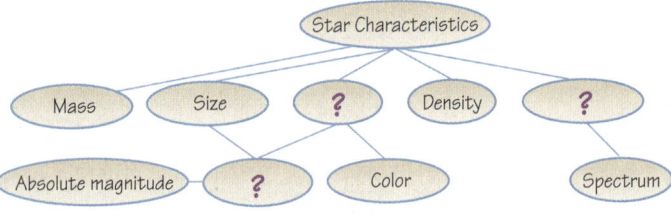

4. What is the connection between a star's surface temperature and its color?

5. **CRITICAL THINKING** Would the sun be more luminous if it were the same size but hotter? if it had the same surface temperature but were more massive? Explain.

6. **MATHEMATICS** How many times brighter than a star with an apparent magnitude of 11.5 is a star with an apparent magnitude of 1.5?

MONITOR AND RETEACH

If students miss . . .

Question 1 Reteach "Constellations." (pp. 617–618) Emphasize that constellations are products of people's imaginations.

Question 2 Reteach "Distances to Stars." (pp. 619–620) Have students make a chart that lists the different units astronomers use to measure distances in space, giving the length of each unit in kilometers.

Question 3 Reteach "Mass, Size, and Temperature of Stars." (pp. 621–622)

Question 4 Point out in the chart on page 622 the relationship between star temperature and color.

SCIENCE & Technology

Telescopes: Windows to the Universe

Although astronomers use many tools, the telescope has increased our understanding of the cosmos the most. As we look through our telescopes, what will we find next?

RADIO TELESCOPES in an array, such as this one in New Mexico, function as one large telescope.

About 400 years ago, the science of astronomy changed profoundly and permanently when Galileo turned a telescope to the moon. Although we think of the telescope as primarily a tool of astronomy, at the time of its invention it was intended to be used at sea.

Galileo, and later Christiaan Huygens, refined the original plan by using lenses that were better made. With better lenses, observers could see better. The earliest telescopes, and many used today, are of a kind called a refracting telescope. Refracting telescopes use lenses both to focus light and to magnify an image for viewers.

In a reflecting telescope, a mirror, rather than a lens, is used to focus light. Usually, a second mirror reflects the light to the lenses of the eyepiece. A reflecting telescope can use an array of smaller mirrors in place of one large objective mirror; this type is called a multiple-mirror telescope.

Today, astronomers observe not only the visible light emitted by distant objects, but also detect emissions throughout the electromagnetic spectrum. For example, astronomers gather valuable information with radio telescopes. Resembling large satellite dishes, these telescopes gather radio waves from distant objects. These telescopes are often used in arrays—arrangements of multiple telescopes that act as one large telescope.

Even after four centuries, telescopes are still opening our eyes to the cosmos. Recent images from the Hubble Space Telescope are allowing us to see more deeply into space than we ever have before. ■

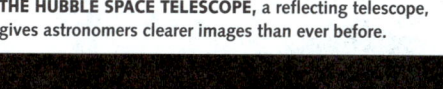

THE HUBBLE SPACE TELESCOPE, a reflecting telescope, gives astronomers clearer images than ever before.

Extension

SCIENCE NOTEBOOK
Telescopes have been critical to our ability to view and learn about our universe. In what ways do you think our perception of the universe would differ had the telescope never been invented?

Discover more about telescopes in space.
Keycode: ES2806

CHAPTER 28 SECTION 2

Discover more about telescopes in space.
Keycode: ES2806

Have students create charts to organize the information they find.

Science and Technology

Refracting vs. Reflecting
Galileo quickly discovered the greatest disadvantage of the refracting telescope: the distortion of colors, to such an extent that stars appeared to be surrounded by rainbow coloring. This color effect has been reduced in today's refracting telescopes, but most astronomers still prefer using reflecting telescopes, which Sir Isaac Newton originally developed. Reflecting telescopes offer the advantages of sharp, high-contrast images, which planet watchers especially appreciate.

Science Notebook
Distant galaxies would be unknown to us, we would have no understanding of stellar evolution, and scientists would not have discovered that the universe is expanding.

MONITOR AND RETEACH

Question 5 Reteach "Luminosity and Absolute Magnitude." (p. 623) Ask: What does a star's luminosity depend on? **the star's size and temperature**

Question 6 Reteach "Apparent Magnitude." (p. 619) Emphasize that increases in magnitude are multiplicative rather than additive. Thus, the difference in brightness between a first-magnitude star and a third-magnitude star is $1 \times 2.5 \times 2.5$, not $1 + 2.5 + 2.5$. Students can use the $[y^x]$ button on a calculator to determine the difference in brightness of two stars based on the stars' apparent magnitudes.

CHAPTER 28 SECTION 3

FOCUS

Objectives
1. Describe the birth of a star.
2. Compare and contrast the life cycle of a star like the sun and of a star more massive than the sun.
3. Describe the remnants of supernovae.

Set a Purpose
Read aloud the key vocabulary terms. Then have students read to answer "What is the life cycle of a star?"

INSTRUCT

VISUAL TEACHING
Discussion
Have students look closely at each axis of the Hertzsprung-Russell diagram. Ask: What is the relationship between the luminosity and temperature of main-sequence stars? *As temperature decreases, luminosity decreases.* Is this true for stars that are not in the main sequence? *no* What is the relationship between the color and temperature of main-sequence stars? *As temperature decreases, color changes from blue, to white, to orange, to red.* Is this true for stars that are not in the main sequence? *yes* Have students describe the stars above and below the main sequence. *Stars above are giants and supergiants, which are more luminous than the main-sequence stars because of their greater size. Stars below are small, dim stars that are near the end of their lives.*
Use Transparency 34.

28.3

KEY IDEA
Stars are born, and they mature, grow old, and die; their lifespan and final form depend on their masses.

KEY VOCABULARY
- main sequence
- giant star
- supergiants
- white dwarfs
- nebula
- planetary nebula
- supernova
- neutron star
- pulsar
- black hole

MAP OF STARS' LIVES The H-R diagram, a plot of luminosity against temperature, shows the stages of stars' lives.

Life Cycles of Stars

Stars are born from great clouds of gas and dust. They mature, grow old, and die. When they die, they may produce new clouds of dust, from which may arise new stars, along with planets that will orbit them. The more massive a star is, the shorter its life will be.

The Hertzsprung-Russell Diagram

The stars in the universe are at different stages in their life cycles. Some are young and hot; others are older and colder. You can get a picture of a star's life from studying the Hertzsprung-Russell (H-R) diagram. The astronomers Ejnar Hertzsprung of Denmark and Henry Norris Russell of the United States developed this diagram independently of each other in the early 20th century. The diagram plots the luminosity of stars against their surface temperatures.

Most stars fall into distinct groups in the H-R diagram because the groups represent stages in the life cycles of the stars. The majority of stars—about 90 percent—are in a band that runs from the upper left of the diagram (high luminosity, high surface temperature) to the lower right (low luminosity, low surface temperature). This band is called the **main sequence,** and the stars in it are called main-sequence stars.

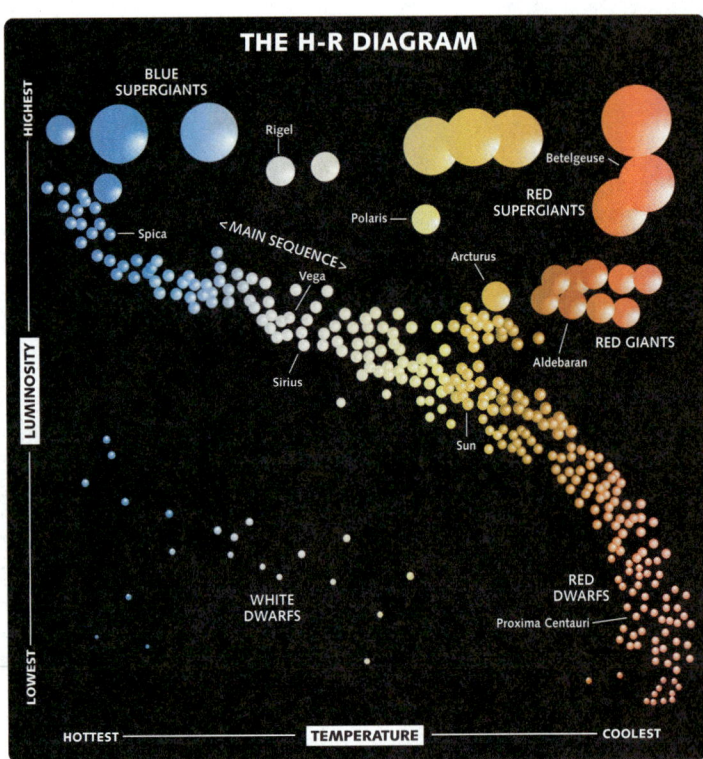

DIFFERENTIATING INSTRUCTION

Reading Support Help students become critical readers. Throughout this section, stars are referred to as having "life" cycles and undergoing "births" and "deaths." Read aloud the first paragraph of the section. Ask students if they can identify phrases that are not intended to be taken literally. *Stars are born . . . They mature, grow old, and die. The more massive a star, the shorter its life will be.* Ask them for ideas on why the paragraph was written in this way. *The writer used a metaphor—likening star cycles to the familiar life cycle of animals—as a way to make the concepts easier to understand.*
Use Reading Study Guide, p. 97.

Main-sequence stars vary in surface temperature and absolute magnitudes. Some are about the same size as the sun, some have diameters about ten times that of the sun, and some have less than one-tenth the sun's. Some main-sequence stars burn for just a million years; other may be able to last billions of years. So what do they have in common? They are actively fusing hydrogen into helium.

As you may recall, larger stars tend to be more luminous. Above the main-sequence stars in the H-R diagram, and therefore of greater luminosity, are the **giant stars** that have diameters from 10 to 100 times greater than that of the sun.

Even more luminous than the giant stars are the **supergiants,** stars with diameters more than 100 times that of the sun. Because of their great size, these relatively cool stars (with surface temperatures between 2500 and 4000°C) are very luminous. Betelgeuse, in the constellation Orion, is a supergiant.

The H-R diagram also includes **white dwarfs,** stars that are near the end of their lives. Such stars were once red giant stars. But the red giant stars lost their outer atmosphere, and now are only the glowing stellar core.

Birth of a Star

A star begins its life in a cloud of gas and dust called a **nebula.** Usually, about 99 percent of a nebula consists of gas, most of which is hydrogen. The remaining 1 percent is a kind of dust made of very tiny grains, with diameters of about one ten-thousandth of a centimeter or less.

A nebula may begin to condense when an outside force, such as a shock wave, acts upon it. The force compresses regions of the nebula, where particles of gas and dust begin to move closer together under the influence of gravity. As these regions become denser, their temperature increases. If the nebula is large enough, parts of it will start to glow. Such large glowing areas are called protostars. As their contraction continues, the protostars become hotter and brighter. Eventually, a protostar's center may become so hot that a fusion reaction begins. When fusion begins, a star is born.

ORION NEBULA A star begins its life in a nebula like this one, great clouds of gas and dust.

CHAPTER 28 SECTION 3

VISUAL TEACHING
Discussion
Direct students' attention to the illustration. Ask: What factor determines which of the two life-cycle courses a star takes? *the initial mass of the star* How does the cloud-like nebula form protostars? *Outside forces lead to compression in regions of the nebula. As the regions become denser and their temperatures increase, they start to glow, forming protostars.* What process marks the transition from protostar to star? *fusion* What remains balanced during the long period when a star's hydrogen is fusing into helium? *The energy produced by fusion is balanced by the force of gravity.* At what point is this balance lost? *when the hydrogen in a star's core is used up*

Death of a Star Like the Sun

Main-sequence stars with a mass similar to the sun's remain about the same size for millions or even billions of years, slowly brightening, perhaps, but with little other change in appearance. During all that time, hydrogen in the core of such as star continues to fuse into helium. The energy produced by the fusion reactions balances the force of gravity that is pulling the star's matter inward. Eventually, however, the hydrogen is used up, and gravity begins to take the upper hand. The hydrogen core shrinks and its contraction produces additional heat, which triggers hydrogen fusion outside the core. The entire star begins to expand.

The star begins to die when its core temperature rises to the point at which helium can fuse into the heavier elements carbon and oxygen. Because the temperature never rises enough for these heavier elements to fuse, a carbon-oxygen core forms. Hydrogen and helium fusion reactions continue in the layers surrounding the core.

Then the gases at the star's surface begin to blow away in abrupt bursts. Eventually, all the gaseous layers are blown away, and only the fiercely hot carbon-oxygen core—a white dwarf—is left. The expelled layers of the star absorb the white dwarf's ultraviolet emissions and give off visible light. The resulting glowing halo of gases is called a **planetary nebula** because 18th-century astronomers thought such halos looked like the disks of planets.

After some 25,000 years, the planetary nebula fades as its gases dissipate into space. Only the white dwarf is left behind.

DIFFERENTIATING INSTRUCTION

Developing English Proficiency Tell students that the word *nebulous* implies something misty or unclear. Ask them how these meanings relate to nebulae in space. *Nebulae are clouds of gas and dust. Thus, nebulae make the stars behind them more difficult to see from Earth.*

628 Unit 7 Space

Death of a Massive Star

Stars that are eight or more times as massive as the sun face a fate different from that of stars similar to the sun in mass. All stars eventually run low on hydrogen. In a star such as our sun, the fusion process would stop with carbon. In a massive star, however, fusion processes continue until iron nuclei are formed. At this point, the star swells to more than 100 times the diameter of the sun, becoming a supergiant.

The formation of iron nuclei does not release energy; instead, it absorbs energy, so the iron core of the star quickly and suddenly collapses. The collapse of the core produces a shock wave that blasts the star's outer layers into space at thousands of miles per second and produces a brilliant burst of light. The explosion results in a **supernova**, a nova far brighter than ordinary—10 to 100 times as bright as our sun. A single supernova can outshine all the other stars in its galaxy for a time.

The most famous supernova of recent times occurred in 1987 in the Large Magellanic Cloud, one of two galaxies closest to the Milky Way. The first supernova to be seen since the development of modern observational equipment, it gave astronomers an unprecedented chance to test their hypotheses about supernovae.

When a large star explodes as a supernova, it produces many elements, including copper, uranium, silver, and lead. These elements are blown away into space as a huge cloud of gas and dust, which mixes with the gas and dust already there. But new elements are not all a supernova leaves behind.

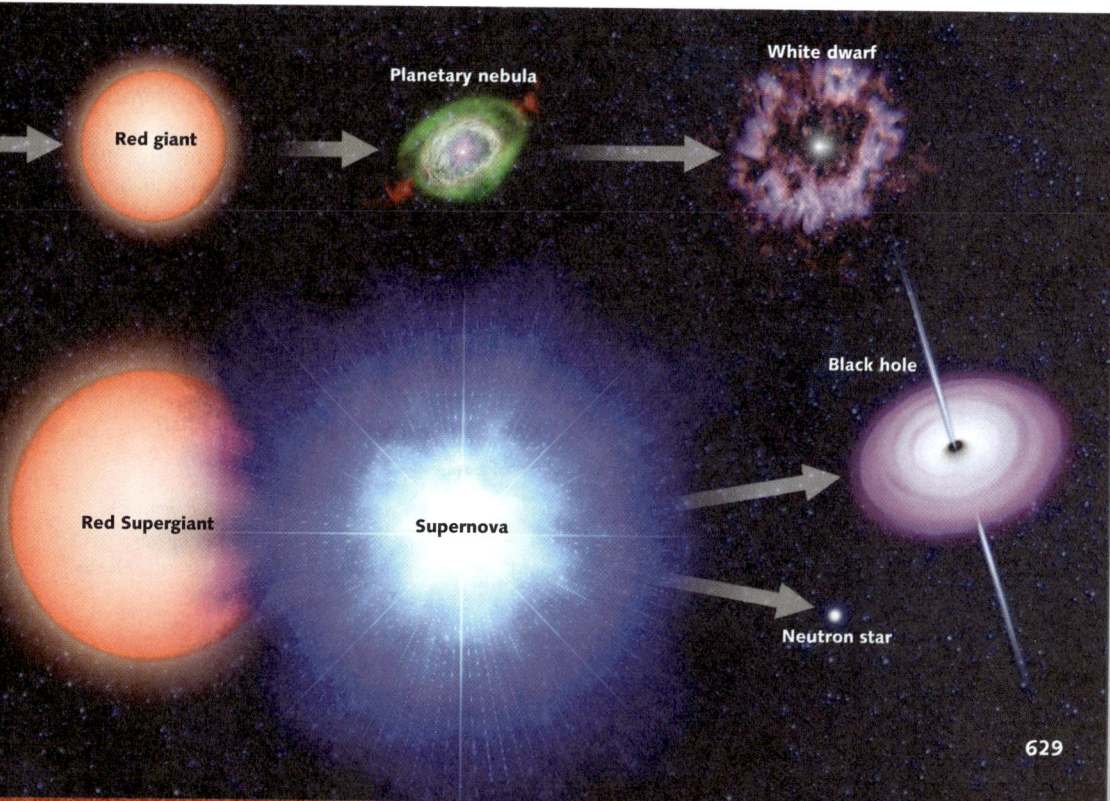

CHAPTER 28 SECTION 3

ASSESS

1. The Hertzsprung-Russell diagram is a plot of star luminosity against star surface temperature and depicts the stages in the life cycles of stars.

2. The force of gravity balances the force of energy from fusion.

3. When a massive star runs low on hydrogen, heavier elements begin to fuse until eventually they produce iron nuclei. The star swells to more than 100 times the diameter of the sun, becoming a supergiant. Because the formation of iron nuclei absorbs energy, the iron core of the star suddenly collapses, creating a shock wave that blasts the layers into space in a brilliant burst of light, called a supernova. Left behind is a black hole (if the original star was at least 15 times more massive than the sun) or a neutron star.

4. It is unlikely that a star with 10 times the sun's mass will live long enough to allow organisms on an orbiting planet to evolve into complex forms.

SUPERNOVA The collapse of a massive star's iron core resulted in this brilliant burst of light.

Remnants of Massive Stars

After a massive star "goes supernova," it leaves behind its core. In some cases, this remnant takes the form of a **neutron star.** This kind of star is so named because of the effect gravity had on its atomic structure.

In an atom, as you may recall, one or more electrons circle a positively charged nucleus. Astronomers think that the gravitational force in a neutron star is so great that each atom's electrons are crushed into the nucleus, overwhelming the natural forces that would keep them apart. The star becomes for the most part a dense mass of neutrons. Astronomers calculate that a typical neutron star is about 20 kilometers in diameter, but trillions of times more dense than the sun.

When first formed, a neutron star spins rapidly, giving off bursts of radio waves as it does so. As it rotates, it emits a beam of radiation along its magnetic axis that sweeps through space like a searchlight. If Earth is in line with this beam, we can detect pulses of radiation as the beam sweeps over us. Astronomers call such a rapidly spinning neutron star a **pulsar** because of the pulses of energy.

Although incredibly dense, neutron stars are not the densest objects in the universe. A **black hole** is the remnant of a star at least 15 times as massive than the sun. Inside a black hole, the mass of 10 suns may be condensed inside an area 30 kilometers wide.

If the gravitational force of black holes is so strong that light cannot escape, how do astronomers identify them? One of the first black holes ever discovered was detected as a strong x-ray source detected in the constellation Cygnus. From their study of this source, astronomers inferred that it consisted of a star orbiting a very massive, yet invisible, companion. They hypothesized that the invisible companion is a black hole that is drawing into itself matter from the visible star. Before they disappear into the black hole, atoms emit powerful x-rays as they are ripped apart by the force of gravity.

Some scientists think that many galaxies have supermassive black holes at their centers. Some black holes, like the one thought to be at the center of the Milky Way, may contain the mass of about 2.6 million suns. Others may have the mass of 1 or 2 billion suns.

28.3 Section Review

1. What does the Hertzsprung-Russell diagram depict?
2. What forces balance each other in a main-sequence star?
3. Describe the end of a massive star's life.
4. **CRITICAL THINKING** Earth was more than 4 billion years old before complex multicellular organisms with hard body parts appeared. Would you expect to find similar life forms on a planet orbiting a star with 10 times the sun's mass? Why?

MONITOR AND RETEACH

If students miss . . .

Question 1 Reteach the H-R diagram. (pp. 626–627) Ask students to identify the three main groups of stars in the H-R diagram. **main-sequence stars, giants and super giants, and white dwarfs**

Question 2 Reteach "The Death of a Star Like the Sun." (p. 628) Have students draw a sketch that shows the balanced inward and outward forces in a main-sequence star.

Question 3 Reteach "Death of a Massive Star." (p. 629) Make a flowchart that shows the sequence of events that lead to the death of a massive star.

Question 4 Remind students that more massive stars have shorter lives. (p. 626) A star much more massive than the sun will turn into a red giant before allowing the evolution of complex organisms.

Galaxies and the Universe

The universe is everything that exists. The observable universe is the term used to refer to everything that we can observe; the observable universe is limited in extent by a combination of the age of the universe and the speed of light. For example, if the universe is 12 billion years old, then light from objects more than 12 billion light-years away has not yet had time to reach us. Astronomers are not sure how old the universe is, but current estimates range from about 10 billion to 20 billion years.

What Are Galaxies?

Without a telescope, you can see several thousand stars in the night sky. You can also see a few hazy patches of light. Small telescopes show thousands more of these patches. Early observers called most of these hazy patches nebulae, from a Latin word for clouds.

Astronomers have studied the hazy patches with powerful telescopes, and have learned that many of them are not nebulae, but are **galaxies**, systems containing millions or even billions of stars. Most estimates place the number of galaxies in the observable universe at 50 billion to 100 billion. Space is so vast that most galaxies are millions of light-years apart.

The galaxy to which the sun belongs is the Milky Way. In it, the sun is one star among hundreds of billions. Every individual star you can see when you look up into the night sky belongs to the Milky Way. The Milky Way is a spiral galaxy. It is shaped like a thin disk with a central bulge. The diameter of the Milky Way is about 100,000 light-years. Its greatest thickness is about 10,000 light-years. The sun is about 26,000 light-years from the Milky Way's center. When looking at the night sky along the plane of the Milky Way's disk, one can see so many stars that this region of the sky looks milky. Observers called this region the Milky Way before they actually knew that it actually was our galaxy.

28.4

KEY IDEA
Billions of galaxies make up the universe, which, according to the big bang model, formed between 10 and 20 billion years ago.

KEY VOCABULARY
- galaxies
- quasar
- big bang model

Examine the Milky Way Galaxy at different scales.
Keycode: ES2808

SPIRAL GALAXIES, such as this one called M51, are named for their shapes, which resemble a spiral or a pinwheel.

CHAPTER 28 SECTION 4

FOCUS

Objectives
1. Tell what a galaxy is and describe the various types of galaxies.
2. Explain the origin of the universe according to the big bang model.

Set a Purpose
Have students read this section to find the answer to "What is the universe made of, and how did it form?"

INSTRUCT

More about...

Galaxies
The earliest astronomers noticed faint patches of light in the night sky. Once telescopes were invented, these fuzzy objects confused astronomers who were searching for comets. The French astronomer Charles Messier had cataloged nearly 100 of these objects by 1781. Updated versions of Messier's catalog are still used today by amateur astronomers. For example, the huge Andromeda Galaxy is M31.

Examine the Milky Way Galaxy at different scales.
Keycode: ES2908

 Visualizations CD-ROM

DIFFERENTIATING INSTRUCTION

Reading Support A concept web can help clarify students' understanding as they read. Have students skim the section, then ask them for ideas on how to set up a concept web for this section. For example, they might begin with the section title encircled in the center of the web, then include a separate strand for "What Are Galaxies," with galaxy characteristics coming off from it. The next strand might include "Types of Galaxies," with the final strand reading "Origin of the Universe." Have students read through the section and suggest items to include on the concept web.
Use Reading Study Guide, p. 98.

CHAPTER 28 SECTION 4

DISCUSSION

Ask: How is the Milky Way Galaxy classified into increasingly larger groups? **groups from smallest to largest are: Local Group, Virgo Cluster, Local Supercluster** The Milky Way Galaxy can be seen from Earth as a milky band running across the sky. How do you think elliptical and irregular galaxies might look from planets within them? **An elliptical galaxy might look like a band of light that is brighter than the Milky Way; the band formed by an irregular galaxy would be much dimmer than the Milky Way if visible at all.**

CRITICAL THINKING

Ask students to explain why elliptical galaxies have few, if any, young stars. **Elliptical galaxies have little interstellar gas and dust from which stars can form.**

SIDE VIEW When a spiral galaxy is viewed edge-on, the thickness of its center is plain to see.

The Milky Way belongs to a group of more than 30 galaxies called the Local Group. The Milky Way's nearest neighbors in the Local Group, the two Magellanic Clouds, can be seen without a telescope from the Southern Hemisphere. Another neighbor, the Andromeda Galaxy, is faintly visible in the Northern Hemisphere sky to the unaided eye. The Milky Way and the Andromeda Galaxy are by far the most massive galaxies in the Local Group. The Local Group and hundreds of other groups surround the Virgo Cluster, which may consist of as many as 2000 galaxies. All together, these make up the Local Supercluster, consisting of some 10,000 galaxies. The observable universe contains large numbers of super-clusters separated by great voids of space in which few or no galaxies are seen.

Types of Galaxies

No two galaxies are exactly alike. Most galaxies can be classified, however, by shape. Many are spiral galaxies, like the Milky Way. Spiral galaxies come in a range of types—from ones with large, bright nuclei of stars and tightly wound spiral arms to ones with very small, dim nuclei and open sprawling arms. The Andromeda Galaxy is a spiral galaxy.

Elliptical galaxies range from nearly spherical to lens-shaped. Their stars are concentrated in their centers, and they have no arms or similar structure. Elliptical galaxies contain far less interstellar gas and dust than spiral galaxies, and they contain few, if any, young stars. The galaxy called M87 is an example of an elliptical galaxy.

A galaxy can also be irregular in shape. Irregular galaxies are much smaller and fainter than spiral and elliptical galaxies, and their stars are spread unevenly. The two Magellanic Clouds are irregular galaxies.

Types of Galaxies

Spiral	Elliptical	Irregular
Pinwheel Galaxy	M87	Large Magellanic Cloud

Active Galaxies

The Milky Way is a normal galaxy; the total energy it emits is the totality of the energy emitted by its component stars. Some galaxies, however, emit far more energy than could be given off by their stars. Scientists call these active galaxies. Some of them emit large amounts of radiation (at radio,

---

DIFFERENTIATING INSTRUCTION

Support for Visually Impaired Students Students might not be able to discern the details of the galaxies in the photographs on page 632. In particular, the spiral arms of the Pinwheel Galaxy and the bulge in the center of the edge-on-viewed spiral galaxy could present difficulty. Enlarge these photographs on a copy machine and use a marker to delineate the features of the galaxies. Read the text describing these galaxies and have students point to places on the photographs corresponding to the descriptions. Alternatively, draw simple diagrams of these galaxies on the board, exaggerating the distinguishing features.

infrared, ultraviolet, x-ray, and gamma-ray wavelengths), and some are highly variable, changing in brightness considerably over a short periods of time.

The current model of active galaxies suggests that such a galaxy may be powered by a supermassive black hole at its center, from which pour forth jets of hot gas in opposite directions at nearly the speed of light. The black hole is surrounded by a disk of gas that is spiraling into it. As it does so, the gas gives off huge amounts of radiation before disappearing inside the black hole.

Scientists currently recognize many phenomena associated with active galaxies, including quasars and blazars. **Quasars** are extremely distant objects, some as far away as 12 billion light-years. A quasar is also extremely luminous, perhaps a hundred or even a thousand times brighter than a normal galaxy. Some scientists have suggested that quasars may be highly active galactic nuclei, burning so brightly that they blot out all the light from the galaxies' stars. A blazar is believed to be an active galaxy that has one of its jets pointed toward Earth, so that observers here on Earth are looking directly into the jet of escaping energy.

Lower-luminosity active galaxies also exist. Astronomers suspect they may be weaker versions of quasars or may be powered by star formation rather than supermassive black holes.

Active galaxies are the subject of much debate among astronomers. The study of these distant objects is one of astronomy's greatest challenges, and astronomers are continually making new discoveries about them, and devising new theories to explain their characteristics.

VISUALIZATIONS
CLASSZONE.COM

Observe some regular, irregular, and very peculiar galaxies.
Keycode: ES2809

CHAPTER 28 SECTION 4

VISUALIZATIONS
CLASSZONE.COM

Observe some regular, irregular, and very peculiar galaxies.
Keycode: ES2809

Have students make sketches of the galaxy shapes they observe.

Visualizations CD-ROM

VISUAL TEACHING
Discussion

Refer students to the illustration of the active galaxy. Ask: What do scientists presently think causes an active galaxy to emit jets of energy? **A supermassive black hole at the center of the galaxy superheats a disk of gas spiraling inward. This superheated material emits huge amounts of radiation before disappearing into the black hole.** Why isn't the black hole visible in the center of the picture? **Black holes have never been "seen," only inferred by the behavior of nearby matter.** Aside from its "active" nature, how would you categorize the galaxy shown? **It is a spiral galaxy.**

ACTIVE GALAXY In this artist's rendering, an active galaxy's jets of gas are clearly seen pouring from its nucleus.

DIFFERENTIATING INSTRUCTION

Challenge Activity Challenge students to develop a demonstration of how a spinning or spiraling object can shoot off jets of material. Suggest they look in school and at home for a source of "spinning" and that they carefully consider safety implications of what they use as jet material. Have students set up their demonstrations outside, wearing safety goggles and clothing protection. Ask them to lead a discussion with their classmates that ensures the class understands what the demonstrations represent.

Origin of the Universe

Scientists cannot say where or how the universe originated. However, by observing stars and galaxies with telescopes and other devices and by experimenting with matter here on Earth, they have been able to put together a model of how it has developed—the **big bang model.** The big bang model explains the history of the universe from a tiny fraction of a second after it came into being up to the present time.

A Model of the Beginning

According to the big bang model, about 10 billion to 20 billion years ago all matter in the universe existed in an incredibly hot and dense state, from which it expanded and cooled, slowly condensing into stars and galaxies. Among the many recent attempts to describe the conditions that followed the big bang model are the inflationary models. In these models, the universe began as a tiny region of space-time that expanded at incredible speed. Then, almost at once, most of the energy of the high-speed expansion materialized—at the same time, through all of space— into high-energy particles of matter and light. This superhot, superdense universe then continued to expand, but at a much slower rate. It is important to understand that the universe did not expand into existing space—space itself was expanding.

The temperature and density of the universe gradually decreased. However, the temperature was so still so high that ordinary atoms of matter could not survive. When an electron and a proton would combine to form a hydrogen atom, the atom would be ripped apart instantly by a collision with high-energy particles. Eventually, after a few hundred thousand years, the temperature of the universe dropped to the point where atoms could survive.

After the universe became cool enough for atoms to form, they began to clump together into clouds of gas, which in time became organized as galaxies. The first stars consisted mostly of hydrogen, with a small amount of helium. No planets like Earth orbited them, because heavier elements, like iron and silicon, had not yet been manufactured inside stars.

But as more and more stars formed, entered the main sequence, grew old, and died, more and more matter was fused into heavier elements and expelled back into interstellar space by supernovas and dying red giant stars. Eventually, our sun and its nine planets formed from this interstellar gas and dust.

Evidence for the Big Bang Model

The first piece of evidence supporting the big bang model is the universe's apparent expansion. The distance between galaxies and groups of galaxies seems to have been increasing with time. This evidence was first announced in 1929 by the astronomer Edwin Hubble, who found redshifts in the spectra of galaxies he studied. The redshifts show that distant galaxies in all directions are receding from Earth faster than nearby galaxies. Because the galaxies are moving apart, Hubble concluded they must have been closer together in the past.

More about...

The Steady-State Theory
In 1920, the steady-state theory was put forth to explain the origin of the universe. Among its proponents was the British astronomer Sir Fred Hoyle, who provided a mathematical theory within the model. It was Hoyle who coined the term "big bang" during a radio lecture. Basically, the steady-state theory states that the universe is always expanding but maintains constant density because new matter is created at a rate equal to the expansion. This implies that there is no beginning or end in time and that galaxies of all possible ages are mixed together in space. However, the theory has been contradicted repeatedly by observational evidence showing that the universe is constantly evolving. Since the discovery of cosmic background radiation in 1964, most astronomers do not accept the steady-state theory.

DIFFERENTIATING INSTRUCTION

Hands-On Demonstration This demonstration can be done as an alternative to the Lab Activity. Use a large balloon, marker, and tape measure to model the movement of galaxies away from each other as the universe expands. On the balloon, mark and number three dots around a central dot, each a different distance from the center. Record the distances of the three dots from the central dot. Blow up the balloon and tie it off. Record the distances of the three dots from the central dot once again. Have students compare the distances traveled by the dots. They should observe that the dot farthest from the central dot traveled the farthest and therefore the fastest. Relate students' observations of the moving dots on the expanding balloon to moving galaxies in an expanding universe.

More support for the model came in 1964, when the engineers Arno Penzias and Robert Wilson discovered radiation apparently left over from the universe's beginning. This radiation is called cosmic background radiation. Although the redshift of this cosmic background radiation has moved beyond the visible part of the spectrum, infrared and radio telescopes can detect this radiation. The temperature of this radiation has been precisely measured by the *Cosmic Background Explorer,* or COBE, a satellite dedicated to exploring the origins of the universe. COBE found its the temperature to be -270.3°C—similar to what the cosmologists (scientists who study the universe) had predicted, taking into account the amount of time that the universe has had to cool. The background radiation varies very little across the sky.

The big bang theory continues to be tested and examined. Astronomers and cosmologists around the world are seeking further evidence to support it, and others are considering alternative ways in which the universe may have reached its present state.

It is impossible to know for certain how the universe began. With the big bang model surviving crucial tests like the ones described above, it remains the best explanation we have about the universe's origin. If one day scientists discover phenomena that cannot be explained by this model, they will have to start looking for a new model of the universe's history.

EDWIN HUBBLE found redshifts in the spectra of the galaxies he studied, evidence of the universe's expansion.

28.4 Section Review

1. What is the difference between the universe and the observable universe?
2. Into what basic shapes can most normal galaxies be classified?
3. How do active galaxies differ from normal galaxies?
4. Copy and complete this concept map concerning what we know about the universe.

   ? → Superclusters → ? → Galaxy → Solar Systems → Small Bodies (?, ?, ?), ? Stars ?, Moons

5. **CRITICAL THINKING** Observations of very distant galaxies show that they are at an earlier stage of development than nearby galaxies. How do these observations support the big bang model?

CHAPTER 28 SECTION 4

ASSESS

1. The universe includes everything that exists; the observable universe includes only that which we can observe. The observable universe is limited in size by a combination of the age of the universe and the speed of light.

2. Most galaxies are either spiral or elliptical.

3. The total energy emitted by a normal galaxy equals the sum of the energy emitted by its component stars. An active galaxy emits far more energy than the sum of energy emitted by its component stars. Active galaxies emit large amounts of energy and/or are highly variable and change in brightness over a short period.

4. from top: Observable universe, Virgo Cluster; from left: Comets, Planets, Sun, Asteroids, Meteoroids

5. Light from distant galaxies takes a very long time to reach Earth. Thus, the more distant the galaxy, the earlier in time we are seeing it. These observations support the big bang model because they suggest that galaxies have evolved through time as the universe has expanded and that the galaxies formed at nearly the same time.

MONITOR AND RETEACH

If students miss . . .

Question 1 Have students reread the first paragraph in this section. (p. 631) Encourage them to speculate about what might exist in the universe that has not been observed.

Question 2 Reteach "Types of Galaxies." (pp. 632–633) Have students make sketches of spiral, elliptical, and irregular galaxies.

Question 3 Refer students to the illustration on page 633. Ask: How is this galaxy different? *The galaxy is releasing energy jets.*

Question 4 Have students review the first paragraph on page 632 and Chapter 27 to recall other kinds of objects in our solar system.

Question 5 Review the first paragraph on page 631. Ask: How long would it take for light from a galaxy that is 8 billion light-years away to reach Earth? *8 billion years* What would the galaxy look like? *the way it did 8 billion years ago*

CHAPTER 28 ACTIVITY

PURPOSE
To gain understanding of the expanding-universe theory by using a balloon model of the universe

MATERIALS
Provide students with a copy of Lab Sheet 28. A copymaster is provided on page 150 of the teacher's laboratory manual.

High-quality latex balloons from scientific supply companies work better than party balloons for this activity.

PROCEDURE

2 Clipping the balloon might prove tricky; advise students to practice before beginning their measurements.

7 The answers recorded in the *rate-of-distance-increase* row should equal the numbers in the *distance-increase* row divided by 8 years.

9 Assuming the balloon expands at the same rate, the distances should increase by the same amount between 16 and 24 years as they did between 8 and 16 years. Therefore, most students should record in the *predicted-distance* row the sum of the distance at 16 years and the distance increase.

CHAPTER 28 LAB Activity

Expansion of the Universe

Many astronomers support the theory that our universe began with an explosion called the big bang. Since the big bang, the universe has continued to increase in size. This theory is supported by observations of distant galaxies, all of which appear to be moving away from our galaxy at enormous speeds. In this activity, you will assume the role of a scientist in the hypothetical universe of Balloonia. Beings living in Balloonia are only able to see along the surface of their balloon universe. By measuring the distance between different points on Balloonia every eight years, you will develop a theory to explain the rate of expansion of Balloonia.

SKILLS AND OBJECTIVES
- **Calculate** the rate of expansion of a hypothetical universe.
- **Compare** the behavior of a balloon universe to our universe.

MATERIALS
- round balloon
- binder clip or paper clip
- fine-tip marker
- graph paper
- piece of string
- metric ruler

Procedure

1 Use a fine-tip marker to mark 13 dots around the surface of the deflated balloon. Mark dots only on the round, inflatable part of the balloon, not on the stem. Be sure to distribute the dots all over the balloon. Label the dots A through M, randomly.

2 Inflate the balloon until its diameter is about 8 centimeters. Pinch and fold the stem over and use a binder clip or paper clip to keep air from leaking out.

3 Use a piece of string to measure the distance in millimeters from point A to each of the other 12 points on the balloon's surface. Copy the data table below to record the results as "Distance at time = 8 years".

4 Unclip the balloon and inflate the balloon until it has a diameter of about 16 centimeters. Re-clip the stem to prevent air from leaking.

5 Use a piece of string to measure the new distance in millimeters from point A to each of the other 12 points. Write the results in the data table in the row "Distance at time = 16 years".

Point on Balloon	B	C	D	E	F	G	H	I	J	K	L	M
Distance (mm) at time = 8 years												
Distance (mm) at time = 16 years												
Distance increase (mm)												
Rate of distance increase (mm/yr)												
Predicted distance at time = 24 years												
Measured distance at time = 24 years												

6 Subtract the distances at time = 8 years from the distances at time = 16 years. Record the result in the row "Distance increase (mm)."

⑦ Assume that the distance measurements were actually taken eight years apart. Compute the average speed that each point is moving away from point A. Record your answer in the row "Rate of distance increase". Hint: average speed = total distance ÷ total time.

⑧ Make a scatter plot of the data from your calculations made during Steps 5 and 7. Label the x-axis with "Distance at time = 16 years". Label the y-axis with "Rate of distance increase". Draw a best-fit line for the points on the scatter plot.

⑨ Using the rates from Step 7, predict how far each point on the balloon would be from point A at a time of 24 years. Assume the balloon will expand at the same rate as it did between times of 8 and 16 years. Record your predictions in "Predicted distance at time = 24 years".

⑩ Unclip the balloon's stem and inflate the balloon so that its diameter is approximately 24 centimeters. Re-clip the stem.

⑪ Use the string to measure the distance between point A and the other points of the balloon. Record the distance in the row "Measured distance at time = 24 years".

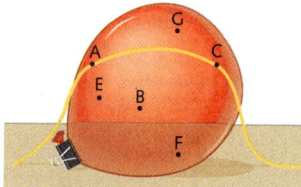
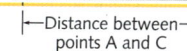
Distance between points A and C

Analysis and Conclusions

1. Calculate the slope of the best-fit line in your scatter plot from Step 8. What are the units for the slope of the best-fit line? What does the slope of this best-fit line represent?

2. Give a few examples of possible sources of error for the predictions you made of any distances from point A at a time of 24 years.

3. This activity was done from the point of view of a two-dimensional being living at point A in Balloonia. Would the expansion of Balloonia appear faster or slower if this being lived at one of the other points?

4. From the point of view of a being living on Balloonia, where is the center of the expansion of the universe? Remember that this being can only see along the surface of the balloon. Where is the center of the Balloonia universe according to an outside observer?

5. Astronomers today theorize that our universe is expanding, similar to the way all points on Balloonia move away from each other as the balloon inflates. If this theory is accurate, where is the center of our universe?

CHAPTER 28 ACTIVITY

ANALYSIS AND CONCLUSIONS

1. On a perfect sphere, the best-fit line would have a slope of $\frac{1}{16}$ yr^{-1} (that is, $\frac{1}{16}$ per year). Students' answers should vary between $\frac{1}{10}$ and $\frac{1}{20}$ yr^{-1}. The units are 1/yr. The slope is a measure of how the rate of change of the distance between two points will change over a given time.

2. Possible sources of error include incorrect measurements of distances, such as the balloon diameter; air escaping from the balloon; the balloon not inflating evenly in all places; and the balloon not being a perfect sphere.

3. The expansion of Balloonia would appear the same no matter where the measurements are taken.

4. Acceptable answers include the center of expansion being everywhere and nowhere. The answer to Question 3 implies there is no center for the expansion on the surface of the balloon. Therefore, a student might infer that every point on the balloon is the center of expansion. Equally valid is the conclusion that since no place is unique in its expansion, no center of expansion exists. An outside observer (one who can perceive all three dimensions) can point to the center of the balloon as the center of the Balloonia universe.

5. The analogy between Balloonia and our universe is that we cannot perceive the center of the universe as a specific point. Distant galaxies all appear to be moving rapidly away from Earth, yet if we switched our vantage point to one of those distant galaxies, we would still observe all distant galaxies as rapidly moving away from us. Therefore, it is impossible to identify a specific point from which the universe is expanding.

Refer students to Safety in the Earth Science Laboratory on pages xxii–xxiii.

CHAPTER 28 REVIEW

VOCABULARY REVIEW

1. An astronomical unit is the average distance from Earth to the sun; a light-year is the distance light travels in a year.
2. Absolute magnitude is a measure of stars' brightness if they were all equidistant from Earth; apparent magnitude is how bright stars would appear when viewed from Earth.
3. A pulsar is a neutron star that is spinning very rapidly; a quasar is a type of active galaxy.
4. Giant stars have diameters 10 to 100 times greater than that of the sun, while supergiants have diameters more than 100 times that of the sun.
5. Parallax is the change in an object's direction due to a change in the observer's position; a parsec is 3.258 light-years.
6. A constellation is a human invention—a pattern of stars in the sky; a galaxy is a natural grouping of stars.
7. A nebula is a cloud of gas and dust; a planetary nebula is the layers of brightly glowing gases around a white dwarf.
8. A white dwarf is the glowing stellar core of a star near the end of the star's life; a black hole is the remnant of a star at least 15 times as massive as the sun.
9. Blueshift is the shift of a star's emission spectra lines toward the blue end of the spectrum, indicating the star is moving toward Earth; redshift is the shift of a star's spectral lines toward the red end of the spectrum, indicating the star is moving away from Earth.
10. A neutron star is the remnant of a massive star that had become a supernova; a main-sequence star is actively fusing hydrogen into helium.
11. An emission spectrum appears in a spectroscope as a series of unevenly spaced lines of different colors and brightness; an absorption spectrum is a continuous spectrum with dark lines.
12. Protostars are large, glowing areas in nebulas where contraction of matter is occurring; a nebula is a cloud of gas and dust.

CHAPTER REVIEW 28

Summary of Key Ideas

28.1 The electromagnetic spectrum includes radio waves, microwaves, infrared, visible light, ultraviolet, X rays, and gamma rays. Astronomers analyze stellar spectra to determine stars' composition and movement.

28.2 Stars differ in mass, size, and surface temperature. Surface temperature affects the color of a star. Hydrogen and helium are the two most abundant elements in stars. Apparent magnitude, luminosity, and absolute magnitude all describe the brightness of stars. Distances in space are measured in astronomical units, light-years, and parsecs.

28.3 The Hertzsprung-Russell diagram plots a star's luminosity against its surface temperature. The diagram's groups of stars represent life-cycle stages of stars. A star's fate depends on its mass. Most stars are main-sequence stars.

28.4 Galaxies contain millions or billions of stars. The big bang model is a hypothesis about the origin of the universe.

KEY VOCABULARY

absolute magnitude (p. 623)
absorption spectrum (p. 614)
apparent magnitude (p. 619)
astronomical unit (p. 619)
big bang model (p. 634)
black holes (p. 630)
Cepheid variables (p. 624)
constellation (p. 617)
Doppler effect (p. 615)
electromagnetic spectrum (p. 612)
emission spectrum (p. 614)
galaxy (p. 631)
giant (p. 627)
light-year (p. 620)
luminosity (p. 623)
main sequence (p. 626)
nebula (p. 627)
neutron star (p. 630)
parallax (p. 620)
parsec (p. 620)
planetary nebula (p. 627)
pulsar (p. 630)
quasar (p. 633)
supergiant (p. 627)
supernova (p. 628)
white dwarf (p. 627)

Vocabulary Review

Explain the difference between the terms in each pair.

1. astronomical unit, light-year
2. absolute magnitude, apparent magnitude
3. pulsar, quasar
4. giant, supergiant
5. parallax, parsec
6. constellation, galaxy
7. nebula, planetary nebula
8. white dwarf, black hole
9. blueshift, redshift
10. neutron star, main-sequence star
11. emission spectrum, absorption spectrum
12. protostar, nebula

Concept Review

13. Why does each star have a unique absorption spectrum?
14. How does the Doppler effect indicate a star's movement toward or away from Earth?
15. How does a star's surface temperature relate to its color?
16. Using the map on page 617, find the Big Dipper in the night sky. What other constellations can you identify?
17. **Graphic Organizer** Copy and complete the Venn diagram below. How are apparent magnitude and absolute magnitude alike? How do they differ?

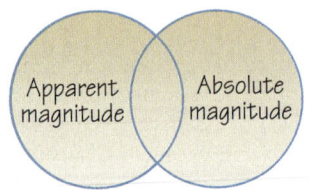

CONCEPT REVIEW

13. Each star contains different amounts of various elements.
14. As the star moves, the wavelengths of the light it emits change, shifting toward the blue end of the spectrum if the star is moving toward Earth and toward the red end if it is moving away from Earth.
15. The hottest stars are bluish-white; the coolest stars are reddish. Stars with "in-between" temperatures are white, yellow, and orange.
16. Point out that the Big Dipper is part of the constellation Ursa Major. Other bright constellations students may find include Cassiopeia, Pegasus, and Cygnus.

Critical Thinking

18. **Hypothesize** A shock wave may be the impetus that causes a nebula to start condensing and form new stars. Write a statement describing what events might produce such a shock wave.

19. **Interpret** Look at the Hertzsprung-Russell diagram on page 626. As the sun ages, how will its position on the diagram change?

20. **Infer** Could we see the sun if it were 32.6 light years from Earth? Explain your answer.

21. **Infer** Planets shine by reflected sunlight. How can planets be brighter than stars?

22. **Communicate** Prepare a presentation describing how a star's mass affects the course and length of its lifespan.

Internet Extension

What Happens as a Star Runs Out of Hydrogen?
Use stellar properties to predict the stages of stars' lives.
Keycode: ES2810

Writing About the Earth System

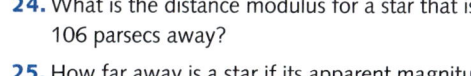 **SCIENCE NOTEBOOK** Look at patterns of constellations in the star maps on pages 714 and 715. Using existing stars, create new constellations based on modern objects. Draw your constellations in your Science Notebook.

Interpreting Graphs

Graph A shows the relationship between the distance to a star in parsecs (1 parsec = 3.26 light-years), and the distance modulus. The distance modulus is the difference between the apparent magnitude (m) of a star and its absolute magnitude (M). Graph B shows the relationship between the period of a cepheid variable and its absolute magnitude.

23. The distance modulus for a star is 10. How far away is the star?

24. What is the distance modulus for a star that is 10^6 parsecs away?

25. How far away is a star if its apparent magnitude is 10 and its absolute magnitude is −10?

26. What is the absolute magnitude of a Cepheid variable star with a period of five days?

27. A Cepheid variable has a period of 50 days. If its apparent magnitude is 0, how far away is the Cepheid?

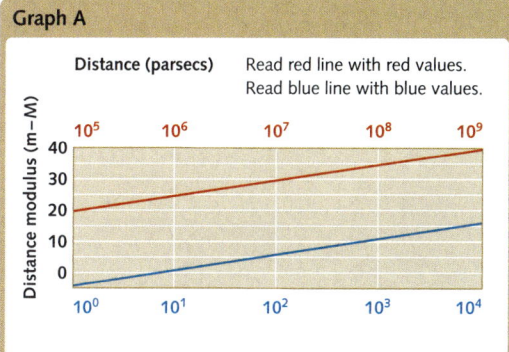

Graph A

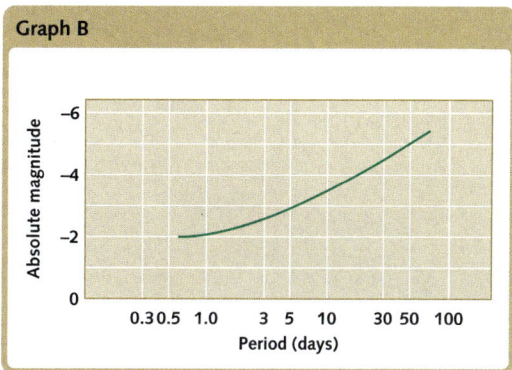

Graph B

WRITING ABOUT SCIENCE

Writing About the Earth System
Constellations will vary, but should include existing stars shown in correct relative positions.

CHAPTER 28 REVIEW

What Happens as a Star Runs Out of Hydrogen?
Keycode: ES2810

📖 *Use Internet Investigations Guide, pp. T96, 96.*

(continued)

17. Diagrams will vary. Where the two circles overlap, students might write the phrases *measure of star brightness* and *the lower the number, the brighter the star*. In the non-overlapping part of the Absolute Magnitude circle, students might write the phrase *depends on star's size and temperature*. In the non-overlapping part of the Apparent Magnitude circle, students might write the phrase *depends on star's absolute magnitude and distance from Earth*.

CRITICAL THINKING

18. Fusion in a massive star forms iron nuclei. The iron core collapses, and a shock wave results.

19. The star's position will move from the main sequence into the area of the red giants, and then will move to the lower left as it becomes a dwarf star.

20. Yes; the sun's absolute magnitude is +4.8, and its apparent magnitude would be +4.8. We can see stars of that magnitude without a telescope.

21. The planets' proximity to Earth makes them appear brighter.

22. Answers should indicate that large-mass stars have relatively short lifespans and end in supernovae, whereas small-mass stars have longer lifespans and end as white dwarfs.

INTERPRETING GRAPHS

Point out to students that the red line in Graph A is a continuation of the blue one below.

23. 10^3 parsecs (1,000 parsecs)
24. 25
25. 10^5 parsecs (100,000 parsecs)
26. −3
27. 10^2 parsecs (100 parsecs)

UNIT 7 FEATURE

▶ Focus

Objective
Explain what scientists can learn about living things by studying bacteria that live in extreme environments of Earth's hydrosphere.

Set a Purpose
Have students read to find answers to the question "What are the characteristics of places in Earth's hydrosphere where extremophiles live?" **Extremophiles are true to their name. They thrive in conditions that are considered extreme for most life as we know it. Such conditions include very high or very low temperatures, high pressure, the absence of light or oxygen, and an overly saline environment.**

UNIT 7 The EARTH SYSTEM

Life in Extreme

Imagine that you are in Antarctica, at the Russian research station Vostok. You are surrounded by ice as far as your eyes can see, continually blanketed in freezing temperatures, and looking out upon a seemingly barren landscape.

It is here at Vostok that the coldest continental temperature on Earth was ever recorded, about −129°F on July 21, 1983. Drilling through thick deposits of ice, you find a hidden freshwater lake more than two miles beneath the surface—a lake that has been sealed off from the rest of world for a million years. This lake, also named Vostok, is the largest of more than 70 subsurface lakes discovered so far on the southernmost continent. The thick layer of ice over it helps to insulate it, keeping the water liquid. Scientists are studying this lake for answers of some of our most intriguing questions—What are the necessary conditions for life? Is there life beyond Earth?

Scientists are looking to Lake Vostok for clues because of the possibility that life forms may dwell there. Scientists have discovered that some life forms not only survive in such extreme environments, but actually thrive in them. These organisms, usually bacteria, are called extremophiles (ihk-STREHM-oh-FYLS).

The discovery of extremophiles has stretched the boundaries of what was once considered to be necessary conditions for life. They have been discovered in extreme environments of intense pressure, cold, heat, and salt, or in the absence of light and oxygen—environments in which most types of life, as we know it, would quickly perish. The organisms that may live in Lake Vostok would be called psychrophiles (sy-kroh-FYLS), meaning "cold-lovers."

Since Lake Vostok is covered by a layer of ice about two miles thick, access to the water is very difficult. Although researchers have drilled through the ice to approximately 100 meters above the liquid surface of the lake, they will not proceed further until they find a way to access the water without contaminating it. In ice samples already taken from the lake, researchers have found an astounding diversity of microbes. There could be communities of microbes which have survived in isolation, living in the lake for all these years.

If life is present in Lake Vostok, it will shed light on the possibility of life existing on other worlds. In particular, scientists compare Lake Vostok to one of Jupiter's moons, Europa. Observations of Europa indicate that there might be significant liquid oceans underneath its frozen surface. Could life survive in such conditions? We may not know until robots land on Europa to examine the surface and drill beneath it. Atmospheric and

PSYCHROPHILIC BACTERIA are found in cold environments around the world, such as Antarctic pack ice.

640 Unit 7 Space

DIFFERENTIATING INSTRUCTION

Reading Support A concept web is a useful tool for clarifying the content of articles such as "Life in Extreme." Ask students to create a concept web in their science notebooks. Suggest that they begin with extremophiles as the topic, then give them time to organize their webs. After they have done so, ask volunteers to recreate their webs on the board. Discuss the webs as a class. A sample web would include two spaces in the second level: one for psychrophiles and another for hyperthermophiles. The third level would include specific habitat information and implications for life elsewhere (on Europa for psychrophiles, on early Earth for hyperthermophiles). Students should revise their concept webs as needed, then fill in the details themselves.

and the ENVIRONMENT

geologic conditions are much different on Europa than on Earth. If life is found in Lake Vostok, scientists will ask again, Could life exist on Europa? With such possibilities in mind, NASA has exploration programs underway for Europa and beyond.

On the opposite side of the spectrum to psychrophiles, are hyperthermophiles, meaning "heat-lovers." Hyperthermophiles thrive in water temperatures 176°F or greater. They are found in natural hot springs, geysers, and near deep-sea thermal vents. Scientists draw connections from these organisms to the earliest forms of life on Earth. Earth's environment was not always as it is today. When the Earth formed about 4.6 billion years ago, there was no oxygen readily available in the atmosphere and conditions were seemingly inhospitable to life. Yet, there is evidence proving the presence of microbial life as far back as 3.8 billion years. In a relatively brief period of time, life bloomed on Earth.

HYPERTHERMOPHILIC BACTERIA thrive in very hot, wet environments, such as the hot pools in Yellowstone National Park.

By studying modern hyperthermophiles, we can search for answers as to how life could develop and survive in such extreme conditions. Our boundaries are continually expanding. We know that all living beings on Earth are composed of molecular combinations of carbon, oxygen, hydrogen, and nitrogen. Scientists continue to examine the question of how life can arise when those elements and others combine. In this search we gain a new understanding and appreciation of life on Earth—and also a perspective that, in the scale of the universe, life may have the potential to be widespread.

UNIT 7 FEATURE

INVESTIGATIONS
CLASSZONE.COM

Could Mars Support Life?
Keycode: ESU701

Use Internet Investigations Guide, pp. T97–T98, 97–98.

Investigation Photograph: This image of Mars was taken by the Hubble Space Telescope and shows two large seasonal dust storms (yellowish green): one above the northern polar cap and the other in the southern hemisphere (lower right).

UNIT INVESTIGATION
CLASSZONE.COM

Could Mars Support Life?
What are the conditions necessary for life? How could we recognize life on another planet? Consider these questions while examining images of Mars's surface and data from past experiments. Explore the chance that Mars supported life in the past, or the potential for Mars supporting life in the present or future.
Keycode: ESU701

Unit 7 The Earth System and the Environment   641

UNIT 7 ASSESSMENT

UNIT 7 STANDARDIZED TEST Practice

ANSWERS

1. d, Hydrogen fuses to form helium.
2. b, The impact theory is widely accepted.
3. c, Copernicus proposed the concept of a sun-centered solar system; Galileo was the first to use a telescope for astronomical observations; Kepler formulated the laws of planetary motion; Newton articulated the law of gravitation.
4. b, Neither Venus nor Mercury has moons.
5. d, Jovian planets are mostly gaseous.
6. c, A rille is a trenchlike valley that runs through maria bedrock.
7. c, A and F are closest in size to the sun, which is a main-sequence star.
8. b, B and C are the two largest and most massive stars and so are most likely supergiants.
9. a, D and E are between the supergiants and main-sequence stars in size and so are most likely giants.
10. c, Stellar masses are expressed as multiples of the mass of the sun.
11. red giant or red supergiant, given that the sun is dying and seems to be vaporizing comets
12. Scientists used a spectroscope to analyze the light emitted by the star, thereby determining the elements in the star's outer layer.
13. The star's oxygen atoms would be bound up in the form of carbon monoxide, thus, no oxygen would remain for forming water molecules.
14. According to the diagram, Sirius lies within the main sequence, with a mass similar to that of the sun and a

Directions (1–10): For *each* statement or question, select the word or expression that, of those given, best completes the statement or answers the question. Record your answer on a separate answer sheet.

1. The sun obtains its energy through nuclear fusion. Which of the following elements are primarily involved in this reaction?
   (a) hydrogen and deuterium
   (b) carbon and oxygen
   (c) helium and xenon
   (d) hydrogen and helium

2. According to currently accepted theory, the moon was formed
   (a) at the same time as Earth and out of similar materials
   (b) when a planet-sized object struck Earth
   (c) independently of Earth, which then captured it
   (d) when a chunk of molten material spun off from Earth early in its formation

3. Which of the following is one of Nicolaus Copernicus's contributions to science?
   (a) He made the first astronomical observations with a telescope.
   (b) He devised three laws of planetary motion.
   (c) He proposed the heliocentric solar system.
   (d) He discovered the law of gravitation.

4. Which planet does not have any moons?
   (a) Pluto   (c) Mars
   (b) Venus   (d) Neptune

5. Which of the following is not a characteristic of all Jovian planets?
   (a) a system of rings
   (b) more than one moon
   (c) a gas surface of hydrogen–helium
   (d) a mantle of molten rock

6. On the still-molten moon, a crust formed on top of a lava river. The lava drained away. Later, the roof of the resulting tunnel collapsed, forming a
   (a) mascon   (c) rille
   (b) mare   (d) highland

Base your answers to questions 7 through 10 on the table below. The table shows the size of several stars.

Star Sizes		
Star	Solar Mass	Solar Radius
A	2.3	2.5
B	20	36
C	14	60
D	3.5	13
E	4	18
F	2	1.5

7. Which stars are most likely main-sequence stars?
   (a) A and D   (c) A and F
   (b) D and E   (d) C and D

8. Which stars are most likely supergiants?
   (a) A and F   (c) A and D
   (b) B and C   (d) C and D

9. Which stars are most likely giants?
   (a) D and E   (c) E and F
   (b) D and F   (d) B and C

10. One solar mass is equal to the mass of
    (a) any giant star   (c) our own sun
    (b) Vega   (d) Earth

Directions (11–21): Record your answers on a separate answer sheet. Some questions may require the use of material in the Appendix. Base your answers to questions 11 through 13 on the newspaper article below.

First Signs of Water Found Outside Our Solar System

According to an article in the journal *Nature*, there is a distant sun dying that seems to be vaporizing a swarm of comets, releasing a huge cloud of water vapor. This discovery is the first evidence that extra-solar planetary systems contain water, which is essential for all known forms of life.

Researchers studied the dying star with the Submillimeter Wave Astronomy Satellite (SWAS), a small radio observatory NASA launched in December 1998. Although SWAS has detected water vapor from a variety of astronomical sources, "What makes the results so unusual is that we have found a cloud of water vapor around a star where we would not ordinarily have expected to find water," said Dr. Gary Melnick of the Harvard-Smithsonian Center for Astrophysics in Cambridge, Mass.

The star, called IRC+10216, is 500 light-years away. It is rich in carbon. "We expect all the oxygen atoms to be bound up in the form of carbon monoxide, with almost nothing left over to form water," Melnick said. "Yet we find substantial concentrations of water vapor around this star; the most plausible explanation for this water vapor is that it is being vaporized from the surfaces of orbiting comets, that are composed primarily of water ice."

11. What kind of star is IRC+10216?
12. How did the scientists know that IRC+10216 is rich in carbon?
13. Why would the scientists expect that the star was vaporizing nearby comets, rather than producing water vapor itself?

To answer questions 14 through 18, refer to the Hertzsprung-Russell diagram on page 626.

14. Based on the diagram, describe the star Sirius, including an approximation of its mass and surface temperature.
15. Describe how main-sequence stars differ from stars such as Rigel.
16. On the diagram, both Betelgeuse and Barnard's Star are shown in red as having surface temperatures of less than 5000°C but more than 2500°C. Why is Barnard's Star so much less luminous than Betelgeuse?
17. Procyon B is a white dwarf star. During its prime, to which star would it have been most similar: Rigel, Betelgeuse, Barnard's Star, or the sun? Explain your reasoning.
18. Of the stars labeled on this diagram, which will likely have the longest life span? Which will have the shortest life span? Explain your answers.

Base your answers to questions 19 through 21 on the seasonal star charts in the Appendix on pages 714–717.

19. In which constellation does the Big Dipper appear? Describe how you could locate it.
20. Name three circumpolar constellations and give a brief description of each.
21. Locate the constellation Orion. Describe its position on each star chart. Is Orion visible during each of the four seasons? Explain.

(continued)

temperature higher than that of the sun.

15. Main-sequence stars are dimmer and much smaller in diameter than stars such as Rigel.

16. Betelgeuse is a much larger star, making it brighter than Barnard's Star, despite their similar temperatures.

17. During its prime, Procyon B would have been most similar to the sun because stars with mass similar to the sun's eventually become white dwarfs. Rigel and Betelgeuse are supergiants, indicating their initial masses were far larger than the sun's. The low temperature and luminosity of Barnard's Star compared to the sun's indicate a far smaller mass than the sun's.

18. The more massive a star is, the shorter its life span will be. Due to their low masses, Proxima Centauri and Barnard's Star will have the longest life spans. Rigel, Polaris, and Betelgeuse, which are high-mass stars, will have the shortest life spans.

19. The Big Dipper appears in the constellation Ursa Major. The two stars in its outer rim always point to Polaris, the North Star.

20. The three circumpolar constellations are: Ursa Minor, which is the Little Dipper; Ursa Major, which is the Great Bear and contains the Big Dipper; and Cassiopeia, which looks like a huge W.

21. Orion, a seasonal constellation, can be best seen in the winter. It is visible above the southern horizon in January. It cannot be seen on the star charts for spring, summer, and fall.

UNIT 8

UNIT OVERVIEW

Students may not be familiar with the view of Earth shown here. We are looking down upon the North Pole. Point out the areas of white in the center of the image (Arctic icecap, Greenland, northern Siberia, and northern Canada). Students may wonder how this image is relevant to studying Earth's past. Discuss how expeditions to collect long cores of material from the icecaps have provided a huge amount data regarding Earth's past climates, the impact of meteorites, and a timeline of ancient volcanic eruptions and earthquakes. Understanding Earth's past can help predict future trends, for example, cycles of shifting temperatures and changes in climate.

VISUAL TEACHING

Chapter 29 Studying the Past

Pangaea was a supercontinent that existed more than 200 million years ago. At first, it broke up into two landmasses: Laurasia and Gondwanaland. Later, each of the landmasses broke up into the continents that exist today. Laurasia broke up into North America and Eurasia; Gondwanaland broke up into Africa, South America, India, Australia, and Antarctica. Students will learn about the methods scientists have used to infer the breakup of Pangaea and other events in Earth's past.

UNIT 8 Earth's History

The Earth System Layers of rock or polar ice on Earth's surface are like pages in a diary; each layer can record an event in Earth's history. But scientists must learn a new language to read such records. This language includes evidence and events seen on Earth today. Examining rock layers is only one way geologists interpret Earth's long and amazing past.

CHAPTER 29 Studying the Past

This computer image shows how Pangaea might have appeared. How do scientists identify Earth's age and the changes it has undergone?

CONNECTIONS TO . . .

Prior Knowledge In previous units, students learned about processes that continuously change Earth. Unit 8 describes how Earth has changed over time as a result of these processes. It also explains the methods scientists use to infer the changes that Earth has undergone over billions of years.

Unifying Themes Since it formed more than 4 billion years ago, Earth has *evolved*. This evolution includes changes in land formations, climate, and organisms. The geologic time scale summarizes Earth's *evolution*, based on evidence preserved in the rock record. As students study the chapters in Unit 8, have them focus on evidence that shows how Earth has *evolved*.

UNIT 8

CHAPTER 30
Views of Earth's Past

How has Earth, including its lifeforms, landforms, and climate, changed over the billions of years since its formation?

(continued)

Answer: Using radiometric dating, scientists have determined that the oldest rocks on Earth formed about 4 billion years ago. They have inferred the past positions of Earth's landmasses from current plate motions, the shapes of the continents, rock layers, and fossils. Scientists have inferred Earth's past climate from cores extracted from trees, the ocean floor, and polar icecaps.

Chapter 30 Views of Earth's Past

Eurypterids, ancient arthropods typically 13 to 43 centimeters (5 to 17 inches) long, were most abundant during the Silurian Period (410–440 million years ago). Paleontologists believe eurypterids were predators that lived in the oceans and used their claws to swim and attack prey. They may have been the ancestors of modern crabs. The last eurypterids probably became extinct about 248 million years ago, during the great Permian extinction. Students will learn about the evolution and extinction of organisms during Earth's history.

Answer: Since Earth's formation, landmasses, mountains, and seas have formed and disappeared. Earth has fluctuated between glacial and interglacial periods. Changes in Earth's climate have forced organisms to adapt to the new conditions or become extinct.

CONNECTIONS TO . . .

National Science Education Standards
- Students will understand that early Earth was very different from Earth today.
- Students will understand that life evolved in response to changes in the Earth system, but also that life caused changes in the Earth system. For example, early organisms released free oxygen into the atmosphere.
- Students will understand how scientists use rock sequences, fossils, and radioactive isotopes to estimate geologic time.

Unit Features
The Earth System and the Environment
Mass Extinction, pp. 690–691

What Caused the Mass Extinction Recorded at the K-T Boundary? Investigate data from images and tables to compare theories that deal with the K-T extinction.
Keycode: ESU801

Standardized Test Practice, pp. 692–693

645

CHAPTER 29 PLANNING GUIDE: STUDYING THE PAST

	Section Objectives	Print Resources
Chapter-Level Resources		Laboratory Manual, pp. 127–130 Internet Investigations Guide, pp. T101–T103, 101–103 Spanish Vocabulary and Summaries, pp. 57–58 Formal Assessment, pp. 85–87 Alternative Assessment, pp. 57–58
Section 29.1 Fossils, pp. 648–649	1. Define fossil. 2. Describe how different kinds of fossils form.	Lesson Plans, p. 99 Reading Study Guide, p. 99
Section 29.2 Relative Time, pp. 650–655	1. Summarize the principles scientists use to determine the relative age of Earth's rocks. 2. Describe three types of unconformities. 3. Identify methods scientists use to correlate rock layers.	Lesson Plans, p. 100 Reading Study Guide, p. 100 Teaching Transparency 35
Section 29.3 Absolute Time, pp. 656–659	1. Explain the process of radioactive decay. 2. Define half-life. 3. Describe how radiometric dating is used to measure absolute time.	Lesson Plans, p. 101 Reading Study Guide, p. 101

INVESTIGATIONS
CLASSZONE.COM

Section 29.2 INVESTIGATION
What Stories Do Rocks Tell?
Computer time: 45 minutes / Additional time: 45 minutes
Students read rock layers to interpret the geologic story of a place.

Section 29.3 INVESTIGATION
How Do Trees Record Time?
Computer time: 45 minutes / Additional time: 45 minutes
Students analyze patterns of tree rings to find out how trees tell absolute time.

Technology/Online Resources

Electronic Teacher Tools
Visualizations CD-ROM
Classzone.com
Online Lesson Planner
Test Generator

Visualizations CD-ROM
Classzone.com
  ES2901 Visualization

Visualizations CD-ROM
Classzone.com
  ES2902 Visualization, ES2903 Investigation,
  ES2904 Data Center

Classzone.com
  ES2905 Investigation

Classroom Management

Gather these materials for minilab and laboratory activities.

Minilab, p. 657
  periodic table of elements
  graph paper

Lab Activity, pp. 660–661
  colored pencils
  graph paper
  tracing paper
  ruler
  Laboratory Manual, Teacher's Edition
    Lab Sheet 29, p.151

Chapter Review INVESTIGATION
How Did the Layers of the Grand Canyon Form?
Computer time: 45 minutes / Additional time: 45 minutes
Students learn to tell the geologic story of the Grand Canyon.

CHAPTER 29

Studying the Past

At a dig site in Orchard, Nebraska, paleontologists excavate the remains of a 10-million-year-old rhinoceros.

How do scientists determine the age of such fossils?

INTRODUCE

Build Interest

Before sharing background information about the photograph with students, allow time for their questions and comments.

Ask questions like the following:

- What parts of the skeleton in the foreground are visible? **skull, part of a leg, ribs, vertebrae**
- What material surrounds the skeleton? **rock**
- Do you think recovering these fossils is a quick or a slow process? **slow** Why? **to avoid damage**

Respond

- How do scientists determine the age of fossils? **through radiometric dating, based on the fact that radioactive elements decay at a known constant rate**
- Do you think fossils found in rock below these rhinoceroses would be older, younger, or the same age as the rhinos? **older, if the rock layers had not been disturbed**

About the Photograph

The fossil rhinoceros in the photograph is part of Ashfall Fossil Beds in northeastern Nebraska. The rhinoceros was quickly buried in volcanic ash, which helped preserve its skeleton. The bones of the skeleton are still joined.

BEYOND THE TEXTBOOK

Print Resources

- Reading Study Guide, pp. 99–101
- Transparency 35
- Formal Assessment, pp. 85–87
- Laboratory Manual, pp. 127–130
- Alternative Assessment, pp. 57–58
- Spanish Vocabulary and Summaries, pp. 57–58
- Internet Investigations Guide, pp. T101–T103, 101–103

Technology Resources

- Visualizations CD-ROM
- Classzone.com
- Online Lesson Planner
- Electronic Teacher Tools
- Test Generator

Unit 8 Earth's History

CHAPTER 29

PREVIEW

▶ **FOCUS QUESTIONS** In this chapter you will study Earth's past and learn more about the key questions below.

Section 1 What is a fossil and what does it reveal about Earth's past?

Section 2 What is relative dating, and how is it used to order past events?

Section 3 What is absolute time and how is it measured?

▶ **REVIEW TOPICS** As you investigate how scientists study Earth's past, you will need to use information from earlier chapters.

- elements and isotopes (pp. 90–93)
- igneous intrusions (p. 125)
- features of sedimentary rocks (pp. 130–131)
- stress and folds (pp. 238–239)
- weatherings (p. 258)

▶ **READING STRATEGY**

QUESTION

Before you begin Chapter 29, read the chapter opener, scan the chapter contents, and read the key idea for each section. In your science notebook, write questions about the material for which you want to find answers in the chapter.

At our Web site, you will find the following Internet support for this chapter.

DATA CENTER
EARTH NEWS
VISUALIZATIONS
- Fossil Formation
- Unconformity Formation

LOCAL RESOURCES
INVESTIGATIONS
- What Stories Do Rocks Tell?
- How Do Trees Record Time?
- How Did the Layers of the Grand Canyon Form?

CHAPTER 29

PREPARE

Focus Questions
Find out what knowledge students have about how scientists study the past. For each focus question, have students consider the section title and then develop two or three additional questions for each section. For example, Section 2: How does relative time differ from absolute time?

Review Topics
In addition to the earth science topics listed, students may need to review this concept:

Physical Science
- Atoms of an element that have different masses are isotopes of that element.

Reading Strategy
Model the Strategy: To help students develop questions for each section, use a version of the following: "While scanning Section 1, I look at the key idea, key vocabulary, red and blue headings, diagrams, photographs, and captions. From this information, I can think of questions such as: What is amber? How many different ways can fossils form? What is a trace fossil? I can write these questions in my notebook, then answer them as I read the section. These notes will help me to remember and study this material, and to prepare for tests." Have students continue to use this strategy for each section.

BEYOND THE TEXTBOOK

INVESTIGATIONS
- What Stories Do Rocks Tell?
- How Do Trees Record Time?
- How Did the Layers of the Grand Canyon Form?

VISUALIZATIONS
- Animation of how fossils form
- Animation of how an unconformity forms

DATA CENTER
EARTH NEWS
LOCAL RESOURCES
CAREERS

Online Lesson Planner

CHAPTER 29 SECTION 1

FOCUS

Objectives
1. Define *fossil*.
2. Describe how different kinds of fossils form.

Set a Purpose
Discuss what students know about fossils from books, news stories, movies, field trips, and other sources. List some of their ideas on the board. Then have them read the section to find the answer to "How do fossils form?"

INSTRUCT

More about...

Fossil Formation
Only a small percentage of all organisms that lived on Earth have become fossils because the necessary conditions for fossil formation are infrequently met. In general, two conditions favor the preservation of an organism as a fossil: quick burial and the presence of hard parts. Since fossil formation usually depends on these conditions, a biased fossil record exists, consisting of an abundance of the remains of organisms with hard parts that lived in areas of sedimentation.

29.1

KEY IDEAS
Fossils form in several ways. Fossils are recognizable remains or traces of organisms that lived in former geologic ages.

KEY VOCABULARY
- fossil
- paleontology
- mold
- cast
- trace fossil

PETRIFIED LOGS These fossilized remains are in the Petrified Forest National Park, Arizona.

Fossils

A **fossil** is any evidence of earlier life preserved in rock. By analyzing fossils, scientists have been able to describe Earth's past. The study of the life that existed in prehistoric times is called **paleontology.** Traces of ancient life, fossils are both the basis for the geologic time scale and an important part of the rock record.

Formation of Fossils

Fossils include shells, bones, petrified trees, footprints, impressions made by leaves, or even burrows made by worms. Fossils may form in several ways, from original remains or from replaced remains, by being preserved in molds or casts, as trace fossils, or in carbonaceous film.

Original Remains

Fossils are rarely the original unchanged remains of plants or animals. Most often, the original remains decay or decompose before rock forms around them. In rare cases, an organism is preserved in its entirety. Large woolly mammoths found frozen in permafrost in Siberia and Alaska are examples of original remains. Also, some prehistoric insects that were trapped in resin—a sticky sap that oozes from trees—were preserved intact when the resin hardened into amber. In most cases, however, the soft body parts of an organism decay, and only the hard parts—such as the bones and teeth of dinosaurs—are preserved.

AMBER This insect is an example of original remains preserved in amber.

Replaced Remains

Some fossils are formed as replaced remains. In such fossils, the soft parts of the plants or animals have decayed, and the hard parts have been replaced by minerals. The replacement usually results from the movement of underground water. Circulating groundwater removes the original organic material from the organism and replaces it with molecules of minerals, such as calcite, silica, and pyrite. The result is an exact copy of the original plant or animal. Petrified wood is a good example of a fossil formed as replaced remains.

Molds and Casts

Fossils may also be preserved as molds and casts. After an organism, or part of an organism, such as a plant, leaf, or insect, is buried in mud or other sediments, its hard body parts become a fossil as the sediments become rock. If the fossil later dissolves out of the rock, a hollow depression in the rock, called a **mold,** results. The mold shows the original shape and surface of the fossil. Minerals may then seep into the mold and fill it, forming a **cast,** or copy, of the original fossil. Molds and casts of shellfish are common fossils. In the rock layers of some areas, there are molds of ferns, leaves, and fish.

DIFFERENTIATING INSTRUCTION

Reading Support Sequencing the steps in a process is a strategy that students can use throughout this chapter to help them understand and remember what they read. Tell students that whenever they come to a process that they don't understand, they should stop and identify the specific steps in order, writing them down if necessary. The information about the formation of replaced remains is a good example of events that can be sequenced. Have students identify each step in that process. Suggest that students who are having difficulty write each step of a process on an index card. They can mix the cards up and then place them in correct order as a means of review.

Use Reading Study Guide, p. 99.

Cast and Mold Formation

① A brachiopod has died and fallen into soft sediments. The hard shell remains after the soft body parts decay.

② The sediments become rock. The shell decays, leaving a mold of the shell.

③ Minerals fill the mold, creating a cast of the original shell.

Trace Fossils

Often, no part of a skeleton or plant survives as a fossil. However, other, more indirect evidence of life may be preserved as a **trace fossil**. Trace fossils include any impressions left in rock by an animal, such as trails, footprints, tracks, burrows, and even bite marks on fossils of trilobites. Scientists infer the existence of many animals from trace fossils. For example, scientists can learn about dinosaurs that lived in a particular area from footprints the animals left behind.

Carbonaceous Films

Sometimes the only fossil trace is a thin carbon film resembling a silhouette. The remains of plants or animals in sediments are affected by high temperature and pressure as additional sediments are deposited. These conditions cause the carbon compounds making up the tissues of animals and plants to undergo chemical changes. This carbonizing process results in a thin film of carbon that details the remains.

29.1 Section Review

1. Identify two ways in which remains may be preserved as fossils.
2. Describe trace fossils and how paleontologists use them to make inferences about the past.
3. **COMPARE AND CONTRAST** Explain how original remains, replaced remains, molds, and casts are similar and how they are different.
4. **PAIRED ACTIVITY** Work with a partner to complete the table with examples of each type of fossil.

Original	Replaced	Molds	Casts

VISUALIZATIONS
CLASSZONE.COM

Observe how fossils can form.
Keycode: ES2901

FOSSIL BRACHIOPODS You can see examples of both molds and casts in this photograph. The molds are hollowed, the "reverse" of the original shell.

Chapter 29 Studying the Past **649**

CHAPTER 29 SECTION 2

FOCUS

Objectives
1. Summarize the principles scientists use to determine the relative age of Earth's rocks.
2. Describe three types of unconformities.
3. Identify methods scientists use to correlate rock layers.

Set a Purpose
Have students read this section to find the answer to "How do scientists determine the relative ages of Earth's rock layers?"

INSTRUCT

More about...

Relative Dating
Nicolaus Steno, a physician working in Florence, Italy, is given credit for being the first to recognize the principle of superposition and to apply it while studying rock layers in the mountains of western Italy. In 1669, the principle was first clearly stated by Steno, who is also given credit for stating the principle of original horizontality.

29.2

KEY IDEA
Scientists use a number of principles and methods to determine the order of past geologic events.

KEY VOCABULARY
- relative dating
- strata
- unconformity
- correlation
- index fossils
- key bed

VOCABULARY STRATEGY
The word *superposition* comes from the Latin words *super* and *pōnere*, which mean, respectively, "above" and "to place." *Superposition* means "the state of being placed above."

Relative Time

Prior to the late 1700s, scientists estimated the age of Earth to be only about 6000 years old. Evidence of geologic processes such as erosion and sedimentation occurring over extremely long time spans, however, led to the realization that Earth must be much older. Scientists then tried to determine Earth's age; by examining layers of rock, they were able to get an idea of geological history. Paleontologists cannot always identify the exact age of a specimen or the exact time of an event. Therefore, they often use a process called relative dating.

Relative Dating

Relative dating is the process of placing events in the sequence in which they occurred. Relative dating does not identify the actual dates on which the events occurred. For example, suppose your family took a trip several years ago. You cannot remember the exact date of the trip, but you know that it took place before you entered fifth grade. In recalling the trip in this way, you are using relative dating. Scientist use relative dating to help them determine the relative ages of Earth's rocks. Relative dating is based on certain principles, or rules, for telling relative time.

The Principle of Superposition

Sedimentary rock forms after particles settle out of a fluid and, over time, are compressed into layers, or **strata**. The Principle of Superposition states that in an undisturbed sequence of sedimentary strata, the oldest rock layer will be at the bottom and the youngest will be at the top. The Principle of Superposition is the basis for all relative dating and is a fundamental concept in studying Earth's history.

Remember that most layers of sediment are deposited in a horizontal position. You can see that the strata shown in step 1, below, have not been disturbed. Thus you can infer that they are in their original horizontal positions.

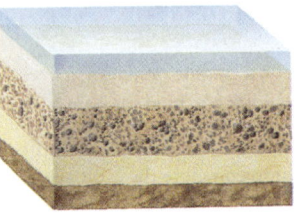

1 According to the principle of superposition, the oldest layer is at the bottom. The youngest layer of sediment, the last deposited, is at the top.

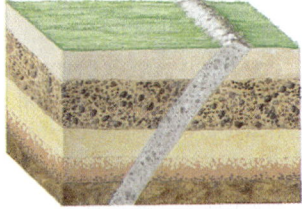

2 The intrusion is the youngest of the layers shown here. It demonstrates the principle of cross-cutting relationships.

DIFFERENTIATING INSTRUCTION

Hands-On Demonstration Demonstrate the principle of superposition. On a desk, stack newspapers in chronological order from oldest on the bottom to youngest on the top. Read aloud the date of each paper as you place it on the desk. Then ask: Where in the stack is the oldest paper? **bottom** the youngest paper? **top** Next, ask students to imagine that each newspaper is a sedimentary layer deposited one after the other. Ask: Which layer was deposited first? **bottom** Which was deposited last? **top** Apply the principle of superposition to describe the relative age of each layer.

The Principle of Cross-Cutting Relationships

Step 2 on page 650 shows a magma intrusion cutting across the horizontal strata. The principle of cross-cutting relationships states that an igneous intrusion is always younger than the rock it has intruded or cut across. Therefore an intrusion is always younger than the surrounding sedimentary layers.

Embedded Fragments

Rocks that are embedded in another rock must be older than the rock in which they are contained. For example, the pebbles in a conglomerate must have existed before the conglomerate formed and therefore are older than the conglomerate.

Gaps in Relative Time

Another way of reading the rock record involves the examination of unconformities. An **unconformity** indicates where layers of rock are missing in the strata sequence; thus, unconformities suggest missing evidence. The processes that cause unconformities all require huge expanses of time and represent large gaps in geologic time. Strata may be missing because they were never deposited. More likely, as when pages are missing from a book, the rock layers may have been deposited and later removed.

Angular Unconformity

Original horizontal sediments are sometimes tilted during uplift. Over time the exposed surface is worn down. An angular unconformity results when younger, flat strata are deposited on top of the older strata.

Disconformity

Sometimes original horizontal sediments are lifted up and the top layers eroded. When they are resubmerged beneath a freshwater or saltwater sea,

Observe an animation showing the formation of an unconformity.
Keycode: ES2902

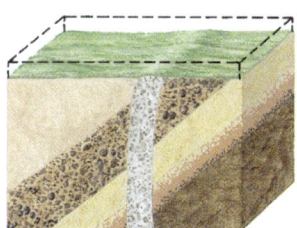

③ Over time, rock layers may tilt and erode. Such changes can cause gaps in the rock record. This demonstrates angular unconformity.

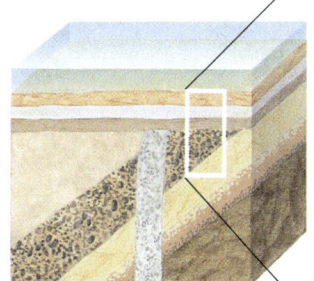

④ An angular unconformity has formed after new sediments were deposited on tilted rock layers.

UNCONFORMITY The horizontal layers at the top of this photograph of an unconformity are younger than the tilted rock layers below them.

CHAPTER 29 SECTION 2

Observe an animation showing the formation of an unconformity.
Keycode: ES2902
Visualizations CD-ROM

VISUAL TEACHING
Discussion

Ask: Why is the top layer blue in diagram 1 and green in diagrams 2 and 3? The blue represents water from which sediment is settling. The green represents grass on the exposed surface. Then ask: In diagram 4, which is the oldest rock layer? the tilted brown layer in the lower right corner Which is the youngest rock layer? the horizontal rock layer on top Have students examine diagram 2 carefully, ask: How many intrusions do you see? two, note a horizontal intrusion (sill) predates the later intrusion to the surface (dike)

Extend

Relate the strata shown to the rock cycle on page 119 of Chapter 6. Weathering and erosion of igneous rock, metamorphic rock, or sedimentary rock could have contributed to the strata shown in diagram 1. The intrusions in diagram 2 are composed of igneous rock. While the layers of rock have been tilted in diagram 3, they have not changed in structure, which would be characteristic of metamorphic rock.

Use Transparency 35.

DIFFERENTIATING INSTRUCTION

Challenge Activity Have students create their own series of 4–6 diagrams representing a particular scenario of geologic events. Suggest they use Chapter 6 as a guide to rock formation, introducing into their scenarios igneous intrusions (p. 125) and conglomerates (p. 128). They can use Chapter 11 to introduce folds (p. 239) and faults (p. 240). They should prepare a report that accompanies their diagrams, explaining the geologic events that led to the particular formations. Then have them present their diagrams to the class and ask the class to use the principles of superposition and cross-cutting relationships to establish the sequence of events.

CHAPTER 29 SECTION 2

Visual Teaching

Interpret Diagrams

Ask: What two formations have been correlated between the two national parks? **Moenkopi formation and Kaibab limestone** Do you think that these rock layers were likely correlated by walking the outcrop? Explain your answer. **No, the areas are widely separated in different states.**

Extend

Ask students if they think the two orange-colored rock layers in the two strata sequences are also matching strata, and to explain their answers. **According to the principle of superposition, the layers are different relative ages, one being above and younger than the Kaibab limestone and the other being below and older. Thus, they cannot be matched.**

CRITICAL THINKING

Describe the following events: In a river environment, a layer of sand and then a layer of silt is deposited. These layers later become sandstone and shale. With time, the rock layers are uplifted and the shale is eroded. On the surface of the remaining sandstone, a layer of glacial till is deposited followed by a layer of sand. Ask: Between which layers does an unconformity occur? **at the boundary between the sandstone and till** What type is it? **a disconformity**

What Stories Do Rocks Tell? Read rock layers to interpret the geologic story of a place. Check your predictions by watching an animated sequences.
Keycode: ES2903

new sediments are deposited. The result is called a *disconformity* because, even though the layers are still horizontal, some layers are missing. It may be hard to identify a disconformity because no folding or tilting has occurred. Nevertheless, a disconformity results in a gap in the rock record.

Nonconformity

A nonconformity results when sedimentary layers have been deposited on igneous or metamorphic rock. If a crystalline rock such as granite is lifted up from deep within Earth's interior, it begins to erode on the surface. If the granite is then submerged under a body of water, sediments are deposited on top of the eroded granite. The resulting boundary is called a nonconformity.

Rock Layer Correlation

Correlation is the matching of rock layers from one area to another. For example, you can see the strata from two locations below. A geologist might wish to know if a layer of limestone in Zion National Park is the same limestone layer as one in Grand Canyon National Park. Geologists use several methods to match, or correlate, rock layers.

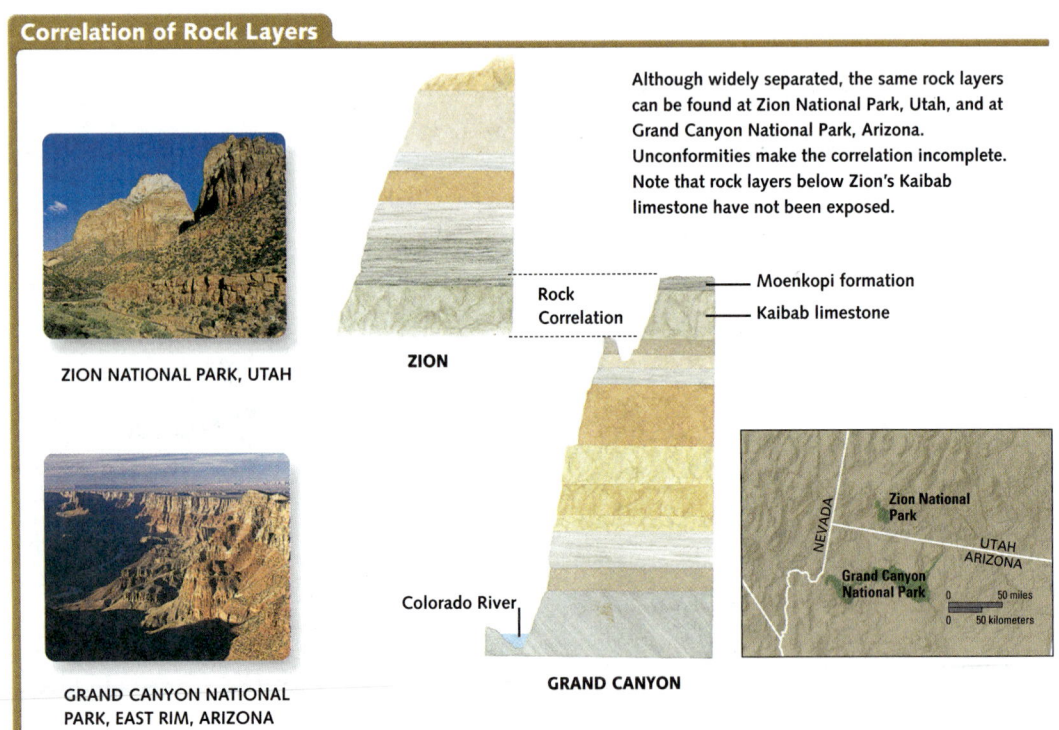

Correlation of Rock Layers

Although widely separated, the same rock layers can be found at Zion National Park, Utah, and at Grand Canyon National Park, Arizona. Unconformities make the correlation incomplete. Note that rock layers below Zion's Kaibab limestone have not been exposed.

ZION NATIONAL PARK, UTAH
GRAND CANYON NATIONAL PARK, EAST RIM, ARIZONA
ZION
GRAND CANYON
Rock Correlation
Moenkopi formation
Kaibab limestone
Colorado River

652 Unit 8 Earth's History

DIFFERENTIATING INSTRUCTION

Developing English Proficiency Students may have difficulty distinguishing the three types of unconformities. On the board, write the terms *angular unconformity, disconformity,* and *nonconformity.* Underline the root *conformity* common to all three terms. Then circle *angular un-, dis-,* and *non-*. Under each term, use colored chalk to sketch the unconformity, erasing to represent erosion. Point out the differences among the three unconformites, and help students to connect the differing prefixes with the differing circumstances represented in the sketches.

Use Spanish Vocabulary and Summaries, pp. 57–58.

Walking the Outcrop

A simple and direct method of correlation is "walking the outcrop." An outcrop is the part of a rock layer that can be seen at Earth's surface. Walking along outcrops is an easy way of finding out whether two rock layers are the same. Walking can be difficult, however, in areas where the outcrop is covered by thick soil or heavy vegetation or where extensive erosion has occurred. Also, this method is not practical when comparing outcrops that are far apart.

Matching Rock Characteristics

In this method, rocks are matched by characteristics such as appearance, color, and composition. For example, a rock that weathers with a distinctive rust color would be easy to recognize and therefore usable for correlation.

Using Index Fossils

One of the best methods of correlating rock layers over long distances is the use of index, or guide fossils. **Index fossils** are the remains of animals that lived and died within a particular time segment of Earth's history. After death, their bodies were deposited in sediments. Most of the animals then became extinct. Therefore, each layer contains fossils unlike those in the layer above or below it. Geologists can use these distinctive fossils to correlate layers of rock from different areas.

An index fossil has four characteristics. First, the fossils are easily recognizable. That is, they are unique in some way that enables scientists to easily distinguish them from other, similar fossils. Second, index fossils are abundant. Third, the fossils are widespread in occurrence. That is, they are found over a broad geographical area. This makes it possible to use them to match rock layers of a particular age over wide areas. Fourth, index fossils are the remains of organisms that existed for only a brief period, so the fossils occur in only a few rock layers. Index fossils found in rock layers in different parts of the world indicate that the layers were formed at the same time.

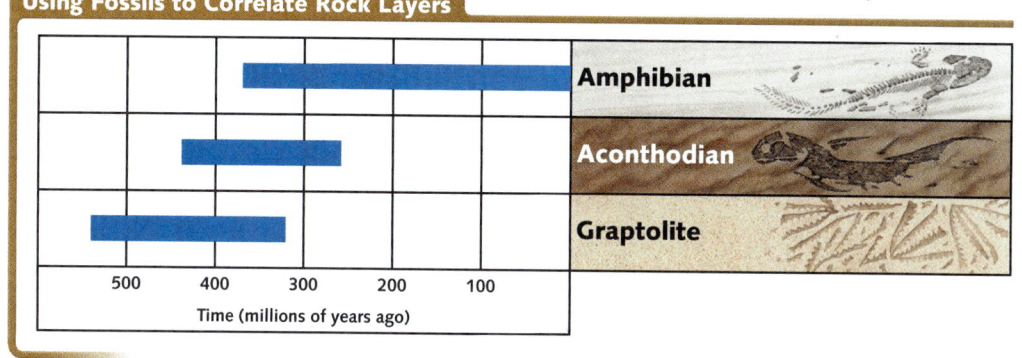

INDEX FOSSILS are rare. For this reason, paleontologists use groups of fossils to correlate rock layers. Fossil amphibians, aconthodians, and graptolites can be used to correlate rock layers.

Scientific Thinking

INTERPRET HISTORY

Using correlation, scientists have determined the relative ages of most of the rocks on Earth's surface. They have worked out a timetable that divides geologic time into units. Refer to this timetable on pages 668–669. Based on the rock correlation chart of Zion and the Grand Canyon, find the relative ages of the area of correlation. Hint: Kaibab is Permian limestone, and the Moenkopi formation is largely Triassic sandstone. Based on information in the timetable, infer what North America may have been like. What life forms were in existence? What geologic forces were present?

CHAPTER 29 SECTION 2

INVESTIGATIONS
CLASSZONE.COM

What Stories Do Rocks Tell?
Keycode: ES2903

Use Internet Investigations Guide, pp. T101, 101.

Visual Teaching

Interpret Diagram

Point out that of the three types of organisms shown in the graph, only amphibians still live on Earth. Ask: When did graptolites live on Earth? **about 540 to 330 million years ago** How old is a rock layer that contains graptolites, acanthodians, and amphibians? **about 370 to 330 years old** Which organism shown lived on Earth for the shortest period of time? **acanthodians**

Scientific Thinking

INTERPRET HISTORY

Most of the rock layers at Zion are younger than those of the Grand Canyon, as indicated in the diagram on page 652. The abundance of corals in the Permian Period (see pages 668 and 669) indicate this area was covered by a shallow sea at that time. Reptiles and land plants were abundant during the Triassic Period, and the sea had receded in this area.

DIFFERENTIATING INSTRUCTION

Reading Support Model for students how they can preview the text material under the red heading "Rock Layer Correlation" to prepare for reading and improve comprehension. Read the first paragraph. Point out that the blue headings that follow discuss different methods of correlating rock layers. By thinking for a moment about each heading, students can begin to construct an understanding of what those methods entail. Say: "By reading the first blue heading, I figure that this method involves walking outside. I'm not sure what an outcrop is, but I'll find out when I read the paragraph." Have volunteers reason aloud similarly for the other headings.

Use Reading Study Guide, p. 100.

CHAPTER 29 SECTION 2

More about...

Fossils and Correlation
A basic principle of geology in terms of correlation of strata is the *principle of faunal succession*. It states that fossil organisms succeed each other in a definite and recognizable order. In other words, the kinds of animals and plants that are found as fossils change predictably through time. As a result, the relative age of rocks can be determined from their fossil content.

ASSESS

1. No; it can only be used to determine the age relative to another rock layer or event, or the sequence in which it formed relative to other rock layers or events.

2. They indicate where layers of rock are missing in the strata sequence.

3. Rock characteristics, index fossils, key beds, or strata sequences in the two locations can be matched.

4. Since dinosaurs were extinct before humans evolved, you would expect dinosaur fossils to be older and thus, according to the principle of superposition, below human fossils in an undisturbed strata sequence.

5. 5.0×10^8, 4.0×10^8, 3.0×10^8, 2.0×10^8, 1.0×10^8

Fossils as Environmental Indicators

In addition to determining relative time and correlating rock layers, fossils are important as indicators of past climate. For example, coral reefs today form only in shallow, warm water between approximately 30° N and 30° S latitudes. A rock containing fossil coral is evidence that the particular area was once covered with shallow, warm water.

Matching Key Beds

A single rock layer that, like an index fossil, is unique, easily recognizable, and widespread is called a **key bed.** Large volcanic eruptions are an excellent source of material for key beds. Volcanic ash and debris are spread over a large area. This material eventually becomes a clay, called bentonite, in sedimentary rock layers. In some areas, beds of bentonite have been used to correlate rock layers. Similarly, dust and debris from the impacts of large meteorites on Earth's surface create key beds. The layer that results from such events represents a single instant in time but is unique and widespread.

Stratigraphic Matching

Suppose you have followed an outcrop of limestone around a canyon wall. You then travel a short distance to a second location, in a small valley, and notice what appears to be the same layer of limestone. You remember that the limestone layer in the canyon was sandwiched between two layers of conglomerate rocks. The limestone in the valley is also sandwiched between conglomerate layers. You can conclude that the two limestone layers represent two outcrops of the same stratum. You have matched not just the outcrop of limestone, but the sequence of the three layers—conglomerate, limestone, conglomerate—from the two locations. This correlating of strata sequences is called stratigraphic matching.

29.2 Section Review

1. Can the principles of relative dating be used to determine the date a rock layer was deposited? Explain your reasoning.

2. How do unconformities represent gaps in geologic time?

3. Describe three methods you could use to correlate rock layers in two distant locations.

4. **CRITICAL THINKING** Would you expect to find dinosaur fossils above or below human fossils in an undisturbed strata sequence? Explain your reasoning.

5. **MATHEMATICS CONNECTION** Use scientific notation to express the dates given on the table on page 653.

MONITOR AND RETEACH

If students miss . . .

Question 1 Explain that *relative* means "in relation to something else."

Question 2 Review "Gaps in Relative Time." (pp. 651–652) Have students define *unconformity* in their own words.

Question 3 Reteach "Rock Layer Correlation." (pp. 652–654) Be sure students can distinguish all the different methods used to correlate rock layers.

Question 4 Reteach "The Principle of Superposition." (p. 650) Ask: Which are older, dinosaur remains or human remains? dinosaur remains

SCIENCE & Technology

Ancient Rocks Provide Clues

A group of geologists working in western Australia dug tiny zircon crystals from ordinary-looking sedimentary rocks. When they measured the ratios of uranium to lead, they found the crystals to be 4.4 billion years old. That's about 130 million years older than any previously discovered Earth crystals!

What is the significance of this information?

Radiometric dating has led to the discovery of Earth materials so ancient that scientists are scratching their heads about what it means. What's more, further tests on the zircon crystals gave evidence that they formed on a solid crust in the presence of water.

Until this discovery, most geologists thought Earth at that time was a molten blob, still millions of years away from forming a crust! If the ancient crystals formed on a crust, geologists may have to change their picture of early Earth.

If water existed that early on Earth, life, too, might have been present far earlier than had been previously thought. At that time, asteroids and meteorites were still slamming into Earth on a regular basis. Some

ZIRCON CRYSTALS that date back 4.4 billion years have forced scientists to reconsider their ideas about Earth's history.

scientists are now wondering if forms of life may have started during a lull in the bombardments. Could more impacts have led to a mass extinction, with all the fossil records having been obliterated? Consider the result of the impact of a single asteroid on the dinosaur population only 65 million years ago.

The zircon crystals found in Australia were tiny fragments of bigger rocks eroded away millennia ago. Radiometric dating of rocks in northwestern Canada has recently revealed a staggering age—around 4.03 billion years old. These ancient metamorphic rocks provide additional clues about the formation of Earth's crust and the continents. ∎

Extension

SCIENCE NOTEBOOK
Certain areas of Earth, such as mountains and valleys beneath the sea, have not been dated. Explain how radiometric dating could be used to explore these areas of Earth.

Learn more about current methods of scientific dating.
Keycode: ES2904

SPECTROMETER Lab workers can use a spectrometer to analyze the chemical composition of rocks and crystals.

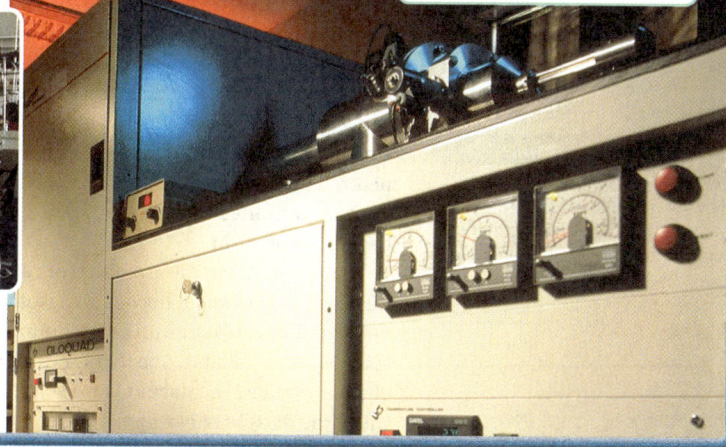

CHAPTER 29 SECTION 2

Learn more about current methods of scientific dating.
Keycode: ES2904

Science and Technology

Dating Zircon
Geologists use a technique called fission-track dating to determine the age of crystals in the mineral zircon. Zircon normally contains a small number of atoms of the radioactive element uranium. These radioactive atoms give off particles that leave tracks in the crystals. The older the crystals, the more of these tracks are visible inside them. Scientists count the number of tracks and, with additional information, calculate the amount of time that has passed since the crystals formed.

Science Notebook
Submersibles or drilling equipment can remove samples of rock from the ocean floor so that they can be sent to labs for dating. This information could tell scientists more about when and how features on the ocean floor formed.

MONITOR AND RETEACH

Question 5 Students should first write out the numbers. Point out that the labels on the bottom of the graph refer to *millions* of years. Refer students to the section on scientific notation in Appendix C, page 731.

CHAPTER 29 SECTION 3

29.3

FOCUS

Objectives
1. Explain the process of radioactive decay.
2. Define *half-life*.
3. Describe how radiometric dating is used to measure absolute time.

Set a Purpose
Read the key idea aloud. Then have students read the section to find the answer to "How can we tell the actual age of a rock or fossil?"

INSTRUCT

More about...

Historical Methods
Some scientists at the turn of the 20th century believed that the salinity of Earth's oceans held the key to determining the age of Earth and its rocks. Assuming that the oceans originally held freshwater, scientists believed they could estimate the total amount of salt in the oceans and the amount of salt transported to the oceans each year, and then use this information to calculate Earth's age. Using this method, John Joly, an Irish physicist, estimated Earth's age at about 90 million years.

KEY IDEA
Scientists use radioactive-dating methods to measure absolute time.

KEY VOCABULARY
- absolute time
- varve
- radioactive decay
- parent isotope
- daughter isotope
- half-life
- radiometric dating

TREE SAMPLES By studying tree rings, scientists can determine a tree's age and the environmental conditions in which it grew, including rainfall and temperature. Somewhat like rock layers, tree rings can be correlated.

INVESTIGATIONS CLASSZONE.COM
How Do Trees Record Time?
Analyze patterns in tree rings to find out how trees tell absolute time.
Keycode: ES2905

Absolute Time

Relative dating principles help scientists place events in a sequence. But those principles cannot help determine the actual dates of events. **Absolute time** identifies the actual dates of the events.

Historical Methods

Before the 20th century, scientists had only a few, limited means of measuring absolute time. One method involved estimating rates of erosion and sedimentation. Unfortunately, such rates are not constant, especially over time periods greater than thousands of years. Therefore only the ages of young geologic features could be accurately measured.

Another method used to measure absolute time was counting tree rings—a method still used today. As you can see on the left, a stump or limb has concentric rings of various widths. Each ring usually represents a single year. The width of a ring depends upon the temperature and the amount of rainfall that year. Each ring in a tree differs from other rings in that tree, yet is similar to the rings formed during that same year in other trees. Therefore, the pattern of rings in one tree can be correlated with that in another tree. By applying this technique to wood in Native American ruins in the southwestern United States, scientists have determined dates back to almost 2000 B.C.

A third method of measuring absolute time is by counting varves. A **varve** is any sediment that is deposited on a yearly cycle. While varves can form in any body of water, they are clearest in glacial lakes formed during an ice age. In these lakes, two distinct layers of sediment were deposited each year—a thick, light-colored sandy layer in summer and a thin, dark-colored clay layer in winter. Like an annual tree ring, each annual varve is distinctive. As a result, the varves of one lake can be correlated with the varves of other lakes. By matching deposits, scientists have determined dates back to 15,000 years ago.

Radioactivity

In the mid-1900s, scientists began using a new tool to measure absolute time—radioactive isotopes. You recall learning about isotopes in Chapter 5. Radioactive isotopes emit or capture tiny particles in a process called **radioactive decay.** The diagram on page 657 shows the three main types of decay: alpha decay, beta decay, and electron capture. In alpha decay, an alpha particle (two protons and two neutrons) is emitted. In beta decay, a neutron becomes a proton and an electron (or beta particle) and the electron is emitted. In electron capture, a proton captures an electron and becomes a neutron.

Each time a particle is emitted or captured, the isotope's atomic number changes and it becomes an isotope of a different element. If this new isotope is radioactive, decay continues. Radiation is released until a stable isotope—that is, one that is not radioactive—forms. The original element is called the **parent isotope,** and the product of the decay is called the **daughter isotope.**

656 Unit 8 Earth's History

DIFFERENTIATING INSTRUCTION

Reading Support Help students master the special vocabulary of science. Point out that definitions are often followed by one or more examples that help clarify their meaning. For example, draw students' attention to the boldface type and definitions of *parent isotope* and *daughter isotope* on page 656. Then point out the examples of a parent isotope, U-238, and a daughter isotope, Pb-206, given in the next paragraph on page 657.
Use Reading Study Guide, p. 101.

656 Unit 8 Earth's History

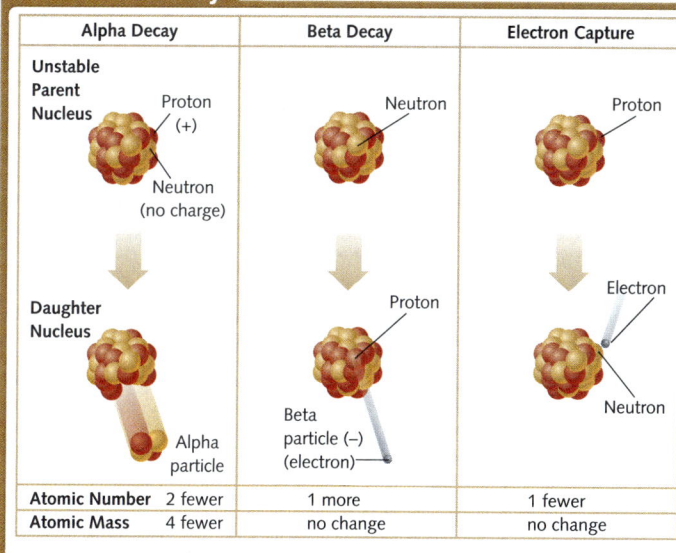

An example of this process is the decay of the radioactive isotope uranium-238 to lead-206. When the parent isotope U-238 gives off an alpha particle, it becomes thorium-234, which is also radioactive. Decay continues, with alpha and beta decay and electron capture, until finally a stable isotope is formed that is not radioactive—the daughter isotope Pb-206.

Half-Life

Radioactive elements decay at characteristic constant rates that are not affected by the passage of time or changes in temperature or pressure. At the moment an igneous rock crystallizes, radioactive elements in the rock begin to decay. The ratio of the amount of radioactive element left in the rock to the amount of stable product can be used to determine the absolute age of the rock.

The rate at which a radioactive element decays is called its half-life. **Half-life** is the time it takes for half the radioactive atoms in a sample to decay to a stable product. Refer to the graph of half-life in the Appendix, page 706. The graph shows how half-life can be used to date material.

After one half-life, half of the atoms of a radioactive element will have decayed to a stable product and half will remain unchanged. During the next half-life, half of the remaining radioactive atoms will decay.

During each half-life, half of the radioactive material in a sample, no matter how small the amount, will decay to a stable product. Half-lives range from a fraction of a second to billions of years. For example, the isotope protactinium-234 has a half-life of about one minute, whereas the isotope uranium-238 has a half-life of 4.5 billion years.

25-Minute Mini LAB

A Decay Path

Materials
- periodic table of elements
- graph paper

Procedure

1. Before becoming Pb-206, U-238 decays to Th-230, as shown below (1 alpha, 2 beta, and 1 alpha decay). Copy this graph on the top half of a sheet of graph paper.

2. The decay events from Th-230 to Pb-206 are 4 alpha, 2 beta, 1 alpha, 2 beta, and 1 alpha decay. Complete your graph. Label each point.

Analysis

Why does the atomic number increase after beta decay? Explain why there might be a pattern to a decay path.

CHAPTER 29 SECTION 3

INVESTIGATIONS
CLASSZONE.COM

How Do Trees Record Time?
Keycode: ES2905

Use Internet Investigations Guide, pp. T102, 102.

Visual Teaching

Interpret Diagrams
Remind students that the atomic number of an element is the number of protons in the nucleus of each atom, the atomic mass is the number of protons and neutrons together.

MINILAB
A Decay Path

Outcome
The labels should be, in order: uranium-238, thorium-234, protactinium-234, thorium-230, radium-226, radon-222, polonium-218, lead-214, bismuth-214, polonium-214, lead-210, bismuth-210, polonium-210, lead-206.

Analysis Answers
- During beta decay, a neutron splits into a proton and an electron. The electron is emitted, leaving the proton. The atomic number goes up by one.
- The atom goes through a pattern of one alpha decay, two beta decays. With alpha decay, the atom loses two protons, two neutrons. To balance positive and negative charges, the atom twice converts a neutron to a proton (beta decay). A lighter isotope of the same element results.

DIFFERENTIATING INSTRUCTION

Hands-On Demonstration To demonstrate half-life, use 16 paper clips to represent atoms of a radioactive parent isotope with a half-life of 1 minute, and pennies to represent atoms of a stable daughter isotope to which it decays. Show students that after 1 minute or 1 half-life, 8, or half, of the parent atoms decay to form 8 daughter atoms by replacing 8 paper clips with 8 pennies. After 2 minutes or 2 half-lives, replace 4 more paper clips with 4 pennies. After 3 minutes or 3 half-lives, replace 2 more paper clips with 2 pennies. Have students predict what will happen after 4 half-lives. One paper clip will be replaced with 1 penny.

CHAPTER 29 SECTION 3

More about...

Carbon 14
The isotope carbon 14 is produced continuously in the upper atmosphere where high-energy nuclear particles shatter the nuclei of gases, releasing neutrons. The neutrons become absorbed by nitrogen gas, causing each nitrogen nuclei to emit a proton. The resulting carbon 14 atoms combine with oxygen to form carbon dioxide gas, which is absorbed from the atmosphere by living things. Carbon 14 has become an essential tool of geologists, archaeologists, and anthropologists for determining the age of artifacts and remains.

CRITICAL THINKING

Say: A scientist picks up a sample of conglomerate. It contains fragments of granite, an igneous rock, but no zircon crystals. What type of rock is this? **sedimentary** What methods might the scientist use to date the sample? **rubidium-strontium or potassium-argon** What method of dating cannot be used? **uranium-lead, because of the lack of zircon** Ask: What does the date determined indicate about the sample? **The resulting date (the age of the granite) will predate the age of the conglomerate.**

Radiometric Dating

Scientists use radioactivity and half-lives of elements to measure absolute time through a technique called radiometric dating. In **radiometric dating,** scientists measure the amounts of a parent and a daughter isotope within a rock or mineral and use the ratio of the two to find the age of the rock. A version of radiometric dating used on organic material, such as preserved remains, can determine the time of an organism's death. Carbon-14 is used to date organic material, and uranium-lead, rubidium-strontium, and potassium-argon, are used to date rocks and minerals.

Radiocarbon Dating

All living things continually take in from their environments both radioactive carbon-14 and nonradioactive carbon-12. These amounts remain constant as long as the organism is living. When an organism dies, however, the amount of carbon-14 in its tissues decreases. The amount of carbon-12 remains the same. Thus the ratio of carbon-12 to carbon-14 can be used to tell when a plant or animal died.

Radiocarbon dating has two limitations. First, the method can only be used to date objects from organisms that once lived, such as logs and bones. Second, because the half-life of carbon-14 (5730 years) is relatively short, the method is limited to dating items that are about 70,000 years old or younger. Radiocarbon is invaluable, though, for dating such items as the wooden tools and skeletons of prehistoric people, and plant and animal remains preserved in deposits from the Ice Age.

Uranium-Lead Dating

The radioactive isotope uranium-238 decays to the isotope lead-206, the stable end product. The half-life of U-238 is 4.5 billion years. This long half-life makes it possible to use this method to date the oldest rocks of Earth's crust. Although uranium is common throughout the world, it occurs only in trace amounts in a mineral called zircon, which is found in some igneous rocks. It is rarely found in sedimentary or metamorphic rocks.

RADIOCARBON DATING Shavings of fossilized reindeer bone will be tested using the radioactive isotope carbon-14.

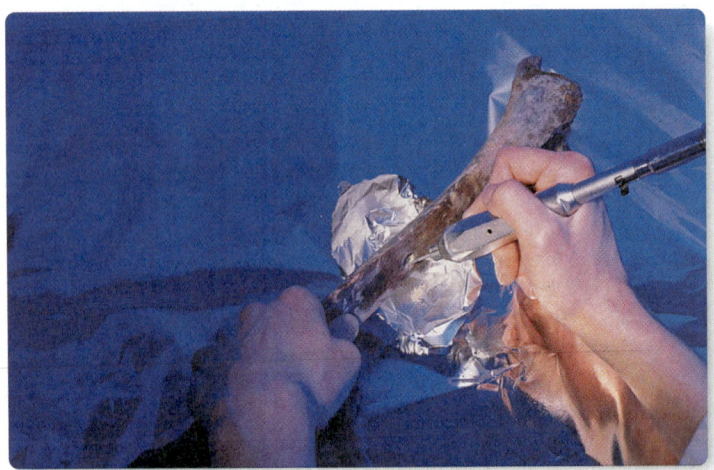

DIFFERENTIATING INSTRUCTION

Challenge Activity Provide students with a hypothetical situation in which scientists find the remains of three Ice-Age animals, such as a woolly mammoth or saber-toothed cat, in different locations. Measurements indicate that the bones of one animal contain 1/4 of its carbon-14. The bones of the other two animals contain 1/8 and 1/32 of its carbon-14. Challenge students to calculate the age of each animal. **$1/2 \times 1/2 = 1/4$, two half-lives = 11,460 years; $(1/2)^3 = 1/8$, three half-lives = 17,190 years; $(1/2)^5 = 1/32$, five half-lives = 28,650 years**

Isotopes Used in Radiometric Dating

Parent Isotope	Decay System	Daughter Isotope	Half-life (years)	Effective Range (years)	Possible Materials for Dating
Carbon-14	Beta decay	Nitrogen-14	5730	100–70,000	Once-living matter (wood, charcoal, bone)
Uranium-238	Alpha decay, Beta decay	Lead-206	4.5 billion	>10 million	Uranium-bearing minerals (zircon)
Rubidium-87	Beta decay	Strontium-87	47 billion	>10 million	Micas, feldspars, metamorphic rocks
Potassium-40	Beta capture	Argon-40	1.3 billion	>50,000	Micas, amphiboles, feldspars, volcanic rocks

Also, because of the long half-life of U-238, the uranium-lead method does not give reliable results for rocks that are younger than 10 million years old. In rocks that are younger, too little of the U-238 will have decayed to measure accurately.

Rubidium-Strontium Dating

A second method of dating rocks involves the decay of rubidium-87 to strontium-87. The half-life of rubidium-87 is about 47 billion years—more than ten times Earth's age! Because of this long half-life, the rubidium-strontium method is best for dating extremely old rocks. Rubidium occurs in common minerals such as feldspars and micas and thus can be used to date almost all igneous rocks. If both rubidium-87 and uranium-238 occur in the same rock, both the uranium-lead and rubidium-strontium methods can be used, and the values obtained can be checked against each other.

Potassium-Argon Dating

Potassium-40, with a half-life of about 1.3 billion years, decays to the element argon-40. Potassium, a very common element, is found in feldspars, micas, and amphiboles. Minerals that can be dated by the potassium-argon method are found in metamorphic and sedimentary rocks as well as in igneous rocks. In some cases, this method can date rocks as young as 50,000 years. Therefore, the method is used to date many rocks that cannot be dated by uranium or rubidium.

29.3 Section Review

1. What are the similarities between tree rings and varves?
2. Describe the three main types of radioactive decay.
3. What is the half-life of carbon-14? In what way does this half-life limit what can be dated by the radiocarbon method?
4. If an igneous rock is more than 3 billion years old, what are the two best methods you could use to find the exact age of the rock?
5. **APPLICATION** If you were to date a piece of petrified wood, would you use carbon-14 or potassium-argon dating? Why?

VISUAL TEACHING
Discussion

Use the table to compare and contrast the four isotopes used in radiometric dating. Make sure students understand that the different half-lives of the different isotopes result in their being effective for use in dating materials of differing ages.

ASSESS

1. Both are methods for determining absolute time and depend on yearly cycles. Each tree ring is distinctive, as is each varve, so both can be correlated.

2. In alpha decay, an alpha particle, consisting of two protons and two neutrons, is emitted. In beta decay, a neutron becomes a proton and an electron, or beta particle, and the electron is emitted. In electron capture, a proton captures an electron and becomes a neutron.

3. 5730 years; Because the half-life is short, the method can only be used to date items that are about 70,000 years old or younger.

4. uranium-lead dating and rubidium-strontium dating

5. carbon 14-dating, because this isotope can be used to date once-living matter such as wood

MONITOR AND RETEACH

If students miss . . .

Question 1 Reteach "Historical Methods." (p. 656) Have students prepare a compare and contrast chart for tree rings and varves.

Question 2 Reteach "Radioactivity." (p. 656) Have student pairs prepare flash cards with the name of each type of decay on one side and its description on the other. Using the cards, partners can take turns testing each other.

Question 3 Reteach "Radiocarbon Dating." (p. 658) Have students compare the half-lives and uses of carbon 14 and U 238 for absolute dating.

Question 4 Review the table of isotopes used in radiometric dating on page 659, emphasizing the ages for which each isotope is effective.

Question 5 Review the table of isotopes on page 659, emphasizing the possible materials for which each isotope is effective in dating.

CHAPTER 29 ACTIVITY

PURPOSE
To graph tree ring growth data and analyze the effect of environmental factors on tree ring growth

MATERIALS
Provide students with a copy of Lab Sheet 29. A copymaster is provided on page 151 of the teacher's laboratory manual.

If necessary, provide students with calculators to help them determine average growth.

PROCEDURE

1. A complete data table of tree ring growth and weather data is in Appendix A on page 707.

2. Students may also use Lab Sheet 29, or prepare their graphs on a computer using a spread sheet program.

4, 6. Students preparing their graphs from scratch may need help choosing an appropriate scale and limits for the y-axis. Refer to Lab Sheet 29. Limits and scale for the graph of precipitation can be −6 to 0 by increments of one; for the graph of temperature, suggest limits of 15 to 45, by fives. Remind students to title all graphs.

ANALYSIS AND CONCLUSIONS

1. In general, there is a negative correlation between growth and temperature; years having the lowest temperatures correlate with large ring widths.

CHAPTER 29 LAB Activity

Deciphering Tree Rings

SKILLS AND OBJECTIVES
- **Plot** tree ring growth data.
- **Analyze** the effect of different environmental factors on growth.
- **Correlate** the relationship between these factors.

MATERIALS
- colored pencils
- graph paper
- tracing paper
- ruler

The study of dating past events and environmental changes by analyzing tree rings is called *dendrochronology*. Before sophisticated technology and dating tools were available, scientists could study tree ring samples to make inferences about climate change and weather conditions, including events such as volcanic eruptions, fires, floods, or droughts.

In order to reconstruct past conditions based on tree rings, scientists must carefully analyze data. For example, tree growth data can be compared to weather data to find possible relationships. When data from one data set has a similar trend to another set, this is called a *positive correlation*. When the data from two sets have opposite trends, this is called *negative correlation*. When the data from different sets do not seem to follow any pattern, the sets are *uncorrelated*. In this activity, you will investigate the relationship between tree rings and past conditions. The experimental data was taken from five trees in a region of Alaska in the years from 1960 to 1994.

Tree Ring Growth and Weather

Year	Tree 1 Growth	Tree 2 Growth	Tree 3 Growth	Tree 4 Growth	Tree 5 Growth	Average Growth	Average Temperature	Total Precipitation
1994	2.54	1.73	3.51	2.39	3.66		−1.39	27.13
1993	2.58	1.59	4.05	3.25	4.15		−1.19	34.77
1992	2.67	2.38						28.68
1991	2.21	2.35						31.83

Procedure

Part A: Calculating Average Tree Growth

1. Using the data in the table, calculate the average growth for each year.

2. Draw a graph similar to the graph shown. Label the x-axis "Year" and the y-axis "Tree ring width (mm)." Choose a colored pencil and plot the data from Step 1 on your graph. Connect your data points to create a line graph. For each axis, choose appropriate limits and scales.

Part B: Correlating Growth and Precipitation

❸ Lay a sheet of tracing paper over the graph from Part A. With the aid of a ruler, trace the outline of the graph setup but not the data from Part A. Copy the x-axis labels, but not the y-axis labels.

❹ Label the y-axis of your traced graph "Total precipitation (cm)." Choose a different colored pencil. Using the grid of the graph from Part A as a guide, plot the precipitation for each year on the traced graph. Be sure to choose an appropriate scale and limits. Connect the data points.

Part C: Correlating Growth and Temperature

❺ Place the graph from Part B aside. Lay a clean sheet of tracing paper over the graph from Part A and trace the outline of the graph setup without the data. Again, copy the x-axis labels, but not the y-axis labels.

❻ Label the y-axis of your traced graph "Average temperature (°C)." Choose an appropriate scale and limits. Using a different colored pencil, plot the temperature data for each year. Connect the data points.

Analysis and Conclusions

1. Overlay the Growth and Temperature graph on top of the original Growth graph. Are these data related? If so, how? Use the terms *positive correlation, negative correlation,* or *uncorrelated* in your answers.

2. Based on your answers to Question 1, predict tree growth during a a warm year, a cold year, and a very cold year.

3. Overlay the Growth and Precipitation graph on top of the other two graphs so that all three are aligned. Is there a relationship between all three sets of data? If so, how are they related?

4. What observations can you make based on all three graphs? Which years had best growth? What were the temperature and precipitation trends during these years? What about the years of poorest growth?

5. What recent trends can you observe? Do you notice trends that show changes in the rate of growth?

6. Based on these trends, can you predict tree growth in 2010? 2050?

7. Based on the experimental data given, is there evidence for global warming or a global warming trend? Explain your reasoning.

Find out more about dendrochronology.
Keycode: ES2907

CHAPTER 29 ACTIVITY

Find out more about dendrochronology.
Keycode: ES2907

(continued)

2. **Possible Response:** Tree growth would be lowest during a warm year, moderate during a cold year, and greatest during a very cold year.

3. Uncorrelated; there is no discernable relationship among growth, precipitation, and temperature.

4. Based on the three graphs, one can conclude that, for this particular sample, growth is negatively correlated to temperature and that precipitation does not appreciably affect growth. Using the criterion of 5 mm of growth or more as "best growth", the years 1964 to 1966 had the most annual growth. The years 1978, 1980, and 1981 exhibited the least amount of growth and are, using the criterion of 1 mm or less, the years of poorest growth.

5. Students may observe the following general trends: since 1960, the average temperature has increased and the average rate of growth has decreased.

6. Students may predict that the average rate of growth will continue to decrease to 2010, but may suggest that 2050 is too far away to make an accurate prediction.

7. The data suggest that the average temperature has increased since 1960. Students may therefore suggest that there is evidence of a global warming trend.

Refer students to Safety in the Earth Science Laboratory on pages xxii–xxiii.

CHAPTER 29 REVIEW

VOCABULARY REVIEW

1. half-life
2. fossil
3. strata
4. correlation
5. index fossils

CONCEPT REVIEW

6. Any three of the following: Original remains: parts of or an entire organism preserved by being frozen or trapped in resin; Replaced remains: an organism's hard parts replaced by minerals, resulting in an exact copy; A mold: an organism buried in sediment that becomes rock later dissolves, leaving a hollow depression; A cast: a copy of a fossil that forms when minerals fill a mold; Trace fossil: an impression left in rock by an animal; Carbonaceous film: thick carbon films formed when high temperature and pressure cause carbon compounds in the buried remains of organisms to undergo chemical changes

7. Yes; they are replaced remains. The wood was replaced by minerals.

8. Both processes are used to determine when events occurred. Relative dating determines the sequence, whereas absolute dating determines the dates.

9. Each indicates gaps in the rock record. An angular unconformity consists of strata tilted by uplift and then overlain by horizontal strata. A disconformity is an area where some rock layers are missing, caused when sediments are lifted up, the top layers are eroded, and then new sediments are deposited. A nonconformity is the boundary between eroded igneous or metamorphic rock and sedimentary layers deposited on top, after the uplifted metamorphic or igneous rock was submerged.

10. The half-life of carbon-14 and the ratio of carbon-12 to carbon-14 are used to date the remains.

11. *Unstable parent nucleus* forms daughter nuclei by *alpha decay, beta decay*, and *electron* capture.

CHAPTER REVIEW 29

Summary of Key Ideas

29.1 Fossils, evidence of earlier life preserved in rock, are the basis for the geologic time scale and are an important part of the rock record. Fossils may occur as original remains, replaced remains, molds and casts, trace fossils, and carbonaceous films.

29.2 Relative dating is used to place past events in order without determining actual dates during which events occurred. Relative dating is also used to determine relative ages of specimens such as rocks or fossils. Relative dating is based on the Principle of Superposition, the Principle of Original Horizontality, and the Principle of Cross-Cutting Relationships. A gap in relative time may be the result of an unconformity in the rock record.

29.3 Absolute time identifies actual dates of events or ages of specimens. Scientists use radioactive dating methods to measure absolute time. Radiometric dating relies on radioactive decay and half-lifes of elements such as carbon, uranium, rubidium, and potassium.

KEY VOCABULARY

absolute time (p. 654)
cast (p. 647)
correlation (p. 650)
daughter isotope (p. 654)
fossil (p. 646)
half-life (p. 655)
index fossil (p. 651)
key bed (p. 652)
mold (p. 646)
paleontology (p. 646)
parent isotope (p. 654)
radioactive decay (p. 654)
radiometric dating (p. 656)
relative dating (p. 648)
strata (p. 648)
trace fossil (p. 647)
unconformity (p. 649)

662 Unit 8 Earth's History

CRITICAL THINKING

12. The oldest rock layer is at the bottom. The youngest rock layer is at the top. The area was once covered by water; volcanic activity later occurred.

13. No; the fossil is not unique, or easily distinguished from other fossils.

Vocabulary Review

Write the term from the key vocabulary list that best completes the sentence.

1. The time it takes for half the radioactive atoms in a sample to decay to a stable end product is the ___?___ .
2. Any evidence of earlier life preserved in rock is a(n) ___?___ .
3. Layers of rock are called ___?___ .
4. The matching of rock layers from one area to another is ___?___ .
5. Scientists correlate layers of rock over long distances by comparing the ___?___ found embedded in the rocks.

Concept Review

6. Describe three types of fossils and how they are formed.
7. Are the trees in Petrified Forest National Park fossils? Why or why not?
8. Compare and contrast relative dating and absolute dating.
9. Describe an angular unconformity, a disconformity, and a nonconformity. What does each indicate about past events?
10. Describe how radiocarbon dating can be used to determine when a plant or animal died.
11. **Graphic Organizer** Copy and complete the concept map below.

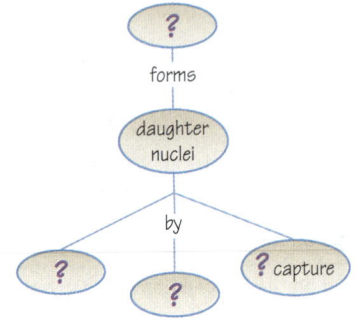

14. Presentations should include alpha and beta decay and electron capture. The unstable parent isotope uranium-238 undergoes alpha decay to become the daughter nucleus thorium-234. A series of daughter isotopes form through alpha and beta decay and electron capture until the formation of the stable daughter isotope lead-206.

CHAPTER 29 REVIEW

Critical Thinking

12. **Infer** A series of rock layers, from bottom to top, is sandstone, shale, limestone, basalt. What can you infer about the rock layer at the bottom? the top? about geologic events in the area?

13. **Evaluate** A fossil being considered for use as an index fossil has been found in large numbers in many parts of the world and seems to be limited to rock layers from the Ordovician period. It closely resembles a Cambrian trilobite. Would this fossil be a good index fossil? Explain.

14. **Communicate** Use diagrams as well as words to prepare a presentation that explains how a parent isotope of uranium becomes the daughter isotope lead via the process of radioactive decay.

15. **Hypothesize** A particular mammal fossil is discovered in identical rock strata in eastern South America and in western Africa. Form a hypothesis that explains how fossils of the same organism, dating from the same time, can be found in such disparate places.

16. **Apply** A table leg was made from the center of a tree that contains all its growth rings. Based on what you have learned about tree-ring dating, can you determine the age of the table? Of the tree at the time it was cut? Explain.

Interpreting Illustrations

The illustration shows a series of rock strata. Review what you know about sediment deposition, igneous intrusions, the Principle of Superposition, and cross-cutting relationships. Examine the illustration carefully, then answer the questions.

17. Which of these layers can you most easily identify as igneous rock? Explain.

18. Considering your answer to question 17, which strata are most likely *not* igneous rock? What types of rock could they be?

19. Would this stratification be possible if all the layers shown were sedimentary rock? Explain.

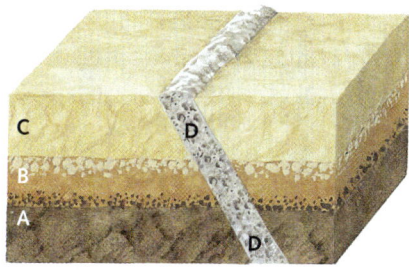

20. Based on your answers to questions 17–19, list the letters of the rock layers in order from oldest to youngest. Explain your answer.

Internet Extension

How Did the Layers of the Grand Canyon Form?
Tell the geological story of the Grand Canyon.
Keycode: ES2906

Writing About the Earth System

SCIENCE NOTEBOOK Rock layers, tree rings, and varves are like pages holding the record of the Earth system's story. Review information about those in your book, the Data Center, and other resources. Then describe in your science notebook how each one records events. Consider cataclysmic events, such as a volcanic eruption, as well as less dramatic events, such as the passage of time. Of rocks, trees, and varves, which one do you think best tells the story of Earth's four spheres—geosphere, hydrosphere, biosphere, and atmosphere? Explain your answer.

How Did the Layers of the Grand Canyon Form?
Keycode: ES2906

Use Internet Investigations Guide, pp. T103, 103.

(continued)

15. The identical rock strata correlate; therefore, South America and western Africa must once have been joined and later were moved apart.

16. You could determine the age of the tree when it was cut by counting growth rings. You could not determine the age of the table. Many years could have passed between the time the tree was cut and when the table was made.

INTERPRETING ILLUSTRATIONS

17. layer D; it crosscuts like an intrusion

18. A, B, and C are most likely sedimentary rock.

19. No; although sedimentary rock may occur at an angle (angular unconformity), it will not crosscut other layers.

20. A, B, C, D; according to the Principle of Superposition, A, B, and C were deposited horizontally and the oldest layer is on the bottom. According to the Principle of Cross-Cutting Relationships, D is younger than the layers it cuts across.

WRITING ABOUT SCIENCE

Writing About the Earth System

Sample Response: Rock layers and the fossils they contain provide clues about Earth's history, including climate and landscape changes and the evolution of organisms. Tree rings record the passage of time, and from them inferences can be drawn about changes in weather and climate, including cataclysmic events. Varves record the passage of geologic time in sediments and provide clues to climate change. Rocks best tell the story of Earth's four spheres, since they are part of the geosphere, are deposited by water from the hydrosphere, include fossil remains of the biosphere, and provide clues about changes in the atmosphere.

CHAPTER 30

PLANNING GUIDE VIEWS OF EARTH'S PAST

	Section Objectives	Print Resources
Chapter-Level Resources		Laboratory Manual, pp. 131–136 Internet Investigations Guide, pp. T104–T105, 104–105 Guide to Earth Science in Urban Environments, pp. 5–12 Spanish Vocabulary and Summaries, pp. 59–60 Formal Assessment, pp. 88–90 Alternative Assessment, pp. 59–60
Section 30.1 The Geologic Time Scale, pp. 666–672	1. Explain how the geologic time scale summarizes Earth's history. 2. Describe some physical and biological changes that have occurred on Earth over time and give reasons for these changes.	Lesson Plans, p. 102 Reading Study Guide, p. 102 Teaching Transparency 36
Section 30.2 The Precambrian and Paleozoic, pp. 673–677	1. Compare and contrast the characteristics of Precambrian time and the Paleozoic Era. 2. Identify major events in early geologic time.	Lesson Plans, p. 103 Reading Study Guide, p. 103
Section 30.3 The Mesozoic, pp. 678–680	1. Identify distinguishing characteristics of the Mesozoic Era.	Lesson Plans, p. 104 Reading Study Guide, p. 104 Teaching Transparency 10
Section 30.4 Earth's Recent History, pp. 681–685	1. Describe characteristics of the Cenozoic Era. 2. Discuss the evolution of humans.	Lesson Plans, p. 105 Reading Study Guide, p. 105 Teaching Transparency 37

INVESTIGATIONS
CLASSZONE.COM

Section 30.1 INVESTIGATION
How Has Life Changed over Geologic Time?
Computer time: 45 minutes / Additional time: 45 minutes
Students analyze fossil and rock records to construct an evolutionary timeline.

Chapter Review INVESTIGATION
Where and When Did Dinosaurs Live?
Computer time: 45 minutes / Additional time: 45 minutes
Students determine the time period and conditions in which a species of dinosaur lived.

664A Unit 8 Earth's History

Technology/Online Resources

Electronic Teacher Tools
Visualizations CD-ROM
Classzone.com
Online Lesson Planner
Test Generator

Visualizations CD-ROM
Classzone.com
 ES3001 Visualization, ES3002 Investigation, ES3003 Career, ES3004 Data Center

Visualizations CD-ROM
Classzone.com
 ES3005 Visualization

Visualizations CD-ROM
Classzone.com
 ES3006 Visualization

Classroom Management

Gather these materials for minilab and laboratory activities.

Minilab, p. 680
 bag of colored beads

Lab Activity, pp. 686–687
 pencil
 paper
 puzzle pieces from two or more puzzle sets
 cardboard box (at least 60 × 60 × 30 cm)
 soil and sand
 small shovels
 small hand trowels
 paintbrushes
 string
 thumbtacks
 masking tape

CHAPTER 30

Views of Earth's Past

Millions of years ago, a dinosaur left tracks on a muddy floodplain. Its fossilized footprints can be seen in what is now Arizona's Painted Desert.

What other types of changes have taken place since Earth was formed, more than 4 billion years ago?

INTRODUCE
Build Interest
Before sharing background information about the photograph with students, allow time for questions and comments.

Ask questions like the following:
- How do you think the feet that made these tracks differ from human feet? **The dinosaur's foot appears to have had only three toes, similar to a bird's foot.**
- How old do you think the tracks are? **The tracks are between 135 and 205 million years old—from the Jurassic Period.**
- What can scientists learn from studying fossils like these? **They can learn about the dinosaur's height and weight, anatomy, movement patterns, and behavior.**

Respond
- What other types of changes have taken place since Earth was formed, more than 4 billion years ago? **Continents have moved, mountains have been built up and worn down, climate has changed, life forms have come and gone, and the landscape has changed.**

About the Photograph
The tracks are of two types of theropods—carnivores from the early Jurassic Period. The tracks are part of the Dinosaur Trackway Site in Arizona's Painted Desert.

BEYOND THE TEXTBOOK

Print Resources
- Reading Study Guide, pp. 102–105
- Transparencies 10, 36, 37
- Formal Assessment, pp. 88–90
- Laboratory Manual, pp. 131–136
- Alternative Assessment, pp. 59–60
- Spanish Vocabulary and Summaries, pp. 59–60
- Internet Investigations Guide, pp. T104–T105, 104–105
- Guide to Earth Science in Urban Environments, pp. 5–12

Technology Resources
- Visualizations CD-ROM
- Classzone.com
- Online Lesson Planner
- Electronic Teacher Tools
- Test Generator

CHAPTER 30

PREVIEW

▶ **FOCUS QUESTIONS** In this chapter you will study Earth's past and learn more about the key questions below.

Section 1 How do scientists organize the major events of Earth's past?
Section 2 How do Precambrian time and the Paleozoic Era differ?
Section 3 What are the major events of the Mesozoic Era?
Section 4 What significant events characterize the Cenozoic Era?

▶ **REVIEW TOPICS** As you investigate Earth's past, you will need to use information from earlier chapters.

- fossils (pp. 130, 648–649)
- fossil fuel formation (pp. 148–150)
- plate tectonics and continental growth (pp. 182–183)
- ice ages (pp. 330–331)
- the composition of Earth's atmosphere (pp. 366–367)
- global warming (pp. 381–382)
- climate change (pp. 474–475)

▶ **READING STRATEGY**

QUESTION

Before you begin Chapter 30, read the chapter opener, scan the chapter contents, and read the key ideas for each section. In your science notebook, write questions about the material for which you want to find answers in the chapter.

INTERNET RESOURCES
CLASSZONE.COM

At our Web site, you will find the following Internet support for this chapter.

DATA CENTER
EARTH NEWS
VISUALIZATIONS
- Geologic Time
- Breakup of Pangaea
- K-T Boundary Asteroid Impact

LOCAL RESOURCES
CAREERS
INVESTIGATIONS
- How Has Life Changed over Geologic Time?
- Where and When Did Dinosaurs Live?

Chapter 30 Views of Earth's Past 665

CHAPTER 30

PREPARE
Focus Questions
Find out what knowledge students have about Earth's past. For each focus question, have students consider the section title and then develop two or three additional questions for each section. For example, Section 2: What was Earth like during Precambrian time? What animals lived during the Paleozoic Era?

Review Topics
In addition to the earth science topics listed, students may need to review these concepts:

Life Science
- A species is the smallest biological classification; it refers to a group of organisms that can successfully interbreed and produce fertile offspring.
- Members of the same species and of different species compete with one another for limited resources.

Reading Strategy
Model the Strategy: Brainstorm a list of questions based on the chapter opener—both the caption and the photograph. Use the questions from Build Interest as a springboard to other questions, such as: In what type of rock are these tracks located? How did people find these tracks? What did the dinosaur that made them look like? Then have students work in small groups to continue brainstorming questions by using the chapter's key ideas, visuals, and headings. Be sure they write their questions in their notebooks, leaving space to answer them as they study the chapter.

BEYOND THE TEXTBOOK

INTERNET RESOURCES
CLASSZONE.COM

INVESTIGATIONS
- How Has Life Changed over Geologic Time?
- Where and When Did Dinosaurs Live?

VISUALIZATIONS
- Animation of changes through geologic time
- Animation of the breakup of Pangaea
- Animation of an asteroid impact at the K-T boundary

DATA CENTER
EARTH NEWS
LOCAL RESOURCES
CAREERS

Online Lesson Planner

Chapter 30 Views of Earth's Past 665

CHAPTER 30 SECTION 1

FOCUS

Objectives
1. Explain how the geologic time scale summarizes Earth's history.
2. Describe some physical and biological changes that have occurred on Earth over time and give reasons for these changes.

Set a Purpose
Have students read the key idea. Then have them read this section to answer the question "What does the geologic time scale show?"

INSTRUCT

More about...

The Geologic Time Scale
The geologic time scale represents the input of dozens of great minds, beginning back in the 17th century. The Danish geologist and antatomist Nicolaus Steno (1638–1686) was one of the first to link Earth's history to the rock record. Steno pointed out that sediments build up layer upon layer and that these layers are sometimes disturbed, indicating a changing Earth. Giovanni Arduino (1714–1795) divided the rock record into four main divisions, based on composition and age. He recognized that the types of fossils found in rock layers changed with the age of the rocks.

30.1

KEY IDEAS
The geologic timetable, based on the rock record, summarizes major events in Earth's history.

The geologic time scale is divided into eons, eras, periods, and epochs.

KEY VOCABULARY
- geologic time scale
- eon
- era
- period
- epoch
- evolution
- natural selection

THE ROCK RECORD Rock layers such as these at Vermillion Cliffs, Utah, provide information about Earth's lengthy past.

The Geologic Time Scale

Historians sometimes name time periods after unique characteristics or remarkable events—the Ming Dynasty, for example. Similarly, geologists divide Earth's past into time periods based on distinguishing traits. Scientists have reconstructed Earth's extensive past by using geological evidence, sophisticated dating techniques, and deductive reasoning. Like the periodic table of elements, the time scale is a useful organizational tool.

Division of Geologic Time

A geologist studying the rocks in an area can use rules, such as the law of superposition, to determine the relative ages of the rocks. By correlating rocks over large areas, geologists have determined the relative ages of most of the rocks on Earth's surface. Over many years, geologists have used rock formations to develop a time scale that divides geologic time into units.

The **geologic time scale** (pages 668–669) is a summary of major events in Earth's past that are preserved in the rock record. Although several slightly different time scales exist, all are based on evidence. Fossils are an important part of the history. In fact, many rock layers have been identified and matched based on the fossils they contain.

Geologic time is divided into eons, eras, periods, and epochs. An **eon** is the longest segment of geologic time. The Archean Eon (ahr-KEE-uhn) is the oldest, beginning with the formation of Earth's crust almost 4 billion years ago. The earliest known rocks formed during this eon. The Proterozoic Eon (PROH-tur-uh-ZOH-ihk) began about 2.5 billion years ago. Rocks from this time contain the earliest fossils, simple organisms that lived in the oceans. No fossil evidence of life on land has been found from this eon.

The most recent eon, the Phanerozoic (FAN-ur-uh-ZOH-ihk), is characterized by signs of visible life. It is subdivided into three **eras.**

DIFFERENTIATING INSTRUCTION

Challenge Activity Tell students that the Archean and Proterozoic eons are often referred to collectively as Precambrian time. Have students research the origin of the word *Precambrian*. *Precambrian refers to all geologic time before the Cambrian Period, which was named for Cambria, the Roman word for Wales. Geologic periods are often named after the area where defining fossils were originally discovered. In this case, certain ancient fossils were first discovered in Wales.*

666 Unit 8 Earth's History

- The Paleozoic (PAY-lee-uh-ZOH-ihk) Era began about 543 million years ago. Rocks formed during the Paleozoic contain fossils of both land and ocean plants and animals.
- The Mesozoic (MEHZ-uh-ZOH-ihk) Era began about 248 million years ago. Dinosaurs thrived during most of this era.
- The Cenozoic (SEH-nuh-ZOH-ihk) Era, the most recent era, began about 65 million years ago and continues today. Events of the era include the last Ice Age and the appearance of humans in the fossil record.

Eras are divided into **periods.** Like eras, periods differ from one another in characteristic plant and animal life; these differences, however, are less dramatic than differences between eras. Some geologic periods are further divided into **epochs** (EHP-uhks). These divisions are briefer, and the distinguishing changes in life are not as great as those between periods.

Changes Through Geologic Time

Since its formation, Earth has undergone many physical and biological changes. These changes have made today's Earth very different from early Earth. The changes were not all dramatic; very small alterations can become significant over millions or billions of years. For example, small, gradual changes in Earth's movements have resulted in a modern year that is days shorter than an early Paleozoic year. Geological events have resulted in the formation or disappearance of landmasses, mountains, and seas. Where corals once grew in warm oceans, you might now find cool mountains. Climatic changes have included the rise and fall of global temperatures.

The atmosphere has also changed significantly over time. Scientists theorize that Earth's first atmosphere was composed mostly of gases released from volcanic eruptions, including carbon dioxide, sulfur dioxide, water vapor, and nitrogen. Although the molecules of some of these gases contained oxygen atoms, the form of oxygen on which human life depends (O_2), was not originally part of the atmosphere. Oxygen (O_2) entered the air with the appearance of photosynthetic organisms, such as algae. The level of oxygen in the atmosphere increased gradually. As the atmosphere changed, so did the numbers and types of organisms. The fossil record shows the appearance, evolution, and disappearance of a great variety of organisms over time.

VOCABULARY STRATEGY

The suffix *-zoic* comes from the Greek word *zōikos,* meaning "pertaining to living beings." "*Protero-*" comes from the word *proteros,* meaning "earlier," and *paleo-* comes from the word *palaios,* meaning "ancient." *Meso-* and *ceno-* come from the words *mesos* and *kainos,* respectively. *Mesos* means "middle," and *kainos* means "new."

Explore events occurring through geologic time.
Keycode: ES3001

CHANGE OVER GEOLOGIC TIME
The fossilized reef organism (left) was found in the Canadian Rocky Mountains in Golden, Canada (right). Geological events over time changed what was once an ocean environment where coral thrived to a mountain environment.

CHAPTER 30 SECTION 1

VISUAL TEACHING

Discussion
Help students understand the organizational principles behind the geologic time scale. Stress that major events usually define the various divisions. Ask: What type of event separates both the Paleozoic from the Mesozoic and the Mesozoic from the Cenozoic? **a mass extinction** Why are the Mesozoic and the Cenozoic sometimes called the "Age of Reptiles" and the "Age of Mammals," respectively? **because reptiles and dinosaurs dominated the Mesozoic and mammals diversified and became much more plentiful during the Cenozoic**

Extend
Have students examine the North American Rock Record in the chart and identify an event that occurred in or near their area. If applicable, have them research a time line of the event. For example, students living in Kentucky can make a time line of the history of the Appalachians.

Use Transparency 36.

CRITICAL THINKING
Point out that the Pennsylvanian and Mississippian periods are collectively called the Carboniferous. Ask: What connection do you see between the name *Carboniferous* and the North American rock record for this time? **Extensive coal-forming swamps formed during the Carboniferous; coal is composed largely of carbon.**

Geologic Time Scale

Eon	Era	Period		MYA* *Millions of years ago*	Epoch
Phanerozoic	Cenozoic "Age of Mammals"	Quaternary		0.01 (10,000 yrs)	Holocene or Recent
					Pleistocene
				2	
		Tertiary	Neogene		Pliocene
				5	
					Miocene
				24	
			Paleogene		Oligocene
				34	
					Eocene
				55	
					Paleocene
				65	
	Mesozoic "Age of Reptiles"	Cretaceous			
				144	
		Jurassic			
				206	
		Triassic			
				248	
	Paleozoic "Age of Invertebrates"	Permian			
				290	
		Carboniferous	Pennsylvanian		
				323	
			Mississippian		
				354	
		Devonian			
				417	
		Silurian			
				443	
		Ordovician			
				490	
		Cambrian			
				543	
Proterozoic					
				2,500	
Archean					
				3,800?	

DIFFERENTIATING INSTRUCTION

Hands-On Demonstration Plot the time line of the geologic time scale on a roll of butcher paper to dramatize the differences in the durations of the various time divisions. You might display the time line along the walls of the classroom.

MYA	Life	North American Rock Record
0.01 (10,000 yrs)	Humans dominant. Domestic animal species develop.	West Coast uplift continues in U.S.; Great Lakes form.
2	Hominids develop. Elephants flourish in North America, then die out.	Ice Age. Raising of mountains and plateaus in western U.S.
5	Hominids appear. Modern horse, camel, elephant develop. Sequoias decline; tropical trees driven south.	North America joined to South America. Sierras and Appalachians re-elevated by isostatic rebound.
24	Horse migrates to Asia, elephant to America. Grasses, grazing animals thrive.	North America joined to Asia. Volcanism in northwestern U.S., Columbia Plateau
34	Mammals progress. Cats and dogs develop and diverge. Elephants in Africa.	Alps and Himalayas forming. Volcanism in western U.S.
55	Pygmy ancestors of modern horse, other mammals. First whales. Diatoms, flowering plants thrive.	Coal forming in western U.S.
65	Many new mammals appear.	Uplift in western U.S. continues.
144	Dinosaurs, ammonites die out. Mammals, birds show new adaptations. Flowering plants, hardwoods rise.	Uplift of Rockies begins. Colorado Plateau raised. Coal swamps in western U.S. Intrusion of Sierra Nevada batholith.
206	Giant dinosaurs. First birds. Conifers and cycads abundant. Earliest mammals.	West-central North America under huge sea. Gulf of Mexico, Atlantic Ocean begin to form as No. America and Africa part.
248	Reptiles thrive. Forests of conifers and cycads.	Volcanism and faulting along East Coast. Palisades of the Hudson River formed.
290	Mass extinction of existing species. Trilobites, seed ferns, scale trees die out. Corals abundant.	Final uplift in Appalachians. An ice age in South America. Salt-forming deserts in western U.S.
323	First reptiles. Many giant insects. Spore-bearing plants, amphibians flourish.	Great coal-forming swamps in North America and Europe.
354	Sharks, amphibians, and crinoids flourish. Seed ferns, conifers abundant.	Extensive submergence of continents.
417	First amphibians; fishes abound. First forests.	Mountain building continues in New England and Canada. White Mountains raised.
443	First land plants and animals (spiders, scorpions). Fishes develop; marine invertebrates thrive.	Salt and gypsum deserts forming in eastern U.S.
490	Marine invertebrates thrive: mollusks, trilobites, graptolites.	Beginning of Appalachian mountain building. Taconic and Green Mountains form. Half of North America submerged.
543	First vertebrates (fish). Many marine invertebrates (first trilobites, shelled animals). Many seaweeds.	Extensive deposition of sediments in inland seas.
2,500	No life on land. Simple marine organisms (algae, fungi, worms). Stromatolites dominant. Other life probably existed, but fossil evidence is lacking.	Great volcanic activity, lava flows, metamorphism of rocks. Formation of iron, copper, and nickel ores.
3,800?		Formation of Earth's crust.

CHAPTER 30 SECTION 1

DISCUSSION

Explain that the Archean and Proterozoic eons were times of extensive volcanic activity. Ask: How did this volcanic activity affect the early atmosphere? **The early atmosphere was made up of gases released during the volcanic eruptions.** Have students identify the volcanic gases in the early atmosphere. **carbon dioxide (CO_2), sulfur dioxide (SO_2), water vapor (H_2O), and nitrogen (N_2)**

CRITICAL THINKING

Write the chemical formulas of the gases in the early atmosphere on the board, then have students reread the last paragraph on page 667. Ask: Given that some of these gases contain oxygen, why isn't oxygen considered part of the early atmosphere? **Oxygen existed in the early atmosphere in combination with other elements; free oxygen atoms were not present until the appearance of photosynthetic organisms.**

DIFFERENTIATING INSTRUCTION

Challenge Activity In 1866, the Royal Society of Edinburgh published Lord Kelvin's paper, "The 'Doctrine of Uniformity' in Geology Briefly Refuted." Lord Kelvin argued convincingly that Earth could not be much older than 100 million years—an estimate that he later lowered to 20 million years. Have students research Kelvin's argument and how it was eventually proved wrong. **Kelvin stated that Earth began as a molten body, then cooled. He calculated the rate of cooling to arrive at the figure of 20 million years. Kelvin assumed, however, that Earth did not generate its own heat—an assumption proved wrong in 1903 by Pierre Curie's discovery that radium compounds constantly give off heat.**

CHAPTER 30 SECTION 1

More about...

Evolution

Scientists can observe a model of evolution in the laboratory by studying organisms with short life cycles, such as fruit flies and bacteria. Bacteria in particular move rapidly through generations, and mutations are common. For example, a population of bacteria may have a few individuals that contain a gene that allows them to resist the effects of an antibiotic. When the deadly antibiotic is introduced to a population of bacteria, all but those few resistant bacteria die. The survivors thrive, multiplying rapidly, and the mutated gene passes on to future generations. The global medical community has become increasingly alarmed about rising bacterial resistance to antibiotics.

How Has Life Changed over Geologic Time? Analyze patterns in the fossil and rock records to construct an evolutionary timeline.
Keycode: ES3002

Evolution

The fossil record indicates that the first organisms were simple in structure. Over the millions of years since their appearance, organisms have developed an astonishing variety in size and structure. The rocks of one place, the Grand Canyon in Arizona, reveal a great deal about the history of such changes. The youngest rocks, near the canyon's top, contain imprints of land reptiles, ferns, and insects. A quarter of the way down the canyon, a sedimentary rock layer contains marine fossils, including fish. Deeper layers contain only a few shells and traces of worms. The oldest rock layers, at the bottom of the canyon, have no fossils at all.

The rock record repeatedly shows the disappearance of organisms and the appearance of different organisms. The evidence indicates a changing, or evolving, pattern of life forms. This process of change that produces new life forms over time is called **evolution.** The theory of evolution provides a scientific explanation for the past and present diversity of life. At one time, most people thought that life forms were fixed and unchanging. However, no theory could account for the fossils of dinosaurs that no longer existed on Earth. Then, in the 1800s, scientists proposed explanations for the changes evident in the fossil record. Charles Darwin, a British naturalist, suggested in 1859 that **natural selection** accounts for the changes that produce new species. The theory of natural selection states that the organisms that survive to produce offspring are those that have inherited

CAREER

Natural History Museum Curator

Where was the first place you saw a dinosaur's skeleton? For many people, the answer is in a natural history museum. The museum's curators most likely spent much time and effort to present that specimen of Earth's history to the public. Museum curators use their imagination and research skills to create exhibits that are both exciting and educational. At the beginning of the process, curators brainstorm ideas for interesting exhibits. Once a topic is chosen, they do extensive research to ensure that each detail is accurate. Curators then put their ideas into action by coordinating the loan or the purchase of exhibit items and by supervising their assembly. Once the exhibit is open to the public, they often provide lectures on the exhibit's topic.

Natural history museum curators usually have at least a bachelor's degree in a specific area of interest, such as anthropology. Aspiring curators also gain a competitive edge by volunteering at local museums and participating in student internships. For positions above entry level, curators often find advanced degrees necessary. One rewarding aspect of the career is the opportunity to learn continuously about fascinating topics and share that knowledge with the public. ■

MUSEUM CURATORS not only educate the public but also preserve valuable scientific material for researchers.

Learn more about a career as a museum curator.
Keycode: ES3003

DIFFERENTIATING INSTRUCTION

Hands-On Demonstration To demonstrate how organisms have evolved over time, show students pictures of skeletal models of the pentadactyl (five-fingered) limbs of vertebrates such as humans, dogs, and porpoises. Point out the differences in the sizes and thicknesses of the bones. Tell students that these differences reflect the different functions of the limbs. Have students compare and contrast the human hand and the porpoise flipper. Both are pentadactyl, but the human hand is made of flexible fingers for grasping, whereas the porpoise's long flipper is adapted for swimming.

the most favorable traits for surviving in a particular environment. Darwin's theory is still the best explanation for most of the existing evidence for evolution.

From his years of observations, Darwin concluded that a species is adapted to its environment because it has evolved gradually for generations. If so, the fossil record should show a series of organisms that undergo small changes over geologic time, eventually resulting in modern life forms. But few organisms have a complete, unbroken fossil record.

Instead, evolution may occur in short, "rapid" bursts. Much of the fossil record shows that some organisms existed essentially unchanged for very long periods. Then, suddenly—in a million years or less—new populations of different but related organisms appear in the fossil record. (Remember, a million years is a relatively short period of geologic time.) There is considerable debate among scientists about whether evolution follows a steady, gradual path of minor changes or is interrupted by short periods of dramatic change.

Scientific Thinking

INFER
Scientists theorize that species which do not adapt readily to a changing environment are more likely to become extinct than those that do adapt. One such environmental change is the introduction of a new species which then competes with the original inhabitants for the same resources—food, nesting areas. What other situations, of either natural or human causes, could lead to such competition?

ADAPTATION Note the different shell structures of these Galapagos tortoises show adaptations to their respective environments. Above the neck, the shell of the brushlands tortoise (right) is higher, which allows the tortoise to reach food higher above the ground.

30.1 Section Review

1. Name the three most recent eras. What is the current epoch?
2. How was oxygen (O_2) introduced into Earth's atmosphere?
3. **CRITICAL THINKING** *Gradualism* and *punctuated equilibrium* are two theories of evolution. Based on what you have learned about evolutionary processes in this section, how would you define *gradualism* and *punctuated equilibrium*? Explain.
4. **GENETICS** Suppose that two populations of rabbits live in an area with a mild climate. One group has brown fur, and the other has white fur. Based on their ability to hide from predators, which group is favored by natural selection? Would that be true if the climate became colder and winters lasted longer? Explain.

CHAPTER 30 SECTION 1

How Has Life Changed Over Geologic Time?
Keycode: ES3002

Use Internet Investigations Guide, pp. T104, 104.

Learn more about a career as a museum curator.
Keycode: ES3003

Scientific Thinking

INFER
An example of a natural cause is the disruption of the habitat of a species due to forest fire or other natural disaster. The species might then move into a new area. An example of a human cause is the accidental introduction of a nonnative species that "hitchhikes" on fruit or other food and is transported to a new habitat.

ASSESS

1. most recent eras: Cenozoic, Mesozoic, and Paleozoic; current epoch: Holocene or Recent
2. by photosynthetic organisms
3. The names of the theories are clues to the processes they describe. Gradualism refers to Darwin's theory that species evolve gradually through a series of small changes. Punctuated equilibrium proposes that evolution occurs in short, rapid bursts following long periods of no change.
4. The brown rabbits are probably favored in a mild climate. Those with white fur would probably be favored after the climate change because they could hide more easily from predators in a snowy environment.

MONITOR AND RETEACH

If students miss . . .

Question 1 Help students analyze the time divisions in the geologic time scale on pages 668 and 669.

Question 2 Review "Changes Through Geologic Time." (p. 667) Ask: What do organisms release during the process of photosynthesis? **oxygen**

Question 3 Reteach "Evolution." (pp. 670–671) Ask: Does the word *gradualism* imply gradual change or rapid change? **gradual**

Question 4 Review the process of natural selection discussed on pages 670 and 671.

CHAPTER 30 SECTION 1

Learn more about endangered species and the efforts to save them.
Keycode: ES3004

Science and Society

The Endangered Species Act
The U.S. Endangered Species Act was originally passed in 1973. It has since been amended, although its main mission of conserving habitats and preserving species remains intact. According to the act, endangered species are those plants and animals that are in danger of becoming extinct within the foreseeable future. Threatened species are those organisms that are likely to become endangered within the foreseeable future. Currently, some 507 species of animals and 737 species of plants are listed as endangered or threatened. The act is administered jointly by the U.S. Fish and Wildlife Service and the National Marine Fisheries.

Science Notebook
Some students may say habitat destruction is most important because many species are affected. Others may point out that all the factors are interrelated and thus all must be addressed equally.

SCIENCE & Society

Biodiversity

What's being done to conserve the vast array of plant and animal species on Earth? Are some locations on Earth more important to that quest than others? As large tracts of Earth's wilderness are being developed by a burgeoning human population, scientists and world leaders are facing the threats to the planet's biological diversity.

How can we best preserve Earth's biological wealth?

Biodiversity is the variety of life on Earth in all its forms, levels, and combinations. Millions of years in the making, Earth's rich biodiversity cannot be taken for granted. Nearly 11,000 years ago, North America held a spectacular array of big animal species, including three species of elephant, saber-toothed cats, giant beavers and wolves, hippo-sized ground sloths, and condors with five-meter-long wingspans. All of these species are now extinct.

Earth has undergone five episodes of mass extinction, probably due to major and rapid changes in the environment that destroyed the habitats for many species. Among the explanations hypothesized for those extinctions are glaciation, oxygen depletion in the oceans, large-scale volcanism, asteroid or comet impacts, and even supernova explosions.

The acronym HIPPO explains the reasons species become extinct today: habitat destruction, introduction of exotic species, population growth, pollution, and overconsumption. Today's extinction rate is alarming enough that some scientists say the planet is experiencing its sixth episode of mass extinction.

ENDANGERED sea turtles are just one species of Earth's biological diversity that faces extinction without preservation of their habitat.

Now, however, people are developing effective new strategies for preserving endangered species. These focus on preserving ecosystems filled with interdependent species. The challenge is how to preserve enough habitats while the human population is growing so quickly. Many international, national, and state environmental laws now mandate the identification and preservation of critical habitats. The U.S. Endangered Species Act has been increasingly focused on habitat conservation. ■

CONSERVATIONISTS in Miami Beach, Florida, relocate sea turtle eggs from a nest on a public beach to a safer site.

Extension

SCIENCE NOTEBOOK
Consider the acronym HIPPO, which explains the factors contributing to species extinction today. Which factor do you think is most important to deal with? Explain why.

Learn more about endangered species and the efforts to save them.
Keycode: ES3004

30.2 The Precambrian and Paleozoic

The majority of Earth's history is encompassed by Precambrian time and the Paleozoic Era. While organisms in these early environments may seem different compared with today's life forms, the organisms that appeared during these times gave rise to all modern life, including humans.

Precambrian Time

Precambrian time includes all geologic time before the start of the Cambrian period in the Paleozoic Era. It is not a unit of geologic time but a common way to refer to the Proterozoic and Archean eons. These eons cover the majority of Earth's past, nearly 4 billion years. The Archean Eon began with the formation of Earth's first rocks, about 3.9 billion years ago, and ended about 2.5 billion years ago with the beginning of the Proterozoic Eon. The Proterozoic lasted almost 2 billion years.

The Precambrian rock record is difficult to interpret. Not only does the record cover an enormous time period, but also Precambrian rocks are often severely bent and folded. The rocks generally lack index fossils, making correlation difficult.

Despite these problems, geologists have determined that processes such as plate movement, erosion, and deposition occurred during the Precambrian. As you may recall from Chapter 8, the craton is the oldest continental rock. Most rocks that have survived since Precambrian time are in the craton, the remains of Precambrian mountains and highlands. An exposed area of the craton is called a **shield.** The North American craton experienced at least four orogenies, or mountain-building episodes, due to plate movements. The last Precambrian orogeny, known as the Grenville Orogeny, occurred about 1 billion years ago when a landmass, possibly South America, collided with eastern North America. Today's Adirondack Mountains of New York were formed by this collision.

Precambrian rocks can be economically important. Iron, copper, gold, silver, uranium, and other minerals are mined from Precambrian rocks. However, Precambrian rocks have relatively few fossils. Many Precambrian

KEY IDEA
Precambrian time is characterized by active plate movement, simple life forms, and glaciation. The Paleozoic Era is noted for a rapid increase in the numbers and types of life forms, extensive swamps and forests, and, at times, climates similar to today's.

KEY VOCABULARY
- Precambrian
- shield
- stromatolite
- trilobite
- graptolite
- eurypterid
- crinoid
- foraminifera

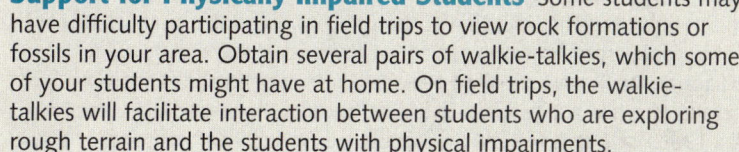

Global map of shield areas (Greenland shield, Canadian shield, Baltic shield, African shield, Indian shield, Brazilian shield, Australian shield)

PRECAMBRIAN SHIELDS About one-fifth of the surface rocks of the continents date back to Precambrian time.

CHAPTER 30 SECTION 2

Scientific Thinking

HYPOTHESIZE
Have students work in small groups to formulate their hypotheses. Students should realize that most modern organisms could not have survived prior to the appearance of Precambrian algae because the atmosphere had no free oxygen. Through the process of photosynthesis, the first algae and plants added oxygen to the air, thus changing its chemical composition. If students have trouble forming hypotheses, have them reread "Changes Through Geologic Time" on page 667.

CRITICAL THINKING

Explain that the rock record indicates that stromatolites were abundant in shallow seas during the Proterozoic. Modern stromatolites, however, are rare. Ask students to infer why. During the Proterozoic, there may have been no organisms that fed on cyanobacteria and disturbed the growing mats that form stromatolites. It is also possible that climatic and atmospheric conditions were more favorable for stromatolites in the Proterozoic than they are in modern times.

rocks are metamorphic or igneous. Igneous rocks do not contain fossils, and metamorphic processes often destroy any plant or animal remains. Also, many Precambrian organisms were tiny or microscopic and lacked the hard shells or skeletons that are more readily fossilized and less easily overlooked. Despite these factors, the first evidence of life is found in Archean rocks.

The African and Australian rocks containing these fossils are about 3.5 billion years old. Some of the early organisms resembled today's bacteria and a blue-green algae called cyanobacteria. Mats of cyanobacteria and trapped sediments formed layered domes or columns, called **stromatolites.** Stromatolites make up the greatest number of

STROMATOLITES such as these (photograph, right) at Hamelin Pool, Australia, are mostly found in intertidal zones or in shallow waters near shore. Precambrian stromatolites probably formed in similar environments. A section through a stromatolite (illustration, above) reveals the layers formed by the cyanobacteria and trapped sediments.

Precambrian fossils, but other Precambrian fossils exist. Some Proterozoic fossils from the Ediacara Hills in Australia resemble modern jellyfish, worms, and sponges.

Although stromatolites are thought to have lived in tropical waters, formations similar to varves and glacial till indicate that ice ages occurred several times during the Precambrian. The first may have taken place early in the Proterozoic. A later glaciation was nearly worldwide, affecting Europe, North America, Africa, China, and Australia.

Paleozoic Era

The Paleozoic Era includes six periods—Cambrian, Ordovician, Silurian, Devonian, Carboniferous, and Permian. This era marks the beginning of an abundant fossil record, in part because of the rapid increase in life forms sometimes called the Cambrian explosion. In addition, many Paleozoic organisms had hard body parts that were readily preserved as fossils.

Several separate continents existed at the beginning of the Paleozoic. The continent that later became North America was on the equator, with what is now the Arctic facing eastward. This continent's climate was warm, with few seasonal changes. In what is now Greenland, tropical plants and animals could be found. However, the continents were moving. By the

Scientific Thinking

HYPOTHESIZE
Could today's animals have survived prior to the appearance of Precambrian algae? Write an hypothesis statement to explain how the first algae and plants contributed to atmospheric change.

DIFFERENTIATING INSTRUCTION

Challenge Activity Have students research and write about the initial failure of the scientific community to recognize that Ediacaran fossils date from Precambrian time. When first discovered in 1946, Ediacaran fossils were classified by scientists as early Cambrian, mainly because they did not believe such life-forms existed in Precambrian time. In the 1960s, scientists realized the fossils were Precambrian, following the discovery of other Ediacaran fossils in strata beneath Cambrian rock.

Ordovician Period, the landmass including today's continents of Africa, Antarctica, South America, Australia, and parts of Asia and Europe had moved to the area over the South Pole.

Cambrian Period

The most commonly preserved Cambrian animal is the **trilobite,** a crablike invertebrate. Another common Cambrian fossil, the brachiopod, resembles a clam. Evidence for Earth's first vertebrates also appears in rocks formed late in the Cambrian. These are pieces of the bony "skin" of ostracoderms (AHS-truh-KOH-durm), a primitive fish.

Evidence from the Burgess Shale, a Cambrian rock formation in Canada, indicates that soft-bodied animals still existed during this period. Here, worms and other invertebrates were preserved in remarkable detail. More than 120 types of animals have been found. One of them is so strange and different from current life forms that it was named *Hallucigenia*.

Little mountain building occurred during the Cambrian. Warm oceans covered much of what is now North America, and marine life flourished. In fact, all Cambrian fossils are of life forms that lived in the oceans. No evidence of Cambrian land plants or animals has been found.

Ordovician Period

Many Ordovician fossils are the same as or similar to common Cambrian invertebrates, but the **graptolite** is a useful index fossil of the Ordovician. Graptolites were tiny animals that lived in colonies, or groups, throughout the world's oceans. Although a few date from other periods, their greatest numbers and distribution occurred during the Ordovician.

As in the Cambrian, all Ordovician animals lived in the ocean. Brachiopods became more numerous than trilobites. Colonial animals called bryozoans appeared. Cephalopods (relatives of the Nautilus), gastropods (snails), and echinoderms (relatives of seastars) were common. Corals and pelecypods (puh-LEHS-uh-pahdz), the group to which clams belong, first appeared.

During the Ordovician, a volcanic island arc similar to modern-day Japan collided with North America, causing the Taconic Orogeny. Vermont's Green Mountains and New York's Taconic Mountains are the remnants of that collision.

TRILOBITES such as this were once common throughout Earth's oceans. This fossil dates from the Cambrian period, over 500 million years ago.

FOSSIL GRAPTOLITES These organisms lived in oceans during the Ordovician period.

CHAPTER 30 SECTION 2

Observe the break up of Pangaea.
Keycode: ES3005

VISUALIZATIONS
CLASSZONE.COM

Observe the breakup of Pangaea.
Keycode: ES3005

Visualizations CD-ROM

DISCUSSION

Explain that not all scientists agree that the Cambrian explosion represents a true increase in life-forms. Some scientists argue that many organisms were present during the Proterozoic but are absent from the rock record because they lacked hard parts and so did not form fossils. Lead students to recognize that after the Cambrian explosion, the fossil record indicates an increase in the number of species (increased diversity), including organisms with more complex body organization and structure and other adaptations to new environments (land).

DIFFERENTIATING INSTRUCTION

Developing English Proficiency Students may encounter many unfamiliar words in this section, such as *Burgess Shale* and *cephalopods*. Visual reinforcement may help some students understand or retain these words. Tell students to identify unfamiliar words and then search the Internet or other sources for images that are linked to the words. For added reinforcement, students can compile a booklet of the unfamiliar words with the accompanying images.

Use Spanish Vocabulary and Summaries, pp. 59–60.

CHAPTER 30 SECTION 2

More about...

Amphibians and Reptiles
Amphibians flourished during the middle Paleozoic. Although they were adapted to living on land, amphibians couldn't venture far from water because their eggs had no covering that would keep them from drying out. According to fossil evidence, reptiles developed amniotic eggs sometime during the Carboniferous Period, perhaps as early as 340 million years ago. An amniotic egg has a leathery shell that prevents water loss. It also has an amnion, or fluid-filled sac, that cushions the growing embryo, as well as sacs that feed the embryo and dispose of waste. Partly because of the development of amniotic eggs, reptiles began to dominate the land by the late Paleozoic.

Silurian Period

An interesting animal common in Silurian oceans was the **eurypterid,** sometimes called a *sea scorpion.* Some eurypterids, which may be distantly related to trilobites, were more than two meters long. Yet these animals are not unique to the period; in general, most Silurian life forms resembled those of the Ordovician. Bryozoans, brachiopods, echinoderms, and corals were all common. The most significant event was the appearance of terrestrial animals. The first land animals included distant relatives of spiders, millipedes, and scorpions.

The Silurian record of land plants is also unmistakable. While photosynthetic algae lived along ocean shores during the Cambrian and Ordovician periods, plants such as club mosses spread over the land during the Silurian. Some of these plants still exist today.

During the late Silurian, the climate of what is now the northern United States became very dry. Shallow seas in eastern North America evaporated continuously, leaving thick beds of rock salt and gypsum in a belt that extends from central New York State to Lake Michigan. The salt deposits near Detroit, Michigan, and Syracuse, New York, are part of this belt.

EURYPTERIDS may have first appeared during the Ordovician. These marine organisms were especially abundant during the Silurian.

Devonian Period

The Devonian Period, often called the Age of Fishes, is known for the appearance of many types of fish. Jawless fish, similar to today's lamprey, and jawed fish covered with heavy plates were common. Some armored fish were giants of the Devonian seas, reaching lengths of up to nine meters. The first fossils of lungfish are also found in Devonian rocks. Before this period ended, a group of fish similar to lungfish developed very strong fins. With these fins, they could crawl out of the water and live briefly on land. These lobe-finned fish may be the ancestors of the first amphibians.

The first forests date to the Devonian Period, during which land plants multiplied in number and variety. Archaeologists have uncovered fossils of types of ferns, giant rushes, and primitive conifers. Trees with scaly bark appeared in the Devonian Period.

Landmass movements during the Devonian include the collision of North America with a continental fragment. These landmasses had been moving toward each other during earlier periods. The collision, known as the Acadian Orogeny, raised mountains from Newfoundland to the Appalachian region.

LUNGFISH Modern Australian lungfish share characteristics with fossils of Devonian lungfish.

Carboniferous Period

The Carboniferous Period is sometimes divided into the Mississippian and Pennsylvanian Periods. Crinoids, or sea lilies, and foraminifera are two common fossils of the Mississippian Period. **Crinoids** (CRY-noydz), which look like plants, are actually invertebrate animals related to sea stars. **Foraminifera** (fuh-RAM-uh-NIHF-ur-uh) are one-celled organisms with tiny calcite shells.

676 Unit 8 Earth's History

DIFFERENTIATING INSTRUCTION

Hands-On Demonstration Show students a clean glass filled with fresh water. Place a drinking straw in the glass and take a deep sip through the straw. Tell students that you have just modeled an important plant adaptation that allowed forests to thrive on dry land. Have students infer what the adaptation was. rigid tubes that transport water upwards through the stem and leaves of a plant Ask: Could a plant without such tubes grow tall? no Explain. Each cell in the plant would have to absorb water directly from its surroundings; thus, all the cells would have to be close to water. The rigid tube could also help provide support, allowing the plant to grow taller.

OKEFENOKEE SWAMP, GEORGIA
When sea levels rose during the Pennsylvanian, freshwater swamps similar to this became inland seas. Coal formation stopped and sediments accumulated, eventually becoming sandstone, shale, or limestone.

The later Pennsylvanian Period is marked by the appearance of reptiles, the first true land vertebrates. Insects increased in number and variety; the Pennsylvanian is sometimes called the Age of Cockroaches. This period is also known for its huge freshwater swamps, formed when interior basins of what is now the eastern United States became flooded. Millions of years later, these swamps became the coal deposits of Pennsylvania, West Virginia, Ohio, Kentucky, Indiana, and Illinois. Parts of the Appalachians are a result of the Allegheny Orogeny of the Carboniferous.

Permian Period

The Permian Period is noted for its dry climate. In addition, a great ice age took place, covering parts of South America, Australia, South Africa, and India with ice. Widespread mountain building caused by continental collisions occurred toward the end of the period. By the time the Permian ended, most continental crust had merged to form the supercontinent Pangaea.

Corals, algae, and sponges thrived in Permian seas. Yet by the close of the Paleozoic Era, nearly half of all known animal groups had become extinct, and the number and diversity of other groups plummeted. The worst losses were among marine invertebrates; trilobites and eurypterids, abundant in earlier periods, became extinct. Among land plants, almost all seed ferns, scale trees, and early conifers of the Carboniferous swamps were extinct by the end of the period. Marine cephalopods and reptiles were among the survivors, however, and would become important in the Mesozoic Era.

30.2 Section Review

1. In what rock formation would you first look for Precambrian rocks?
2. Create a time line that shows one major event for each Paleozoic period.
3. **CRITICAL THINKING** If igneous and metamorphic rocks did not exist, where would you expect to find Archean fossils? Explain.
4. **BIOLOGY CONNECTION** Many of today's oxygen-dependent organisms could not have survived in the Archean atmosphere. Yet organisms similar to the earliest life forms still exist today. Describe an environment in which you might find such organisms.

CHAPTER 30 SECTION 2

DISCUSSION

Explain that many scientists theorize that climatic changes coupled with a marine regression, or lowering of sea level, caused the Permian mass extinction. Point out, however, that marine regressions have occurred throughout Earth's history without causing major mass extinctions. Discuss why the Permian regression was so devastating. Ask: Which organisms did the mass extinction hit hardest? **marine invertebrates** Tell students that many marine invertebrates lived on the continental shelf. Ask: If sea level lowered, what happened to the continental shelf? **It became exposed.** How did this affect marine habitats? **The number and diversity of marine habitats became reduced.** Lead students to understand that the existence of only one continent—Pangaea— worsened the problem because no alternative habitats existed.

ASSESS

1. a shield
2. **Sample Time Line:** Cambrian, 543 MYA, Cambrian explosion; Ordovician, 490 MYA, Taconic Orogeny; Silurian, 443 MYA, appearance of terrestrial animals; Devonian, 417 MYA, first forests; Carboniferous, 354 MYA, many continents submerged; Permian, 290 MYA, mass extinction
3. in the oldest, possibly deepest, layers of sedimentary rock, formed from sediments in shallow seas
4. **Sample Response:** in low-oxygen environments such as a hot spring

---

MONITOR AND RETEACH

If students miss . . .

Question 1 Review "Precambrian Time." (p. 673) Ask: What is the craton? **the oldest continental rock** What is a shield? **an exposed area of the craton**

Question 2 Refer students to the geologic time scale on pages 668 and 669 and have them reread the paragraphs about the Paleozoic periods on pages 675–677.

Question 3 Reteach "Precambrian Time." (pp. 673–674) Ask: What are stromatolites? **mats of cyanobacteria and trapped sediments that form layered domes or columns in shallow waters near shore**

Question 4 The Archean was a time of extensive volcanic activity and little free oxygen. Ask: Where might similar conditions be found today? **volcanic vents, hot springs, hydrothermal vents**

CHAPTER 30 SECTION 3

30.3

FOCUS

Objective
Identify distinguishing characteristics of the Mesozoic Era.

Set a Purpose
Have students read this section to answer the question "What key events occurred during the Mesozoic Era?"

INSTRUCT

More about...

Pangaea
The ocean floor that existed during the formation of Pangaea is gone, destroyed by subduction. Thus, Pangaea's formation must be pieced together using other clues. For example, scientists study ancient mountain belts, such as the Appalachians and Urals, that mark the collision boundaries of paleocontinents. In addition, fossils and rocks indicate the locations of ancient seas and other features and offer evidence of climatic conditions, which help scientists pinpoint where the paleocontinents formed.
Use Transparency 10.

KEY IDEA
The Mesozoic Era is characterized by the rise and fall of dinosaurs, the breakup of Pangaea, and the development of a wide variety of plant life.

KEY VOCABULARY
- dinosaur
- ammonite

VOCABULARY STRATEGY
The word *dinosaur* comes from the Greek words *deinos*, meaning "monstrous," and *sauros*, meaning "lizard."

VISUALIZATIONS
CLASSZONE.COM
Observe an animation of an asteroid impact at the end of the Cretaceous Period.
Keycode: ES3006

CYCAD Modern cycads, such as this tree in Walpoua Forest, New Zealand, resemble those common during the Mesozoic.

The Mesozoic

The Mesozoic Era began about 248 million years ago and ended 65 million years ago. The era is divided into the Triassic, Jurassic, and Cretaceous periods.

The climate was mild during much of the Mesozoic. Some evidence indicates that the poles were free of glacial ice and the ocean surface temperature there was 10°C or warmer. For example, forests grew in polar regions, and coral grew in what is now Europe.

The era may be best known for its **dinosaurs.** Although scientists are debating exactly what types of animals dinosaurs were, there is no question that they were the dominant form of animal life at the time. Dinosaurs lived on all continents, but many excellent fossil sites in the western United States and Canada indicate that these areas probably were particularly favorable environments for dinosaurs.

As interesting as it is, the rise and fall of the dinosaurs was not the only Mesozoic event. Other groups of animals and plants continued to evolve, and new species appeared. Some of these groups, like the dinosaurs, were extinct by the end of the era; others continued to flourish into the Cenozoic Era.

Triassic Period

The Triassic Period lasted from about 248 million to 206 million years ago, during which dinosaurs made their first appearance on land. Some of the first dinosaurs were about the size of a cat. Many dinosaurs were small, moved swiftly, and walked on their hind legs. Some groups of dinosaurs became adapted to life in the seas during the Triassic, and marine reptiles developed in a variety of sizes and shapes. The largest ichthyosaurs (IHK-thee-uh-SAWR), reptiles that resembled dolphins, were over 15 meters long. Some long-necked plesiosaurs (PLEE-see-uh-SAWRS), which lived in oceans of the late Triassic, were nearly five meters in length.

Cephalopods, relatives of squid, were common in the Ordovician Period, but the cephalopods called **ammonites** are an important index fossil of the Triassic. Many types of ammonites occurred worldwide during the Triassic and then almost became extinct by the end of the period. The survivors gave rise to even more varieties in the Jurassic and Cretaceous Periods.

Triassic plants, including tree ferns, spore-bearing ferns, and rushes, showed few changes from the plants of the Permian Period. Plants ranged in size from smaller ferns to the midrange cycads and the taller conifers. Forests of cycads, trees with cones and palmlike leaves, and cone-bearing conifers were common. Well-known remains of Triassic conifers can be found in Arizona's Petrified Forest National Park.

During most of the Triassic, almost all of Earth's land was joined as the supercontinent Pangaea. Starting late in the period, faulting and igneous activity in Europe, North and South America, and Africa began to split Pangaea apart. The northern part, Laurasia, included what are now North America and Eurasia. The southern part, Gondwanaland, held the remaining continents. In the area that is now the United States, several areas of geologic interest formed during the Triassic. New Jersey's Palisades Sill and the igneous rocks of other East Coast locations formed.

678 Unit 8 Earth's History

DIFFERENTIATING INSTRUCTION

Reading Support Have students make a concept web to help them organize the information presented in this section. The first level of the web should be the Mesozoic Era. The next level should indicate the periods of the Mesozoic (Triassic, Jurassic, and Cretaceous), followed by significant events that occurred during each period.
Use Reading Study Guide, p. 104.

678 Unit 8 Earth's History

Jurassic Period

The rock record of the Jurassic Period, spanning from about 206 million to 144 million years ago, is remarkable for its dinosaur fossils. During this period, large dinosaurs were not uncommon, and the number of types of dinosaurs increased. Large plant eaters such as *Brachiosaurus* (BRAY-kee-uh-SAWR-uhs) were 20 meters or more in length. *Allosaurus* (A-luh-SAWR-uhs), a meat eater, and armored, plant-eating dinosaurs such as *Stegosaurus* (STEHG-uh-SAWR-uhs) also lived during this period.

Flies, grasshoppers, and other insects that change form (as when a caterpillar becomes a butterfly) first appeared during the Jurassic. Other animal groups appearing in the Jurassic include the first true mammals—tiny, rodentlike creatures—and the first animal generally recognized as a bird. Although a fossil of a feathered animal dating from the Triassic has been discovered, scientists are debating its relationship to two Jurassic fossils. The birdlike *Protoavis*, the older of the two, had some bone structures similar to those of dinosaurs. *Archaeopteryx* (AHR-kee-AHP-tur-ihks), the younger Jurassic bird fossil, had teeth, wings, and feathers and could probably fly.

During the Jurassic, fernlike plants of earlier times declined. However, mosses, cycads, and conifers were abundant. The ginkgo, an ancient tree with fleshy, fruitlike seeds, was widespread at this time.

Several present-day water bodies formed during the Jurassic Period. The South Atlantic Ocean between South America and Africa began to open. The Indian and the North Atlantic oceans had formed by the close of the Jurassic. India separated from Antarctica and Australia and began to move toward Asia. In North America, a huge sea covered much of the continent's west and center. The Morrison Formation, a widespread rock unit famous for dinosaur bones, formed in what is now the Rocky Mountains.

SIZE COMPARISON A Jurassic *Brachiosaurus*, a Triassic *Coelophysis*, and a school bus. The dinosaurs were from different time periods, and all dinosaurs were extinct before modern humans evolved.

BRACHIOSAURUS Late Jurassic

COELOPHYSIS Late Triassic

CHAPTER 30 SECTION 3

VISUALIZATIONS
CLASSZONE.COM

Observe an animation of an asteroid impact at the end of the Cretaceous Period.
Keycode: ES3005

Visualizations CD-ROM

VISUAL TEACHING
Discussion
Tell students that dinosaurs can be classified into two main groups: ornithischians and saurischians. Ornithischians were herbivores; many walked on two legs, but some walked on four. Saurischians can be further divided into two subgroups: sauropods and theropods. Sauropods were huge herbivores that walked on four legs; theropods were carnivores that walked on two legs. Ask: What type of dinosaur was *Brachiosaurus*? **sauropod** How do you know? **It ate plants, walked on four legs, and was huge.** Is it possible to classify *Coelophysis* based on the illustration? **no** Explain. **You would need to know its eating habits.** Tell students that *Coelophysis* was a carnivore, and then have them classify the dinosaur. **theropod**

CRITICAL THINKING
Tell students that the collisions of ancient landmasses created long mountain belts. These belts split apart when Pangaea broke up. Ask: If part of one such belt lies in North America, where would you expect to find the other part? **Eurasia** Explain. **The ancient continents that formed the core of North America and Eurasia collided to form Laurasia, then split apart.**

DIFFERENTIATING INSTRUCTION

Challenge Activity Have students make models showing the breakup of Pangaea over the course of the Mesozoic Era. Models should show that at the beginning of the Mesozoic, Pangaea was a large landmass in the middle of a single large ocean, the Panthalassa. About 200 million years ago, rifts began to form and great slabs of the supercontinent began drifting apart. By the early Jurassic, the Tethys Sea, the forerunner of the Mediterranean, was contracting, while the Atlantic Ocean was opening up. As the Jurassic came to a close, South America split from Africa, giving rise to the South Atlantic. The rough outlines of modern continents were present by the end of the Cretaceous.

CHAPTER 30 SECTION 3

MINILAB
Continental Isolation's Effect on Species

Management
- This Minilab introduces the concepts of speciation and genetic isolation. You may want to explore these ideas further with students.
- Step 2, using 15 beads, will lead to a more uniform, predictable outcome that conforms to the overall frequency of colors in the bag.

Materials
The bag should have at least 60 beads evenly divided among at least 3 different colors.

Analysis Answers
- The trials that sample 5 beads will lead to very divergent fur-color outcomes for each trial; therefore, the distribution of fur color would not be the same on all continents.
- fewer; the large samplings will tend to reflect more accurately the overall distribution of genes for the whole population prior to isolation.

ASSESS

1. Cretaceous, Jurassic, Triassic
2. During the Triassic, dinosaurs appeared on land. They increased in size and diversity during the Jurassic. The largest dinosaurs lived during the Cretaceous. Dinosaurs became extinct at the end of this period.
3. **Sample Entries:** Triassic: terrestrial dinosaurs evolve, Pangaea begins to break up; Jurassic: first mammals and birds, India separates from Antarctica and Australia; Cretaceous: flowering plants, mass extinction, continents appear much as today
4. the structure and unique characteristics of the species, its similarities to other species or fossils, the location of its fossils in the rock record
5. No; ammonites are index fossils of the Triassic, which began about 250 MYA.

20-Minute Mini LAB
Continental Isolation's Effect on Species

Materials
- bag of colored beads

Procedure
1. Randomly take out 5 beads and set aside. Do this four times: keep each set separate. Record the colors in each set.
2. Repeat Step 1 with 15 beads.

Analysis
As the continents separated, their species evolved based on the genes available from the individuals present. Here, each bead color represents a fur color gene, and each set represents the individuals of the species isolated on a continent. Would the distribution of fur color be the same on all continents? Would evolution based on a large sampling lead to more or fewer similarities among species on different continents? Explain.

MAGNOLIA, a flowering deciduous plant of the Mesozoic. Flowers produce seeds. Many food crops on which humans depend—such as zucchini and oranges—come from flowering plants.

Cretaceous Period

The Cretaceous is the final period of the Mesozoic Era, lasting from about 144 million to 65 million years ago. The largest dinosaurs lived during this period, often exceeding 25 meters in length; one plant eater found in Argentina may have been over 45 meters long. *Tyrannosaurus* may no longer be the largest-known meat eater of the Cretaceous; recent findings reveal that *Carcharodontosaurus* had a larger skull.

Evergreen conifers are found throughout the Mesozoic fossil record, but the appearance of flowering plants is perhaps the greatest event. Deciduous trees eventually crowded the forests. First came magnolia, sassafras, and fig. Oak, maple, birch, and other modern trees developed later. The sequoia, ancestor of California's giant redwoods, also appeared.

By the end of the Mesozoic, the South Atlantic had become a major ocean and the continents appeared much as they do today. However, Australia and Antarctica were still joined, as were North America and Eurasia. Significant geological events in North America include the birth of the Rocky Mountains and the re-elevation of the Appalachians. The Dakota Sandstone, the famous aquifer of the Great Plains, also dates to the Cretaceous.

The Mesozoic Period ends with a mystery. What caused the mass extinction of much of the life on Earth, including the dinosaurs? Scientists estimate that over 50 percent of Earth's plant and animal groups were wiped out. Many hypotheses have been proposed to explain the mass extinction, including a change in climate, the rise of mammals, a drop in global sea level, and massive volcanic eruptions. The most widely accepted explanation is that a large asteroid struck Earth 65 million years ago near the present-day Yucatán Peninsula in Mexico. Dust from the impact could have blocked sunlight for years. Land plants and marine plankton that needed sunlight to survive would have died, starving the animals that fed on them.

30.3 Section Review

1. List the Mesozoic periods in order, from most recent to oldest.
2. Write a brief summary of how dinosaurs changed from the beginning of the Triassic to the end of the Cretaceous.
3. Create a table that summarizes the major changes to life forms and landmasses that occurred over the three periods of the Mesozoic.
4. **CRITICAL THINKING** Debate about the origins of a group of animals—birds, for example—is relatively common. What factors can contribute to the debate over when a species "first appears"?
5. **GEOLOGY** During fieldwork, a geologist finds several ammonite fossils in a rock layer said to be over 300 million years old. Do you agree with the date assigned to the rock layer? Explain your answer.

MONITOR AND RETEACH

If students miss . . .

Question 1 Have students find the sentences on pages 678–680 that tell when each period of the Mesozoic Era started and ended.

Question 2 Have students read about the Triassic, Jurassic, and Cretaceous Periods. (pp. 678–680) Tell them to take notes as they reread the paragraphs that discuss dinosaurs.

Question 3 Refer students to the concept web they created for this section. See Reading Support on page 678 in this Teacher's Edition for guidance.

Question 4 Review "Jurassic Period." (p. 679)

Question 5 Review "Triassic Period." (p. 678) Ask: What is an index fossil? an abundant, widely distributed fossil used to correlate or date rock layers

Earth's Recent History

The Cenozoic Era began 65 million years ago. Geologists divide the era into three periods. The oldest, the Paleogene, lasted about 41 million years; the Neogene, about 22 million years. The most recent period, the Quaternary, spans from about 2 million years ago to the present. Some sources refer to the Paleogene and the Neogene as the Tertiary. Either method further divides the periods into epochs. They are, from oldest to most recent, the Paleocene, Eocene, Oligocene, Miocene, Pliocene, Pleistocene, and Holocene. The Cenozoic is the most recent era, and more is known about it than any other era. Scientists have information about significant events that characterize each epoch.

The early Cenozoic climate was warm and humid, much like the climate of the Mesozoic. However, global temperatures steadily decreased as the era progressed. By the beginning of the Quaternary Period, great sheets of ice covered about one-fourth of all land.

Life in the Cenozoic is characterized by the rise of mammals. Tiny mammals that survived extinction at the end of the Mesozoic Era evolved into today's familiar animals. Modern plants also evolved. Plate movements that began with the Mesozoic breakup of Pangaea continued through the Cenozoic, bringing the continents to their current locations. The appearance and disappearance of land bridges (such as the one connecting the Americas) affected the distribution of land and marine organisms.

The Paleogene and the Neogene

Nearly all major mountain ranges present today, including the Rockies, the Alps, and the Himalayas, existed or began to form during the Paleogene and Neogene Periods. The Himalayas, today's highest mountains, were pushed up as India began to collide with Asia during the Oligocene Epoch. The Alps were formed as the African Plate pushed into the Eurasian Plate. The Appalachian and Rocky mountains were worn down and raised again in this era, and the Colorado Plateau was raised a number of times. During the plateau's last uplift, the Colorado River carved out the Grand Canyon. Faulting created the fault-block mountains of the Nevada Basin and Range and the Sierra Nevadas.

30.4

KEY IDEA
The Cenozoic Era, which includes Earth's most recent history, covers the time from 65 million years ago to the present. The era includes the history of humans.

KEY VOCABULARY
- hominid
- bipedal

CLIMATE AND LANDFORMS Mountains dating from the Cenozoic, including the Himalayas shown here, are covered by only a fraction of the snow and ice that was present during the Pleistocene.

CHAPTER 30 SECTION 4

FOCUS

Objectives
1. Describe characteristics of the Cenozoic Era.
2. Discuss the evolution of humans.

Set a Purpose
Have students read the section to learn the answer to the question "What significant events occurred during Earth's recent geologic history?"

INSTRUCT

More about...

Land Bridges
During times of extensive glaciation, global sea level falls as water is locked in ice. Scientists theorize that a drop in sea level occurred sometime during the late Pleistocene, exposing a land bridge across the narrow Bering Strait between Asia and North America. Early humans from northeast Asia likely crossed this bridge to colonize North America, and later—moving over the Isthmus of Panama—South America. During the Pleistocene, sea level rose and fell numerous times, and animal migration occurred in both directions across the Bering Strait. Food probably triggered the migration of humans; they followed game into new territory.

DIFFERENTIATING INSTRUCTION

Challenge Activity Tell students that Mount Everest in the Himalayas is the world's tallest mountain. Have them find out the latest measurement of the height of Mount Everest and compare it to its measured height from about 50 years ago. How can they account for the difference? In 1999, measurements found Mt. Everest's height to be 8850 meters—2 meters taller than its measured height in 1954. The difference could be the result of the Indian Plate moving northward into Asia, more accurate measurements, or a combination of both.

Developing English Proficiency The names of the periods and epochs of the Cenozoic may be difficult for students to read. List the names on the board and pronounce them. Next to each name, write the years during which that period or epoch lasted. You might keep the list on the board to add the names of organisms that first appeared during the different periods.

CHAPTER 30 SECTION 4

VISUAL TEACHING
Discussion
Have students identify modern animals that could be relatives of those shown in the illustrations of the Green River Basin on pages 682 and 683. **rhinoceros, elephant, horse, mountain lion, hyena** Point out that crocodiles have changed little since the Eocene. Ask: Which type of animal shown in the illustration most likely evolved first? **crocodile** How do you know? **The crocodile is a reptile; the rest of the animals are mammals. Reptiles evolved before mammals.**

During the Paleogene, North America looked much as it does today. Only the Atlantic and Gulf coastal plains and parts of California were submerged. These areas were covered and uncovered by seawater several times during the Cenozoic Era. What is now the western United States was volcanically active during the end of the Paleogene. During the early Neogene, lava flows built up the Columbia Plateau in present-day Washington, Oregon, Idaho, and California. In the area that now includes Yellowstone National Park, lava and ash buried whole forests of trees several times. Later, minerals in groundwater petrified these trees.

Paleogene climates were similar to those of the Mesozoic, but temperatures dropped significantly by the end of the Eocene. In the Antarctic, ice sheets began a massive buildup through the last two epochs of the Paleogene. Global temperatures fluctuated throughout the Neogene until a sharp period of cooling near the end of the period.

When the Paleogene Period began, the warm, humid climate favored the growth of tropical plants, even in what is now the northern United States. Palm, fern, fig, and camphor trees were common. Cypress and sequoia grew as far north as Greenland and northern Canada. As global temperatures dropped during the Neogene, tropical plants were driven south. Grasses that could survive cooler climates thrived. Some of these grasses were the wild ancestors of today's wheat, corn, barley, rye, oats, and rice. The appearance of grasses may have triggered the almost explosive evolution of grazing animals during the Neogene.

Cenozoic mammals have left an extensive fossil record. With the extinction of the dinosaurs at the end of the Mesozoic, mammals began to increase in number and variety. Early meat eaters called creodonts were among the first mammals of the Paleogene. Although creodonts are extinct, some animal groups that first appeared in the early Cenozoic are still present today. In almost every case, the first animal of each group to appear was much smaller than the modern animal descended from it. For example, the first horses to appear in the Eocene Epoch were about the size of large cats.

Spiders, centipedes, scorpions, and insects such as butterflies, bees, ants, and beetles continued to thrive throughout the era. Birds evolved that were similar to those of today. Modern forms of the horse, camel,

PALEOGENE PERIOD, 65 mya–24 mya
The Green River Basin of the Eocene epoch was much warmer than it is today. Animals shown here (clockwise from bottom, left) include Smilodectes, Uintatherium, Palaeosyops, crocodile, Mesonyx, and Orohippus.

682 Unit 8 Earth's History

DIFFERENTIATING INSTRUCTION

Reading Support Help students find patterns in the text. Under the subheading "The Paleogene and the Neogene," the first topic discussed is geologic processes, followed by climate, then life-forms. Ask: What topic would you expect to be discussed first under the subheading "Quaternary Period?" **geologic processes** Have students scan the text to confirm their answers. Encourage them to look for patterns as they take notes. This will make it easier for students to locate important concepts as they read.
Use Reading Study Guide, p. 105.

and elephant are just a few of the animals that first appeared in the Neogene.

Throughout the Paleogene and Neogene, the oceans were home to nearly the same invertebrate animals as today. Sponges, corals, starfish, and sand dollars were common. Mollusks such as clams, mussels, and snails also thrived. Many fish of these periods were similar to those of the late Mesozoic. For example, sharks and rays were still abundant.

Quaternary Period

The Quaternary Period covers the time from about 2 million years ago to the present. It is divided into the Pleistocene and Holocene Epochs. The Quaternary is a brief segment of geological time and so is marked by relatively minor geologic activity. In South America, the existing Andes Mountains were raised even higher as the Nazca Plate was subducted under the South American Plate.

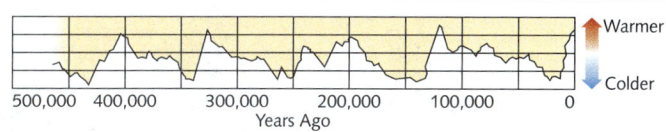

Global Cooling and Warming of the Quaternary

When ice builds up during an ice age, global temperatures drop. This can lower sea levels and alter the levels of isotopes in sea shells. These and other data are used to predict relative changes during the last 450,000 of the Quaternary period.

During the Quaternary Period, the climate underwent cycles of temperature changes that resulted in the forming and thawing of glacial ice. In fact, the Pleistocene, first epoch of the period, is sometimes called the Great Ice Age, or Ice Age. Global temperatures had fallen by the start of the Quaternary. Ice covered large areas of North America, Europe, Asia, and all of Greenland and Antarctica. Only about one-tenth of that area remains covered today. The Pleistocene Epoch came to an end when the last ice sheets disappeared from North America, Europe, and Siberia about 10,000 years ago. That time marks the beginning of the Holocene Epoch.

QUATERNARY PERIOD, 2 mya–present The Green River Basin experienced much cooler temperatures during the Pleistocene epoch. Animals shown here (left to right) include mammoth, giant ground sloth, and Smilodon.

Scientific Thinking

HYPOTHESIZE

Scientists theorize that many factors contribute to ice ages, periods of global cooling. These factors may include variations in Earth's orbit, continental movements, ocean currents, and the atmosphere.

The level of atmospheric carbon dioxide (CO_2) may be a key factor. Given what you know about the role of atmospheric CO_2 in Earth's heat balance, hypothesize how a graph of CO_2 levels might compare with the temperature graph below. Would the "peaks and valleys" be the same? Different? In what way?

CHAPTER 30 SECTION 4

Scientific Thinking

HYPOTHESIZE
Before students form their hypotheses, review the concepts of greenhouse effect and global warming, discussed in Chapter 17. Ask: What is the greenhouse effect? **the natural heating of Earth's surface that occurs when greenhouse gases in the atmosphere trap heat** What is global warming? **an increase in global temperatures caused by increases in certain greenhouse gases** What gas contributes heavily to global warming? **carbon dioxide** Following this discussion, students should hypothesize that a graph of CO_2 levels would be similar to the temperature graph on this page. Some students may suggest that the peaks and valleys on the graph of CO_2 levels would occur before those on the temperature graph, indicating a cause-effect relationship. Changes in temperature would lag slightly behind changes in CO_2 levels.

Critical Thinking

Have students examine the graph on this page. Ask: What pattern do you observe? **Temperatures appear to peak on a 100,000-year cycle.** Based on the pattern, when would you expect the next "valley" on the graph? **about 50,000 years from now**

DIFFERENTIATING INSTRUCTION

Support for Hearing-Impaired Students Use plenty of visual cues when teaching so that students can follow along. Before discussing a topic, write the relevant page numbers on the board or raise your book and point out the text or image being discussed. For example, when reading the caption of the illustration on page 683, point out the mammoth, ground sloth, and Smilodon. You may want to ask students themselves for tips on how to facilitate their learning experience.

CHAPTER 30 SECTION 4

More about...

Lucy
There are five generally accepted *Australopithecus* species: *A. anamnesis*, *A. afarensis*, *A. africanus*, *A. boisei*, and *A. robustus*. Lucy belongs to the species *Australopithecus afarensis*. She represents a tremendous find because more than 40 percent of her complete skeleton was recovered—a rare event for such an ancient fossil. The bones indicate that the hominid was approximately 1 meter tall. Based on their observations of a separate *A. afarensis* fossil found in South Africa, scientists think Lucy could grip a branch with her feet much like a chimpanzee. Thus, Lucy was bipedal and could climb easily.

CRITICAL THINKING

Ask: Why do you think Lucy was given that name? Students might reason that Lucy was the name of the discoverer of the skeleton. Paleontological finds are often named after the discoverer. In this case, however, Lucy was named after The Beatles' song "Lucy In the Sky with Diamonds" which was often playing on tape at the expedition site.

As temperatures cooled late in the Cenozoic, tropical plants died off except in equatorial areas. By the Quaternary Period, these plants had disappeared from western North America. Only two types of the previously widespread sequoia trees are found there today: the giant sequoias and the redwoods. Scientists believe that the significant changes in climate throughout the Cenozoic Era may have led species to adapt more quickly to their environment. This change might have increased the rates of evolution and extinction. Many mammal species that may have competed for resources with human ancestors became extinct toward the end of the Ice Age. Their extinction could have led those early humans to broaden their territory, migrating to other continents and regions.

REDWOOD Cooler global temperatures affected the distribution of Quaternary plant life. Once widespread, trees like this naturally fallen redwood are now found only in California and Oregon.

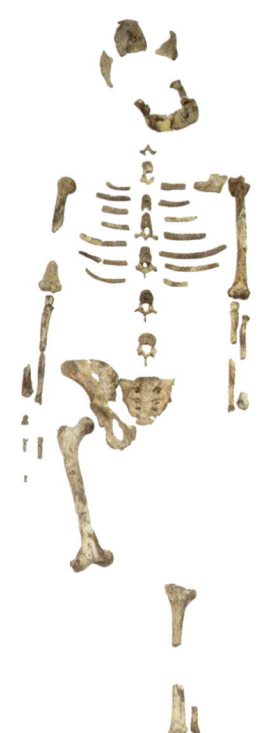

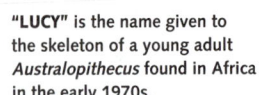

"LUCY" is the name given to the skeleton of a young adult *Australopithecus* found in Africa in the early 1970s.

Rise of Humans

Hominid is the general name for a modern human or a recent humanlike ancestor. Scientists differentiate hominid fossils from similar apelike fossils by looking for several characteristics. For example, modern humans typically have larger brains than do apes. A human brain occupies a volume of about 1300 cubic centimeters. Another distinguishing trait is that humans are **bipedal**; that is, they walk upright on two legs.

The fossil record contains extensive traces of early human life. Thousands of fossils, consisting of jaws, teeth, skulls, and other bones, have been found. The oldest is probably about 4.5 million years old, although more recent finds may eventually prove to be as old as 6 million years. The oldest generally accepted hominid is *Australopithecus* (aw-STRAY-loh-PIHTH-ih-kuhs). These hominids had apelike brains and humanlike jaws and were bipedal. The fossil record beyond 4 million years ago is not perfectly clear, but it seems *Australopithecus* dates from 5 million years ago and possibly more.

In the 1960s and 1970s, hominid fossils were found with a brain size of about 700 cubic centimeters—much larger than that of *Australopithecus*. These hominids, named *Homo habilis*, were able to make and use simple tools. They probably lived from about 2 million to 1.5 million years ago.

Another hominid, *Homo erectus*, had a brain volume of about 1000 cubic centimeters. The species lived around 1.5 million years ago and was likely the first hominid to control fire.

DIFFERENTIATING INSTRUCTION

Challenge Activity Have students find current reports on DNA sequence research and determine which two groups are more closely related to humans—chimpanzees or gorillas. Humans and chimpanzees are more closely related than are humans and gorillas. Studies indicate that humans and chimpanzees differ by an average of 1.5 percent at the nucleotide level, compared to a 2.1 percent difference between humans and gorillas.

Support for Visually Impaired Students Some students may have difficulty reading the map on page 685. Hang a large world map on the wall or bulletin board. Have pairs of students trace the paths of hominid migration on the map. Make sure students with visual impairments are partnered with students with good eyesight.

Hominid Migration Routes

HOMINID MIGRATION is difficult to determine, and the facts are constantly reviewed and debated. A map such as this is revised frequently to reflect new information gathered through research in laboratories and the field.

Hominid fossils from the last 400,000 to 300,000 years are called *Homo sapiens,* the group to which modern humans belong. Early *Homo sapiens* differed from those of today. For example, a group of *Homo sapiens* known as Neanderthals had shorter, more powerfully built skeletons than do most modern humans. Another group of early *Homo sapiens,* Cro-Magnons, had skeletons almost identical to those of modern humans.

Tracing human evolution from the fossil record is difficult. However, it is certain that humans have been on Earth a relatively short time on the geologic time scale. For example, dinosaurs roamed Earth for about 160 million years, whereas hominids have been around only for the last 2 million to 3 million years. The contrast is much greater when viewed from the history of Earth, which extends back 4 billion to 5 billion years. If the span of your arms, from left fingertip to right fingertip, represents Earth's entire geologic time scale, human existence is represented by the outermost edge of your right fingernail.

30.4 Section Review

1. Identify a significant event—of climate, geology, or life—for each Cenozoic period. Be sure to identify the name of the period.
2. Why is the Cenozoic also known as the Age of Mammals?
3. List four hominid species in chronological order.
4. **CRITICAL THINKING** Suppose that climatic changes of the Cenozoic are cyclic. Predict the pattern of climatic change for the next 65 million years. What changes in plant life might occur?
5. **PAIRED ACTIVITY** The use of simple tools and fire influenced the course of human evolution. With a partner, brainstorm three other technological advances that have shaped human development.

CHAPTER 30 SECTION 4

VISUAL TEACHING
Discussion
Point out that the map of hominid migration is based on the locations of hominid-fossil findings. Ask: Where did hominids probably originate? **Africa** How do you know? **The oldest hominid fossils were found in Africa.** Describe the pattern of hominid migration shown in the map. **From Africa, hominids migrated to East Asia and Australia, then to Europe, and finally to the Americas.**

Use Transparency 37.

ASSESS

1. **Sample Response:** Paleogene: first horses; Neogene: fluctuating global temperatures with sharp cooling near the end of the period; Quaternary: rise of humans
2. because mammals dominated that era
3. *Australopithecus, Homo habilis, Homo erectus, Homo sapiens*
4. The climate would be warm and humid, then steadily decrease over the course of the next 65 million years. An ice age would occur near the end of the period; then temperatures would gradually increase. Plant life would change from tropical vegetation to grasses to hardy plants that could withstand cold climates.
5. **Sample Response:** domestication of animals, agriculture, writing

---

MONITOR AND RETEACH

If students miss . . .

Question 1 Review "The Paleogene and the Neogene" (pp. 681–683) and "The Quaternary Period." (pp. 683–684) Have students make a chart that lists significant climatic, geologic, and biologic events for each period.

Question 2 Refer students to the illustration on page 682. Ask: How can you tell the illustration is a scene of animal life during the Cenozoic, not the Mesozoic? **The illustration shows a number of large mammals, which did not exist during the Mesozoic.**

Question 3 Reteach "Rise of Humans." (pp. 684–685) Ask: Which lived first: *Homo sapiens* or *Homo erectus*? *Homo erectus* Which hominids preceded *Homo erectus*? *Australopithecus,* followed by *Homo habilis*

Question 4 Have students reread the second paragraph on page 681.

Question 5 Explain that any significant technological advance, whether ancient (irrigation) or modern (computer), impacts human development.

CHAPTER 30 ACTIVITY

PURPOSE
To simulate an archaeological dig and analyze the findings

MATERIALS
Provide 6–10 puzzle pieces per box (per lab group). Small, colored interlocking plastic blocks can be used instead of puzzle pieces. Shallow, square boxes work well, but boxes of other shapes and sizes are acceptable. The amounts of soil and sand needed depend on the size of the box.

PROCEDURE
To save time and materials, create one dig site using a master layout. Have students work in small groups; each group can excavate one "fossil" or layer. All groups should observe and record the entire excavation. Assess the accuracy with which students record their findings.

ANALYSIS AND CONCLUSIONS
1. Answers will vary depending on the selection of puzzle pieces buried in the boxes.
2. The accuracy of recordings will vary. Recording the coordinates of the location of the fossils allows scientists to more clearly reconstruct the environment of an area during a given period of geologic history. Knowing the layers in which fossils are found can also be used to help date the fossils. Lastly, past geologic disturbances can be traced when scientists locate fossils in layers in which they are not usually found.

CHAPTER 30 LAB Activity

Fossil Excavation

SKILLS AND OBJECTIVES
- **Design** and **implement** the layout of a dig site.
- **Construct** a dig site simulating a fossil excavation.
- **Diagram** the locations of the excavation findings on a chart.
- **Analyze** the contents of a dig site.

MATERIALS
- pencil
- paper
- puzzle pieces from two or more puzzle sets
- cardboard box (at least 60 × 60 × 30 cm)
- soil and sand
- small shovels
- small hand trowels
- paintbrushes
- string
- thumbtacks
- masking tape

Paleontology is the study of forms of life from the geologic past through the analysis of plant and animal fossils. Paleontologists investigate all aspects of fossils, including their structure, how they relate to existing plants and animals, and the locations where they are found. Finding fossils is a challenge due to their rare nature. Most plants and animals that lived in the past decomposed before forming any fossils. Fossils that exist today result from being preserved in optimal conditions.

Once fossils are located, trained professionals must excavate them. Using simple tools such as picks and chisels, researchers painstakingly remove the fossil from its surrounding rock and catalog it for their records. The specimens then go to a laboratory for further study. In this activity, you will have the opportunity to practice techniques used by paleontologists and researchers on excavation digs.

Procedure

Part A: Prepare a Dig Site

1. With your lab partner(s), design the layout of your dig site. Your site should have at least three layers, with puzzle pieces (the fossils) placed randomly on the top of each layer. Prepare a sketch of each layer, similar to those shown below, to indicate the location of the fossils.

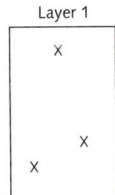 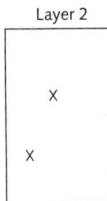 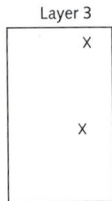

Each X represents the location of a puzzle piece.

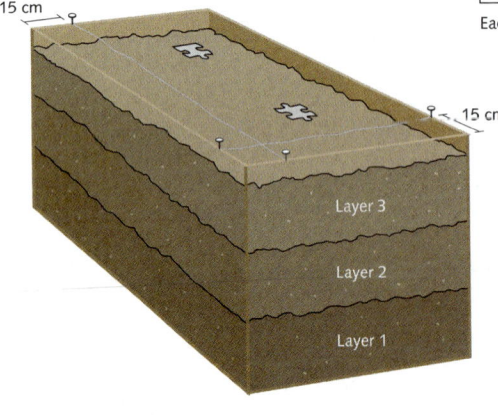

2. Using a small shovel, place sand or soil into the box to a depth of 10 cm. Use the trowel to make it compact. Taking puzzle pieces from different sets, position them according to your design for Layer 1. Press them into the sand or soil, making them level with the top of the layer.

3. Repeat Step 2 for Layers 2 and 3, alternating between layers of sand and soil, so that each layer is distinct. Cover over the exposed puzzle pieces of the top layer, and exchange your dig-site box with another group.

686 Unit 8 The Earth's History

CHAPTER 30 ACTIVITY

Learn more about paleontological fieldwork and expeditions.
Keycode: ES3009

Part B: Excavate a Dig Site

4. Prepare a location grid by attaching pieces of string over the top of the dig-site box at 15-cm intervals. Secure the string to the edge of the box with thumbtacks. (See the diagram at the bottom of page 686.)

5. Use masking tape to label the string grid as shown at right.

6. Prepare a paper grid like the one shown at right for each layer.

7. Using the trowel, begin to carefully remove the top layer of soil or sand, until you locate a fossil. Use the paint brush to fully uncover the fossil, taking care not move it out of position. Assign the fossil a number for identification, and note its location, using the grid coordinates. Record the information on your paper grid.

8. Remove the fossil and brush away any remaining loose dirt or sand. Using masking tape, tag it with its identification number.

9. Continue to excavate the dig site. Compare your results with the preparers' original design to ensure that you have located all fossils.

Analysis and Conclusions

1. Considering all of your fossil finds, can you predict which puzzle set each piece came from? How much of the total puzzle illustration can you reconstruct from your findings? Do any of the pieces fit together?

2. Comparing your paper grids to the original dig-site designs, how accurately did you record the site of each piece? Why is it necessary to record the coordinates of the location where fossils are found?

3. Did you find any pieces from the same puzzle in different layers? Would you expect to find fossils from the same animal or era in different rock layers? How might fossils from one era end up with fossils from another, both being found in the same rock layer?

4. Similar to the layers of sand and soil you used, fossils are found almost exclusively in sedimentary rock formations. Why do you think that is?

5. Predict what conditions would be best for preserving fossils. Explain your reasoning.

6. What types of logic or tools would you use to date fossils found in an excavation dig?

Learn more about paleontological fieldwork and expeditions.
Keycode: ES3009

(continued)

3. Answers will vary depending on the arrangement of the puzzle pieces. Occasionally, fossils from the same animal or era are found in different layers. Fossils from different eras might be found in the same rock layer if the area experienced past tectonic activity, such as mountain building. Accept all reasonable answers.

4. The processes that form igneous and metamorphic rocks—intense heat and pressure—destroy fossils.

5. Fossils tend to form in areas that have low rates of weathering, high rates of deposition for rapid burial, and few scavengers to consume remains. In addition, fossils that are original remains tend to be better preserved in dry and/or cold climates with low rates of decomposition.

6. Types of logic include information based on the law of superposition, index fossils, rock correlations, and key beds. Data from varves and tree rings can also be used. Tools that can be used to date fossils include radioactive decay methods and radiocarbon dating.

Refer students to Safety in the Earth Science Laboratory on pages xxii–xxiii.

Courtesy of PBS series, *Newton's Apple.*
www.tpt.org/newtons

CHAPTER 30 REVIEW

Vocabulary Review

1. Precambrian
2. stromatolites
3. trilobites
4. eurypterid
5. bipedal

Concept Review

6. It helps scientists to organize Earth's vast geologic history.
7. Evolution might occur in short bursts or gradually over generations.
8. Coal forms from the remains of plants and animals that lived in swamps. There was no life on land during the Precambrian.
9. Fossils of a tropical organism might be found in an area that now has a temperate or polar climate.
10. The *Phanezoic* Eon includes the *Cenozoic* Era, which includes the *Paleogene Period, Neogene Period,* and *Quaternary Period.* The Quaternary includes the *Pleistocene Epoch* and the *Holocene* or *Recent Epoch.* How students choose to organize the rest of the concept map may vary, but their maps should include the following: the Mesozoic Era and its Cretaceous, Jurassic, and Triassic Periods; the Paleozoic Era and its Permian, Carboniferous (Pennsylvanian and Mississippian), Devonian, Silurian, Ordovician, and Cambrian Periods.

Critical Thinking

11. There would be far fewer differences between plant and animal life in North America and life in other continents. Life-forms would have evolved on one landmass. Migration and species distribution would have

CHAPTER REVIEW 30

Summary of Key Ideas

30.1 The geologic time scale summarizes the major events of Earth's history as revealed in the rock record. Changes over the past 3.8 billion years include the formation of Earth's crust, the appearance and disappearance of landforms, atmospheric changes, and the appearance and evolution of life forms. Natural selection accounts for changes in species.

30.2 Precambrian time includes the Archean and Proterozoic Eons and covers nearly 4 billion years of Earth's history. Precambrian time is characterized by active plate movement, simple life forms, and glaciation. The Paleozoic Era covers about 295 million years and marks the beginning of an abundant fossil record.

30.3 The Mesozoic Era covers about 180 million years of Earth's history. It is characterized by the rise and fall of dinosaurs, the breakup of Pangaea, and the development of a variety of plant life.

30.4 The Cenozoic Era began about 65 million years ago and includes Earth's most recent events. Life in this era is characterized by the appearance of mammals, including humans.

Key Vocabulary

ammonite (p. 676)
bipedal (p. 682)
crinoid (p. 672)
dinosaur (p. 676)
eon (p. 664)
epoch (p. 665)
era (p. 664)
evolution (p. 670)
eurypterid (p. 672)
foraminifera (p. 672)

geologic time scale (p. 664)
graptolite (p. 671)
hominid (p. 682)
natural selection (p. 670)
period (p. 665)
Precambrian (p. 669)
shield (p. 669)
stromatolite (p. 670)
trilobite (p. 671)

688 Unit 8 Earth's History

Vocabulary Review

Write the term from the key vocabulary list that best completes the sentence.

1. Together, the Archean and the Proterozoic Eons make up what is known as ___?___ time.
2. The most common fossils from the Precambrian are ___?___ .
3. The crablike ___?___ are important fossils of the Cambrian Period.
4. The sea scorpion, or ___?___ , is an index fossil of the Silurian Period.
5. Hominids such as Neanderthals were ___?___ .

Concept Review

6. Explain in what way(s) the geologic time scale is a useful tool.
7. Describe two ways evolution might occur.
8. Why are there no Precambrian coal beds?
9. How might index fossils provide evidence of a change of climate in a particular area?
10. **Graphic Organizer** Copy and complete the concept map below. Then continue the map, including all missing eras, periods, and epochs of this eon.

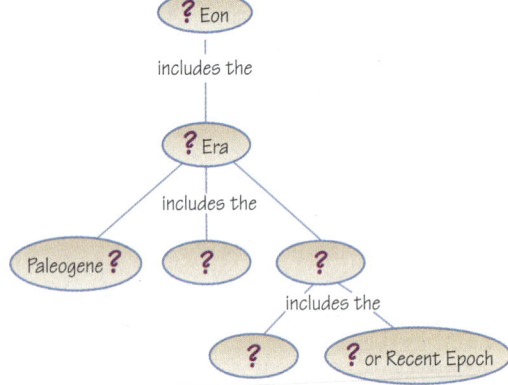

Critical Thinking

11. **Hypothesize** How might plant and animal life in North America appear today if Pangaea had not broken apart? Consider the roles of evolution, migration, and species distribution in your answer.

12. **Communicate** Imagine a swampy forest of the Pennsylvanian Period, then envision the exact same spot during the Cambrian Period. Describe how and why the views would differ.

13. **Hypothesize** Given what you learned in this chapter about climatic changes, write an hypothesis stating whether you expect global temperatures to remain the same, decrease, or increase over the next 100,000 years. Explain your reasoning.

14. **Draw Conclusions** Siberia and Alaska are separated by the Bering Strait, but evidence indicates that humans migrated to North America from Asia about 20,000 to 35,000 years ago. How was such migration possible without boats?

Internet Extension

Where and When Did Dinosaurs Live? Use geologic evidence to determine the time period and conditions in which a species of dinosaur lived.
Keycode: ES3008

Writing About the Earth System

 SCIENCE NOTEBOOK How did the state in which you live appear during the Jurassic Period? the Pliocene Epoch? Choose a time from the Phanerozoic Eon and then use a variety of resources to find information about your state's geologic past. Use the information to describe your area at that time. Note any significant differences in the spheres of the Earth system.

Interpreting Charts

The chart shows the occurrence of fossils in the rock record. Letters representing certain species are placed in the period during which that species is found in the rock record.

15. During which geologic period(s) did the trilobites of species A live?

16. What is the range of geologic periods for armored fish? for armored fish of species M?

17. Two brachiopods, species N and O, are listed. Neither can be used to identify a rock from the Mississippian. Why?

18. What is the probable age of a rock layer that contains species D and species K?

19. List the fossils (including the letter of any species) that could occur in a rock of Ordovician age.

Time Distribution of Fossils

	Cam	O	S	D	Carb	P	Tri	J	Cret	Tert	Q
Trilobites A, B, C	A	B		C							
Ammonoids D				D							
Crinoids E, F			E	F							
Graptolites G, H	G	H									
Dinosaurs I								I			
Eurypterids J, K		J	K								
Mammals L										L	
Armored Fish M				M							
Brachiopods N, O		N	O								

WRITING ABOUT SCIENCE

Writing About the Earth System
Students should include information about the fossils, climate, and geology of their area during a time from the Phanerozoic and compare this information to present conditions. Encourage students to incorporate illustrations, diagrams, and photographs in their reports.

CHAPTER 30 REVIEW

Where and When Did Dinosaurs Live?
Keycode: ES3008

📖 *Use Internet Investigations Guide, pp. T105, 105.*

(continued)
been limited by climate and relatively small geographic barriers, such as rivers and mountains, rather than large geographic barriers, such as oceans.

12. During the Pennsylvanian Period, the swampy forest would have been full of plant and animal life, including spore-bearing plants, amphibians, reptiles, and insects such as cockroaches. During the Cambrian Period, the area could have been under a warm ocean teeming with marine life. If not, the spot was likely barren and rocky. There is no evidence of life on land during the Cambrian.

13. If the cyclic trends indicated by the graph on page 683 continue, temperatures might continue to rise for a few centuries, but will eventually drop to ice-age levels. Some students may predict a continuous increase due to human activities that contribute to global warming.

14. An ice bridge existed across the Bering Strait. Humans used this ice bridge to cross into North America from Asia. Alternately, sea levels drop during ice ages; a land bridge could have formed, allowing the migration.

INTERPRETING CHARTS

15. Cambrian
16. Silurian to the beginning of the Carboniferous; Devonian
17. They became extinct before the Mississippian.
18. Devonian, between 417 and 354 million years
19. trilobites (B), crinoids, graptolites (G, H), eurypterids, brachiopods

UNIT 8 FEATURE

Focus

Objective
Describe changes in Earth's hydrosphere, atmosphere, and geosphere that may have led to mass extinctions of species.

Set a Purpose
Have students read to find answers to the question "What were the consequences and possible causes of each of the mass extinction episodes described in the article?"
65 mya: consequences—dinosaur extinction, rise of mammals and insects, diversification of surviving species; possible causes—asteroid collision with Earth, massive volcanic eruptions.
440 mya: consequences—many marine invertebrates perished; possible cause—massive glaciation lowered sea levels and destroyed continental-shelf habitats.
360 mya: consequences—marine life perished, land plants survived; possible causes—glaciation, salinity, meteorites.
250 mya: consequences—80–96% of all Earth species perished; possible causes—formation of Pangaea, glaciation, volcanoes.
213 mya: consequences—killed reptiles and amphibians, rise of dinosaurs; possible causes—climate change, increased rainfall.

The EARTH SYSTEM

Mass Extinction

Imagine attending the Super Bowl, then returning after the game to an empty stadium. During the game the noise was deafening. Now you hear only a few distant voices. Extend that lonely sensation to the entire Earth and you may get a sense of the affect of a mass extinction.

Every day you experience the biological wealth of a planet crowded with some 12.5 million species. At least five times in Earth's past, however, the stadium, so to speak, has emptied. Mass extinctions have obliterated as much as 96 percent of the planet's land and sea species. The survivors repopulated the planet, sometimes changing the nature of life on Earth.

The most recent mass extinction happened 65 million years ago, wiping out the dinosaurs. Their fossils are abundant worldwide and it is clear that dinosaurs dominated the Mesozoic landscape. However, these animals disappear suddenly from the fossil record, along with 85 percent of all other species existing at that time. One hypothesis blames a deadly collision of an asteroid into Earth. Other evidence blames massive volcanic eruptions. Whatever the cause, surviving species multiplied and diversified. Mammals were among the survivors, although it might surprise you to learn that insects, not humans, now dominate Earth.

We don't know why mass extinctions occur, but the types of species affected can hint at a cause. For example, most Ordovician life existed in the ocean shallows. Around 440 million years ago, more than 100 marine invertebrate families perished. Scientists hypothesize that their continental shelf habitat was destroyed when massive glaciation lowered sea levels.

Nearly 360 million years ago a late Devonian Period extinction hit marine life, but left land plants untouched. Why 70 percent of Earth's species perished remains a mystery. Was the cause glaciation? Changes in the ocean's chemistry salinity? Meteorite strikes?

BRACHIOSAURUS and all other dinosaurs became extinct in the late Cretaceous.

690 Unit 8 Earth's History

DIFFERENTIATING INSTRUCTION

Reading Support Have students summarize the information in "Mass Extinction." Suggest that they create headings for each of the mass extinction episodes discussed in the article. Below each heading, they can fill in the details of the time frame, affected species, and suspected causes. Ask students if the article contains important information that does not fit into this summary format. Solicit suggestions as to where such information might be included. You can suggest students create an additional heading, such as "General mass extinction information."

and the ENVIRONMENT

THE PERMIAN EXTINCTION eliminated trilobites from Earth.

CONSERVING BIODIVERSITY in rainforests may help ensure that life continues after a mass extinction.

The greatest mass extinction occurred 250 million years ago at the end of the Permian Period when 80 to 96 percent of Earth's species vanished. Pangaea was formed during that period. Was that a factor? Did glaciation or volcanic activity alter climates?

The mass extinction of the late Triassic Period, about 213 million years ago, killed reptiles and amphibians, and allowed dinosaurs to flourish. Climate change and increased rainfall are suspected reasons for this extinction.

It is difficult to pinpoint the cause of these extinctions for several reasons. The fossil record is incomplete. Accurately dating fossils is difficult as well. Also, fossils form only under certain conditions, so a lack of fossils could mean either an extinction or the absence of fossil-forming conditions. Despite these difficulties, researchers have noted that mass extinctions occur roughly every 26 million years. Are mass extinctions the result of a single, cyclic catastrophe? Perhaps Earth runs into a deadly cloud of comets every 26 million years. If so, we needn't worry. The next event isn't due for 10 million years.

Nearly 99 percent of the 5 billion species that ever existed are extinct. The vast majority disappeared not in mass extinctions, but in extinctions caused by minor environmental changes. Many of today's extinctions, however, are the result of human activity—habitat destruction and the introduction of alien species, for example. Despite the fact that human activity threatens so many species, our global community holds a unique position—we have the ability to preserve Earth's biodiversity.

What Caused the Mass Extinction Recorded at the K-T Boundary? Mass extinctions have occurred throughout Earth's history. Perhaps the best-known took place between the Cretaceous and the Tertiary time periods, marked by the K-T boundary. More than half of all plants and animals living on Earth at that time became extinct. Investigate data from images and tables to compare theories that explain what happened at the K-T boundary. *Keycode:* ESU801

UNIT 8 FEATURE

What Caused the Mass Extinction Recorded at the K-T Boundary?
Keycode: ESU801

Use Internet Investigations Guide, pp. T106–T107, 106–107.

Investigation Photograph: The hammer tip points to the K-T boundary as seen in clay deposits located in Alberta, Canada.

UNIT 8 ASSESSMENT

ANSWERS

1. c, The other choices are not in the correct order.

2. b, Flowering plants first appeared during the last period of the Mesozoic Era, the Cretaceous Period.

3. a, See the chart on pages 668 and 669.

4. b, The other choices are studies of current life, minerals, and rocks.

5. a, The evidence can be shells, bones, petrified trees, footprints, impressions made by leaves, or burrows made by worms.

6. b, The rapid increase in life forms occurred later, in the Paleozoic Era.

7. c, Relative dating places events in sequence.

8. c, Species of grazing mammals experienced explosive evolution after grasses developed.

9. d, The current climate in Europe is too cold for corals.

10. d, The Alps formed when the African plate pushed into the Eurasian plate; the Himalayas formed when India collided with Asia.

11. c, The half-life of carbon-14 is 5730 years.

12. a, The Precambrian Era ended 570 million years ago.

13. b, Two half-lives of uranium-238 is 9 billion years.

14. 0.55 million years, or 550,000 years

15. Scientists use potassium-argon dating. Fossils could also be dated by rock layer correlation or by matching index fossils, key beds, and stratigraphy.

16. An ice age followed the Pliocene; it probably changed the temperatures and precipitation patterns in the area.

UNIT 8 STANDARDIZED TEST Practice

Directions (1–13): For *each* statement or question, select the word or expression that, of those given, best completes the statement or answers the question. Record your answer on a separate answer sheet.

1. The correct order, from the greatest to the smallest segment of geologic time, is
   (a) eon, period, epoch, era
   (b) era, eon, epoch, period
   (c) eon, era, period, epoch
   (d) epoch, era, eon, period

2. Flowering plants first appeared in the
   (a) Paleocene Epoch (c) Triassic Period
   (b) Cretaceous Period (d) Quaternary Period

3. The periods of the Cenozoic Era include the
   (a) Quaternary, Neogene and Paleogene
   (b) Holocene and Pleistocene
   (c) Cambrian and Silurian
   (d) Triassic, Jurassic, and Cretaceous

4. The study of prehistoric life is called
   (a) biology   (c) mineralogy
   (b) paleontology (d) geology

5. Any evidence of life preserved in rock is called
   (a) a fossil   (c) original remains
   (b) preserved remains (d) a carbonaceous film

6. Which is not characteristic of Precambrian time?
   (a) stromatolites
   (b) abundant fossils
   (c) lava flows
   (d) formation of the first rocks

7. The dating method which determines the order of events, but not the actual dates of the events, is called
   (a) absolute dating   (c) relative dating
   (b) radiometric dating (d) radiocarbon dating

8. Which animal group benefited most from the evolution of grasses during the Cenozoic Era?
   (a) hominids   (c) grazing mammals
   (b) dinosaurs (d) nest-building birds

9. Which is evidence for a mild Mesozoic climate?
   (a) glaciers in Africa (c) palm trees in Florida
   (b) orogenies in Asia   (d) corals in Europe

10. Two mountain ranges that formed during the Cenozoic Era are the
    (a) Adirondacks and Appalachians
    (b) Appalachians and Urals
    (c) Urals and Alps
    (d) Alps and Himalayas

Base your answers to questions 11 through 13 on the table below, which provides information about isotopes used in radiometric dating.

Isotopes Used in Radiometric Dating

Parent Isotope	Half-life (years)	Effective Range (years)
Rubidium 87	47 billion	> 10 million
Uranium 238	4.5 billion	> 10 million
Potassium 40	1.3 billion	> 50,000
A	5730	100 – 70,000

11. The parent isotope labeled A above is
    (a) nitrogen–14 (c) carbon–14
    (b) lead–206   (d) argon–40

12. Which method is best for dating igneous rocks formed during Precambrian time?
    (a) rubidium-strontium (c) uranium-lead
    (b) potassium-argon   (d) radiocarbon

13. Given one gram of uranium–238, how much will be left after 9 billion years?
    (a) $\frac{1}{2}$ gram (c) $\frac{1}{8}$ gram
    (b) $\frac{1}{4}$ gram (d) $\frac{1}{16}$ gram

Directions (14–23): Record your answers on a separate answer sheet. Some questions may require the use of material in the Appendix. Base your answers to questions 14-17 on the reading passage below.

Idaho During the Pliocene Epoch

The bluffs that rise above the Snake River in Idaho comprise the Hagerman Fossil Beds. The sediment layers date from 3.7 million years old at river level to 3.15 million years old atop the bluff. The bluff's sediments include river sands, thin shale layers deposited in ponds, clay flood deposits, and occasional volcanic deposits. Radioactive elements such as potassium 40 in the volcanic ashes allow scientists to determine the age of the fossils. The beds preserve the large numbers of fossils needed to study past climates and ancient ecosystems. No other site preserves such varied species from the Pliocene Epoch.

When significant environmental change occurs, a population may adapt, migrate, or become extinct. Hagerman fossils illustrates each response as the environment changed from a wetter grassland savanna (with over twice today's ten inches of yearly precipitation) to today's drier high-desert conditions. Hagerman's beaver and muskrat adapted; they are similar to today's species. Other fossils indicate that llamas migrated from this area to South America, and camels and horses crossed the Bering Land Bridge to Eurasia. Large herbivores such as ground sloths and mastodons became extinct, as did the animals that preyed on them, the sabre-tooth cats and hyena-like dogs.

14. How many years-worth of strata are in the bluff?

15. How do scientists determine the age of the fossils at this site? What other methods could they use to date the fossils?

16. What major climatic event occurred during the epoch following the Pliocene and how might it have affected the environment at this site?

17. How might the changed climate have caused species in this area to migrate or become extinct?

Base your answers to questions 18 through 23 on the illustration to the right, which shows a series of rock layers. Each layer is identified by a letter. The types of rock are identified in the key.

18. Which layer is the oldest? Which layer is the youngest? Explain your reasoning.

19. Describe the principle of cross-cutting relationships. Which layer or layers demonstrate that principle?

20. Identify the rocks that make up layers D, E, and F, and order them from oldest to youngest. What is the relative grain size of the layers? What can you infer about geological events in this area from the rock types and their sequence?

21. Describe the location and identify the type of unconformity shown in the illustration.

22. Can the relative ages of layers D, E, F, and G be determined with certainty? Why or why not?

23. Suppose that after examining these strata you travel several miles away and find strata that appear to match layers B, C, and D. What is this correlation of strata sequences called? What techniques might you use to correlate the strata?

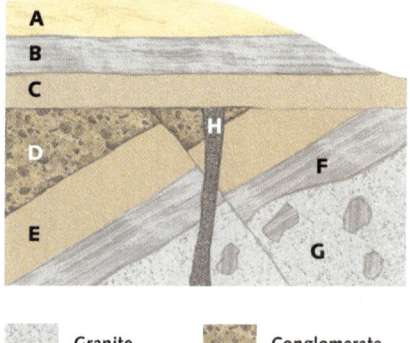

UNIT 8 ASSESSMENT

(continued)

17. Each species had optimal temperatures and diets that depended on established precipitation patterns. When the climate changed, the temperatures and precipitation patterns might no longer have been favorable, causing species to migrate to areas with favorable conditions. Species that were unable to migrate or adapt became extinct.

18. Layer F, shale, is the oldest; layer A, limestone is the youngest. Layer G, granite, is not the oldest because it contains pieces of the shale. For the granite to contain the shale, it must be younger than the shale. Layer A is the youngest, as it is on top of all the other layers (principle of superposition).

19. The principle of cross-cutting relationships states that an igneous intrusion is always younger than the rock it intruded or cut across. Layer H demonstrates this principle.

20. From oldest to youngest, the rock layers are F, shale; E, sandstone; D, conglomerate. The grain size becomes larger as the layers become younger; thus, shale has the finest particles, conglomerate the coarsest. This could indicate that a shoreline was changing (a sea was retreating). It is also possible that the land was rising so that streams flowed faster and carried larger particles farther from shore.

21. The diagram shows an angular unconformity, where horizontal layers A, B, and C were deposited on top of layers that had been tilted.

22. No; the only certainty is that layer G formed after layer F. Layer G could have formed after the deposition of layer E or after the deposition of layer D.

23. Stratigraphic matching; techniques include walking the outcrop, matching rock characteristics, and using index fossils. A key bed can also help correlate strata.

Appendix Contents

APPENDIX A REFERENCE TABLES

International System of Units	696
Topographic Map Symbols	697
Periodic Table of the Elements	698
Properties of Common Minerals	700
Scheme for Sedimentary Rock Identification	702
Scheme of Metamorphic Rock Identification	702
Scheme for Igneous Rock Identification	703
Inferred Properties of Earth's Interior	703
Dew Point Temperatures	704
Relative Humidity	704
Properties of Earth's Atmosphere	705
Pressure Scale	705
Properties of Water	705
Specific Heats of Common Materials	706
Solar System Data	706
Half-life Graph	706
Equations	706
Tree Ring Growth and Weather Data	707

APPENDIX B ATLAS

Physical United States Map	708
Physical World Map	710
Earth's Tectonic Plates	712
Seasonal Star Maps	714
Topographic Map: Harrisburg, PA	718

APPENDIX C SKILLS HANDBOOK

Reading and Study Strategies	719
Laboratory Techniques: Measurement Skills	722
Laboratory Techniques: Error Analysis	724
Mathematics	

Algebra Skills

Fraction Operations	726
Fractions, Decimals, and Percents	728
Ratio, Rate, and Proportion	730
Scientific Notation	731
Units of Measure	731

Geometry Skills

Area and Volume	732
Circumference	733
Angles and Arcs	733

Graphing Skills

Graphing Skills and Scatter Plots	734
Bar Graphs and Histograms	735
Line Graphs and Circle Graphs	736

REFERENCE TABLES AND MAPS IN CHAPTERS

Water Cycle	13
Characteristics of Earth's Layers	73
Common Elements of Earth's Crust	96
Mohs Scale of Hardness	106
Rock Cycle	119
Some Common U.S. Renewable Resources	144
Some Common U.S. Nonrenewable Resources	145
World Mineral Resources	146
Some U.S. Nonrenewable Energy Resources	148
Locations of Earthquakes and Volcanoes	173
Breakup of Pangaea	182
North American Craton	185
Interpreting an Earthquake Travel-Time Graph	218
Earthquake Risk in the United States	225
P and S Waves Inside Earth	228
North American Soils	266
Ice Coverage During the Pleistocene Epoch	330
Principal Gases of Dry Air	366
Mean Sea Level Temperatures in Asia and Australia	377
Common Air Pollutants	378
Global Precipitation Map	405
Three-Celled Model of Planetary Wind Belts	423
Global Surface Winds	424
North American Air Masses	437
Kinds of Fronts	440
Fujita Tornado Intensity Scale	448
Saffir-Simpson Hurricane Scale	452
Weather Symbols	457
World Climate Zones Map	470
Antarctic Temperatures for Past 420,000 Years	474
Salts in Sea Water	495
Surface Ocean Currents	532
Surface and Deep Ocean Currents	536
The Sun's Layers	573
The Electromagnetic Spectrum	613
Comparisons of Star Sizes	621
Stellar Color and Temperature	622
Apparent Magnitude vs. Absolute Magnitude of Stars	623
Luminosity and Temperature of Stars	626
Isotopes Used in Radiometric Dating	659
Geologic Time Table	668

Appendix A: Reference Tables

International System of Units (SI)

Some Base Metric/SI Units of Measurement

Quantity	Name	Symbol
length	meter	m
mass	kilogram	kg
volume	liter	l
temperature	kelvin*	K

Examples Using the Meter

Name	Symbol	Equivalent
kilometer	km	1000 m
meter	m	1 m
centimeter	cm	0.01 m
millimeter	mm	0.001 m

Metric System Prefixes

Prefix	Symbol	Multiples
kilo	k	1000
hecto	h	100
deka	da	10
deci	d	0.1 $\left(\frac{1}{10}\right)$
centi	c	0.01 $\left(\frac{1}{100}\right)$
milli	m	0.001 $\left(\frac{1}{1000}\right)$

Metric Conversion Table

Length

1 m = 39.37 in.
1 m = 3.280 ft
1 m = 1.093 yd
1 m = 0.00062 mi
1 cm = 0.393 in.
1 km = 0.62 mi

1 inch (in.) = 0.0254 m or 2.54 cm
1 foot (ft) = 0.3048 m or 30.48 cm
1 yard (yd) = 0.9144 m or 91.44 cm
1 mile (mi) = 1609 mi or 1.609 km

Area

1 square meter (m^2) = 1550.0 $in.^2$
1 m^2 = 10.76 ft^2
1 m^2 = 1.19 yd^2

1 inch ($in.^2$) = 0.0254 m^2 or 2.54 cm^2
1 foot (ft^2) = 0.3048 m^2 or 30.48 cm^2
1 yard (yd^2) = 0.9144 m^2 or 91.44 cm^2
1 mile (mi^2) = 1609 mi^2 or 1.609 km^2

Volume

1 liter (L) = 1.06 qt
1 L = 33.9 fluid ounces (oz)

1 quart (qt) = 0.95 L
1 cubic inch ($in.^3$) = 16.38 cm^3

Mass

1 kg = 2.204 lb
1 kg = 35.374 ounces (oz)

1 pound (lb.) = 0.4536 kg
1 lb = 453.6 g

Temperature *

* Even though kelvin is the SI base for temperature, the unit "degree Celsius" is most often used in your study of Earth system science. The scale at the right shows the relationship between temperature in degrees Fahrenheit (°F), degrees Celsius (°C), and kelvin (K).

$°C = \frac{5}{9} \cdot (°F - 32)$ $K = °C + 273$

0°C or 32°F = freezing point of water
100°C or 212°F = boiling point of water
37°C or 98.6°F = normal human body temperature
20°C or 68°F = room temperature

Temperature

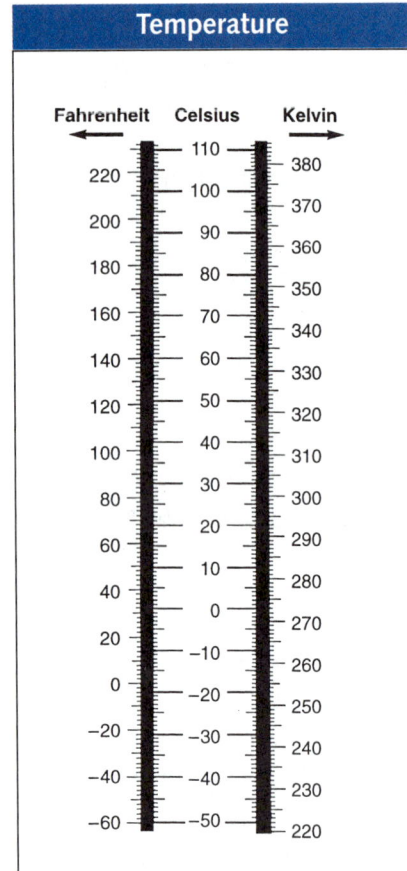

Key to Topographic Map Symbols

Primary highway, hard surface
Secondary highway, hard surface
Light-duty road, hard or improved surface
Unimproved road
Trail
Railroad: single track
Railroad: multiple track
Bridge
Drawbridge
Tunnel
Footbridge
Overpass—Underpass
Power transmission line with located tower
Landmark line (labeled as to type)

Dam with lock
Canal with lock
Large dam
Small dam: masonry—earth
Buildings (dwelling, place of employment, etc.)
School—Church—Cemeteries
Buildings (barn, warehouse, etc.)
Tanks; oil, water, etc. (labeled only if water)
Wells other than water (labeled as to type)
U.S. mineral or location monument—Prospect
Quarry—Gravel pit
Mine shaft—Tunnel or cave entrance
Campsite—Picnic area
Located or landmark object—Windmill
Exposed wreck
Rock or coral reef
Foreshore flat
Rock: bare or awash

Benchmarks
Road fork—Section corner with elevation
Checked spot elevation
Unchecked spot elevation

Boundary: national
State
county, parish, municipio
civil township, precinct, town, barrio
incorporated city, village, town, hamlet
reservation, national or state
small park, cemetery, airport, etc.
land grant
Township or range line, U.S. land survey
Section line, U.S. land survey
Township line, not U.S. land survey
Section line, not U.S. land survey
Fence line or field line
Section corner: found—indicated
Boundary monument: land grant—other

Index contour
Supplementary cont
Cut—Fill
Mine dump
Dune area
Sand area

Intermediate contour
Depression contours
Levee
Large wash
Distorted surface
Gravel beach

Glacier
Seasonal streams
Water well—Spring
Rapids
Channel
Sounding—Depth curve
Dry lake bed

Intermittent streams
Aqueduct tunnel
Falls
Intermittent lake
Small wash
Marsh (swamp)
Land subject to controlled flooding

Woodland
Submerged marsh
Orchard
Vineyard
Areas revised since previous edition

Mangrove
Scrub
Wooded marsh
Many buildings

Reference Tables 697

Appendix A: Reference Tables

Periodic Table of the Elements
(based on $^{12}_{6}C + 12.0000$)

To learn more about reading the periodic table, turn to page 92.

Key
- 6 — Atomic number
- C — Element symbol
- Carbon — Element name
- 12.01 — Atomic mass

1A	2A							
1 **H** Hydrogen 1.008								
3 **Li** Lithium 6.941	4 **Be** Beryllium 9.012							
11 **Na** Sodium 22.99	12 **Mg** Magnesium 24.31							
19 **K** Potassium 39.10	20 **Ca** Calcium 40.08	21 **Sc** Scandium 44.96	22 **Ti** Titanium 47.88	23 **V** Vanadium 50.94	24 **Cr** Chromium 52.00	25 **Mn** Manganese 54.94	26 **Fe** Iron 55.85	27 **Co** Cobalt 58.93
37 **Rb** Rubidium 85.47	38 **Sr** Strontium 87.62	39 **Y** Yttrium 88.91	40 **Zr** Zirconium 91.22	41 **Nb** Niobium 92.91	42 **Mo** Molybdenum 95.94	43 **Tc** Technetium (98)	44 **Ru** Ruthenium 101.1	45 **Rh** Rhodium 102.9
55 **Cs** Cesium 132.9	56 **Ba** Barium 137.3	57 **La** Lanthanum 138.9	72 **Hf** Hafnium 178.5	73 **Ta** Tantulum 180.9	74 **W** Tungsten 183.9	75 **Re** Rhenium 186.2	76 **Os** Osmium 190.2	77 **Ir** Iridium 192.2
87 **Fr** Francium (223)	88 **Ra** Radium 226.0	89 **Ac** Actinium (227)	104 **Rf** Rutherfordium (261)	105 **Db** Dubnium (262)	106 **Sg** Seaborgium (263)	107 **Bh** Bohrium (264)	108 **Hs** Hassium (265)	109 **Mt** Meitnerium (268)

58 **Ce** Cerium 140.1	59 **Pr** Praseodymium 140.9	60 **Nd** Neodymium 144.2	61 **Pm** Promethium (145)	62 **Sm** Samarium 150.4
90 **Th** Thorium 232.0	91 **Pa** Protactinium (231)	92 **U** Uranium 238.0	93 **Np** Neptunium (237)	94 **Pu** Plutonium (244)

						8A
3A	4A	5A	6A	7A		2 **He** Helium 4.003
5 **B** Boron 10.81	6 **C** Carbon 12.01	7 **N** Nitrogen 14.01	8 **O** Oxygen 16.00	9 **F** Fluorine 19.00	10 **Ne** Neon 20.18
13 **Al** Aluminum 26.98	14 **Si** Silicon 28.09	15 **P** Phosphorus 30.97	16 **S** Sulfur 32.07	17 **Cl** Chlorine 35.45	18 **Ar** Argon 39.95

28 **Ni** Nickel 58.69	29 **Cu** Copper 63.55	30 **Zn** Zinc 65.38	31 **Ga** Gallium 69.72	32 **Ge** Germanium 72.59	33 **As** Arsenic 74.92	34 **Se** Selenium 78.96	35 **Br** Bromine 79.90	36 **Kr** Krypton 83.80
46 **Pd** Palladium 106.4	47 **Ag** Silver 107.9	48 **Cd** Cadmium 112.4	49 **In** Indium 114.8	50 **Sn** Tin 118.7	51 **Sb** Antimony 121.8	52 **Te** Tellurium 127.6	53 **I** Iodine 126.9	54 **Xe** Xenon 131.3
78 **Pt** Platinum 195.1	79 **Au** Gold 197.0	80 **Hg** Mercury 200.6	81 **Tl** Thallium 204.4	82 **Pb** Lead 207.2	83 **Bi** Bismuth 209.0	84 **Po** Polonium (209)	85 **At** Astatine (210)	86 **Rn** Radon (222)
110 **Uun** Ununnilium (269)	111 **Uuu** Unununium (272)	112 **Uub** Ununbium (277)						

63 **Eu** Europium 152.0	64 **Gd** Gadolinium 157.3	65 **Tb** Terbium 158.9	66 **Dy** Dysprosium 162.5	67 **Ho** Holmium 164.9	68 **Er** Erbium 167.3	69 **Tm** Thulium 168.9	70 **Yb** Ytterbium 173.0	71 **Lu** Lutetium 175.0
95 **Am** Americium (243)	96 **Cm** Curium (247)	97 **Bk** Berkelium (247)	98 **Cf** Californium (251)	99 **Es** Einsteinium (252)	100 **Fm** Fermium (257)	101 **Md** Mendelevium (258)	102 **No** Nobelium (259)	103 **Lr** Lawrencium (260)

Appendix A: Reference Tables

Properties of Some Common Minerals

The minerals are arranged alphabetically, and the most useful properties in identification are printed in italic type. Most minerals can be identified by means of two or three of the properties listed below. In some minerals, color is important; in others, cleavage is characteristic; and in others, the crystal shape identifies the mineral.

Name and Chemical Composition	Hardness	Color	Streak	Type of Cleavage	Remarks
Amphibole (complex ferromagnesian silicate)	5–6	*Dark green to black*	Greenish black	Two directions at angles of 56° and 124°	Vitreous luster. Hornblende is the common variety. Long, slender, six-sided crystals. *Black with shiny cleavage surfaces at 56° and 124°.*
Apatite (calcium fluorophosphate)	5	Green, brown, red, variegated	White	Indistinct	Crystals are common as are granular masses; vitreous luster.
Beryl (beryllium silicate)	8	*Greenish*	Colorless	None	*Hardness, greenish color, six-sided crystals.* Aquamarine and emerald are gem varieties. Nonmetallic luster.
Biotite mica (complex silicate)	2.5–3	Black, brown, dark green	Colorless	*Excellent in one direction*	Thin elastic films peel off easily. Nonmetallic luster.
Calcite (CaCO₃)	3	Varies	Colorless	Excellent, three directions, not at 90° angles	*Fizzes in dilute hydrochloric acid. Hardness.* Nonmetallic luster.
Chalcopyrite (CuFeS₂)	3.5–4	*Golden yellow*	Greenish black	None	*Hardness and color distinguish from pyrite.* Metallic luster.
Copper (Cu)	2.5–3	*Copper red*	Red	None	Metallic luster on fresh surface. Ductile and malleable. Sp. gr. 8.5 to 9.
Corundum (Al₂O₃)	9	Dark grays or browns common	Colorless	None, parting resembles cleavage	Barrel-shaped, six-sided crystals with flat ends.
Diamond (C)	10	Colorless to black	Colorless	Excellent, four directions	Hardest of all minerals.
Chlorite (complex silicate)	1–2.5	*Greenish*	Colorless	*Excellent, one direction*	Nonelastic flakes, scaly, micaceous.
Dolomite (CaMg(CO₃)₂)	3.5–4	Varies	Colorless	*Good, three directions, not at 90°*	Scratched surface fizzes in dilute hydrochloric acid. Cleavage surfaces curved.
Feldspar (Potassium variety)(silicate)	6	*Salmon pink, and red are diagnostic; may be white and light gray*	Colorless	Good, two directions, 90° intersection	Hardness, color, and cleavage taken together are diagnostic.
Feldspar (sodium plagioclase variety)(silicate)	6	White to light gray	Colorless	Good, two directions, about 90°	*If striations are visible, they are diagnostic.* Nonmetallic luster.
Feldspar (calcium plagioclase variety)(silicate)	6	Gray to dark gray	Colorless	Good, two directions, about 90°	*Striations commonly visible;* may show iridescence. Associated with augite, whereas other feldspars are associated with hornblende. Nonmetallic luster.
Fluorite (CaF₂)	4	Varies	Colorless	*Excellent, four directions*	Nonmetallic luster. In cubes or octahedrons as crystals and in cleavable masses.
Galena (PbS)	2.5	*Bluish lead gray*	Lead gray	Excellent, three directions, intersect 90°	*Metallic luster.* Occurs as crystals and cleavable masses. *Very dense.*
Gold (Au)	2.53–3	*Gold*	Gold	None	Malleable, ductile, *dense.* Metallic luster.
Graphite (C)	1–2	Silver gray to black	Grayish black	Good, one direction	Metallic or earthy luster. *Foliated, scaly masses common. Greasy feel, marks paper.* This is the "lead" in a pencil (mixed with clay).

700 Appendix A

Properties of Some Common Minerals (cont.)

Name and Chemical Composition	Hardness	Color	Streak	Type of Cleavage	Remarks
Gypsum (hydrous calcium sulfate)	2	White, yellowish, reddish	Colorless	Very good in one direction	Vitreous luster. *Can be scratched easily by fingernail.*
Halite (NaCl)	2–2.5	Colorless and various colors	Colorless	Excellent, three directions, intersect at 90°	*Taste,* cleavage, hardness.
Hematite (Fe$_2$O$_3$)	5–6 (may appear softer)	*Reddish or silvery*	*Reddish*	None	Sp. gr. 4.9 to 5.3. Metallic or earthy luster
Kaolinite (hydrous aluminum silicate)	2–2.5	White	Colorless	None (without a microscope)	Dull, earthy luster. Claylike masses.
Limonite (group of hydrous iron oxides)	4–5.5	*Yellowish brown*	*Yellowish brown*	None	Earthy, granular. Rust stains.
Magnetite (Fe$_3$O$_4$)	5.5–6.5	*Black*	*Black*	None	Metallic luster. Occurs in eight-sided crystals and granular masses. *Magnetic. Sp. gr. 5.2.*
Muscovite mica (complex silicate)	2–2.5	Colorless in thin films; yellow, red, green, and brown in thicker pieces	Colorless	Excellent, one direction	*Thin elastic films peel off readily.* Nonmetallic luster.
Olivine (iron magnesium silicate)	6.5–7	*Yellowish and greenish*	White to light green	None	Green, glassy, granular.
Opal (hydrous silica)	5–6.5	Varies	Colorless	None	Glassy and pearly lusters, conchoidal fracture.
Pyrite (FeS$_2$)	6–6.5	*Brass yellow*	Greenish black	None	*Cubic crystals* and granular masses. Metallic *luster.* Crystals may be striated. *Hardness important.*
Pyroxene (complex silicate)	5–6	Greenish black	Greenish gray	Two, nearly at 90°	Stubby four or eight-sided crystals. *Augite* is a common variety. Nonmetallic.
Quartz (SiO$_2$)	7	Varies from white to black and colors	Colorless	None	Vitreous luster. Conchoidal fracture. Six-sided crystals common. Many varieties. Very common mineral. *Hardness.*
Serpentine (hydrous magnesium silicate)	2.5–4	*Greenish (variegated)*	Colorless	Indistinct	Luster resinous to greasy. *Conchoidal fracture.* The most common kind of asbestos is a variety of serpentine.
Sphalerite (ZnS)	3.5–4	Yellowish brown to black	White to yellow	*Good, six directions*	Color, hardness, cleavage, and resinous luster.
Sulfur (S)	1.5–2.5	*Yellow*	White to yellow	Indistinct	Granular, earthy.
Talc (hydrous magnesium silicate)	1	White, green, gray	Colorless	Good, one direction	*Nonelastic flakes, greasy feel.* Soft. Nonmetallic luster.
Topaz (complex silicate)	8	Varies	Colorless	One distinct (basal)	Vitreous. *Crystals commonly striated lengthwise.*
Tourmaline (complex silicate)	7–7.5	Varies; *black* is common	Colorless	Indistinct	*Elongated, striated crystals with triangular-shaped cross sections are common.*

Appendix A: Reference Tables

Scheme for Sedimentary Rock Identification

TEXTURE	GRAIN SIZE	COMPOSITION	COMMENTS	ROCK NAME	MAP SYMBOL
Clastic (fragmental)	Pebbles, cobbles, and/or boulders embedded in sand, silt, and/or clay	Mostly quartz, feldspar, and clay minerals; may contain fragments of other rocks and minerals	Rounded fragments	Conglomerate	
			Angular fragments	Breccia	
	Sand (0.2 to 0.006 cm)		Fine to coarse	Sandstone	
	Silt (0.006 to 0.0004 cm)		Very fine grain	Siltstone	
	Clay (less than 0.0004 cm)		Compact; may split easily	Shale	

CHEMICALLY AND/OR ORGANICALLY FORMED SEDIMENTARY ROCKS

TEXTURE	GRAIN SIZE	COMPOSITION	COMMENTS	ROCK NAME	MAP SYMBOL
Crystalline	Varied	Halite	Crystals from chemical precipitates and evaporites	Rock Salt	
	Varied	Gypsum		Rock Gypsum	
	Varied	Dolomite		Dolostone	
Bioclastic	Microscopic to coarse	Calcite	Cemented shell fragments or precipitates of biologic origin	Limestone	
	Varied	Carbon	From plant remains	Coal	

Scheme for Metamorphic Rock Identification

TEXTURE		GRAIN SIZE	COMPOSITION	TYPE OF METAMORPHISM	COMMENTS	ROCK NAME	MAP SYMBOL
FOLIATED	MINERAL ALIGNMENT	Fine	MICA, QUARTZ, FELDSPAR, AMPHIBOLE, GARNET, PYROXENE	Regional (Heat and pressure increase with depth)	Low-grade metamorphism of shale	Slate	
		Fine to medium			Foliation surfaces shiny from microscopic mica crystals	Phyllite	
		Fine to medium			Platy mica crystals visible from metamorphism of clay or feldspars	Schist	
	BANDING	Medium to coarse			High-grade metamorphism; some mica changed to feldspar; segregated by mineral type into bands	Gneiss	
NONFOLIATED		Fine	Variable	Contact (Heat)	Various rocks changed by heat from nearby magma/lava	Hornfels	
		Fine to coarse	Quartz	Regional or Contact	Metamorphism of quartz sandstone	Quartzite	
		Fine to coarse	Calcite and/or dolomite		Metamorphism of limestone or dolostone	Marble	
		Coarse	Various minerals in particles and matrix		Pebbles may be distorted or stretched	Metaconglomerate	

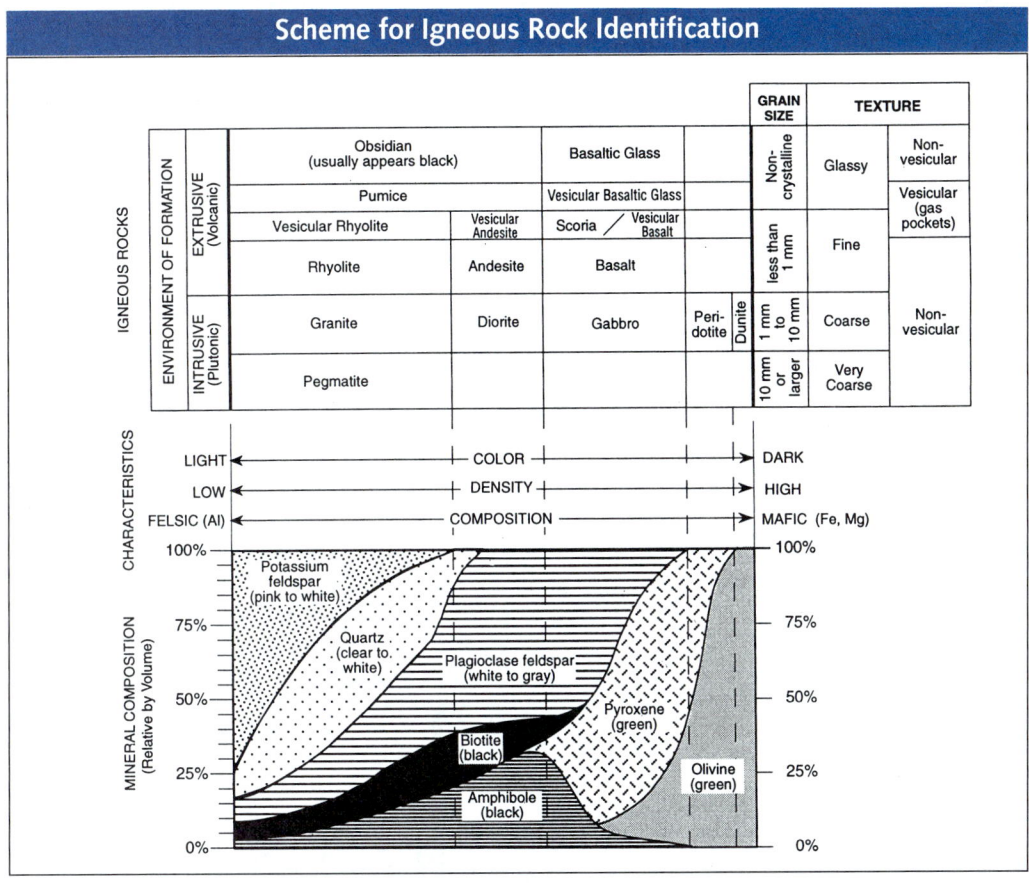

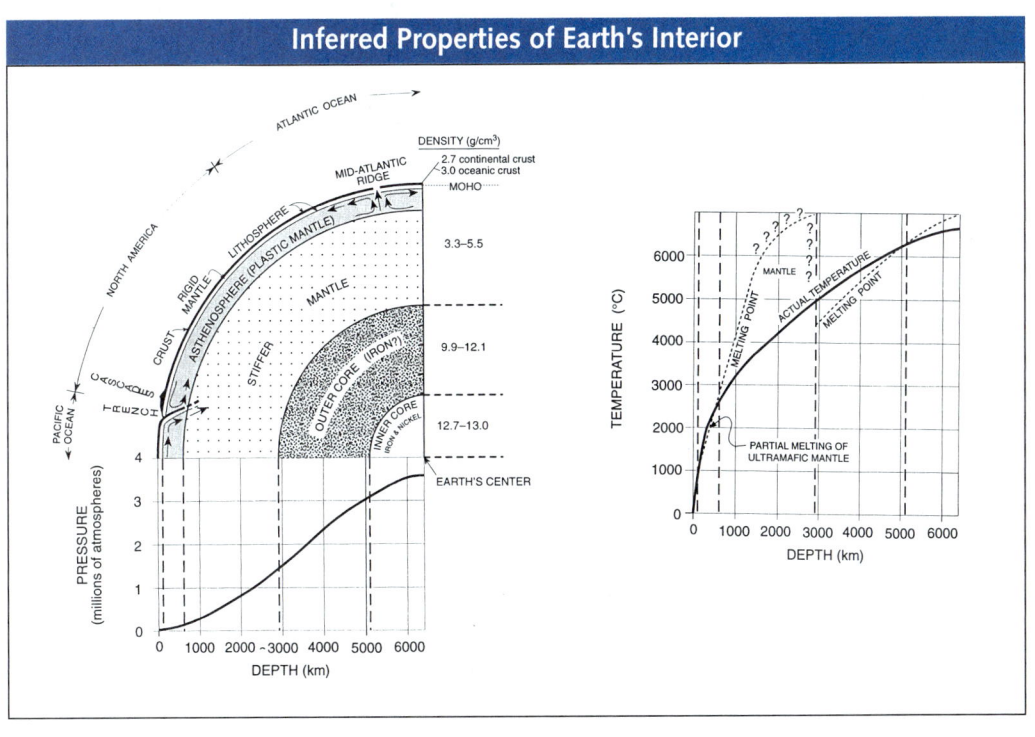

Reference Tables 703

Appendix A: Reference Tables

Dewpoint Temperatures (°C)

Dry-Bulb Temperature (°C)	Difference Between Wet-Bulb and Dry-Bulb Temperatures															
	0	1	2	3	4	5	6	7	8	9	10	11	12	13	14	15
−20	−20	−33														
−18	−18	−28														
−16	−16	−24														
−14	−14	−21	−36													
−12	−12	−18	−28													
−10	−10	−14	−22													
−8	−8	−12	−18	−29												
−6	−6	−10	−14	−22												
−4	−4	−7	−12	−17	−29											
−2	−2	−5	−8	−13	−20											
0	0	−3	−6	−9	−15	−24										
2	2	−1	−3	−6	−11	−17										
4	4	1	−1	−4	−7	−11	−19									
6	6	4	1	−1	−4	−7	−13	−21								
8	8	6	3	1	−2	−5	−9	−14								
10	10	8	6	4	1	−2	−5	−9	−14	−28						
12	12	10	8	6	4	1	−2	−5	−9	−16						
14	14	12	11	9	6	4	1	−2	−5	−10	−17					
16	16	14	13	11	9	7	4	1	−1	−6	−10	−17				
18	18	16	15	13	11	9	7	4	2	−2	−5	−10	−19			
20	20	19	17	15	14	12	10	7	4	2	−2	−5	−10	−19		
22	22	21	19	17	16	14	12	10	8	5	3	−1	−5	−10	−19	
24	24	23	21	20	18	16	14	12	10	8	6	2	−1	−5	−10	−18
26	26	25	23	22	20	18	17	15	13	11	9	6	3	0	−4	−9
28	28	27	25	24	22	21	19	17	16	14	11	9	7	4	1	−3
30	30	29	27	26	24	23	21	19	18	16	14	12	10	8	5	1

Relative Humidity (%)

Dry-Bulb Temperature (°C)	Difference Between Wet-Bulb and Dry-Bulb Temperatures															
	0	1	2	3	4	5	6	7	8	9	10	11	12	13	14	15
−20	100	28														
−18	100	40														
−16	100	48														
−14	100	55	11													
−12	100	61	23													
−10	100	66	33													
−8	100	71	41	13												
−6	100	73	48	20												
−4	100	77	54	32	11											
−2	100	79	58	37	20	1										
0	100	81	63	45	28	11										
2	100	83	67	51	36	20	6									
4	100	85	70	56	42	27	14									
6	100	86	72	59	46	35	22	10								
8	100	87	74	62	51	39	28	17	6							
10	100	88	76	65	54	43	33	24	13	4						
12	100	88	78	67	57	48	38	28	19	10	2					
14	100	89	79	69	60	50	41	33	25	16	8	1				
16	100	90	80	71	62	54	45	37	29	21	14	7	1			
18	100	91	81	72	64	56	48	40	33	26	19	12	6			
20	100	91	82	74	66	58	51	44	36	30	23	17	11	5		
22	100	92	83	75	68	60	53	46	40	33	27	21	15	10	4	
24	100	92	84	76	69	62	55	49	42	36	30	25	20	14	9	4
26	100	92	85	77	70	64	57	51	45	39	34	28	23	18	13	9
28	100	93	86	78	71	65	59	53	47	42	36	31	26	21	17	12
30	100	93	86	79	72	66	61	55	49	44	39	34	29	25	20	16

Selected Properties of Earth's Atmosphere

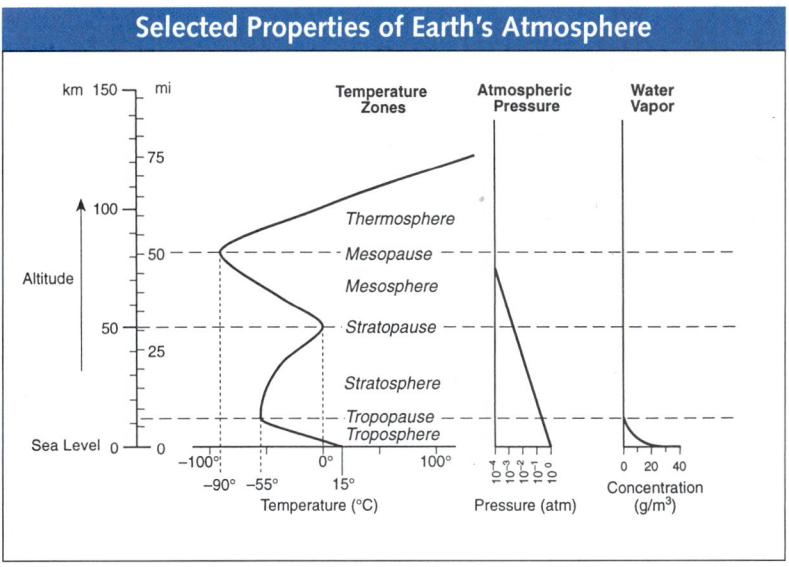

Properties of Water

Energy gained during melting	80 calories/gram
Energy released during freezing	80 calories/gram
Energy gained during vaporization	540 calories/gram
Energy released during condensation	540 calories/gram
Density at 3.98°C	1.00 gram/milliliter

Graph of Half-Life

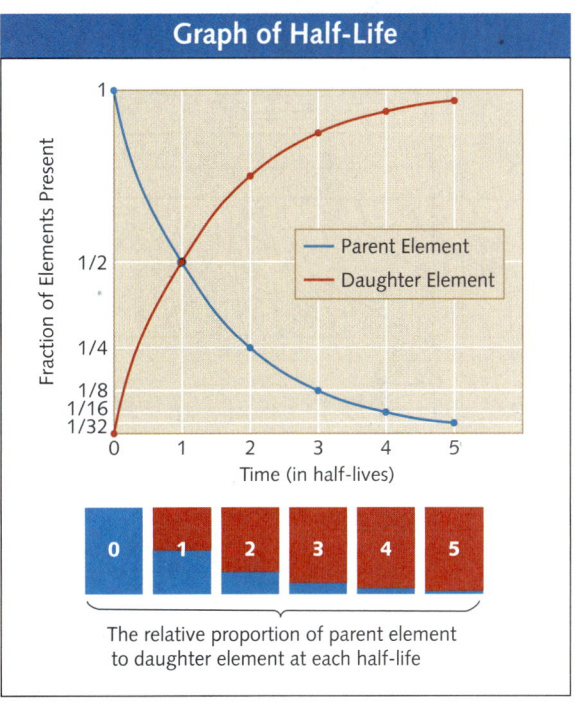

The relative proportion of parent element to daughter element at each half-life

Pressure

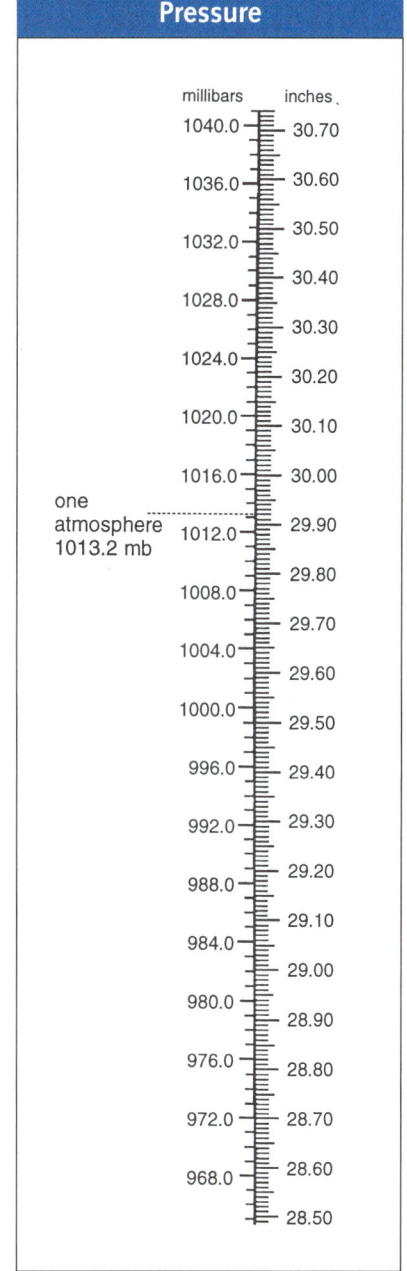

one atmosphere 1013.2 mb

Appendix A: Reference Tables

Solar System Data

Object	Mean Distance from Sun (millions of km)	Period of Revolution	Period of Rotation	Eccentricity of Orbit	Equatorial Diameter (km)	Mass (Earth = 1)	Density (g/cm^3)	Number of Moons
SUN	—	—	27 days	—	1,392,000	333,000.00	1.4	—
MERCURY	57.9	88 days	59 days	0.206	4,880	0.553	5.4	0
VENUS	108.2	224.7 days	243 days	0.007	12,104	0.815	5.2	0
EARTH	149.6	365.26 days	23 hr 56 min 4 sec	0.017	12,756	1.00	5.5	1
MARS	227.9	687 days	24 hr 37 min 23 sec	0.093	6,787	0.1074	3.9	2
JUPITER	778.3	11.86 years	9 hr 50 min 30 sec	0.048	142,800	317.896	1.3	16
SATURN	1,427	29.46 years	10 hr 14 min	0.056	120,000	95.185	0.7	18
URANUS	2,869	84.0 years	17 hr 14 min	0.047	51,800	14.537	1.2	21
NEPTUNE	4,496	164.8 years	16 hr	0.009	49,500	17.151	1.7	8
PLUTO	5,900	247.7 years	6 days 9 hr	0.250	2,300	0.0025	2.0	1
EARTH'S MOON	149.6 (0.386 from Earth)	27.3 days	27 days 8 hr	0.055	3,476	0.0123	3.3	—

Equations

Percent deviation from accepted value: $\text{deviation (\%)} = \dfrac{\text{difference from accepted value}}{\text{accepted value}} \times 100$

Eccentricity of an ellipse: $\text{eccentricity} = \dfrac{\text{distance between foci}}{\text{length of major axis}}$

Gradient: $\text{gradient} = \dfrac{\text{change in field value}}{\text{distance}}$

Rate of change: $\text{rate of change} = \dfrac{\text{change in field value}}{\text{time}}$

Specific Heats of Common Materials

MATERIAL		SPECIFIC HEAT (calories/gram • C°)
Water	solid	0.5
	liquid	1.0
	gas	0.5
Dry air		0.24
Basalt		0.20
Granite		0.19
Iron		0.11
Copper		0.09
Lead		0.03

Series of Rock Layers

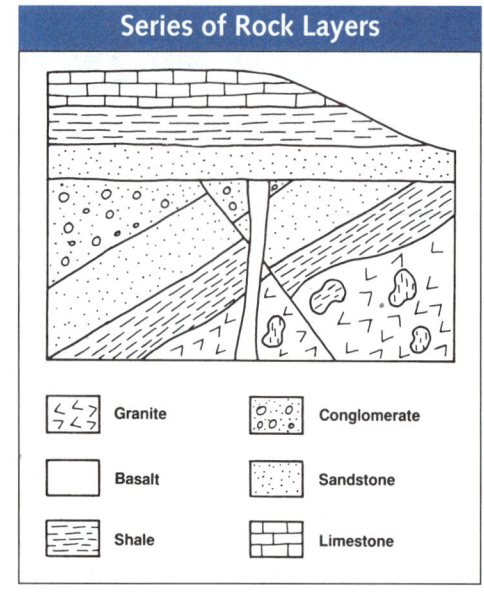

Granite, Conglomerate, Basalt, Sandstone, Shale, Limestone

Tree Ring Growth and Weather Data

Year	Tree 1 Growth (mm)	Tree 2 Growth (mm)	Tree 3 Growth (mm)	Tree 4 Growth (mm)	Tree 5 Growth (mm)	Average Growth (mm)	Average Temperature (°C)	Total Precipitation (in.)
1994	2.54	1.73	3.51	2.39	3.66	?	−1.39	27.13
1993	2.58	1.59	4.05	3.25	4.15	?	−1.19	34.77
1992	2.67	2.38	3.75	3.13	3.78	?	−2.97	28.68
1991	2.21	2.35	3.99	3.19	3.23	?	−2.54	31.83
1990	1.49	1.19	2.67	1.01	1.97	?	−2.2	39.52
1989	1.93	0.91	2.81	1.49	2.79	?	−2.88	24.97
1988	1.4	1.03	2.31	1.76	1.41	?	−0.66	24.49
1987	1.59	0.96	2.45	1.97	3.93	?	−0.83	19.25
1986	2.38	1.83	3.02	2.93	5.01	?	−2.69	28.37
1985	2.7	1.61	3.27	3.43	4.97	?	−2.47	29.77
1984	2.69	1.96	2.19	3.1	3.71	?	−2.71	29.31
1983	2.56	0.97	1.36	3.04	2.78	?	−1.66	26.67
1982	2.25	1.41	1.42	2.35	2.38	?	−2.71	30.89
1981	1.16	0.42	0.58	0.79	0.69	?	−0.87	25.53
1980	1.04	0.39	0.58	0.69	0.61	?	−0.64	20.22
1979	1.66	1.09	1.23	1.48	1.18	?	−2.94	30.66
1978	1.14	0.6	0.97	1.05	0.88	?	−2.54	25.45
1977	1.51	1.33	0.86	1.25	1.22	?	−0.86	26.72
1976	1.39	1.15	1.93	1.56	1.92	?	−4.04	17.55
1975	1.41	1.12	1.85	1.51	1.76	?	−2.67	23.02
1974	1.07	0.68	1.95	1.34	1.43	?	−3.48	16.87
1973	1.63	1.55	3.09	2.25	2.26	?	−2.65	29.46
1972	2.37	2.34	4.08	3.38	4.23	?	−4.39	25.78
1971	2.24	3.13	5.34	3.75	4.04	?	−3.89	36.17
1970	1.73	1	1.74	2.56	1.63	?	−1.03	22.4
1969	1.4	1.24	2.28	2.89	3.03	?	−3.56	20.5
1968	3.11	3.54	5.47	2.99	6.4	?	−2.21	21.49
1967	2.28	2.16	4.87	5.96	5.31	?	−3.9	43.56
1966	3.43	2.74	5.73	8.07	6.44	?	−5.29	26.38
1965	3.46	3.38	6.54	8.06	6.94	?	−4.85	20.02
1964	3.53	3.41	5.25	7.94	5.85	?	−4.29	24.74
1963	3.63	3.74	4.8	7.05	5.51	?	−5.8	35.65
1962	3.74	2.86	4.28	5.27	4.86	?	−3.92	16.61
1961	3.31	3.17	3.75	4.18	3.29	?	−3.25	30.15
1960	2.26	2.29	2.95	2.84	1.99	?	−2.91	22.2
1959	2.08	2.21	1.92	1.89	1.98	?	−5.03	25.96
1958	1.57	1.06	0.79	1.55	1.12	?	−1.53	15.44
1957	2.67	1.83	4.43	3.9	3.06	?	−3.54	20.65
1956	2.89	2.49	5.07	5.05	4.52	?	−4.92	28.91
1955	2.45	2.02	4.39	3.6	3.27	?	−4.03	36.14
1954	1.87	1.92	1.7	2.47	1.7	?	−4	22.68
1953	2.06	1.55	2.63	2.11	1.74	?	−1.98	24.94
1952	2.88	2.27	4.36	3.2	2.7	?	−3.8	23.16
1951	2.06	1.4	3.1	2.4	1.63	?	−4.43	23.7
1950	3.51	2.76	5.79	4.02	5.18	?	−2.68	24.21

Appendix B: Atlas

Physical United States Map

Physical United States Map

Appendix B: Atlas

Physical World Map

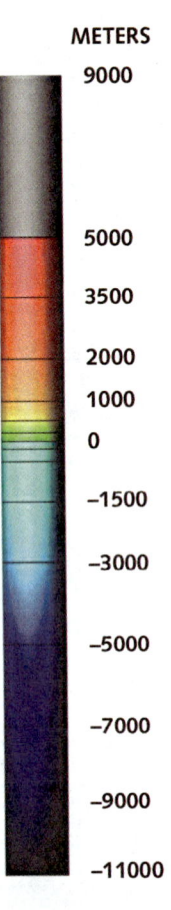

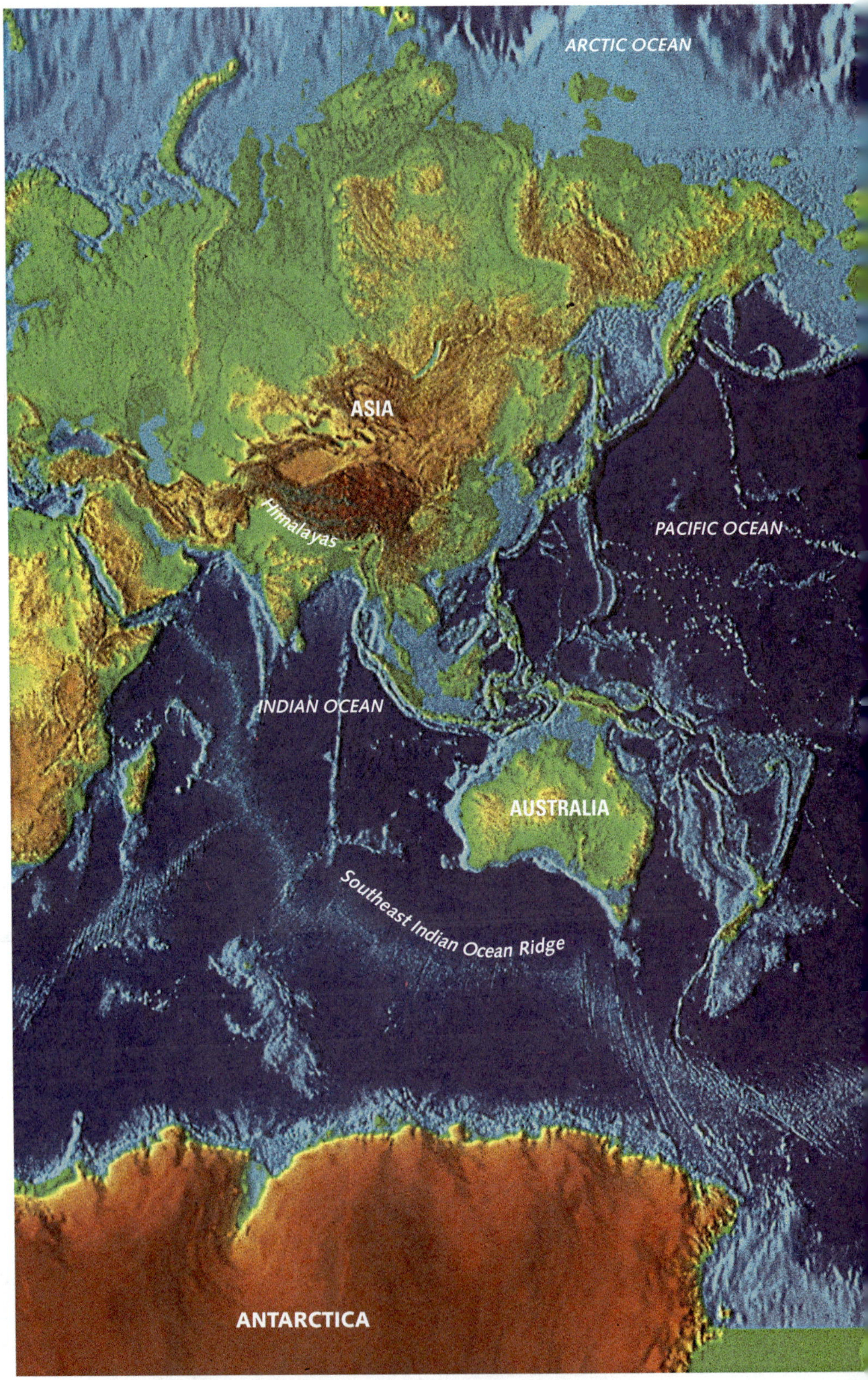

This image was generated from a combination of satellite altimetry data, ship-based data, and land-based data.

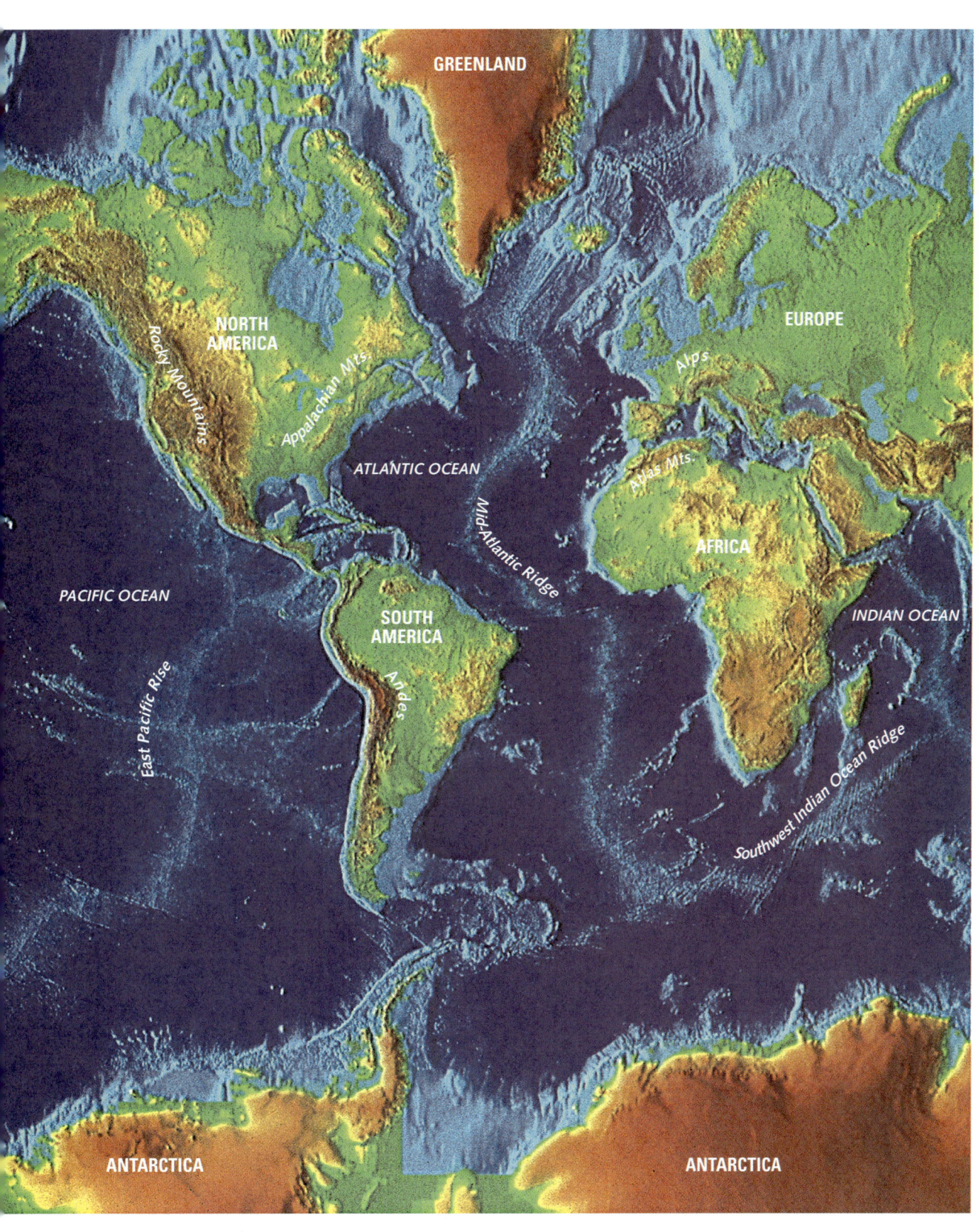

Physical World Map

Appendix B: Atlas

Earth's Tectonic Plates

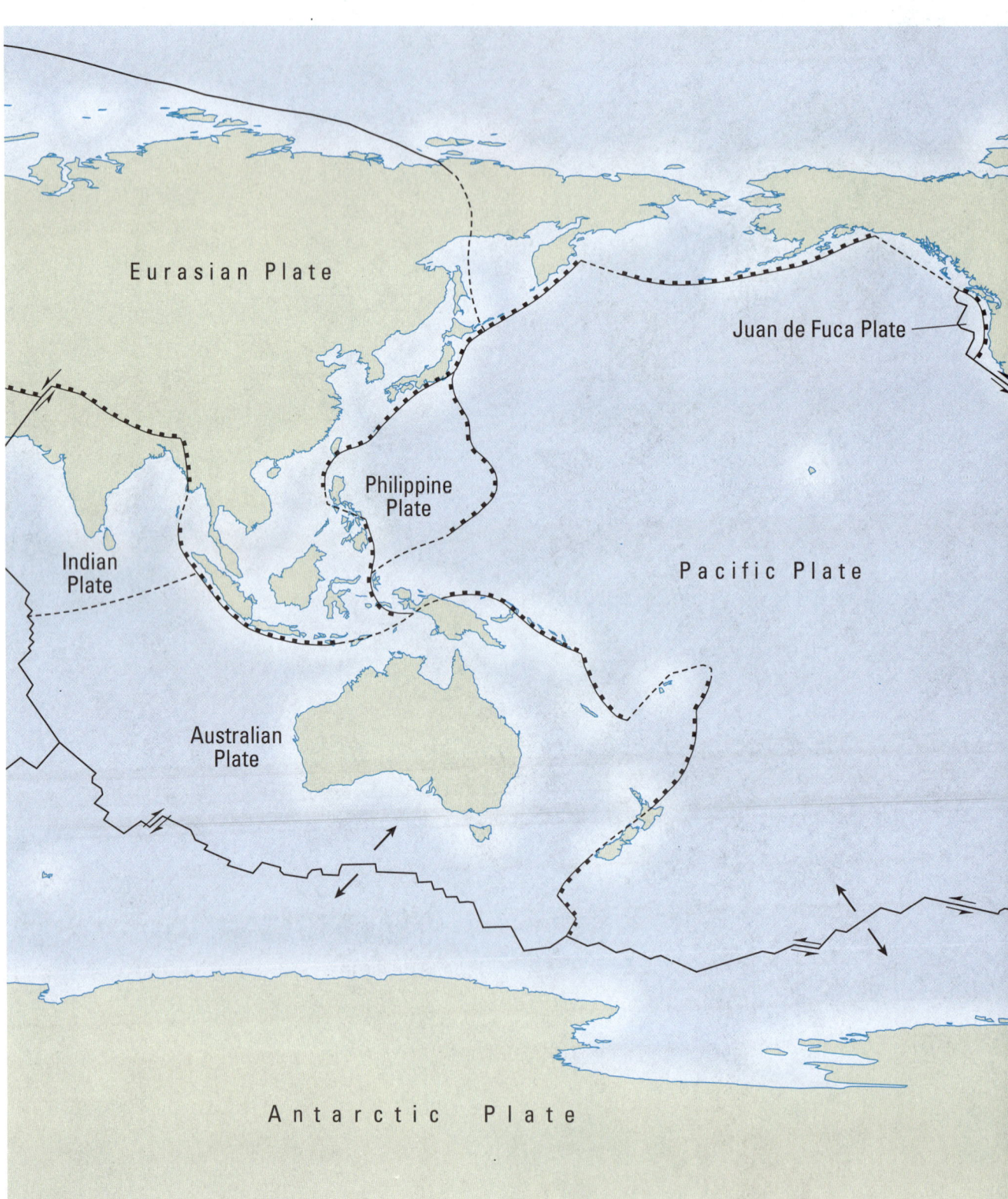

Earth's Tectonic Plates

Appendix B: Atlas

Seasonal Star Maps

WINTER SKY to the NORTH, *January 15*

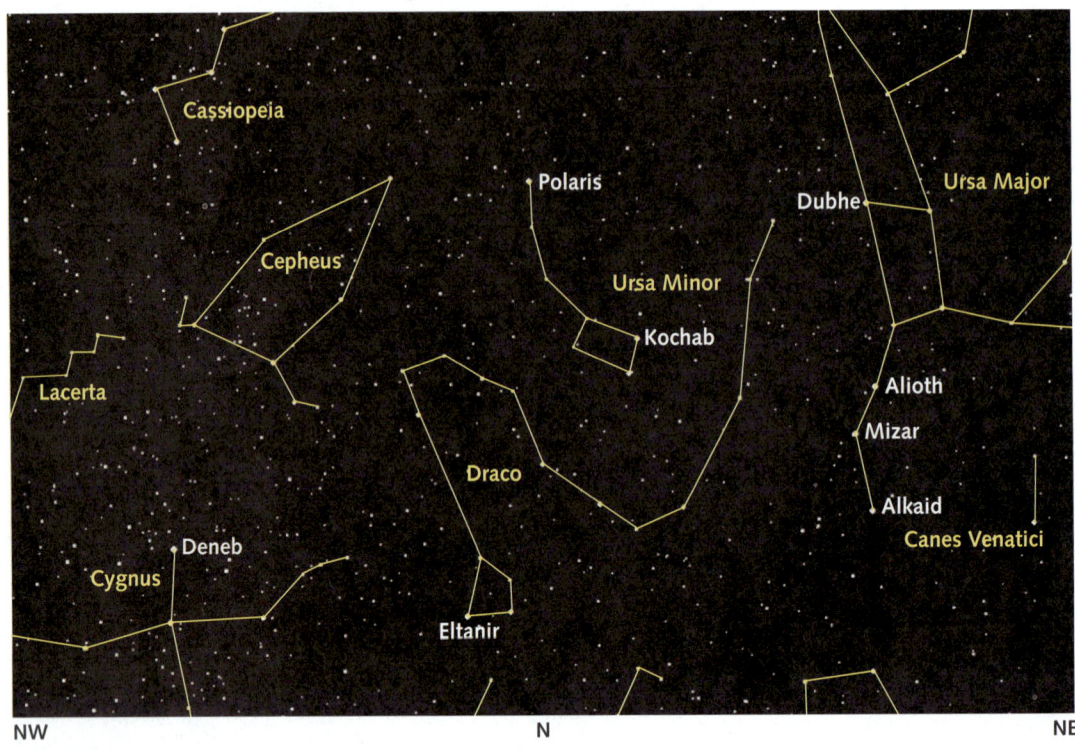

WINTER SKY to the SOUTH, *January 15*

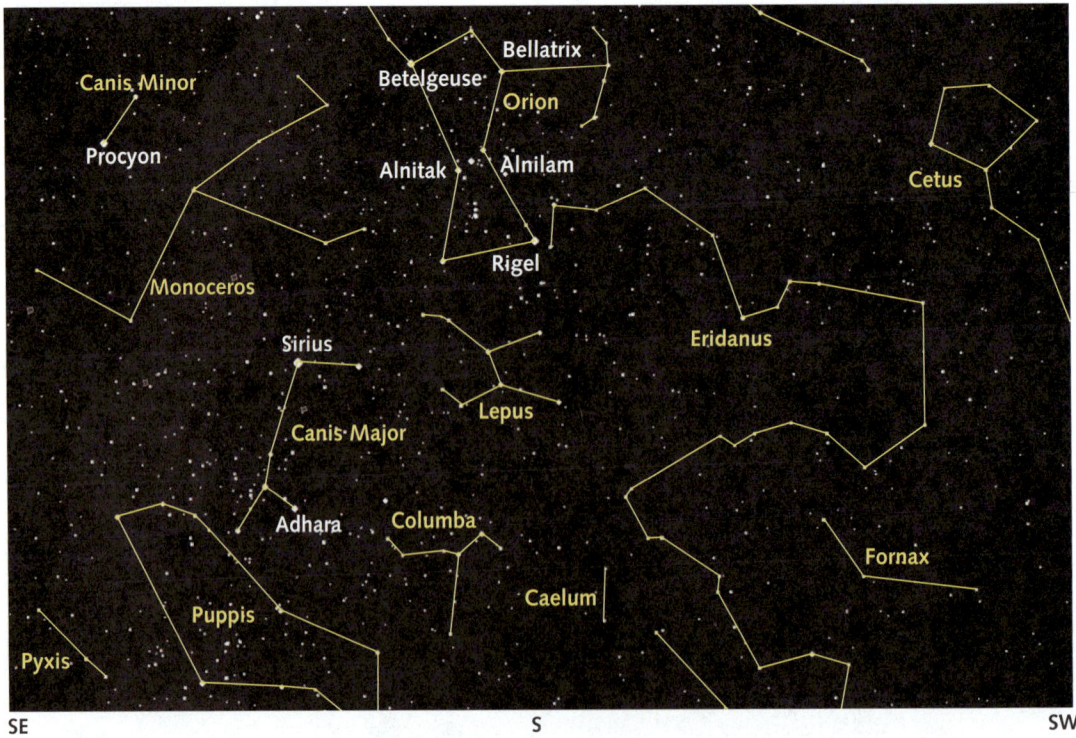

SPRING SKY to the NORTH, April 15

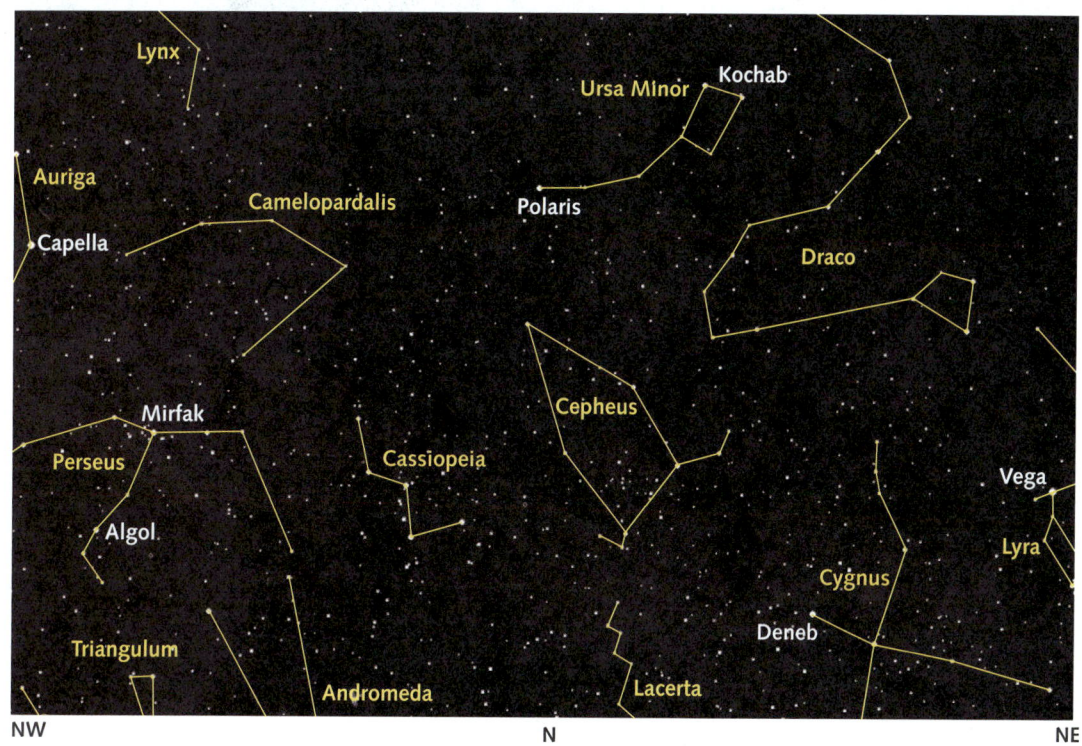

SPRING SKY to the SOUTH, April 15

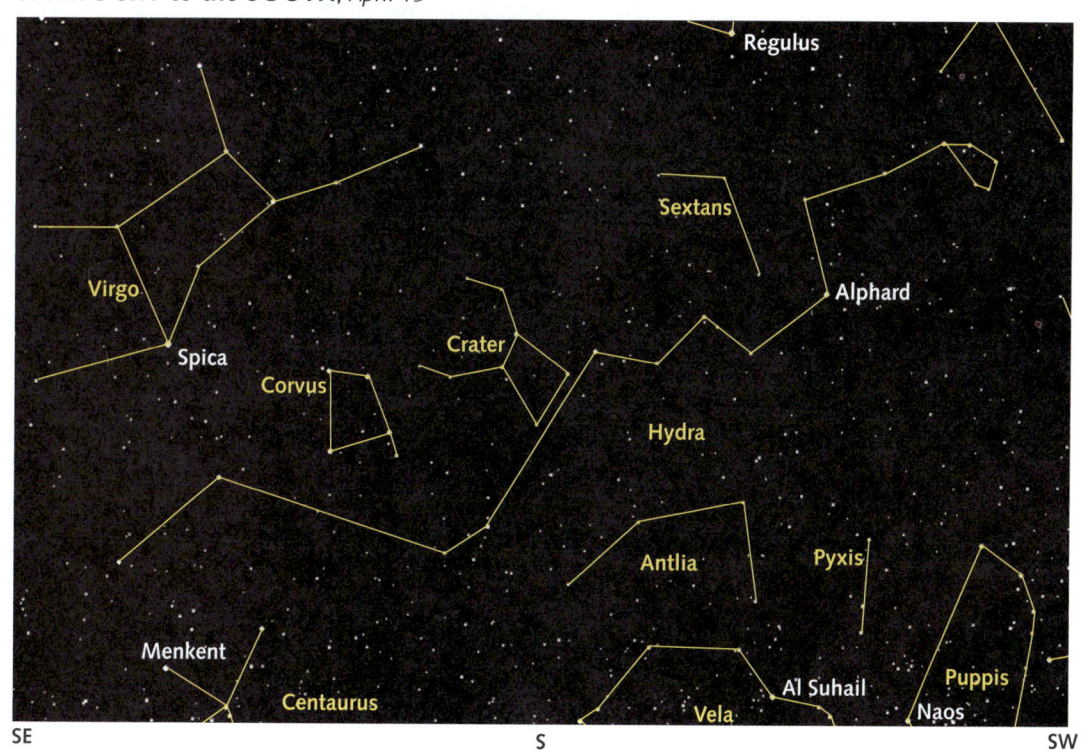

Appendix B: Atlas

Seasonal Star Maps

SUMMER SKY to the NORTH, *July 15*

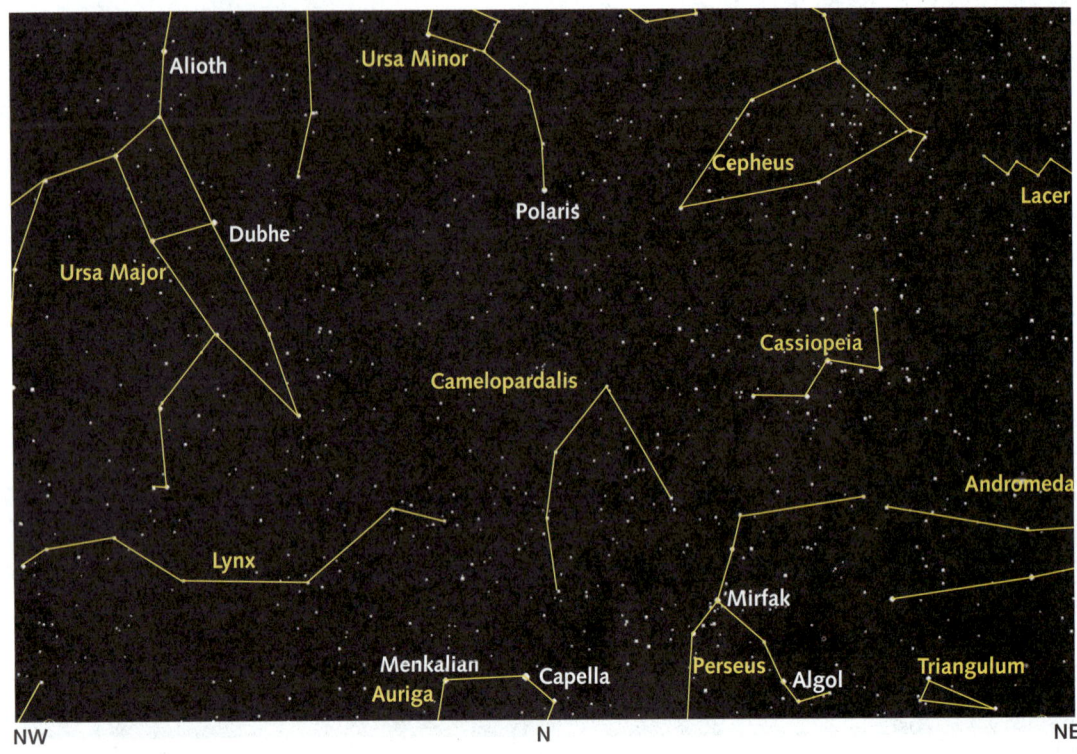

SUMMER SKY to the SOUTH, *July 15*

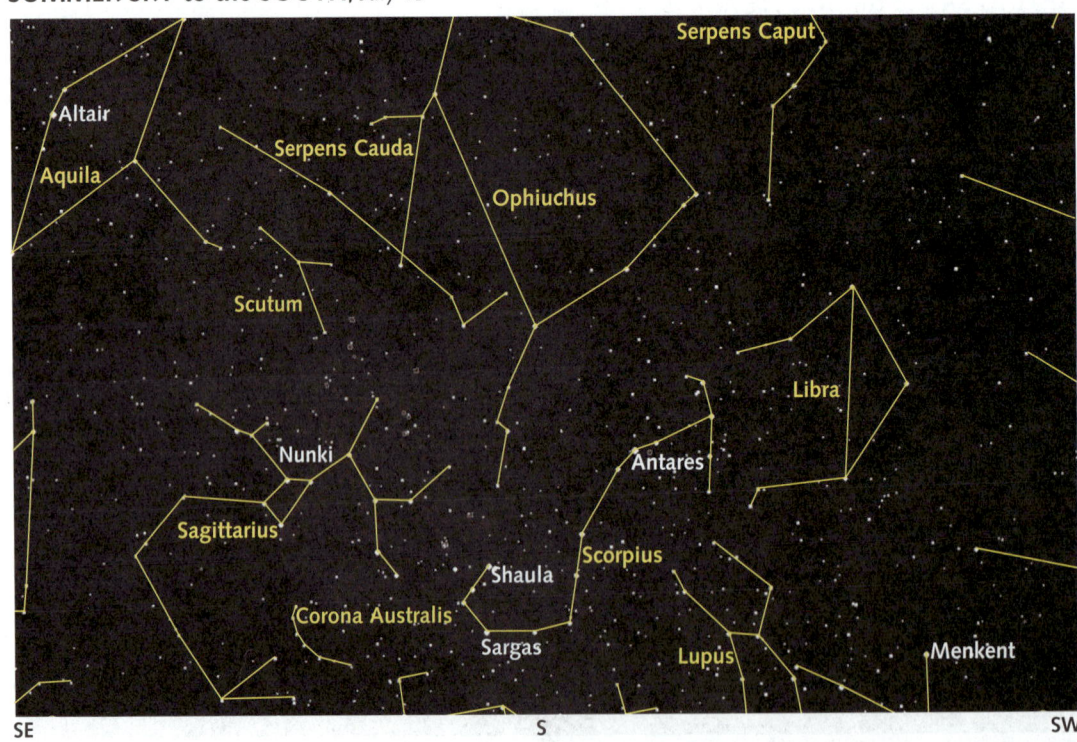

AUTUMN SKY to the NORTH, October 15

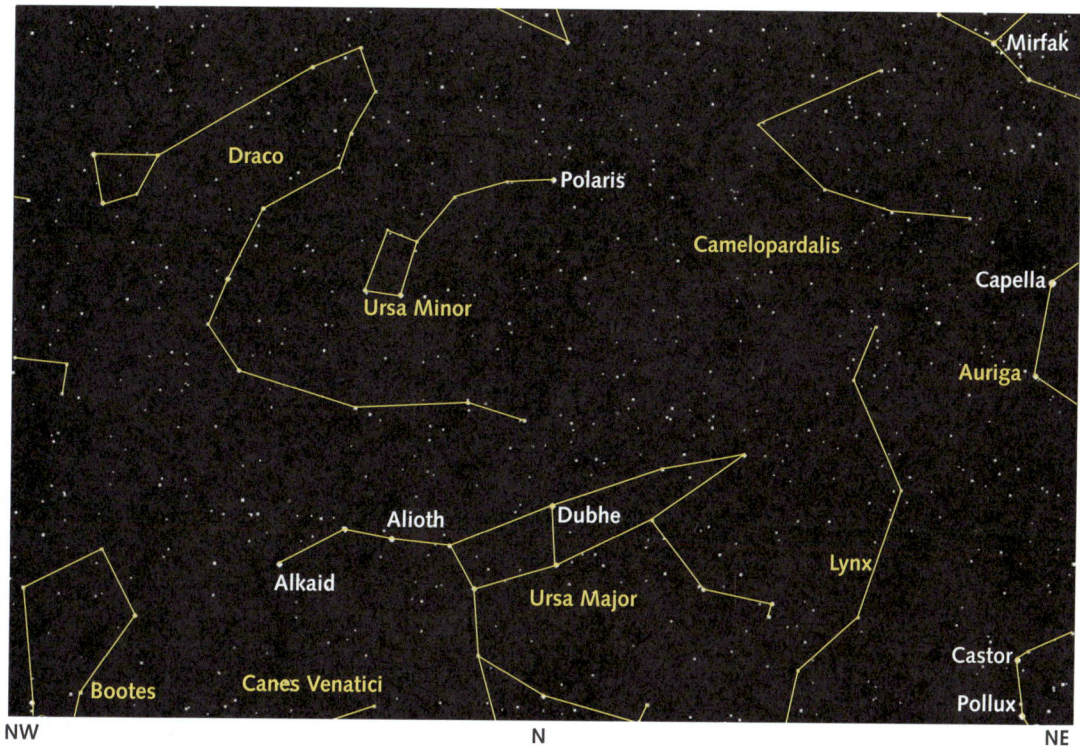

AUTUMN SKY to the SOUTH, October 15

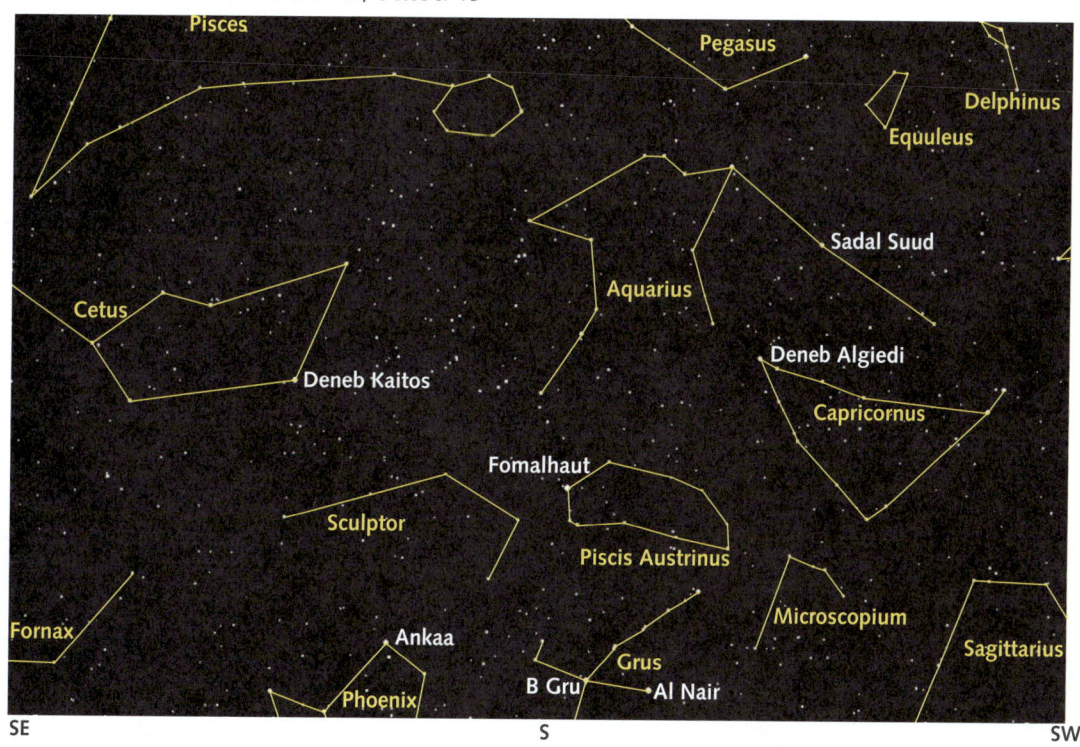

Seasonal Star Maps

Appendix B: Atlas

Topographic Map: Harrisburg, PA *(15 minute series, partial)*

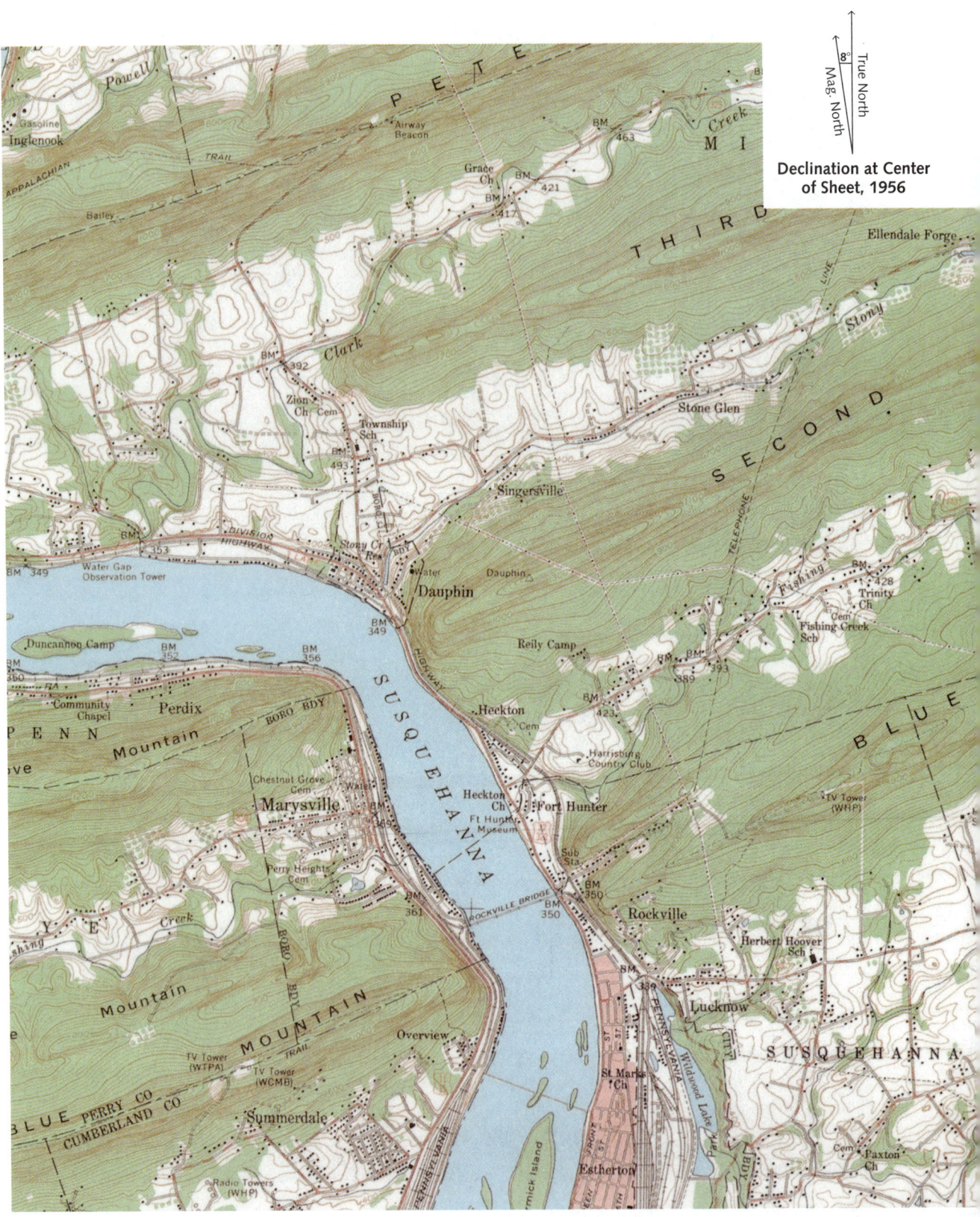

Appendix C: Skills Handbook

Strategies for Reading Science

To get the most from reading science textbooks and science articles, you need to be an active reader. There are specific strategies that you can use as an active reader to help you learn new information and make sense of important concepts. Some strategies are most effective when used before reading, while others are most effective during or after reading.

At any stage of reading, using a **Science Notebook** is valuable.

BEFORE YOU READ

PREVIEW Skim through the section you are about to read. Identify where it starts and ends. Read the headings, because they can help you identify the main ideas. Look closely at any diagrams, illustrations, and photographs. Read all the captions. Some students even read the questions in the Section Review so that they have an idea of what they are expected to learn in the section. Such a preview prepares you for successful reading.

After you have previewed the section, do one or more of the following activities in your Science Notebook:

- List what you already know about the topic that the section presents.
- List any predictions or expectations you have for what you might learn in this section.
- List what you want to find out about the topic.

QUESTION As you preview the chapter, formulate questions about the contents and record them in your Science Notebook. The questions you ask yourself can help you set a purpose for reading.

SET A PURPOSE Each section begins with a statement the presents the Key Idea. You can use this Key Idea to set a purpose for reading. For example, the Key Idea of a section might be "The composition of magma affects how explosive a volcanic eruption will be." Already, from reading this one statement—even if you knew nothing about volcanoes before—you can figure out two things: (1) that there must be a variety of magmas and (2) that some volcanic eruptions are more explosive than others. Now turn that Key Idea into a question: In your Science Notebook write, "How does the type of magma determine how explosive a volcanic eruption will be?" Make it your goal to find the answer to this question as you read the section.

Appendix C: Skills Handbook

WHILE YOU READ

LOOK FOR DESIGN FEATURES Pay attention to any text that is set in a box or treated differently. Notice, for example, the words or phrases that are boldfaced or highlighted in some way. Such design features are supposed to draw your attention to key words and concepts and help you recognize what is important.

USE TEXT ORGANIZERS Chapter and section titles, headings, and subheadings all provide a framework that helps you to identify and organize main ideas.

DEVELOP YOUR VOCABULARY Whenever you come across a word that you do not know, take the time to look up that word in the dictionary. You will benefit from developing the habit of writing down and looking up new words. You may even want to dedicate a section of your Science Notebook to new words. These words may not be "science" words, but they are important to your understanding of new material. So, do yourself a favor—find out what you don't know!

CONNECT As you read the section, make connections between what you are reading and what happens in your day-to-day life and in the "real world" as you experience it. Making such connections while learning new information helps you to make sense of the world around you.

CHECK YOUR UNDERSTANDING Remember, reading science is not a passive activity. To learn, you need to do more than let your eyes roam over the pages. Stop every now and then to ask yourself, "Am I getting this?" When you read a paragraph, see if you can restate the information in your own words.

STUDY THE VISUALS The photographs, diagrams, illustrations, charts, and graphs, all provide you with succinct information. Often times you might discover that the visuals present information in a way that helps you to grasp a concept more easily. Be sure to read the captions and pay attention to the labels.

RECORD YOUR THOUGHTS Use your Science Notebook to keep track of new information that you find interesting or ideas that you believe are important.

AFTER YOU READ

REVIEW Look over the section. Pause at each heading to ask yourself what you know about it.

SUMMARIZE Use your Science Notebook to answer that question you wrote when you Set a Purpose.

REFLECT What did you learn? Take the time to consider how reading the section has affected your prior knowledge or what you thought you knew about the topic. Does this new knowledge change the way you understand the world around you? How so?

CLARIFY One way to help you retain new information is to share it with someone else. When you have to explain a process or a concept to someone else, you end up making an extra effort to make that information clear. Hence, the process of "retelling" deepens your own understanding.

WHAT IS A SCIENCE NOTEBOOK?

In this textbook, you will find many references to the Science Notebook— consider it your tool for learning. You can turn any spiral notebook or binder with loose leaf pages into your own Science Notebook:

- a place where you can record your thoughts and observations
- a space for you to reflect on what you are learning as you read.

Sometimes your teacher might assign activities for you to keep in your notebook. Sometimes the textbook will direct you to your notebook.

You can use your Science Notebook:

- for keeping notes
- for organizing information
- for writing assignments
- for recording answers to section and chapter reviews in one place so that you can readily prepare for tests.

More importantly, the practice of keeping a Science Notebook enables you to deepen your understanding of important concepts. Thinking and writing about ideas — about what you know as well as what want to know — makes you an active, vibrant learner.

Appendix C: Skills Handbook

Laboratory Techniques: *Measurement Skills*

Throughout your study of earth science, you will be using equipment to help you observe and measure earth materials in the laboratory. To use the laboratory equipment safely and effectively, you must first become familiar with its appearance, function, and methods of use.

METRIC RULER

Length is measured with a metric ruler or meterstick. A meterstick is divided into 100 centimeters (cm). Each centimeter is divided into 10 parts, or 10 millimeters (mm). You will often measure length to the nearest tenth of a centimeter—that is, to one of the ten divisions within the centimeter.

When measuring length with a metric ruler or meterstick, do not start measuring from the very end of the meterstick, because the end is often imperfect. Instead, measure from the 1-centimeter mark on the ruler. Lay the ruler flat, and make sure it does not move between the time when you line it up and the time when you take the measurement. Whenever you line the ruler up to a mark on the page, read the ruler from directly above the mark on the page. Otherwise, the slanted perspective will cause your reading to be incorrect. After you have made your measurement, remember to subtract 1 centimeter from your reading, since you started from the 1-centimeter mark on the ruler and not from the end.

BALANCE

Mass is measured using a balance. A triple-beam balance is a common type of laboratory balance. It has a pan at one end and three beams, with riders, extending away from the pan and ending in a pointer.

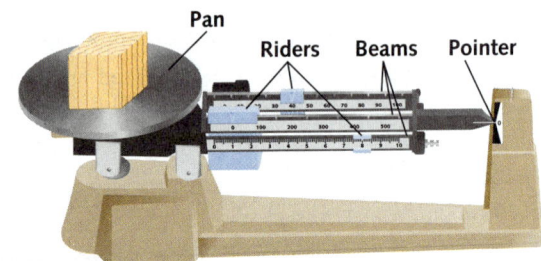

To find mass using a triple-beam balance, place the object to be measured on the pan. Starting with the largest rider, move the riders away from the pan, one notch at a time, until the pan is balanced and the pointer is at zero. In the laboratory exercises in this manual, you will often measure mass to the nearest tenth of a gram—that is, to the smallest division on the front beam of the balance.

THERMOMETER

A thermometer is used to measure temperature. Laboratory thermometers, unlike the thermometers you may use at home, are not shaken down before they are used. When measuring temperature, allow the thermometer to stabilize (stop changing) before taking the reading. Read the level of the liquid in the thermometer at eye level.

GRADUATED CYLINDER

The volume of a liquid is measured with a graduated cylinder. Place the graduated cylinder on a level surface and pour in the liquid. Read the volume of the liquid with your eye at the level of the surface of the liquid. If you try to read the volume from some other angle or level, your reading will be incorrect. If the cylinder is made of glass, the surface of the liquid forms a curve, or meniscus. In this case, the volume of the liquid is read from the bottom of the meniscus.

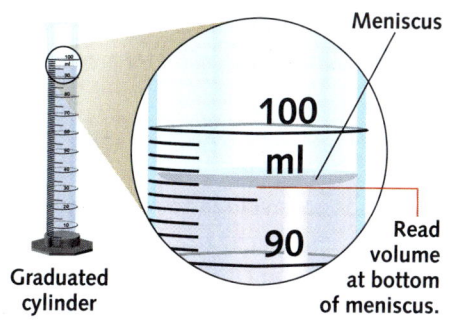

A graduated cylinder can also be used to find the volume of an object. Add water to the cylinder to a convenient level and note the volume. Carefully slide the object you wish to measure down the side of the cylinder and into the water. The change in water level is the volume of the object.

STOPWATCH

A stopwatch is used to determine the duration of an event. Most stopwatches have a button for starting and stopping the watch and another button for resetting the watch to zero. Some stopwatches read to the nearest tenth of a second and others read to the nearest hundredth.

ELECTRONIC PROBES

New technology makes it possible for electronic instruments to make more numerous and accurate measurements than it is possible for a person to make. Many electronic devices are available for use in a classroom laboratory setting. One such instrument is an electronic probe. The probe is connected to a computer or a graphing calculator through a data collection interface. The probe is inserted into a sample or pointed towards an energy source, while the computer or calculator displays and stores the data.

One advantage of using this technology to collect data is that it allows you to collect large amounts of data very quickly and efficiently. Graphing the data is made simpler because the data is being stored directly in a computer or inside a graphing calculator. The data collection interface also allows you to use several different types of probes simultaneously. If used properly, the electronic instruments make measurements that are much more accurate than measurements judged by the human eye.

If you decide to use an electronic device to record data, be sure to keep the computer and calculator from getting wet. Also, do not clean your probe devices with napkins or paper towels; rinse the sensors with distilled water and, if necessary, dry with a lens-cleaning tissue or similar wipe.

Appendix C: Skills Handbook

Laboratory Techniques: *Error Analysis*

The word *error* has a different meaning from its everyday definition when used in science. Scientific error is actually a measure of "experimental uncertainty." The ability to minimize errors and to predict how they might affect an experiment's outcome will greatly increase the usefulness of your results.

THE NATURE OF MEASUREMENTS

Examine the first illustration below. How long is the pencil: 9 centimeters? 10 centimeters? The markings on the ruler determine how accurately you can measure the pencil's length. The smallest unit of measure you can reliably read with this ruler is one centimeter. An acceptable answer, therefore, is 9 centimeters, with an uncertainty of ± 0.5 centimeter. This answer indicates that the length of the pencil is between 8.5 centimeters and 9.5 centimeters.

Now consider the second illustration below. Because the ruler has millimeter markings, you can give a more precise answer. A reasonable measurement of the pencil's length is 9.3 centimeters, with an uncertainty of about ± 0.05 centimeter.

The second answer, 9.3 centimeters, has two significant figures. If you could use a ruler with hundredth-of-a-centimeter marks, you could measure more closely, to three significant figures. The number of significant figures reveals the degree of uncertainty in a measurement.

SCIENTIFIC ERROR

You might think of a scientific error as a mistake that occurs while you perform an experiment. However, misunderstanding a procedure, reading scales incorrectly, or mishandling equipment are considered blunders, not errors. If you make a blunder, make a note in your data and redo the part of the procedure that was affected.

An error is an uncertainty about measurements that can result from limitations of the equipment, or of the person performing the measurement, and from environmental factors.

Scientific errors are classified by the way they affect experimental measurements. A systematic error is an error that shifts a measurement in a certain predictable direction. For example, a triple-beam balance that is calibrated incorrectly will cause measurements to be either all higher or all lower than the correct value. Sometimes large systematic errors are detectable, and adjustments can be made. Small systematic errors may not be detectable, and cannot necessarily be avoided.

A random error is one that shifts a measurement in an unpredictable direction. Their effect can be minimized by averaging the results of multiple trials.

PRECISION VERSUS ACCURACY

Experimental errors can affect the precision and accuracy of measurements. Precision describes the refinement and consistency of measurements. For example, the ruler showing the millimeter markings has higher precision than the ruler showing only centimeters. Another indicator of precision is the variation between multiple trials for the same measurement. Just because measurements are precise does not mean that they are accurate. Accuracy describes the difference between measurement and an actual value. The dart boards below illustrate the difference between precision and accuracy.

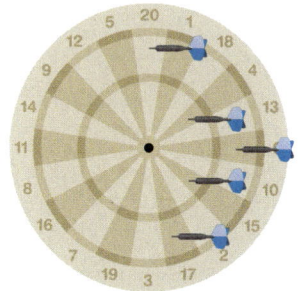

low precision

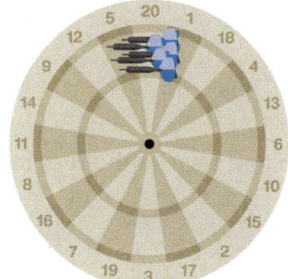

precision, but not accuracy

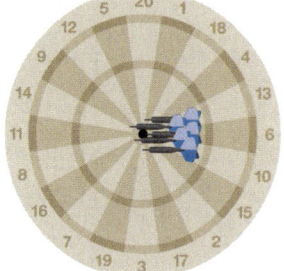

precision and accuracy

CALCULATING ERROR

Scientists use statistical methods to analyze error in their experiments. When the actual value of that which you are measuring is known, you can use the formula below to calculate the percent error of your measurement:

$$\text{Percent Error} = \frac{\text{Experimental Value} - \text{Actual Value}}{\text{Actual Value}} \times 100$$

MINIMIZING ERROR

Before you begin an experiment, read the procedure and analysis carefully. Pay attention to steps where potential error might occur. Also try to anticipate how a certain error might affect your final results. Remember that error can result from many sources throughout the experiment.

Skills Handbook 725

Appendix C: Skills Handbook

Mathematics: *Algebra Skills*

FRACTION OPERATIONS

To add or subtract two fractions with the same denominator, add or subtract the numerators.

Example

$$\frac{3}{5} + \frac{4}{5} = \frac{3+4}{5}$$ Add numerators.

$$= \frac{7}{5}, \text{ or } 1\frac{2}{5}$$ Simplify.

Example

$$\frac{7}{10} - \frac{2}{10} = \frac{7-2}{10}$$ Subtract numerators.

$$= \frac{5}{10}$$ Simplify.

$$= \frac{5}{2 \cdot 5}$$ Factor numerator and denominator.

$$= \frac{1}{2}$$ Simplify fraction to lowest terms.

To add or subtract two fractions with different denominators, write equivalent fractions with a common denominator.

Example

$$\frac{3}{5} + \frac{5}{6} = \frac{18}{30} + \frac{25}{30}$$ Use the lowest common denominator (LCD), 30.

$$= \frac{18 + 25}{30}$$ Add numerators.

$$= \frac{43}{30}, \text{ or } 1\frac{13}{30}$$ Simplify.

To add or subtract mixed numbers, you can first rewrite them as fractions.

Example

$$3\frac{2}{3} - 2\frac{1}{4} = \frac{11}{3} - \frac{9}{4}$$ Rewrite mixed numbers as fractions.

$$= \frac{44}{12} - \frac{27}{12}$$ The LCD is 12.

$$= \frac{44 - 27}{12}$$ Subtract numerators.

$$= \frac{17}{12}, \text{ or } 1\frac{5}{12}$$ Simplify.

To multiply two fractions, multiply the numerators and multiply the denominators.

Example

$$\frac{3}{4} \times \frac{5}{6} = \frac{3 \times 5}{4 \times 6}$$ Multiply numerators and multiply denominators.

$$= \frac{15}{24}$$ Simplify.

$$= \frac{3 \cdot 5}{3 \cdot 8}$$ Factor numerator and denominator.

$$= \frac{5}{8}$$ Simplify fraction to lowest terms.

Appendix C

Two numbers are **reciprocals** of each other if their product is 1. Every number except 0 has a reciprocal.

$5 \times \frac{1}{5} = 1$, so 5 and $\frac{1}{5}$ are reciprocals.

$\frac{2}{3} \times \frac{3}{2} = 1$, so $\frac{2}{3}$ and $\frac{3}{2}$ are reciprocals.

$1\frac{1}{4} \times \frac{4}{5} = \frac{5}{4} \cdot \frac{4}{5} = 1$, so $1\frac{1}{4}$ and $\frac{4}{5}$ are reciprocals.

To find the reciprocal of a number, write the number as a fraction. Then interchange the numerator and the denominator.

Example   Find the reciprocal of $3\frac{1}{4}$.

Solution $3\frac{1}{4} = \frac{13}{4}$ Write $3\frac{1}{4}$ as a fraction.

$\frac{13}{4} \longrightarrow \frac{4}{13}$ Interchange numerator and denominator.

▶ The reciprocal of $3\frac{1}{4}$ is $\frac{4}{13}$.

✓ **CHECK** $3\frac{1}{4} \times \frac{4}{13} = \frac{13}{4} \times \frac{4}{13} = \frac{13 \times 4}{4 \times 13} = 1$

To divide by a fraction, multiply by its reciprocal.

Example $\frac{3}{4} \div \frac{5}{6} = \frac{3}{4} \times \frac{6}{5}$ The reciprocal of $\frac{5}{6}$ is $\frac{6}{5}$.

$= \frac{3 \times 6}{4 \times 5}$ Multiply numerators and denominators.

$= \frac{18}{20}$ Simplify.

$= \frac{2 \cdot 9}{2 \cdot 10}$ Factor numerator and denominator.

$= \frac{9}{10}$ Simplify fraction to lowest terms.

Example $2\frac{1}{2} \div 4\frac{1}{6} = \frac{5}{2} \div \frac{25}{6}$ Write mixed numbers as fractions.

$= \frac{5}{2} \times \frac{6}{25}$ The reciprocal of $\frac{25}{6}$ is $\frac{6}{25}$.

$= \frac{5 \times 6}{2 \times 25}$ Multiply numerators and denominators.

$= \frac{30}{50}$ Simplify.

$= \frac{3 \times 10}{5 \times 10}$ Factor numerators and denominators.

$= \frac{3}{5}$ Simplify fraction to lowest terms.

Appendix C: Skills Handbook

FRACTIONS, DECIMALS, AND PERCENTS

Percent (%) means "divided by 100."

$53\% = 53 \text{ divided by } 100 = \frac{53}{100}$

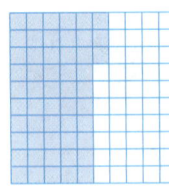

The grid contains 100 squares.
53 of the 100 squares are shaded.
53% of the squares are shaded.

To write a percent as a decimal, move the decimal point two places to the left and remove the percent symbol.

Example

a. $85\% = 85 = 0.85$
b. $3\% = 03 = 0.03$
c. $427\% = 427 = 4.27$
d. $12.5\% = 12.5 = 0.125$

To write a percent as a fraction in lowest terms, first write the percent as a fraction with a denominator of 100. Then simplify if possible.

Example

a. $71\% = \frac{71}{100}$
b. $10\% = \frac{10}{100} = \frac{1}{10}$
c. $4\% = \frac{4}{100} = \frac{1}{25}$
d. $350\% = \frac{350}{100} = \frac{7}{2} = 3\frac{1}{2}$

To write a decimal as a percent, move the decimal point two places to the right and add a percent symbol.

Example

a. $0.93 = 0.93 = 93\%$
b. $1.47 = 1.47 = 147\%$
c. $0.025 = 0.025 = 2.5\%$
d. $0.005 = 0.005 = 0.5\%$

To write a decimal as a fraction in lowest terms, first write the decimal as a fraction with a denominator of 100. Then simplify if possible.

Example

a. $0.73 = \frac{73}{100}$
b. $0.25 = \frac{25}{100} = \frac{1}{4}$
c. $0.05 = \frac{5}{100} = \frac{1}{20}$
d. $2.75 = 2\frac{75}{100} = 2\frac{3}{4}$

It is simple to write a fraction as a percent if the denominator of the fraction is a factor of 100. If not, divide the numerator by the denominator.

Example

a. $\dfrac{17}{25}$ ⟶ 25 is a factor of 100, so write $\dfrac{17}{25} = \dfrac{17 \cdot 4}{25 \cdot 4} = \dfrac{68}{100} = 68\%$.

b. $\dfrac{1}{8}$ ⟶ 8 is not a factor of 100, so divide: $1 \div 8 = 0.125 = 12.5\%$.

c. $\dfrac{1}{6}$ ⟶ 6 is not a factor of 100, so divide: $1 \div 6 = 0.1666\ldots \approx 0.167 = 16.7\%$.

Equivalent percents, decimals, and fractions		
$1\% = 0.01 = \dfrac{1}{100}$	$33\dfrac{1}{3}\% = 0.\overline{3} = \dfrac{1}{3}$	$66\dfrac{2}{3}\% = 0.\overline{6} = \dfrac{2}{3}$
$10\% = 0.1 = \dfrac{1}{10}$	$40\% = 0.4 = \dfrac{2}{5}$	$75\% = 0.75 = \dfrac{3}{4}$
$20\% = 0.2 = \dfrac{1}{5}$	$50\% = 0.5 = \dfrac{1}{2}$	$80\% = 0.8 = \dfrac{4}{5}$
$25\% = 0.25 = \dfrac{1}{4}$	$60\% = 0.6 = \dfrac{3}{5}$	$100\% = 1$

A fraction shows what part one number is of another. You can first write a fraction to determine what percent one number is of another.

Example What percent of 120 is 48?

Solution First write a fraction that compares 48 to 120: $\dfrac{48}{120}$.

$\dfrac{48}{120} = 0.4$ Write the fraction as a decimal.

$= 40\%$ Write the decimal as a percent.

▶ 48 is 40% of 120.

To find a percent of a given number, first write the percent as a decimal or as a fraction. Then multiply.

Example What is 75% of 160?

Solution $75\% = \dfrac{3}{4}$ Write 75% as a fraction.

$\dfrac{3}{4} \times 160 = 120$ Multiply.

▶ 75% of 160 is 120.

Skills Handbook

Appendix C: Skills Handbook

RATIO, RATE, AND PROPORTION

A ratio compares two numbers using division. If a and b are two quantities measured in the same units, then the **ratio of a to b** is $\frac{a}{b}$. The ratio of a to b can be written in three ways: a to b, $a:b$, and $\frac{a}{b}$.

Example Write the ratio 7 to 15 in three ways.

Solution 7 to 15   7 : 15 $\frac{7}{15}$

Example Write the ratio 12 to 60 in lowest terms.

Solution 12 to 60 $= \frac{12}{60}$ First write the ratio as a fraction.

$\qquad\qquad\qquad = \frac{1}{5}$ Then write the fraction in lowest terms.

If a and b are two quantities measured in different units, then the **rate of a per b** is $\frac{a}{b}$.

A **unit rate** is a rate per one unit of a given quantity. To determine a unit rate, write the rate with a denominator of 1.

Example A car traveled 648 miles using 18 gallons of gas. Find the unit rate in miles per gallon.

Solution $\frac{\text{miles}}{\text{gallons}} = \frac{648}{18} = \frac{36}{1}$

▶ The unit rate is 36 miles per gallon.

Two ratios, a/b and c/d, are said to be in **proportion** if $a/b = c/d$. You can use proportions to solve problems involving ratios and rates.

Example A car traveled 648 miles using 18 gallons of gas. If the car were to travel at the same rate, how far would it travel on 20 gallons of gas?

Solution $\frac{\text{miles}}{\text{gallons}} = \frac{648}{18} = \frac{?}{20}$

$? = \frac{648 \text{ miles} \times 20 \text{ gallons}}{18 \text{ gallons}} = 720 \text{ miles}$

SCIENTIFIC NOTATION

Values that are used in scientific calculations can be very large. One way to make these numbers more manageable for calculations is to use scientific notation. In scientific notation, a value is expressed as a number between 1 and 10 multiplied by a power of ten in exponential form.

Example The distance from the sun to the nearest star, Proxima Centauri, is about 40 trillion kilometers. Express this distance in scientific notation.

Solution 40 trillion kilometers = 40,000,000,000,000 km = 4.0×10^{13} km

Example The size of a pebble is measured to be 25 ten-thousandths of a meter across. What is the size of the pebble in scientific notation?

Solution 25 ten-thousandths of a meter = 0.0025 m = 2.5×10^{-3} m

UNITS OF MEASURE

Sometimes it is helpful to convert a measurement to a more convenient unit of measure. Unit conversions may be done by using metric prefixes or by changing the base unit of measure.

Example The size of a pebble is measured to be 0.0025 m across. What is the size of the pebble in millimeters?

Solution There are 1000 millimeters for every meter.

$$0.0025 \text{ m} = 00.25 \text{ m} \times \frac{(1000 \text{ mm})}{(1 \text{ m})}$$

$$= 2.5 \text{ mm.}$$

Example One light-year is equal to 9.5×10^{12} kilometers. The distance from the sun to Proxima Centauri is 4.0×10^{13} km. Express this distance in light-years.

Solution

$$4.0 \times 10^{13} \text{ km} = 4.0 \times 10^{13} \text{ km} \times \frac{(1 \text{ light-year})}{(9.5 \times 10^{12} \text{ km})}$$

$$= 4.2 \text{ light-years.}$$

Appendix C: Skills Handbook

Mathematics: *Geometry Skills*

AREA AND VOLUME

The area A of a figure is the number of square units enclosed by the figure.

Example

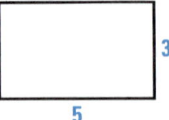 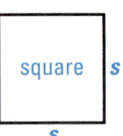

$A = 5 \times 3$ $A = \text{length} \times \text{width}$ $A = \text{side} \times \text{side}$
$= 15$ square units $= \ell \times w$ $= s \times s$
                                      $= \ell w$ $= s^2$

Example Find the area of a square with sides 15 inches long.

Solution $A = s^2 = 15^2 = 225$

▶ The area is 225 square inches (in.^{2}).

The volume of a solid is the amount of space contained by the solid. Volume is measured in cubic units.

One such unit is the cubic centimeter (cm^3). It is the amount of space contained by a cube whose length, width, and height are each 1 cm.

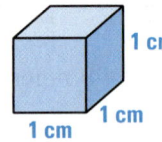

Example

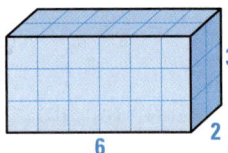 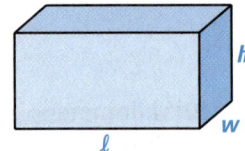

$V = 6 \times 2 \times 3$ $V = \ell \times w \times h$
$= 36$ cubic units

Example Find the volume of a box with length of 8 feet, width of 5 feet, and height of 9 feet.

Solution $V = \ell \times w \times h = 8 \text{ ft} \times 5 \text{ ft} \times 9 \text{ ft} = 360 \text{ ft}^3$

▶ The volume is 360 cubic feet (ft^3).

ANGLES

An angle consists of two different rays with a common endpoint. Angles are measured in degrees. You can use a protractor to measure and draw angles.

Example Line up the protractor's 0° line with one side of the angle as shown. Read the measure of the angle where the other side crosses the protractor. The measure of the angle is 60°.

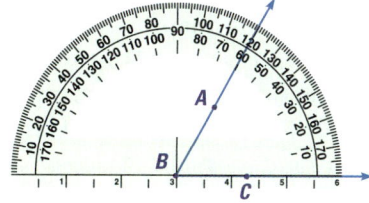

CIRCUMFERENCE AND ARC LENGTH

The circumference of a circle is the distance around the circle. To find the circumference C of any circle, multiply the diameter d by the number π. You can also find the circumference when you know the radius r.

$$C = \pi d$$
$$C = 2\pi r$$

The irrational number π is approximately equal to 3.14. Many calculators have a π key that approximates the value of π.

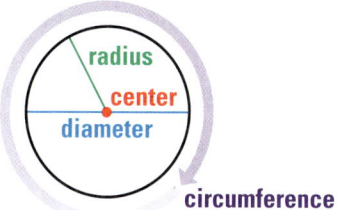

Example

Find the circumference of a circle with diameter 37 cm.

$C = \pi d$	Write formula.
$C = \pi(37)$	Substitute.
$C \approx 3.14(37)$	Use 3.14 for π.
$C \approx 116.2$ cm	Multiply.

Find the circumference of a circle with radius 5 m.

$C = 2\pi r$	Write formula.
$C = 2\pi(5)$	Substitute.
$C \approx 31.4$ m	Use a calculator.

An arc is a portion of a circle. The measure of an arc is given in degrees, as shown. An arc length is a portion of the circumference of a circle. You can use the measure of an arc to find its arc length. For example, the arc shown measures 90°. There are 360° in a circle, so the arc length is $\frac{XX}{XX} = \frac{90°}{360°}$, or $\frac{X}{X} = \frac{1}{4}$ the circumference of the circle.

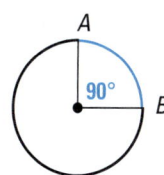

Example Give the measure of arc MN and find its arc length.

Solution The measure of arc MN is 100°.

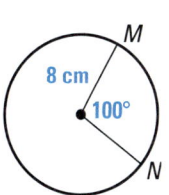

▶ Arc length of $MN = \frac{100°}{360°} \cdot$ circumference

$$= \frac{100°}{360°} \cdot 2\pi r = \frac{100°}{360°} \cdot 2\pi(8) \approx 14 \text{ cm}$$

Skills Handbook 733

Appendix C: Skills Handbook

GRAPHING SKILLS AND SCATTER PLOTS

A scatter plot compares two sets of data that change relative to each other. When making a scatter plot from a table of data, follow these general rules:

❶ Decide which quantity to plot on each axis. Sometimes, one quantity is being changed deliberately by you, and the other quantity is being measured after each change. The quantity that you are changing is called the controlled quantity and is usually placed on the horizontal axis.

❷ Choose a scale for each axis. Each scale should do two things. First, all of your data should lie between the highest and lowest numbers on your scale. Second, the scale should be easy to read. Each square of graph paper should represent convenient numbers such as 0.2, 1, 5, etc.

❸ Plot your data. To plot a data point, find the value on the scale of the horizontal axis. Imagine a line that rises vertically from that place on the scale as shown below. Now find the value on the vertical axis, and imagine a line that moves horizontally from that value. The point where these two imaginary lines intersect is where the data should be plotted.

❹ Draw a line or curve to show the general trend of the data. Often the data points will appear to fall along a straight line or a smooth curve. However, since measured data always contains some error, measured data seldom fall exactly on a straight line or curve. Rather than connect each data point, draw a straight line or curve that best fits the data points. A good best-fit line may pass through some of the points, but also has points above and below the line.

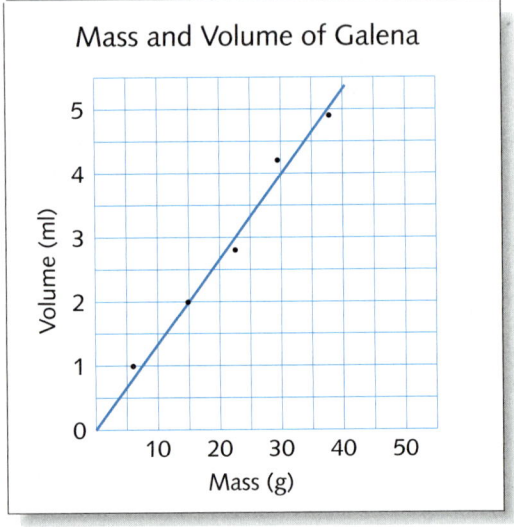

❺ Make predictions by interpolating or extrapolating. The best-fit line is your prediction of the relationship between the two quantities you measured. If you wish to make a prediction of a future measurement, simply find the value of the controlled quantity on the horizontal axis, and find the corresponding value on the vertical axis given by the best-fit line.

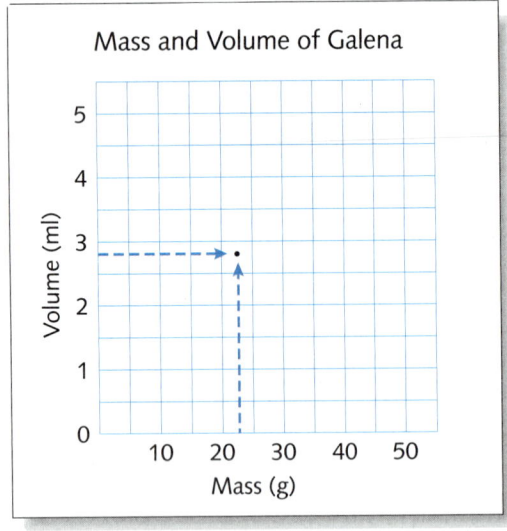

| Mass and Volume of Galena ||
Mass (g)	Volume (ml)
6.2	1.0
15.0	2.0
22.7	2.8
29.6	4.2
38.0	4.9

BAR GRAPHS AND HISTOGRAMS

A bar graph can be used to display data that fall into distinct categories. The bars in a bar graph are the same width. The height or length of each bar is determined by the data it represents and by the scale you choose.

Example In 1998, baseball player Mark McGwire hit a record 70 home runs. The table shows the locations to which the home runs were hit. Draw a bar graph to display the data. ▶ Source: *The Boston Globe*

Field location	Number of home runs
left	31
left-center	21
center	15
right-center	3
right	0

Solution

❶ First, choose a scale. Since the data range from 0 to 31, make the scale increase from 0 to 35 by fives.

❷ Draw and label the axes. Mark intervals on the vertical axis according to the scale you chose.

❸ Draw a bar for each category.

❹ Give the bar graph a title.

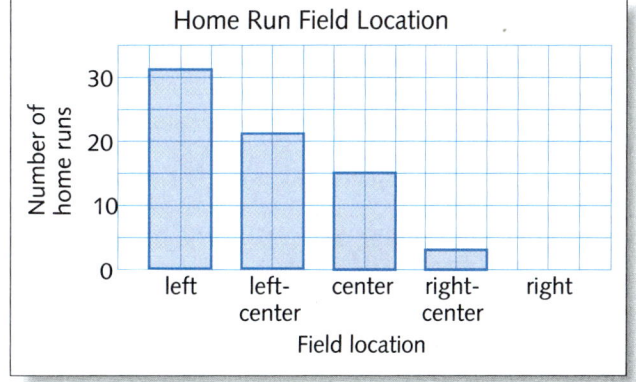

A histogram is a bar graph that shows how many data items occur within given intervals. The number of data items in an interval is the frequency for that interval.

Example The table shows the distances of McGwire's home runs. Draw a histogram to display them.

Distance (ft)	Frequency
301–350	4
351–400	24
401–450	27
451–500	11
501–550	4

Solution Use the same method you used for drawing the bar graph above. However, do not leave spaces between the bars.

Since the frequencies range from 4 to 27, make the scale increase from 0 to 30 by fives.

Draw and label the axes. Mark intervals on the vertical axis and draw a bar for each category. Do not leave spaces between the bars.

Give the histogram a title.

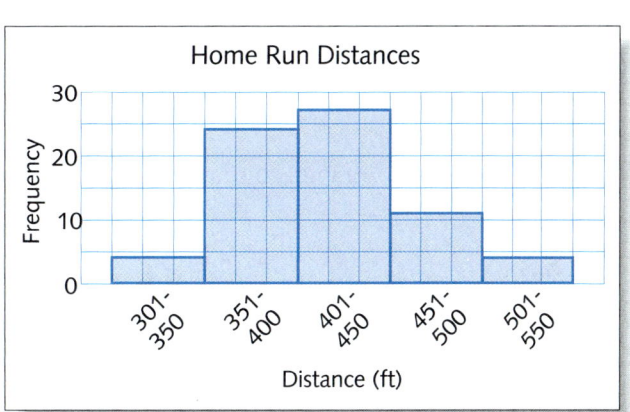

Skills Handbook

Skills Handbook

LINE GRAPHS AND CIRCLE GRAPHS

A line graph can be used to show how data change over time.

Example A science class recorded the highest temperature each day from December 1 to December 14. The temperatures are given in the table. Draw a line graph to display the data.

Date	1	2	3	4	5	6	7	8	9	10	11	12	13	14
Temperature (°F)	40	48	49	61	24	35	34	42	41	40	22	20	28	30

Solution

1. Choose a scale.

2. Draw and label the axes. Mark evenly spaced intervals on both axes.

3. Graph each data item as a point. Connect the points.

4. Give the line graph a title.

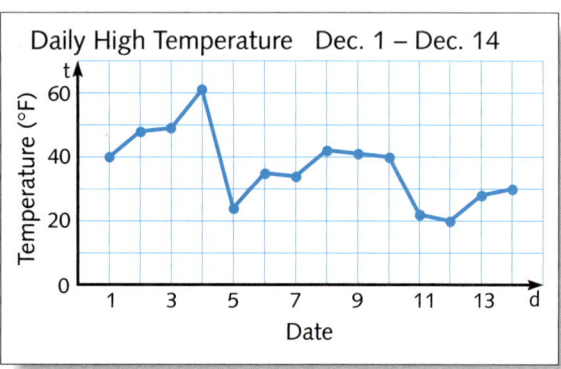

A circle graph can be used to show how parts relate to a whole and to each other.

Example The table shows the number of sports-related injuries treated in a hospital emergency room in one year. Draw a circle graph to display the data.

Related sport	Number of injuries
basketball	56
football	34
skating/hockey	22
track and field	10
other	28

Solution

1. To begin, find the total number of injuries.

   $56 + 34 + 22 + 10 + 28 =$ **150**

   To find the degree measure for each sector of the circle, write a fraction comparing the number of injuries to the total. Then multiply the fraction by 360°. For example:

   Football: $\frac{34}{150} \cdot 360° = 81.6°$

2. Next, draw a circle. Use a protractor to draw the angle for each sector.

3. Label each sector.

4. Give the circle graph a title.

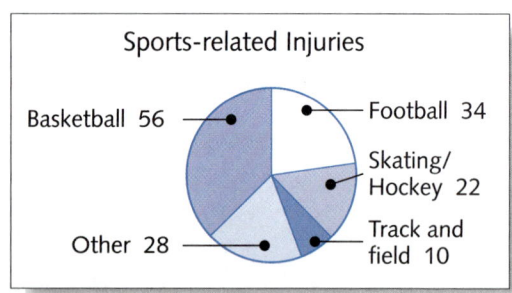

736 Appendix C

Glossary

Word Roots and Prefixes Science is full of complex words, many of which may seem difficult to understand and remember. Yet you can unlock the meaning of many words, both scientific and everyday, by understanding the meaning of the parts that make up the word. Words have three parts: the prefix, the root, and the suffix. Words do not always have prefixes or suffixes, but they all have roots.

English draws words from many other languages. As you may have noticed as you have used the vocabulary strategies in this text, many of the words you have been reading come from two ancient languages, Latin and Greek. The list below expands upon what you have learned in the vocabulary strategies. It names the word part, tells of its origin, and then provides sample words or phrases.

Alt- comes from the Latin word *altus,* meaning "high." Altitude, altimeter, altostratus.

Aster- is from the Greek word, *asteroeidēs,* which means "starlike." Asteroid.

Astheno- is from the Greek word that means "weak," *asthenēs.* Asthenosphere.

Ap- means "far" and is drawn from the Greek prefix *apo-.* Apogee, aphelion.

Astro- comes from the Greek word for star, *astron.* Astronomy, astronomer.

Baro- comes from the Greek word for weight, *baros.* Barometric, barometer.

Batho- is a variety of the prefix *bathy-,* which is drawn from two Greek words: *bathus,* meaning "deep," and *bathos,* meaning "depth." Batholith, bathyscaphe.

Bi- comes from three Latin words, *bis* and *bi-,* which mean "twice," and *bīnī,* "two by two." Bicycle, bisect, bimonthly.

Bio- comes from the Greek word *bios,* which means "life." Biology, biosphere.

Centric is based in the Greek word *kentron,* the center of a circle. Heliocentric, geocentric.

Cline comes from the Greek word *klīnein,* which means "to lean." Incline, syncline.

Com- and *con-,* which both mean "with" or "together," come from the same root, the Latin word *com,* "with." Composite, converge.

Cycl- comes from the Greek word *kuklos,* meaning circle or wheel. Cyclone, cycle, epicycle.

Epi- comes from the Greek word for "upon," *epi.* Epicenter, epicycle.

Equi- is derived from the Latin prefix *aequi-,* which is from the word *aequus,* meaning "equal." Equator, equinox.

Geo- comes from the Greek word *geō,* which in turn came from the word *gē,* meaning "earth." Geology, geocentric.

Helio- comes from the Greek word *helios,* meaning "sun." Aphelion, perihelion, heliocentric.

Hemi- is derived from the Greek word *hēmi,* which means "half." Hemisphere.

Hydro-, which means "water," comes from the Greek words *hudro* and *hudr,* which are from *hudōr.* Hydrosphere, hydrometer.

Iso- comes from the Greek word *īsos,* which means "equal." Isobar, isotherm, isocline.

Both *lith* and *litho* are from the Greek *litho,* which means "stone." Lithosphere, batholith.

Meso- comes from the Greek word for "middle," *meso.* Mesozoic, mesosphere.

Meta comes from the Greek word for "after," *meta.* Metamorphic.

Meter comes from the French word *mètre,* which in turn came from the Greek word *metron,* both of which mean "measure." Altimeter, seismometer, meter.

Morphic has its beginnings in the Greek word for "shape," *morphē.* Metamorphic, morph.

Peri- is from the Greek word for "near," *peri.* Perigee, perihelion.

Photo- derives from the Greek word *phōs,* which means "light." Photosphere, photosynthesis.

Proto- is from the Greek *protōs,* which means "first." Protoplanet, protostar.

Rego- is from the Greek word for blanket, *rhēgos.* Regolith.

Seismo- comes from *seiein,* the Greek verb that means "to shake," which became the Greek word for earthquake, *seismo.* Seismograph, seismometer.

Sub- is a prefix that comes from the Latin word for "under," *sub.* Subduction, submarine.

Terre- comes from the Latin word *terra,* meaning "earth." Terrestrial, terrane.

Thermo- comes from Greek *therme,* which means "heat," which in turn came from *thermos,* which means "warm" or "hot." Thermocline, thermosphere, thermometer.

Word Roots

Glossary

A

aa A solidified basaltic lava flow on land that has formed with rough, jagged surfaces. (p. 200)

abrasion The wearing away of rock material by grinding action. (p. 259)

absolute magnitude The measure of how bright a star would be it were located 10 parsecs from Earth. (p. 623)

absolute time The method of recording events that identifies the actual date of an event, such as when a rock formed. (p.656)

absorption spectrum A continuous spectrum crossed by dark lines produced when light passes through a non-incandescent gas. (p. 614)

abyssal hill One of a series of small rolling hills on the ocean floor that occur next to continental margins and oceanic ridges. (p. 517)

abyssal plain A large flat area on the deep sea floor; composed of sediments that originated mostly on continents. (p. 517)

acid rain Rainwater that contains unusually high amounts of acids that can be traced back to pollutants, including sulfur dioxide, nitrogen compounds, and carbon dioxide. (p. 261)

active continental margin A continental margin at which an oceanic plate is subducting under a continental plate, characterized by the presence of a narrow continental shelf and a deep-sea trench. (p. 515)

aftershock A smaller earthquake that follows a larger earthquake. (p. 223)

air mass A large body of air in the lower troposphere that has similar characteristics throughout. (p. 436)

air pollutant Any airborne gas or particle that occurs at a concentration capable of harming living things or disrupting the functioning of the environment. (p. 378)

air pressure The force air exerts per unit area. (p. 414)

alluvial fan A sloping triangular deposit of sediment located where a mountain stream meets level land.

altitude The height of an object in the sky above the horizon.

amphibole A family of complex silicate minerals that tend to form long, needlelike crystals.

anemometer An instrument that measures wind speed.

anticline An upfold in rock layers. (p. 239)

anticyclone An area of high pressure

aphelion The point in a planet's orbit when it is farthest from the sun.

apogee The point farthest to Earth in the moon's orbit. (p. 542)

apparent magnitude The measure of how bright a star appears to be to an observer on Earth. (p. 619)

aqueous solution A homogeneous mixture of two or more substances in which the solvent or dissolving medium is water. (p. 493)

aquifer A permeable layer of rock and sediment that stores and carries groundwater in enough quantities to supply wells. (p. 304)

arête A sharp divide that separates to adjoining cirques.

artesian formation An arrangement of a permeable layer of rock sandwiched between two layers of impermeable rock. (p. 304)

asteroid A solid, rocklike mass that revolves around the sun; most asteroids are found between the orbits of Mars and Jupiter. (p. 603)

asthenosphere The partially melted layer of the mantle that underlies the lithosphere (p. 73)

astronomical unit The average distance between Earth and the sun, about 150 million kilometers. (p. 619)

astronomy The study of the universe.

atmosphere The gaseous envelope of air surrounding Earth, made up of a mixture of about 78 percent nitrogen, 21 percent oxygen, and 1 percent other gases and water vapor ; one of the four spheres of the Earth system. (p. 8)

atom The smallest part of an element that has all the properties of that element.

atomic number The number of protons in the nucleus of an atom. (p. 91)

aurora A glow in the night sky produced in the upper atmosphere by particles from the solar wind interacting with Earth's magnetic field. (p. 575)

autumnal equinox Start of fall in the Northern Hemisphere, occurring on or about September 22 each year when the noon sun is directly over the equator; one of the two days each year (with the vernal equinox) when day and night are of equal length in both hemispheres. (p. 83)

axis of rotation An imaginary straight line through Earth between the North Pole and the South Pole, on which Earth rotates; it is tilted 23.5° from the plane of Earth's orbit.

B

backwash A gentle current of water that runs down a beach slope under an oncoming wave.

barometer An instrument that measures atmospheric pressure.

base level The level of the body of water into which a stream flows. (p. 288)

batholith A large mass of igneous rock exposed by erosion at Earth's surface; forms the core of many mountain ranges. (p. 126)

beach The area of shore between the high-tide level and the low-tide level. (p. 350)

Beaufort scale A scale that relates wind speed to its effects on land and sea.

bed load Sand, pebbles, and boulders that are moved along the bed of a stream and that are too heavy to be carried in suspension. (p. 284)

bench mark A marker in the ground indicating the exact elevation above sea level.

big bang model The theory holding that the universe originated from the instant expansion of an extremely small agglomeration of matter of extremely high density and temperature. (p. 634)

biogenous sediments Sea-floor sediments that are composed mostly of the shells and skeletons of living things. (p. 524)

biosphere All living organisms in the Earth system and their environments; one of the four spheres of the Earth system. (p. 10)

black hole The final life stage of an extremely massive star, with a gravitational field so intense that not even light can escape. (p. 630)

black smoker A hot spring on the ocean floor vent where hot water (380°C) comes up through chimney-like vents; metallic sulfides precipitate when the hot water comes into contact with the cold ocean water (2°C). (p. 502)

blizzard A winter storm characterized by winds exceeding 56 kilometers per hour, temperatures of 27°C or lower, and falling or blowing snow that reduces visibility. (p. 453)

blowout A small, shallow depression formed by deflation.

body waves Waves of energy that travel from the focus of an earthquake through the material of Earth's body; P waves and S waves are two types of body waves. (p. 215)

boiling point The temperature at which a solution turns from liquid to gas. (p. 494)

buffer An aqueous solution that acts to stabilize changes in pH. (p. 494)

C

caldera A large, crater-shaped basin formed after the top of a volcano collapses. (p. 204)

calving The process by which a block of a glacier breaks off and falls into the sea to form an iceberg. (p. 322)

capacity A measure of the total amount of sediment a stream can carry. (p. 285)

capillary action A phenomenon whereby groundwater rises because the water molecules are attracted to soil particles. (p. 302)

carbon cycle The biogeochemical circulation of carbon through the Earth system. (p. 14)

carbonate A nonsilicate mineral that has as its major building block one carbon atom covalently bonded to three oxygen atoms. (p. 110)

carbonic acid A weak acid formed when carbon dioxide dissolves in water.

cartographer One who makes maps. (p. 44)

cast An object created when minerals seep into and fill a mold, forming a copy of the original fossil. (p. 648)

cavern A large underground chamber. (p. 311)

cementation The process by which minerals precipitate out of solution to fill the spaces between sand grains, pebbles, or other rock particles and bind the fragments together. (p. 128)

cepheid variable A variable star that brightens and dims regularly, or pulses, and whose distance can be determined from its period of pulsation. (p. 624)

Glossary

chemical weathering The breakdown or decomposition of rock that takes place when minerals are changed into different substances. (p. 258)

chlorofluorocarbons Gases containing chlorine, fluorine, and carbon atoms that break down the ozone layer and weaken protection from the sun's ultraviolet waves.

chromosphere The inner layer of the sun's atmosphere, located above the photosphere and below the corona. (p. 573)

cinder cone A cone-shaped volcano formed from lava fragments that have been ejected from a volcanic vent. (p. 202)

cirque A semicircular basin found at the head of a glacial valley formed by a valley glacier. (p. 324)

cirrus High-altitude clouds that are thin, feathery tufts of ice crystals. (p. 397)

clastic rocks Sedimentary rocks formed from fragments of other rocks.

cleavage The tendency of a mineral to split along planes of its crystalline structure where bonds are weakest. (p. 100)

climate The characteristic pattern or course of the weather that an area has over a long period of time. (p. 466)

climate controls The conditions that determine the climate of an area, including but not limited to latitude, elevation, topography, and prevailing winds. (p. 467)

closed system A system in which energy can enter or leave, but matter cannot. (p. 5)

cold front The boundary between an advancing cold air mass and the warm air mass it is displacing. (p. 440)

cold-core ring An offshoot stream from an ocean current that forms a ring with a core of cold water. (p. 535)

collision boundary A convergent boundary where two continents have come together and are welded into a single, larger continent. (p. 178)

comet A mass of rock, ice, dust, and gas traveling around the sun, usually in a highly eccentric orbit. (p. 602)

competence A measure that describes the maximum size of the particles a stream can carry. (p. 285)

composite volcano A volcano made of layers of hardened lava flows and pyroclastic materials. (p. 203)

compound A substance that contains atoms of two or more elements that are chemically combined. (p. 93)

condensation The change from water vapor to liquid water. (p. 390)

condensation level The altitude at which water vapor begins to condense. (p. 398)

condensation nuclei Microscopic particles on which water vapor condenses to form cloud droplets. (p. 394)

conduction The transfer of heat or another form of energy from one particle of a substance to another. (p. 369)

conservation The protection, restoration, and management of natural resources. (p. 156)

constellation A group of stars that appear to form a pattern in the sky. (p. 617)

contact metamorphism The process of rock formation that results when hot magma forces its way into overlying rock and changes that rock.

continental drift A hypothesis that Earth's continents move on Earth's surface. (p. 172)

continental glacier A large sheet of ice that covers a large part of a continent. (p. 319)

continental margin The underwater part of the continental crust, which includes the continental shelf and the continental slope. (p. 514)

continental rise A gently rolling undersea slope between a continental slope and an abyssal plain. (p. 514)

continental shelf The flat or gently sloping submerged part of the continent; extends from the shoreline out to the continental slope. (p. 514)

continental slope The steep boundary between the continental shelf and the continental rise. (p. 514)

continuous spectrum A spectrum that contains all colors or wavelengths. (p. 614)

contour interval The difference in elevation between two consecutive contour lines. (p. 54)

contour lines Lines on a topographic map showing elevation of land above or below sea level, all points connected by a line having the same elevation. (p. 53)

convection The transfer of heat energy in a liquid or gas through the circulation of currents of heated particles within the substance. (p. 369)

convergent boundary A boundary between two plates that are moving toward each other, or converging. (p. 177)

cooperation The act of working together toward a common goal; sharing scientific data and information through professional organizations and conferences, and in scientific journals. (p. 28)

coral atoll A ring-shaped coral island surrounding a central lagoon. (p. 521)

coral Tiny, colonial sea animals that form a large part of Earth's reefs. (p. 500)

core sampling A technique that involves the use of a special hollow drill to remove cylindrical samples from sediment or rock, so they can be analyzed for certain properties or features. (p. 511)

Coriolis effect The effect of Earth's rotation that causes the deflection of moving objects toward the right in the Northern Hemisphere and toward the left in the Southern Hemisphere. (p. 419)

corona The outermost layer of the sun's atmosphere, located above the chromosphere. (p. 573)

correlation The matching of rock layers from one area with those of another area. (p. 652)

countercurrent A surface ocean current that flows in the opposite direction of the wind-related current. (p. 535)

covalent bond The bond formed from the sharing of electrons.

crater A bowl-shaped depression on the surface of a moon or a planet, usually caused by the impact of a meteorite. (p. 556)

craton The ancient core of a continent, which is tectonically stable. (p. 185)

creep The slow, often imperceptible, movement of soil down a slope. (p. 268)

crevasse A great fissure or crack in a glacier. (p. 321)

crust The very thin outer layer of Earth above the mantle, composed of a rigid layer of lighter rocks that can extend 65 kilometers at its deepest point. (p. 73)

crystal A solid substance in which the atoms or ions are arranged in an orderly pattern that repeats over and over again. (p. 98)

crystal shape The pattern a mineral's ions or atoms form if there is enough time and room to grow.

cumulus Thick, fluffy clouds with flat bases, formed by vertically rising air currents. (p. 397)

cycle The physical or chemical processing of Earth materials that repeats over time, for example, the water cycle, carbon cycle, and rock cycle. (p. 13)

cyclone Any counterclockwise movement of air.

D

daughter isotope An element that is the product of radioactive decay. (p. 656)

deep-ocean trench A long, narrow, steep-sided trough that runs parallel to continental margins or to volcanic island chains. (p. 518)

deferent An orbit along which the center of a planet's epicycle moved around Earth in Ptolemy's geocentric model of the solar system.

deficit The condition in which stored soil water is gone and the need for moisture is greater than the rainfall. (p. 306)

deflation The removal of loose rock particles by the wind. (p. 341)

deform The changing of a rock's shape by heat, friction, stress, and pressure. (p. 134)

delta A fan-shaped deposit that forms when a river flows into a quiet or large body of water, such as a lake, an ocean, or an inland sea. (p. 285)

density The ratio between the mass and the volume of a substance.

density current A subsurface current that is heavier or more dense than the surrounding water. (p. 536)

deposition The process by which materials are dropped, such as sand or silt by a stream; also, the process by which frost forms when water vapor condenses as a solid. (p. 283)

desert pavement A surface of pebbles and boulders left behind after sand and silt have been blown away. (p. 341)

Glossary

desertification The removal of soil by wind of rain in areas left exposed by the removal of plant covers.

dew Water vapor that condenses on a surface as a liquid when the air is saturated.

dew point The temperature at which saturation occurs and condensation begins; a measure of the amount of water vapor in the air. (p. 393)

diatom A one-celled plant with a delicate, thin shell made of silica. (p. 499)

dike An igneous intrusion that cuts across rock layers and is formed when magma intrudes into vertical or nearly vertical fissures in bedrock.

discharge The volume of water that passes a certain point in a stream or river in a given amount of time. (p. 281)

divergent boundary A boundary between two lithospheric plates that are moving apart. (p. 176)

divide The higher land that separates one drainage basin from another. (p. 280)

doldrums A rainy equatorial belt of low air pressure and slowly rising air, characterized by calm or windless weather.

dome mountain A nearly circular folded mountain. (p. 244)

Doppler effect The apparent change in the wavelength of radiation or of sound in which there is relative motion between the source and the receiver.

drainage basin All the land that drains into the river either directly or through its tributaries; a watershed. (p. 280)

drumlin A long, smooth, canoe-shaped hill that is usually found in groups, shaped by an advancing glacier. (p. 327)

dry-adiabatic lapse rate The rate at which unsaturated air cools as it rises, about 10°C for every kilometer. (p. 398)

dust storm A storm with strong, steady winds that lift great amounts of silt and clay from topsoil. (p. 340)

E

Earth system science The study of Earth materials and processes subdivided into a group of four interconnected spheres of activity: atmosphere, geosphere, hydrosphere, and biosphere. (p. 5)

earthflow The downslope movement of a mass of earth materials that have been saturated with water. (p. 269)

earthquake The shaking of Earth's crust caused by a release of energy. (p. 214)

echo sounding A system that uses transmitted and reflected sound waves to measure distances to the ocean floor; sonar. (p. 510)

elastic-rebound theory The theory that earthquakes occur when the stress building up between two lithospheric plates overcomes the force of friction, causing the plates to suddenly move, release energy, and then snap back to their former shapes.

electromagnetic radiation Energy radiated in the form of a wave, resulting from the motion of electric charges and the magnetic fields they produce. (p. 612)

electromagnetic spectrum A continuum depicting the range of electromagnetic radiation, with the longest wavelength at one end and the shortest at the other. (p. 612)

electron A negatively changed particle that spins around the nucleus of an atom.

element A substance composed of atoms that are chemically alike and that cannot be broken down into simpler parts by ordinary chemical or physical means. (p. 90)

elliptical orbit An ellipse-shaped orbit with two foci.

El Niño A warm surface current in the Pacific Ocean that contributes to a short-term warming trend.

emission spectrum A spectrum consisting of individual lines at characteristic wavelengths produced when light passes through an incandescent gas; a bright-line spectrum. (p. 614)

energy cycle The movement of energy into and out of the Earth system. (p. 16)

environment All of the resources, influences, and conditions at or near Earth's surface. (p. 144)

epicenter The point on Earth's surface directly above the focus of an earthquake. (p. 214)

epicycle A small orbit along which a planet traveled in Ptolemy's geocentric model of the solar system.

742 Glossary

epoch A subdivision of a geological period on the geologic time scale.

equator The imaginary line dividing Earth's surface into Northern and Southern Hemispheres; establishes 0° latitude. (p. 46)

era A major division of geologic time.

erosion The removal and transport of materials by natural agents such as wind and running water. (p. 268)

erratic A large boulder that has been transported into an area by a glacier. (p. 327)

esker A long, winding ridge formed when sand and gravel fill meltwater tunnels beneath a glacier. (p. 327)

evapotranspiration The rapid cycling of water vapor into the atmosphere by evaporation from Earth's surface or transpiration from plant leaves. (p. 13)

evidence The material or data from which conclusions can be drawn and by which proof can be established. (p. 28)

exfoliation The peeling of surface layers from exposed bedrock. (p. 259)

F

false-color image A computer image that assigns distinctive colors to different wavelengths of light reflected from a distant object, delineating surface features; used frequently in satellite images of Earth from space. (p. 49)

fault A break in the lithosphere along which movement has occurred. (p. 214)

fault-block mountains Mountains formed from blocks of crust that have been faulted and tilted at the same time. (p. 245)

feldspar A family of the most common and abundant of all minerals formed by silica tetrahedrons that are joined by ions of aluminum and other metals.

felsic A type of magma rich in silica that forms light-colored igneous rock containing minerals such as quartz and feldspars. (p. 121)

fetch The length of open water over which wind blows steadily. (p. 344)

firn Partially compacted and refrozen snow which has yet to become a glacier. (p. 318)

fjord A long, deep, steep-sided bay formed when the sea floods a glacial trough. (p. 353)

flash flood A sudden flood, usually caused by intense, heavy rainfall. (p. 291)

flood The phenomenon whereby a river overflows its banks. (p. 290)

floodplain A wide, level area that borders a river and is covered by its water during a flood. (p. 290)

focus The point at which the first movement occurs during an earthquake. (p. 214)

folded mountains Mountains formed when two plates carrying continental crust collide, folding the rocks and earth with great force. (p. 243)

fossil The remains, impression, or any other evidence of life from another geologic age preserved in rock. (p. 130)

fossil fuel A nonrenewable fuel that formed from the remains of organisms that lived long ago; coal, petroleum, and natural gas. (p. 148)

fracture The property of a mineral that describes an irregular pattern of breakage in a direction other than along cleavage planes (p. 105)

fracture zone An area of ocean ridges that are broken by transform faults. (p. 520)

freezing point The temperature at which a solution turns from liquid to solid. (p. 494)

freezing rain Supercooled raindrops that freeze instantly when they hit a solid surface. (p. 403)

front The boundary that separates dissimilar air masses. (p. 439)

frost Water vapor that has condensed on a surface as a solid when the temperature is at or below 0°C.

frost wedging A mechanical weathering process in which water freezes in the cracks of a rock and wedges it apart. (p. 258)

fusion The combining of the nuclei of lighter elements to form a heavier element. (p. 572)

G

galaxy A group of millions, or even billions, of stars held together by gravity. (p. 631)

geocentric A model of the solar system that has Earth at the center. (p. 577)

Glossary

geology The study of the earth's surface and interior.

geosphere The rocks, mountains, lithospheric plates, and other physical features of Earth, except for water; one of the four spheres of the Earth system. (p. 9)

geothermal energy Heat energy that originates from within Earth and drives the movement of Earth's tectonic plates. (p. 16)

geyser A hot spring that intermittently shoots columns of hot water and steam into the air. (p. 305)

giant star A large star with great luminosity and a diameter 10 to 100 times greater than that of the sun. (p. 627)

gibbous A phase of the moon in which almost all of the bright half of the moon faces Earth. (p. 563)

glacial valley A U-shaped valley formed by glacial erosion. (p. 324)

glacier A large mass of ice and snow that exists year-round and moves under the influence of gravity. (p. 318)

gradient The slope or incline of an area of land or of a stream. (p. 281)

granules Individual cells that are the tops of gas columns that form below the sun's photosphere.

gravitation The force of attraction between any two objects with mass; the strength of the attraction is proportional to their masses and inversely proportional to the distance between them. (p. 580)

greenhouse effect The absorption and retention of the sun's radiation by a planet's atmosphere, resulting in an increase in surface temperature.

groundwater Water that enters and is stored in the ground. (p. 300)

gully A miniature valley formed by erosion from heavy rains.

guyot A flat-topped seamount. (p. 520)

H

hail Small, nearly spherical pieces of ice made up of concentric layers formed by the successive freezing of layers of water. (p. 403)

half-life The time it takes for half the atoms in a sample of a radioactive element to decay to a stable end product. (p. 657)

hardness The resistance of a mineral to scratching.

headward erosion The process by which land is worn away at the head of gully or stream valley. (p. 287)

heat The total kinetic energy of all of the particles of a substance. (p. 370)

heliocentric A model of the solar system that has the sun at its center. (p. 578)

hemisphere An equal division Earth's surface into northern and southern halves or eastern and western halves. (p. 46)

high-pressure area An area in which the barometric pressure is greater than that of the surrounding air. (p. 416)

horn A pyramid-shaped peak formed where three or more cirques meet.

horse latitudes Belts of high air pressure and very dry descending air, located at about 30° north and south of the equator and characterized by calm or very light wind conditions.

hot spot An area of volcanic activity that results from a plume of hot solid material that has risen from deep within Earth's mantle. (p. 197)

hurricane A large, rotating storm of tropical origin with sustained winds of at least 119 kilometers per hour. (p. 450)

hydrogen bonding A chemical bond in which the hydrogen atom of one molecule is attracted to the negatively charged atom of another molecule. (p. 493)

hydrogenous sediments Sea-floor sediments, such as manganese nodules, that form when chemical reactions cause minerals to crystallize from seawater. (p. 525)

hydrolysis The chemical reaction of water with other substances. (p. 260)

hydrosphere All water in the Earth system—gaseous (water vapor), solid (snow and ice), and liquid (rain and water); one of the four spheres of the Earth system. (p. 9)

hygrometer An instrument that measures relative humidity.

hypothesis A tentative explanation for an observation or phenomenon, developed from available information and used as a basis for testing. (p. 28)

I

ice age An extensive period of glaciation. (p. 330)

ice cap A dome-shaped glacier that covers an area smaller than 50,000 square kilometers. (p. 320)

ice front The end of a glacier. (p. 321)

igneous rock Rock formed by the cooling and hardening (crystallization) of magma; one of three types of rock in the rock cycle. (p. 118)

impermeable Describes a rock material through which water does not pass easily.

index fossil The fossilized remains of organisms that lived and died within a particular time segment of Earth's history and that can be used to correlate rock layers. (p. 653)

inner core The solid innermost layer of Earth, composed of iron and nickel under extremely high pressure and temperature. (p. 72)

inner planets The four planets nearest the sun (Mercury, Venus, Earth, and Mars), separated from the outer planets by the asteroid belt. (p. 588)

inquiry The examination of a matter to learn more about it. (p. 28)

insolation The solar radiation (energy from the sun) that reaches Earth. (p. 372)

International Date Line The imaginary line placed at roughly 180° longitude where the new calendar day begins, moving east to west. (p. 78)

intertropical convergence zone The low-pressure zone at the equator, where the winds from the Northern and Southern Hemispheres converge. (p. 425)

ion An electrically charged atom or group of atoms. (p. 94)

ionic bond The force of attraction between oppositely charged ions that holds them together.

ionosphere The portion of the thermosphere between about 90 and 500 kilometers above Earth, where the air is highly ionized due to the effects of the sun's ultraviolet rays. (p. 372)

island arc A volcanic island chain that forms along a deep-ocean trench. (p. 518)

isobar A line on a weather map that joins points having the same barometric pressure. (p. 416)

isotherm A line drawn on a weather map through places having the same atmospheric temperature at a given time. (p. 374)

isotope Any of two or more forms of the same chemical element that differ in atomic mass. (p. 92)

J

jet stream A high-altitude air current forms a narrow band of very strong westerly winds flowing above the middle latitudes; travels up to and beyond 400 kilometers per hour (approximately 250 miles per hour) at altitudes of 15 to 25 kilometers (about 10 to 25 miles). (p. 420)

joint A crack or break in the bedrock along which no movement has occurred. (p. 241)

Jovian planets The gaseous planets Jupiter, Saturn, Uranus, and Neptune, which are much larger but less dense than the inner or terrestrial planets.

K

kame A small, cone-shaped hill of stratified sand and gravel formed at a glacial front by meltwater carrying sediment off the glacier's surface. (p. 328)

karst topography Topography characterized by sinkholes, sinkhole ponds, lost rivers, and underground drainage; forms in areas with bedrock made of calcite, dolomite, or other minerals that dissolve easily. (p. 311)

kettle A bowl-like hollow found in deposits of glacial outwash; formed when a large block of ice was left behind by a glacier melted. (p. 328)

key bed A single, widespread rock layer that is unique and easily recognizable; used to correlate rock layers. (p. 654)

L

laccolith A dome-shaped mass of intruded igneous rock.

lahar A fast-moving mudflow that occurs when the heat associated with a volcanic eruption melts snow and ice on top of a volcano. (p. 203)

Glossary

landslide The rapid movement of a mass of bedrock or loose soil and rock down the slope of a hill, mountain, or cliff. (p. 268)

latitude East-west lines parallel to the equator used to measure distance in degrees north and south, from 0° at the equator to 90° north and south at the poles. (p. 46)

lava Magma that reaches Earth's surface. (p. 200)

lava plateau A plateau formed from basaltic lava pouring from a crack or fissure in Earth's surface. (p. 204)

law A generalized statement about how the natural world behaves under certain conditions and for which no exceptions have ever been found. (p. 32)

lightning A discharge of electricity from a thundercloud to the ground, to another cloud, or to another spot within the cloud itself. (p. 446)

light-year The distance that light travels in one year, about 9.5 trillion kilometers. (p. 620)

liquefaction A temporary state in which loose soil and rock materials take on the property of liquid, often as a result of severe ground-shaking. (p. 222)

lithosphere The outer shell of the Earth consisting of the crust and uppermost portion of the mantle. (p. 73)

load The eroded rock and soil materials that are transported downstream by a river. (p. 284)

loess Wind-deposited sediment consisting of fine, silt-sized particles. (p. 342)

longitude North-south lines running between the poles, used to measure distance in degrees east and west of the prime meridian, from 0° at the prime meridian to 180° east and west. (p. 46)

longshore current An ocean current that slows parallel to the shoreline.

Love wave An earthquake wave that travels along Earth's surface.

low-pressure area An area in which the barometric pressure is lower than that of the surrounding air. (p. 416)

luminosity The brightness of a star. (p. 623)

lunar eclipse The eclipse that occurs when the moon passes into Earth's total shadow, or umbra, preventing sunlight from reaching the moon; it occurs only at the full-moon phase. (p. 564)

luster The property of a mineral that describes the quality or appearance of light reflected from its surface. (p. 105)

mafic Type of magma rich in iron and magnesium and low in silica; forms dark-colored igneous rock containing minerals such as hornblende, augite, and biotite. (p. 121)

magma The hot molten rock that forms beneath Earth's surface. (p. 118)

magnetic declination The angle by which the compass needle varies from true north; magnetic variation. (p. 56)

magnetic field An area in which the motion of charged particles creates a magnetic force, such as the field of magnetic force generated by the movement of fluid in Earth's outer core. (p. 74)

magnitude The measure of the amount of energy released in an earthquake. (p. 220)

main sequence A star that is at the point in its life cycle in which it is actively fusing hydrogen nuclei into helium nuclei; also, the band of the Hertzsprung-Russell diagram depicting such stars. (p. 626)

mantle convection A process by which heat from Earth's inner and outer cores is transferred through the mantle. (p. 180)

mantle The thickest of Earth's layers, located between the outer core and Earth's crust, composed mostly of compounds rich in iron, silicon, and magnesium. (p. 72)

map A flat, two-dimensional representation of Earth's surface and features. (p. 44)

map scale On a map, the comparison of distance units used with actual distances on Earth's surface; may be expressed as a ratio, a fraction, or in a scale bar. (p. 46)

maria Extensive dark areas on the moon that represent great basins and level plains. (p. 558)

mascons Areas of higher gravity that exist over lunar maria; short for "mass concentrations." (p. 558)

mass movement The downslope transportation of large masses of earth materials by gravity. (p. 268)

mass number The sum of the numbers of protons and neutrons in an atom. (p. 92)

matter Anything that has mass and volume.

meanders Broad, looping bends in a river. (p. 290)

mechanical weathering The breakdown of rock that takes place when a rock is broken into smaller pieces of the same material without changing its composition. (p. 258)

meridian An imaginary half-circle that runs in a north-south direction from the North Pole to the South Pole; a longitude line.

mesosphere The layer of the Earth's atmosphere that extends from the stratosphere to the thermosphere, characterized by decreasing temperatures. (p. 371)

metal An element that loses electrons easily to form positive ions. (p. 94)

metamorphic rock Rock that has undergone chemical or structural change due to the effects of heat and pressure; one of three types of rock in the rock cycle. (p. 118)

metamorphism The process by which a rock's structure or composition is changed by pressure, heat, and moisture. (p. 133)

meteor The light made by a meteoroid as it passes through Earth's atmosphere. (p. 604)

meteor shower A large number of meteors entering and burning up in the atmosphere; meteor showers take place when Earth passes through a debris field. (p. 604)

meteorite The part of a large meteoroid that survives its trip through the atmosphere and strikes Earth's surface. (p. 604)

meteoroid A rocky or icy fragment that travels through space. (p. 556)

meteorology The study of processes that govern Earth's atmosphere. (p. 436)

mica Soft silicate minerals with flat, shiny flakes that are found in many rocks, including granite and gneiss.

micrometeoroids Tiny rock fragments no larger than sand grains that travel through space. (p. 557)

middle latitudes The area between about 30 degrees and 60 degrees latitude, where surface winds are determined by the locations of transient high- and low-pressure systems. (p. 424)

mid-ocean ridge A long chain of mountains with a central rift valley that is located along a divergent boundary on the ocean floor. (p. 174)

millibar A unit used to measure air pressure; 34 millibars equals 1 inch of mercury.

mineral A naturally occurring inorganic solid with a distinct chemical composition and crystalline structure. (p. 96)

mineral deposit A deposit that is left behind when groundwater that contains minerals cools or evaporates. (p. 309)

mineralogy The study of minerals and their properties. (p. 104)

mixed layer The surface layer of ocean water. (p. 497)

model A simplified representation of an object, process, or phenomenon, used as the basis for further study or investigation. (p. 5)

Mohorovičić discontinuity (Moho) The boundary between Earth's crust and mantle.

moist-adiabatic lapse rate The rate at which saturated air cools as it rises; varies from about 5°C per kilometer to about 9°C per kilometer. (p. 398)

mold A hollow depression in rock formed when a fossil dissolves out of the rock; shows the original shape and surface of the fossil. (p. 648)

molecule A group of atoms linked together by chemical bonds. (p. 93)

monsoon A wind of southeast Asia and the Indian Ocean that changes direction seasonally, from the northeast from November until March and the southwest from April until October, when it is accompanied by heavy rains. (p. 427)

moon Any natural satellite of a planet.

moraine A deposit of till left behind when a glacier retreats. (p. 326)

mountain A large mass of rock that rises a great distance above its base. (p. 236)

mudflow The downslope movement of water that contains large amounts of suspended clay and silt. (p. 269)

Glossary

N

natural levees Elevated ridges along a river's bank that are formed by the deposition of the river's sediment load. (p. 290)

neap tide A tide of small range occurring at the quarter phase of the moon.

nebula A large cloud of gas and dust in space. (p. 627)

nekton Free-swimming marine life, such as fish, reptiles, whales, squid, and jellyfish. (p. 501)

neutron One of three basic atomic particles, with a mass slightly greater than that of a proton but no electrical charge.

neutron star The superdense remains of a massive star that collapsed with enough force to push all of its electrons into the nuclei they orbit, resulting in a mass of neutrons. (p. 630)

nonmetal An element that gains electrons easily to form negative ions. (p. 95)

nonrenewable resource A resource that exists in a fixed amount or is used up faster than it can be replaced in nature. (p. 144)

normal fault A fault where the hanging wall moves down with respect to the footwall. (p. 240)

O

occluded front The front that is formed when a cold front overtakes a warm front and displaces it upward in an area of low pressure. (p. 441)

ocean current A continuous flow of water along a broad path in the ocean. (p. 532)

oceanography The scientific study of the ocean and seas. (p. 490)

oozes Fine lime or silica muds found on the ocean floor. (p. 523)

open system A system in which there is a free exchange of both energy and matter between the system and its surroundings. (p. 6)

orbit The path of a revolving body such as that followed by Earth as it revolves around the sun.

ordinary well A well that is dug or drilled down to the water table. (p. 303)

ore mineral The valuable mineral, a metallic element, that can be separated from the rock in which it is found. (p. 145)

outer core The layer of Earth's interior located between the inner core and mantle, composed of iron and nickel in a liquid state. (p. 72)

outer planets The five planets farthest from the sun (Jupiter, Saturn, Uranus, Neptune, and Pluto) separated from the inner planets by the asteroid belt. (p. 588)

outwash Sediment deposited in front of a glacier by streams of meltwater. (p. 326)

outwash plain A broad, stratified, gently sloping deposit of sediment formed beyond the terminal moraine by streams from a melting glacier. (p. 327)

oxbow lake A crescent-shaped body of water formed when sediments deposited by a river cut off a meander from the river. (p. 290)

oxidation The chemical reaction of oxygen with other substances. (p. 261)

oxide A mineral consisting of a metal element combined with oxygen. (p. 111)

ozone A molecule of oxygen that consists of three oxygen atoms. (p. 371)

P

P waves Body waves that squeeze and stretch rock materials as they pass through Earth; also known as compressional waves or primary waves. (p. 215)

pahoehoe Solidified basaltic lava flow on land that has formed with smooth, ropelike surfaces. (p. 200)

paleontology The study of the life that existed in prehistoric times. (p. 648)

Pangaea The name of a hypothetical landmass consisting of all the continents welded together, which evidence indicates existed about 250 million years ago. (p. 183)

parallax The apparent shift in one object's position relative to another caused by a change in the location of the observer. (p. 80)

parallel An imaginary circle going east to west around the earth, parallel to the equator; a latitude line.

parent isotope The original element that will, after radioactive decay, become an isotope of a different element with a different atomic number. (p. 656)

parent material The rock material from which a soil is formed. (p. 264)

parent rock The original rock material that forms metamorphic rock. (p. 133)

parsec A unit of measurement used to describe distances between celestial objects, equal to 3.258 light-years. (p. 620)

passive continental margin A continental margin that does not occur along a plate boundary. (p. 236)

peer review The process by which one's work or research is reviewed by experts in the field to evaluate the validity of the work. (p. 30)

penumbra The area of partial shadow surrounding the darkest part of the shadow of the Earth or moon. (p. 564)

perigee The point closest to Earth in the moon's orbit. (p. 542)

perihelion The point in a planet's orbit when it is closest to the sun.

period The time it takes for one full wavelength to pass a given point (p. 345); or, a subdivision of a geologic era. (p. 667)

permeability The rate at which water or other liquids passes through the pore spaces of a rock. (p. 301)

phases The daily changes in the moon's appearance as it is viewed from Earth. (p. 563)

photosphere The visible surface of the sun. (p. 573)

phytoplankton Microscopic plants that float freely in the ocean's surface waters, make up the base of the ocean's food chain, and are the primary energy source for the marine ecosystem. (p. 499)

pillow lava Lava that cools underwater, taking on a distinctive pillow-like shape as it hardens. (p. 200)

planetary nebula A halo of gases that is formed by the expelled layers of a star's atmosphere. (p. 628)

plasma A state of matter consisting of charged particles—positively charged ions, and negatively charged electrons. (p. 572)

plate tectonics The theory that the lithosphere is made of plates that move and interact with each other at their boundaries. (p. 172)

pluton Intrusion of magma into Earth's crust, creating igneous rock formations such as dikes, sills, laccoliths, volcanic necks, and batholiths. (p. 125)

polar front The boundary at which air flowing away from the polar regions collides with warmer air from the lower latitudes. (p. 423)

polarity The differences in electrical charge at the ends of a molecule; for example, a water molecule has a positive charge near the hydrogen atoms and a negative charge near the oxygen atom. (p. 493)

porosity The percent of a material's volume that is pore space. (p. 300)

porphyry Rocks containing conspicuous crystals surrounded by a fine-grained mass.

pothole Deep oval or circular basins cut into a stream or river bed by abrasion from swirling sand, pebbles, and small boulders. (p. 283)

precipitation Any form of water that falls from a cloud to Earth's surface, such as rain, snow, sleet, and hail. (p. 402)

pressure gradient The rate at which atmospheric pressure declines over a given distance. (p. 416)

prevailing winds Winds, such as the trade winds and polar easterlies, that usually blow from the same direction. (p. 425)

prime meridian The imaginary line dividing Earth's surface into Eastern and Western Hemispheres, established as 0° at Greenwich, England; the starting point for standard time zones. (p. 78)

projection A representation of the spherical Earth on the flat plane of a map. (p. 44)

proton A positively charged particle in the nucleus of an atom.

pulsar A distant neutron star that emits rapid pulses of light and radio waves instead of steady radiation. (p. 630)

pyroclastic flow A dense, superheated cloud of gases and pyroclastic materials that moves rapidly downhill from an erupting volcano. (p. 201)

pyroclastic material Solid rock fragments that are ejected during a volcanic eruption. (p. 200)

pyroxenes Silicate minerals that have cleavage surfaces that meet nearly at right angles.

Glossary

Q

quasar A very distant, extremely luminous celestial object that scientists consider to be a type of active galactic nuclei. (p. 633)

R

radar A method of detecting distant objects and recording their features and properties by analysis of electromagnetic waves reflected from their surfaces. (p. 48)

radiation The transfer of energy through space by electromagnetic waves. (p. 369)

radio telescope An instrument that picks up radio waves emitted by objects in space.

radioactive decay The process by which radioactive isotopes emit or capture tiny particles; includes alpha decay, beta decay, and electron capture. (p. 656)

radiometric dating A technique used to measure absolute time in which the amounts of a parent and daughter isotope within a rock or mineral are measured and the ratio used find the age of a rock. (p. 658)

rain gauge An instrument for measuring the amount of precipitation.

rawinsonde A small balloon-carried weather observatory, which carries a radio transmitter that sends out signals about temperature, pressure, and relative humidity.

ray A bright streak of shattered rock and dust that radiates from a lunar crater. (p. 560)

recharge The refilling of soil water supply at times when plants need little moisture. (p. 306)

recycle To collect and reuse materials from waste. (p. 156)

refraction The change in the direction of a wave as it passes from one substance to another of different density; the bending of water waves as they reach shallow water. (p. 345)

regional metamorphism The process of rock formation that results from large areas of rocks being under intense heat and pressure of mountain-building movements.

regolith A grayish-brown mixture of small rock pieces and fine particles that covers the moon. (p. 560)

relative dating The process of placing events in the sequence in which they occurred; does not identify actual dates. (p. 650)

relative humidity A comparison of the actual amount of water vapor in the air with the maximum amount of water vapor that can be present in air at a given temperature and pressure. (p. 392)

renewable resource A resource that can be replaced in nature at a rate close to its rate of use. (p. 144)

reserve The known deposits of a mineral in ores that are worth mining at the present time. (p. 146)

residual soil Soil whose parent material is the local bedrock beneath it. (p. 264)

retrograde motion A periodic backward, or westward, loop made by a planet in front of the background of constellations.

reverse fault A fault where the hanging wall moves up with respect to the footwall. (p. 240)

revolution The movement of one body around another, such as the Earth in its orbit around the sun. (p. 80)

ridge push A force that is exerted by cooling, subsiding rock on the spreading lithospheric plates at a mid-ocean ridge. (p. 181)

rift A crack or opening in Earth's crust. (p. 176)

rift valley A deep valley at a point where lithospheric plates are moving apart, such as at a mid-ocean ridge. (p. 176)

rille A trenchlike valley running through the bedrock of lunar maria, believed to have formed after the cave-in of the roof of a tunnel that had transported lava. (p. 558)

river system A river and all of its tributaries. (p. 280)

rock A naturally formed group of minerals bound together; can consist largely of one mineral or several different minerals in varying quantities. (p. 118)

rock cycle A repeated series of events by which rock gradually and continually changes between igneous, sedimentary, and metamorphic forms. (p. 119)

rock flour Fine sand and silt formed by the crushing of rock beneath a glacier.

rock-forming mineral A specific group of minerals known to form rocks.

rotation The turning of a body, such as Earth, on its axis. (p. 75)

S

S waves Body waves that cause particles of rock material to move at right angles to the direction in which the waves are traveling; also known as shear waves or secondary waves. (p. 215)

Saffir-Simpson scale The 1-to-5 scale used to rate a hurricane's intensity and estimate potential property damage and flooding. (p. 453)

salinity The measure of the dissolved salts in water. (p. 495)

salinization A soil condition caused by the evaporation of irrigation water, which leaves too much mineral matter on the soil's surface for crops to grow. (p. 271)

sand dune A wind-formed mound or hill of sand. (p. 342)

sandbar A ridge formed by ocean currents depositing sand near the shore. (p. 351)

satellite A smaller body orbiting around a larger body; a natural satellite is also called a moon.

saturated The condition in which the air is holding as much water vapor as possible at a given temperature and pressure. (p. 391)

scientific inquiry Investigation of a natural phenomenon by observing, asking questions, forming a hypothesis, gathering data, testing the hypothesis, and sharing what has been learned. (p. 29)

seamount A cone-shaped undersea mountain of volcanic origin that rises high above the ocean floor. (p. 520)

sediment Solid particles such as weathered rock fragments, plant and animal remains, or minerals that settle out of solution onto lake and ocean bottoms. (p. 118)

sedimentary rock Rock formed by the compaction and cementing of layers of sediments; one of three types of rock in the rock cycle. (p. 118)

seismic gap An area along a seismically active fault where no earthquake activity has occurred over a long period of time. (p. 225)

seismogram The recording of an earthquake made by a seismograph. (p. 217)

seismograph An instrument that detects and records waves produced by earthquakes. (p. 217)

shadow zone The wide area around Earth on the side opposite an earthquake's focus where neither P waves or S waves are received.

shield The exposed area of the oldest rocks, or craton, of a continent.

shield volcano A shield-shaped volcano with a broad base and gently sloping sides that is made of basaltic lava. (p. 202)

silica tetrahedron A grouping of one silicon ion and four oxygen ions that forms the basic building block of silicate. (p. 100)

silicate Any mineral that has as its building block a tetrahedron of silicon and oxygen. (p. 100)

sill A sheet of intrusive igneous rock forced between rock layers parallel to the rock layers it intrudes.

skeptical To be questioning; inquisitive. (p. 28)

slab pull A force at a subduction boundary that the sinking edge of the subducting plate exerts on the rest of the plate. (p. 181)

sleet Clear ice particles formed when raindrops freeze before they reach the ground. (p. 403)

slope The steepness of a landscape, calculated as the change in elevation divided by the distance covered. (p. 55)

slump A mass movement in which a block of land tilts and moves downhill along a surface that curves into the slope. (p. 268)

snow line The lowest elevation that permanent snows reach in summer. (p. 318)

soil Loose, weathered rock and organic material in which plants with roots can grow. (p. 264)

soil depletion The process by which soil gradually becomes so lacking or depleted in nutrients that it can no longer grow a crop. (p. 271)

soil fertility The ability of soil to grow plants. (p. 271)

soil horizon A soil layer with physical and chemical properties that differ from those of adjacent soil layers. (p. 264)

soil profile A cross section of soil layers that displays all soil horizons. (p. 264)

Glossary

solar eclipse The eclipse that occurs when the moon passes between the sun and Earth and the moon's shadow strikes Earth's surface. (p. 564)

solar energy Energy emitted by the sun. (p. 16)

solar flare A sudden outburst of energy that rises up in areas of sunspot activity.

solar prominence A huge flamelike arch of material that occur in the corona of the sun.

solar system The sun and its family of orbiting planets, moons, and solar system debris.

solar wind A constant stream of electrically charged particles that is blown out from the sun in all directions. (p. 575)

solution The state in which mineral matter dissolved from bedrock is carried in a river.

specific gravity The ratio of the weight of a substance to the weight of an equal volume of water; property used to identify minerals. (p. 106)

specific humidity The amount of water vapor in the air at a given time and place; expressed as the number of grams of water vapor per kilogram of air. (p. 391)

spreading center The area where lithospheric plates are moving apart.

spring A small stream of water whose source is groundwater that has reached the surface. (p. 303)

spring tide A tide of large range occurring at the new moon and full moon.

squall line A line of thunderstorms that occurs ahead of a front, often preceded by strong winds. (p. 446)

standard time zones Areas roughly defined by twenty-four 15° sections of longitude, each centered on a time meridian that establishes the hour of the day. (p. 77)

station model A compact expression of weather information for an area, including temperature, dew point, weather conditions, wind speed and direction, barometric pressure, and cloud cover. (p. 457)

stationary front The boundary between two dissimilar air masses, neither of which is displacing the other, usually resulting in cloudy weather and mild temperatures. (p. 441)

stock A large igneous intrusion, similar to a batholith, but with an exposed surface area of less than 100 square kilometers.

storm surge The rapid rise in water level along the coast as a hurricane or other tropical storm approaches. (p. 451)

strata The layers of sedimentary rock that form after particles settle out of a fluid and are compressed over time. (p. 650)

stratification The arrangement of layers of sedimentary rock. (p. 130)

stratosphere The layer of the Earth's atmosphere that extends from the troposphere to the mesosphere; concentrations of ozone cause temperatures within the stratosphere to increase with altitude. (p. 370)

stratus Clouds that form in low, horizontal layers. (p. 397)

streak The property of a mineral that describes its color in powdered form. (p. 105)

stream piracy The diversion of the upper part of one stream by the headward growth of another stream. (p. 288)

striations Long, parallel scratches left on rocks and bedrock by glacial movement. (p. 323)

strike-slip fault A fault where the rocks on opposite sides of the fault plane move horizontally. (p. 241)

subduction boundary A convergent boundary where an oceanic plate is plunging beneath another, overriding plate. (p. 177)

submarine canyon An undersea gully that cuts through the continental shelf and continental slope. (p. 515)

subsoil The B-horizon of soil; contains clay and iron oxides washed from the topsoil. (p. 265)

sulfide A mineral consisting of a metal element combined with sulfur. (p. 111)

summer solstice The first day of summer in the Northern Hemisphere, which occurs on or about June 21 each year when the noon sun appears to reach its most northern point in the sky. (p. 82)

sunspots Dark areas on the sun's photosphere that result from variations in the sun's magnetic field. (p. 574)

supercell A very large, single-cell thunderstorm with particularly strong updrafts (p. 446)

supercooled water Water that has cooled below 0°C without freezing.

supergiants The most luminous, most massive stars, with diameters greater than 100 times the diameter of the sun. (p. 627)

supernova The brilliant burst of light that follows the collapse of the iron core of a massive star. (p. 629)

surface current An ocean current that generally flows in the upper 1000 meters of the ocean and is primarily driven by the wind. (p. 532)

surface waves Earthquake waves that travel along Earth's surface; Love waves and Rayleigh waves are two types of surface waves. (p. 216)

surplus The condition of having rainfall greater than the need for moisture when the soil is already saturated. (p. 306)

suspension A state in which rock materials carried by a river are stirred up and kept from sinking by the turbulence of stream flow. (p. 284)

syncline A downfold in rock layers. (p. 239)

system A naturally occurring group of objects or phenomena that share matter and energy, for example, the four spheres of the Earth system. (p. 5)

T

talus Cliff rock fragments that have been weathered loose and pulled down by gravity. (p. 268)

technology The application of scientific discoveries to meet human objectives. (p. 28)

temperature The measure of the average kinetic energy of the atoms or molecules in a substance. (p. 369)

temperature inversion An increase in temperature with an increase in altitude; occurs when a layer of cold air is trapped beneath a layer of warm air. (p. 380)

terrane A large block of lithospheric plate that has been moved, often over a distance of thousands of kilometers, and attached to the edge of a continent. (p. 186)

terrestrial planets The planets (Mercury, Venus, Earth, and Mars) with rocky crusts and dense mantles and cores; the inner planets.

terrigenous sediments Sea-floor sediments that came from eroded minerals that had comprised continental rocks. (p. 523)

theory An explanation or model based on observation, reasoning, and experimentation, especially one that has been tested and confirmed as a general explanation for phenomena that has been observed. (p. 32)

thermocline The layer of ocean directly beneath the mixed layer, in which the temperature decreases rapidly. (p. 498)

thermosphere The layer of the Earth's atmosphere above the mesosphere, characterized by increasing temperature with altitude. (p. 371)

thrust fault A reverse fault in which the fault plane dips 45° or less from the horizontal. (p. 240)

thunderstorm A storm with rain, lightning, and thunder, and sometimes hail and tornadoes. (p. 445)

tidal energy Energy created by the gravitational pull of the sun and moon on Earth's oceans. (p. 16)

tidal range The difference in level between low tide and high tide.

tides The periodic rise and fall of the ocean surface due to the gravitational pulls of the moon and sun. (p. 541)

till Unsorted and unstratified rock material deposited directly by glacial ice. (p. 326)

time meridian A line of longitude exactly divisible by 15° on which each standard time zone is roughly centered. (p. 77)

topographic map A map that uses contour lines and symbols to show the surface features of a particular area, including natural features like mountains, valleys, bodies of water, as well as human-made features like bridges, buildings, and roads. (p. 53)

topography All natural and human-made surface features of a particular area. (p. 53)

topsoil The A-horizon of soil; contains organic material, or humus, that forms from decayed plant and animal materials. (p. 265)

tornado A violent, rotating column of air that extends down from dark clouds and moves overland in a narrow, destructive path. (p. 447)

Glossary

toxic waste A poisonous chemical waste that must be disposed of with extreme care.

trace fossil Any indirect evidence of life preserved as an impression in rock; trails, footprints, tracks, burrows, and bite marks. (p. 649)

trade winds Steady winds that blow across the equator, from about 30° north latitude to about 30° south latitude. (p. 425)

transform boundary A boundary between two plates that are sliding past each other. (p. 178)

transported soil Soil that formed from parent material left by winds, rivers, or glaciers or soil that itself was moved from its original location. (p. 264)

tributary A stream that runs into another stream or river. (p. 280)

troposphere The lowest layer of Earth's atmosphere, characterized by decreasing temperature with altitude. (p. 370)

tsunami A large ocean wave that results from an underwater earthquake, landslide, or volcanic eruption. (p. 223)

turbidity current An undersea landslide of mud and sand, triggered by earthquakes or by gravity, that speeds down the continental slope carving out channels. (p. 516)

U

umbra The darkest part of the shadow cast by the moon or by Earth. (p. 564)

unconformity The layer or layers of rock missing from a strata sequence. (p. 651)

upwelling The vertical movement of cold, deep ocean water toward the surface. (p. 538)

usage The condition where plants draw water from the soil at times when the need for moisture is greater than the rainfall. (p. 306)

V

valley glacier A long, slow-moving, wedge-shaped glacier that moves within valley walls. (p. 319)

variable stars Stars that change in brightness at regular periods or cycles.

varve Any sediment that is deposited on a yearly cycle. (p. 656)

ventifact A rock with surfaces that have been abraded by wind-blown sand. (p. 341)

vernal equinox Start of spring in the Northern Hemisphere, occurring on or about March 21 each year when the noon sun is directly over the equator; one of two days each year (with the autumnal equinox) when day and night are of equal length in both hemispheres. (p. 83)

viscosity A substance's resistance to flow. (p. 199)

volcanic neck The solidified lava filling the central vent of an extinct volcano. (p. 270)

volcanic rock Igneous rock that forms from cooled lava, or from volcanic dust and ash.

volcano An opening in Earth's crust through that molten rock, gases, and ash erupt; also, the landform that develops around this opening. (p. 194)

W

waning The decreasing of the visible amount of the moon's illuminated surface, from full moon to new moon. (p. 563)

warm front The boundary between an advancing warm air mass and the colder air it is displacing. (p. 441)

warm-core ring An offshoot stream from an ocean current that forms a ring with a core of warm water. (p. 535)

water budget Describes the income and the spending of water in a region. (p. 306)

water cycle The continuous circulation of water through the hydrosphere as solid, liquid, or gas. (p. 13)

water mass A body of water with distinct properties based on where it originated. (p. 495)

water table The surface below which the ground is saturated with water. (p. 302)

water vapor An invisible gas formed when water reaches 100°C or above and evaporates. (p. 390)

watershed All the land that drains into the river either directly or through its tributaries; a drainage basin. (p. 280)

wave height The vertical difference between a wave's high point, or crest, and its low point, or trough. (p. 344)

wavelength The distance between two successive wave crests. (p. 344)

waxing The increasing of the visible amount of the moon's illuminated surface, from new moon to full moon. (p. 563)

weather The state of the atmosphere at a given time and place.

weathering The breakup of rock due to exposure to processes that occur at or near Earth's surface. (p. 258)

white dwarf The remnant of a giant star that has lost its outer atmosphere; the glowing stellar core. (p. 627)

winter solstice The first day of winter in the Northern Hemisphere, occurs on or about December 21 each year when the noon sun appears to reach its most southern point in the sky. (p. 83)

Z

zenith The point in the sky directly above the observer.

zooplankton The microscopic animals found in ocean water, which eat phytoplankton or smaller zooplankton. (p. 500)

Index

Italic page numbers preceded by *i* are references to illustrations. Those preceded by *m* refer to maps, and those preceded by *c* refer to charts and graphs.

A

aa, 200, *i 200*
abrasion, 259, 283, 341
absolute dating, 656–659
absolute magnitude, 623, *i 623*, 624
absorption spectrum, 614–615, *i 614*, *i 615*
abyssal hill, 517
abyssal plain, 517
acid rain, 155, 261, 379, 407
active galaxy, 632–633, *i 633*
adiabatic lapse rate, 398
aeration, zone of, 302
aftershock, 223, 227
air. *See also* atmosphere.
  gases in, *c 366*
  pollution of, 155, 378–380, *c 378*, *c 382*, 383, 407
air mass, 436, 444
  front and, 439, 440, 441
  types of, 437–438, *i 437*
air pressure, 414–416, *c 705*
  weather and, 441–444
  wind and, 417, 420, 423, 424, 425, 427–429, 444
albedo, 17–18
alluvial fan, 286
altitude, 82
ammonite, *i 130*, 678
amphibole, 110, *c 700*
  in igneous rocks, 124
  molecular structure of, *i 101*
andesite, 124
andesitic magma, 199, *c 199*
anemometer, 418
aneroid barometer, 415
angular unconformity, 651, *i 651*
animals
  in carbon cycle, 14–15
  in Cenozoic Era, 681, 684–685
  conservation of, 672
  fossils of, 130, 648–649
  mechanical weathering by, 259
  in Mesozoic Era, 678–680
  in ocean, 500–502
  in Paleozoic Era, 675–677

Antarctic Circle, 82, *i 82*, 83
anthracite, 149
anticline, 239
  as oil trap, 150
apatite, *c 700*
aphelion, 81
apogee, 542, 562
Apollo, 556, 558, 559, 561
apparent magnitude, 619, *c 623*
aqueous solution, 493–494
aquifer, 304
  High Plains, 308, 359
  perched, 303
Archean Eon, 666, 673
Arctic Circle, 82, *i 82*, 83
arête, 324
Ariel, 601
artesian formation, 304, *i 304*
artesian spring, 304
artesian well, 304
ash, 123, 201
asteroid, 588, 603, *i 603*, 680
asthenosphere, 73, 180, 194
astronomical unit, 574, 619–620
atmosphere, 8. *See also* air; weather; wind.
  balance in, 367–368
  composition of, 366, 667
  heat and, 369–373
  human impact on, 378–382
  ozone in, 4–5, 371, 380–381
  structure of, 370–372, *i 371*, *c 705*
  water in, 390–406, 416
atomic mass, 92
atomic number, 91
atoms, 90
  bonding of, 93–95
  isotopes of, 92–93
  structure of, 91
augite, 109
  in igneous rocks, 121
  weathering of, 262
aurora, 372, *i 372*, 575–576, *i 575*
  on Jupiter, 595
autumnal equinox, 83
axis, 76, 331, 475

B

back swamp, 291
barchan, 342
barograph, 415, *i 415*
barometer, 414–415
Barrera, Enriqueta, 26, *i 27*
barrier island, 352, *i 352*
basal slip, 321
basalt, *i 122*, 124
  on moon, 558
basaltic magma, 199, *c 199*
base level, 288
batholith, 126
beach, 350
  currents and, 347
  erosion of, 347, 348
bed load, 284
beryl, *c 700*
big bang model, 634–635
biodiversity, 672
biogenous sediment, 524–525
biosphere, 10
biotite, 109, 121, 124, 262, *c 700*
bituminous coal, 149
black hole, 630, 633
black smoker, 502
blazar, 633
blizzard, 453, *i 453*
block, 201
blowout, 341, *i 341*
blueshift, *i 615*, 616
body wave, 215
bond
  covalent, 93, *i 93*
  hydrogen, 493
  ionic, 94–95, *i 94*
  metallic, 95
Brahe, Tycho, 578, *i 579*
breaker, 346–347, *i 346*
buckminsterfullerene, 103
buffer, 494

C

calcareous ooze, 524
calcite, 110, *i 110*, *c 700*
  chemical test for, 107
  chemical weathering of, 261, 263
  deposition by groundwater, 131, 309, 310, 311
  double refraction and, 107

as gangue mineral, 145
  in sedimentary rocks, 128, 129, 131
caldera, 204
Callisto, 600, *i 600*
calving, 322, 523, *i 523*
Cambrian Period, 675
canyon, 287
  on Mars, 592
  submarine, 515–516, *i 516*
capillary action, 302
capillary fringe, 302
capillary water, 301
cap rock, 304
carbon
  atomic structure of, 91, *i 91*
  "buckyball" molecules of, 103
  crystal structures of, 102, *i 102*
  isotopes of, 93, *i 93*
carbonaceous film, 649
carbonate, 110
carbon cycle, 14–15, *i 14–15*
carbon dioxide, 14, 15, 332, 366, 367–368, 372, 373, 381, 475, 476, 590
carbonic acid, 260–261, 310–311, 332
Carboniferous Period, 676–677
careers
  Air Force meteorologist, 400
  earthquake engineer, 221
  electromagnetic geophysicist, 187
  environmental consultant, 368
  hydrologist, 282
  land surveyor, 51
  marble sculptor, 136
  marine biologist, 500
  mission specialist, 596
  natural history museum curator, 670
  recycling technician, 158
  research vessel officer, 524
  science writer, 622
  solar physicist, 576
  tropical rain forest ecologist, 12
  wastewater treatment plant operator, 301
  weather observer, 456
Carson, Rachel, 62–63
cast, 649
cave. *See* cavern.

cavern, 261, *i 261*, 310–311, *i 310*
cementation, 128
Cenozoic Era, 667, 681–685
Cepheid variable, 624
chalcopyrite, *c 700*
Charon, 598, *i 598*
chemical sedimentary rock, 128
chert, 131
chinook, 405
chlorite, *c 700*
chlorofluorocarbon, 4–5, 380, 381
chromosphere, 573
cinder cone, 202, *i 203*
cirque, 324
cirrostratus, 396, 444
cirrus, *i 396*, 397, 444
clastic sedimentary rock, 127–128, *c 128*
clay
  as product of weathering, 260, 261, 262, 263
  in shale, 128
  in soil, 265
cleavage, 100, 105
climate, 466–467
  changes in, 331–332, 474–477, 482–483, 549
  erosion and, 270
  fossils as indicators of, 654
  human effect on, 476
  oceanic effects on, 548–549
  soil and, 264, 265
  weathering and, 263
  zones, 469–473, *c 469*, *m 470–471*
clock, 79
closed system, 5
cloud
  constituents of, 396
  formation of, 398–401
  insolation and, 376
  types of, 396–397, *i 396–397*
coal, 96, 129, 148–149, *i 149*
cold-core ring, 535
cold front, 440–441, *i 440*, 444
collision boundary, 178
comet, 602, *i 602*, *c 603*
  meteor shower and, 604
composite volcano, 203, *i 203*
compound, 93, 95
compression, 238–239
computer

mapmaking and, 50, 52
  seismographic network and, 227
  as tool of scientist, 36–37
  weather forecasting and, 459
concretion, 131
condensation, 390, 393–395
condensation level, 398, 401
condensation nucleus, 394, 406
conduction, 369
conglomerate, 128, *i 128*
  metamorphosis of, *i 133*, 135
conservation, 156, 158–159
  of groundwater, 306–307, 308
  of plant and animal species, 672
  of soil, 272–273
  of water, 359
constellation, 577, 617–619
contact metamorphism, 134, *i 134*
continental arctic air mass, 437
Continental Divide, 280
continental drift, 172
continental glacier, 319, 325
continental margin, 236–237, 514–516, *i 514*
continental polar air mass, 437
continental rise, 514
continental shelf, 514
continental slope, 514
continental tropical air mass, 438
continuous spectrum, 614, *i 614*
contour line, 54, 55, 56
convection
  in mantle, 180–181, *i 180*
  as transfer of energy, 369
convergent boundary, 177–178, *c 179*
  mountain formation at, 236–237, 243
Copernicus, Nicolaus, 578
copper, *c 700*
coral, 501, 521, 675, 676
coral atoll, *i 520*, 521
coral reef, 501, 503, *i 503*, 521
core
  of Earth, 72, 228
  of sun, 573
Coriolis effect, 75, 419–420, *i 419*, *i 420*, 422, 423, 451, 532, 538, 539

Index

corona, 573, 575
correlation of strata, 652–654, *i 652*
corundum, *c 700*
cosmic background radiation, 635
countercurrent, 535
covalent bond, 93, *i 93*
Crater Lake, 204, *i 204*
craters
  on Callisto, 600
  on Earth, 605, *i 605*
  on Mars, 591
  on Martian satellites, 599
  on Mercury, 589
  on moon, 556–557, 560
  on Uranian satellites, 601, *i 601*
  on Venus, 590
craton, 185, 673
creep, 268
Cretaceous Period, 680
crevasse, 321, *i 321*
crinoid, 676
cross-cutting relationships, principle of, 650
crust, 73, *c 96*, 118, 229
crystals, 98–102, 122
cubic system, *c 99*
cumuliform cloud, 396
cumulonimbus, 397, *i 397*, 399, 445
cumulus, 397, *i 397*, 398
current
  density, 536–538
  shoreline, 347
  surface, 532–535, *m 532*
  turbidity, 516, 517, 536–537
  upwelling, 538–539, 540
cycad, 678, *i 678*

D

dam
  to create reservoir, 291, 292
  for flood control, 292
  hydroelectric, *i 151*
  natural, 291
  removal of, 293
Darwin, Charles, 670–671
daughter isotope, 656, 658
deep-sea trench, 177
deflation, 341
deformational metamorphism, 134
Deimos, 599, *i 599*
delta, 285–286, *i 286*
density, 72
  of Ganymede, 600
  of Io, 599
  of minerals, 102, 106–107
  of moon, 557
  of Saturn, 596
  of seawater, 494, 499, 537–538
  of stars, 621
  of Titan, 601
  of water, 492
density current, 536–538
deposition
  by glaciers, 326–328
  by groundwater, 309–310, 311
  on ocean floor, 519, 523–525
  by streams and rivers, 127–128, 285–286
  by wind, 342–343
desert pavement, 341, *i 341*
deuterium, 93
Devonian Period, 676
dew, *i 393*, 394
dew point, 393, *c 704*
diabase, 124
diamond, *i 90*, *c 700*
  crystal structure of, 102, *i 102*
  hardness of, 106
  luster of, 105
diatom, 499, *i 499*
dike, 125
dinosaurs, 678, 679, *i 679*, 680
diorite, *i 97*, 124
disconformity, 652
divergent boundary, 176, *c 179*
  earthquake depth at, 214
  volcanic activity at, 196
divide, 280
dolomite, 110, *i 110*, *c 700*
dome mountain, 244
Doppler effect, 449, 615–616
Doppler radar, 449
drainage basin, 280
dripstone, 311
drizzle, 403
drumlin, 327, *i 327*
dry-adiabatic lapse rate, 398
dry climates, 470–471
dust storm, 340, *i 340*

E

Earth
  as closed system, 6–7
  cycles in, 13–18
  formation of, 70–71
  heat of, 16, 73–74, 152–153, 180, 305
  interior of, 72–73, *i 72*, *c 73*, 228–229, *c 703*
  magnetic field of, 74
  past of, 182–187, 648–659, 666–685
  revolution of, 80–83
  rotation of, 75–76, *i 75*, 77–78, 419, 422, 618
  size and shape of, 71–72
  spheres of, 8–12
  tilt of, 76, 82–83, *i 82*, 427, 475
earthflow, 269
Earth Observing System, 49
earthquakes
  causes of, 214–215
  damage caused by, *i 216*, *i 220*, 222–224, *i 222*, *i 223*, *i 224*, 227
  Earth's interior structure and, 228–229
  epicenters of, 214, *i 214*, 219
  as evidence for plate tectonics, 173
  foci of, 214–215, *i214*
  at fracture zones, 520
  locating, 218–219
  magnitudes of, 220
  predicting, 225–226
  risk of, 224–225, *m 225*
  waves produced by, 215–218, 222, 228–229
Earth system science, 4–7
  environment and, 62–63, 164–165, 250–251, 358–359, 482–483, 548–549, 640–641, 690–691
echo sounding, 510–511, *i 510*
eclipse
  lunar, 564, *i 564*
  solar, 564–565, *i 565*
electromagnetic radiation, 612–613
  timekeeping and, 79
electromagnetic spectrum, 612, 613–616, *i 613*

electron, 91, 92
  atomic bond and, 93–95
elements, 90
  classification of, 92
  in Earth's crust, c 96
elliptical galaxy, 632, i 632
El Niño, 468, 503, 540, 548, 549
emission spectrum, 614, i 614, i 615
energy budget, 16–17
energy cycle, 16–17
energy level, 91, 93
energy resource, 148–153, 154–156
  conservation of, 156
environment, 144
  Earth system and, 62–63, 164–165, 250–251, 358–359, 482–483, 548–549, 640–641, 690–691
  exploitation of resources and, 154–158
eon, 666
epicenter, 214, i 214, 219
epoch, 667
equinox, 83, i 83
era, 666–667
erosion
  absolute time and, 656
  by glaciers, 323–325
  by groundwater, 310–311
  headward, 287, 288
  landforms and, 270
  on moon, 557
  soil, 272–273
  by streams and rivers, 283, 287, 288–289
  by waves, 347, 348, 349
  by wind, 340–341
erratic, 327, i 327
esker, i 326, 327
Europa, 599–600, i 600, 640–641
eurypterid, 676, i 676
evaporation, 390
  density currents and, 538
  mineral deposition and, 97, 129, 309
  in water cycle, 13
evapotranspiration, 13, 306
Everest, Mount, 242
evolution, 670–671
exfoliation, 259–269, i 269
extinction, 677, 680, 690–691

extremophile, 640–641
extrusive igneous rock, 121, 122–123

F

fault, 214, 240–241
  deformational metamorphism and, 134
fault-block mountains, 245
feldspar, 109, c 700
  cleavage of, 105
  as gangue mineral, 145
  in igneous rocks, 121, 123, 124
  weathering of, 110, 262
felsic magma, 121
felsite, 124
firn, 318–319
fission, 150, i 151
fissure spring, 304
fjord, 353
flash flood, 291
flint, 131
flood, 291–292
floodplain, 290–291, i 290
fluorite, c 700
fluorescence, 107
focus
  of earthquake, 214–215, i 214
  of elliptical orbit, 579
fog, 394–395, i 395
fold, 239, i 239
folded mountains, 239, 243, 244
footwall, 240
foraminifera, 676
Forde, Evan B., 26–27, i 27
fossil, 130, 182, 648–649, 653, 654, 666
fossil fuel, 148–150, 155
Foucault pendulum, 75
fracture, 105–106
fracture zone, 176, 520
freezing rain, 403
front, 439, i 439
  thunderstorm and, 446
  types of, 440–441
frost, 394
frost wedging, 258–259, i 258
Fujita scale, 448, c 448
fumarole, 305
fusion, 31, 70, 572, i572, 627, 628, 629

fusulinid, 186–187

G

gabbro, i 122, 124
Gaia hypothesis, 63
galaxy, 631–633
galena, c 700
Galileo Galilei, 581, i 581
gangue, 145
Ganymede, 600, i 600
geocentric model, 577, 578
geode, 131, i 131
geographic information systems (GIS), 50, 52
geologic time scale, 666–667, c 668–669
geosphere, 9
geothermal energy, 16, 17, 152–153, 156, 196, 305
geyser, 305, 309
  deep-ocean vent as, 519
geyserite, 309, i 309
giant, 627, 628
GIS. See geographic information systems
glaciers
  deposition by, 326–328
  erosion by, 323–325
  evidence for climate change in, 477
  fjords and, 352–353
  formation of, 318–319
  ice ages and, 330–331, 332–333, 674, 683
  movement of, 321–322, i 322
  transport by, 323, 523
  types of, 319–320
Global Positioning System, 50, 242
global warming, 381–382, 476, c 476, 549
gneiss, 137, i137
gnomonic projection, 45
gold, c 700
Gondwana, 183, 678
graben, 245, i 245
granite, i 118, 123, i 123, i 137
granodiorite, 125
graphite, 102, c 145, c 700
graptolite, 675, i 675
gravity
  black holes and, 630

Index

mass and, 90, 621
nebular condensation and, 70, 627
neutron stars and, 630
Newton's law and, 580
ridge push and, 181
solar system's formation and, 70
stellar equilibrium and, 573, 628
transport of weathered materials by, 268–269
variations of moon's, 558
Great Lakes, 332–333
greenhouse effect, 372–373, 381–382, 590
groundwater, 300–305
 conservation of, 306–307, 308
 erosion and deposition by, 131, 309–311
 pollution of, 307
Gulf Stream, 534, 535, *i 535*
gully, 287
guyot, 520
gypsum, 129, *c 701*
 weathering of, 262–263

H

hail, 403, *i 403*
half-life, 657, *c 705*
halite, 94, 107, 129, *c 701*
 cleavage of, 100
 crystal structure of, 98, *i 98*
Halley's comet, 602, *i 602*
hanging valley, 324, 353
hanging wall, 240
hardness, 102, 106
hard water, 309
Hawaiian Islands, 197–198, 250
headward erosion, 287, 288
heat
 of common minerals, *c 706*
 of Earth's interior, 16, 73–74, 152–153, 180, 305, 502
 on land and in water, 376
 movement of, 369, 536
 temperature and, 369–370
heat budget, 372–373, *i 373*, 474
heliocentric model, 578
helium, 572, 620, 627, 628, 629
 atomic structure of, 91, *i 91*
hematite, 111, *i 111*, *c 701*
 streak of, 105
Hertzsprung-Russell diagram, 626–627, *i 627*
hexagonal system, *c 99*
highland climates, 473
High Plains aquifer, 308, 359
high-pressure area, 416
 upper-air flow and, 443
 weather associated with, 444
 wind and, 417, 420, 423–425, 427–428, 444
hominid, 684–685
horn, 324
hornblende, *c 104*, 110
 in igneous rocks, 121, 123
 weathering of, 262
horst, 245, *i 245*
hot spot, 194, 197–198
hot spring, 176, 305
Hubble, Edwin, 634, *i 635*
humidity, 391–393
 of air mass, 436, 437–438
 air pressure and, 416, *i 416*
humid tropical climates, 472
humus, 265
hurricane, 450–453, *i 450*, 454
Hutton, James, 62
hydroelectric power, 151
hydrogen
 atomic structure of, 91, *i 91*
 isotopes of, 93
 on Jovian planets, 588
 in stars, 572, 620, 627, 628, 629
hydrogen bond, 493
hydrogenous sediment, 525
hydrogen sulfide, 502
hydrolysis, 260
hydrosphere, 9
hyperthermophile, 641
hypothesis, 28, 29, 30

I

ice age, 330–333, 674, 683
ice cap, 319
ice front, 321–322
ice stream, 329
ice wedging, 258–259, *i 258*, 323
iceberg, 322, 523
igneous intrusion, 125–126, *i 125*, 651
igneous rocks
 continental growth and, 186
 formation of, 121–123
 groups of, 123–125, *c 124*
 identification, scheme for, *c 703*
 magnetic record in, 174–175, 182
impact crater. *See* craters.
index fossil, 653
inner core, 72
insolation, 372–373, 374–376, *i 374*
International Date Line, 78, *m 78*
International System of Units (SI), *c 696*
intertropical convergence zone, 425, 428, 472
intrusive igneous rock, 121–122
Io, 207, *i 207*, 599, *i 599*, *i 600*
ion, 94–95
 in atmosphere, 372
ionic bond, 94–95, *i 94*
ionosphere, 372
iron
 atomic bonds in, 95
 in meteorites, 604
 ores of, 110–111
irregular galaxy, 632, *i 632*
island
 barrier, 352, *i 352*
 coral, 521
 as open system, 6
 volcanic, 177, 195, 197–198, 518, 520, 521
island arc, 177, 195, 518
isobar, 416, 417, *m 417*, 420, 458
isotherm, 377
isotope, 92, 93
 parent and daughter, 656, 658

J

jet stream, 420–421
joint, 241
Jovian planets, 588, 594
Jupiter, 595, *i 595*
 satellites of, 207, 599–600, 640–641
Jurassic Period, 679

K

kame, 328

kaolinite, 110, c 701
karst topography, 311
Kepler, Johannes, 578–579, 580
kettle, 328
key bed, 654

L

laccolith, 126
lagoon, 351
lahar, 203, 269, i 269
lake
  cirque, 324
  kettle, 328, i 328
  moraine-dammed, 328
  oxbow, 290, i 291
land breeze, 429, i 429
Landsat, i 5, 49
landslide, 268
lapilli, 201
latitude, 46
  insolation and, 375
Laurasia, 183, 678
lava, i 121, 122, 200
lava plateau, 204
law, scientific, 32
levee, 290, 291, 292, i 292
light, 612–616
lightning, 446, i 446
light-year, 620
lignite, 149
limonite, c 701
limestone, 110, 129, i 135
  acid rain and, 379
  caverns in, 261, 310–311
  metamorphosis of, 135
  sinkholes in, 311
liquefaction, 222–223
lithosphere, 73, 194
  plates of, 173, 175, 176–181, 214
Little Ice Age, 482–483
local metamorphism, 134
loess, 342, i 342
Loihi Seamount, 198, i 198
longitude, 46
longshore current, 347, i 347, 351
lost river, 311
Love wave, 216
low-pressure area
  life cycle of, 441–442
  upper-air flow and, 443
  weather associated with, 444
  wind and, 417, 420, 423–425, 427–428
low-water landscaping, 19
luminosity, 623, 626
luster, 105

M

mafic magma, 121
magma
  contact metamorphism and, 134
  formation of, 194
  igneous-rock formation and, 118, 119, 120, 121–123
  mineral formation and, 97, i 97
  on moon, 556–557
  types of, 199, c 199
magnetic field
  of Earth, 74, 174, 575
  of Ganymede, 600
  of Jupiter, 595
  of Neptune, 597
  of Saturn, 596
  of sun, 574
  of Uranus, 597
magnetite, 107, 111, i 111, c 701
magnitude
  of earthquake, 220
  of star, 619, 623, 624
main-sequence star, 626–627
manganese nodule, 525, i 525
mantle, 72–73, 118
  convection in, 180–181
  seismic waves in, 228, 229
map, 44
  orientation of, 47
  projections, 44–45
  scale of, 46–47, 57
  temperature, 376–377
  topographic, 53–57, c 697
  weather, 376–377, 416, i 417, 457–458, c 457, i 458
mapmaking, 48–52, 329, 510, 511
marble, 110, 135, i 135
maria, 206, 558
maritime polar air mass, 438
maritime tropical air mass, 438
Mars, i 582, 591–592, i 591
  retrograde motion of, i 577, 578
  satellites of, 599
  space probes and, 593
  volcanic activity on, 206–207, 592
mascon, 558
mass, 90, 621
mass extinction, 677, 680, 690–691
mass number, 92
matter, 90
meander, 290
Mercator projection, 44
Mercury, 589, i 589
mercury barometer, 414, i 414
meridian
  prime, 78
  time, 77
mesocyclone, 447
mesosphere, 371
Mesozoic Era, 667, 678–680
metal, 94
  as resource, 145, 147, 156, 158
metallic bond, 95
metamorphic rocks
  formation of, 118, 120, 133–134
  identification, scheme for, c 702
  types of, 134–137
metamorphism, 133–134
meteor, 604
meteor shower, 604
meteorite, 604, i 604
meteoroid, 556, 604
meteorology, 436
mica, i 105, 109–110
  in igneous rocks, 123
  luster and cleavage of, 105
  metamorphism of shale and, 135–136
  molecular structure of, i 101
  weathering of, 262
micrometeoroid, 557
mid-ocean ridge, 174–175, 180, 181, 519–520
  volcanic activity at, 194, 196, i 196
Milky Way, 631–632
millibar, 415
mineralogy, 104
minerals
  composition of, 96

Index

crystalline structure of, 98–102
formation of, 97
in groundwater, 309–310
groups of, 108–111
identifying properties of, 104–107, *c 700–701*
reserves of, 146
as resources, 144–147, *m 146*, 154
rock-forming, 104
specific heats of, *c 706*
mineral spring, 310
mining, 147, 154, 156
Miranda, 601, *i 601*
Mississippian Period, 676
Mohorovičić discontinuity, 229
Mohs scale, 106, *c 106*
moist-adiabatic lapse rate, 398–399
moist mid-latitude climates, 472–473
mold, 648–649, *i 649*
molecule, 93
moment magnitude, 220
monoclinic system, *c 99*
monsoon, 427–428, *i 427*, 533
moon. *See also* satellite (natural).
  eclipse of, 564, *i 564*
  geologic history of, 556–557
  layers of, *i 557*
  orbit of, 562
  phases of, 563, *i 563*
  properties and features of, 557–560
  solar eclipse and, 564–565
  theories of formation of, *i 32–33*, 556
  tides and, 541, *i 541*
  volcanic activity on, 206
moraine, 323, *i 323*, 326–327, 328, 331
motion, laws of, 579, 580
mountains, 236
  climate and, 473
  formation of, 238–241
  glaciers and, 318, 319
  location of, 236–237
  lunar, 559
  plate tectonics and, 178, 238, 239, 243, 673, 675, 676, 677, 681, 683
  precipitation and, 404, 405

types of, 243–245
wind and, 405, 429
mud cracks, 130–131
mud pot, 305, *i 305*
mud volcano, 305
mudflow, 269, *i 269*
muscovite, *i 101*, 109, *c 701*

N

nanotube, 103
natural gas, 149, 150
natural selection, 670–671
nebula, 627, *i 627*
  planetary, 628
nebular hypothesis, 70
nekton, 501–502
Neogene Period, 681–683
Neptune, 597, *i 597*
  satellites of, 601
NEPTUNE project, 522
neutron, 91, 92–93
neutron star, 630
névé, 318
Newton, Isaac, 580, *i 580*
Niagara Falls, 289, *i 289*
nimbostratus, 397, 444
nodule, 131
nonconformity, 652
nonmetal, 95
nonrenewable resource, 144, *c 145*, 148–150, *m 148*, 154–155
normal fault, 240, *i 240*
North Star, 76, 618
nuclear power, 150, *i 151*, 154–155
nucleus
  of atom, 91, 92–93
  condensation, 394, 406

O

Oberon, 601
obsidian, *i 123*, 124, 132, *i 132*
occluded front, 441
ocean
  composition of water in, 495, *c 495*
  currents in, 347, 532–539
  effects on weather and climate, 548–549
  life in, 499–502

pollution of, 502, 521
salinity of water in, 495, 496
study of, 490–491
temperature zones in, 496–498, *c 497*
tides in, 16, 17, 541–543
ocean floor, 512, *m 512–513*
  at continental margins, 514–516
  hydrothermal vents on, 176, 502, *i 502*, 519
  in ocean basins, 517–521
  sediments on, 476–477, 511, 523–525
  study of, 510–513, 522
ocean wave, 344–349
oceanography, 490–491
oil. *See* petroleum.
oil shale, 150
olivine, *i 101*, 110, *c 701*
  in igneous rocks, 124, 125
  molecular structure of, *i 101*
Olympus Mons, 206, *i 206*, 592
ooze, 523, 524–525
opal, *c 701*
open system, 6
orbit (Earth's), 76, 80–81
orbital plane, 76, *i 81*
Ordovician Period, 675
ore, 145, 147
organic sedimentary rock, 128
original horizontality, principle of, 650
orthoclase, 109, *i 109*
  in igneous rocks, 121, 123
orthorhombic system, *c 99*
outer core, 72, 228–229
outwash, 326, 327, 328
outwash plain, 327, 331
oxbow lake, 290, *i 291*
oxidation, 261
oxide, 111
oxygen
  in atmosphere, 11, 366, 667
  chemical weathering and, 261
ozone, 4–5, 371, 380–381

P

pahoehoe, 200, *i 200*
Paleogene Period, 681–683
paleontology, 648
Paleozoic Era, 667, 674–677

Pangaea, 183, 677, 678
parallax, 80, *i 81*, 620, *i 620*
parent isotope, 656, 658
parent rock, 133
parsec, 620
peat, 149
peer review, 30
Pennsylvanian Period, 676, 677
penumbra, 564
perched aquifer, 303
perigee, 542, 562
perihelion, 81
period, 667
periodic table of elements, 92, *i 92, c 698–699*
permeability, 301
petrified wood, 309
petroleum, 149–150
  conservation of, 156
pH, 379, *c 379*, 494
Phanerozoic Eon, 666–667
Phobos, 599, *i 599*
phosphorescence, 107
photosphere, 573, 574
photosynthesis, 11, 14
phyllite, 136, *i 136*
phytoplankton, 14–15, 499, 500, 539
pillow lava, 200
Pinatubo, Mount, 205
plagioclase, 109, *i 109*
  in igneous rocks, 123, 124
planetary nebula, 628
planetesimal, 70, 71
planets. *See also names of individual planets.*
  apparent movement of, 577
  inner, 588–592
  laws of motion and, 579, 580
  outer, 588, 594–598
  satellites of, 599–601
plants
  in Cenozoic Era, 681, 684
  conservation of, 672
  fossils of, 130, 648–649
  mechanical weathering by, 259
  in Mesozoic Era, 678–680
  in ocean, 499
  in Paleozoic Era, 676–677
  photosynthesis of, 11, 14
plasma, 572
plastic flow, 321
plate tectonics, 172–187

climate change and, 332, 475
mountain building and, 178, 238, 239, 243, 673, 675, 676, 677, 681, 683
plunge pool, 283
Pluto, 588, 597–598, *i 598*
pluton, 125
plutonic dome mountain, 244, *i 244*
polar climates, 469
polar front, 423
polar-front jet stream, 421, *m 421*
Polaris, 76, 618
polarity, 493
pollution
  air, 155, 378–380, *c 378, c 382*, 383, 407
  water, 154, 155, 156, 307, 502, 503, 521
polyconic projection, 45
porosity, 300, *i 300*
porphyry, 122
potassium-argon dating, 659
pothole, 283, *i 283*
Precambrian time, 673–674
precipitation, 402–406
  air masses and, 437, 438
  fronts and, 439, 440, 441
pressure gradient, 416, 417, 421, 428, 429
prevailing wind, 423, 425, 427
prime meridian, 46, 78
Proterozoic Eon, 666, 673
proton, 91, 92–93
psychrometer, 392
psychrophile, 640
Ptolemy, 578
pulsar, 630
pumice, *i 123*, 124
P wave, 215, *i 215*, 218, 228–229
pyrite, 111, *i 111, c 701*
  luster and streak of, 105
pyroclastic flow, 201
pyroclastic material, 200–201, *i 201*
pyroxine, 109, *c 701*
  in igneous rocks, 124, 125
  molecular structure of, *i 101*

Q

quartz, 108, *i 108, c 701*

color, luster, and fracture of, 105
  crystal structure of, 98, *i 98*
  as gangue mineral, 145
  in igneous rocks, 121, 123, 124
  in sedimentary rocks, 128, 131
  weathering of, 262
quartzite, 135, *i 135*
  weathering of, 263
quasar, 633
Quaternary Period, 683–684

R

radar, 48, 449
radiation
  cosmic background, 635
  electromagnetic, 612–613
  as transfer of energy, 369
radioactive decay, 656–657, *i 657*
radiocarbon dating, 658
radiometric dating, 655, 658–659, *c 659*
radiosonde, 455–456
rain, 402, 403, 404
  acid, 155, 261, 379, 407
  artificial production of, 406
  freezing, 403
  measuring, 403
rain forest, 267, 472, *i 472*
rapids, 288
Rayleigh wave, 216
recycling, 156, 158, 159, 164–165
redshift, *i 615*, 616, 634
reef. *See coral reef.*
refraction
  of light waves, 107, 613
  of ocean waves, 345, *i 346*
  of seismic waves, 228–229
regional metamorphism, 133–134
regolith, 560
relative dating, 650–654, *c 706*
relative humidity, 392, *c 393, c 705*
renewable resource, 144, *c 144*, 151–153, 155–156
reserve, 146
residual soil, 264
resource
  conservation of, 156, 158–159,

Index

272–273, 306–307, 308
energy, 148–153, 154–156
mineral, 144–147, *m 146*, 154
nonrenewable, 144, *c 145*,
    148–150, *m 148*, 154–155
renewable, 144, *c 144*,
    151–153, 155–156
soil as, 271–273
retrograde motion, 577, 578
reverse fault, 240, *i 240*
revolution
  of Earth, 80–83, 331, 475, 618
  of moon, 557, 562
rhyolite, *i 123*, 124
rhyolitic magma, 199, *c 199*
Richter magnitude, 220
ridge push, 181
rift, 176, 196
rift valley, 176, 519
rille, 558
rip current, 347
ripple marks, 130, *i 131*
river
  channel of, 282
  deposition by, 285–286
  discharge of, 281
  erosion by, 283, 287, 288–289
  flooding of, 291–292
  floodplain of, 290–291, *i 290*
  gradient of, 281
  lost, 311
  rapids and waterfalls in,
      288–289
  system, 280, *i 280*
  transport by, 284–285, *i 284*,
      523
  valley of, 287–288
  velocity of, 281
roche moutonnée, 323, *i 325*
rock, 96, 118
  igneous, 118, 119, 120,
      121–126
  metamorphic, 118, 120,
      133–137
  sedimentary, 118, 119–120,
      127–131
rock cycle, 119–120, *i 119*
rock flour, 323
rock-forming mineral, 104
rotation
  of Earth, 75–76, *i 75*, 77–78,
      419, 422, 618
  of Jupiter, 595

of Mercury, 589
of moon, 557
of Neptune, 597
of Saturn, 596
of sun, 574
of Uranus, 597
of Venus, 589
rubidium-strontium dating, 659

S

Saffir-Simpson scale, *c 452*, 453
Sagan, Carl, 26, *i 26*
St. Helens, Mount, 199, 203,
    250–251, *i 250–251*
salinity, 495, 496
salinization, 271, *i 271*
San Andreas Fault, 178, *i 178*
sand
  on beaches, 347, 350
  in soil, 265
  windblown, 340, 341, 342–343
sandbar, 347, 351–352
sand dune, 342–343, *i 343*
sandstone, 128, *i 128, i 135*, 263
  metamorphosis of, 135
Santa Ana, 405
Sargasso Sea, 534
satellite (artificial)
  mapmaking and, 49, 329, 511
  as tool of scientist, 37
  for tracking plate movements,
      184, *i 184*
  weather forecasting and, 455,
      468
satellite (natural). *See also*
    moon.
  of Jupiter, 599–600
  of Mars, 599
  of Pluto, 598
  of Saturn, Uranus, and
      Neptune, 601
saturation, zone of, 302
saturation vapor pressure, 391
Saturn, 596, *i 596*
  satellites of, 601
schist, *i 97*, 136–137, *i 136*
scientific literacy, 34
scientist, 26–27. *See also*
    careers.
  methods of, 29–33
  thinking of, 28
  tools of, 35–37

scoria, 124
sea. *See* ocean.
sea breeze, 417, *i 417*, 428, *i 428*
seamount, 198, 520
seasons, 82–83, 375
  constellations and, 618–619,
      i 618
  on Mars, 591
  winds and, 427–428
sediment
  absolute time and, 656
  continental growth and, 185,
      186
  glacier-borne, 326–328, 331,
      523
  at passive continental margin,
      237, *i 237*
  river-borne, 127–128, 280, 281,
      284–286, 290, 523
  rock formation from, 118, 119,
      120, 127–129
  on sea floor, 476–477, 523–525
  sorting of, 127–128, *i 127*
  wave-borne, 351–352
  wind-borne, 340, 342–343
sedimentary rocks
  features of, 130–131
  formation of, 118, 119, 120,
      127–129
  identification, scheme for,
      c 702
  strata of, 130, 650–652
seismic gap, 225–226
seismic wave, 215–218, 222,
    228–229
seismogram, 217–218, *i 217*,
    i 218, 220
seismograph, 217, *i 217*
seismographic network, 227
serpentine, *c 701*
shale, 128, *i 128, i 136*
  concretions in, 131
  metamorphosis of, 135–136
  oil, 150
  weathering of, 263
shear stress, 238–239
shield, 673, *m 673*
shield volcano, 202, *i 202*
shoreline
  erosion of, 349
  features of, 349–352, *i 350–351*
  types of, 352–353
  waves and currents at, 345–347

SI. *See* International System of Units.
silica
  deposition by groundwater, 309
  magma viscosity and, 199
  in sedimentary rocks, 128, 131
silicates, 100, 108–110
  molecular structure of, *c 101*
silica tetrahedron, 100, *i 100*, 108
siliceous ooze, 524–525
sill, 125–126, *i 125*
Silurian Period, 676
sinkhole, 311, *i 311*
slab pull, 181
slate, 135–136, *i 136*
sleet, 403
slump, 268
smog, 380
snow, 403, 453
  glaciers and, 318–320
  measuring, 404
snow line, 318, *c 318*
sodium chloride, 495
soil
  composition of, 265
  erosion and conservation of, 272–273
  formation of, 264–265
  mass movements of, 268–269
  on moon, 560
  problems threatening fertility of, 271
  in rain forests, 267
  residual, 264
  surface mining and, 154, 156
  transported, 264
  types in North America, *m 266*
soil depletion, 271
soil horizon, 264–265, *i 264*, *c 265*
soil profile, 264, *i 264*
solar energy, 16, 17, 332
solar flare, 575, 576
solar power, 152, 156, 157, 383
solar system, *i 588–589, c 706*. *See also* sun.
  history of models of, 577–580
  objects in, 588–605
  origin of, 70–71, *i 70–71*
solar wind, *i 574–575*, 575–576
solstice, 82, *i 82*, 83

specific gravity, 106–107
specific humidity, 391
spectroscope, 613, 614
spectrum, 612, *i 612*, 613–616, *i 613*
sphalerite, *c 701*
spiral galaxy, 631, 632, *i 632*
spring, 303
  artesian, 304
  hot, 176, 305
  mineral, 310
squall line, 446
stalactite, 311
stalagmite, 311
standard time zone, 77
star
  apparent movement of, 577, 618
  distance of, 619–620
  early observations of, 617–619
  elements in, 620, *c 620*
  fusion in, 70, 572, 627, 628, 629
  life cycle of, 626–630, *i 628–629*
  luminosity of, 623
  mass and size of, 621, *c 621*
  movement relative to Earth of, 615–616
  temperature of, 621–622, *c 622*, 623, 626
  variable, 623–624
stationary front, 441
station model, 457, 458
storm, 445–453
  dust, 340, *i 340*
  winter, 453
storm surge, 451–452
strata, 650–654
stratification, 130
stratiform cloud, 396, 400
stratigraphic matching, 654
stratocumulus, 396–397
stratosphere, 370–371
stratus, *i 396*, 397, 444
streak, 105
stream. *See* river.
stream piracy, 288
striations
  on feldspar cleavage surface, 109
  glacial, 323, *i 323*
strike-slip fault, 241, *i 241*, 520

stromatolite, 674, *i 674*
subduction boundary, 177
  active continental margin and, 515
  earthquake depth at, 214–215
  volcanic activity at, 194, 195, *i 195*, 518
subsidence, 306–307
subsoil, 265
sulfide, 111
sulfur, *c 104, c 145, c 701*
summer solstice, 82
sun
  altitude of, 82
  eclipse of, 564–565, *i 565*
  energy of, 16, 17, 332
  as energy resource, 152, 156, 157, 383
  features on, 574–576
  formation of, 70
  fusion in, 572
  insolation from, 372–373, 374–376
  layers of, 573, *i 573*
  size of, 572
  spectrum of, 614
  tides and, 542, *i 542*
sunspots, 574, *i 574*
  climate and, 475
  ionosphere and, 372
  solar flares and, 575
supercell, 446
supergiant, 627, 629
supernova, 629, *i 630*
Superposition, Principle of, 650
surface current, 532–533, *m 532*, 535
  warm/cold, 534
  wind and, 533, *i 533*
surface wave, 216
S wave, 215, *i 215*, 218, 228–229
syncline, 239
system, 5–6

T

talc, *c 701*
talus, 268, *i 268*
tar sand, 150
tectonic dome mountain, 244, *i 244*
telescope, *i 37*, 625
temperature, *c 696*

Index

of air mass, 436, 437–438
air pressure and, 416, 422, 427–429
heat and, 369–370
maps of, 376–377
in ocean, 496–498
of stars, 621–622, *c* 622, 623, 626
variations in land, 374–377
temperature inversion, 380, 444
tension, 238–239
terrane, 186–187
terrigenous sediment, 524
tetragonal system, *c* 99
theory, 32
thermocline, 498
thermodynamics, 17
thermohaline circulation, 549
thermometer, 370
thermosphere, 371
three-cell circulation model, 422–425, *i* 423
thrust fault, 240, *i* 240
thunderstorm, 445–446, *i* 445, 449
tidal energy, 16, 17
tide, 16, 541–543
  as power source, 151
till, 326, 328, 331
tillite, 331
time
  absolute, 656–659
  Earth's rotation and, 77–78, *i* 77
  measuring, 79
  relative, 650–654
time meridian, 77
time zone, 77
Titan, 601
Titania, 601
topaz, *c* 701
topographic map, 53–57
  symbols, *c* 697
topsoil, 265
tornado, 447–449, *i* 447
tourmaline, *c* 701
trace fossil, 649
trade wind, 425, 426, 468, 533
transform boundary, 177, *c* 179
  active continental margin and, 515
  earthquake depth at, 214
transition zone, 229

trans-Neptunian object, 602
transpiration, 13
transported soil, 264
travertine, 310, *i* 310
tree-ring dating, 656
trench, 518
Triassic Period, 678
tributary, 280, 291
triclinic system, *c* 99
trilobite, 675, *i* 675
tritium, 93
Triton, 601
Tropic of Cancer, 82, *i* 82
Tropic of Capricorn, *i* 82, 83
troposphere, 370, 420–421
trough, 443, *i* 443
tsunami, 223, 345
tuff, 123
turbidity current, 516, 517, 536–537

U

ultramafic igneous rock, 125
umbra, 564
Umbriel, 601
unconformity, 651–652
universe, 631
  origin of, 634–635
upwelling, 538–539, *i* 539, 540
uranium
  energy levels in atom of, 91
  as energy resource, 150
  isotopes of, 93
  radioactive decay of, 657
  radioactive minerals of, 107
uranium-lead dating, 658–659
Uranus, 597, *i* 597
  satellites of, 601

V

valley
  glacial, 321, 324
  hanging, 324, 353
  rift, 176, 519
  river, 287–288
  wind and, 429
valley glacier, 319, *i* 319
  erosion by, 324, *i* 324
variable star, 623–624
varve, 656
ventifact, 341, *i* 341

Venus, 590–591, *i* 590
  volcanic activity on, 207, 590
vernal equinox, 83
volcanic island arc, 177, 195, 518
volcanic neck, 126, 270, *i* 270
volcanoes, 194–198, *i* 195, 202–204
  climate and, 475
  early atmosphere and, 366, 667
  effects on biosphere, 250–251
  as evidence for plate tectonics, 173
  extraterrestrial, 204–205
  formation of, 194–198, *i* 195
  as landforms, 202–204, 244–245
  materials erupted by, 199–201
  predicting eruptions of, 205
  types of, 202–204, *i* 202–203

W

warm-core ring, 535
warm front, *i* 440, 441, 444
water. *See also* groundwater; ocean; river.
  in atmosphere, 390–406, 416
  chemical weathering and, 260–261
  density of, 492, *c* 705
  energy of, *c* 705
  as energy resource, 151, 155
  hard, 309
  landscaping and, 19
  management of, 358–359
  mechanical weathering and, 258–259
  molecular structure of, 493
  polarity of, 493
  pollution of, 154, 155, 156, 307, 502, 503, 521
  solutions in, 493–494
  states of, 390, *i* 391
water budget, 306
water cycle, 13, *i* 13
water mass, 495
waterfall, 288–289, 324
watershed, 280
water table, 302–303
wave. *See* light; ocean wave; seismic wave.

wavelength, 344, 612–613, 616
weather, 436–453. *See also* climate.
  forecasting, 449, 455–459, 468
  modification of, 406
  oceanic effects on, 548–549
weathering, 258
  chemical, 258, 260–261
  factors affecting rate of, 262–263
  mechanical, 258–259
  by streams and rivers, 283
  by wind, 340–341
Wegener, Alfred, 172
well, 303, 304
white dwarf, 627, 628
wind
  air pressure and, 417, 420, 423, 424, 425, 427–429, 444
  chinook, 405
  Coriolis effect and, 75, 419–420, *i 420*, 422, 423
  as energy resource, 152, 155–156, 383
  erosion and deposition by, 340–343
  friction and, 420–421
  global patterns of, 422–425
  local, 428–429
  measuring, 418
  ocean currents and, 533, *i 533*, 538–539
  ocean waves and, 344
  Santa Ana, 405
  seasons and, 427–428
  trade, 425, 426, 468, 533
windmill farm, 152, *i 152*
wind vane, 418
winter solstice, *i 82*, 83
winter storm, 453

Y

yazoo tributary, 291

Z

zenith, 82
zooplankton, 500–501

Acknowledgements

Cover Photography
wraparound image Adam Jones/Dembinsky Photo Associates.
back cover *top to bottom* Marc Meunch/Corbis; courtesy, NASA/GSFC; courtesy, NASA; Norbert Wu Photography.

Photography
i Adam Jones/Dembinsky Photo Associates; **iv** Digital Stock; **v** David R. Frazier; **vi** *t* Digital Stock; *m* Timothy G. Laman/National Geographic Image Collection; **vii** *top to bottom* AFP/Corbis; Alfred Pasieka/Science Photo Library/Photo Researchers, Inc; Bob Krist/Corbis; Owen Seumptewa/Native Shadows; **viii** *top to bottom* Krafft/Photo Researchers, Inc.; REZA; Chamoux/Liaison Agency, Inc.; **ix** *top to bottom* Kunio Owaki/Corbis Stock Market; Mark Burnett/Photo Researchers, Inc.; Bill Bachman/Photo Researchers, Inc.; Joel Bennett/Peter Arnold, Inc.; Lee Foster/Bruce Coleman, Inc.; **x** *top to bottom* Michael P. Gadomski/Photo Researchers; Inc.; Photri, Inc.; AFP/Corbis; Ralph H. Wetmore/Stone; William J Herbert/Stone; **xi** *top to bottom* Jan Hirsch/Science Photo Library/Photo Researchers, Inc., ©2000 MBARI; Francois Gohier/Photo Researchers, Inc; **xii** *mt* NASA/Photodisc; *mb, b* courtesy, NASA; **xiii** *t* Annie Griffiths Belt/Corbis; *m* Jeff Greenberg/Photo Researchers, Inc.; *b* Sandy Felsenthal/Corbis; **xiv** *tl* Cary Wolinsky; *cr* Dave G. Houser/Corbis; *br* Peter Wilson/Corbis and Royal Greenwich Observatory/Science Photo Library/Photo Researchers, Inc.; **xv** *t* Jeff Greenberg/PhotoEdit; *inset* Reuters NewMedia/Corbis; *b* Mark Moffett/Minden Pictures; **xvi** *t* Science VU/Visuals Unlimited, Inc.; *b* Russell D. Curtis/Photo Researchers, Inc.; **xvii** Pat O'Hara/Corbis; **xviii** *tl* courtesy, NASA; *bl* Earthweek: Earth Environment Service; *br* courtesy, Robert Crippen/JPL/NASA; **xx** *both* courtesy, USGS; **xxi** *t* courtesy, NASA; *c, b* NOAA/National Climatic Data Center; **xxii** David Young-Wolff/PhotoEdit.

Unit 1
xxiv, 1 *background* courtesy, NASA; **1** *t* courtesy, NOAA; *b* courtesy, NASA; **2, 3** Digital Stock Royalty Free; **4** NASA/Corbis; **5** courtesy, NASA; **7** Steve Striddand/Visuals Unlimited, Inc.; **8** Corbis; **9** *c* Digital Stock; *b* Michael Marten/Science Photo Library/Photo Researchers, Inc.; **10** Gerald & Buff Corsi/Visuals Unlimited, Inc.; **11** Digital Stock; **12** Mark Moffett/Minden Pictures; **14** *t* J.B.Diederich/Picturequest; *b (inset)* Andrew Syred/Science Photo Library/Photo Researchers, Inc.; **15** Gary Retherford/Photo Researchers, Inc ; **16** *l* Photodisc, Inc. *m* Ragnar Sigurdsson/Stone; *r* Digital Stock; **18** Ross Frid/Visuals Unlimited, Inc.; **19** *c* Tom Bean Photography; *b* Mark E. Gibson/Visuals Unlimited, Inc.; **24, 25** Timothy G. Laman/National Geographic Image Collection; **26** James P. Blair/National Geographic Image Collection; **27** *t* courtesy, Dr. Eriqueta Barrera/University of Akron; *b* courtesy, NOAA; **30** Photoworks; **31** Leif Skoogfors/Corbis **34** *bl* Roger Ressmeyer/Corbis; *c* Jerry Cooke/Corbis; *tr* Patrick Ramsey/International Stock Photo; **35** Jonathan Blair/National Geographic Image Collection; **36** *l* David Parker/Science Photo Library/Photo Researchers, Inc.; *r* Nick Caloyianis/National Geographic Image Collection; **37** *l* Mark Burnett/Picturequest; *r* Roger Ressmeyer/Corbis **42, 43** courtesy, NASA; **48** Hulton Archive Photo; **51** Rick Poley/Visuals Unlimited, Inc.; **53** James L. Amos/National Geographic Image Collection; **55** *b* Digital Stock; **62, 63** *b* courtesy, NASA; **62** *t* Bettman/Corbis; **63** *cl* Eric and David Hosking/Corbis; *c* Galen Rowell/Corbis; *br* © Woods Hole Oceanographic Institute.

Unit 2
66, 67 *background* ESA/TSADO/Tom Stack & Associates; **66** *t* Gary Schultz; **67** *bl* Bob Krist/Corbis; *t* Jeff Scovil; *br* Ed Kashi/Corbis; **68, 69** AFP/Corbis; **75** David R. Frazier; **79** *tr* Robert Holmes/Corbis; *bl* Peter Wilson/Corbis; *br* Royal Greenwich Observatory/Science Photo Library/Photo Researchers, Inc.; **88, 89** Alfred Pasieka/Science Photo Library/Photo Researchers, Inc.; **90** Jeff Scovil; **94** *t* Martyn F. Chillmaid/Science Photo Library/Photo Researchers, Inc.; *m,b* Charles D. Winters/Photo Researchers, Inc.; **97** *both* Breck P. Kent; **98** Breck P. Kent; **99** *top to bottom* Breck P. Kent; Mark A. Schneider/Dembinsky Photo Associates; *all others* Jeff Scovil; **101** *top to bottom; (1, 2)* Mark A. Schneider/Dembinsky Photo Associates; *(3, 4)* Jeff Scovil; *(5, 6)* Breck P. Kent; **103** *t* Photo Researchers, Inc.; *b* Courtesy of Richard Smalley; **104** *left to right;* Mark A. Schneider/Dembinsky Photo Associates; E.R. Degginger/Dembinsky Photo Associates; Breck P. Kent; Mark A. Schneider/Dembinsky Photo Associates; **105** E.R. Degginger/Color-Pic, Inc.; **106** William K. Sacco; **108** *l* Courtesy of Jessica Haigh; *c* Breck P. Kent; *r* Ken Lucas/Visuals Unlimited, Inc.; **109** *l* Breck P. Kent; *r* Ed Degginger/Color-Pic, Inc.; **110** *t* Mark A. Schneider/Dembinsky Photo Associates; *b* Jeff Scovil; **111** *l, c* Mark A. Schneider/Dembinsky Photo Associates; *r* Jeff Scovil; **116, 117** TomBean/Corbis; **118** *clockwise from top* Breck P. Kent; Jeff Scovil; Breck P. Kent; Jose Manuel Sanchis Calvetee/Corbis; Stephen Frisch; **119** *clockwise from top;* Yann Arthus-Bertrand/Corbis; Breck Kent; Jeff Scovil; Vittoriano Rastelli/Corbis; Breck P. Kent; **121** Krafft, Explorer/Photo Researchers, Inc.; **122** *t* Jeff Scovil; *b* Breck P. Kent; **123** *tl, tr* Breck P. Kent; *cr* Robert Pettit/Dembinsky Photo Associates; *b* Mark A. Schneider/Dembinsky Photo Associates; **125** Tom Bean Photography; **126** Mark E. Gibson; **128** *l* Breck P. Kent; *c* Lee Boltin/Boltin Picture Library; *r* E.R. Degginger/Color-Pic, Inc.; **129** *l* Scott T. Smith/Dembinsky Photo Associates; *r* Bob Krist/Corbis; **130** *l* Tom Bean Photography; *r* Breck P. Kent; **131** *l* Tom Bean Photography; *r* Lee Boltin/Boltin Picture Library; **132** *t* Visuals Unlimited, Inc.; *b* A.J. Copley/Visuals Unlimited, Inc.; **133** Charles C. Plummer; **135** *tl* Breck P. Kent; *tr* Robert Pettit/Dembinsky Photo Associates; *bl* Dusty Perin/Dembinsky Photo Associates; *bc* Breck P. Kent; *br* Breck P. Kent; **136** *left to right (top)* E.R. Degginger/Color-Pic, Inc.; Breck P. Kent; Lee Boltin/Boltin Picture Library; Breck P. Kent; *b* Hubert Stadler/Corbis; **137** *l* The Natural History Museum, London; *r* Breck P. Kent; **142, 143** Jim Corwin/Photo Researchers, Inc.; **144** Paul A. Grecian/Visuals Unlimited, Inc.; **145** Francois Gohier/Photo Researchers, Inc.; **149** *all* E.R.

Degginger/Color-Pic, Inc.; **151** Porterfield/Chickering/Photo Researchers, Inc.; **152** *t* Russel D. Curtis/Photo Researchers, Inc.; *b* Craig Miller/Photo Researchers, Inc.; **153** E.R. Degginger/Color-Pic, Inc.; **156** Jeff Greenberg/PhotoEdit; **157** *t* Owen Seumptewa/Native Shadows; *b* Pat Armstrong/Visuals Unlimited, Inc.; **158** *t* Spencer Grant/Photo Edit; *b* Joseph Sohm/Chromosohm Inc./Corbis; **159** Trent Steffler/David R. Frazier; **164, 165** *b* Liz Hymans/Corbis; **164** Ken O'Donoghue; **165** *t* Dave G. Houser/Corbis; *br* Adrian Arbib/Corbis.

Unit 3

168, 169 *background* NASA/GSFC/NOAA/USGS; **168** *br* K. Segerstrom/USGS; **169** *t* Reuters/Alcir Salvala/TimePix; **170, 171** Tony Waltham/Geophotos; **172** A.J. Copley/Visuals Unlimited, Inc.; **176** Mary Plage/Bruce Coleman Inc.; **177** *l* Rona Photography/Bruce Coleman Inc.; *r* Baron Wolman/Stone; **178** *t* Takehide Kazami/Peter Arnold, Inc.; *b* Dewitt Jones/Corbis; **184** courtesy, NASA; **186** *l* Winifred Wisniewski/Corbis; *r* C.C. Lockwood/Animals Animals; **187** courtesy, USGS. Photo by David B. Westjohn; **192, 193** Krafft/Photo Researchers, Inc.; **194** Steve Kaufman/Corbis; **196** Bernhard Edmater/Photo Researchers, Inc.; **199** *l* Douglas Pebbles/Corbis; *c* courtesy, USGS; *r* Eric and David Hosking/Corbis; **200** *l* D. W. Peterson/Corbis; *r* courtesy, USGS; **201** *l* courtesy, D.E. Wieprecht/USGS; *c* courtesy, USGS; *r* courtesy, J.P. Lockwood/USGS; **202** Mike Zens/Corbis; **203** *l* Galen Rowell/Corbis; *r* David Meunch/Corbis; **204** Greg Vaughn/Tom Stack & Associates; **205** *t* B. Diederich/Contact Press Images/PictureQuest; *b* Sid Balatan/Black Star Publishing/PictureQuest; **206, 207** *all* courtesy, JPL/NASA; **212, 213** REZA; **214** courtesy, NOAA/NGDC; **216** *l* DPA/RAJ/The Image Works; *r* Reuters NewMedia Inc./Corbis; **217** Reuters NewMedia Inc./Corbis; **220** Marshal Lockman/Black Star; **221** Roger Ressmeyer/Corbis; **222** *b* Mark Downey/Liaison Agency Inc; **223** Mark Downey/Liaison Agency, Inc.; **224** *both* courtesy, EQE; **227** courtesy, USGS; **234, 235** Piere Beghin/Liaison Agency Inc.; **236** Kent & Donna Dannen/Photo Researchers, Inc.; **239** G.R. Roberts/Photo Researchers, Inc.; **242** *b* Chamoux/Liaison Agency Inc; *t* Peter Russell; The Military Picture Library/Corbis; **245** C. Weaver/Ardea London Limited; **250** P. La Touretts/VIREO; **250, 251** John Marshall/Stone; **251** *br* David H. Harlow/USGS; *b* Pat O'Hara/DRK Photo.

Unit 4

254, 255 *background* Nigel Press/Stone; **254** *l* Zandria Meunch Beraldo/Corbis; *r* Kunio Owaki/Corbis/Stock Market; **255** *bl* Adam Jones/Dembinsky Photo Associates; *t* Susan Ruff; *br* Annie Griffiths Belt/Corbis; **256, 257** Kunio Owaki/Corbis/Stock Market; **258** Todd Gipstein/Photo Researchers, Inc.; **259** Kenneth W. Fink/Ardea London Limited; **260** Breck P. Kent; **261** Ray Juno/Corbis/Stock Market; **262** Laurence Parent Photography, Inc.; **264** Jeffery Howe/Visuals Unlimited, Inc.; **265** courtesy, USDA Photo Library; **267** *tr* M. Edwards/Peter Arnold, Inc.; Gary Braasch/Corbis; **268** Betty Derig/Photo Researchers, Inc.; **269** courtesy, USGS; **270** Breck P. Kent; **271** courtesy, Tim McCabe/USDA/Natural Resources Conservation Service; **278, 279** Cameron Davidson/Stone; **281** *l* Laurence Parent Photography; *r* Mark E. Gibson; **282** Mark Burnett/Photo Researchers, Inc.; **283** Photo Researchers, Inc.; **284** Mark E. Gibson; **285** Mark E. Gibson; **287** Tom Bean Photography; **289** Ray Juno/Corbis/Stock Market; **291** *t* Jeff Lepore/Photo Researchers, Inc.; *b* Cameron Davidson/Stone; **292** E.R. Degginger/Color-Pic, Inc.; **293** *t* Robert F. Bukaty/AP Wide World Photos; *b, inset* Stephen Brooke, Inc.; **298, 299** Bill Bachmam/Photo Researchers, Inc.; **301** Greg Pease/Stone; **305** David Cavagnaro/Visuals Unlimited, Inc. **306** courtesy, USGS; **307** Joseph Sohm;ChromoSohm Inc./Corbis; **308** Phillip Gould/Corbis; **309** Laurence Parent Photography, Inc.; **310** James L. Amos/Corbis; **311** G. R. " Dick" Roberts; **316, 317** Joel Bennett/Peter Arnold, Inc.; **319** E.R. Degginger/Color-Pic, Inc.; **320** Barbara Gerlach/Visuals Unlimited, Inc.; **322** Tom Bean Photography; **323** *t* S.J. Krasemann/Peter Arnold, Inc.; *b* Carr Clifton/Minden Pictures; **324** *t* Francis Gohier/Ardea London Ltd.; *b* Mark E. Gibson; **326** Tom Bean Photography; **327** *l* Tony Waltham/Geophotos; *r* Douglas Faulkner/Photo Researchers, Inc.; **328** Glen M. Oliver/Visuals Unlimited, Inc.; **329** *both* courtesy, NASA/Goddard Space Flight Center, Scientific Visualization Studio. Canadian Space Agency, RADARSAT International, Inc.; **338, 339** Minden Pictures; **340** Corbis; **341** *t* Tom Bean Photography; *b* Cleet Carlton; **342** Phil Schermeister/Corbis; **348** *t* courtesy, Wave Dispersion Technologies; *b* Annie Griffiths Belt/National Geographic Image Collection; **349** Lee Foster/Bruce Coleman Inc.; **350** *l* Ed Collacott/Stone; *r* G. R. "Dick" Roberts; **351** Phil Degginger/Color-Pic, Inc.; **352** Jim Wark/Peter Arnold, Inc.; **358** *l* Richard T. Nowitz/Corbis; *b* Jonathan Smith/Cordaiy Photo Library Ltd./Corbis; **358, 359** Steve Kaufman/Corbis; **359** Lowell Georgia/Photo Researchers, Inc.

Unit 5

362, 363 *background* Earth Imaging/Stone; **362** *bl* Provided by the SeaWiFS Project, NASA/Goddard Space Flight Center, and ORBIMAGE; *tr* Corbis; **363** *t* Provided by the SeaWiFS Project, NASA/Goddard Space Flight Center, and ORBIMAGE; *bl* Madhusudan B. Tawde/Dinodia.com; *br* Stefano Cellai/Corbis/Stock Market; **364, 365** W. Lynn Seldon Jr.; **366** Michael P. Gadomski/Photo Researchers, Inc.; **368** Bob Wiatrolik/Illinois Environmental Protection Agency; **372** Pekka Parviainen/Dembinsky Photo Associates; **374** Robert Pickett/Corbis; **378** Jim Wark/Peter Arnold, Inc.; **379** Seth Resnick/Liaison Agency Inc.; **380** *l* Barnabus Bosshart/Corbis; *r* Reuters NewMedia Inc./Corbis; **383** *t* Jeff T. Green/AP Wide World Photos; *b* Jockel Finck/AP Wide World Photos; **388, 389** Phil Degginger/Color-Pic, Inc.; **390** Craig Tuttle/Corbis/Stock Market; **393** Pat O'Hara/Corbis; **394** Wayne Wastep/Stone; **395** Nancy Rotenberg/Animals Animals/Earth Scenes; **396** *l* Macduff Everton/Corbis; *r* Joyce Photographers/Photo Researchers, Inc.; **397** *l* Phil Degginger/Color-Pic, Inc.; *r* John Skye Chalmers/Corbis/Stock Market; **400** courtesy, U.S. Air Force Photo; **403** Howard B. Bluestein/Photo Researchers, Inc.; **406** Photri, Inc.; **407** *t* Ted Wood/Black Star; *bl* © NYC Parks Photo

Acknowledgements

Archive/Fundamental Photo; *br* © Joseph P. Sinnot/Fundamental Photo; **412, 413** Bill Ross/Corbis; **415** *t* Sam Ogden/Photo Researchers, Inc.; *b* JPL/NASA; **418** Itsuo Inouye/AP Wide World Photos; **425** NASA/Goddard space flight Center/Photo Researchers, Inc.; **426** *t* North Wind Picture Archives; *b* Herman Moll/Osher Map Library; **428** AFP/Corbis; **434, 435** Tina Gerson/L.A. Daily News/Liaison Agency, Inc.; **436** Pete Saloutos/Corbis/Stock Market; **438** Philip James Corwin/Corbis; **444** Plymouth State College Weather Center; **446** Ralph H. Wetmore/Stone; **447** Corbis; **448** Reuters NewMedia/Corbis; **450** courtesy, Laboratory for Atmospheres, NASA/Goddard Space Flight Center; **453** Don Heupel/AP Wide World Photos; **454** *t* courtesy, U.S. Air Force Photo; *cl, b* courtesy, NOAA; **455** *both* National Climatic Data Center; **456** Jonathan Kannair; **464, 465** Lisa Merrill/Danita Delimont Agency; **466** *l* AFP/Corbis; *b* Dave G. Houser/Corbis; **468** *tr* Commander Emily B. Christman/NOAA; *b (both)* JPL/NASA; **469** Tom Stewart/Corbis/Stock Market; **470** James Randklev/Stone; **471** Stefano Cellai/Corbis/Stock Market; **472** James Martin/Stone; **473** William J. Herbert/Stone; **475** courtesy, NASA; **477** Micheal/Sewell/Peter Arnold, Inc.; **482** *bl* Werner Forman Archive/Art Resource; **482, 483** *center across* Corbis; **483** *bl (both)* courtesy, Earth Sciences and Image Analysis, NASA-Johnson Space Center; *r* Courtesy, Dorothy Hall and Janet Chien, NASA Goddard Space Flight Center, and Ron Beck, EROS Data Center.

Unit 6

486 Stuart Westmorland/Corbis; **487** *b* Morton Beebe, S.F./Corbis; **488, 489** D. Sweeney/NOAA, TAO Project Office, Dr. Michael J. McPhaden, Director; **490** North Wind Picture Archives; **491** Woods Hole Oceanographic Institute; **492** Robert Estall/Corbis; **494** Richard T. Nowitz/Corbis; **495** Roger Wood/Corbis; **497** Robert Yin/Corbis; **498** David Doubilet; **499** Jan Hinsch/Science Photo Library/Photo Researchers, Inc.; **500** Douglas Faulkner/Photo Researchers, Inc.; **502** *both* courtesy, NOAA; **503** *t* Roger Grace/Environmental Images; *b* Susan Crane; **508, 509** © 2000 MBARI (Monterey Bay Aquarium Research Institute); **520** Yann Arthus-Bertrand/Corbis; **522** *t* Ken Smith/Scripps Institute of Oceanography; *b* Center for Environmental Visualization, University of Washington; **523** Hans Strand/Stone; **524** Maria Brown/National Marine Sanctuaries, Sustainable Seas Expeditions; **525** Woods Hole Oceanographic Institute; **530, 531** Michael St. Maur Shiel/Corbis; **534** Ralph A. Clevenger/Corbis; **535** Image courtesy Liam Gumley, MODIS Atmosphere Team, University of Wisconsin-Madison CIMSS; **538** Bruno Barbey/Magnum/Picturequest; **540** Bate Littlehales/Corbis; **543** *both* Francois Gohier/Photo Researchers, Inc.; **548** Cary Wolinsky; **548, 549** STR/Reuters 1998; **549** *b* James A. Sugar/Corbis.

Unit 7

552, 553 *background* courtesy, NASA; **552** USGS/JPL/NASA; **553** *tr* NASA/PhotoDisc, Inc.; *b* NASA/AURA/Goddard Space Flight Center/Corbis; **554, 555** NASA/Photodisc; **558, 559** *b* NASA/Artville; **558** *l* courtesy, NASA; **559** *r* courtesy, NASA; **560** NASA/Corbis; **561** *b* NASA/Photodisc; **564** Roger Ressmeyer/Corbis; **565** *top to bottom* NASA/PhotoDisc; NASA/Corbis; John Bova/Photo Researchers, Inc.; **570, 571** NASA/Photodisc; **574** Science VU/Visuals Unlimited, Inc.; **575** NASA/Corbis; **576** courtesy, Emilie Drobnes/NASA; **581** *r* Scala/Art Resource, New York; **586, 587** courtesy, NASA/USGS; **589** *t* NASA/Photodisc; *b* courtesy, NASA; **590** *t* Science and Society Picture Library; *b* courtesy, NASA; **591** NASA/JPL/Malin Space Science Systems; **592** courtesy, NASA/USGS; **593** *b* courtesy, NASA/JPL; **595** University of Arizona/JPL/NASA; **596** *t* courtesy, NASA/USGS; *b* courtesy, NASA/Johnson Space Center Houston, Texas; **597** *t* Heidi Hammel (MIT), and NASA; *b* courtesy, NASA; **598** courtesy, Eric Carlson/NASA; **599** courtesy, University of Arizona/NASA; **600** *t* courtesy, NASA/JPL/PIRL/ University of Arizona; *b (all)* courtesy, NASA/DLR; **601** courtesy, NASA/USGS; **602** *l* NASA/Corbis; *r* Jerry Lodriguss/Photo Researchers, Inc.; **603** courtesy, JPL/NASA; **604** *l* Tom McHugh/Science Source/Photo Researchers, Inc.; *c* Science and Society Picture Library; *r* Detlev Van Ravenswaay/Science Photo Library/Photo Researchers, Inc.; **605** PhotoDisc, Inc.; **610, 611** AURA/STSCI/NASA; **612** Galen Rowell/CORBIS; **613** *left to right;* Max-Planck–Institute for Radio Astronomy/Science Photo Library/Photo Researchers, Inc.; Corbis; Photodisc, Inc.; NASA/Photodisc; **617** Pekka Parianinen/Science Photo Library/Photo Researchers, Inc.; **619** John Chumack/Photo Researchers, Inc.; **622** STR/Reuters News Picture Archive; **625** *t* PhotoDisc, Inc.; *b* NASA/PhotoDisc, Inc.; **627** NASA/PhotoDisc, Inc.; **630** NASA/AURA/Goddard Space Flight Center/Corbis; **631** NASA/Corbis; **632** *t* National Optical Astronomy Observatories/Science Photo Library/Photo Researchers, Inc.; *bl* NASA/Corbis; *bc, br* Royal Observatory, Edinburgh/Science Photo Library/Photo Researchers, Inc.; **633** NASA/Corbis; **635** Carnegie Institute of Washington; **640, 641** *b* Wolfgang Kaehler/Corbis; **640** Professor N. Russell/Science Photo Library/Photo Researchers, Inc.; **641** *tr (inset)* Alfred Pasieka/Science Photo Library/Photo Researchers, Inc.; *tr* Scott T. Smith/Dembinsky Photo Associates; *br* AFP/Corbis.

Unit 8

644, 645 *background* Thomas Ligon/Photo Researchers, Inc.; **645** Photo Researchers, Inc.; **646, 647** Annie Griffiths Belt/Corbis; **648** *l* Francois Gohier/Photo Researchers, Inc.; *r* Jeff J. Daly/Visuals Unlimited, Inc.; **649** Laurence Pringle/Photo Researchers, Inc.; **651** Biophoto Associates/Photo Researchers, Inc.; **652** *t* P.Wysocki/Explorer/Photo Researchers, Inc.; *b* John R. Foster/Photo Researchers, Inc.; **655** *t* Mark A. Schneider/Dembinsky Photo Associates; *bl* Dean Conger/National Geographic Image Collection; *br* Deep Light Productions/Science Photo Library/Photo Researchers, Inc.; **656** James A. Sugar/Corbis; **658** Photo Researchers, Inc.; **664, 665** Tom Bean/Corbis; **666** Dembinsky Photo Associates; **667** *both* Jonathan Blair/Corbis; **670** Phil Degginger/Color-Pic, Inc.; **671** *l* Inga Spence/Visuals Unlimited, Inc.; *r* Jeff Greenberg/Photo Researchers, Inc; **672** *t* Reuters NewMedia Inc./Corbis; *b* Jeff Greenberg/PhotoEdit; **674** Brian Rogers/Visuals Unlimited, Inc.; **675** *t* Sinclair Stammers/Science Photo Library/Photo

Researchers, Inc.; *b* Martin Land/Science Photo Library/Photo Researchers, Inc.; **676** *t* Photo Researchers, Inc; *b* Jean Paul Ferrero/Auscape; **677** Patricia Caulfield/Photo Researchers, Inc.; **678** E.R. Degginger/Color-Pic, Inc.; **680** Geoff Bryant/Photo Researchers, Inc.; **681** Tim Hauf R/Visuals Unlimited, Inc.; **684** *l* John Reader/Science Photo Library/Photo Researchers, Inc.; *r* Wolfgang Kaehler/Corbis; **690** *l* Sandy Felsenthal/Corbis; **690, 691** *tr* Kevin Schafer/Corbis; **691** *tr* Mark A. Schneider/Dembinsky Photo Associates; *br* Francois Gohier.

Illustration

Earth Vistas (background), 419, (background), 427
Susan Biebuyck 626
Chris Forsey/MCA 72, 180, 280, 284, 404
The Granger Collection, New York 579, 580, 581
David A. Hardy xxiv–1 (background), 6, 50, 70–71, 102 (l), 371, 439, 440 (b), 554–555 (background), 586–587 (background), 588, 589, 594
Gary Hincks 176–177, 202–203, 302–303, 326–327, 514, 518–519, 520
Interactive Visualization Systems 487
MapQuest.com, Inc. 44, 45, 46, 146, 148, 172, 173 (all), 182 (all), 183 (source: Future Maps by Christopher R. Scotese, PALEOMAP Project, University of Texas at Arlington, www.scotese.com), 185, 188, 219, 225, 275, 266 (source: USDA), 295, 308, 330, 333 (all, source: U.S. Army Corps of Engineers/Great Lakes Commission), 377 (all), 405, 417 (t), 421, 458, 460, 461, 470–471, 532, 533, 536, 652, 673, 682 (l), 685, (r)708–709, 712–713.
Janos Marffy 125, 127, 149, 197 (all), 239, 286, 288, 289, 290, 599, 649, 650, 651, 652 (l), 663, 674
Davis Meltzer 501
NASA vi, 424, 511
NASA Goddard Space Flight Center 381
NASA Ames Home Page 561
National Geographic Holdings, 2001 (url: www.topo.com) 54, 55
National Geographic Image Collection 549 (both)
National Geographic/Scripps Institute of Oceanography/NOAA/NESDIS/NODC 169
NOAA: Image courtesy of Captain Albert E. Theberge, NOAA Corps (ret.) 198
NOAA/NGDC (National Geophysical Data Center) viii, 175 (both), 351, 486–487 (background), 512, 513, 517, 710–711
P.E. Olsen 644
Precision Graphics 36, 58, 85, 107, 112, 138, 139, 150 (all), 160 (all), 175, 204 (all), 230 (all), 240, 274, 313, 334, 354, 355, 367, 384, 385, 391, 393, 398, 417 (b), 422, 423, (all) 428, 429, 452, 474, 476, 478, 504, 510 (all), 521 (all), 526, 527, 544, 545, 606, 637, 657, 660, 686, 687
Matthew Pippin xxiv, 13, 14-15, 154–155
Mike Saunders 258, 262, 396–397
Raymond Smith 653, 679, 682, (r) 683
Space.com Canada, Inc. 80 (both), 714–717 (all)
Raymond Turvey 174, 214, 238 (all), 240 (all), 241, 244, 245, 300, 343, 346, 347
USGS (U.S. Geological Survey) 707
USGS: Stacey Tighe/The University of Rhode Island/USGS 168, 181
Julian Baum/Wildlife Art Ltd. xii, 32, 33, 34, 553, 556, 573, 574–575, 628, 629
Richard Bonson/Wildlife Art Ltd. 5, 6, 8, 272, 273, 321, 325 (all), 350–351, 353 (all)
Peter Bull/Wildlife Art Ltd. 97, 133, 134
Matthew Frey/Wood Ronsaville Harlin 237, 243 (all)
Rob Wood/Wood Ronsaville Harlin 122, 123, 194, 195, 196, 237, 260 (all), 304, 310, 557
Wood Ronsaville Harlin 53, 516, 539

Text Acknowledgments

Chapter 7, page 150: Adapted from "A Model Oil Well." Reprinted by permission of the Seeds Foundation, St. Albert, Alberta, Canada.

Chapter 8, pages 188–189: "Comparison of Nuclear Sites with Known Volcanoes and Earthquake Activity" (Map). Courtesy of Thomas Worthley, University of Connecticut, Cooperative Extension.

Chapter 10, pages 230–231: Adapted from "Earthquakes: Why Do Earthquakes Happen?". Courtesy of PBS series, Newton's Apple (www.tpt.org/newtons).

Chapter 15, pages 334–335: Adapted from "Glacier Climbing: What Is a Glacier and How Does One Move?". Courtesy of PBS series, Newton's Apple (www.tpt.org/newtons).

Chapter 20, pages 460–461: Two maps of surface weather conditions from the Department of Atmospheric Sciences (http://ww2010.atmos.uiuc.edu/). Courtesy of Dr. Mohan Ramamurthy, Department of Atmospheric Sciences, University of Illinois–Urbana/Champaign.

Acknowledgements

Chapter 21, pages 478–479: Adapted from "Greenhouse Effect Activity" by Angela K. Hancock, appearing on the NASA: Classroom of the Future web site. Reprinted by permission of the author.

Unit 6 Assessment, page 551: Excerpt from "Is Bleaching Coral's Way of Making the Best of a Bad Situation?" by Ben Harder, from *National Geographic News*, July 25, 2001. Reprinted by permission of the National Geographic Society.

Chapter 25, pages 566–567: Adapted from "Impact Craters," from *Exploring the Moon: A Teacher's Guide with Activities*. Reprinted by permission of the National Aeronautical and Space Association (NASA), Earth Sciences for Education.

Chapter 27, pages 606–607: Data generated using Jovian Moon Events, version 3.0 High Accuracy. Courtesy of Dr. Dan Bruton, Physics and Astronomy, Steven F. Austin State University, Nacogdoches, Texas.

Chapter 29, pages 660–661: Adapted from "Tree Rings: A Study of Climate Change". Reprinted by permission of Science Applications International Corporation (SAIC).

Chapter 30, pages 686–687: Adaptation of "Dinosaur I: Finding and Dating". Courtesy of PBS series, Newton's Apple (www.tpt.org/newtons).

Appendix A, page 698: "Periodic Table of the Elements," from *World of Chemistry* by Steven S. Zumdahl, Susan L. Zumdahl, and Donald J. DeCoste. Copyright © 2002 by Houghton Mifflin Company. All rights reserved.

From *Earth Science Reference Tables*. Courtesy of the New York State Education Department.